Image and Logic

Frontispiece. Magnetic detector cartoon. Source: SLAC.

Image and Logic

A Material Culture of Microphysics

Peter Galison

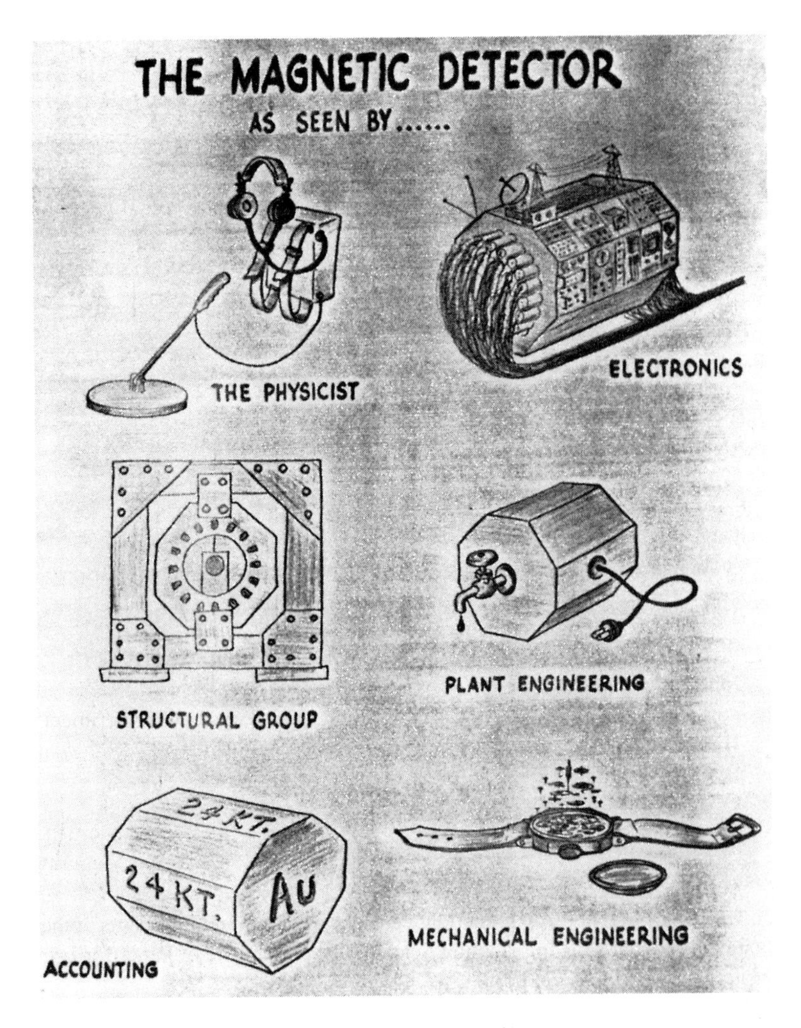

The University of Chicago Press • Chicago and London

Peter Galison is Mallinckrodt Professor of the History of Science and of Physics at Harvard University. He is the author of *How Experiments End,* published by the University of Chicago Press, and coeditor of *The Disunity of Science: Contexts, Boundaries, and Power.*

The University of Chicago Press, Chicago 60637
The University of Chicago Press, Ltd., London

Printed in the United States of America
06 05 04 03 02 01 00 99 98 97 1 2 3 4 5
ISBN: 0-226-27916-2 (cloth)
ISBN: 0-226-27917-0 (paper)

Library of Congress Cataloging-in-Publication Data
Galison, Peter Louis.
Image and logic : a material culture of microphysics / Peter Galison.
p. cm.
Includes bibliographical references and index.
ISBN 0-226-27916-2 (cloth : alk. paper). — ISBN 0-226-27917-0 (pbk. : alk. paper)
1. Microphysics—History—20th century. 2. Physics—Experiments—History—20th century. 3. Physics—Instruments—History—20th century. I. Title.
QC173.4.M5G35 1997
539.7′7—dc21 96-52177
CIP

♾ The paper used in this publication meets the minimum requirements of the American National Standard for Information Sciences—Permanence of Paper for Printed Library Materials, ANSI Z39.48-1984.

Who is "the experimenter" whose activities we have been discussing? Rarely, if ever, is he a single individual. . . . The experimenter may be the leader of a group of younger scientists working under his supervision and direction. He may be the organizer of a group of colleagues, taking the main responsibility for pushing the work through to successful completion. He may be a group banded together to carry out the work with no clear internal hierarchy. He may be a collaboration of individuals or subgroups brought together by a common interest, perhaps even an amalgamation of previous competitors whose similar proposals have been merged by higher authority. . . .

The experimenter, then, is not one person, but a composite. He might be 3, more likely 5 or 8, possibly as many as 10, 20, or more. He may be spread around geographically, though more often than not all of him will be at one or two institutions. . . . He may be ephemeral, with a shifting and open-ended membership whose limits are hard to determine. He is a social phenomenon, varied in form and impossible to define precisely. One thing, however, he certainly is not. He is not the traditional image of a cloistered scientist working in isolation at his laboratory bench.

—A. M. THORNDIKE, Brookhaven National Laboratory, 1967

Contents

Figures

Preface

This is a book about the machines of physics. Out of the experimental apparatus come the delicate track images that have launched, backed, and challenged the abstractions of unified field theories—pictures that, as symbols of science, have graced the covers of a hundred books. My goal is not to begin with the tracks and position them within arguments for great experimental discoveries, such as those of the psi, the omega minus, or the positron. Nor is my goal to retell, once again, the long march of theories of matter from atoms to quarks. Instead, I want to expose the practices presupposed by these images, to peer into all that grubby, unplatonic equipment that lies such a long way from Lie algebras and state vectors. I want to get at the blown glass of the early cloud chambers and the oozing noodles of wet nuclear emulsion; the insistent hiss of venting nitrogen gas from the liquifiers of a bubble chamber; the resounding crack of a high-voltage spark arcing across a high-tension chamber and leaving the lab stinking of ozone; the silent, darkened room, with row after row of scanners sliding trackballs across projected bubble chamber images; the late-night computer screens flashing with the skeined complexity of rotating and disappearing tracks; the one remaining iridescent purple line cutting across the background of a terminal. Pictures and pulses—I want to know where they came from, how pictures and counts got to be the bottom-line data of physics. The Göttingen number theorist Edmund Landau, on hearing of something too applied for his taste, looked over in disgust and pronounced it *Schmieröl*—engine grease. But that is what I want to know about: how all these machines, these gases, chemicals, and electronics, came to make facts about the most theoretically articulated quadrant of nature. Historically, historiographically, philosophically, this book is a back-and-forth walk through physics to explore the site where engine grease meets up with experimental results and theoretical constructions.

These machines have a past. To walk through the laboratories of the twentieth century is to peruse an expanse of history in which physics has played many parts. Over here, film for atomic physics, X-ray film out of boxes destined for medicine; over there, a converted television camera rewired as part of a spark chamber. In this corner a piece of preparatory apparatus for a hydrogen bomb, in that a cannibalized bit of computer. Around you in the 1950s the structure of mutable, industrial-style laboratories introduced to physics in the wartime scramble to ready nuclear weapons and radar. Shaped by the exigencies of industry and war, but also shaping the practices of both, the machines of physics are part of a wider technological material culture—neither below it, nor above it.

The embedding of physics machines in material culture is never just about machines by themselves. A nuclear emulsion plate of the 1940s was also a means of preserving individual work high in the Alps. A bubble chamber of the 1950s came with an industrial and military notion of how to partition teamwork in the laboratory. When the early spark chamber designers put together their instruments, they were in many instances trying to gain control, both epistemic control and workplace control, over the process of experimentation on their most elusive targets. By following the all-too-material culture of the laboratory, I want to get at what it meant to be an experimenter and to do an experiment across the century—for the very category "experiment" will not stay still.

The particular laboratory machines that I focus on are detectors, those objects large and small that mediate between the microworld and the world of knowledge. There are other machines: from those that dig the vast tunnels of the accelerator, to the accelerator beam technology itself. If I begin the subtitle of this book with an indefinite article, it is to leave space for such inquiries. But detectors have a singular place, lying as they do at the intersection of epistemology, physics, technology, and the social structure of physics. For physicists, these devices have a magnetic fascination: they fix what it is like to *be* an experimenter. One of the great designers, Carlo Rubbia, once put it this way: "Detectors are really the way to express yourself. To say somehow what you have in your guts. In the case of painters, it's painting. In the case of sculptors, it's sculpture. In the case of experimental physics, it's detectors. The detector is the image of the guy who designed it" (quoted in Alvarez, "Bubble Chambers" [1989], 299). In many ways, and at many points in history, I am after the material culture of the laboratory as the "image" of the designer. At the same time, I am after the reciprocal: the experimenters themselves—their relations and their practices—as the image of the machine at which they work. Imaging works both ways.

One final remark on the title concerns "microphysics." By tracking instruments, not theories, we pick out different parts of physics. "Particle physics" appropriately designates postwar physics but fails to capture the situation earlier. Experimentalists moved between cosmic rays, alpha particles, atoms, even raindrop formation. "Microphysics" names the assortment of entities studied through chambers and counters without imposing our later classification.

While it is meant to stand alone, *Image and Logic* was written as part of a trilogy. I organized *How Experiments End* (1987) around the experimental procedures, the modification of apparatus and sorting of data that separate effect and background. There, my goal was to keep the narrative gaze on the establishment of several specific effects in order to probe how the standards of demonstration shifted over time. In part that approach meant looking "up" to theory—to unravel just which parts of atomic, nuclear, or particle physics theory mattered to the experimentalists, and how. But it also meant looking "down" to the apparatus, to the ways in which cloud chambers, counters, and bubble chambers were used in bringing those particular experimental arguments to a close.

In *Image and Logic* I invert the perspective and caution about too hierarchical a conception of the relation of instruments, experiments, and machines. Instruments become central, and experimental demonstrations of particular effects recede. Instead of attending to the continuity of experimental topic—such as the existence or nonexistence of neutral currents—I bring out a continuity of the devices of physics and the changing experimental life that surrounded them. How does the pictorial (image) tradition of cloud chambers become that of nuclear emulsions and bubble chambers? How does the counting (logic) tradition of Geiger-Müller counters carry over to spark chambers and wire chambers? And finally, how do these two material and epistemic traditions hammer out an intermediate category of experimentation that borrows from both?

Unfortunately, the word "tradition" might suggest a static, totally autonomous, and transtemporal set of practices. Not here. The early cloud chambers and emulsions are grounded not only in the ultimate constitution of matter but in volcanoes, thunderstorms, and photography. And while the bubble chamber image borrows much from the practices of cloud chamber photography, this newer chamber has slid far from its natural historical roots, tied more to the factory and wartime production. Finally, the centralized factory itself begins to fragment into the dispersed, multinational collaborations of the 1980s and 1990s, and the material culture of the laboratory shifts once again. Experiments centralize—pulled into ever larger collaborations working on ever vaster pieces of equipment. At the same time they disperse, through the diffusion of construction, data processing, and experimental analysis. Computers and simulations cease to be merely substitutes for mechanical parts, they come to stand in a novel epistemic position within the gathering of knowledge—not quite a piece of empirical machinery, and not quite one with theoretical apparatus.

How Experiments End was a move toward a dynamic history of the culture of experimentation as it shifted across the twentieth century. *Image and Logic* is aimed at the historical development of the material laboratory and the world of work that surrounded it—at the always shifting culture that surrounds the knowledge machines of physics. The last in this trio of books will be approached similarly: there I will want to shift the focus one last time, toward the changing location of theoretical physics, as it is constituted—and reconstituted—between the computer, mathematics, condensed matter physics, and cosmology.

Underlying *Image and Logic,* and the larger three-part project of which it is part, is a sense of the barrenness of a series of dichotomies. We bounce back and forth between untenable positions. Science, it is sometimes said, must follow a set of prescribed rules: algorithms for discovery, replication, verification, or confirmation. The practice of the physical sciences never works quite that simply, so the opposite vision comes into favor. Science is really, at root, about interests, the surface reflection of more basic forces, be they economic, psychological, or social. Or, more philosophically, responding to the often exaggerated empiricist claims for knowledge from "raw" experience comes the counterclaim that experimental knowledge should not be considered autonomous at all, so heavily is it "theory laden." Yet elsewhere, one side claims that replication in science demands no more than the unambiguous execution of published procedure. The other side ripostes that all replication is grounded on unarticulated (and even unarticulable) tacit knowledge that can *only* be disseminated through the bodily transport of personnel or complete machines.

Lost in such dichotomies is a sense that what holds physics together is neither a single, unified, deductive or inductive apparatus, nor a Potemkin village of rationality hiding the raw exertion of competing interests. Physics is a complicated patchwork of highly structured pieces: instrument makers thoroughly versed in the manipulation of gases, liquids, and circuitry; theorists concerned with the coherence, self-consistency, and calculability of the behavior of matter in their representation of matter most finely divided; and experimenters drawing together instruments into combinations in pursuit of novel effects, more precisely measured quantities, or even null results. But between and among these subcultures of physics lie substantial border territories, and it is only by exploring the dynamics of those border regions that we can see how the whole fits together. Between theoretical nuclear physics and the art of cloud chamber construction is the art of interpretation, borrowing bits of theoretical ideas, pieces of craft knowledge about films and optics, and portions of experimental knowledge. Neither "theory ladenness" nor "theory independence" gets at the working out of interpretive practices that makes it possible to narrate the series of decays, materialization, and scattering events from their "fossil" photographic record.

It is surely true, as authors from Michael Polanyi, to Thomas Kuhn, to Harry Collins have insisted, that there are moments when individuals cannot spell out rules for replication. That should be occasion, not to stop the inquiry, but to ask: Why not? What pieces of practice will not fit the public discourse of science, and why? Sometimes the movement of machine knowledge may be impeded because no one knows what portion of a complicated procedure is efficacious and what is superfluous. At other times laboratory procedures get blocked for reasons of state, at other times for reasons of industrial propriety or competitive advantage. There are even times when instruments *do* sail smoothly into replication without the necessity of sending copies of the device or experts with tacit *Fingerspitzengefühl.* But at border crossings, the movement of machine knowledge typically takes place not all of a piece; engineers from different laboratories meet to share tricks about gaskets and seals, about computer analyses and simulations, about chemistry, cryogenics, and optics. Sometimes the parts can be yanked bodily from off-the-shelf devices, at other times a published report is good enough to allow college freshmen to make a spark chamber or a high-temperature superconductor a thousand miles from its site of first production.

The point is that the dispersion of instrument knowledge follows no univocal pattern. More constructively, one could say that the diversity of technical and laboratory traditions suggests that knowledge diffusion is going to depend crucially on the volatile intersection arenas of physics—the sites where chemical engineer and nuclear physicist meet, where electronic engineer and mathematician work on the computer—in short, on the "trading zone" between groups with different technical traditions. For it is there that we can begin to capture the inevitably incomplete, but essential coordination between the different subcultures of physics.

In the early 1970s, as work came to completion on an extraordinary new particle detector at the Stanford Linear Collider, George Lee, an engineer and amateur cartoonist, sketched the frontispiece illustration. To the Accounting Department, the SPEAR Magnetic Detector was a pile of gold—quite a bit of gold by the standards of the time; to Plant Engineering, the hybrid device was a substantial problem in water cooling and electrical feeding; to Electronics there were the delicate strands of instrument wire, and the powerful conduits to the solenoid. This was not a world of vanishing technical assistants, like those artisanal helpers who wrapped wire for Maxwell, or assisted Rutherford in counting scintillator flashes; this was a laboratory where the physicists played a leading role, but where everyone knew that the engineers were by no means mere extras. Amid such a coordinated cacophony of construction, physicists could be lampooned by likening them to bewildered scavengers, walking parks and beaches in the hopes of culling a gold ring from the discarded flip tops and bent nails. The humor of the cartoon—and the reality of physics, I would argue—resides in its startling diversity in how the machine looked from the vantage points of the various subcultures of physics. More seriously, the cartoon prompts the question posed by any piece of complex instrumentation: How do these subcultures come together without homogenizing their distinct identities? In November 1974, not long after the cartoon was drawn, the scavenger's earphones were ringing like mad—here as elsewhere the data tapes that grounded the resulting discovery of the charmed quark have to be understood as resting in no small measure on the coordination of these separate spheres.

Even this penciled sketch is but a partial presentation of the multitude of worlds within physics; there were other worlds beyond. Left out are the different university and national groups participating in large experiments, not to speak of the theorists, phenomenologists, administrators, and industrialists; there are computer programmers simulating runs and figuring out how to acquire, store, and sort the data; there are postdocs running shifts. Somehow, out of it all, comes an argument. This picture of science fits badly into the narrowly construed rationality of the algorithmic, and equally badly into the image of an unreasoned struggle by opposing forces to divvy up the territory of knowledge. Physics as a whole is always in a state of incomplete coordination between extraordinarily diverse pieces of its culture: work, machines, evidence, and argument. That these messy pieces come together as much as they do reveals the presence, not of a constricted calculus of rationality, but of an expanded sense of reason.

This book has been such an immense part of my life for so long that I cannot imagine its ever having been written without uncounted discussions with colleagues and students, especially Keith Anderton, Alexi Assmus, Richard Beyler, Mario Biagioli, Allan Brandt, Cathy Carson, Nancy Cartwright, Jordi Cat, Hasok Chang, Lorraine Daston, Arnold Davidson, Clifford Geertz, Howard Georgi, Ian Hacking, Carl Hoefer, Gerald Holton, Myles Jackson, Martin Jones, David Kaiser, Cheryce Kramer, Timothy Lenoir, Everett Mendelsohn, Naomi Oreskes, Matt Price, Hilary Putnam, Simon Schaffer, Joan Scott, David Stump, Patrick Suppes, Emily Thompson, Jonathan Treitel, and Norton Wise.

I had grant support from the National Science Foundation (through research grants and through the Presidential Young Investigator Award), the Center for Advanced Study in the Behavioral Sciences, the Pew Foundation, the Mellon Foundation, the American Institute of Physics, and the Howard Foundation. Without them I would have had neither the time nor the mobility to find the widely scattered sources that made the writing of this book possible.

Specifically, the archival roots of this book are in two types of sources: papers held in archives and institutional settings and "private" papers, by which I mean collections still in the hands of particular physicists and engineers. To both groups I owe an extraordinary debt. On the institutional side, I would like to thank the librarians and staff in charge of the following collections: Albert Einstein Papers, Princeton University Archives; American Institute of Physics, Niels Bohr Library; Bundesministerium für Unterricht, Österreiches Staatsarchiv, Vienna; Harvard University Archives; Rare Books and Special Collections, Princeton University Libraries; University Library, University of Bristol; Archives and

Records Office, Lawrence Berkeley Laboratory; Archives, Brookhaven National Laboratory, Upton, New York; Cavendish Laboratory, University of Cambridge; Churchill College Library, University of Cambridge; University of Edinburgh; National Archives—Pacific Sierra Region, San Bruno, California; Special Collections and University Archives, Stanford University; Fermilab Archives, Fermilab, Batavia, Illinois; Trinity College, University of Cambridge; Hagley Museum and Library, Wilmington, Delaware; Institut für Radiumforschung, Vienna, Austria; University of Wisconsin Archives; Manuscript Division, Library of Congress; American Philosophical Society Library, Philadelphia; Los Alamos National Laboratory Archives, Los Alamos, New Mexico; Niels Bohr Archives, Niels Bohr Institute, Copenhagen; Public Record Office, London; Institute Archives, Massachusetts Institute of Technology; National Archives—New England Region, Waltham, Massachusetts; Science Museum Library, South Kensington, London; Stanford Linear Accelerator Center Archives, Stanford, California. In the acknowledgments to the individual chapters listed below, I thank individuals who gave me unstintingly of their time and (crucially) allowed access to everything from key notebooks to e-mail. Abbreviations for all these sources can be found listed after chapter 9.

Each chapter had a distinct set of source materials—from the Royal Society's collection of C. T. R. Wilson's papers in chapter 2 to the quickly fading SSC laboratory of chapters 7 and 9. In addition to the institutional collections listed above, for help with chapter 2 I would like to thank C. T. R. Wilson's daughters, J. I. M. Wilson and R. H. Wilson, who so generously shared with me photographs, letters, and other documents left by their father. It is a particular pleasure to thank Alexi Assmus, who worked with me almost a decade ago on cloud chamber history, first while she was an undergraduate student of mine at Stanford, and then while a graduate student at Harvard. That collaboration led to the publication of an early version of the cloud chamber chapter, "Artificial Clouds, Real Particles," in D. Gooding, T. Pinch, S. Schaffer, eds., *The Uses of Experiment* (Cambridge: Cambridge University Press, 1989; excerpts reprinted with the permission of Cambridge University Press). Several of the master builders of the cloud chamber era spoke to me at length about technique, including Carl D. Anderson, Pierre Auger, L. Leprince-Ringuet, George Rochester, Robert Thompson, and J. G. Wilson. My thanks go as well to Paul Forman, Gerald Holton, Jane Maienschein, S. Della Pietra, Simon Schaffer, and Roger Stuewer, all of whom had helpful comments on various drafts of this material over these many years.

Sources for chapter 3 came from many places, but of special help were Laurie Brown, Mrs. Fowler, Mrs. Gardner, Leopold Halpern, G. P. S. Occhialini, C. O'Ceallaigh, Mrs. Powell, David Ritson, George Rochester, J. Rotblat, Roger Stuewer, and Cecil Waller. Chapter 4 is grounded almost entirely on institutional archival holdings at the University of California, Berkeley, Harvard University, the University of Wisconsin, Stanford University, Princeton University, and MIT—all of these are more fully referenced above. I also benefited from discussions with Robert Hofstadter, D. Kevles, A. Needel, and John Wheeler, as well as from the MIT-Harvard Conference on the Military and Science, which led to the publication of "Physics between War and Peace," in E. Mendelsohn, P. Weingart, and M. Roe Smith, eds., *Science, Technology and the Military* (Boston: Reidel, 1988; excerpts reprinted by permission of Kluwer Academic Publishers). My collaboration with Rebecca Lowen, then a graduate student at Stanford, and Bruce Hevly, a postdoc working with me also at Stanford, on the history of SLAC led to "Controlling the Monster," in Galison and Hevly, eds., *Big Science: The Growth of Large-Scale Research* (Stanford, Calif.: Stanford University Press, 1992), bits of which I have referred to here as well.

The shift in scale associated with the building of the big bubble chambers is accompanied, not surprisingly, by a radical increase in paper. And without the extraordinary generosity of Donald Glaser and Luis Alvarez in making their notebooks, memos, correspondence, and reports available to me, nothing like chapter 5 could have been written. In addition to their papers, and a variety of collections housed in the LBL archives, I benefited from many discussions with and sources offered by Mrs. Alvarez, H. R. Crane, Jack V. Franck, John Heilbron, Paul Hernandez, Roger Hildebrand, P. V. C. Hough, W. Kerr, Henry Lowood, Donald Mann, Darragh Nagle, David Rahm, Robert Seidel, R. Shutt, Pete Schwemin, Lynn Stevenson, Alan M. Thorndike, and Bruce Wheaton. Early portions of chapter 5 appeared as "Bubble Chambers and the Experimental Workplace," in P. Achinstein and O. Hanaway, eds., *Observation, Experiment, and Hypothesis in Modern Physical Science* (Cambridge, Mass.: MIT Press, 1985), and "FORTRAN, Physics, and Human Nature," in M. J. Nye, J. Richards, and R. Stuewer, eds., *The Invention of Physical Science* (Boston: Reidel, 1992).

In regard to the writing of chapter 6 on the myriad of logic devices, I am pleased to thank Gerald Abrams, Otto Claus Allkofer, Adam Boyarksi, Martin Breidenbach, Georges Charpak, William Chinowski, Conversi, Bruce Cork, James Cronin, David Fryberger, P.-G. Henning, Lillian Hoddeson, Leon Lederman, Massonnet, Martin Perl, Gerson Goldhaber, Robert Hofstadter, Robert Hollebeek, J. A. Kadyk, Harvey Lynch, Piet Panofsky, John Rees, Burton Richter, Michael Riordan, Michael Ronan, Bruno Rossi, Melvin Schwartz, Roy Schwitters, John Scott Whitaker, Perez-Gomez, Martin Deutsch, Bouclier, A. Gozzini, M. S. Fukui, B. Maglic, S. Miyamoto, and William Wenzel. Parts of chapter 6 appeared in Peter Galison, "Bubbles, Sparks, and the Postwar Laboratory," in *Pions to Quarks: Particle Physics in the 1950s,* edited by L. M. Brown, M. Dresden, and L. Hoddeson, 213–51 (Cambridge: Cambridge University Press, 1989; excerpts reprinted with the permission of Cambridge University Press).

By the time we get to the hybrid detectors of chapter 7, we are in the era of e-mail and mass duplication of memoranda—the very size of the collaboration demands a circulation of information, multiply produced. And it is just this multiple production of literary (and not so literary) traces that found its way into the centralized collections at LBL, forming the documentary basis of this chapter. I am grateful, in addition, to Elliott Bloom, Howard Georgi, Paul Hernandez, Jay Marx, David Nygren, and Mike Ronan for many discussions about the TPC, and to Moshe Safdie for access to papers and conversations about the planning of the SSC. George Trilling provided electronic copies of governance documentation for SDC; I am grateful to Gene Fisk, Paul Grannis, and Hugh Montgomery for help with the D0 documentation; Melissa Franklin was my source of information about CDF; and Gigi Rolandi was kind enough to communicate with me about the running of ALEPH. On chapter 8, my thanks go to Sara Foster, who worked on these matters with me while she was a graduate student at Stanford, and to John Bahcall, Foster Evans, Cuthbert Hurd, and Nicholas Metropolis for enormously helpful discussions. Some material in chapter 8 also appears in *The Disunity of Science: Boundaries, Contexts, and Power,* edited by Peter Galison and David J. Stump, and is used here with the permission of the publishers, Stanford University Press; © 1966 by the Board of Trustees of the Leland Stanford Junior University, all rights reserved.

The methodological considerations of chapters 1 and 9 were discussed in Peter Galison, "Multiple Constraints, Simultaneous Solutions," *PSA 1988* 2 (1988): 157–63, and Peter Galison, "History, Philosophy, and the Central Metaphor," *Science in Context* 2 (spring 1988): 197–212 (excerpts reprinted with the permission of Cambridge University Press).

Manuscript preparation on a project of this size is an enormously difficult task, and I want to give great credit to Pam Butler, who worked extraordinarily hard at getting straight several thousand footnotes, bibliographical entries, and a couple of hundred figures. Richard Beyler, Michael Gordin, Barbara Kataoka, Sherri Roush, and Karen Slagle also spent many hours on this book, and I want to give them special thanks.

My greatest debt is to Caroline Ann Jones. Her incisive comments led to revisions in every chapter, and every section of this book has benefited from them. Luckily for me, her extraordinary insights and well-chosen criticisms were coupled with an unstinting moral support. This book is for her, for Sarah, and for Sam.

1 Introduction: Image and Logic

1.1 A Material Culture of the Laboratory

In 1964, some of the world's leading experimental physicists gathered in Karlsruhe, West Germany, to discuss the radical changes then underway in their profession. Where an earlier generation of experimenters saw the design, construction, and use of apparatus as defining features of their identity, now much of this activity was permanently out of the physicists' hands. Where physicists once collated data in notebooks and analyzed them with slide rules, the computer had now taken over much of this work—storing, processing, even analyzing information and delivering it in publishable graphic form. Huge teams of physicists clustered around barn-sized bubble chambers at centralized accelerators. Throughout the laboratory, the relations among computer programmers, experimenters, instrument makers, and engineers were utterly in flux. In a taped discussion from the Karlsruhe conference, Lew Kowarski, an influential physicist from the major European particle accelerator, CERN, waxed enthusiastic about these changes, extolling the virtues of assembling millions of pictures and tackling them with automatic techniques. The audience was stunned. One physicist, obviously pained, asked about the danger present in automation: mightn't it squelch innovation? Another confessed how "frightened" he was to hear that "in a few years . . . one would not go to start a new experiment but one would just go into the archives, get a few magnetic tapes[,] and start to scan the tapes from a new point of view[—]that would be the experiment." Knowledge from archives—what could be worse?

The assembled experimenters were right to be stunned. Their profession, and their daily lives, were being transformed as the meaning of the term "experiment" altered before their eyes. They faced fundamental questions: What would count as a demonstration in physics? Was building, running, and modifying apparatus necessarily part of experimenting? What role would statistical argumentation play? Would a computer simulation—*should* a computer simulation—substitute for an experiment? As the dreaded "archives" loomed, even the venue of experimentation changed, both physically and sociologically. Many could remember vividly an experimental life conducted in rugged mountain cosmic ray huts on Pic-du-Midi, or in small groups huddled around a handful of cloud chamber pictures. Now many found themselves physically banned from what Kowarski so provocatively called the site of "the kill," the interaction of a charged-particle beam on a target and the boiling hydrogen of the bubble chamber that tracked the remnants. Physicists whose epistemic and personal raison d'être had required *control* over experimental apparatus now found themselves on the outside looking in. Nothing less than their notion of an experiment was at stake.

These struggles over control took place in the design, appropriation, use, and rejection of instruments—the machines of physics. And it is among these machines, on the shop floor so to speak, that the action of this book takes place. Lowly instruments, the everyday physical equipment that clutters (and in part defines) the laboratory, constitute our subject. The reader may well ask how mere hardware could compete in interest with dramatic claims about the ultimate constituents of the universe, the origin of all things, or the final unification of all forces. I will argue that laboratory machines can command our attention if they are understood as dense with meaning, not only laden with their direct functions, but also embodying strategies of demonstration, work relationships in the laboratory, and material and symbolic connections to the outside cultures in which these machines have roots. It is by means of such a broader and deeper exploration of tubes, tapes, and tracks that we can get at the material culture of a discipline. Who counts as an experimenter, what counts as an experiment? Those are the guiding questions of this book.[1]

1. "Laboratory studies" has progressed to the point where it simply is not possible to list all the important books and articles in the field. The following, however, will serve as a representative set of such inquiries and can be consulted for further references. Several collections of essays might serve as an overview: Achinstein and Hannaway, *Observation* (1985), includes a mixture of historical and philosophical essays on experimentation, as does Gooding, Pinch, and Schaffer, *Uses of Experiment* (1989). See also James, *Development of Laboratory* (1989); Lenoir and Elkana, *Science in Context* 2 (1988): 3–212; Ophir et al., *Science in Context* 4 (1991); and Van Helden and Hankins, *Osiris* 9 (1994): 1–250. Lynch and Woolgar's *Representation* (1990) includes valuable references (and articles) on ethnomethodological approaches to laboratory studies; see also Coleman and Holmes, *Investigative Enterprise* (1988); Levere and Shea, *Nature* (1990); Horwich, *World Changes* (1993); and Buchwald, *Scientific Practice* (1995).

Hacking, *Representing* (1983), and Cartwright, *Laws* (1983), are two influential philosophical works that argue for a mitigated realism about experimental entities, whereas Franklin's *Neglect* (1986) and *Fifth Force* (1992) take a Bayesian approach in arguing for a realism of a different sort. On instrumentation and philosophy, see Ackermann, *Data* (1985); Ihde's *Instrumental* (1991) grapples with the difficult problem of combining Continental and Anglo-American writings on instrumentation and experiment.

I have found enormously helpful several central contributions to the literature on experimentation in the

Of the myriad of machines strewn about the laboratory, the focus here will be on detectors, the devices large and small that carry the microphysical world into the empirical wing of physics. These are the mediators between the production of phenomena and the production of evidence. Or, looked at spatially, our inquiry will begin with the set of machines most immediately under physicists' control, as opposed to the wider infrastructure in which research takes place. In this respect, it is useful to introduce a few terms. The "outer laboratory" will designate the macroenvironment that encloses the experimental physicist. In early-twentieth-century Germany, the outer laboratory would refer, for example, to a physical institute such as the Physikalisch-Technische Reichsanstalt of pre–World War II Berlin;[2] in early-twentieth-century America, the term might refer to university laboratory buildings such as Palmer Hall at Princeton and Jefferson Laboratory at Harvard or to a loosely directed institute such as the Carnegie Institute for Terrestrial Magnetism. And in the big accelerator laboratories, the construction, upkeep, and running of the accelerator itself is at least once removed from the detector. There is no doubt that the outer laboratory has an impact on the individual experimental physicists who work there (and we will refer to those effects at several crucial points), but such broader institutions do not determine the bench-top environment of physicists' laboratory life. It took

---

early modern period: Hannaway, "Libavius," *Isis* 77 (1986): 585–610; Shapin, "Experiment," *Isis* 79 (1988): 373–404; Shapin and Schaffer, *Leviathan* (1985); and Heilbron, *Early Modern* (1982). All are key entry points into this literature.

For the twentieth-century physical sciences, see Galison, *Experiments* (1987); Galison and Hevly, *Big Science* (1992), which focuses on military, industrial, national, and university-based large-scale research. Holton's *Thematic* (1988) and *Scientific Imagination* (1978) both include important essays on experiment-theory relations; his study of Millikan and Ehrenhaft, "Subelectrons" (1978), has proved significant in its exploration of theoretical presuppositions. Heilbron and Seidel, *Lawrence* (1989)—like the CERN volumes of Hermann et al., *CERN I* (1987) and *CERN II* (1990)—concentrates on a major institution as a way to define its scope. Traweek, *Beamtimes* (1988), is the best example I know of an anthropological study of a laboratory and is unique in its exploration of gender and cross-cultural issues in modern physics. Collins's *Changing Order* (1985) contains his influential sociological studies of replicability. Pickering has written extensively on the social constructivist account of experiment-theory relations—see, e.g., *Quarks* (1984), "Quark," *Isis* 72 (1981): 216–36; and "Monopole," *Soc. Stud. Sci.* 11 (1981): 63–93. For a detailed study of early atomic physics experimentation, see Stuewer, *Compton Effect* (1975).

A good introduction to some of the best recent literature on nineteenth-century physical experimentation can be found in Schaffer, "Manufactory" (1989); Smith and Wise, *Energy* (1989); Buchwald, *Scientific Effects* (1994); Gooding, *Experiment* (1990); Olesko, *Physics* (1991); and Cahan, *Institute* (1989).

For the biological sciences and experimentation, Kay, "Electrophoresis," *Hist. Phil. Life Sci.* 10 (1988): 51–72, and Lenoir, "Electrophysiology," *Hist. Stud. Phys. Bio. Sci.* 17 (1986): 1–54, are particularly useful because they focus on instrumentation. Though it has often been (unwisely) neglected in broader philosophical considerations of demonstration, fieldwork has played a central role in the empirical sciences: see Maienschein's *Transforming Traditions* (1991) and "Experimental," *Stud. Hist. Bio.* 6 (1983): 107–27; Latour and Woolgar, *Laboratory Life* (1987); Latour's *Pasteurization* (1988) and *Science in Action* (1987); and Holmes, *Lavoisier* (1985).

While there is a smaller literature on geology and geophysical experimentation, I have learned an enormous amount from Rudwick's *Devonian* (1985) and "Visual," *Hist. Sci.* 14 (1976): 149–95; Greene, *Geology* (1982); and Secord, *Victorian Geology* (1986). Of particular interest because it contrasts the historical epistemic role of fieldwork with that of the geophysical laboratory is Oreskes, "Rejection," *Hist. Stud. Phys. Bio. Sci.* 18 (1988): 311–48.

2. On the Physikalisch-Technische Reichsanstalt, see Cahan, *Institute* (1989).

ecologists many years to recognize that in an ecological system the *micro*environment can differ markedly from the *macro*environment; given this distinction, it makes sense to launch an inquiry into the wind, temperature, and chemical conditions in the millimeter above a leaf, and to acknowledge their significant deviations from the ambient conditions of the forest. Similarly, in a laboratory, the immediate surroundings of the working physicist may differ radically from the broader institutional world; the microenvironment of the physicist is what I term the "inner laboratory."

The inner laboratory, which will be our principal site of inquiry, includes the working physicist's material culture and experimental practices.[3] These might include the tools on the bench, the methods of calculation, and the roles of technicians, engineers, colleagues, and students. As the later part of this story unfolds at the end of the twentieth century, the mutation of laboratory life undermines even these locational designations: the distinction between "outer" and "inner" loses its meaning as the major particle physics laboratories delocalize through an extended network of computer analysis and control. Despite the eventual porosity of the boundary between inner and outer laboratories, however, there is a separation in space and practice that divides the two. And inside the inner laboratory of atomic, nuclear, and particle physics, I will argue, we can follow two broad classes of detection devices that link the microscopic world with the world of our senses. One class—the image-making devices—produces pictures, while the other—the logic devices—produces counts. This evidentiary or epistemic classification is not arbitrary; it is bound to the building of one device out of another, through the specific, often quite material, details of how instruments like the cloud chamber or bubble chamber are made and used.

The material culture of physics that circulates around the detector within the inner laboratory is our starting point. But if our study of instrumentation returns again and again to the space within laboratory walls, it will not and cannot remain there. With roots in the Victorian fascination with Krakatoa, thunderclaps, and lightning, the cloud chamber was before all else a site for reproducing nature in the wild; Cecil Powell's emulsions emerged from seismological observer networks and commercial film laboratories. Postwar electronics were based in wartime radar research; Luis Alvarez's bubble chamber borrowed equipment directly from the thermonuclear bomb effort from Eniwetok to Boulder, Colorado, and Los Alamos, New Mexico. Understanding the origin

3. Anthropologists and archaeologists use the term "material culture" in a wide variety of ways, from the study of objects taken by themselves (so to speak) to the analysis of objects along with their uses and symbolic significance. I share with some recent authors in this field a concern with the analysis of objects as "encultured" (or to use Nicholas Thomas's useful term, "entangled"). This is a large literature, but some of the most intriguing anthropological writing can be found in Thomas, *Entangled Objects* (1991); Pearce, *Museum* (1989); Appadurai, *Social* (1986); and Ferguson, *Historical* (1977). A more recent collection of essays (Lubar and Kingery, *History* [1993]) takes factory machines, landscape architecture, and household items and shows how one might go about using them in the writing of history.

of Monte Carlo simulations takes us back to nuclear weapons research and the advent of the electronic computer. Marietta Blau, the now-forgotten Viennese inventor of the nuclear emulsion, drew her original material from ordinary X-ray dental film. The genealogy of these instruments helps to explain how they became certified as legitimate keys to the domain of the subvisible. And it is through an understanding of the original and sustaining technologies, often far outside the usual narrative (from molecule to quark by way of atom and nucleus), that we can answer the historian's insistent question: Why were *these* instruments created, reproduced, and used in particular ways in particular places?

By bringing instruments front and center, we get a different history, a history only awkwardly classed under the old rubrics "internal intellectual history" and "external sociological history." Of course, the history of instruments must be a technical history, a history (for example) of the chemistry of nuclear emulsions, the coordination of mechanical and electrical engineering in a hybrid detector, or the craft-based blowing of glass bulbs for cloud chambers. But the history of instruments is necessarily also part labor history, part sociology, and part epistemology. It is a history that is inseparable from individuals' search for a way of working in laboratories sited squarely in a particular culture—here of Victorian Scotland, there of wartime Los Alamos. It is also a history of twentieth-century microphysics, but a nonstandard one, written from the machines outward not from the unifying theories inward. It is, in the end, a history—refracted through specific material objects—of the elusive historical dynamic that generates and sustains the meaning of "experiment" over time.

To get at the shifting meaning of "the experiment" and "the experimenter," we will need to empathize with the anxiety that physicists have often felt at their loss of control over the rapidly expanding laboratory, at the complex confrontation of physicist with engineer, and at the often productive tensions between experimenter and theorist. For it is the historically specific awkwardness felt by experimental physicists in their relation to others in the laboratory (safety supervisors, engineers, technicians, scanners, government inspectors) that signals a shift in the experimenters' identity.

It has been a long, irregular, and often broken road between a time when it was unthinkable that a physicist be anything *but* someone who built equipment, designed procedures, manipulated experiments, wrote up results, and analyzed them theoretically to a time when it would be a matter of near-universal consent that someone could count himself (or, more rarely, herself) as an experimenter while remaining in front of a computer screen a thousand miles from the instrument itself. These alterations in practice contradict the notion that there is a single, unitary concept of experiment. Experiment and experimenter are bound together; their meanings necessarily change together.

Seen through the old and rigid philosophical notion of "*the* experimental method," the instability of "experiment" might seem surprising—but it should not. Instruments and experiments have never been undisputed notions, not in the early modern era, and not since. In their remarkable book *Leviathan and the Air-Pump,* Steven Shapin and Simon Schaffer used the air pump to confront two very different visions of what it might mean to produce an empirical demonstration in the early modern period.[4] On the side of absolute resolution, Shapin and Schaffer located Hobbes, whose picture of resolving a dispute in natural philosophy was to invoke deductive arguments from first principles and demonstrations, like Toricelli's mercury tubes, that wore self-evidence on their sleeves. Hobbes opposed what he considered the contrived and isolated laboratory world of the air pump that Boyle had worked so hard to create: a site where qualified gentlemen would leave their philosophical and religious disputes at the door and stand as witnesses to the execution of experiments in air pumps and evacuated chambers. In different ways, both visions (Hobbes's and Boyle's) were partial solutions to the larger problem of dispute resolution in post-Restoration England. While Hobbes thought political and natural philosophical assent had to come through acquiescence to the absolute authority of a political principal or philosophical principle, Boyle and his allies had in mind an assent that would bracket off divisive first principles and resolve issues of "fact." In Gresham House gentlemen could check their Aristotle and Descartes at the door, and politicians could leave their Bibles outside Parliament; in each instance, the chaos of unresolvable conflict could be avoided. Experiments, as they were understood among the followers of Boyle, would resolve phenomenological questions: What happens in an inverted glass jar when an air pump is attached?[5] Much more can be seen in the meaning of Toricelli's tubes and Boyle's pumps than the motion of mercury and feathers.

One way of reading the Hobbes-Boyle dispute is that Boyle won and, with his victory, established a "modern" experimental science and with it "modern" science, full stop (Shapin and Schaffer write in summary, "in this book we have examined the origins of a relationship between our knowledge and our polity that has, in its fundamentals, lasted for three centuries").[6] While nothing in their primary thesis hinges on the permanency of the experimental method as construed in the seventeenth century, the view that the early modern period set up an unchanging notion of experiment appears in a diverse literature. Richard Westfall, who depicted the origins of experimentation very differently from Shapin and Schaffer, concurred about the durability of the new method: "By the end of the 17th century, the scientific revolution had forged an instrument of investigation that it has wielded ever since. A good part of its success lay in developing a method adequate to its needs, and since that time the example of its

4. Shapin and Schaffer, *Leviathan* (1985).
5. Shapin and Schaffer, *Leviathan* (1985).
6. Shapin and Schaffer, *Leviathan* (1985), 343.

success has led to ever broadening areas of imitation."[7] In contrast, I suggest that we see Hobbes, Boyle, Bacon, Galileo, and others as articulating aspects of experimentation that were enormously influential but drop the assumption that these aspects are frozen in the past; I suggest that the dispute over what empirical knowledge ought to be like has continued, productively, for three hundred years, fought largely over the nature of experimentation. Indeed, far from stabilizing, the meaning of "experiment" continues to be challenged, with no end in sight.

Even in the final years of the twentieth century, experimentation and its instruments appear as contested categories: Who counted as the experimenter in the $700 million detector planned for the Superconducting Supercollider, or in the only slightly smaller activities at CERN during the 1990s? Is the activity of the software designer for a subassembly of the detector "experimentation"? Is a doctoral dissertation written about computer simulations of a device that has never been built a professional qualification for an experimental physicist? Is a single observation sufficient to secure an event's existence? Is the scanner who discovers an important event an experimenter? Do statistical demonstrations count as proof of a new effect? As the opening quotation of this book suggests, the physical, temporal, and epistemic boundaries of "experiment" have been and remain in flux.

At one level, this book is an exploration of the tensions that have persistently trailed the changing life of experimentation. It is the cultural meaning of things close at hand that I am after—read in the fine rain of C. T. R. Wilson's cloud chamber, in Robert Hofstadter's heavy gun-mounted spectrometer, in the quasi-real computer space of the simulation, and in the Time Projection Chamber's delicate array of image-making charged coupled sensors. It is amid these intimate bits of machines, data, and interpretations that the categories of experiment and experimenter are embodied: defined, dismantled, and reassembled.

1.2 The Central Metaphor

In 1961, in his book *The Structure of Science,* the philosopher Ernest Nagel insisted that an experimental law has "a life of its own, not contingent on the continued life of any particular theory" that might justify or explain it.[8] By this autonomy, Nagel meant that there was an observational substratum that functioned as the foundation for a hierarchy of levels, passing through bridge principles, and culminating in theory. High theory might offer new connections among observational terms, but there must be a determinate sense to an empirical law that is statable without reference to any particular theory. When Ian Hacking revived

7. Westfall, *Construction* (1977), 116.
8. Nagel, *Structure of Science* (1961), 87.

the slogan "experiments have a life of their own" 20 years later in *Representing and Intervening,*[9] he had in mind something quite different: that there is a momentum and motivation to experimentation that is not enslaved to theory; experiments have goals that differ from merely checking theories (though they certainly do that too). Experimenters do their work to explore new domains of phenomena, to try out new equipment, to gain added precision for physical constants, among other tasks. On the basis of the robustness of experimentation and the objects that it studies, Hacking concluded that though he could not accept a realism about theories—the entities posited in high-level accounts of physics, biology, and astrophysics—he could (contra Bas van Fraassen's revised instrumentalism) plump for a realism about experimental entities.[10]

There is a further resonance and a more basic meaning that I wish to scan in the striking motto "experiments have a life of their own." The life associated with experimentation is not the life affixed to theorizing. Relations to industry, to material objects, and to standards of right reasoning, methods of argumentation, concepts of elegance, uses of heuristics, and forms of apprenticeship are all different. The life of an experimenter involves knowing (for example) the properties, costs, and uses of materials. A young precision bubble chamber experimenter in the 1980s knew not only about heavy-quark decays but about the tough transparent plastic lexane—about its manufacturers, about its optical qualities, about its tensile, shear, and temperature-resistant strength, in short about the many properties that made lexane ideal for canopies on military jet fighters and justified its existence, but made it equally well suited for holding vats of bubble chamber liquid. An emulsion physicist in the 1950s might well have attended conferences about film with industrial workers more interested in photography or colloidal chemistry than pions. A colliding beam experimenter needed to learn not only about rare decays and Higgs searches but about computer and electronic problems associated with an environment of intense radiation. Indeed, the only other workers interested in such environments were electrical engineers, as they prepared radiation hardening for war on a nuclear battlefield—the effects of high bursts of radiation on delicate circuits led into a technical world altogether unknown to the particle theorist.

Even the social skills of the experimenter and theorist come into play differently. As a National Academy of Sciences report summarized in 1972,

> the well-trained high-energy experimentalist is likely to acquire some experience in such diverse disciplines as theoretical physics, electronics, solid state, cryogenics, electrical engineering, computer science, information theory, cost accounting, or even psychology! He has been educated in the search for new and unusual methods for solving a technical problem and in the ways to stretch existing tech-

9. Hacking, *Representing* (1983), 150.
10. Van Fraassen, *Scientific* (1980).

niques to the utmost, as well as getting the most out of available funds. He is disciplined, as required for a large-scale group effort, in all the troublesome aspects of technical teamwork.[11]

With the exception of the bits of theoretical physics and some computer programming, these were skills and disciplines radically different from those of the aspiring theorist, whose skills (in 1972) clustered around group theory, particle phenomenology, and quantum field theory. As the report emphasized: "[I]n contrast to the experimentalist, the theorist must be prepared to work by himself. It is characteristic of theoretical physics that it is a relatively lonely game even though it is often carried out in a large laboratory."[12] The intellectual and social world of the experimenter is different from that of the theorist. Arguments take place in different physical and social settings, with different standards of demonstration.

There are both historiographical and philosophical consequences of taking seriously the distinctness of these subcultures within physics.[13] Historically, the significance of these partially separate lives is that—once one abandons "observation" or "theory" as the basis for a univocal account—no single narrative line can capture the physics of the twentieth century, even within a single specialty. If we follow the practices of instrumentation, we find ourselves crossing disciplinary boundaries at different points than would be suggested by a restriction to theory. Cutting the disciplinary matrix by the subject matter of particles and their theories, the cloud chamber and bubble chamber appear and disappear as contributions to the study of "fundamental" physics. But if we attend to the practices surrounding these instruments per se, we end up tracking Wilson (and his cloud chamber) into meteorology, photography, and steam engines and Alvarez (and his bubble chamber) from nuclear weapons work into high energy

11. National Academy of Sciences, *Physics 2* (1972), 114.

12. National Academy of Sciences, *Physics 2* (1972), 114.

13. My use of the notion of "subculture" in this book has two aims. First, I want to indicate that experimenters, instrument makers, and theorists as groups each have a certain degree of autonomy—to call them different "cultures" or different "forms of life" would, however, suggest that these groups are totally autonomous, fully isolated, and self-contained. To call experiment or theory "subdisciplines" would also miss the emphasis I want to put on the irreducible entanglement of the way physicists want to work with the kinds of instruments they design and the modes of demonstration they employ; "subculture," in the sense I use it, carries technical, symbolic, and social dimensions. It is crucial to my argument that, for all their differences, experimenters and theorists have striven, often against enormous difficulty, to work out specific points of exchange that each judged crucial; where they have endeavored jointly, these various subcultures have constituted a historically specific culture of physics. There is an immense and contentious literature in anthropology revolving around the issue of whether culture does or ought to capture a "total system of human action." This is not the place to pursue the various positions taken on the issue; an entry to the debate may be found in the following sources: Clifford, *Predicament of Culture* (1988); Geertz, *Interpretation of Cultures* (1973); Gupta and Ferguson, "Beyond 'Culture,'" *Cult. Anthro.* 7 (1992): 6–23; Geertz's *Local Knowledge* (1983) and "Culture War," *N.Y. Rev. Books* 42 (1995); Obeyesekere, *Work of Culture* (1990); Sahlins's *Islands of History* (1985), "Goodbye *Tristes*," *J. Mod. Hist.* 65 (1993): 1–25, *How "Natives" Think* (1995), and "Culture Not Disappearing" (1995); Thomas's "Against Ethnography," *Cult. Anthro.* 6 (1991): 306–22, and *Entangled Objects* (1991); Kuper, "Culture," *Man* 29 (1994): 537–54.

physics. Even where the cloud chamber and the bubble chamber overlap, they do so more because they share practices associated with their design, operation, and characteristic demonstration strategies than because they are both used to resolve the same particular puzzle within physics. As a result, the break points of the past, usually given by theoretical breakthroughs and experimental discoveries, occur at different times when viewed from the perspective of changing instrumental techniques.

To the everyday life of the experimenter, the introduction of the cloud chamber or bubble chamber meant both more and less than the full specification of the quark model of the nucleons. The quotidian physics of the laboratory and the theory room often do not move in lockstep. Consequently, our instrumental practice map has different boundaries than maps drawn to capture theoretical techniques or even experimental programs.

There are maps for every occasion—mineral maps, linguistic maps, political maps, religious maps, demographic maps—that these need not coincide in their boundaries is as evident as the latest war. To the historian, the choice of periodization (discontinuity in time) roughly parallels the cartographer's choice of map type (discontinuity in space). And, by and large, inquiries into the conceptual development of physics have taken as given two basic types of mapmaking. On one side is the quest for an entity map, a periodization of the past that defines its continuities and discontinuities around ontology, around the specification of what entities physics says exist. On this map we find prominently displayed the discovery of oxygen, the atom, the electron, the nucleus, the quark, the neutrino. Principally (but not exclusively), mapmaking on these lines highlights the history of experimental programs—for example, the work of Rutherford, leading ultimately to his famous alpha-scattering experiments that made his case for the nucleus. Obviously such accounts cannot be given in the absence of theory—arguments about energy conservation play a crucial role in motivating the search for the neutrino—yet the emphasis in a history of the entities of physics is more on laboratory practice than on theory practice. At the other extreme is a cartography of theory: Think, for example, of the histories of gravitation, from Galileo through Newton to Einstein, and perhaps beyond. Or of the many histories of quantum theory that cite particular results from the laboratory (blackbody radiation, spectra, Davisson-Germer electron diffraction) but focus on a narrative that runs from Bohr and Sommerfeld through Schrödinger and Heisenberg to the statistical interpretation by Born. Both entity maps and theory maps have had their heyday in the development of the history and philosophy of science; these choices have shaped how we understand instruments and experiments.

In part because of its association with a positivist philosophy of science that took "observation" or "experience" to be the source of all secure knowledge, the history of physics went through what Gerald Holton has called an "experimenticist" stage in which entity maps had priority: an interpretation of history that depicted the development of the discipline as proceeding along inductive, observation-based lines.[14] In this view, the Michelson-Morley experiments were the culmination of a sequence of laboratory ventures designed to measure the earth's motion in the ether, and special relativity takes the historical stage as the inevitable climax of a long string of experimental results. As philosophy (or, more precisely, as a tendency within a popularized version of logical positivism), experimenticism certainly made history simple: instruments were tied to experimental programs, and experimental programs were ordered into sequences capped (often only provisionally) by the construction of a theory. The central metaphor was the solid, aggregative observational "basis" upon which the more delicate structures of theory were necessarily built.

Beginning in some sense with Quine's holism in the 1930s, and burgeoning in the 1960s, the philosophical reaction against the logical empiricists' obsession with observation came hard. Paul Feyerabend, Russell Hanson, Mary Hesse, Thomas Kuhn, and others sought to stand the positivist program on its head: instead of starting with observation and treating theory as its superstructure, the antipositivists took theory as primary and observation as subordinate. In many ways the antipositivists—using a theory map this time—successfully countered the move to advance a protocol or neutral language of observation that would stand outside of all theories and adjudicate among them. No such neutral ground, they argued, could exist.[15]

Both logical positivists and antipositivists had an overriding concern with observation, and for good cause. Observation, the use of the senses to gain knowledge, was the mainstay not just of the Vienna Circle's brand of positivism but of a longer and deeper tradition with roots grasping both Comtian positivism and British empiricism. The philosophical stakes were as high as the old debate over the relation between knowledge gained through the senses and knowledge present already in the mind or shaped through categories already in the mind. Positivist and antipositivist therefore shared an abiding interest in the nature of perception, for example in the dynamics and lessons of Gestalt psychology.[16] Their conclusions, however, went in different directions.

From the argument against an observation-based, theory-free "protocol language" the antipositivists concluded that a break in theory, in the objects and

14. Holton, *Thematic* (1988), 293–98.

15. Kuhn, *Scientific Revolutions* (1970); Feyerabend, *Realism* (1981), chaps. 2–7; see also Feyerabend's *Free Society* (1978), 65ff.; Hesse, "Theory" (1974); and Hanson, *Positron* (1963).

16. E.g., Carnap, in his *Logical Structure* (1967), sec. 67, writes: "Modern psychological research has confirmed more and more that, in the various sense modalities, the total impression is epistemically primary, and that the so-called individual sensations are derived only through abstractions, even though one says afterward that the perception is 'composed' of them: the chord is more fundamental than the individual tones, the impression of the total visual field is more fundamental than the details in it, and again the individual shapes in the visual field are more fundamental than the colored visual places, out of which they are 'composed.'" Kuhn, in *Scientific Revolutions* (1970), uses the Gestalt shift as a representation of the changing image of observations under different conceptual schemes.

laws, would precipitate a disruption all the way through the entire practice of a science. *Philosophically* (antipositivists contended), this rupture undermined the possibility of a neutral sideline from which to view theory change: there could be no Archimedean point from which to view two sides of an epistemic break. *Historically,* the total rupture suggested that interest lay more in the dynamics of theory change because theory thoroughly conditioned the possibility of observation. This historico-philosophical picture depended on what I will call a "block periodization" in which the practice of physics cleaves all the way through from a specific fracture at a single moment of history. If the history of science is block periodized, there is no stepping outside the fray to "pure observation"—only a brute choice remains. Take observation-plus-theory within Ptolemy's geocentrism or observation-plus-theory within Copernicus's heliocentrism. No neutral territory can survive between the paradigms of Newton's absolute space and Einstein's relativistic spacetime, nor between the paradigms of Priestley's phlogiston and Lavoisier's oxygen. Without doubt, this conception of "theory contaminated" or "theory laden" observation was a fruitful counterweight to the dream of a sense-data language that can exist without theory. One central metaphor had replaced another.

The appeal of this picture of science as a collection of island empires, each under the rule of its own system of validation, can be measured by its ubiquity, even in the work of historians, philosophers, and sociologists whose views are otherwise distinct from or opposed to one another's. As will be discussed more completely in chapter 9, Carnap's "frameworks," Quine's "conceptual schemes," Lakatos's "programmes," Kuhn's "paradigms" (among others) all performed a crucial task in the 1960s: they stood against the view that knowledge can be obtained and secured piecemeal, from morsels of observation. More recently, in the 1970s and 1980s, some sociologists of science such as Barry Barnes, Harry Collins, Trevor Pinch, and Andrew Pickering adopted the Kuhnian picture of incommensurable paradigms while according it an underlying sociological mechanism to explain why the paradigm is adopted, sustained, and eventually dismantled. As Collins and Pinch put it: "The special quality of revolutionary transitions arises out of the all-inclusive quality of paradigm commitment." Members of different communities (Collins and Pinch say elsewhere) frequently hold to different paradigms. Quoting Albert Weimer approvingly they conclude: "[T]here can be no 'simultaneous' research in conflicting paradigms: once the world is seen from a new point of view, it will take the 'Gestalt Switch' to retrieve the old perspective and like the alternative perceptions of the Necker cube, they cannot both be had at once."[17] Expressing the same commitment to

17. Collins and Pinch, *Frames* (1982), 160, 155. Collins and Pinch conclude their study of the "psychokinetic revolution" ("revolution" in the Kuhnian sense): "What has been demonstrated, with specific reference to science, is that the idea of paradigm incommensurability, even when interpreted in its most radical Winchian/Wittgensteinian sense, has a part to play in the understanding of contemporary science" (1984–85). As Collins

the totality of the paradigm shift, Barnes contended that "[i]n actual historical cases, as a new paradigm arises so there is an associated transformation of the entire conceptual fabric. What has to be evaluated are two alternative frameworks of discourse and activity. There is a reconstruction of the whole pattern of both."[18] Pickering too specifically invoked Kuhn to tie experiments and instruments to the conceptual schemes that ruled over them, stressing the "flexibility" of the empirical as it conformed to theories that came first: "[T]he possibility of such tuning [of instrumental practices to accord with conceptualizations] implies a flexibility of the empirical base of science beyond that recognized in the philosophical notion of the theory-laden nature of observation languages; a flexibility which itself implies the possibility of a radical incommensurability between the 'natural worlds' of societies subscribing to different conceptual frameworks—each society tuning its instrumental practice to phenomena appropriate to its conceptual framework."[19] In such block-periodized accounts, the historical picture of "all-inclusive" breaks underpinned one particular variant of scientific relativism grounded in incommensurable "ways of life" or paradigms that pass each other like ships in the night.

My own sense is that the assumption of block periodization is not needed for any of the more recent projects in the study of science. Social constructivists could drop the picture of the *coperiodization* of theoretical and laboratory practices, and the block relativism that is supposed to follow from it; they could nonetheless maintain their productive vision of scientific work and standards as the assembly of a wide range of cultural resources. Philosophers could hold to the idea that conceptual schemes, frameworks, and paradigms might be useful ways of capturing the different ways of speaking about, for example, electrons among Maxwellians or electrons among Lorentzians or electrons among late-nineteenth-century experimenters. We need not, in short, careen between the view that talk about physics is composed of atomistic bits of language and the view that scientific language only comes in whole worldviews.[20] The picture presented here is that the language and practice of experimentation, instrumentation, and theory are distinct, but linked—and in interesting ways. Above all,

and Pinch see it, their sociological application of Kuhn applies to science insofar as "a universal rationality" fails to provide "bridges" between paradigms. "If evidence of radical cultural discontinuities can be found in contemporary science, and that is the claim of this work, then the scope of the sociology of knowledge is very wide indeed." (Their quotation of Weimer is from Weimer, "History of Psychology," *Sci. Stud.* 4 (1974): 367–96, on 382.) Similarly, Pickering, summarizing his fine book *Quarks* (1984), argues, "I have analyzed the emergence of the new physics against the background of the old, and manifestations of incommensurability are stamped across the transition between the two regimes. The old and new physics constituted, in Kuhn's sense, distinct and disjoint worlds" (409). He goes on to specify that the choice of theories "required a simultaneous choice between the different [experimental] interpretive practices . . . which determined the existence or nonexistence" of an important microphysical phenomenon (the neutral current).

18. See Barnes, *Kuhn* (1982), 67.

19. Pickering, "Monopole," *Soc. Stud. Sci.* 11 (1981): 63–93, on 88–89.

20. For a sustained attack on holism, see Fodor and Lepore, *Holism* (1992).

we do not need to assume that experimental, theoretical, and instrumental practices all change of a piece.

1.3 Multiple Constraints, Simultaneous Solutions

Instead of depicting the practices of instrumentation, experimentation, and theory as changing synchronously, I want to leave open the possibility that each has its own tempo and dynamics of change. Put in short form: *The periodizing breaks of the various subcultures of physics are intercalated, not necessarily coincident.* Breaks in, say, theoretical practice may occur during a time of continuity in instrumental practice. Ruptures in experimental practice—such as the introduction of the cloud chamber—need not coincide with changes in theories of microphysics. In a broad sense, this book is written toward two ends: First, to make a historical and philosophical place for these partly autonomous subcultures—in particular, to follow the ways in which the material culture of the laboratory moves through different branches of physics, indeed far outside of physics, often without the direct guidance of particular theoretical agenda. Second, I want to address the problem of how these various subcultures do, in fact, interact; how, seen through historical specificity, do the subcultures form anything worth calling a culture as a whole?

In order to characterize the partial autonomy of instrument making, theorizing, and experimenting, we need a way to characterize their internal structures. As a start, we can ask where the difficulties lie: Where and when for a given subculture of physics is it difficult to proceed? Suppose you are trying to design a detector for a major particle physics experiment. What constraints bind you? In 1984, much of the American high energy physics community was preoccupied with possible detector designs for the Superconducting Supercollider. One study group listed its reasons for moving the magnet (which would deflect particles and so help fix their energies) outside the calorimeter (which measured the energy of showers of charged particles). The group's first reason suggests that the constraints of physics were the only bounds within which it had to work: "[Moving the magnet would lead to] no deterioration of calorimeter performance with respect to energy resolution, shower spreading or, especially, hermeticity." Taken in isolation, this reasoning indicates that the basic problem set facing the experimenters was internal: the decision to move the magnet was made because of technical demands issuing from heavy-quark decays—precision of energy measurements, the ability to seize all the particles generated in a cascade (shower) of interactions, and the need to account for all the particles generated in the interaction. But another set of constraints from the same list dealt with the cost and availability of uranium within the whole network of the nuclear, industrial, and military cycle of production and consumption. This argument, taken alone, implies the opposite of the first conclusion: the detector design followed from external socioeconomic considerations. A third constraint issued from the domain of engineering: the structural arrangement of components could be greatly simplified by moving the magnet out; here is grist for the technological determinist's mill. But none of these reductive schemes—*internalism* (design for physics reasons alone), *externalism* (design for socioeconomic reasons alone), or *technological determinism* (design for engineering reasons alone)—captures the physicists' routine experience of working within a multiplicity of constraints, any one of which can be violated, but at a cost. If the energy resolution of the detector was inadequate, the resulting ambiguity had to be resolved elsewhere and with difficulty; if the cost of uranium rose, the number of channels would have to change, or some other compensatory budget cutting would have to be endured.[21]

Experimenters, then, frequently find themselves in the position of the mathematician who is searching for a curve that satisfies a number of different constraints—only here the constraints are not all within the domain of science (narrowly construed). And as in mathematics, it is not unusual to find that more than one curve passes through the filters in place. That more than one solution is possible rules out blind determinism (or historical functionalism): it is not the case that the existing constraints always and in general fix a laboratory move unequivocally. But to push the mathematical analogy further, it is by no means guaranteed in advance that *any* curve will satisfy all the imposed demands simultaneously. For example, as we will see in some detail in chapter 5, Donald Glaser, the inventor of the bubble chamber, desperately sought a way to make his new detector both an image device (producing pictures of high resolution) and a logic device, "triggerable" in just those cases where the particles had certain properties (energies, angles, etc.). Cloud chambers had been triggerable for 20 years. But despite the efforts of tens of groups all over the world over several decades, no one could combine the electronics and the hydrodynamics to make such a triggerable image device: the constraints of the properties of liquids, heat, and technology made it impossible. The particular intraexperimental constraints associated with triggering were simply not compatible with the constraints associated with the practice of building bubble chambers.

This possibility of no solution is of particular importance when theorists

21. For more on this example, see Galison, "Philosophy," *J. Phil.* 185 (1988): 525–27. In this respect, I am sympathetic to the effort Michel Callon and Bruno Latour have made to stress the heterogeneity of elements constitutive of the production of scientific work. See their accounts of scallop harvesting (Callon, "Sociology of Translation" [1986]) and of the consensus formation around the work and name of Pasteur (Latour, *Pasteurization* [1988]). As Jardine (*Scenes of Inquiry* [1991], 185ff., esp. n. 10) aptly points out, there is a significant correspondence between the use I have made of "long-term constraints" and Latour's use of "givens." The difference in emphasis between my picture of multiple constraints and the Callon-Latour actor/network theory is perhaps this: Their work emphasizes the effectiveness and elasticity of the leveraged power that comes of creating alliances and enlisting other groups toward a predetermined end. My work tends to focus on the material (and nonmaterial) obstacles that shape and delimit action in the sphere of science over time.

and experimenters meet. For it is when these two very different subcultures clash that one sees the most dramatic failure of the central metaphor of block periodization: physics as divided into self-validating island empires.

Theorists have their own constraints, typically different from those of instrument makers or experimenters. Some of these may be frankly practical. A theory with predictions that cannot be calculated is of little help, and calculability has, for some time, been a constraint on theory building. In different epochs such constraints have included the restriction of theoretical forms to the linear, the analytic, the covariant, the natural, or the renormalizable. Some theoretical constraints may be aesthetic—restriction of representation to that which can be expressed in compact, symmetric, or homogeneous symbolic form. But whatever their origins, constraints shape the theorist's positive research program. Intratheoretical constraints, like intraexperimental constraints, can in different circumstances allow multiple, single, or no solutions. For example, the original hope for a grand unified theory (GUT) was that such an account of the strong, weak, and electromagnetic forces would satisfy a mathematical constraint (involve only a simple gauge group), a theoretical constraint (be renormalizable), and various phenomenological constraints (e.g., be consistent with the measured weak-mixing angle—roughly the ratio of electric and weak force strengths—and predict a measurable proton decay). True, it was (logically) possible to weaken or modify these restrictions and carry on. But the minor theoretical industry that grew up around GUTs in the 1970s collapsed in the 1980s as it became evident that the concurrent satisfaction of these clear constraints left no path free and clear.

Two points about constraints are important to note here. First, the constraints operating within the theoretical and experimental subcultures are quasi-autonomous, and not *absolutely* separate. It is impossible even to formulate many of the experimental questions being asked without many aspects of theory. Conversely, theory can depend deeply on aspects of experimental knowledge—even string theory, the most rarified of physical theories begins with some basic assumptions about the nature of gravity. That said, the constraints of theory and experiment are often distinct enough that they can clash—to the point where one or the other has to be given up. My favorite instance of such incompatibility came in the early 1970s when a leading experimenter, after years of effort devoted to showing that weak neutral currents did not exist, came up hard against the results of his own experiment. Laboratory constraints, issuing from a variety of hardware moves, data analysis, simulations, and calculations, left him no way to maintain the older theory. He wrote to his colleagues: "I don't see how to make these effects go away."

Second, theoretical (and experimental) constraints play a constructive as well as a restrictive role. Because constraints restrict moves, they shape the theorist's positive research program—giving a problem-domain form, structure, and direction. In the following remarks, Steven Weinberg describes his own reaction to renormalizability, the metatheoretical constraint that requires a theory to make predictions to all orders of accuracy with a fixed and finite set of parameters as input:

> I learned about renormalization theory as a graduate student, mostly by reading Dyson's papers. From the beginning it seemed to me to be a wonderful thing that very few quantum field theories are renormalizable. Limitations of this sort are, after all, what we most *want;* not mathematical methods which can make sense out of an infinite variety of physically irrelevant theories, but methods which carry constraints, because these constraints may point the way toward the one true theory. In particular, I was very impressed by the fact that quantum electrodynamics could in a sense be *derived* from symmetry principles and the constraint of renormalizability; the only Lorentz invariant and gauge invariant renormalizable Lagrangian for photons and electrons is precisely the original Dirac Lagrangian of [quantum electrodynamics].[22]

Because constraints are used by physicists to construct theories, devices, and experiments ("what we most want"), it is crucial *not* to see constraints as purely negative. Indeed, theoretical constraints are often inscribed into notational rules themselves. Rules about matching indices in general relativity render virtually automatic the preservation of relativistic invariance, for example. Other more subtle notational devices serve to police the conservation of a host of quantities, from probability to electric charge or angular momentum. Many of the more abstract quantities conserved in the theory of particular interactions (isospin, charm) are a long way from being directly measurable. While these boundary signs (e.g., no laws of strong interaction without conservation of isospin) may say "ne plus ultra" to a speculative line of thought, they also function positively to suggest solutions to particular problems.

Block periodization obscures the diversity of these theoretical practices, their distinctness from the myriad of obstacles that experimenters or instrument makers meet in their fields of action. But if on historical grounds it is hard to accept block periodization schemes, one can understand their seductiveness on more abstract philosophical grounds. Each of many block representations (neo-Kantian *Weltanschauungen,* Carnapian frameworks, Kuhnian paradigms, Lakatosian programs) divvied science into conceptual schemes. Many authors within science studies continue to exploit such strategies because they believe block representation challenges a naive progressivism in which each new theory in science subsumes and explains the prior one. In more recent times, the island empire view has found a home among those who use it as the foundation for an attack on metaphysical realism and a defense of one species of relativism. As Trevor Pinch put it, Barry Barnes, David Bloor, and Harry Collins "all advocate

22. Weinberg, "Conceptual," *Rev. Mod. Phys.* 52 (1980): 515–23, on 517.

some form of relativism" based on the notion of "paradigm incommensurability"; all are engaged in what Pinch calls study of a "radical Kuhnian" type.[23] How (this argument for relativism goes) can science have any purchase on "truth" or "the world" if each knowledge block (the symbiotic, self-reinforcing amalgam of theory, experiment, observation, and instrument) is unbridgeably isolated from and incommensurable with its forerunner and successor?

In the world of positivism, theories are demoted to mere summaries of observations. In the world of antipositivism, the reverse is true. Referring to instruments, Gaston Bachelard's remarks in *Epistémologie* have become a rallying cry against the autonomy of instrumentation and for the all-pervasiveness of theory: "An instrument, in modern science, is truly a reified theorem." Bachelard goes on to say that hypotheses and instruments are thoroughly coordinated. An apparatus like Millikan's oil drop experiment or Stern and Gerlach's space quantization experiment is, on Bachelard's reading, "conceived directly as a function of the electron or the atom."[24] Sweeping statements like these trade on a lack of precision about the relation of theory to experiment. While it is true that Millikan designed his oil drop experiment in order to measure the charge of the electron (he was already committed, as Holton has shown,[25] to the discreteness of electric charge), his device was, at root, a cloud chamber, variants of which had been and subsequently were used (by Ehrenhaft, among others) to prove that electrons did not carry quantized charge. Indeed, the cloud chamber, as we will see in the next chapter, cut across a spectacular variety of views about particulate matter, and across wide swaths of natural science more generally. It was a piece of machinery built on the Victorian obsessions with dust, clouds, rain, and some form of particulate Cantabrigian ions—the idea of an ion enabled the Wilson cloud chamber. But Wilson's chamber was certainly not a reified theory of the ion or electron. Conceptual scheme talk homogenizes the practices of theory and the practices of experiment and exaggerates the determination of instrument and experiment by theory. Instead, for both philosophical and historical reasons, we need a picture of physics that can capture the *partial* autonomy of laboratory practices from the details of theoretical commitments without making either an absolute basis for the other.

We will proceed here without the prior assumption that physics divides

23. Pinch, " Conservative and Radical," *4S Newsletter* 7, no. 1 (1982): 10–25, on 10–11; Pinch also includes Cox under the rubric "radical Kuhnian" and notes that all invoke incommensurability between paradigms as a reason to eschew ranking work on either side of a break as better or worse.

24. See Bachelard, *Epistémologie* (1971), 137–38. Bachelard's remarks here are actually more subtle (and interesting) than the circularity argument to which the "reified theory" comments have been appended. After the material cited, Bachelard goes on to criticize Vaihinger for the fictionalism of his *"Philosophie des als ob,"* which ignored, e.g., the material practices that embodied atomism. Therefore, at least one plausible reading of Bachelard would make him not an advocate of the self-referentiality of theory, in virtue of the theory-ladenness of apparatus, but rather a gently materialist opponent of a certain stripe of neo-Kantian idealism.

25. Holton, "Subelectrons" (1978).

into self-contained and self-stabilizing blocks. Instead of presupposing that the subcultures of physics are coperiodized, we will take their relation to one another as a contingent matter to be explored in the specific historical situation in which each is embedded.[26] In fact, far from anticipating that coperiodization is the rule, we might well expect the histories of the different subcultures each to have their own breaks, with the breaks coinciding only rarely. This intercalated periodization would depict the history of the discipline as a whole as an irregular stone fence or rough brick wall rather than as adjacent columns of stacked bricks. And, just as the offsets between joints in a brick wall give the wall much of its strength, it is this intercalation of diverse sets of practices (instrument making, experimenting, and theorizing) that accords physics its sense of continuity as a whole, even while deep breaks occur in each subculture separately considered.

1.4 Image and Logic

Instead of constructing a theory-dominated account, or collecting stories of isolated experimental discoveries, the goal here is to demonstrate the deep continuity of experimental practice through an analysis of the instruments of modern physics. I will follow two competing traditions of instrument making reflected in this book's title (*image* and *logic*). One tradition has had as its goal the representation of natural processes in all their fullness and complexity—the production of images of such clarity that a single picture can serve as evidence for a new entity or effect. These images are presented, and defended, as *mimetic*—they purport to preserve the form of things as they occur in the world. Particles make tracks of bubbles in superheated hydrogen, or water drops in supersaturated vapor, or chemically altered emulsions on photographic plates, and these records ipso facto recreate the very form of invisible nature. Because this ideal of representation relies on the mimetic preservation of form, I will call it "homomorphic."

Against this mimetic tradition, I want to juxtapose what I have called the "logic tradition," which has used electronic counters coupled in electronic logic circuits. These counting (rather than picturing) machines aggregate masses of data to make statistical arguments for the existence of a particle or effect. The logic tradition gives up, or in some cases explicitly rejects, the sharp focus its image-making competitors bring to bear on individual occurrences. In its place, the logical relations between certain circumstances are determined: the particle did not penetrate iron plate 3 but did pass through iron plates 4, 5, and 6. Because this statistical mode of registration preserves the logical relation among events, I will call it "homologous" representation.

26. The term "embedded" is used differently by other authors; cf. Wise, "Mediating Machines," *Sci. Con.* 2 (1988): 77–113.

The mimetic representation is well girded against the charge that something is missing from its account but vulnerable to the accusation that it has located a fluke or oddity. The statistical approach, by contrast, consciously sacrifices the detail of the one for the stability of the many. In many instances, both approaches can yield the same information, for example the disclosure of a new particle. In such cases, physicists invoke what might be called a metaphysical version of the ergodic theorem: information about a single event rendered with full detail is in all relevant ways equivalent to information deduced from partial details about many events of the same class. (In statistical mechanics, studying the behavior of one bucket of gas for a long time is equivalent to examining many identical buckets for shorter periods of time.) If you have seen one anti-muon-neutrino, you have seen them all; if you have seen bits and pieces of a million lambdas, you can know all about any given one. Abstractly, then, the image tradition (with its homomorphic representations) evolves in tension with the logic tradition (with its homologous representations). Concretely, I will focus on instruments designed to pursue the microphysical entities of physics: cloud chambers, nuclear emulsions, and bubble chambers on the image side; counters (Geiger-Müller, Cerenkov, and scintillation), spark chambers, and wire chambers on the logic side; and electronically produced images, an image-logic hybrid (see figure 1.1):

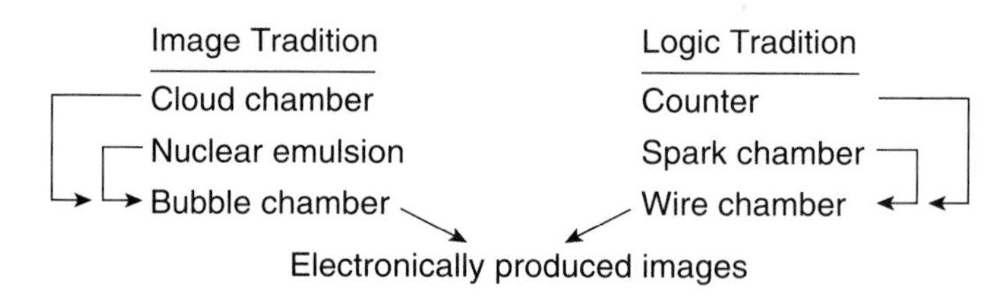

Figure 1.1 Image and logic—schematic.

When the cloud chamber was first turned to the world of particles in the early twentieth century, it was hailed as one of the great instruments of all time. The physicist E. N. Andrade proclaimed in 1923 that "the future historian of science will probably record as one of the outstanding features of the period in which we live the ability, so rapidly acquired during the last decade or so, to deal with single atoms and small numbers of atoms."[27] First and foremost, this ability was exemplified by Wilson's chamber and the new electronic counters capable of registering the passage of individual charged particles. When electronic counters came into their own in the late 1920s and early 1930s, they seemed ready to take over from the image tradition; the cloud chamber and its image tradition

27. Andrade, *Atom* (1923), 285. Andrade counts Aston's mass spectroscope among the devices able to manipulate small numbers of atoms.

offspring, the nuclear emulsion, staged a comeback in the early postwar years, only to be threatened by the innovative electronics produced by the Manhattan and Radar Projects. In the 1950s and 1960s, Glaser and Alvarez not only revived the image tradition with the massive hydrogen bubble chamber; they made it dominant. The image tradition seemed to have entered the age of industrial-scale research and triumphed—this time for good. But the logic tradition had another card to play. Exploiting the microelectronic revolution of the late 1960s and early 1970s, counter physicists introduced a host of new electronic detectors with such stunning resolving power that they began to approach the persuasive clarity of the image tradition, leading some experimenters to dub various devices "electronic bubble chambers."

The new instruments (drift chambers and time projection chambers coupled to powerful computers) had begun to meld together the data-sorting capabilities of the logic tradition with the singular detail and inclusiveness of the image tradition. So it was that in the early 1980s the two traditions fused, with the production of electronically generated, computer-synthesized images. It was just such an electronic "photograph" that heralded the discovery of the W and Z particles in 1983—the first time a single electronic detection of an event had ever been presented to the wider physics community as compelling evidence in and of itself.[28]

In arguing for the endurance of the traditions designated in figure 1.1, I have in mind three levels of constantly reinforced continuity. First, there is *pedagogical continuity:* Not long after C. T. R. Wilson built his first cloud chamber, Robert Millikan learned the technique by studying Wilson's work and used a variant of it to perform his oil drop experiment. Millikan taught the technique to future cloud chamber master Carl Anderson; Anderson then instructed Donald Glaser, the inventor of the bubble chamber. Not surprisingly, many of Glaser's students went on to work in the enormous bubble chamber groups of the 1960s. More generally, one can follow "family trees" of students on each side of the image and logic divide.

Second, there is *technical continuity.* The day-to-day laboratory skills continue from cloud chamber to bubble chamber, from cloud chamber to emulsion, and from emulsion to bubble chamber. All three involve track analysis, photographic skills, and micrometry. Similarly, on the logic side is an unbroken cluster of electronic skills, including the use of high voltages, logic circuit design, and gas discharge physics. Once we recognize these (interrelated) continuities of pedagogy and technique, it becomes less surprising to find in the image tradition, for example, a significant flow of "raw" data in the form of track photographs from bubble chamber collaborations to emulsion groups in the early 1960s. In contrast, skills did not transfer easily across the image-logic divide.

28. Arnison et al., "Experimental," *Phys. Lett. B* 122 (1983): 103–16.

Both individuals and groups of physicists found it difficult to drop cloud chamber or bubble chamber work and to take up the soldering iron, vacuum tube, or circuit board. The difference between the two skill clusters is marked, as the following report from the early 1970s made clear: "If it is a counter spark chamber group with which he [a Ph.D. student] is associated, he may help in the design and development of equipment, and he may participate in computer interfacing and software development. In a bubble chamber group, he may spend a great deal of his effort on developing programs or procedures for sorting out the information contained in the pictures."[29] Because the writing of analysis programs and picture analysis had so little to do with breadboard circuit fabrication, crossover between the two traditions faced a hurdle by the 1930s and a solid wall by the mid-1960s.

Finally, there is what I will call *demonstrative* or *epistemic continuity*—continuity within each tradition in the characteristic form of argumentation. On the image side resides a deep-seated commitment to the production of the "golden event": the single picture of such clarity and distinctness that it commands acceptance. To name but a few such instances: Anderson's picture of the positron in 1932, Alvarez's and his collaborators' shots of the cascade zero and their muon-catalyzed fusion in the 1950s, Brookhaven's capture of the omega minus in the 1960s, and the Gargamelle (bubble chamber) team's picture of the single-electron neutral-current event in the 1970s. The last example is perhaps the best illustration of the demonstrative force that a single well-structured picture can carry. Despite years of statistical data accumulated by some members of this big collaboration, it was this picture, projected on lecture hall screens, reproduced in journals and textbooks, and literally carried around the world, that swayed many physicists into believing, for the first time, in the physical reality of neutral currents.[30] Not every significant result in the image tradition has been a golden event. Many rested on the compilation of thousands of carefully scanned pictures. Nonetheless, the golden event represented a characteristic form of demonstration entirely unavailable to the logic tradition (until the 1980s). As such it was an emblem, a banner of the validity of the approach—if an individual pho-

29. National Academy of Sciences, *Physics 2* (1972), 114.

30. On the positioning of this single event within the demonstration structure of Gargamelle's case for neutral currents, see Galison, *Experiments* (1987). Latour rightly emphasizes the important rhetorical power that accrues to the image makers by virtue of the simultaneously stable and transportable character of photographs (his "immutable mobiles," Latour, "Visualization," *Know. Soc.* 6 (1986): 1–40); along related lines, see Lynch, "Discipline," *Soc. Stud. Sci.* 15 (1985): 37–66. As I will stress throughout this book, however, visualization is a contested form of demonstration as well as a contested form of laboratory work. Within Gargamelle, e.g., many members were *not* persuaded by the single electron picture; a similar division can be found within the American experimental group that competed with Gargamelle. More generally (see, e.g., chapter 5 below) there was—and remains—an epistemic and work organizational vision of experiment that refuses pictures in favor of statistical, formal, and other nonvisual strategies of demonstration.

tograph could be waved in front of a packed lecture hall, who could doubt the force of a demonstration built on 100,000 such pictures?

The golden event was the exemplar of the image tradition: an individual instance so complete, so well defined, so "manifestly" free of distortion and background that no further data had to be invoked. (We will need to explore in detail how this transparency was created.) Because of the extraordinary difficulty of obtaining a perfect photograph, the harvest of cloud chamber and emulsion (or even early bubble chamber) photographs was meager. It was simply impossible to gather the tens or even hundreds of thousands of instances amassed overnight by a Geiger-Müller counter. But when it was feasible (in the 1950s and 1960s) to produce statistical arguments with enormous collections of bubble chamber photographs, the statistical argument built (so its defenders could claim) on a foundation in which individual pictures could be examined one by one.

In contrast, the logic tradition relied fundamentally on statistical demonstrations. The typical logic argument is illustrated by the 1930s cosmic ray experiments on the penetration of matter by charged particles: one Geiger-Müller counter was placed above a gold brick and one below. The experimenters then determined the particle penetration of the gold by counting the number of coincident firings of the two counters. Since there will always be some random coincident firings of the counters due to "accidental" arrivals of different particles (e.g., members of a shower of cosmic ray electrons), the experiment relied *intrinsically and inescapably* on the excess of joint firings over the calculated accidental rate. Any single coincident firing of the two counters meant nothing. Indeed, it is because these electronic selection mechanisms are based on logical relations (and/ or/ if/ not/ then/) that I have chosen to name the tradition as I have.

Through such logical relations (embodied by electronic sorting) other claims were made about the microworld, ones altogether as far-reaching as the famous golden events. During the late 1930s, one path to the muon (by Curry Street and E. C. Stevenson) was principally made through coincidence and anticoincidence counters—these showed, statistically, that there exists a particle with energy and charge equal to those of an electron that penetrates much farther through lead than an "ordinary" electron does. In 1953, Frederick Reines and C. L. Cowan similarly based their experimental argument for the free neutrino entirely on statistical measurements from a sophisticated array of counters wired in coincidence and anticoincidence around a large vat of liquid scintillator. While Reines and Cowan took their particles from a reactor, other logic tradition experimenters drew theirs from accelerators. Simultaneously competing against and collaborating with image tradition work, Owen Chamberlain, Emilio Segrè, and their Bevatron colleagues at what would become Lawrence Berkeley

Laboratory (LBL) deployed Cerenkov counters and scintillation counters to hunt down the antiproton using precision timing and coincidence circuits (1955). While striking images of the antiproton followed shortly thereafter, in the first instance counter-generated statistical significance—not image-produced golden events—bolstered the discovery claim. Even in the late 1950s, small-scale logic tradition experimental equipment still had a role to play: Chien-Shiung Wu and her collaborators conducted a series of experiments leading, in 1957, to the experimental case for parity violation. They too worked with a completely nonvisual apparatus, one that made fundamental use of electron counters measuring the angular distribution of beta particles from oriented nuclei. If parity was violated the rate of particles emitted in some direction θ (relative to the polarization of the nuclei) would differ from the rate in the direction $180^\circ - \theta$. Wu detected that asymmetry, and parity fell. One can continue in this vein, for even the shortest of lists of logic tradition triumphs would include James Cronin and Val Fitch's measurement of charge-parity (CP) violation, Mel Schwartz, Leon Lederman, and Jack Steinberger's 1962 spark chamber discovery of a second species of neutrino, and the 1970s Harvard-Wisconsin-Pennsylvania-Fermilab spark chamber measurement of the weak neutral current.[31]

Image machines and logic machines—each had their victories, their Nobel Prize–winning successes. At the same time, neither tradition was able to claim a privileged hold on truth for any length of time, and neither held unique sway (for long) over the physics community. Each found its own form of argumentation persuasive and judged the competition to be faulty in certain respects: image physicists argued that logic physicists could be fooled because they missed essential details of the physical process. Only through the vividness of an image could the causal elements be traced all the way through a process. Ceding the force of such track-by-track arguments, logic physicists nonetheless remained dubious about demonstrations predicated on a handful of events. By the 1960s, any logic experimenter could point to a myriad of examples of unrepeated (unrepeatable?) prints that had beckoned experimenters and theorists alike down blind alleys. In the words of one experimenter in the logic tradition (about a singular picture he had taken), "[A]nything can happen once."[32]

31. These issues will be discussed in detail and with further references in chapter 6. On the muon, see Galison, *Experiments* (1987), chap. 3. On Reines and Cowan's discovery of the neutrino, see Reines and Cowan, "Free Neutrino," *Phys. Rev.* 92 (1953): 830–31; and Seidel, "Hunting of the Neutrino" (1996). On the antiproton, see Chamberlain et al., "Observation," *Phys. Rev.* 100 (1955): 947–50; Chamberlain, "Discovery" (1989); and Piccioni, "Antiproton Discovery" (1989). On parity violation, the original article is Wu et al., "Parity Conservation," *Phys. Rev.* 105 (1957): 1413–15; see also Franklin, "Discovery and Nondiscovery," *Stud. Hist. Phil. Sci.* 10 (1979): 201–57. On CP violation, the original article is Christenson et al., "2π Decay," *Phys. Rev. Lett.* 13 (1964): 138–40; see also Cronin, "CP Symmetry Violation" (1981); Fitch, "Discovery" (1981); Franklin, "Discovery and Acceptance," *Hist. Stud. Phys. Sci.* 13 (1983): 207–38; Danby et al., "High-Energy Neutrino," *Phys. Rev. Lett.* 9 (1962): 36–44; A. Benvenuti et al., "Observation of Muonless," *Phys. Rev. Lett.* 32 (1974): 800–803; and Aubert et al., "Further Observation of Muonless," *Phys. Rev. Lett.* 32 (1974): 1454–57.

32. See Galison, *Experiments* (1987), 123.

To the image experimenters, the passivity of their systems of registration was a virtue: theoretical presuppositions would not enter if the emulsion dutifully recorded all that passed through it and the unprejudiced eye of the camera array flashed away at every bubble chamber expansion. For these experimenters, there was always something suspect about the logic tradition's highly selective "cuts" of the data before recording ever took place. Could one really know, after the fact, whether these selections had avoided the structuring of data beyond that which was intrinsic? By contrast, the logic experimenters professed horror at their competitors' passivity. As the logic physicists saw it, the image physicists had given up being experimenters when they removed themselves from the real-time manipulation of the apparatus. How else could one truly know what was happening? How could one eliminate the spurious without the ability to get results, modify the setup, and promptly run the experiment again? At every level the traditions clashed: golden event versus statistical demonstration, the objectivity of passive registration versus the persuasiveness of experimental control, vision versus numbers, and photography versus electronics.

One way to think of the clash of image and logic traditions is through the epistemic distinction between what Hartry Field has called "world→head reliability" and "head→world reliability."[33] Perfect world→head reliability obtains when we have a certain inner state only if a particular external situation holds. So if there is a rock in the world in front of me, I know about it. By contrast, perfect head→world reliability holds when it is the case that if I think there is a rock, there is. When a bubble chamber is working properly it can be seen to aim at world→head reliability: by imprinting on film every track that crosses a sensitive volume, the image device, ideally at least, ensures us awareness of every event. The nightmare of overselection is avoided. But looked at from the logic tradition, this ecumenical registration is problematic. Consider the rare event—often the apple of the experimenter's eye. For experimenters to assure themselves that they are not being deceived by accidents of the apparatus, angle of photograph, or distortion of the recording liquid, emulsion, or gas, they want to be able to extract the interesting events from the background. They want to manipulate these occurrences—make the particles pass through a steel plate or be deflected by electromagnetic fields. Selection and manipulation are what we use to persuade ourselves that what we take to be the case is, in fact, the case. The logic tradition is willing to sacrifice inclusiveness for head→world reliability.

In my view each tradition of instrument making has captured something crucial about empirical knowledge and embedded it into the long-term fabric of physics. At the same time neither tradition has epistemic priority. Indeed,

33. On head→world and world→head reliability, see Field, "'Narrow' Aspects" (1990), esp 106. For a work of great interest that develops these ideas in a different context from that treated here (Godfrey-Smith focuses on evolutionary biology and its philosophy), see Godfrey-Smith's "Signal, Decision, Action," *J. Phil.* 88 (1991): 709–22, esp. 710–712, and "Indication and Adaptation," *Synthese* 92 (1992): 283–312, esp. 286–87, 302–3.

understanding the strength and weakness of each helps elucidate the incessant drive on each side alternately to defeat and encompass the other. In attending to the productive tension between the two, I am after not so much a transtemporal philosophy *of* machines as a historically specific philosophy *in* machines.

The power of these traditions of knowledge-making machines is visible in various ways. In addition to the continuity of personnel, technique, and mode of argumentation, one can look at three categories of evidence. First, there is a continuity of shared classification within the two traditions. Just after the (imaging) bubble chamber was introduced, physicists classified articles on it under the rubric "cloud chambers." Similarly, conferences on (logical) counters regularly included discussions of other logic tradition devices, such as spark chambers, and conferences on spark chambers included discussions of the next logic tradition innovation, wire chambers. Strikingly, during the early years of the spark chamber (1957–62) there were no separate entries under "spark chamber" in the definitive reference *Physics Abstracts.* To pull these articles from the database required searching under "counters," more specifically "counters, spark."[34]

Second, quantitative studies of the high energy physics literature reveal an overwhelming likelihood that someone who has published in one tradition will remain in that tradition for subsequent publications. Barboni, for example, has classified several thousand papers in weak interaction physics, distinguishing "optical" instruments (emulsions, cloud chambers, and bubble chambers) from "electronic" ones (spark chambers and counters). The probability that a physicist employs the same detector type in two consecutive publications, or changes detector type, is shown in table 1.1, where Barboni's categories have been relabeled "image" and "logic." Though the study of weak interactions is but one of the subspecialties within particle physics, it is clear that the diagonal (same device) correlations are far higher than the off-diagonal (different device) pairings: 0.94 versus 0.06. Physicists remain within their traditions.

On the basis of the correlation data alone, there remains the logical possibility that physicists typically wrote, say, 17 papers in one tradition, switched, then wrote another 17 in the other tradition (and continued oscillating between image and logic in this way). Such behavior would lead to a net contribution of 50% of their papers on each side of the image-logic divide, while still maintaining the 94% correlation between sequential papers. While logically allowable by this particular set of data, this oscillation is manifestly not the case historically. Consider the careers of some of the most active principal investigators within particle experimentation in the last 75 years. Not only did they coauthor a total of several thousand articles, but in the process they charted instrument paths that were taken by hundreds of collaborators and students.

Take Enrico Fermi, perhaps the last major contributor to both theory and

34. See Institution of Electrical Engineers, *Physics Abstracts,* for the years before 1962.

Table 1.1 Autocorrelation of Subsequent Publications within and between Image and Logic Traditions

	Publication 2		
Publication 1	Image	Logic	Total
Image	2,011 (0.94)	130 (0.06)	2,141
Logic	85 (0.06)	1,254 (0.94)	1,339
Total	2,096 (0.60)	1,384 (0.40)	3,480

Source: Modified from Barboni, Differentiation (1977), 190.

Note: Barboni labels his detectors "optical" and "electronic." I avoid these labels because of the hybrid image-producing electronic detectors that arose in the 1970s, which belong to both the image and logic traditions but would not rightly be described as electronic and optical.

experiment to work in the domain of individual particles and fundamental forces. With considerable restraint, the editors of his collected works selected some 270 major papers; of the hundreds of experimental inquiries, none were conducted with cloud chambers, emulsions, or bubble chambers. This was true despite the fact that Fermi's work ranged over nuclear forces, resonances, the Raman effect, fission, and a host of other topics. Another exemplar of the logic tradition, Frederick Reines, published his first nonclassified paper in 1945 on a photoelectric hygrometer. From that time, through his work on the enhanced atomic bomb, the hydrogen bomb, the first detection of the free neutrino, and a series of underground and underwater neutrino detectors that joined particle physics to astrophysics, Reines remained within the logic tradition. In principle, could Reines have prosecuted neutrino physics through bubble chambers during the 1960s? Yes, and it was done (e.g., by the long sequence of propane chambers in France and CERN). Nevertheless, of Reines's roughly 316 publications (up to 1995), only one can be classed in the image tradition: a 1950 *Reviews of Scientific Instruments* paper that analyzed low energy neutron spectra using nuclear emulsions. And that paper was in fact a counter-based test of the validity of nuclear emulsion as a measuring device for neutron and proton recoil near the low energy limit of emulsions.[35]

35. Understanding the precise role of the emulsion in experiments is crucial. E.g., in the famous discovery of the antiproton at Berkeley in 1955, Segrè's collaboration contained two subgroups: a "counter subgroup" (as Owen Chamberlain called it) consisting of Chamberlain, Segrè, Clyde Wiegand, and Thomas Ypsilantis and an "emulsion subgroup" (E. Amaldi, G. Goldhaber and R. Birge, and their respective collaborators). First, the combined collaboration published counter-based evidence for the antiproton, then, shortly thereafter, they clinched

Logic experiments characterize the over 200 physics articles that Carlo Rubbia has written or cowritten. Taking his Ph.D. in Pisa (1958), Rubbia did his thesis work on a new breed of fast coincidence circuits, spark counters, and high-voltage pulses. His investigations have taken him over many different fields of physics (cosmic rays, weak interactions, strange-particle decays, quantum chromodynamics, solar neutrinos, among others), but his career is unified by a steadfast commitment to electronic detectors. Beginning with tabletop counter hodoscopes, moving through ever-larger spark and wire chambers, the detectors culminated in the massive hybrid electronic detectors at CERN, such as UA1, that drew the image tradition into the logic tradition. Aside from occasionally analyzing some bubble chamber results (and once, in 1959, designing a counter-based pressure-measuring device for bubble chambers), Rubbia kept at arm's length from the dominant (photographic) image devices of the 1950s and 1960s. He conducted no work with nuclear emulsions or bubble chambers and collaborated in only a single set of cosmic ray measurements that combined cloud chambers and scintillators, a foray of four papers in 1958–59.[36] One could, in similar fashion, follow the (principally) logic orientations of Robert Hofstadter, Samuel Ting, Burton Richter, Mel Schwartz, Leon Lederman, and Jack Steinberger.[37]

---

their argument with a series of four articles based on emulsion results (see Chamberlain, "Discovery" [1989]; and Goldhaber, "Early Work" [1989]). In all, Chamberlain's list of publications included 97 logic experiments stretching from 1946 through the mid-1990s and crossing easily from various counters, to spark chambers, to the Time Projection Chamber. Aside from the four antiproton emulsion pictures, Chamberlain's work falls completely in the logic camp. Note that Segrè himself was also almost entirely a logic physicist. Nonetheless, he coauthored the set of emulsion-based evidence for the antiproton discussed above, work on an antiproton event in a propane bubble chamber, and a couple of miscellaneous papers using emulsions earlier in his career. The rest of his career takes place within the counter tradition.

36. See Galison, *Experiments* (1987), 202–6.

37. Robert Hofstadter began his career doing spectroscopic experiments in chemical physics using sophisticated coincidence circuits; he then turned to Na I counters, Cerenkov counters, and a variety of other logic techniques. Of his 112 listed papers that were not reviews, all fell within the logic tradition—none were on cloud chambers, bubble chambers, or nuclear emulsions. Samuel Ting is best known for his work leading to the codiscovery of the *J*/psi particle in 1974, but his work with electronic detectors goes back to his very first experiments following his Ph.D. in 1962. Over several years he used spark chambers of various types to explore strong interactions, especially in pion-proton events. Later, he turned to proton beams impacting on beryllium to produce heavy particles that would, in due course, yield electron-positron pairs detectable through spectrometers and counters. These techniques took Ting through a variety of different physics problems ranging from tests of quantum electrodynamics through his Nobel Prize–winning work of 1974 in which the 3.1 GeV resonance was measured through precision assessments of its decay products. Though scale, topic, and even laboratories changed over some 30 years, the electronic orientation of his work did not: he moved without much break from tabletop spark chambers to the massive L3 detector at CERN, never shifting to experiments involving cloud chambers, emulsions, or bubble chambers. Sharing the Nobel Prize with Ting, Burton Richter also exhibited a strong loyalty to the logic tradition. Throughout his career, Richter moved among detectors, accelerators, and physics. Papers on gas counters, Cerenkov counters, magnetic chambers, and storage rings alternated with explorations of a variety of physics topics centered on tests of quantum electrodynamics in his early career and then moving to weak and strong interactions. A handful of streamer chamber articles (these chambers were modified spark chambers that, by using weaker electric fields, allowed the capture of more detailed particle trajectories than ordinary spark chambers)

On the image side, the loyalty to instrument class was equally powerful. After inventing the cloud chamber, Wilson experimented on electrical discharge phenomena and the mobility of ions, but this work never invoked the electronic infrastructure of the logic tradition and was dominated by his cloud chamber work. In turn, his student Cecil Powell, who developed the nuclear emulsion after working on Wilson's cloud chamber for his 1928 Ph.D., spent a few years working with electrical experiments on the mobility of ions in gases (1929–35). From 1938 until his death in 1969, Powell worked entirely within the visual tradition, producing over 75 papers, mostly using nuclear emulsions. Like Wilson, Powell never used a coincidence detector, scaler, timer, or logic discriminator of any sort. Also trained in cloud chamber techniques, Donald Glaser's adherence to the visual tradition was as complete. Though he once toyed (unsuccessfully) with the construction of an electrical track-producing device, the whole of his career in physics fell into the image tradition, either through single cloud chambers, double cloud chambers, diffusion cloud chambers, or the bubble chamber he invented in the early 1950s. Even Glaser's initial work in biology was marked by the visual techniques he carried over from the track-following tradition. Continuity within traditions is also apparent in other, institutionalized forms. Again and again, one sees cloud chamber or emulsion collaborations become bubble chamber groups, or cloud and emulsion physicists who lack the infrastructure to build a bubble chamber acquire bubble chamber film to scan.[38]

---

were probably the closest he ever came to working with a visual device. Like Ting's work, Richter's 350 or so papers (before 1995) contain no crossovers to cloud chamber, emulsion, or bubble chamber physics. One can continue in this way: Martin Perl, who started his career working with Glaser and was best known for his work on the tau lepton, did a handful of bubble chamber experiments at the very beginning of his career (1956–58) but went on to publish more than 250 papers in the logic tradition. Mel Schwartz similarly began his career with a few years of bubble chamber experiments (his Ph.D. was in 1959 and his last image experiment in 1962); he then switched entirely and permanently to the logic tradition from 1962 through 1988. Leon Lederman wrote roughly 22 papers in the image tradition between his Ph.D. (1951) and 1959, at which point he too permanently joined the electronic world. Of that generation of particle experimenters Jack Steinberger stands out as the most ambidextrous between the two traditions. Having begun life as a theorist, Steinberger, more than others, moved back and forth between image and logic. Steinberger may have been especially flexible because he worked more as an interpreter, binding experiment to theory, than as an experimenter enmeshed in the details of hardware.

38. Some critical examples: Ralph Shutt brought his cloud chamber skills to the United States in the 1930s and, with A. Thorndike, led Columbia-Brookhaven to its position as one of the strongest visual collaborations in the postwar era. The team first pushed the traditional cloud chamber into a high-pressure cloud chamber facility of impressive size and then launched a long series of bubble chamber projects culminating in the discovery of the crowning resonance, the omega minus. On early BNL chambers, see chapter 5; see also R. Shutt, interview by the author, 23 December 1983; and A. Thorndike, interview by the author, 22 December 1983. At CERN, we find a similar story: leading French physicists, including L. Leprince-Ringuet, his student Charles Peyrou, and Peyrou's student André Lagarrigue, were deeply immersed in cloud chamber work at the École Polytechnique and then elsewhere as they began to play increasingly central roles in the development of French and CERN-based bubble chambers; see Galison, *Experiments* (1987), 139–50; see also L. Leprince-Ringuet, interview by the author, 14 May 1986; and C. Peyrou, interview by the author, 14 July 1984. Indeed, almost every collaboration on the list of principal bubble chamber groups in 1968 can be traced immediately to a previously existing cloud or emulsion group. For references to the major bubble chamber groups in 1967 and before, see the extremely useful technical

For all these reasons—pedagogical continuity, technical continuity, and epistemic continuity—it strikes me as useful to take this different cut through history. Let us put aside, for a moment, the structuring of our history of physics through the discovery of individual particles or effects. Let us step back from the history of weak interactions, of strong interactions, or even of theory. At the level of material culture there is a large-scale structure to the history of experimentation, the long-term history of image and logic.

Indeed in those relatively rare instances where an individual or a group did cross over to the competing tradition—and there surely are important examples of this— argumentative strategies of the previous tradition often carried over into the new. For example, bubble chamber physicists who switched to electronic detectors often continued to look for golden events. We will see evidence of this "epistemic hybridization" in chapters 6 and 7 when the remnants of the Alvarez bubble chamber team joined forces with electronic (logic) groups at the Stanford Linear Accelerator Center (SLAC).[39] But hybridization is apparent elsewhere as well. It is no accident, for example, that when David Cline (trained in the delicacies of scanning for individual rare events) switched to a spark chamber experiment in the early 1970s, he wanted a few golden events in hand before he signed on to the existence of weak neutral currents. When Martin Perl chose to search for single, especially persuasive instances of the tau particle in the Mark I experiment, he too was drawing on earlier experience with bubble chambers and the image tradition's golden events.

Each side of the image-logic divide has repeatedly tried to emulate the other's advantages, and the competition has pushed both technologies further. In 1952, Glaser wanted to build a bubble chamber that could be deployed selectively, the way cloud chambers could, so that only interesting cosmic ray events would "take pictures of themselves." Even 20 years later, it was still only a dream of bubble chamber physicists to design an electronically triggerable chamber. In the mid-1960s, Soviet physicists designed a series of high-resolution "streamer chambers"—a modification of a spark chamber that produced a track of light without actually sparking. For some physicists this was the "electronic bubble

work edited by Ralph Shutt, *Bubble and Spark Chambers* (1967), on cryogenic bubble chambers chap. 3, 146–50, and on heavy-liquid chambers, chap. 4, 166–67.

The distinction between image and logic left other institutional footprints. At CERN, the Track Chamber Committee, which ran the bubble chambers, was absolutely distinct from the Electronic Experiments Committee not only in type of experiment but in the amount of organizational structure and degree of centralized control that each found appropriate; see Hermann et al., *CERN II* (1990), chap. 8, esp. 465ff.

39. A particularly striking example of the "crossover" phenomenon is the case of one of the collaborations whose work first supported the existence of weak neutral currents. David Cline, a leading bubble chamber analyst of rare weak interaction decays, had joined forces with a big spark chamber group. As the team struggled to balance competing claims for the effect as real or artifactual, Cline began to study the individual spark discharges as if they were bubble chamber photographs, looking for the golden event that would carry the day; see Galison, *Experiments* (1987), chap. 4, esp. 225–28.

chamber" they had been looking for because it combined the ability to trigger in response to the event with the ability to register many tracks traveling in almost any direction. To its advocates' disappointment, the streamer chamber had little impact within the field, unable to compete successfully with either bubble chambers or spark chambers.

Yet a third attempt to find an electronic bubble chamber began in the 1970s, with the development of computer-generated images using data generated by wire chambers. By this time there had been changes in both the inner and outer laboratories. In the outer laboratory, colliding beams were coming into operation, so instead of a beam blasting out particles in a narrow cone aimed at a target, now two counterrotating beams slammed into each other, spewing particles in every direction. Collecting these many spatially dispersed particles demanded larger, more complex detectors. In the inner laboratory, even in fixed target colliders, spark chambers had grown massive and, by virtue of that growth, had begun to lose their status as havens from the "factory" work structure of the big bubble chamber groups. With the creation of groups such as the SLAC-LBL Solenoidal Magnetic Detector (later Mark I) and even more so with the Time Projection Chamber (TPC), the electronic instrument had recreated, and exceeded, the intricate, hierarchical, inner structure of its image competition. Indeed, these electronic groups absorbed the bubble chamber groups and, like a satisfied python, displayed the outline of what they had devoured.

One can only understand the design of many of these electronic image experiments by viewing them as collaborations between former bubble chamber groups (looking for distinctly visual evidence such as golden events) and wire chamber groups (looking for electronically manipulable evidence that could be selectively extracted).[40] This synthesis, which was at once sociological, technical, and epistemic, is typical of the mid- to late 1970s, as bubble chamber groups across the country began to close down, their members finding refuge in a very new style of electronic experimentation. The *W* and *Z* discovery of 1983 was an emblem of this binding together of the two traditions. It had become routine to extract a mere handful of golden events from a billionfold background.

1.5 Overview: Instruments and Arguments

One advantage of treating instrumentation, experimentation, and theory as quasi-autonomous subcultures is that one is forced to confront the strikingly different contexts in which each is embedded. It is hard to find a better illustration of this diversity than the cloud chamber, the subject of chapter 2. This extraordinary tabletop device is usually depicted as an instrument of atomic, nuclear,

40. Treitel, "Tau," *Centaurus* 30 (1987): 140–180.

or particle physics. Its origins, however, could scarcely have been more distant. C. T. R. Wilson's work began with his immersion, as a very young man, in the Victorian fascinations with weather and photography. In chapter 2, I trace Wilson's course through his education in natural history, his field work in meteorology, and his eventual arrival at the Cavendish Laboratory. The cloud chamber itself was a result of the fusion of these several sets of practices: on the one side, Wilson's long-standing fascination with the reproduction of natural phenomena was grounded in a rich subfield of experimental meteorology; on the other side, Wilson's interest in electrically catalyzed precipitation was squarely located in the theory of charged atomic particles that dominated research at the Cavendish.

In the first instance the cloud chamber was just that—a space where clouds were reproduced. Rooms for the reproduction of nature were hardly unusual. The camera obscura and the camera lucida took light from three-dimensional objects and focused it as images on a flat surface. Others had exploited a myriad of techniques to make miniature representations of clouds and the optical phenomena seen after Krakatoa. This trope of bringing the complex exterior into the manageable interior was a kind of mimetic experimentation; experimentation with the goal of capturing nature in its full complexity, rather than risking distortion through the analytical dissection of phenomena into "simpler" parts. From Wilson's notebooks it becomes clear that these meteorological origins shaped the categories by which he viewed his phenomena ("clouds," "precipitation," "fogs," and "atmospheric optical phenomena"). Indeed, it seems quite impossible to separate his work into later fields of particle physics and meteorology; rather, his work is better conceptualized as condensation physics, a conjoint subfield that was at once analytic and natural historical.

The cloud chamber's glory days were vividly captured by a new literary form that came into being just before the Second World War: the cloud chamber "atlas." These glossy oversized volumes, patterned on the medical atlases produced so copiously since the late nineteenth century, reproduced exemplary photographs—each one its own golden event. By appropriating this presentational form, physicists accomplished two things: they instructed initiates in the technique of cloud chamber research, and exploiting the image tradition, they helped create the category of "particle physics," distinct from the preexisting fields of atomic, nuclear, and cosmic ray physics. The objects of such collections were the defining entities of the basic processes and particles of the new science of particle physics. More generally, atlases of this sort, along with analogous volumes for physiology, anatomy, astronomy, and ornithology assisted in forming—and popularizing—the notion of scientific objectivity.[41]

By the 1920s, the cloud chamber was already hailed as a transformative instrument of physics, its ability to yield individual photographs of particles, un-

41. Daston and Galison, "Objectivity," *Representations* 40 (1992): 81–128.

rivaled. The extension of this instrument into the nuclear emulsion is the subject of chapter 3. At first sight, nothing could be simpler: Boxes of ready-made photographic plates were exposed to radioactive materials or cosmic rays. Then, as particles skittered across the emulsion surface, they induced chemical changes in the emulsion, blackening the film on development. With microscopic analysis, the physicist could measure these tracks and deduce properties of the incident particles. Looking more closely at these film boxes, not only can we see the discovery of a host of new particles, we can track a shift through three fundamentally different stages of twentieth-century experimentation, each in a different setting.

Scene 1 finds Marietta Blau, a young woman physicist, working at the Institut für Radiumforschung, never quite integrated into the most prestigious research teams, and attempting, on the margins of the institute, to make dental X-ray film sensitive to protons. With the help of a young woman student, and the inspired use of pinakryptol yellow, Blau achieved a chemical wonder. Suddenly her film revealed nuclear phenomena no one else could see—longer tracks, proton tracks, and high energy nuclear "stars." But just as she reached and began to pass the edge of known processes, the Nazis entered Vienna, and she fled for her life—to Scandinavia, Mexico, and eventually Brookhaven. Emulsions were the instrument of the outsider because they were so cheap, a laboratory-in-a-box that aided her through the quasi isolation of Vienna and through her years on the run.

Nuclear emulsions, scene 2, opens in the late 1930s as Cecil Powell, one of Wilson's few cloud chamber students, began a careful reading of Blau's papers. While Powell was setting up his Bristol laboratory for a series of cloud chamber experiments, he began to speculate about methods by which he could use the film as a *direct* recording instrument for tracks—without the chamber. Starting at the point where Blau had been forced to stop, Powell began building a new kind of laboratory around this novel technology. In place of Blau's solitary lab-in-a-box, Powell established a cottage industry, one in which a coordinated staff of physicists was supplemented by a crew of women who became full-time "scanners" of film as the exposed plates arrived from mountaintops or cyclotrons. As the war began, Powell declined to work directly on the Tube Alloy (British atomic bomb) Project but trained others who did. Because of the wartime application of the emulsion method (e.g., in sorting out neutron dynamics), the postwar British government became an enthusiastic sponsor of emulsion research. In the late 1940s, Powell and his colleagues, many of them cloud chamber veterans, raised the use of emulsions to a fine art, discovered a bevy of new particles and particle decays, effectively launched experimental particle physics, and exported their famous emulsions to laboratories from CERN to Berkeley.

The development of emulsion techniques brought anxiety as well as triumph to the laboratory. Part of the physicist's bargain, in making the new technique work, was a redistribution of control over the detector, and with it a redistribution of scientific authorship. To chemical companies, physicists ceded the design and manufacture of the photographic plates. They sacrificed knowledge of the chemical composition and technique of production in return for a device that worked far better than anything they could cook up in the back rooms of their laboratories. Then, to teams of women scanners, the physicists handed over the search for tracks on freshly exposed plates. In return for efficiency, the physicists gave up the "discovery moment" that traditionally had fallen to the first observer of a new phenomenon. To theorists such as Abraham Pais, Robert Oppenheimer, Murray Gell-Mann, and many others, the experimenters deferred on many aspects of interpretation. In return for the theorists' understanding of the increasingly complex world of particle phenomenology, the experimenters lost some of their role as interpreters of their own findings.

Finally, scene 3 opens as Powell and his emulsion collaborators begin, in the mid-1950s, to launch multinational balloon flights with huge emulsion stacks. These create a kind of big science society as the emulsion is parceled out to the participating groups for analysis. Moreover, the majority of this last phase of emulsion work abandoned cosmic rays in favor of the increased flux of the cyclotron; experimenters bade farewell to the romantic autonomy of mountain-top labs and seafaring cosmic ray expeditions. Instead came the identity cards, safety protocols, shift work, and factory-style organization inaugurated in the postwar era. Scene 3 was, in fact, the closing one for emulsions as the technique moved to CERN, Brookhaven, Berkeley, and beyond. In the late 1950s, the emulsion physicists encountered a different kind of marginality—not Blau's kind, not that of the Europeans looking over their shoulders at the Americans, not even that of the loss of control over discovery and fabrication. The final anxiety was watching the emulsions themselves give way as a field to the industrial-sized bubble chambers that began to dominate microphysics.

Before proceeding with the bubble chamber (in chapter 5), we need to step back in time again. Though cosmic rays and the British use of nuclear emulsions take us from the 1930s through the war to the late 1950s, we will have seen the effects of the war on a too-restricted field. Chapter 4 shifts the field of view to the wartime laboratories. Traditionally, World War II has been accorded two roles in the historiography of modern physics: as the cause of increased federal funding of peacetime physics after 1945 and as the goad forcing an illustrious corps of refugee scientists to emigrate from Nazi Germany and Fascist Italy to the United States and Great Britain. While both roles are important, the argument of chapter 4 is that the impact of the hostilities began early during the war, and went deeper, shaping not just the personnel, budget, and national prominence of physics but the fabric of laboratory practice and instruments at its core. Chapter 4, then, is about the imaginative restructuring of the ways that physicists conceived of the laboratory; it is about the link between physicists' wartime experiences in the radar and atomic bomb projects and the new picture they formed of how research would proceed—the picture of a collaborative, factory-scale effort that would displace the ideal of individual and small-group work.

For many physicists, the lesson of the war concerned the outer laboratory. They left their military assignments with the view that collaboration in a centralized, mission-directed facility was an extraordinarily effective way of making weapons *and* doing physics. The Metallurgical Laboratory under Enrico Fermi, the Los Alamos Scientific Laboratory under Robert Oppenheimer, and the MIT Radiation Laboratory under Lee DuBridge, all provided examples of the new style of research. It was this image of the future that served postwar leaders like H. D. Smyth, Victor Weisskopf, and John Slater as they began to plan a collaborative laboratory for postwar physics. Their plans were realized in Brookhaven National Laboratory, and Brookhaven became an institutional template for laboratories across the United States, Europe, and the Soviet Union.

I have referred to the impact of this new centralized laboratory as a transformation of the "outer laboratory" because that phrase captures the bounds of change in the late 1930s and 1940s. This was a time when physicists took their cosmic ray experiments out of shacks in the Alps and Andes and set them more or less intact in front of an accelerator. No better illustration of this transport exists than the Berkeley emulsion work under Eugene Gardner. When Berkeley's fabulous new cyclotron opened, its proprietors could not repeat the demonstration of the pion until one of the Bristol team arrived with the particular film and processing procedures used at Bristol. Even as accelerator laboratories grew in size, it was not uncommon to find the old nuclear emulsions, counters, and cloud chambers perched in front of the latest beam line at Brookhaven, Berkeley, or CERN. Only later, as we will see in chapter 5, with the advent of the bubble chamber in the 1950s, would there be a radical restructuring of the *immediate* environment of the physicist—the microenvironmental work space of the experiment itself. Then and only then would experimental work be broken up into subtasks and a rationalized work structure. This final shift transformed the inner laboratory as well.[42]

In chapter 4, we see how the massive infusion of war-surplus electronic equipment, together with war-generated plans for novel devices (new vacuum tubes, low-noise amplifiers, electronic counters, pulse-height analyzers), reshaped the daily routines of physicists and structured the kinds of projects they undertook. As the wheels of declassification began to turn in 1946–47, the

42. Since the focus of this book is on the immediate environment of the physicists, I will discuss the development of accelerator technology only insofar as the accelerator laboratories served as models for or otherwise induced changes in work around the inner laboratory. Fortunately, there are several excellent institutional studies on accelerator laboratories on which I will draw. A few of the more recent are Heilbron and Seidel, *Lawrence* (1989); Hermann et al.'s *CERN I* (1987) and *CERN II* (1990); Westfall, "Fermilab," *Phys. Today* 42 (1989): 44–52; and Hoddeson, "KEK," *Soc. Stud. Sci.* 13 (1983): 1–48.

extent of the wartime innovations in particle detectors and electronics became known. The extraordinary burst of new instruments (especially detectors) and novel experiments after the war resulted in large measure from this multileveled transfer of hardware, physics goals, and—just as important—work organization from weapons to civilian laboratories. World War II, far from forcing physics into a holding pattern, provided a crucial impetus for the electronics of the logic tradition. In addition, as I will show in chapter 5, the war offered a new organizational form of experimental life that would utterly transform the image tradition.

If the cloud chamber was Powell's jumping-off point for emulsion detectors, it was equally Glaser's for bubble chambers. In chapter 5, we follow Glaser's painstaking efforts to imitate the cloud chamber while substituting the phase change from liquid to gas for the cloud chamber's phase change from gas to liquid. In his theory, design, and application of the apparatus, Glaser patterned the bubble chamber on the cloud chamber. Indeed, Glaser fully expected his device to be used in the same individual-oriented work organization as the cloud chamber. His bubble chamber was to be applied to cosmic rays, and it was the splendid isolation of the cosmic ray experimenter's life that appealed to him. By contrast, Glaser shuddered at the "factory-like" atmosphere of the big accelerators, and he saw the bubble chamber as just the instrument to "save" cosmic ray physics by making it competitive with the centralized outer laboratories at Berkeley and Brookhaven.

There is a deep irony to Glaser's invention. Instead of rescuing cosmic ray physics, the bubble chamber killed it. Because the chamber could not be triggered by cosmic rays, the experimenter would have to know when to make the device sensitive—that is, would have to know *in advance* when particles would be arriving. Since this knowledge is impossible in nature—and always possible in the accelerator environment—the bubble chamber was perfect for the big machines and useless for cosmic rays.

Opposites attract. In 1952, as Glaser was displaying his tabletop device, Alvarez was leaving a canceled nuclear weapons project and searching in the civilian sector for a way to exploit what he had learned about large-scale research. Better preparation for big physics would have been hard to find. During the war, Alvarez had served both in the Radiation Laboratory at MIT and in the intense development program of the implosion bomb at Los Alamos. He, like many of his wartime colleagues (discussed in chapter 4), came to revel in the ambition and scale of such scientific industrial enterprises. Now at Berkeley he sought to recreate that atmosphere in the inner laboratory of the bubble chamber.

As chapter 5 relates, the bubble chamber was a perfect vehicle for Alvarez's expansion plans, and he soon scaled the chamber up: from the size of a marble to the size of a building; from hundreds of dollars to millions of dollars; and from a single experimenter to teams of nearly a hundred physicists, scanners, and technicians. Perhaps most important, Alvarez routinized and automated the computer-aided scanning of the hundreds of thousands of photographs that came spewing out of the bubble chamber operation. Because the bubble chamber produced enormously complex pictures, however, Alvarez judged that they were far too complicated to be *fully* automated, and so he integrated squadrons of human scanners and data processors into the routine. But Alvarez's view about how pictures should be processed was not the only one. A competing program (known as the Flying Spot Device) centered largely but not exclusively at CERN aimed at obviating the human element in data processing. Where Alvarez saw a uniquely human capacity for recognizing patterns, many of his European competitors saw a bottleneck, an idiosyncratic obstacle in the path of pattern recognition systems that would be fast and uniform. In the tension between the two views there is more than mere disagreement over data processing: we will find different technological cultures inextricably tied to conflicting ideals of experimentation.

To throw the Berkeley approach to big physics into sharper relief in one further dimension, chapter 5 follows a bubble chamber that worked out disastrously: the Cambridge Electron Accelerator (CEA) chamber, which exploded in the summer of 1965. Stanley Livingston, who ran the Harvard-MIT CEA, was horrified by the industrial ethos of his competitors, and again and again asserted his prerogative to conduct research in the open atmosphere of an older style of physics. He saw safety officers, managers, and sequenced flowcharts as impediments to academic inquiry, and he did not hesitate to say so. Then, when the CEA bubble chamber blew up, the historical lesson—the conclusion drawn by the Atomic Energy Commission and other administrative bodies—was that the disaster proved the necessity of adopting a rigorous industrial form of work. Chapter 5 follows this incident in detail because the response to perceived danger played a crucial (if now forgotten) role in the establishment of a new laboratory culture.

Together, construction methods, administration, data analysis, and safety concerns reshaped the laboratory and the physicists who occupied it. New hierarchies were established within the experimental group; subgroups sprouted with the mission of preparing data reduction software; other teams specialized in the analysis of particular physical problems, yet others in hardware operations. As responsibility for the running of the apparatus shifted to highly trained cryogenic and structural engineers, experimenters came to regard their goal as the analysis of data. Soon this task was shared with other image tradition (cloud chamber and emulsion) groups scattered across the world—groups that had never laid eyes on the Berkeley chamber. Once again, the meaning of the term "experiment" had changed and along with it the nature of scientific authorship.

Both Glaser and Alvarez earned Nobel Prizes for their innovations. In the early 1960s, however, Glaser left particle physics for biology, largely because he felt that the Berkeley operation had cut him off from the kind of solitary work that had drawn him to physics in the first place. More surprising, Alvarez, though he flourished for over a decade and a half in the industrial environment he helped create, also eventually found it oppressive. At the end of the 1960s, he returned to cosmic ray physics, a branch of the discipline marginalized in no small measure because of his own earlier success.

It is all too easy to celebrate the image tradition as the only path to experimental demonstration. Bubble chamber pictures grace the covers of hundreds of books, from physics textbooks to histories and philosophies of science. But despite this decorative dominance, images have always been contested throughout science—in mathematical physics and mathematics, for example—not just in experiment. When Cauchy and Lagrange voiced their suspicion of diagrams in mathematics, it was because they suspected the visual would import physical (and unrigorous) arguments where logical deduction alone should be.[43] Elsewhere in mathematics the struggle has appeared again: in the battle over intuitionism in the 1920s and in Hilbert's famous dictum that, properly understood, geometry could be done with "chair" and "table" standing in for "line" and "point." In the late 1940s and 1950s, the theoretical physicist Julian Schwinger was horrified by Richard Feynman's invocation of diagrams; Schwinger thought that, by obscuring the underlying field theoretical calculations, Feynman's diagrams degraded physics and impeded the advancement of theory. Later, general relativity theorists divided on the importance of geometry: Some, like John Wheeler, advocated a vastly increased role for geometry—general relativity was, Wheeler believed, destined to become a branch of "geometrodynamics." Steven Weinberg, by contrast, argued vigorously that geometry was leading physicists astray—properly understood, general relativity was a particular branch of quantum field theory, and the intuitions suggested by differential geometry hid the underlying dynamics.[44] But our focus, as usual, is on the instrumental practices that embody these issues. Chapter 6 is about the *anti*visual traditions within the laboratories: the iconoclasm of the logic tradition.

Even in their heyday, the image-producing cloud chamber physicists did not have the laboratory to themselves: the logic tradition had begun. In a certain sense, the search for homologous rather than homomorphic demonstrations began with electroscope measurements at the turn of the century, or even earlier. But the competing electronic counter tradition began in earnest in 1928 after a great number of frustrating dead ends—and Walter Müller's intervention. It was then that Hans Geiger's *Spitzenzähler* became the "Geiger-Müller" counter, a usable instrument for nuclear and cosmic ray physics. Chapter 6 tracks the

43. See Daston, "Physicalist," *Stud. Hist. Phil. Sci.* 17 (1986): 269–95.

44. On the differing interpretations of the role of geometry, see Galison, "Rereading" (1983).

development of the nonvisual, often antivisual, logic tradition that flourished with the proliferation of these counters.

From the very first pulses, the problem of distinguishing "real" from "spontaneous" discharges haunted the design of counting apparatus and was the source of serious controversy.[45] After extensive attempts to improve Geiger's counter, Müller finally tied the spontaneous discharges to the presence of cosmic radiation. Not long afterward, W. Bothe and W. Kolhörster vastly improved the process of separating effect from artifact when they began using the coincident firing of two counters to signal the passage of a particle through both. But their cumbersome photographic recording of pulses, which left sheets of film drying from the ceiling like clothes in a laundry, offended the aesthetic sensibility of Bruno Rossi, who designed a fully electronic system that pulsed only when both counters fired. Since Rossi's circuits could be wired to any number of counters in any combination of coincidence and anticoincidence, the electronics played the role of logical connectives, for example, $(A \wedge B) \wedge \sim(C \vee D)$. As a result of these innovations, by the start of World War II the counter was an instrument fully capable of catching up to Wilson's cloud chamber. By the war's end, the art and science of electronics had advanced beyond anyone's expectations.

If the bubble chamber groups resembled factories, with leaders and corporate hierarchies, the spark chamber groups were (at least initially) self-consciously democratic. Spark chamber experimenters wanted to build an inner laboratory around a set of instruments and work practices that gave individual scientists control over their experiments, control that the experimenters felt had been sacrificed in the big bubble chamber groups. The spark chamber appeared in early forms in Japan, England, South Africa, and Italy, as well as in the United States—reflecting, in part, the wide dispersal of electronics innovations (available through the laboratories, factories, and the surplus war matériel of World War II). In contrast to the bubble chamber, which has had a fairly unambiguous lineage of invention, the spark chamber multiplied everywhere in a myriad of forms.

No single national history, no focused institutional study, no biographical structure can capture the dispersal of the electronic research on detectors. It grew up in Pisa, around the discarded radar equipment of American troops as they pulled out after the war. It continued in occupied Germany in reconstructed laboratories with personnel recently released from Allied custody. It flourished in Japan with a melange of home-built and imported electronic gear. Important work occurred too in the Soviet Union, in Britain, and in the more opulent postwar laboratories of the United States. Passed back and forth, sometimes with

45. For a fascinating account of the Cambridge-Vienna controversy over artificial disintegration, see Stuewer, "Disintegration" (1985).

only the exchange of preprints, the technology of building fast spark counters and, later, spark chambers finally wended its way to Berkeley in 1960.

Culminating the work of the late 1950s was a new "discharge chamber" built by two Japanese physicists, Fukui and Miyamoto, that could display the rough position of tracks by sparks. At Berkeley, the Japanese innovation fell into grateful hands. Alvarez's fantastic image-filled bubble chamber had left the logic physicists almost out of work. For with each stunning success of the big bubble chamber team, it seemed that not only the technology but the work organization of the individual counter experimenters became more and more out-of-date. Alvarez wrote in September 1957 to Edwin McMillan, director of the laboratory, arguing that the counter groups should close up shop. In high energy physics, the counter experimenter "can hardly get started in a serious investigation of the much more common inelastic processes if he does not have a visual detector, and preferably a bubble chamber." For Alvarez, "counters appear to me to be on the way out as precision instruments in high energy physics."

With the arrival of the Japanese discharge chambers, however, the counter physicists acquired a new weapon. They soon modified the device to produce a localizable, visible spark between plates when a particle traversed the intervening space. In a matter of months, new devices sprang up at universities across the country. The first one at Princeton was the work of a handful of college sophomores, a long way from the demanding engineering staffs at the Berkeley bubble chambers. For many logic physicists, however, the necessity of photographing the visible sparks was repulsive; they wanted a fully electronic system that never passed through the stage of photographic reproduction. Indeed, these logic experimenters reacted with the same force as Heisenberg did to Schrödinger's formulation of a wave mechanics. The analogy has thus far been ignored because historians have been lovingly attentive to theoretical aesthetic judgments while turning a deaf ear to the aesthetics of instrument design.

Two interrelated, and ultimately inseparable, forces thus combined in the production of the spark chamber: One force was the logic experimenters' *social* desire to recapture control over their workplace and avoid the hierarchical teams of the "bubblers." The other, more ethereal force was the logic physicists' *epistemic* commitment to eliminating the visual. In the design of the new electronic detectors the social and the intellectual were welded together: the desire was for a way of experimenting in which control was possible—control over how one worked in the laboratory, over the disposition of the apparatus, and over the generation and analysis of data.

Chapters 6 and 7 describe the recent synthesis of image and logic traditions in particle physics. It should be mentioned that a synthesis between these opposing traditions has taken place in fields well beyond particle physics. For example, in the astronomy program of the Space Telescope, the photography-based astronomers viewed their electronic detector competitors as having no device of adequate resolution and sensitivity. Conversely, space scientists with electronic predilections were gravely concerned that photographic emulsions would fog because of penetrating radiation or, worse yet, require potentially disruptive human intervention. Robert Smith and Joseph Tatarewicz traced this story and showed that the two sides eventually built their way to a solution with the adoption of charged coupled devices (CCDs), which produced electronically generated pictures of very high resolution.[46]

The parallel with particle physics is not spurious. In this case—and in earth-imaging methods, CAT scanning of the human body, and nuclear magnetic resonance imaging—a great deal of shared technology underpins the new methods. It is beyond the scope of this book to trace these astronomical, geophysical, and medical instrument histories in the same detail as the physics case. Nonetheless, I would argue that the tension between analog technical knowledge and digital technical knowledge is a deep one and that this division has cut across disciplinary boundaries. The weaving together of the two traditions in the last few decades represents a previously hidden unifying trend in an age of scientific specialization. Homologous and homomorphic representations have coalesced.

In physics, the epistemic shift toward a merger of the two forms of representation brought with it a new scale of work. The vastly increased size and complexity of detectors such as PEP-4 in the late 1970s guaranteed that many aspects of control would be lost. No longer was it possible, or even conceivable, to vary an apparatus the way Wilson could move around electrodes, or the way Powell could alter (or have altered) the chemical composition of his emulsions. Even the proper design of a piece of apparatus exceeded the capability of analytic calculations.

Image and logic traditions found their confluence in specific devices, perhaps none more famous than the Mark I at SLAC. This experiment was one of two that produced the *J*/psi particle in November 1974, but even before then the Mark I stood as an embodied combination of the two very different strategies for producing evidence. From Berkeley, a significant number of Alvarez's old bubble chamber team joined forces with a SLAC team that had always been committed to electronic logic devices and never produced an image machine. Together, they built an electronic device that made pictures—and employed the skills and epistemic practices of both traditions. Coordinating these different approaches became crucial as heterogeneity of practices and increased scale made instrument building an ever more complicated process.

46. Smith and Tatarewicz, "CCDs," *Proc. IEEE* 73 (1985); see also Lynch and Edgerton, "Aesthetics" (1988). Much interesting work has been done on the history and sociology of images within scientific texts; an excellent introduction to this work can be found in Lynch and Woolgar, *Representation* (1990), where the authors explore the complicated chains of references that images invoke to do their work, including procedures, other images, and series of images, one modified into the next; see also Fyfe and Law, *Picturing Power* (1988).

As an increasing fraction of the high energy physicists joined large colliding beam detector groups, questions of scale and diversity came to the experimental community with ever greater force. Even compared with groups the size of Alvarez's, these new collaborations were large—experiments increased another factor of 10. Instead of 20 to 30 physicists working with 20 or so engineers and technicians and another 20 to 30 scanners, the huge hybrid experiments of the 1980s carried 100 to 200 physicists and a correspondingly large number of engineers and technicians. By the early 1990s, detector groups at Fermilab, others at the planned Superconducting Supercollider, and yet others at the CERN-based Large Electron Positron Collider (LEP) ranged well over 500 physicists, some over 800. Chapter 7 brings us into this multiply hybrid world through a study of the Time Projection Chamber.

Considered in isolation the TPC is extraordinary. It is a cylindrical volume of gas bathed in electric and magnetic fields that guide the ionized tracks, intact, to the end caps, where the information contained in the tracks is digitized and sent to a computer. It is, in a sense, the purest realization of the dream of an electronic bubble chamber—no wires disturb its interior volume. At the same time, the TPC when considered as deployed is hybridized with ancillary detectors in such a way as to make it the large-scale embodiment of all that has gone before: calorimeters, wire chambers, image-making capacities, sophisticated electronic triggering, large-group structures, and the massive use of computers. But the idea of aggregation alone does not capture the change in scale of work accompanying the move from million-dollar detectors to hundred-million-dollar devices. In a crass sociometric sense, hybrid detectors like the TPC and those that were destined for the Superconducting Supercollider signaled the end of the line: one such device employed 800 physicists; with an American high energy community of some 2,500 experimenters, the detector groups are within striking distance of constituting the entire community.

Beyond numbers, the complexity of the organizational structure of a device like the TPC/PEP-4/PEP-9 facility built for SLAC by an international collaboration raises one final set of issues about the meaning of "experiment" and "experimenter." The boundaries of the detector are ill defined. Because the Department of Energy wanted to open the machine to the wider community of physicists, the device itself was called a "facility" not an "experiment." This label meant that the proprietary relationship between the builders of the device and their product was severed; after a prescribed interval, other groups and individuals were to take over. Or, even more concretely, several different groups with different leadership structures and memberships had "facilities" such as PEP-4 and PEP-9 (PEP stands for the Positron Electron Project, a colliding beam facility at SLAC), two different "experiments" sharing the same microphysical events. What in one sense was a purely political problem (how to distribute data) was at the same time a physics problem (how to reconstruct the phenomenon in question). Governance, too, came to borrow from the surrounding culture: managerial skill was borrowed from business seminars at Asilomar, and wide-ranging debates circulated through the collaboration on the nature of leadership. Should the collaboration be subject to a benign dictatorship? To a Senate-like representative structure drawing equally from the different institutions? Or should House of Representatives–style proportional representation serve to make final decisions?

If the construction, employment, and management of the hybrid facility broke with the past, so did the results of the experiment, as will be apparent from our study of TPC/PEP-4. In the era of bubble chambers, theorists played an essential role, but one that could be circumscribed. A theorist like Gell-Mann could propose a particle along with its quantum numbers, and experimenters could make it one of their targets. Conversely, particles could emerge from the scanning process, and theorists would work to integrate it into their schemes. With the TPC and other detectors of the 1980s and 1990s, the boundary between theory and experiment became less sharp, principally because quantum chromodynamics (QCD) was a notoriously difficult theory out of which to extract phenomenological predictions. Which models should the experimenters use? And when the model was compared with data, were they testing the model or QCD itself? Understanding QCD experimentally became one of the longest lasting, least dramatic, and most important components of physics in the 1980s and early 1990s. Though "confirming" a theory would earn no trips to Stockholm, it was a prerequisite to recognizing any new phenomena. Without understanding the background there would be no foreground. The socio-epistemological consequences of this blurred experiment-theory line were deep. Unlike the situation only a few years earlier, experimental groups began to need their "own" theorists, theorists whose work lay, not in the outer speculative realms, but close enough to a machine's specificity that they could begin to put data into a form comparable to the output of models from theory. On the theory side too, one begins to see "phenomenologists," whose work was designed to generate experimentally testable consequences of QCD, its variants, and its competitors. Schematically, instead of the simple comparison

$$\text{theory} \leftrightarrow \text{experiment}$$

one might have a string of comparisons:

$$\text{experiment} \leftrightarrow \text{experimentalist's theory} \leftrightarrow \text{theorist's phenomenology} \leftrightarrow \text{theory}$$

This change will be apparent from the discussion of internal experimental meetings in the TPC/PEP-4 collaboration that aimed to link experiment to models to theory.

Finally, the assembly of the massive hybrid experiment precipitated a major reconstruction in the notion of scientific authorship. The distribution of results in the collaboration was governed, not by an individual, but by a group, appropriately named SCAT (Standing Committee on the Assignment of Talks). Which participating experimenters could talk, where they could talk, and what they could talk about had become an issue of pressing concern. For it was feared that with a group of 200 or 400 or 800 physicists all speaking for the experiment, the solidity (and propriety) of the experimental results would melt into air. Taken together, these debates over construction, ownership, deployment, and dissemination brought the high energy physics experiment into a new and, to many, terrifying epoch.

Looked at with hope, this complex collective was a utopian community of communities, one in which every participant had access to the vast data bank summarizing experimental runs. Like Jean-François Lyotard's postmodern fantasy of a society in which every constituent had access to information,[47] the collaboration allowed every member—from the least experienced graduate student to the most senior principal investigator—to jack into the resources of the Data Summary Tapes and produce original research. Looked at with pessimism, the experiment was a nightmare inversion of everything for which physics had once stood. In a collaboration of 800 on a billion-dollar device, the individual experimenter necessarily lost his or her control over building, maintaining, adjusting, or altering the apparatus. Given the bureaucratic, physical, and financial constraints, the experimenter could no longer select or quickly modify the goals of an experiment. Even understanding the subtleties of QCD and its ties to experimental output became more difficult—the experimenter struggling to make a sector of a subdetector work was often hard-pressed to follow the data analysis. To maintain some coherent presentation to the outside world, the large collaborations put a tight lid on what the individual experimenter could write, publish, or even say in public. As a final insult, the experimenter had, in many instances, lost contact with the physical world entirely and worked with simulations instead. Both bright-side and dark-side views of the changing practice of physics tell us something of the conjoint epistemic and social structure of the laboratory; experimentation with 800 colleagues is itself an unproven social and epistemic experiment full of tension.

After the bubble chamber—but before, in, and after wire chambers like the TPC—is the computer. At first no more than a faster version of the electromechanical calculator, the computer became much more: a piece of the instrument, an instrument in its own right, and finally (through simulations) a stand-in for nature itself. In chapter 8, the last historical chapter, we come face to face with the computer, and more specifically with the origins of the "artificial reality" that lies in simulated physical systems. In a nontrivial sense, the computer began to blur the boundaries between the "self-evident" categories of experi-

47. Lyotard, *Postmodern Condition* (1984).

ment, instrument, and theory. Understanding these developments, the subject of chapter 8, takes us back not only to the early days of computation but to the radical changes demanded in weapons design as Los Alamos moved from atomic to hydrogen bombs.

During World War II, questions that arose about the design of the atomic bomb pressed at the limit of what could be handled using hand or electromechanical calculators. After the war, as attention shifted toward the hydrogen bomb, the Manhattan Project approached a Rubicon of calculational difficulty. First, there was no equivalent of Enrico Fermi's controlled fission under the stadium at the University of Chicago where so many of the measurable (though not calculable) parameters of fission were quantified. In short, there was no cold fusion. Second, it was almost immediately apparent that the physics behind the hydrogen bomb would vastly exceed the difficulties associated with the Hiroshima and Nagasaki bombs. Faced with the inability to measure or calculate, physicists, mathematicians, and electrical engineers embarked on the full-scale development of a combined system of computer hardware and software (*avant la lettre*) to conduct simulations.

This tertium quid between experiment and theory was, as one physicist put it, "a new way of life in nuclear science," and it prompted intense self-questioning. Was the Monte Carlo simulation an experiment? Was its execution a proper activity for the experimenter trained to probe with the oscilloscope, to measure track angles, to explore photographs, or to wire logic circuits and phototubes?

By a "proper activity," I mean to suggest that a fundamental questioning was taking place: recurrent doubts about professional identity, about the pleasure of being an experimenter, and about the forcefulness of claims about nature made on the basis of simulations. Some physicists were comfortable with the new mode of argumentation, others ill at ease. A change was afoot that was both epistemic and moral, and for many, the alienation from what they considered the classical activity of experimentation crossed the usual boundaries between private reactions and public argumentation.

The implications of the debate went deep. To some members of the growing community of simulators, the Monte Carlo method was just another way to approximate the "true" subject of physics: the continuous relations of fundamental entities reflected in the mathematics of partial differential equations. What was new was a point of view that inverted this relation. Instead of regarding differential equations as the furniture of Plato's heavenly abode, the new computer vision saw the Monte Carlo itself as the key to reality. In this "stochasticist" vision, the world was inherently discrete and stochastic; it was the partial differential equation that was a mere approximation and the Monte Carlo that, by imitating the randomness of the world, truly stood for nature. There is a metaphysics to computation, and the redefinition of "experimenter" and "experimentation" was—at one and the same time—metaphysical, sociological, and eminently practical. Manifestations of the anxiety associated with simulations appeared from the inception of the method. Some tried to assimilate the activity into theory, citing its removal from the "hard" reality of solder and circuit: How, after all, could an experiment exist inside a computer? Others saw simulations as essentially experimental: How could a procedure that never gave the same response twice be anything else? My goal is neither to elevate nor to denigrate the simulation, but rather to indicate the growing role it played in experimental demonstration, and its significance in problematizing the notions of "experiment," "instrument," and "theory." There is an age-old tradition of opposing headwork to handwork, ideas to experiences, and rationalism to empiricism. The growing netherland of simulations makes those radical oppositions ever more difficult to sustain.

1.6 The Context of Context

Guiding the inquiry of chapters 2 through 8 is the argument that the subcultures of physics are diverse and differently situated in the broader culture in which they are prosecuted. But if the reductionist picture of physics-as-theory or physics-as-observation fails by ignoring this diversity, a picture of physics as merely an assembly of isolated subcultures also falters by missing the felt interconnectedness of physics as a discipline. To sail between the Scylla of exaggerated homogeneity and the Charybdis of mere aggregation, I repeatedly use the notion of a trading zone, an intermediate domain in which procedures could be coordinated locally even where broader meanings clashed. The ninth and final chapter of this book returns to these questions in greater depth to explore the dynamics of these exchange processes.

Subcultures trade. Anthropologists have extensively studied how different groups, with radically different ways of dividing up the world and symbolically organizing its parts, can not only exchange goods but also depend essentially on those trades. Within a certain cultural arena—what I call in chapter 9 "the trading zone"—two dissimilar groups can find common ground. They can exchange fish for baskets, enforcing subtle equations of correspondence between quantity, quality, and type, and yet utterly disagree on the broader (global) significance of the items exchanged. Similarly, between the scientific subcultures of theory and experiment, or even between different traditions of instrument making or different subcultures of theorizing, there can be exchanges (coordinations), worked out in exquisite local detail, without global agreement. Theorists and experimenters, for example, can hammer out an agreement that a particular track configuration found on a nuclear emulsion should be identified with an electron and yet hold irreconcilable views about the properties of the electron, or about philosophical interpretations of quantum field theory, or about the properties of films.

The work that goes into creating, contesting, and sustaining local coordination is, I would argue, at the core of how local knowledge becomes widely accepted. At first blush, representing meaning as locally convergent and globally divergent seems paradoxical. On one hand, one might think that meaning could be given sentence by sentence. In this case the global sense of a language would be the arithmetical sum of the meaning given in each of its particular sentences. On the other hand, the holist would say that the meaning of any particular utterance is only given through the language in its totality. There is a third alternative, namely, that people have and exploit an ability to restrict and alter meanings in such a way as to create local senses of terms that speakers of both "parent" languages recognize as intermediate between the two. The resulting pidgin or creole is neither absolutely dependent on nor absolutely independent of global meanings.

Translation, our usual linguistic picture of meaning shift, is frequently associated with a holistic view of meaning. In the translation view of cross-boundary talk, terms are so interconnected that the translation of "electron" from the culture of the theorist to the culture of the experimenter, or from Stoney to Lorentz, is so deeply tied to all other possible utterances and associations that people who speak "Stoneyan" or "Lorentzian" must speak past one another. Rather than depicting the movement across boundaries as one of translation (from theory to experiment, or from military to civilian science, or from one theory to another), it will prove useful to think of boundary work as the establishment of local languages—pidgins and creoles—that grow and sometimes die in the interstices between subcultures.[48]

In this view, exchanges between the subcultures of physics and between each of these subcultures and the broader embedding culture are part of the same problem. Electrical engineers must speak to structural engineers on the TPC; electrical engineers must speak to experimental physicists; theorists building radar at the MIT Radiation Laboratory must communicate with radio engineers;

48. In working out the idea of a trading zone, I have benefited enormously from the work of anthropological linguists, who have incorporated into anthropology a sophisticated taxonomy of pidgins and creoles and an analysis of the development and uses to which these tongues have been put. Of particular value to me have been Highfield and Valdman, *Creole* (1980); Mühlhäusler, *Pidgin* (1986); Foley, "Language" (1988); Hymes, *Pidginization* (1971); Romaine, *Pidgin and Creole* (1988); Dutton, *Police Motu* (1985); Todd, *Pidgins and Creoles* (1990); and Bickerton, *Roots* (1981). Additional sources are cited in chapter 9.

The idea of "trading zones" is developed in various places, including Galison, "The Trading Zone" (1989); and Galison, "Contexts" (1995). Of all the literature on scientific exchange, with respect to the notion of the trading zone I find most congenial the work of Star and Griesemer's "'Translations' and Boundary Objects," *Soc. Stud. Sci.* 19 (1989): 387–420, where they use the history of ecology to show how certain objects (such as fossils) could participate simultaneously in the different "views" held by collectors, trappers, administrators, and others—and yet have some element of continuous identity. The notion of cooperation through heterogeneity is key for their project and mine. As I will argue below, however, I urge first that we drop the notion of "translation" and replace it by the establishment of local exchange languages. Second, as the term "trade language" suggests, more can be shared between languages than the nouns designating specific objects: we will frequently be concerned in the context of experimentation with locally shared *procedures* and *interpretations* as well as objects.

and the Einsteinian theorist must come to terms (so to speak) with the Newtonian theorist. We must be willing to admit that the theorist of 1952 may have seen the electron as a quantum field theoretical object spending part of its time surrounded by a sea of other ephemeral particles while the experimenter of 1952 may have had no such notion and that, nonetheless, the theorist and experimenter shared many structured assumptions about the electron's behavior: charge, spin, mass, and related knowledge about ionization patterns in cloud chambers, films, and bubble chambers. I am interested throughout this book in the process by which the "deep" and global ontological problems of what an electron "really" is can be set aside while these local exchanges get worked out. In other words, bracket off for now the startling contrast between the way the Newtonian employed "time" or "space" in the late seventeenth century and the way the Einsteinian used these terms in the early twentieth century. Look instead at the way Poincaré, Bucherer, Einstein, Kaufmann, and Lorentz coordinated their use of these terms in the particular laboratory setting of particular experiments. Just how these exchanges take place is a question to be tackled empirically. One consequence of this view is that the (by now) classical problem of incommensurability between pre-Einsteinian and Einsteinian physics is not to be treated in isolation. The problem is considered here as a particular instance of local exchange between two theoretical subcultures. More generally, I want to treat the movement of ideas, objects, and practices as one of local coordination through the establishment of pidgins and creoles, not by invoking the metaphor of global translation and its philosophical doppelgänger, the conceptual scheme.

Anthropological linguists have studied hundreds of these remarkable language pockets from the northwest American coast to the western coast of Africa. Some *pidgins,* like the Papuan Koriki Hiri Trade Languages, are composed of no more than a few hundred words and are designed to coordinate a highly specific exchange of goods.[49] *Extended pidgins,* including Tok Pisin (formerly New Guinea Pidgin), are characterized by a significantly larger lexicon and more flexible syntax than trade languages—while such pidgins may originate in trade, some grow to play a wider communicative role between two or more "natural" languages. Finally, *creoles,* some of which are expanded versions of pidgins, are languages powerful enough to support the range of poetic, metaphorical, metalinguistic, and referential work that people using them demand of a first language: people can grow up in a creole. All of these linguistic systems, to one degree or another, facilitate local communication between communities of what would otherwise be mutually incompatible languages while preserving the separateness of the parent languages.[50]

49. Dutton, "Rare Pidgins" (1983).

50. The span of interlanguages is actually more finely categorized than indicated here. E.g., Peter Mühlhäusler includes prepidgin continua (varieties not yet stabilized), minimal pidgins, pidgins, extended pidgins, initial

Because the picture of physics sketched here is one of distinct but coordinated subcultures, the notion of an interlanguage is a useful decentered metaphor. In different forms the same kind of question arises: How should we think about the relation of theorists to theorists, of theorists to experimenters, of physicists to engineers, of chemists to physicists, of image instrument makers to logic instrument makers, and of the myriad of detector subgroups within a hybrid experiment one to the other? To homogenize these various groups artificially is to miss their distinct ways of going about their craft; to represent them as participating in isolated conceptual schemes "translating" back and forth is to shut our eyes to the productive, awkward, local coordination by which communities, machines, and knowledge get built. Consider three aspects of the interlanguage.

Locality.—Pidgins and creoles are specific to the uses to which they are put and the languages they connect. As such, they are emphatically not global lingua francas. Analogously, the working out of a interlanguage between experimenter and theorist is specific, not the realization of a universal protocol language. In 1944, Gregory Bateson commented on New Guinean Tok Pisin, envisioning the overlap between European and indigenous groups as a "third culture" different from each participating group: "[N]either [group's] philosophy of life . . . crops up in the neutral and special fields of Pidgin English conversation. It is not that democracy and private enterprise could not be described in Pidgin, it is just that, in fact they are not."[51] It might have been possible for the theorists Einstein, Lorentz, and Poincaré and the experimenter Kaufmann to have discussed the metaphysics of time as they worked out a way to coordinate theory and data bearing on the experimental examination of their conflicting theories, it is just that—following Bateson—they did not. It seems to be part of our general linguistic ability to set broader meanings aside while regularizing different lexical, syntactic, and phonological elements to serve a local communicative function. So too does it seem in the assembly of meanings, practices, and theories within physics.

Diachrony.—When translation is used as a philosophical metaphor for the relation between different theories, paradigms, or conceptual schemes, time drops out of the equation in favor of the structural difficulties associated with holistic meaning. By contrast, pidgins and creoles are time variant, sometimes shifting dramatically over a few decades, and it is this temporality that is needed to describe the shifting links between the subcultures of physics. The study of pidgins and creoles launches questions about language creation, expansion, contraction, and extinction. We will see, for example, the way the postwar interfield

creoles, creoles, and the postcreole continuum, where a creole drifts in various forms into a superordinate language. See Romaine, *Pidgin and Creole* (1988), 116; and Mühlhäusler, *Pidgin* (1986).

51. Bateson, "Pidgin English," *Trans. N.Y. Acad. Sci. II* 6 (1944): 137–41, on 139, cited in Mühlhäusler, *Pidgin* (1986), 84.

of *ionographie* drew together colloidal chemists and nuclear physicists around the registration of particle tracks in nuclear emulsions. But after an extraordinary efflorescence following World War II, the skill cluster surrounding these photographic plates died away as the mode of experimental work shifted from the tabletop to the bubble chamber factories of the 1950s. To anticipate another example: Monte Carlo simulations began near the end of World War II as a technique applied to thermonuclear weapon design and expanded in the 1960s to become a kind of pidgin between electrical engineers, physicists, airplane manufacturers, applied mathematicians, and nuclear weapons designers, eventually evolving into a defining problem area of computer science. Finally, though outside the focus of this book, it may be useful to think of links between other disciplines and physics—such as chemical physics and physical chemistry—as pidgins that gradually emerged into full disciplinary languages (creoles) rich enough to "grow up" in professionally.[52] Translation metaphors of scientific thought exclude history; the dynamic of interlanguage change keeps history front and center.

Contextuality.—Unlike the study of pure language and translation, the task of investigating interlanguages cannot avoid the historical and sociological circumstances under which these languages expand, contract, differentiate, and combine. War—to point to the most extreme circumstance—throws people of different languages together, and not surprisingly, a host of interlanguages have evolved during periods of conflict: specific coordinative languages emerged during World War II, in Korea, and in Vietnam.[53] For similar reasons, through the press of imposed, coordinated wartime work, interaction between physicists and engineers brought about a new way of thinking about microwaves and nuclear fission during World War II—and eventually changed the way both physicists and engineers practiced their professions. But war is not the only sociohistorical shaper of language.[54] The distribution of power between two communicating groups frequently plays a role, with the dominant group providing the lexicon and the less powerful group, in reduced form, the syntax. Another phenomenon is that of *relexification*—the replacement of the lexical structure in

52. The starting point of any such work would surely be the excellent book by Servos, *Physical Chemistry* (1990). The Ionists (van't Hoff, Arrhenius, and Ostwald) created their field as a localized set of problems—reaction kinetics, equilibrium, and affinity—borrowing the mathematized formalism from physicists and combining it with procedures and explanatory mechanisms that were distinctly chemical. One would then track the expansion of this field against the very different alignment of the two parent disciplines that gave rise to physical chemistry (Pauling et al.). Another inquiry into an interfield, though not one attached to physics, is Kohler's disclipinary history, *Biochemistry* (1982).

53. An example: In Vietnam a pidgin (Annamite French Jargon) that began in the French garrisons of Cochin China was spoken in Saigon and other sites of major French military presence until the French rout at Dien Bien Phu. By the 1970s it was almost extinct. See Reinecke, "Tây Bôi" (1971).

54. On sociolinguistic evolution of pidgins and creoles see, e.g., Valdman, *Pidgin and Creole* (1977), sec. 4; Mühlhäusler, *Pidgin* (1986), chap. 3; and Alleyne, introduction to Valdman and Highfield, *Theoretical Orientations* (1980). My favorite sociohistorical account of a pidgin is Dutton, *Police Motu* (1985).

a worked-out pidgin by the lexical content of the language of a newly dominant group. More generally, the context dependence of the development of pidgins and creoles—when a pidgin stabilizes, when it declines, and when it expands into a creole—raises important questions about the fate of practices lying between established fields of scientific inquiry.

This book is a history of instruments, but it is a history that cannot remain hermetically sealed. The laboratory draws together what at first may seem to be a Borgesian miscellany: meteorology, nuclear weapons, chemistry, volcanology, business strategy, and photography, along with other fields, practices, and materials. The image of physics as a densely connected map of distinct cultures bound by interlanguages offers an alternative both to a picture of crazy-quilt fragmentation and to one of homogeneous unification. But no metaphor is perfect, and the linguistic structure of pidgins and creoles should not obscure the focus here on material objects and their use. It is not that I want to explore the language that we use to talk *about* machines, but rather that I want to expand the notion of language to include the disposition of laboratory objects. Stated differently, we are familiar with asking about the changing meanings of theoretical objects: Physicists, philosophers, and historians have long asked, for example, how the quantum-theoretical wave shifts its meaning from the old quantum theory, through Schrödinger's wave mechanics, to Max Born's statistically interpreted wave function. Since the meaning of the quantum psi also connects, however partially, to issues outside of physics narrowly construed, a full unpacking of the meaning of the psi draws simultaneously on the technical uses of wave functions *and* on the broader range of meanings associated with notions of determinism, causality, and probability. Here I want to accord laboratory objects and associated practices a similarly embedded and changing character—spark gaps, films, and computer simulations have meanings as dense as electric fields, wave functions, and spacetime. The meaning of instruments should include not only what we say about them but the often unspoken patterns of their functional location with respect to other machines, patterns of exchange, use, and coordination. Like theoretical concepts, these knowledge-producing machines acquire meaning through their use within the physical laboratory and, in complex ways, through their material links to machines far from physics. The meaning of photographic apparatus, statistical counters, or computer simulations cannot be confined to microphysics—each is tied to sites far from the laboratory floor. To capture the movement of machines within different subcultures of laboratory practice, to get at the creation of more or less autonomous machine practices, I will frequently refer to interlanguages as a way of capturing the relation between the uses of detectors and theory, but also to get at the relation between different subcultures, between instrument-making traditions, and between instrument making and technical uses of these machines away from physics altogether. In order to emphasize that it is not just a matter of sharing objects between traditions but of the establishment of new patterns of their use, I will often refer to *wordless pidgins* and *wordless creoles:* devices and manipulations that mediate between different realms of practice in the way that their linguistic analogues mediate between subcultures in the trading zone.

By one reading, then, this book aims to show how two subcultures of instrumentation formed a wordless creole: On one side is the image tradition of photographic practices aiming at noninterventionist objectivity—producing homomorphic representations of nature. On the other side is the logic tradition of electronic practices aiming at manipulative persuasion—producing homologous representations of nature. Out of the two, I will argue historically, sociologically, and philosophically, came a wordless pidgin that evolved into a wordless creole: electronically produced images, homologous in source and homomorphic in presentation, a powerful system in which generations of physicists could then "grow up."

It is my hope that considerations like those sketched here will have some purchase in the study of the everyday practice of physics and at the same time bring out some of the philosophical and metahistorical difficulties of its study: experimental and instrumental traditions, intercalated periodizations, trading zones between subcultures, multiple constraints and simultaneous solutions, and work organization as a dynamic force in the growth and practice of modern physics. Instruments contain much more than springs, circuits, and film, and they are certainly not (pace Bachelard) merely reified theories. Instruments embody—literally—powerful currents emanating from cultures far beyond the shores of a master equation or an ontological hypothesis. To probe these currents, I want these specific, *sited* instruments to be emblematic of the changing nature of experimental work: these histories of things—cloud chambers, bubble chambers, wire chambers, nuclear emulsions—function allusively. The machines sit squarely in the laboratory and yet always link the laboratory to other places and practices. How?

One picture of the ways that instruments link one laboratory to another has been explored intriguingly by a group of sociologists of science following Polanyi and Kuhn on the notion of "tacit knowledge," the unarticulated craft aspect of scientific work. Collins's original argument was against those sociometrically inclined sociologists of science who thought relevant communities of scientists could be identified purely by tracking the movements of published information. Pointing out the many subtleties of information transfer, especially selective competitive secrecy, Collins and his collaborators could show that in at least some instances (such as the building of a certain type of laser) only those laboratories that acquired personnel from the originally successful laboratory managed to replicate the event they sought. "The point is that the unit of knowledge cannot be abstracted from the 'carrier.' The scientist, his culture and skill are an integral part of what is known."[55] This phenomenon was taken to be an instantiation of Kuhn's paradigm (alternatively expressed as a "form of life"), in which knowledge was not or could not be divided into articulated algorithmic procedures and therefore could not be conveyed through publications alone.[56]

As part of the antipositivist reaction to an overemphasis on the algorithmic or protocol vision of scientific action, Collins's work was apt and persuasive: experimentation was not and could never be purely a cookbook affair (he calls this picture "enculturational" as opposed to the "algorithmical" conception of scientific practice).[57] And indeed, there is no doubt that there were instruments and effects the replication of which required the movement of personnel and sometimes objects. For example, recall that it was only when a member of the Bristol team of emulsion experts arrived in Berkeley that the Berkeley team succeeded in making its emulsions exhibit the tracks of pions. But even in such cases the transfer was not totalistic, hardly a "way of life." Indeed, what is noteworthy about the Bristol-to-Berkeley transfer is the difference in the patterns of laboratory life. Bristol exemplified the postwar European cosmic ray effort, with individual workers, mountaintop laboratories, and small, cottage-industry scanning teams. Berkeley was, by the late 1940s, the quintessential factory-laboratory, in which the emulsions were exposed, not on a windswept alpine summit, but in a shift-run cyclotron. It is the stripped-down partiality of the shared procedures that interests me—the very opposite of a gestalt shift. I want to attend to the local, specific coordinative moves that made it possible for two such different scientific cultures to agree about emulsions, processing, and the analysis of tracks.

Appropriate to the totalistic tacit model of knowledge diffusion are the maps of transfer that have become common within the sociology of science. Arrows designate the transfer of people and things from a central location to outlying regions. Looking for all the world like the armies of a colonial power, these arrows spread out from London, Cambridge, and Paris as instruments and people are moved lock, stock, and barrel to new sites. Appropriate to the picture we seek is something much less centralized, much more partial. Chemical engineers share certain common understandings with emulsion physicists; emulsion experimenters share bits and pieces of interpretive strategies with theoretical physicists. Our map is not one of conquering armies or spreading disease

55. Collins, "TEA Set," *Sci. Stud.* 4 (1974): 165–86, on 183.

56. "[I]f it is taken that an important component of scientists' knowledge is 'tacit knowledge' (Polanyi, 1958) then the transfer of that knowledge to the scientist is likely to have been as invisible as the knowledge itself. More generally, if an actor's knowledge comprises his 'form of life,' and a scientist's knowledge comprises his paradigm, then the way that they came by that knowledge, or even elements within that knowledge, is unlikely to be properly investigated through means designed to explore information" (Collins, *Changing Order* [1985], 171).

57. Collins, *Changing Order* (1985), 57.

launched from one site and dominating all it encounters. Instead the limit zones between the various industrial, military, experimental, instrumental, and theoretical pieces of physics have complex and local boundary conditions: not the mathematically thin boundaries of topology, but the substantive, irregular, often productive boundaries that Peter Sahlins so effectively describes in the Pyrenees.[58] There are enclaves, and intermediate communities, jointly administered sectors, borderland languages, and autonomy movements at the margin. Experimental physics too is bounded in unpredictable and productive ways, drawing in turn on computer manufacturers, cryogenic experts, radar engineers, emulsion chemists, electrical engineers, and army-surplus gun mounts.

The very partiality of exchange between sites of knowledge makes possible the productive contact with these extraordinarily diverse parts of the world in which physics is embedded. There is no doubt that there were instruments and effects the replication of which required no movement of personnel and objects—vastly important instances in which scientist-to-scientist "craft skill" exchange does not figure at all. The spark chambers of 1960 never demanded the kind of careful craft transfer that is seen as an illustration of the inalienable importance of tacit knowledge. The reason is, in part, that the technologies—from Geiger counters to radar—that made the transfer from one laboratory to another needed only the most restricted bits of technique. Cork and Wenzel saw Fukui and Miyamoto's article on the discharge chamber in the scientific literature and within weeks had built one. There were no telephone calls, no transshipment of devices, and no personal visits. Indeed, across the United States, from Berkeley to Princeton, and many places between, groups built spark chambers with practically no difficulty—in physics classes for undergraduates and in the preparation rooms of the big accelerators.

In many different ways this book is a working out of the following observation: pieces of devices, fragments of theories, and bits of language connect disparate groups of practitioners even when these practitioners disagree about their global significance. Experimenters like to call their extractive moves "cannibalizing" a device. Televisions, bombs, computers, radios, all are taken apart, rearranged, and welded into the tools of the physicist. And the process can be inverted: instrumentation from physics becomes medical instruments, biological probes, and communication apparatus. Geiger-Müller counters were cannibalized to make the first electronic logic units for a computer, but pieces of the computer were soon stripped out for use in particle detectors. There is no unique direction, no requirement that the move be a platonic one from technology through experiment and then to the ethereal reaches of theory. Nor do we live in a Comtian world in which high theory always cascades down through experi-

58. Sahlins, *Boundaries* (1989).

ments to instruments and finally to the prosaic details of telephones, computers, and engines.

Because the argument here is for a connectivity of physics achieved through partial moves of coordination between theory, experiment, instrument making, and technology, there is no unique way to separate "science" from "context." Take the link between physics and the early computer just alluded to (and analyzed more fully in chapter 6). What is the context here? Should we say that the counter tradition in cosmic ray and nuclear physics formed the context of the development of the computer? Or should we say that computer technology formed the context for the development of instrumentation in nuclear and particle physics? When scale enters the picture, the demarcation of technological context from science is even more difficult. Do we say the MIT Radiation Laboratory grew up in the "context of industry" and was therefore "like" a factory? No, the Radiation Laboratory became part of an industrial system of production—a $2 billion project is not an epiphenomenon of wartime industry, even by the standards of war production. When the multinational collaborations of the TPC and the Solenoidal Magnetic Detector set up their finances, administration, and data production systems, these efforts began to shape the world around them. They too are wrongly regarded as islands of science in a sea of context. Methods of distant-site control, e-mail networks, and teleconferencing pioneered by these projects provided templates for multinational production; leaders of the collaborations attended management seminars with leaders of other multinational corporations. One final example: In the early 1990s, the Superconducting Supercollider was used by architect Moshe Safdie as a model for the design of new planned cities. Now what does one say: that the experimental instrument is the context for our urban life? Science-in-context mutates into science-as-context. With such shifting boundaries between science and context, we would do well to see "context" as a historically specific concept and *not* as a self-evident, transhistorical category of analysis.

Because these bottles of gas and racks of circuit boards participate so heterogeneously in the wider culture of Victorian mimetic experimentation, or postwar industrial production, or military projects, I am not comfortable treating the history of instrumentation as "case studies"—at least in the way that "case study" is frequently understood in science studies. This practice too often presupposes a relation of *typicality* between the episode being studied and a gamut of scientific procedures purportedly like it. One studies the incorporation of a company in business school to learn how companies, in general, are incorporated; or one studies the progression of a case of gangrene to know how to diagnose and treat the pathology in the future. The case study was imported into the history of science in the years after World War II by the president of Harvard, James Bryant Conant. Some history of history of science may help here in distinguishing between the use of local history as typical and the more located and emblematic use invoked in this book.

The "case system" (as it was called long before the turn of the century) had its origins in the Harvard Law School, where, in the 1870s, Dean C. C. Langdell boosted its usefulness as a pedagogical tool for the teaching of law in general, and contract law in particular. "Law," Langdell argued in the first (1871) edition of his *Selection of Cases,* "considered as a science, consists of certain principles or doctrines. To have such a mastery of these as to be able to apply them with constant facility and certainty to the ever-tangled skein of human affairs, is what constitutes a true lawyer; and hence to acquire that mastery should be the business of every earnest student of law." It was not that the doctrines were fixed in axioms centuries earlier and then applied. Rather (according to Langdell) current legal doctrines could be understood only through the cases that embodied them.[59] Properly culled and systematized, a choice set of cases would reveal that the "number of fundamental legal doctrines is much less than is commonly supposed."[60] Or as William A. Keener, dean of Columbia College Law School, put it: "The case system . . . proceeds on the theory that law is a science and, as a science, should be studied in the original sources."[61]

The original sources were, by Keener's lights, adjudged cases and not the distillations of textbook writers. Adjudged cases to the lawyer were the direct analogues of specimens to the mineralogist. Why read about the rocks when you can have them under your fingers and before your eyes? Here and throughout the literature of the 1890s on the case method, one finds a Protestant ethos[62]—James Carter, a leading figure of the New York Bar, advocated a return to pure text unsullied by the vagaries of the commentators or the interpolation of intermediaries.[63] Formerly, Carter continued, we went to text books for our law, but

59. Langdell, *Selection of Cases* (1879), viii.

60. Langdell, *Selection of Cases* (1879), ix. The case system was not without its detractors. Attacked on some sides for straying from clearly enunciated principles, the case study appeared to its enemies to encumber the profession with precedent over principle. Yale law professor Edward J. Phelps advocated a school library emptied of the too many books that burdened its shelves.

61. Keener, "Methods," *Yale Law J.* 1 (1892): 144.

62. Religious imagery emerged explicitly in the dramatic nineteenth-century defense of the case system by Carter:

> What is this thing which we call 'law,' and with the administration of which we have to deal? Where is it found? How are we to know it? It is not found in that code which was proclaimed amid the thunders of Sinai! It is not immediately and directly found in the precepts of the gospel. It is not found in the teachings of Socrates or Plato, or Bacon. It is alone found in those adjudications, those judgments which from time to time, its ministers and its magistrates are called upon to make in determining the actual rights of men. (Carter, quoted in Keener, "Methods," *Yale Law J.* 1 [1892]: 147)

63. Sir Frederick Pollock, an English lawyer, is quoted by Keener: "One of the first and greatest fallacies besetting law students is to suppose that law can be learned by reading *about* the authorities. Professor Langdell's method . . . strikes at the root of this" (quoted in Keener, "Methods," *Yale Law J.* 1 [1892], 147). Keener's emphasis on law as science was put in even stronger terms by Langdell himself as the case master reviewed the history of

such secondhand knowledge can never be worthy of science. Our goal must be "to study the great and principal cases in which are the real sources of the law, and to extract from them the rules which, when discovered, are found to be superior to all cases."[64]

The search for original cases and the "superior" rules that would emerge from them spread far outside legal practice.[65] Wallace Donham, dean of the Harvard Business School from 1919 to 1942, was trained at the law school in the heady days of the case system's early and enthusiastic reception. Where law and business parted ways was in the contingent matter of the availability of ready-made cases—law faculty simply reached for their shelves, while professors of business needed to create a new literary species—the business case book.[66]

Long before World War II, then, Conant had before him the twin exemplars of the case system, one from law and the other from business; by direct analogy with the business case book, the history of science case book was invented.[67] Following his World War II experience as director of the National Defense Research Council, Conant came to believe in the necessity of making

---

the method: "[It] was indispensable to establish at least two things; first that law is a science; secondly, that all the available materials of that science are contained in printed books." Only if law was indeed a science would it merit teaching at a university and not in a pure apprenticeship as a handicraft. "We have also constantly inculcated the idea that the library is the proper workshop of professors and students alike; that it is to us all that the laboratories of the university are to the chemists and physicists, all that the museum of natural history is to the zoologists, all that the botanical garden is to the botanists" (*Anniversary* [1887], 97–98, cited in Sutherland, *Law at Harvard* [1967], 175). We therefore have a fine hermeneutic circle: the legal case study is modeled on the physical and chemical laboratory, while some years later, the physical and chemical laboratory comes to be represented in a literary form designed around the legal case study.

64. Carter, quoted in Keener, "Methods," *Yale Law J.* 1 (1892): 147.

65. To a twentieth-century reader, the legislative, constructive function of law and the adjudicative function are distinct and equally important. But to Langdell, indeed to most American lawyers in 1870, the function of law, and therefore of cases, would have been primarily adjudicative. Legislative and administrative regulation was not a widespread part of the law—certainly less than it became in the decades to follow. See Sutherland, *Law at Harvard* (1967), 177.

66. To Donham, the case method stood squarely in the legal and cultural tradition of Anglo-American thought. Unlike French or Spanish law, Donham emphasized, English law was grounded on the doctrine of *stare decisis,* in which the written case decisions of the past shape, and instantiate, the law. Just as the recording of cases allowed English common law to break the arbitrariness of local law, Donham argued in 1925, business needed to universalize its procedures by itself adopting the case system. The chaos of local law that ruled in England before the common law, Donham contended, "is exactly the same situation that we have [in the world of business] where practically every large corporation is tightly bound by traditions which are precedents in its particular narrow field and narrow field only. The recording of decisions from industry to industry [enables] us to start from facts and draw inferences from those facts; [it] will introduce principle . . . in the field of business to such an extent that it will control executive action in the field where executive action is haphazard or unprincipled or bound by narrow, instead of broad precedent and decision" (W. Donham, transcript of talk to the Association of Coll. School of Business Committee Reports and Other Literature, 5–7 May 1925, Harvard University Graduate School of Business Administration Dean's Office Files, Wallace Brett Donham, 1919–42, Baker Library, Harvard Business School, box 17, folder 10, 62). To Donham, law, business, and economics all were open to treatment by the case system. All (he contended) "may broadly be considered sciences based in part on precedents and customs and in part on natural and economic laws" (W. Donham, "Business Teaching," *Amer. Econ. Rev.* 12 [1922]: 53–65, on 55).

67. Donham mentions the case method to Conant in, e.g., his letter of 18 February 1936 and accompanying memorandum, "The Theory and Practice of Administration," 15 February 1936, Harvard University Graduate

science available to the citizenry. Turning back to the case system, Conant wanted to convey the "special point of view" peculiar to scientists, one that transcended biology, physics, or chemistry. In particular, by examining what he envisioned as the "simpler" days of pre-twentieth-century science, this point of view could be elucidated, and it was to this end that he dedicated a third species of case book, the *Harvard Case Histories in Experimental Science.* Introducing the now-famous collection of inquiries, Conant wrote: "Some of the cases in this series present the work of men who lived over three hundred years ago; others are drawn from the eighteenth and nineteenth centuries; a few may involve twentieth-century discoveries. But irrespective of their dates, the examples presented illustrate the methods of modern science."[68] Langdell on law, Donham on business, and now Conant on experimental science, all three took it as fundamental that students could learn their disciplines only through confrontation with "original" cases; then and only then could the inductive method extract underlying principles that cut across time and, in the case of science, define "modern science."[69]

It is precisely to the denial of a rigid experimental method, fossilized forever in the early modern period, that this book is devoted. It is not just that the twentieth century studies fluctuations in the quantum vacuum state instead of birds in the vacuum jar. Rather, I am interested in the changing sense of what it means to be an experimenter (or an experiment), and it is for this reason that I focus throughout the book on just those difficult moments of transition in the practices of experimentation and the changing locations of experiment. I begin these histories with the material objects that populate the laboratory and ask simple questions: Where does the liquid hydrogen come from? Why is Wilson taking pictures of clouds? Out of which machine did Gozzini pull his microwave generator? Building on these stripped-down objects and transferred material, one begins to re-outline the activity in a way not obvious from the final experimental conclusions taken alone. And in the links of practices and pieces of machines one can see other, less visible transfers, appropriations, and adaptations: pieces of shared administrative structure, organization of architectures, and the handling of data. How, I want to know, does it come to pass that a single photograph of a set of tracks can count as an experimental demonstration—even when another "similar" photograph may not be produced for months, even years? How is it that the management of an experimental team comes to supplant the act of discovery as the crucial feature of experimentation? When and why does data

School of Business Administration Dean's Office Files, Wallace Brett Donham, 1919–42, Baker Library, Harvard Business School, box 37, folder 35.

68. Conant and Nash, *Harvard Case Histories* (1950–54), 3.

69. On recent use of the notion of a case study, see, e.g., Lynch and Woolgar, *Representation* (1990), 3, where the editors write of the "Strong Programme": "Bloor and Barnes developed several research policies which they and their colleagues exemplified with socio-historical case studies."

analysis come to be seen as the essential feature of experimentation—even when the experimenter is more or less completely excluded from the operation of the instrument? How does the changing division of work in the laboratory between engineer and physicist reshape what counts as an experiment? How do computer simulations alter (or expand) the meaning of experimentation?

By returning over and again to the changing site of experimentation—the Cavendish, Bristol, alpine summits, Berkeley, Los Alamos, Waxahachie—these and other questions acquire a generality that is nonetheless fixed in history. It was not an accident that Monte Carlo methods prospered just at the time that programmable electronic computers and a particular set of engineering and weapons problems entered the scene. I want to explore the configuration of events that made possible a certain view of simulations and their place in scientific demonstration. While some of these issues are purely mathematical, some concern the technological difficulties of computation, and others are rather philosophical issues tied to the ideals of scientific reasoning. But my analyses of these historically specific, yet philosophical, issues about the nature of an experiment are not "case studies" in the sense that they expose the universal structure of "modern science."

If the term "case study" implies to the reader no more than a detailed study of scientific work, then the chapters that follow are case studies. But if, as I suspect, the term now carries far more meaning than this minimal reading implies, the matter must dealt with up front. Imagine a book entitled *A Case Study in European History: France.* This made-up title strikes me as immensely funny, not because it purports to be a detailed study of an individual country (there are many important national histories), but because it encourages the reader to imagine a homogeneous class of European countries of which France is an instance. The absurdity rests on the discrepancy between the central and distinctive position we accord France in history and the generic position we must assume France occupies if we wish to treat it as a "case."

In studying instruments a similar problem arises: the cloud, bubble, spark, and wire chambers cover both too wide and too diverse a swath of the history of recent physics to be "instances" of instruments sub specie aeternitatis. Yet it is my hope that through these particular instruments the changing texture of experimentation can be conveyed; if my approach is successful, these chapters will convey the often vertiginous shifts from the late Victorian cloud chamber, to the postwar factory-like bubble chamber, to the multinational executive-board-governed $100 million TPC.

Because of this diversity of experimental work, I hesitate to see, as Conant did, an idealized version of experimentation as a model of human decision making; even where Conant's broad claim has been abandoned by subsequent analysts of science, the picture of microhistory as "typical" frequently persists. In looking for a case study to offer clues to a universal form, structure, or dynamic of "modern science," we perhaps too easily transfer an epistemological precept from science to history itself: For the physicist, to have probed the structure of one kaon is to have studied them all, but it does not then follow that for the case study historian, to unravel the discovery of oxygen is to know some essential element of the discovery of all other objects of natural philosophical inquiry. It is in this sense that I am uncomfortable with the definite article that begins the title of Kuhn's great work, *The Structure of Scientific Revolutions,* and equally so with more recent sociological claims that modern experimental science can be studied, with essentially the same results, from the seventeenth century to supercollider simulations in the early 1990s. But to deny typicality is often seen as a doomed retreat to history as detail-for-its-own-sake. The problem is not unique to the history of physics—the very project of history must contend with the problem of how to make the particular stand for the general, and how to limit claims for such a stance.

Carlo Ginzburg devoted a full book to the tribulations of Menocchio, a sixteenth-century miller hauled before the Inquisition for his heretical dairy cosmogenesis: life appears in the universe spontaneously like worms in cheese.[70] But what could Menocchio have been typical of? Millers? Inquisitional targets? Cosmogenical heretics? Menocchio's universe is hardly a representative belief system, even among the class of millers located within an afternoon's walk of his home. Yet challenges to typicality like these seem to miss the point of Ginzburg's project. In quantitative social history, it has made good sense to focus on specific attributes (family size, age at marriage, life expectancy, income). By sacrificing the qualitative aspects of values and meaning, it is possible to bring measurable data to the fore. But trying to read microhistorical cultural history as if it were a weak form of cliometrics mismatches the lines of connection between specificity and generality. Ginzburg, as I understand him, wants to convey a sense of the different ways that elite and peasant cultures read in the sixteenth century, the various ways they manipulated the symbolic and the literal, and the discrepant values they affixed to the stories they told. To draw from a different episode in cultural history, out of the inquisitorial records of one Jacques Fournier the historian Emmanuel Le Roy Ladurie extracted the tale of the tiny town of Montaillou in the south of (modern day) France.[71] Could these documents, this town, these 25 peasants, be the basis for a "representative" case study? Hardly. Of the poor men and women hauled before a court in Pamiers, one finds a motley crew of Albigensians, as little representative of society (in a narrow demographic sense) as they could be. If the absence of ready-made typicality needs one final example, I cannot imagine a less "typical" story of everyday life in the sixteenth-century Pyrenees than Natalie Davis's account of imposture, love, and

70. Ginzburg, *Cheese and the Worms* (1983).
71. Ladurie, *Montaillou* (1979).

jurisprudence in *The Return of Martin Guerre*.[72] Ginzburg, Ladurie, and Davis are surely after something broader than the troubles of their handful of characters, but the connection between the specific and the general is not metonymic: the small in their studies is not a core sample of the large. What characterizes their histories is a use of the local and specific to explore beyond the proximate by probing the uses of symbols, meanings, and values as expressed within the constraints of particular cultures.[73]

The seven histories that follow this introduction similarly are not "typical"—we are not going to find even one other physicist "like" Alvarez—no one else flew in the chase plane over Nagasaki, created the largest bubble chamber group in the world, and eventually withdrew from particle physics because he found the routine unbearable. No one else is "like" Marietta Blau as she used her "day job" skills to parlay dental X-ray film into a fundamental tool of modern physics: even the margins are singular. But this book does aim beyond the particular instruments discussed. The bubble chamber takes us to an understanding among physicists in the 1950s of the importance of data analysis as the defining activity of experimentation, of the power and limits of statistical argumentation, of the value and opprobrium attached to the novel event. This book is a brief for *mesoscopic history,* history claiming a scope intermediate between the macroscopic (universalizing) history that would make the cloud chamber illustrative of all instruments in all times and places and the microscopic (nominalistic) history that would make Wilson's cloud chamber no more than one instrument among the barnloads of objects that populated the Cavendish Laboratory during this century. The chapters each aim to capture both literally and allusively the characteristic ways of working in experimental physics across several decades. That the chapters are roughly chronological yet not purely linear in their sequence is inevitable. The cloud chamber does not drop off the face of the earth with the invention of the nuclear emulsion, and to maintain narrative continuity in following the instruments, we will necessarily pass many times through the war years: to tell the story of emulsions in wartime Europe is to tell a history very different from the massive transformation of instruments in the war laboratories of the United States.

72. Davis, *Martin Guerre* (1983).

73. Levi put it nicely in his essay, "On Microhistory" (1991): "[All] social action is seen to be the result of an individual's constant negotiation, manipulation, choices and decisions in the face of a normative reality which, though pervasive, nevertheless offers many possibilities for personal interpretations of freedoms. . . . In this type of enquiry the historian is not simply concerned with the interpretations of meanings but rather with defining the ambiguities of the symbolic world, the plurality of possible interpretations of it and the struggle which takes place over symbolic as much as over material resources" (94–95). In the history of laboratory material culture the heterogeneity of constraints may be even more pronounced than that found in everyday life. There are not only the widely based obstacles of social structure, material properties, and work organization but also the specific barriers originating in the symbolic domain, ranging from conservation laws and mathematical constraints to rules of thumb.

The chapters of this book, like the Medieval and Renaissance histories I have cited, are grounded in the local. But I resist the designation "case study" because I do not believe that there is a set of defining precepts that can be abstracted from these or other studies to "experiment in general" (or, for that matter, "theory in general" or "instruments in general"). Experimentation is so rich and varied—from the replication of natural phenomena in the laboratory to the fixing of fundamental constants, from metaexperimental statistical demonstrations to demonstrations based on a single prize photograph—that I have no interest in pronouncements about putatively time-transcendent principles of laboratory action. The passage of time changes too much about the nature of the laboratory and experimentation for a "typical" experiment to capture the "principles" of experimentation the way Conant hoped to in his case books. And the history of experimentation is not over; those who build instruments, run experiments, and produce theory will continue to expand, contract, and reconfigure these activities in ways linked to earlier practice but not necessarily reproducing any "essence" of experimentation. To one generation the idea of an experiment on a computer was absurd; to the next it was natural. Experimentation is not captured by procedure alone, even the expanded sense of procedure that includes the protocol-escaping *Fingerspitzengefühl* or tacit knowledge that has so captivated commentators from Polanyi to more recent practitioners of social studies of knowledge.[74] Beyond bench skills, experimentation draws on and alters broader cultural values—values that emphasized moral and scientific exactitude in the early nineteenth century,[75] the imitation of natural phenomena in late Victorian times, managerial acuity in the 1930s, factory production and military organization in the postwar bubble chamber era, and more recently, the information-based delocalization of the laboratory in the 1980s and 1990s that can be seen to fit into a wider cultural debate about aspects of postmodernism.[76]

The stories that follow are about physics. There is no understanding the texture of this history without grasping debates over the nature of mesons, nucleons, hadronization, and quarks. But the stories are also about the culture of experimental science in the twentieth century and the complexities of its connection to the wider cultural spheres of theory, industry, warfare, professional identity, and philosophical inquiry. Perhaps one should see the following chapters as parables that work by evoking particular epochs of experimentation rather

74. On tacit knowledge, see Polanyi, *Personal Knowledge* (1958); and Collins, *Changing Order* (1985), esp. 56–58, 70–71.

75. On the espousal of exactitude as a laboratory and moral value, see the very thorough and interesting book by Olesko, *Physics* (1991), 381–82, 450, 460. Wise and Smith have developed the notion of "methodological mediators" to characterize the links between industrial culture and scientific practice. Along these lines, they focus on "force" and "work" as connective tissue between Victorian society and electrodynamic theory, the telegraph, and the steam engine; see Wise, "Mediating Machines," *Sci. Con.* 2 (1988): 77–113; Smith and Wise, *Energy* (1989); and Wise, *Values of Precision* (1995).

76. See Galison and Jones, "Laboratory, Factory, and Studio" (forthcoming).

than as cases that work by being representative. But unlike Aesop's these parables allude to *changing* values and meanings as they are read into and out of the knowledge machines we call instruments. The clashing goals of the American cyclotroneers and the European cosmic ray mountaineers, the tabletop science of Donald Glaser and the industrial-scale ambitions of Luis Alvarez—indeed the stories throughout this book—are pinned to the laboratory bench but always refer elsewhere. Necessarily this is sited, not typical, history, and its aim is to evoke the mesoscopic periods of laboratory history, not a universal method of experimentation.

My question is not how different scientific communities pass like ships in the night. It is rather how, given the extraordinary diversity of the participants in physics—cryogenic engineers, radio chemists, algebraic topologists, prototype tinkerers, computer wizards, quantum field theorists—they speak to each other at all. And the picture (to the extent one simplifies and flattens it) is one of different areas changing over time with complex border zones that sometimes vanish, coalesce, and even burgeon into quasi-autonomous regions in their own right.

2 Cloud Chambers: The Peculiar Genius of British Physics

2.1 Cloud Chambers and Reality

When E. N. da C. Andrade surveyed atomic physics in 1923, noting that future historians would see the cloud chamber as one of the lasting accomplishments of the age, he was less sanguine about the durability of theoretical models. These he believed would come and go. Each conceptualization captured features of a restricted field of inquiry, but none (Andrade judged) could be pushed too hard. Optics invoked "ellipsoids of optical elasticity," the kinetic theory fielded small elastic spheres, quantum theory and the wave theory advanced entities in manifest contradiction with each other. Under the fruitful regime of imaginative work, he believed, these tensions were not at all a bad thing; it was far better for new theories, in their first youth, not to introspect excessively. Later, in their dotage, there would be ample time for the peaceful reconciliation of respectable consistency and textbook security. "Whatever may be the fate of the theories which have been so inadequately exposed in this book, whatever modifications or mishaps they may meet, the experimental facts which led to their formation, and those others to whose discovery they in their turn gave rise, will remain as definite knowledge, to form a lasting ornament to an age otherwise rich in manifold disaster and variety of evil change."[1] Despite the depths of felt disaster, the crisis of early quantum theory and the crisis of the Great War left the strong foundation of experimental results intact.

To Andrade, "The triumph of the atomic hypothesis is the epitome of modern physics."[2] So it has been for many from the late nineteenth century to the late twentieth. Philosophers of every stripe returned again and again to the

1. Andrade, *Atom* (1923), 295.
2. Andrade, *Atom* (1923), 1.

question of the reality of atoms; indeed, debates over the "real" existence of the atom stood at the origin of the twentieth-century debate over scientific realism. Consider the young Moritz Schlick (later leader of the Vienna Circle), who in 1917 protested: "[T]he strictly positivist picture of the world seems to be unsatisfactory on account of a certain lack of continuity. In narrowing down the conception of reality in the above sense, we tear, as it were, certain holes in the fabric of reality, which are patched up by mere auxiliary conceptions. The pencil in my hand is to be regarded as real, whereas the molecules which compose it are to be pure fictions." Only realism could patch these rents in the tissue of belief; only realism could reconcile the "antithesis, often uncertain and fluctuating," between the real and the working hypothesis.[3] Historically, the molecule (or atom) was the quintessential theoretical entity. Because the cloud chamber purported to display particulate matter piece by piece, realists and fictionalists alike found they had to confront these wispy cloud tracks. Many agreed that the best support for an experimentally grounded realism below the threshold of microscopy lay with this dramatic, if simple device.

The cloud chamber, by making the subvisible world visible (as in figure 2.17, below), entered the struggle over the status of "theoretical entities" during every one of the last seven decades. Those asserting the reality of the microphysical celebrated the visualization of particulate collisions, dissociations, and scatterings. Percy Bridgman, in his *Logic of Modern Physics* (1932), put stock above all in operations, procedures that (he believed) defined the valid concepts vital to any legitimate description of the world. "In searching for such new experimental operations it seems to me that by far the greatest promise for the immediate future is offered by improvements in our powers of dealing with individual atomic and electronic processes, such as we now have to a limited extent in the various spinthariscope methods of counting radioactive disintegrations, or Wilson's β-track experiments. In this self-conscious search for phenomena which increase the number of operationally independent concepts, we may expect to find a powerful systematic method directing the discovery of new and essentially important physical facts."[4] To Bridgman, as to Andrade, the importance of the Wilson chamber lay in its ability to display individual processes, directly, and not through a long, complicated, and indirect chain of inference. Unlike instruments that measured molecular properties in the aggregate, the cloud chamber singled them out, gave them, as Bridgman so often stressed in other contexts, an "operational" meaning—a meaning tied to features of the sensible world.[5] For

3. Schlick, *Philosophical Papers* (1979), 265–66.

4. Bridgman, *Logic* (1927), 224.

5. Elsewhere Bridgman wrote of the new experimental discoveries, in particular of Brownian motion and the Wilson cloud chamber tracks, "the convenience of the atomic picture has become so overwhelming that we have discarded the alternative point of view completely, and speak of the situation in different terms, as when it is often stated that all these facts have proved the *reality* of atoms." In fact, the Wilson chamber photographs, along

somewhat different philosophical reasons, Sir Arthur Eddington in 1939 had similar concerns about the subvisible world. Like Bridgman, he was troubled by the gap between the phenomena we observe and the theoretical entities that are the subject of physical laws and theories. The cloud chamber promised to bridge this gap. As far as Eddington was concerned, we "actually do count electrons in a Wilson chamber, where their tracks are made visible."[6] Tracks and electrons blurred together; the faint cloud trajectories bound the elusive reaches of theoretical objects to the perceptible. This visibility extended our sight, our sensible world, down to the atom itself: "We can almost see protons and electrons in a Wilson chamber; we can almost see mass being conserved. We do not actually see these things; but what we do see has a very close relation to them."[7]

Andrade, Bridgman, and Eddington used what they took to be the similarity of cloud tracks to underlying particle trajectories to dispense with the common view that atoms and molecules were "useful fictions" not to be considered fully real.[8] This "almost seeing" that made the cloud chamber an extension of our sense of sight is what I mean by the homomorphic form of evidence that becomes characteristic of the image tradition. With these palpable tracks, there appears to be no need for the "inference" of statistics characteristic of the logic tradition—one is (nominally) simply extending sight. Such a generalized argument for microphysical reality and against atomistic agnosticism began long before quantum mechanics. However, it and the cloud chamber assumed added significance when such physicists and philosophers as Pascual Jordan, Werner Heisenberg, and Henry Margenau confronted the philosophically problematic foundation of quantum mechanics.[9]

with allied developments, actually define what "the operational meaning of *reality* is" (Bridgman, *Physical Concepts* [1952], 22).

6. Eddington, *Physical Science* (1939), 175.

7. Eddington, *Physical Science* (1939), 134.

8. On debates over the reality of the atom, see Nye, *Question* (1984), xiiiff., and her *Molecular Reality* (1972).

9. Pascual Jordan (who had already made fundamental contributions to the new quantum theory) hoped in the 1930s and 1940s to unify quantum physics and biology through the processes that began at the scale of the microscopic and ended with the macroscopic. Here again, by illustrating the micro-macro connection, the cloud chamber played a crucial explanatory role. As Richard Beyler has noted, Jordan originally used what he called his "amplifier theory" to account for the way in which (macroscopic) living beings were "directed" or "steered" by quantum events within the cell nucleus; see Beyler's "Physics" (1994) and "Targeting the Organism," *Isis* 87 (1996): 248–73. This steering mechanism would, at least in principle, Jordan hoped, avoid both a vitalistic independence of biology from physics *and* any simpleminded mechanistic reductionism. (Occasionally, as Norton Wise has observed, Jordan's cellular leadership principle made substantive contact with his less than microcosmic nazistic proclivities; Wise, "Pascual Jordan" [1994].)

Among the interpretations of quantum mechanics that Jordan rejected was the notion that quantum "decisions" (for an electron to be in one eigenstate rather than another) were in any way tied to our conscious or mental states. Amplification processes in the cloud chamber, in which an electron track was "particle-like" (and not wavelike), illustrated the way in which amplifiers-cum-recording mechanisms fixed a quantum state without reference to anything supraphysical—no mental processes, no psychic phenomena, no consciousness. According to Jordan, this purely physical process of ionization and droplet coalescence is identical with the "quantum decisions" over

The question of the existence of molecules was put into focus earlier by the kinetic theory of gases, in which the observable properties of gases are deduced from the assumption that gases are made of molecules, even though an individual molecule has no observable effect. To the combative philosopher of science Herbert Dingle, the physics of the molecular theory of gases unreasonably abandoned the age-old philosophical ideal of coming to terms with the "nature of a real, external, material world"[10] and in its place offered merely useful relations among observables. It was, he contended in a lecture reprinted in *Nature* (1951), "a betrayal of the true mission of physicists according to the accepted philosophy." Whereas earlier physicists had been "dedicated to the investigation of reality," now the discipline had fallen to probing "mere appearances." In this fallen state, the ultimate constituents of nature, the molecules, had become "counters . . . dummies . . . useful conceptions," and while Dingle considered such fictionalism an inescapable part of modern physics, he wrote of it in tones of saddened resignation.[11] Shocked by Dingle's reluctant instrumentalism, Max Born retorted that the kinetic theory had much to recommend it beyond merely recapitulating a phenomenal description of gases, the kinetic theory's prediction of the specific heat of monatomic gases being just one example. But Born's ultimate argument was the cloud chamber. There, he contended, one hit bedrock. How could one stand face to face with track photographs while maintaining that microphysics had halted at fictionalism? Born: "Here the evidence of the reality

---

which his overly mystical contemporaries had spilled so much ink; Jordan, "Measurement," *Phil. Sci.* 16 (1949): 269–78, on 271. In a footnote (272), Jordan then elaborated by quoting himself from an unspecified source: "Ein Lichtquant, welches ein Silberkorn in der Photoplatte entwickelbar gemacht, oder welches durch Ionisierung eines Moleküls die Bildung eines Nebeltröpfchens in der Wilsonkammer eingeleitet, oder welches endlich in einem Zellkern eine Mutation zustande gebracht oder ein Bakterium getötet hat—dies Lichtquant hat sich lokalisiert, hat einen bestimmten Ort angenommen, unter Verzicht auf seine komplimentäre erscheinungsmöglichkeit als räumlich ausgedehnter Wellenzug." See also Jordan's *Geheimnis* (1941), 89, and *Erkenntnis* (1972), 156.

Throughout the 1950s, virtually all of the major philosophers and physicists who wrote on the philosophy of science figured the cloud chamber into their discussions of measurement, observability, and realism; see Margenau's *Physical Reality* (1950) and Heisenberg's *Wandlungen* (1945). Heisenberg wrote in *Wandlungen:* "As far as the track of an electron in a Wilson cloud chamber can be investigated, the laws of classical mechanics can be applied to it. Classical mechanics does predict the correct track of the electron. But if, without observation of its track, the electron is reflected at a diffraction grating, [classical mechanics fails]" (translated in *Philosophical Problems* [1952], 42; see also 43, 46). Heisenberg, along with many others, used Wilson's chamber to exemplify the full force of the wave-particle duality. Inquiries like Millikan's oil drop experiment, Margenau insisted, made electrons stand motionless, "which obviously no self-respecting wave would do." That feat alone would seem to exclude the emerging picture of the wavelike nature of the elementary charge. "But certainly the Wilson cloud chamber cannot be dismissed without mention. . . . It is difficult to conceive how a wave, uncollimated and free to spread in space, could make narrow tracks such as those observed in a cloud chamber" (Margenau, *Physical Reality* [1950], 318). For Margenau and Heisenberg, the cloud chamber served, not to refute fictionalism per se, but to discredit the more specific attempts to minimize the physical reality of complementarity. Once again, this time even more elaborately, the cloud chamber made experimentally accessible aspects of nature previously only representable through theory.

10. Dingle, "Philosophy," *Nature* 168 (1951): 630–36, on 634.

11. Dingle, "Philosophy," *Nature* 168 (1951): 630–36, on 633.

of molecules is striking indeed, and to speak of a 'dummy' producing a track in a Wilson chamber or a photographic emulsion seems to me—to say the least—inadequate."[12]

Where, Born demanded, is the difference between this and the everyday reality that we accept as the standard against which microphysics is to be compared? Following the long history of physicists' armed metaphors, Born shot back at Dingle: "You see a gun fired and, a hundred yards away, a man breaking down. How do you know that the bullet sticking in the man's wound has actually flown from the gun to the body?" No one actually sees or could see the bullet fly through the air, unless high-speed photographic apparatus of the type Ernst Mach had invented had been installed. Conceding that he could not defend the bullet story against a determined skeptic (e.g., someone prepared to claim that the bullet in flight is merely posited to preserve the laws of mechanics), Born did insist that the putative divide between medium-sized and very small objects could not serve as the basis for a principled distinction. A defender of the dichotomy would be forced, for example, to exclude the "existential evidence" of a cloud chamber alpha track, which can be seen, while allowing the bullet, which cannot.[13]

Born's remarks make clear that there is something deeply valued about the visual character of evidence: you *see* a gun fired, you *see* a man breaking down, but while no one *sees* the bullet fly through air, you can *see* the cloud chamber alpha track. Just this imprinted preference for visual evidence struck Stephen Toulmin in his *Philosophy of Science* (1953):

> To a working physicist, the question "Do neutrinos exist?" acts as an invitation to "produce a neutrino," preferably by making it *visible*. If one could do this one would indeed have something to show for the term "neutrino," and the difficulty of doing it is what explains the peculiar difficulty of the problem. For the problem arises acutely only when we start asking for the existence of *sub-microscopic* entities, i.e., things which by all normal standards are invisible. In the nature of the case, to produce a neutrino must be a more sophisticated business than producing a dodo or a nine-foot man. . . . [C]ertain things are, however, generally regarded by scientists as acceptable—for instance, cloud chamber pictures of α-ray tracks, electron microscope photographs or, as a second-best, audible clicks from a Geiger counter. They would regard such striking demonstrations as these as sufficiently like being shown a live dodo on the lawn to qualify as evidence of the existence of the entities concerned. And certainly, if we reject these as insufficient, it is hard to see what more we can reasonably ask for: if the term "exists" is to have any application to such things, must not this be it?[14]

12. Born, "Physical Reality," *Phil. Q.* 3 (1953): 139–49, on 142.
13. Born, "Physical Reality," *Phil. Q.* 3 (1953): 139–49, on 142.
14. Toulmin, *Philosophy of Science* (1953), 136. J. J. C. Smart quotes (or rather misquotes) extensively from Toulmin in "Theoretical Entities" ([1956] 1973), 95.

At this moment one has reached the end: not even the forensic evidence of everyday macroscopic life can compare with the evidence of a cloud chamber. As Born put it, if existence is not demonstrated by these photographs, then what conceivable occurrence would count? If not here, where?

When W. V. O. Quine turned to the cloud chamber a few years later, it was seemingly for a different, and rather more subtle, philosophical end. His "Posits and Reality" (1956) turned on the thesis that the directness of the link between sense impression and theoretical inference would vary over time—what was indirect evidence at one moment could be restructured into a direct link as the practices of science shifted. "Many sentences even about common-sense bodies rest wholly on indirect evidence; witness the statement that one of the pennies now in my pocket was in my pocket last week. Conversely, sentences even about electrons are sometimes directly conditioned to sensory stimulation, for example, via the cloud chamber." Quine's point is that statements about any kind of body gain significance only within a "collectively significant containing system," and so the ontological status of a particle might move from indirect to direct. While some people might use this "slack" within science to claim that a theoretical entity such as an electron had a more dubious status than a couch, Quine disagreed: why, he asked, should one ascribe full reality to everyday objects for which there exists no complete system of discourse at all? Positing subvisible entities has brought benefits to us, including the added simplicity of physical theory, the fecundity for further testing offered by the hypothetical entities, and the added scope of physical theory. "Benefits [of this sort]," Quine concluded, "are what count for the molecular doctrine or any, and we can hope for no surer touchstone of reality."[15] On this view, electrons may well turn out to be *better* established than the room-sized objects against which they are usually, so deprecatingly, contrasted. Like Born, Quine looked at cloud chamber tracks and saw the best argument for existence, not a weak cousin to everyday sights and sounds.

Observation sentences could not live on their own, Quine taught. And throughout the 1960s, this lesson was repeated over and again, though increasingly often with a relativist slant that was not Quine's own. The positivist doctrine was inverted: now the cloud chamber became the target of choice for anyone arguing against observation as bedrock. In perhaps the best, most sustained version of the new *anti*positivist doctrine, Russell Hanson used his *Concept of the Positron* (1963) to defend the view that tracks in a cloud chamber that are now "read" as evidence for the positron were earlier "read" out of physical significance altogether. As Hanson concluded: "Whenever seen, such tracks were discounted as 'spurious,' or as 'dirt effects.' Certainly no experimental physicist before late 1932 made any such track his prime object of study. Part of the

15. Quine, "Posits and Reality" ([1955] 1973), 161.

function of [Hanson's *Positron*] will be to understand why this is so, why such tracks were always overlooked, underevaluated, or explained away."[16] Theory gave meaning to pair-creation tracks just as surely as a gestalt image of a duck or rabbit gave meaning to lines and squiggles on a page. Hanson contended that there was literally no raw data before theory made it into evidence for pair creation; the tracks in many instances could hardly be seen.

In 1970, Mary Hesse dubbed the view that observation and theory were inextricably enmeshed, one with the other, the "network model"; she too invoked the cloud chamber. Unlike many of her contemporary antipositivists, however, Hesse scrupulously distinguished scientific practice (which often treated observations or experiments as highly distinct from theory) from a priori claims that there was something about observation sentences that rigidly distinguished them from theoretical statements. Consequently, she was ready, as Quine was, to concede that "examples are thinkable where highly theoretical descriptions would be given directly: 'particle-pair annihilation' in a cloud chamber." Hesse, following Quine and Hanson, rejected the notion that it is possible to withdraw to a pure observation language from theory-laden descriptions. Even if one replaces "particle-pair annihilation" with "two white streaks meeting and terminating at an angle," the second does "not show that [it is] free from lawlike implications of [its] own, nor even that it is possible to execute a series of withdrawals in such a way that each successive description contains fewer implications than the description preceding it."[17] Hesse's point is that there is no distinction between observation and theory in kind—there is no safe haven to which one can retreat that is free of any reference to theoretical laws. True, it may be possible to find temporary shelter from any particular theory (e.g., of pair creation), but that possibility does not imply that the language of cloud chamber track geometry can be freed of *all* theory.

Bas van Fraassen was therefore building on a long and illustrious tradition when, in 1980, he attacked the notion that the cloud chamber made the electron "observable." Van Fraassen's starting point was a challenge to the very notion of "observability." Like "portability," he contended, the term refers to our human capacities (and not to some entirely abstract definition). We could say that there is nothing special about the bounds of what it is within our strength to lift—in other words, that "in principle" the Empire State Building is portable (were we giants)—but this would be to bend the term "portability" out of all recognizable similarity with standard usage. Similarly, "observability" does not allow for our having the eyes of eagles (or electron microscopes): "observability" refers to what we, as humans, can do under normal circumstances. The point is not that instruments are never useful as stepping stones to observability that would hold

16. Hanson, *Positron* (1963), 139.
17. Hesse, "Observation Language" ([1970] 1974), 23–24.

under better conditions. After all, van Fraassen readily acknowledged that "a look through a telescope at the moons of Jupiter seems to me a clear case of observation, since astronauts will no doubt be able to see them as well from close up."[18] Cloud chamber tracks are *not* like this: the electron or alpha particle is not observable by humans under any conditions; therefore, van Fraassen argued, it occupies a very different place in our epistemic hierarchy from that held by objects humans can observe. We carry an everyday realism about things we can see and feel, but our relation to the microworld is one that he concluded is inescapably instrumental: the cloud chamber reveals regularities of phenomena that are observable. That is all. It does not offer us a direct view of the subvisible in any way "like" or "almost like" our seeing a building, dodo, or man.[19] By now, it will come as no surprise that realists like Alan Musgrave responded to van Fraassen by invoking the cloud chamber in other ways, claiming, for example, that to speak of "detecting" a particle by the device was already to believe it to be "true that the object really exists."[20]

There you have it. The image of the cloud chamber has flown like a banner above almost every realist and antirealist crusade. To Bridgman, the American operationalist, the Wilson chamber was just the procedure needed to give meaning to utterances about the subvisible world of the atom—more, it provides an instance of what we mean by the term "exist," full stop. To Toulmin and Born, the cloud chamber was the very symbol of realism; Wilson's tracks are the final station of the argument against a division between ordinary objects and the population of the submicroscopic. As Born put it, "[I]f the term 'exists' is to have any application to such things, must not this be it?" To Jordan, Heisenberg, and others, cloud chamber pictures illustrated the spacetime half of the complementarity relations—how much more like a realistically interpreted classical particle could one get than the detailed footprints of an alpha particle skittering through the clouds? For Quine, the cloud chamber's images exemplified the most direct evidence for theoretical entities that we were likely to get, and in combination with the rest of our physics, our account of the microworld

18. Telescopes are not, however, to be analyzed in the same terms as microscopes since "the purported observation of micro-particles in a cloud chamber seems to me a clearly different case—if our theory about what happens there is right. . . . Suppose I point to [a contrail that looks like a cloud chamber track] and say: 'Look, there is a jet!'; might you not say: 'I see the vapour trail, but where is the jet?' Then I would answer: 'Look just a bit ahead of the trail . . . there! Do you see it?' Now, in the case of the cloud chamber this response is not possible. So while the particle is detected by means of the cloud chamber, and the detection is based on observation, it is clearly not a case of the [p]article's being observed" (van Fraassen, *Scientific* [1980], 16–17).

19. A closely related point was made many years ago by Nagel, *Structure of Science* (1961), where the cloud chamber occurs in the course of a longer argument intended to dismiss the realist-antirealist debate as "a conflict over preferred modes of speech" (152).

20. E.g., Musgrave, "Realism" (1985), 205–6. It should be emphasized once again that there are many kinds of realist and antirealist arguments that do not depend on the cloud chamber. The point here is that more than any other experimental apparatus, the cloud chamber has formed the locus classicus of the debate for many decades and many (contradictory) philosophical positions.

suggested a pragmatic realism. Full-fledged antipositivists, à la Hanson and Hesse, invoked the cloud chamber photograph as the best example of "neutral data" but went on to argue that even statements about these photographs could never truly be independent of theory. Against Born, van Fraassen celebrated theoretical links among tracks without conceding an inch to the reality of any entity ultimately causing the tracks, while his opponents saw realism embedded in the very notion of cloud chamber detection.

A vast spectrum of views among physicists surrounds the cloud chamber, a similar variety among philosophers. Yet all these accounts share the view that the cloud chamber, perhaps better than any other instrument, instantiates the production of direct evidence for the subvisible world of microphysics. The chamber revealed the positron and the muon to Carl Anderson and allowed George Rochester and C. C. Butler to "see" a new class of "strange" particles. John D. Cockcroft and Ernest T. S. Walton used the device to demonstrate the existence of nuclear transmutation. Indeed, for generations of cosmic ray physicists, and then briefly for accelerator physicists, the cloud chamber gave concrete meaning to the wide range of new particles whose discovery inaugurated the field of particle physics. Cloud chambers were the prototypes for the spate of later detectors that we will examine in the following chapters, including the high-pressure chamber, sensitive nuclear emulsions, and most important, the bubble chamber.

Given the physical and philosophical background, it comes as a shock to find that this quintessential particle detector had its origin in a time, place, and subject utterly removed from the scattering, production, and disintegration of particle physics. But by transporting ourselves away from particle physics, out of the postwar laboratory, and back to an era of Victorian meteorology, we can begin to reconstruct the process by which this "cloud room" moved from the storm-drenched hills of Scotland to become, as Lord Rutherford called it, "the most original and wonderful instrument in scientific history."[21] Only through the lens of Victorian experimentation, through the fascination with the wholesale reproduction of the great forces of nature, can one see the origin of the image tradition. My hope is that this genealogy will help us understand what these vivid pictures have come to be for so much of our century, what Samuel Johnson's rock was for his: the final word in evidence.

2.2 The Romance of Re-creation

Despite the uses to which his device was later put, Charles Thomson Rees Wilson, the creator of the cloud chamber, cannot possibly be considered a particle physicist.[22] From his earliest work in 1895 to his last ruminations on

21. "Professor C. T. R. Wilson" [Obituary], *Times* (London) 16 November 1959, 16.

22. There are only a few articles written about C. T. R. Wilson, who lived from 1869 to 1959. The most complete is Blackett, "Charles Thomson Rees Wilson," *Biog. Mem. F.R.S.* 6 (1960): 269–95; Turner has a short

thunderclouds when he had turned ninety, Wilson was riveted by the phenomena of weather. Even J. J. Thomson, one of his greatest admirers, commented that Wilson's experimentation on fogs was not "a very obvious method of approaching transcendental physics."[23] By "transcendental physics" Thomson meant analytical research into the basic structure of matter that shunned complex but mundane problems like fog and rain. Though the eventual use of the cloud chamber has led atomic, nuclear, and particle physicists to appropriate Wilson as one of their own, his life's work is incomprehensible outside the context of weather. One must come to terms with the dust, air, fogs, clouds, rain, thunder, lightning, and optical effects[24] that held the rapt attention of Wilson and his nineteenth-century contemporaries in order for the invention of the cloud chamber to make historical sense.

The historical reconstruction of mundane as well as transcendental physics is necessary to set Wilson's cloud chamber work in its properly Victorian context. That world will emerge as a startling source of modern physics, revealing in Wilson's work a coherence that is entirely lost when it is divided into "meteorology" on the one hand and "physics" on the other. Lying between or perhaps in the intersection of the two domains is an area better called "mimetic

article, "Wilson," *Dict. Sci. Biog.* 14 (1981): 420–23. Wilson himself left two retrospective articles, "Ben Nevis," *Weather* 9 (1954): 309–11, and "Reminiscences," *Not. Rec. Roy. Soc. London* 14 (1960): 163–73. His Nobel Prize Lecture, in 1927, gives a general overview of his work ("Making Visible Ions" [1965]). In addition, Wilson wrote chapter 7, "1899–1902," of *History of the Cavendish* (1910). Crowther discussed Wilson in two of his books, but without much detail: *Scientific Types* (1968), 25–55, and *Cavendish Laboratory* (1974), 126–75, 213–24. Tomas, "Tradition" (1979), has a provocative thesis that uses art-historical techniques to fix the authorship of certain chambers; Tomas does not use Wilson's notebooks. For a modern explanation of the cloud chamber, see J. G. Wilson, *Cloud-Chamber Technique* (1951). Meteorologists recall Wilson as an important figure in their discipline; see, e.g., Halliday, "Some Memories," *Bull. Amer. Met. Soc.* 51 (1970): 1133–35. The most important source for the cloud chamber, however, is Wilson's own laboratory notebooks, which will constitute the core of this study. They are referenced below.

23. Thomson, *Recollections* (1937), 416. This language was not unique to Thomson, as is evident when Airy wrote Stokes in April 1867: [The Mathematical Tripos] "gave no adequate encouragement or assistance to men of the second degree, who might by proper direction of their studies become men highly educated as philosophers though not so transcendent as the first" (quoted in David B. Wilson, "Experimentalists," *Hist. Stud. Phys. Sci.* 12 [1982]: 325–71, on 338).

24. Several different meteorological optical effects will be mentioned in this paper. The terminology gets a bit confusing, so a short description of each phenomenon is given here for future reference. A *corona* is a series of concentric colored rings formed by the diffraction of light from the sun or moon when it hits water droplets in clouds or high fogs. A *halo* is also a series of colored rings, produced, however, by refraction through ice crystals. Although these two terms are explicitly defined by meteorologists, they are used inconsistently by different observers. A *glory* consists of colored rings that are seen around the shadow that an observer casts on a cloud or a fogbank. It is caused by the diffraction of light by water droplets and can occur with a corona. The entire phenomenon of the shadow and rings is sometimes called a *Broken Spectre.* A *Bishop's Ring* is not due to water droplets but instead is due to the interaction of light with solid particles in the atmosphere. Bishop's Rings were often seen after the Krakatoa eruption of 1883 but are very rare in normal circumstances; they depend on there being a large number of particles in the atmosphere and therefore are usually seen after a catastrophic explosion. See Whipple, "Meteorological Optics" (1923), 3:527–29.

experimentation," a term that will designate the attempt to reproduce natural physical phenomena, with all their complexity, in the laboratory.

It is with mimetic experimentation that Wilson's work begins, and the evolution of his thought and his machine must be understood as a continuing dialogue between general theories of matter held at the Cavendish Laboratory and particular demonstrations of remarkable natural phenomena. The Wilson cloud chamber is the material embodiment of this conversation between the analytic and the mimetic and, as such, became the foundation for the hundred-year reign of the tradition of image-making devices. Exploring the origin of the cloud chamber will offer insight into the intersection and subsequent transformation of the material cultures of both meteorology and matter theory. By following the movements of this specific instrument we will be able to study the formation and disintegration of "condensation physics," a transitory subfield of physics. Then, as the device moves away from condensation phenomena per se, an analysis of the subsequent construction and deployment of cloud chambers will expand our picture of the historical roots both of particle physics and of physical meteorology.

For the Victorian imagination, the extremities and rarities of nature held an endless fascination. Explorers ventured to the ends of the empire, to the deserts, jungles, and icecaps. Painters and poets tried to capture the power of storms and the grand scale of forests, cliffs, and waterfalls. And both artists and scientists recognized a tension between the rationalizing, lawlike image of nature proffered by the natural philosophers and the irreducible, often spiritual aspect of nature presented by their contemporaries in the arts.[25] There was a similar split in science itself between an abstract, reductionist approach to the physical world and a natural historical approach that authors from Goethe to Maxwell had dubbed the "morphological" sciences.[26] Of these sciences, Goethe took particular joy in meteorology, for "atmospheric phenomena can never become strange and remote to the poet's or to the painter's eye."[27] Up through the eighteenth century, there had been no systematic classification of clouds. Then in 1802–3, a British chemist, Luke Howard, presented a classification system that he modeled on Linnaean taxonomy.[28] Through Goethe, Howard's system entered the cultural mainstream.

In front of a small philosophical society, Howard sorted clouds according to a "methodical nomenclature": "cirrus," "cumulus," and "stratus." Howard chose Latin for his system because he thought Latin would capture the universal

25. Much of the debate was fueled by reforms in scientific instruction. See Super, "Humanist" (1977).

26. *Encyclopaedia Britannica,* 9th ed., s.v. "physical sciences" (by J. C. Maxwell): "What is commonly called 'physical science' occupies a position intermediate between the abstract sciences of arithmetic, algebra, and geometry and the morphological and biological sciences." Goethe, quoted in Merz, *European Thought* (1965), 212–13.

27. Goethe, quoted in Badt, *Constable's Clouds* (1950), 17.

28. Howard, "Modifications of Clouds," *Phil. Mag.* 16 (1803): 97–107; 344–57; 17 (1803): 5–11.

validity of his scheme. Moreover, in contrast to chemists who used Greek terms to represent invisible chemical entities, Howard wanted to classify clouds "by [their] visible characters, as in natural history."[29] No doubt this affinity with natural history appealed to Goethe. When he discovered Howard in 1815, the poet was deeply impressed with the new way of seeing clouds: "I seized on Howard's terminology with joy," Goethe announced, for it provided "a missing thread."[30] From Goethe the Dresden school of painters learned to view clouds differently; the art historian Kurt Badt surmises that it was Luke Howard's expanded work of 1818–20 that triggered John Constable's astonishing cloud studies of 1821–22 (though this has been disputed; see figures 2.1 and 2.2).[31] Whether or not Constable's studies were inspired by the new meteorology, it is clear that he avidly followed the popular work of Thomas Forster, *Researches about Atmospheric Phaenomena* (1815), mixing theories of the weather with Forster's own observations. Constable challenged some passages and marked others of particular interest to him, including the following bit of Forster: "[O]n the barren mountain's rugged vertex, in the uniform gloom of the desert, or on the trackless surface of the ocean, we may view the interesting electrical operations which are going on above, manifested in the formation and changes of the clouds, which bear water in huge masses from place to place, or throw it down in torrents on the earth and waters; and occasionally creating whirlwinds and water spouts; or producing the brilliant phaenomena of meteors and of lightening; and constantly ornamenting the sky with the picturesque imagery of coloured clouds and golden haze."[32] Meteorology, given cultural weight through this kind of evocative popular science, fostered cloud studies in painting, in poetry, and later in photography. Clouds became a central figure in romantic thought.

For Luke Howard the study of clouds was much more what Goethe would call "morphological" than "abstract." Howard never felt at ease with mathematics or the newer, more mathematized forms of chemistry.[33] At the end of the nineteenth century, the historian Theodore Merz commented on this dual aspect of systematic thought; he stressed that the "abstract sciences" (e.g., optics,

29. Howard, "Modifications of Clouds," *Phil. Mag.* 16 (1803): 97–107, on 98.

30. Goethe, quoted in Badt, *Constable's Clouds* (1950), 18.

31. Louis Hawes has vigorously disputed the central role of Howard's meteorology in the development of Constable's sky studies. Hawes argues instead that "environmental conditions," such as Constable's work in his father's windmill, along with the cloud depictions by earlier painters played a more important role; see Hawes, "Constable's Sky," *J. Warburg Courtauld Inst.* 32 (1969): 344–65. It seems, however, that some of Hawes's claims have themselves run into trouble. For example, he writes that Constable "nowhere mention[s] Howard or his terminology" (346). Three years after Hawes's article appeared, Constable's copy of Forster's *Atmospheric Phaenomena* (1815) came to light, and it includes, among many other notations, references to Howard's meteorology; see Thornes, "Constable's Clouds," *Burlington Mag.* 121 (1979): 697–704.

32. Thornes, "Constable's Clouds," *Burlington Mag.* 121 (1979): 697–704, on 698.

33. At Goethe's request, Luke Howard composed a brief autobiographical statement that is reproduced in Goethe's collected works: see Howard, "Luke Howard an Goethe" ([1822] 1960); on chemistry and mathematics, see 824–25.

Figure 2.1 Luke Howard's clouds (1802–3). Luke Howard's classification system, widely disseminated by Goethe, launched an aesthetic, popular, and scientific fascination with clouds in Victorian times. Here, *a* exhibits cirrocumulus up close and in the distance; *b* shows a light cirrostratus (just before rain) and a dark cirrostratus (in twilight); and *c* displays "mixed" and "distinct" cumulostratus. Note: Captions in original are misnumbered. Source: Howard, "Modifications of Clouds," *Phil. Mag.* 16 (1803): 97–107, plate on 64.

Figure 2.2 Constable's cloud study (1821–22). John Constable's wildly popular representations of clouds were generally perceived to have broken with previous, formal means of depicting them. His paintings of dramatic sky phenomena brought meteorological explorations front and center for Victorian culture. Reproduced here is Constable's cloud study: "Looking South." 21st Sept. 1822. 3 o'clock afternoon, Brisk wind at East, warm and fresh. Source: Constable, *Study* (1822): No. B1981.25.116, Yale Center for British Art, Paul Mellon Collection.

mechanics, electricity, and magnetism) involved either "literally a process of removal from one place to another, from the great work- and store-house of nature herself, to the small workroom, the laboratory of the experimenter," or a process of removal "carried on merely in the realm of contemplation."[34] The morphological sciences had, almost by definition, a place in nature itself.

But abstraction was the true goal of physics, according to the leading practitioners of matter physics, not classification of the "countenance of the sky" or the earth. William Thomson (Lord Kelvin), for example, allowed natural history a merely preliminary role: "[I]n the study of external nature, the first stage is the description and classification of facts observed with reference to the various kinds of matter. . . . [T]his is the legitimate work of Natural History. The establishment of general laws in any province of the material world, by induction from facts collected in natural history, may with like propriety be called Natural Philosophy." For the "abstract" or "natural philosophical" investigator, the goal of experimentation was to extract the universal law from the particular

34. Merz, *European Thought* (1965), 200.

description—and to thereby achieve a more "transcendent" truth than could be obtained by excessive attention to special phenomena. The success of the natural philosophical approach was present for all to see in Maxwell's theory of electrodynamics and Kelvin's theory of heat. Both took certain known phenomena and gave them a mechanical, dynamical basis.[35]

But despite the manifold benefits of the "chemical and electrical laboratories with the calculating room of the mathematician on the one side, and the workshop and factory on the other,"[36] Victorian scientists realized that there were times when the abstract scientific method was inadequate. Difficulties arose because the natural philosopher who exploited analysis and abstraction exclusively was "forcibly reminded that he [was] in danger of dealing not with natural, but with artificial, things. Instances are plentiful where, through the elaboration of fanciful theories, the connection with the real world has been lost."[37] When the natural philosophers invented a dynamical basis for phenomena, they ran the risk of inferring from the success of a model the existence of possibly spurious entities put forward in the model.

Opposing the "one-sided" working of abstract science lay another ideal of investigation, embodied in the morphological sciences. These sciences were motivated, as Merz somewhat rhapsodically put it, by "the genuine love of nature, the consciousness that we lose all power if, to any great extent, we sever or weaken that connection which ties us to the world as it is—to things real and natural: it finds its expression in the ancient legend of the mighty giant [Antaeus] who derived all his strength from his mother earth and collapsed if severed from her."[38]

The morphological scientist "look[s] upon real things not as examples of the general and universal, but as alone possessed of that mysterious something which distinguishes the real and actual from the possible and artificial."[39] Explaining that he is borrowing and extending Goethe's term, Merz included the large-scale study of landscape—mountains and valleys, glaciers, land and water, stratification of rocks, and formation of clouds—under the rubric "morphological" sciences.[40] Alexander von Humboldt, the great explorer and measurer of natural phenomena like wind and air pressure, was in Merz's eyes the leading advocate of this new line of investigation: Humboldt, Merz told his readers, "may be called the morphologist of nature on the largest scale."[41]

35. Thomson, "Introductory Lecture" (1846), quoted in Smith and Wise, *Energy and Empire* (1989), 121–22; and *Encyclopaedia Britannica*, 9th ed., s.v. "physical sciences" (by J. C. Maxwell).

36. Merz, *European Thought* (1965), 202.

37. Merz, *European Thought* (1965), 201.

38. Merz, *European Thought* (1965), 202.

39. Merz, *European Thought* (1965), 203.

40. Merz, *European Thought* (1965), 219.

41. Merz, *European Thought* (1965), 226.

Susan Faye Cannon, writing in the 1970s, was thus echoing Merz when she identified the "great new thing in professional science in the first half of the 19th century" as "Humboldtian science, the accurate, measured study of widespread but interconnected real phenomena" such as "geographical distribution, terrestrial magnetism, meteorology, hydrology, ocean currents, the structures of mountain-chains and the orientation of strata, [and] solar radiation." By their commitment to work on location, the Humboldtians opposed "the study of nature in the laboratory or the perfection of differential equations."[42] The laboratory dealt with artificially isolated phenomena—not nature—and differential equations disembodied the variegated reality of actual things. Geologists held just such an antilaboratory view; Mott Greene reports that two of the most prominent early-nineteenth-century geologists were "resistant to the idea that laboratory testing could recreate or duplicate natural conditions."[43] According to Martin Rudwick, fieldwork (not laboratory work) was "the mark of the true [nineteenth-century] geologist; its sometimes arduous nature was the test of his apprenticeship and the badge of his continuing membership in the 'brethren of the hammer.'"[44]

Later in the nineteenth century the legacy of this precise, field-based investigation was a scientific interest in phenomena as they "really occurred" in the world. But instead of shunning experiment, late-nineteenth-century morphological scientists began to use the laboratory to reproduce these same natural occurrences. The morphologists strove to make laboratory versions of real phenomena with all the richness the cyclones or glaciers had in nature. By *re*-creating nature in the controlled world of the laboratory, the scientists hoped to discover the physical processes underlying the natural world. It is this attempt to imitate nature in the small that I will refer to as the "mimetic tradition"; when the morphological sciences enter the laboratory, mimesis becomes their characteristic form of representation.

E. Reyer, a geologist writing in 1892 from Vienna, heralded the beginning of a new experimental physical geology. In the past, Reyer wrote, workers had given up either because quantitative experiments seemed impossible or because experiments had been unable to imitate (*nachbilden*) natural conditions. Now, by reproducing these phenomena at least partially, much could be learned: huge masses of lava could be replicated with the flow of limited quantities of pulpy substances; chalk banks and soil sediments could be adequately represented by clay, plaster, and other suitably malleable or rigid materials; Reyer similarly found it possible to model the successive stages of deformation leading to fold

42. Cannon, *Science in Culture* (1978), 105, and throughout chap. 3.

43. Greene, *Geology* (1982), 53, referring to Hutton and Werner on the problem of duplicating temperature and pressure effects on material of the subcrust.

44. Rudwick, *Devonian* (1985), 41.

mountains. Should the program succeed, Reyer wrote, "we will have to assign the long-ignored geological experiment a deep significance."[45] Mimesis preserves the morphological geologist's ideal of capturing nature in toto but does so through imitation, not the hammer.

During this same period, much of meteorology, like geology, welcomed experiment only if it could imitate nature without gross distortion. An advocate of this morphological approach was John Aitken; as we will see, it was on Aitken's miniature cloud building that Wilson most liberally drew. Aitken was born at Falkirk, Scotland, on 18 September 1839, son of the head of an established legal firm. After studying engineering at the University of Glasgow, Aitken served an apprenticeship in Dundee, followed by three further years with the shipbuilders Napier and Sons in Glasgow. Almost immediately, his career as a marine engineer ended with a breakdown in his health. In part inspired by William Thomson's lectures on natural philosophy, the young engineer transferred his manual skills to building a laboratory and workshop, outfitting it with lathes, blowpipes, and glasswork. Throughout his life Aitken pursued mimetic experiments. For example, he designed a large-scale demonstration model of vortices in order to present the dynamics of cyclones and anticyclones to the Royal Society of Edinburgh.[46]

To explain the circulation of ocean waters Aitken built a trough with glass sides, filled it with water, and used the motion of tracer dyes to track the effects of air jets (miniature windstorms).[47] When seeking to account for the motion of glaciers, Aitken noted that freezing water in a laboratory glass tube offered a "mimic representation of glacier motion."[48] Aitken imitated nature without offering any reductionist analysis. He proceeded by *making* a whole cyclone—not by deducing the cyclone's dynamics from the "application" of mathematical principles of the primitive interactions of matter. Indeed, Aitken's collected papers contain not a single equation. Nothing could be further from the ideals of analytical (transcendental) experimentation so pervasive at the Cavendish.

C. T. R. Wilson's fascination with science was grounded, like Aitken's, in the remarkable, real phenomena of the natural world. Charles was born in 1869, the eighth and youngest son of John Wilson, a Scottish sheep farmer. John died when Charles was four, leaving Charles's mother, Ann Wilson, to care for her own three children and four stepchildren. Taking all seven with her, she moved to Manchester where the Wilsons managed a precarious existence. Charles's stepbrother, William, helped to support the family from Calcutta where he

45. Reyer, *Geologische* (1892), 3–5, on 5.

46. Aitken, "Notes" ([1900–1901] 1923) and "Dynamics" ([1915] 1923); biographical details on Aitken can be found in Knott, "John Aitken's Life" (1923).

47. Aitken, "Ocean Circulation" ([1876–77] 1923), 25–26.

48. Aitken, "Glacier Motion" ([1873] 1923), 4.

worked as a businessman. Although William had never received what he considered an adequate education, he was determined that Charles should have one.[49]

At 15 Charles entered Owens College, then part of Victoria University in Manchester, in order to prepare for a medical career. The relatively new institution had equipped itself for science teaching by drawing on middle-class manufacturers to sponsor its scientific and technological facilities.[50] Before he could begin studying medicine, Wilson had to complete a course of study that included lectures in botany, zoology, geology, and chemistry. On graduating with his B.Sc. in 1887, and after spending an additional year studying philosophy, Latin, and Greek, Wilson won a scholarship at Cambridge. William Wilson, obviously proud of his younger brother, wrote from India in January 1888 that "[o]ne of the pleasures of my life was imparted to me on Sunday morning when I heard that you had been successful at Cambridge. I was *very* pleased to hear it, and hope the acquisition of this scholarship may eventuate in much real advantage to you. Now that the impetus of success has set in I expect it won't expend its energy until it finds you seated in the presidential chair of the British Association!"[51]

Charles completed Cambridge's Natural Science Tripos in 1892, keen to pursue science but fearful that he would be unable to support the other members of his family. One route that appeared open was the vocation of a mapper: "I felt I might be of use as an explorer as I had some knowledge of a wide range of sciences and powers of endurance tested on the Scottish hills."[52] It was a career entirely in keeping with the Humboldtian, morphological tradition—exploration in the service of science, the attempt to add precision to knowledge of the variety of nature by examining it in situ and reproducing it to scale. While at Owens, Wilson had spent his school vacations exploring the Scottish countryside, his eye opened by a trip to the North High Corrie in Arran (an island off the West Coast). There he was "strongly impressed with the beauty of the world. . . . [I]n Manchester I spent all my spare time looking for and studying beetles and pond-life which I also learned to love."[53]

Like many Victorians, Wilson and his brothers took up photography. The depiction of nature must have appealed to him, as he specialized in pictures of landscapes and clouds, from which his mother would often paint (see figures 2.3 and 2.4).[54] In Calcutta, William was devoting his "spare evenings . . . to

49. For biographical details on Wilson's childhood and youth, see Blackett, "Charles Thomson Rees Wilson," *Biog. Mem. F.R.S.* 6 (1960): 269–95, on 269–70; and Wilson, "Reminiscences," *Not. Rec. Roy. Soc. London* 14 (1960): 163–73, on 163–64.

50. See Thackray, commentary in "Physical Science," by Sviedrys, *Hist. Stud. Phys. Sci.* 2 (1970): 127–51, on 148.

51. William Wilson to C. T. R. Wilson, 24 January 1888, CWP.

52. Wilson, "Reminiscences," *Not. Rec. Roy. Soc. London* 14 (1960): 163–73, on 165.

53. Wilson, "Reminiscences," *Not. Rec. Roy. Soc. London* 14 (1960): 163–73, on 164.

54. J. Wilson (Wilson's daughter), interview by the author, 10 September 1986.

Figure 2.3 Early Wilson cloud photograph I (ca. 1890). Taken by C. T. R. and George Wilson. Source: Courtesy of Miss Jessie Wilson, CWP, rephotographed from originals by the author.

photography. I have got my enlarging lantern and have been experimenting."[55] About the same time, Wilson and his brother George reported to their elder sibling on their own first steps in the new art. William guided his novices by mail, counseling them even in the details of their selection of a lens: "But I must hasten to give you my opinion of your first pictures. Your exposure has perhaps been full, but your development has been first rate, I think, judging from the pictures. You have printed them very well. I would strongly recommend you to soak your negatives in every case in a saturated solution of alum after they have been fixed. . . . I would advise you to keep a supply of Manchester or Wratten and Wainwrights [instead of Ilfords] beside you for first class pictures."[56]

Thus, long before C. T. R. Wilson turned his camera on the microphysical world, he had used it to recreate the natural world, especially the crags, cliffs,

55. William Wilson to C. T. R. Wilson, 28 November 1887, CWP. We would like to thank Miss J. Wilson for making this and other documents available to us.

56. William Wilson to C. T. R. Wilson, [1887?], CWP.

Figure 2.4 Early Wilson cloud photograph II (ca. 1890). Taken by C. T. R. and George Wilson. Source: Courtesy of Miss Jessie Wilson, CWP, rephotographed from originals by the author.

and clouds of Scotland. Later, Wilson took his camera with him on his hikes to Ben Nevis and elsewhere and, alongside scientific notes, inscribed his notebooks with the circumstance of each individual exposure.[57] The immense popularity of such amateur nature photography was a notable feature of Victorian Britain, and in general British photography occupied "a stylistic and conceptual midpoint between French and American photography of the nineteenth century." Where American photographers were for the most part scientists or entrepreneurs and the French tended to come from the ranks of painters, the Victorian amateurs "compromise[d] between these two extremes . . . successfully

57. CWnb A21, e.g., 9–16 April 1907. Wilson's lab books (CWnb) are housed at the Royal Society, London. They have been indexed in Dee and Wormell, "Index," *Not. Rec. Roy. Soc. London* 18 (1963): 54–66. Dee and Wormell have divided Wilson's notebooks into two groups: A, which deals with condensation phenomena and includes the notes on the development of the cloud chamber, and B, which treats the earth's electric field and thunderstorms; as we shall see, the division between the two categories is rather artificial. Hereinafter, references to the Wilson notebooks will simply be by letter, number, and date of entry.

manag[ing] to blend emotional evocation with an objective assertion of sheer physical fact."[58]

The British desire to reproduce "sheer physical fact" had no better object in the 1880s than the effects caused by the violent eruption of Krakatoa on 26 and 27 August 1883. Sounds of the explosion echoed through Rodriguez and Diego Garcia, respectively 3,080 and 2,375 miles from the volcano. Windows burst and walls cracked a hundred miles distant.[59] Filtered by the staggering mass of particles shot into the upper atmosphere, strange optical phenomena appeared around the world. From Honolulu an observer saw a "peculiar lurid glow, as of a distant conflagration, totally unlike our common sunsets";[60] Herr Dr. A. Gerber from Glöckstadt recalled in the first volume of *Meteorologische Zeitschrift* how "[t]he sailors declared, 'Sir, that is the Northern Lights!' and I thought I had never seen Northern Lights in greater splendour. After 5 minutes more the light had faded . . . and the finest purple-red rose up in the S.W.; one could imagine oneself in Fairyland."[61] Reports from Italy, Ohio, Switzerland, Portugal, India, Japan, and Australia cascaded into newspapers; both unknown observers and renowned scientists posted notices to all the major scientific journals. Even Hermann von Helmholtz in Berlin took the time to report on "cloud-glow" and the remarkable illusion of green clouds.[62]

Photographers, artists, and scientists all tried to capture the extraordinary visual events: mere black and white photographs could not do justice to the display. William Ascroft drew the extraordinary Bishop's Ring (corona) he saw on 2 September 1884 (figure 2.5), and at 10-minute intervals he drew quick pastels of the yellow-green sunset over the Thames (figure 2.6).[63] When in 1886, the Royal Society published its comprehensive report on the events of Krakatoa, they included Ascroft's sunset sketches as scientifically useful. Throughout the years following the eruption, debate raged over whether the sun's strength was changing, whether the effects really were correlated with the eruption, and if so what mechanisms could account for them. Even if no explanation was absolutely persuasive, at least scientists could mimic this extraordinary event; Karl Kiessling attempted to reproduce the dramatic coronae of late 1883 with a "diffraction chamber" filled with dust, ordinary air, and filtered vapor.[64]

Aitken was well prepared to contribute to the Krakatoa debate. In 1883 he was continuing the important work he had begun with his article "On Dust,

58. Millard, "Images" (1977), 23–24.

59. Judd, "Volcanic Phenomena" (1888), 27.

60. Rollo Russell and Archibald, "Optical Phenomena" (1888), 153.

61. Rollo Russell and Archibald, "Optical Phenomena" (1888), 157.

62. Rollo Russell and Archibald, "Optical Phenomena" (1888), 171.

63. Zaniello, "English Sunsets" (1981), locates the Krakatoa sunsets as a meeting point for scientific and artistic concerns in Britain during this decade.

64. Kiessling's experiments are described in Rollo Russell and Archibald, "Optical Phenomena" (1888).

Figure 2.5 Ascroft pastel (1884). William Ascroft's pastels of the atmospheric aftereffects of Krakatoa were widely known, not least through their inclusion as the frontispiece to the monumental book *Eruption of Krakatoa,* Symons (1888). This pastel shows the reddish brown corona around the sun known as the Bishop's Ring. Ascroft recorded this scene on 2 September 1884, sunset 7:35 P.M. Source: Ascroft Sketch, Science Museum Library, South Kensington.

Fogs, and Clouds," which demonstrated the role of dust in nucleating fog and cloud droplets;[65] his observations on the extraordinary sunsets followed as a natural sequel. To avoid Britain's cloud cover, Aitken voyaged to the south of France, where he repeatedly witnessed the white glare of the daytime sun, the yellow-orange-red sequence of colors on the western horizon at sunset, and then the brilliant afterglows that emerged some 15 and then 30 minutes after sunset.[66] Even a decade later Aitken was still struggling to understand those cataclysmic events by reproducing the Krakatoa green sunsets in his laboratory using electrified steam: "The colours produced by such simple materials as a little dust and a little vapour are as beautiful as anything seen in nature, and well repay the trouble of reproducing them."[67]

Victorian England thus was fascinated with all kinds of reproductions of the dramatic in nature: through painting, poetry, photography, and even laboratory recreation. In addition, there were immediate, practical issues at stake. Weather affected transportation, fishing, public health, military affairs, agriculture, and communication.[68] Aitken, among others, frequently stressed this practical side of

65. Aitken, "Dust, Fogs, Clouds" ([1880–81] 1923), 34–64.

66. Aitken, "Remarkable Sunsets" ([1883–84] 1923), 123–24.

67. Aitken, "Cloudy Condensation" ([1892] 1923), 283.

68. E.g., an international conference on the study of weather at sea was suggested by Lieut. M. F. Maury of the U.S. Navy in 1853 and the (British) Board of Trade established a Meteorological Department in 1854; see Shaw, "Meteorology," *Nature* 128 (1931): 925–26. The Scottish Meteorological Society had close relations with fisheries organizations, which substantially subsidized the society's research; see, e.g., Scottish Meteorological Society, "Report," *J. Scot. Met. Soc.* 7 (1884): 56–60, on 57. Also, as Sir Ernest Wedderburn mentions, in "Scot-

Figure 2.6 Ascroft pastel (1885). Ascroft's sketch of 3 September 1885 illustrates the "Amber Afterglow with Crepuscular Rays," 7:10 P.M.; sunset 6:42. Source: Ascroft Sketch, Science Museum Library, South Kensington (53A5).

meteorology. For example, he knew that the dust-filled industrial output of England produced fogs that could endanger its citizenry: "All our present forms of combustion not only increase the number and density of our town fogs, but add to them evils unknown in the fogs which veil our hills and overhang our rivers."[69] Such evils, Aitken asserted, force our attention to the importance of dust in the origin of clouds: "As our knowledge of these unseen particles increases, our interest deepens, and I might almost say gives place to anxiety, when we realize the vast importance these dust particles have on life, whether it be those inorganic ones so small as to be beyond the powers of the microscope, or those larger organic ones which . . . though invisible, are yet the messengers of sickness and of death to many—messengers far more real and certain than poet or painter has ever conceived."[70]

tish," *Q. J. Roy. Met. Soc.* 74 (1948): 233–42, on 235: "Amongst [the society's] subsidiary objects were . . . the bearing of meteorology on public health, agricultural [*sic*] and horticulture, the alleged periodical recurrence of wind storms, and the general laws regulating atmospheric changes, the discovery of which might lead to a knowledge of the coming weather."

69. Aitken, "Dust, Fogs, Clouds" ([1880–81] 1923), 49–50.

70. Aitken, "Small Clear Spaces" ([1883–84] 1923), 84.

Industrial fogs, Victorian exoticism, and the pragmatic demands of transport combined to bolster a worldwide establishment of weather networks, observatories, and professional meteorological societies.[71] Significantly, Wilson's path would lead him to an observatory located on the peak of Ben Nevis. Ben Nevis, in Northern Scotland, a few miles south of the Caledonian Canal, is the highest mountain in the British Isles. Impressed by the new style of meteorological research pioneered in Germany and the United States, many Britons had enthusiastically supported the establishment of an observatory. With the help of devoted amateurs and private subscriptions, the Meteorological Council publicized their interest in founding an observatory to assist in tracking "vertical meteorological sections of the atmosphere."[72] One Clement L. Wragge volunteered to take weather readings from Ben Nevis by himself to show the value of careful observation. A "king among eccentrics," Wragge assigned Christian names to cyclones, and even edited his own journal, *Wragge—For God, King, Empire and People.*[73] On the heels of the wide publicity given to Mr. Wragge, contributions were made by such diverse benefactors as Her Majesty and the Worshipful Company of Fishmongers, London. These funds allowed the Ben Nevis Observatory to open in 1883. Until it was closed in 1904, resident observers sent daily weather information by telegraph to England. For years Wilson's research was shaped by his experiences at this distant meteorological outpost.

2.3 Mountaintop Glory, Laboratory Ion

Wilson's career as a professional physicist began haltingly.[74] After graduating from Cambridge in 1892, he stayed on, demonstrating at both the Cavendish and Caius Chemical Laboratories. Hoping to obtain a Clerk Maxwell fellowship (he did not succeed), Wilson wrote to J. J. Thomson in November 1893 about his work on the distribution of a substance in solution that was kept hot on the top and cold on the bottom: "Very few experiments appear to have been made on the subject, and it seems of considerable importance in connection with theories of solution and osmotic pressure." The aspiring physicist "would determine the concentration of different parts of the solution optically."[75] In his later work Wilson used theoretical calculations of vapor pressures to try to understand the condensation of water vapor, and he pursued the problem experimentally with an optical method (the cloud chamber). His earlier work with nonequilibrium systems

71. For an excellent bibliography of meteorology, and for further general sources and, in particular, studies of individual national weather systems, see Brush and Landsberg, *Geophysics* (1985).

72. Report of the Council of the Scottish Meteorological Society for 1875, quoted in Paton, "Ben Nevis," *Weather* 9 (1954): 291–308, on 292.

73. Paton, "Ben Nevis," *Weather* 9 (1954): 291–308, on 294.

74. See details in Wilson, "Reminiscences," *Notes Rec. Roy. Soc. London* 14 (1960): 163–73, on 166.

75. Wilson to Thomson, 8 November 1893, CWP.

Figure 2.7 Exterior, Ben Nevis Observatory (1885). Wilson worked at Ben Nevis in autumn 1894 and summer 1895; the electrical and condensation phenomena he witnessed became guiding themes for the entirety of his scientific career. Both became natural wonders that he sought again and again to mimic within the confines of the laboratory. Source: *Graphic* 36 (1885): 638, University of Cambridge Library.

contributed to his interest in thermodynamic instability—the fundamental feature of cloud condensation. But the immediate catalyst for Wilson's work came from his stint in the fall of 1894 at the Ben Nevis Observatory.

By the early 1890s, the mountaintop observatory was flourishing and its staff welcomed volunteers to work on Ben Nevis during the easy observing periods of summer and fall (see figures 2.7 and 2.8). After graduating from Cambridge, Wilson's affection for the mountains led him to the small station many times. He made his first trip on 8 September 1894, which began as a cloudless day. Soon a thick haze embraced patches of fog that gradually became continuous. The next evening, at 9 P.M., the observers sighted first one lunar corona, then another at 10 P.M., and at least seven more during the next two weeks. On the fifteenth the logbook records: "Solar fogbow & glories at 16^{H} & Lunar Corona at 23^{H}." Just hours before Wilson descended from the heights on 22 September, light and clouds performed spectacularly—even the dry tone of the logbook rises, in the remark that "some beautiful Triple Lunar Coronae were seen this morning through thin passing fog."[76]

On descending from the station, Wilson, in tune with Aitken and so many contemporaries, wanted to mimic the wonders of nature: "In September 1894 I

76. Ben Nevis Observatory logbook, 15 and 22 September 1894, Meteorological Office, Edinburgh. I would like to thank Marjory Roy for making this material available to me.

Figure 2.8 Interior, Ben Nevis Observatory (1885). Source: *Graphic* 36 (1885): 638, University of Cambridge Library.

spent a few weeks in the Observatory . . . of Ben Nevis. . . . The wonderful optical phenomena shown when the sun shone on the clouds surrounding the hill-top, and especially the coloured rings surrounding the sun (coronas) or surrounding the shadow cast by the hill-top or observer on mist or clouds (glories), greatly excited my interest and made me wish to imitate them in the laboratory."[77]

Returning to the mountain peak just nine months later (June 1895), Wilson recorded that he "[w]alked along Spean Bridge road. Saw lightning lighting up mist in big corrie & heard thunder rolling in that direction." The power of the electrical storm was evident: "Saw the damage done by lightning yesterday. Telegraph instrument fused at various places, including part of one of the steel keys. . . . Boxes on shelf above [lightning] arrester thrown on to floor." After recording the events of a few days mostly spent hiking in the hot, blazing sun, Wilson noted on 26 June an abrupt alteration of weather. "Mist suddenly began to pour down a gully between B[en] Nevis & Carn Dearg, & afterwards hid upper part of cliff. . . . Heard continual muttering of thunder in distance. Walked along ridge to top of Carn Mor Dearg. After a minute or two there, suddenly felt St. Elmo in my hair, & in my hand on holding it up. Ran down into corrie. Bright lightning & loud thunder."[78]

77. Wilson, "Making Visible Ions" (1965), 194.
78. CWnb, A21, 19, 20, and 26 June 1895.

The optical and electrical phenomena that Wilson witnessed during those two trips to Ben Nevis set the outlines of his lifelong scientific goals. Meteorological optics and atmospheric electricity remained central to Wilson's work until his death in 1959.

Between demonstrating physics and tutoring students, young Wilson had earned a living in Cambridge but been unable to find the time to do any research. To improve the situation, he had tried teaching for a few months at the Bradford Grammar School, but this left him no freer than before. Returning to Cambridge, he considered himself lucky to land a job demonstrating physics to medical students at the Cavendish. "With this I had just enough to live on, a connexion with the Cavendish, and at last time to do some work of my own just when I had ideas which I was impatiently waiting to test."[79]

The few months he spent teaching young grammar school students were not directly productive, but his dramatic encounter with meteorological effects at Ben Nevis left him with a burning desire "to reproduce the beautiful optical phenomena of the coronas and glories I had seen on the mountaintop."[80] With this goal in mind, Wilson began a notebook in which he recorded speculations as well as research notes. Immediately preceding the notes on his first "cloud" experiment, we find a page of questions he has addressed to himself. Under a section labeled "Cloud Formations & c[etera]" he pondered:

> 1 Are the rings of corona and glory formed simultaneously in same cloud, equal in radius? Try monochromatic light.
> (2) Conditions of formation. When best formed. (In dusty or tolerably pure air or in presence of soluble substances.[)]
> (3) Are ice particles ever formed instead of water drops. (Halos & c.)
> 4. Are coronae & c. formed when one liquid separates out in milky form from another. (Mixture of ether and water allowed to cool & c.) Also are halos ever formed when crystalline precipitates are formed. Applications.[81]

On what tradition did Wilson draw when he decided to perform these experiments to reproduce clouds, glories, and coronae in his laboratory? Other scientists were working on similar problems; several had even tried to reproduce clouds and fogs. Early in his notebook Wilson reviewed the quite extensive literature on cloud formation. Aitken, Jean Paul Coulier, and Robert von Helmholtz (the son of Hermann) had all performed experiments to see if they could get water vapor to condense—either by using an air pump or an india rubber ball to expand saturated air in a glass vessel, or by watching a steam jet abruptly expand as it escaped from a nozzle.[82] The theoretical justification for this

79. Wilson, "Reminiscences," *Notes Rec. Roy. Soc. London* 14 (1960): 163–73, on 166.

80. Wilson, "Reminiscences," *Notes Rec. Roy. Soc. London* 14 (1960): 163–73, on 166.

81. CWnb A1, preceding entries dated March 1895.

82. Aitken, "Dust, Fogs, Clouds" ([1880–81] 1923); Coulier, "Nouvelle propriété," *J. Pharm. Chim.*, 4th ser., 22 (1875): 165–73, 254–55; and Helmholtz, "Dämpfe und Nebel," *Ann. Phys. Chem.* 27 (1886): 508–43.

method is as follows: an adiabatic expansion of saturated gas lowers the temperature[83] of the gas, causing supersaturation that can lead to condensation. In the term "supersaturated air," supersaturation is usually defined to be the state in which the ratio of actual vapor pressure to the equilibrium vapor pressure above a *flat body of water* is greater than one. The obvious way to study condensation is with supersaturated vapors, but it is not always easy to precipitate the liquid. Under certain circumstances it is possible to have supersaturation without condensation; this occurs, for example, if the vapor is not over a flat body of water.

Among Wilson's predecessors who used some form of expansion of air to investigate condensation Aitken was the most important for Wilson. Also a Scotsman, Aitken initiated research work at Ben Nevis that involved equipment that Wilson would take as a model for his own experiments. The remarkable similarity between Aitken's and Wilson's instruments, and Wilson's reference only to Aitken in his first published paper, suggests looking more closely at the material culture of Aitken's meteorology.[84]

Aitken's first conclusion from his cloud experiments was that dust particles acted as nuclei for water droplets in supersaturated air. Without dust there was no condensation under what Aitken considered "normal" conditions. In 1880 he pointed out that "[d]usty air—that is, ordinary air, gives a dense white cloud of condensed vapour."[85] He decided to use his dust chamber in a purposeful way, by designing an instrument to measure the number of particles in the air. "Powerful as the sun's rays are as a dust revealer, I feel confident we have in the fog-producing power of the air a test far simpler, more powerful and delicate, than the most brilliant beam at our disposal."[86] Aitken's motivations for studying dust were nominally meteorological; he believed in the "possibility of there being some relation between dust and certain questions of climate, rainfall, etc."[87] But dust was clearly much more than a nucleation site for rain. As Aitken repeatedly stressed, he hoped to settle the "great fog question," the problem of town fogs, whose "increased frequency and density . . . [is] becoming so great as to call for immediate action."[88] Dust and fog were both signs and symptoms of power, both industrial and natural, and for the Victorians these motes carried a cultural meaning far beyond their physical content.

By 1888 Aitken had a method of counting dust particles in the air based on his observation that they were a source of condensation in supersaturated air.

83. In practice, the expansion was quick enough to give a good approximation to an adiabatic system (no heat exchange with the outside). Therefore, the gas obeys the equation $pV^{\gamma} =$ constant, where γ is the ratio of specific heats c_p/c_V. Using the ideal gas equation we obtain a fall in temperature as the air expands, $T_1/T_2 = (V_2/V_1)^{\gamma-1}$.

84. Wilson, "Formation of Cloud," *Proc. Camb. Phil. Soc.* 8 (1895): 306.

85. Aitken, "Dust, Fogs, Clouds" ([1880–81] 1923), 35.

86. Aitken, "Dust, Fogs, Clouds" ([1880–81] 1923), 41.

87. Aitken, "Dust, Fogs, Clouds" ([1880–81] 1923), 41.

88. Aitken, "Dust, Fogs, Clouds" ([1880–81] 1923), 48.

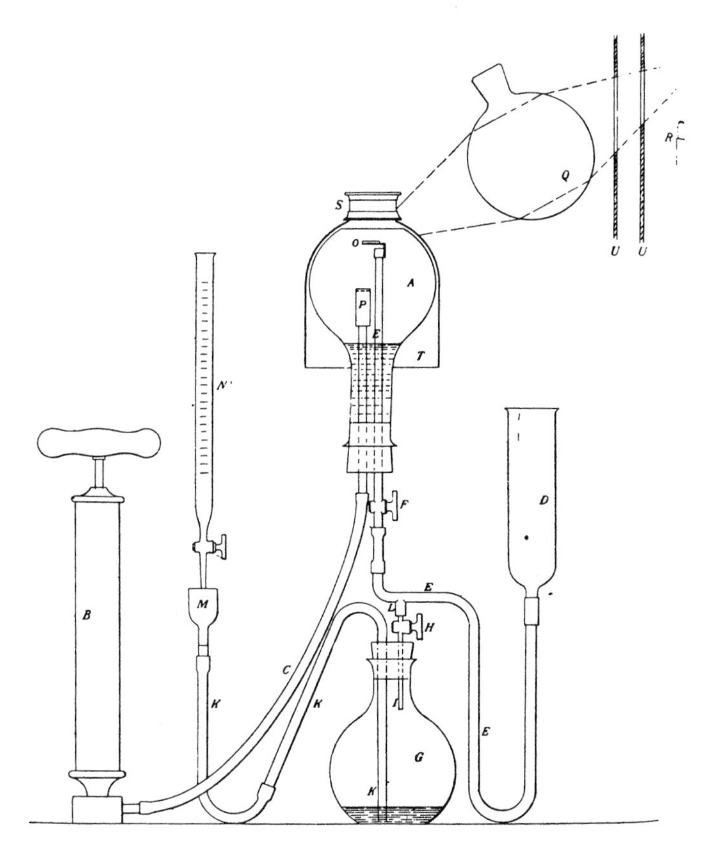

Figure 2.9 Aitken's dust chamber (1888). Aitken's dust chambers were simultaneously a means for the measurement of dust in a given volume of air and a means of imitating the effects of dust on condensation. In both respects, the dust chamber became a model for the cloud chamber—and, in a sense, its antithesis once Wilson began arguing for an electrical rather than a particulate source of condensation. Source: Aitken, "Dust Particles" ([1888] 1923), 190.

His instrument is reproduced in figure 2.9. The condensation occurs in the ordinary glass flask *A*, the receiver, and the water droplets (each surrounding a single dust particle) are counted by means of the compound magnifying glass *S*. The glass flask *G* contains the air to be tested, which is kept saturated by water in the flask. *D* is a cotton wool filter that clears ordinary air of dust particles. Air from *G* and *D* is mixed so that "too much dusty air [is not] sent into the test receiver at one time, or the drops will be too close for counting."[89] Aitken wanted to ensure that the dust particles would be far enough apart that each particle would

89. Aitken, "Dust Particles" ([1888] 1923), 193.

serve as a center of condensation and all would be counted. After this mixture of air is in receiver *A* and stopcock *F* has been closed, the experimenter makes one stroke of the pump (*B*), while watching stage *O* very carefully. As the air in *A* expands, it supersaturates; condensation occurs on the dust particles that then fall onto *O*. Counting is facilitated by a square grid (on *O*) where the rulings are 1 millimeter apart.

Aitken spent five years (1889–94) counting dust particles in British and Continental air with a pocket version of his counter. In a three-part paper published during this period, "On the Number of Dust Particles in the Atmosphere of Certain Places in Great Britain and on the Continent, with Remarks on the Relation Between the Amount of Dust and Meteorological Phenomena," Aitken compared measurements taken on Ben Nevis with those from a low-altitude station at Kingairloch.[90] Mr. Rankin, the observer who took the readings on Ben Nevis, wrote that two dust counters were bought for the observatory in 1890, "one, a portable form of the instrument mounted on a tripod stand, for use in open air; the other, a much larger form, for use in the laboratory. . . . Both instruments were made from plans and specifications prepared by Mr. Aitken."[91] The last measurement recorded on Ben Nevis was in 1893, just a year before Wilson arrived for his first visit of September 1894. Wilson certainly saw the Aitken apparatus on his visit; the observatory was small and crowded and there was not much to do on the mountain after the sun went down.

After his short stay in Scotland, Wilson began to study the optical effects of clouds. By condensing water vapor in an expansion apparatus similar to Aitken's (see figure 2.10), Wilson was able to produce the rings of color that had delighted him on Ben Nevis. Though Wilson obviously derived his method from Aitken's work, their procedures differed crucially. Instead of mixing purified and dusty air, Wilson filtered *all* the air that entered the receiver. Figure 2.10, reproduced from the first page of Wilson's notes on condensation experiments, displays an apparatus that tests only filtered air; there is no valve system, such as Aitken's, for mixing pure and dusty air. Air enters receiver *V* through the cotton wool filter *F* and is kept saturated by water in *V*. A pump (*P*) evacuates *R*, which then is allowed to come into contact with receiver *V* when expansion is desired.[92]

There is a profound puzzle lodged in that cotton wool filter. If Wilson wanted to make fogs, why was he removing the dust that Aitken had so painstakingly demonstrated to be the condensation nuclei? Surely, if Wilson's only motivation was to reproduce the natural phenomena of clouds he would have used ordinary air, not air that was specially prepared for laboratory purposes. This

90. Aitken, "Dust Particles in Certain Places" ([Part I, 1889–90; Part II, 1892; Part III, 1894] 1923).

91. Rankin, "Dust Particles," *J. Scot. Met. Soc.* 9 (1891): 125–32, on 125. The dust counter remained at the observatory until at least 1901. In a deposition of the property at the observatory the dust counter is specifically mentioned; see McLaren et al., "Memorandum," *J. Scot. Met. Soc.* 12 (1903): 161–63, on 162.

92. CWnb A1, 26 March 1895.

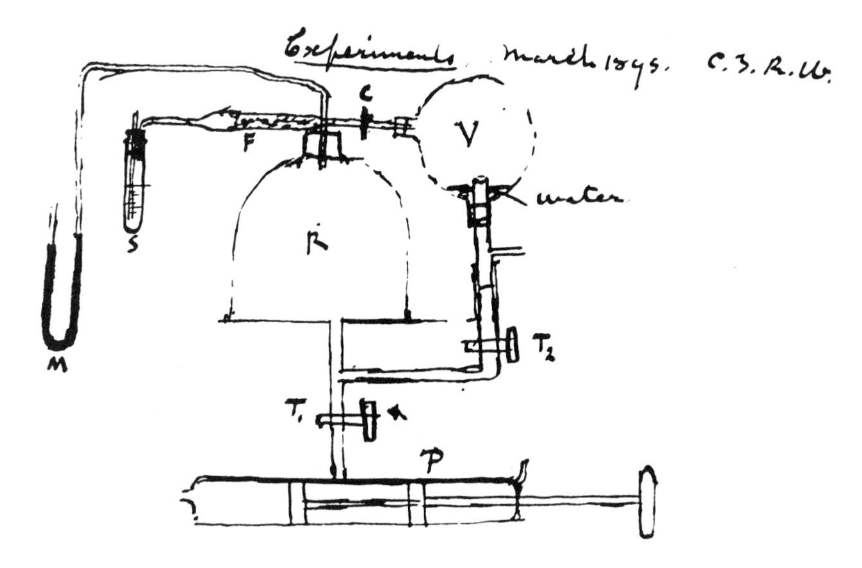

Figure 2.10 Wilson's cloud chamber apparatus (1895). Source: CWnb A1, March 1895.

"artificial" aspect of his experiments signals a significant departure from the mimetic tradition in which his work was initially embedded.

Was there any reason to suppose such filtered air could produce droplets? Several different observers had suggested that electrified air—even without dust—might be capable of nucleating rain. First, it was generally known that as steam hisses out of a nozzle it expands, supersaturates, and condenses. Though they could not explain *why,*[93] R. von Helmholtz, Aitken, Carl Barus, and Kiessling had all found they could increase that condensation by electrifying the jet.[94] Second, Aitken and others had noted (again without knowing why) that a high expansion would produce condensation even in supposedly purified air.[95] After

93. Aitken reasoned that electricity would prevent smaller drops from coalescing into larger drops because of electrostatic repulsion and that therefore the number of the drops would remain large, promoting dense condensation. But, as in the other 600 pages of Aitken's collected papers, he offers no quantitative analysis or formal derivation. Aitken, "Cloudy Condensation" ([1892] 1923), 258–59.

94. In his notes, Wilson referred to Helmholtz, "Dämpfe und Nebel," *Ann. Phys. Chem.* 27 (1886): 508–43 (CWnb A1, preceding entries dated March 1895), and in a later paper, to Barus, Kiessling, and Aitken (Wilson, "Condensation," *Phil. Trans. Roy. Soc. A* 189 [1897]: 265–307).

95. Aitken, "Dust Particles" ([1888] 1923): "We see from this experiment that an expansion of 1/50 is nearly, perhaps quite sufficient to cause condensation to take place on even the smallest particles in the air tested; from which we may conclude that the showers which unexpectedly took place from time to time in the experiments described, where high expansions were used, were not due to the presence of extremely small particles which had become active with the high degree of supersaturation" (201). Aitken got rid of this problem by pumping slowly. He believed that he was eliminating the "shock" of expansion which caused unwanted condensation (202–3). Actually, because his expansion was slow and therefore did not approximate an adiabatic expansion as

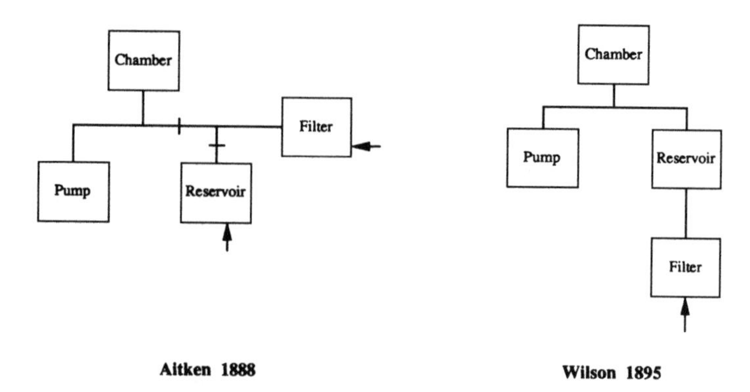

Figure 2.11 Schematic comparison, dust chamber and cloud chamber. Representations of the Aitken and Wilson apparatuses depicted in figures 2.9 and 2.10. In the Aitken apparatus (*left*), air from the outside enters through the reservoir and is mixed with a certain amount of filtered air coming through the filter. The filtered air serves merely to keep the air in the chamber from becoming too dusty. By contrast, in Wilson's apparatus *all* the air must pass through the filter before it can enter the reservoir, the main chamber, or the pump system. Thus, Wilson had to find it plausible that water vapor might condense on dust-free air, a situation only imaginable with the enabling notion of the ion.

satisfying himself that he could eliminate this bothersome effect in his dust counter, Aitken dropped the matter. Wilson was aware of both phenomena,[96] though it would be years before he would be certain of the connection between the two.

Thus, suspecting that there might be several ways for droplets to form, Wilson filtered the air to identify the best conditions in which to make the "wonderful" optical effects, coronae and glories: "When [are the optical effects] best formed[?] (In dusty or tolerably pure air . . . [?])"[97] Soon, the notebooks reveal, he began experimenting *exclusively* with filtered air. For Wilson, dust became just an annoyance to be removed; in contrast, dusty air was Aitken's prize specimen, and condensation in purified air was a background problem to be eliminated and then forgotten. Wilson created artificial laboratory conditions in order to "dissect" nature; Aitken wanted to remove the artificiality from his experiment to imitate nature as it was. It is at first puzzling that Wilson used an apparatus almost identical to Aitken's, yet the two fastened on mutually exclusive phenomena—one on dust, the other on purified air. Wilson's apparently simple move of the filter F (see the schematic figure 2.11) hides a profound shift both in

well as a fast expansion did, the supersaturation was not as large as with the quicker motion, and the unexpected condensation did not take place.

96. CWnb A1, preceding entries dated March 1895.

97. CWnb A1, preceding entries dated March 1895.

material culture and in conceptual structure. The source of this change in traditions was the scientific program of the Cavendish Laboratory.

2.4 Analysis and Mimesis

Through J. J. Thomson and his collaborators, Wilson learned the Cavendish style of analytic physics, a mode of work and tradition of instrumentation altogether different from the mimetic mode to which Wilson was so committed. Thomson, appointed to the Cavendish chair in 1884, was a firm believer in analytic solutions to problems, using explanatory entities far from the visible manifestations of nature: "[T]he principal advances made in the Physical Sciences during the last fifty years . . . [have] intensif[ied] the belief that all physical phenomena can be explained by dynamical principles and [stimulated] the search for such explanations."[98] As a student at Cambridge from 1888 to 1892, and as a researcher at the laboratory starting in 1895, Wilson absorbed much of the matter physicist's style of physics. In addition, the Cavendish was unique in its firm programmatic commitment to the assumption that electric charge came in discrete bits (ions). Wilson borrowed heavily from his Cambridge colleagues; ion physics (or, more specifically, the idea of ions in general) had been essential to his invention of the cloud chamber. In return, Wilson's work became an integral part of the Cavendish investigations into the electrification of gases. Eventually, under Rutherford, his instrument became the laboratory's primary research tool. The cloud chamber, developed under one research program, would itself (in its role as a piece of experimental equipment) establish the boundaries of another, completely different research project. Instruments will not stay put.

Evidence of the effect of the Cavendish research program on its students is easy to find. As a Natural Sciences Tripos student (Wilson took the new laboratory course rather than the more formal Mathematical Tripos), Wilson faced questions such as "Give some account of the phenomena observed in the neighbourhood of the negative electrode when an electric discharge passes through a tube containing gas at low pressure."[99] Interest in this type of discharge was central to the basic thrust of Cavendish research, which sought to understand the structure of matter through the investigation of cathode rays. One of the program's great triumphs followed not many years later when Thomson, in 1897, began arguing that the electron was a subatomic particle.[100]

Since the middle of the nineteenth century when it became possible to produce a moderate vacuum, physicists across Europe had been exploring the

98. Thomson, *Applications of Dynamics* (1888), 1.

99. David B. Wilson. "Experimentalists," *Hist. Stud. Phys. Sci.* 12 (1982): 325–71, on 346–47.

100. Much of the following summary of early atomic research is taken from Heilbron, "Atomic Structure" (1964), especially the first two chapters. Heilbron summarizes: "[T]he 'great discoveries' [of X rays, radioactivity, and the corpuscular electron] developed mainly out of experimental investigations of the phenomena accompanying the discharge of electricity through rarified gases" (59).

effects of electrical discharges. German physicists led the effort to probe the invisible rays emitted from a hot cathode contained in a tube with rarified gas. At first, attention focused on light and dark bands in the tube; as higher and higher vacuums were achieved, scientists saw the glowing in the tube dwindle until, at very low gas pressure, light only appeared on the walls of the tube. Since the rays seemed to originate from the cathode (rather than the anode), they were named cathode rays.[101] In 1879, William Crookes demonstrated that magnets deflected the rays, which led him to identify the rays as streams of charged particles.[102] Crookes thus added support to a view, held intermittently since Faraday's work in the 1830s, that electricity comes in discrete "atoms." Opposing the particulate view were scientists who believed that all electromagnetic effects, including charge itself, were deformations of the ether.[103] By the 1880s English, though not Continental scientists, had decided in favor of the corpuscular nature of electricity. One of the chief proponents of the ion picture was J. J. Thomson at the Cavendish.

Thomson began his study of the discharge of electricity through gases in 1886,[104] and soon most of the laboratory was devoted to this problem. From 1893 to 1895, the year Wilson arrived at the Cavendish, fully half of the papers published by Cavendish scientists were concerned with the discharge of electricity.[105] As a discharge effect possibly explicable in terms of ions, Aitken's electrified steam experiments would have seemed relevant to Thomson. And in 1893, two years before Wilson published his paper on condensation in the absence of dust,[106] Thomson provided a theoretical justification for the growth of a drop in the presence of a nonuniform electric field.[107]

In particular, Thomson first argued that because surface tension puts a drop under pressure, the equilibrium pressure surrounding the drop is very high for small drops. Quantitatively, the presence of a drop of radius r increases the equilibrium vapor pressure by a factor of $1/r$, promoting evaporation. The smaller the drop, the greater this effect. As Thomson put it, in terms readily understandable to his class and culture, "It is evident that this property makes the growth of drops from smaller ones of microscopic dimensions impossible, for these small

101. Wheaton, *Tiger* (1983), 5.

102. Wheaton, *Tiger* (1983), 6.

103. The best source for a discussion of Maxwellian charge concepts and the related electrodynamics is Buchwald, *Maxwell to Microphysics* (1985), e.g., 23, 38ff. Locating these developments within other contexts are two works: Bruce Hunt, *Maxwellians* (1991); and Harman, *Energy* (1982), 89–103.

104. Thomson's first paper explicitly concerned with the discharge of electricity in gases was "Electric Discharge," *Proc. Camb. Phil. Soc.* 5 (1886): 391–409.

105. *History of the Cavendish* (1910), 297–98. I have taken the number of papers published on electrical discharge phenomena from the list (given by year) of all Cavendish papers provided at the end of the volume.

106. Wilson, "Formation of Cloud," *Proc. Camb. Phil. Soc.* 8 (1895): 306.

107. Thomson, "Steam-Jet," *Phil. Mag.*, 5th ser., 36 (1893): 313–27. Much of the theoretical work in this paper was taken from Thomson's earlier book *Applications of Dynamics* (1888), chap. 11, "Evaporation," 158–78.

drops would evaporate and get smaller, and the smaller they get the faster will they evaporate. They are in the position of a man whose expenditure increases as his capital decreases, a state of things which will not last long."[108]

Thomson next argued on thermodynamic grounds that a nonuniform electric field *decreases* the vapor pressure as $1/r^2$. Energy considerations provide a simple way to understand this: Because the dielectric constant of water is about 80 and that of air about one, the energy in the electric field decreases when the charge is surrounded by water. Assuming the air acts as a thermal reservoir, condensation of water is favored since the Gibbs free energy of the system tends to a minimum. Because surface tension increases equilibrium vapor pressure by $1/r$ and the presence of charge decreases it by $1/r^2$, for small drops the electric field wins and the drop can grow.

Thomson's research program, by providing a quantitative model of the physics of ionic condensation in dust-free air, allowed Wilson to consider experiments that would have seemed pointless without Thomson's detailed scheme of ion drop formation. A week after Wilson began his cloud experiments, he explicitly used Thomson's formula to calculate the magnitude of the electric charge he suspected he might be seeing in his chamber. "If nuclei be present in [the] shape of small electrified drops of radius 2 times 10^{-7} [centimeters] each charged with atomic charge, we can calculate magnitude of this charge necessary to neutralize effect of S. T. [surface tension]."[109] As Wilson turns his thoughts to the world of subvisible ions, one sees the impact of Cavendish research on his programmatic goals.

During the remainder of 1895 Wilson used his cloud chamber daily. Each morning he would fill his chamber with air and make several expansions to remove dust, a method he found more satisfactory than using a cotton wool filter. Then for the rest of the day he used this purified air to determine expansion ratios for condensation. (The expansion ratio was defined to be V_2/V_1, where V_2 was the volume after expansion and V_1 the volume before.) On 3 April alone, he made 115 runs.[110] In a one-page paper, the only article he published in 1895, Wilson stated the critical expansion ratio given an initial temperature of 16.7°C: $V_2/V_1 = 1.258$.[111] Such precision would become important to Wilson later while investigating the new "rays" that were soon to be discovered on the Continent.

As he developed the cloud chamber, Wilson never left meteorology, constantly interjecting weather questions into his notebook. During spring 1895 he speculated on the significance of his experiments in a section labeled

108. Thomson, *Recollections* (1937), 416.

109. CWnb A1, 22 April 1895. Wilson is directly citing Thomson's article "Steam-Jet," *Phil. Mag.*, 5th ser., 36 (1893): 313–27.

110. CWnb A1, 3 April 1895.

111. Wilson, "Formation of Cloud," *Proc. Camb. Phil. Soc.* 8 (1895): 306.

"Meteorological." Beginning with optics, Wilson claimed that the "[e]xistence of coronae shows uniformity in size of drops in the clouds showing them." The drops, Wilson speculated, might oscillate in a cloud: as drafts carry them up into regions of high supersaturation they grow bigger until their weight causes them to descend, thus reaching an area of lower saturation where they begin to evaporate until they are light enough to begin the cycle again.[112] In addition, Wilson began to think that weather phenomena might be affected by ions and wondered, "*When drops are suddenly formed do they throw off small electrified drops or free ions?* (Thunderstorms & c.) In either case air should be left electrified."[113]

During the New Year's celebration of 1896 the world was thrilled by news of Röntgen's discovery of X rays. The unique photographic properties of the rays created great excitement among both scientists and nonscientists; Röntgen was even called to Potsdam to give Kaiser Wilhelm a demonstration.[114] At the Cavendish, experiments with X rays began at once. Ernest Rutherford, a new research student at the laboratory, wrote on 25 January 1896, "The Professor [Thomson] of course is trying to find out the real cause and nature of the waves, and the great object is to find the theory of the matter before anyone else, for nearly every Professor in Europe is now on the warpath."[115] Thomson used the new rays in conjunction with a well-known experiment at the Cavendish. He and Rutherford watched the effect that Röntgen rays had on the passage of electricity through gas.[116] Soon they discovered that the rays enhanced conduction and explained this by suggesting that Röntgen rays produced ions in the gas as they passed through it.

When Wilson heard about Röntgen's vision, he was anxious to shine the new rays into his cloud chamber. He borrowed an X-ray tube from Ebeneezer Everett,[117] Thomson's assistant, and upon turning it on his chamber was delighted to find that "no effect is produced by the X-rays unless the expansion is great enough to produce condensation in any case. When it is sufficient to cause condensation without the rays, they produce a very great increase in the number of the drops."[118] Wilson had carefully determined the expansion ratio that caused condensation around nuclei produced by Röntgen rays and found that exactly the same expansion ratio condensed vapor about nuclei naturally present in dust-free air. Therefore, Wilson reasoned, "[i]t seems legitimate to conclude that when the Röntgen rays pass through moist air they produce a supply of nuclei of the same kind as those which are always present in small numbers, or at any rate of

112. CWnb A1, 27 March 1895.

113. CWnb A1, 30 March 1895; Wilson's emphasis.

114. Keller, *Atomic Physics* (1983), 57.

115. Rutherford to Mary Newton (fiancée), 25 January 1896, quoted in Keller, *Atomic Physics* (1983), 57.

116. Thomson and Rutherford, "Passage of Electricity," *Phil. Mag.*, 5th ser., 42 (1896): 392–407.

117. "Wilson of the Cloud Chamber," transcription of radio interview with Wilson on Scottish Home Service, 16 February 1959, 8, item D57, EAP.

118. CWnb A2, 17 February 1896.

exactly equal efficiency in promoting condensation."[119] Quantitative experiment was the tool that allowed Wilson to identify the kinds of nuclei as one and the same because precisely computed expansion ratios were needed to make this comparison. Such a move would have been inconceivable in the qualitative style that marked the work of John Aitken.

Becquerel's discovery of uranium rays in March 1896 added yet another dimension to Wilson's experiments. Wilson found that, like X rays, uranium rays increased condensation at the established expansion ratio of $V_2/V_1 = 1.25$. By the second half of the year 1897, Wilson was willing to go to press with the assertion that "[t]he electrical properties of gases under the action of Röntgen rays and Uranium rays point to the presence of free ions."[120]

Wilson's patience and precision in determining expansion ratios led to a modification in the ion theory of condensation. By 1898 he could show that precipitation in the cloud chamber fell into four regimes bounded by three expansion ratios. Below $V_2/V_1 = 1.25$ there is no condensation in dust-free air, between 1.25 and 1.31 there are distinct "rain" drops, at 1.31 there is a sudden increase in the number of drops, and finally, above 1.37 a dense fog is seen. Wilson supposed that the fog was formed by the statistical aggregation of water molecules because of extremely high supersaturation; no nuclei were needed. He did not, however, understand the reason for the existence of the *two* lower expansion ratios. In March 1898 he speculated, in his notebook, that the difference in condensation in the two domains could be explained by the quantity of charge acting as precipitant, for a larger charge would promote condensation at a lower supersaturation. He thought that two sets of ions might exist, one with a charge twice that of the other: the double charge was strong enough to pull in water at 1.25, while the single charge would only precipitate water at the ratio 1.31. At this stage Wilson drew his ion models from chemistry. His doubly charged carrier (ion) was an oxygen atom. He did not suggest that particles smaller than atoms carried charge.[121]

By 7 July 1898 Wilson was considering another reason for the two different expansion ratios. His notebook entry of that day began with Thomson's idea: "J. J. T. suggests that if the expansion required to catch positive and negative ions is different (say less for the negative than for the positive) gas would be left charged if expansions were only sufficient to catch the negative but not the positive." Wilson saw immediately that "[t]his would have obvious meteorological application if atmospheric air were ionised to even very small extent."[122] What did Wilson mean by this? It was generally known that the earth is negatively charged and that a potential gradient exists between the earth and the ionosphere.

119. Wilson, "Röntgen's Rays," *Proc. Roy. Soc. London* 59 (1896): 338–39, on 339.

120. Wilson, "Condensation," *Proc. Camb. Phil. Soc.* 9 (1897): 333–38, on 337.

121. CWnb A8, 4 March 1898.

122. CWnb A8, 7 July 1898.

(On an ordinary day this gradient is 100 volts per meter, amounting in total to a change in voltage of 400,000 volts.)[123] If there were no external forces acting on this system, such a huge potential difference would discharge the earth in approximately half an hour. In 1898 no mechanism was known that could maintain that voltage gap.

Wilson was quick to realize that if negative ions were more likely to cause condensation, rain would bring down negative charge to the earth, thus maintaining this fair-weather gradient. Immediately, he began to speculate on meteorological consequences of Thomson's suggestion. Under a section heading, "Condensation problems (suggested during previous work) with *meteorological* bearings," Wilson asked himself: "[1] [A]re there any free ions in ordinary moist air in absence of external influences? . . . [2] Are there any agents . . . capable of ionising [air]? . . . [3] Are negative ions . . . more easily caught than positive ones?" He concluded that the "three questions are all connected with the theory of atmospheric electricity suggested on the previous page. That is that the ground is kept negative by the fall of rain, each raindrop containing one negative ion."[124] Another of Wilson's notebooks for this period offers a detailed discussion of balloon experiments that measure the potential gradient accurately and records ideas for laboratory experiments "to find probable nature of the carriers of atmospheric electricity."[125]

Between 7 and 23 January 1899 Wilson carried out the experiments that confirmed Thomson's hunch about the different expansion ratios for negative and positive ions. Wilson used a modified cloud chamber, one with a brass wire down the center that could be held at a constant potential. His conclusions were inescapable: "This showed difference plainly, and to some extent the distribution of the ions in the tube appeared to be made visible; such fogs as were obtained with EMF [electromagnetic field] arranged to drive positive ions outwards, being concentrated around the wire, while the much more marked fogs obtained with the EMF in the other direction extended throughout the tube. The expansions available were scarcely sufficient to catch positive ions."[126] In notebook entries like this one, we see both the Cavendish transcendental, analytic language (ions) and the natural historical, morphological, mimetic language (fogs). Ions, while invoked in both natural philosophical and natural historical projects, were stripped of the detailed physical and chemical attributes that they had in "pure" physical and chemical work.

By this time Wilson's Cavendish colleagues were also using the cloud chamber. Most notably, J. J. Thomson resolved to establish a value for the charge of the electron, *e*, and thus add support to his growing conviction that the electron

123. Feynman, Leighton, and Sands, *Feynman Lectures* (1963–65), 2:9-1–9-2.

124. CWnb A8, 7 July 1898.

125. CWnb B1, 4 October 1898.

126. CWnb A3, 7–23 January 1899.

was a fundamental particle. Just a year before, in 1897, he had claimed that cathode rays were streams of elemental particles[127] and had shown that the charge-to-mass ratio, e/m, was a constant for these particles. This was no guarantee, however, that e and m were separately constant, and some workers in the field judged it insufficient proof of the existence of an electron.[128] At the Cavendish, however, there were few doubters, and work to determine accurately the value of e started immediately.

Thomson's first attempt was crude, but it established a method that Robert Millikan later exploited in his famous oil drop experiments.[129] Thomson was able to determine ne, the number of ions times the electron charge, from the measurement of current through gas (a procedure he had been using for a long time); he only had to determine n to get a value for the electron charge. Quoting Thomson: "The method I have employed to determine n is founded on the discovery made by Mr. C. T. R. Wilson."[130] Thomson's method was to take the gas for which ne had been determined and subject it to expansion. Assuming that each water droplet caused by the expansion contained a single ion, n would be equal to the number of droplets. By observing how quickly the cloud caused by the expansion of the gas in a cloud chamber fell under the influence of gravity, he was able to estimate the number of drops. Take the mass of water in the air to be fixed. (The greater the number of drops, the smaller each drop must be.) Stokes's law then predicts that the cloud falls more slowly for smaller drops. These early experiments gave the value of 7.3×10^{-10} electrostatic units for the electron charge.[131]

Wilson adopted Thomson's falling cloud method. In a more complicated version of the experiment to test whether negative charge induces condensation before positive, Wilson placed three vertical brass plates in his chamber, creating two regions.[132] The middle plate was grounded, the left plate was kept at a positive potential, and the right at a negative potential. Since oppositely charged ions were drawn into different parts of the chamber, Wilson could determine whether negative charges differed from positive in their ability to condense water vapor. After making an expansion of 1.25, he "counted" the number of charges in the two sides of the chamber using Thomson's method and found that negative charge did precipitate water but positive charge did not.

Soon H. A. Wilson, also at the Cavendish, joined Thomson in his determination of e. Together, they added electrically charged plates (set horizontally,

127. Thomson, "Cathode Rays," *Phil. Mag.*, 5th ser., 44 (1897): 293–316.

128. Holton, "Subelectrons" (1978). This paper provides a detailed discussion of Millikan's different methods of measuring the electron charge, e. Also on Millikan's oil drop experiments, see Franklin, "Oil Drops," *Hist. Stud. Phys. Sci.* 11 (1981): 185–201.

129. Holton, "Subelectrons" (1978).

130. Thomson, "Charge of Electricity," *Phil. Mag.*, 5th ser., 46 (1898): 528–45, on 528.

131. Thomson, "Charge of Electricity," *Phil. Mag.*, 5th ser., 46 (1898): 528–45, on 542.

132. Wilson, "Comparative Efficiency," *Phil. Trans. Roy. Soc. London A* 193 (1899): 289–308.

not vertically as in C. T. R.'s experiments) to Thomson's original experiment and compared the fall of clouds under the action of an electric field with that under no field. This comparison allowed them to obtain an estimate of e without resorting to counting n, the number of droplets. Eventually Millikan, in America, embraced the Thomson–H. A. Wilson method, modifying it with a far more powerful electric field until he could observe individual drops, either balancing them or using Stokes's law to calculate their charge while they rose or fell.[133]

It is clear that C. T. R. Wilson's work was assimilated into the heart of Cavendish research. On one hand, many of his experiments were interpretable only in terms of ion physics, and Thomson was quick to suggest ionic explanations of cloud chamber phenomena. On the other hand, Thomson and other Cavendish physicists used Wilson's chamber to bolster a major premise of ion physics, the existence of an elemental charge. Yet Wilson hardly contributed to these experiments on matter. Why not? To understand why Wilson's attention fell elsewhere, we need to specify more completely the problems and goals of Wilson's self-defined subfield. What, in short, was the relation of laboratory to atmospheric phenomena? At stake was the nature of mimesis itself.

It was in the link between laboratory and sky that Wilson's identification of condensation nuclei with electrical ions ran into trouble. In 1899, before his experiment on positive and negative ions, Wilson had been able to demonstrate conclusively what he could only suggest in 1897: condensation nuclei were ions. To prove his point, Wilson had applied an electric field to his chamber after it had been exposed to X rays or other sources of ions, but before the expansion was made. He expected the field to sweep away the electrical ions, and to his satisfaction, the plates did, in fact, diminish the condensation. At the time, Wilson had felt confident enough to state categorically: "This behaviour of the nuclei proves them to be charged particles or 'ions.'"[134] Unfortunately, Wilson's identification procedure soon encountered two striking anomalies. Ultraviolet light produced condensation, but an electric field had no effect on droplet formation. These surprises worried Wilson tremendously. Making matters worse, Wilson found that his electric fields seemed powerless to sweep away the nuclei continuously present in dust-free air. Both effects threatened the ion explanation of dust-free condensation.

Thus, when Wilson turned to the experiments with positive and negative ions, he again tried to prove that ions caused condensation both in filtered air and in air exposed to ultraviolet light. But even with a stronger electric field and a more sophisticated method of determining whether there was any effect, Wilson failed completely. An electric field produced no noticeable diminution of

133. Holton, "Subelectrons" (1978), esp. 42–43. Franklin points out that few drops were actually observed in suspension; see Franklin, "Oil Drops," *Hist. Stud. Phys. Sci.* 11 (1981): 185–201.

134. Wilson, "Condensation Nuclei," *Proc. Roy. Soc. London* 64 (1899): 127–29, on 129.

condensation. He now faced an almost impossible choice: The enormously productive ion hypothesis appeared to be refuted by two clear experimental results. Could these be evidence of two different kinds of ion? In other words, might there be essentially different explanations of (1) "the slight rain-like condensation which takes place, when V_2/V_1 lies between 1.25 and 1.38, in the absence of all radiation, as well as the much denser condensation produced by the same expansions when the air is exposed to weak ultra-violet light," and (2) "the apparently similar condensation produced in air ionised by Röntgen rays." Perhaps the "normal air" condensation and ultraviolet-induced condensation had nothing to do with ions at all. Wilson agonized over the conundrum: "There is, however, the difficulty of the unlikelihood of two entirely different classes of nuclei being so exactly identical in the degree of supersaturation necessary to cause water to condense on them. The apparent existence of a second coincidence (an increase of the number of drops when V_2/V_1 exceeds 1.31) is still harder to explain on this view."[135]

Wilson's only explanation for why an electric field could not remove the ionic nuclei was to suggest that the ions were an artifact of the abrupt expansion process itself. Since the expansion occurred *after* the electric field was applied, Wilson's conjecture would explain why the field had no effect. Wilson never seemed too happy with this ad hoc account. To resolve the conflict between experiment and theory Wilson modified the hardware—he tried to create a continuously operating rain chamber that would not require sudden expansions.[136] It was a short-lived attempt, however, and soon Wilson realized that he would have to give up the cloud chamber altogether if he wanted to find out whether there were ions present in air, in the absence of external agents. Why?

Wilson's primary purpose in designing the cloud chamber had been to reproduce faithfully atmospheric phenomena, yet it now threatened to be a repository of mere laboratory artifacts. Two problems threatened the link between the chamber and the earth's atmosphere. First, when used on unmolested air, the device caused condensation—but the nuclei did not respond to an electric field. Were the ions produced by the machine itself? Or were they nuclei of a kind that would not respond to electric fields? Second, while Wilson could show that Thomson's ions (generated, e.g., by X rays) *could* cause condensation, he could not show that the real atmosphere contained any such ions. Thus, Wilson feared that he had entirely lost contact with the authentic conditions of nature—dust-free condensation might have nothing to do with fog, rain, or atmospheric optics. It was the nightmare of the morphological physicist; Antaeus had been cut from Earth.

135. Wilson, "Comparative Efficiency," *Phil. Trans. Roy. Soc. London A* 193 (1899): 289–308, on 305.

136. CWnb B2, June 1900.

In a paper written in 1901, Wilson explained his quandary and described the electroscope he would now adopt as his preferred instrument: "After much time had been spent in attempts to devise some satisfactory method of obtaining a continuous production of drops from the supersaturated condition, I abandoned the condensation method, and resolved to try the purely electrical method of detecting ionisation. Attacked from this side the problem resolves itself into the question, Does an insulated-charged conductor suspended within a closed vessel containing dust-free air lose its charge otherwise than through its supports, when its potential is well below that required to cause luminous discharges?"[137] Constrained on one side by his commitment to a morphological, mimetic physics and on the other by his commitment to ion theory and his new instrument, he was stuck. It seemed the cloud chamber had carried Wilson to a dead end. If the images Wilson produced were not homomorphic—if they failed to capture nature "out there" in the lightning and thunderstorms—then the cloud chamber was nothing at all.

2.5 Trailing the Invisible

In desperation Wilson turned to an entirely different class of devices—electroscopes—which, though they offered no means of visually reproducing atmospheric phenomena, could at least monitor atmospheric electricity continuously without creating artifacts. To save his condensation program Wilson had to discover whether the ions he had studied in the cloud chamber really existed in the atmosphere. Having built an electroscope, Wilson turned to the well-known problem of leakage. Two German physicists, J. Elster and H. Geitel, had shown that an electrified body lost its charge even in dust-free air; moreover, it did so in night as in day, as much for positive as for negative charges on the electrodes of the apparatus, and the leak did not seem to depend on the voltage.[138] Wilson confirmed these findings and, while ceding Geitel priority, added two original contributions: the rate of leakage was roughly proportional to the pressure, and at atmospheric pressure, in the language of the Cavendish the leakage corresponded to the production of 20 ions/cc/sec.[139] Wilson toyed briefly with the suggestion that the vessel walls might be radioactive. When that possibility did not stand up to experiment, Wilson had to know whether the production of ions in dust-free

137. Wilson, "Ionisation," *Proc. Roy. Soc. London* 68 (1901): 151–61, on 152.

138. Wilson first became familiar with Hans Geitel's work on spontaneous leakage from Geitel's article, "Elektrizitätszerstreuung," *Phys. Zeit.* 2 (1900–1901): 116–19, which Wilson received from Geitel on 28 November 1900; see C. T. R. Wilson to H. Geitel, 28 November 1900, Staatsbibliothek zu Berlin–Preussischer Kulturbesitz. Geitel apparently beat Wilson to the discovery of spontaneous leakage by a few days. In a letter to Rutherford, Wilson explicitly rejected extraterrestrial particles as a source of this ionization: "[T]he ionisation does not seem to be due to very penetrating rays which have traversed our atmosphere" (C. T. R. Wilson to E. Rutherford, 20 April 1901, ERP. Add. 7653, microfilm copy deposited at AIP).

139. Wilson, "Ionisation," *Proc. Roy. Soc. London* 68 (1901): 151–61, on 153.

air was "due to radiation from sources outside our atmosphere, possibly radiation like Röntgen rays or like cathode rays, but of enormously greater penetrating power."[140] Lugging his electroscope into a deep rock tunnel owned by the Caledonian Railway, Wilson found no evidence that his instrument leaked more slowly underground than it did on the surface. Later observers accounted for Wilson's failure to detect cosmic rays by the earth's radioactivity, but at the time he saw no choice but to agree with Geitel that the ineradicable ionization was caused by "a property of the air itself."[141]

For three years Wilson pursued his radioactive air hypothesis by studying the radioactivity brought to earth by rain and snow,[142] while his Cavendish colleagues exploited the chamber to explore more central questions of "transcendental physics"—such as Thomson and H. A. Wilson's use of the chamber to determine the unit of elemental charge. In 1903 in a short review article on atmospheric electricity for *Nature,* C. T. R. Wilson wrote, "It is quite conceivable that we may be driven to seek an extra-terrestrial source for the negative charge of the earth's surface."[143] But Wilson never offered concrete arguments for the existence of cosmic rays. Instead, he concentrated his efforts on atmospheric electricity, speculating that the ionization might come from meteorological processes, such as the asymmetrical scattering of charges during the splashing of raindrops.[144] A decade later, Victor Hess took three electroscopes on a balloon ascent and proved that the ionization effects did not decrease with height as they should have were the radiation emanating from the earth's crust.[145] It may be that Wilson's nondiscovery of cosmic radiation reflected his preoccupation with meteorology, a worldview that kept his mind's eye fastened on the sublunary sphere.

While Thomson pushed the cloud chamber unambiguously toward reductionist natural philosophy, Wilson's concerns circled around his self-constructed complex of condensation problems: atmospheric electricity, rain, hail, fog, ion properties, and atmospheric optics. His notebooks record a constant movement back and forth between questions of ionic charge and the nature of atmospheric phenomena. Again and again he returned to these core problems: his notes from 1908, for example, include a carefully lettered section entitled, "Theory of

140. Wilson, "Ionisation," *Proc. Roy. Soc. London* 68 (1901): 151–61, on 159.

141. Wilson, "Ionisation," *Proc. Roy. Soc. London* 68 (1901): 151–61, on 161.

142. Wilson, "Radio-Active Rain," *Proc. Camb. Phil. Soc.* 11 (1902): 428–30; "Further Experiments," *Proc. Camb. Phil. Soc.* 12 (1902): 17; and "Radio-Activity from Snow," *Proc. Camb. Phil. Soc.* 12 (1903): 85.

143. Wilson, "Atmospheric Electricity," *Nature* 68 (1903): 102–4, on 104.

144. In Wilson's "Atmospheric Electricity," *Nature* 68 (1903): 102–4, on 104, he seemed to favor two purely meteorological explanations for the source of atmospheric electricity: (1) the differential condensation on positive and negative charges could bring down the negative charges while leaving the positives; (2) raindrops, with their charges polarized in an ambient electric field, could hit other drops on their way down, driving one kind of charge up and the other down.

145. See Steinmaurer, "Erinnerungen" (1985), 22.

Corona."[146] From storms to the laboratory, his concerns could not be split into the purely analytic and the purely mimetic. For example, Wilson's working theory of rain formation involved several stages. First, water vapor condensed around negative ions—just as it did in the cloud chamber. If these small drops rose more slowly than the updraft of supersaturated air, the drops would rapidly grow, perhaps providing the mechanism by which a cumulus cloud metamorphoses into a cumulonimbus. Wilson speculated that a thin cloud cap might form "suddenly over the head of the cumulus and sink rapidly into it. . . . The drops formed on the ions may themselves fall through the lower cloud and reach the ground as rain." In addition, the separation of positive ions (left in the air) from the negative ones (borne downward by rain) would account for the intense electric fields found in thunderstorms.[147] Such concerns led him back to work on cloud chambers with electrical fields.

Conversely, Wilson's concern with the real condition of nature drove him out of the laboratory. To test whether beta rays from the atmosphere might be charging the earth, he set his electrometer under the "natural screening" of trees, or under sections of sod; he also used wire screens outdoors: "*Jan 16* [1909]. Experiments in garden. Cambridge. Electrometer placed on table under bare apple trees. Electrical field too small to detect. . . . Cloudless sky. [F]air breeze."[148] Apparently to study the very small drops that would initiate thunderstorms, Wilson came back to condensation experiments in order to count drops when they were very numerous. He set up a Nernst lamp and inserted an eyepiece but was distracted by the coronae that the new apparatus displayed.[149] When Wilson resumed his rain experiments in March 1909, he wanted to know whether his ion-condensed droplets grew into raindrops. At least since his reading of meteorologist Osborne Reynolds's work, he had known that coalescence was a strong candidate for the growth mechanism of both rain and hail. Simple inspection of the ice crystals' cross sections demonstrated that hail could not be frozen rain. Furthermore, in both rain and hail the heat of condensation is too great for the growth to be attributed entirely to condensation. Again, the notebooks: "Do ordinary cloud particles coalesce spontaneously? If not, do they when rather larger drops fall through the cloud[?] Effect of electrical field[?]" Using a vertical tube, Wilson hoped to make visible the process of drop formation and growth.[150]

Obviously persuaded that condensation alone would not suffice to produce real rain, Wilson intensified his efforts to *see* the drop formation process itself.

146. CWnb A9, preceding entry dated April 1908.

147. CWnb A9, April 1908, section entitled "Atmospheric Electricity & Condensation on Negative Ions."

148. CWnb A9, 16 January 1909.

149. CWnb A9, 16 January 1909. Notes describing these experiments are between entries dated 16 January and 6 February 1909.

150. CWnb A9. Wilson then took careful reading notes on Reynolds, "Raindrops and Hailstones," *Mem. Lit. Phil. Soc. Manchester*, 3rd. ser., 6 (1879): 48–60.

As he speculated on droplet dynamics in his notebook, Wilson recalled the astonishing high-speed pictures of drops and splashes recently publicized by Worthington.[151] Worthington's book *A Study of Splashes* had just appeared in 1908 and held a twofold fascination for Wilson. First, the photographer had brilliantly exploited the light of sparks that endured mere *millionths* of a second to capture on film events ordinarily too fleeting for a human observer (see figure 2.12). This high-speed method offered Wilson the technical means to reveal the elementary processes of condensation and coalescence. Second, Worthington's photographically "frozen" splashes might indicate the mechanism that separated positive and negative particles, causing the electrical gradient in rain clouds. Again Wilson used photography to record the laboratory reproduction of meteorological cloud phenomena, though now at a more fundamental level. Wilson soon adapted Worthington's spark illumination system for his own use. In a section of his notebook labeled "*On drop counting methods,*" he reasoned that the high-speed camera would enable him to count droplets more accurately. "Methods depending on instantaneous photography of drops immediately after their production are superior to those in which drops falling through an illuminated layer are counted—the possibility of photographing the drops being presupposed."[152] It was a crucial and original step away from the falling cloud methods of H. A. Wilson and Robert Millikan; by May 1909 Wilson had produced his first successful negatives.

As was so often the case, within his own special science of condensation physics Wilson oscillated between thunderstorms and atoms. This time his technical success with the photography of rain formation led him from meteorology into ion physics. His notebooks display an attempt to use a permanent film record to improve Thomson and H. A. Wilson's ionic charge experiments.[153] But despite some inventive schemes recorded in the notebook, Wilson never advanced the cause of charge determination and soon returned to the photography of his laboratory clouds.

The day before Christmas 1910, Wilson sat down to summarize the "[p]resent state of work on expansion apparatus." He recorded his satisfaction with newly discovered tricks of the trade; in particular, he admired a procedure for coating the inside of the glass bulb with a thin layer of gelatin. This coating prevented the glass from fogging as water vapor deposited in a dewlike form and offered a conducting surface for use in electrical experiments. Experimental skills like these could be used in conjunction with a variety of chambers. As he noted, the "simple flash form" chamber would provide a stage for the

151. Worthington, *Splashes* (1908); see CWnb A9, section marked "Atmospheric Electricity and Condensation on Negative Ions," immediately following entry dated April 1908.

152. CWnb A9, between entries dated 10 April and 17 July 1909, probably May 1909.

153. CWnb A9, 13 September 1909 and 16 January 1910.

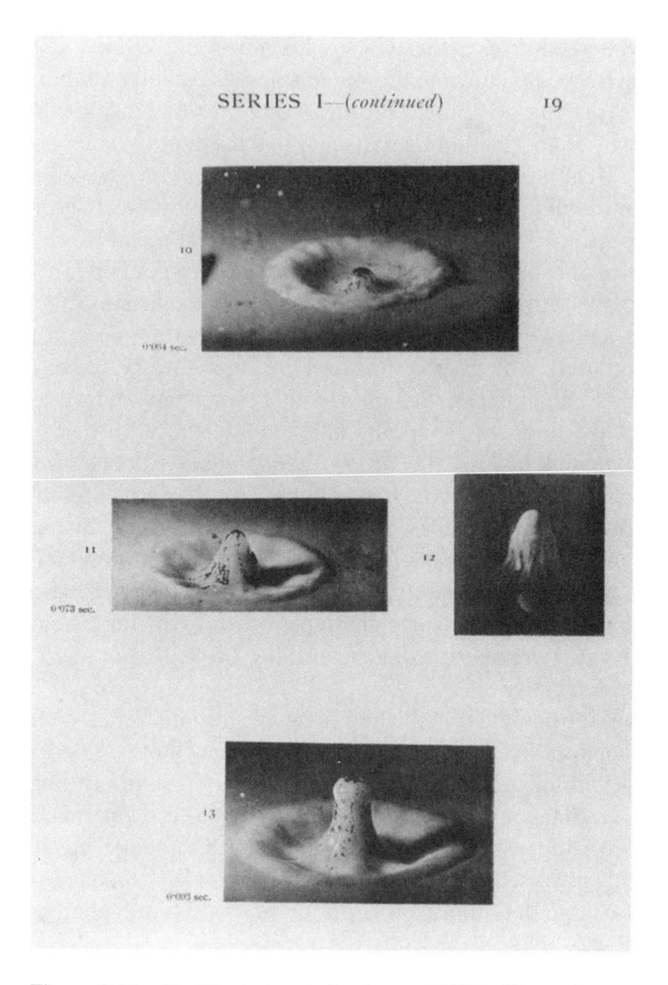

Figure 2.12 Worthington's splash pictures (1908). These pictures impressed Wilson by the novel form of photography they offered (nanosecond illumination by spark). Source: Worthington, *Splashes* (1908), 19; in Wilson's unpublished work, see CWnb A9, "Atmospheric Electricity," immediately following entry dated April 1908.

photographing of individual drops condensing around "spontaneously" produced ions. A second chamber would sport an electrode down the center and would measure the effect of fields on spontaneously produced ions—and, with luck, help find the electron charge, e. Yet a third chamber is listed simply with "applications:—Tracks of α or other rays. . . . Assuming that the difficulties of photographing the drops were overcome, one might see tracks of α rays marked out

Figure 2.13 Wilson's cloud chamber for tracks (1911). Wilson used this chamber (modified in various ways) for decades. The chamber itself is the glass cylinder to which are attached coiled electrical leads for "clearing" residual ions. Below the chamber is an expansion cylinder that rides above a pan filled with water (to keep the air saturated). Before the chamber is triggered, the large glass ball to the right is evacuated by a pump; when the trigger is activated, the air below the plunger rushes into the glass ball, and the plunger falls quickly. This sudden drop of pressure in the cloud chamber lowers the temperature, and the air within the chamber is suddenly in a supersaturated state. Source: Courtesy of Cavendish Laboratory, University of Cambridge.

by narrow core of drops formed on positive ions with more scattered cloud of those formed on negative ions round [the positive ions]."[154]

Wilson approached the most elementary condensation process with all the means at his disposal. Pursuing this goal, he designed an electrometer that he hoped might reveal the presence of a charge of no more than "about 30 ions." On the very next page of his notebook, in the entry of 18 March 1911, Wilson returned to the cloud chamber (see figure 2.13), hoping to bring his photographic vision ever closer to the condensation around a single ion. Wilson continued to scribble: "Cloud chamber exposed to X rays (from above). Nernst lamp illumination. Cloud was discontinuous, showing numerous knots. Are these cross sections of tracks of rays? They showed up without field (exposure for short time only) Next expansion (without rays) showed uniform cloud (on residual nuclei)

154. CWnb A9, 24 December 1910.

Figure 2.14 First golden event (1911). One of the distinguishing features of cloud chamber events was the possibility of examining physical effects that either were not repeated or were exceptionally rare. It is striking that the very first golden event was identified as such so soon after Wilson first saw tracks at all. Both his familiarity with the registration of unusual meteorological phenomena and his prior engagement with photography no doubt contributed to his interest in individual events. Source: CWnb A9, 29 March 1911.

without any such knots. Expansion apparatus working perfectly. Field was due to single secondary cell."[155]

Was the striking first result an artifact? Wilson increased the electric field to encourage drop formation; days later, there was no doubt: "[R]ays showed up much better. They were extremely sharply defined. . . . Look[ed] at from above. The individual rays were seen, in many cases as extremely fine lines, chiefly radiating from aperture—but in some cases running in other directions (secondary rays)."[156] Photography, by then second nature to Wilson, lent credence to the phenomena. From the splash and drop formation studies, he knew the photographer's art could probe where the eye could not. Within days, Wilson had the knack of photographing tracks; on 29 March 1911 he recorded that he could produce images of beta rays taken "as before." Almost simultaneously, he successfully found the footprints of X rays and used gamma rays from radium to produce tracks even through a lead plate.

Later that March day, Wilson had enough of a feel for the "ordinary" images of radioactive processes to record an extraordinary one: "On one occasion in addition to ordinary thread-like rays, one large finger-like ray was seen, evidently a different form of secondary ray—giving rise to enormously more ionization than even ordinary [alpha] ray" (see figure 2.14).[157]

In mid-April 1911, Wilson submitted his first track paper, "On a Method of Making Visible the Paths of Ionising Particles through a Gas." There he exhibited his photographs of alpha rays and the "clouds" that precipitated on X-ray–produced ions. In a register at once natural philosophical and natural historical

155. CWnb A9, 18 March 1911.
156. CWnb A9, 20 March 1911.
157. CWnb A9, 29 March 1911.

Wilson wrote: "The photograph [conveys] but a poor idea of the really beautiful appearance of these clouds."[158] It was clear that X rays ionized by producing secondary, charged corpuscles. But still "undecided" was the question of whether an X ray was a "continuous wave front, or is itself corpuscular as Bragg supposes, or has in some other way its energy localised . . . in the manner suggested by Sir J. J. Thomson."[159]

Although Wilson would have little to do with the theoretical development of quantum theory, his first cloud chamber pictures thus entered directly into the controversy between wave and particle interpretations of X and gamma rays. The two figures in this controversy, which had come to a head in 1907, were William Henry Bragg and Charles G. Barkla.[160] At the time, the accepted explanation of these rays was J. J. Thomson's pulse theory. According to Thomson, "bundles of energy" (or pulses) varied in size depending on the type of rays; for example, gamma rays were ether pulses that occupied a much smaller volume than the pulses of X rays. Bragg challenged Thomson's theory, proposing instead that X rays and gamma rays were streams of neutral particles.[161] He postulated that these penetrating neutral particles were composed of a "neutral pair" of oppositely charged particles; in some cases the two constitutive entities were none other than the alpha and the beta.[162]

Barkla rejected Bragg's neutral pairs; in fact, his experiments showed that X rays were scattered, in agreement with Thomson's theory. (The intensity of the scattered rays varied as $1 + \cos^2 \theta$, where θ is the scattering angle.) In addition, Barkla found that he could polarize the rays. Although Bragg found confirmation of his theory in J. P. V. Marsden's experiment with gamma rays, which showed that there was more scattering in the forward direction than in the backward (and thus challenged Barkla's results),[163] there was little reason to believe in the completely unprecedented neutral-pair theory over the well-established wave theory.

158. Wilson, "Ionising Particles," *Proc. Roy. Soc. London A* 85 (1911): 285–88, on 286.

159. Wilson, "Ionising Particles," *Proc. Roy. Soc. London A* 85 (1911): 285–88, on 287–88.

160. For an account of the controversy, see Stuewer, "Corpuscular Theory," *Brit. J. Hist. Sci.* 5 (1971): 258–81; Wheaton, *Tiger* (1983), 71–103; and Caroe, *Bragg* (1978), 53–63. Bragg was led to this theory by his realization that the absorption coefficient for particulate scattering would be different than that for waves. As the particles passed through matter, they would slow down, making collisions with other atoms more likely, and thus, the absorption coefficient would increase. For homogeneous waves incident on matter, this would not occur—absorption would be constant. In 1904, Bragg was able to show that gamma rays were not absorbed at a constant rate and therefore could not be traditional waves. By 1906, he had extended this reasoning to hard X rays and declared that both types of waves were corpuscular. This view further allowed Bragg to explain why secondary electrons produced by ionizing X rays had high velocities, inconsistent with the pulse theory. If the rays were particulate, such concentrated energy transfers would be a necessary consequence. See Wheaton, *Tiger* (1983), 81–87.

161. Stuewer's *Compton Effect* (1975), 6–23, and his "Corpuscular Theory," *Brit. J. Hist. Sci.* 5 (1971): 258–81.

162. See Wheaton, *Tiger* (1983).

163. Bragg and Madsen, "Gamma Rays," *Phil. Mag.*, 6th ser., 15 (1908): 663–75.

Bragg was delighted when Wilson was able to photograph "tracks" of X rays and gamma rays. Naturally, the neutral ray itself was invisible, but branching off from its path were numerous tracks of beta rays. The tracks were easily identifiable and were similar to ionization caused by known corpuscular rays (e.g., by beta rays themselves). Barkla and others who believed in the pulse theory had expected diffuse ionization. Wilson recognized the importance of his work for Bragg, and in late April 1911, Wilson assured Bragg that "[t]he distribution of cloud when the expansion was made while the cloud chamber was exposed to γ-rays—was quite in accordance with your views. There appeared to be no ionisation apart from that due to corpuscular β-rays starting from the walls of the vessel. The cloud was confined chiefly to perfectly straight fine lines running from side to side."[164] Rutherford also encouraged Bragg: "I am sure you are highly delighted at the way things are turning out in favour of your views."[165]

Neither Bragg nor Barkla seems to have been aware of Einstein's 1905 hypothesis that light came in "quantum bundles"; even if they had been, it is questionable whether they would have seen the connection between his work and theirs.[166] Although Bragg eventually gave up his neutral-pair theory, at least part of his belief in the corpuscularity of X rays and gamma rays was vindicated; but then so were Barkla's convictions that the objects were wavelike. By 1912, Bragg was calling for a theory that recognized a fundamental duality in the rays.[167] Other theoretical issues were also settled in Wilson's glass-enclosed world. In a paper of 7 June 1912, Wilson contended that his pictures were thoroughly consistent with Rutherford's predictions of nuclear scattering: Wilson interpreted his pictures as showing that a few alpha rays underwent sudden, large-angle scattering, with the vast majority suffering a gentler bending from multiple deflections. Moreover, improvements in the chamber allowed the cloud master to capture on film even the fastest beta particles, to render even individual ions visible, and to record tracks without distortion.[168] Wilson was particularly impressed by the astonishing match between alpha tracks that Bragg had drawn based on theory (figure 2.15) and the cloud tracks Wilson himself had just produced in his chamber (figure 2.16). Extraordinarily articulated beta tracks could only add to Wilson's conviction that he had indeed made the invisible visible (see figure 2.17). Just as Wilson's chamber could carve out a trading zone between mimetic and analytic experimentation, so too could it define a border area between experiment and theory. All at once, Wilson's chamber was all the rage.

164. Wilson to Bragg, 23 April 1911, ARI.

165. Rutherford to Bragg, 9 May 1911, quoted in Caroe, *Bragg* (1978), 55.

166. Stuewer, *Compton Effect* (1975), 23–24. On the slow reception of Einstein's light quantum, see, e.g., Kuhn, *Black-body* (1978).

167. See Stuewer, *Compton Effect* (1975), esp. 69; Stuewer, "Corpuscular Theory," *Brit. J. Hist. Sci.* 5 (1971): 258–81, on 273–80; and Wheaton, *Tiger* (1983), 165–66.

168. Wilson, "Expansion Apparatus," *Proc. Roy. Soc. London A* 87 (1912): 277–90.

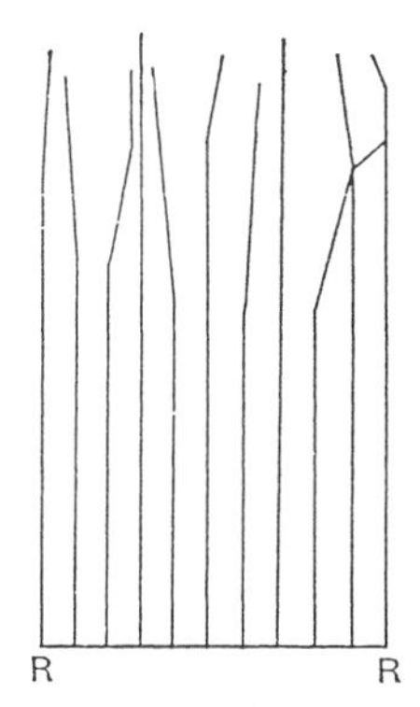

Suggested forms of the paths of a particles projected upwards from radium at R R.

Figure 2.15 Bragg, theoretical alpha tracks (1911). When Bragg drew these tracks, he knew nothing of Wilson's new device. But judging from Wilson's comments not much later, the striking visual similarity between the cloud chamber tracks and these theoretical ones added to Wilson's conviction that he had been able to "photograph" this previously hidden, atomic domain of nature. Source: Bragg, *Radioactivity* (1912), 34.

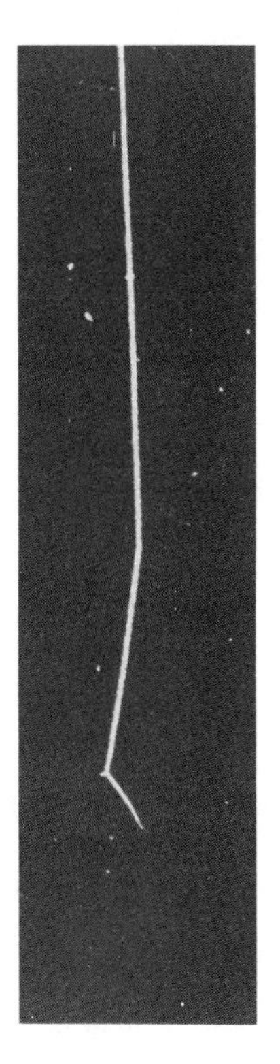

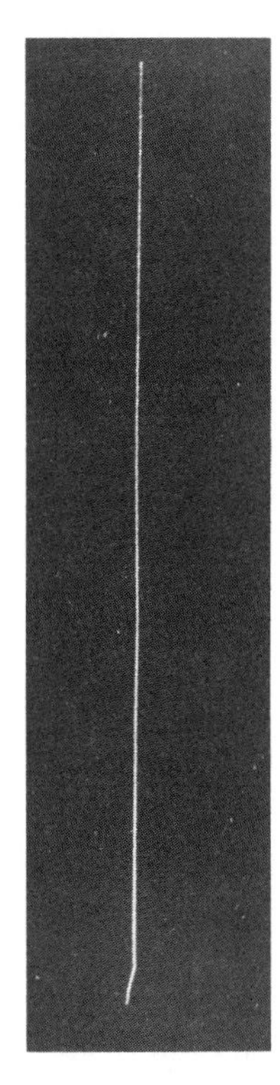

Figure 2.16 Wilson, cloud chamber alpha tracks (1912). Here is visible what Wilson identified as the complete track of an alpha particle from radon and the end of an alpha track showing two single scatterings. Source: Wilson, "Expansion Apparatus," *Proc. Roy. Soc. London A* 87 (1912): 277–90; reproduced in Gentner, Maier-Leibnitz, and Bothe, *Atlas typischer Nebelkammerbilder* (1940), figs. 14 and 15 on 21.

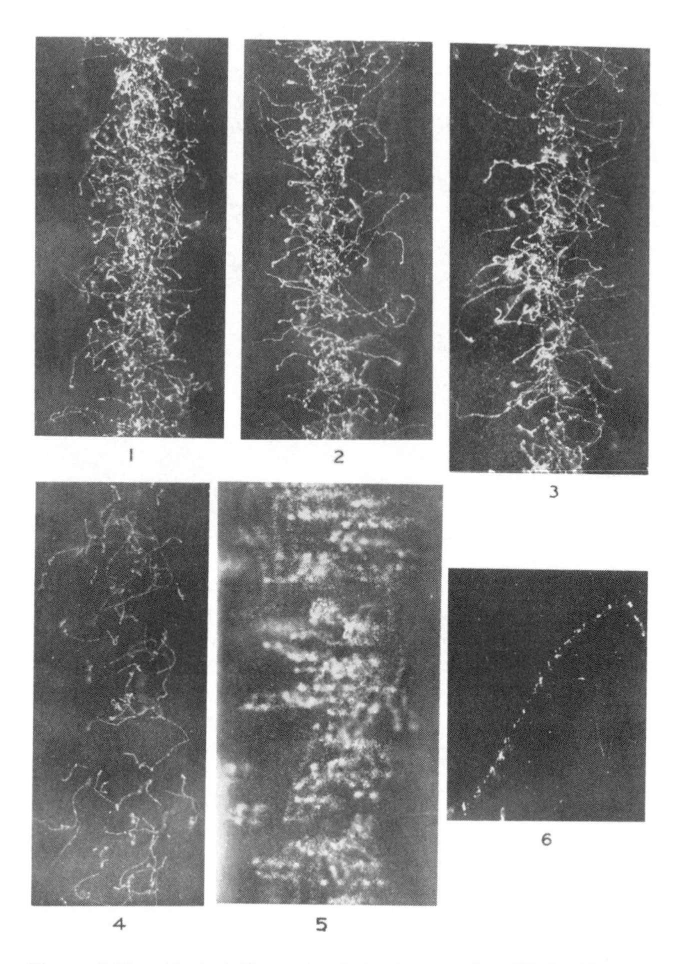

Figure 2.17*a* Typical X-ray clouds, high expansion (1912). Source: Wilson, "Expansion Apparatus," *Proc. Roy. Soc. London A* 87 (1912): 277–90, pl. 8.

Perhaps nothing illustrates better the speedy reception of Wilson's new device than the startlingly fast production of commercial versions of the chamber. Not only did the Cambridge Scientific Instrument Company, Ltd., immediately swing into action in making a marketable instrument, just a year later (1913) they issued a separate publication titled "The Wilson Expansion Apparatus for Making Visible the Paths of Ionising Particles." For the bargain price of £20, the aspiring ion measurer could have a Cloud Expansion Apparatus, and a mere

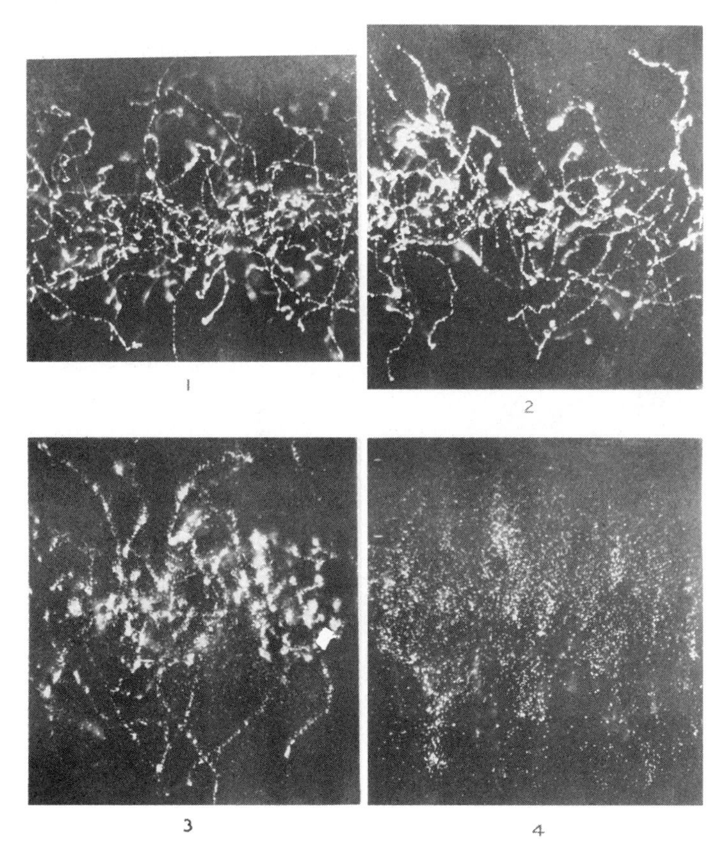

Figure 2.17*b* Typical X-ray clouds, low expansion (1912). Source: Wilson, "Expansion Apparatus," *Proc. Roy. Soc. London A* 87 (1912): 272–290, pl. 9.

£115 would have the vacuum gauge and support at your door in no time (see figure 2.18).[169]

Wilson's cloud chamber work soon came to occupy center stage at the Cavendish, although the few students that Wilson himself supervised found it difficult to work with him.[170] P. M. S. Blackett, James Chadwick, Norman Feather, J. D. Cockcroft, and E. T. S. Walton—all working under the watchful eye of Rutherford—used the chamber to probe the newly discovered atomic and subatomic realms.

169. Cambridge Scientific Instrument Co., *Wilson Expansion Apparatus* (1913), 5.
170. Halliday, "Some Memories," *Bull. Amer. Met. Soc.* 51 (1970): 1133–35, on 1133–34.

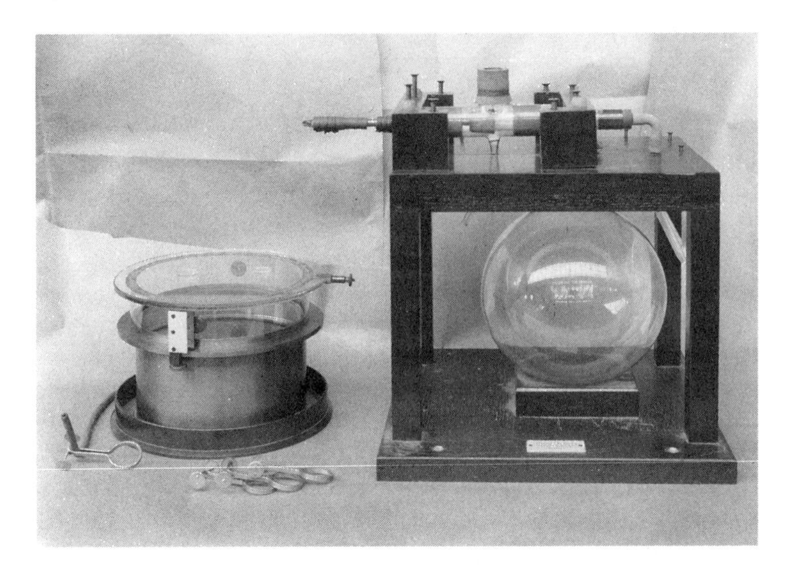

Figure 2.18 Commercial clouds (1913). In June 1913, the Cambridge Scientific Instrument Co. issued a catalog for their multiply produced chamber. Not only was it possible to buy a chamber, the enthusiast could also purchase a set of lantern slides of alpha rays, X rays, and a close-up of one track showing individual ions. Source: Cambridge Scientific Instrument Co., Ltd., Archives, University of Cambridge Library.

In 1919, Rutherford discovered that nitrogen and other light elements were disintegrated on the impact of fast alpha particles; the result was the production of very fast protons. Rutherford made his discovery using the scintillation method, but evanescent flashes of light told him little about what was happening during the actual disintegration process. As Blackett later remarked, "What was more natural than for Rutherford to look to the Wilson cloud method to reveal the finer details of this newly discovered process."[171] T. Shimizu, a Japanese physicist working at the Cavendish, modified the chamber so that he could quicken the expansions and take many more photographs. Shimizu also arranged the cameras to extract stereoscopic pictures from the cloud trails. Before he could do any more, however, he had to return to Japan, and the project fell to Blackett.

Blackett raved about his new work: "Provided by Rutherford with so fine a problem, by C. T. R. Wilson with so powerful a method, and by Nature with a liking for mechanical gadgets, I fell with a will to the problem of photographing some half million alpha-ray tracks."[172] The problem was to find a few golden

171. Blackett, "Cloud Chamber Researches" (1964), 97. Some interesting comments on the stylistic continuity of later cloud chambers can be found in Tomas, "Tradition" (1979), 32–48.

172. Blackett, "Cloud Chamber Researches" (1964), 98.

events; most collisions between the alpha particles and nitrogen nuclei would be elastic, but a few would not and these would show Blackett and Rutherford what was happening in the supposed disintegration. After photographing 400,000 tracks and carefully examining them all, Blackett identified 18 anomalous tracks that did not correspond to elastic collisions. These were evidence of the transmutation of nitrogen. Blackett discovered that the process was not one of disintegration, but one of "integration"; only two tracks were seen after the interaction occurred, meaning that the alpha particle was absorbed as the proton was ejected. The resulting nucleus was a heavy isotope of oxygen. Since all Rutherford could know from his scintillation experiments was that alpha particles infrequently caused nitrogen nuclei to emit protons—he could not "see" the actual interaction—he had assumed it was a disintegration process.[173] Only the cloud chamber could provide a *visual* representation of the transmutation process itself and give physicists the chance to discover the intricacies of the exchange.

After his first resounding success with the cloud chamber, Blackett went on to make the instrument the centerpiece of his life's work. In 1931, he and a visiting Italian physicist, G. P. S. Occhialini, began work on a counter-controlled chamber, a chamber that would expand when an electronic counter-coincidence circuit signaled that a cosmic ray had shot through (see figure 2.19).[174] This apparatus, capable of making particles "photograph themselves," gave cosmic ray physicists a hope of studying rare and new phenomena. Many more physicists went on to use the chamber, a good portion of them at the Cavendish. Working under Rutherford was James Chadwick, who discovered the neutron in 1932. Right next door, John Cockcroft and Ernest Walton were producing accelerated protons with a high-voltage apparatus, and by 1933 they were shooting them at lithium, boron, carbon, and deuterons. They too "saw" into the atom by photographing the events in a cloud chamber.

Rutherford, Thomson's successor as director of the Cavendish, steered the laboratory into a new channel of physics research. The goal was to understand the nature of subatomic matter. It was not completely different from Thomson's goal of explaining the ionic structure of nature, in fact it followed from it, but it was both broader in scope and finer in detail. Neutral particles were admitted into the game, and the strategy was to try to smash and bombard the more complex manifestations of matter until the simplest sprayed forth, bringing with them the secret of the atom. Wilson's cloud chamber was indispensable for this task; having been developed under the aegis of ionic physics, it now led the way to the subatomic.

Wilson, however, neither led nor followed his Cavendish colleagues, withdrawing ever more into weather phenomena. During the heyday of Cavendish

173. On Rutherford, see Stuewer, "Satellite Model," *Hist. Stud. Phys. Sci.* 16 (1986): 321–52.

174. Blackett and Occhialini, "Corpuscular Radiation," *Nature* 130 (1932): 363.

Figure 2.19 Blackett and Occhialini's counter-controlled cloud chamber (1932). Perhaps the most significant advance in cloud chamber technology after Wilson's creation of the device was a cloud chamber triggered, not randomly, but by a set of preprogrammed counters. With Blackett and Occhialini's device it became possible, for example, to trigger the device just when particles passed through the chamber and a thick plate of lead. After the discovery of rare events, the ability to let events "photograph themselves" became an essential part of the cloud chamber craft. Source: Courtesy of Cavendish Laboratory, University of Cambridge.

cloud chamber atomic physics, Wilson's isolation was as physical as it was scientific; he moved his apparatus to the Solar Physics Laboratory, geographically separate from the Cavendish. There he found the solitude he needed to begin the detailed study of thunderstorms that would culminate in his final paper at the age of 90—the first paper to give a dynamical explanation of thunderstorms.

2.6 Objectivity and the Tutored Eye

With its successes of the 1920s and 1930s in hand (including the discovery of the positron, the neutron, the muon, nuclear transmutation, and the Compton effect), Wilson's chamber had come to dominate experimental inquiry into nuclear physics. In addition to the growing body of lore about the construction of the device itself—expansion, lighting, magnetic field, temperature control—a visual language began to emerge around the analysis of pictures. Nowhere is this more evident than in the introduction of a novel forum for physical evidence: the cloud chamber atlas.

Atlases date back to the sixteenth century, when Gerard Mercator published his map of the world. Astronomical maps became commonplace in the eighteenth century, and through these various kinds of big maps, the term "atlas" soon designated a particularly large size of paper. But it was in the nineteenth century that the oversized sheets lent their name to a much wider species of literature, the genre of systematic presentation of scientific pictures. There were atlases of every sort, from the pathology of the human body to African plants, from atlases of fossils to ones of micrographs and X rays.[175]

For the early part of the nineteenth century, atlas makers argued that "truth to nature" in their depictions could be achieved only by genial intervention on the part of the author. Some thought the author's obligation was above all to find the most perfect specimen of a given object: the best possible example of a flower, fossil, or diseased organ. Others sought an ideal altogether different from the instances of (real) nature, while yet others pursued a kind of mean among the many samples examined. Roughly in the middle of the nineteenth century, a change occurred, and a visual notion of objectivity emerged that broke significantly with the older conception of truth to nature. Now the goal was not to reconstruct, idealize, or approximate the form behind a wealth of actual objects (no longer were atlas makers after an *Urpflanze* behind ordinary plants), but rather to exert superhuman effort to remove the author from the process of depiction. Self-abnegation became the order of the day, and atlas makers assiduously tried to avoid imposing idealization, imaginative artistic aids, or indeed any form of "personal interpretation." Of particular relevance to our concerns here, the X-ray atlases by skilled clinicians like Rudolf Grashey (*Atlas typischer Röntgenbilder vom normalen Menschen* [1905] and *Atlas chirurgisch-pathologischer Röntgenbilder* [1908]) were designed to delineate the boundary between normal and pathological with pictures. When faced with a new case, the radiologist could use these pictures to decide whether he or she actually faced something new, or merely a variation on the normal. Explicit judgment was to be removed from the process of depiction in the name of objectivity, and the trained eye shifted or extended from author to audience.[176]

Building primarily on the medical atlas tradition, W. Gentner, H. Maier-Leibnitz, and W. Bothe (from the Institute of Physics at the Kaiser Wilhelm Institute for Medical Research in Heidelberg) put together their *Atlas of Typical Cloud Chamber Pictures* (1940).[177] It included pictures of beta rays in magnetic fields, of Anderson's extraordinarily clear glimpse of a positron, and of

175. Daston and Galison, "Objectivity," *Representations* 40 (1992): 81–128.

176. Daston and Galison, "Objectivity," *Representations* 40 (1992): 81–128, on 105–7.

177. Gentner, Maier-Leibnitz, and Bothe, *Atlas typischer Nebelkammerbilder* (1940).

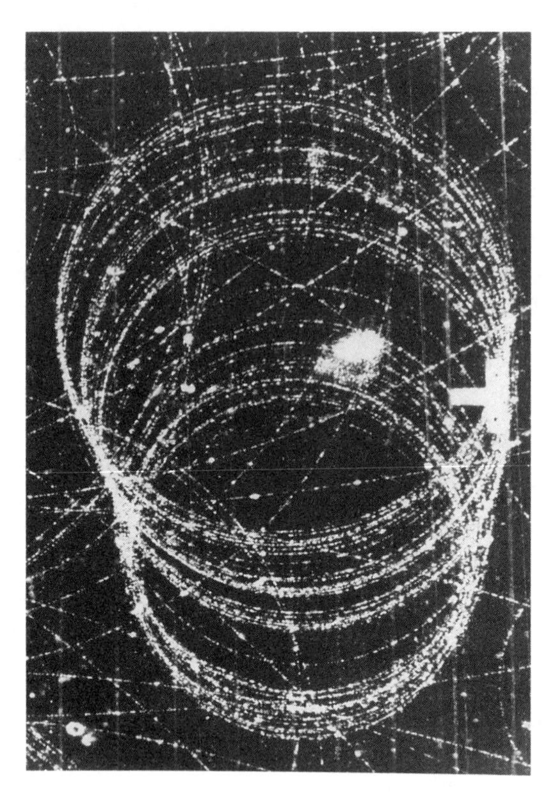

Figure 2.20 Radiation of fast electron (1940). At the arrow (*center bottom*), an electron enters the picture with an energy of 16.9 MeV, its positive partner lost. As it spirals some 36 times in a magnetic field, it moves slowly toward the stronger portion of the field, causing an upward drift. By the end of the visible track, the electron's energy has sunk to 12.4 MeV. By measuring the length of track (1030 cm), the energy loss by ionization can be calculated to be 2.8 MeV, leaving 4.5 MeV unaccounted for. This energy deficit "must . . . be caused" (according to the atlas) in part by bremsstrahlung, the emission of a photon by an accelerated charged particle. And, in fact, a sudden change of curvature can be seen in the seventeenth circle. Source: Gentner, Maier-Leibnitz, and Bothe, *Atlas typischer Nebelkammerbilder* (1940), fig. 43 on 51.

photoelectrons from X rays and others scattered by gamma rays. There were electron pairs produced by gamma rays, bursts of alpha rays, and of course, the by-then famous pictures of Wilson's first alpha-scattering events. In the latter sections of the book were pictures of protons being hit by neutrons and, perhaps most topically, the just-discovered splitting of the uranium nucleus. Together, these images were supposed to tutor the eye, to train the growing cadre of cloud chamber physicists to recognize the typical, and therefore to underscore the new. By the second edition (1954), Gentner, Maier-Leibnitz, and Bothe had added to their atlas a host of cloud chamber pictures taken at accelerators (see, e.g., figure 2.20). Together these volumes became standard reference works for cloud chamber particle hunters.

From high in the atmosphere at mountaintop observatories to the protection of underground laboratories, cloud chambers became, in the late 1940s and early 1950s, the means to the discovery of new particles. So successful was the cloud chamber, and its close cousin the nuclear emulsion that we will encounter in the next chapter, that in 1951 Blackett reported (in the introduction to another cloud chamber atlas): "All but one of the now known unstable elementary particles have been discovered by one or other of these techniques [cloud chambers or nuclear emulsions]."[178] Both methods produced visual evidence, giving "us pictures of what single particles do." Because the skills involved in identifying characteristic tracks soon became a quasi-autonomous branch of experimental science, leaders in the field responded enthusiastically to solicitations for pictures to fill the atlases. Aside from the intrinsic public relations appeal of the picture books, one stated function of these cloud chamber atlases was to facilitate the learning of pattern recognition skills that were a prerequisite for discovery. In the foreword to the one atlas edited by George Rochester and John G. Wilson, Blackett put its purpose succinctly: "An important step in any investigation using [the visual technique] is the interpretation of a photograph, often of a complex photograph, and this involves the ability to recognise quickly many different types of sub-atomic events. To acquire skill in interpretation, a preliminary study must be made of many examples of photographs of the different kinds of known events. Only when all known types of event can be recognised will the hitherto unknown be detected."[179]

In writing this foreword, Blackett could not help but have in mind the recent events at his own laboratory in Manchester. On 15 October 1946, Rochester and Clifford C. Butler inspected their daily harvest of cloud chamber photographs (see figures 2.21–2.23) and to their surprise found the unusual forked track in the lower right-hand quadrant of figure 2.23. The two young physicists reasoned this way: the small amount of ionization left by the arms of the V ruled out the possibility that they were protons. A glance at the stereoprojections indicated that they truly began in a single point, so the two tracks' common origin could not be a mere illusion of superimposed tracks. If the arms of the V came from a collision process involving the nucleus (nuclear disintegration, pair production, nuclear scattering), more such events ought to have been observed near the ample source of nuclei present in the 5-centimeter-thick lead plate that traversed the chamber. Throughout the winter of 1946–47 Blackett, Butler, and Rochester struggled over the picture, hesitating to go to press with but one instance. On 23 May 1947, Rochester and Butler found a second V, this one a candidate, not for the decay of a new neutral meson, but of a previously unknown charged one. Their announcement came in December 1947 at the Institute for Advanced

178. Blackett, "Foreword" (1952). For other examples of cloud chamber atlases see Gentner, Maier-Leibnitz, and Bothe, *Atlas typischer Nebelkammerbilder* (1940), and the same authors' revised edition, *Atlas of Typical Photographs* (1954).

179. Blackett, "Foreword" (1952).

Figure 2.21 *V*-particle chamber (1947). Clifford Butler is adjusting the cloud chamber used in his and George Rochester's discovery of the first *V* particle (later known, as a class, as the "strange particles"). Source: Courtesy of George Rochester.

Studies in Dublin,[180] and on the basis of these two pictures, they published their discovery in *Nature.*[181]

Even theorists quite close to the Manchester team were unsure how to proceed. W. Heitler, for example, was stymied: "[A]s to the theoretical side I do not wish to make any comments yet. I feel it somewhat futile to speculate at the

180. L. Jánossy to Rochester, 6 November 1947, GRP: "By the way, regarding rest mass of the primary, I find without assumptions as to mass etc. of the secondaries (not even assuming that they are equal) the following inequality $M \geq 2p/c$, where M is the mass of the primary and p is the *transverse* component of the momentum of the secondary"; Rochester to Jánossy, 12 November 1947, GRP.

181. Rochester and Butler, "Elementary Particles," *Nature* 160 (1947): 855–57.

Figure 2.22 Analyzing the *V* particles (1947). Butler is retroprojecting cloud chamber images onto a screen where their length, curvature, and ionization can be measured. Source: Courtesy of George Rochester.

present shape about the genealogy of all the many particles that have been discovered now (most of which have only been observed once). Evidently, the field is for the experiment now to find out all the particles that exist and the modes of decay they undergo."[182] In a similar vein, John Wheeler wrote from Princeton that while he was not, himself, among the doubting Thomases, there were others with less faith: "As usual, on things of this kind, there are certain skeptics. In this case the skepticism generally takes the form of suggesting that the pair of positive and negative mesons created below the lead plate was due to the impact of a neutron on a nitrogen or oxygen nucleus."[183]

Two and a half long years passed with no further sightings, though physicists spared no effort in their search. It was a time when groups from around the world were announcing the existence of a myriad of particles, each with great fanfare; some went on to enter the pantheon of fundamental entities, but a great many vanished unceremoniously into the cluttered refuse heap of artifacts. The Manchester group itself began to plan an elaborate search from a mountain

182. Heitler to Rochester, 23 November 1947, GRP.
183. Wheeler to Rochester, 9 December 1947, GRP.

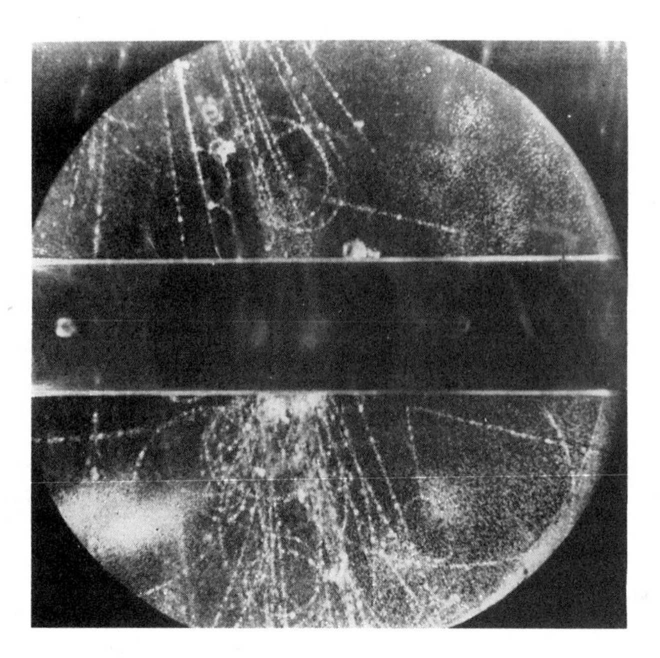

Figure 2.23 First *V* particle (1946). Just below the lead plate, in the lower right-hand quadrant, an inverted V extends to the lower right. Rochester and Butler measured the momentum of the upper particle as 300 MeV/*c* and determined its charge to be positive. The other (lower) particle has a negative charge (if it is moving downward) or a positive charge (if it is moving upward). From past experience, they argued, the two tracks would have to be closer together to be an electron-positron pair. The two tracks cannot be a two-track star; if they were, there should be a visible recoiling nucleus at the apex. Finally, conservation of momentum excludes a pion decaying into an electron, or a muon decaying into an electron. Since this argument eliminates the only plausible alternatives, Rochester and Butler concluded that this event had to be a photographic record of a novel phenomenon: the decay of a previously unknown neutral meson. Source: Rochester and Wilson, *Photographs* (1952), pl. 103 on 102.

observatory, when Carl Anderson wrote Blackett on 28 November 1949 with welcome news: "Rochester and Butler may be glad to hear that we have about 30 cases of forked tracks similar to those they described in their article in *Nature* about two years ago, and so far as we can see now their interpretation of these events as caused by new unstable particles seems to be borne out in our experiments."[184] Speculating that their own group's failure to have found any

184. Anderson to Blackett, 28 November 1949, file B48, PBP.

more V's might be attributable to their using an insufficiently thick lead absorber over the chamber, Blackett responded in December 1949: "Rochester, Butler and I are extremely pleased to hear you have succeeded in confirming the existence of some peculiar forked tracks of the general type that they found a year or two ago. We have been getting rather worried at not finding any more. Rochester and Butler have taken a lot more photographs with the same chamber and have obtained many more beautiful photographs of penetrating showers, but without finding another case."[185] It was on the heels of this confirmatory evidence that Rochester and J. G. Wilson put together their atlas, with relatively few examples of better known phenomena and correspondingly "many of high energy nuclear interactions, and in particular of the recently discovered *V* particles."[186]

While the Manchester discovery of *V* particles may have been the proximate cause of publication of the new atlas, it was by no means its only goal. The new entities were but two of a proliferating zoo of such objects, and physicists needed a guide both to what had been seen and to how to see properly. As Blackett put it, "Such new events may be extremely rare, and it is important that, when found, the chamber record should prove to be technically suitable for accurate measurement."[187] "Accurate measurement" meant recognizing artifacts as well as genuine effects, and the Manchester atlas included a substantial section on wrong results, as warnings against going astray. For example, figures 2.24 and 2.25 illustrate artifacts that are possible in the vicinity of a metal plate. Figure 2.24 shows that inadequate background condensation could make a track seem to disappear near the plate, while figure 2.25 illustrates the distortion introduced by a temperature differential between the plate and the chamber walls. Part of Rochester and Butler's defense of their first *V* particle depended precisely on showing that they had not fallen victim to the illnesses of figures 2.24 and 2.25, in other words, that the forked track did not come from a lost track just below the plate. As the atlas had it: "Gross distortion near to a metal plate . . . must as a rule be taken as an indication of distortion throughout the section of chamber concerned."[188] Such photographs were irredeemably compromised. Imposing technical standards on the community was part and parcel of the atlas's stated purpose.

Blackett and his colleagues directed the atlas beyond workers in the field, "to all students of the physics of elementary particles, even if they are not themselves engaged in original investigation, [because] the book should prove an invaluable means of acquiring physical understanding." This understanding was not to be one based on understanding the full complement of high theory: there is no mention of Yukawa's exchange forces mediated by mesons, nothing

185. Blackett to Anderson, 5 December 1949, file B48, PBP.
186. Rochester and J. G. Wilson, *Photographs* (1952), viii.
187. Blackett, "Foreword" (1952).
188. Rochester and J. G. Wilson, *Photographs* (1952), 8.

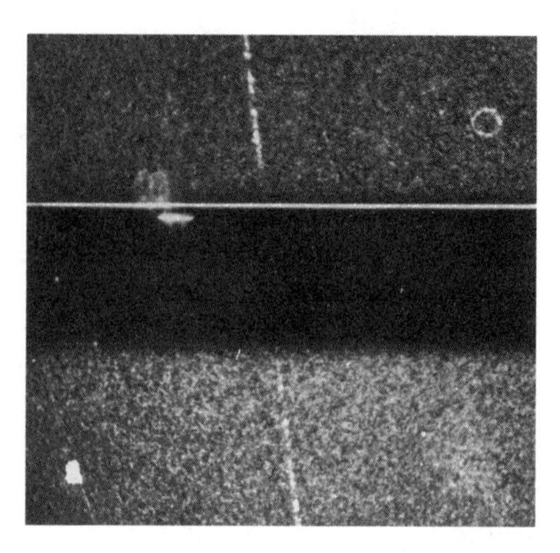

Figure 2.24 Artifacts near a plate, variable saturation (1952). Learning to use the cloud chamber (and its visual evidence) meant being able to identify the spurious as well as the real. In this shot, the authors write, the chamber had a too-low supersaturation. Close to the plate, the supersaturation level was so low as to impede track formation, creating the illusion that the track ended or began a few millimeters above the plate. The very possibility of golden event as evidence demanded that artifacts be identified picture by picture. Source: Rochester and Wilson, *Photographs* (1952), pl. 9 on 8.

substantial about quantum electrodynamics (QED) or the apparatus that QED utilized (Feynman diagrams, Lagrangian field theory, the field-theoretic vacuum). Quite the contrary. Blackett assured his readership that the atlas "must surely help us to make clear that this world of sub-atomic events is one which can be easily visualised and understood without the aid of complicated mathematics or the mastery of deep theories. If one asks why [some of] these [complicated] events happen, one may be led into the subtle intricacies and uncertainties of modern fundamental theoretical physics, but if the experimenter contents himself with asking how they happen, then these pictures, and the attached commentaries, are an ideal guide to the world of energetic elementary particles."[189]

I would argue that the division between experimenter and theorist (which had clearly existed for many years) was widened by the creation of a zone that mixed the mundane theory used in characterizing interactions with the assorted track phenomena. For now it was all any group could do to design, maintain, and exploit their detectors, get the pictures, locate the new phenomena, and begin to analyze them. In October 1952, Clifford Butler visited Anderson's Caltech laboratory and wrote Blackett with detailed, revealing descriptions about the American operation. Most striking to the British physicist was the crafted simplicity of the American chambers:

189. Blackett, "Foreword" (1952).

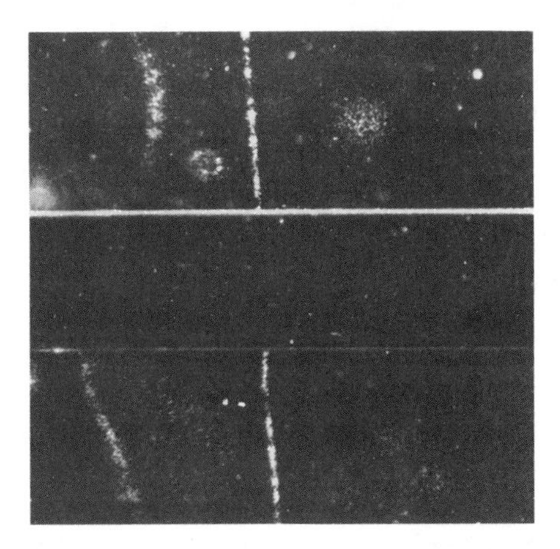

Figure 2.25 Artifacts near a plate, temperature distortion (1952). If the temperature of the plate is significantly different from that of the chamber walls, the gas can settle into inhomogeneous states. In the picture, Rochester and Wilson taught the initiate, the top half of the chamber is problematic. Not only is there a manifest (and artifactual) curvature near the lower side of the plate, but under careful viewing conditions the top and bottom of the track do not meet—which means that the whole curvature of the track is, as the authors put it, "suspect." Again, as in figure 2.24, evidence had to be assessed event by event. Source: Rochester and Wilson, *Photographs* (1952), pl. 10 on 8.

> The chambers are of light construction—fabricated from brass sheet and chrome plated inside. Only one side window is provided—the opposite side is chrome plated and acts as the mirror. The inside of the chamber is very free of obstructions—no velvet is used at all. The solid piston is made of aluminum which is anodized black and lacquered. The edges of the piston come only about 1 cm from the side walls of the main chamber box. The piston consists of a flat plate with a simple ribbed back 1″ thick. When the chambers are in the magnet the piston is pushed forwards by a rather heavy metal arrangement which is operated by air.

Of the whole, Butler reported, "I think the situation may be summarized by saying that there are no radically new techniques here at all. They have evolved a very simple design of chamber which is very clean." No glue, no velvet, no cements—in short, nothing to create distortions. In its work structure, the laboratory reflected its priorities: "It is clear . . . that the senior people—Anderson, Leighton and Cowan . . . [—] have spent most of their time on the design and construction of the apparatus and the analysis of the data has been done by research students."[190] As we will see in the chapters to come, this delegation of data analysis became ever more common in the postwar years.

Around the world, a few experimenters, Anderson perhaps foremost among them, had acquired a reputation as scrupulous cloud chamber experimenters in

190. Butler to Blackett, 14 October 1952, file B52, PBP.

large measure by being masters of the laboratory instruments. This society of picture experts included Robert Thompson at Indiana, Pierre Auger and Louis Leprince-Ringuet in Paris, Dimitry Skobeltzyn in the Soviet Union, and Blackett and his group at Manchester. Their reputations meant that, increasingly, pictures by these people carried considerably greater weight than those by other experimenters. But one of the more remarkable features of cloud chamber photographs was that their value was independent of provenance; because they promised to reveal experimental error, cloud chamber pictures traveled. Many, but not all, sources of distortion could be attributed to such features as an irregular condensation background. In part because of this self-sufficient feature of cloud chamber pictures, they were widely exchanged, published, stored, and reanalyzed by groups far distant from the original photographic site. The same could not be said of the less self-contained "raw data" that emerged from many of the electromechanical experiments so characteristic of late-nineteenth-century physics (or even of the raw data of the counter experiments we will turn to in chapter 6). The Manchester and Caltech cloud chamber groups, for example, sent pictures back and forth. Clearly, Anderson believed that the new discoveries of this subculture of experimenters had far outpaced anything theory or theorists could explain: "The photo of the double decay which Butler sent us is extraordinarily interesting. It is still true that neither theory nor imagination can keep up with facts."[191]

When a Blackett, an Anderson, or a Rochester uses the term "theory," contrasting it with "facts," one must read with care, especially when considering their remarks in light of the philosophical comments regarding cloud chambers that introduced this chapter. In particular, there is never the slightest hint from the experimenters that they believe it possible to extirpate theory altogether from the tracks—that is, there is no suggestion that one can withdraw from the theoretically "loaded" or "contaminated" analysis to one of "pure" sensory experience. In fact, "interpretation" (as used by the experimenters themselves) is integral to the experimenter's task; without it the job is not done. Even the "picture book" atlases included interpretive remarks. As Rochester and Wilson put it in the introduction to their atlas: "Any record of cloud chamber application would be misleading which did not adequately stress the element of critical interpretation which is necessary if photographs are to yield the full information contained in them. With this in view, we have tried to make the legends to photographs illustrate the process of interpretation as well as the accepted description of the central phenomenon."[192] I take it to be significant that Blackett, like Rochester and Wilson, speaks of "interpretation," rather than either "theory" or "measurement"; his choice of words presumably reflects a judgment that no royal road

191. Anderson to Blackett, 10 August 1950, file B48, PBP.
192. Rochester and J. G. Wilson, *Photographs* (1952), viii.

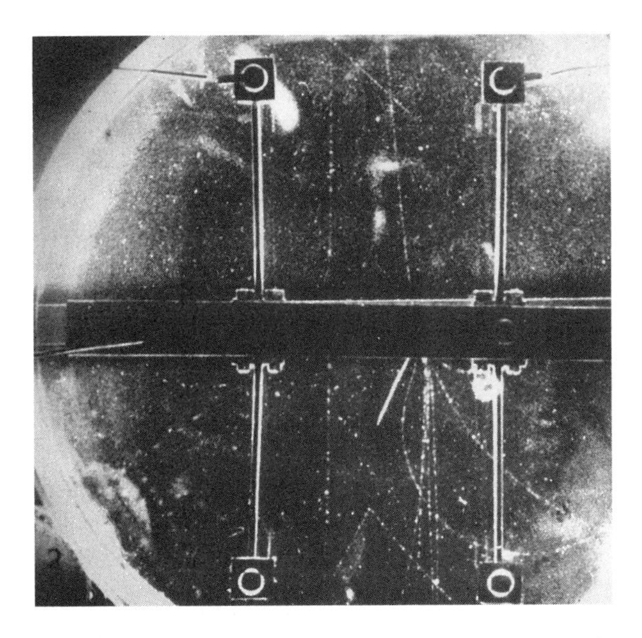

Figure 2.26 Fast neutral *V* particle (1952). "These photographs are further examples of the decay in flight of fast V^0-particles. . . . Because of the chamber distortion it is not possible to state the momenta of the secondary particles. However, since the ionization is near minimum, the velocity of the particles must be almost *c*, and the momenta are probably high. The V^0-particle on [this plate] clearly originated outside the chamber, presumably in a nuclear interaction in the lead." Source: Rochester and Wilson, *Photographs* (1952), pl. 106 on 105.

runs from cloud tracks to an identification of particle identities and behaviors. It is this possibility of multiple understandings that one finds in Blackett's response to the letter Anderson had written announcing the Caltech group's confirmation of the *V* particle's existence: "I myself was fairly convinced of the reality of the phenomena," Blackett confided, "though naturally other possible interpretations were not inconceivable."[193] Interpretations would differ occasionally, and the atlases duly reported such divergences. For example, the Rochester and Wilson atlas reproduced a photograph taken by Anderson's group, one that purportedly showed fast neutral *V* particles (see figure 2.26). This previously unpublished plate contained an inverted V track in the lower center section of the plate. Two "interpretations" were suggested: One argued that the forked

193. Blackett to Anderson, 5 December 1949, file B48, PBP.

track was the result of a V^0 particle originating outside the chamber, presumably in the lead just above it. Or, Rochester and Wilson suggested, it is possible that the V^0 originated in the lead bar that bisects the chamber. In defense of the second interpretation, the authors pointed out that the left-hand track of the fork is of higher momentum than the right-hand one (thinner, less dense droplets meaning less ionization, meaning a faster particle had left it). This argument makes it plausible (by momentum conservation) to track the neutral particle back to the (inferred) interaction in the plate (the interpolated interaction point from which the shower of other tracks emerges).[194]

How much theory is included in "interpretation"? Quite a bit, it turns out. The whole point of analyzing a picture like figure 2.26 is to deduce the properties of a track that cannot be seen—the putative source of the forked secondaries that can be observed. Relativistic physics figures in the ionization calculations that are frequently used; relativistic mechanics underlies the conservation laws. "Interpretation" also involves a significant knowledge of possible distorting effects (temperature gradients in the gas, contamination, irregular expansion of the gas, optical problems, and the effects of the scattering properties of different gases). By contrast, there is nothing in Rochester and Wilson's atlas, or in any of the other cloud chamber atlases published between the 1930s and the 1950s, that pays any significant attention to theory at the level of the Dirac equation, the meson exchange theories of Yukawa, Heisenberg, and others, or to the postwar development of QED. These were the "subtle intricacies," the "deep theories," that introduced "uncertainties" Blackett wished to avoid.

The tension between what satisfied a theorist and what satisfied an experimenter is written over and again, in correspondence, in conferences, and at all the intersection points of the two subcultures. In 1937, for example, two American groups (Anderson's and Curry Street's) came to the conclusion that there were particles like electrons in charge and interaction properties, but heavier in mass.[195] Blackett's group too saw the unusual objects and quickly entered into correspondence with a skeptical audience of theorists. Paul Dirac was quite blunt: "After Shankland [who had published experimental claims of the failure to confirm the photon interpretation of the Compton effect][196] I feel very sceptical of all unexpected experimental results. I think one should wait a year or so to see that further experiments do not contradict the previous results, before getting worried about them."[197] Rudolf Peierls, another of the leading theorists of the 1930s, also voiced doubt, to which Blackett replied: "I understand your lingering

194. Rochester and J. G. Wilson, *Photographs* (1952), pl. 106, discussed on 104.

195. Galison, *Experiments* (1987), chap. 3.

196. Shankland, "Failure," *Phys. Rev.* 49 (1936): 8–13. A cloud chamber argument against the Shankland work soon appeared, see Crane, Gaerttner, and Turin, "Cloud Chamber Study," *Phys. Rev.* 50 (1936): 302–08.

197. Dirac to Blackett, 12 February 1937, file B137, PBP.

doubts about the results, in view of the radical conclusions that they lead to! But really there is no doubt at all as to the main results." He then rehearsed the arguments about curvature and ionization, concluding, "I think you would be convinced if you looked at the photos."[198] Possibly. But "looking" at photographs, or more precisely being convinced by them, could well be considerably easier for an experimenter who knew how to read the clues of self-validation in ways that a theorist did not. (Peierls was eventually persuaded, but even the photographs were not enough to squelch lingering theoretical objections.)[199]

Theorists—Dirac and Peierls among them—had their own constraints, and indeed their own criteria of acceptance. The gap between the two subcultures was quite visible even when experimenters spoke among themselves. Focus, for example, on the type of theory that experimenters integrate easily into intra-experimenter discourse and the type of theory that is kept extrinsic. One example: In 1948, Robert Thompson conducted some cloud chamber experiments that he thought were consistent with the assumption that the meson decays into an electron and a neutral particle "neutretto" of about 90 electron masses. When Thompson wrote to fellow experimenter Bruno Rossi, Rossi responded sympathetically, agreeing that the results were consistent with the neutretto hypothesis. Rossi then reasoned as follows: Because the electron had the intrinsic angular momentum (spin) of 1/2, the meson would have to have a spin of 1/2 whether the neutretto was spin 1/2 or spin 1 (by addition of angular momenta). This, in turn, meant that when the meson is produced in the collision of two nucleons, a second spin-1/2 particle must also be produced. This second particle must have very small mass (it was probably a neutrino) since the experiments at Berkeley would have taken more energy to produce mesons if this unseen particle had a significant rest mass and it had to be produced simultaneously. "The theorists," Rossi added, "seem to feel that if a neutrino is emitted when a negative heavy meson is absorbed by a nucleus, it would carry with it a large fraction of the rest energy of the meson," and that would mean that less energy would have been left to disrupt the nucleus. This conclusion, Rossi believed, was inconsistent with the experimental observation of disintegrating nuclei. "However, I do not know how strong the theoretical argument is about the share of energy taken up by the neutrino."[200] It is clear that to Rossi, at this time, relativistic quantum mechanics (which fixed absolutely the share of neutrino energy) was something of a black box. But as this example indicates, experimenters' "interpretive" reasoning is anything but "theory free": it includes a rich combination of quantum mechanics, conservation laws, experimental induction, and even hypothetical particles such as the neutretto. In the trading zone between experimenters and theorists,

198. Blackett to Peierls, 20 May 1937, B138, PBP.

199. Peierls to Blackett, 25 May 1937, file B138, PBP.

200. Rossi to Thompson, 20 July 1948, file "Rossi," RTP.

relativistic quantum mechanics is left aside. A trading language in terms of particles, their energies, and motions was assembled in the late 1940s, a pidgin that functioned by employing neither the high theory of the Dirac or Yukawa equations nor the experimental details of expansion chambers and optics.

Considerations like these about "critical track interpretation" make it possible to return to the issues of periodization and the relation of theory to experiment and, more specifically, to the long-term significance of Wilson's cloud chamber. For around the cloud chamber grew a set of interconnected construction and interpretive skills that cut across both theoretical and experimental research programs. To the cloud chamber worker, it was possible to pass from research on nuclear disintegrations to work on electrodynamic showers, or from penetrating particles to the newly discovered *V* particles. Across all of these domains lay the skill clusters of measuring angles, measuring ionization density, calculating trajectories of unseen particles, sorting out the signs and masses of the constituent entities, and reconstructing the kinematics of unseen neutrals. But while this package of interpretive skills was manifestly independent of high theory (the details of meson, Dirac, or quantum field theory), this independence can in no way be conflated with the logical positivists' dream of finding a protocol language divested of *all* theory.

The interpretive lore of the cloud chamber functions as a trading language in the sense that it bound together the various subcultures of microphysics. Anyone in the larger culture of physics can make out statements like "Below the plate, a few electrons, notably a characteristic pair to the left of the photograph, are distributed in a way that is consistent with development from the decay photons of a neutral meson."[201] But it is not necessarily the case that these comments would be integrated into the discursive practices of experimenter and theorist in the same way. Electrons, for the cloud chamber experimenter, were inseparably bound to characteristic visual appearances—of the tight spirals of delta rays, of the striking offset helices associated with synchrotron radiation, of the minimum ionizing tracks, of a particle with specific mass, spin, and charge. To many theorists of the 1950s, the electron was embedded in quite a different set of practices: principally in the calculational and conceptual structure of the Feynman-Schwinger-Tomonaga formulation of QED. For the theorist using QED, an individual electron was an entity with no fixed and recoverable identity as it passed some of its life as virtual versions of other objects, with no fixed charge and mass because those quantities were momentum dependent; the whole representational scheme was based formally on the new calculus of renormalization, not a whiff of which could be scented in the atlases of cloud chamber photographs. Where these two subcultures could (and did) meet was in the language of what Blackett and others called "interpretation."

201. Rochester and J. G. Wilson, *Photographs* (1952), 98.

In the next chapters, we will see how the creation of the skill cluster surrounding the cloud chamber could constitute another kind of trading zone, one that reached to other instrument systems: physicists trained in cloud chamber construction, cloud chamber theory, and cloud chamber photographic analysis moved easily to emulsion groups and bubble chamber groups; emulsion physicists likewise transferred their abilities to the analysis of bubble chamber film. By creating an example of a visual, single-particle detector, Wilson had done far more than improve an existing instrument. His achievement was not merely on the order of a more sensitive galvanometer, a better source of electromotive force, or a new engine for the measurement of time and distance. The cloud chamber created a new *class* of physical evidence by concomitantly ushering in a new visually based language. In the narratives that surrounded individual pictures, in the stories that could be told about the history of a microphysical occurrence, was the possibility of a novel mode of demonstration: the decisive golden event heralded by Wilson and then by the famous snapshots of Anderson's positron and Rochester and Butler's V's.

2.7 Cloud Chambers and the Ideals of Experimentation

Both historically and philosophically the cloud chamber lies at a crossroads. The device issued from two distinct branches of natural science: the "analytic" tradition, located at the Cavendish, and the very different "morphological" tradition, located in the field sciences such as geology and meteorology. C. T. R. Wilson straddled the two, and for almost two decades succeeded in creating a hybrid subject at which he excelled. This trading zone, condensation physics, was a domain that drew its practitioners, devices, techniques, theoretical entities, and goals from both branches of knowledge (see figure 2.27).

During the period 1895–1910, both morphological and analytic physics were transformed by Wilson's work. In the early nineteenth century, meteorology, along with geology, was predicated on accurate classification. Successful research drew its strength from scrupulous analysis in the field. Humboldt set the tone with his wide-ranging explorations and systematic use of careful data taking. Through his extensive use of isobaric and isothermal charts, he lifted meteorology from the chaos of conflicting local reports to reveal an order only visible on scales of tens, even hundreds of miles. A similar, field-based effort around the globe led to a widely accepted stratigraphy and its accompanying geological history of the earth. Cloud physics could only begin after Luke Howard's remarkable classification.

At the time Wilson came of scientific age, a second transformation of the morphological sciences was under way, one in which in situ explorations could be complemented by in vitro investigations. For decades the morphologists had held the laboratory in some suspicion—after all, the great gains of recent decades

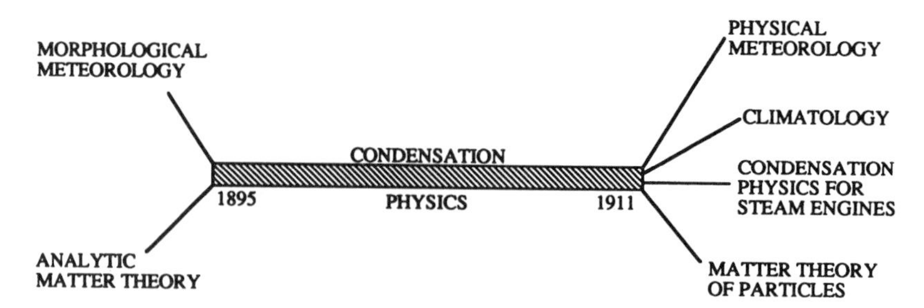

Figure 2.27 Creation and disintegration of condensation physics. Though the category "condensation physics" never rivaled "atomic physics" as a theoretical entity, the term binds together practices cutting across the practical (design of steam engines), the geological (volcanic eruptions), the matter theoretical (nature of ions and radiations), and the meteorological (origin of rain). Brief as its existence was, this transient grouping of phenomena makes sense of Cecil Powell's work leading to the nuclear emulsion.

had been made by looking at nature as it really was, not as it was artfully constituted in the glass, chemicals, and springs of the workshop. To achieve legitimacy in the eyes of the morphological scientists, the new experimentation of the 1880s and 1890s had to promise to connect to the real world. And it was here that experiments on synthetic lava flows, miniature glaciers, and scaled-down mountain building had a role to play.

John Aitken, whose work stood most immediately in Wilson's view, was an exemplar of this mimetic transformation of the morphological sciences. His artificial cyclones, ocean currents, and glaciers were just such attempts to recreate rather than dissect nature. The causal structure of the miniature cyclone was supposed to be *just like* the vast entity that could destroy towns and flood fields. Most important, Aitken's work with the dust chamber could recreate some of the most striking meteorological phenomena of the times. In a controlled, but faithful fashion, the chamber could produce the fogs that had descended threateningly on England's industrial cities; it could make rain; and most striking, the chamber could reproduce the amazing green sunsets, the afterglow of Krakatoa. For Wilson, Aitken's and others' related work provided the basic material culture for his entry into science. It was from Aitken that Wilson drew the mechanical design of the cloud chamber: the pump, reservoir, filter, valves, and expansion mechanisms were all taken directly from the dust chamber. Indeed, Wilson's desire to reproduce clouds meshed well with Aitken's own fascination with the recreation of dramatic aspects of nature. At the same time, Wilson's youthful enthusiasm for the photographic imitation of nature, the building of models, the drawing of maps, and the taxonomic collection of insects flowed smoothly into his mature interest in the production and reproduction of clouds. The morphological combined with the laboratory to yield the mimetic; the mimetic-photographic joined with the analytic to yield the image tradition.

When Wilson went up to Cambridge, he entered a community with ambitions far from the morphological or mimetic. Under J. J. Thomson, Cavendish physics held classification and reproduction to be "mere" natural history. For Thomson, the Cavendish and a handful of other physics centers promised a new physics, one that would *explain* the descriptive features of nature dynamically, in terms of more basic, more simple units. It would do so by taking nature apart, not by imitating it.

Thomson himself had contributed significantly to knowledge of the ion—the charged particle that, in aggregate, constitutes bulk matter. Using electrified gases, vacuum pumps, and electrometers, the ion physicists had by the late 1880s and early 1890s begun to articulate some of the entity's properties. Thomson and his colleagues could produce and manipulate the ion using electric and magnetic fields; ultimately, their work led to the further specification of the notion of an electron and the atomic models of early quantum theory.

Though Wilson would *not* follow the Cavendish route toward purely matter-theoretical goals, the ion figured in his work from the very first page of his notebook on clouds. Even while he desperately wanted to retrieve in the laboratory coronae and St. Elmo's fire, he imbibed with his Cambridge education the tradition of ion physics. Emphatically, this did not mean he was committed to particular, rigid beliefs about the nature of the ion. It did mean that for each new phenomenon, he would search for an ion-theoretical account. Wilson was not building atomic models, and though he would use the broad idea of an electric particle, he had no use for explanations of spectroscopic lines, chemical combination, or virtually any other element of what John Heilbron has called Thomson's "program."[202]

In the tension between the morphological and the analytic, the cloud chamber was born and raised. In good Thomsonian fashion Wilson used X rays to make rain nuclei, just as Thomson exploited X rays to make conduction ions. Wilson built charged parallel plates to move condensation nuclei just as Thomson had installed them to displace conduction ions. But in contrast to Thomson, Wilson then directed his energies toward the reproduction of phenomena that had drawn him into physics in the first place: thunderstorms, coronae, and atmospheric electricity.

Was Wilson's cloud chamber research of 1895–1910 meteorological, or was it "fundamental" physics? Is it reasonable to assess meteorology as a practical inspiration that launched Wilson's purer science? Or should we instead deduce that the meteorology was no more than pure Cavendish ion physics applied

202. Heilbron, "Lectures" (1977), 52–63.

to the complicated, great outdoors? In light of what we have seen, both the "application" and "inspiration" accounts appear thoroughly inadequate; a creole is not a mere application of one language to another, it is a working set of linguistic practices drawing from both parent languages and, significantly, extending beyond them.

Wilson wanted to know how condensation worked in the world. Thus, his work certainly had ties to the more general Cavendish program—there was a shared concern with the nature of the ion, its charge, and the relation between positive and negative ions. But it was also tied to physicists altogether outside that tradition. Wilson drew heavily on the work of the American meteorologist Carl Barus, on the dust master John Aitken, on the student of high-speed splashes Worthington, and on the many others who had experimented with droplets, hailstones, steam jets, and splashing water. Perhaps the most striking evidence of Wilson's participation in the morphological tradition was his abandonment of the cloud chamber when he suspected that it was generating condensation nuclei that were not normally present in the atmosphere. For Wilson the morphological physicist, this discrepancy with the real atmosphere was an artifact of the most subversive kind.

The notion of a hybrid practice drawing on two previously distinct areas is explanatory in a positive and a negative sense. *Positively,* Wilson drew on technologies from both predecessor fields, pulling the dust chamber from meteorology and the electric deflection plates from matter physics. Perhaps most important (and most subtle), Thomson's ion served as an *enabling notion:* the idea of particulate electricity, not the detailed theory embedded in atomic models, made possible the rearrangement of the cotton filter permitting experimentation on dust-free air. In a strong sense Wilson used the notion of an ion to transform the dust chamber into the cloud chamber but did so without the full set of assumptions made by Cavendish physicists about the nature of the ion.

Simultaneously, Wilson's morphological commitments exerted a crucial *negative* or constraining influence in the definition of condensation physics. Meteorology imposed boundaries on what Wilson would consider reasonable experimentation. When Wilson obtained results that appeared not to hold in the bulk of the earth's atmosphere, he abandoned his cloud chamber research. When the determination of the charge *e* exceeded in accuracy what Wilson needed to address real atmospheric condensation, he lost interest. And when the spontaneous discharge of electroscopes revealed cosmic rays, Wilson observed the research from a respectful distance. Ion theory too imposed constraints, for Wilson expected—or, more precisely, demanded—that his experiments be reconcilable with Cavendish experiments on conduction ions. Other physicists working at the same time as Wilson did not feel their work to be under such constraints. For all these reasons, there was a brief moment in the history of the physical sciences, from 1895 to 1911, when condensation physics was a trading zone, participating in both fields at once: it was an investigation into the fundamental electrical nature of matter *and* an exploration of the formation of clouds.

Condensation physics as a creolized overlap of meteorology and ion physics ended abruptly in 1911. In March of that year, Wilson turned the cloud chamber toward its role as a detector. With that move, the physics community seized upon the chamber-as-detector and consequently relegated its earlier functions—its ability to produce all the phenomena that principally intrigued Charles Wilson—to the background (though in the next chapter we will see at least one important—and continuing—application of condensation physics). There was no longer any useful sense in which ions, coronae, fogs, and rain raised the same questions as problems in the theory of matter.

As the knotty clouds blended into the tracks of alpha particles and the "thread-like" clouds became beta-particle trajectories, the sense and meaning of the chamber changed. Previously, Wilson had an audience among the matter physicists for his question: How do droplets form? Now, in the water trails, the matter physicists had other, more pressing questions: What are the energies of the gamma rays that produce electrons? How do alpha particles scatter? What does the scattering of a particle imply about it and its target? With the addition of magnets, and then of triggerable expansions, the chambers began to reveal new particles, their masses and interaction properties to be deduced from the curvature and density of the curved trajectories. For hunters, snow tracks tell about their quarry, not the quality of the snow.

Condensation physics as an interfield disintegrated. Meteorology and the physics of matter diverged, but Wilson's work had permanently changed both. When contributors to the ninth edition of the *Encyclopaedia Britannica* reviewed meteorology in 1895, they identified the founding of scientific meteorology with the publication of Humboldt's *Isothermal Lines.*[203] Geographical variation and periodic temporal variation of climate and weather were the substance of the field. Essentially, the subject amounted to the systematic compilation of daily, weekly, monthly, and yearly variations of temperature, pressure, wind, rain, and cloud cover. Sixteen years later the eleventh edition began by dividing the subject into the statistical study of the atmosphere, known as *climatology,* and the theoretical and physical study of meteorology. The latter embraced both the macroscopic, hydrodynamical processes of bulk air and the microscopic explanations of processes that are "mostly comprised under thermodynamics, optics and electricity." None of the analytic schemes are given the same prominence as C. T. R. Wilson's theory of dust-free condensation, which the encyclopedia presents as the "first correct idea as to the molecular processes involved in the formation of rain" and atmospheric electricity.[204] Meteorology thus passed from

203. *Encyclopaedia Britannica,* 9th ed., s.v. "meteorology" (by A. Buchan and Balfour Stewart).
204. *Encyclopaedia Britannica,* 11th ed., s.v. "meteorology" (by Cleveland Abbe).

pure climatology to a split discipline; it embraced a large-scale descriptive function and a dynamical explanatory component that would be known as physical meteorology.

Cavendish-style physics too was transformed. After the cloud chamber the subatomic world was suddenly rendered visualizable—and consequently took on a reality for physicists that it never could have obtained from the chain of inferences that had previously bolstered the corpuscular viewpoint. In this faith in direct reproduction we find the powerful legacy of the mimetic tradition: by the time Wilson made his photographic claims for the reality of electron motions beneath the cloud inscriptions, he had a long tradition behind him, a tradition of replication in the form of a series: artificial fog mapped real fog; artificial atmospheric optics mapped real atmospheric optics; artificial rain mapped real rain; until, in the ultimate term, he had cloud tracks mapping out the motions of the real, but invisible, alpha and beta rays.

Soon the cloud chamber and its close photographic relation, the nuclear emulsion, became the defining instruments of a new field: particle physics. The cloud chamber experts, scientists including Anderson, Auger, Blackett, Leprince-Ringuet, Rochester, Skobeltzyn, and Thompson, all succeeded in large part because they came to have a craft familiarity with tracks themselves. What grounded this craft familiarity? In part, one could argue, it rested on the explanation of the track-forming process—an explanation that came down to vapor pressures and Gibbs free energy. But it is clear that this explanatory scheme was, in practical circumstances, secondary to the *visual recognition* of certain track patterns. After a modest amount of training, it was possible to sort particle types, to interpret processes, even to recognize distortion, all from the tracks themselves. I suspect that it is in no small measure this quasi independence of interpretation from the details of any underlying theory that gave the cloud chamber pictures their power to persuade. By sidestepping a long chain of inferences, the cloud chamber pictures give us the sense, as Eddington put it, that we can "almost see" particles. It is not so much that the cloud chamber fulfills our standard of realism at the macroscopic level as that the cloud chamber sets the gold standard of existence. This, I take it, is the sense of Born's saying a cloud chamber picture is better evidence than being an eyewitness to a shooting, or Bridgman's asserting that such a picture defines "what the operational definition of reality is."

In November 1927 George P. Thomson wrote a congratulatory letter to Wilson on his receipt of the Nobel Prize: "Your work has always seemed to be the *beau ideal* of an experiment, carrying such immediate and complete conviction, and making real and visible what was before only, after all, a theory. It is a perfect example of the peculiar genius of British physics."[205] Thomson,

205. Thomson to Wilson, November 1927, CWP.

I suspect, was right. But the peculiar genius of British physics was not merely an abstract excellence at a universal challenge. It was a peculiarly Victorian enterprise that Wilson had set out on, a gathering-in of the powers of nature. The camera lucida and the camera obscura could only reproduce images. Wilson's camera nebulosa did more: it was designed to remake the lightning, haloes, rainstorms, and fog that so captivated his contemporaries. It should be said that Wilson eventually failed in his attempt to persuade others that clouds formed around ions—later meteorologists determined rather unanimously that dust and sea salt capture much more water than electrical particles. His laboratory clouds therefore never transformed the study of storms. But with artificial clouds Wilson had created a language of tracks that has lasted almost a century; by so doing he made particles real.

3 Nuclear Emulsions: The Anxiety of the Experimenter

3.1 Introduction

After C. T. R. Wilson's death in 1959, Cecil Powell reflected on his teacher's life work. Powell's memorial essay covers Wilson's important physics discoveries: the cloud chamber, condensation phenomena, atmospheric electricity, and thunderclouds. But read carefully, the essay reveals as much about Powell as it does about Wilson. For Powell, Wilson's scientific life came to stand for more than the evident scrupulous care in experimentation. Wilson's solitary persistence in his quest for a glass chamber capable of withstanding sudden changes in pressure, his search for precision cloud chamber photographs, and his delight in a certain "style" of highly crafted experimentation all impressed Powell greatly. Unlike Rutherford, who forged ahead to important results heedless of craft precision, Wilson's search for decisive experiments, to Powell, stood for "the same kind of aesthetic satisfaction as a mathematical demonstration carrying complete conviction."[1] Wilson's care, self-sufficiency, persuasiveness, and craftsmanship became a fixed point against which Powell's later experience in physics could be contrasted. "To me," Powell mused, "[Wilson's] qualities seem to gleam like the banners of a vanished age."[2] Even Powell's choice of heraldic imagery signaled how vastly physics had changed in the quarter-century of Powell's career. One goal of this chapter is to pursue this shift. Nuclear emulsions—the chemical "device" on which Powell set his career—will be the emblem of this evolution, marking the transformation of physics from an individual pursuit to the joint effort of a hierarchical team of both "skilled" and "unskilled" workers.

1. Powell, "C. T. R. Wilson" (1972), 361.
2. Powell, "C. T. R. Wilson" (1972), 368.

I want to use this history of the nuclear emulsion not only to study the development of techniques for the registration of elementary particles but to probe the shifting ideal of what it meant to be a skilled experimenter in the mid-twentieth century.

Powell's rise through the ranks of physicists to his receipt of the Nobel Prize in 1951 is inextricably bound up with his transformation of the laboratory team. By the time of his award he was head of a large, efficient experimental group with significant resources, crucial industrial connections, and a worldwide reputation. To locate Powell's work in the rapidly changing experimental ethos of the mid-twentieth century and to fill out a broader history of the emulsion method, it is useful to see Powell in relation to three other figures, as if we were following the art-historical method of contrasting pairs of slides. First pair of contrasting images: Powell's bustling lab and his successful career stood worlds apart from another extraordinary—though now utterly forgotten—innovator in the emulsion technique, the Viennese physicist Marietta Blau. Years before Powell's triumph, Blau essentially created the field of emulsion physics by using film to record tracks directly and by developing the techniques needed to make commercially available film sensitive to the passage of individual protons. It was only then that emulsions became useful as a fundamental tool for nuclear physics, and it was from Blau's work that Powell began his inquiries. Though Blau too was nominated for the Nobel Prize (in her case unsuccessfully), our story shows how differently her trajectory is embedded in the events of the 1940s. Marginalized at her native Austria's sophisticated laboratory (the Institut für Radiumforschung) for reasons of sex and religion, she spent years working without a proper laboratory, literally in the corridors of the institute. Then, fleeing the Nazis, Blau struggled from country to country trying to use her emulsions to reestablish a disrupted life and career in nuclear physics.

Our second contrast: Contemporary with Powell at the height of his career was Pierre Demers, a Canadian physicist-chemist whose work in *making* emulsions provides another useful counterpoint to Powell. Emulsion fabrication, it turns out, is one of the most secretive, delicate, and intuitive of chemical tasks, and producing sufficiently reliable and sensitive emulsions was beyond the grasp of almost everyone in the widely dispersed community of nuclear scientists. Indeed, one of the great accomplishments of Powell and his British colleagues was their recruitment (with government support) of the industrial giants Ilford and Kodak to provide them with ever more sensitive film. Adopting Ilford and Kodak as contractors had a price, however: the physicists neither made their own tool of inquiry nor even had access to aspects of its inner workings. Demers was one of a mere handful of *individuals* who could make, with their own hands, a film competitive with the mass-produced output of these huge corporations. Understanding how physics looked to the physicists who chose the industrial path—and contrasting their vision with the view from Demers's position—throws further light on the early era of postwar particle experiments.

A third contrast comes as we position Powell's Bristol experimenters against the group of emulsioneers that formed around Eugene Gardner after the war at Lawrence's Berkeley laboratory. At first, the emulsion method seemed almost impossible to transport across the Atlantic; the Berkeley scientists struggled through an immensely frustrating period in which, despite their access to the most powerful accelerator in the world, they could not replicate the Bristol discoveries, much less make new ones. Then, with the help of imported matériel and personnel, the Bevatron succeeded at registering pions. Berkeley promptly rose to become overwhelming competition to the European style of small-group, cosmic ray work. But even accelerator-based emulsion physics could not survive long. By the mid-1950s, the nuclear emulsion technique itself was limited to very particular applications, as the bulk of particle physics was reoriented toward the massive bubble chambers of Berkeley, Brookhaven, and CERN.

In different ways, and for different reasons, each of the nuclear emulsion efforts just mentioned—Blau's, Powell's, Demers's, Gardner's—was marked by a certain fragility. At the same time, the film itself was epistemically fragile, rife in a hundred ways with possibilities of deception. Particles appearing in physicists' reports often had to be denied a few weeks later. Different entities emerged only to be amalgamated, like the evening and morning stars, into a single object. Just as often, one object would splinter after further analysis into several different particles. The anxiety precipitated by these elusive entities was profound. Whether it was the Berkeley team's frustration at not being able to produce the new pion in emulsions, or the Bristol team's fear that its pioneering work on the pion was predicated on faulty assumptions about film grain, the anxiety of the emulsion experimenter was acute. To resolve these problems, the physicists brought in nonexperimenters—scanners, theorists, chemists—to stabilize the method and their results. And while the expertise of each of these collaborating groups helped, control inevitably spun further from the physicists' own hands.

For all its fragility, the simple, portable emulsion method offered a marginalized experimenter like Blau a lifeline to the physics community; carrying a cyclotron around with her was not on offer. Nor was she the only one to hold on to emulsions as a link to physics when world events foreclosed more elaborate apparatus. After World War II, European physicists were surrounded by the rubble of their laboratories, while their American colleagues enjoyed unprecedented material plenty. It was then that the more established European nuclear physicists came to embrace the relatively cheap and easily transferred technique as a means to compete with the burgeoning empire of American accelerator physics. In a sense, Powell occupied a sociological midpoint: He and the well-structured laboratory he led, with its links to chemists, theorists, and scanners, was vastly more elaborate than anything a Blau could muster by herself. But compared with the factory-like American accelerator centers, Powell's lab was a cottage industry.

Embedded in the rapidly moving events of the late 1940s, the emulsion discoveries of kaon decays and hyperons became more than new data points. They were claims by a community of physicists desperate to preserve a place in a changing discipline—a place for European physics in general, and a place for small groups of autonomous researchers more specifically. For several years, the emulsion method remained the key to survival. Its practitioners largely created the subdiscipline of particle physics, and it kept both alpine cosmic ray stations and European physics alive; the emulsion was a precarious medium in an anxious time. Physics, like the world, was in flux as Europe staggered toward its convulsion, through World War II, and then into an unsure postwar era. It would be a hopeless task to construct an isolated history of the emulsion technique, as if it were disconnected from the high stakes that surrounded it. From the microscopic physical properties and difficult chemical practices entailed by the interaction of films and particles, we will repeatedly be drawn outward; and from the outer events, we will be drawn back inward again.

3.2 Marietta Blau: Between Nazis and Nuclei

If Wilson's life as a cloud chamber experimenter stood as a shining example, his principal instrument also presented a model for other instruments, among them the nuclear emulsion. One can see the links between cloud chambers and nuclear emulsions quite strikingly by looking at the eclectic and desultory way that film intersected with the history of nuclear physics before Wilson found his tracks in 1911. When Wilson announced his track-recording cloud chamber, everything changed.

In 1839, Daguerre produced his astonishing images on silver that had been exposed to iodine vapor, and Fox-Talbot invented the first negative-positive process using paper soaked in silver nitrate and sodium chloride. Almost immediately after these photographic innovations became known, experimenters began probing the reaction of light-sensitive compounds to other substances, and enthusiasts found responses to chalk, marble, cotton, and feathers. Though it was known that uranium salt fogged film, even through paper, this circumstance did not become salient to the community until 1896. That year, as the German scientist Wilhelm Conrad Röntgen probed the world with X rays, Antoine Henri Becquerel began work that established the radioactivity of uranium.[3] Even then, physicists saw film as a means of studying radioactivity rather than of photographing individual particles—typical questions were: How thick a barrier could alpha rays cross? or What was the rate of alpha emission?

3. Powell, Fowler, and Perkins, *The Photographic Method* (1959), 11.

For example, consider the 1909 work of S. Kinoshita at the University of Manchester, where he had the active support of Rutherford. Years before, Rutherford had found that the photographic action of alpha particles ceased when an aluminum barrier increased beyond a particular thickness. In these experiments the film was deployed on the model of a scintillation screen—perpendicular to the particles' trajectories. Kinoshita continued Rutherford's work until he could show that individual alphas could cause discrete marks on the film. But Kinoshita never laid the film at a tangent to the particles, and he never reproduced a photograph of the marks he observed.[4] Indeed, when Kinoshita sought thick emulsions (he inaugurated the use of stacks of emulsions to increase the effective thickness), it was not to ensure a three-dimensional track, but rather to guarantee that no alpha particle would penetrate the screen of film without encountering a grain of silver halide. In this way, he concluded, "the photographic method can be applied for counting α-particles with considerable accuracy."[5]

Within a few months of Kinoshita's work, Wilson's tracks appeared in the *Proceedings of the Royal Society,* and immediately, Kinoshita's "counting device" began its life as a track recorder. Wilhelm Michl, for example, began his 1912 paper with a discussion of Wilson's success at finding tracks in his cloud chamber.[6] Maximilian Reinganum, too, began to probe individual tracks only after Wilson's artificial clouds had revealed full particle trajectories. Kinoshita himself, when he came back to the problem in 1915, had reconceptualized his project; his new title was "The Tracks of the α Particles in Sensitive Photographic Films."[7]

Historical continuity, it seems, is not to be found here in the attempts to resolve a particular theoretical question. These discussions do not turn around the development of the new quantum theory. There is no talk of electron orbits, the constitution of atoms or molecules, or the energy levels of matter. Nor do questions of continuity or discontinuity hinge on a specific experimental object. Instead, continuity lies at the level of laboratory practice, of the characteristic deployment of material objects. In some ways, the emulsion before Wilson resembles the dust chamber Wilson inherited from Aitken. There, the ion theory, or more precisely the idea of ions in general, served as an enabling notion that made it plausible to move the stopcocks into a configuration forbidden in dust chamber design. The emulsion encountered a different kind of enabling notion, one grounded not in theory but in instrument design itself. Film was associated first with a scintillation screen and then with a cloud chamber; the transforma-

4. Kinoshita, "Photographic Action," *Proc. Roy. Soc. London A* 83 (1910): 432–53.

5. Kinoshita, "Photographic Action," *Proc. Roy. Soc. London A* 83 (1910): 432–53, on 453.

6. Michl, "α-Teilchen," *Sitzungsb., Akad. Wiss. Wien, Math.-naturwiss. Kl., Abt. IIa* 121 (1912): 1431–47; Michl, "Photographischen Wirkung," *Sitzungsb., Akad. Wiss. Wien, Math.-naturwiss. Kl., Abt. IIa* 123 (1914): 1955–63.

7. Kinoshita, "Tracks," *Phil. Mag.* 28 (1915): 420–25.

tion came from *within* the domain of instruments—not from the exigencies of theory or experiment.

But while work on emulsions continued, the technique remained at the outskirts of atomic physics. Even Rutherford—who certainly knew of Kinoshita's and others' progress—exploited primarily electronic counters and the ever-present scintillation screen. Both techniques, however, had their disadvantages as Rutherford well knew: flashes were hard to record reliably, and Geiger-Müller tubes took a black art to build and were fickle to operate. As Roger Stuewer has shown so nicely, between 1923 and 1924 the maddening scintillators were behind one of the most bitter and hard-fought controversies in atomic physics. The start of the feud was Rutherford's claim that he had disintegrated the nucleus of nitrogen using RaC alpha particles. Within a short time, Rutherford and his laboratory were in a pitched battle with Swedish scientist Hans Pettersson and Austrian physicist Gerhard Kirsch, both at the Institut für Radiumforschung in Vienna: at stake was more than a result about nitrogen, for the two sides had drastically different understandings of the nature of the atomic nucleus.[8] The battle was both experimental and theoretical. Rutherford and his collaborator James Chadwick argued that the nucleus held a proton in satellite orbit—and they reported experiments to sustain the claim. According to their account, collisions of an orbiting proton with an incoming alpha particle would release the proton, though the likelihood of such an encounter was relatively small. Pettersson and Kirsch, by contrast, contended that the incoming alphas were exploding the target nuclei, that this was supported by the collision fragments, that there were no orbiting protons, and that this was true for all elements. Begun quietly, the British-Austrian engagement escalated dramatically, each challenging the other's instruments, experiments, and interpretations.

Applying their by-then well-established scintillation screen to the problem, the Cantabrigians finally closed the dispute by showing, both to their and to their opponents' satisfaction, that the observers in Vienna were not reliably recording the scintillations flashing before their eyes. It is hardly surprising that, having lost this epochal dispute, at an immense cost to his reputation, Pettersson should search for a counting method less vulnerable to the training and skill of his observers. Though Pettersson's own involvement with emulsions ceased at this point, his assignment of the problem to a young physicist, Marietta Blau, proved of crucial importance.

Born in Vienna in 1894, Blau grew up in a prosperous Jewish family that had made its mark in Viennese high culture by founding the foremost music publishing company in Europe. Blau received her Ph.D. in 1919 with a thesis on ray physics—on the absorption of gamma rays.[9] Following her doctorate, she

8. Stuewer, "Disintegration" (1985).

9. Blau, self-description (in German), five-page typescript, LHP.

moved to Berlin in 1921, taking a position with a company that manufactured X-ray tubes (Röntgenröhrenfabrik Fürstenau)—her tasks involved electrotechnical and spectral analysis. This job was followed by a stint as an "*Assistentin*" at the Institute for Medical Physics at the University of Frankfurt (am Main), where she worked and published papers on X-ray physics. Primarily, however, her assignment was to instruct doctors in the theoretical and practical bases of radiology.[10] This border zone between medicine and physics brought Blau much nearer the realm of nuclear physics than it may at first appear. For not only did she carry over a deep knowledge of ray physics and film from medicine to physics (at times even exploiting standard dental X-ray film to do nuclear physics), she took with her a lasting commitment to the persuasive power of the image and the concomitant close analysis of artifacts that accompanied the establishment of real effects amid the visual noise.

As we saw in chapter 2, the tradition of radiology had, through the genre of the scientific atlas, carried over directly from the invisible rays of Röntgenology to those of the subatomic domain. Marietta Blau's career trajectory underscores that link. Not only did she publish extensively in journals of photography, she contributed to joint projects such as a multiauthored 1931 volume on the physical-medical border area (*Zehn Jahre Forschung auf dem physikalisch-medizinischen Grenzgebiet*).[11] In that volume, after summarizing her own research, she emphasized the continuity of the long history of photography as an aid to the study of radioactivity; less obviously, she insisted that the examination of radioactivity would contribute reciprocally to the development of the photographic process itself.[12]

For a decade and a half, between 1923 and 1938, Blau's investigations were centered at the Institut für Radiumforschung in Vienna and at the Second Physical Institute. Despite these solid surroundings, she was always peripheral (and by and large unpaid). When the issue of her getting a *Dozent* position arose, her brother later recalled, a professor told her that to be a woman and a Jew was just too much.[13] To make ends meet, in the years before 1937 she taught *Praktikumsunterrichten* and occasionally worked for other institutes, including the oceanographical institute in Göteborg (where Pettersson was based), the Röntgentechnische Versuchsanstalt in Vienna, and a variety of photographic and precious metal enterprises.[14] In the interstitial zone between scientific and commercial ventures, her work in physics continued. At Pettersson's suggestion, Blau began

10. Blau, self-description (in German), five-page typescript, LHP; Blau, "Curriculum Vitae," ca. 1937, in Meyer-Blau correspondence, IfR. On X rays, see, e.g., Blau and Altenburger, "Über eine Methode," *Z. Phys.* 25 (1924): 200–214.

11. Blau, "Photographische Untersuchungen" (1931).

12. Blau, "Photographische Untersuchungen" (1931).

13. Otto Blau to Leopold Halpern, 22 January 1977, LHP.

14. Blau, "Curriculum Vitae," ca. 1937, in Meyer-Blau correspondence, IfR.

to explore the possibility of finding protons and smashed atoms using photographic emulsions. Finally, in 1925, she succeeded in detecting the fragments of atoms hit by alpha particles, including the thinner, harder-to-find tracks of protons.[15] These experiments were followed in 1926 and 1927 by a series of experiments in which Blau bombarded aluminum with alpha particles in order to measure the nuclear fragments that would emerge. Unfortunately, with a weak radioactive source (the only kind available to her), she had to settle for the very lowest energy particles.[16] It was clear that if she was going to make fast protons visible (as opposed to the much more heavily ionizing nuclear fragments or slow-moving protons), she would have to improve both the emulsion and the development process that would bring out the narrow tracks.[17]

For a variety of reasons, perhaps in part because of the encouragement offered by Stefan Meyer, the head of the institute, and in part because the boundary between physics and chemistry had, as a field, been more open to women physicists, the Institut für Radiumforschung became a mecca for women exploring the complex of fields surrounding nuclear physics, radiochemistry, and radiophysics. Meyer brought in, among others, Blau and Berta Karlik; Blau was then able to coauthor papers with or supervise the dissertations of at least five other women in the years 1930–37 alone: Elizabeth Rona, E. Kara-Michailowa, Hertha Wambacher, Stefanie Zila, and Elvira Steppan.[18]

In mid-1932, Blau began a longer collaboration with Wambacher, continuing Wambacher's thesis topic on densensitizers in an effort to improve the photographic method. Their first important success occurred in the fall of that year, when they were able to exhibit the recoil protons from unseen neutrons (neutrons having just been discovered by Chadwick).[19] Blau and Wambacher's result was, on the face of things, bizarre and counterintuitive. When the photographic plates were soaked in a photographic *de*sensitizer, the organic dye pinakryptol yellow, beta rays and gamma rays were clearly less able to leave an imprint. But plates so desensitized seemed to register the same number of alpha tracks, and

15. Blau, "Photographische Wirkung," *Sitzungsb., Akad. Wiss. Wien, Math.-naturwiss. Kl., Abt. IIa* 134 (1925): 427–36; and Blau, "H-Strahlen aus Paraffin und Aluminium," *Z. Phys.* 34 (1925): 285–95.

16. Blau, "Photographische Wirkung II," *Sitzungsb., Akad. Wiss. Wien., Math.-naturwiss. Kl., Abt. IIa* 136 (1927): 469–80; and Blau, "H-Strahlen aus Paraffin und Atomfragmenten," *Z. Phys.* 48 (1928): 751–64.

17. Blau, self-description (in German), five-page typescript, LHP.

18. E.g., Blau and Rona, "Anwendung der Chamié'schen Methode," *Sitzungsb., Akad. Wiss. Wien., Math.-naturwiss. Kl., Abt. IIa* 139 (1930): 275–79; Blau and Kara-Michailova, "Über die durchdringende γ-Strahlung," *Sitzungsb., Akad. Wiss. Wien., Math.-naturwiss. Kl., Abt. IIa* 140 (1931): 615–22; Wambacher, "Untersuchung der photographischen Wirkung," *Sitzungsb., Akad. Wiss. Wien., Math.-naturwiss. Kl., Abt. IIa* 140 (1931): 271–91; Zila, "Ausbau der photographischen Methode," *Sitzungsb., Akad. Wiss. Wien., Math.-naturwiss. Kl., Abt. IIa* 145 (1936): 503–14; Steppan, "Zertrümmerung von Aluminium," *Sitzungsb., Akad. Wiss. Wien., Math.-naturwiss. Kl., Abt. IIa* 144 (1935): 455–74.

19. See Blau and Wambacher, "Neutronen II," *Sitzungsb., Akad. Wiss. Wien., Math.-naturwiss. Kl., Abt. IIa* 141 (1932): 615–20; cf. Blau, "Méthode Photographique," *J. Phys. Radium,* 7th ser., 5 (1934): 61–66.

the size of the blackened grains *increased*—at least in large-grained emulsions. Indeed, for protons and alpha particles, Blau noted a marked increase in recognizable series of grains.[20] To secure their results, Blau and Wambacher compared their recoil tracks with those registered under neutron bombardment both in the Wilson chamber and in scintillation experiments. The striking similarity in outcomes among the methods legitimated (in their view) the new method. They then directly compared, on the same plate, the result of a neutron source placed at one end and a proton source on the other. Recoil protons could be observed beginning at a variety of distances from the neutron source, whereas in the proton case they began directly in front of the source. Compiling the range of the recoil protons, Blau and Wambacher could then find at least a rough energy distribution for the initial neutrons. In Blau and Wambacher's hands, the photographic plate promised both a new quantitative and a new qualitative way to explore the emission of neutrons.[21]

With a grant from the Federation of Women Academics of Austria, Blau went first to Göttingen. But when, in April 1933, Marie Curie offered Blau the use of strong radioactive sources at the Institut du Radium in Paris, it was an offer Blau could not refuse. Making quick use of the concentrated polonium that she received, Blau continued her emulsion studies, now on a neutron beam produced by alpha particles hitting beryllium.[22] By this time, it was quite clear, Blau emphasized, that the choice of plates was a subtle matter. Choose too sensitive a film and the observer will be overwhelmed with "parasitic" effects. Fine grains might be an advantage, but *too* fine a grain made individual grains indistinguishable under the microscope, which lowered the precision of range and ionization measurements. Moreover, it appeared to Blau that the very fine grains simply would not register the passage of a single alpha particle: Lippmann plates, for example, often failed to register a single hit by an alpha. While the *Röntgenzahnfilm* Agfa may have been superb at recording the insults of a decaying tooth, it was less good at seizing the relativistic proton. And as the energy of the incident protons increased—and their ionizing power decreased—they became harder and harder to detect. Only the mysterious effects of pinakryptol yellow could materialize the image, and there, as Blau put it, "we are totally in the dark as to how to explain the apparent sensitivization by a desensitizer."[23]

In 1934, Blau returned to Austria to push forward her neutron studies with

20. On pinakryptol yellow, see Wambacher, "Untersuchung der photographischen Wirkung," *Sitzungsb., Akad. Wiss. Wien., Math.-naturwiss. Kl., Abt. IIa* 140 (1931): 271–91.

21. Blau and Wambacher, "Neutronen II," *Sitzungsb., Akad. Wiss. Wien., Math.-naturwiss. Kl., Abt. IIa* 141 (1932): 614–21; and Blau and Wambacher, "Neutronen," *Anzeiger, Akad. Wiss. Wien, Math.-naturwiss. Kl.*, 9 (1932): 180–81.

22. Blau, "Méthode Photographique," *J. Phys. Radium*, 7th ser., 5 (1934): 61–66, invitation from Curie on 62.

23. Blau, "Méthode Photographique," *J. Phys. Radium*, 7th ser., 5 (1934): 61–66, on 66.

Wambacher but confronted the thinness of the emulsions—tracks simply up and left the plate before they had deposited enough of a track to allow a full measurement. The photographic giant Ilford obligingly began to thicken the plates so that more of the inclined tracks could be traced in their entirety. But now difficulties arose as the thicker plates created new problems in the dark room: problems of homogeneity in drying, sensitivity, and processing that would haunt the method for many years.[24] Nor was this all. Since 1931, Blau had sought to understand why the latent image of the tracks themselves seemed to fade; that is, between the time of the exposure and the time of development, the image spontaneously vanished into the chemicals.[25] And despite Ilford's help, the thinness of the films continued to put geometric constraints on film capture of particle motion.[26]

When it came time to apply the method to the specific problem of determining neutron energies, other challenges to the photograph's legitimacy arose. In 1935, specifically targeting Blau, H. J. Taylor rejected the very possibility of using a photographic emulsion to estimate the energy distribution of neutrons and protons freed in nuclear disintegrations: "[T]he uncertainties introduced by this method of investigation are considered, and the conclusion is reached that the method is unsuitable for determining the detailed distribution of neutron energies." Using the radioactive material thorium-C′ (ThC′), Taylor measured 60 alpha tracks and found a 20% spread between the shortest and longest alpha tracks. From this determination, and the fact that perfectly registered tracks should all be the same length, he concluded that tracks in general would carry a 20% error in length, catastrophically high for the quantitative legitimacy of the new method.[27] Blau and Wambacher shot back that the size of the error Taylor reported reflected his use of low energy alpha rays, which produced shorter tracks. The error in the emulsion method was associated with the variation in the length of track missing near the end of the trajectory. If alpha tracks were longer (as the paths of the high energy protons in disintegration experiments generally were), then the percentage error was much smaller than Taylor had reckoned.[28] With a budget near zero, Blau struggled with a handful of students and her ex-student Wambacher to stabilize a deeply insecure method, one with fading images, inconsistent tracks, distorted trajectories, and published opposition to

24. Blau, self-description (in German), five-page typescript, LHP.

25. Blau, "Bericht über die Entdeckung," *Sitzungsb., Akad. Wiss. Wien, Math.-naturwiss. Kl., Abt. IIa* 159 (1950): 53–57. Blau's worries about the problem of fading go back at least to 1931; cf. Blau, "Abklingen," *Sitzungsb., Akad. Wiss. Wien., Math.-naturwiss. Kl., Abt. IIa* 140 (1931): 623–28.

26. Blau and Wambacher, "Längenmessung," *Sitzungsb., Akad. Wiss. Wien., Math.-naturwiss. Kl., Abt. IIa* 146 (1937): 259–72. Blau and Wambacher recognized the geometric difficulty as the greatest disadvantage of the method but argued, by example, that the thickness was sufficient for their purposes.

27. Taylor, "Tracks," *Proc. Roy. Soc. London A* 150 (1935): 382–94; and Taylor and Dabholkar, "Ranges," *Proc. Roy. Soc. London* 48 (1936): 285–98; the authors qualified their objection somewhat.

28. Blau and Wambacher, "Längenmessung," *Sitzungsb., Akad. Wiss. Wien., Math.-naturwiss. Kl., Abt. IIa* 146 (1937): 259–72, on 264–65.

its validity. But instability in the emulsion was nothing compared to the political precariousness of the world outside (and sometimes within) the institute walls.

Blau's collaboration with Wambacher must have been fraught; for in Wambacher Blau had chosen an ardent Nazi as her laboratory partner. Indeed, the entire circle of institute experimenters working on and around emulsions had formed an alliance with the still-secret fascist movement. Throughout Blau's collaboration with Wambacher during the 1930s, it appears to have been no great mystery to those in the laboratory that Wambacher was extremely close personally to Georg Stetter, an active Nazi and powerful figure in the Austrian scientific community after the Anschluss; at the same time she was collaborating with a second Nazi, Gerhard Kirsch, on emulsions. Kirsch had been a leader of a *Keimzelle* (roughly "seed group") of the National Socialist Teachers League at the University of Vienna from 1933 to 1937,[29] and Stetter had joined the National Socialist Teachers League in 1932, taking up (secret) membership in the NSDAP during June 1933.[30] In 1938, asked about his "previous activities for the N[ational] S[ocialist] Movement," Stetter replied that for years he had been involved with "Private propagand[a], support of the N[ational] S[ocialist] Students League . . . preparation of a reserve short-wave station, etc."[31] Rounding out the leading Nazi triumvirate in the laboratory, Stetter worked with Gustav Ortner.[32] Together, Stetter and Ortner had shown that the photographic method was picking up "all" the alpha tracks by comparing the results obtained to those found under similar conditions using the better established electrical method.[33] This scientific legitimation of her new method (by her political enemies) was clearly important to Blau—she cited it, almost as a touchstone, throughout the thirties. So it was, that while Blau never worked personally with Kirsch, Stetter, and Ortner, she was tied pedagogically and then collaboratively to Wambacher, and through her to a set of affiliated colleagues whose political dedication to Nazism was early, deep, and enduring. With scientific reference to those who threatened her own existence, she hoped to ensure the survival of the images delicately engraved in silver on photographic plates.

Amid the threats to both her and her method's existence, Blau had a break. With the help of the man usually credited with discovering cosmic rays, Victor

29. Kirsch, questionnaire, 20 May 1940, Personalakte Kirsch, BmUP, AdR. With some pride, Kirsch testified that he had been a member of "the first NSDAP in Austria from 15 Nov. 1923 until its demise, [and] since March 1934 [a member] of the V[aterländischen] F[ront]."

30. Stetter questionnaire, 11 May 1938, fols. 269, 272, on fol. 269v, Personalakte Stetter, BmUP, AdR.

31. Stetter questionnaire, 11 May 1938, fols. 269, 272, on fol. 269v, Personalakte Stetter, BmUP, AdR.

32. Ortner had become a member of the Vaterländischen Front in March 1934 (Confirmation, Office of V.F., 16 December 1935, fol. 163, Personalakte Ortner, BmUP, AdR), paid his dues to the National Socialist Teachers League from 1934–35 onward, and taken on full and official Nazi Party membership in November 1937. Together Ortner and Wambacher had made, inter alia, an energy determination of neutrons; Kirsch and Wambacher, "Geschwindigkeit," *Sitzungsb., Akad. Wiss. Wien., Math.-naturwiss. Kl., Abt. IIa* 142 (1933): 241–249.

33. E.g., Ortner and Stetter, "Korpuskularstrahlen," *Z. Physik* 54 (1929): 449–70.

F. Hess, Blau and Wambacher were able to send their new emulsions to the 2,300-meter peak on Hafelekar (near Innsbruck) for a five-month exposure, ending in June 1937. On first examination, they found proton tracks of a length (and therefore an energy) far in excess of what anyone else had observed, some extending as far as the equivalent of 6.5 meters of penetration through air. Considering that Japanese and American teams had recently launched emulsion-bearing balloons into the stratosphere showing tracks of about 1 meter (air equivalent), the two women's results were astounding. But Blau and Wambacher's more salient result was what I consider to be the first emulsion golden event (analogous to Wilson's event depicted in figure 2.14), which bore little resemblance to anything previously seen. The phenomenon was this: on the emulsion there appeared several "contamination stars" (several tracks emanating from a point) with tracks leading from them that were longer than any Blau and Wambacher had ever seen in the laboratory. Could this, they wondered, be a new radioactive decay? Or was it merely a lessening of the braking power of the emulsion? The method was not yet secure enough a base on which to erect, by way of a single golden event, a major scientific claim. A week later, they found another star that was unambiguously clear of any irregularity in the emulsion and that manifestly could not be associated with any known decay.[34] With four such events in their collection, Blau and Wambacher sent a 25 August 1937 paper to *Nature*.[35]

Blau and Wambacher's golden event "star" consisted of nine branches, of which only one could be identified as an alpha particle. Two others were protons with ranges of 11 and 30 centimeters, and the rest were protons with larger energies that penetrated the emulsion. Interpreting the event was difficult, but two features were clear. First, the particle causing the star had to be from the cosmic rays because its energy far exceeded those from known radioactive decays. Second, the destroyed nucleus must have been one of the heavier elements of the emulsion because it had to have begun with a charge of at least 9; the most likely candidates were bromine and silver. Third, the authors pointed out that other physicists had found a single case of disintegration in a cloud chamber, but no one previously had ever actually seen the center of disintegration. Immediately, Blau and Wambacher began trying new experiments, adding thin layers of different materials above the emulsion to see whether stars formed differently when other nuclei were the targets, exposing film at different altitudes to probe the effects of cosmic radiation at varied heights, and sending emulsion samples high into the atmosphere with balloons (see figures 3.1 and 3.2).[36]

34. Blau, "Bericht über die Entdeckung," *Sitzungsb., Akad. Wiss. Wien, Math.-naturwiss. Kl., Abt. IIa* 159 (1950): 53–57.

35. Blau and Wambacher, "Disintegration Processes," *Nature* 140 (1937): 585.

36. Blau, "Bericht über die Entdeckung," *Sitzungsb., Akad. Wiss. Wien, Math.-naturwiss. Kl., Abt. IIa* 159 (1950): 53–57; Blau and Wambacher, "Disintegration Processes," *Nature* 140 (1937): 585; and Blau and Wambacher, "II. Mitteilung," *Sitzungsb., Akad. Wiss. Wien, Math.-naturwiss. Kl., Abt. IIa* 146 (1937): 623–41.

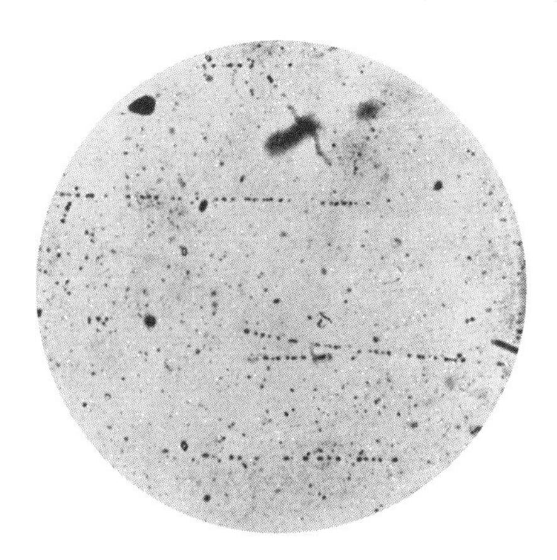

Figure 3.1 Blau and Wambacher, star (1937). Blau and Wambacher's first golden event was a nuclear "star" with eight tracks, found in emulsion exposed on the Hafelekar near Innsbruck. Because of the angles of the tracks, only some of them could be put into focus simultaneously. The emulsion itself was 70 microns thick. Source: Blau and Wambacher, "Disintegration Processes." Reprinted with permission from *Nature* 140 (1937): 585, © 1937 Macmillan Magazines Limited.

For one brief moment Blau was at the peak of her career. She had a film prepared in such a way and exposed at such a height as to give her an advantage over virtually all competitors. The golden moment ended abruptly with the rapidly deteriorating political situation. Suddenly Wambacher, one-time student and subordinate, had the upper hand over her former advisor.

On Friday 11 March 1938, the Germans entered Vienna. Blau fled first to Oslo, at the invitation of E. Gleditsch (to work at the Institute for Organic Chemistry). Then, desperate to find another home and to rescue her mother from Vienna,[37] she began exploring the possibility of getting to Mexico. She was, as she wrote in one only partly preserved letter, "obviously ready to do not only scientific work, but whatever is needed for the country"—which might include, for example, geological work, spectral studies of ores, or even further work on X rays.[38] With the recommendation of Einstein,[39] Blau moved (in November 1938) to Mexico City, where she became a professor of physics at the Polytechnic School. From there, as she told Einstein, she gave a series of successful lectures at a provincial university (in Morelia). There, it seems, she deeply impressed

37. Blau to Einstein, 10 June 1938, AEP, 52-606-1, 2.

38. See fragmentary letter, Blau-[?], n.d., AEP, 52-610-1, 2.

39. Blau wrote to Einstein on 10 June 1938 (presumably from Oslo) requesting that Einstein help her with the Mexican appointment (AEP, 52-606-1, 2). Einstein then sent a letter directly to Ingeniero Batiz (n.d., probably also June 1938 though later dated in typed insertion as 1939):

> My dear Ingeniero Batiz, It was a great pleasure for me to learn that the eminent Austrian scientist Dr. Marietta Blau has got a call to Mexico. I am sure that her large abilities will be useful for the instruction and research activities of your country; for we all know her to be a very intelligent and energetic personality who is trained in many fields and has excellent teaching abilities. Due to the great interest of my colleagues and of myself in her I take the liberty of asking you for information concerning the kind of her position and duties. You would oblige me very much personally and at the same time would give me the possibility of informing my colleagues who take special interest in the researches and fate of Dr. Blau. With kindest personal regards, Yours very sincerely, [Einstein].

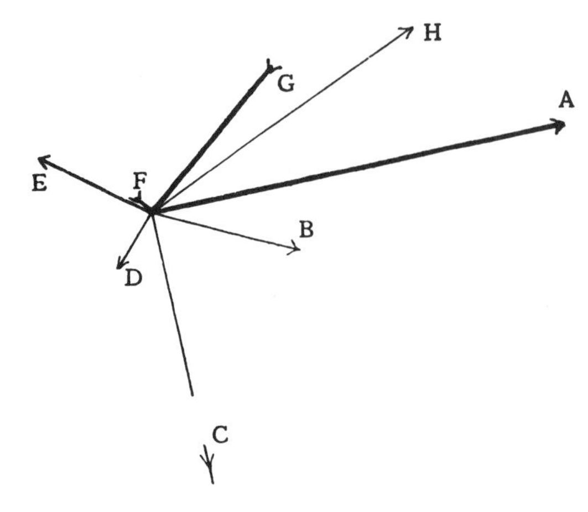

Figure 3.2 Blau and Wambacher, schematic star (1937). "Reading" figure 3.1, Blau and Wambacher offered the following schematic interpretation: Thick lines indicate a comparatively large number of grains per unit length of the track. An interrupted line means that the track is too long to be reproduced on the same scale. The arrows indicate the direction from the surface of the emulsion to the glass. Source: Blau and Wambacher, "Disintegration Processes." Reprinted with permission from *Nature* 140 (1937): 585, © 1937 Macmillan Magazines Limited.

the rector of the school, whose pride in the institution had been manifest when he went to the United States to purchase the makings of a laboratory. For want of a scientist, the equipment still stood in cases, and the rector was pleased to invite Blau to set it up. Excited by the prospect of working again, Blau had first to finish a stint teaching in Mexico City, no trivial matter since each week she had some 24 hours of lectures to deliver. Unfortunately, by the time she finished, the apparatus in Morelia had vanished, only to reappear shortly thereafter in a pawn shop. Stymied by events and frustrated by three years of isolation from physics, she pleaded for an intervention from Einstein so that she might at least do something in geophysics or a related field. "If one would simply let me work," she pleaded, "I could prove, at least to the best of my abilities, that an emigrant can be more than a useless burden."[40]

In May 1944, Blau moved north to New York City, where she took a position with the Canadian Radium and Uranium Corporation and the International Rare Metals Refinery.[41] With the war over, the newly established Atomic Energy Commission (AEC) set Blau up at Columbia University as a research physicist; the AEC then moved her in 1950, to their Brookhaven National Laboratory, which was just then turning to high energy work. Among other results that Blau

40. Blau to Einstein, AEP, 54-835-1, 2. Einstein did oblige in a letter of 24 June 1941 to the Mexican ambassador to the United States: "I am taking the liberty of drawing your attention to a case very close to my heart. Since three years my colleague, the physicist Dr. Marietta Blau, lives in Mexico City. Officially she holds a position on the Institute for Technology there; but the trouble is that she has not had until now the opportunity for useful work. I know Miss Blau as a very capable experimental physicist who could do valuable service to your country. She is an experienced investigator in the field of radio-activity and cosmic rays." Einstein concluded, "I should be very much obliged to your Excellency if you would be good enough to bring this fact to the attention of your government" (English original, AEP, 54-837).

41. Blau and Dreyfus, "Multiplier Photo-Tube," *Rev. Sci. Inst.* 16 (1945): 245–48; and Blau and Feuer, "Radioactive Light Sources," *J. Opt. Soc. Amer.* 36 (1946): 576–80.

obtained there with her emulsions was the first demonstration that accelerator-produced mesons could produce more mesons.[42] In further work, she contributed to the design of scintillation counters, while her main research continued to be with emulsions.[43] Although Brookhaven initially seemed congenial, even that resting spot did not last, as personal friction with some staff members, coupled with dire financial difficulties and health problems, led Blau back to Vienna in 1960.[44] Constantly on the move, Blau was nowhere long enough to become fully visible to the world of physics.

Wambacher and Stetter did well indeed during the war and continued to publish in the field of photographic emulsions. According to a friend and colleague of Blau's, Leopold Halpern, after the war Blau told him that as she was leaving Germany from Hamburg in 1938, her zeppelin was forced down and all her scientific notebooks stolen by the Gestapo. Apparently the Nazis knew precisely the object of their search, dismissing her once they had found the crucial papers. What happened to those notebooks has never been clear. Blau, according to Halpern, believed that they ended up back in the hands of her old colleagues at the Institut für Radiumforschung in Vienna.[45] Although we may never be able to confirm this, we can know something of Wambacher's attitudes in the years of Nazi rule. Writing to the Nazi membership division, Wambacher was incensed:

> I have been an applicant for party membership since June 1934. . . . [A]s you can tell by inquiring to Circle IX, my status as party applicant has been in good order even when it was illegal [to join the party]. . . .
>
> In May 1940 I suddenly received a form according to which my application request had been refused! The form consisted of a printed slip, evidently used for many cases, and contained no indication of the grounds for my refusal; I was "free" then in 1940, once again to apply that is to say together with all those characterless, previously hostile rifraff who exploited the economic "boom" of 1940!
>
> According to word of mouth in the *Ortsgruppe* there is a large number of

42. See, e.g., Blau, Caulton, and Smith, "Meson Production," *Phys. Rev.* 92 (1953): 516–17.

43. Blau, "Grain Density," *Phys. Rev.* 75 (1949): 279–82; Blau, Caulton, and Smith, "Meson Production," *Phys. Rev.* 92 (1953): 516–17, on meson-meson production; and Blau and Caulton, "Inelastic Scattering," *Phys. Rev.* 96 (1954): 150–60, on the scattering of negative pions at 500, 750, and 1300 MeV.

44. Date of return from Blau, self-description (in German), five-page typescript, LHP.

45. According to Halpern: "Dieses Luftschiff machte in Hamburg von Skandinavien kommend Station. Blau, die in Skandinavien zugestiegen war fuehrte ihre gesamten Arbeiten mit sich. Nachdem das Luftschiff in Hamburg anlegte kamen Maenner an Bord die sie aufforderten mit ihrem gesamten auszusteigen. Die Leute oeffneten fachkundig die Koffer und nahmen, ohne sich um den Rest ihrer Habe zu kuemmern die wichtigsten photographischen Platten sowie die Aufzeichnungen ueber ihre zukuenftigen Forschungsplaene an sich. Danach liessen sie sie wieder einsteigen. Blau meinte es bestuende kein Zweifel, dass die Aktion vorbereitet war und die Leute genau das Wesentliche kannten. Alles andere wurde ihr belassen. Blau berichtete mir dann, dass spaeter die in Hamburg konfiszierten Ergebnisse ihrer Arbeiten in gemeinsamen Veroeffentlichungen von Stetter und Wambacher erschienen sind und dass Stetter gerade die Experimente durchfuehrte die sie in ihren Arbeitsplaenen beschrieben hatte" (Halpern, "About the Life of Marietta Blau," 16-page typescript, LHP).

upstanding illegal Nazis being subjected to this shocking and totally unnational-socialistic treatment. Obviously I have not let myself be subjected to this *Schweinerei* and turned first to the bureau for membership in Vienna; there I received the verbal information that my case was in no way unique. . . .

Of course I realize that in the war bureaucratic matters go slowly and so I waited from September 1940 to now, January 1942! I have therefore in no way put the urgency of my personal business above the war! But if your office continues to function at all during the war, then it finally ought to start to function and . . . in an exemplary and conscientious manner!

I insist that you finally take care of my circumstance in a way appropriate to a national socialist office and I intend to pursue these matters through all stages of appeal right up to the chancellery of the Führer, especially since I see that I am not alone. . . . Heil Hitler! Dr. Hertha Wambacher[46]

With the Nazis' many successes, right-thinking National Socialist physicists had two ordinary and two extraordinary chairs to fill in Vienna, positions freed, as one report put it delicately, "by the departure of the Jewish professors Meyer, Ehrenhaft, Przibram and Kottler."[47] (Blau's departure from her minor position left few spoils.) It was a bit awkward bureaucratically to appoint docents from the same university to such major positions, and Ortner, for example, had hardly been a rising star. Ranked as "extraordinary assistant" since 1924, Ortner's docentship was renewed periodically through 1934, a long enough span that some explanation was demanded: he was, as one official explained it, in "full mastery of complicated radioactive measurement methods," while Professors Przibram and Jaeger had been overburdened.[48] Faint praise. Stetter likewise had been an "extraordinary assistant" for quite some time; at the Second Physical Institute since November 1922, he was renewed for the next 15 years or so. With Przibram's dismissal, the Nazi star had risen, and Stetter was called to fill, provisionally, Przibram's chair.[49] Major promotions were in the offing, even if Nazi officials recognized that internal promotions into professorships had not

46. Wambacher to NSDAP, Reichsleitung München, Amt für Mitgliedschaftswesen, 24 January 1942, AdR.

47. Commission members were Dean Christian, Flamm, Franke, Haschek, Kirsch, Rector Knoll, Kralik, Marchet, Mayrhofer, Prey, Schweidler, Späth, Stetter, Dworzak, and Ortner, who met in May 1938 to consider the structure of physics at the university and the institutes. Haschek, Kirsch, Stetter, and Ortner were absent for discussion of personnel matters; Report, n.d. (8 June 1938), fol. 139, Personalakte Ortner, BmUP, AdR. An excellent secondary source for information on the fate of academics dismissed from the Institut für Radiumforschung and other Austrian institutions of higher learning can be found in Stadler, *Vertriebene Vernunft II* (1988), especially Wolfgang Reiter, "Das Jahr 1938."

48. Dean Schweidler to [unidentified], n.d. (ca. March 1936), fol. 162v; further renewed as assistant, 1 June 1938–31 May 1940, Dean, Phil. Fak., to BmU, 18 December 1937, fol. 145. Personalakte Ortner, BmUP, AdR.

49. Stetter's assistantships: Draft of BmU to Dean, Phil. Fak., 24 October 1934, fol. 277v (for 1 November 1934–31 October 1936); draft of BmU to Dean, Phil. Fak., 16 October 1936, fol. 273v (for 1 November 1936–31 October 1938). On the provisional appointment as extraordinary professor in 1938, see MikA-IV to Rector, Univ. Vienna, 9 December 1938, fols. 310–11. Personalakte Stetter, BmUP, AdR.

been a done thing in the *Altreich.* Now, however, in the deadening prose of the bureaucracy: "There is available in Vienna a professionally highly qualified succession, which also is quite advanced in years, which unquestionably deserves consideration. If one were to now, corresponding to existing custom, pass over the proposed Austrians and call those from the *Altreich,* there would arise the danger that the aforementioned Austrians, who frequently because of their National Socialist attitude were excluded from any appointments [as professors] in the old Austria, would be placed behind a younger, scientifically in no way better qualified workforce from the *Altreich.*"[50] Kirsch too had been professionally slow to advance before the German army entered Vienna. He had been an assistant at the Institut für Radiumforschung from October 1921 to 1928, then promoted to Oberassistent at the Second Physical Institute, and finally, in 1932, made "titular extraordinary Professor."[51] After some discussion back and forth, Stetter was proposed for *Ordinarius,* Kirsch for *Ordinarius,* and Ortner for *Extraordinarius* professorships. Kirsch, by official lights, emerged as a "worthy fighter [for] the cause."[52] The Academy resolved simultaneously to name Ortner chair of the Institut für Radiumforschung.[53]

By the end of the war, Stetter, Kirsch, and Ortner were in full power. Marietta Blau, by contrast, had moved some ten times, lost all her scientific papers and notebooks, and still had no clear path to a permanent position. Wherever she went, the relatively inexpensive emulsions gave her a kind of replaceable laboratory, from the mountains of Austria to the halls of the Institut für Radiumforschung, from Mexico City to Brookhaven. Unlike a cyclotron or even an electronic apparatus, the emulsion was an instrument well suited, however awkward its images, to those on the margin. With a box of Ilford halftone photographic plates, the use of a microscope, and some desktop chemicals, atomic disintegrations, neutron dynamics, and radioactive decays could be studied and brought to the pages of *Nature, Zeitschrift für Physik,* or the *Mitteilungen aus dem Institut für Radiumforschung.* But while the method worked (to a certain extent) for the nomad, institutions still held the power to locate evidence amid the noise, and to orchestrate the cross-checking of novelties among different observers. Only in the structured environment of a new kind of laboratory could the anxiety of the experimenter be allayed. Cut off from her former colleagues in Vienna, Blau was not yet part of a network of collaborators in film production, film testing,

50. Copy of Christian to MikA, 8 June 1938, fol. 138, Personalakte Ortner, BmUP, AdR.

51. Information form, 16 June 1939, and Kirsch, questionnaire, 20 May 1940; both from Personalakte Kirsch, BmUP, AdR.

52. Draft of Plettner, MikA-IV to REM, 9 December 1938, fols. 115–16, Personalakte Ortner, BmUP, AdR.

53. General secretary, Akad. der Wiss., to MikA-IV, 17 December 1938, fol. 109. Dual appointment confirmed by REM, February 1939, Dames, REM, to MikA-IV, 11 February 1939, fol. 105; officially made on 28 September 1939, effective 1 October 1939, Zschintzsch, REM, to Ortner, 28 September 1939, fol. 92. Personalakte Ortner, BmUP, AdR.

and interpretation of events. In Britain, Cecil Powell began to create that new kind of collaborative laboratory, assembling experimenters, chemists, scanners, and links to theorists into a powerful new consortium: it was, as we will see, nothing less than a new kind of experimental physics. Isolated, Marietta Blau spun ever further from the center.

3.3 A Miniature Cloud Chamber

C. T. R. Wilson had only a handful of students; they all went on to build cloud chambers. All had high standards of craftsmanship, but Cecil Powell's had already been honed by being the son of a master craftsman.[54] For Powell, the cloud chamber was the first of many devices that "allowed far-reaching conclusions to be reached, with some confidence, on the basis of a single observation."[55] To his colleagues, Powell's leaning toward the photograph seemed only natural given his general approach to the natural world. Theorist Victor Weisskopf found it

> characteristic of Cecil Powell that he should choose the simple method of taking pictures of nature as his way of doing physics. It is the direct approach, it brought him into direct connection with nature. He looked at things and he saw something interesting. His greatest discoveries were made when he observed the effects of cosmic rays. He used nature's own high energy beams; they were freely available and came from the open skies. Here we find a characteristic trait of Powell's approach to natural phenomena. He was a naturalist at heart; he took pictures of cosmic rays like taking pictures of animals, only in a much more sophisticated way.[56]

Weisskopf's summary is apt. Like Wilson, whose mimetic work began in the photographic reproduction of natural phenomena of the Scottish Highlands, Powell situated himself as a collector and imitator of nature. Only by understanding the Wilsonian style of physics can we make plain the contrast between Powell's early work and the emulsion physics that followed. In the divide between the well-defined macrophysics of condensation on one side and the ephemeral microphysics of particles on the other, we can track the roots of the experimenter's anxiety.

Like Wilson's, Powell's earliest scientific work centered on condensation. In Wilson's laboratory, Powell struggled to build an all-glass chamber in order to study the condensation process at different degrees of expansion. Not only was the chamber extremely difficult to assemble using soda glass and emery powder, but once completed it was still so fragile that the device would sometimes

54. O'Ceallaigh, "Powell's Group," *J. Phys.* 43 (1982): C8-185–C8-189, on C8-186.
55. Powell, "Autobiography" (1972), 20.
56. Weisskopf, "Life and Work" (1972), 1.

implode on the first expansion.[57] Once it functioned, however, it gave Powell the means to push forward the old core of Wilson's endeavor: condensation physics. Although condensation no longer functioned as a central problem for the theory of matter, it reemerged in the world of engineers, who could use the cloud chamber to design steam engines.

Already in 1915 Hugh Callendar had used cloud chambers to explore the thermodynamic theory of the steam turbine, in particular the discharge of steam by nozzles.[58] The issue was practical since condensation during the steam's compression in and expulsion from the cylinders would affect the efficiency of an engine. Real engines entered the scene, not abstractions. Callendar and Nicolson did their physics on full-size engines made by the Robb Engineering Company of Amherst, Nova Scotia. Experiments were delicate: one had to probe the temperature of steam using sensitive platinum thermometers in the cylinder of the steam engine. Theory depended on whether one was to consider "dry" or "wet" (supersaturated) steam. Here Wilson's experiments entered because they showed that under suitable decompression, water vapor would condense whether or not there were dust nuclei present. Such supersaturated steam released heat upon condensation and had additional frictional properties.[59]

Although Wilson had not come to the problem through the steam engine, condensation as a subject of investigation was as much a part of steam engine study as it was of ionic physics or meteorology. Such inquiry immersed Powell in the literature about the pressure distribution in the turbines of the S.S. *Mauretania* (see figure 3.3) during his detailed determination of the "cloud limit line" in his heat-pressure graphs of 1925.[60] Three years later, in 1928, Powell used the cloud chamber to explore more generally the effect of temperature on supersaturation.[61] And finally, completing the cycle, Powell spent some four years studying the mobility and interactions of ions in gases. This interest was perhaps a natural one for someone emerging from the Cavendish interested in the underlying mechanisms not only of condensation but of the constitution of matter more generally.[62]

What Krakatoa had been to Aitken and his contemporaries in the 1880s, Mt. Pelée (on Martinique) was to Powell (see figure 3.4). On 8 May 1902, the island's volcano had erupted with an extraordinarily powerful jet of steam, and rocks and ash were shot into the atmosphere and hurtled down the slopes of the mountain, killing over 20,000 people. Only one man (a convict in the town jail)

57. Powell, "Autobiography" (1972), 20.

58. Callendar, "Steady Flow," *Proc. Inst. Mech. Eng.* (1915): 53–77.

59. Callendar, "Steady Flow," *Proc. Inst. Mech. Eng.* (1915): 53–77; and Callendar and Nicolson, "Condensation," *Min. Proc. Inst. Civ. Eng.* 131 (1898): 147–268.

60. Powell, "Supersaturation" ([1925] 1972).

61. Powell, "Condensation Phenomena" ([1928] 1972).

62. See, e.g., Tyndall and Powell, "Mobility of Ions" ([1930] 1972).

Figure 3.3 SS *Mauretania,* engine room (1907). Instruments wander far from their points of origin. Begun in meteorology, the cloud chamber migrated, with Powell, through the engine room of the *Mauretania.* Displayed here is the erecting shop at the Sallsend Works, in which one sees three turbines from the *Mauretania:* one astern, one low-pressure ahead, and one high-pressure ahead turbine. Powell used what he had learned about condensation from Wilson's laboratory to study the dynamics of the highly compressed steam in these engines. Source: Taken from *The Cunard Turbine-Driven Quadruple-Screw Atlantic Liner "Mauritania,"* by Mark D. Warren (1907), pl. 18, fig. 128, published by Patrick Stevens, Ltd., 1987. Reprinted with permission.

survived, and even he died shortly thereafter. Noticing that the frequency and power of volcanic eruptions were increasing alarmingly in the Lesser Antilles, and watching with dismay as Pelée again erupted between 1930 and 1932, the Royal Society and Colonial Office formed a seismic expedition to their possessions in the British West Indies. Frank A. Perret, one of the world's leading volcanists, wrote in the summer of 1934 that the newly opened fumarole gas vents, long thought to be extinct, had awakened. These emissions, coupled with local tremors and earth shocks, were disquietingly reminiscent of the years prior to the catastrophic explosion of 1902.[63]

Powell joined the Royal Society expedition, which was to center its efforts on Montserrat in the Caribbean. His reasons were undoubtedly several. The *nuée ardente* itself, as a natural phenomenon, was a large-scale version of the condensation phenomena to which he had devoted much of his career. Most vol-

63. Perret, *Pelée* (1935), 11.

Figure 3.4 Explosion of Pelée (1935). No natural phenomenon so dramatically illustrated the *nuée ardente,* the massive condensation that accompanies certain explosive volcanic eruptions. Here, in the explosion of Pelée in Martinique, the example is striking. Powell was dispatched urgently by the Royal Society to set up a system of monitors to watch over the British island of Montserrat. Source: Perret, *Pelée* (1935), 7.

canoes erupt primarily upward; trapped gas, having been compressed to extreme pressures, breaks the cork formed by lava above it and rockets the lava up and out of the volcano. As their name suggests, *nuée ardentes* (also known as *Glutwolken* or *nuages denses*) involve "magnificently spectacular convolutions of vapor and ash, [along with] the swiftly unfolding cauliflower conformations of finer dust which mount upward."[64] These displays are produced by masses of active lava, the particles of which spew gas, sometimes explosively. Because the lava is broken apart by gas, the liquid is not coherent and cascades down the side of the volcanic mountain with little friction and a speed far exceeding more usual lava flows.[65]

Since the proper measurement of earthquakes required observers spread out not only over time but over space, no single site would suffice. After a stay in Montserrat that lasted from 24 March to 24 July 1936, Powell reported to the Royal Society: "Our present knowledge of the seismic-volcanic phenomena in these islands is so limited that the establishment of a quite modest equipment of observing instruments might well lead to discoveries of considerable value. The

64. Perret, *Pelée* (1935), 5.
65. Perret, *Pelée* (1935), 84–88.

organization which has been established in Montserrat could be regarded as a first step in the establishment of instruments at a number of points in the island chain and as an experiment in the maintenance of instruments in the hands of untrained observers and volunteers" (see figure 3.5).[66] Aside from nourishing Powell's fascination with unique occurrences in nature, the Montserrat expedition gave him experience in the organization of "untrained observers," a scientific managerial style that he exploited for the rest of his career.

Elsewhere in the same report, Powell reiterated the importance of using such "unskilled" workers, a theme with lasting implications for the whole of postwar physics:

> The knowledge necessary for a real understanding of the fundamental processes which are occurring can only be obtained by extended observations at a number of points in the Caribbean arc. Valuable information could be obtained in the first instance by a quite modest investigation. If the experiment of leaving the instruments in the care of untrained observers in Montserrat is successful, then similar equipment might be installed in other British islands in this region. In this way an organization could be built up which might, in time, be able to predict the nature and extent of any threatening catastrophe and which would make important contributions to the science of geophysics.[67]

Powell's use of "unskilled" observers began in his search for seismic activities; in addition, the Royal Society expedition served to reinforce his already strong commitment to the power of individual images to reveal underlying dynamics. For in a sense the seismograph offered a "picture" of a particular and never-quite-repeated earth movement, potentially a golden event, portrayed not in a photograph but in the spiking trace of a needle. Evidence for such an occurrence, such as the Wiechert seismograph results of figure 3.6, provided the observers with the information they needed.[68] Through the careful coordination of numerous observers, Powell hoped hundreds of lives might eventually be saved. That understood, it was not with geophysics that he wanted to cast his lot.

With recent developments in nuclear physics by Enrico Fermi, Lise Meitner, Otto Hahn, and others, it became clear to Powell that nuclear experimentation could not wait. On returning to Bristol, he immersed himself in the subject and by 9 February 1938 began to ponder how he could exploit the cloud chamber to maximum advantage: "The question whether the proton-proton forces are attractive or repulsive depends upon the absolute determination of large angle scattering probabilities. The experiment could be done by making expansion chamber photographs of long range protons in hydrogen. Advantages are large energies available and unambiguous counts. Can one obtain sufficient protons?

66. Powell, "Expedition to Montserrat" ([1937] 1972), 146.
67. Powell, "Expedition to Montserrat" ([1937] 1972), 148.
68. Powell, "Expedition to Montserrat" ([1937] 1972), 138.

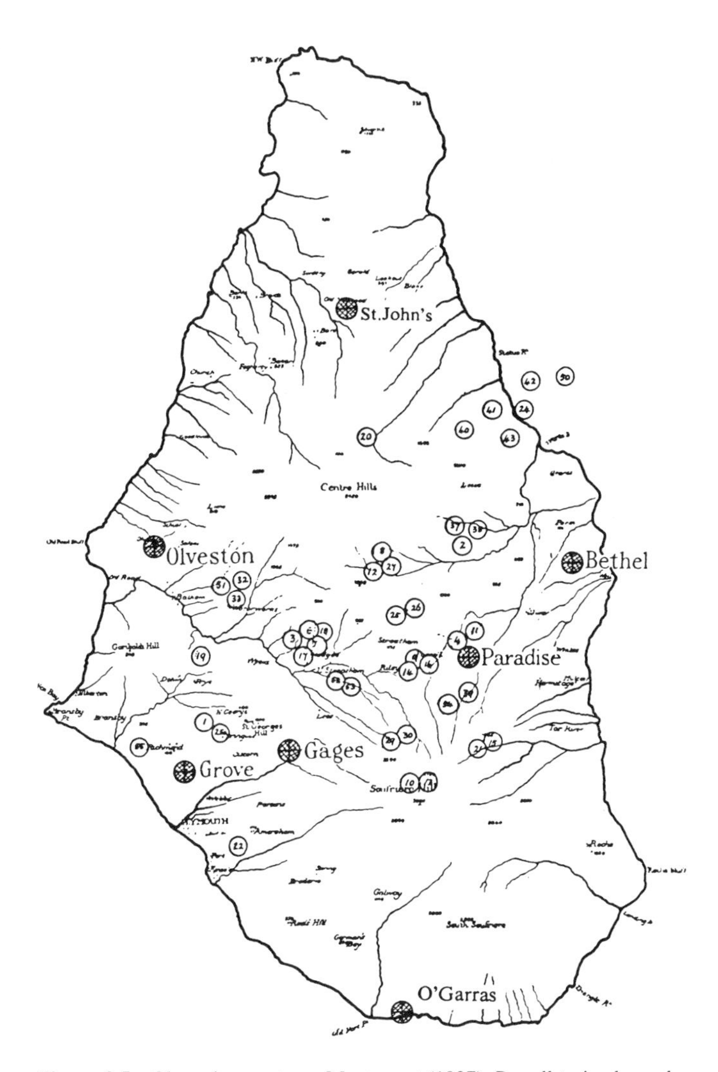

Figure 3.5 Observing posts on Montserrat (1937). Powell trained a cadre of "unskilled" observers to cover Montserrat with observing stations, both to advance geophysics and to warn of impending catastrophe. In this map, the team indicated the positions of the Jaggar (seismographic) stations and suggested epicenters. Jaggar stations shown with shaded plus; epicenters with circled number. Source: Powell, "Expedition to Montserrat," *Proc. Roy. Soc. London A* 158 (1937): 479–94, fig. 10 on 485.

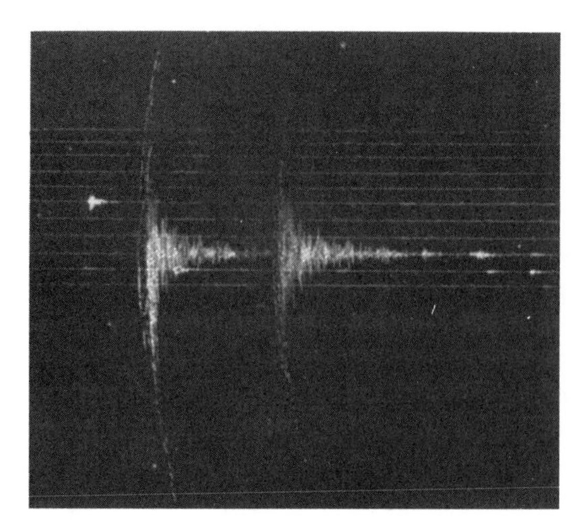

Figure 3.6 Seismograph from Montserrat (1937). Golden events were not restricted to cloud chamber photographs. Here Powell and his team reported a seismographic record of a single event recorded on 22 June, 21 hours, 52.3 minutes. Powell argued that the event shown must have been local as there was no resolution of the S (surface waves) and the longitudinal P (pressure waves). Source: Powell, "Expedition to Montserrat," *Proc. Roy. Soc. London A* 158 (1937): 479–94, pl. 24, fig. 5, after 488.

Large chamber probably better than present one and high pressures."[69] Powell's reference to "unambiguous counts" is undoubtedly a comparison with the competing method of exploiting counters, where no individual count could be identified unambiguously with a real, as opposed to spurious, occurrence. With the cloud chamber, by contrast, each single scattering event would leave a full track, and the determination of the number of particles scattering by a given angle would be more evident. Making a large chamber would be necessary to ensure a target with a sufficient number of protons from which the beam protons could scatter; operating at high pressure served the same function.

Powell's concern that he could obtain a sufficient number of scattering events continued. On 19 February 1938, he commented in his notebook: "Increase stopping power of gas increases probability of collision but not probability in a given energy range. If possible to obtain 100 tracks/collision then one fork for every 60 photographs. For 1000 forks, 10,000 photographs."[70] Even at a high pressure, the number of photographs required was daunting. Then, amid the calculations, on 12 March 1938, Powell carefully wrote: "Hitler marches into Austria. We shalln't be doing this much longer. Birmingham yesterday. Oliphant says condensed discharge unsatisfactory and that it's better to have continuous discharge with a shutter. We shall see."[71]

Work alternated with politics, and 21 May brought new trepidation: "Rumoured massing of German, Hungarian + Czech armies."[72] There followed

69. Entry opposite text marked 9 February 1938, CPnb.
70. 19 February 1938, CPnb.
71. 12 March 1938, CPnb.
72. 21 May 1938, CPnb.

several exchanges with colleagues about building the larger cloud chamber that Powell required. But construction difficulties left Powell stumbling over one impediment after another. Then, on 2 June 1938: "Heitler suggests investigation of cosmic ray bursts by the photographic plate method. [There follow two references to Blau's pre-Anschluss work.][73] Why not shoot boron or nitrogen protons into film tangentially and investigate range and also, possibly, scattering. Might be possible to get space orientation of grains with high power microscope by focussing."

Powell was pushing the cloud chamber to its limit in every way. He struggled to augment its stopping power—increasing size, bolstering the pressure of the hydrogen gas. But at the same time his work within the visual tradition of the cloud chamber imposed constraints on the kind of evidence that would satisfy him: counters with their "ambiguous" individual data points would not do. Thus from the start, even *before* it had been shown to be possible, Powell wanted the emulsion to recreate the central feature of the cloud chamber: the ability to display the three-dimensional spatial orientation of the particle tracks. With proper calibration the focus setting of the microscope could reveal the track's depth within the thin emulsion layer of the film.

For Powell and other physicists of the 1930s, the spirit of cloud chamber work was embodied in the new emulsion technology. In 1939, Powell called the emulsion "equivalent to a continuously sensitive high-pressure expansion chamber" and repeated the claim the next year with the following amplification: "We believe it possible to press the analogy [of the emulsion to the Wilson chamber] further in the sense that the spatial orientation of a track in the emulsion can be determined."[74] Though Powell at this time remained innocent of most previous work on emulsions, the leading Russian expert, A. Zhdanov, had argued along similar lines just a few years before: "Thanks to [various] improvements in photographic method, we obtained roughly the same approximation to a particle's trajectory that is given by a cloud chamber with a picture a thousand times smaller of the same phenomenon."[75] In the years that followed, one finds the analogy pursued in other ways: stacks of emulsions become "emulsion chambers."[76] References like these indicate that despite the physical differences between the production of a photographic track and a cloud track, the *visual* structure of

73. Blau and Wambacher, "Vorläufiger Bericht," *Sitzungsb., Akad. Wiss. Wien, Math.-naturwiss. Kl., Abt. IIa* 146 (1937): 469–77; Blau and Wambacher, "II. Mitteilung," *Sitzungsb., Akad. Wiss. Wien, Math.-naturwiss. Kl., Abt. IIa* 146 (1937): 623–41. Note: the notebook incorrectly refers to the second as pp. 621ff. As discussed above, the first of these reported on the high energy protons and (indirectly) neutrons that were observed in the first analysis of pictures from the Hafelekar. The second is the full report on high energy stars.

74. Powell and Fertel, "High-Velocity Neutrons" (1939 [1972]), 153; Powell, "Further Applications" ([1940] 1972), 159.

75. Jdanoff, "Traces des Particules," *J. Phys. Radium* 6 (1935): 233–41, on 234. Note: Jdanoff and Zhdanov are different transliterations from the Russian.

76. Zhdanov, "Quelques problèmes" (1958), 233.

both made them seem continuous with one another even at the beginning of emulsion work.

For the first year of his work on emulsions, Powell and his associates moved smoothly back and forth between the Wilson expansion chamber, which neared completion, and the newer method. Yet he could not keep his mind on the old technique much longer. After noting the status of the cloud chamber, Powell turned back, in his notebook entry of 18 June 1938, to the problem of making the emulsion play the role of a miniature cloud chamber:

> It seems possible that the photography technique for investigating nuclear reactions might become very valuable. It has a number of important advantages:
>
> (a) It is extremely simple to make the exposures and an enormous amount of information can be contain[ed] in a single small piece of plate.
>
> (b) There is no associated gear and it is therefore possible to make experiments at high potential. Thus with our set a proton source of great power could be run at earth potential with the set giving negative polarity. Alternately the tube can be made with the two ends at + and − with the pumping at earth in the middle. The chief difficulty is digging the information out of the film. With scattering experiments using an expansion chamber one can certainly get 100 tracks per photograph and a fork will be easily distinguishable. With the emulsion method it *may* be possible to look over large numbers of tracks in the field of view of the microscope with low powers (200 diameters) but it may be necessary to examine each track in succession. For absolute values of the scattering the expansion chamber results enable one to estimate the number of tracks by counting the tracks on a relatively few photographs and estimating a mean but with the emulsion method the individual tracks must be counted. The results with the α tracks suggest that it will certainly be possible to obtain spacial orientations by the focussing method. The depth of focus is of the order of a few wave lengths only.[77]

Using the precision focus of the microscope would help resolve the depth coordinate of a track, but to use the emulsion fully as a three-dimensional recorder, Powell needed thicker film, on the order of 300 microns rather than the halftone's 70–100 microns. But thicker films required new processing techniques to avoid distortions during development and drying. If Powell could succeed at making such films and providing a viable method of extracting the track information, the method could be exceedingly valuable in probing nuclear disintegrations (such as $B^{11} + H^{1}$ or $N^{14} + H^{2}$), where a mere handful of good pictures could give all the information one desired. Like Blau, Powell glimpsed the fecundity of the emulsions, but at the same time, he felt the elusiveness of the objective image.

Returning more systematically to the photographic literature, Powell spent late June and early July 1938 reading more about previous work on the emulsion

77. 18 June 1938, CPnb.

technique by H. J. Taylor and Maurice Goldhaber, who used specially prepared film to confirm the possibility of disintegrating boron nuclei by slow neutrons ($B^{10} + n^{1} \rightarrow Li^{7} + He^{4}$).[78] Simultaneously, he redesigned the new cloud chamber. But if prospects were brightening for the new emulsion, political events were not. On 14 September 1938, Powell recorded that "Chamberlain flys [*sic*] to Hitler and throws all peace forces into confusion. Reestablishes the failing Fascist prestige. Next move will be to try and establish 4-power pact and thus isolate Soviet Union."[79] As if in sympathy, the cloud chamber project faltered. A leak appeared and then vanished overnight. There was a catastrophic flood of the laboratory, and a variety of other instrumental failures meant that Powell could do practically no useful work on tracks in his cloud chambers. Perhaps in desperation at the absence of productive beam time, at the end of September 1938, Powell took absorbers and ordinary halftone photographic plates (ordinarily used to make lantern slides) to the laboratory roof in search of cosmic rays.[80]

If cosmic rays were the quarry, the place to hunt them was not on a Bristol roof, as Powell soon determined, but at the mountain observatory at Jungfraujoch in Switzerland. Dispatched to expose plates there, one of Powell's colleagues returned on 20 December 1938, and despite the tedium of microscopic examination, Powell was clearly delighted with the haul:

> *Tracks on Plates*
>
> (1) Thorium α particle stars (5 components).
>
> (2) Stars with long range components from cosmic rays.
>
> (3) Individual tracks—long—from cosmic rays—sometimes associated.
>
> (4) Stars with short components showing large angle scattering.

With these images as a resource, Powell set out his program:

> The following must be done:
>
> (1) Measure number of grams/cm in tracks from α particles and from high energy protons obtained with H[igh] V[oltage] set.
>
> (2) Compare with tracks of type (4) which may be heavy electrons.
>
> (3) Establish association of type (3).
>
> (4) Confirm tracks (1) by length measurement.
>
> (5) Determine absorption coeff[icient] in Pb of primary radiation responsible for (2) + (3).[81]

Suddenly Powell's carefully enunciated agenda was obsolete. On 1 February 1939, the nuclear physics community received the most astonishing news: the nucleus had been split. Powell, after underlining the words "uranium transmutation" four times, turned his attention entirely to the problem of fission:

78. Taylor and Goldhaber, "Detection," *Nature* 2 (1935): 341.
79. 14 September 1938, CPnb.
80. 30 September 1938, CPnb.
81. 20 December 1938, CPnb.

> *Uranium Transmutation*
> Reported from Columbia, N.Y., that Uranium is disintegrated by slow neutrons into Ba + Kr etc and that Meitners trans uranium series are wrongly identified. Range of 100MV barium nucleus in air is about 2–3 mms too small for our photographic plates to detect satisfactorily. Might try a plate flooded with uranium.[82]

Other suggestions and attempts followed in quick succession. Could one use the photographic method with plates containing fluorescent substances that would make the tracks distinctive? Could one load the plate with a finely divided uranium phosphor or with a uranium salt along with an organic molecule phosphor? Would it be conceivable, as Neville Mott contended, to exploit a scintillation screen of a uranium phosphor that "might show bright flashes under the bombardment of slow neutrons"?[83] Could the photographic method reveal gamma rays by recording the disintegration of deuterons?

Most significant, though, Powell immediately began to employ the emulsion method to analyze individual events, as was clear from the following entry on 12 March 1939:

> We have observed a star on the Jungfrau plates with one component of air equivalent 14 cms, in which the number of grains/cm is four times that in the proton tracks. The observation of grain distance in cases which we believe to be Th[orium] α is not distinguishable in individual cases from proton tracks. The particle is therefore probably a heavy particle of great energy and charge. It is further distinguished from the lighter particles in that the ionisation falls off with distance from beginning of the track and at the end is similar to that of an α particle or proton. This may be explained as due to the decrease in Z_{eff} as the velocity decreases by "capture + loss" since for heavy particles these processes are important at much greater energies than in the case of the α particle or proton. Energy at which capture of electrons into K shell takes place is proportional to Z^3 roughly.[84]

On 15 March 1939 Powell recorded: "German troops go into the rest of C[zecho]-S[lovakia]. The consequences of Munich are becoming clear." A few days later: "Took C[osmic] R[ay] plates off roof for development. . . . Rumoured off of pact to the S[oviet] U[union]."[85] Full legitimation appears to have required Powell to bring the new detector technology head-on against the more established technique. To do this he chose to use the emulsion to repeat some cloud chamber experiments already conducted by T. W. Bonner and W. M. Brubaker at Caltech. The two Pasadena physicists had sent deuterons into a high-pressure cloud chamber filled with methane and indirectly measured the energy distribution of the emitted neutrons. They could do so because, when the

82. 1 February 1939, CPnb.
83. 3 February 1939, CPnb.
84. 12 March 1939, CPnb.
85. 15 March 1939, CPnb; 19 March 1939, CPnb.

neutrons hit a proton or helium nucleus, the charged particle took on most of the neutron's energy and left a visible trail.[86] In this way, they could examine neutrons from beryllium, boron, and carbon and produce a statistical distribution of the number of neutrons at each given energy—this application is a good illustration of the ways in which cloud chamber data could be used without invoking golden events. Powell and his collaborator G. E. F. Fertel wanted to replicate the cloud chamber work, only this time with an emulsion.

First, however, Powell and Fertel had to calibrate the new instrument. Using a beam of 600 KeV deuterons from the Bristol high-tension generator, he bombarded a boron target, sending the resulting protons through a mica window tangentially into an Ilford halftone plate. Graphing the number of protons with a given range against the range, Powell and Fertel could then compare their "spectrum" with that obtained by well-established counter techniques. In this case, the process being measured was

$$^{10}B_5 + {}^2H_1 \rightarrow {}^{11}B_5 + {}^1H_1;$$

that is, boron isotope of mass number 10 plus deuteron goes to boron isotope of mass number 11 plus proton. Now the mass of the reactants (on the left) is greater than that of the reaction products on the right; the difference (by $E = mc^2$ and taking into account the kinetic energy of the incident deuterons) emerges as kinetic energy distributed among the reaction products. The spectrum of proton ranges is a series of bumps (roughly like figure 3.7 below), and the longest range (highest energy) bump corresponds to elastic scattering. Bumps of lesser range (and so lesser energy) correspond to the excited energy states of the boron nucleus.

To Powell and Fertel's delight, the counter and emulsion peaks matched; this agreement gave them the means to calibrate the stopping power of the Ilford plate. The idea was this: By matching an easily identifiable peak in the emulsion against the same peak in a counter experiment, Powell could say that a proton that left a track of a certain number of microns in the emulsion (they actually measured at first in "eye piece divisions") corresponded to a proton penetrating a measured distance in air as determined by the counter. Since the stopping power of air was known, this correspondence told Powell the stopping power of the emulsion—some number of MeV per micron of track. With the emulsion now calibrated against the counter, Powell was ready to try his hand at reproducing the recent cloud chamber results of Bonner and Brubaker.[87]

To study neutrons, Powell and Fertel employed the same basic setup as for protons: they used 600 KeV accelerated deuterons to hit various light elements

86. Bonner and Brubaker, "Disintegration," *Phys. Rev.* 50 (1936): 308–14.

87. For details, see Powell and Fertel, "High-Velocity Neutrons" ([1939] 1972); and Powell, "Photographic Plate" ([1942] 1972).

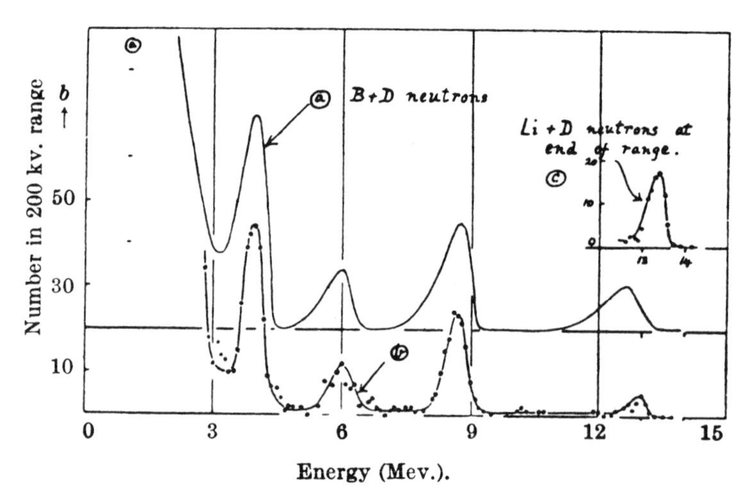

Figure 3.7 Powell and Fertel vs. Bonner and Brubaker (1939). The cloud chamber was both the model for the photographic method and its direct competition. In mid-1939, Powell and Fertel took just 3 square inches of film and in four hours of exposure had data they judged superior to anything the cloud chamber physicists could laboriously compile. Curve *a* plots the energy of a neutron produced in collisions of a deuteron (D) on boron (B); data come from measuring the recoil tracks of protons hit by the (unseen) neutrons. Caltech physicists used some 1,000 tracks to produce this curve, harvesting them from 11,000 stereoscopic pairs. In curve *b*, D + B neutrons are measured by the photographic method. Curve *c* displays the high energy end of the energy spectrum of neutrons emitted from Li + D and measured by the photographic method. This peak (Powell and Fertel argued) had not been correctly produced by the Caltech cloud chamber physicists. In short, as early as 1939, Powell was willing to correct the cloud chamber by means of emulsions. Source: Powell and Fertel, "High-Velocity Neutrons." Reprinted with permission from *Nature* 144 (1939): 152, © 1939 Macmillan Magazines Limited.

then let the emitted neutrons pass tangentially into an Ilford plate. Simple kinematics showed that if a neutron "knocked on" a proton within 5° of the neutron's line of flight, more than 99% of the neutron's energy would be conveyed to the proton. So measuring the energy of knocked-on protons with tracks along the neutron flight line would accurately yield the energy of the original neutron. The results of deuterons on boron ($^{11}B_5 + {}^2H_1 \rightarrow {}^{12}C_6 + {}^1n_0$) pleased Powell, as did a sharply delineated peak corresponding to deuterons hitting deuterons in the target:

> Tonbridge. 7 April 39. The last three weeks have been a period of almost continuous excitement. The first neutron plates to be examined were those observed with a boron target. About eighty tracks with angles within 5° of the direction of the incident neutron beam were examined. They showed a very sharp peak in energy of

> about 2½ MV, evidently from the D + D reaction. The peak was a good deal better than that obtained by Bonner and Brubaker. Instead of taking some hours and stereoscopic pairs the analysis was made in about three hours or less.

The comparison of emulsion and cloud chamber results, shown in Powell and Fertel's diagram (figure 3.7), illustrated how tracks from a several-minute emulsion exposure on 3 square centimeters of photographic film could get sharper peaks than many months of cloud chamber work. Powell continued, his confidence now at the point where he was willing to *correct* Bonner and Brubaker on the basis of the emulsion work:

> The lithium plates were then examined. The energy spectrum of these neutrons has been deduced by B[onner] + B[rubaker] by taking about 20,000 stereoscopic pairs of high pressure expansion chamber photographs from which 1500 suitably directed proton tracks were chosen. From measurements on one of our plates on 400 particles it is certain that their distribution is seriously in error, and that there is considerable fine structure in the spectrum. It is probable that the three-particle explosion does not occur and that the disintegration always proceeds in steps.[88]

Before submitting their paper for publication two months later (June 1939), Powell and Fertel eliminated their claim to a "serious" error correction but still vaunted the "considerably higher" resolving power of the chamber over a much shorter and more convenient exposure.

As far as Peter Dee (like Powell, one of C. T. R. Wilson's students) was concerned, Powell's case was solid before the neutron paper appeared in print; both Dee and another colleague "[a]ccept results, and want to use method."[89] As before, plans for the future slipped without ado between accelerator physics at various cyclotrons and cosmic ray studies at the Jungfrau, or Mt. Elmus.[90]

When the plates arrived from the mountains in late 1939, Powell turned again to the problem of processing the information: "Microscope stage fitted with large scales with verniers for accurate mapping. Champion, London and I all do same area counting, measuring and mapping all tracks. Depth gauge satisfactory. First 120 tracks completed. Very satisfactory." And a few weeks later, in mid-November 1939, "Over 500 tracks now measured. Champion + I about equal in speed."[91] Slowly, the number of observers swelled: at the end of 1939, the group purchased its fourth binocular microscope,[92] which Powell reported was "of the highest quality" since good optics were necessary to "reduce the nervous strain on the observer to a minimum."[93]

88. 7 April 1939, CPnb.
89. 7 June 1939, CPnb.
90. 7 June 1939, CPnb.
91. 27 October 1939, CPnb; 17 November 1939, CPnb.
92. 17 November 1939, CPnb.
93. Powell, "Further Applications" ([1940] 1972), 163.

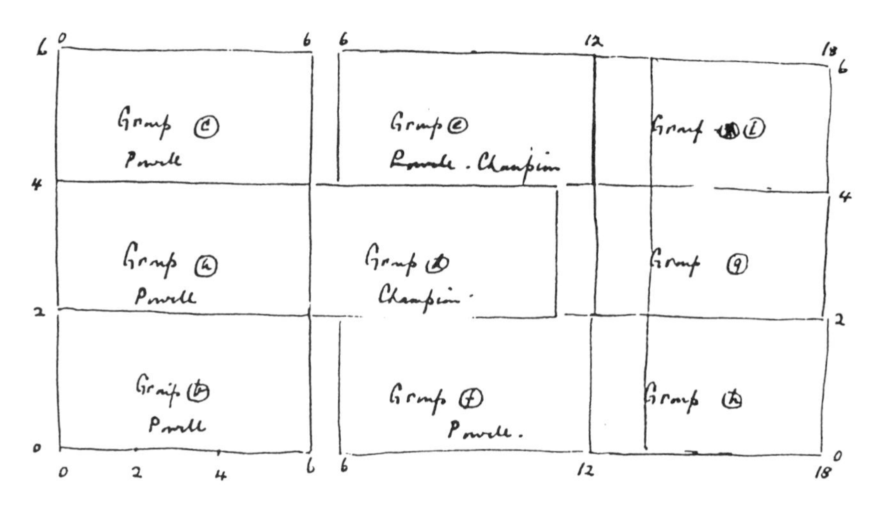

Figure 3.8 Space of the emulsion (1939). As work on the photographic method burgeoned, Powell began to subdivide the film into territories assigned to each of the team's subgroups. In a sense, the work organization of the laboratory was mapped onto the emulsion. Source: 17 November 1939, CPnb. Extracts from the Cecil Frank Powell Archives are reproduced by kind permission of the Librarian of the University of Bristol.

Tracks, microscopes, observers, and physicists all multiplied. To cope with the rush of information, Powell began to divide the emulsion into sectors, assigning each sector to a different group of scanners. Thus, the space of the film was expanded in all three dimensions: by thickening the emulsion and exploiting the focus mechanism of the microscope, the film became "equivalent" to a cloud chamber. By expanding the film in its *x-y* or lateral dimensions, and then dividing it into sections, Powell came to recreate in the emulsion the social space of its analysis; for each group held responsibility for its own sector of the plate, as illustrated in figure 3.8.

With the scanning procedure and exposure methods now in place, Powell was prepared in January 1940 to undertake an experiment not previously attempted by another group. For Powell the suspension of the psychological predisposition to concur with what was known would reinforce the legitimacy of the new method: "The first experiments were made on neutrons of which the energy spectrum had already been investigated. It was, therefore, of importance to apply the method to an investigation in which previous work could not serve as a guide able to prejudice the experimental observations. Adequate checks were applied in the previous work and it is believed that the results are free from this objection, but nevertheless an independent experiment was desirable."[94] By

94. Powell, "Further Applications" ([1940] 1972), 158.

publishing his results before the cloud chamber competition had readied its conclusions, objectivity would be guaranteed. The race was to determine the spectrum of neutrons produced from fluorine when the fluorine was bombarded by deuterons. By the invitation of Dee and J. D. Cockcroft, Powell would exploit 900 KeV deuterons from the Cambridge 1 MV generator. Using 4 square centimeters of Ilford halftone 100-micron-thick plate, he obtained 1,700 tracks in 2 hours of exposure and 60 hours of scanning. "We understand," Powell added, "that the same spectrum has been investigated by Dr. T. W. Bonner using the expansion chamber, and that his results will be published shortly. We have not seen his results, and it will thus be possible to make a completely independent check on the validity of the photographic method."[95] While this additional check added to the persuasive force of the emulsion method, the physics issues of the day lay, not with fluorine, but in the splitting of the atom.

In the aftermath of the discovery of nuclear fission, nuclear physics had acquired a new urgency. War threatened England, the Germans had advanced into Czechoslovakia, and American physics was leaping ahead with cyclotrons; Powell and his colleagues in England worried about the future of their small-scale enterprise. On 10 January 1940, Powell set down the results of his most recent discussions with James Chadwick, the discoverer of the neutron and a powerful figure in Britain's scientific establishment:

> (1) According to Walker Lawrence has the money (£500,000) to build an 100″ cyclotron. To be built in the side of a hill, with control panel 1/2 mile away. Power requirements of the order of 2500 kw.
>
> (2) Proposes to make a defined neutron beam to do tumour work. Slow neutron beam with tissues impregnated with boron.
>
> (3) Chadwick understands that nuclear physics will make revolutionary changes in the whole economy. He says that Rutherford frequently came back from London saying that somebody on this or that committee had said that it was absurd to be playing about with the nucleus, and that he ought to devote himself to classical physics. — as if nature were "merely" classical + that "quantum" physics was really a fashion which would die out.
>
> (4) Chadwick says that the basis of this is that the manufacturers want *cheap* assistants to turn out better insulators and better lavatory pans.
>
> (5) Another reason for resistance to nuclear research is that some of the older people are entrenched in classical physics and don't see why they should sweat to learn modern physics.
>
> (6) Unless the allocation of money to physical and other sciences in this country is much increased we shall begin to fall rapidly behind the developments in the U.S.A. and Russia.[96]

95. Powell, "Further Applications" ([1940] 1972), 158–59.
96. 10 January 1940, CPnb.

During early April 1940, Chadwick—by then deeply involved with the British atomic bomb project, Tube Alloy—wrote to Powell "saying that he hadn't realised we were interested in inelastic scattering and that this [was] the problem for which he wished to learn the photographic technique."[97] Chadwick's desire was not surprising; Powell was deeply involved in using the emulsion technique to determine neutron spectra, and it was the inelastic scattering of slow neutrons that induced nuclear fission in the heavy elements. Four days later, Chadwick proposed a "thorough" collaboration in the experiments. What exactly Powell knew about the bomb project at this stage is unclear. What is evident is that he knew fission was at issue, as on 9 May 1940 when he "again called [the laboratory director] A. M. T[yndall]'s attention to the importance of fission from the point of view of potential power generation and the importance of maintaining 'pure' nuclear research."[98]

During June 1940, Powell and Chadwick planned three spheres of collaboration: (1) a determination of the energy distribution of fission neutrons using thermal neutrons, (2) a determination of absolute neutron + heavy nuclei cross sections for neutrons of different energies, and (3) measurement of the absolute intensity of neutron beams using a carbon and paraffin wax absorber.[99] At this point, Bristol was designated a "protected area" and all German refugees, including Heitler, had to leave for Wells some fifteen miles south of the city. Powell and his wife cycled over to see the displaced refugees and later brought several microscopes along with drawing materials. Soon, however, "all male refugees interned. Destination unknown."[100]

Presumably, some of the women refugees in Wells were still scanning film. (As women who had become part of the analysis system, they apparently presented no security problem.) When one of the British women, a Miss Lennard, returned from her holiday, Powell immediately had an ortholux microscope sent to her door so that she could work out of her home. With Alan Nunn May—by then a steady collaborator—Powell brought two additional microscopes to T. S. Walton to start work. But just three days later, on 6 July 1940, Miss Lennard called in sick, and Powell despondently recorded that his team of trained observers had been completely dispersed.[101] This team consisted entirely of women, employed to scrutinize the emulsion sheets under the microscopes. Essentially, Powell's attitude toward them remained that of an employer, organizing an interchangeable workforce, as his friend R. L. Mercer's diary made clear when he recorded Powell's request for "three more microscopes and three girls."[102]

97. 8 April 1940, CPnb.
98. 9 May 1940, CPnb.
99. 4 June 1940, CPnb.
100. Entries marked June 1940, CPnb.
101. 3 July 1940 and 6 July 1940, CPnb.
102. Diary of R. L. Mercer, 28 May 1939, quoted in Frank and Perkins, "Powell," *Biog. Mem. F.R.S.* (1971): 541–55, on 546.

Two days after the final dispersal of his observers, May and Powell went to Tyndall to plead for special assistance in the nuclear emulsion work. Powell outlined their position on 8 July 1940:

> (1) That the solution of any specific problem is always closely bound up with general progress on the subject. The continuance of work in nuclear physics in general is therefore of importance.
>
> (2) Apart from its contribution to the solution of specific problems the photographic method is so powerful that, if properly applied, it could completely transform the present rate of progress in nuclear physics. There is a very great probability that with advances in the subject, technical advances, at present unforeseen, will become possible; advances of a character to completely change the whole basis of power production.
>
> (3) We consider it reasonable to suggest that on the basis of our experience, and with the technical resources available for nuclear investigation, we could keep a team of twenty observers occupied in the examination of plates. Such an organization would be able to carry out investigations at something like fifty times the speed possible with previous experiment techniques.
>
> (4) At the present time we are in a better position to take advantage of these technical possibilities than anybody else. This advantage will not be retained if no facilities are made available.[103]

The wartime significance of new sources of power and weaponry was not abstract, even if Powell's notes only hinted at the connection. More days than not Powell could see the Luftwaffe passing directly over the laboratory on its way to bomb South Wales. On 9 July 1940, Powell recorded that he and his associates had built "a splinter proof shelter outside the back-door."[104] Inside the laboratory, work on neutron scattering for Chadwick and Tube Alloy continued unabated;[105] then the diary switched back a few days later to the events just outside the window: "Returned from five days in Tonbridge. In Southburgh on 18th saw flight of about 130 German aircraft flying to attack London. A most formidable array which failed to penetrate the London defences." Then back again to nuclear physics on 22 August 1940, as Powell wrote to the British photographic firm Ilford, asking them to construct thicker plates so that his particle tracks would not leave the emulsion before coming to a stop.[106]

But aside from his allusions to the Blitz, including the first large-scale air attack on Bristol itself on 24 November 1940, Powell's political side comments all concern the battles of the Soviet Union: "[21 August 1941] Reported that the Russians have destroyed Dneprograd. Workers called upon to defend Leningrad. . . . [22 September 1941] Leningrad and Odessa still held. German

103. 8 July 1940, CPnb.
104. 9 July 1940, CPnb.
105. 11 August 1940, CPnb.
106. 22 August 1940, CPnb.

defeats in the Ukraine. No news of battles round Smolensk. . . . [6 December 1941] Recapture of Rostov. . . . [25 February 1942] Important Russian successes in region of Staraya Rusya." Powell's pro-Soviet sympathies were clear: there are no such intense, dense entries about Pearl Harbor or even Dunkirk.

3.4 Wartime

Powell refused to join Tube Alloy on moral grounds but cooperated with the effort in a variety of ways. May and Powell, for example, completed one major report that went into the bomb project papers, entitled "A Preliminary Report of an Investigation of the Scattering of High Energy Neutrons in Uranium Oxide and in Lead," undated but probably produced in 1941.[107] In it, the authors made use of plates exposed at the Liverpool cyclotron, where the Chadwick team had directed a collimated beam of deuterons at a target then further collimated the emerging neutrons through a cylindrical hole in a block of paraffin. When the neutrons hit a scatterer (uranium or lead), the emerging neutrons, both scattered and transmitted, were then measured by the protons they deflected as the neutrons rammed into photographic plates deployed in a circle around the scatterer. By measuring the diminution of the number of recoil tracks in the plates due to the presence of the scatterer, it was possible for Powell and May to determine the total cross section of neutron-uranium and neutron-lead scattering (see figure 3.9).

As promised to Chadwick, May and Powell delivered several properties of neutron scattering. Their main result was a confirmation of work by the American experimenter Robert Bacher,[108] in which Bacher had shown that neutrons scattered from lead at very small angles; the new results added precision to those previously published. By working outside of the circle of classified work (Powell did not have access to Manhattan District results), his photographic plate work existed in a kind of twilight secrecy. It was itself classified, yet it was located in an intellectual environment quasi-disconnected from the work at Chicago, Los Alamos, and elsewhere. Internally, there were other difficulties. May and Powell pointed to the problems they encountered with observation errors due to inconsistent photographic development associated with the depth of the emulsions themselves. Future work, they advised, would do well to sacrifice the geometric advantage of thickness and use thinner, more reliable emulsions. In addition, all plates ought to be processed together, avoiding any "idiosyncrasies" in the conditions of development.[109]

107. May and Powell, "A Preliminary Report of an Investigation of the Scattering of High Energy Neutrons in Uranium Oxide and in Lead," n.d. [probably produced in 1941], AB4/98, PRO; old ref. BR 97, PRO.

108. Bacher, "Elastic Scattering," *Phys. Rev.* 55 (1939): 679; and "Elastic Scattering," *Phys. Rev.* 57 (1940): 352.

109. May and Powell, "A Preliminary Report of an Investigation of the Scattering of High Energy Neutrons in Uranium Oxide and in Lead," n.d. [probably produced in 1941], AB4/98, PRO; old ref. BR 97, PRO.

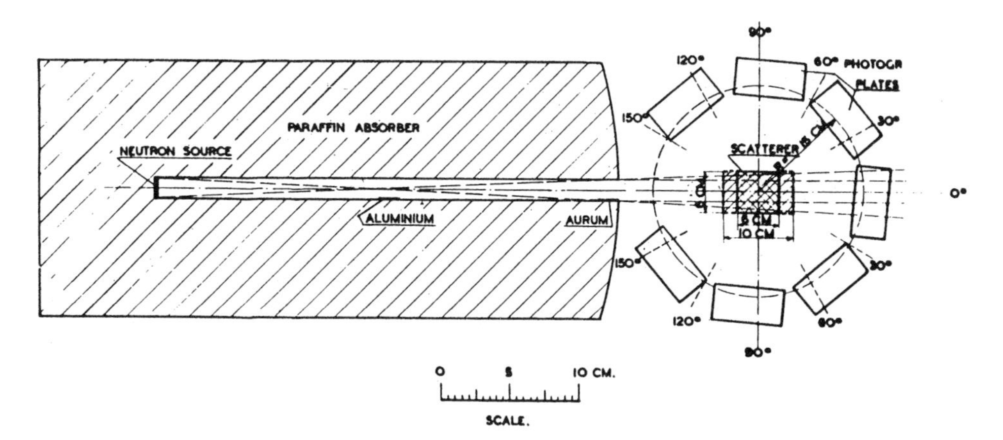

Figure 3.9 May and Powell, "uranium" (1941). Powell's wartime efforts hovered in the shadowland between the Tube Alloy (atomic bomb) project, which he refused to join officially, and "pure physics" of great interest to the Chadwick team. In the 1941 experiment depicted here, Alan Nunn May and Powell measured the total cross section of the uranium neutron, a critical quantity for bomb design and construction. Source: AB4/98, PRO; old ref. BR 97, PRO.

Other groups at the George Holt Physical Laboratories at the University of Liverpool pushed ahead with the photographic method for the duration of the war. Powell contributed by inviting Tube Alloy physicists to Bristol for training, including Joseph Rotblat, who helped lead the Liverpool emulsion effort.[110] In July 1942, the Liverpool progress report indicated considerable work on tracks produced by fission fragments and by fission neutrons; their eventual goal was to determine whether the neutrons were emitted at the instant of fission or later.[111] Frustration soon followed, as it became evident that energy spectrum results were filled with artifacts—due, in part, to the badly collimated incident beam of protons, which meant that the protons had utterly heterogeneous energies. Since the protons were not of constant energy, the observed energy spectrum of emitted neutrons told the team nothing.[112]

Other problems intervened as well in recording the emission spectrum of neutrons. Now initiated into the ways of film, Rotblat continued to study the proton tracks scattered by the emerging neutrons, and Rotblat and Chadwick

110. J. Rotblat, interview by the author, 18 September 1990.

111. Progress report for July 1942, University of Liverpool, AB1/95, PRO; old ref. 2/9/6, PRO.

112. Monthly report for August 1942, dated 9 September 1942, University of Liverpool, AB1/95, PRO; old ref. 2/9/6, PRO.

launched further studies along the same lines by supplying similarly exposed plates to Powell at Bristol for confirmation. Yet other plates went to D. L. Livesey and B. B. Kinsey at Cambridge, again to check the results that were still not stable. Reporting in July 1944, the Cambridge duo invoked the cross-check as essential to their method: "In order to minimise the effect of personal idiosyncrasies on these measurements, several observers have taken part in this work, and we give here a summary of their results [comparing Cambridge and Bristol measurements]."[113] One difficulty became immediately apparent: the energy distribution based on low energy tracks (those below 1.5 MeV) did not agree among the different laboratories. At fault was the dependence of the identifiability of the low energy tracks on the grain density in the plate; how each observer set the lower limit of observability amid the fog was to a certain extent "the choice of the observer." In other measurements (concerning the scattering of neutrons), the vastness of the microscopic domain simply became unbearable: "It is possible that this [neutrons scattered from neighboring objects] is due to tracks being missed owing to the tedium of searching large areas of plate."[114] Once again, no conclusive results were available on the question of a possible low energy peak in the neutron spectrum, as both the report and cover letter to Chadwick made clear: "The trouble is that this maximum, if it occurs, lies too close to the lower limit of detection by photographic plate. I am doubtful myself whether more measurements would now improve the accuracy of the neutron spectrum determination in this region."[115] The subjective variations in the scanning process undercut the stability of these essential results; without a way to secure greater robustness, the emulsion method was useless.

Determined to find the energy of fission neutrons, Rotblat, Thomas Pickavance, S. Rowlands, J. R. Holt, and Chadwick undertook a systematic effort to plot the spectrum. To produce the fission neutrons, the Liverpool team used a beam of 2.84 MeV protons to hit a target of LiOH on a molybdenum sheet. This process produced primary neutrons of energy just under 1.2 MeV. After passing through a block of paraffin wax to slow the neutrons (slow neutrons are better at causing fission), the neutrons hit a 2.9-kilogram block of U_2O_5. It is this last collision that causes fission and the emission of the fission neutrons. Rotblat's team then attempted to measure the energy of the fission neutrons using three different methods: measurements of proton recoil tracks in photographic plates, comparisons of the absorption of fission neutrons in water and thermal neutrons produced in lithium by protons, and finally, studies of the absorption of fission

113. Livesey and Kinsey, "Some Further Measurements of Energy Spectrum," 27 July 1944, AB4/501, PRO; old ref. BR 471, 2/9/6, PRO.

114. Livesey and Kinsey, "Some Further Measurements of Energy Spectrum," 27 July 1944, AB4/501, PRO; old ref. BR 471, 2/9/6, PRO.

115. Livesey and Kinsey, "Some Further Measurements of Energy Spectrum," 27 July 1944, AB4/501, PRO; Kinsey to Chadwick, 3 August 1944, AB4/501, PRO.

neutrons in paraffin. The emulsion method was a direct measure of the energy of individual neutrons. The second method involved the insertion of a series of manganese detectors into a water tank, each at a different distance from the neutron source. Rotblat and his colleagues then compared the shape of the absorption curve produced by thermal neutrons (those emerging from the lithium target under proton bombardment) to fission neutrons. In the third method, the team again measured the average energy of the fission neutrons, this time by placing a Geiger-Müller counter on the far side of an absorber from the source, at any given absorber thickness. The number of recoil protons picked up by the counter provided a measure of the fraction of neutrons energetic enough to get at least that far; increasing the absorber thickness allowed only the most energetic neutrons to penetrate. Rotblat himself undertook the photographic plate work and, after painstakingly sorting out some 242 proton recoil tracks, produced an energy spectrum from 0.7 to 7.3 MeV, with an average fission neutron carrying about 2.1 MeV, a result confirmed by the water method (average energy, 2 MeV) and the paraffin method (2.5 MeV). Finally, the collaboration hoped to put to rest a suspicion voiced several times over the preceding months that there was a great abundance of very low energy neutrons (in the range of a few KeV to a few hundred KeV). Were this the case, the insertion of even a very thin slab of paraffin, on the order of 0.5 centimeters, should have significantly depleted the number of counts on the Geiger-Müller counter. There was no such drop in counts. Photographic plates confirmed the results: there was no low energy peak.[116]

From the energy spectrum studies, the Liverpool group members turned to other problems. Through the remainder of the war, they exploited the emulsion technique to detect fission products and neutrons—adding periodic improvements in technique, including the soaking of emulsions in uranyl salts (salts of uranium, mercury, and other metals intensified the track images and so made scanning with low-power microscopes faster)—and the registration of neutrons emitted following the nuclear capture of thermal neutrons.[117]

By the end of the war, the photographic plate had moved from the novel, not quite reliable, instrument of 1938 to a promising new source of information about many aspects of nuclear physics. Nuclear excitation levels no longer had to be confirmed with older means, as was evident from the report that Livesey

116. Rotblat, J., T. G. Pickavance, S. Rowlands, J. R. Holt, and J. Chadwick, "The Energy Spectrum of the Fission Neutrons produced in Uranium by Thermal Neutrons," n.d., AB4/48, PRO; old ref. BR 47, PRO.

117. George Holt Physics Laboratory, University of Liverpool: monthly reports for January 1944, April 1944, June 1944, AB1/95, PRO; progress reports, Liverpool University (Professor Chadwick), 2/9/6, PRO. See also Feather, "Report on Work Carried out for the Directorate of Tube Alloys Research of the D.S.I.R.," June 1944, AB1/93, in progress reports, Cambridge University (Feather and Bretscher), DTA 2/9/4, PRO, where Feather reported that satisfactory fission tracks had finally been produced in photographic emulsion, and that the D.S.I.R. had succeeded in impregnating uranyl nitrate solution into the emulsion. On the intensification of tracks by metal salts, see Powell et al., "New Photographic Emulsion" ([1946] 1972), 212–13.

and D. H. Wilkinson completed at Cambridge in May 1945. For the first time, they reported, the emulsion method could be used to determine accurately the complete spread of neutron energies and angles from the DD reaction (a beam of deuterons accelerated by the Cambridge High Tension set incident on a deuteron target). Even in the low energy range 2–4 MeV, the emulsion method performed well.[118] While Livesey and Wilkinson conceded that "the best way for attacking these . . . problems simultaneously is probably the cloud chamber method," they agreed with Powell's conclusion of some years earlier that the cloud chamber investigation "would be most excessively tedious." Beyond mere tedium, the cloud chamber presented too sparse a target to be able to give an accurate determination of the total flux of neutrons. Furthermore, aside from confirming its "well-known advantages and labour saving properties" in answering the particular set of questions about DD reactions, the authors also responded affirmatively to the question of whether the emulsion method was capable of precise measurements of neutron energies. Indeed, the authors had enough confidence to challenge the applicability of the Bohr compound nucleus model to ^{4}He, and to throw grave doubt on resonances previously claimed for the reactions they had examined. The one place where Livesey and Wilkinson endorsed the Bohr theory was in the low energy region; emulsions provided evidence for the "partial correctness" of the Bohr account.[119]

While no one could claim that the photographic emulsion played a central role in the development of the atomic bomb, the method had emerged as a significant means of studying the complexities of neutron dynamics in uranium collisions. By enlisting Rotblat, Chadwick, Pickavance, and many others, the circle of emulsion cognoscenti had widened considerably, and the government-created Photograph Emulsion Panel had tied together the physics community, the burgeoning atomic energy establishment, and the film industry in a mutually supportive chain. The emulsion method had proved itself in physics applications that would have been highly awkward (if possible at all) using the more traditional methods of the Wilson chamber or the Geiger-Müller counter. Finally, in the months following Hiroshima and Nagasaki, association with the Allied development of nuclear weapons hurt no one's influence in the postwar scientific reconstruction.

Back at the Institut für Radiumforschung, wartime research progressed rather differently. After Blau's forced departure, Wambacher continued to hunt high energy nuclear stars in cosmic-ray-exposed emulsions, especially by using the method of grain counting. (Range was not helpful given that the high energy protons she observed exited the film.) Most of her work proceeded in a fashion

118. Livesey and Wilkinson, "A Photographic Plate Study of Neutrons from the DD Reaction," AB4/681, PRO, 1–19; old ref. BR 638, May 1945, PRO.

119. Livesey and Wilkinson, "A Photographic Plate Study of Neutrons from the DD Reaction," AB4/681, PRO, 1–19 on 18; old ref. BR 638, May 1945, PRO.

largely unconnected with contemporary issues in nuclear physics. With great care she composed a 54-page paper in which she plotted, inter alia, the number of nuclear stars produced per day per unit area as a function of the emulsion's height in the atmosphere; she also counted the number of particles found per star, measured the lengths of tracks emerging from the stars, and identified the highest energy found deposited in a star.[120] Work that Wambacher did in 1943 with Stetter was more closely tied to contemporary efforts in cosmic ray physics: together they examined the height variation of the number of nuclear stars in terms of the penetrating radiation (muons) and the nonpenetrating radiation (electron-photon showers) sorted out in the late 1930s by Anderson, S. H. Neddermeyer, Street, and Stevenson.[121] Ortner too used the war years to follow up on nuclear-emulsion-based exploration of cosmic rays. After strategically deleting Blau's name from any footnotes, Ortner went on to compare the cosmic ray penetration of nuclei with Heisenberg's 1937 work.[122]

The contrast with the Bristol-Liverpool-Cambridge Tube Alloy project is stark: Where Powell had systematically calibrated the emulsions against existing instruments (both cloud chambers and counters) and then turned the new tool toward the detailed analysis of the excited states of accelerator-produced as well as cosmic ray nuclei, emulsions in Austria remained an unsystematic appendage of cosmic ray research. Ortner and Stetter moved back and forth between instruments, never gaining full control over the ever-variable films. Whereas Powell guided the emulsion to a strategically important intersection point between nuclear weapons development, the promise of power generation, the photographic industry, and the cyclotron laboratory, Wambacher, Stetter, Ortner, and Kirsch left the technique more or less as it was at the time of the Anschluss. In some abstract and counterfactual sense, the Institut für Radiumforschung could have been the world center for emulsion research; in the real world, it never came close.

The end of the war had predictable consequences for Stetter, Wambacher, Ortner, and Kirsch. All four of the emulsion physicists, members of the Nazi party, were booted unceremoniously from office according to paragraph 10 of the Verbotgesetz. Stetter's 1933 NSDAP membership cost him his professorship on 4 August 1945, retroactive to 6 June.[123] Now came some powerful hermeneutic work. Whereas in his May 1938 questionnaire, Stetter testified that he had

120. Wambacher, "Kernzertrümmerung," *Sitzungsb., Akad. Wiss. Wien, Math.-naturwiss. Kl., Abt. IIa* 149 (1940): 157–211; cf. Wambacher and Widhalm, "Kurzen Bahnspuren," *Sitzungsb., Akad. Wiss. Wien, Math.-naturwiss. Kl., Abt. IIa* 152 (1943): 173–91.

121. Stetter and Wambacher, "Absorption," *Sitzungsb., Akad. Wiss. Wien, Math.-naturwiss. Kl., Abt. IIa* 152 (1943): 1–6. On the physics leading to the discovery of the muon, and the sorting out of shower phenomena, see Galison, *Experiments* (1987), esp. chap. 3.

122. Ortner, "Höhenstrahlung," *Sitzungsb., Akad. Wiss. Wien., Math.-naturwiss. Kl., Abt. IIa* 149 (1940): 259–67.

123. Draft of SVUEK to Stetter et al., 4 August 1945, fols. 250–51, Personalakte Stetter, BmUP, AdR.

been a party member for years before the Anschluss, now events were reinterpreted: "The written application for membership in the Party was filled out by me approximately May 1938 and only considerably later . . . answered with the issuing of the provisional member's card." And again, "My management of my office was absolutely correct and my political attitude and bearing, already quite moderate in 1938, completely changed in the subsequent years."[124] More elaborate explanation then went to the Philosophical Faculty dean in October 1945 in a form worth reprinting:

> I was urged to join the NSDAP already before 1933, but I turned this down, since as a scientist I wanted nothing to do with party politics. Now, one evening in June 1933, I met an acquaintance, Dr. Holoubek, who had done scientific work one or two years previously in the Physical Institute. He again encouraged me to join, and I finally gave my agreement. . . . One generally hoped then for an improvement in the difficult economic situation of certain groups in the population through the intervention of the NSDAP. I also gave the aforementioned [person] a modest monetary contribution (10 or 20 schillings). Soon thereafter came the ban of the Party, and Dr. H. left Vienna, so that in the following years I did not hear anything more from him. This verbal agreement did not lead to any Party activity, so that there already one cannot speak of an illegal Party membership.[125]

Moreover, Stetter insisted, he had tried to defend Meyer and H. Thirring and criticized the Hitler Youth and NS Students League for depriving students of real work. Surely, he reckoned, there was a "Jewish style" in physics, but it is "laughable" to dismiss results simply because of this. While on his earlier report, Stetter had advertised his propaganda for the Nazi cause, now he let it be known that he never made political appeals at the institute. Not only that—he had let the statue of the Kaiser, Franz-Josef, stand despite orders to the contrary. "Leaving the Party would have had my dismissal as a result and would have helped no one more than a perhaps radical successor."[126] In an addendum of April 1946, Stetter pointed out his valuable service in guarding the institute as he dealt with the importuning American occupiers.[127]

Slowly the wheels of the Bureau of Truth turned. In late 1947, the Registration Agency promoted Stetter to the category of minor infractions, the so-called *minderbelasteten* persons; Stetter used the occasion to have his dismissal

124. Stetter to Ebert, plenipotentiary of Vienna universities, 25 September 1945, fol. 212, Personalakte Stetter, BmUP, AdR.

125. Stetter to dean, Phil. Fak., 17 October 1945, fols. 213–14; essentially the same letter, with an addendum, is Stetter to BmU via rector, Univ. Vienna, 10 April 1946, fol. 207, copy fol. 221, Personalakte Stetter, BmUP, AdR.

126. Enclosure in Stetter to dean, Phil. Fak., 17 October 1945, fols. 215–17, quote on fol. 217; essentially the same, with an addendum, is Stetter to BmU via rector, Univ. Vienna, 10 April 1946, fols. 208–10, copy fols. 222–24, Personalakte Stetter, BmUP, AdR.

127. Enclosure in Stetter to BmU via rector, Univ. Vienna, 10 April 1946, fols. 208–10 on fol. 210; copy fols. 222–24, Personalakte Stetter, BmUP, AdR.

reviewed and to restore his pension.[128] Next step: the District Captaincy (Bezirkshauptmannschaft) of Zell am See removed Stetter from the registry of Nazis with the explanation that Stetter's application for membership in 1933 had not been acted upon because the party was banned; since his temporary party card of 1938, number 6,105,101, which he claimed never to have received expired in 1940, the expiration itself could be judged as "voluntary departure [from the party] in 1940."[129] Having been promoted to a *minderbelastete* person, then removed from the Nazi roll altogether, Stetter could, in 1953, finally advance to a professorship, which he did with the aid of the Federal Education Ministry. Some bits of the past popped up now and again—one article in *Informationen für alle* on 9 July 1954 reminded its readers that Stetter had been somewhat "alienating" during his directorship of the institute—for example, "he declined to take over the room and furniture used by his deceased predecessor, because this man was a Jew."[130] By 1962, Ortner was nominating Stetter for honors.

Sparing the reader the details, Ortner too was dismissed and went on to challenge the dismissal, on the grounds that he had not really understood where his dues were going, and like Stetter, he protested over the technical consistency of his party memberships.[131] Rehabilitated, Ortner was lifted back through the ranks of the Austrian system. Evidently, the U.S. State Department was not thrilled with this development, especially (and confidentially) because, the State Department reported, Ortner had apparently been "used by the Russians for a time in Moscow. Then [the Austrian file continues] he was active for some years in Cairo, which poses certain riddles [for] the Americans."[132] American misgivings about a Nazi nuclear physicist's consulting activities in Moscow and Cairo

128. Stetter to LEDRRÖ via dean, Phil. Fak., 23 December 1947, fol. 203. He was then retroactively installed to the civil service as of 18 February 1947 but not reappointed to the Austrian universities; see LEDRRÖ to Stetter, 6 April 1948, fol. 202, Personalakte Stetter, BmUP, AdR.

129. Copy of decision by Bezirkshauptmannschaft, Zell am See, 18 October 1950, fol. 183, Personalakte Stetter, BmUP, AdR.

130. Chancellor's Office to Ernest Kolb, Bundesminister, BmU, 16 December 1952, fols. 127–28; official appointment made on 27 April 1953, Kolb, BmU, to Stetter, 6 May 1953, fol. 387v. Copy of article in *Informationen für alle,* no. 25, 9 July 1954, fol. 98, Personalakte Stetter, BmUP, AdR.

131. Ortner's plea: "Already several years before 1938 I gave monetary contributions to a friend from the profession upon his request, which [contributions], according to his statements, were to be channeled to members of the group of teachers who, in consequence of their political position, had been penalized and thereby, along with their families, come into need. I gave 10 sch. monthly during the period from 1934 to 1938, in any event not regularly, since according to my memory at times no contributions were asked from me. I can see no political activity therein, rather only a charitable action. A connection with the NSDAP or one of its organizations did not, in any case, exist for my part. First after the Umbruch in March 1938 was I informed that these contributions . . . were channeled to the NS Teachers League. . . . Only after March 1938 did I apply for Party membership and my member's card gives entry into the Party in May 1938" (Ortner to SVUEK, 13 August 1945, fols. 79 and 79v, Personalakte Ortner, BmUP, AdR). The Staatspolizei thought differently, saying that Ortner knew perfectly well what was happening to his money and that his story about the dates of his membership was not credible; Dürmeyer, Staatspolizei to SVUEK with report, 15 October 1945, fols. 82–84, Personalakte Ortner, BmUP, AdR.

132. Matsch, Bka-AA, to BmU, 9 December 1955, fol. 451, Personalakte Ortner, BmUP, AdR. The BmU then asked Matsch to get ambassador Gruber to facilitate the granting of the visa that Ortner needed. The Bka-AA

notwithstanding, Ortner ascended into the international atomic energy establishment, visiting sites at Harwell, Argonne, Battelle Memorial Laboratory, Brookhaven, Atomics International, Berkeley, and General Atomics, as a sample. By 1960, Ortner was head of a new institute and an *Ordinarius* Professor of Technical Physics at the Technische Hochschule.[133] Under the Verbotgesetz, Kirsch was dismissed as of 8 June 1945; on 23 July 1947 the Bundesminister commuted that ruling to a retirement.[134]

Not surprisingly, perhaps, the only real casualty in the rehabilitation of the Institut für Radiumforschung's Nazi circle was its least powerful member—and only woman—Hertha Wambacher. When the war ended she was 42 years old; thrown out of her post and forbidden from teaching, Wambacher published only one article over the next five years ("Microscopy and Nuclear Physics").[135] She died in 1950 without any significant recognition, her brief obituaries penned by Stetter, Thirring, and Ortner. Stetter and Thirring concluded: "A tragic fate led this pair of women [Wambacher and Blau], blessed with a true spirit of research, to be thrown from their life's course through the weight of external circumstances in our unstable time. The fruits of their investigations and discoveries are now left for other, more fortunate followers."[136] Of these "fortunate followers," none—as Stetter and Thirring readily acknowledged—were as significant as Powell in Bristol and his collaborators at Ilford and Kodak.[137]

3.5 Kodak, Ilford, and the Photographic Emulsion Panel

Without examining their role in the atomic bomb project, it is impossible to understand the resources suddenly available to the British emulsion physicists just

agreed to do this; to Gruber they wrote "that everything must be undertaken to prevent that this matter, so important for Austria, is not scotched by malice" (Personalakte Ortner, BmUP, AdR).

133. On Ortner's rehabilitation, see 11 December 1954 Phil. Fak. vote (48 yes, 1 no, 2 abstentions) to renew Ortner's *venia legendi,* dean, Phil. Fak., to BmU, 1 February 1955, fol. 406. The report points out that the law of 30 July 1949 suffices to give amnesty to Ortner, BmU evaluation, n.d. (1955), fols. 405, 409, on 409, Personalakte Ortner, BmUP, AdR. Austria wanted to participate in the new UN International Atomic Energy Commission. A commission recommended that Ortner serve as a general consultant on atomic energy, especially on reactors, for the BmU, and as representative to a "Reactor Training School" at Argonne Laboratory; see B. Karlik, IfR, to F. Hoyer, BmU, 4 May 1955, fols. 396–97, enclosed commission report, 28 April 1955, fol. 398, Personalakte Ortner, BmUP, AdR. Ortner himself was keen to continue the research on neutrons that he and his colleagues had begun before 1945—work on electron amplifier tube measurements of charged particles, radioactive indicator techniques, and the photographic method of measuring corpuscular radiation. "The activity of the Vienna group was recognized by the then government [the Nazi government] by the foundation of a new organization under the title 'Four Year-Plan Institute for Neutron Research'" (the four-year plan being under Goering's direction; Ortner to Adalbert Meznik, BmU, 26 August 1955, fols. 415–16, on 415v, Personalakte Ortner, BmUP, AdR).

134. Kirsch's dismissal, Skrbensky, SVUEK, to dean, 4 August 1945; dismissal commuted to retirement, Hurdes, Bundesminister, BmU, to Kirsch, 23 July 1947, Personalakte Kirsch, BmUP, AdR.

135. Wambacher, "Mikroskopie und Kernphysik," *Mikroskopie* 4 (1949): 92–110.

136. Stetter and Thirring, "Hertha Wambacher," *Act. Phys. Austriaca* 4 (1950): 318–20; see also Ortner, "Wambacher," *Nature* 166 (1950): 135.

137. Stetter and Thirring also point to Occhialini's laboratory in Brussels: Stetter and Thirring, "Hertha Wambacher," *Act. Phys. Austriaca* 4 (1950): 318–20; see also Ortner, "Wambacher," *Nature* 166 (1950): 135.

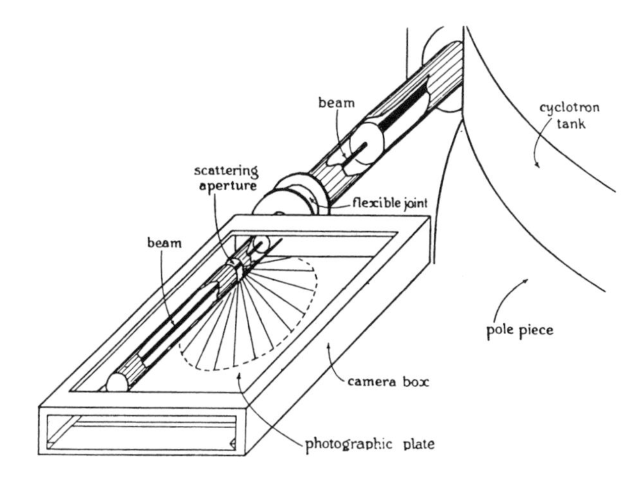

Figure 3.10 Powell's "camera" (1944). While Powell and his group relied heavily on cosmic rays as their source of particles, the cyclotron was vital for the war effort. This perspective drawing shows disposition of the components of the "camera" used to record the passage of particles scattered from the target. Source: Chadwick et al., "Investigation" ([1944] 1972), 173.

after the war. Hiroshima guaranteed that emulsions—and the laboratory practices that accompanied them—would not be forgotten in the new environment of cyclotrons and atomic power. Even before hostilities ceased, Powell had routinized the procedure for photographing particles in the cyclotron (see figure 3.10). Indeed, by January 1945, he was already pressing Ilford to develop new emulsions, a request they deferred (at least publicly) until the end of the war.[138] But, when the war did end, a panel was established by the Cabinet Advisory Committee on Atomic Energy (Powell was on a powerful scientific subcommittee of the Ministry of Supply) to encourage the development of new and more sensitive film. Representing the universities, Powell served on this Photographic Emulsion Panel, as did Livesey (still at Cambridge), Donald Perkins (from Imperial College London), and May (King's College, London University). Unlike Powell, May had immersed himself in the atomic bomb project, first in England, and then, from 1943 through 1945, in Canada, where he was privy to manufacturing and design decisions. Sometime in 1943, he began spying for the Soviet Union, and among the materials he shipped to Moscow were samples of enriched U^{235}. May's name was handed to the Canadian authorities by the Russian cipher clerk Igor Gouzenko. Under watch from the moment of his return to England on

138. 28 January 1945, CPnb.

16 September 1945,[139] May kept up appearances as best he could, continuing to participate both in his laboratory work and on the panel. Rotblat, having emerged from the war with new bona fides, was asked to head the panel, while L. D. Chelton held up Ilford's end of the collaboration (Kodak joined later).

The Emulsion Panel faced three great problems: fog from gamma rays, variation in the number of grains per track, and fatigue associated with scanning under the microscope. "Most of these disadvantages," minutes of the first (21 November 1945) meeting concluded, could be countered "by developing emulsions with a finer grain and smaller spacing. If the spacing were made small enough, the effect of straggling would be considerably reduced, and also the personal strain largely lifted." This question of "personal strain" was of crucial importance since all of the work proposed required thousands of tracks to be scanned. Along these lines, the report continued, "[t]he removal of background fog by the use of desensitizers and the elimination of γ-rays fog by making the emulsions insensitive to γ-rays would eliminate the errors due to counting fog grains as part of the track, and it would greatly reduce the skill required by the observers."[140] Indeed, when Ilford succeeded in producing a more sensitive emulsion under the panel's contract in 1945, Powell and his coauthors reiterated the social necessity of eliminating background grains. Otherwise, he warned, "it is sometimes difficult to decide which [grain] is the beginning of a proton track, especially when it has a large angle of dip. This imposes a nervous strain on the observer which increases the difficulty of the measurements and prevents the employment of unskilled personnel."[141] Background grains and the unskilled women scanners were incompatible.

At its first meeting, the panel recognized that universities were ill equipped to develop the new emulsions themselves; the work required "facilities not available normally in a University Laboratory," and it was therefore "more expedient to entrust this work to experts on photographic emulsions." Even processing the plates struck the physicists as outside their bailiwick, and this too they turned over to Ilford. Before doing so, however, the physicists had to coordinate their efforts with those of the emulsion makers by lending Ilford some 50 millicuries of polonium so that its technicians could investigate the differences in tracks between protons (which the polonium emitted) and common alpha emitters. Using the loaned equipment, and microscopes that they would be authorized to purchase by the Ministry of Supply, Ilford would be under contract to start looking for an emulsion with small grains, one that was insensitive to light and gamma rays. The firm would aim to produce emulsions of sensitivity to various substances

139. See, e.g., Hyde, *Atom Bomb Spies* (1980), esp. 27–69, date of arrival in England on 27.

140. "Photographic Emulsion Panel," GRP.

141. Powell et al., "New Photographic Emulsion" ([1946] 1972), 209.

and to find new ways to load test substances into the emulsion, as well as new techniques to develop the film.[142]

Rotblat remembers May, Powell's wartime collaborator, acting "peculiarly" at the second panel meeting; behavior perhaps understandable in light of the close—and tightening—surveillance May had been under for some months.[143] On 4 March 1946, officers of the Civil Secret Service and the British Military Secret Service seized May in his King's College laboratory, charging him with espionage.[144] A few days later, he confessed: "I gave . . . very careful consideration to [the] correctness of making sure that development of atomic energy was not confined to [the] U.S.A. I took the very painful decision that it was necessary to convey general information on atomic energy [to the Soviet Union] and make sure it was taken seriously."[145] Powell apparently destroyed most of his papers—if he reacted strongly to May's arrest, I have found no trace of it.

Minus one member, the panel continued to press for new emulsions. Ilford delivered the goods, at least in part, by the third panel meeting of 7 May 1946. By then, the photographic firm had in hand a set of new emulsions in four grain sizes (A, B, C, and D), each available in three sensitivities, 1, 2, and 3. Importantly, Ilford's chemists had increased to 8:1 the ratio of silver bromide to gelatin, improving sensitivity strikingly.[146] Powell and his collaborator Occhialini waxed enthusiastic about the new film as soon as it arrived. First, it cut the background due to gamma rays. Second, by eliminating much of the uncertainty about the endpoints of tracks, the new film significantly narrowed the "wings" of the bumps in spectra such as those in figure 3.7; Powell and Occhialini therefore could claim an increase in the power of the method to separate nuclear excitations of closely spaced energy levels.[147]

But the culmination of the Emulsion Panel's work came in March 1948, when Kodak, assiduously guarding the secret of its brew, announced the production of its NT4 emulsion, a film so sensitive that it could exhibit the tracks of *any* charged particle, including the elusive trail of a highly relativistic electron (see figure 3.11).[148] Ilford too had a new electron-sensitive film, G5, which it presented to the panel a few months later.[149]

142. "Report on Meeting of the Photographic Emulsion Panel, 21 November 1945," GRP.

143. J. Rotblat, interview by the author, 18 September 1990.

144. *New York Times,* 5 March 1946, 2.

145. See, e.g., Hyde, *Atom Bomb Spies* (1980), 78.

146. See Rochester and Butler, "Nuclear Emulsions," *J. Phys.* 43 (1982): C8-89–C8-90.

147. Powell et al., "New Photographic Emulsion" ([1946] 1972), 209–10.

148. Berriman, "Electron Tracks," *Nature* 161 (1948): 432; Herz, "Electron Tracks," *Nature* 161 (1948): 928–29; and most conclusive, Berriman, "Recording," *Nature* 162 (1948): 992–93.

149. "It was found possible to increase the grain size of the G.5 emulsion and, by close attention to all stages of its manufacture, to increase its sensitivity whilst keeping background density within an acceptable limit. Sensitivity to particles at minimum ionization has thus been achieved and several large batches of the new material

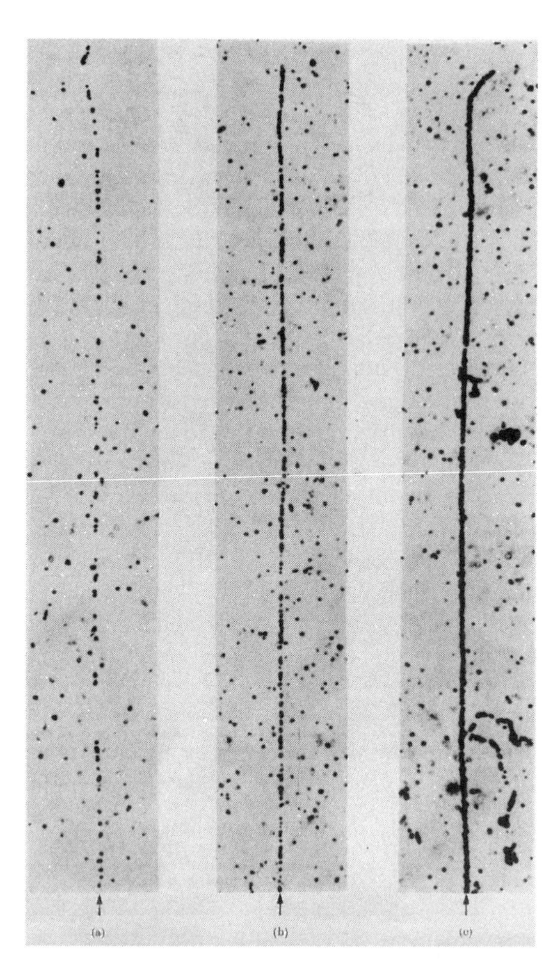

Figure 3.11 Track endings in old and new films (1939–49). Track *a* is recorded in an Ilford halftone plate produced in 1939 for general photography. Track *b* is the Ilford C2 emulsion, made by Glassett and Waller in 1946; it was able to record passing particles if they had an ionization at least four times the minimum (i.e., four times the minimum ionization produced by a fast, light particle with the elementary charge, *e*). Track *c* is the Kodak NT4 of 1949, the first to be sensitive enough to record the minimum ionizing track of an electron and so produce a record of all particles. Source: Powell, Fowler, and Perkins, *The Photographic Method* (1959), pl. 1–13 on 31.

Despite the welcome these emulsions received in the physics community, coordination with an industrial firm was a complicated affair. All information from the scientific side was to flow toward Ilford; in particular, information from Pierre Demers,[150] a Canadian emulsion expert who had worked on the bomb project, would be handed over to Ilford. Ilford, by contrast, was a private company and under no obligation to reciprocate. Indeed, as the physicists recognized

have been supplied to nuclear physicists. The emulsion is still not completely reproducible and the causes of this need to be discovered and eliminated" (Report of work carried out under Ministry of Supply contract 6/atomic/2/CF9A (Con.2A3)287/EMR/1411, 2 September 1949, Ilford Research Laboratory CW/JEP, GRP).

150. See Demers's important paper "Improved Tracks," *Phys. Rev.* 70 (1946): 86, which built on the work of Zhdanov (Jdanoff, "Traces des Particules," *J. Phys.* 6 [1935]: 233–41).

from the outset, "A clause stipulating secrecy would probably have to be imposed with regard to unpublished work."[151] A few years later, when Demers published the most detailed public statement of the methods of emulsion making, he began his discussion with explicit mention of the fog that shrouded everything to do with rapid and sensitive emulsions: "The fabrication of the fastest negative emulsions is surrounded by the greatest secrecy, a fact which is explained simultaneously by industrial necessity and by the nature itself of the techniques employed. . . . These procedures are empirical, they hide tricks of the trade and craft recipes (*des tours de main et des recettes de métier*)."[152]

Since L. D. Chelton from Ilford was present at panel meetings, it seems reasonable to assume that this silence about process was a sine qua non of his firm's involvement. Emulsion making is a very secretive business; even forty years after the war, wartime emulsion fabrication techniques were still considered to have proprietary value, and Charles Waller, one of the leading emulsion experts from Ilford, remained hesitant to talk in detail about the production techniques used in those early years.[153]

In 1987, however, Waller did receive permission from Ilford to point to British patent 580,504, which for decades had been on the public record but had received scant attention. Previously, to separate silver halide from the by-products of its formation, chemists had used two principal methods. In one, the silver halide in gelatin is centrifuged—an awkward, expensive procedure that often left unusable clumps of silver halide. More common was the method of setting the gelatin suspension by chilling, shredding the resulting mass, and washing it with water. Waller and Duncan Pax Woosley's secret improvement, a chemical means of concentrating the silver halide, was first provisionally specified on 3 July 1944 and then issued as a patent on 29 May 1945. Essentially, their idea was to add a "detergent," a surface-active sodium alkyl sulphate or sodium alkylaryl sulphonate. This anion soap precipitates the gelatin in a soft cohesive layer, from which the liquid phase can be decanted—in other words, the concentrated silver halide is left on the bottom of the vessel as a sediment, and the water, along with various salts, can simply be poured away. As the patent reads: "The invention thus affords a means for forming a silver halide suspension in the protein vehicle freed from unwanted salts or other substances which may have been present in the original suspension."[154]

Subsequent emulsion fabrication steps can then proceed by using the more concentrated silver halide and gelatin mixture. By 1945, Ilford had adopted the

151. "Report on Meeting of the Photographic Emulsion Panel, 21 November 1945," GRP.

152. Demers, *Ionographie* (1958), 29.

153. C. Waller, interview by the author, 5 September 1985.

154. Cecil Waller and Duncan Pax Woosley, Complete Specification, patent 580,504: "Improvements in or relating to the Production of Photographic Silver Halide Emulsions," Patent Office (London: Courier Press for His Majesty's Stationary Office, 1946), 6. See also the discussion in Waller, "British Patent 580,504" (1988).

Waller and Woosley concentration technique for most film emulsions. When Powell contacted Ilford in the autumn of 1945, Waller exploited the new process to create a concentrated version of the old halftone emulsion, now dubbed C2. Ilford had these plates in Powell and Occhialini's hands by November 1945, and they went immediately into field use.[155]

The exchange of goods and specifications between physicists and industrial chemists did not, however, mean that the two groups were working as a single homogeneous group. Quite the contrary. Even *after* Waller and his colleagues at Ilford had created the C2 emulsion, the chemists were thoroughly unfamiliar with the basic physical concepts of "electron sensitivity" and "minimum ionization." Once they learned that the physicists needed emulsions that could detect relativistic electrons, though, the chemists set to work on two fronts: to increase the grain size of the silver halide from 0.15 to 0.3 microns and to boost its sensitivity by inserting traces of gold and labile sulphur. Producing the concentrated, sensitized emulsion was more than the research laboratory could handle, and the eventual product, emulsion G5, came from a joint effort of the large-scale production facilities of Ilford's Plate Factory and the smaller scale commitment of the research branch.[156] For their part, the physicists were in no position to reproduce, much less innovate in, the altogether distant realm of colloidal chemistry, anion soaps, and ph-sensitive gelatinous media.

Without the resources of Kodak or Ilford, individual nuclear physicists and chemists had a hard row to hoe if they wanted to make their own instruments. One of the few who mastered the emulsion chemistry and did experiments with the resulting film was Demers, working with only a handful of assistants at the University of Montreal. Apparently, Demers perfected some of his remarkable emulsions as early as 1945, films comparable in quality to those fabricated at the big corporations. Only after the military security of the atomic bomb project was relaxed could he publish, revealing a method that employed two jets, one of silver nitrite and a second of potassium bromide. Details matter in this business, and Demers specified how the precipitated emulsion is forced down to the bottom of a glass bottle, whence "it comes down as long noodles or shreds which are washed thoroughly in cold running water. . . . Washing lasts two to four hours in two stainless steel pots of 5 to 7.5 [liters]. . . . A good rule is as follows. A noodle is melted between closed hands. Then the hands are rubbed, half opened, and are smelled. A very small trace of alcohol is then easily detected. When the smell of alcohol disappears, washing is usually halfway or two thirds through."[157] In Demers's hands, the emulsions were an enormous asset, and he worked productively not only in the design and fabrication of film but in its exploitation.

As Demers insisted, every stage of emulsion preparation was infused with

155. See the discussion in Waller, "British Patent 580,504" (1988).

156. Waller, "British Patent 580,504" (1988), 58.

157. Demers, "Cosmic Ray Phenomena," *Can. J. Phys.* 32 (1954): 538–54, on 541.

craft knowledge. It turns out, for example, that the gelatin best suited for emulsion work must have thiosulfates or related substances. The bones of calves served well because calves eat wild mustard (*Brassica vulgaris*) and other crucifers containing the sulfur compounds; pigs tend to avoid these particular herbs. And even within the class of calf-originated gelatin, one can only hope to identify empirically the specific type containing the perfect balance ("*un dosage heureux*") of sensitizers, moderators, and retarding agents. Since, at least as of 1955, it was impossible to synthesize active gelatin, such properties were of more than purely bovine alimentary interest.[158] Into this "happily" composed gelatin is then introduced the silver bromide (one method among several being the two-jet method), which is responsible for the primary sensitivity of the emulsion.

Why is the whole process so ferociously difficult, so devoid of the "straightforward" procedures to which physicists were accustomed? Demers put it this way: In a gas or a liquid, substances are independent of their mode of preparation. This is, however, not the case for crystals, as is well known from the properties of tempered steel, for example. Semiconductor crystals are even more sensitive to their history of preparation; the quantum mechanical function depends on the distribution and nature of defects in the crystal. In a certain sense, the emulsion resembles these crystalline structures since, like them, the silver bromide crystals depend essentially on their preparation. But the emulsion is far more subtle even than a semiconductor not only because of the intrinsic complexity of the gelatin but because of the crucial (but subtle) chemical interaction that occurs at the surface between the silver bromide crystals and the surrounding gelatin. It is impossible to start with a finished emulsion, decompose it into bromide, silver, and gelatin, then begin again with precipitation and maturation and obtain the same photographic properties. Ultimately unlike the semiconductors and luminescent materials studied by physicists, as far as Demers was concerned, the emulsion lay in its own incompletely understood science of colloids.[159] Emulsions in their *chemical* construction were a long way from the elegant quantum theory of photographic action that Mott and Gurney had developed in 1937.[160] (In fact, as far as I can tell, the quantum theory of the latent image played absolutely no role in the development of nuclear emulsions.) Add to the craft colloidal knowledge a heavy overlay of industrial secrecy, and you have an obstacle few physicists could surmount.

Demers's knowledge of the fabrication of photographic chemistry was astounding for someone working outside of the major film companies, though he was not the only individual to produce high-sensitivity emulsions.[161] He was,

158. Demers, *Ionographie* (1958), 34–35.

159. Demers, *Ionographie* (1958), 52–54.

160. Gurney and Mott, "Photolysis of Silver Bromide," *Proc. Roy. Soc. London A* 164 (1938): 151–67.

161. For further references to other papers about emulsions sensitive to minimum ionizing particles, see, e.g., Demers, *Ionographie* (1958), 17.

however, unique in both successfully making new, highly sensitive emulsions and applying them productively to physics problems. In chapter 1, I introduced the notion of a disciplinary map drawn according to clusters of practice rather than shared ontologies. Such a practice map might be thought of as grouping together the wide collection of activities subsumed under the category "condensation physics," which formed the core of C. T. R. Wilson's work. Here, a similarly conceptualized practice map is drawn for us by Demers himself, in his attempt to capture the connecting thread of some 4,500 articles he had assembled on the use of image records made by ions, a subject that cut in complex ways across cosmic ray physics, geology, and biology: "It seems reasonable to see in this set of knowledge domains a particular science, in the same way that photography or acoustics are defined by their modes of detection, a sensitive plate or an ear. For this science, we propose . . . the name *Ionography.*"[162] Ionography would be the science of materialized trajectories in solids. Track-making particles could be captured that came from the cosmos or were released from nuclei. Looked at from the point of view of a detector, Demers asserted, ionography would shed light on nuclear and cosmic forces; looked at from the point of view of a process, detection would inform us about the theory of crystalline and atomic forces. Demers's image of the "new science" is reproduced in figure 3.12.

Proper pieces of ionography are on or in the ellipse; domains "more or less attached" to the new science are outside the ellipse. Thus, under the "detection aspect" of the science are the nuclear reactions involving alpha, beta, or gamma interactions, electron collisions, proton interactions, antiproton production, and meson manufacture—as well as "geochronology" (using plates to date rocks), hygiene, fission, reactors, and cosmic rays. Under the rubric "process aspect" one finds the dynamics of gelatin and emulsions themselves and the chemicophysical study of the silver bromide colloidal system. While Demers's name, "ionography," did not leave a lasting impression on the physicist's lexicon, his category clearly reflected the location that emulsion physicists had within the culture of scientific work. For, by the early 1950s, it had become common for emulsion physicists to meet with emulsion chemists and cite photographic work far from the pion-nucleon interaction. Both physicists and chemists could agree on the complex procedures of scanning and calibrating the plates, but each group would interpret the results into different matrices of commitments. To the physicist the plate told about the particles, while to the emulsion chemist the particles revealed properties of the plate. The *aspect détection* and the *aspect processus* coalesced in the trading zone of *ionographie.*

Despite widespread admiration for Demers's chemicophysical accomplishment, it should be clear that one feature of his program did not come to fruition: no major group of physicists anywhere in the United States or Europe took up

162. Demers, *Ionographie* (1958), 8.

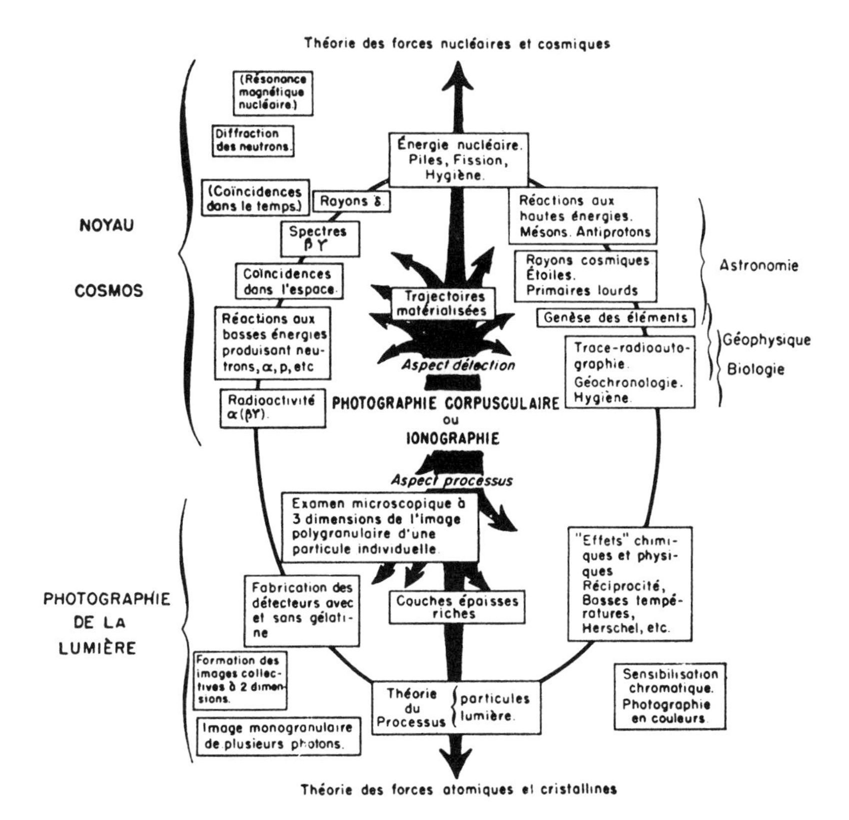

Figure 3.12 Demers, ionography (1955). Ionography, Demers wrote, "is the science of trajectories materialized in solids." This figure is Demers's representation of the cartography of this new science. In the "northern hemisphere" is the "detection aspect" of ionography; in the south is the "process aspect." North, one finds the full range of forces at work in the nucleus and the cosmos from element genesis to the spectra observed of alpha, gamma, and beta ray—but also the less "fundamental" sciences of geochronology and hygiene. To the south are the myriad of chemical and photochemical processes at work in the emulsion medium, from gelatinous and nongelatinous emulsions to theories of colloidal media. Orbiting around the whole are nuclear magnetic resonance, neutron diffraction, color photography, and other ancillary scientific fields. Taken as a whole, this picture illustrates how different the layout of scientific knowledge can look when viewed from the perspective of the laboratory instrument, rather than from that of experiment or theory. Items that are theoretically distal can become proximate (geochronology and element genesis, e.g.). Source: Demers, *Ionographie* (1955), 9.

the Demers procedure for *making* emulsions. Why? It was certainly not because physicists like Powell thought poorly of Demers's product. In fact, in his 1946 paper lauding the new Ilford film, Powell noted that Demers's new film gave even greater detail than Ilford's.[163] The reason for not making emulsions à la Demers lay elsewhere. One technical distinction between Demers's emulsions and Kodak's and Ilford's electron-sensitive films was that the Kodak and Ilford products both had relatively large grain size. This, in turn, had important consequences for the laboratory as workplace. While small grains were relatively difficult to scan and required high-power microscopes, the large-grain commercial films were more easily adapted to the laboratory structure long advocated by Powell. According to George Rochester, who served on the Emulsion Panel, "the fact that an emulsion like G5 had a relatively large grain size of 0.3 μm allowed Powell to introduce mass scanning by a large team of ladies (sometimes facetiously called 'Cecil's Beauty Chorus'!) with low-power microscopes, and thus to obtain relatively quickly a wealth of data for his large team of scientists: It is unlikely that mass scanning could have been employed with finer grained emulsions such as, for example, Demers's emulsions."[164]

At the same time, there was a vast distance between the professional competence of the physicist trained in atomic (and, more recently, nuclear) physics and the expertise of the colloidal chemist. It was one thing to serve on a panel in which a trading zone could be established between a Powell or Rotblat on one side and Waller or Chelton on the other. It was quite another for the physicists to retrain in the dark arts of silver bromide, gelatin, and colloidal gold. Thus, for reasons of craft knowledge as well as for reasons of laboratory work structure, emulsion physicists never took up Demers's mixture of physics and emulsion chemistry. But their chemical abstinence was not without its price. Almost immediately, the physicists had to grapple with the consequences of their isolation from the inner structure of their chosen instrument. Without knowing precisely how the films were manufactured, they did not, could not, know in detail how their apparatus worked.

3.6 Scanning and Spread of the Method

Some 18 months after receiving the new Ilford plates (that is, sometime in mid- to late 1947), Powell and Occhialini were sufficiently taken with the promise of the new method that they decided to produce a book similar to the cloud chamber atlases: *Nuclear Physics in Photographs* (1947). (The muse who suggested such a publication to them, P. Rosbaud, had also initiated several of the cloud chamber volumes.) At one level, Powell and Occhialini argued, the equipment was so simple, so standardized, that any school laboratory could serve as a site of nuclear inquiry. An appendix specified the development procedure: 33 minutes at 65°F, stop-bath immersion for 10 minutes in glacial acetic acid in water,

163. Powell et al., "New Photographic Emulsion" ([1946] 1972).

164. Rochester and Butler, "Nuclear Emulsions," *J. Phys.* 43 (1982): C8-89–C8-90, on C-90.

fixation until the opaqueness has disappeared and the plate has become translucent (not more than 30 minutes), and a wash for an hour under running water. This was a method that was easily transferable, the book emphasized, in its structure, style, and content, one not dependent on massive infusions of either money or theory.[165]

To bolster the legitimacy of the visual approach to nuclear physics, Powell and Occhialini opened their atlas with a homage to Maxwell, and through him, to the virtue of the visualizable as distinct from the formal mode of reason:

> There are . . . some minds which can go on contemplating with satisfaction pure quantities presented to the eye by symbols, and to the mind in a form which none but mathematicians can conceive.
>
> There are others who feel more enjoyment in following geometrical forms which they draw on paper, or build up in the empty space before them.
>
> Others, again, are not content unless they can project their whole physical energies into the scene which they conjure up. They learn at what rate the planets rush through space, and they experience a delightful feeling of exhilaration. They calculate the forces with which the heavenly bodies pull on one another, and they feel their own muscles straining with the effort.
>
> To such men momentum, energy, mass are not mere abstract expressions of the results of scientific enquiry. They are words of power, which stir their souls like the memories of childhood.
>
> For the sake of persons of these different types, scientific truth should be presented in different forms, and should be regarded as equally scientific, whether it appears in the robust form and the vivid colouring of a physical illustration, or in the tenuity and paleness of a symbolical expression.[166]

Paleness of symbolic expression has no place in the image tradition. For Powell, Maxwell's scientific ecumenism created legitimacy for his own commitment to the visualization of the subatomic realm through the cloud chamber and now through the emulsion method. Indeed, the book was written with no more than an elementary use of mechanics and electrostatics, simple scattering laws, for example. By virtue of this reduced level of mathematics and the inherent simplicity of the method, Powell hoped that nonscientific readers would be brought into the study of nuclear physics. But more generally, Maxwell served as an icon of British physics, indeed of British culture more generally, and as a voice of support for the nontheoretical character of Powell's work (even outside of his popular writing).

Powell and Occhialini continued: "In making a useful contribution to their subject, amateur astronomers have overcome much greater technical difficulties than those which have to be met in producing good photo micro-graphs."[167] In a

165. Powell and Occhialini, "Appendix A" (1947).

166. Powell and Occhialini, *Nuclear Physics in Photographs* (1947), title page.

167. Powell and Occhialini, *Nuclear Physics in Photographs* (1947), v–vi.

burst of idealism, Powell and Occhialini imagined amateur nuclear physicists spotting new decays in the way that amateur astronomers had plucked comets and asteroids from the heavens. Conceding that to the present (mid-1947) emulsions had not competed seriously with the Wilson chamber or electrical counters, the production of Ilford's Nuclear Research emulsions promised a great future. Because of the stunning results that had been found in the previous year and a half, the authors hoped that their atlas would serve physicists as well as nonphysicists by presenting the wide possibilities offered by the study of tracks of heavy charged particles.

Alas, the private side of physics is less simple: As Powell had realized in the first few days of his interest in nuclear emulsions, extracting information from the gelatin would be as hard as recording it in the first place. There were problems in the variability of the emulsion, in the exposure conditions, in distortion through development, drying, scanning, and interpretation. Just months after the end of World War II, he wrote to Chadwick, emphasizing the difficulty of reading the photographs in a reliable way: "The most important technical problem, in my view is to establish a team of observers and a routine of measurements in order to increase the speed at which results can be obtained. I think it would be a great advantage to have two teams in different places so that independent checks could be applied. My experience in training people for this work shows that some unskilled people can be employed successfully."[168]

"Unskilled people" in this context meant, by and large, women who were not physicists (just as the unskilled shock-recording observers had not been geophysicists). Their job would be to find specified topologies of events, record the position in the emulsion, and then pass the film to a physicist or physics student, who would make the measurements and label the process. So identified with Powell was the employment of women scanners, that in his biographical memoir for the Royal Society, F. C. Frank and Donald Perkins attributed to him the invention of the term "scanning girl": "A vital innovation necessary for the successful prosecution of the researches with the emulsion method, was the creation of teams of girls to perform the tedious examination of the emulsions by means of high-power microscopes, for events of interest." Led by Powell's wife, Isobel, it was these "girls" who found the first pion track and most of the significant physical processes that were displayed by the Bristol group over the next years. (Somewhat later, in the age of the bubble chamber, one sees that such a division of family labor was not at all rare: Donald Glaser's wife became a scanner; Luis Alvarez's wife led the bubble chamber scanners at Berkeley for many years.) "Powell," his colleagues and obituary writers continued, "soon convinced everyone that it was possible to train young women, with no formal

168. Powell to Chadwick, 16 October 1945, CPP, Chad IV 2/10.

knowledge of physics, to perform this exacting work with expertise and meticulous accuracy."[169]

In some respects, the choice of women for the job of scanning was not innovative. For decades it had been a common European and American assumption about women's nature that they were specially suited to such "meticulous," "tedious," and "exacting" work. One sees women in the employ of astronomers, as star counters, or in Stefan Meyer's Vienna laboratory as counters of flashes on the scintillator screen.[170] During World War II, women served as "computers" grinding out numerical work on mechanical calculators. But in other respects, as we will see directly, the women scanners in Powell's laboratory were doing something different, as Frank, Perkins, and Rochester rightly insisted.

From 1947 to 1957, Powell's group employed about 20 scanners at any one time in an employment arrangement that was widely copied across Europe and the United States. By 1957, female scanners had become so characteristic of the discipline that they could be found throughout the Soviet Union and in virtually every laboratory in Europe and across the United States. So widespread had the practice become that inquiries started to be made by health officials about the conditions under which the women worked. Powell responded to one such inquiry this way:

> For the last ten years we have employed about 20 observers, who work all day on microscopes, scanning plates exposed to nuclear particles. We have had no serious trouble to the eye-sight of any of the people we have employed, although from time to time, some observers find the work strenuous and we try to give them alternative employment. In general, it is important to recognise that there is nervous strain attached to this work and that an observer benefits greatly by not being kept too long on the job, at any one time. With this in mind, we give 20 minutes break in the morning at 10-30 and a similar break in the afternoon at 3-30. It is important that the observer should not feel stressed. I suppose the total working week amounts to about 35 hours. . . . We don't find any considerable loss of efficiency in the conditions we have established.[171]

In the early years of the method, the scanning women were deliberately included in the investigation and granted a kind of quasi-scientific authorship. Each of the discovery photographs—a mosaic of images from each layer of the emulsion stack—bore the name of the scanner who discovered the event: Isobel Powell, Mrs. I. Roberts, Mrs. W. J. van der Merwe, Marietta Kurz, and Mrs. B. A. Moore, among others. With time, the removal of the women from the discovery process became routine, and by the early 1950s, it was rare to find the scanner's

169. Frank and Perkins, "Powell," *Biog. Mem. F.R.S.* 17 (1971): 541–55, on 549.

170. On Meyer's laboratory, see Stuewer, "Disintegration" (1985).

171. Powell to Wilkinson, 7 February 1957, CPP, E61.

name mentioned at all. Indeed, one commentator, Walter Barkas, remarked that physicists should exercise extreme caution in the use of scanners, beginning with precise instructions from the physicist to his employees, in order to "ensure the integrity of the data." Such detailed supervision was only possible if the physicist acquainted "himself with the problem by doing a considerable amount of scanning" and then wrote out detailed instructions, including "scanning biases that may intrude." Instead of seeing advantage in keeping the scanners informed about the physics goals and the expected topologies of interesting events, by the late 1950s, Barkas and presumably others attached epistemic importance to *denying* knowledge to these women: "[I]t is necessary to remember psychological factors. It is important that the scanner be unaware of the result that is expected in a measurement or observation. It is very human to try to obtain the answer that pleases. Some of the false 'discoveries' made in emulsion experiments may have had such a cause."[172] Scanners therefore took the blame when the experiment went amiss, an extraordinary circumstance given that (ostensibly) nothing could be published without an elaborate sequence of measuring and interpreting that deeply involved the physicist.[173]

But clearly something troubled the workers as they grappled with the idea of authorship, given that it was the women scanners who, in the first instance, found the interesting events. Even the language is problematic: in the earliest papers under each mosaic is printed an attribution of the form "Observation by Mrs. I. Powell"; in later ones this information is abbreviated to the form "observer: W. J. van der Merwe." In letters, the physicists write about events "found" by a particular observer, reserving "discoveries" for the physicists themselves. The transition in grammar shifts the specificity of this observer finding that event to an observer "on watch" as an event presented itself. But the often narrow gap between discovery and observation is perhaps most starkly revealed in some of the accounts by the Bristol group of the first of the pion tracks. One of Powell's greatest supporters—the director of the H. H. Wills Laboratory, A. M. Tyndall—wrote a memoir of the Powell group in 1961. There he argued that the measure of Powell's achievement should not focus on the pion at all, but rather on the general orientation of nuclear emulsion work. He reasoned this way:

> Future historians in assessing Powell's contribution to physics will probably think of him primarily as the discoverer of the π-meson: whereas I would lay the emphasis on his experimental insight and manipulative skill in creating the emulsion technique which others had failed to do with originally the same material. It was he who overcame the difficulties in making it a quantitative tool. Once this was done, who it was that discovered something new was purely a matter of chance.

172. Barkas, "Data Handling" (1965), 68.

173. Steven Shapin has a fascinating discussion of the attribution of experimental misfiring to the actions of otherwise invisible technicians in Boyle's inquiries. See Shapin, *Social History* (1994), 389–91.

> Indeed many of the rare events, including the π-meson itself, were first observed by a scanner with no scientific training reporting a track of appearance different from anything *she* had previously seen.[174]

By shifting "discovery" away from the physicist and to the scanner, Tyndall would seem to be denying altogether any authorial weight to Powell qua physicist. Yet while assigning the role of discovery to the women scanners, Tyndall restores the primary role to the invention of the method, and it is here that Powell's "manipulative skill" is to be lodged. With discovery denigrated in this way to a "matter of chance" requiring no "scientific training," it is natural that in their *Photographic Method,* Powell, Fowler, and Perkins ordinarily simply deleted the women's names altogether from the mosaic captions. As the title of the work makes plain, the general method has superseded in importance any specific discoveries.

But there are other issues at stake in the pion discovery, and it is to the wider sphere of events surrounding 1947 that we now turn. One of the first qualitatively novel results obtained with the new nuclear emulsions was the interpretation of several events as slowly moving charged particles, of mass intermediate between those of an electron and a proton, approaching and then disintegrating a nucleus. Donald Perkins submitted the first such event to *Nature* on 8 January 1947, using Ilford's new B1 film that had been exposed in an RAF plane at 30,000 feet. The star had four tracks (see figure 3.13). In this picture, Perkins identified the track labeled H^3 as a triton (proton plus two neutrons) because of the rate of change of grain density along the track; the two protons (labeled H^1) were the by-then-standard footprints of a proton. Unlike these three particles, which slow down as they *leave* the center, the track labeled sigma (σ) decelerated as it *approached* the center. Perkins interpreted the picture as a particle of intermediate mass slowing down and being absorbed by a nucleus that then exploded into the three fragments, the triton and two protons. On this interpretation, the remaining nucleus is observed recoiling between the triton and sigma tracks.[175]

Could Perkins's sigma track be the trace of an electron? This Perkins ruled out because the track's ionization was too high; indeed, he and his colleagues were desperate for a film that *could* display the tracks of electrons. At the same time, a proton is ruled out because the scattering is too large—plotting the number of degrees of scattering against range in microns (related to the particle's energy), it was clear that the sigma was too easily deflected to be a proton. Using scattering formulae for the emulsion, Perkins estimated the sigma mass at between 100 and 300 electron masses; a corroborative check based on a rough

174. Tyndall, appendix to Frank and Perkins, "Powell," *Biog. Mem. F.R.S.* 17 (1971): 555–57, on 557, emphasis added.

175. Perkins, "Meson Capture," *Nature* 159 (1947): 126–27.

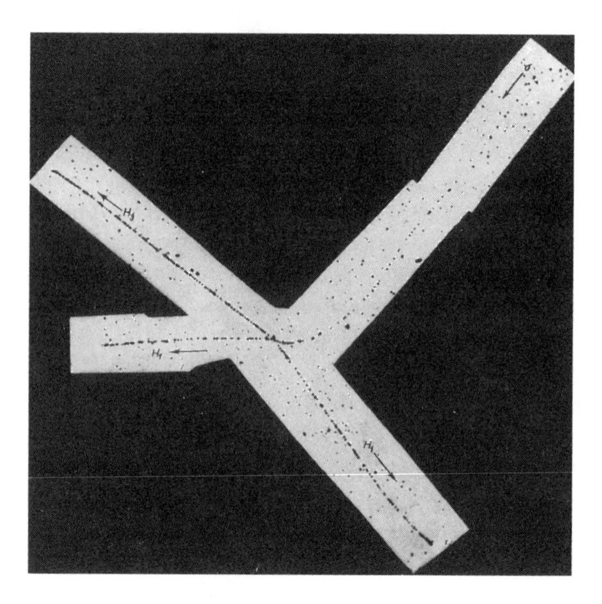

Figure 3.13 Perkins, "meson capture" (1947). First observation of a disintegration by a negative meson. H_1 designates a proton (two tracks), H_3 a triton (proton with two neutrons), and σ a new particle of intermediate mass. Since the new particle's ionization track appears to be denser near the center than farther away, Perkins argued that the sigma was slowing down as it approached the intersection of the four tracks. All of the other particles appear to be leaving the center because their tracks are denser away from the center. Source: Perkins, "Nuclear Disintegration." Reprinted with permission from *Nature* 159 (1947): 126, © 1947 Macmillan Magazines Limited.

nuclear physics calculation (meson annihilating a carbon, oxygen, or nitrogen nucleus in the gelatin) put the mass between 120 and 200 electron masses.[176]

At Bristol too there were new particles. Occhialini, after an imposed exile during the war, returned to the lab and took the new plates to the French Pyrenean observatory at Pic-du-Midi. Developed and set under the microscope, the plates revealed a myriad of new decays; Powell later rhapsodized that it was as if "we had broken into a walled orchard, where protected trees had flourished and all kinds of exotic fruits had ripened in great profusion."[177] One scanner found an event that appeared to be a meson entering into a nucleus and disintegrating it, a find followed by a series of other similar events. Publishing in

176. Perkins, "Meson Capture," *Nature* 159 (1947): 126–27.
177. Powell, "Autobiography" (1972), 26.

the same 1947 volume of *Nature* that carried Perkins's sigma track, Powell and Occhialini reported on six "meson" tracks (tracks of particles that they, like Perkins, identified as having a mass intermediate between those of an electron and a proton) with grain density (energy loss) increasing as it approached a point from which other tracks issued. Their mass determination was based on the following procedure. First, they focused beams of protons each of a given energy into the film in order to determine the range, *R*, versus energy, *E*, curve for a particular emulsion. That curve yielded *dE*/*dR* (energy loss per unit of track length) as a function of energy. Since they knew the original proton energies of these test runs, they could correlate the energy loss per length with the measured decline in grains per length. From then on, they could take any particle of unit charge that stopped in the film and use grain counts to determine how much energy it had lost in the film and therefore what its energy must have been at the beginning of its track. They then applied this method to the meson candidates found in the cosmic-ray-exposed film. Working backward from the stopping point of the meson in question (and assuming that the new particle was of unit charge), Powell and Occhialini obtained the initial energy, *E*, of the new particle and added this to their knowledge of its measured range, *R*. Looking up in their cyclotron-generated calibration tables how far a proton of similar mass could have gotten, they showed that the new object's range was far shorter—a difference explained if the six new particles were lighter than protons (from 100 to 230 electron masses).[178]

Perkins's article appeared (on 25 January 1947) while Powell and Occhialini's paper was still in proof (it appeared on 8 February 1947); for Powell, the coincidence of their results legitimated not only the specific event type, but the method itself: "The observed difference in the grain spacing of the meson tracks, in the B1 and C2 emulsions employed in the two experiments, is in good accord with expectations based on the known recording properties of the two types. The agreement between the results of observers in two different laboratories, working entirely independently with different experimental material, is a definite proof of the reliability of the photographic method in its present stage of development."[179] This global certification of the method, which seems to find expression over and again, reflects the felt precariousness of procedures. Already, we have seen the photographic method certified when it produced results coincident with those of Bonner and Brubaker, when it generated results previously

178. Occhialini and Powell, "Nuclear Disintegrations" ([1947] 1972), 224. One other possibility existed: that the meson stopping and the disintegration apparently "caused" by it were actually two distinct events, accidentally superimposed on top of one another. Based on other observations of the frequency of disintegrations, the authors concluded: "The probability, in any one case, that a charged particle, unrelated to the star, has, by chance, come to the end of its range within 1 micron of the disintegrating nucleus, is less than 1 in 10^5. We must therefore conclude that the particle entered the nucleus and produced a disintegration with the emission of heavy particles."

179. Occhialini and Powell, "Nuclear Disintegrations" ([1947] 1972), 227.

unknown by others (confirmed by Bonner), when it produced roughly coinciding results during the war (in experiments by the Bristol and Cambridge groups), and when it produced independent, but matching, golden events. As we will see, this anxiety about variability that had dogged the emulsion method since Blau's time, and continued through the war, persisted in different ways year after year.

As with cloud chambers, arguments based on grain density were not universally accepted. In addition, the period of exposure of the emulsions to cosmic rays was typically a half-year to accumulate a large number of tracks. But in that time the latent image of the undeveloped Nuclear Research emulsion could fade, leading to the (false) conclusion that the particle was light and not massive. In an effort to avoid the fading problem, Powell and Occhialini restricted their exposure to six weeks, but even in that interval the older tracks would fade, compromising the persuasiveness of their claims to have recorded a meson. They therefore leaned on an alternative argument: the number of small-angle coulomb-scattering events of the track exceeded that of a proton of similar range and therefore militated in favor of their hypothesis that the particle was significantly lighter than a proton.[180]

Perkins, Occhialini, and Powell all saw their discoveries as helping to resolve a then-current conundrum. For several years it had been known that the negative meson (muon, a.k.a. mu-meson or heavy electron) could annihilate a nucleus. In addition, theorists since Yukawa in the 1930s had sought a particle of roughly this mass to mediate the nuclear force. The problem was that evidence was rapidly accumulating that, despite its felicitous mass and charge, the muon could not be what Yukawa had ordered. In 1940, Tomonaga and Araki had argued that negative "Yukawa particles," because they interacted so strongly, should fall into the nucleus before they could decay, and they offered a quantitative prediction of just how fast the negatives should be captured. Clinching the matter was a series of elegant experiments, including some conducted by Conversi, Pancini, and Piccioni while hiding in a basement during World War II: they showed definitively that the penetrating mesons (muons) were *not* absorbed as rapidly as Tomonaga and Araki said they should be.[181] The experimenters had a meson (the muon found by Anderson, Neddermeyer, Street, and Stevenson), and the theorists had a meson (the Yukawa particle). But constraints imposed by the Italian experimenters on the lifetime of their negative meson and the

180. Occhialini and Powell, "Nuclear Disintegrations" ([1947] 1972).

181. See Tomonaga and Araki, "Slow Mesons," *Phys. Rev.* 58 (1940): 90–91; Rassetti, "Disintegration," *Phys. Rev.* 60 (1941): 198–204; Conversi, Pancini, and Piccioni, "Decay Process," *Phys. Rev.* 68 (1945): 232; Conversi, Pancini, and Piccioni, "Disintegration of Negative Mesons," *Phys. Rev.* 71 (1947): 209–10. Further references and more details on these counter-based experiments may be found in Brown and Hoddeson, *Birth of Particle Physics* (1983), esp. chapters by S. Hayakawa, O. Piccioni, and M. Conversi.

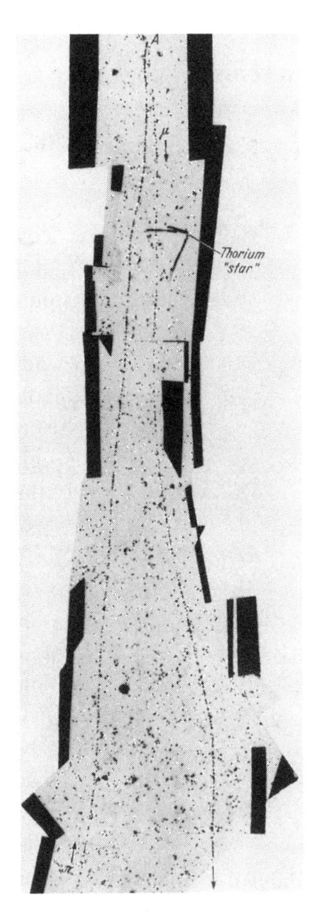

Figure 3.14 One meson decaying into another (1947). First observations of the decay of a pion. As in figure 3.13, one track increases in density as it approaches the vertex, while the other decreases—Powell and his group concluded that the left track (labeled π) was moving up and the right track (labeled μ) was moving down. Since both tracks varied in their ionization density much more than a proton would over the same distance, the team concluded that they were both of lighter mass (and so more easily slowed). Source: Lattes et al., "Charged Mesons." Reprinted with permission from *Nature* 159 (1947): 695, © 1947 Macmillan Magazines Limited.

constraints imposed by the Japanese theorists on the lifetime of their negative meson were strictly incompatible. The experimenters' muon was not the Yukawa particle.

Just a few months after Occhialini, Powell, and Perkins had published their first meson pictures, a scanner in Powell's laboratory, Marietta Kurz, found a particularly striking event among the cosmic rays: what looked like one meson decayed into another (see figure 3.14). The exact sequence of events is not clear—one physicist thought recognition was a matter of seconds, whereas another remembers it this way: "I think she [Kurz] had it under the microscope for a day or two before she convinced people that actually a meson decayed into another and convinced them to look into her microscope. So the thirty seconds you mentioned only occurred after an initial delay!"[182] This discovery along with another meson-to-meson decay formed the centerpiece of Powell and his collaborators' first publication on the intermediate-mass particles. Their conclusion was that there were two different mesons of different mass. Though they hesitated to provide a detailed account of the decay, the authors thought the new meson decay might "contribute to the solution" of the discrepancy between the long-lived, penetrating mesons and those responsible for nuclear forces.[183]

Despite the best efforts of the emulsion physicists, their method continued to have severe limitations when it came time to be quantitative. In particular, while they continued to use the grain-counting method, the emulsion physicists felt obliged to defend the procedure and cautioned the reader about its accuracy. By way of securing their results, they argued that grain counting reproduced accurately the masses of test protons injected into the film at known calibration energies. The danger of an unknown number of grains being lost to fading could be controlled (Powell and his collaborators claimed) by checking the grain deposition of protons emerging from nuclear "stars"; any mesons in a given plate would nominally be faded by the same percentage as the calibration protons. But all these methods were valuable only as averages—individual events, and the mesons were surely that, could not be counted on to obey the laws of large numbers. Grain counting showed that the new particles were mesons (i.e., less ionizing than protons and more ionizing than electrons), but it would not suffice as a way to calculate the meson mass. Of the 65 mesons that had been found to have tracks terminating in the emulsion, only two produced a single, light secondary particle: the one Kurz had found and a second one discovered by another scanner, I. Roberts. As far as Powell and his associates could tell, however, the two mesons had, within statistical error, the same mass.[184]

That conclusion changed in Lattes, Occhialini, and Powell's *Nature* articles of 4 and 11 October 1947; by accumulating more events from cosmic ray film exposed in the Bolivian Andes and at the Pic-du-Midi, they could show that the secondary meson had a fixed range (and therefore a fixed energy), implying that the decay of the primary meson was two-body (see figure 3.15). Using grain-counting techniques, they gave $m_\pi/m_\mu = 2$ as the mass ratio of pion to muon (Powell so dubbed them in this article for the first time), but the authors pointed to several sources of error. There was the statistical error introduced by counting grains. But equally vexing were difficulties that "arise, for example, from the fact that the emulsions do not consist of a completely uniform distribution of silver halide grains. 'Islands' exist, in which the concentration of grains is significantly higher, or significantly lower than the average values, the variation being

182. Foster and Fowler, *40 Years* (1988), 51.

183. Lattes et al., "Charged Mesons" ([1947] 1972), 216.

184. Lattes et al., "Charged Mesons" ([1947] 1972).

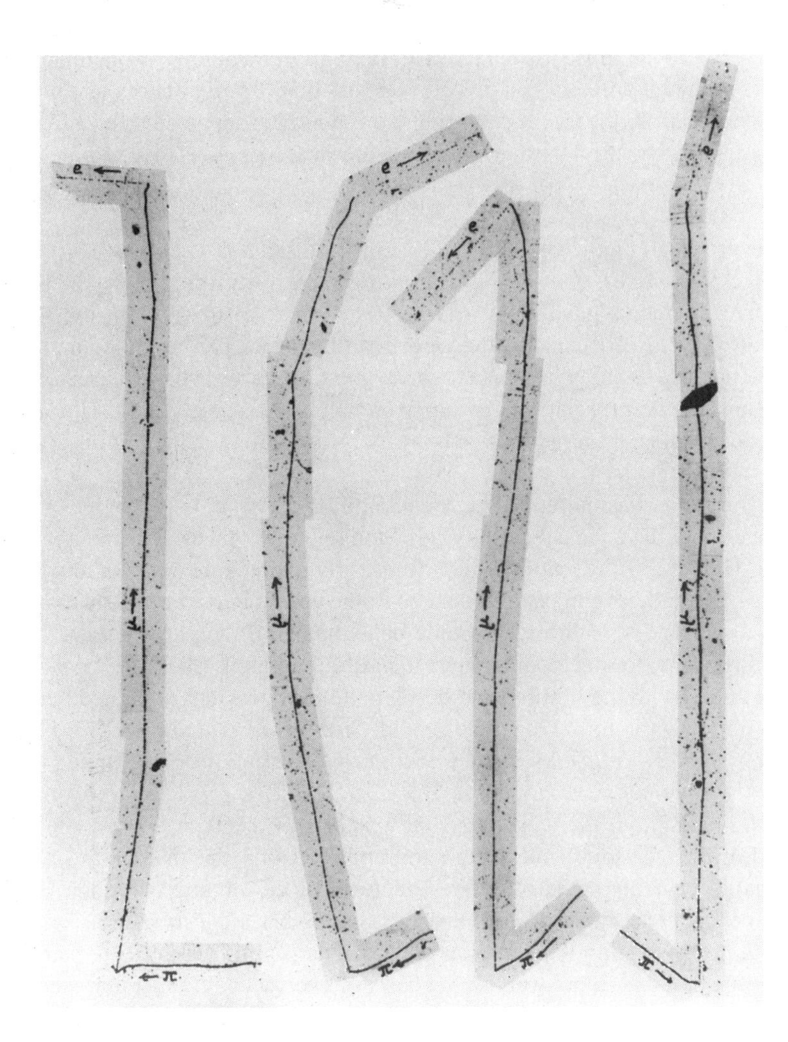

Figure 3.15 Constant range of muon in pion decay (1950). Each of these four mosaics contains a meson (here, for the first time, dubbed the "pion") decaying into a second, lighter meson ("muon"). If the pion decays essentially from rest into two objects (the muon and a single unseen partner), one would expect that the muon would always emerge with the same energy. And indeed, as indicated in this figure from 1950, Powell demonstrated that the muon tracks were of equal length (hence equal energy). At the end of their range, the muons each decay into an electron (and another unseen neutral partner). Source: Powell, "Mesons," *Rep. Prog. Phys.* 13 (1950): 350–424, plate opposite 384.

much greater than those associated with random fluctuations." Nonetheless, the authors judged it unlikely that m_π/m_μ was less than 1.5. By energy and momentum conservation, the muon emerging from a pion at rest would have to be accompanied by a neutral partner. If that partner was a photon (for example), then:

$$c^2 m_\pi = c^2 m_\mu + E_\mu + h\nu.$$

Simple arithmetic indicated that for the range of muon masses between 100 and 300 electron masses, one obtains a pion-to-muon mass ratio below 1.45. Since their measurements gave them a larger ($m_\pi/m_\mu = 2$) mass ratio, the authors committed themselves against the view that the unseen partner of the muon was a light particle—either a neutrino or a photon. Instead, they plumped for a heavy unseen neutral particle (of approximately the same mass as the muon) that would yield a m_π/m_μ ratio of about 2, corresponding to their grain-counting observations.[185]

But there was more. Lattes, Occhialini, and Powell could show, in many instances, that the primary meson originated in the "explosive" disintegration of nuclei. Unlike negative pions, which frequently disintegrated nuclei, the muons did not. The authors conjectured that, as Bethe and Robert Marshak had only recently suggested, this difference could be explained if the pion interacted much more strongly with nucleonic matter than did the muon. These new data could be combined with the lifetime and decay pattern of previously studied penetrating radiation if the following picture held: *Strongly interacting pions were produced in nuclear collisions; these decayed to lighter, weakly interacting muons and an unseen heavy neutral in a two-particle decay. Pions could be Yukawa particles and muons the penetrating mesons of prewar fame.*[186]

Powell's meson-to-meson photographs became the talk of the physics community. Having heard of the new pictures of the summer, Norman Feather (an experimenter) wrote to Powell on 19 September 1947 to say that he could rule out the possibility that the pion produced an (unobserved) e^+e^- pair at the same time and that, moreover, on the basis of energy and momentum conservation, he agreed with Powell that one could exclude the emission with the muon of a photon, a neutrino, or a single electron as the nonmuon product. The only possibility, he concluded, was that the muon emerged opposite a neutral pion of essentially the same mass as the charged pion, as its twin.[187]

Powell responded on 24 September 1947 with a slightly guarded re-endorsement of that conclusion: He would bet, but not too heavily, against a light partner to the muon. In his words he was "not yet prepared to exclude . . .

185. Lattes, Occhialini, and Powell, "Observations" ([1947] 1972).

186. Lattes, Occhialini, and Powell, "Observations" ([1947] 1972). Note: the authors cite a prepublication version of Bethe and Marshak's "Two Meson" paper.

187. Feather to Powell, 19 September 1947, CPP, FEAT 18/7.

that mass ratio, [pion/muon], is sufficiently low to allow process to be explained as a spontaneous decay, in which a photon or a particle of small rest-mass provides the momentum balance. We think this is very unlikely, however, and the values obtained hitherto suggest the emission of a neutral particle with a mass of the same order of magnitude as that of the μ-meson. A final decision will be possible only when further examples, giving favourable conditions of measurement have been found."[188] What was clear was that all the secondary particles (the muons) seemed to travel approximately the same distance through the emulsion before coming to rest. As mentioned above, similar ranges meant similar energies, and if one decay product was monoenergetic (the muon), then the parent decay (of the pion) must be two-body.

That same day Dee wrote to Powell with congratulations and yet another endorsement of Powell's view that the neutral was heavy:

> The plate work has obviously become a major technique & largely due to you. I want to put someone on it as we shall clearly need it. So far we've been too busy. At our colloquium today when the Harwell talks were described I asked why you don't assume that the $\pi \rightarrow \mu$ transition is not via the emission of a neutral particle with mass about equal (say) to μ i.e. as a rough explanation π = say 416 $m_0 \rightarrow$ two particles each 200 m_0 one charged one neutral & each having about 8 mc^2 kinetic energy. I feel you must have considered this but according to our party you didn't explain why you don't assume what seems a simple explanation i.e. (an 8 MeV binding energy of the two particles!!?) So I decided to write & ask you if only to prove to my "young" that they shouldn't be afraid to ask simple questions! But perhaps they weren't listening.[189]

Each of Powell's modes of contact with scanners, emulsion makers, and fellow experimenters had its own dynamics, each its own concerns and vocabulary. Different yet was Powell's contact with theorists. Just as the colloidal chemists had concerns that were in good measure distinct from those of the emulsion physicists, theorists had characteristic modes of reasoning about mesons far from the pragmatic, informal mode of argumentation that Powell favored. One of Powell's students, David Ritson, recalled that Powell hardly ever invoked the details of even nonrelativistic quantum mechanics, much less the intricacies of quantum field theory or meson dynamics.[190] Powell's correspondents gauged their missives accordingly. For example, when theorist Christian Møller wrote to Powell on 11 August 1947 (in the midst of excitement over the meson tracks) he was silent about the real content of his theory of nuclear forces, which had to do with the attractive pseudoscalar meson coupling and an inserted term to cancel the characteristic $1/r^3$ divergence. Indeed, Møller mentioned nothing of the inner

188. Powell to Feather, 24 September 1947, CPP, FEAT 18/7.
189. Dee to Powell, 24 September 1947, CPP, E7.
190. David Ritson, interview by the author, 1 March 1990.

workings of his theory—its motivation, its formal development, its relation to other theoretical approaches to meson theory. Instead, the letter held to a language restricted to lifetimes, masses, and momenta. Under various assumptions, Møller adumbrated an assortment of combinations of particles that might emerge from the decaying pion, and stressed above all the pi/mu mass ratio (m_π/m_μ), which he set at 1.3.[191] Powell responded in kind, for his part omitting the kind of detail that other experimenters would want to hear about technique or the types of laboratory uncertainty encountered. Instead, Powell simply held the door open for Møller's theory by refusing to rule out $m_\pi/m_\mu = 1.3$, adding that a newly discovered track would soon add to the empirical database.[192]

3.7 The Transfer of Knowledge

The problem of the transfer of knowledge was complex, on many levels. Even extremely good laboratories—such as the one run by Peter Dee—could not reproduce the spectacular results obtained by Powell. The only cure seemed to be through the exchange of personnel, in Dee's case by sending an emissary of the laboratory to study with the masters.[193] As Dee arranged with Powell for the apprenticeship of one of Dee's young colleagues, he put her task this way: "Of course, I should like her to see how your people prepare, expose, develop and examine plates, but, no doubt, she could make her own opportunities for doing most of these things if you put her for part of the time on some routine observations or assistance. I hope you will see my point of view. I am sure that she will learn more in one month from being surrounded by your expert workers in these techniques than from any amount of reading and hearsay."[194]

At Berkeley's 184-inch accelerator, Eugene Gardner's group was equally frustrated. Beaten to the meson, one of the cyclotron's reasons for existing, by Bristol, the Lawrence Radiation Laboratory team found itself in late 1947 and early 1948 shaken by the fleeting results emerging from the plates. Gardner drafted a letter to Kodak on 5 January 1948 that made no secret of his annoyance both with their product and the secretive fashion in which the film company presented its results to the physics community. The issue, however, was obviously touchy, and he began by remarking that while it might not be "possible to discuss reproducibility in a published article, then at least we should come to an agreement among ourselves about what the situation is." Gardner continued:

191. Møller to Powell, 11 August 1947 and 3 September 1947, NBA.

192. Powell to Møller, 8 September 1947, NBA.

193. In this respect, the legitimation of the emulsion resembles that of the air pump in seventeenth- and eighteenth-century Europe. See Shapin and Schaffer, *Leviathan* (1985).

194. Dee to Powell, 29 July 1948, CPP, E7. On the sociological nature of the exchange of craft expertise in science and replication, see Collins, *Changing Order* (1985), esp. chaps. 2–4.

> Now I don't think that Eastman [Kodak] should apologize for this lack of reproducibility. Most of the types of plates have been developed very recently, and if they can't be made with a constant sensitivity, I don't think that Eastman would lose face by announcing that they can't. The point is that if people are led to believe that the plates are uniform from batch to batch, when in reality the plates aren't uniform, then workers {are going to} may lose confidence in the whole technique. The situation now is that Eastman says nothing about lack of uniformity, but workers in the field do discuss it.[195]

As an example, Gardner cited a recent article by R. A. Peck, Jr., who had written in the *Physical Review* that "[u]ntil there is some assurance of the reproducibility of Eastman NTB plates [Kodak's equivalent of Ilford's all-purpose C2 plate for heavy and medium particles] it seems necessary to repeat the calibration procedure for every new batch received."[196] Calibration was no triviality. It involved, among other things, exposing plates to beams of protons of known energy and then determining the range in film. Gardner added, "This paper by Peck leaves me with the rather unhappy feeling that I don't know just what it is that varies from batch to batch, or how much. It is easy to say that workers should calibrate their own plates, and that this may be possible in some cases. It seems to me, however, that most laboratories would not have the facilities to {make a range-energy calibration} do calibrations."[197]

No doubt the Berkeley emulsion group's anxiety was exacerbated because Peck's irreproducibility was not even their main concern (grain density inhomogeneity was). How many other variations were there that the physicists had not yet spotted? "If someone [presumably from Kodak, who should know] would clarify the situation by stating specifically what varies from batch to batch, and what does not, we might find that the plates are a lot better than most people think, and that there are types of experiments for which the variations don't matter." Perhaps trying the appeal of company pride, Gardner then added his clincher: "I won't say this outside school, but the Ilford plates that we have used are extremely good from the point of view of reproducibility. In fact, we have never noticed any variation from batch to batch. . . . In spite of the fact that the Ilford plates are really rather uniform, the British workers are careful to call attention to possible sources of error."[198] The most obvious example, Gardner insisted, was Lattes, Occhialini, and Powell's *Nature* article quoted above, where they circumscribed the error associated with nonuniform "islands" of silver halide in Ilford film.[199] Frank "statements of this kind," Gardner insisted, "tend to increase confidence of other workers in these photographic materials"[200]—and Kodak had

195. Gardner to Julian Webb, 5 January 1948, EGP; braces indicate deleted segment.
196. Peck, "Calibration," *Phys. Rev.* 72 (1947): 1121.
197. Gardner to Webb, 5 January 1948, EGP; braces indicate deleted segment.
198. Gardner to Webb, 5 January 1948, EGP.
199. Lattes, Occhialini, and Powell, "Observations" ([1947] 1972), 237.
200. Gardner to Webb, 5 January 1948, EGP.

not provided the information for him to do the same in his new work on alpha and deuteron tracks in Kodak's NTA film.

When Lattes arrived at Berkeley bearing the secrets of the trade, one of his first acts was to review the procedures of Gardner's group for developing the film. It was obvious that something was not right; just what remained a mystery. In May 1947, Rotblat (who headed the Emulsion Panel in Britain) wrote to Pickavance, then at Berkeley, where the letter was copied and distributed. Rotblat was as dismayed as Gardner about the film: "Unfortunately, the plates which Gardner sent us are very difficult to investigate. For one thing, since the plates were irradiated in the tank, the actual direction of the beam is not well defined; in fact we find that most of the particles enter the emulsion from the glass and not from the top. This makes it impossible for us to follow the track over its entire length. Secondly, there is a large background of various other particles, probably coming from the top and bottom of the dees. . . . This may be due to processing, but the result is that only few plates of the batch are suitable for examination." And later, "I am surprised to hear that the processing suggested by us was not good. There must be something wrong with the developer they are using. Can you find out what developer they have used and also other details of processing? All our plates are processed according to the instructions I sent to Gardner and they come out alright."[201]

Over the next few months, paper flew back and forth between England and Berkeley over the right processing directions. En route, things misfired. On 23 September 1946, Waller suggested using the developer D-19 for 20 minutes, diluted 1:2 with water; on 27 June 1947, someone copied directions out of the "Photographic Emulsions for Nuclear Research" handbook indicating that D-19 should be used for 10–20 minutes diluted 1:2; on 9 April 1947, laboratory notes prescribe D-19 for 4 minutes (no dilution specified); in March 1948, instructions read 35 minutes at 1:3. "Take the camel's hair brush," the guide advised, "and 'paint' each plate while submerged, and do this every minute or so. A ferrotype tin makes a good cover for the tray if you want to turn the safelight up brighter in order to eat a lunch, mix chemicals, or clean the place up. . . . People eager to scan plates just exposed frequently ask when they'll be ready, and it's a good idea to be ready with a quick answer. . . . [I]f there's no interest in possible mesons in the emulsion don't dilute the D-19 and develop as little as 4 minutes."[202]

Somewhere in the shuffle, the protocol was altered. As Powell reported to a theorist at the beginning of March 1948, the error was soon evident: "Apparently, Lattes, when he got there, found that they [Gardner and his colleagues]

201. Excerpts from Rotblat to Pickavance, n.d. [24 May 1947?], two-page typescript, EGP.

202. "Successful Developing Procedure for Mesons. Ilford C-2 Plates, March 1948," EGP.

had been developing their plates for 4 instead of 40 minutes. When they adopted the normal [that is to say, Bristol] procedure, they found many mesons, some of which, at any rate, produce disintegrations at the end of their range in the emulsion."[203] It was a small point of explicit procedure, not an arcane bit of tacit knowledge barely communicable in ordinary language. No, this was a bit of photographic lore that transferred awkwardly to a physics laboratory rich in electronics and accelerator knowledge, but ill prepared for this kind of technology.

With the new protocol in effect, everything changed: "We have observed tracks which we believe to be due to mesons in photographic plates," Gardner and Lattes announced on 12 March 1948. They showed the same scattering pattern and change in grain density as a function of energy that Lattes, Occhialini, and Powell had noted the previous year. Qualitatively, the tracks were distinct enough that a "practiced observer" could spot them on sight. Quantitatively, the Berkeley group had a vast advantage over their cosmic ray competitors: by sending the mesons through a magnetic field and calculating the resultant radius of curvature, the mass could be determined to within strict limits, $m_\pi = (313 \pm 16)m_e$. Now that they could produce pions at a rate some 100 million times that available in cosmic rays, they expected that "the rate of progress in this field can be greatly accelerated."[204] Rather sheepishly, Gardner wrote Kodak that Mr. Lattes "has come here to work for a year and convinced me that the electron tracks can be detected. Thus I look to retract my objections."[205]

Berkeley's discovery was both a blessing and a curse for the Bristol group. On one hand, Gardner's group now had unambiguously confirmed the existence of a strongly interacting particle of intermediate mass. On the other hand, the mass ratio of pion to muon as determined by the Berkeley physicists ($m_\pi/m_\mu = 1.33 \pm 0.02$) was irreconcilable with the Bristol value near 2. This discrepancy was not a trivial matter: it meant, for example, that the entire interpretation of the pion decay had to be overhauled. Where Bristol's value seemed to imply a neutral particle of muonic mass as the unseen partner, the Berkeley result definitively ruled that possibility out and pointed in the direction of a neutral massless or light partner to the muon when the pion decayed. Facing up to the crisis that seemed to threaten the mass determination methods used by Bristol, Powell's group published a response in which they acknowledged that the "recent experiments at Berkeley suggest that the true value is [$m_\pi/m_\mu =$] 1.33 $\pm$ 0.02, a result which throws serious doubt on the reliability of the method based on grain-counts."[206] Looking back, Brown, Camerini, Fowler, Muirhead, Powell, and

203. Powell to Møller, 5 March 1948, NBA.

204. Gardner and Lattes, "Production of Mesons," *Science* 107 (1948): 270–71.

205. Gardner to Webb, 1 March 1948, EGP.

206. Brown et al., "Cosmic Radiation," *Nature* 163 (1949): 47–51 and 82–87, on 82, repr. in Powell, *Selected Papers* (1972), 265–75, on 270.

Ritson pointed to two sources of error as their undoing: fading of the latent image and variation of development with depth.[207] The uncertainty engendered by these errors forced Powell and his team to rely on contemporaneous tracks—pions that decayed into muons—so that both tracks would be subject to the same development and the same latent image fading. This restriction meant, however, that only relatively short pion tracks were available for measurement (less than 400 microns long), giving the grain-counting method a large statistical uncertainty in these cases.[208] The Bristol authors then made use of the conclusion from their earlier work to assert that the particle giving rise to the stars was the π^-; this identification gave a new set of long pion tracks on which m_π could be measured. Then the authors assumed that all the lighter meson tracks were either π^+ or π^-, again yielding far longer pion tracks than the original restriction to pairs of tracks in which one meson nominally decayed into another. With the wider data set and longer tracks the new ratio m_π/m_μ was calculated to be 1.33 ± 0.05, in splendid agreement with the Berkeley accelerator team's figure. The price of conformity was dear: Bristol could match Berkeley only by giving up the original Bristol interpretation of a pion decaying into a heavy neutral and a muon. Moreover, the entire episode left many people in the field suspicious of the stability of the grain-counting method. Anxiety over the method retroactively undercut the security of the full range of the Bristol team's results.

Immediately, the Powell group had to reestablish the method not just to backstop their older work. They had to secure the procedure in order to certify their latest find: a complicated event found by Mrs. W. J. van der Merwe (figure 3.16) that they interpreted as a single heavy meson decaying into three pions. Published on 15 January 1949, the new object, dubbed the tau (τ), was something of a sensation.

Not everyone acquiesced. Experimenters demurred because of residual doubts about the method. And J. Robert Oppenheimer, leader of one of the most influential theoretical groups in the world, was not sanguine about admitting such an entity into the particle zoo. When no additional instances of the tau decay were seen over the course of 11 months, Oppie wrote Powell in November 1949 to say, "We have . . . been very much pleased at the lingering death of the τ meson.

207. On the difficulties the grain method had encountered, see Brown et al., "Cosmic Radiation" ([1949] 1972). The problem of uneven development of the emulsion was yet another very serious instability in the emulsion method. Two important techniques were developed to stabilize the process. In the first, Dilworth, Occhialini, and Payne ("Thick Emulsions," *Nature* 162 [1948]: 102–3) proposed that the film be soaked in a cold developer. Then, when the developer had thoroughly and evenly permeated the emulsion, the emulsion would be warmed up and the developing process would proceed evenly throughout the material. In the second technique, Blau and de Felice ("Two-Bath Method," *Phys. Rev.* 74 [1948]: 1198) proposed that development be executed in two baths. The first held the developing agents but no alkali. As in the Dilworth-Occhialini-Payne method, the emulsion could be evenly soaked; placing the film in the alkali solution then began the actual development evenly. See Rotblat, "Photographic Emulsion Technique," *Prog. Nucl. Phys.* 1 (1950): 37–72, on 48–49.

208. Brown et al., "Cosmic Radiation" ([1949] 1972).

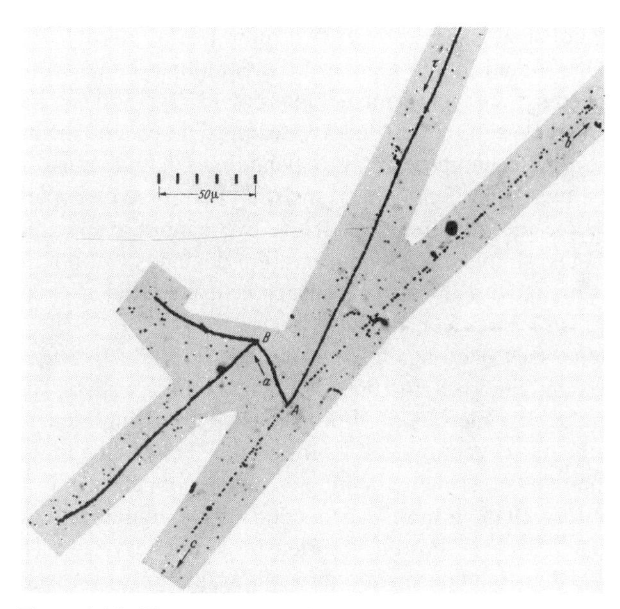

Figure 3.16 First tau photograph (1949). Using Kodak's new NT4 emulsion, which could record even minimum ionizing particles, Brown, Camerini, Fowler, Muirhead, Powell, and Ritson snagged this complicated decay. According to the authors' interpretation, particle *a* is identified as a pion, with its characteristic nuclear decay at point B. By means of grain counting and scattering, particle *b* was classed as a meson of mass 285 ± 30 m_e (and so probably a pion); the final decay product (*c*) was also identified as a pion. The particle labeled τ, entering from above was clearly approaching the center; if one assumed (this was not altogether obvious) that it was decaying at A into three pions (*a*, *b*, and *c*), then its mass could be estimated as somewhat less than 1000 m_e. Interpreted—and reinterpreted—many times over, this candidate tau event became the first in a series of events that, taken together, formed the basis for the claim that the *K* mesons had a decay mode into three pions. Coupled with the argument that the *K* mesons could also decay into *two* pions, one had the famous tau-theta puzzle that led to the discovery of parity violation in the weak interactions. Source: Brown et al., "Cosmic Radiation." Reprinted with permission from *Nature* 163 (1949): 47–51, 82–87, © 1949 Macmillan Magazines Limited.

The novelty in this domain of physics ought to be a more surprising and less expected novelty than that wretched structure ever seemed to offer."[209] Nor was Oppenheimer's protest unique. In 1951, C. O'Ceallaigh, visiting Powell's laboratory

209. Oppenheimer to Powell, 18 November 1949, ROP.

from Cork, published an analysis of two more heavy, unstable particles in the emulsions. One, found by the scanner Miss Stradling, he identified as a kappa particle, dubbed κ_1, with an estimated mass of 1320 $\pm$ 170 m_e; the second, observed by Isobel Powell, he labeled κ_2 as a particle of similar mass. Were the two the same? O'Ceallaigh hesitated, calculating what the mass of the particle must be if the two were identical and insisting that more examples were necessary to complete the interpretation. If the two kappas are identical, however, then, as he notes, simple energy conservation demands that both cases (κ_1 and κ_2) involve a muon with at least two unseen neutral particles. Next, he asks, are κ_1 and κ_2 the same particle as the tau?[210]

Here one sees once again a certain, often productive tension between experimenters and theorists. In 1951, Bethe had publicly advocated the identification of all the new mesons of similar mass with a single one; this, he contended, was the most "economical assumption."[211] O'Ceallaigh demurred: "We believe . . . that the experimental evidence, while not decisive, is against such a view."[212] By July 1953, at the Bagnères de Bigorre conference, that discussion remained contentious, as Powell reported in a summary talk. There was evidence for a tau that could decay into three pions; there was evidence for a kappa that could decay to a muon of varying energy, which implied that it was accompanied by at least two unseen neutrals; there was evidence for a chi (χ) that could make a pion plus one unseen neutral since the pion always had the same energy. Were these the same particle with different decay modes ($\tau = \kappa = \chi$)? Was the tau equivalent to the kappa, and the chi distinct? So was there one new particle? Two particles? Three particles? The masses appeared to be similar, but even that result was not secure: the number of certain processes was small, and in addition, systematic errors loomed large. Were the laboratories using the same scanning techniques? Were they calibrating their emulsions similarly? Were the track lengths long enough to get statistically significant measures of mass?[213] What was needed, Powell asserted, were observations of "greater statistical rate" on the masses of the particles and the products into which they decayed.[214]

Reasoning about the existence of particles used statistics in different ways. The simplest, brute fact of increasing numbers of events was one—let us call this external statistics. By the time of the Bagnères de Bigorre conference there

210. O'Ceallaigh, "Modes of Decay," *Phil. Mag.* 42 (1951): 1032–39.

211. Bethe, Copenhagen Conference, 1951, quoted in O'Ceallaigh, "Modes of Decay," *Phil. Mag.* 42 (1951): 1032–39.

212. O'Ceallaigh, "Modes of Decay," *Phil. Mag.* 42 (1951): 1032–39, on 1038.

213. "The scattering-range method of mass determination contains possible systematic errors due to the uncertainty of our knowledge of the scattering constant and the energy-range relation. The situation would be improved if workers in different laboratories were to use a standardised method of measurement" (O'Ceallaigh, discussion following Powell, "Recapitulation" [1953], 224). On the track length needed, see O'Ceallaigh, discussion following O'Ceallaigh, "Determination" (1953), 128; and the discussion below on grain gaps.

214. Powell, "Photographic-Post-Discussion" (1953).

were, as M. Menon and O'Ceallaigh reported, some 60 *K* particles (by Menon's definition, unstable particles of mass around 1,000 m_e that decayed into a charged particle and other neutrals) in Bristol alone and an additional 20 in other laboratories. These data could be compiled to come to conclusions about masses and decay products.[215] A second kind of statistical argumentation (call it internal statistics) is used, not between events, but within a single event. For example, O'Ceallaigh would measure the spacing between developed grains on the plate. He would then plot the number of gaps of a given length (number of pairs of sequential grains separated by a given distance) versus separation. As theoretically expected (for an even distribution of silver halide in the emulsion), this plot formed an exponential with a decay constant—and this constant is related to the particle's mass. To calibrate the method, the experimenter could study the grain gap distribution for a particle of known mass—such as the muon. Then the experimenter could use statistical methods (least squares analysis) to find the best-fitting curve passing through the grain gap plot of an unknown particle and compare it with the muon's calibration constant.[216] Using methods like this one, particles received ever more accurate masses.

Nonetheless, events were not consistent, and particles continued to appear and disappear, sometimes becoming identical to one another, and at other times branching off into ontological independence. Given this world of ephemeral existence, it is no surprise that the editors of the Bagnères de Bigorre proceedings chose as their motto: "The particles described in this conference are not entirely fictitious and every analogy with the particles really existing in nature is not purely coincidental."[217]

For many of the image physicists assembled at Bagnères de Bigorre in 1953, the dependence on statistical argumentation was acutely uncomfortable. Everything in the image tradition seemed to cry out against it: where, people seemed to ask, was the straightforward and persuasive event that would "carry complete conviction"? The great experimenter P. M. S. Blackett captured this feeling when he devoted part of his concluding remarks at the conference to an indictment of the insufficiently careful attention paid both to systematic and to random error. "I am reminded of one of the numerous stories of Lord Rutherford. One day a research student brought him some experimental results and embarked on a rather long and somewhat complicated discussion of their accuracy in the light of the statistical theory of error. Rutherford got more and more impatient and eventually burst out, 'Do, goodness sake, forget about the theory of error and go back into the laboratory and do the experiment again!'"[218] In

215. Menon and O'Ceallaigh, "Observations" (1953).

216. O'Ceallaigh, "Determination" (1953).

217. *Congrès International sur le Rayonnement Cosmique* (1953).

218. Blackett, "Closing Remark" (1953), 291.

their talks, both French and British summary speakers highlighted the awkward instability that characterized not only the elementary particles they studied but the forms of demonstration that such particles required. In his closing remarks, L. Leprince-Ringuet, like Powell, insisted that "we will need to pursue the possibility of exchanging documents, between specialized laboratories, in order to obtain a mutual standardization (*contrôle reciproque*). I recall a conversation with Powell in Copenhagen, where Powell was ardently wishing for the exchange of documents and physicists in order to establish a coherence among the methods of observation, measures of errors and the collection of experimental parameters. It is good that one particle that appears aberrant be measured not only in one single, but in many laboratories; we all want this, and I am persuaded that more and more it will be done."[219]

3.8 Visual Statistics: The Dalitz Plot

Powell's insistence on standardization between laboratories aimed above all to quiet the chaotic appearance and disappearance of new particles, to still an ontology gone amok. For Powell, however much he might prize the golden event, there seemed to be nothing for it but to use statistics when the golden events conflicted. In particular, getting a statistical grasp of particle masses was crucial to the establishment of the basic objects of the world: when Powell said the tau and kappa were "equivalent," he would systematically add "or that $m_\tau \equiv m_\kappa$," meaning that the masses were averaged over the results of many laboratories' work. Similarly, at a Bagnères de Bigorre summary talk, Bruno Rossi uneasily announced that kappa and tau were equivalent then a few minutes later, during discussion, reckoned that there were good arguments for separating them. Another few exchanges and Rossi was back to wondering whether perhaps there were explanations to do with the different nuclear absorption of the two particles that might allow them to be identified after all. This kind of debate ran through the host of conferences devoted to cosmic rays during the first half of the 1950s. Into the fray stepped the theorists.

To a large extent, nuclear theorists and emulsion physicists were speaking different languages in the early 1950s—the theorists were preoccupied with topics like renormalization in QED and the possibility of constructing a satisfactory theory of meson-carried nuclear forces. The gap is visible in letters written by theorists to Powell, almost always stripped of the authors' principal concerns and restricted to the narrowest of questions about decay patterns. For a moment, I would like to focus on a powerful theoretical intervention to mediate between theorist and experimenter: here the theorists brought to the conversation a form of statistical reasoning about taus that was neither between grains

219. Leprince-Ringuet, "Discours de clôture" (1953), 288.

within tracks nor simply an averaging over masses and production ratios. More specifically, I want to concentrate on the theorists' categorization of the mesons by properties (spin and parity) that had long been central to theoretical work but virtually absent in the analytical practices of emulsion experimenters. This work—by theorist Richard Dalitz—illustrates in detail how theorists played a role in steadying discussion about which particles were which. At the same time, these "Dalitz plots" illustrate a sophisticated statistical method applied to a *visual* method of experimentation, a method that, as we will see, carried over into the age of bubble chambers. Finally, the Dalitz plot provides an elegant illustration of a set of practices that could constitute a trading zone between experiment and theory.

That theorist and experimenter did not always see eye-to-eye in mid-1953 is clear. When École Polytechnique theorist Louis Michel stood to talk at Bagnères de Bigorre, he began: "Most of you are quite suspicious against theorists and prefer, to their advice, the answer of Nature. You are right." But in subcultural self-defense, Michel went on to insist that—among physicists—no one ought work without the "best proved theoretical laws"; no one, he continued, would violate momentum, energy, or charge conservation. At least at that time and place, these were considered to be bedrock constraints for experimenters; I know of no instance where Powell and his team even contemplated violating them. Michel's point was that there were other constraints, theorists' constraints, that fundamentally reshaped which particles could be considered equivalent. There were constraints of nucleon number: if a proton (nucleon number 1) hit a neutron (also nucleon number 1), then the outgoing particles must also sum to a nucleon number of 2. At the same time, conservation of angular momentum was implied by classical physics. More generally still, special relativity and quantum mechanics together demanded conservation of parity and conservation of (quantized) angular momentum. Addressing a recurring theme of the conference—the question of the identity of the theta with the tau—Michel concluded that the two might be "corresponding particles" if both were $J^P = 2^+$, where J was the total angular momentum of the final states and P was the total parity (positive if all spatial variables were switched as if in a mirror image, and negative if the quantum mechanical wave function changed sign).[220] Dalitz addressed the question of spin and parity in a different way.

In 1948–49, Dalitz had been a young theorist working as a research assistant to Neville Mott at Bristol. Though he was not an experimenter, he was frequently in touch with the band of emulsion hunters and remained in contact with them over the following years. Through his links with Bristol's "fourth floor" (Powell H.Q.), by late January 1953 Dalitz knew that the Bristol film analysts were characterizing positive tau decays by their decay into three pions, two

220. Michel, "Absolute Selection Rules" (1953).

positive and one negative.[221] Preparing for the Bagnères de Bigorre meeting in July 1953, Dalitz had taken the tau data and put them in a form unlike any previous representation: his idea was to use the energies of the tau decay products to constrain the spin and parity of the tau by studying the decay:

$$\tau^+ \rightarrow \pi^+ \pi^+ \pi^-.$$

It is worth following Dalitz's representational scheme in some detail as an illustration of how a theorist—using, not experimenters', but theorists' constraints—could sort the data differently and so divide the particle ontology differently from his colleagues of the bench.

Dalitz's representation worked this way:[222] Inside an equilateral triangle he inscribed a circle (see figure 3.17*a*). Each individual tau decay into $\pi^+ \pi^+ \pi^-$ would have a unique point, P, to represent it. (The usual charts with special comments on each event now vanished.) By elementary euclidean geometry, for any P, the sum of the three perpendiculars, PL, PM, and PN, is the height of the triangle: PL + PM + PN = CY, and since CYV is a 30-60-90 triangle, the height (CY) is $\sqrt{3}/2$ times the side, YV. The point P is fixed by choosing the perpendiculars to stand for the pion kinetic energies; since the two positive pions (π^+) are indistinguishable, it is possible without loss of generality to set PM = T_1 (by definition one positive pion's kinetic energy) less than PL = T_2 (by definition the other pion's kinetic energy). This simplifying assumption puts all points P in the right-hand side of the diagram. Dalitz then chose PN = T_3 to represent the energy of the negative pion (π^-). Using momentum conservation and simple geometric arguments, Dalitz showed that the point P is always inside the circle in figure 3.17*a*, that is, that no one of the three pions can have energy too much larger than the other two. Thus, P is always in the right-hand semicircle.

Fermi's "Golden Rule" (from nonrelativistic quantum mechanics) tells us that for three-body decay, the probability of a decay with given energies (T_1, T_2, and T_3 above, which determine the point P) is proportional to the square of the matrix element which describes the transition from the initial state of the tau particle to the final state of the three pions. If the matrix element does not depend on the particular distribution of energy among the three pions (as would be approximately the case if the decay products have spin zero and no mutual rotation), then we would expect the tau decay points P to fall evenly throughout the allowed right-hand Dalitz plot semicircle.

221. Powell, "Discussion," *Proc. Roy. Soc. London A* 221 (1954): 277–420.

222. See Dalitz's contribution at Bagnères de Bigorre, "Modes of Decay" (1953), and his very useful retrospective account of the sequence of cosmic ray conferences in "Strange Particle Theory," *J. Phys.* 43 (1982): 195–205; also see his articles "Decay of τ-Mesons," *Phys. Rev.* 94 (1954): 1046–51; and "Analysis of τ-Meson Data," *Phil. Mag.* 44 (1953): 1068–80. See also the very useful discussions of the Dalitz plot in Perkins, *Introduction* (1987), 130–38; and Cahan and Goldhaber, *Experimental Foundations* (1989), 52–59.

To investigate what would happen if the spin were nonzero, Dalitz let the angular momentum of the $\pi^+\pi^+$ system be called L (which had to be even since the particles were identical); he dubbed the angular momentum of the negative pion relative to the center of mass of the $\pi^+\pi^+$ dyad, l. Then J (total angular momentum) equals the vector sum $L + l$. The parity of the tau is given by the product of the intrinsic parity of the pion (determined in previous experiments to be -1) and the parity of the orbital wave function, $(-1)^J$. Since there are three pions in the decay, the tau parity is deduced to be:

$$P(\tau) = (-1)^3(-1)^J = (-1)(-1)^{L+l} = (-1)^{l+1} \quad (3.1)$$

as L is even. For a decay to two pions (as in the theta decay) the parity is $(-1)^l$, so Dalitz could immediately conclude that for $J = 0$ and parity odd (pseudoscalar case), a tau decaying to three pions would be allowed, while a tau decaying to two pions would be forbidden. For $J = 0$ and even parity (scalar case), the reverse would be true, a situation of no interest since the whole point was to model the tau-to-three-pion process.

What about higher spin assignments for the tau, which led to interactions among the decay pions? These, Dalitz argued, would reveal themselves through deviations from a homogeneous distribution on the Dalitz plot—a method simple, qualitative, and visually apparent. Suppose, as became manifest during 1954 and 1955, that in the vicinity of the south pole of figure 3.17*c*, there was no dramatic depletion of events. This would mean that there were events in which the negative pions have energy near zero, a statement equivalent to saying that there was no negative pion angular momentum with respect to the positive pion pair ($l = 0$). In turn, this implied that $(-1)^{J+1} = (-1)^{L+1}$ and, because L is even, that the tau must be in the sequence $0^-, 2^-, 4^-, \ldots$ Since the theta decay was parity positive, come what may, it seemed that the tau-to-three-pion decay was incompatible with tau to two pions: the two particles could not be the same. Incompatible, that is, unless parity was not conserved—but that is a different and longer story that functions inversely to this one: how new theoretical constraints joined, rather than parted, previously established entities.[223]

Dalitz's talk about pseudoscalars, vectors, and pseudovectors was a language that, however familiar to theorists, was not a weapon in the usual armamentarium of the assembled experimenters, some of the most accomplished of their day. Bruno Rossi, for example, reviewed the Bagnères de Bigorre conference results, specifically addressing the question of whether the theta, tau, kappa, and chi were to be identified. His detailed and thoughtful analysis *never* invoked the symmetry constraints so dear to the theorists' hearts. Rossi's world of constraints was (in July 1953)—as Louis Michel implied—one of masses, decay

223. See Franklin, "Discovery and Nondiscovery," *Stud. Hist. Phil. Sci.* 10 (1979): 201–57.

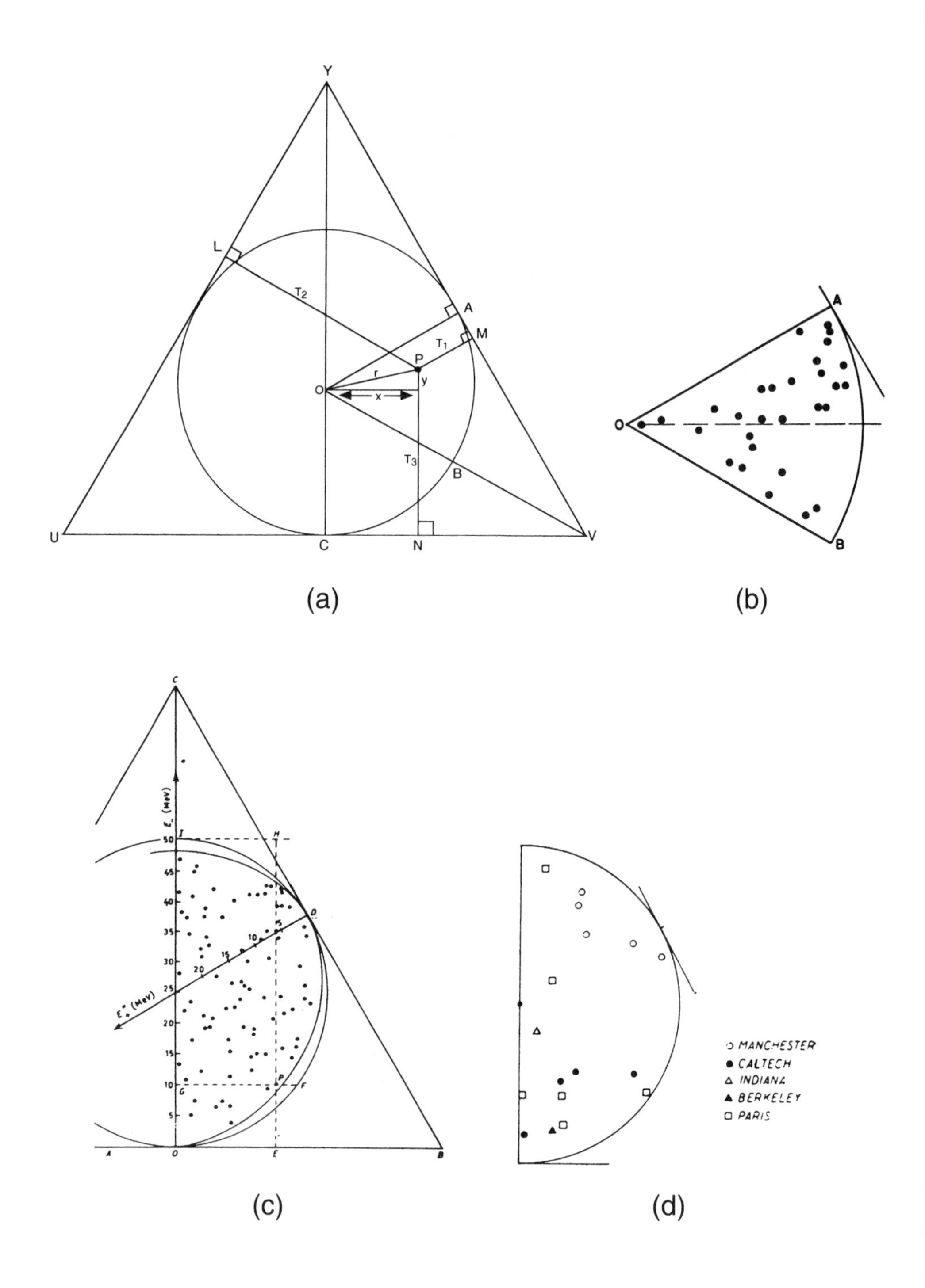

Y
L
T2
A
M
T1
P
r
O
y
x
T3
B
U
C
N
V
(a)
A
O
B
(b)
C
E_+ (Mev)
E_+^* (Mev)
I
H
D
G
F
A
O
E
B
(c)
MANCHESTER
CALTECH
INDIANA
BERKELEY
PARIS
(d)

ratios, energies, and the multiplicities of decay patterns. The same could be said of Blackett's and Powell's closing comments. For the experimenters, the coincidence between the 970 m_e mass for the tau and the 971 m_e mass for the theta zero was prima facie evidence that the two particles were identical. For a theorist like Dalitz, the fact that one was parity even and the other parity odd set them apart. By 1956, experimenters were adopting the Dalitz plot as a standard item in their toolkit. E. Amaldi concluded on its basis (see figure 3.17*c,d*) that the "τ and θ mesons must be different particles."[224] Different constraints meant parsing the world differently.

The need to represent as many taus as possible on the Dalitz plot added to Leprince-Ringuet's call for *contrôle reciproque* between the laboratories. At the Padua meeting of April 1954, a "tau committee" was established to coordinate data. By the Pisa meeting in 1956, Amaldi could offer detailed charts of all the world's taus, contributed variously from Bern, Brookhaven, Berkeley, Bombay, Bristol, Bethesda, Paris, Göttingen, Milano, Padova, Rochester, Rehovoth, Roma, Torino, and Madison. But still Amaldi complained that "the agreement between the results of various laboratories is not satisfactory." There might still be differences in the "practical details of the calibration" or it might be that the procedure itself was no good (using at-rest pions decaying to muons as a calibration of emulsion stopping power).[225]

Aside from illustrating how theorists and experimenters can divide the world differently (and eventually come to resolve that difference), the Dalitz chart

224. Amaldi, "Report on τ-Mesons," *Nuovo Cimento* 4 Suppl. (1956): 179–215, on 207.
225. Amaldi, "Report on τ-Mesons," *Nuovo Cimento* 4 Suppl. (1956): 179–215, on 196.

Figure 3.17 (opposite page) Dalitz's plot (1953). Richard Dalitz took individual photographs, from both emulsions and cloud chambers, and plotted them in a revisualized map. These plots therefore constituted a "second visualization" in which significant theoretical conclusions could be extracted by eye as one inspected the clustering (or equidistribution) of points. Though a theorist, Dalitz immersed himself in the visual practices of Powell's Bristol laboratory. Through these plots, he established a trading zone, bringing previously theoretical categories (such as spin-parity relations) to experimenters and rendering cloud chamber and emulsion work accessible to theorists. (*a*) Dalitz plot geometry (see text). Unlike magnetized cloud chambers, emulsions did not always unambiguously yield the charge of the meson, and without charge information, there was an ambiguity as to the sector of 3.17*a* into which the points should fall. Dalitz put them all in the right-center sector (see 3.17*b*), and in February 1954 he judged them "not yet inconsistent with a random distribution." Figure 3.17*c* prepared by Amaldi, 3.17*d* by Dalitz, both in 1957. They show emulsion data for tau-meson decays from emulsion and cloud chamber pictures, respectively. Sources: 3.17*a* after, and 3.17*b* from, Dalitz, "Decay of τ-Mesons," *Phys. Rev.* 94 (1954): 1046–51, on 1051; 3.17*c* and 3.17*d* from Amaldi, "Report on the τ-Mesons," *Nuovo Cimento* 4 Suppl. (1956): 179–215, on 206.

raises two important points about the nature of demonstration and scientific authorship. There is no doubt that Powell and many of the other emulsion physicists found the clarity of an argument based on a single, sharply defined event to be of enormous demonstrative force. This predilection should not, however, obscure the manifold uses of statistics when individual events clashed. Along with the turn toward statistics (and the concomitant shift toward aggregating the fruits of many laboratories) came a shift in scientific authorship. For while a golden event interpreted by an O'Ceallaigh, a Powell, or a Perkins left the scientific author clearly identified, the Dalitz plot did not. *By definition,* the demonstration of the chart had to do with the equidistribution of points in the semicircle, and those points came from unspecified physicists and scanners in 15 laboratories in eight countries. The signature of the individual scientific author vanished just as the characteristic trace of individual events collapsed into Dalitz plot points. *With respect to the demonstration of the spin-parity of the tau, no single point meant anything at all.* The (temporary) separation of the tau from the theta was "authored" by the community as a whole.

3.9 Instability and the Army of Science

As Amaldi indicated, by 1956 it was clear both that the emulsion method needed coordination between laboratories and that the method had *still* not fully stabilized. Reproducibility remained problematic, and workers in the field knew it. At an international conference at CERN, one worker complained about yet another difficulty arising from the experimenters' isolation from the emulsion makers; this time the accusing finger pointed at Ilford:

> [T]he other day I was talking to Mr. Ehrlich of Ilford Limited, and I told him that we take great care to keep the emulsion packages well sealed, so that we can be sure that the water content is always that which corresponds to the relative humidity of 50% at which the emulsions are dried and packed at Ilford's. "Oh," said Mr. Ehrlich, "but we don't have the time to let the emulsions reach equilibrium." . . . So the water content of the emulsions may vary from batch to batch, or even package to package and pellicle to pellicle. This is something I had not thought of before.[226]

He then noted that the nuclear emulsion had not shed its early responsiveness to humidity, which caused fading, altered emulsion density, and precipitated shrinking. Even the film's sensitivity, one of its characteristic features, was "not a very well defined quantity." Like the tracks by Reinganum a half-century before, "our tracks . . . are subject to distortion that can be confused with scattering, and like all the emulsion workers . . . we have to be very careful in our processing if we

226. Herz, "Measurements" (1965), 84.

are to obtain reproducible results and uniform sensitivity." The emulsion physicist elaborated:

> The word "reproducible" brings me to a point which is rarely made, perhaps because it is so obvious, but which is of great importance to all our work. It is that there are very few people indeed who know how to make nuclear research emulsions in large quantities in a reproducible fashion. That it is possible to obtain good results in such a messy medium is indeed a miracle, and we should never forget to give credit to the manufacturers, especially to the Ilford group of Dr. Waller and Mr. Vincent and, now, Mr. Ehrlich, and to Professor Bogomolov, who have succeeded in giving us a product of almost constant properties in which precision measurements can be made.
>
> In all our work we rely on the Principle of Uniformity of Nature; that is, we believe that if the same experiment is performed several times in exactly the same way, then the results will be the same. However, I do not believe that there is also a Principle of Uniformity of Nuclear Emulsion, except by the grace of the manufacturers. Today there are physicists who have spent an appreciable part of their working lives investigating the properties of this medium. It is true that important results have come out of such work, but I think it is useful to remind ourselves from time to time that nuclear emulsion is not a phenomenon of nature; it is a very powerful tool that we must apply to justify its existence.[227]

Even the grace of manufacturers, then, could not underwrite the Principle of Uniformity of Nuclear Emulsions. As Berkeley learned the hard way, all the little side techniques—even of processing—had to be honed in the day-to-day practice of the laboratory. Just as the physicists ceded some of their earlier functions to the scanners, they were simultaneously delivering others—the fabrication and even analysis of their instrument's inner workings—to the film companies. Nor did changes in the experimenter's practice end there.

One group leader from the University of Bern, F. G. Houtermans, had his wife travel to Bristol to learn the method and then began negotiations over how to acquire the exposed plates.[228] As Powell wrote back, "[Y]ou might like to adopt the procedure of about half a dozen laboratories who send us money and we provide them with exposed stacks. Commonly we also develop the plates, but that need not be an invariable rule. If you like to come to a similar arrangement you should let us know what size of stack you want. . . . I think it reasonable to suppose that if you sent us £500 we should be able to supply you with a stack of 40 stripped emulsions, 8″ × 6″ flown at 85,000 ft or over for 4 hours at this latitude."[229] While Houtermans declined this particular offer (he had already

227. Herz, "Measurements" (1965), 82.

228. Houtermans to Powell, 21 January 1954, CPP, E18.

229. Powell to Houtermans, 13 July 1954, CPP, E18.

subcontracted for flight exposure by the Office of Naval Research in Texas), he did inquire about the famed ability of one of Powell's associates to cut the emulsion stacks for distribution to the various laboratories.[230]

Similarly, in correspondence with Feather in 1954, Powell offered to expose Feather's plates in the Bristol balloon flights: "[I]t occurred to me that if you are interested in experiments with plates it might be an advantage for you to work with stripped emulsions. We are shortly to make flights from Cardington, and have arranged to expose and process stacks for Copenhagen, Oslo, and the Japanese Universities. They give us £300 each and we do our best to provide them each with a stack of 40 stripped emulsions exposed above 85,000 ft. for 3 hours and processed."[231] The so-called G-stack brought a large number of laboratories into a centralized collaboration, involved laboratories without great financial resources in particle physics, set the stage for international cooperation at CERN, and created a source for particle physics that could, in the trying postwar conditions of Europe, compete with the great accelerators of the United States. The British David with a microscope and film pitted against the Goliath of Berkeley was a matter of some discussion, even among the Americans. One Californian, having visited the Bristol group in late 1948, reported back to Berkeley, including the latest plates to be passed on to Oppenheimer: "I am enjoying my work in England tremendously and I am learning a lot. Mainly, how the British are able to turn out so many fundamental discoveries with such limited budgets. It is all a matter of thinking the problem out carefully before tearing into it which we in the States tend not to do."[232]

For the Europeans, the extension of the cosmic ray apparatus through the inexpensive emulsion medium bought time. For Powell, the various cosmic ray laboratories had to band together, in much the same way that late-nineteenth-century surveyors of the earth's magnetic field had; no doubt Powell also called upon his own experience in setting up the observer network at Montserrat. As he introduced a major session at the Bagnères de Bigorre conference in July 1953, Powell began by invoking Maxwell: "[T]he scattered forces of science are converted into a regular army."[233] For Powell that army would consist not only of the various emulsion laboratories but of myriad allied cloud chamber groups spread from the École Polytechnique to India.[234] Together, they would march

230. Houtermans to Powell, 22 March 1955, CPP, E19.

231. Powell to Feather, 11 May 1954, CPP, E15.

232. Thomas Coor to David Bohm with plates for Oppenheimer, 10 December 1948, ROP.

233. Powell, "Recapitulation" (1953), 221.

234. It was evidently a metaphor he liked. It occurs frequently in his writing; for example, just a few years later in his lecture celebrating the International Geophysical Year (1957–58): "The cosmic radiation provides us with a source of particles of much greater energy than any we can generate at the present time with the big machines. . . . In the past ten years, notably at Manchester and Bristol, such studies have led to the discovery of many different kinds of unstable elementary particles which we refer to as mesons and hyperons. In a sense, investigators of the cosmic radiation are the advance guard of the great scientific army involved in the study of nuclear

against the anxiety of instability that plagued any group foolish enough to foray out on its own.

3.10 Chased by Machines

But there was another form of instability, one vastly more threatening than the physical variability of emulsions: would the field of cosmic rays itself survive? From standardization between laboratories, Leprince-Ringuet turned in his wide-ranging 1953 speech at the Bagnère de Bigorre conference to a subject that, he confessed, was dear to his heart. Others among the *cosmiciens* would sympathize, but perhaps most of all the *anciens*. What, Leprince-Ringuet asked, was the future of cosmic rays? Must we toil for a few new results, or would we do better to turn it all over to the machines? Reciting with pleasure the sequence of new detection devices that had kept the cosmic ray physicists in business—cloud chambers, counter-controlled cloud chambers, emulsions, emulsion stacks—he took pride in the cosmic ray physicists' past accomplishments. But it might not be enough: "We must go quickly, we must run without slowing our rhythm: we are being chased . . . we are being chased by the machines!" Already, he reported, Brookhaven was at that very moment making a neutral *V* meson and would be producing ever more of them. "We know that we could perfectly well take six months of vacation in the countryside and come back to find Brookhaven with the truth about those problems that we have been examining on super protons, super neutrons, and all the hyperons that we have been studying these last few days." At stake was more than priority. A way of life for these physicists was on the table, and accelerator physicists held better cards. Leprince-Ringuet concluded with a metaphor: "I think we are a little bit in the position of a group of alpinists who scale a mountain, a mountain of extraordinary height, maybe almost indefinitely high, and we are clambering in ever more difficult conditions. But we cannot stop to sleep because coming from below, from beneath us, is a rising tide, an inundation, a flood which builds progressively and forces us to climb ever higher. It is evidently a not very comfortable position, but isn't it lively and of marvelous interest?"[235]

Leprince-Ringuet's cri de coeur held both a figurative and literal meaning since the cosmic ray physicists were indeed forced ever higher into the alps and into the stratosphere to increase their yield of intriguing events. Others echoed his concern. P. Cüer introduced the second international emulsion conference by commenting that the "exceptional importance of the French delegation in difficult national circumstances well demonstrates that . . . they can remain active in

processes" (Report on International Geophysical Year and the Bristol Laboratory, CPP, B11; no author [but certainly Powell], no title).

235. Leprince-Ringuet, "Discours de clôture" (1953), 290.

one of the most modern scientific domains where ingenuity can still play a big role when pitted against the abundance of material means."[236] These were brave words. But even in the dreaded American accelerators, the nuclear emulsion technique, like the latent images it held, was beginning to fade. Giant bubble chambers offered similar stopping power with vastly greater effective volumes and a far finer discrimination in time.

In the alpine cosmic ray station lay a romanticized image of what it meant to be an experimenter (see figure 3.18). Gathered around Leprince-Ringuet's depiction—or ideal representation—are, once again, the banners of a now-departing age. Americans were taking over where Europeans once were; material means superseded ingenuity, and factory-like conditions supplanted the isolated explorer. But to complete the image of the hut, more needs to be added: the camera must pull back to capture the County of Essex Ilford laboratories, the teams of scanners peering through microscopes, the distant theorists corresponding with the emulsion experimenters, and the allied laboratories cross-checking each other's results. Still the central authority—in all senses—remained with the director of research, and a physicist like Powell (or Leprince-Ringuet or Blackett) could well feel the threat of the impending invasion of machines.

By 1965, the brave purveyors of cosmic ray emulsions had lost their race; the alpine hut had surrendered to the accelerator control room. Gathering at CERN, experts on the subject from around the world reported on the latest developments, but the technique had begun to feel its domain restricted, hemmed in both by the newer visual technique of the bubble chamber and by the competing electronic methods that were garnering more and more attention. When Bernard Gregory began the closing remarks of the conference, he warned the assembled devotees that his was to be an honest, rather than truly summary, set of remarks. First, he apologized for not having attended most of the meeting, only to add that as far as he was concerned "technical conferences are necessary evils, since what we really want is to do the physics with the technique and not really discuss the technique as such." Still, he allowed, such "technical" assemblies were needed to secure the reliability of the results sought by "experimental physicists." It was a move evidently dividing instrumentalists from experimenters.[237]

Gregory next acknowledged some recent progress with the film. The discovery of the sigma plus in emulsions was significant, but any increase in the accuracy of measuring it properly might cost more pain in scanning than it was worth. Small-angle proton-proton scattering was a useful result as well, but even those inquiries had a dim future: "I think that one should consider that counter or spark chambers have now managed to develop their precision in measurements, so that they can match the precision of emulsions; moreover, they have

236. Cüer, "Introduction" to Demers, *Photographie* (1959), 9–18, on 9.
237. Gregory, "Closing Talk" (1965), 107.

Figure 3.18 Jungfraujoch Observatory, Switzerland (1954). Typical of the high-altitude observatories was that on the Jungfraujoch in Switzerland—it was here, for example, that Brown et al. exposed the film that revealed the tau event of 1949 (figure 3.16). The contrast between this alpine redoubt and the factory structure of the Lawrence Berkeley Laboratory at the same time was striking, sometimes painful, to the European cosmic ray physicists. Source: S. Korff, ed., *The World's High Altitude Research Stations* (New York: Research Division, College of Engineering, New York University, 1954), plate opposite 30.

the enormous advantage of being able to collect their data directly on a computer and give the answer almost right away. Therefore, in spite of these good results recently obtained using emulsions, we are not able to foresee a very great future along this line."[238] Gregory's damningly faint praise must have been hard on the

238. Gregory, "Closing Talk" (1965), 108.

assembled audience. The study of hyperfragments, which he lauded as a useful pursuit with emulsions, was to be supported primarily because it used so little machine time.

While wishing the conferees well in their efforts to study the sigma plus, Gregory pronounced what amounted to a gentle burial speech: "[W]e should consider that this beautiful technique, which has given so many very good results in elementary particle physics in the last years, does not seem to be one which, at this time, should be considered as one of the leading and growing types of techniques around a large accelerator."[239] These words closed the fifth international conference on nuclear emulsions. Never again would emulsion workers assemble on such a scale. Physicists pursuing images of the particle world had turned to the bubble chamber and to their even newer electronic rivals.

3.11 Conclusion: The Anxiety of the Experimenter

The nuclear emulsion had its heyday between 1935 and 1955. In those two decades the method served as the basis for the establishment of a new subdiscipline, "particle physics." As Powell and his colleagues pointed out, by 1959 there were 20 particles in the zoo.[240] Though the "moment of discovery" of the electron is itself arguably undefinable, the authors assign it to the fluorescent screen technique, while they put the neutron down to the ionization chamber. Of the remaining 18 particles, 15 were located using visual techniques: 7 with the cloud chamber (e^+, μ^+, μ^-, K^0, Λ^0, Ξ^-, and Σ^-), 6 with nuclear emulsions (π^+, π^-, anti-Λ^0, Σ^+, K^+, and K^-), and 2 in the bubble chamber (Ξ^0 and Σ^0). Only three had been found using purely electronic techniques (anti-n, anti-p, and π^0): the power of the image to establish the reality of particles remained dominant. But beyond its contribution to the burgeoning field of particle physics, emulsion images offered the devastated countries of Europe the opportunity to participate in physical research when new and expensive accelerators lay beyond their reach. For a brief and fragile moment, the method allowed small-scale physics to survive in an age of accelerators, away from the vast material means of the Americans, as Cüer and Leprince-Ringuet were quick to emphasize. Of course, once the accelerator physicists could adapt the method to their machines, the cosmic ray physicists' days were numbered.

But from the beginning, as I have argued throughout, the photographic method was fragile in another sense: its results were quirky and evanescent. At times purported particles appeared once then vanished. Two particles thought distinct turned out to be one; one particle on closer examination became two. Images on Blau and Wambacher's emulsions disappeared altogether; kappas,

239. Gregory, "Closing Talk" (1965), 110.

240. Powell, Fowler, and Perkins, *The Photographic Method* (1959), xiv.

taus, and chis came together, broke apart, and eventually merged into the kaon. The interpretation of the pion decay shifted dramatically. The causes of this fragility were many, leaving the practitioners anxious on four fronts, as control over their work was dispersed to others.

First, their covenant with the industrial chemists left the particle hunters on shaky ground. True, in the decades that followed the era of emulsions, physicists grew accustomed to a certain distance from the design and function of their instruments: it came to seem natural that electrical, cryogenic, and structural engineers would offer physicists black boxes that they could deploy to their own ends. But for the emulsion physicists, the past (or an image of the past) cast a golden light over the physics that had just passed from view: gone, it was thought, was an era when physicists could supervise—if not participate in—most aspects of instrument production. The shock of exclusion was great. And nowhere was the abruptness of the boundary so visible as in the contractual isolation of the emulsion chemists from the physicists. Exclusion carried with it a vulnerability to real or perceived variance in the experimental results. Whether the emulsion physicist worried about fluctuations in density, humidity, or grain density; whether variations arose from unknown rapidity of fading, from the passing effects of pinakryptol yellow, or from inhomogeneous development or drying, the loss of control meant instability of results. It was an instability that manifested itself in Gardner's declaration and retraction of the pion confirmation, in Powell's group announcement and retraction of a high value for the ratio m_π/m_μ, in fears about the validity of grain counting, and in the sometimes productive, sometimes troubled, private communications in the trading zone between physicists and chemists.

Second, the installation of the women scanners, like the collaboration with the chemists, offered both an opportunity and a threat to the physicists. There is no doubt that without the 20-strong team of scanners at Bristol, the number of interesting new events found would have plummeted. Yet, as we saw, the new reliance on the women came at the price of another kind of uncertainty. If, as many of the physicists believed, the women scanners had a natural proclivity to work meticulously even at intrinsically uninteresting work, might not this virtue collide with the scanners' equally natural tendency to want to please, to produce exactly those results that the physicists wanted? Just as some physicists worried that turning over responsibility for the emulsions to the chemists made their results more vulnerable, so others believed that handing the job of finding interesting phenomena to the women removed that much more control from the physicists, adding to their sense of anxiety about their results. The complexity of the dynamics between physicist and scanner, precisely around the "objectivity" problem, further suggests that women executed a more formative role in the laboratory than merely providing cheap labor. As one emulsion physicist, Barkas, wrote (in a comment cited earlier), "[I]t is important that the scanner be *unaware* of the result . . . expected"; only the unskilled could live in that state of objective ignorance. As the physicists at the time clearly understood, both when they exalted it and when they worried about it, scanning was difficult work, and good scanners were not easily replaceable.[241]

Third, with theorists overstepping themselves to interpret the tiny silver tracks, experimenters felt further destabilized. As theoretical ideas and expectations fluctuated, Powell and other emulsion experimenters found themselves hedging bets, assuring each of the theoretical camps (one thinks here of Powell writing to Christian Møller) that the new result might not be incompatible with the latest pseudoscalar meson theory. Theory was an uneven constraint on the experimenters' interpretation—sometimes experimenters would buck the theorists' expectations, and at other times, the theorist's disapproval left them unwilling to argue too hard for a new result.

Fourth and finally, collaboration among the physicists themselves was not without its tensions, as the task of film exposure, observation, and interpretation was distributed over many individuals (and groups). Who would count as an author? What would count as "adequate" statistical demonstration? Though, by and large, the distribution of plates to other groups, and the joint efforts of the G-stack collaboration, worked to bind different groups together, the exchange of the plates as property raised its own set of difficulties about who an author was and how responsibility ought be partitioned. In the summer of 1948, for example, Powell drafted a letter to Yves Goldschmidt-Clermont, reflecting on the difficulties of their joint efforts. Some time earlier, Powell and Goldschmidt-Clermont had apparently established a common project (with Ugo Camerini, H. Muirhead, and later David Ritson) to study the absorption by ice of the radiation responsible for nuclear stars.[242] According to a draft letter remaining in the Powell papers, Powell contended that he agreed with M. G. E. Cosyns, one of Goldschmidt-Clermont's collaborators, when Cosyns had argued early in the collaboration that "there is no particular virtue in having made an exposure," and consequently "that any novel features observed in the plates, outside the scope of the original purpose of the experiment should be treated as the particular property of the observer who found them. I understood this to mean that he wished some freedom of action for you in relation to observations not connected with the defined object of the investigation, and whilst it clearly limited the scope of the collaboration, I sympathised with his view."[243] Now Powell wanted to publish autonomously (on the pion lifetime), and apparently, as a result, there were some hard feelings. As had occurred in the other branches of collaboration, the trade of control for resources offered both rewards and difficulties: in each of many ways

241. On women's work in astronomy, see Rossiter, *Women Scientists in America* (1982), esp. 54–55.

242. See Camerini et al., "Slow Mesons" ([1948] 1972).

243. Draft letter Powell to Goldschmidt-Clermont, n.d. [probably August 1948], CPP, notebook 22.

the category of the individual experimenter-author came in for revision. With the multilaboratory collaboration of the much larger G-stack experiment, the category of the individual experimenter took an even more severe battering. With the loss of control came anxiety.

I take the experimenter's anxiety seriously. It is important not merely because of the increasing discomfort the experimenter experienced as responsibility was distributed over chemists, scanners, theorists, and other physicists, but because the tension signaled a reformulation of the very definition of the experimenter-author. A principal component of the new experimentation was the team, and its leaders. As many of his contemporaries recognized, Powell's role was not as the individual discoverer in the older sense; that role had splintered into tens of parts. Powell was not an emulsion chemist, he was not the one who did the main scanning work, and he did not do the fundamental theoretical or interpretive work. Indeed, the original method he used to fix the pion-to-muon mass ratio (grain counting) soon came under fire, and Powell's interpretation of the decay did not last a year. Powell was hardly the classical "discoverer" of the pion; he was recognized, instead (both then and subsequently), as the organizer of the team, the manager of a new research group in which others' expertise was marshalled and assembled into important physics.

In this new world, there was little place for the solitary worker. The individual who tried to put together chemical expertise and the inquiry of experimental physics, a Blau, a Demers, or a Zhdanov, never had a central role in the Bristol laboratory. By working in concert with a team of other physicists and the largest photographic corporations in the world, Powell was able to standardize his approach, identify gaps in the procedure, and by so doing, at least partially stabilize a highly unstable mode of experimentation. It is no accident that Blau could contribute to the production of the first emulsions but not to the establishment of the myriad of unstable particles that followed: isolation from the new materials, the reciprocal standardization, and the coordination of results was simply incompatible with the new forms of demonstration.

As time passed, teamwork and interdisciplinary collaboration loomed ever larger as the significant milepost of Powell's accomplishment in the history of experimental particle physics. Blau's own contributions, recorded in the first words of the first entry on emulsions in Powell's notebooks, receded as the years passed. In a sense this should not surprise us. Her work was far from center stage, conducted at the margins of a well-known but turbulent Austrian laboratory, already convulsing under political tension in the 1930s. At the peak of her career, Blau had been tossed out of the community of nuclear physicists by forces far beyond her control.

As for Blau herself, her career spiraled away from the visible laboratories of Columbia and Brookhaven. After several years in Coral Gables, Blau's eye troubles and financial problems drove her back to Austria to await an operation.[244] Several physicists tried to gather funds for her. Schrödinger put her up for the Schrödinger Prize (which she won), and twice for the Nobel Prize, to no avail.[245] And Halpern, through Otto Frisch, tried to convince the big film companies to grant her a sinecure in recognition of the minor industry of nuclear emulsions that she helped create. Ilford responded this way: "[O]n the assumption that from all sources a worthwhile contribution to her future welfare will be forthcoming[,] Kodak Limited and ourselves would each be willing to contribute £100 per annum."[246] Frisch then wrote back to Halpern to confirm the arrangement. The £200 "did not sound very much," he admitted, "and Powell suggested that I should see if it could be increased by invoking some of the finer points in British tax law. That turned out not to be possible."[247] Halpern tried one last time to get the firms to take on Marietta Blau in a consultantship of some sort. They declined. "Her eyesight has improved after an operation but the heart seems to have become so weak that she cannot even carry books from the library to her home. . . . I somehow feel, there is hope that I can perhaps persuade her to accept an agreement about some activity for the firm, even if she recognizes it as more than a semi-bogus."[248] Halpern was wrong. Her pride still with her, Blau thanked Frisch in 1964 for his kind efforts and reiterated her gratitude to Ilford. But in the end she declined despite her poverty: "[F]or various reasons I believe that it [the consulting job with no real duties] could not be done in this way. I also want to thank all colle[a]g[u]es who have thought about, how to help me. In this connection I wish to tell you that I do not suffer any material hardship because one of my brothers is well situated and very willing to help me, if I can not manage alone . . . yours Marietta Blau."[249] Blau's use of emulsions in her struggle with marginalization was extreme, but in various forms, the emulsion provided the same lifeline to others—to Leprince-Ringuet as he climbed away from the deluge of American accelerator physics, and eventually even to Powell's well-appointed laboratory as Berkeley came on line with its factory of mesons. Yet somehow Blau's fate remains both specific and emblematic of what it meant to be a woman, a Jew, and a solitary physicist fleeing the collapsing world of Nazi Austria. She died five years after her last letter to Frisch, poor and virtually unknown outside the small world of the first-generation emulsion physicists.

Among the emulsion physicists there was a deeply ambivalent attitude toward the status of individual or golden events. On one hand, from the very beginning

244. Halpern to Frisch, 3 August 1963, OFP.
245. Moore, *Schrödinger* (1989), 479–80.
246. Waller to Powell, 1 May 1964, OFP.
247. Frisch to Halpern, 24 June 1964, OFP.
248. Halpern to Frisch, 29 June 1964, OFP.
249. Blau to Frisch, 5 September 1964, OFP.

of the method, the nuclear emulsion was displayed as a technique in which single pictures could carry demonstrative force. On the other hand, the reliance on confirmations was highlighted as partly constitutive of the discovery. Even in the Emulsion Panel's earliest deliberations we saw how the participants pressed for coordination among laboratories because of the "peculiar" nature of the instrument. We saw this reliance again when Powell expressed his delight over the simultaneous discovery (by the Bristol team and by Perkins) of a particle of mesonic mass emerging from a nuclear star. This, Powell said, was "definite proof of the reliability of the photographic method." And yet again, we encounter the need for confirmation in Powell's request to Chadwick for *two* teams of observers in order to secure objectivity. This ambivalence is manifest in Powell, Fowler, and Perkins's discussion of the problem of individual events in their 1959 magnum opus, *The Photographic Method:*

> [A] single event is commonly not sufficient finally to establish the existence of a particular process; for this two or more are commonly necessary. A review of the first events of this kind observed in photographic emulsion shows, however, that in most of them, an original interpretation was subsequently maintained, and that rarely was an event which appeared to correspond to an important new physical process found to be due to a chance juxtaposition of unrelated events. This result is consistent with the extreme improbability of precise associations of events of a kind which simulate real processes in the conditions of exposure commonly employed, and is a tribute to the great power and discipline of the critical scientific method.[250]

In its ambivalence, the paragraph reflects a deep conundrum of the emulsion worker. First, the paragraph reports that single events are commonly *not* sufficient to establish the existence of a process but then goes on to argue that most such "first events" were sustained and that the interpretive stability both reflected a natural fact (the extreme improbability of artifactual simulation) and a social fact (the great power and discipline of scientific method).

The social dimension of the maintenance of the golden event, however, goes beyond the commitment to method. For if a single event is *not* sufficient as evidence, then the "discovery" cannot reside in the restricted space and time of an individual observer and may not even reside in one single laboratory. That is, to the extent that one does not take the individual κ_1 event published by O'Ceallaigh in January 1951 to be, per se, persuasive evidence, then that event, its discovery, and O'Ceallaigh's own role are altered. For over the few years following O'Ceallaigh's publication, his photograph was reinterpreted (and republished) as exhibiting a different decay (kaon to a muon and one neutral instead of kappa to a muon and at least two neutrals);[251] more important, that particular photograph

250. Powell, Fowler, and Perkins, *The Photographic Method* (1959), 422.
251. Powell, Fowler, and Perkins, *The Photographic Method* (1959), 310–11.

was combined explicitly with a host of other decays to become evidence for the existence of a kaon with a variety of decay modes. In a deep sense, the status of the individual golden event is inseparably bound up with the notion of individual authorship in experimental discovery. Even before institutionalized teamwork, *contrôle reciproque* combined authors. This coordination between laboratories was first exploited in mass averages, then took a more subtle turn with the analysis of the Dalitz plot. Comparing Perkins's meson capture (figure 3.13) with Amaldi's tau plot (figure 3.17*d*), we can ask: What or who has the author become? The tau committee? The 15 laboratories? Dalitz, as theoretical inventor of the representational scheme? Theorists take over a fundamental role in interpretation, scanners in the first moment of "discovery," and film companies in instrument fabrication. Taken together, these various forms of diffusion rewrote what it meant to be an individual experimenter.

One response to the anxiety of diffusion is an attempt to render the past like the present: even in the heroic days of the founding of modern science, collaboration had been necessary—even celebrated. And indeed, immediately following the section of their book on the probability of unrelated events' mimicking a novel process, Powell, Fowler, and Perkins offered a quotation from Thomas Sprat's *History of the Royal Society of London for the Improving of Natural Knowledge* (1667).[252] In context, the quotation serves to remind the reader not only that the scientific endeavor is necessarily social insofar as the individual worker cannot have all the virtues needed for the task but that this was always so: "If I could fetch my Materials whence I pleas'd, to fashion the Idea of a perfect Philosopher; he should not be all of one Clime, but have the different Excellencies of several Countries." The ideal experimenter would have the "Industry, Activity, and inquisitive Humor of the Dutch, French, Scotch, and English in laying the Ground work, the Heap of Experiments." To this should be added the "cold, and circumspect, and wary Disposition of the Italians and Spaniards" to meditate upon these empirical results, and only afterward fly to the dominion of Speculation. But since these qualities are hardly ever to be found in one worker, only a "publick Council" would do by mixing these various dispositions. Sprat's admonition to pursue joint work could then serve as a historical precedent from the golden age of modern science, a talisman against the charge (or fear) that science had lost its bearings as it wandered into the managerial mode.

Now some Speculation. The history of emulsions is a history of anxiety: anxiety about the loss of control over the instrument, ceded to the chemists; anxiety about the loss of discovery, now held by the scanners; anxiety about the loss of the individual experimenter, now subordinated to the team. It is an anxiety about the disappearance and reappearance of particles, about the fading of the

252. Powell, Fowler, and Perkins, *The Photographic Method* (1959), 422.

emulsion, about the problems connected with stabilizing a terrifyingly unstable gelatinous chemical mass in its production, utilization, storage, processing, and interpretation. And not least, it was an anxious, final (if doomed) attempt to forestall the Americanization of particle physics. Before the European experimenters' eyes lay a vision of what physics had been in a receding and, it must be said, partly mythologized past. It was a place where individual workers did everything from washing test tubes to building apparatus, running the device and taking data to the literary assembly of the published paper. This is the vision that Powell had in mind when he spoke of Wilson's qualities gleaming "like the banners of a vanished age." The Wilsonian virtues seemed endangered by the very technology that underpinned Powell's career: Wilson's self-sufficiency replaced by teamwork, Wilson's craftsmanship by commercial chemistry. Perhaps most of all, the quintessence of Wilson's search for decisive experiments brought to Powell "the same kind of aesthetic satisfaction as a mathematical demonstration carrying complete conviction." As if in self-reproach, Powell worked ceaselessly to piece together the photomicrographic track mosaics in imitation of the crystal-sharp cloud chamber photographs with which Wilson had startled the world a half-century earlier. Such reconstructions were an effort that Powell's contemporaries and coworkers often found puzzling. Why, they asked, make pictures that in any case were mere projections of the geometric reconstructions that were needed in three dimensions? To understand the power these photomontages had for Powell is to understand the Cavendish tradition Powell had learned from Wilson: the drive to see and with sight seize the phenomena of nature—volcanoes, clouds, and now particles. These pasteboards stand as symbolic as well as real reassemblies of the fragmented picture of nature. But to Powell and his contemporaries the new instrument was all too clearly *not* a simple repetition of the cloud chamber—at one level cheaper and easier to deploy, at another infinitely more volatile.

Anxiety, issuing from the loss of control in the laboratory, was productive. For with each move to stabilize the method, nuclear emulsions became more capable of sustaining claims for the existence and properties of new particles.[253] At each moment, the film appeared to be unstable: at one moment the photographic plate appeared to be selective in what particles it would reveal; at another it was obscured with fog, distorted by development, or uneven in drying. Reliability was threatened by the chemical and physical inhomogeneity of a plate or a batch of plates, and by other difficulties in scanning or interpreting the photomicrographs. Without cease, the struggle to stabilize the emulsion method was a response to the

253. Harold Bloom, the literary theorist, argued in his *Anxiety of Influence* that poetry cannot be seen as the suppression of anxiety about the influence of tradition, but is rather the embodiment of responses to that anxiety. In physics, too, the practices of the emulsion laboratory evolved to cope with the specific anxieties of distortion, misinterpretation, and authorship. Bloom, *The Anxiety of Influence* (1973).

anxiety of instability. Anxiety and the material, theoretical, and social responses to it were eventually constitutive of the method itself.

Necessarily, the resulting moves were heterogeneous: strategies for organizing the workplace and strategies for selecting grain size were inextricably connected. Strategies for hiring Ilford and Kodak could not be separated from strategies for defending the reliability of the film. To contribute to ionography was to make a move in chemistry, a move in physics, and a move in the experimental workplace.

By the time ionographic tracks had themselves faded from the scene, the emulsion method had ushered in both the field of particle physics and a redefinition of the experiment and of the experimental author. Read carefully, these microphotographs tell more than the decay pattern of the pion; they embody the transformation midcentury physicists felt as a deep, sometimes painful, mutation in what it meant to be an experimental physicist.

4 Laboratory War: Radar Philosophy and the Los Alamos Man

4.1 Laboratory War

Three hundred and fifty years ago, Galileo introduced the notion of mechanical relativity by invoking the experience of sea travel:

> Shut yourself up with some friend in the main cabin below decks on some large ship, and have with you there some flies, butterflies, and other small flying animals. Have a large bowl of water with some fish in it. . . . The fish swim indifferently in all directions; the drops fall into the vessel beneath; and, in throwing something to your friend, you need throw it no more strongly in one direction than another. . . . You will discover not the least change in all the effects named, nor could you tell from any of them whether the ship was moving or standing still.[1]

The imagery and dynamics of ships permeate Galileo's works, at once rhetorically tying the new mechanical physics to the modern navigational achievements of early-seventeenth-century Italy and providing an effective thought-experiment laboratory for the new "world system."

Three centuries later, when Albert Einstein was struggling to overthrow Galilean-Newtonian physics, he too chose an image of contemporary transport as the vehicle for his radically new *Gedankenexperimenten.* Now the railroads, symbol of the success of German technology and industry, replaced the Galilean sailing vessel. As Einstein put it, no optical experiment conducted in a constantly moving train could be distinguished from one performed at rest.[2]

Yet a third image of transport appealed to the American physicist Richard Feynman as he assembled a synthetic picture of quantum mechanics and relativity

1. Galilei, *Two Chief World Systems* (1967), 186–87.
2. See, e.g., Einstein's popularization of relativity first published in 1916: *Relativity* (1961).

in the years after World War II. The positron, as he saw it, could be viewed as an electron moving backward in time. Then the simultaneous creation of an electron-positron pair—ordinarily seen as involving two distinct paths—could instead be viewed as a single, continuous track: the positron travels backward in time until it reaches the moment of creation, whereupon it becomes an electron moving forward in time. Feynman conveyed his vision with a vivid metaphor: "It is as though a bombardier flying low over a road suddenly sees three roads and it is only when two of them come together and disappear again that he realizes that he has simply passed over a long switchback in a single road."[3]

Every age has its symbols, and Feynman's was as telling as Galileo's. Young American physicists of the 1940s and 1950s had seen their discipline recrystallize around the twin poles of the radar and the atomic bomb. The venerated B-29 bomber carried both—an apt symbol, therefore, of the fruits of their labor. As Feynman's choice suggests, the bomber served as a perfect vantage point from which to view the new theoretical and experimental physics.

Of course, the effect of the war on the development of physics in general, and microphysics in particular, goes far beyond a passing metaphor. Indeed, the problem is that the consequences are too great, and too varied, to be treated comprehensively in any one place. For the effects of the war permeated every aspect of postwar history. In the history of physics the realignment of the structure of the discipline included a thoroughgoing overhaul of the institutional structure of government-supported science. From the National Science Foundation to the Atomic Energy Commission and the Office of Naval Research, no aspect of science funding remained unchanged.[4] Among the war's consequences was a profound reconfiguration of relations between the academic, governmental, and corporate worlds, especially as physicists began contemplating the funding necessary for the construction of atomic piles, larger accelerators, and new particle detectors. Further, the war forged many collaborations and working groups among scientists that continued smoothly into the postwar epoch. And finally, the war left stockpiles of astonishing quantities of surplus equipment that fed the rapidly expanding needs of postwar "nucleonics"—the study of a broadly construed nuclear physics, situated at the nodal point of research problems surrounding cosmic rays, nuclear medicine, quantum electrodynamics, nuclear chemistry, and the practical imperatives of industry and defense.

Above all, one cannot ignore the new relation of university physics to military affairs that in a sense began, rather than ended, in the skies over Japan in August 1945. Suddenly, academic physicists could negotiate with high-ranking officials from the navy, the air force, and the army to acquire new machines. At

3. Feynman, "Positrons," *Phys. Rev.* 76 (1949): 749–59.

4. On World War II scientific organizations and their effects on research, see Stewart, *Scientific Research* (1948); Kevles, *The Physicists* (1978); and Dupree, "Organization" (1972). On the postwar contributions of one important agency, the ONR, see Schweber, "Mutual Embrace" (1988).

the same time, the military became an active participant in the shaping of postwar scientific research, through university contracts, the continuation of laboratories expanded during the war, and the establishment of new basic research programs under the aegis of individual armed services. Projects of joint civilian and military interest were lavishly funded, offering physicists the chance to think about exploring cosmic rays, not three or four, but a hundred miles above the earth's surface. Where a handful of technicians had once been sufficient to aid the physicists as they constructed new instruments, now the physics community began a deep new alliance with the various branches of scientifically informed engineers.

This chapter, addressing the impact of wartime research on postwar experimental and theoretical physics, can only begin to sketch some of these ramifications, drawing a few of the lines along which such a history of physics between war and peace might advance. I will not treat, for example, the vicissitudes in the dramatic careers of scientist-politicians such as Vannevar Bush, James Conant, or J. Robert Oppenheimer; the establishment and internal politics of funding organizations; or the alterations in industrial physics research policy. Instead, my goal is to peer into the impact of wartime science on the quotidian proceedings of physics itself, and into the experience of physicists in their research capacity. How did physicists in the war begin to reimagine their future discipline? How would physicists and engineers interact around experiments and instruments? How would theory function within the discipline? Having examined the evolution of the nuclear emulsion method—the European instrument of physics par excellence in the immediate postwar epoch—we now regloss the war, rereading it from the site-specific material culture of the American wartime laboratory.

To do this I have chosen to focus on six exemplary physics departments: those at Harvard, Princeton, Berkeley, Stanford, Wisconsin, and MIT. In many ways, these six illustrate different facets of war in the laboratory: Harvard, MIT, and Princeton, all private East Coast institutions, were powerfully involved at every level of the radar and atomic bomb projects. Berkeley, as the incubation site for large-scale cyclotron research in the 1930s, provides a West Coast perspective along with its private university neighbor Stanford, where microwave research was king. The University of Wisconsin, which already had a strong physics department before the war, was also transformed by the war. At the same time, we will see in this midwestern experience some of the frustration felt as goods and personnel began moving elsewhere near the end of hostilities. Each university had its own trajectory, shaped in part by different war experiences and earlier patterns of research. Yet the six sites had much in common: each had to confront a sudden expansion, search for a new relationship between theorists and experimenters, forge a novel alliance between experimenters and engineers, and face the difficulties that accompanied the move into the epoch of large-scale, centralized, and cooperative research. To get at these themes, we will begin with the shock of the new in radar and atomic bomb projects as the very different cultures of physicists and engineers came into contact; we then turn to the war experience at these six university-based laboratories. This history provides the basis for understanding the material continuity of war and postwar research (even when subject of inquiry changed radically), and the new meaning of "cooperative research."

Together these changes transfigured the physicists' approach to research, radically altered the instruments that populated their laboratory, and shifted their image of themselves in the university and on the world stage: the culture of physics was not what it had been. In the most visible and dramatic fashion the war provided concrete examples of scientific accomplishments, though it remains an open question whether the lessons drawn from that experience were actually the ones responsible for the physicists' success. But in a sense that I will develop below, the major weapons systems—radar, atomic bombs, rockets, and proximity fuses—formed *guiding symbols* that inspired the strategy of much postwar research. Needless to say, one can find earlier examples of one aspect or another of large-scale research: the great philanthropically funded telescopes, Ernest Lawrence's growing array of cyclotrons in the 1930s, and institutes of physics in Europe come immediately to mind—monumental telescopes cost millions of dollars, cyclotrons took several people to operate, and at certain European institutes, state and "pure" scientific concerns shared a roof. Indeed, especially at West Coast universities such as Stanford and Berkeley, the 1930s had already seen the establishment of joint endeavors involving physics and electrical engineering. But despite the importance of such successes as the cyclotron and the klystron, before the war there were no physics achievements born of such large physics-engineering efforts that made the continuation of such centralized big research projects seem inevitable. During the Second World War, however, the large-scale collaboration between physicists and engineers on electromagnetic and nuclear-physics-based weapons systems provided just such imperatives. In part as a result of these powerful projects, between 1943 and 1948 key segments of the American physics community came to accept a mutation in the ideal of the physicists' work and workplace. One after another, physics departments began to conceive of a style of orchestrated research that has come to dominate contemporary investigations in high energy physics and wield growing influence in other domains including synchrotron radiation, space science, laser physics, and plasma physics. These changes were not universally heralded as advantageous. As we will see here and in the subsequent chapters, the issue of *control* recurred over and again—control over the apparatus, control over the research agenda, and control over the increasingly industrialized and rationalized workplace itself. Changes in the hardware, its material impact and its concomitant symbolic meaning, came together with changes in the self-definition of the experimenter.

4.2 Rad Lab/Met Lab: Physicists and Engineers

Though over the years it has received but an infinitesimal fraction of the attention heaped on the atomic bomb project at Los Alamos, MIT's Radiation Laboratory arguably had a greater effect on the postwar development of physics. Located within a major research university (rather than a southwestern mesa), the laboratory, restructured by the war, became a model closer to the heart of research activity.

A proper account of American radar research would reach back into the 1930s; such a history would encompass work at Stanford, Bell Laboratories, the Naval Research Laboratories, and a host of other establishments. But however far the Americans had progressed, nothing quite prepared them for the shock of meeting a team of British scientists who in September 1940 brought over their magnetron (a radar tube built in strict secrecy and capable of producing microwave radiation far in excess of anything the Americans had been able to make). By October 1940, the U.S. National Defense Research Council (NDRC), working with the British delegation, and led by Sir Henry Tizard and a group of high-level American scientists (including Ernest O. Lawrence, Vannevar Bush, and Karl Compton), began to lay out plans for a major, central, civilian-run laboratory. Formed around a core of individuals including Alfred L. Loomis (a New York lawyer and scientist), E. L. Bowles (from MIT), and Lee A. DuBridge (named scientific director of the laboratory), the enterprise took its name—the Radiation Laboratory—as a pun on the Berkeley Radiation Laboratory, which was devoted to "useless" nuclear physics. Based at MIT, the laboratory at the end of 1940 employed about 30 physicists, three guards, and a secretary. As the laboratory grew over the next year, it began to produce prototypes of different radar systems, such as gun-laying radar and air interception systems. Personnel moved easily back and forth across the several specialized research divisions.[5]

The Japanese destruction of Pearl Harbor on 7 December 1941 precipitated a stark debate over the organization and mission of the Rad Lab. Suddenly, what had been an ad hoc arrangement of subgroups and committees came into question, offering the historian a view of participants' conceptions of how they could and should interact. In the crisis of the moment, the different cultures of industry and academia, of engineering, theory, and experiment, met. And the relations revealed looked very different than those sketched in the workings of the Emulsion Panel in England just a few years later.

5. Guerlac, *Radar* (1987), 253–303.

In March 1942, Frederick Dellenbaugh, a research associate in electrical engineering at MIT (and a vice president of Raytheon), sketched out his own understanding of how the laboratory should function. Historically, he began, there were at least four types of organizations: governmental, educational, business, and military. "The first two," he quipped, "have been developed to prevent change or at least are extremely inflexible."[6] As his experience at Raytheon must have made clear, "the last two . . . seem to have much in common."[7] Older military and older industrial structures both established authority in "a straight line." Bell Telephone Laboratories could stand for such an arrangement on the industrial side. Napoleon, Dellenbaugh contended, was the first to put a staff organization to work—though his failure to do so in the navy led to disastrous defeat in Egypt. Dellenbaugh's extraordinary military-industrial history continued, as he related how the staff system finally triumphed in the Franco-Prussian war, when Van Moltke reversed Napoleon's military gains by completing his bureaucratic innovation. Americans followed suit at the turn of the century when Elihu Root, then secretary of war, restructured the U.S. Army to protect its island possessions. It came as no surprise to Dellenbaugh that it was Root, the leading legal advisor to industry, who went on to rebuild the military. But once again, Dellenbaugh warned, Germany had taken the organizational offensive. Though their government was totalitarian, their small-unit autonomy in both the war machine and in industry lent the Nazis extraordinary power. Their trick lay in coordinating the many small units.[8]

Now it was high time for physicists and engineers to learn the German lesson: "In considering the Radiation Laboratory, it might seem that a modern military organization would prove a good pattern." Strategy, Dellenbaugh pronounced, is the analogue of administration—tactics, the analogue of execution. The laboratory director or commander-in-chief should have an associate director acting as "chief of staff." Below the chief of staff should be the staff officers who execute the director's orders. One of the Rad Lab's current failings, according to Dellenbaugh, was precisely the absence of such a chief of staff. Also following the model of the military, the laboratory should have an assistant chief of staff to direct "operations" just as in the army and navy, though at the Rad Lab, operations meant research and development.

In a sense, Dellenbaugh was at one and the same time trying to create a vocabulary, a bureaucracy, and with it an ethos of scientific engineering work that would reenact his understanding of the history of military-industrial action.

6. Dellenbaugh, "Subject: Staff Organization," Group VIII-Engineering, 13 March 1942, file "Reorganization," box 59, RLP.

7. Dellenbaugh, "Subject: Staff Organization," Group VIII-Engineering, 13 March 1942, file "Reorganization," box 59, RLP.

8. Dellenbaugh, "Subject: Staff Organization," Group VIII-Engineering, 13 March 1942, file "Reorganization," box 59, RLP.

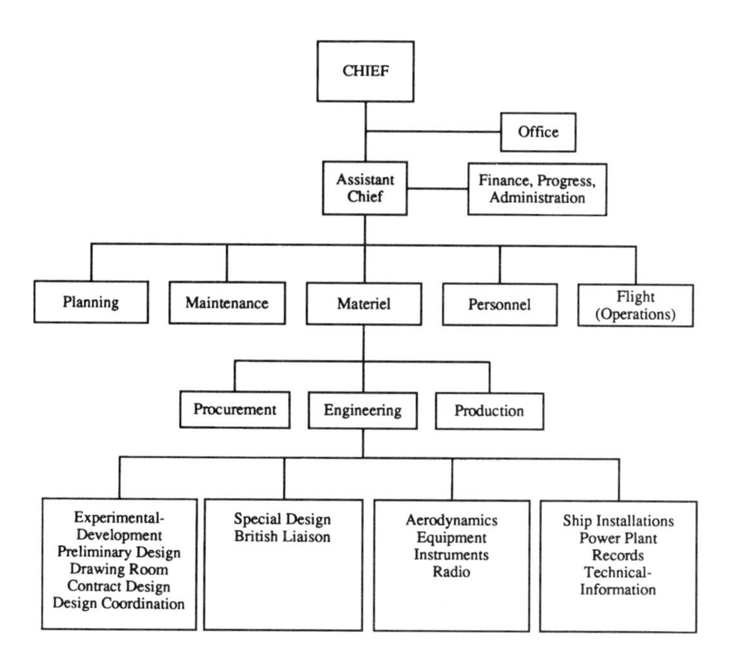

Figure 4.1 Staff structure, Bureau of Aeronautics, U.S. Navy (1942). Redrawn from Dellenbaugh's hand sketch. Source: Dellenbaugh, "Subject: Staff Organization," Group VIII-Engineering, 13 March 1942, file "Reorganization," box 49, RLP.

Assimilated deep into "Rad Lab culture," or "Rad Lab philosophy" as it was later known, physicists, engineers, and laboratory administrators all began to employ a lexicon that began with these elements and learned to move back and forth freely across the divide:

strategy::administration
tactics::execution
commander-in-chief::laboratory director
chief of staff::associate director
staff officers::staff officers
assistant chief of staff::[needed position at Rad Lab]
operations::research and development tasks

Dellenbaugh's reconceptualization of the laboratory went further. After sketching representations of the army and navy staff structures (see figures 4.1 and 4.2), and suggesting the analogue for the Rad Lab (see figure 4.3), Dellenbaugh

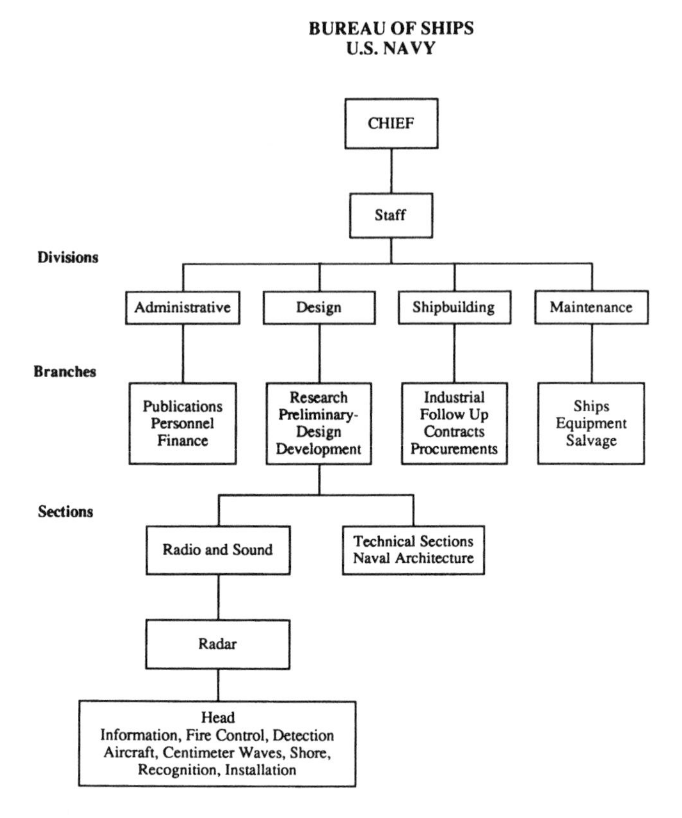

Figure 4.2 Staff structure, Bureau of Ships, U.S. Navy (1942). Redrawn from Dellenbaugh's hand sketch. Source: Dellenbaugh, "Subject: Staff Organization," Group VIII-Engineering, 13 March 1942, file "Reorganization," box 49, RLP.

closed by insisting that "[t]oo much attention to doing, and too little attention to directed planning is a common criticism of business organizations."[9] The Rad Lab needed greater direction.

Few quarreled with Dellenbaugh's injunction to implement directed planning. But there was little agreement as to how that administrative structure ought to function. On one side were those advocating a "vertical" organization, structured around complete weapons systems. Among such advocates were those responsible for airborne radar; they feared that without total control over the complete system, little attention would be paid to their need for lightweight

9. Dellenbaugh, "Subject: Staff Organization," Group VIII-Engineering, 13 March 1942, file "Reorganization," box 59, RLP.

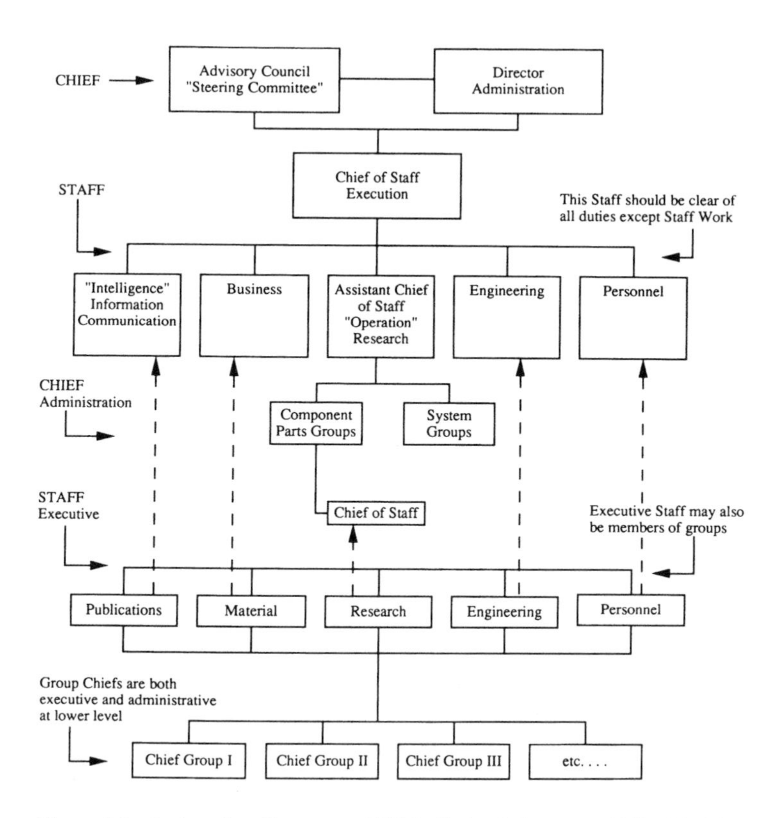

Figure 4.3 Projected staff structure, MIT Radiation Laboratory (1942). Modeled directly on Dellenbaugh's reading of military and industrial organization (see text and figures 4.1 and 4.2), his plan for the Rad Lab is here redrawn from a faint hand sketch. Source: Dellenbaugh, "Subject: Staff Organization," Group VIII-Engineering, 13 March 1942, file "Reorganization," box 49, RLP.

constituent devices, including aluminum waveguides, aluminum chassis, and magnesium parts.[10] Material objects—these building blocks of microwave devices—were inseparable from these questions of administrative and conceptual control. The most extreme vertical proposal came from E. G. ("Taffy") Bowen, who had helped develop radar in Britain, where such work was conducted on an

10. L. A. DuBridge, interview with H. E. Guerlac in presence of T. A. Farrell, Jr., 20 December 1945, file "Interviews," box 59, RLP.

industrial model. On 30 January 1942, Bowen argued to the Steering Committee of the Radiation Laboratory that the effort should be divided into three distinct laboratories, one each for land, sea, and air radar systems. Each laboratory would support a staff of 1,000, and assign approximately 125 to each of the following divisions: (1) Research, (2) Development, (3) Engineering Design, (4) Small Production, (5) Installation, (6) Training, (7) Maintenance, and (8) Operator Training. Out of each laboratory would come some 50 working units of any given system, along with the trained personnel to operate and maintain them.[11]

The danger of an approach like Bowen's, Louis Ridenour countered, was that as the engineering problems grew, they would invariably displace research and development. Even large companies faced such realignment under the pressure of production, and in Ridenour's view, "we have already seen this same process occurring, in a microcosmic way, within the Radiation Laboratory itself . . . as we attempted to design and build and install five and fifteen and fifty equipments of a given sort." In his view, it made no sense to produce a local oscillator or indicator for an air application, then another for sea, then another for land. If it is a good one, "it will be required to float and to fly as well as to sit on a trailer." An AGL-1 system for the air was simply a modified XT-1 (SCR-584) ground-based fire control radar. "The tripartite division may be a good one for engineering design, installation, and the like, but it is indefensible as regards research and development." For Ridenour, whether one plumped for components or for systems, the key was to isolate the research from the engineering. At stake—as Ridenour's choice of metaphor below suggests—were more than mere organizational niceties. Preservation of the purity and autonomy of research was a necessary precondition for the success of the radar effort: "It is quite possible that the greatest service the Radiation Laboratory can perform in the war effort is to keep itself entirely free from being prostituted by production, in order to have its men available for continued research on components and development of systems. Further, it is questionable whether any system [that] can be properly designed by the engineers of an industrial company should be designed by the Radiation Laboratory."[12] Engineering and production values (according to Ridenour) squarely opposed flexible, autonomous research.

Dellenbaugh disagreed: "[E]ngineering in the Laboratory is a definite function, and . . . there should be an engineering executive upon the staff of the Director to coordinate the different degrees of engineering activities." The amount of engineering was such an essential part of the laboratory's activities (upward of 50%) that it would be counterproductive to segregate it. In fact, eight months

11. Lewis N. Ridenour, "Memorandum on the Place of the Radiation Laboratory in the War Effort," 2 February 1942, file "Reorganization," box 59, RLP.

12. Lewis N. Ridenour, "Memorandum on the Place of the Radiation Laboratory in the War Effort," 2 February 1942, file "Reorganization," box 59, RLP.

earlier Dellenbaugh had recommended that engineering liaison members be appointed to each laboratory division, and this diffusion of engineering expertise was put into effect. These "engineering nuclei" could now be exploited to facilitate the increasingly close links with industry and production. More generally, the laboratory would soon need to establish standards, introduce test methods, and adopt "commercial practise" in order to move knowledge smoothly from breadboard to bomber. To Dellenbaugh such tasks were not the "prostitution" they appeared to his colleague; indeed, he quoted approvingly a gendered definition of engineering, laden with a very different conception of the power relations between the pure and applied that ran this way: "Engineering is Science's handmaid following after her in honor and affection, but doing the practical chores of life, concerned with the useful and material."[13]

Milton G. White was all for the useful and the material, but he wanted the "handmaid" to wait until the "basic elements" of the system were ready. As head of the laboratory's Division 5 (Transmitter Components),[14] he spoke for the autonomy of the component groups. True, he allowed, each system could develop its own components, but "[a] far swifter and more painless method of spreading new circuits and general Radar philosophy would be to group together such basic elements as are to be found in most complete systems. This is the historical beginning of the laboratory and a very sound one."[15] Speaking with the voice of a physicist steeped in university research, he insisted that "fundamental work involving a fairly long programme can only be carried on by a group which enjoys some measure of permanency and freedom from deadlines set by the arrival of an airplane, boat or truck. This fundamental work must go on if any marked improvement in performance characteristics of existing apparatus is to be achieved." Of course, such autonomy could not come at the expense of seeking "out and destroy[ing] the enemy," but the new tactical systems could be engineered with the fundamental components as their "black-boxed" constituent parts. For the system to function, White recognized, physicists could not leave their component amplifiers or local oscillators in the state of a mere breadboard. Some workers in each group had to have as their primary occupation and interest the production of a reliable piece of machinery. "For want of a better term," White added, "we can call these men engineers." It was a distinction of considerable importance: "One should not expect the physicist, untrained as he usually is in the details of manufacturing methods, to handle the later stages of development. He won't like it and, worse, will probably bungle it. On the other hand the

13. Dellenbaugh to White, "Functions of Engineering Group," 14 March 1942, file "Reorganization," box 59, RLP.

14. Division leaders are given in Guerlac, *Radar* (1987), 293, 295.

15. White, "A Proposal for Laboratory Organization," n.d. [probably January–February 1942], file "Reorganization," box 59, RLP.

development engineer will uncover many weaknesses in the design, he will find certain materials unobtainable and requiring substitutes, he will see ways of simplifying and reducing manufacturing time."[16] Sometimes the engineering critique will prompt the (physicist) inventor to further innovations; at other times the engineers themselves will rectify the problem.[17] White, Ridenour, Dellenbaugh—all struggled to create a new modus vivendi between engineering and physics.

Neither engineer nor physicist, however, would suffice when it came to understanding how the device worked in the brutal conditions of combat. There, White argued in early 1942, one needed the expertise of yet another type of worker: the technician. For "[e]ven after a unit has passed through these stages [of fundamental design and engineering] and finally emerges as a handsome black crackle box the job is not done. By the time the second or third unit has reached the trusting consumer the first box will be a smoldering heap. Someone forgot the primitive fact that if there is any combination of controls which will ruin the apparatus it will be ruined. So now service and repair becomes a component group function." Should the component groups fail to provide adequate servicing, the fighting groups will. And if the repair function were to slip out of the purview of the component group, it would "isolate the research and development groups from some of the ugly facts of life." Following this "philosophy," White believed, would ensure that when those first breakdowns occurred, "there will be on hand the component man who first designed the apparatus. This man will, in conjunction with the project engineer, run down all difficulties on the spot."[18]

Such simultaneous collaboration and friction between the cultures of engineering and physics was by no means restricted to the Rad Lab: in laboratories across the country, physicists had to face the "facts" of engineering life. The Metallurgical Laboratory at Chicago, to choose an example from the atomic bomb project, was also rife with boundary problems. Du Pont, charged with administering the Met Lab (and producing bomb quantities of plutonium), grounded its war work on its successful prewar manufacture of nylon.[19] In particular, Du Pont's "industrial culture" included the assumption that the company, rather than the physicists, knew how to scale properly from laboratory to factory. Let us track, this time, the view from both the physicist's and the engineer's perspective.

16. White, "A Proposal for Laboratory Organization," n.d. [probably January–February 1942], file "Reorganization," box 59, RLP.

17. White, "A Proposal for Laboratory Organization," n.d. [probably January–February 1942], file "Reorganization," box 59, RLP.

18. White, "A Proposal for Laboratory Organization," n.d. [probably January–February 1942], file "Reorganization," box 59, RLP.

19. Hounshell, *Science* (1988); see also his "Du Pont" (1992).

Eugene Wigner, head of the Met Lab's "theory group," was frustrated and angry about his inability to accelerate the construction of a new and better pile, code-named P-9 and designed for plutonium production. In a polemical memo of 20 November 1942 entitled "On the Relation of Scientific and Industrial Organizations in the Uranium Project,"[20] Wigner set out his views. It was a plea for engineers to acknowledge the absolute centrality of the theorist's vision: "We must not forget that this process [nuclear fission] exists at present only in the minds of scientists, no 'practical' man has any experience with it. In this case, it is not the experience which creates the theory but the theory which creates the experience." He proffered two explanations as to why he and his associates had not been brought more centrally into decision making: The first was that the Du Pont engineers were simply biased against foreigners (Wigner was Hungarian and many of his colleagues were also foreign born). More probably, Wigner reckoned, the reason was the far deeper prejudice against scientists as being unable to handle "practical matters." This belief (he contended) made no sense: "Our process [fission] is not something like the tanning of leather which has been known since the time of ancient Greece and to which science has made only relatively small contributions. In such a case a very serious argument can be put up for leaving the practical man in the leading position. . . . On the contrary, our process has been invented by, and is known to, only scientists, and every step to put others in charge is artificial and deleterious."[21] While it was acceptable to have engineers work on the development of reactors and the production of nuclear explosives, it was (according to Wigner) absolutely wrong to grant them sole responsibility. Wigner's complaints continued over the next months: he was excluded from key decisions regarding P-9, Du Pont was trying to monopolize nuclear industry, and Du Pont was slowing the project for its own interests.[22]

Arthur Compton, the much-acclaimed experimental physicist whom Vannevar Bush had put in charge of chain reaction and weapons theory, had, in turn, vested Wigner with responsibility for the theory of the chain reaction. In October 1942, Compton urged the establishment of "direct contact" between theory, engineering, physics, and chemistry—"coordination" was the order of the day as he made clear in figure 4.4.[23] But for all his respect for the theorist, Compton became increasingly alarmed by Wigner's statements. Eventually, Compton shot off a tough memo on 23 July 1943, reminding Wigner that Du Pont was charged, not just with producing a few bombs, but with making enough fissile material

20. Wigner, "On the Relation of Scientific and Industrial Organizations in the Uranium Project," 20 November 1942, box "Manhattan Project," EWP.

21. Wigner, "On the Relation of Scientific and Industrial Organizations in the Uranium Project," 20 November 1942, box "Manhattan Project," EWP.

22. Compton to Wigner, 23 July 1943, box "Manhattan Project," EWP.

23. Compton to Moore and Allison, "Coordination of Engineering and Research," 22 October 1942, box "Manhattan Project," EWP.

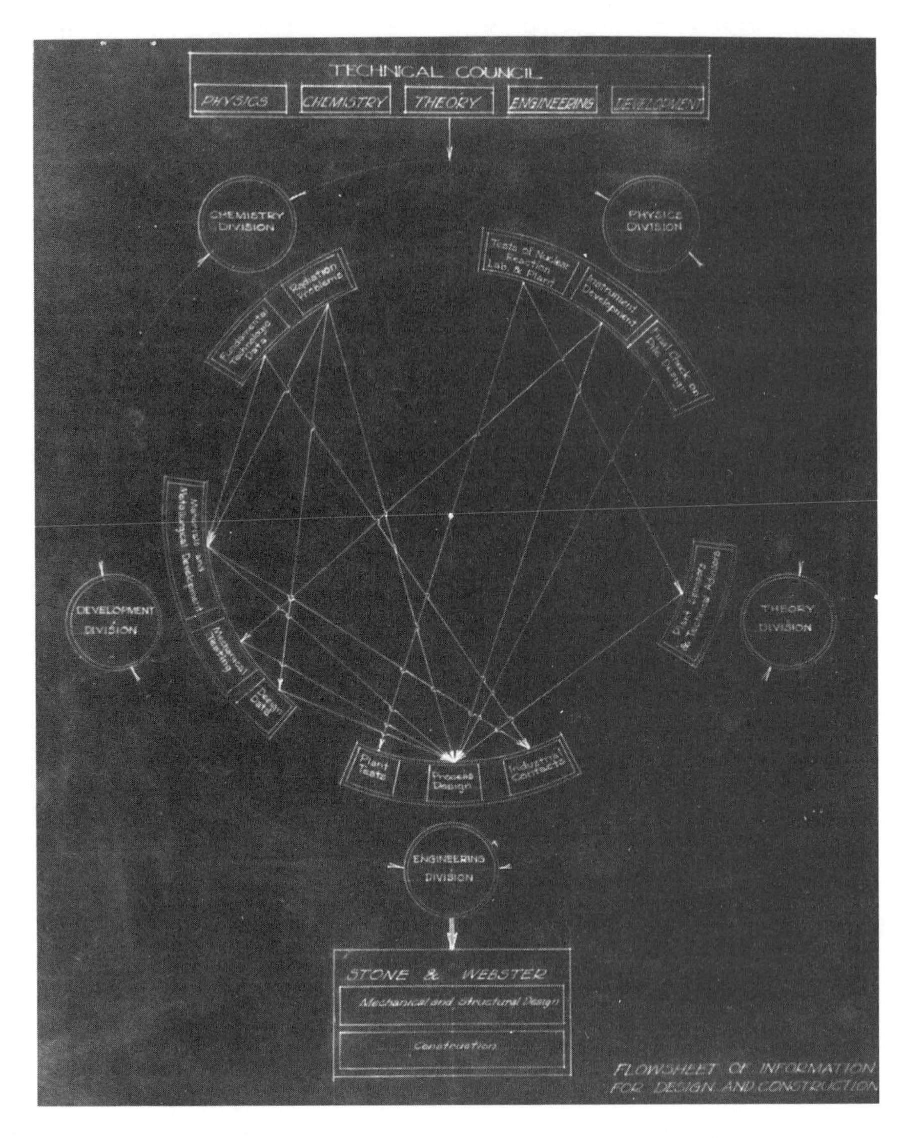

Figure 4.4 Atomic bomb project, information flowsheet (1942). In the plutonium project, Compton assigned the Theory Division a central role in the distribution of ideas and data that would shape the billion-dollar effort. The flow of information would travel from the Physics Division ("Tests of Nuclear Reaction Lab. & Plant") to the Theory Division and from there to the Engineering Division ("Process Design"), followed immediately by Stone & Webster's "Mechanical and Structural Design and Construction." Similarly, the Physics Division was in direct contact with the Development, Engineering, and Theory Divisions. It is these kinds of complex exchange relations that created new techniques, new local languages, and new sites of interaction—reshaping along the way how physicists saw themselves and their discipline. Source: Compton to Moore and Allison, 22 October 1942, box "Manhattan Project," EWP. Papers of Eugene Wigner, Manuscripts Division, Department of Rare Books and Special Collections, Princeton University Libraries.

for mass production. He pressed Wigner to cooperate with industry: "The nation has confidence in its scientists. . . . We now find ourselves with unusual power in our hands. If we cooperate with industry and show our adaptability to war conditions, with the objective of the common interest continually before us, we shall take an ever more responsible place in shaping the world."[24]

The issues at hand were substantive: Should the project go ahead with dry shielding or wet shielding around the reactor? How would geometrical modifications for thermal expansion shift the projected yield of penetrating gamma and neutron radiation? In general, Wigner favored building the P-9 pile first, then modifying it later. Only through trial experiment, he believed, could one ascertain what worked, what failed, and what could be improved. Furthermore, since nuclear physics was best left to the nuclear physicists, it would be time-consuming and ultimately counterproductive to try to teach all that was needed to the engineers. This approach struck Crawford H. Greenewalt, Du Pont's engineer and project manager for the plutonium project, as absolutely backward: he wanted to teach the engineers what they needed to know, and to work through a proper design the first time. After one of their many painful exchanges, Greenewalt wrote to Wigner in March 1943: "I counter that our philosophy is the philosophy of success while yours is the philosophy of failure. We are convinced that the pile, as we shall build it, will work satisfactorily and that there will not be much reason to deviate from this pattern in later piles."[25] The pilot plant at Clinton, Tennessee, went critical in November 1943, and the full-scale plant in Hanford, Washington, did so in September 1944.[26]

In the development of plutonium production methods, Greenewalt took it as axiomatic that intermediate stages, or "semiworks," had to be established as the company sought to expand operations. These semiworks mediated between the test tube and the factory; they aimed, not at proof-of-principle, but at perfecting modes of engineering for the production line. For many physicists on the Manhattan Project, such stepwise increments in scale appeared luxuries that delayed a program that could not help but work as it expanded. If physical law conflicted with engineering practice, it was the engineers who had to yield (so the physicists believed). Just as physicists in the early Rad Lab found it hard to imagine that it would be anything but efficient to have one transmitter for use in air, on land, or at sea, so the Met Lab physicists balked at the idea that pure research would encounter obstacles when scaled to industrial proportions. Declassified portions of Greenewalt's wartime diaries capture the difficulty he had coping with the Chicago physicists. While the physicists found Greenewalt intrusive, from Greenewalt's perspective in December 1942, the first difficulty was

24. Compton to Wigner, 23 July 1943, box "Manhattan Project," EWP.
25. Greenewalt to Wigner, 29 March 1943, box "Manhattan Project," EWP.
26. Hoddeson et al., *Critical Assembly* (1992), 38.

his *lack* of power over the scientists with whom he had to work: "The arrangement I'm to work under is most difficult. I have no authority over the Chicago crowd—but am to see to it by diplomacy and pleading that they do the right things at the right time and don't chase too many butterflies. What an assignment! Fortunately Compton is a swell guy so it may work out O.K."[27] More contact over the next few days only reinforced Greenewalt's assessment of the quasi autonomy of the two communities, an autonomy that manifestly disturbed a man used to the more univocal authority structure of a large industrial concern. As he put it on 22 December 1942, "It is obvious that I couldn't successfully 'boss' the physicists. This can only be done by Compton for whom they all have a great respect. I outlined for him some of the major points on which we need data promptly."[28]

Nor was this friction generated purely by the combination of two different and independent hierarchies. The physicists clearly had their own sense of the character of their efforts as distinct from that of the engineer. Again, Greenewalt: "C[ompton] has peculiar ideas as to the difference between 'scientific' and 'industrial' research. I started some missionary work to convince him that the difference was more of terminology than actuality but will have to do more."[29] Not surprisingly, indigenous peoples of physics remained restive, and Greenewalt's missionary work never succeeded. Despite his protestation that there was merely a verbal difference between the groups, a few days later he was reeling with anger at the inadequate leadership and structure of the physicists' work on the air-blast cooling mechanism for the pile. Words, actions, and values, it would seem, were not so easily split. Greenewalt: "No mechanism yet devised for unloading and sorting, no flow sheet, operating manual or program. No clear idea as to what Du Pont is expected to do—Hell! The first thing to do is to work out an operating organization. . . . I believe we *must* infilter pile design in spite of the fact that we aren't very welcome."[30]

From the other side, Eugene Wigner, Enrico Fermi, and Leo Szilard all wanted the opposite—to "infilter" the Du Pont design structure, insisting that the engineers bring even details of design plans to them, because (as Greenewalt put it) "some small point might violate physical principles."[31] While the physicists continued their exploration of the nuclear physics of the pile, especially the calculation of the neutron multiplication constant k, Greenewalt was worrying about finding group leaders with sufficient industrial research experience, "freezing" the pile design before procurement and development plans were set, and initiating work on scaled-up fabrication techniques. Again and again, Greenewalt

27. Greenewalt diaries, 19 December 1942, HML.
28. Greenewalt diaries, 22 December 1942, HML.
29. Greenewalt diaries, 28 December 1942, HML.
30. Greenewalt diaries, 28 December 1942, HML.
31. Greenewalt diaries, 5 January 1943, HML, "principals" [*sic*] in original.

recorded the clash of cultures: "More arguments with Wigner and Szilard on transfer of information."[32] Wigner got angry enough to bring his complaints to President Truman; Greenewalt, for his part, insisted on the difference between laboratory conditions where "everything works well" and the results at the semiworks. He demanded that physicists calculate under *operating* conditions, underlining the word "operating"—as when he urged Alvin Weinberg to take into account the temperature of the reflector when calculating the temperature dependence of k.[33]

Though they differed on how alignment was to be achieved, all of the advocates of reorganization had to grapple with the problem of bringing the different cultures of experimental physicists, engineers, and technicians together. The problem was expressed differently in the different war projects, sometimes as a need for a "Los Alamos man," sometimes as search for a "radar philosophy." Nominally, everyone involved in a particular project was at work on the same piece of hardware. In the multiply "infiltered" laboratory, however, the guiding principles of design often clashed. Would one build to set out generalizable physical properties that could be widely incorporated? Was one building to produce working breadboard models? Did one want to divide work horizontally (along components) or vertically (along systems)? Did design have to include production? Operating conditions? Combat stresses? Though the struggles at first glance appear local to the Radiation Laboratory, to Los Alamos, or to the Metallurgical Laboratory, the debate has ramifications that extend far into the postwar epoch. For in the confrontation of different work structures, design principles, relations to industry, and understandings of scale came a transformation of both physicists and engineers. And while it is often (perhaps too often) told how engineers needed to learn the physics, the crucial point here is that the physicists, though they might declaim about purity and prostitution, began to learn how to think like engineers. Whatever expressions were used, the wartime projects had forced a rearrangement of the role that physics played, not only in terms of visibility to generals and politicians, but in terms of the right relation of these various subcultures to one another. Sometimes a machine is not just a machine. It was in the material culture of the wartime laboratory, in designing reactors, radar sets, and weapons that physicists learned this most fundamental lesson about how physics might look when peace came.

4.3 War in the Universities

Our first need is for a closer analysis of how the expansion and redefinition of laboratory culture affected physics departments, and to achieve this we need to

32. Greenewalt diaries, 6 January 1943, HML.

33. Greenewalt diaries, 28 July and 12 August 1943, HML.

dispose of the myth that changes in civilian scientific planning began only after the guns of World War II had ceased firing. For it was during the period 1943–45 that physicists and administrators first debated and set in motion the coming boom. Across the country, from Berkeley to Harvard, pressure to think about postwar expansion began at the top.[34] It would be possible to continue the analysis beyond the six departments considered here. One would naturally turn to the University of Chicago, where Fermi's first reactor stood; to Caltech, with its extensive work on rockets; to Johns Hopkins, where the proximity fuse was developed; the list goes on. But with Harvard, Princeton, Berkeley, Stanford, Wisconsin, and MIT we will see both variety and certain common threads of the reconceptualization of physics at war.

4.3.1 Harvard

At a meeting on 20 November 1944, the president and fellows of Harvard College agreed to create a panel whose task was to direct the expansion of physics, chemical physics, and engineering. Selecting representatives from the various physical sciences at Harvard, President James Conant established a Committee on the Physical Sciences.[35] As the head of the NDRC, a driving force in shaping scientific war research, Conant had arrived at a clear conception of the shape of postwar science. It was an image shared by several of his Harvard colleagues, as was evident at the committee's very first assembly. The physicist Edwin C. Kemble lobbied for his discipline in these terms: "*[T]he war* has given a great boost to physics. It has stressed the importance of physics to industry and national defense and has underscored the usefulness of men trained in pure physics when emergency requires that they turn to applications." Consequently, he argued, the university needed to expand into new fields, get the best personnel, and enlarge the staff for instructional purposes. Kemble conveyed the department's desire to keep its existing staff in nuclear physics, "the [field] of greatest interest," to add "top caliber" theoretical physicists—for example, Julian Schwinger, Hans Bethe, or Harvey Brooks—and, of course, to augment its total budget with funds for mechanical assistance and construction costs.[36]

To justify the augmentation of theoretical physics, Kemble composed a memorandum on 9 December 1944 to the Physical Sciences Panel which began, under the rubric "presuppositions," by contending that there would be a "nationwide acceleration in the growth of the science of Physics as a result of war

34. On the postwar work at Berkeley, see Seidel, "Accelerating Science," *Hist. Stud. Phys. Sci.* 13 (1983): 375–400.

35. Conant to Kemble, 28 November 1944, summary of meeting of 20 November 1944, box 3, RHP. Already in the summer of 1943, Conant was deeply involved in postwar planning for military-scientific relations, though the early speculations did not lead directly to accepted policy. See the excellent book on the military's expectations for their postwar condition, Sherry, *Postwar Defense* (1977), 137–38.

36. File cards in handwriting of E. C. Kemble, note cards, RHP.

emphasis." As Kemble saw it, the growth would take place in two areas. *Solid state physics* ("the field of properties of matter in bulk") demanded the efforts of a new, more powerful contingent of theoretical physicists, who would be masters of quantum mechanics, statistical mechanics, and chemical thermodynamics. For Kemble, the need for more theory was illustrated by his colleague Percy Bridgman, whose superb investigation into the high-pressure domain had nonetheless "undoubtedly fallen short of its maximum potentialities since, to date, he has worked without steady and effective collaboration from theoretical physicists."[37] Just as the physicists had to learn to work with engineers, experimenters had to find a means of coordinating their work with theorists. Trading zones marked both contact sites.

Beyond solid state physics lay *nuclear physics,* which was, by Kemble's lights, "the most spectacular field in physics today." It was there that "the riddle of the physicist's universe is found," amid the cosmic rays, mass spectrographs, cyclotrons, and forms of radioactivity. And as the war was making perfectly clear, medical and chemical applications were "manifold" and were accompanied by the even more tantalizing "possibility of unlocking stores of atomic energy [which added] urgent significance to the investigations." Exactly this combination of purely intellectual benefits and hoped-for practical consequences characterized the physics community's justification for expansion. In brief, the intellectual argument positioned atomic physics as a stepping-stone to nuclear physics.[38]

Speaking for his department, Kemble argued that since World War I, the focus of physicists' concern had shifted away from understanding atomic structure and the structure of simple molecules. Previously, Harvard's own efforts had to a large extent been devoted to spectroscopy of all kinds; but now problems of this type "have largely been solved," and "work in this field operates against a law of diminishing returns." These moribund lines of inquiry were being replaced by the more alluring problems of nuclear and solid state physics. Both of these new domains demanded deeper cooperation between experimental and theoretical workers, to handle "the increasingly abstract and complex character of present-day physical theories. This quality is the result of the intensive search for more powerful means of attack on problems of a more and more difficult character. As one consequence of the complexity of the theory, the most brilliant experimental physicist is in need of theoretical collaboration to an extent previously unknown."[39] Already during the 1930s, these changes were in motion. An

37. Kemble, "Panel: Physical Sciences. Memorandum on Proposals for the Development of the Department of Physics," 9 December 1944, RHP.

38. Kemble, "Panel: Physical Sciences. Memorandum on Proposals for the Development of the Department of Physics," 9 December 1944, RHP.

39. Kemble, "Panel: Physical Sciences. Memorandum on Proposals for the Development of the Department of Physics," 9 December 1944, RHP.

increasing number of the "more gifted young men" were choosing theoretical physics. Oppenheimer's presence at the Berkeley Radiation Laboratory was an example for all to see of the theorist's usefulness in joining the skills of experimental and theoretical physicists; the young theorist's contribution "has been of crucial importance in the meteoric rise of that laboratory to its place as principal center for nuclear investigations in this country." Calculations of nuclear dynamics, analysis of cosmic ray phenomena, and analysis of electromagnetic showers among electrons and photons all played essential roles in the interpretation of experiment.

Unspoken—but undoubtedly understood—was Oppenheimer's masterful guidance of the Manhattan Project, in particular his nurturing of a powerful theoretical group under the leadership of Hans Bethe. Their fundamental role was well known among those with security clearance: an implosion design without theory was unimaginable, and almost every facet of the project interacted with Bethe's team, from critical mass and efficiency calculations to studies of blast effects and neutron diffusion.[40] Now contemplating the future, the Harvard physicists hoped Schwinger or Bethe could enliven postwar theoretical life in Cambridge.[41]

Acquiring theorists was, however, only part of a much larger pattern of growth. By the end of December 1944, Kemble was sure enough of the expansion plans to write Conant: "My personal acquaintance with the state of engineering arts at the time of the last war and at the present time convinces me that it is imperative for our future national safety that the scientific bases of engineering practice shall have far more intensive study than heretofore. We must have a much increased number of analytical engineers with brains and advanced training if we are to hold our own in the technological race."[42] As 1944 drew to a bloody close, the "technological race" began shifting from a contest against the Germans to the postwar, post-Axis world. Kemble added that the universities could best aid in the competition by prosecuting work in "pure science," which would yield more "in relation to the investment" and attract the best "quality of the men . . . likely to . . . execute it."[43] Here, the Harvard physicist embraced the central justifications for expansion that would be repeated over and over again during the following decades: connections between fundamental research and teaching, industrial spin-offs, and military preparedness.

The "modest" investment Kemble had in mind was hiring Julian Schwinger and Edward M. Purcell; adding $25,000 for operating expenses, $30,000 per

40. Hoddeson et al., *Critical Assembly* (1992), esp. 179–84.

41. Kemble, "Panel: Physical Sciences. Memorandum on Proposals for the Development of the Department of Physics," 9 December 1944, RHP.

42. Kemble to Conant, 22 December 1944, RHP.

43. Kemble to Conant, 22 December 1944, RHP.

year for operating the cyclotron, $30,000 per year for the physics of metals, and $10,000 per year for electronics research; making further appointments in electronics, mechanical engineering, and aeronautical engineering; and constructing a new electronics building at a cost of $100,000, a new mechanical engineering building at $500,000, a wind tunnel at $75,000, and an interdisciplinary science center at between $1.5 and $2 million.[44] Clearly, Kemble's plans extended far beyond nuclear or "fundamental" physics and embraced a picture of a much-enlarged program for both physics and engineering. A few days later, Kemble wrote to his physicist colleagues Curry Street and Kenneth Bainbridge, asking them to consider how their respective research areas might participate in the expansion.[45]

A boosted physics budget made possible both the new machines and the larger role for theoretical physics. Throughout the American university system, the growth of physics was also fueled by a spectacular jump in the number of students. During the war, the armed forces had called upon physics departments to teach thousands of conscripts the elements of physics so they could cope with a new generation of technical war equipment, especially radar, radio, rockets, and navigational aids. Providing instruction to so many students often taxed universities' already depleted resources, but wartime instruction also presented schools with an unprecedented opportunity to expand their clientele. And after the war, the G.I. Bill funded these and other students as they came back to the universities in droves.

Temporary physics programs had formed during the war at Harvard and Bowdoin in electronics and communication, at MIT in radar, at Los Alamos, and at other institutions as well. At Harvard alone, 5,000 students passed through the preradar electronics course. After demobilization the faculty expected that large numbers of veterans from that course would return to further their physics education. As E. L. Chaffee put it on 9 December 1944, the Officer War Training Courses put "Harvard in an advantageous position for attracting students after the war."[46]

Attract them, it did. In July 1946, the department's annual report looked back a year before at a department in which "the instruction and research activities of the Department were at low ebb. The Army Specialized Training Program had been abandoned, and the Navy V-12 program greatly curtailed." Never, in recent years, had the department had so few students. By the second semester of 1945–46 that had changed: there were more students than in any

44. Kemble, "Tentative Summary of Proposals before Physical Sciences Panel," RHP.

45. Kemble to Street, 30 November 1944, RHP; Kemble to Bainbridge, 30 November 1944, RHP.

46. Chaffee, "Expansion of Research and Instruction in the Cruft Laboratory," RHP. We still have little systematic information on the effects of this teaching: Were there lasting effects on the physics curriculum? Where did the students go after the war—to academic pursuits? industrial positions? military assignments?

Table 4.1
Physics Enrollments, Harvard and Radcliffe, 1944–47

Year	Concentration	Enrollment
1944–45		
	Physics	37
	Electronic physics	23
	War service physics (abandoned during the year)	3
	Physics and chemistry	1
	Radcliffe	17
Total		81
1945–46		
	Physics	81
	Electronic physics (abandoned during the year)	15
	Physics and chemistry	1
	Radcliffe	7
Total		104
1946–47		
	Physics	197
	Electronic physics	11
	Physics and chemistry	2
	Radcliffe	9
Total		219

prewar year.[47] The number of physics concentrators catapulted beyond anything anyone had seen (see table 4.1). Aside from the staggering increase in physics concentrators, these figures say something significant about the number of women in physics. In July 1943, the Federal Security Agency, U.S. Office of Education, wrote to the deans of higher education institutions specifically to emphasize that given the intense military need for physicists, "It seems particularly appropriate to encourage women to study physics." Women, the memo continued, would profitably pursue the subject, preparing themselves for "important service in war industries and in the various women's auxiliaries and branches of the armed forces."[48] Federal concern was quickly mirrored in the universities—in the 1943–44 annual physics department report, E. C. Kemble bemoaned the low percentage of women in physics, mathematics, and chemistry. A startling jump in physics concentrators among women had occurred in 1943, a change he

47. Harvard University, Department of Physics, "Annual Report of the Department of Physics," 1 July 1945–1 July 1946, HUA. I would like to thank Katherine Sopka for making these reports available to me.

48. Federal Security Agency, U.S. Office of Education, to Deans of Colleges of Arts and Sciences, July 1943, folder "War probs-misc. corresp.," box 40, UWA.

ascribed to a combination of interest generated by the war and to the inclusion of a full-time instructor who could give the women students a degree of attention "previously unavailable." Such success, Kemble believed, had to be furthered by even greater efforts to provide individualized help. The number continued to climb in 1944–45 to a peak (shown in table 4.1) of 17 out of 81 concentrators. With the huge influx of male students after V-J Day, the efforts to attract women appear to have been largely abandoned, the women students were mixed in with the men, and the percentage of women fell precipitously from 7/104 to 9/219.[49] In short, physics simply reflected the displacement women underwent in professional jobs of every description immediately after the war. With the massive return of (male) veterans, the physics department turned to other problems.

Aside from introducing promising students to the university, Chaffee remarked that "[t]he war-training courses have provided us at no expense with a very considerable amount of laboratory equipment. There is insufficient space in the Cruft Laboratory even to store this equipment, [to] say nothing of setting it up for instruction."[50] Instructional equipment could be supplemented by war-surplus apparatus suitable for research. Chaffee noted that Harvard's antiradar work and other war projects would offer "an unusual opportunity to purchase advantageously some very valuable equipment from the OSRD [Office of Scientific Research and Development] projects and perhaps some equipment from the war training courses. . . . We should be prepared to purchase a considerable amount of this equipment. There will [be] available machine tools, obtainable at much reduced prices from the same sources, and I believe we should purchase a considerable amount of this machinery both to increase our present shop facilities and to replace some outmoded and worn machine tools."[51] With this new equipment, Chaffee expected that physicists would be able to exploit their newly developed capability to generate microwave signals "by methods which have worked but which are not understood." Such applications included detection of molecular resonances, pulse systems of communication, and high-frequency heating. As Purcell rather overmodestly put it: "[Y]ou didn't have to be too smart to design an experiment with the extraordinary resources offered by the new electronics."[52]

Surplus equipment came from many war sources and went to a wide spectrum of users. Connections established between the scientific and defense

49. Harvard University, Department of Physics, "Annual Report of the Department of Physics," 1943–44, 1944–45, 1945–46, and 1946–47, HUA.

50. Harvard University, Department of Physics, "Annual Report of the Department of Physics," 1943–44, 1944–45, 1945–46, and 1946–47, HUA.

51. Harvard University, Department of Physics, "Annual Report of the Department of Physics," 1943–44, 1944–45, 1945–46, and 1946–47, HUA.

52. Edward M. Purcell, interview by the author, 26 May 1987. Purcell discovered the 21-centimeter line and won the Nobel Prize for his coinvention of nuclear magnetic resonance techniques—both exploited Rad Lab techniques.

communities multiplied, yielding benefits for scientists long after V-J Day. Since they were on contracts from the Office of Naval Research (ONR), as late as 1960 physicists at Brookhaven could acquire large armor plating, originally intended for cruisers, to use in neutrino experiments.[53] On the West Coast, Robert Hofstadter collected naval gun mounts on which he could perch his magnetic spectrometer.[54] Overseas (as we will see in chapter 6), A. Gozzini's experiments with surplus pulse-generator circuits and microwave equipment led to his development, with Marcello Conversi, of the "flash tubes" that played such an important role in cosmic ray physics and in subsequent work on the spark chamber.[55] Phototubes, crucial for scintillation devices, had been much improved and exploited during the war for noise generation in radar countermeasures. And in England (as we saw in chapter 3), the great change in emulsion preparation techniques during the atomic bomb project set the stage for postwar emulsion-based particle physics. This type of continuity between wartime and postwar work ran deep; beyond the formation and evolution of administrative organizations such as the OSRD and its accompanying contract system, an essential consequence of the war work was the carryover of physicists' techniques, equipment, and collaborations into the late 1940s.

With the promise of new physicists, new students, and new equipment, departments could embark on major new research projects. Harvard—along with many other universities—was determined to restart its cyclotron program on a much larger scale. Planning to accelerate both electrons and protons, the Committee on Nuclear Physics met for the first time in January 1946. They wrote: "From the point of view of physics this program represents a vigorous and progressive plan which should enable Harvard to compete favorably for financial support and, in addition, enhance its attractiveness as a center of research in the nuclear field."[56]

When the Harvard physicists turned from war work to accelerator work, they brought with them experience from the Manhattan Project (Kenneth Bainbridge), the Radar Project (Edward M. Purcell, Wendell Furry, Curry Street, and Julian Schwinger), Radar Countermeasures (Roger W. Hickman and John Van Vleck), the Interservice Radio Propagation Laboratory (Harry Rowe Mimno), NDRC Ordnance Sections (Percy Bridgman), and the Underwater Sound Laboratory (Roger W. Hickman and Frederick "Ted" Hunt). Many of these men served

53. M. Schwartz, interview by the author, 20 October 1983.

54. Hofstadter, Fechter, and Helm to commandant, Mare Island Naval Shipyard, 9 February 1952, RHofP, SUA.

55. See Galison, "Bubbles" (1989).

56. "Proposal for Nuclear Physics Program at Harvard," submitted 2 February 1946 and enclosed with minutes of the first [11 January 1946] meeting of the Committee on Nuclear Physics [later Committee on Nuclear Sciences], file "Index and Records," box "Committee on Research and Nuclear Sciences, Records 1946–1951," HUA.

on Harvard's newly established committee. After selling their old cyclotron to the government for $250,000 and getting a commitment of $590,000 from Harvard and $425,000 from the navy, the committee members could begin to plan an 84-inch cyclotron.[57] Their physics program included proposals to produce 25 MeV deuterons and 50 MeV alphas in order to explore the nature of the proton, to produce high energy neutrons, and to extend the wartime fission experiments to elements lighter than uranium and thorium. With accelerated electrons on tap, the planning committee hoped that the cyclotron could also be used to pursue radiation therapy, the photodisintegration of nuclei, and the formation of electromagnetic showers.[58]

Building plans called for two separate structures linked by a corridor. One, standing on heavy foundations and surrounded by 3-foot-thick concrete walls, would enclose the cyclotron. A 20-ton crane would shuttle equipment back and forth, while primary power, mechanical support, photographic services, and electronic equipment would be staged in the second building. Almost $500,000 was promised by the government, a host of new, contract-supported "research fellows" entered the scene, and new hires included two mechanical engineers, one electrical engineer, one junior physicist, two machinists, a technician, two draftsmen, and several graduate students. These resources were augmented by the immediate support of the military: the Watertown Arsenal, for example, would machine some 750 tons of iron for the magnet frame. In this massive infrastructure, Curry Street, for example, saw the opportunity to continue with measurements of meson masses and lifetimes, work he had touched on in an earlier and less expensive era of $500 cloud chambers and mountaintop expeditions. In scale and scope it was truly a new era.[59] The ever growing squadron of theorists would take over the space previously occupied by Aiken's Automatic Sequence Controlled Calculating machine—Aiken and his machine had moved up to the new Computation Laboratory.[60] Here, in the very physical distribution of activities of the laboratory, one can track the shifting conception of physics: an industrial scale of experimentation, a new quasi-autonomous theory group, and a sizable

57. Buck to Hickman, 5 October 1946, file "Miscellaneous Financial Material and Correspondence," box "Committee on Research and Nuclear Sciences, Records 1946–1951," HUA; "Proposal for Nuclear Physics Program at Harvard," submitted 2 February 1946 and enclosed with minutes of the first [11 January 1946] meeting of the Committee on Nuclear Physics [later Committee on Nuclear Sciences], in file "Index and Records," box "Committee on Research and Nuclear Sciences, Records 1946–1951," HUA.

58. "Proposal for Nuclear Physics Program at Harvard," submitted 2 February 1946 and enclosed with minutes of the first [11 January 1946] meeting of the Committee on Nuclear Physics [later Committee on Nuclear Sciences], in file "Index and Records," box "Committee on Research and Nuclear Sciences, Records 1946–1951," HUA.

59. See Harvard University, Department of Physics, "Annual Report of the Department of Physics," 1946–47, HUA; on Street's prewar meson work, see Galison, *Experiments* (1987), chap. 3.

60. Hickman, "Annual Report Concerning the Laboratories Associated with the Department of Physics," 1 July 1946–1 July 1947, attached to "Annual Report of the Department of Physics," 1946–47, HUA.

computation laboratory. All of these visible branches had thick, deep roots in the infrastructure of the scientific war.

4.3.2 Princeton

At Princeton, as at Harvard, planning for the postwar expansion began long before euphoric crowds descended on Times Square. On 4 January 1944, a somewhat overoptimistic John Wheeler wrote to H. D. Smyth that he trusted the war would soon be over and he could return to physics shortly. He then went on to formulate a "Proposal for Research on Particle Transformations," of which the "ultimate purpose" was "[t]o determine the number of elementary particles, the transformations between them, the combinations which they permit, the nature of their interactions, the relation between these particles and the existing theories of pair formation, electromagnetism, gravitation, quantum mechanics and relativity." He had witnessed an immensely successful collaboration of theory and experiment at the University of Chicago's Met Lab, culminating in the achievement of fission in December 1942, and he clearly saw joint work as the wave of the future. Wheeler began: "*Plan.* Effective progress calls for the collaboration of experiment and theory. The interaction between the two will be most fruitful, I believe, when in addition both approaches are combined in a single institution, under the same leadership."[61] He, like his Harvard counterpart Kemble, argued that such collaboration ought to be the model for physics research; the lesson they were drawing was one being taken to heart at institutions across the United States.

Wheeler needed help to solve the main theoretical problems. These included the use of action-at-a-distance theories to eliminate the self-energy difficulty in quantum electrodynamics, the classification of relativistic quantum field theories, and the theoretical exploration of positronium. There were also such "associated experimental" problems as nuclear meson capture, the masses of cosmic ray particles, and the gamma-ray production of mesons. Like Kemble at Harvard, Wheeler counseled building the theoretical side of the Princeton department by hiring "[t]hree theoretical assistants of the type of Feynman or Jauch," along with some experimenters "of the type of Luis Alvarez or Bob Wilson." Of course the 33-year-old Wheeler would "[w]elcome collaboration of interested older members of staff," and again grounding his recommendation on his Chicago experience, he suggested that experimenters should "call on experienced electronics men for the design and development of instruments." The program promised, as Wheeler continued to insist after the war, the possibility of sources

61. Wheeler to Smyth, 4 January 1944, file "Postwar Research," box "Physics Department Departmental Records, Chairman, 1934–35, 1945–46, no. 1," PUA.

of energy many times more powerful than all known nuclear reactions, with "obvious implications for the problem of national defense."[62]

As ambitious as young Wheeler's plans may have sounded initially, even before the war's end the physics department had begun to address the Princeton administration in a new, more confident tone. With the successes of proximity fuses, radar, and then the atomic bombs, physicists—who for years had occupied a decidedly secondary place within the university—swelled with pride, evident as they spoke to colleagues and administrators. The physics department captured this tone in a draft of its departmental report:

> The end of the war finds the department in a praiseworthy but embarrassing condition. The record of the members of the department in war work is laudable, so much so that many of them, particularly in the younger group, are receiving very attractive offers from other institutions and from industry. Such offers are not only attractive in terms of salary but are usually backed by promises of large expenditures for apparatus and equipment. The university must choose between going ahead vigorously, capitalizing [on] the fine record of this department during the war or letting its physicists drift away to such a degree that it may take a generation to restore the department. The first course will require money for men and for equipment, a great deal of money, but it offers a magnificent opportunity, completely in the tradition of the university. We have never been in a better position to push forward in the field of fundamental physical research.[63]

Indeed, their position *was* entirely unprecedented.

Physicists across the continent were attracted by the promise of the new technology and science for advancing the physics of nucleons and mesons, but developments in "fundamental" experimental physics at Princeton took on a particular cast. In part this reflected an imaginative style of work that John Wheeler had developed before the war, but the echoes of what he had seen in the Metallurgical Laboratory can be heard in his ideas for postwar research. As early as June 1945, Wheeler had written a proposal on the future of physics research that gave three goals for the postwar epoch. First, though he voiced doubts about some of their features, he advocated the development of accelerator sources for particles. Among these, he mentioned Luis Alvarez's latest plans to build a linear electron accelerator; Wheeler judged that one would want at least a 5 GeV proton accelerator in order to produce pairs of mesons. Second—and here he saw a real payoff in physics—he hoped that the department would seize upon cosmic rays as the primary domain in which to search for answers to basic

62. Wheeler to Smyth, 4 January 1944, file "Postwar Research," box "Physics Department Departmental Records, Chairman, 1934–35, 1945–46, no. 1," PUA.

63. "Report to the President 1944–45," box, "Physics Department Chairman's Correspondence 1942–43, 1943–44, no. 16," PUA.

questions because there alone could one find the high energy collisions needed to probe the subnuclear domain. For the cosmic ray project, Wheeler suggested using Flying Fortresses to hoist experiments and experimenters into the upper atmosphere.[64]

The idea of enlisting bombers in the study of high-altitude cosmic radiation had several appealing aspects. It would, for example, alleviate the costs of such exploration for the universities while leaving control over research apparatus entirely in the hands of the physicists. As for its physics justification, Wheeler noted that only by reaching far into the sky could one study particles with something like 10^{17} electron volts—vastly more energy than the 5 GeV envisioned for the accelerator—and therefore exhibit the multiple-meson production process that interested him. "This plan calls for army transportation of equipment up to 10 tons to altitudes of the order of 40,000 feet. Research money would in this way be freed for research itself, and for research of a most effective kind."[65]

Finally, Wheeler felt that a survey of the entire field of ultranucleonics was of the highest priority. From what had been learned by cosmic ray studies, he suspected that it might be possible to transform matter directly and completely into energy on the model of protons being transformed into mesons in the upper atmosphere: "Discovery [of] how to release the untapped energy on a reasonable scale might completely alter our economy and the basis of our military security. For this reason we owe special attention to the branches of ultranucleonics—cosmic ray phenomena, meson physics, field theory, energy production in supernovae, and particle transformation physics—where a single development may produce such far-reaching changes."[66] To reach these dramatic goals, physicists had to make their needs known, and here the survey would play a vital role. It would offer workers in postwar physics "a prospectus of long-range objectives" and gain financial support for fundamental physics by making research public and "by demonstrating that scientists in free association can show more vision and judgment on research planning than any centralized government authority." Not least, the ultranucleonic survey would "uncover lines of investigation of evident present or future value to the country's war power."[67] Once again, the wider world impinges on the agenda of abstract inquiry.

On 15 August 1945 (local), Japan surrendered. World War II was over. But while crowds rejoiced, Wheeler uneasily contemplated the future and composed

64. Wheeler, "Three Proposals for the Promotion of Ultranucleonic Research," 15 June 1945, copy to Smyth, box "Physics Department Departmental Records, Chairman 1934–35, 1945–46, no. 1," PUA.

65. Wheeler, "Three Proposals for the Promotion of Ultranucleonic Research," 15 June 1945, copy to Smyth, box "Physics Department Departmental Records, Chairman 1934–35, 1945–46, no. 1," PUA.

66. Wheeler, "Three Proposals for the Promotion of Ultranucleonic Research," 15 June 1945, copy to Smyth, box "Physics Department Departmental Records, Chairman 1934–35, 1945–46, no. 1," PUA; see also Wheeler, "Particle Physics," *Amer. Sci.* 35 (1947): 177–93, on 170, 172, 174.

67. Wheeler, "Three Proposals for the Promotion of Ultranucleonic Research," 15 June 1945, copy to Smyth, box "Physics Department Departmental Records, Chairman 1934–35, 1945–46, no. 1," PUA.

a letter to Edward Teller: "Dear Edward, with the conclusion of the war today my work here will soon reach its conclusion. . . . What I can now do most effectively is, I believe, fundamental research. But I do not feel quite at ease to do so over the next five year period." It seems that not long before, Teller had spoken with Wheeler at Los Alamos about his fears for postwar national security and hopes for the completion of a weapon a thousand times more powerful than the atomic bomb—the hydrogen bomb. Having been invited to participate in this research, Wheeler recalled, "[b]y that time it was becoming obvious that if I join you in the work . . . it would be a matter of getting ready for the next war, not of prosecuting this one. And it is just this matter of getting ready for the next war about which I am concerned." For security reasons, Wheeler then lapsed into a conceit:

> Here is a group of men absolutely isolated on an island. They have got into a fight. Two groups of men with quite different ways of doing things have teamed together to try to put down the troublemakers. Our group has learned to put together a bow and arrow. By that means we have put an end to the fight. Our ally is observant. Now that the fight is over he has gone back and is spending part of the time behind his wall. We know that some of his men would get delight out of building a bow and arrow of their own. We suspect that some of his men would not hesitate to use that bow and arrow on us if someday we happened to get into a disagreement on who is get the pears from that fine looking pear tree over there. For some reason or other the two former allies don't seem to be able to get together to turn the bow and arrow weapon over to a custodian whom both can trust. . . . Some people in our group say, "So what" and are making plans to go fishing. I'm one of the people who feels that if we're going to get into an armament race, we'd better start now, and we'd better try to build the best weapon we know how to build—a machine gun which will outmode the bow and arrow.[68]

As Wheeler knew, Teller had already been working on the machine gun (superbomb) since before the Manhattan Project had even broken ground at Los Alamos. He now asked Teller what priority the work ought to have, what priority it in fact had, and how he (Wheeler) could help. "If the next war is not to come for fifteen or twenty years, I believe fundamental particle research is the field in which I can make the most useful contribution; but if it's to come in 5 or 10 years, I'm inclined to think I ought to do something on the machine gun."[69] In a month, Wheeler returned to Princeton, leaving open exactly how his time would divide between elementary particle physics and thermonuclear weapons. When the call to hydrogen arms did come for Wheeler—in late 1949—he responded promptly, first through work at Los Alamos, and then in 1950 with a Princeton-based effort designated "Matterhorn B."[70] But as Wheeler's island

68. Wheeler to Teller, 12 August 1945, A-84-019, LAA.

69. Wheeler to Teller, 12 August 1945, A-84-019, LAA.

70. Galison and Bernstein, "In Any Light," *Hist. Stud. Phys. Bio. Sci.* 19 (1989): 267–347.

parable made clear, even before the war came to an end, for a great variety of reasons, his thoughts had turned to fusion, a far more complete annihilation of matter than mere fission.

In June 1945, three months earlier, by which time he certainly knew of Teller's "machine gun," Wheeler had begun to plan the future of the Princeton physics department, centering on the transformation of mass into energy: he wanted a new accelerator program, a new, far-better-funded cosmic ray program, and a theoretical "ultranucleonics" program in which theory would probe new kinds of particle transformations. With Wheeler's V-J Day letter in mind, his speculations about the ability of ultranucleonics to alter the "basis of our military security" gain a more concrete basis. For though fusion would still technically be "nucleonics," its vast superiority to fission made it a kind of stepping-stone into a new world.

Renaissance navigators, Wheeler told a symposium audience in November 1945, could delude themselves about how much of the distance around the globe they had traveled. Not so modern physicists. "The relation of Einstein between mass and energy is our sextant." That famous formula $E = mc^2$ not only promised the possibility of the conversion of any kind of mass into energy, it also provided an absolute measure of how far we were along the path to the total destruction of matter. The science of the atomic bomb and the pile was coming to completion; now attention must turn to ultranucleonic transformations in which neutrons and protons are not merely rearranged but "reduced to lower entities or altogether destroyed." Here was the real point of cosmic ray investigations that soon became a $200,000 per year program at Princeton: "The possibility of the complete conversion of matter to energy is suggested by present incomplete information on the production of particles of lower mass by or from protons in the upper atmosphere of the earth. Discovery of how to release the untapped energy . . . might completely alter our economy and the basis of our military security. . . . Other nations have not neglected this work during the war. [By "other nations," Wheeler meant the Soviet Union.] We must prepare to resume it vigorously."[71]

In his theoretical work, Wheeler turned to calculations as part of his program. He explored the dynamics and annihilation rates for what he called "polyelectrons"—atoms composed of bound states of positrons and electrons. By October 1946, he had results.[72] His hope was that these polyelectrons, their excited states, and their decay into the vacuum could explain not only the mass spectrum of the newly discovered mesons but also the decay times and products. Even more than fusion, electron-antielectron annihilation advanced the Wheeler program toward pure matter-to-energy conversion. Flush with the success of

71. Wheeler, "Problems and Prospects," *Proc. Amer. Phil. Soc.* 90 (1946): 36–47, on 36–37.

72. Wheeler, "Polyelectrons," *Ann. N.Y. Acad. Sci.* 48 (1946): 219–38.

these polyelectron calculations, Wheeler continued in 1946 to press for a further exploration of what he began calling "transformation physics." "Nuclear explosions," with its double meaning, now was elevated to a separate category of inquiry. Individual nucleons, preeminently the neutrons, were seen in cloud chambers to disintegrate into a myriad of other entities. How, Wheeler asked, did the cross section of nuclear explosion vary with incoming photon energy? How did the cross section of nuclear explosion change with atomic number? What were the number, type, and energy of the outgoing decay products?[73] Only by answering these sorts of questions (some of which were standard and some of which were Wheeler's own) could the navigator know when the limits of nucleonic transformations had been broached.

If the physics and engineering of nuclear weapons left their imprint on the day-to-day planning of postwar research at Princeton, the enemy's war matériel did too. From Nazi Germany had come the dreaded "vengeance weapons": the V-1 "buzz bomb" and the V-2 guided missile. These bombs had required an engineering project of immense scope, costing over $3 billion—fully one and a half times the resources put into the Manhattan Project. The V-2 was brought into the war relatively late, but starting in September 1944, the Germans successfully launched more than 3,000 V-2's, killing almost 10,000 people in England. Soon, however, the Allies began advancing across Europe, and Wernher von Braun retreated from his headquarters at Peenemünde with some 4,000 workers to the V-weapon production facility situated in the concentration-camp complex of Dora-Nordhausen in Thuringia. Installing themselves in the Harz mountains, the V-2 workers successfully evaded the approaching Russian army, eventually surrendering themselves to an American garrison. Under a secret mission code-named "Overcast," the American army shipped the Nazi scientists to the United States to continue missile development work at several sites. The Germans arrived in October 1945, and soon the Peenemünde team was split up into groups, working with American industry to produce a variety of rocket types.[74]

From 16 April 1946 to 19 September 1952, 64 V-2's were launched at White Sands. The first failed three and a half miles into the air when a fin ripped off and the rocket was destroyed; the next launch, on 10 May 1946, successfully rose to 71 miles.[75] For physicists at Princeton, the capture and installation of the German rocket team offered an immediate opportunity. In November 1945—a month after von Braun and his associates were brought to White Sands—M. H. Nichols jotted an interoffice memo to Smyth suggesting that the department propose to study optical and electrical phenomena in the upper regions of the earth's

73. Wheeler, "Particle Physics," *Amer. Sci.* 35 (1947): 177–93, on 185–86.

74. Ordway and Sharpe, *Rocket Team* (1982); see also the important article by Hunt, "U.S. Coverup," *Bull. Atom. Sci.* 41 (April 1985): 16–24.

75. Ordway and Sharpe, *Rocket Team* (1982), 353–54.

atmosphere. At the same time, the Princeton group could explore cosmic rays and neutron densities. All this was made possible by "new advances in rocket technique as well as progress here at Princeton and elsewhere in the field of radio telemetering from aircraft and missiles," which would "make possible an extension of present data to regions as high as 500,000 feet."[76]

For some time, Wheeler had seen cosmic ray physics, not accelerator physics, as the primary vehicle for understanding the "transformations" of elementary particles. In a memo of January 1946, he stated that "cosmic ray research will take on an even more important role in physics in the next few years," and he advocated immediately setting up a joint experimental and theoretical research group: "Inasmuch as the V-2 firings will not last indefinitely, and inasmuch as the experienced researchers are becoming scarcer every day, it appears that some action be taken as soon as sound decisions can be made." Presumably addressing himself to the department chairman, Wheeler stressed that they would be needing four to six assistants with experience in experimental electronics, nuclear physics, or cosmic rays, as well as experienced cosmic ray and nuclear physics experimenters: "[R]esearch in physics is starting fresh, and . . . new techniques and new vehicles are now available"—and so it behooved the department to search out consultants from among the best universities, institutes, and weapons laboratories.[77]

Wheeler himself headed a navy-funded project that would handle the telemetric transmission of cosmic ray data from the German team's missiles. Begun on 1 January 1945 for other purposes, the navy grant had been extended in March 1946 and was to cover the development of telemetry equipment, while at the same time serving to study cosmic ray showers and the properties of "mesotrons" through the design, fabrication, and operation of cloud chambers and Geiger counters to be mounted on the V-2's. In July 1946, D. J. Montgomery reported to the navy that the Princeton V-2 expedition had arrived at White Sands and was making final tests for what they hoped would be the 100-mile-high Princeton Shot on 6 August. At this point $335,000 had already been allotted, with $250,000 more to be shared over the next two years between chemistry and physics.[78]

Cooperation with the military remained close. Military and elementary

76. Nichols to Smyth, 26 November 1945, file "Postwar Research," box "Physics Department Departmental Records, Chairman 1934–35, 1945–46, no. 1," PUA.

77. Wheeler, "Program in Cosmic Rays," January 1946, file "Postwar Research," "Physics Department Departmental Records, Chairman, 1934–35, 1945–46, no. 1," PUA.

78. Montgomery, "Annual Report on Project Assisted by Outside Funds," 23 July 1946, file "A-475 Wheeler," box "Physics Department Laboratory and Research Files, 1929–54," box 1 of 5, PUA. See also "Elementary Particle Projects as of 6 May 1946," in same file; among physics goals listed in this report were determination of total cosmic ray intensity, meson production, neutron intensity, multiply charged particles at rocket altitudes, study of radio propagation in ionosphere, telemetry tests, pressure-temperature studies, and coordination with the Schein group's ground cloud chamber and balloon tests.

particle problems were interspersed with planning and designing the mission. Both physicists and strategic planners needed a comparison of "Lark" and Naval Research Laboratory (NRL) telemetering systems, especially with regard to the reliability, intensity of signals, and freedom from disturbances of each system. Both civilians and uniformed personnel had to study radio signal propagation in the ionosphere by transmitting and receiving signals from the missile. In addition, the physicists could use the high-altitude flight to measure cosmic ray intensity, to distinguish primary cosmic ray electrons from primary protons, and to measure the neutron productivity in the atmosphere as a function of altitude.[79] Such explorations directly continued some of the Princeton group's wartime accomplishments in telemetry. In fact, at least one member of the staff, Walter Roberts, wanted to maintain close connection with the work on guided missiles at the Johns Hopkins Applied Physics Laboratory. As Wheeler assured his readers, this connection would ensure a "satisfactory liaison" between the Princeton laboratory and weapons development efforts.[80]

During the period from 1945 to the early 1950s, the liaison between civilian and military nucleonics functioned well—from both parties' perspective. The ONR liberally funded civilian science, and in return the scientists moved easily back and forth between nuclear physics, cosmic rays, and weapons problems. Princeton's nuclear physicist Milton White, for example, was pleased to report on some recent Princeton instrumentation work that seemed perfectly suited for transfer to the military sector. The laboratory had perfected a new, simple, rugged, and reliable scintillation counter, and White lost no time in alerting the ONR to possible defense applications of the new device: "If the U.S. Government has need of α-particle counters, either in connection with plutonium plants, or atomic bombs, there should be set in motion a program for further engineering and quantity manufacture. I can visualize an eventual need of many thousands of counters; if this is correct then our contract with the Navy will already have given the government more than a fair return on the money thus far allocated."[81] White added only that he hoped the Princeton researchers could be spared the engineering details, an echo of the 1942 Rad Lab debate in which he had argued powerfully for the physicist to be exempted from "the details of manufacturing methods." Zones of collaboration bound experimenter to theorist, weapon designer, and engineer; much could be shared around the local

79. Wheeler, "Appendix IV—General Survey of the Princeton Project Program—Cosmic Rays and Telemetering," 28 August 1946, file "A-475 Wheeler," box "Physics Department Laboratory and Research Files, 1929–54," box 1 of 5, PUA.

80. Wheeler, "Appendix IV—General Survey of the Princeton Project Program—Cosmic Rays and Telemetering," 28 August 1946, file "A-475 Wheeler," box "Physics Department Laboratory and Research Files, 1929–54," box 1 of 5, PUA.

81. White to Liddel, Nuclear Physics Section, ONR, 11 July 1947, file "761," box "Physics Department Laboratory and Research Files, 1929–54," box 5 of 5, PUA.

concerns of a scintillation counter or a telemetered rocket launch. But the distinct identity of the physicists and their subculture remained.

4.3.3 Berkeley

On the West Coast, physicists at Stanford and Berkeley had no intention of being spared the engineering details. The style of research in the West was different from that in the East: it was more tightly bound to engineering, and it drew more liberally from philanthropic and industrial sources. Such entrepreneurial physics had brought Berkeley's E. O. Lawrence international fame for his big accelerators, funded through private donors. As engineering accomplishments, Lawrence's accelerators were unrivaled, though he was less successful at drawing deep physics from them. As Robert Seidel has shown, World War II saw a substitution of federal for philanthropic funds, and the important assignment that Lawrence's laboratory direct the electromagnetic separation of U^{235}. Lawrence was as well prepared to begin large-scale research as anyone and soon was supervising a dramatically bigger laboratory with a wartime expenditure of $692,000 *per month.* By the middle of 1944 the Berkeley Radiation Laboratory held a total working population of 1,200 scientists, engineers, and technicians; indeed, once their war work began in earnest the number of engineers at the laboratory never dipped below 60.[82]

At Oak Ridge, where the Berkeley workers were setting up their accelerators to separate the highly fissionable U^{235} from U^{238}, Lawrence endured months of intense frustration because of difficulties extracting the U^{235}. To bolster the flagging effort, he began in the summer of 1944 to lobby for 10 new isotope separation facilities, leading General Leslie Groves (who was in charge of the Manhattan Project) to some cautious thinking about spending $7–$10 million. A year later Lawrence began arguing for the rapid expansion of nonweapons facilities, including Luis Alvarez's planned linear accelerator and Edwin McMillan's projected synchrotron. A few months after the war's end, Groves authorized $250,000 in surplus radar sets for the linear accelerator, $203,000 in surplus capacitors for the synchrotron, $630,000 for construction in the laboratory, and $1.6 million for six months of operating expenses. Building on earlier experience, engineering and physics had grown together at Berkeley, to make the university one of the models of postwar physical research. In fact, the Berkeley Radiation Laboratory became the pacesetter for much of the AEC's development of regional laboratories.[83]

Lawrence had brought big science to Berkeley in the 1930s; in a sense he helped to create the category "big science." Before World War II, the scale of

82. Seidel, "Accelerating Science," *Hist. Stud. Phys. Sci.* 13 (1983): 375–400.

83. Seidel, "Accelerating Science," *Hist. Stud. Phys. Sci.* 13 (1983): 375–400; see also Seidel, "Big Science," *Hist. Stud. Phys. Sci.* 16 (1986): 135–75.

the operation itself had done much to alter the outer laboratory—more physicists and engineers than ever before crowded together to build a single piece of apparatus. Their fates were consequently bound together. World War II surely reinforced these trends, and Lawrence's effort at electromagnetic isotope separation was one of the exemplars of large-scale science. The transformation of the inner laboratory, however, was not one initiated by Lawrence. Even Eugene Gardner's postwar emulsion work left the immediate site of experimentation unchanged from cosmic ray inquiries. Changes in the inner laboratory came, as we will see in chapter 5, with Alvarez's importation of collaborative research into the inner experimental space itself. It was then, and only then, that the new work practices and technology came all the way down to the creation, analysis, and interpretation of the physical data. Because this is a turning point in experimental practice, it demands a more extensive analysis. We therefore defer further discussion of the Berkeley Radiation Laboratory to chapter 5.

4.3.4 Stanford

Stanford, like its Berkeley neighbor, had successfully linked engineering and physics before the war. While Lawrence and his team were building ever-larger cyclotrons, the Stanford physicists were binding electrical engineering to physics as they learned to manipulate microwaves. William Hansen had set the character of that collaboration at Stanford with his stunning development of the "rhumbatron," which set electrons in an oscillatory dance by creating electromagnetic resonances within a copper cavity. Although the device was quickly superseded as a particle accelerator, it formed the core of the klystron, a powerful microwave tube that the Varian brothers designed and deployed in airplane navigation and locating systems. Soon the Sperry Gyroscope Company was underwriting much of the joint physics and electrical engineering efforts. With the help of their electrical engineer leader, Frederick Terman, Stanford's electrical engineering department built a myriad of radio communications systems around their jewel, the klystron. Gradually, the Stanford engineers transformed the klystron from a fascinating, isolated tube to a standardized component within a whole gamut of microwave circuits.[84]

On a technical level, the microwave klystron-based research continued unabated into the early years of World War II: weapon innovations included instrument landing systems and Doppler radar. But dramatic changes quickly accompanied the increased pace, scope, and funding of laboratory work. Already in April 1942, Paul Davis (Stanford's general secretary) was writing Terman that "[t]here are many things that could be done under the pressure of the present

84. On Stanford's early combination of physics and engineering, see Leslie and Hevly, "Steeple Building," *Proc. IEEE* 73 (1985): 1169–80.

war situation that will be more difficult to achieve in peace times"—including ambitious plans for electrical engineering. In August 1942, Stanford issued its "Proposal to Organize the Stanford Resources for Public Service," which focused on how to organize "a vastly augmented program of service on a contractual basis."[85]

After listing suggested projects (from surveys of mineral and industrial resources to the creation of psychological warfare tunes, such as "Marching Civilization"), Stanford's August proposal turned to the effects of war work at the university itself. A radical increase in contractual research would provide an opportunity to reorganize faculty administration and to improve the physical plant for more effective war and postwar work. Substantial contracts would bring federal war priority, keeping faculty on campus, creating interdisciplinary research, and engaging a new cadre of talented students who would stay on after the war. Moreover, working on contracts would make Stanford students known to public and private agencies and might contribute to the long-term development of the West. But above all, government-sponsored research would rocket Stanford to a position comparable to Harvard, Chicago, Caltech, Berkeley, and Columbia by bringing in "substantial additional income."[86]

Ironically, while Stanford did greatly expand during the war, the expansion's great advocate, Frederick Terman, spent the war years on the East Coast as director of the Radio Research Laboratory in Cambridge (RRL), a facility to produce radar countermeasures. As the new laboratory grew into a powerful organization, Terman became ever more conscious of the models that Harvard, MIT, and RRL itself presented for postwar Stanford. He was also deeply impressed with many of the administrators from Harvard, especially with his neighbor William Henry Claflin, Jr., Harvard's treasurer.[87] Terman was quite keen for Stanford's general secretary to speak with Claflin, and the Stanford engineer-administrator soon sought and arranged a meeting between Donald Tresidder, Stanford's president, and Claflin. By December 1943, Terman had concluded that "[t]he years after the war are going to be very important and also *very critical ones* for Stanford. I believe that we will either consolidate our potential strength, and create a foundation for a position in the West somewhat analogous to that of Harvard in the East, or we will drop to a level somewhat similar to that of Dartmouth, a well thought of institution having about 2 per cent as much influence on national life as Harvard."[88] Terman then presented a plan to "lick" Caltech by equaling it in the physical sciences; "after all they are only a

85. Davis to Terman, 18 April 1942, SC 160, 1:1:2, FTP; "Stanford Resources," typescript, 24 August 1942, SC 160, 1:1:2, FTP.

86. Typescript, "Public Service," 24 August 1942, SC 160, 1:1:2, FTP.

87. Terman to Davis, 23 August 1943, SC 160, 1:1:2, FTP.

88. Terman to Davis, 29 December 1943, SC 160, 1:1:2, FTP; quoted, in part, in Leslie and Hevly, "Steeple Building," *Proc. IEEE* 73 (1985): 1169–80, on 1176.

specialized school, and Stanford is a complete university." In part, Terman wanted a "technical institute" that would create a joint identity among scientific and engineering fields. This alliance would aid in attracting and placing students, raising money, and creating an identity based on special western areas of strength. One such field was the characteristically western oil industry, which would link geology, heat transfer, and chemical engineering with the radio industry and accompanying research. Moreover, Terman argued, the competition was softening: Caltech had become "smug," leaving "cracks in its armor" by not developing electrical engineering, and Harvard, Yale, Columbia, and Princeton had thus far slighted the applied sciences in favor of natural philosophy and the humanities.[89]

The codevelopment of electrical engineering and physics was the hallmark of Stanford physics. Indeed, the commitment to scientific entrepreneurship had begun during the mid- to late 1930s, as Bruce Hevly and Stuart Leslie have shown,[90] but it now took a new turn. The push to create an overlap area—a trading zone—between physics and engineering accelerated under the pressures of war and the beckoning opportunities that would follow the conflict. In the short term, the need was to create a site, a huge new Microwave Laboratory that would draw simultaneously from physics and engineering. Hansen wrote the physics department chairman pro tem, Paul Kirkpatrick, on 6 November 1942, explaining that when peace came, the microwave lab would draw further staff and equipment from both the physics and the electrical engineering departments. "At this point, this laboratory goes out and gets a government contract for some microwave job. There should be no difficulty in doing this. With this job will come a priority. . . . Then you start spending money and also, of course, doing the job." Hansen hoped to establish as many attractive fellowships as could be filled, and then to order "equipment—of a sort that will be useful after the war." This should include "machine tools, measuring equipment, . . . books and any other things that can be used to generate apparatus or research." Played well, the plan would guarantee that "even if we don't have a dime after the war, good physics can be done."[91]

Almost exactly a year later, in November 1943, Hansen elaborated on his initial scheme. It had already been "obvious" before the war that physicists would play a crucial role in industry and that radio engineering (including electronics) was in for rapid expansion. But "[w]hile these trends were obvious before the war, the war has both accelerated and called attention to them. This is especially easy to notice . . . in the field of radar. . . . The result will be that, after the war, all major universities will be forced to offer strong instruction in these two branches of science." Because it could strengthen and link the two domains,

89. Terman to Davis, 29 December 1943, SC 160, 1:1:2, FTP.

90. See Leslie and Hevly, "Steeple Building," *Proc. IEEE* 73 (1985): 1169–80.

91. Hansen to Kirkpatrick, 6 November 1942, SC 160, 1:1:7, FTP.

the Microwave Laboratory would prove essential. Modeled on the Berkeley Radiation Laboratory, it would remain under the control of the physics department, although the director would control the budget. Indeed, by establishing the laboratory with university funds, the inevitable private support would not dictate the direction of work.[92]

The Stanford Microwave Laboratory itself reflected the successful alliance in 1943 between Felix Bloch, Hansen, Kirkpatrick, and Terman (still in Cambridge, Massachusetts). Its mission was to consolidate prewar discoveries (such as the klystron) and serve as a magnet for postwar students and postwar money. Intellectually, it would seamlessly enclose work in microwave waveguides, antennas, and tube design while extending earlier research on the acceleration of electrons for use in experimental nuclear physics.[93]

Before they even had a roof over their heads, Hansen, Edward Ginzton, and their students started a microwave-powered linear accelerator, the Mark I. Their physics goal was the production of electrons of a billion electron volts. At the same time, both had a stake in the military function of microwave technology, and between the two worlds stood the klystron and its associated technologies. When the navy set aside funds for the linac in 1946, their bounty included a grab bag of radar matériel that fit together: S-band spectrum analyzer, standing-wave meters, synchroscopes, modulators, radar transmitters, and S-band echo boxes and wavemeters. For the Stanford physicists who had plied their trade at MIT's Rad Lab, at Harvard's Radar Countermeasures Laboratory, or with Sperry Gyroscope, these were components of a long-familiar system. These components, and their characteristic modes of use, served as a kind of materialized creole binding work together from two different universes: the classified world of radar and antiradar lodged in the Microwave Laboratory and the open research housed in the budding program in electron-nuclear physics.[94]

The hybrid practice cluster surrounding microwave tubes gained the kind of stability that "ionography" had in the visual-photographic domain. Both established a local pidgin, so to speak, that grew up through informal notes and then textbooks into a full-fledged creole of microwave technique binding engineering to physics. But unlike ionography, which functioned at a relatively small scale even after the intervention of Kodak, Ilford, and Britain's Ministry of Supply, microwave physics was conducted at a scale unimaginable before the war. One need only think of the establishment of the DEW-line (the Distant Early

92. Hansen, "Proposed Micro-Wave Laboratory at Stanford," 17 November 1943, SC 160, 1:1:8, FTP.

93. For a much more complete examination of the struggle at Stanford to maintain control over different varieties of sponsored research (industrial, military, and governmental), see Galison, Hevly, and Lowen, "Controlling" (1992), esp. 55–58 for World War II.

94. Task Order IV, Contract N6Oori-106, 12 June 1946, SC 126, box 1, WHP, SUA; cited in Galison, Hevly, and Lowen, "Controlling" (1992), 58–59.

Warning System), designed to protect all of North America with a radar shield, to see that the stakes (and money) vested in centimeter waves were enormous.

That the binding together of physics and engineering would change the very meaning of the term "physicist" was clear to the participants. David Webster, former chair of the Stanford physics department, felt burned by the close constraints of the department's previous alliance with Sperry Gyroscope. Warning Hansen in February 1943, he wrote, "I still can't see how you, as well as the [Sperry royalty] money, can fail to be swallowed up by E.E. [Electrical Engineering] and lost to Physics if we go into this plan. If you want to, even so . . . and putting the microwave lab under E.E. from the start, I can see your idea even though I would regret it. . . . We hired ourselves out to Sperry once, as a result of our desire to develop a military gadget. In view of the present uses of our gadget, I am glad we did. . . . But a choice after the war, with no military duty, is something else." To Webster, the military, the industrialists, and the engineers all threatened the autonomy, and therefore the definitional existence, of the physicist: "It may be argued that it is better for us all to become engineers with lots of money to spend than to remain physicists with little. That, possibly, is a matter for each individual to decide for himself." Sperry money was surely an enticement, but one that would come with powerful chains. Webster: "[I]f Sperry pays us anything beyond royalties, it can soon begin to gently hint that if only we would direct the royalty-fed researches into more useful lines, and if only we would not publish just yet—'Will you come into my parlor?' said the spider to the fly."[95]

Felix Bloch, one of the great small-scale experimenters before the war, had come through his experience in the war laboratories to favor collaboration between physicist and engineer. Responding to Webster (he had seen Webster's February 1943 letter to Hansen), Bloch wrote in March 1943: "[I]t is certainly funny to think how we two have exchanged our positions; the time is not so far back when you went all out for applied physics (klystrons, airplanes), even to the extent of joining with the interests of a commercial enterprise, while I snobbishly maintained the principle of 'l'art pour l'art' for physicists. . . . Right now I am gladly using 'l'art pour the war.'" Responding to Webster's "parable of the fly" Bloch asserted that even flies must know what spiders are, by nature, inclined to do. "If a fly forgets this major fact and follows into the parlor, it must be because it feels either honored by or hopes to profit from the invitation, or both, and I wouldn't shed a tear for the obvious result. I believe we can have perfectly happy flies buzzing around in the microwave lab; they just have to remember about spiders." This said, Bloch thought the flies should be able to

95. All of these quotations from Webster to Hansen, 13 February 1943, SC 126, 4:40, WHP. Part of the letter is referenced in Leslie, *Cold War* (1993), 166. (Note: Leslie accidentally elides two distinct parts of the letter.)

politely decline dangerous entreaties. To forbid flies to visit the spiders because they might be eaten, however, went entirely too far, restricting academic freedom.[96] Flies and spiders on the West Coast, handmaidens and prostitutes on the East Coast—both sets of metaphors signalled the turmoil in border zones that suddenly joined physics to industry, military, and engineering.

If anything tempered the gung-ho entrepreneurial spirit of Stanford's physicist-engineers, it was, as Webster repeatedly insisted, their chastening prewar conflicts with Sperry Gyroscope Company. In exchange for powerful industrial sponsorship, many of the staff had found themselves forced again and again to turn from their concern with physics to the rearguard defense of patents. Instead of moving from one piece of innovative work to its sequelae, it became necessary to spend weeks varying the original design, imagining patent-busting alternatives and plugging possible loopholes with the appropriate technical lines of defense.[97] Fearing loss of control to industrial forces, even Hansen—the prime mover in the Microwave Laboratory—jotted down his worries on 27 September 1944:

> Some help should be obtainable from the outside, e.g. Sperry.
> What can we sell them?
> Not our soul.
> Consulting.
> Students.
> Some research.
> Patents—new.
> Patent help.
> Scientific advertising.
> Terman thinks they will pay, and so do I, but how much and for what are questions.[98]

In the struggle to define just what the physicist's "soul" was lies the shifting boundary of the experimenter as author, as builder, as individual.

By the war's end in August 1945, the Microwave Laboratory had been in gestation for nearly three years. And with the atomic bomb project passing from top secret to a national obsession, the physics department scaled up its requests. The internal doubts about the identity of the mid-1940s were coupled to a visage of unblinking authority. Now, in October 1945, discussions of physicists and physics took on a new, assertive tone, unheard before the war. Of Norris Bradbury, whom Kirkpatrick wanted to lure back to Stanford: "He is thus the head of

96. Bloch to Webster, 23 March 1943, SC 126, box 4, folder 40, SUA.

97. Galison, Hevly, and Lowen, "Controlling" (1992), esp. 47–55.

98. Draft of Hansen to Tresidder, 27 September 1944, SC 126, 4:41, WHP; cited in Galison, Hevly, and Lowen, "Controlling" (1992), 57.

the group that changed all human history, a continuing group whose power to effect such changes is by no means exhausted."[99] Even Bradbury, it had become clear, had to struggle to keep the staff together at Los Alamos—and the salaries across the country had simply skyrocketed. As an indication of the precipitous rise, Stanford offered one physicist a salary of $3,750 only to see it met immediately by a $6,000 counteroffer from the University of Chicago. In fact, of a sample of 14 physicists (almost all between 31 and 41 years old), the *average* salary was $8,460, with a high of $15,000 and a low of $6,000. "Whether one likes it or not [this salary level] reflects the present pronounced bull market in physicists, which has naturally resulted from a short supply and a heavy demand."[100] While salaries were just one indicator of their new status, discussions like these took place at practically every university.

As never before, the physics department at Stanford could address a memo to the president of the university that bluntly asserted in its first paragraph: "It appears to be the manifest destiny of this department to expand." Unanimously, the Stanford physicists brought forth several "considerations": ROTC and NROTC students were going to be taking physics courses, as were soldiers in the army's Officers Training Program; the Consulting Committee on Undergraduate Studies was considering requiring new physics courses; the federal government was probably going to subsidize science students, and "physics will be the first science affected"; and the microwave program was going to draw students, and indeed the whole "war record of physics is causing students to plan careers in this science." Whereas before the war about one freshman per year indicated the he wanted to major in physics, the figure now stood at 11 and would surely rise further with the influx of veterans. On the basis of these facts, the department welcomed "a chance to enlarge the department" from six, to eight or nine on its permanent staff.[101]

Enlargement by 50% would cost money, and as Hansen had advocated during the war, the source would be government-contracted research. Clearly this benefited the universities, but how so the government? Colonel O. C. Maier of the Air Technical Service command put it concisely to Stanford's president in January 1946. The Army Air Force, Maier wrote, had two purposes in continuing close cooperation between universities and the military: "We would not only get a good deal of our work accomplished by capable personnel, but in addition build up a pool of trained engineers and scientists, who could be of assistance in War Department research in case of emergency."[102] The choice of tasks spanned the gamut of microwave research: propagation in low, high, ultrahigh,

99. Kirkpatrick to Tresidder, 1 October 1945, SC 158, 1:4, DTP.
100. Kirkpatrick to Tresidder, 1 October 1945, SC 158, 1:4, DTP.
101. Kirkpatrick to Tresidder, 6 November 1945, SC 151, 25:1, DTP.
102. Maier to Office of the President, Stanford University, 11 January 1946, SC 158, B1:4, DTP.

very high, and microwave frequencies; "Modulation Systems (including accurate timing for radar applications); Pulse Modulators; Broadbanding of Antennas and Circuit Elements; Magnetrons and Klystrons; Research on Millimeter Waves; 3-dimensional Radar Data Presentation; Beacon Communication; Moving Target Indicator Research; Random Polarization Jammers; Flight Computers; Navigation Systems; Loran Research; [and] New Radar Systems."[103] And this was just one prospective offer from one branch of the armed services—supplemented significantly by the AEC, which soon took over contracting from the Manhattan Project. Within a few years, workers at Stanford would be moving easily between classified, applied research and the open domain of academic studies.

All the way up and down the ladder of the physicists, the new access to authority, resources, and collective action had a flip side of resistance, hesitation, and fear before the transformation in the nature of what it meant to be an experimenter. These issues, especially those surrounding the surrender of control over research autonomy, continued to arise throughout the mid-1940s and beyond. Misgivings did not, however, alter the move toward increased sponsorship. Indeed, in the end, worries about patents and industry may have hastened the flight into government and military research, which at least initially promised vastly more autonomy, or at least a different kind of constraint, than industry. After the war, the spider appeared in the guise of the navy or the AEC rather than Sperry. One project followed helter-skelter on the tail of another: electron accelerators grew like Topsy, mushrooming to such a size that even some of their advocates began to beg off. Bloch had, by 1954, become acting director of CERN, from which position he looked back at Stanford with trepidation. To Hofstadter (who was at that point boosting the newest 1-mile-long linear accelerator project dubbed "M" for Monster and later SLAC), Bloch wrote:

> 10^7 bucks means administrators, personnel managers, foremen, auditors, navy contact men, prolonged sojourns in Washington, etc . . . etc . . . , in short a fantastic bureaucratic and political rigmarole which—I regret to say—I can altogether too well envisage in my present position.
>
> 1 mile means building contractors, city ordinances, powerlines, excavations, etc . . . , in short vast problems of "dirt" in both senses of the word.
>
> 10^2 klystrons with their attachments mean mass production—and I think that sufficiently characterizes this aspect of the thing.[104]

Britain's John Cockcroft, a veteran of both accelerators and the bomb project, had urged Bloch to shift such a project off the back of the university. And

103. Maier to Office of the President, Stanford University, 11 January 1946, SC 158, B1:4, DTP.
104. Bloch to Hofstadter, 8 December 1954, SC 303, 1:12, FBP.

that was Bloch's ultimate hope: that the "superaccelerators" could be kept out of the university, leaving something, somewhere, free of the "security-hunted" national AEC laboratories. It was all very well to say that one would have a "nice little 'spigot'" that would turn on the electrons when they were needed. "But the actual beast with which one would have to deal could not be so easily handled."[105] As the project shifted away from the "university style" and toward an "industrial laboratory," Hofstadter too broke ranks and abandoned the Monster.[106]

These are debates over control—control over the pace and direction of research that came up against the proprietary and legal demands of patent work in the 1930s; control over the dissemination of results against the constraints of secrecy during the war and thereafter; control over research against the importuning gestures of the ONR and the AEC; and finally control over the laboratory against the sheer size of teams, apparatus, and bureaucracy and the polysemic "dirt" of big science. To navigate these shoals, to maintain their "soul," as Hansen put it, the physicists did what they could to maintain autonomy in the sphere of their research. But acquiring the matériel they coveted, while avoiding the loss of research prerogatives, became an ever more elusive ambition.

4.3.5 Wisconsin

Though the Wisconsin department was founded at the same time as the University of Wisconsin in 1849, the term "physics" first arose as a professorial title there in 1878, and the focus of activity until the late 1890s was on teaching rather than research. Like many American physics departments, Wisconsin's began to expand in the 1930s.[107] It was then that the school first began to recruit theorists well versed in the new quantum theory: John Hasbrouck Van Vleck (at Wisconsin 1929–34), Eugene Wigner (at Wisconsin 1935–38), and Gregory Breit (at Wisconsin 1935–46). Together, the experimenters and theorists formed a powerful alliance, one that made the department a prime target for recruitment to the radar and atomic bomb efforts.

Indeed, Wisconsin played an active role in formulating the projects themselves. When war came, Breit lobbied hard for an atomic bomb project, even before the Manhattan Project began its official existence, and went on to play a powerful role in the drive for a nuclear weapon. The NDRC asked Wisconsin's Joseph McKibben to head a research program to measure atomic properties of tamper materials with an electrostatic, pressure-insulated high-voltage generator developed by Raymond George Herb. Herb himself spent the war years at

105. Bloch to Hofstadter, 8 December 1954, SC 303, 1:12, FBP.

106. Hofstadter to Schiff, 8 September 1958, SC 220, 3:15, LSP.

107. On the growth of physics in the United States during the 1930s and the relation of this expansion to economic and political changes, see Kevles, *The Physicists* (1978), esp. chaps. 17–19.

the Radiation Laboratory at MIT working on radar. Alfred O. Hanson headed a Los Alamos group on the four-million-volt atom smasher, McKibben on the two-million-volt generator. H. B. Wahlin was a consultant at the Metallurgical Laboratory, while Julian Mack was in Los Alamos after March 1943, running the optics and photographic group with former students F. E. Geiger and D. M. Livingston.[108]

Very rapidly, the experience of war work began to alter the Wisconsin physicists' image of the future of their home institution in the postwar era. And like so many other physicists in the midst of unprecedented resources and interdisciplinary collaboration, Wisconsin's Ragnar Rollefson reported back in 1944 that his university would do well to consider recruiting from the new generation of experimenters. Luis Alvarez, Rollefson judged, was probably the best of the new crop but would be hard to get. Especially interesting would be B. Chance, Edward Purcell, or J. L. Lawson, all near 30 years old. In the hothouse environment of the war laboratory, deference to seniority collapsed. Any of these physicists, according to Rollefson, would figure among the very best at the Rad Lab, regardless of age or rank, and would be the equal of an Alvarez. His message: get young men from the Radiation Laboratory, appoint them immediately, and do so at a high rank.[109]

When American bombers loosed the nuclear bombs on Japan, the physicists could not contain themselves. One long, and in many ways horrific, poem circulated through the department, rehearsing the state of physics since Einstein but culminating in the events of 6 and 9 August 1945. It spoke volumes about the ethos of being a physicist at that moment—the simultaneous insecurity and arrogance, the pride and doubt felt by young men who spent the war in a laboratory when millions of their contemporaries were fighting and dying in Europe and the Pacific. One small and disturbing excerpt:

> The college professor, you think, is a dreamer,
> But see the shellacking he gave Hiroshima.
> The people that thought these fellows were wacky
> Were lucky they didn't live near Nagasaki.[110]

108. Krohn, "Release on the Contributions Made on the Atom Bomb Project by Faculty Members and Former Students of the University of Wisconsin Physics Department," folder "A," 7/26/2, UWA. On Wahlin and Mack, see also Leonard Ingersoll, "University of Wisconsin Department of Physics," folder "Dean M. H. Ingraham," box 2, 7/26/10, UWA.

109. Rollefson to Leonard Ingersoll, 12 January 1944, folder "R," box "Correspondence, 1945–46," 7/26/2, UWA.

110. "When Einstein came out with his World Revelations," typescript poem, n.d. [probably 1945], box 5, 7/26/10, UWA.

Suddenly the Wisconsin physicists were on radio programs, giving lectures on the bomb,[111] and they wasted no time approaching the university authorities for more of everything. According to Leonard Ingersoll, the long-time departmental chair, "We expect a large increase of activity in the field of physics beginning at once and continuing for many years. This will mean more undergraduate students, more research activity and—perhaps most notably of all—more graduate students to supply the tremendously expanding needs of industry for physicists."[112] The department chairman reacted with some shock to the changes around him. "Harvard is putting $400,000 into a new physics research program, Pittsburgh $250,000, Illinois $1,350,000, Chicago has plans involving several million dollars, and Cornell is asking for some millions for physics expansion. Washington is doubling its physics staff, and MIT is apparently paying almost any salary to get brilliant young physicists."[113] Wisconsin, by contrast, had stagnated in the mid-1930s and now, the chairman told his troops, would need to expand: "*[E]ven the most modest of plans should involve a 30% increase.*"[114]

A month after the end of the war already seemed too late to begin the hiring spree. On 27 September 1945, Mack wrote Ingersoll from Los Alamos, where he saw before him "a wholly unprecedented concentration of first-rate physicists." Because Wisconsin had already procrastinated past the selection of the "top of the cream," Mack reported, "I have been increasingly impressed in the last few weeks with the lateness of the hour. I am convinced that if we want any of the best young men now available, we must act within a relatively few days unless we are going to follow the suggestion made by Gregory, to wait a year or so."[115] Further delay struck Mack as foolish: in a year all the good young people from the weapons laboratories would have jobs at the best universities in the country and once settled would be unwilling to move. Even institutions that previously were no threat to Wisconsin were moving fast with their sights high. Illinois was advancing with "Kerst's millions," and Washington University, armed with Arthur Compton as president, was putting together what amounted to new chemistry and physics departments. And when Los Alamos itself had congressional authorization, it too would enter the fray with offers to any of the

111. See, e.g., Mack to Leonard Ingersoll, 4 January 1946, 7/26/13/1–2, JMaP: "I note that you have down for May 1st 'Social Implications of the Atomic Bomb.' I am curious as to whether you have a speaker already scheduled for this. This is a matter with which you probably know many of us have been seriously concerned. If no speaker has as yet been arranged for, it might be possible for me to have a 'big name' speaker from this project. If the intention is to confine the lecture schedule to members of the department, I would be willing to try getting up such a talk. . . . I have some strong convictions on the subject which, incidentally, are the orthodox convictions of the local religion."

112. Ingersoll, "Department of Physics Plans," 29 October 1945, box 1, 7/26/1, UWA.

113. Ingersoll, "Department of Physics Plans," 29 October 1945, box 1, 7/26/1, UWA.

114. Ingersoll, "Department of Physics Plans," 29 October 1945, box 1, 7/26/1, UWA.

115. Mack to Ingersoll, 27 September 1945, folder "Physics," box 3, 7/1/2–3, UWA.

"young men not already committed elsewhere." Pleading to his university that they had to act within a matter of days, Mack again went over the list of colleagues from the war laboratories, any one of whom would add luster to Wisconsin, the nuclear physicists Bob Wilson, Norman Ramsey, and John Manley foremost among them. In the youngest group Edward Creutz, Kenneth Greisen, Boyce D. McDaniel, and Henry H. Barschall stood out.[116]

Not only the content of physics but its work practices had changed radically: "We have all seen the growth in importance of team research." This growth had immediate consequences for the choice of new faculty: "If the department wants, and can very soon definitely offer as a prospect, a teamwork program involving a major construction problem, as for instance on further Van de Graaf work or a billion-volt accelerator, Creutz becomes very important." Or again, later in the same letter: "[I]f we feel ourselves carried along by this trend," then a particular order would be naturally dictated in recruitment efforts—some of the returning scientists had absorbed the new collaborative, large-scale ethos, others had not.[117] Big machines and the physicists that prospered with them came in together.

By November 1945, Julian Mack, still at Los Alamos to pursue nuclear physics research that he could not do in Madison, wrote in desperation to Ingersoll: "[T]he critical weeks of decision" had passed; Wisconsin, Mack lamented, had missed its main chance. To bring the university administration to its senses, he rehearsed the astronomical salaries accepted by men he had recommended just a few weeks before. Creutz, for example, had taken an associate professorship at Carnegie Tech for $5,500; Greisen, an assistant professorship at Cornell for $4,500; Hanson, an assistant professorship at Illinois for $4,000; Ramsey, an associate professorship at Columbia for $5,000; and Wilson an associate professorship at Harvard for $6,300. The list continued to grow. A graduate student, rumor had it, had just taken up a half-time assistantship at MIT for $3,800.[118] Of this group, it was Barschall who eventually joined the Wisconsin faculty. Hugh Richards, another product of Los Alamos, signed on as an assistant professor, E. E. Miller came from the Radar Project to Wisconsin, and others, ranging from the theorist F. T. Adler (an assistant professor) to the senior geophysicist L. F. Slichter, returned to Madison with weapons work of other kinds.[119]

Students at Wisconsin flocked to physics. Before the war, the department enrolled some 30 graduate students; now they were 75 strong, crammed into evening classes lasting till after 10 P.M. Navy contracts brought in five additional men, and the laboratories were filled to overflowing.[120] In August 1946,

116. Mack to Ingersoll, 27 September 1945, folder "Physics," box 3, 7/1/2–3, UWA.

117. Mack to Ingersoll, 27 September 1945, folder "Physics," box 3, 7/1/2–3, UWA.

118. Mack to Ingersoll, 19 November 1945, folder "Dept Mtg," box 2, 7/26/10, UWA.

119. Ingersoll, "The First Hundred Years of the Department of Physics of the University of Wisconsin," typescript, n.d. [probably 1947], 7/26, UWA.

120. Ingersoll to Woodburn, Building Committee, 18 April 1946, folder "W," 7/26/2, UWA.

amid this explosive growth, Ingersoll put together his biennial report for the department: "As in the case of all other departments of the university the past biennium has seen the reconversion from a war time to peace time footing, but perhaps in no other case has the change been so profound. Two years ago . . . we had 72 people on our payroll, including 25 undergraduates. Among them were a 19 year old sophomore girl and a 69 year old geologist, and in between about everyone on the campus who knew enough physics to help us."[121] With 2,000 students, largely enlisted men, civilian research had taken a low priority. Now, with the war over, the eclipse of civilian research had passed: "Herb alone has some 20 men working . . . under him."[122] The war record spoke for itself (so Ingersoll assured the authorities): Breit had played a role in the theory group at Los Alamos, Herb had headed a powerful division at the Radiation Laboratory, and his high-voltage generator had served Los Alamos. Rollefson's work on radar was still too secret to be explained; Mack's Los Alamos efforts and Wahlin's consultations at the Metallurgical Laboratory had been crucial. New teaching strategies would allow undergraduate engineers to take an "option in physics," and graduate schemes along the same lines were already in the pipeline.[123] As at Harvard, Princeton, Stanford, MIT, and many other institutions, engineering and physics had aligned themselves in ways that transformed each.

And, again, as in many other laboratories, the physicists of Wisconsin began to prepare for a major expansion of theoretical physics: "[W]hile the demands of industry for pure theorists may be limited in the future as in the past, there can be no doubt that it will expect men to be able to combine theory and experiment in their work to gain a larger insight into what is taking place. This war has given a striking demonstration of the tremendous possibilities of the pencil as a research tool."[124] To advance the cause of pencils, the department assembled a "Proposal to Build up Theoretical Physics Facilities at the University of Wisconsin." Salaries, spiraling upward, would have to push even beyond the level of the new crop of experimenters, brushing $10,000. They were sweetened by a special perquisite, understandable to anyone, even a theorist: "Optional attendance at our department meetings," stated the proposal. Who could resist an offer like that? Especially since the new theorists would be given the privilege of "short circuit[ing] chairman and department committee" to deal directly with the dean or president. But, more seriously, the war had altered the relation of

121. Ingersoll, "University of Wisconsin Department of Physics Biennial Report 1945–46," 7 August 1946, folder "Dean M. H. Ingraham," box 2, 7/26/10, UWA.

122. Ingersoll, "University of Wisconsin Department of Physics Biennial Report 1945–46," 7 August 1946, folder "Dean M. H. Ingraham," box 2, 7/26/10, UWA.

123. Ingersoll, "University of Wisconsin Department of Physics Biennial Report 1945–46," 7 August 1946, folder "Dean M. H. Ingraham," box 2, 7/26/10, UWA.

124. Ingersoll, "University of Wisconsin Department of Physics Biennial Report 1945–46," 7 August 1946, folder "Dean M. H. Ingraham," box 2, 7/26/10, UWA.

theory to experiment in a lasting way; this new order of things was to be communicated to the new recruit—"[a]ssurance of theoretical physics being given the same consideration as experimental in future appointments, both major and minor."[125]

Beyond expansion and the realignment of experimentation with engineering and theory, the war radically shifted the work of the department in ways that lasted beyond the termination of hostilities: Herb, Barschall, and Richards, for example, continued the nuclear physics they had undertaken at Los Alamos. More subtly, the priorities of the expanding atomic establishment reoriented the work of faculty members even when their interests might have pointed elsewhere. One such parabolic story concerns Julian Mack.

Mack was an expert spectroscopist, a wizard at the gamut of photographic techniques and nuclear dynamics needed to probe the hydrogen atom. His war work having drawn to a close, Mack had a clear sense of where his research should go and why. How to proceed was another question altogether, as he confided to Herb on 1 May 1946:

> So far as I am concerned my single objective in immediate post war research is to do the best that I can toward a definitive study of the hydrogen alpha fine structure problem. I am currently lacking in equipment and assistance. . . . In order to obtain facilities for doing the hydrogen problem I am willing to commit myself to work on the structure of heavy nuclei which is of considerable interest in itself. Although I should not have made any moves toward a commitment in this field this spring had it not been for the fact that the equipment needed for the two problems is largely identical, and I can expect the Navy to show more concern at present over heavy atoms than over the validity of the Dirac theory.[126]

Sadly enough for Mack, the navy's concern took precedence over his investigations of Dirac's theory of the relativistic electron. Six months later, in June 1946, Mack was exploring the spectroscopy of plutonium, still working at Los Alamos, and feverishly negotiating with the Metallurgical Laboratory in Chicago for the transfer of classified materials (heavy nuclear isotopes) to Madison.[127]

By the time Mack got around to the hydrogen atom, in February 1947, he had not yet adapted the spectacular new microwave instruments to the problem, but rather struggled to resurrect his old (prewar) atomic beam apparatus. Perhaps the earlier material culture beckoned as a refuge from the infrastructure of war work. In any case, betting on prewar instrumentation was a strategy soon blocked. Through the grapevine of student gossip, Mack learned that Willis Lamb at Columbia had begun an intensive inquiry into the same atomic test of

125. "Proposal to Build up Theoretical Physics Facilities at the University of Wisconsin," n.d. [probably first half of 1946], folder "Dean M. H. Ingraham," box 2, 7/26/10, UWA.

126. Mack to Herb, 1 May 1946, 7/26/13/1–2, JMaP.

127. Farrington Daniels to Mack, 6 June 1946, 7/26/13/1–2, JMaP.

the Dirac theory. Writing in May 1947 to Lamb, Mack in a mixture of collegiality and nervousness said, "If I am right as to what you are doing, our work overlaps partly but not enough so that one's success would eliminate the usefulness of the other's work—since your object is to get one interval with great accuracy, and mine, to get the whole structure of the $n = 2$ and $n = 3$ complexes, but far less accurately."[128] Lamb replied that he was not yet sure whether his accuracy would be superior to Mack's and assured Mack that the Wisconsin work would be well suited for other atoms as well as hydrogen, whereas his own work was dedicated only to the hydrogens and perhaps helium. Lamb continued: "Since Rabi saw you, we think we have detected effects due to metastable hydrogen, and have obtained r.f. [radio frequency] flops in the region of 3.2 cm wavelength in rather low magnetic fields. If real, this would indicate that the $S_{1/2}$-$P_{3/2}$ separation is less than the 10,900 m.c. [megacycles] predicted by theory by about 1200 to 1400 m.c."[129]

Unfortunately for Mack, as Lamb's letter explained, Lamb had already made an extraordinary discovery in the domain to which Mack was about to apply himself.[130] Dirac's extremely successful relativistic quantum theory of the electromagnetic field was wrong, but just barely. Lamb had found a tiny difference in energy between two states of the hydrogen atom, $2S_{1/2}$ and $2P_{3/2}$, which the Dirac theory predicted would have the same energy. This result, the Lamb splitting, had a brilliant success over the next few years as an experimental inspiration for the new theory of quantum electrodynamics created by Feynman, Schwinger, Tomonaga, and Dyson.[131]

Enclosed with Lamb's 12 May 1947 letter was a preprint (by Lamb and Retherford) that made it clear just how different his apparatus was from Mack's. In the first paragraph, the Columbia physicists announced that "[t]he great war time advances in microwave techniques in the vicinity of three centimeters

128. Mack to Lamb, 9 May 1947, 7/26/13/1–2, JMaP.

129. More precisely, the idea was this: The Dirac theory predicted that two atomic states $2S_{1/2}$ and $2P_{1/2}$ have the same energy level ($2S_{1/2}$ designates the second principal energy level of the hydrogen atom, with orbital angular momentum zero and a total angular momentum of 1/2; $2P_{1/2}$ is also the second principal energy level of the hydrogen atom, with orbital angular momentum one and a total angular momentum of 1/2). In the Dirac theory, the fact that these two levels have the same total angular momentum and the same principal energy level suffices to make them degenerate—the same—in energy. The $2P_{3/2}$ state, however, is not degenerate with $2P_{1/2}$ because $2P_{3/2}$ has a higher total angular momentum (3/2 instead of 1/2), and the coupling between the spin and orbit components of angular momentum generates a difference in energy well known to spectroscopists (about 10,900 megahertz). Quantum field theory introduced an additional energy gap (subsequently known as the Lamb shift) due to fluctuations in the electromagnetic field of the hydrogen nucleus (a single proton). For even though the average value of the fluctuating field is zero, the jiggling of the electron adds energy to it. Because this newly predicted energy shift exhibited the quantum fluctuations of the electrodynamic field, it helped to legitimate the exploration of the new quantum electrodynamics. In this letter, Lamb was reporting that his observations supported the theoretical prediction of this additional energy gap between the $2S_{1/2}$ and $2P_{3/2}$ of about 1200 megacycles.

130. Lamb to Mack, 12 May 1947, 7/26/13/1–2, JMaP.

131. On the Lamb shift and the postwar development of quantum electrodynamics, see Schweber, *QED* (1994).

wave-length make possible the new use of new physical tools for a study of the $n = 2$ fine structure states of the hydrogen atom."[132] By pumping microwave energy into the atom, Lamb could induce atomic transitions with a precision that was unimaginable before the routines of radar engineering had become part of the experimenter's set of habitual practices. The contrast between Lamb and Mack illustrates, dramatically, two ways to convert from war to peace—in this case both with relatively small scale equipment. Mack chose, after the Japanese surrender, to continue, at least for a while, his spectroscopic work on the weapons materials themselves. When he returned to "peaceful" activities, he left both the plutonium and war apparatus behind. Lamb, by contrast, switched away from the direct *subject* of war making to the hydrogen atom. At the same time, he continued the *practices* associated with the war-generated apparatus of microwave physics. Within months, it was apparent that Mack had bet on the wrong horse. Those, like Lamb, who had carried over wartime nuclear and microwave techniques had located a resource for research that brought them success beyond anyone's dreams.[133]

4.3.6 MIT

Microwave radar—and World War II—came to MIT long before the attack on Pearl Harbor. In September 1940, a secret British mission, bringing knowledge and equipment from their radar program, arrived in the United States to meet

132. Lamb and Retherford, "Microwave Method," *Phys. Rev.* 72 (1947): 241–43; and Lamb, "Hydrogen," *Rep. Prog. Phys.* 14 (1951): 19–63.

133. This said, it would be wrong to suggest that pressure from the government ceased once the war ended; the Cold War had its own dynamic. To give one example: With the Soviet atomic bomb test of August 1949 ("Joe 1"), the start of the American hydrogen bomb program in January 1950, and the start of the Korean conflict in June 1950, the AEC pressed for more cooperation from the scientists it supported. K. S. Pitzer, research director of the AEC, wrote to Ragnar Rollefson, ex-chair of the Wisconsin department, in a letter dated 7 May 1951 (folder "A," 7/26/2, UWA): "The increased tenseness of the international situation and the movement of this nation into a state of emergency have raised the very real question of the optimum utilization of scientific manpower. . . . The probable long-range nature of the present emergency, coupled with the lessons learned as a result of World War II, have pointed up to us the desirability and necessity of a strong and continuing basic research program in the universities, foundations, and government laboratories."

Such an effort, Pitzer emphasized, would "assure a continuing flow of competent young scientists into industry, university, and government laboratories, and the armed forces." From those not moving to the defense laboratories, Pitzer pressed for an "increase" in "their contribution to the defense effort by associating themselves more closely with the activities of the Atomic Energy Commission, and other defense agencies, and by orienting their research activities so that they will have a more direct bearing on the problems of national defense." Finally, Pitzer offered added resources for those who "reoriented" their research toward defense problems. More students and faculty would need to be given Q (nuclear weapons) clearance.

Barschall (who succeeded Rollefson as chair) responded on 19 May 1951 (folder "A," 7/26/2, UWA): "Several members of our department would be willing and anxious to carry out work which would be of direct benefit to the program of the Atomic Energy Commission. As far as the application of atomic energy to military purposes is concerned, I have been in close contact with the developments in this field and the program which is being carried on in the field of neutron physics [at Wisconsin] has taken into account the needs of the Los Alamos Laboratory." For more on the ways that Cold War demands shaped the research choices that physical scientists made, see Forman, "Quantum Electronics," *Hist. Stud. Phys. Bio. Sci.* 18 (1987): 149–229; and Schweber, "Mutual Embrace" (1988).

with their American counterparts. Enormously impressed with the British progress on the magnetron, the key component of the British radar system, the NDRC resolved to press ahead with an American laboratory. On 17 October 1940, the Microwave Committee (in charge of NDRC radar research) let MIT President Karl Compton know that they believed MIT was a perfect site for a wartime radar laboratory. Compton immediately concurred, granting the new research center complete independence. The MIT physics department was neither consulted nor informed about the Radiation Laboratory.[134]

Nevertheless, members of the MIT physics department rapidly became key staff members of the Rad Lab and, with colleagues drawn from across the country, began to develop the art of microwave engineering, a high-pressure, astonishingly successful amalgam of electromagnetic theory, empirical fiddling, engineering lore, and industrial coordination. The total NDRC microwave appropriation (virtually all of which went to the Rad Lab) went from $1.7 million in 1941, to $31.7 million in 1943, to $51.4 million in 1945.[135] Such expenditures dwarfed the research funding previously known at the physics department or, for that matter, at any American physics department before the war. And wartime resources left their mark. While working within the Rad Lab's productive structure, physicists like John Slater soon began to think about continuing microwave work after the war. Indeed, as Slater later recalled, even during the war it was obvious that physics would expand in peacetime, and equally clear in what directions: "[T]he two fields of electronics, as exemplified in radar, and of nuclear structure as applied in the atomic bomb, were bound to lead to greatly accelerated research and application, and greatly increased numbers of students and opportunities for their employment."[136]

The decision to focus postwar research on microwaves and nuclear structure must have come early in the war, since by August 1943 the MIT Corporation's executive committee had already begun setting aside funds to strengthen microwave electronics in the hopes that $50,000 of university seed money would attract further funds from industry and government for later expansion.[137] Slater, as chairman of the physics department, was determined to preserve what he could of the impressive Rad Lab structure that he and his colleagues saw growing around them; their efforts came to a head in an effort to keep their physics department colleague Julius Stratton from being lured away by Bell Laboratories. As Slater put it in an August 1944 letter to President Compton, MIT should

134. On the establishment of the Rad Lab, see Guerlac, *Radar* (1987), 243–308; John Slater, "History," mimeographed version ("part 2"), 27, JSP. See also Leslie's fine essay, "Profit and Loss," *Hist. Stud. Phys. Bio. Sci.* 21 (1990): 59–85, which traces other laboratories at MIT modeled on the RLE and LNSE and follows these latter two laboratories after the period considered here; and Schweber, "Cornell and MIT" (1992), esp. 165–75.

135. Guerlac, *Radar* (1987), 658.

136. Slater, "History of the M.I.T. Physics Department 1930–1948," 35, JSP.

137. "Report of the President, Action of the MIT Corporation Executive Committee, October 1943," [24 August 1943], cited in Ehrmann, *Past, Present and Future* (1974).

make a dramatic institutional effort in electronics, an effort Slater hoped would tempt Stratton to stay: "[The new laboratory] should certainly be a cooperative project of the Physics and Electrical Engineering Departments . . . set up on as large a scale as seems feasible."[138] Thinking big seemed entirely reasonable, for long before Hiroshima, Slater foresaw that "the war [had] brought physics to the attention of the public, of the industry and government, as had never happened before. It was obvious that this would result in a greatly expanded interest in physics after the war, just as the first world war focussed attention on chemistry."[139]

Building on this enthusiasm for physics, and the resources MIT commanded as host to the Rad Lab, Slater saw to it that "before the close of hostilities, [the Institute] determined to push electronics as one of its post-war ventures."[140] Lest there be any doubt as to the model for the new laboratory, Slater even located the planned institution physically in the "present space of the Radiation Laboratory" scheduled to be closed at the end of the war. To staff the reincarnated laboratory, Slater seized the opportunity to hire some senior staff members from the Rad Lab, and by suitably modifying the category of "research associate," MIT retained a number of junior Rad Lab staff as well. Slater reasoned that not only would such a peacetime electronics laboratory allow MIT to compete effectively with Harvard's Cruft Laboratory, it would also make intellectual room for crucial joint physics-engineering research into microwaves and their associated devices, including triodes, velocity modulation tubes, and magnetrons. Of all the potential projects, Slater found the most interesting lay in the exploration of shorter and shorter wavelengths, using the harmonics of nonlinear elements such as crystals to make wavelength measurements at several millimeters. Such technology would allow electronics to be extended into the infrared, inaugurate a new branch of supersonic acoustics through piezoelectric methods, and foster a host of other new research programs.[141] As one writer put it, Slater had made "the future real" by enunciating a broad and novel physics program, gesturing competitively up the Charles toward Harvard, tying the program to Stratton, and associating the planned institution with the already extant Radiation Laboratory.[142]

Funding for the new laboratory soon started to flow. MIT raised its internally generated ante by an additional $250,000 over five years, still hoping for industrial funds;[143] by February 1945 it had become clear that large-scale federal support of the laboratory would be forthcoming; on 1 July 1946 the Signal

138. Slater to Karl Compton, 23 August 1944, folder 10, JSP.

139. Slater, "History of the M.I.T. Physics Department 1930–1948," 35, JSP.

140. Slater, "History of the M.I.T. Physics Department 1930–1948," 38, JSP.

141. Slater to Karl Compton, 23 August 1944, folder 10, JSP; discussed in Ehrmann, *Past, Present and Future* (1974), II-100–103; and Slater, "History of the M.I.T. Physics Department 1930–1948," 40, JSP.

142. Ehrmann, *Past, Present and Future* (1974), II-103.

143. On the $250,000, see "Minutes of the Corporation Executive Committee," 12 September 1944, cited in Ehrmann, *Past, Present and Future* (1974).

Corps, Army Air Force, and ONR began a $600,000-per-year contract with the new laboratory. Like other laboratories trying to secure "Los Alamos men," MIT was eyeing possible leaders from the Rad Lab before they became otherwise engaged. Indeed, so great were the possibilities of expansion under the new alliance among university, government, and military that one member of the staff began to worry that the MIT physics department might split into two parts: the first component following the Rad Lab, supported by federal money, and the second funded by tuition, endowment, and industrial funds in the prewar tradition.[144]

The transfer of space and personnel was soon complemented by the shift of equipment. Even a cursory examination of the halls of the laboratory revealed that it possessed not only an extraordinary wealth of standard equipment but also "vast quantities of apparatus designed in the laboratory, built to its specifications by manufacturers all over the country, and not obtainable at any price in the open market."[145] But as soon as the NDRC began to plan the closing of the Rad Lab, other birds of prey began circling, looking for the best equipment for service laboratories located in sites around the United States. Through skilled bureaucratic maneuvering, Slater's nascent Research Laboratory of Electronics (RLE) managed to protect what its founders considered their birthright; when the new RLE began operations, it did so with an extraordinary amount of the best electronic equipment in existence and a mandate to continue—still under army and navy support—the high-frequency research pioneered in the radar work of the Second World War.[146] At least for the first years, as Stuart Leslie notes, it was central to the RLE's mission to keep close ties to the military and to ensure that MIT research was at the forefront of electronic science and engineering. The liaison was often based on techniques of interest to the military but remained central to civilian science as well. Out of MIT's wartime accomplishments came a myriad of microwave applications and, soon after, their renowned program in plasma physics.[147]

If the RLE was to serve as a postwar Rad Lab, the MIT physicists still needed a postwar nuclear physics laboratory to serve as their Los Alamos, and here the war left them in a poorer condition. True, before the war MIT had possessed the valuable high-voltage investigations of Robert Van de Graaff and John Trump, but the Institute had not sent a large contingent to Los Alamos or Chicago. Jarrold Zacharias and others at MIT were determined to remove that deficiency by transferring skills, instruments, information, problems, and

144. On federal support, see "Memorandum of Conversation between K. T. Compton, J. A. Stratton, P. M. Morse, and J. C. Slater," 3 February 1945, and on possible dangers "Memorandum from Compton to Hazen," 10 March 1945; both cited in Ehrmann, *Past, Present and Future* (1974), II-110–111. See Slater, "History of the M.I.T. Physics Department 1930–1948," 44, JSP, for yearly budget and contract arrangements.

145. Slater, "History of the M.I.T. Physics Department 1930–1948," 43, JSP.

146. Slater, "History of the M.I.T. Physics Department 1930–1948," 43–44, JSP.

147. Leslie, "Profit and Loss," *Hist. Stud. Phys. Bio. Sci.* 21 (1990): 59–85.

personnel from Los Alamos to MIT. In what soon became a recruiting mission, Zacharias arrived at Los Alamos just after Trinity and remained through Hiroshima, Nagasaki, and the surrender of Japan. While on the mesa, he succeeded in tempting two of the New Mexico laboratory's most experienced and active physicists to join the MIT team: Victor F. Weisskopf in theoretical physics and Bruno Rossi in experimental cosmic rays. Because planning had begun during the war, by late September 1945, the university-wide MIT Nuclear Committee was ready to hold a formal meeting, in part to create the "nucleus" of a program at MIT to attract later subsidies by industry and government. As in the wartime programs, the committee sought to combine expertise from a host of departments, including metallurgy, radiochemistry, physics, and several engineering specialties. Several of these departments had already made offers to alumni of the bomb project, and the physicists were in close pursuit of five senior figures "all now at Los Alamos accumulating valuable experience."[148]

The committee was asking MIT for a minimum of $100,000 per year for five years to support the nuclear effort, and a similar amount for the physics department.[149] At about the same time offer letters had been sent to several junior researchers; by 9 October 1945, the physics department alone had made twelve offers to younger Los Alamos people, including three assistant professors, one instructor, and eight research associates.[150] The Laboratory of Nuclear Science and Engineering (LNSE) was officially founded in November 1945, directly on the model of the RLE. Like the RLE, the LSNE began with seed money from the Institute, but these funds were soon used to prime the larger military pump that delivered hundreds of thousands of dollars the first year and much more after that. Unlike the RLE, its nuclear sibling did not have a cornucopia of instrumentation waiting for it; the "navy connection" made up for that insofar as possible by securing "large amounts of surplus materials."[151]

In addition to money, laboratory space, shop facilities, equipment, research orientation, and personnel, *information* from the war projects was desperately sought by laboratories across the United States. For example, at the very first meeting of the MIT Nuclear Committee, on 26 September 1945, one physicist reported on the declassification procedures that would soon disgorge almost all of the results from new advances in instrumentation; this release was to be followed in quick succession by a wealth of data on "fundamental physics."[152] Such information would play a key role in the building of a nuclear pile, one of the centerpieces of the nuclear program—and itself a direct continuation of the Chicago Met Lab research begun in wartime.

148. "Minutes—Meeting #1 of M.I.T. Nuclear Committee, 9 October 1945," B: SL2p, JSP.
149. "Minutes—Meeting #1 of M.I.T. Nuclear Committee, 26 September 1945," B: SL2p, JSP.
150. "Minutes—Meeting #1 of M.I.T. Nuclear Committee, 9 October 1945," B: SL2p, JSP.
151. Slater, "History of the M.I.T. Physics Department 1930–1948," 47, JSP.
152. "M.I.T. Nuclear Committee, 26 September 1945," B: SL2p, JSP.

The transformation of physics at MIT on the wartime model was profound, and it carried a price. Throughout the archival record one sees fragments of what must have been longer dialogues about the proper relation of "basic research" to "process development." Should the two occur in the same laboratories? Some contended that a dual-function laboratory was needed for the best and most efficient production of weapons; others wanted complete openness in their research. On the whole, however, physicists emerging from the wartime projects tended to argue that the weapons-related ("process") programs could not possibly be conducted in the open, leaving some categories of basic research classified by virtue of their proximity (physical or intellectual) to the more secret investigations. Most discussions took this latter view for granted and argued principally over the proper location for the fence.[153] For many of the physicists at MIT, such questions did not dampen enthusiasm; the newfound techniques and support for physics appeared to be an unmitigated blessing. For others, at Stanford, for example—and one thinks here of Hansen, Bloch, and Webster—the cascade of wealth and equipment brought with it a style of work and a web of constraints that impinged on their conception of a physicist.

From planning that began early in the war, the Research Laboratory of Electronics had sprung intact from the MIT Radiation Laboratory like Athena from the head of Zeus. Using the RLE as a model, MIT then established its hugely successful Laboratory of Nuclear Science and Engineering. Together, these two laboratories—both founded within a year of the end of the war—served as guiding symbols for the postwar proliferation of university-based, government-funded, interdepartmental institutes that characterized the new face of science at MIT and elsewhere for decades to come.

4.4 Continuity of Technique, Discontinuity of Results

On many levels, then, physics began to change during, not after, the war. Nonetheless, there is a natural tendency among physicists and historians to overlook the continuity between wartime weapons development work and postwar research, and to reach back before the war to points of common peacetime research. The difficulty may stem from an understandable focus solely on results, ignoring the planning, expectations, techniques, practices, work structure, and material culture of the discipline. Contributing to the physicists' inclination to elide the effects of war on research is the preponderance of theorists among those who have narrated the discipline's history.

It may also be that war-postwar continuities are slighted because, after the war, the physics community found itself divided over the opportunities and

153. "Minutes–Meeting #1 of M.I.T. Nuclear Committee, 5 December 1945," B: SL2p, JSP.

hazards of weapons research. Physicists walked a tightrope, using government funds to build the machines and teams they needed, but at the same time trying to reestablish a domain of work free from the constraints of a too closely directed and supervised research. The struggle to maintain that independence also contributed to a vision of the history of physics as skipping lightly over the war years.

But whatever the source of this hesitancy in tracing the continuity of wartime and postwar research, we have inherited a broken narrative. Let me illustrate this point by focusing on the work of the long-productive physicist Bruno Rossi. Rossi, an important contributor to cosmic ray physics before the war, to the war effort itself, and subsequently to high energy physics, offered the following recollection: "In 1939, a systematic investigation of air showers was initiated by Auger and his collaborators. Their work, still carried out by means of Geiger-Müller counters, produced results of very great significance. However when, in the late 40's, air shower work was resumed, it became clear that, in order to substantially advance and refine these studies, more sophisticated kinds of detectors were needed."[154] If attention is paid only to the specific results of air shower research, Rossi's comment ("air shower work was resumed") makes perfect sense. Air shower work in the United States did, in fact, halt during the war and resume in the mid-1940s. But instead of halting our historical inquiry at that point, let us descend to a "lower level" of analysis—that is, to the instruments and techniques of the work in question.

Many of the instruments developed after 1945 to detect air showers were fundamentally linked to war work. In Rossi's case, this link was abundantly clear because he, with H. Staub, wrote the book on the subject. Their volume *Ionization Chambers and Counters* (1949) was produced for the National Nuclear Energy Series, Manhattan Project Technical Section. It summarized the advances in electronics and detectors that issued from the radar and bomb projects. Starting in July and August 1943, Staub had directed a Los Alamos team in charge of improving counters, and Rossi had led a group intended to improve electronic techniques. In September 1943, the two groups were merged into a single Experimental Physics Division group P-6, the detector group, under Rossi.[155]

Roughly speaking, their task was to design and implement detector systems that could determine the type, energy, and number of particles emerging from a variety of interactions. Their principal mission was to develop ionization counter systems that functioned in four stages: a first device detected the particle by producing a small current; a second amplified the current; a third separated the signal from unwanted noise; and a final instrument counted and recorded the total number of pulses. Physicists from the two big war projects had improved

154. Rossi, "Cosmic Ray Techniques," *J. Phys.* 43 (1982): 82.

155. Hawkins, Truslow, and Smith, *Project Y* (1983), 90.

electronic instrumentation in all four areas—detection, amplification, discrimination, and counting.

An ionization counter works as follows: A charged particle passes through a gas that is contained between two parallel plates at different voltages. Along the particle's track it ionizes gas atoms; the electrons wander toward the positive plate, and the ions toward the negative plate. If the field is not too strong, the charge deposited on one of the plates is equal to the number of ions produced. When these charges arrive at the collecting plate, the current that they produce can be amplified; the shape and height of this current can then be used to determine the charge and energy of the incoming particle.

Rossi's immediate postwar contribution to physics involved the development of fast timing circuits for cosmic rays. His work of the late 1940s built directly on the wartime timing circuits that he had used to link ionization chambers in tests of the Los Alamos "Water Boiler" reactor. For that "Rossi experiment," as it became known, the Italian physicist set a neutron detector to register the presence of a chain reaction inside the reactor. Using a fast coincidence circuit, the experimenter could count the number of other neutrons emitted during a brief period after the start of fission. In this way Rossi and his coworkers determined the period between the emission of prompt neutrons (those simultaneous with the fission event) and that of delayed neutrons.[156]

By 1947 military authorities had declassified not only Rossi's electronic contributions but a compendious batch of 270 Los Alamos technical reports. Immediately, journals on instrumentation brimmed with the new information. Even a cursory look at the 1947 volume of *Reviews of Scientific Instruments* indicates the depth of interest in the instrumentation that had been developed in the weapons projects.

Consider just one example from each of the four stages of measurement mentioned above. One way to find a neutron's energy was to scatter it from a hydrogen nucleus inside an ionization chamber. The recoiling hydrogen nucleus, since it is charged, ionizes other particles in the gas; these, in turn, cascade toward the negative plate. The pulse is then proportional to the number of ions, which is proportional to the energy of the recoiling proton, which is a good estimate of the energy of the original fast neutron. A variety of such "proportional counters" issued from the Manhattan Project, including sensitive ones that could measure the energy of neutrons traveling in a particular direction.[157] Signals from devices like these could then be analyzed. The simplest device only registered a pulse if its height came above a certain set level. In more sophisticated instruments developed at other laboratories (e.g., by the Chalk River group), separate

156. Hawkins, Truslow, and Smith, *Project Y* (1983), 90.

157. Coon and Nobles, "Hydrogen Recoil," *Rev. Sci. Inst.* 18 (1947): 44–47.

channels were activated by pulses of varying energy. Such "pulse-height analyzers" gave an immediate energy spectrum; they were (and are) essential instruments in postwar nuclear physics.[158] Finally, once the pulses emerged from the discriminator they needed to be counted. Here too a great deal of progress was made during the war. One such device that was designed at Los Alamos to be used with a wide variety of detectors was the "Model 200 Pulse Counter," which, like the other devices just described, was made public in 1947.[159] Instruments like these were among the bonds connecting weapons work with postwar "basic" research.

At MIT the bridge between war and peace afforded by instrumentation was evident. For nuclear physics, the highest priorities were the building of a cyclotron, a 5 MeV electrostatic generator, a several-hundred-MeV electron accelerator, and a small experimental nuclear pile. Accompanying these developments were the many detection devices (such as those just mentioned), and a host of electronic techniques and devices that issued from the war. For the participants in the MIT expansion, it was this domain of instrumentation that would link nuclear physics to the new electronics laboratory; together the joint effort of several laboratories would "carry on into peace time the collaboration between the Radiation Laboratory and the Los Alamos laboratory which has been so fruitful during the war."[160] Instrumentation thus provided the core of those projects "of most immediate interest"; for through these devices physicists could generate and detect the neutrons that were necessary prolegomena to the building of piles, the construction of model power plants, and the eventual generation of electricity.[161]

Physicists from the Manhattan and Radar Projects disseminated their work at other institutions as well. One expert on electronics, William C. Elmore, prepared a series of Saturday lectures that he delivered at Princeton in spring 1947. The Department of Physics mailed nearly 300 copies of the lectures to physicists all over the United States, and many of the Princeton physicists made quick application of the techniques.[162] Robert Hofstadter, to offer one example, recalled

158. Before the war there were essentially two ways to obtain an energy distribution. One could record the pulses photographically with an oscillograph, which was cumbersome and required large amounts of film. Or one could employ a counting circuit with a discriminator that would record the number of counts N above an energy amplitude E; the resulting "bias curve" (N vs. E) then had to be reduced after the experiment by taking $N(E) = dN/dE$ and plotting this quantity against E. See Freundlich, Hincks, and Ozeroff, "Pulse Analyser," *Rev. Sci. Inst.* 18 (1947): 90–100. As of February 1947, descriptions of the other devices still had not been published; e.g., their citation of Sayle, British Project Report, January 1944.

159. Higinbotham, Gallagher, and Sands, "Model," *Rev. Sci. Inst.* 18 (1947): 706–15.

160. Slater to Captain R. D. Conrad, Naval Office of Research and Invention, "A Program for Basic Research on Atomic Power," 24 October 1945, JSP.

161. Slater to Captain R. D. Conrad, Naval Office of Research and Invention, "A Program for Basic Research on Atomic Power," 24 October 1945, JSP.

162. Princeton University, Committee on Project Research and Inventions, "Proposal for Continuation of Research Project in Proton-Nuclear Reactions for the Year 1948–49," 5 March 1948, box 5 of 5, PUA.

that his own work on the inorganic scintillation detector was strongly shaped by Elmore's talks. The next year Elmore published an expanded version of his lectures in the journal *Nucleonics* as a four-part article, "Electronics for the Nuclear Physicist." According to the author, the series constituted in part a "commentary on electronic instruments designed at the Los Alamos Scientific Laboratory, and now employed extensively at various university laboratories."[163]

These examples are only a sample of the variety of ways in which wartime ionization detectors, fast electronics, discriminators, and scalers were used after the war. Experimenters used their wartime expertise to design devices for experiments with X rays, electrons, positrons, neutrons, protons, gamma rays, and fission fragments. Perhaps more influential than any of these developments were the microwave techniques that played crucial roles in accelerator technology after the war: waveguides, transmission lines, klystrons, molecular beams. In addition, there were the benefits of the Radiation Laboratory efforts—better low-noise amplifiers, lock-in amplifiers, microwave oscillators, which profoundly shaped nuclear magnetic resonance techniques, radio astronomy, and microwave spectroscopy. In a certain sense, the story of the continuity of Rossi's techniques is already familiar from chapter 3. Nuclear emulsions are usually described in the way Rossi recalled air shower work: early efforts to detect "pure" nuclear physics phenomena were interrupted by the war. Then, the story goes, an Emulsion Panel was established (no explanation of why) to reanimate the pre-1940 research. But such a broken narrative leaves out the neutron work in Tube Alloy that not only radically widened the circle of emulsion cognoscenti but also altered equipment and created the conditions under which Ilford, Kodak, the Ministry of Supply, and the university groups were willing to devote significant human and other resources to the emulsion problem.

It would take us too far astray to speak of the myriad of other war-bred technologies that led to calculating machines, computers, and many aspects of computer programming, though some of these topics will arise in chapters 5, 6, and 8. What is clear is that wartime research profoundly transformed the material as well as the social culture of physics. Here was "matériel" culture.

4.5 Collaboration, Work Organization, and the Definition of Research

Thus far our attention has been fixed principally on skills and the instruments of physical research. But the war left another legacy, one alluded to in the spate of new research apparatus, and in the redeployment of surplus war matériel that

163. Elmore, "Electronics," *Nucleonics* 2 (1948): no. 2, 4–17; no. 3, 16–36; no. 4, 43–55; no. 5, 50–58; on no. 2, 4. Together, Elmore and Matthew Sands wrote a book, *Electronics* (1949), that was widely distributed and translated into several languages.

formed such an important basis for experimental work. The war taught other lessons about the nature of research that left an indelible stamp on the physicists who participated in the massive programs at the Chicago Met Lab, the MIT Rad Lab, Berkeley, Oak Ridge, Hanford, and Los Alamos; we therefore return to these various sites with an eye to the growth of "cooperative" research. One such "lesson" concerned large-scale research organized upon complex managerial lines. But "managerial" is not quite the right term because the restructuring of experimentation altered fundamental aspects of the relation of physicist to apparatus. To see how the notion of war-based "cooperative research" entered into planning for physics after the war, we return to some of the sites visited earlier. Begin at Princeton, a few weeks after D-day. Henry Smyth sat down to sketch a proposal for a new kind of physics laboratory, one not devoted to weapons production, but which could duplicate the scientific engineering successes already in hand from the various wartime enterprises.

Smyth titled his July 1944 effort "A Proposal for a Cooperative Laboratory of Experimental Science," and the document reflected on the vast changes facing physics: "The war," he wrote, "has now reached the stage where it is desirable to make plans for the postwar period and the period of transition. The complete disruption of the normal activities of the universities and, in particular, of the scientific groups in the universities leaves the whole condition of science in this country highly fluid."[164] Smyth's remark strikes at a central, and often ignored point: change was facilitated to a large extent because the traditional structures of research, leave time, personnel, teaching, and interdepartmental boundaries had been radically altered. While "normal activities" were suspended, deeper and faster mutations could be imposed on the system than would have been possible in peacetime. As Smyth noted, the direction of those mutations would shape the definition of a physicist, and of physical research:

> Forty years ago the physicist working on a research problem usually was largely self-sufficient. He had available a certain number of relatively cheap instruments and materials which he was able to assemble himself into an apparatus which he could operate alone. He then accumulated data and interpreted and published them by himself. Most of the special apparatus that he needed he himself constructed with his own hands. He was at once machinist, glassblower, electrician, theoretical physicist, and author. He instructed his students in the various techniques of mind and hand that were required, suggested a problem, and then let the student work in the same fashion under his general supervision.[165]

164. Smyth, "A Proposal for a Cooperative Laboratory of Experimental Science," 25 July 1944, file "Postwar Research 1945–46," box "Physics Department Departmental Records, Chairman, 1934–35, 1945–46, no. 1," PUA.

165. Smyth, "A Proposal for a Cooperative Laboratory of Experimental Science," 25 July 1944, file "Postwar Research 1945–46," box "Physics Department Departmental Records, Chairman, 1934–35, 1945–46, no. 1," PUA.

But even before the war, Smyth pointed out, physics had been growing more complex. Large laboratories had begun adding specialized technicians to their staffs, including glassblowers and machinists. Even graduate students came to depend on these technicians. Devices such as grating spectrographs dwarfed in size and complexity the simple tabletop devices that had previously been sufficient. Thus it was that "a certain amount of cooperation in the use of such installations had to be worked out. But even twenty years ago research problems were largely individual." Only between 1930 and 1945 had predominantly individual research and authorship dwindled, as equipment grew larger, more expensive, and hard enough to handle that it came to require a team of workers to run. Of all such devices, the cyclotron was the most dramatic example,[166] costing in some cases more than the combined prewar physics budgets of 50 laboratories.

For Smyth such developments held dangers, as well as promises. If every university aspired to build a cyclotron or a betatron, the costs could prove overwhelming. There was a danger that only a few elite institutions would be left in command of research and that consequently the "background of strength in science which [had] grown up so successfully in the country in the past twenty years" would be weakened. Only cooperative research, he felt, could salvage the situation by consolidating various universities and "other institutions" into centralized enterprises.[167] In a February 1945 revision of his document, Smyth added that such big projects should not "oppress the individual scientists. Such installations must be the servants, not the masters of the research man."[168]

The OSRD experience of physicists carried three lessons, Smyth argued: First, "fundamental science" was important in solving problems that mere "experts" could not handle. "The moral which is to be drawn from this experience is that the ultimate technological strength of the country, even for military purposes, rests on men trained in fundamental science and active in research on fundamental problems of science."[169] Second, the war demonstrated the benefits that could accrue from "large cooperative research enterprises." Finally, the third lesson was that this cooperation would extend not only to other scientific fields such as chemistry, biology, and medicine but also to the deep links between

166. Smyth, "A Proposal for a Cooperative Laboratory of Experimental Science," 25 July 1944, file "Postwar Research 1945–46," box "Physics Department Departmental Records, Chairman, 1934–35, 1945–46, no. 1," PUA.

167. Smyth, "A Proposal for a Cooperative Laboratory of Experimental Science," 25 July 1944, file "Postwar Research 1945–46," box "Physics Department Departmental Records, Chairman, 1934–35, 1945–46, no. 1," PUA.

168. Smyth, "A Proposal for a Cooperative Laboratory of Experimental Science," revised version, 7 February 1945, file "Postwar Research 1945–46," box "Physics Department Departmental Records, Chairman, 1934–35, 1945–46, no. 1," PUA.

169. Smyth, "A Proposal for a Cooperative Laboratory of Experimental Science," revised version, 7 February 1945, file "Postwar Research 1945–46," box "Physics Department Departmental Records, Chairman, 1934–35, 1945–46, no. 1," PUA.

physicists and engineers.[170] The latter was an alliance with roots dating from before the war, but which bore fruit only in the wartime efforts. Although for security reasons Smyth passed over it, the obvious reference of this section of his proposal is to the Manhattan Project; at the time the revised proposal was written, in February 1945, Los Alamos was only a few months from detonating its first nuclear weapon.

The laboratory of Smyth's dreams clearly evoked a reality that lay, still secret, in the New Mexican desert. He figured 300 square feet per physicist, about 100 physicists, and about 15-foot ceilings, along with 5,000 square feet for large installations. This added up to 525,000 cubic feet, which at $0.70 per cubic foot would have cost $367,500. Industrial production provided the architectural prototype: "The laboratory should be essentially of factory-type construction, capable of expansion and alteration. Partitions should be nonstructural." And with a democratic flourish Smyth appended his intention that "[p]anelled offices for the director or any one else should be avoided."[171]

Soon Smyth found his thoughts echoed in his colleagues' memoranda. One physicist was "genuinely concerned that in the Atlantic coast region we have some possibilities in this field [of cooperative nuclear research] and that all of the government support is not thrown to those other sites which have a prior claim, of course, because of existing facilities."[172] Wheeler also reacted with enthusiasm, in December 1945, to the idea of a multiuniversity collaborative enterprise and had no doubt that universities were owed support by the federal government:

> Any one familiar with work on nuclear physics and its applications to military and peace time uses is aware that progress in this field in the United States has now dropped to a very low rate. Scientists are leaving to go to laboratories where they can have conditions of freedom appropriate for independent investigations. The country is losing out because it hasn't been able to work out a system suitable to enlist the participation of the scientists. In addition to this problem of applying science in the country's service, there is also the problem of what the country can do to replenish the scientific capital on which it drew so heavily during the war. The universities paid in years of peace for the fundamental research of which the government took advantage in time of war. The universities need and can rightly call for government support in the future.[173]

170. Smyth, "A Proposal for a Cooperative Laboratory of Experimental Science," revised version, 7 February 1945, file "Postwar Research 1945–46," box "Physics Department Departmental Records, Chairman, 1934–35, 1945–46, no. 1," PUA.

171. Smyth, "A Proposal for a Cooperative Laboratory of Experimental Science," revised version, 7 February 1945, file "Postwar Research 1945–46," box "Physics Department Departmental Records, Chairman, 1934–35, 1945–46, no. 1," PUA.

172. Watson to Smyth, 23 June 1945, file "Postwar Research 1945–46," box "Physics Department Departmental Records, Chairman, 1934–35, 1945–46, no. 1," PUA.

173. Wheeler, draft of "Proposal for Cooperative Laboratory," 11 December 1945, file "Postwar Research 1945–46," box "Physics Department Departmental Records, Chairman, 1934–35, 1945–46, no. 1," PUA.

Such support should come in the form of engineering assistance and equipment, Wheeler argued. In addition to the cooperative work he advocated in cosmic ray physics, Wheeler favored a system in which the government financed and ran a facility where university researchers could bring their cloud chambers or magnetic spectrographs, make some measurements, and return to their home institutions. A self-administering center would therefore burden academics no more than necessary. For support Wheeler looked to the men and institutions that had built such laboratories in the past: the Manhattan Engineer District, Vannevar Bush, and industrial concerns.[174]

John Slater at MIT strongly endorsed the incipient regional laboratory in a proposal of 9 February 1946. Acknowledging the precedent of Berkeley's Radiation Laboratory—that "impressively large institution"—he nonetheless emphasized the novelty of the postwar environment: "With the war . . . and the undertakings of the Manhattan District, nuclear research has entered an altogether new order of magnitude." Only "extremely large" teams of researchers and correspondingly large installations could carry out such tasks, institutions necessarily larger than "even the largest university organization." According to Slater, only three possible outcomes presented themselves: (1) such research could be dropped; (2) the government could carry out the mission, separated from the universities; or (3) the research could be executed by "large cooperative laboratories, in which the universities will have their part."[175]

Option 1, presented for rhetorical purposes only, "would be the defeatist outcome, which we do not have to consider." Option 2 repelled Slater and his colleagues; they had been uncomfortable under military control whenever they had encountered it during the war, and they had little faith that the military could conduct high-level scientific technological research. Even more important, the education of the next generation of scientists could not take place outside of the universities. Not surprisingly, this left the cooperative laboratory. Imitating the MIT Radiation Laboratory in scale, the nuclear laboratory would also model its organizational structure. For example, Slater argued that the new laboratory would have at first "a very strong permanent skeleton organization, with direction on as large a scale as was the M.I.T. Radiation Laboratory." At the same time, the funding structure would also take the Rad Lab as its template: "It would seem that [the cooperative laboratory's organization] would be much more flexible if it were not a government laboratory, but were instead a private laboratory operating under government contract, as M.I.T. operated the Radiation Laboratory under O.S.R.D. contract during the war."[176]

174. Wheeler, draft of "Proposal for Cooperative Laboratory," 11 December 1945, file "Postwar Research 1945–46," box "Physics Department Departmental Records, Chairman, 1934–35, 1945–46, no. 1," PUA.

175. Slater, "Proposal for Establishment of a Northeastern Regional Laboratory for Nuclear Science and Engineering," 9 February 1946, JSP.

176. Slater, "Proposal for Establishment of a Northeastern Regional Laboratory for Nuclear Science and Engineering," 9 February 1946, JSP.

When Slater wrote to Leslie Groves to enlist his support, Slater emphasized that the cooperative laboratory's backers were almost all veterans of the Manhattan District or of other large war projects. Many of these scientists and engineers, he emphasized, had held significant posts in the administration of these weapons laboratories; it was an experience that had uniquely prepared them to run such a laboratory. And because they had sponsored the extensive OSRD projects, MIT, Princeton, Harvard, and the other universities in the consortium backing the new laboratory "understand the problems of organization and personnel, of financial responsibility, of large scale research and development, of security, of relationships to government bodies and the armed services, which would be concerned in a large laboratory of nuclear science and engineering."[177]

Over the next half-year the Princeton and MIT physicists combined forces with others in the Northeast to draft a proposal for a nucleonics laboratory that would cost about $2.5 million (it soon increased to $15 million, then to $22 million, and finally to $25 million). Blessed with support from Groves and the Manhattan Project, the planning staff of the budding Brookhaven National Laboratory recruited the building and management expertise of Hydrocarbon Research Inc. In addition, and to the consternation of Oak Ridge, the planners explicitly resolved to crib experience and information from the proven facilities there.[178]

Unlike reactors, which were obviously too big for most universities, cyclotrons hovered for the next few years at the boundary between being too big for universities and too small to merit their own cooperative laboratories. Indeed, Brookhaven's initial attempts to get its accelerator division funded were unsuccessful. Later, when synchrocyclotrons appeared, the Brookhaven reactor laboratory became a prototype for collective research at the accelerator, and the model was soon extended elsewhere in the United States and then to Europe as a template for CERN. War laboratories thus provided the managerial models, the technical expertise, and even the personnel for the establishment of postwar collaborative laboratory work.

Concern about the impact of a big cyclotron on university physics is apparent in the case of Princeton—as it was at Stanford. Milton G. White reported to Smyth in December 1945 that many members of the department were keen to find a place for themselves at a cyclotron facility but remained a bit apprehensive about the nature of the research that awaited them: "Dicke is leaning heavily toward elementary particle physics, but not too anxious to press for high energy if the engineering must come out of his hide. He wants to help get the cyclotron under way and then work on some simple interactions." Therefore, to get the accelerator program "back on its feet," White wanted "*very* much to

177. Slater, "Proposed Letter to General Groves," 18 February 1946, JSP.

178. For more on the origin of Brookhaven, see the fine article by Needell, "Nuclear Reactors," *Hist. Stud. Phys. Sci.* 14 (1983): 93–122.

acquire someone who would attend to moving, wiring, redesign problems." In addition, he wanted one of those sought-after types, "a man from Los Alamos."[179]

More generally, White foresaw the need to create new positions for the changed environment of the large-scale laboratory, ones outside the traditional academic hierarchy from assistant professor to tenured full professor. "On the one hand," White reported, "we find physics research going in the direction of complex equipment requiring a supporting staff of highly competent, broadly trained physicists, engineers, chemists and administrative personnel; while on the other hand we have the customary university policy of regarding all scientific employees as likely candidates for academic positions." Instead of hunting for the "well rounded man" appropriate to academia, White advocated a Division of Research that would hire specialists with soft money provided in part by endowment but significantly supplemented by industrial and government funds.[180]

White was concerned about on-campus accelerator physics, whatever might come of proposals for the cooperative laboratory. And expected costs for the cyclotron were high: from $100,000 to $300,000. Size would also be significant since forecasts called for at least 75,000 square feet. Using Smyth's estimate quoted earlier, a plant of this dimension would run at least another $825,000. In all, White forecast expenses of around $100,000 per year for the next five years, and even this sum was exclusive of the building and power requirements of the accelerator: "No crystal ball is required to outline the trend in high energy physics—the trend is up! In not more than six months it should be possible to settle on the part we wish to play in high energy physics, and having settled this we must pick some one accelerator scheme and back it for all we are worth."[181] Faced with such enormous costs, White advocated a cooperative nuclear physics laboratory, with funding in large part to be provided by private industry—citing, for example, the Monsanto Company.[182] Unfortunately for White's plans, Monsanto declined.

4.6 War and the Culture of Physics

For years we have treated the history of physics as if it simply stopped between 1939 and 1945; only the movement of refugee scientists, bomb building, and the

179. White to Smyth, 20 December 1945, file "Postwar Research 1945–46," box "Physics Department Departmental Records, Chairman, 1934–35, 1945–46, no. 1," PUA.

180. White to Smyth, 6 May 1946, file "Postwar Research 1945–46," box "Physics Department Departmental Records, Chairman, 1934–35, 1945–46, no. 1," PUA.

181. White to Smyth, 20 December 1945, file "Postwar Research 1945–46," box "Physics Department Departmental Records, Chairman, 1934–35, 1945–46, no. 1," PUA.

182. White to Thomas, Monsanto Chemical Company, 18 September 1945, file "Postwar Research 1945–46," box "Physics Department Departmental Records, Chairman, 1934–35, 1945–46, no. 1," PUA.

federal administration of a dramatically larger science budget have commanded attention. But the conflict had deeper implications for twentieth-century physics, and to understand these, we must look to the techniques and practices of the discipline. In this chapter I have followed five lines of continuity: the transfer of technology, the transfer of support, the realignment of physics and engineering, the new relation between theoretical and experimental physics, and the reorganization of the scientific workplace.

Technological transfer consisted, in part, of the invention of new devices: Novel accelerator technology, such as the klystron technology that figured powerfully in the development of the series of linear electron accelerators leading to SLAC;[183] strong focusing—a technique for controlling particle beams—also emerged out of wartime concerns. Cosmic ray physics benefited from new access to captured V-2's and facilitated access to high-altitude military flights. The new electronic technology of counters, timers, amplifiers, and pulse-height analyzers all contributed to the postwar burgeoning of the physics of nuclei and particles. As we saw in the case of Rossi, this was true for many aspects of cosmic ray detection; in chapter 6 we will find that the same holds for a myriad of postwar electronic detectors that found their way into accelerators as well. Wartime improvements gave a new lease on life to nuclear emulsions, especially on the British end of the bomb project, and the list does not end there. Microwave systems proved crucial to postwar work beyond microphysics: experimental solid state physics made significant use of the new techniques, as did radioastronomy (although the attachment to radar technique sometimes proved a disadvantage when pitted against longer wavelength detection).[184] But beyond pure invention, the war increased the industrial production of high-performance components. In turn, this capacity made available to the experimenter tools that had previously required custom manufacture. Finally, technological transfer occurred at the most literal level—great storehouses of equipment and hundreds of millions of dollars' worth of machinery that the federal government shipped directly from war-designated activities to the civilian sector as surplus. From the lone researcher picking up a microwave generator in Europe to a new generation of students exposed to state-of-the-art devices at the best-funded universities in the United States, this infusion of tools and machining equipment transformed the scope and capacities of postwar research.

The financial support for the discipline of physics had also changed completely: wartime government support of research and development multiplied by a factor of 10 during the war, from \$50 million to \$500 million, and after a

183. On the development of the linear accelerators at Stanford, see Galison, Hevly, and Lowen, "Controlling" (1992).

184. Brian Pippard's work in skin measurements of microwave penetration came directly out of wartime radar work; see Hoddeson et al., *Crystal Maze* (1992), 214–15; and Sullivan's work on the history of radioastronomy, e.g., his "Early Years" (1988), esp. 336–38.

brief hiatus during the year or two following the Allied victory, those levels resumed their rise.[185] After Hiroshima, the Manhattan Project continued to underwrite many activities, and when it finally closed shop, its sponsoring activities were quickly taken over by OSRD, the ONR, and then by the AEC. Most important, at each of the universities discussed here (Harvard, Princeton, Berkeley, Stanford, Wisconsin, and MIT) the war trained academic physicists to think about their research on a new scale, invoking a new organizational model. Not only did physicists envision larger experiments than ever before, they now saw themselves as *entitled* to continue the contractual research that both they and the government had seen function successfully during the war. This way of thinking molded both the continuation of wartime projects and the planning for new accelerators and national laboratories. Across the country, with occasional strong dissent, and with different emphases in different regions, the physics community strove to reenact the trilateral collaboration among government, university, and private enterprise.

The expanded institutional base of research permitted more complex relations between theorists and experimenters, and between physics and engineering. These collaborative relations were firmly established at the huge project centers for scientific warfare: in the Metallurgical Laboratory of Chicago, in the vastly augmented Berkeley Radiation Laboratory, at Harvard's Radio Research Laboratory, in the rocket plants of Caltech, at the MIT Rad Lab, at Oak Ridge, and at Hanford. So it was, when the war ended, that a Brookhaven, a Stanford Microwave Laboratory, or a rejuvenated Berkeley Radiation Laboratory naturally assumed a style of physics that elevated the role of theorists in the shaping of research programs, while keeping scientific engineering front and central.

Some of these wartime weapons laboratories maintained their identity as such—Hanford, Oak Ridge, and Los Alamos foremost among them. Others branched out: Argonne began as an early reactor site spun off from Fermi's work at the University of Chicago, its first heavy-water–moderated reactor went critical in 1944, and in 1946 (under the Atomic Energy Act) the laboratory became one of the primary and permanent national laboratories to conduct research in nuclear power.[186] Similarly, Brookhaven National Laboratory emerged from the Manhattan Project, launched in 1946 after the long series of programmatic attempts (traced earlier in this chapter) to fashion a Northeast interuniversity laboratory. Though both Argonne and Brookhaven formed around reactors, the former was propelled more by Met Lab personnel and the latter by emeriti of the MIT Rad Lab.[187] It was through these new alliances between the universities, the

185. Owens, "Federal 'Angel,'" *Isis* 81 (1990): 188–213, chart on 212; and Owens's extremely helpful (unpublished) "Bush, Science in War."

186. Mozley, "Change in Argonne," *Science* 173 (1971): 30–38.

187. The founding of Brookhaven is discussed lucidly by Needell, "Nuclear Reactors," *Hist. Stud. Phil. Sci.* 14 (1983): 93–122, esp. 95–100.

AEC, and the remnants of the Manhattan Project that the first great particle physics laboratories were founded.

Finally, and perhaps most important, the war changed physicists' mode of work, in the process redefining what it meant to be an experimental physicist or to do an experiment in microphysics. As Smyth put, 40 years earlier the physicist had been at once machinist, glassblower, electrician, theoretical physicist, and author. A few years after the war all that had changed: consolidating a prewar trend, the new breed of high energy physicists were no longer taught to be both theorists and experimenters—they were to choose one path or the other. In the place of physicist-craftsmen arose a collaborative association among theoretical and experimental physicists and engineers of accelerator, structural, and electrical systems.

One consequence of these interrelated transformations was a marked shift in rhetoric. The new, often triumphalist tone of having changed "human history" broadcast pride in the physicists' wartime contribution; it also signaled defensiveness, as the scientists struggled to justify government funding while avoiding tight restrictions on the prosecution and dissemination of their research. Physicists' desire for control over their research confronted government secrecy, military pressure toward applications, and ultimately the raw infrastructural complexity of large-scale research. The inevitable tension found expression in a mordant language of preserving one's soul against the temptation of riches, of hapless flies and beckoning spiders, of physicists tangling with a Monster run amok, of *l'art* versus dirt (of administration and excavation), of purity and corruption. Suddenly patterns of command, production schedules, and organizational charts were tools they had to confront, not alien life forms from the distant worlds of the military and business. The physicists' war years, from 1940 to 1945, brought out these tensions as never before.

Nonetheless, my guess is that these alterations in work structure would not, alone, have precipitated such apocalyptic talk among some of the physicists. The issue of control goes deeper: control over the workplace—the ability to manipulate and alter the apparatus—was closely tied to the prewar definition of experimentation. Indeed, since the time of Francis Bacon, control of nature had been fundamental to the definition of experiment and experimenter. The anxiety that Dicke voiced at being forced into engineering, the disgust Ridenour expressed over the prostitution of the Rad Lab to production values, and the fear that Bloch, Hofstadter, and Webster communicated about the loss of control (and of their "souls") were all, to a certain degree, responses to the recasting of time-honored professional identities. At the peak of the Project M/SLAC debate in 1955, Hofstadter wrote to Bloch that his "enthusiasm for [the accelerator] has flagged considerably because I realized that it is soul destroying in the sense that one's personality (in physics) disappears in the merging that goes into its fabrication. And I have found that I do not merge well."[188] Even those who supported the "new physicist" understood that it implied a radical shift in the nature of the practitioner, as Samuel Goudsmit, director of the Brookhaven Cosmotron, wrote in a 1956 internal memorandum:

> In this new type of work experimental skill must be supplemented by personality traits which enhance and encourage the much needed cooperative loyalty. Since it is a great privilege to work with the Cosmotron, I feel that we now must deny its use to anyone whose emotional build-up might be detrimental to the cooperative spirit, no matter how good a physicist he is. . . . I shall reserve the right to refuse experimental work in high energy to any member of my staff whom I deem unfit for group collaboration. I must remind you that it is, after all, not you but the machine that creates the particles and events which you are investigating which such great zeal.[189]

It is not just that experimenters built machines in the image of their aspirations—experimenters were being reconstructed by the machines on which they worked.

This perceived assault on the individuality of the scientific subject—the existence of an *author* of experimentation—is, I take it, the wellspring of the passion. Many experimenters saw the new modality of research as threatening the very foundation of their separate existence as physicists. Their debate about change was also about survival—all at once, of soul, individuality, professional autonomy, and experiment. There was a new type of experimenter, and given the difference in character between a physicist who could "merge" and one who could not, it is perhaps no surprise that the Wisconsin department had two differently ordered priority hiring lists depending on whether they were to pursue a team player or an individualist. As the team grew, older functions split off from the concept of "experimentation," as both sides of the dyad experiment/experimenter shifted.

Of course this fragmentation of the experiment/experimenter did not occur all at once, and new requirements such as managerial expertise, engineering skills, and computer usage accreted to the notion of the experiment as the old ones fell away. What we see as we track the process here, and in subsequent chapters, is the pain of the experimental author, challenged as an individual, reconfigured professionally, all the while having to develop new epistemic strategies of demonstration, yet losing certain forms of control. We saw this anxiety in the emulsion groups as they dispersed both functions and control. It was present in the development of large bubble chamber groups in the 1960s, in the creation of large electronic collaborations in the 1970s, and in the formation of 100-, 200-, and 500-Ph.D. teams at American and European colliding beam facilities in the 1980s. And it arose again in the detector groups planned for the

188. Hofstadter to Bloch, 13 March 1955, SC 303, series I, box 6, folder 4, FBP.

189. Quoted in Heilbron, "Historian's Interest" (1989), 52.

ill-starred Superconducting Supercollider until its cancellation in 1993, and in those destined for the large hadron collider at CERN. "Real experimental physics" has always been just fading behind the horizon. Each generation looks back to the one before and sees the experimental author intact and integral. Before now, says the experimenter (but the "now" keeps slipping forward), being an experimental physicist was an autonomous affair, specialization hadn't yet spoiled the inquiry. Here I would say the physicists are right. They are right in the sense that experimentation *is* always changing: the tasks that define it are always being splintered off, reconfigured, and supplemented. The integral "personality (in physics)," as Hofstadter called it, *is* always merging.

Changes in the material culture, organization, and goals of physics therefore went far beyond new turns of phrase, giving rise to a new style of research in nuclear and particle physics. Schematically, it is useful to think of this "industrialization" of university accelerators and national laboratories as having occurred in two stages. The first stage, discussed here, involved an upheaval in the laboratory environment in which nuclear physicists worked. This transformation of the "outer laboratory" began with Lawrence's prewar forays into big science but became the norm of nuclear physics only in the years 1943–48. In the following decade, the change of scale in physics would reach even further into the conduct of experimental high energy physics. With Alvarez's massive hydrogen bubble chambers, which we will explore in chapter 5, the inner laboratory, the microenvironment of the experimentalist's measuring and calculating devices, grew like the outer laboratory to industrial size.[190]

When Henry Smyth requested a laboratory with "factory-type construction" using nonstructural partitions, and a director's office without paneling, he was making a straightforward architectural request. But in those plans were other, less visible architectures. From the war physicists had inherited a new sense of mission-directed, team-executed research that required a new human architecture as well—specialization and collaborations with well-defined leaders (as is evident from remarks like Wheeler's as he speculated on postwar physics). These directors were to lead interdisciplinary nuclear physics programs that, with their movable partitions, could shift priorities as new instruments or questions appeared. Gone were the days when Palmer Hall at Princeton or Jefferson Laboratory at Harvard could devote small rooms purely and permanently to acoustical or magnetic research. And of course the new research would find its natural place in the lavishly funded regional laboratories where—as in the war projects—university, industry, and government would work together.

At the same time, one senses in the postwar architectural plans an apprehension about the new physics—a concern that the leaders should not isolate

190. For more on the "inner" and "outer" laboratories and an extended discussion of the transformation of the inner laboratory in the 1950s, see Galison, "Bubble Chambers" (1985); and Galison, "Bubbles" (1989).

themselves or stifle the working physicist. In many of the postwar physics planners one feels a deep tension, social and intellectual, between the power available through collaboration and the ideological commitment to individual research. It is a tension only incompletely resolved through the Los Alamos model of large-scale, hierarchical teamwork where the physicist could argue with anyone up the line. Even the director of the huge National Accelerator Laboratory would later write an autobiographical essay entitled "My Fight against Team Research."[191] When Smyth sat down to plan a regional laboratory, he assumed there had to be a director's office—but without paneling.

Plans for physics after the war constituted more than a shift in research priorities; they were simultaneously reflective of the wartime projects and determinative of the future direction of big physics. Whether physicists turned for guidance to the Met Lab or the Rad Lab, they were constructing a new culture by supplanting the guiding symbols of research. No longer could the image of laboratory work come from the precision interferometry studies of an Albert Michelson. Now, the cultural symbol of physics would originate in a Los Alamos, a Brookhaven, or a National Accelerator Laboratory: these sites became trading zones in which engineers, physicists, chemists, and metallurgists composed a new idiom and practice of experimental research.

In speaking of the guiding role of cultural symbols, I have in mind something similar to the role Clifford Geertz accords symbols—agreed-upon programs for future action, not single-valued icons like those on a typewriter key.[192] It is in this more robust sense of the term that the weapons projects were symbols. In the context of America in the mid-1940s, the Manhattan Project was far more than an indicator of the usefulness of physics; it was a prescription for the orchestration of research. As a representation of how technical, physical, military, and political activities could coalesce, the wartime laboratory became the site for a mutation in the culture of physics.

Beyond the experience of war itself, and beyond the technical and organizational details of weapons physics, was the euphoria that surrounded the impending, and then actual, victory—a victory that American scientists felt was substantially theirs. In this time of exaltation the scientists participated in the self-confidence of the larger society. Across the country, one found American prowess contrasted to the decadence of a divided and collapsing Continent. The belief among scientists, and held in many segments of society, was that the torch of arts and sciences had been passed from Europe to the United States. This triumphalism has been well documented in the arts. Already in spring 1941, one could read in the *Kenyon Review* of the "conviction that the future of the arts is in America. . . . [T]he center of western culture is no longer in Europe. It is in

191. R. R. Wilson, "Team Research" (1972).

192. Geertz, *Interpretation of Cultures* (1973), esp. 44–49.

America." Or, as a letter to the *New York Times* put it: "Under present circumstances the probability is that the future of painting lies in America." By 1946, such sentiments had spilled into the mainstream, as the *Encyclopaedia Britannica* acknowledged: "During the year 1946 more than ever before, a significant realization was experienced by many. . . . U.S. art is no longer a repository of European influences." [193]

The end of World War II, following so immediately the detonations at Hiroshima and Nagasaki, had altered American physicists' sense of their place in the world and, at the same time, the cultural meaning of their enterprise as seen from the outside. The impact of these events on German physicists, who before the war had worn the mantle of leadership, was sharp. On 6 August 1945, many of the leading German physicists were gathered near Cambridge at Farm Hall, imprisoned by the Allies, for interrogation on their work surrounding nuclear physics. Shortly before dinner, their captors informed Otto Hahn that the BBC had announced the detonation of the Hiroshima bomb. "Completely shattered by the news," as the interrogator reported, Hahn went back to his colleagues and, unbeknownst to him, to hidden microphones taping every word. Erich Bagge, Kurt Diebner, Max Von Laue, Paul Harteck, Werner Heisenberg, Carl Friedrich Weizsacker, Walther Gerlach, Horst Korsching, Karl Wirtz, and the rest listened with astonishment to Hahn, who turned abruptly to the assembled physicists to say, "If the Americans have a uranium bomb then you're all second-raters. Poor old Heisenberg." Heisenberg retreated into denial: the detonation at Hiroshima was merely a trick, nothing but high explosives. As they talked through the problem, however, denial gave way to guilt. How, this elite of German physicists began to ask, had they failed? It was immediately clear that isotope separation must have required an extraordinary collaboration among the American physicists, teamwork of a type and scale unknown in their own laboratories. The following exchange, in my view, acknowledged a turning point in the history of physics, a shift from the Munich or Berlin *Ordinarius* professor surrounded by his *Assistenten* to the focused, team-run, industrial-scale laboratories that symbolized sites of victory in the United States:

> KORSCHING: [This] shows at any rate that the Americans are capable of real cooperation on a tremendous scale. That would have been impossible in Germany. Each one said that the other was unimportant.
> GERLACH: You really can't say that as far as the uranium group is concerned. You can't imagine any greater cooperation and trust than there was in that group. You can't say that any one of them said that the other was unimportant.
> KORSCHING: Not officially of course.
> GERLACH: (*Shouting*) Not unofficially either[!] Don't contradict me[!] [194]

193. Guilbaut, *Modern Art* (1983), 63, 65, 124.
194. Frank, *Operation Epsilon* (1993), 70–71, 75–76.

Korsching a bit later reemphasized the point, insisting that "[t]he Americans could do it better than we could, that's clear." Hahn acknowledged that it was a "tremendous achievement without parallel in history," after which Gerlach left for his bedroom where, sobbing, he portrayed himself to Heisenberg and Harteck as having betrayed his country through failure.[195]

It was a moment of extremes—American euphoria, German desolation. Harry Smyth found the same disintegration in European physics as others found in Old World art. Thinking of science overseas, he started to write of its "disap[pearance]" (he crossed that out), then its "probable disappearance" (crossed out again), and finally put it this way: "[There is a] certain decay of Europe as a source of discoveries in fundamental science. In the past we have been relatively strong in applied science, weak in basic science. We could afford this because we could depend on Europe for basic science. We can do so no longer."[196] In physics, as in the arts, an explosion of self-confidence accompanied the rise of American power. American science entered the postwar era with a sense of national, and international, mission. But with the headiness came anxiety—as physicists saw their global power expanding, they felt their local control slipping away. As we look at the clean, black crackle boxes of World War II, we can see more than radar components; these material objects are the embodiment of a new picture of experimentation, one saturated with history. To understand these now chipped, heavy, and obsolete metal cases is to understand an era when new and often disputed meanings were imputed to "experiment" and "experimenter."

The physicist's postwar mission was conducted in turbulent years stuck, in many senses, between war and peace.[197] As World War II mutated into the Cold War, physicists confronted new styles of experimentation and new tensions over the meaning of experiment and scientific authorship. As we turn to the bubble chamber, we will see the struggle for autonomy and control continue after the stresses of total war had ended.

195. Frank, *Operation Epsilon* (1993), 77, 79.

196. [Smyth], "The Plans of the Department of Physics for the Next Five Years," 20 December 1945, box "Physics Department Departmental Records, Chairman, 1934–35, 1945–46, no. 1," PUA.

197. While the organizational features of the World War II weapons projects endured into postwar "pure" science, the extraordinary political consensus that bound the civilian physics community to the defense physics establishment did not. During the 1950s, for many reasons, the two scientific groups began to bifurcate. Some of these reasons were institutional—the slow decline of the General Advisory Committee, the rising capability of weapons laboratories outside universities; some were political—splits over the hydrogen bomb, the ABM system, the role of secrecy, the Cold War; and some were physical—high energy physics decisively split from nuclear physics both theoretically (e.g., current algebra and field theory) and experimentally with the exploitation of devices that were useful in one field but not in the other (e.g., bubble chambers). This is not to say that the two communities should be seen as completely autonomous. Links remained through advisory panels, students, funding sources, and, most of all, shared technologies. But in the decades following the atomic bomb, the nature of the connection between the civilian and military scientific establishments changed from one of joint enterprise to one of shared resources. These issues will have to be pursued elsewhere.

5 Bubble Chambers: Factories of Physics

PART I: Making Machines

5.1 The Physics Factory

Skip forward, for a moment, from the immediate postwar years to 17 February 1959, as engineers and physicists from the Lawrence Radiation Laboratory (LRL) began final installation of their 72-inch bubble chamber. Their immediate task was to maneuver 520 liters of highly dangerous liquid hydrogen, held at −411°F, into steel tanks coupled to precision optics, powerful compressors, and a massive refrigeration system, with the whole bathed in a strong magnetic field.[1] Engineering memos circulated through the laboratory to prepare everyone for possible disaster. Should a failure occur, remote sensors would signal the director of operations in the control room, and he would instantly issue the already scripted command over the laboratory paging system: "HYDROGEN DETECTED IN HIGH BAY COMPRESSOR ROOM AREA. . . . ALL PERSONNEL CLEAR THE HIGH BAY AND COMPRESSOR ROOM. ASSEMBLE IN SHOP. WAIT FOR ALL CLEAR." The words would echo off the reinforced, blast-resistant concrete walls, and the team would begin shutdown procedure, flicking switches across their panel mounts and clamping down the intricately woven valve system. If fire had already broken out, a new, even more urgent message would send physicists, engineers, and technicians scrambling for cover: "EXPLOSIVE GAS LEAKAGE BUILDING 59. CLEAR AREA OF PERSONNEL." The shift leader himself would then have just three minutes to calibrate the portable explosimeter, climb to the high bay to verify the danger readings, and, if possible, reset detection instruments. As this choreography of crisis

1. Lawrence Radiation Laboratory, *72-Inch* (1960).

response operations indicates, this was a different work world from the tabletop experiments of the 1930s.[2]

Just 48 hours after these safety regulations were promulgated, bubble chamber installation began, and Chief Engineer Paul Hernandez began a taped, shift-by-shift log. Deep inside the cavernous halls of the Berkeley cyclotron laboratory, cranes lowered the huge steel chamber into the main vacuum tank, and vacuum carts began to suck air from the chamber. Day-shift technicians secured bolts on the beam end of the window. Friday 20 February 1959, the team reviewed the nitrogen shield that insulated the hydrogen chamber, checking that the steel would be sufficiently secure. Over the next days the joints and gaskets were examined one final time and final tubing connections were verified. Procedure governed every step of the 18-hour Final Assembly and Testing Sequence. Even the tiny amounts of dust gathered on reflective surfaces threatened to interfere with operation and had to be eliminated. On Thursday 26 February, formal operating procedures using a "crew chief" began, and all members of the round-the-clock team received instructions to check the operations log as they came on shift. Day, swing, and graveyard shifts followed one another without a break. Leaks had to be repaired and standardized operations instituted on every phase of work—from how to leave a valve in a manifestly open position to the maximum permissible cooldown rate. By 6 March the chamber temperature was recorded at 160° above absolute zero, and by Sunday 8 March, the frigid centerpiece of the instrument was down to 105° above the bottom of the scale. Even years of training did not fully prepare the team for the complexity of its operation; by 9 March the chief engineer insisted that the team would be needing more qualified crew chiefs to keep this physics factory open and productive 24 hours a day, seven days a week.[3]

Bubbles, the constituents of the particle tracks that were the physicists' grail, first appeared in nitrogen tests on Tuesday 10 March, and the optical system, the pressurized pulsing system, and camera systems all passed muster. The crew started introducing hydrogen gas on 16 March with full security procedures in effect. From the fragmentary remains of notes recorded on Thursday 19 March, it is clear that the chief engineer was concerned that no one individual had yet grasped the whole of this delicate and dangerous operation; only through far more training could any one person master the valves, regulators, welding techniques, and gasket types. Bubbles in the hydrogen showed up at last on 7 April, and the crew wanted a better view. Cleaning the windows was an operation with its own jitters: whoever did it would be face to face with enough flammable hydrogen to pulverize and incinerate the cleaner, his colleagues, and most of the laboratory. On the tape the next day: "The operation is a hazardous one, and it is

2. Smits, "Hydrogen Bubble Chambers Operation," UCRL Engineering Note 4313-03 M1, box 4, book 137, HHP.

3. Hernandez, "First Hydrogen Run February, 1959," box 4, book 137, HHP.

sort of like looking down the barrel of a cannon since the only protection is the lucite window between us and the 5″" glass. Windows cleaned, film production mounted to five reels per day and the recorded particle tracks passed muster with the physicists. By Wednesday 22 April, the team leader, Luis Alvarez, and his deputy, Don Gow, reviewed the chamber operation and gave their blessing to both the engineering and the physics. Finally, on Monday 27 April 1959, the chief engineer leaned over his control room microphone to issue his final recorded remarks for this experiment: "The run is now complete."[4]

I will argue that with these words, more than the run was complete. The industrial grade of experimentation had now closed down even the small space of autonomy left to the physicist when he or she put a cloud chamber or emulsion pack in an accelerator. The gigantism of this 1959 scene represented exactly what had been most feared by physicists like Leprince-Ringuet when he bemoaned the American onslaught on European cosmic ray expeditions. By the close of business at LBL on 27 April, any imaginary refuge from teamwork seemed foreclosed, both in the hills of Berkeley and in the Alpine peaks.

Experimental physicists confront nature through instruments, their daily work largely determined by the character of the apparatus. In high energy experimental physics there was a radical transition in that character between 1939 and 1959, a transition we followed in the war laboratories and the plans that issued from them. To appreciate the depth of that shift, one need only think for a moment about the scene just described with its crew leaders, chief engineers, graveyard shifts, cranes, checklists, loudspeakers, and integrated technological systems. Contrast it with the postwar use of nuclear emulsions, such as the work by Eugene Gardner we examined in chapter 3; Gardner used the accelerator, but his work with emulsions was altogether continuous in structure and practice with the cosmic ray work being pursued in Europe. Typically, the physicist worked by him or herself, or perhaps with a few collaborators, on a tabletop device (like a cloud chamber) made in a machine shop or, in the case of nuclear emulsions, bought from Ilford or Kodak. The physicist designed, partly built, used, and modified the apparatus and eventually took the data, reduced them, and brought them to publication. Problems, such as the photographic difficulties Gardner encountered, could be solved by the concentrated effort of a visitor whose knowledge of cosmic ray work carried over completely into the still-sheltered inner laboratory. Individual craftsmanship continued to figure prominently in the immediate postwar years—one thinks here of the extraordinary cloud chamber virtuosos like Robert Thompson or Carl Anderson, as they manipulated hand-built instruments that cost perhaps thousands, occasionally tens of thousands, of dollars.

Into this protected preserve of the inner laboratory—the relatively quiet

4. Hernandez, "First Hydrogen Run February, 1959," box 4, book 137, HHP.

microecological niche of the physicist—slammed the massive bubble chambers. There was no separate sphere for the individual worker in Alvarez's 72-inch monster, inaugurated in the scenes just described. By the 1960s, the total number of people involved in running experiments on this machine reached a hundred or so, divided into a wide assortment of semiautonomous subgroups. Specialists devised software and hardware for data reduction; engineers handled aspects of safety, design, and construction; lay scanners encoded raw data into Dalitz and effective mass plots. Apparatus alone cost millions of dollars—without accounting for the cost of building, maintaining, and running the accelerators. Changes had occurred in almost every respect, from the kinds of physics questions being asked to the instruments and work structure that shaped routine tasks. Closely allied to these developments were fundamental debates over the nature of experimentation.

One way to explore experiment is through the comparison of detailed case studies of discovery. We could take on a particular bubble chamber discovery—the cascade zero, say—and examine it on the model of historical studies of discoveries made with other devices, including the experimental argument for the positron, the muon, the tau, the neutrino, charge-parity violation, or neutral currents.[5] Case studies offer great detail about the interaction of many experimental and theoretical concerns all brought to bear on a single complex of physics questions. For all their virtues, however, case studies by their very focus on the resolution of a specific particle or interaction point away from the continuity in experimental practices that transcends particular puzzles of physics.

That continuity, as we saw in the transition from wartime to postwar physics, may reside more in a tradition of material culture than in specific physics problems. C. T. R. Wilson pursued condensation physics back and forth across the boundary of meteorology and the physics of the atom, just as the microwave physicists slid easily from radar to the resonances of the nucleus. Marietta Blau passed from medical X rays to images of the fragmenting nucleus. What linked these practices were not theoretical approaches, or even similar kinds of physics questions. What linked them were traditions of image making as reflected in the material culture of the experimental apparatus as it evolved in time. Quite generally, experimenters seem to follow clusters of practices through varied subjects. I want to follow the machines that hold these practices together.

In particle physics, by the late 1950s, virtually all experiments were attached to accelerators, and the move from cosmic ray stations to accelerator laboratories marks one of the most significant long-term trends of the discipline.

5. Some of the historical case studies on experimental physics after 1930 are Hanson, *Positron* (1963); Galison, "Muon," *Centaurus* 26 (1983): 262–316; Purcell, "Nuclear Physics" (1964), 121–33; Heilbron and Seidel, *Lawrence* (1989); Franklin, "Discovery and Nondiscovery," *Stud. Hist. Phil. Sci.* 10 (1979): 201–57; Galison, "Neutral Current Experiments," *Rev. Mod. Phys.* 55 (1983): 477–509; Pickering, "Phenomena," *Stud. Hist. Phil. Sci.* 15 (1984): 85–117; and Franklin, "Discovery and Acceptance," *Hist. Stud. Phys. Sci.* 13 (1983): 207–38.

Studies of individual accelerator laboratories therefore have a crucial role to play. They can depict the history of high energy physics within a framework that complements case studies organized around the resolution of specific questions.[6] Histories such as those of LBL and CERN have offered us a view of the institutional and technical changes that have transformed the budget and the infrastructure of the postwar high energy physics community. Through these studies we can see how, along with a remarkable number of experimental discoveries, the move to accelerator laboratories was accompanied by changes in the experimenter's laboratory life. As the accelerators grew in size, the experimenter became progressively more distant from the machines, no longer wielding direct control over parts of the experimental apparatus. More and more specialists were needed to supervise the scheduling, the maintenance, the construction, and the operation of the accelerators themselves.

There is no doubt that the move from the cosmic ray physics discussed in chapters 2 and 3 to accelerator-based experimentation was a shock to the physicists' self-image. At Bagnères de Bigorre in 1954, Leprince-Ringuet had undoubtedly spoken for many of his European colleagues when, in closing one of the greatest cosmic ray conferences, he conveyed a mixture of resignation and fear, recoiling in horror before the rising tide of American accelerator work. Racing in front of that foreign horde of accelerators he described was the solitary cosmic ray physicist, grappling to escape ever higher into the alpine peaks. As Leprince-Ringuet realized, when accelerator laboratories took over the discipline, physicists lost some of the control they previously had over their experiments. Suddenly, they had to compete for beamtime, not only asking for approval from the accelerator management but negotiating when and for how long they could run. The quality of the beam (its intensity, its energy, its spread in momentum and energy) determined what experiments could be conducted. Yet, as the technology of beam construction and control increased, physicists had less and less involvement with the accelerator side of the laboratory. This first postwar transformation, the transformation of the outer laboratory, was a profound shock. A second, the complete restructuring of the inner laboratory, was even greater.

In the transition at the end of the war, planning the outer laboratory was the primary concern. From Stanford to Brookhaven, accelerator building in vast multidisciplinary laboratories was in full swing during the late 1940s and the early 1950s. Nonetheless, certain features of the experimenter's work remained stable, or at least changed more gradually. In particular, despite the growth of

6. Some of the historical accounts of accelerator development are the following: Livingston, "Early History," *Adv. Elect. Electron Phys.* 50 (1980): 1–88; Goldsmith and Shaw, *Europe's Giant Accelerator* (1977); Baggett, *AGS 20th Anniversary* (1980); Hoddeson, "KEK," *Soc. Stud. Sci.* 13 (1983): 1–48; Seidel, "Accelerating Science," *Hist. Stud. Phys. Sci.* 13 (1983): 375–400; Heilbron and Seidel, *Lawrence* (1989); and Kevles, *The Physicists* (1978). Two major accelerator history projects are currently underway, one at CERN and one at LBL.

the accelerator, the particle detector continued to fall almost completely under the experimenter's control. Whether it was a cloud chamber, an emulsion stack, or a counter array, the detector continued as the proximate source of the data needed for experimental demonstrations. It was altogether natural, for example, for Marietta Blau to move to Brookhaven and load her emulsions into the accelerator's pion beam, and for Lattes to slide from European emulsion physics to Gardner's side at Berkeley, and then to return to Europe. In addition, the detectors still offered the experimenter many of the same technical challenges typical of cosmic ray research before the war. Could one make faster electronics, more distortion-free cloud chambers, or more sensitive emulsions? For many experimenters, the skills, physics questions, and apparatus that they had acquired when building detectors for cosmic ray physics (or for radar or for the atomic bomb) could be applied to accelerator-based physics. Lattes's pilgrimage to the Gardner group at the Lawrence Radiation Laboratory in Berkeley was but one in a long series of productive transfers between cosmic rays and accelerators; one could cite the movement of cosmic ray cloud chamber groups into accelerator beams at Berkeley, Brookhaven, and later CERN as further examples.

The stage was thus set for a second painful mutation in practice that took place during the 1950s and the 1960s, a change that mirrored in the inner laboratory developments previously confined to the outer laboratory. It was then, as physicists began to construct the first large-scale detectors, that the growth in scale begun during the move to accelerators was repeated within the immediate experimental workplace around the detectors. The consequences of this second step for the physicist's laboratory life were even sharper than those of the first. It was one thing for engineers and physicists to collaborate in the postwar construction of accelerators. It was quite another when the experimenters could no longer build, operate, or modify the large detection devices on which their experiments depended without extensive collaborations among physicists and engineers. Finally, the new generation of detectors altered the kind and quantity of data that had to be analyzed. And soon, the massive teamwork that marked the production of data came to characterize their analysis as well. It is to this revolution in laboratory life, the industrialization and rationalization of the inner laboratory, that this chapter is devoted.

The removal of the physicist from the apparatus, the specialization of tasks, the increased role of computation, and the establishment of hierarchical collaborations have become hallmarks of high energy physics experiments. Such changes raise a host of questions, many of which we have touched on in earlier chapters, although in different contexts. How has the role of the experimental physicist changed? What part of laboratory activity counts as "the experiment"? In what ways have the criteria of experimental demonstration been altered? What parts do computer programming and engineering play in the experimental workplace? What counts as scientific authorship? Since the 1950s, many types of detectors have undergone the expansion that provokes these questions, but the first, the one whose pictures have become a symbol of particle physics, was the bubble chamber.

The bubble chamber proved ideal for use with accelerators and useless for cosmic rays. It is therefore often described as if it had been invented for the big machines. However, I will argue here that despite the bubble chamber's eventual use, its origin lay not with accelerators but in the tradition of cloud-chamber cosmic ray research. This is true on several levels: The bubble chamber borrowed many of its technical features from the cloud chamber; both involve compressors, optics, and hydraulics. Skills of data analysis for the two devices had much in common, including the recognition of track patterns, the apparatus of data reduction and analysis, and in many cases, even the work structure of tasks divided among (mostly women) scanners. In the early stages of the bubble chamber's development, the underlying processes of bubble formation and drop formation were thought to be fundamentally similar. Finally, the physics goals toward which Donald Glaser aimed his bubble chamber work grew out of cloud chamber research. Beyond technical considerations, Glaser hoped to preserve a relation between experimenter and experiment that was present in his early cosmic ray work and markedly absent in large detector groups based at accelerators. A major concern of this chapter is the relation of the organization of experimental work to the content of instrument use and design.

During the 1950s and the 1960s the bubble chamber served in the establishment of an enormous number of new entities. The η, the ω, the Ξ^0, the $Y_1^*(1385)$, the $K^*(890)$, the $Y^{*0}(1405)$, the $\Xi^*(1530)$, and the $\Omega^-(1672)$ are but a few of the many particles and resonances discovered by bubble chamber physicists (subscripts indicate the particle's isospin, the asterisk shows that it is an excited state, and the number in parentheses is the mass in MeV). These discoveries played an essential role in the development and confirmation of Murray Gell-Mann and Yuval Ne'eman's "eightfold way" and thereby paved the way for the quark model. Contributing to these successes were bubble chamber groups from across the United States, Europe, and the Soviet Union, including those at Argonne, Brookhaven, CERN, the Institute for Theoretical and Experimental Physics (ITEP), the Rutherford Laboratory, the Deutsches Elektronensynchrotron in Hamburg (DESY), SLAC, the Cambridge Electron Accelerator (CEA), not to speak of chambers in Rome, Bologna, Oxford, and Dubna.[7]

Though many groups eventually applied the bubble chamber to important matters of physics, it was a young University of Michigan experimenter, Donald Glaser, and his collaborators who first developed the instrument, and it was the

7. See Pickering, *Quarks* (1984), for a discussion of the role of the bubble chamber particle discoveries in the development of SU(3) as a classification scheme. More specifically, there is a fine overview of the Lawrence Radiation Laboratory in Heilbron and Seidel, *Lawrence* (1989).

group at Berkeley, under the direction of Luis Alvarez, that first constructed and productively used the large hydrogen bubble chambers. In a technical sense, Alvarez and his collaborators continued Glaser's work. The Berkeley team began by recreating, step by step, Glaser's invention. But the style of their work could not have been more different. Where Glaser sought to preserve his tabletop, individually run experiment, Alvarez brought in a team of structural, cryogenic, and accelerator engineers to work with physicists, postdocs, and film scanners. Alvarez's image of data reduction was in the same spirit. But where he sought to insert a sizable team of human scanners, both physicists and nonphysicists, at the center of an interactive system of automation, competitors hoped to dispense with the human altogether, making data reduction a kind of robotic auxiliary to the experimenter's primary concerns.

The result of these upheavals in daily laboratory life was profound. What counted as an experiment in 1968 was not the same as what had counted as one in 1953. This chapter takes the material culture of the bubble chamber physics laboratory as evidence for and an instantiation of that transformation. Along the way we will turn to the consequences of that transformation for the inseparable cluster of categories that lie at the heart of this book: experiment, experimental demonstration, and experimenter.

5.2 Tabletop Reality

Questions about the origin and the nature of cosmic rays played an unexpected role in physics at the California Institute of Technology for three scientific generations. In pursuit of his idiosyncratic theory about the origin of cosmic rays, Robert A. Millikan stumbled across several of their important properties, such as the peak in their intensity in the upper atmosphere.[8] His student Carl Anderson discovered the positron while on an assignment to examine other experimental consequences of his adviser's theory.[9]

The basic technique Anderson exploited was to examine the energy distribution of charged cosmic rays by measuring their curvature in a magnetic field as revealed by their tracks in a cloud chamber. Some 20 years after Anderson finished his thesis, he set his graduate student Donald Glaser the task of extending to high energies the cloud chamber measurements of sea-level muon momenta. With great care Glaser succeeded.[10] One difficulty was that very energetic muons were hardly bent by the magnetic field strengths then available. Worse, as one

8. Galison, "Muon," *Centaurus* 26 (1983): 262–316, on 263–76; Kargon, "Birth Cries" (1981), 309–25; and Seidel, "Physics Research" (1978), chap. 7.

9. Galison, "Muon," *Centaurus* 26 (1983): 262–316, on 271–72; and, more extensively, in Galison, *Experiments* (1987), chap. 3; and Seidel, "Physics Research" (1978), 287–91.

10. Glaser, "Momentum Distribution" (1950).

Figure 5.1 Glaser's double cloud chamber (1950). A powerful electromagnet was set between two cloud chambers. By determining the angle of track deflection between the upper and lower chambers, Glaser could find the energy of very high energy muons. The bubble chamber grew out of an attempt to improve this apparatus for the study of these muons and the then recently discovered strange particles. Source: Glaser, "Momentum Distribution" (1950), fig. 3b on 43.

increased the magnetic fields, the heat generated by the electromagnet caused convection currents in the chamber, ruining the track. The key to the device Glaser and his coworkers used lay in doubling the instrument. By employing two cloud chambers separated by a strong magnetic field (figure 5.1), they could measure the angle of deflection, if not the elbow of the curve itself.[11]

Even though the basic device was partly assembled before Glaser's arrival,[12] his assignment forced him to become familiar with many details of the

11. Glaser, "Momentum Distribution" (1950), 3–4.
12. Glaser, interview by the author, 4 March 1983.

operation of the cloud chamber: high-pressure devices for expansion and recompression, photographic techniques, track analysis methods, and the theory of droplet formation.[13] The project was, despite its success, "rather humdrum"; its main advantage, Glaser decided, was that it was likely to be over quickly. Meanwhile, the apprentice to the cloud chamber and his graduate student roommate, Bud Cowan, promised themselves that at least once each week they would try to think of something original to break the monotony of their routine thesis work.[14]

Several problems threatened the usefulness of the cloud chamber. While straightforward momentum measurements could be undertaken in the cloud chamber, the gas offered relatively few targets with which an incoming particle could interact. This meant that most particles simply passed through the chamber. If interactions did occur, they often did so in the dense walls of the chamber, where they could not be seen. Only a rare particle was obliging enough to interact in a place in which the vertex could be seen and the decay tracks studied. To compensate for this, as we have seen, physicists had deployed nuclear emulsions. Here the sensitive volume was a solid and therefore offered many more targets and better stopping power, but there were problems of a different sort. First, as Blau, Powell, and the other emulsion advocates had painfully discovered, film collects tracks registered over the entire lifetime of the emulsion, from its creation until its development. It was consequently difficult to disentangle tracks made at different times. Second, to make film sensitive to events not coplanar with the emulsion, physicists had to place the plates in expensive and awkward stacks; otherwise it was impossible to tell whether a particle stopped or simply left the emulsion's surface. Finally, there were the many sources of distortion we saw in chapter 3: latent image fading, variable composition of the emulsion, physical deformation in development and drying, among others.

As for the double cloud chamber: while it could measure high-momentum particles, it could not produce enough data, and it was barely capable of catching the most exciting events of the late 1940s. These, as we saw in chapter 2, came from Manchester, where in 1947 George D. Rochester and C. C. Butler culled two inexplicable events from some 5,000 cloud chamber pictures. Both shots seemed to depict the decay of a heavy particle into two other particles with a total mass on the order of 500 MeV.[15] For over a year, no one, including Rochester and Butler, could repeat the filming of these elusive *V* particles, or "pothooks" (named after the shape of the tracks into which the neutral particle decayed). Finally, by removing a cloud chamber to a mountaintop, Carl Anderson, R. B. Leighton, and collaborators were able to achieve a better view of the

13. Glaser, "Momentum Distribution" (1950), 8–11.
14. Glaser, interview by the author, 2 November 1983.
15. Rochester and Butler, "Evidence," *Nature* 160 (1947): 855–57.

pothooks—they could photograph, on average, one per day.[16] Analysis of the events became the priority in Anderson's group. On his blackboard in 1948 and 1949, everyone could see Anderson's admonition: "What have we done about the pothooks today?"[17]

When Glaser finished his thesis in 1950, he hoped finally to be able to do something about the pothooks by building a better detection device. Glaser later remembered his desire to work on his own as his prime consideration in deciding among job offers from Columbia, MIT, Minnesota, and Michigan.[18] Although the University of Michigan had experimental groups centered on the synchrotron (the group of H. R. Crane), the cyclotron (the group of Bill Parkinson), and cosmic rays (the group of Wayne Hazen),[19] Glaser "characteristically stated that he did not wish to join one of the existing research efforts . . . but preferred to pursue an independent course of research."[20] After all, this was the work pattern that Glaser had seen used with striking success by Anderson, his adviser at Caltech. Anderson had been working alone in 1932 when he had found the positron, and a few years later, collaborating with a single coworker, he made the case for a heavy electron (dubbed the muon). As he continued his cosmic ray program, Anderson successfully produced paper after paper, working in small collaborations around certain general lines of inquiry, well into the late 1940s (when Glaser was his doctoral student). Michigan was willing to allow Glaser the freedom to continue this Caltech style of cosmic ray work. And so, in 1949, upon his arrival in Ann Arbor as an instructor, he began a two-year period of independent exploration, looking into a great variety of mechanisms in the hope of designing a new visual detector for use with cosmic rays. Like Powell, Leprince-Ringuet, and many other European cosmic ray physicists, Glaser lived in horror of the advance of the big machines that already loomed in the Berkeley hills.

Dick Crane remembered the contemplative silence of Glaser's early period in Ann Arbor this way: "I took the credit for recruiting him, but for the next year or two I wondered whether I had been very smart, because there were few external signs of activity." And Paul Hough, who was there, recalls Crane and others leaning very hard indeed on Glaser to have him help with their recalcitrant synchrotron. Glaser refused absolutely, and he confided in Hough that he had been told, as a result, that he ought to think about looking for another job.[21]

16. Seriff et al., "Cloud Chamber Observations," *Phys. Rev.* 78 (1950): 290–91.

17. Glaser, "Elementary Particles" (1964), 530.

18. Glaser, interview by the author, 2 November 1983.

19. For an idea of the kind of work Hazen, Crane, and Parkinson were doing, see Parkinson and Crane, "Final Report" (1952); and Hazen, Randall, and Tiffany, "Vertical Intensity," *Phys. Rev.* 75 (1949): 694–95.

20. Rahm, "Donald A. Glaser" (1969), 1.

21. Crane, manuscript of unpublished colloquium talk to University of Michigan physics department, 1977. I thank H. R. Crane for making this available to me. Paul Hough, e-mail to the author, 11 December 1995.

The focus of Glaser's attention remained cosmic rays; the research teams associated with Van de Graaff machines had not appealed to him as a new graduate student in 1947, and the synchrotron team repelled him in 1950. "There was," he has said, "a psychological side to this. I knew that large accelerators were going to be built and they were going to make gobs of strange particles. But I didn't want to join an army of people working at the big machines. . . . I decided that if I were clever enough I could invent something that could extract the information from cosmic rays and you could work in a nice peaceful environment rather than in the factory environment of big machines."[22] The sequence of choices Glaser made—to work in cosmic ray physics, to avoid joining an accelerator team, to plunge into tabletop tinkering rather than team-based research—all put him at right angles to the postwar enthusiasm for a new scale of physics. The years just after the war remained a time of flux, as physicists like Felix Bloch and Robert Hofstadter agonized over whether they could keep their "souls" under the demands of highly coordinated research.

Support for Glaser's first research at Michigan came from the Michigan Memorial-Phoenix Project for a "Study of the High Energy Cosmic Rays" from June 1950 to January 1951 and totaled a paltry \$750.[23] By the time the short grant expired, Glaser had little to show. Nonetheless, in December 1950, the physics department extracted \$1,500 from the School of Graduate Studies for Project R 250, "Investigation on Cosmic Ray Mesons"—\$1,000 for a research assistant and \$500 for supplies.[24] Fortunately for the historian, in May 1951 Glaser had to justify even these small initial grants.[25] There was complete continuity between the physics goals of his thesis project—exploring the upper end of the known energy spectrum of cosmic rays—and the new project. "The immediate goal of the research carried on under Phoenix Project No. 11 was to develop a method for measuring the momentum of cosmic ray particles of very high energies, perhaps up to a thousand billion electron volts." The measurements were of "interest for their bearing on the question of the nature and origin of cosmic rays."[26]

Studies of the origin and nature of cosmic rays represented the continua-

22. Glaser, interview by the author, 4 March 1983.

23. Glaser, University of Michigan, "Application for Grant from Research Funds," 6 October 1953, MMPP. See under "Previous Grants for Research Purposes 1. Phoenix #11."

24. Sawyer to Glaser, 28 December 1950. Copy is in files of E. F. Barker, Department of Physics, University of Michigan.

25. Glaser, University of Michigan, "Report on Research Project," Project #11, 23 July 1951, MMPP.

26. Glaser, University of Michigan, "Report on Research Project," Project #11, 23 July 1951, 1, MMPP. Years later, after seeing these forgotten proposals, Glaser wrote: "It is embarrassing to admit it—but I really wasn't interested in high energy sea-level muons after my thesis. What really interested me was the 'pothooks.' I must have asked for money based on high energy measurements because I thought that was the area in which I had demonstrated competence" (Glaser to author, 6 March 1984).

tion of his pedagogical lineage—his teacher Anderson and Anderson's teacher Millikan both had struggled over these questions. Even Glaser's new instrument, with its two visual detectors separated by a powerful electromagnet,[27] was to function with a detector above and below an electromagnet in much the same way as the old double cloud chamber he had used for his thesis work. But if the deflection of cosmic rays was adequate in the double detector, the existing detectors themselves were not. All extant detectors, Glaser figured, were based on the catalysis of an instability, be it chemical (nuclear emulsions), electrostatic (Geiger-Müller counters, amplified scintillation counters), or thermodynamic (cloud chambers).[28] In an effort to make a new visual device that could record tracks, he set out a list of all instabilities he could think of and how they could be exploited in a fast, dense detector. One by one, he began to build new detectors. Most failed.

One attempt involved letting the ion tracks polymerize a monomer. One can polymerize acryle nitrile; the monomer is soluble in certain fluids, whereas the polymer is not. Glaser's hope was that when a charged particle sped through the solute, the path of ionization would polymerize some of the molecules, leaving a solid precipitate. In Glaser's words, "the 'total fantasy' was to be able to lift out a 'solid Christmas tree of tracks' and measure them at leisure." These tracks then could be extracted, or at least photographed. This try was an utter failure, resulting in much brown liquid and no tracks.[29]

Glaser's second try was inspired by some work he had seen J. Warren Keuffel perform when they were both graduate students at Caltech.[30] One problem with ordinary Geiger-Müller counters was the rapid falloff of the field as one moved away from the center wire. This variation meant that an avalanche precipitated near the center wire developed much more quickly than one starting near the walls. To improve the resulting time resolution, Keuffel tried making a counter with parallel planes instead of concentric cylinders. As a passing observation on the eighth page of his 1948 thesis, Keuffel noted that "the discharge is localized," posing "obvious possibilities as a means of determining the path of the particle."[31] It would take at least nine more years before this "obvious" possibility could be realized as a track-following device, but meanwhile Glaser began an abortive attempt to do so. When the glass electrode was coated with a conducting but transparent medium, the sparks could be localized to within a few tenths of a millimeter. However, the conduction layer was chemically unstable,

27. Glaser, University of Michigan, "Application for Grant from Research Funds," 6 October 1953, MMPP. See under "Previous Grants for Research Purposes 1. Phoenix #11."

28. Glaser, interview by the author, 4 March 1983; Glaser to author, 6 March 1984.

29. Glaser, interview by the author, 4 March 1983.

30. Keuffel, "Parallel-Plate" (1948); and Glaser, interviews by the author, 4 March and 2 November 1983.

31. Keuffel, "Parallel-Plate" (1948), 8.

and the device was useless after a day or two of operation. Equally unpleasant was the tendency of the spark to wander around the plate, ruining attempts at sharp photography.[32]

By late February 1951, the spark counter, like the polymerizing Christmas tree project, was certified as a lost cause.[33] Fortuitously, at about this time there was a resurgence of interest in another detector: the diffusion cloud chamber. In an expansion cloud chamber a sudden enlargement of the gas volume causes a temperature drop, supersaturating the gas, which precipitates around the charged tracks. By contrast, in a diffusion chamber the gas supersaturates as it approaches the cooled floor of the container. Recommending the device was its continuous sensitivity—since no expansion is needed, the device is never "dead."[34] "[I]t is hoped," Glaser wrote to the granting agency, "that such a chamber, which operated reliably for long periods of time, may allow a position determination with accuracy sufficient for the present experiment [on high energy cosmic rays]."[35]

The indefatigable diffusion cloud chamber had long intrigued Glaser. In graduate school he and Cowan had speculated about its possible uses. Shortly after Glaser's departure from Caltech, Cowan had begun work on such a device; he published an often cited review article on the subject in 1950.[36] With money from the Phoenix Project, Glaser hired David Rahm as a research assistant to assist in building one.[37] Designing the chamber was fairly straightforward: A 12- to 18-inch glass cylinder would be filled with ethanol, which would drift down toward a dry-ice–cooled plate at the bottom.

Once again disappointment struck. Glaser and Rahm soon convinced themselves that the instrument might make a fine demonstration device but could never be precise enough for cosmic ray energy determination. After several other failures, Glaser abandoned chemical and electrostatic devices. His questions narrowed to one: Could a thermodynamic device be produced, analogous to the cloud chamber, in which bubbles rather than droplets formed at the site of an ion? In other words, instead of supersaturating a gas with a vapor one would superheat a liquid (heat it above its boiling point by holding it under pressure).

32. Glaser, University of Michigan, "Report on Research Project," Project #11, 23 July 1951, MMPP. Glaser, interview by the author, 4 March 1983; Rahm, interview by the author, 15 November 1983.

33. Glaser, University of Michigan, "Report on Research Project," Project #11, 23 July 1951, 2, MMPP.

34. The continuously sensitive cloud chamber was invented by A. Langsdorf, Jr., as part of his doctoral work at MIT; see Langsdorf, "Cloud Chamber," *Rev. Sci. Inst.* 10 (1939): 91–103. Work resumed on the device in 1950; see Cowan, "Cloud Chamber," *Rev. Sci. Inst.* 21 (1950): 991–96; and Needels and Nielsen, "Cloud Chamber," *Rev. Sci. Inst.* 21 (1950): 976–77. Related work at Brookhaven began to be published in 1950; see also Miller, Fowler, and Shutt, "Operation," *Rev. Sci. Inst.* 22 (1951): 280. The Brookhaven group went on to progressively larger and higher-pressure devices.

35. Glaser, University of Michigan, "Report on Research Project," Project #11, 23 July 1951, 2, MMPP.

36. Glaser, interview by the author, 2 November 1983; and Cowan, "Cloud Chamber," *Rev. Sci. Inst.* 21 (1950): 991–96.

37. Glaser, interview by the author, 2 November 1983.

Success this time depended on a prior question: Is the energy deposited by an ion in a liquid sufficient to cause a vapor bubble to grow? With this question, Glaser opened "Book 1" of two notebooks still surviving from his early work by writing "Study of the Formation of Bubbles in Liquids (Theory and Literature)."[38] Slowly and carefully, Glaser alternately translated (from German), annotated, and summarized a book by Max Volmer, *The Kinetics of Phase Formation.*[39] The simplest phase change induced by an ion results when the ion condenses a liquid out of a vapor.

As we know from chapter 2, in 1888 J. J. Thomson had explained how water tended to coalesce around ions because the dielectric constant of liquid water was much larger than that of air.[40] He pointed out that the electrostatic effect added a term to the potential energy of the system equal to $\Delta V = (1/2)(e^2/r\epsilon)$ for a droplet of charge e and radius r. Thus, a region of high electric field will tend to pull in a dielectric in order to minimize the total energy of the system.[41] Thomson's result would seem to imply that a charged bubble would collapse much faster than it would without the charge, a result that seemed to preclude the possibility of ion-induced bubble tracks in a liquid. Volmer discussed the issues of electrical fluctuations on bubble formation, and Glaser read on avidly.

In an undated list of queries preserved in one of the early notebooks, Glaser wondered, "Is an electrically-modified fluctuation theory sensible as [Volmer] does it, or should we consider multiple charges as a way of getting the same type of effect."[42] This question sheet seems likely to have been written just before Glaser calculated the effect of multiple charges on a bubble's pressure. Though in retrospect the physics community (including Glaser) repudiated the electrostatic repulsion theory, it is essential historically, for two reasons. First, it provided for Glaser a theoretical bridge from the Wilsonian principles of a cloud chamber to a rough plausibility argument for a bubble chamber. That is, the multiple charge theory enabled Glaser to proceed where the relevant physical theory (constructed by analogy with Thomson's drop formation) seemed to announce ne plus ultra, bubbles would never form. Second, the electrostatic theory gave a rough quantitative prediction for the conditions of temperature and pressure under which the device would work and therefore helped shape its design.[43]

38. DGnb1, 5, DGP.

39. Volmer, *Phasenbildung* (1939).

40. Thomson, *Applications of Dynamics* (1888), 164–66, esp. 166: "We should . . . expect an electrified drop of rain to be larger than an unelectrified one."

41. Thomson, *Applications of Dynamics* (1888), 165.

42. Glaser, "Cloud Chamber Droplet Counts," n.d., loose in DGnb2, DGP.

43. Actually, any mechanism that deposits between 10 and 1000 eV will give roughly the correct conditions for bubble growth, so the fact that the electrostatic theory "worked" was not (in retrospect) a stringent test of its validity. As will be discussed below, the actual mechanism for energy transfer to the bubble is by the heat deposited by the penetrating ionizing particle. See Peyrou, "Bubble Chamber" (1967), esp. 27–38.

If we ask the question so popular in antipositivist philosophy of science, Was the experimental apparatus built using theory? the answer is clearly yes. But this question is badly formed. For the issue is not whether theory entered, but which theory intruded and how it functioned. Here, in the creation of the bubble chamber, as in that of the cloud chamber, the "theory" employed was more than anything else the simple existence of ions of no special structure and the energy considerations that made it possible to consider phase changes around them. Let us turn to Glaser's Volmer-based electrostatic argument.[44]

We have the energy E of a bubble with a surface charge ne distributed over a surface of $4\pi r^2$ and a surface tension of $\sigma(T)$, where T is the temperature and $\epsilon(T)$ is the dielectric constant of the liquid:

$$E = 4\pi r^2 \sigma(T) + \frac{(ne)^2}{2\epsilon(T)r}. \tag{5.1}$$

From E we can deduce the outward force, F:

$$F = \frac{dE}{dr} = 8\pi r \sigma(T) - \frac{(ne)^2}{2\epsilon(T)r^2}. \tag{5.2}$$

It is assumed that the surface tension and the dielectric constant are roughly independent of r. Then the pressure P is given by

$$P = \frac{F}{\text{area}} = \frac{2\sigma(T)}{r} - \frac{(ne)^2}{8\pi\epsilon(T)r^4}. \tag{5.3}$$

Setting $dP/dr = 0$ and solving for r yields r_0, the radius for which the pressure P is maximal. Inserting r_0 into equation (5.3), we deduce $P_{\max}$, the maximum value of the combined electrostatic repulsion and surface tension contraction. It follows that the condition for a bubble to grow to macroscopic size is

$$P_\infty - P > \frac{3}{2}\left(\frac{4\pi}{(ne)^2}\right)^{1/3} [\sigma(T)]^{4/3}[\epsilon(T)]^{1/3} = P_{\max}, \tag{5.4}$$

where P_∞ is the vapor pressure of the liquid and P is the (mechanically adjustable) applied pressure. Many qualitative features are immediately discernible. Bubble

44. The first dated reference to the electrostatic theory is from June 1952: Glaser, "Some Effects" *Phys. Rev.* 87 (1952): 665. The actual inequality involving P, $\sigma(T)$, $\epsilon(T)$, and ne first appears in Rahm, "Progress Report," August 1952, DGP. In many papers (e.g., Glaser and Rahm, "Characteristics," *Phys. Rev.* 97 [1953]: 474–79) Glaser indicates that the theory predated his first experiments on superheated liquids. These trials, he states, were "guided by a detailed physical model of the mechanism by which ionization could nucleate bubble formation" (474).

formation is facilitated by increasing the expansion, decreasing the surface tension, decreasing the dielectric constant, or increasing the charge. In particular, the expression suggests operating at a high liquid temperature, which would augment P_∞, while reducing surface tension. (At the thermodynamic critical point the surface tension vanishes.)

Volmer's book offered Glaser a detailed review of the kinetic theory of droplet formation as well as a parallel development of bubble nucleation in liquids. The close analogy is stressed but not exaggerated, for there is one essential difference between the bubble and the droplet. At the surface of the droplet the vapor pressure depends on the curvature of the liquid gas surface. But did this apply to a bubble as well? Glaser opened his notebook entries on Volmer by addressing just that section of Volmer's book in which the question of vapor pressure occurs.[45] The answer is that the pressure inside a bubble does not depend on curvature and is equal simply to the vapor pressure at the surface of the liquid. After skipping back and forth between chapters in Volmer to understand this surprising fact, Glaser looked up and translated W. Döring's article of 1937 in which the result was first proved.[46] Consider two bubbles (both of a gas G insoluble in the liquid L) of different sizes. If the partial vapor pressure of L differs between the large bubble and the small bubble, then a small capillary tube with a semipermeable membrane could transport vapor L from one bubble to the other while blocking the flow of gas G. The liquid would evaporate from the wall of the bubble with high vapor pressure and condense in the bubble with low pressure. It would be a perpetual motion machine. Similarly, the capillary could connect a bubble to the surface, and the argument can be extended mutatis mutandis. Therefore, bubbles of any size must have vapor pressures equal to the saturated vapor pressure of the liquid. "The mistake [of supposing the vapor pressure to depend on the radius] is often made," Glaser noted.[47] (Droplets are unlike bubbles because the droplet's radius-dependent surface tension determines the liquid pressure. The vapor pressure depends on this liquid pressure, not on the geometry per se.)

Immediately after discussing Döring's argument, Glaser used Volmer to compare the theory of bubble formation with experiment. However, he did so with a cautionary remark: "Tensile strength hasn't been done very well and for superheating Wismer and Co. are principally quoted."[48] K. L. Wismer, a chemist at the University of Toronto, had, with his collaborators, studied how long a

45. DGnb1, 15–18, DGP; Volmer, *Phasenbildung* (1939), 156–57.

46. Döring, "Berichtigung," *Z. Phys. Chem. B* 36 (1937): 292–94; trans. in DGnb1, 93–95, DGP.

47. DGnb1, 95, DGP.

48. DGnb1, 87, DGP. The works cited by Volmer, *Phasenbildung* (1939), 163, are Wismer, "Pressure-Volume," *J. Phys. Chem.* 26 (1922): 301–15; and Kenrick, Gilbert, and Wismer, "Superheating," *J. Phys. Chem.* 28 (1924): 1297–1307.

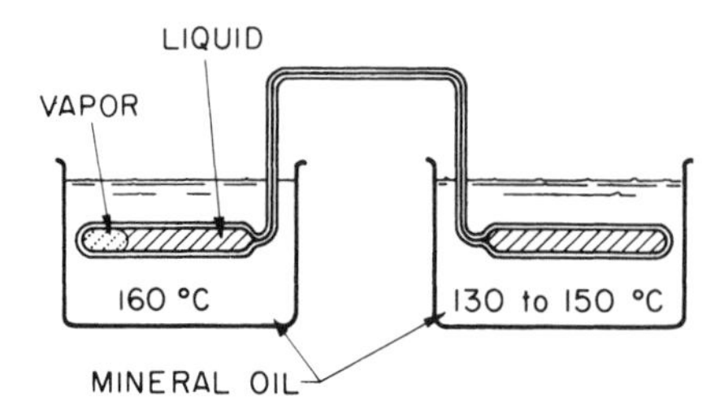

Figure 5.2 Capillary tube experiment, left side (1954). In Glaser's apparatus, a bubble of vapor diethyl ether in a capillary tube was heated to 160°C in an oil bath. The resulting pressure on the liquid diethyl ether was about 22 atmospheres, high enough to prevent the liquid from boiling even though it is above its normal boiling temperature (135°C). Source: Glaser, "Progress Report," *Nuovo Cimento* 11 Suppl. (1954): 361–68, fig. 2 on 366.

capillary tube of diethyl ether heated to 140°C could last before exploding. As Glaser reported from Volmer, this was typically about one second.[49] Volmer ascribed the inevitable explosion to local fluctuations due to dirt or imperfections in the glass surface,[50] but when Glaser arrived at the Michigan chemistry library to look up Wismer, Gilbert, and Kenrick's paper, he found something much more exciting than dirty glass: not all of Wismer and Kenrick's capillary tubes had exploded almost instantaneously. They were puzzled by their results; Glaser was thrilled. What the physical chemists had found was that the superheated material sometimes burst out of its tube in a few seconds but at other times waited for well over a minute.[51] Wismer attributed the "capriciousness in the results" to the imperfections of inadequately polished glass walls. When Glaser plotted the results for a sample run with a temperature of 130.5°C, the points formed a Poisson distribution with an average waiting time of 60 seconds.[52] From his thesis Glaser knew the sea-level cosmic ray flux. A brief calculation yielded the average time between cosmic ray particles for a capillary tube of the size used by the chemists: 60 seconds.[53] The crucial test remained: Would a radioactive source induce boiling?

Glaser connected two glass tubes by a glass capillary, partly filled the assembly with diethyl ether liquid, and dipped the contraption into two oil baths held at differing temperatures (see figure 5.2).[54] The higher temperature (and

49. DGnb1, 87, DGP.

50. "Die Keimbildung, d.h. die durch lokale Schwankungen bedingte Bläschenbildung setzt stets vorher ein und lässt den überspannten Zustand zusammenbrechen" (Volmer, *Phasenbildung* [1939], 165).

51. Kenrick, Gilbert, and Wismer, "Superheating," *J. Phys. Chem.* 28 (1924): 1297–1307, on 1298–99.

52. Data given in Kenrick, Gilbert, and Wismer, "Superheating," *J. Phys. Chem.* 28 (1924): 1297–1307, on 1304.

53. A very useful account of the early work of Glaser and Rahm is found in Rahm, "Development" (1956). A reference to Glaser's cosmic ray calculation appears on 4. There is a fascinating parallel between these events and the development of the Geiger-Müller tube. For many years, Geiger, Rutherford, and Müller had despaired of making a useful counter because when the voltage was turned up to make the device sensitive, wild collisions ("wilde Stösse") were recorded that did not seem to be due to impurities in the air. Great effort was expended cleaning the walls of the tube before it was realized that cosmic rays were being seen. See Trenn, "Geiger-Müller-Zählrohres," *Deut. Mus., Abh. Ber.* 44 (1976): 54–64; and Trenn, "Geiger–Müller Counter," *Ann. Sci.* 43 (1986): 111–35.

54. Glaser, "Progress Report," *Nuovo Cimento* 11 Suppl. (1954): 361–68, on 366.

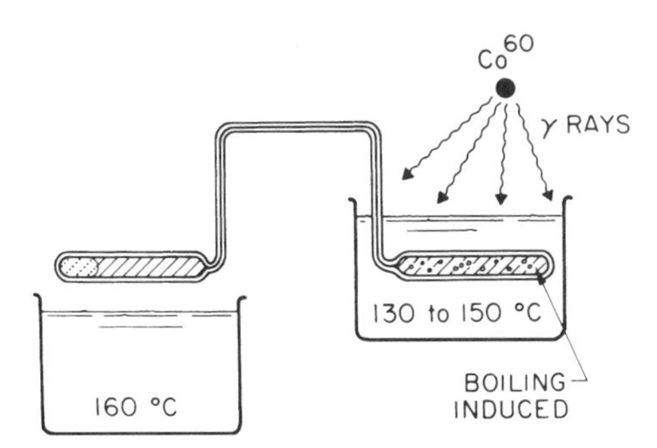

Figure 5.3 Capillary tube experiment, right side (1954). When Glaser removed the left side of the capillary tube from its hot bath, the gas bubble cooled, and contracted, leaving the liquid on the right under less pressure. On exposure to a radioactive source, violent boiling ensued. Source: Glaser, "Progress Report," *Nuovo Cimento* 11 Suppl. (1954): 361–68, fig. 2 on 366.

therefore the higher vapor pressure) on the left-hand bulb forced the liquid to fill the right-hand bulb. Boiling did not start in the right-hand bulb because the liquid was under too great a pressure. The ban on boiling was suddenly lifted when the hotter capillary tube (the left-hand one in figure 5.3) was removed from its bath, because the vapor pressure dropped, making the liquid superheated. Without a radioactive source, boiling would occasionally not begin for several minutes. By contrast, late one night at the lab, Glaser enlisted a graduate student to open a lead box of cobalt 60 that was 30 feet away, and the boiling began violently and suddenly. In a brief letter to the editor of *Physical Review* received 12 June 1952, Glaser reported his radioactively induced boiling.[55] Before prosecuting the work any further, Glaser left on a previously arranged European trip.[56] His student, David Rahm, continued on the project by investigating the "temperature, pressure and ionization dependence of bubble formation in superheated liquids."[57] To create a high vapor pressure over the hot liquid diethyl ether, Rahm built a hand-cranked hydraulic piston-cylinder apparatus. The pressurized ether in the glass bulb was potentially explosive, so the whole device was covered by a large shield with clear plastic observation ports.

For two days after assembly, Rahm's every trial fizzled. As soon as the pressure was released, boiling would set in. Eventually Rahm got the system to work properly and was able to record the waiting time as a function of temperature and pressure. The empirical data on the critical pressure change, together with Glaser's electrostatic theory, made it possible to calculate the number of charges on a bubble. Solving for n in equation (5.4) with standard values of $\sigma(T)$ and $\epsilon(T)$ gave the number of charges responsible for the bubble growth as

55. Glaser, "Progress Report," *Nuovo Cimento* 11 Suppl. (1954): 361–68, on 366; Glaser, "Bubble Chamber," *Sci. Amer.* 192 (February 1955): 46–50, on 47–48; and Glaser, "Some Effects," *Phys. Rev.* 87 (1952): 665. The graduate student was Noah Sherman.

56. Rahm to Warnow, American Institute of Physics, 22 September 1976, DGP.

57. Rahm, "Progress Report," August 1952, DGP. See also the cover letter to Warnow (Rahm to Warnow, American Institute of Physics, 22 September 1976, DGP). This work is summarized briefly in Rahm, "Development" (1956), 1, 5.

$n = 37.5$. Looking back on this period of development, such calculations might appear irrelevant: no feature of the multiple charge explanation survived more than a few years. At the time, however, these equations were central to Glaser's understanding how the elementary bubble was being formed.

Modifications of Rahm's experiment were possible until the glass broke. Throughout the summer the goal was clear, as Rahm emphasized in his report: "Bubble Chamber—a large bulbed system (3/4″ diameter × 1-1/4″) was used with the hope that tracks could be observed. The only thing that could be seen, however, was a single large bubble which almost always appeared to form on the walls of the bulb. . . . Further work on more carefully cleaned bulbs should be carried out."[58]

Scrupulous bulb cleaning became the order of the day when Glaser returned from Europe. Using Rahm's hand-cranked compressor,[59] some new bulbs, and a high-speed movie camera, he resumed the search for tracks. From the notebooks we know that on 14 October 1952, Glaser shot 300 feet of Super XX Panchromatic 16 mm film at 8,000 frames per second. Under "conclusions" he wrote the following (see figure 5.4):

> 1) bubbles can grow in ~1 msec from sizes invisible with the optics used to diameters ~1 mm. . . .
> 4) Twice apparently very faint tracks of 4 or more bubbles were seen to precede a grown globe, but their faintness makes it possible to assume that these were dirt effects.
> 5) More light is needed to see smaller bubbles.[60]

Four days later, on 18 October, with better illumination, Glaser recorded the following under stills from his bubble chamber movies (see figure 5.5):

> *Conclusions* Tracks *can* be photographed
> *Plans* Next one must attempt to observe counter-controlled expansions.[61]

On reflection, Glaser must have decided to take an intermediate step, for on 20 October he began to try a counter-controlled flash while maintaining a random expansion.[62] Some bubbles showed up on the photographs, but no tracks. Under "Conclusions" he recorded that the sensitive time of the chamber was too short to produce useful results given that the counting rate of cosmic ray particles was only four or five per hour. He therefore decided to press on with the attempt to make a counter-controlled expansion.[63] Glaser enlisted an old family

58. Rahm, "Progress Report," August 1952, sec. 5, DGP.
59. Rahm to Warnow, American Institute of Physics, 22 September 1976, DGP.
60. DGnb2, 6–7, DGP. The movies were first reported publicly in Glaser's "Progress Report," *Nuovo Cimento* 11 Suppl. (1954): 361–68, on 366.
61. DGnb2, 7, DGP.
62. DGnb2, 11, DGP.
63. DGnb2, 11–13, DGP.

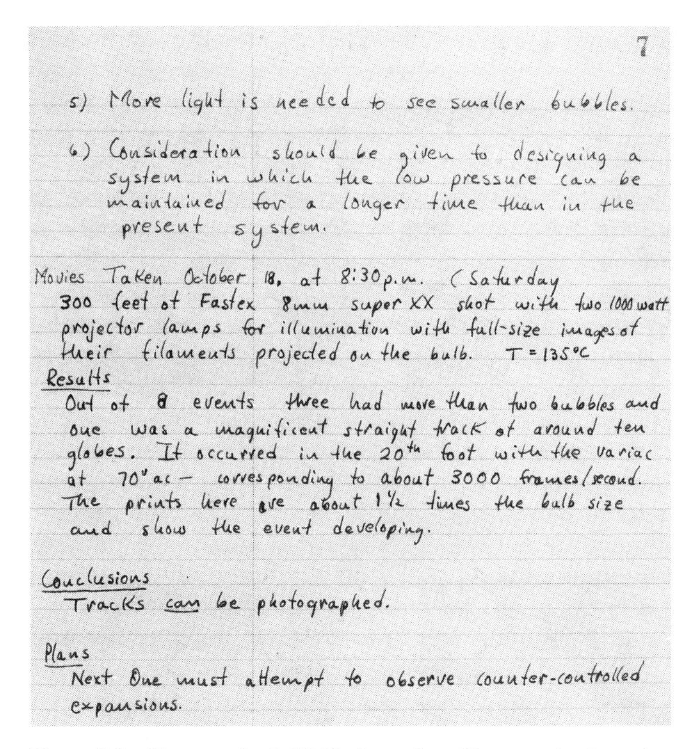

7

5) More light is needed to see smaller bubbles.

6) Consideration should be given to designing a system in which the low pressure can be maintained for a longer time than in the present system.

Movies Taken October 18, at 8:30 p.m. (Saturday
300 feet of Fastex 8mm super XX shot with two 1000 watt projector lamps for illumination with full-size images of their filaments projected on the bulb. T = 135°C

Results
Out of 8 events three had more than two bubbles and one was a magnificent straight track of around ten globes. It occurred in the 20th foot with the variac at 70v ac — corresponding to about 3000 frames/second. The prints here are about 1½ times the bulb size and show the event developing.

Conclusions
Tracks can be photographed.

Plans
Next One must attempt to observe counter-controlled expansions.

Figure 5.4 Glaser notebook (1952). Pages from Glaser notebook on photographs from the 1 cc bubble chamber. Source: DGnb2, 7.

radio, a 1930 Colson sporting a loudspeaker with a very large magnet.[64] With a big vacuum tube to deliver a hefty current, the magnet could be made to release a valve very quickly. (That this was dubbed a "pop valve" needs no further explanation.) The work to this point had cost a total of $2,000. At the end of November 1952, the Phoenix Project doubled the yearly stipend to $3,000 for "New Methods of Detecting Ionizing Radiation by Its Effect on Phase Changes."[65] (Even then, as the grant title suggested, Glaser's inquiry was not focused on the specific liquid-to-gas phase change.) By 18–19 December 1952, a prototype (Geiger-Müller) counter-triggered bubble chamber was ready, pop valve and all. Glaser wrote: "53 pictures were taken of which 30 were at a sensitive time. Of these, 20 should have been good, but none were."[66] The failure was followed by

64. Glaser, interview by the author, 4 March 1983.

65. Sawyer to Glaser, 21 November 1952. Copy is in files of E. F. Barker, Department of Physics, University of Michigan.

66. DGnb2, 17, DGP.

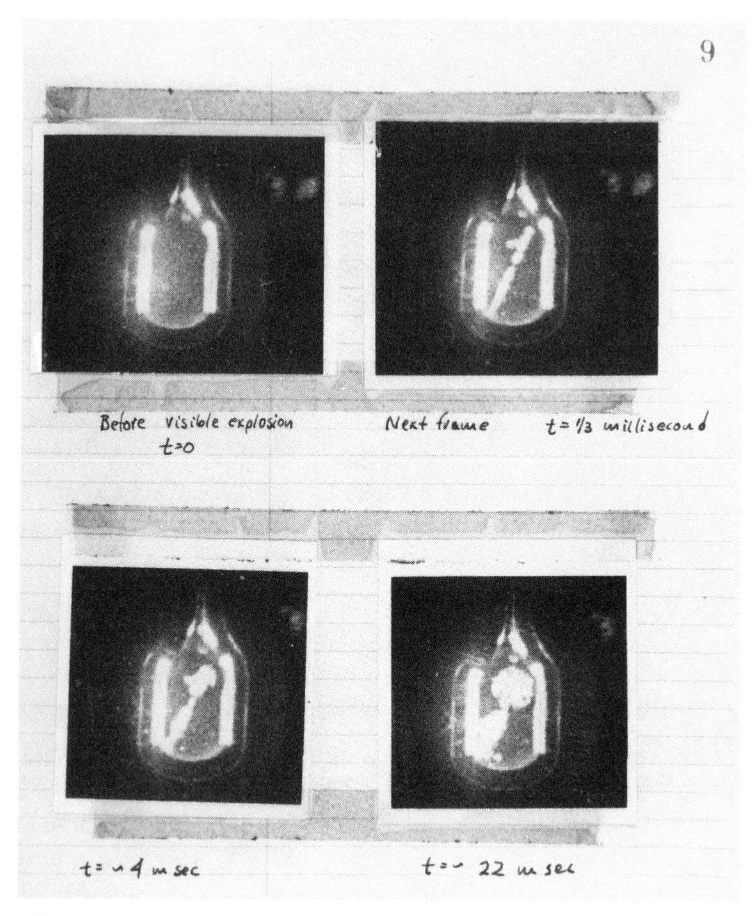

Figure 5.5 Bubble chamber movies (1952). Glaser first filmed distinct tracks on 18 October 1952, using a high-speed 16 mm camera. Source: DGnb2, 9.

a return to random expansions, amid much frustration.[67] Why the immediate fascination with counter control?

Above all, Glaser did not want to work on the big accelerators: "I wanted to help save cosmic-ray physics," he reminisced later.[68] To rescue his old specialty he needed to make the method adaptable to counter control—otherwise the bubble chamber was fated to be wed to the predictable pulses of the accelerator. Or perhaps we should state this the other way round: if an accelerator-based detector was the goal, counter control would not have been the first improvement

67. Glaser, interview by the author, 4 March 1983.
68. Glaser, interview by the author, 4 March 1983.

needed for the new device. More than any pronouncement about intentions, this early insistence on counter control spoke to Glaser's picture of the bubble chamber as an element of cosmic ray physics. A work structure was built into the hardware.

As befit the importance of creating a self-registering device modeled on the cosmic-ray cloud chamber, many attempts at counter control followed. Glaser and Rahm tried injecting CO_2 into the diethyl ether to slow the formation of bubbles; the delay would have given the expansion mechanism time to function before track formation. No luck.[69] Another attempt began in April 1953 when Rahm and Glaser tried to exploit the "plink" heard when the violent eruption of boiling begins. To tap this signal the physicists attached a phonograph pickup to the side of the glass chamber. The method worked, though not entirely satisfactorily; by the time the sound arrived at the pickup the bubbles had grown too large for precision measurement.[70] By the fall of 1953, they had a new method in which a photoelectric cell would trigger the flash when the optical disturbance of the bubbles indicated that a particle had passed through. Again, the resulting photographs were not spectacular.[71]

Even though Glaser was unable to rescue cosmic ray physics with a new counter-controlled detector, his small, basic bubble chamber was ready for presentation. On Saturday 2 May 1953, in session XA of the Washington meeting of the American Physical Society, he presented a 10-minute account of his results[72] to a somewhat depleted audience, it being the last day of the conference.[73] "My first paper," Glaser later wrote, "was scheduled by the secretary of the American Physical Society, Karl Darrow, in the Saturday afternoon 'crackpot session.'" Shortly after the conference, Glaser submitted a letter to *Physical Review.* It was promptly returned "on the ground the [he] had used the word bubblet which is not in Webster."[74] When the letter was received without the offending word on 20 May 1953, it contained the outline of the charge repulsion theory of bubble growth and a description of the counter-controlled flash, random expansion bubble chamber.[75] This published letter served as the point of departure for the many experimental groups that soon began to construct prototype chambers.

In light of Glaser's appearance in the "crackpot session," the surprising feature of his audience was that two very attentive listeners hung on every word of the bubblet producer's ideas. Luis Alvarez, from the University of California,

69. Glaser, interview by the author, 4 March 1983.

70. DGnb2, 37, DGP; see also Glaser and Rahm, "Characteristics," *Phys. Rev.* 97 (1953): 474–79, on 475.

71. DGnb2, 55–61, DGP; Glaser, "Progress Report," *Nuovo Cimento* 11 Suppl. (1954): 361–68, on 367; Glaser and Rahm, "Characteristics," *Phys. Rev.* 97 (1953): 474–79, on 475.

72. Glaser, "Possible 'Bubble Chamber,'" *Phys. Rev.* 91 (1953): 496.

73. Glaser, interview by the author, 4 March 1983. Darragh Nagle remembers only himself and the other scheduled speakers in the audience (Nagle, interview by the author, 28 November 1983).

74. Glaser to Weber, 17 August 1970, DGP.

75. Glaser, "Bubble Chamber Tracks," *Phys. Rev.* 91 (1953): 762–63.

Berkeley, had met Glaser by chance at a reception the day before his talk; Darragh Nagle, from the University of Chicago, was one of the few people who had remained to hear the talk itself.[76] Unlike Glaser, neither had the slightest interest in saving cosmic ray physics—indeed, both wanted nothing more than to use the bubble chamber to make full use of accelerator programs at their respective institutions. They had in mind a physics of a very different kind from that of the hand-cranked compressor, the 1-inch chamber, and the loudspeaker pop valve. And if the world of Jungfraujoch, Pic-du-Midi, and Lake Arrowhead vanished, it was of little concern; their universe of physics lay in the accelerators of Berkeley, Chicago, and Brookhaven.

5.3 Between the Scales

It was Glaser's fantasy of a hydrogen-filled bubble chamber that intrigued Nagle and the Alvarez group. Nagle was just finishing collaborative work that had used liquid hydrogen as a target for the Chicago pion beam. If the target could double as a visual detector, a whole new realm of accurate interaction measurements would be possible.[77] The old collaboration had mastered the art of handling a liquid hydrogen target 3 inches in diameter and 7.5 inches long and had discovered some excited proton states (among other things).[78] Clearly, hydrogen, with its most simple nucleus, was the medium in which to do interaction experiments, even if high stopping power and decay questions were best addressed with heavier and more complicated nuclei. And elementary particle accelerator physics was quintessentially the science of the simplest interactions.

Returning to Chicago after the American Physical Society meeting, Nagle suggested to Roger Hildebrand that they work together on the hydrogen bubble chamber project. Since Glaser was also interested in making a hydrogen chamber, he began to collaborate with them, visiting them several times to participate in design sessions.

The switch to liquid hydrogen presented several problems, two of which stood out above the others. By weight, hydrogen gas is more explosive than dynamite, so safety features were needed far more than with diethyl ether. Second, whereas diethyl ether boils at atmospheric pressure at about 130°C, hydrogen does so at −246°C. Cryogenic expertise was thus required for the project to succeed. The low-temperature physicists Earl Long and Lothar Meyer at Chicago had worked with the hydrogen target group. Together, Nagle and Hildebrand, consulting Long and Meyer, began to repeat on nitrogen the radiation sensitivity

76. Alvarez, interview by the author, 7 March 1983; Nagle, interview by the author, 28 November 1983.

77. Nagle, interview by the author, 28 November 1983.

78. Anderson et al., "Negative Pions," *Phys. Rev.* 85 (1952): 934–35; Fermi et al., "Ordinary," *Phys. Rev.* 85 (1952): 935–36; and Anderson et al., "Positive Pions," *Phys. Rev.* 85 (1952): 936.

experiments Glaser had performed on diethyl ether. (Long had become quite an expert on hydrogen cryogenics; he had been consulting on liquid hydrogen problems for the H-bomb project since his days with Teller early in World War II.) On 4 August 1953, the very first series of runs worked. No tracks appeared, but superheated nitrogen clearly was sensitive to a radioactive source. Two days later, hydrogen followed suit.[79]

Over the next weeks, the Chicago physicists continued to explore the properties of their chamber. How should it be kept cold? How could the pressure be properly regulated? Above all, how could the bulb be kept exquisitely clean?[80] Hildebrand recalls late that summer trying (unsuccessfully) to reach Glaser in Europe, asking him about their collaboration.[81] In the event, the paper was submitted by Hildebrand and Nagle and received by *Physical Review* on 21 August 1953.[82] Glaser decided to continue his work separately.

Despite their initial success in showing radiation sensitivity, the Chicago group still had no tracks. Toward this end, they turned in February 1954 to the building of a new chamber.[83] Most of the work centered on improving the expansion system, building bigger chambers, and improving the photography (see figure 5.6). Meanwhile, Hildebrand's graduate students Irwin Pless and Richard Plano launched a simultaneous effort to obtain accelerator-generated meson tracks in a pentane chamber. This work succeeded.[84]

By the summer of 1953 certain physical properties of bubble chambers had become obvious, above all that the chamber was sensitive to track formation for only several microseconds. This augured poorly for cosmic ray applications, since by the time a counter control could effect an expansion the tracks could no longer be made visible. On the other hand, accelerators emitted particles at controllable times, so the chamber could be set up to expand at a propitious moment. Indeed, the short sensitivity time now became an advantage by excluding older (and therefore irrelevant) tracks. Glaser had hoped his invention would "allow physicists to work independently or at least in small groups." But now, he admitted, "I was trapped. It was just what the accelerators needed and it wasn't useful for cosmic rays."[85]

So it was that when Glaser came back to the United States from Europe in the fall of 1953, his attention turned to the Brookhaven National Laboratory

79. Hildebrand and Nagle, "Bubble Chamber Log," August 1953–November 1955, RHilP.

80. Hildebrand and Nagle, "Bubble Chamber Log," August 1953–November 1955, 1–8, RHilP.

81. Hildebrand, interview by the author, 17 February 1983.

82. Hildebrand and Nagle, "Operation," *Phys. Rev.* 92 (1953): 517–18.

83. Hildebrand and Nagle, "Bubble Chamber Log," August 1953–November 1955, 13, RHilP.

84. Preliminary results: Pless and Plano, "Study," *Phys. Rev.* 99 (1955): 639–40; Plano and Pless, "Negative," *Phys. Rev.* 99 (1955): 639. Pless's dissertation was published as "Proton-Proton," *Phys. Rev.* 104 (1956): 205–10.

85. Glaser, interview by the author, 2 November 1983.

Figure 5.6 First hydrogen bubble chamber (1954). Here Darragh Nagle adjusts his and Roger Hildebrand's cryogenic chamber. Source: Courtesy of Darragh Nagle.

cyclotron, where he and Rahm hoped to put the new detector through its paces. Arrangements were made in late 1953 for the Michigan physicists to travel to the Long Island laboratory. Knowing all too well the difficulty of handling liquid hydrogen, Alan Thorndike, from the Brookhaven Cloud Chamber Group, wrote to Glaser: "We have no intention of getting into the liquid hydrogen business, but think that a hydrogenous liquid such as ether would make a very valuable device for studying rare nuclear events, such as *V*-particle production, at the Cosmotron."[86] There is some retrospective irony in this remark, since Thorndike was to spend the better part of the next 30 years involved in the development of data analysis from hydrogen chambers. For the moment, however, building somewhat bigger versions of Glaser's device was the goal. Even here, Glaser cautioned Thorndike, there were difficulties: "My first reaction toward the possibility of making a successful ten-liter chamber is that it probably won't be feasible. One liter seems ambitious in view of my present knowledge of the problem."[87]

86. Thorndike to Glaser, 4 December 1953, ATP.
87. Glaser to Thorndike, 14 December 1953, ATP.

In the first experiments at Brookhaven, Glaser and Rahm propped the chamber on a table in front of a crack in the Cosmotron's radiation shielding. Photography was undertaken with Glaser's personal 35 mm camera. By capturing several π-μ-e events in the first roll of film, they impressed their sponsors enough for them to pursue the combination of accelerator and bubble chamber. When more funding was sought from the NSF, the ONR, and the AEC, three goals figured prominently. First, a much improved means of data analysis was needed to take advantage of the precise pictures now being copiously produced. Second, a heavy-liquid bubble chamber seemed promising because it could offer high stopping power (making decays easier to study and photons more likely to convert into observable e^+e^- pairs). Third, Glaser hoped that he could build a hydrogen chamber. Part-time salaries for Glaser, Martin Perl (a new faculty member), a secretary, and four research assistants, added to the salaries for a full-time postdoc and a full-time machinist-technician, came to about $25,000. Equipment, supplies, and machining ran about the same.[88]

The highest priority was accorded to the construction of a liquid xenon chamber, a project that inter alia required most of the world's xenon supply. Here, Glaser recalled, was one last "unique niche that I could [fill] at Michigan without access to all this high technology and large engineering staffs."[89] With a working chamber now successfully demonstrated with an accelerator, funding (and xenon) flowed freely from the AEC. The first physics objectives centered on the chambers with high stopping power, which promised results on the decay and production of strange particles without necessitating cryogenic engineering.

In the process of perfecting their new chamber, Perl, Glaser, and J. L. Brown succeeded in demolishing the old electrostatic theory of bubble formation—the enabling theory that had opened the conceptual possibility of building bubble chambers. In this respect, success followed from failure. When the xenon chamber was completed, absolutely no tracks were to be seen. At about this time, Glaser and his collaborators learned from Los Alamos that gaseous xenon scintillated in the optical frequencies. This fact suggested that the energy that might have been exploited to form bubbles was escaping in the form of ordinary light. If it could be captured in the liquid, the chamber might be rescued. Happily, when tiny quantities of a quenching agent (ethylene) were mixed with xenon, tracks began to appear. (Apparently, by collisions of the second kind, the ethylene molecules were deexciting the xenon atoms before the latter could emit optical light. Instead of radiating away, energy was being converted into kinetic energy and internal molecular excitation of the ethylene.) Since even very small

88. Glaser, "Study of High Energy Nuclear Interactions Using Bubble Chambers and Further Development of the Bubble Chamber Technique," grant proposal to AEC Physics Division, with cover letter to Trese, 6 December 1954, DGP.

89. Glaser, interview by the author, 2 November 1983.

quantities of the absorber did the trick, the experiment suggested that bubble production was probably due to local heat trapping rather than electrostatic repulsion.[90] A more comprehensive treatment of the "heat spike theory" was worked out by Frederick Seitz, whose interest in bubble formation had begun in the days of the Manhattan Project, when it was feared that radiation-induced bubbles in reactor cooling water might lead to a meltdown. Inspired by Glaser's work, Seitz described in 1958 how a charged particle would cause local heating of the superheated liquid along its path, much as a hot needle would induce a long sharp line of nucleation.[91] The mutual repulsion of multiple charges, it seemed, had nothing to do with bubble formation.

In short, during the transitional period of the mid-1950s, both Glaser's group and the Chicago team succeeded in adapting a number of heavy-liquid bubble chambers for accelerator experiments. Though costs for both groups mounted into the mid-five figures, the style of work did not shift radically around Glaser's xenon and the Chicago-based heavy-liquid chambers; much like the Gardner nuclear emulsion group at the Berkeley accelerator, the older work structure around the detector was grafted in its entirety to the newer outer laboratory of the accelerator. The switch to hydrogen chambers, by contrast, was much more difficult. Liquid hydrogen alone cost the Chicago group a sizable fraction of its 1954–55 budget,[92] and that did not include the costs of additional safety conditions imposed by the dangerous substance. Nonetheless, with funding from the ONR[93] and the AEC they were able, by late 1956, to move from their successful demonstration of the radiation sensitivity of liquid hydrogen to the production of tracks of pion-proton collisions in hydrogen.[94] But the honor of first demonstrating a track-producing hydrogen bubble chamber went to the rapidly growing Berkeley collaboration. We return to those hills.

5.4 Physicist and Engineer

To a great extent, Glaser's experience in the early 1950s as an independent researcher using makeshift, low-budget equipment was out of step with the times. Too young to have been on the scene during the big wartime projects on radar and the atomic bomb, he had not been exposed to the very different style of research we saw in chapter 4. The war-expanded teams at Berkeley, Los Alamos,

90. J. L. Brown, Glaser, and Perl, "Liquid Xenon," *Phys. Rev.* 102 (1956): 586–87.

91. F. Seitz, interview by the author, 29 November 1983; and Seitz, "Theory," *Phys. Fluids* 1 (1958): 2–13.

92. Nagle, budget proposal, 28 May 1954, sent to Wright, Nuclear Physics Branch, Division of Physical Sciences, Office of Naval Research, and to George Kolstad, AEC Physics Division, DNP. I thank Nagle for making these documents available to me.

93. Anderson, "High Energy Accelerator Research," 1 February 1955 to 31 January 1957, dated 13 December 1954, DNP; also see letter, Ebel, USN Nuclear Physics Branch, to Anderson, 7 December 1954, DNP.

94. Nagle, Hildebrand, and Plano, "Scattering," *Phys. Rev.* 105 (1957): 718–24.

MIT's Radiation Laboratory, Oak Ridge, and the Metallurgical Laboratory in Chicago had established a new mode of physics research, lavishly funded by government and integrated horizontally among many physicists and vertically among engineers, administrators, students, and technicians. As we saw in the last chapter, Ernest Lawrence had initially anticipated a much reduced Radiation Laboratory at Berkeley.[95] By the time peace came in August 1945, however, the Radiation Laboratory, the government, and the military all agreed that laboratory activity should be continued at a high level, and not just at Berkeley. Encountering the new government largesse, one physicist, Arthur Roberts, was moved to something like verse:

> How nice to be a physicist in 1947,
> To hold finance in less esteem than Molotov does Bevin,
> To shun the importuning men with treasure who would lend it,
> To think of money only when you wonder how to spend it,
> Oh,
> Research is long,
> And time is short
> Fill the shelves with new equipment,
> Order it by carload shipment,
> Never give
> A second thought,
> You can have whatever can be bought.[96]

And if treasure and physics went together, nowhere did they converge in such amounts as at Berkeley. As Robert Seidel put it, "By accelerating science along the path of crash programs for large-scale technological development in science, the Radiation Laboratory helped set the tone of modern big science."[97]

Lawrence's style of big physics was congenial to Luis Alvarez. Not only had Alvarez seen the building of the large, philanthropically supported accelerators, he had also participated in both of the two great scientific war efforts: radar and the atomic bomb. Each project was budgeted at around $2 billion; one learned in such an environment to plan on a large scale. It was in the midst of the war work of winter 1941 and spring 1942 that Alvarez made the great shift in his "laboratory lifestyle,"[98] as did so many others that we tracked in chapter 4: Wheeler, Dellenbaugh, White, Kemble, Terman, and Bloch. In the three years between 1942 and 1945, Alvarez had been "indoctrinated" with "radar philosophy"

95. Seidel, "Accelerating Science," *Hist. Stud. Phys. Sci.* 13 (1983): 375–400, on 377.

96. Roberts, "How Nice to Be a Physicist," typescript, 27 April 1947, LAP.

97. Seidel, "Accelerating Science," *Hist. Stud. Phys. Sci.* 13 (1983): 375–400, on 399–400.

98. For further discussion of Alvarez's work on the radar project, see Guerlac, *Radar* (1987), sec. C, pt. 1; and Alvarez, *Adventures* (1987), 86–110. Alvarez's work on the atomic bomb is summarized in Alvarez, *Adventures* (1987), 111–142; and Hawkins, Truslow, and Smith, *Project Y* (1983), 112, 128, 203–6.

and then, having moved to the southwest during the latter part of the war, had become the quintessential "Los Alamos man." On 6 August 1945, Alvarez flew in the observation plane *Great Artiste* for the Hiroshima mission.[99] Before the war, Alvarez's work on cosmic rays and nuclear physics had been done alone, in a small group, or with a few graduate students; afterward, for most of his career, he would hold line jobs in large research organizations.

The transition in Alvarez's scientific work structure began in the early stages of his radar efforts at the MIT Radiation Laboratory. There, during the fall of 1941, he was made group leader of JBBL (Jamming Beacons and Blind Landing), and then head of Division 7 (Special Systems). In these capacities he and his group concentrated on three projects: the GCA (Ground Controlled Approach radar), which allowed planes to make blind landings, the MEW (Microwave Early Warning radar) later used to direct the air armada to the Normandy coast on D Day, and the Eagle Blind Bombing System, which was employed in strategic attacks all over Europe. By 1943 most of the physics-related radar problems had been solved, and Alvarez transferred to the Manhattan Project, going first to Fermi's reactor at Chicago and then to Los Alamos, where he joined the team working on the implosion device. The implosion group grew quickly, and in June–July 1944 it underwent a reorganization that put Alvarez at the head of group E-11. This group's assignments were, first, to exploit the gamma-emitting radiolanthanide to follow the course of imploded material during tests and, second, to investigate electric detonators for the imploding charge.[100]

When the war ended, the Berkeley Radiation Laboratory budget was extended in new directions: Alvarez made use of $250,000 worth of surplus radar sets in his linear accelerator, $200,000 worth of surplus capacitors went to McMillan's synchrotron, and some $10 million was destined for the Bevatron.[101] One of the scientific benefits of the governmental largesse was the training and cultivation of an extraordinary group of engineers who enjoyed an unprecedented degree of autonomy. An example of the new and powerful contribution of engineers in physics facilities was the design of the Bevatron by the engineer William Brobeck.

Each of these three projects—radar, the atomic bomb, and the linear accelerator—placed Alvarez at the nexus of scientific, engineering, and management concerns. Certainly each alone would have been ample schooling in the execution of large-scale physics. But there was a fourth project that, had it been pursued, would have been bigger than the three previous ones put together.

Debate within the secret atomic establishment of the United States came to a head in the late 1940s. Before the summer of 1949, the General Advisory

99. See Alvarez, *Adventures* (1987), 143–44.

100. Guerlac, *Radar* (1987), on 293, 386–94, 497–506; and Johnston, "War Years" (1987), 55–71; on the implosion program, see Hoddeson et al., *Critical Assembly* (1993), esp. chaps. 8, 9, 14–16.

101. Hawkins, Trower, and Smith, *Project Y* (1983), 112.

Committee (GAC), the highest advisory panel to the AEC, recommended that the AEC not put priority on the thermonuclear weapon. Slow deliberation ended in the flash of the Russian atomic bomb explosion of August 1949, and Alvarez and Lawrence became personally and passionately involved in the often ferocious secret dispute over whether the thermonuclear weapon was needed for national defense. Both were among the most ardent of the H-bomb advocates.[102] On 31 January 1950, to their relief, President Truman gave his approval, and the crash program began.

Lawrence surmised that the weapons program would soon require increased supplies of tritium (for the fusion bomb) and fissionable material (for a variety of atomic bombs). As Alvarez saw it, his linear accelerator design could be scaled up to an enormous neutron foundry capable of producing both kinds of material. Had it been completed, this Materials Testing Accelerator (MTA), headed by Alvarez, would have borne a price tag of $5 billion.[103] Despite a great deal of planning, only the front end of one of the projected machines was built; it alone cost over $20 million. By 1952, the entire enterprise was obviated by the discovery of rich natural sources of uranium in the Colorado plateau; in August 1952 it was decided to postpone Alvarez's MTA indefinitely.[104] Alvarez played a central role in much of the MTA work, an experience that (in addition to providing even more expertise in large-scale planning and management) solidified his ties with the AEC.

Before the MTA project ended formally, Alvarez began to reimmerse himself in fundamental physics. Samuel Allison, a physicist at the University of Chicago, wrote to Alvarez in September 1951 inviting him to summarize "your" recent work at an international conference.[105] Responding to the query, Alvarez queried Allison about his use of the indexical "you": "I am slightly confused as to whether you mean 'you' or 'you all.' I doubt that I would have anything that would interest the group, since I have been an engineer for the past year and a half and haven't spent much time in the laboratory myself. If you mean 'you all' I could mention some of the highlights of the work here at Berkeley. . . . I am looking forward to the conference as a chance to catch up on what is going on in the world of physics, since I feel a bit out of touch myself."[106] As Alvarez "caught up" with particle physics, his experience in defense work on the boundary between physics and engineering served him well. Indeed, the design engineering, planning, and management of the Berkeley bubble chamber work were integral

102. On the H-bomb debate and Alvarez's role in it, see Galison and Bernstein, "In Any Light," *Hist. Stud. Phys. Bio. Sci.* 19 (1989): 267–347, and further references contained therein.

103. See "Atomic Energy Commission Objectives of the MTA Program: Report by the Director of Reactor Development," 1952, General Correspondence, document no. AB-2153; DC 8990, LBL. For an excellent account of MTA, see Heilbron and Seidel, *Lawrence* (1989), 66–75.

104. Heilbron and Seidel, *Lawrence* (1989), 69.

105. Allison to Alvarez, 6 September 1951, LAP.

106. Alvarez to Allison, 11 September 1951, LAP.

to its construction and execution at every stage from the choice of personnel to the character of everyday life in the Berkeley hills.

Glaser's presentation thus caught Alvarez at a time when engineering was in his immediate past and accelerator physics on his present agenda. Immediately after the Washington meeting at which Glaser introduced the new detector, Alvarez resolved to set his lab the goal of building very large hydrogen chambers.

Work on the new instrument began at Berkeley much as it had at Michigan and Chicago: on a very small scale. And, as at Chicago, the hydrogen chamber had the highest priority. On 5 May 1953, Lynn Stevenson underscored the word *hydrogen* twice on the second page of the Berkeley group's notebook on bubble chambers. Then came data from the International Critical Table, along with plots of isotherms.[107] Alvarez later put it bluntly: "Who wanted nucleons tied together in a ball of junk? Still we thought we'd start from where Don [Glaser] had gotten to."[108] On these grounds, Stevenson began constructing a small diethyl ether chamber. On 18 May 1953, the prototype, Mark I, was ready. Stevenson noted: "Crappy chamber (Not evacuated). At $t = 70$ bubbles formed."[109] The next machine, Mark II, produced bubbles soon after a radioactive source was inserted. On 20 May 1953, Mark III entered the world with a whimper: "[G]lass slipped out of O rings and we evaporated bulb ether good night. P.S. The reason this chamber might not have worked was the presence of minute glass particles from the filter."[110] Dirt, dreaded by cloud chamber builders and scrupulously avoided by Glaser, was often the accused party. Mark IV showed promise for a while on 22 May 1953—the source gave "a definite effect." The next step was to photograph the bubbles using pickup-triggered strobes. Sadly, wrote Stevenson, "[t]here was no success in viewing bubbles." All Stevenson got was "a splitting headache ~2 atmospheres of ether vapor in the sinuses."[111] So it continued through June, July, and August 1953.

While Stevenson's headaches with prototype ether chambers continued, a young Berkeley accelerator technician, John Wood, bypassed the Chicago group, Glaser, and the senior physicists at Berkeley. Working from the Hildebrand-Nagle design published in August, Wood produced a hydrogen chamber that, by October 1953, was the center of attention; it was therefore a technician, not a physicist, who authored the first paper on a track-producing hydrogen chamber.[112] Then, more frustration evidenced in the notebooks: "Attempts were made to superheat the liquid and get bubble formation produced with a radium source. . . . There was no effect when radiation was incident on the chamber."

107. BCL 1, 2–5, LBL.
108. Alvarez, interview by the author, 7 March 1983.
109. BCL 1, 24, LBL.
110. BCL 1, 25, LBL.
111. BCL 1, 27, LBL.
112. Wood, "Bubble Tracks," *Phys. Rev.* 94 (1954): 731.

Under "Suggested improvements," the log book concludes that pressure and thermal gauges are needed and that the experimenters should "replace present chamber with a 'cleaner one.'"[113]

Glaser's experiment on very smooth glass chambers had set a standard (an operational constraint) for everyone—without "cleanliness," it was universally assumed, the spontaneous bubbling initiating at the walls would ruin any attempt to produce tracks. Thus, Stevenson admonished himself in mid-October to make a "cleaner" chamber. More work was done to get control of temperature. On 25 January 1954, the chamber was tried again. The experimenters varied the delay between expansion and flash,[114] and 50 tracks were photographed. The first of these Polaroid photographs are preserved in Stevenson's first notebook.

While staring at these early pictures, Alvarez realized that his group had accidentally stumbled across the most essential feature of all.[115] True enough, boiling was beginning near the walls. Nonetheless, tracks near the center of the chamber were clearly visible. Thus, the failure to make a chamber as clean as Glaser's precipitated the Berkeley group's first great triumph. If accidentally "dirty" walls did not preclude track formation, neither would purposefully imperfect walls. A letter by Wood to *Physical Review* put it this way: "We were discouraged by our inability to attain the long times of superheat, until the track photographs showed that it was not important in the successful operation of a large bubble chamber."[116]

In this very specific sense, the craft tradition accompanying cloud chamber and early bubble chamber construction carried with it a powerful constraint on detector design. Instrument designers like Glaser had imbibed respect for optical-quality glass with mother's milk. The incident with Wood's edge-boiling picture illustrates that constraints can be powerful yet not absolute: surrounded and preceded by industrial-scale design, the Berkeley team spent more than half a year battling to maintain the purity and cleanliness of the glass walls. By the time of Wood's letter, the team was prepared to recognize that this barrier could, in fact, be crossed: purity was not needed.

Immediately, work began on larger bubble chambers in which the walls were not glass but industrial-grade metal.[117] Had it not been possible to use dirty walls, bubble chambers would have forever remained much too small to study highly energetic events. Even the 2.5-inch chamber designed by Doug Parmentier and Pete Schwemin (two other Berkeley technicians)[118] would have required such thick glass walls as to have made the device impractical to maintain

113. BCL 1, 30, LBL.
114. BCL 1, 31, LBL.
115. Alvarez, interview by the author, 7 March 1983.
116. Wood, "Bubble Tracks," *Phys. Rev.* 94 (1954): 731.
117. Schwemin's chamber was first ready for testing on 9 March 1954; BCL 1, 39, LBL.
118. Parmentier and Schwemin, "Liquid Hydrogen," *Rev. Sci. Inst.* 26 (1955): 954–58.

at the right temperature. Instead, the chamber was brass, 1 inch long, 2.5 inches in inside diameter, and capped with glass ends. It worked, and the pace quickened. On 29 April 1954, the first tracks showed up in the 2.5-inch chamber; by August, two 4-inch chambers were built, roughly on the same lines as the 2.5-inch one.[119] All the results were presented in December 1954 at the American Physical Society meeting in Berkeley.[120]

From the very beginning of bubble chamber work at Berkeley, technicians and engineers played an essential role. In fact, none of the early papers were authored by physicists. Wood, Schwemin, and Parmentier were all technicians, and they worked with a fair amount of autonomy, though they consulted frequently with the physicists.[121] The establishment from the outset of close ties between engineering and scientific personnel characterized work at the Rad Lab, differentiating it from other American laboratories, and even more so from European institutions such as the cosmic ray groups and the new European laboratory CERN. The archival legacy of the physics-engineering coordination at Berkeley is a collection of thousands of "engineering notes," the first of which was filed in the engineering records of the laboratory and dated 13 January 1955.[122]

By late 1954, it was clear that Berkeley, at least for some time, was going to dominate bubble chamber technology. The Chicago and Brookhaven groups looked west. Alan Thorndike wrote to Alvarez from the Brookhaven Cloud Chamber Group requesting a chance to work with the Alvarez group. Thorndike added that at Brookhaven, "all the effort [is] going into the pentane-filled chamber approach." "[B]ut I think we all feel," he continued, "the liquid hydrogen filling, which you have been pioneering, is the thing that will really pay off."[123] For similar reasons, Darragh Nagle put in for some time in the Berkeley lab.[124] Roger Hildebrand spent summer 1954 with the Berkeley group and wrote back glowing reports of the expanding project. On 27 June 1954, the remaining Chicago group (Darragh Nagle, Irwin Pless, and Richard Plano) got notice from their representative that the Berkeley group was designing a bubble chamber "4 inches diameter × 3 inches deep!"[125] Hildebrand suggested that his colleagues not "change any of our plans except to think hard about *big* chambers (for the future) (they work)."[126] Later tips that were passed along included the

119. Alvarez, "Recent Developments" (1972), 251–52.

120. Parmentier et al., "Four-Inch," *Phys. Rev.* 58 (1955): 284.

121. Alvarez, "Recent Developments" (1972), 250–51.

122. Stevenson, "Bubble Chamber Development," UCRL Engineering Note 4311-14, file M1, LAP.

123. Thorndike to Alvarez, 4 November 1954, LAP; see also Alvarez to Thorndike, 16 November 1954, LAP; Thorndike to Alvarez, 23 November 1954, LAP.

124. Nagle to Alvarez, 23 November 1954, LAP; Alvarez to Nagle, 13 December 1954, LAP.

125. Hildebrand to Nagle, Pless, and Plano, 27 June 1954, DNP. I thank Nagle for making this and the letters of 20 August and 9 September 1954 available to me.

126. Hildebrand to Nagle, Pless, and Plano, 27 June 1954, DNP.

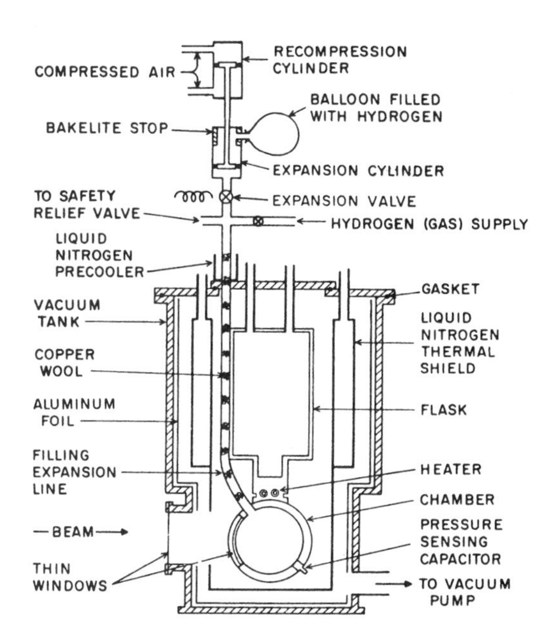

Figure 5.7 Four-inch LRL bubble chamber (1960). The 4-inch liquid hydrogen bubble chamber was too small to use at Bevatron energies but could produce physics in the Radiation Laboratory's 32 MeV proton linear accelerator and the 340 MeV electron synchrotron. It was still possible for the 4-inch chamber to be built by technicians in consultation with physicists. Source: Lawrence Radiation Laboratory, *72-Inch* (1960), 11. Courtesy of Lawrence Berkeley Laboratory, University of California.

use of a liquid hydrogen level indicator and methods of sealing the glass-metal interface.[127]

Alvarez happily reported to Hildebrand in February 1955 that the 4-inch chamber was working in a reproducible fashion and that attention was moving to the physics of stopping positive and negative K mesons in hydrogen (see figure 5.7).[128] However, while the physics was extremely interesting, expansion continued. Shortly after the 4-inch chamber first gave tracks, construction began on a 10-inch chamber (see figure 5.8). Completed in 1956, the 10-inch chamber revealed tracks on its first run, though it stumped the Berkeley designers why the images came only in the top few centimeters of liquid. At fault were the tracks themselves. It seemed that the heat of vaporization was cooling the deeper liquid with the bubbles floating to the top. When the chamber recompressed, the hot vapor bubbles were squashed out of existence, depositing their heat in the top of the chamber. Subsequent decompressions therefore occurred in an environment in which the lower part of the chamber was too cool to boil while the upper part worked perfectly. To counter the temperature gradient, the Alvarez group installed a much faster recompression system that flattened the bubbles

127. Hildebrand to Nagle, 9 September 1954, DNP; Hildebrand to Nagle, Pless, and Plano, 20 August 1954, DNP.

128. Alvarez to Hildebrand, 18 February 1955, LAP.

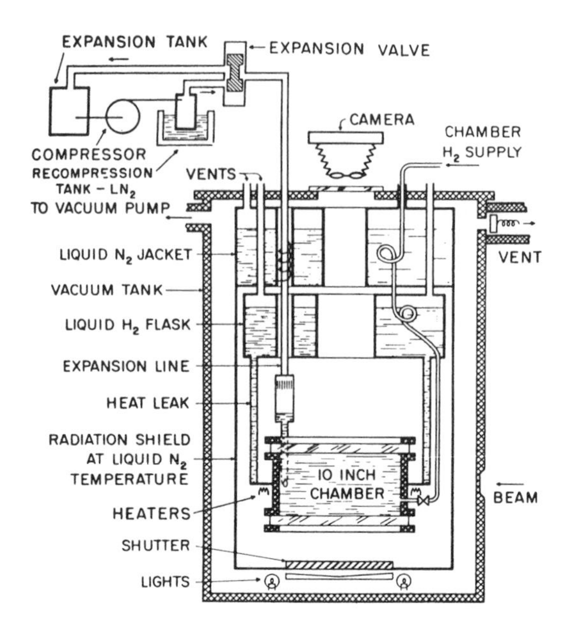

Figure 5.8 Ten-inch LRL bubble chamber (1960). The 10-inch chamber marked the cutoff point for the older model of physicist-technician collaboration. The scale, costs, and physical dangers had now increased to the point where engineers became a permanent part of the construction team. Data production also surged, and the 10-inch produced some 600,000 pictures from turn-on in 1956 to its dismantling the next year. Source: Lawrence Radiation Laboratory, *72-Inch* (1960), 14. Courtesy of Lawrence Berkeley Laboratory, University of California.

before they could rise, and reinforced the temperature homogeneity with a cool plate placed at the high point of the chamber. Over 600,000 pictures issued from the 10-inch chamber before its magnet was cannibalized for its descendent—the 15-inch chamber that came on-line in 1957.[129]

Long before the 10-inch chamber had been turned on, the Alvarez team began the 15-inch chamber. Designed to avoid a significant practical limitation of the 10-inch chamber, the 15-inch device allowed the lights to be replaced without taking the machine completely apart (in figure 5.9, the lights on the 15-inch can be seen located within the vacuum jacket of the chamber). Illumination would be simple: the 15-inch would have only one window, with light shining in through that window and the reflected light from the bubbles emerging through the same portal. This scheme was fine, almost. If the reflector sent the light back directly to where it came from, the pictures would be entirely gray. To avoid this washout, Alvarez's team used optical reflectors (dubbed "coat hangers") to reflect direct light back in the direction of the lights, while light scattered from the bubbles would be reflected off to the side and into the waiting camera lens (see figures 5.9 and 5.10).[130] Expansion at LRL came head over heels. The 2.5-inch chamber had ceded to a 4-inch, next to a 10-inch, and finally to a 15-inch; then all eyes turned to a chamber so vast that it dwarfed any extant or planned detec-

129. Lawrence Radiation Laboratory, *72-Inch* (1960), 13–14.
130. Lawrence Radiation Laboratory, *72-Inch* (1960), 16–17.

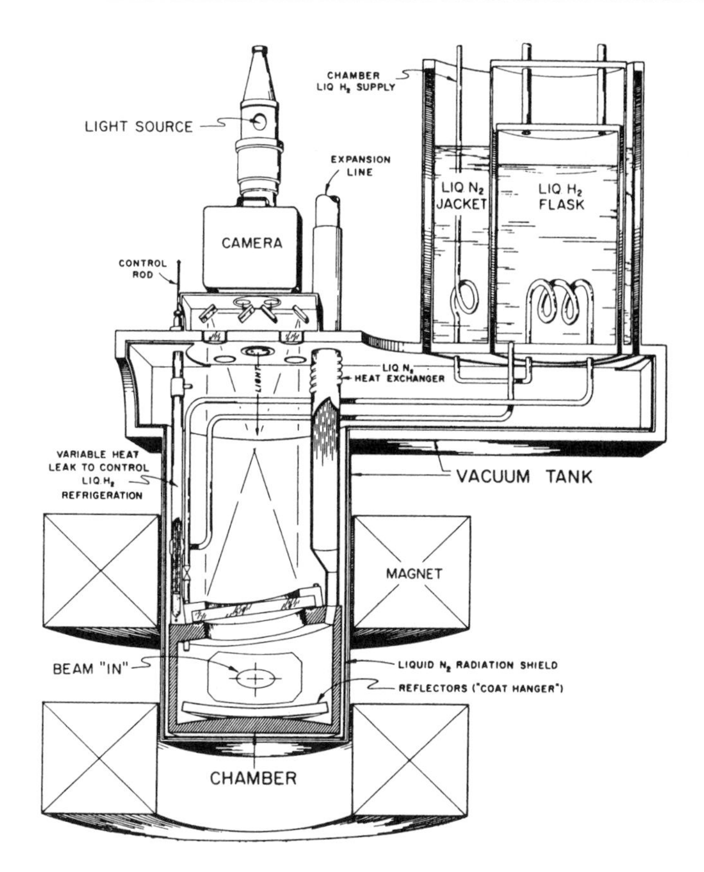

Figure 5.9 Fifteen-inch LRL bubble chamber (1960). Completed in 1957, the 15-inch chamber used parts from the 10-inch and presented many of the design features to be incorporated into the already-planned 50-inch machine (later restructured as the 72-inch bubble chamber). Source: Lawrence Radiation Laboratory, *72-Inch* (1960), 15. Courtesy of Lawrence Berkeley Laboratory, University of California.

tor. Just two months after the 4-inch chamber first made pictures, that is in April 1955, Alvarez began drafting plans for a much larger instrument, 50 × 20 × 20 inches in sensitive volume.[131] Building devices as big as the 10-inch bubble chamber required two important changes in physicists' lives from the earlier, smaller machines. First, physicists and their usual complement of technicians were joined by engineers in the design and construction of the chamber. Second, the physicists, now unable to cope with the stream of data emerging in the

131. Alvarez, "The Bubble Chamber Program at UCRL," 18 April 1955, stenciled typescript, 9, LAP.

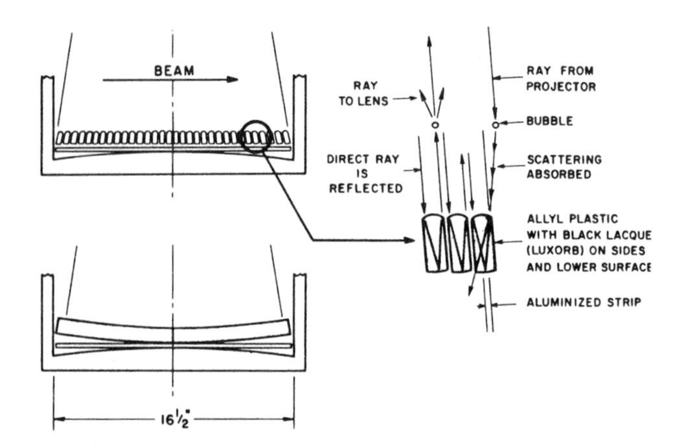

Figure 5.10 Optical reflectors ("coat hangers") (1960). The problem of "washout" was solved by the design of unique plastic reflectors called "coat hangers." Each coat hanger has a narrow aluminized reflecting strip on the bottom at the focal point of the optically polished (upper) elliptical refracting surface. Nonoptical surfaces are painted black to minimize scattered light. Rays from the light source focus on the aluminized strip and return through the liquid unattenuated. However, scattered light coming from other sources (e.g., light that has scattered in bubbles before reaching the reflectors) does not focus on the strip and is instead absorbed in the black surfaces. This retrodirective system prevents reflections (double images) of bubbles. Because the reflected rays return directly to the light source and only those reflected rays that are scattered by bubbles go into the camera lenses, tracks appear light against a dark background. This dark-field illumination gives pictures of excellent contrast. Source: Lawrence Radiation Laboratory, *72-Inch* (1960), 16. Courtesy of Lawrence Berkeley Laboratory, University of California.

thousands of photographs, began to delegate some of the preparation of hardware and software for the processing of a vastly increased volume of data.

By 1955, with only a 4-inch chamber actually working, but 10-, 15-, and 50-inch devices on paper and in various stages of construction, the bubble chamber program had grown large enough that its administration required both managerial expertise and extensive funding. Neither the one nor the other had been acquired in the course of cosmic ray research. For Alvarez, as for many of his contemporaries on the Manhattan and Radar Projects, it was the war that had taught them how to run a large scientific engineering project. For example, in structuring his staff Alvarez put Don Gow in a role that was "not common in physics laboratories, but . . . well known in military organizations"—he made him chief of staff.[132] More specifically still, the March 1942 reorganization of the Rad Lab had revolved around the issue of centralized "operational" control. As we saw in chapter 4, scientific administrators like Frederick Dellenbaugh had contended after Pearl Harbor that the single greatest failing of the pre-reorganized Rad Lab had been the absence of the civilian equivalent of a chief of staff (see figures 4.1, 4.2, and 4.3). Furthermore, some of the cryogenic hardware and much of the engineering expertise came through military connections, primarily via the Atomic Energy Commission–National Bureau of Standards (NBS) cryogenic laboratory in Boulder, Colorado. This laboratory's first and primary task had been to provide large quantities of liquid hydrogen for the hydrogen bomb project. Toward this end, the laboratory produced two identical hydrogen compressors: one to be left in Boulder for cryogenic experimentation and one to be shipped to the Pacific atolls for the Eniwetok "wet" (i.e., liquid) hydrogen bomb test.[133] With the equipment at Boulder, engineers and physicists explored the convenience and safety of storing and handling the low-temperature liquids.[134] For example, stainless steel, high-nickel steels, and aluminum alloys were tested down to 20°K for tensile and fatigue strength. Other programs sought to explore high-vacuum techniques and methods of transferring the liquefied gases.[135] Yet another group of workers began to study the thermodynamics of liquefied gases; enthalpy, entropy, and the Gibbs function all needed to be known to provide the large-scale refrigeration and production of the cryogenic liquids. None of these had been studied adequately over the wide range of conditions that were required.

Even before the detonation of the first lithium-6 deuteride H-bomb (a solid fusionable fuel which obviated the use of liquid hydrogen), the NBS laboratory's primary mission was finished, and the staff turned its attention to other applications.[136] Liquid oxygen was needed for propulsion in rocket testing, and portable cryogenic systems were required for various other air force airplane and missile uses, such as aerodynamic control surfaces, landing gear, communications equipment, cabin pressurization, and trajectory control.[137] None of these other projects had the priority of the H-bomb, and by 1955, the H-bomb had no further use for the massive production and control of liquid hydrogen. The physicists and engineers who by then had become adept at handling this difficult substance could turn to other tasks. But whether its final destination was to be pacific or not, liquid hydrogen presented dangers that haunted every aspect of the experimenter's work life.

132. Alvarez, "Recent Developments" (1972), 258–59.

133. Donald Mann, interview by the author, 15 December 1983.

134. Brickwedde, "A Few Remarks" (1960), 1–4.

135. "Research Facilities," in Timmerhaus, *Advances* (1960), 1:10.

136. Donald Mann, interview by the author, 15 December 1983.

137. Lieberman, "E.R.E.T.S." (1960), 2:225–42; Hohmann and Patterson, "Cryogenic Systems" (1960), 4:184–95.

5.5 The Meaning of Laboratory Danger

In the spring of 1955, the Berkeley bubble chamber group began consulting with the experts in cryogenics at Boulder. Dudley Chelton, Bascom Birmingham, and Donald Mann all traveled to Berkeley from Boulder to apply their engineering skills to the rapidly expanding needs of the particle physicists.[138] Ordinarily, safety issues would arise in discussions among Alvarez, Gow, and Alvarez's trusted lieutenant, the mechanical engineer Paul Hernandez; these would then be addressed in detail by Hernandez working with the NBS personnel. Some typical problems: How rapidly could the glass windows be cooled without fracturing? How often could they be recooled without breaking? Should excess hydrogen gas be vented or burned? These questions became significant during the construction of the 10-inch chamber and remained so for all larger devices. Other joint testing projects included investigating the boiling, burning, and detonation of hydrogen (see figure 5.11).[139] To cope with these threats, Hernandez structured the engineering defense in three categories: blocking one failure from causing another, separating air from hydrogen, and designing equipment to withstand internal explosions.[140]

It is essential to understand just how frightening these large quantities of liquid hydrogen were. The danger affected the decision of some laboratories not to build hydrogen chambers at all, and in those laboratories that did proceed the threat of explosion played a crucial role in the partitioning of work among engineers and physicists, as well as in the daily routine and affect of experimentation: it is part of understanding what it meant to be an experimenter to know that standing in front of the chamber window felt like "looking down the barrel of a cannon." Worries about a possible catastrophe stemmed from many sources. Many participants in the bubble chamber work remembered the May 1937 Hindenburg disaster,[141] and more proximately, there was the dramatic but far less destructive accident in the operation of the 10-inch chamber, as Hernandez reported in 1956: "The 180 psig [pounds per square inch, gauge] rupture disc between the expansion line and the vent system failed causing a sudden drop of pressure in the bubble chamber. . . . About 4 liters of the 8 liters [of hydrogen] flashed almost instantaneously. The gas is believed to have reached supersonic speed for a part of the time as it traveled through the vent pipe. The gas literally screamed traveling down the vent line. When the hydrogen gas reached the flame it burned with a big 'whoosh,' estimated less than a second, with a flame esti-

138. Donald Mann, interview by the author, 15 December 1983.

139. Hernandez, "Safety" (1960), 2:336–50; Chelton, Mann, and Hernandez, UCRL Engineering Note 4311–14, file M33, cited in Hernandez, "Safety" (1960), 2:344.

140. Hernandez, "Safety" (1960), 2:346–49.

141. Donald Mann, interview by the author, 15 December 1983; Alvarez, interview by the author, 7 March 1983.

Figure 5.11 Explosion testing at NBS. Many of the components for the 72-inch hydrogen chamber had to be tested for behavior under conditions of high pressure and low temperature. Here, behind a concrete blast wall, an engineer at the National Bureau of Standards in Boulder, Colorado, tests a Joule-Thomson heat exchanger. Source: Courtesy of Donald Mann.

mated from 10 to 20 feet high."[142] Fortunately, in this case the venting system worked and no one was hurt.

Considerably less fortunate were those working on the 40-inch liquid hydrogen chamber at the Cambridge Electron Accelerator (CEA) in the early morning hours of Monday 5 July 1965. One of the bubble chamber windows failed. Hydrogen detonated and burned, killing one man, injuring six others, and causing over $1 million worth of damage.[143] For the high energy physics community as a whole, the tragedy also contained a final warning, justified or not, against any attempt to import the unfettered work style of small-scale physics into the dangerous, quasi-industrial laboratory that was coming to dominate the new era. In the place of a personal laboratory, where tinkering and individual authorship

142. Hernandez, "Safety" (1960), 2:341.
143. Livingston, "Semi-Annual Report," CEAL-1031, 13 July 1966, 2–3, CEAP.

à la Glaser was the norm, a shop floor more like Berkeley's had to be created. With the laboratory felt to be a dangerous place, new constraints fell into place about who could be there, what people could do, and how they could do it.

The CEA had its origins in the Harvard Cyclotron, a prewar device appropriated for use at Los Alamos, then replaced and refurbished in larger form at the war's end. Arguing that neither Brookhaven (as a regional center) nor the western accelerators could provide the needed access to physicists in the Boston area, the Harvard Nuclear Physics Committee proposed a 4–6 GeV alternating gradient synchrotron to be financed primarily by the federal government. Cambridge, the committee members contended, contained far and away the largest community of physicists with "neither a meson-producing cyclotron nor a billion-volt electron accelerator." Making use of the strong focusing principle introduced by Ernest Courant, Milton Stanley Livingston, and Hartland Snyder, accelerating electrons to these energies appeared to be well within the domain of technical possibility.[144] From early in the construction process tensions arose between the AEC's "industrial" demands and the proclivity of the Harvard and MIT physicists to work in an older, more independent research style. Already in March 1957, Livingston blasted the AEC's New York office for encumbering him and his staff with accounting and planning formalities he thought unnecessary and burdensome ("arbitrary bureaucratic limitations"): "We cannot escape the conclusion that your reporting and authorization requirements are based on 'industrial' type construction and cost accounting procedures which we consider completely inappropriate. We have done our best to inform your staff of the reasons why such procedures do not represent the most effective and economical methods for administration of the design and assembly of a research-type instrument such as the CEA accelerator."[145]

More than accounting was at stake. Livingston protested that he would need to allow for "continuous modification of design of certain . . . components." Prototypes and models would need to be produced and altered as the team progressed—so too would new materials, subassemblies, and even the laboratory and administrative facilities of the institution. Above all, Livingston insisted,

> [a] great deal of the success attendant upon this research-development-type procedure is due to the flexibility in use of both personnel and equipment, which must not segregate then into rigid ruts of production-type operations. . . . Further pressure from the AEC to force us to modify our internal procedures would greatly

144. See "Proposal for Strong-Focussing Synchroton," Joint Accelerator Committee (Livingston, Ramsey, Street, and Zacharias), 6 October 1952, and "Recommendations for a Five Billion Volt Electron Accelerator," Harvard Nuclear Physics Committee (Ramsey, Bainbridge, Pound, Hickman, Purcell, Preston, Schwinger, and Street), 7 June 1954, CEAP.

145. Livingston to Eisenbud, manager, New York Operations Office, AEC, 5 March 1957, CEAP.

> handicap our progress. We are confident that, if your staff people had expended as much time and effort in the direction of finding out how we do operate as they have spent telling us how we should operate, they would have observed that our normal procedures provide completely adequate controls and safeguards.[146]

For Livingston the twin moments of control—control over the experiment as a part of the ethos of research and control as a way of work life—were thoroughly intertwined. "Industrial"-style work would (Livingston contended) threaten the success of the inquiry into the subatomic world.

Not surprisingly, the friction over independence from AEC regulations continued into the operation of the accelerator. In particular, the AEC wanted safety procedures, manuals, and contingency plans, features by then standard at Berkeley. The CEA, by contrast, insisted on a philosophy that left the senior investigators in charge, with some supplemental help from the CEA staff. In practical terms, this meant that there were surveys of fire protection, industrial safety, and radiation and cryogenics safety, but that the whole program was never formally or centrally structured. Unlike Berkeley's Radiation Laboratory, for example, there was never a single individual or committee in charge of the whole facility. Individual researchers never had to submit their equipment for review by the CEA staff. And while the staff would raise issues about hazards on the laboratory floor at the weekly meetings, such reviews were informal.[147]

For several more years, the CEA and AEC vied over the type of safety procedures and the details of the safety program. By mid-1964, the New York office seemed resigned to the less formal style of the CEA. As one internal report stated: "CEA policy is opposed to the issuance of organization charts. Safety responsibility and lines of authority appear to have been assigned but are not committed to writing."[148] One improvement seemed to be the adoption of the requirement that all but very specialized equipment used by visiting scientists would be fabricated by laboratory staff, thus ensuring better safety control. But at the same time the increasing complexity and number of experiments operating on the floor was making control more difficult. In their last report before the accident, the New York AEC office recorded internally on 30 March 1965 that "[a]lthough the appraisal team could hope for a formal delineation of safety responsibility and an organization chart, this is a concept which is abhorrent to the administration [of CEA], and in view of the fact that the safety objectives are

146. Livingston to Eisenbud, manager, New York Operations Office, AEC, 5 March 1957, CEAP.

147. Wasser to Cummings, 20 December 1960, cited in AEC, New York Operations Office, "Investigation of Explosion and Fire, Experimental Hall, Cambridge Electron Accelerator, Cambridge, Massachusetts, July 5, 1965," typescript [1966], 10–11, CEAP.

148. Bootman and Weintraub to Breslin, 19 June 1964, cited in AEC, New York Operations Office, "Investigation of Explosion and Fire, Experimental Hall, Cambridge Electron Accelerator, Cambridge, Massachusetts, July 5, 1965," typescript [1966], 12, CEAP.

achieved by their own less formal organization, there is no point in pressing this issue."[149]

Several days before the explosion, liquid hydrogen began to condense in the bubble chamber for the first time at 2 A.M. on 2 July 1965. For the next several days the operation continued around the clock. By the end of the second shift on the fourth of July, the chamber was 80% full, and the crew began to circulate gaseous hydrogen to further cool the inner parts of the chamber and in order to speed the filling. All seemed well. Quickening its flow, the liquid hydrogen continued to pour into the new detector. At 1:15 A.M. on 5 July, the crew had the chamber condensing hydrogen at its design rate. J. Szymanski, an assistant professor at Harvard, gave a statement of what followed:

> At 3 A.M. the hourly log was taken and all was normal. . . . I estimate it was about 3:15 A.M. when the fill rate started to decrease slowly and again leveled off. . . . I then again went to the purifier to adjust the bleed rate. When I returned to the control panel, which was only several seconds later, namely the time it takes to walk from the purifier to the control panel, I noticed a high flow rate (~12 CFM [cubic feet per minute]) through the flow meter. This was entirely unexpected. I quickly noticed that for the first time during our entire cooldown and fill period the hydrogen fill pressure (and consequently Bubble Chamber pressure) was rising for unexplained reasons. . . . I immediately shut off the fill valve . . . to the Bubble Chamber while reading the gauge indicating the fill pressure. The fill pressure was still rising and reached a value of 80 PSI. I did not consider 80 PSI as extremely serious at that instant since all the peripheral systems are capable of easily handling such a pressure. At this point I turned to check the pressure in the Bubble Chamber to make sure that it was not rising excessively. I never did see the Bubble Chamber pressure gauge.[150]

From witnesses and the evidence of debris, the accident investigation team reconstructed the events that followed. Deep within the bubble chamber, the inner beryllium window had shattered along a microscopic imperfection in its surface. Splintering outward, the inner window fragments blasted open the outer beryllium window accompanied by the pressure wave of the expanding hydrogen (see figure 5.12).[151] Within half a second, the laboratory floor was bathed

149. Weintraub and Glauberman to Rizzo, 30 March 1965, cited in AEC, New York Operations Office, "Investigation of Explosion and Fire, Experimental Hall, Cambridge Electron Accelerator, Cambridge, Massachusetts, July 5, 1965," typescript [1966], 12, CEAP.

150. Statement by J. Szymanski, 5 July 1965, cited in AEC, New York Operations Office, "Investigation of Explosion and Fire, Experimental Hall, Cambridge Electron Accelerator, Cambridge, Massachusetts, July 5, 1965," typescript [1966], 21–22, CEAP.

151. AEC, New York Operations Office, "Investigation of Explosion and Fire, Experimental Hall, Cambridge Electron Accelerator, Cambridge, Massachusetts, July 5, 1965," typescript [1966], CEAP; on the sequence of window failure, see the calculations by Copland and Croeni, Research Physical Metallurgists, Albany Metallurgy Research Center, U.S. Bureau of Mines, 52–55, CEAP; and Borch, Process and Materials Development Division, Lawrence Radiation Laboratory, University of California, 68–69, CEAP.

Figure 5.12 Ruptured beryllium windows, CEA (1965). One of the perennial fears in bubble chamber design had been that the windows would fail under the constant stresses of thermal and mechanical shock. According to the AEC accident report, a likely point of origin for the catastrophic CEA accident of July 1965 lay in imperfections on the surface of the beryllium lens. Source: U.S. Atomic Energy Commission, "Investigation" (1966), fig. 17, HUA.

with some 400 liters of turbulent, burning hydrogen. Ignited when the outer window failed, the fire burned wherever the hydrogen and air were mixed. Seconds later, a fierce explosion ripped through the laboratory, strong enough to blow the 31,000 square foot laboratory roof 10 feet into the air. As it crashed back down, roof material cascaded onto the floor and began to burn, raining down hot tar. Now other areas erupted in flames as the soft soldered joints melted in the tubes that linked large quantities of liquid petroleum gas, as well as other combustibles. Worsening the fire, one of the building's vents was capped shut; others functioned poorly, concentrating gases within the building (see figure 5.13).[152]

152. For the accident model deemed most probable, see AEC, New York Operations Office, "Investigation of Explosion and Fire, Experimental Hall, Cambridge Electron Accelerator, Cambridge, Massachusetts, July 5, 1965," typescript [1966], published as TID-22594, appendix III, CEAP; and some details from the typescript that were later deleted in the published version. On hot tar from the roof, see 25; on vents, Hernandez, project engineer, LBL, see 75–77.

Figure 5.13 Outside view, CEA fire (1965). This picture, taken by a passerby, shows the CEA conflagration already fully engaged. Source: U.S. Atomic Energy Commission, "Investigation" (1966), fig. 14, HUA.

A. Goloskie, an MIT technician, was working at a bench west of the chamber. He was slammed down backward and saw orange and yellow flames, as well as a fireball near the ceiling; another fireball hovered at the height of the chamber. His eardrum damaged, he wandered, dazed, through the corridors until he was escorted out by less injured technicians from the accelerator staff.[153] One graduate student's clothes caught fire, and he sustained serious burns leaving him in critical condition. Others working in the laboratory received second-degree burns. Worst hit were a graduate student, F. L. Feinberg, whose clothes began burning, leaving him with third-degree burns over much of his body, and A. C. Reid, a 19-year-old MIT technician, who was burned over much of his body in addition to sustaining serious internal injuries. Fifteen days later, Reid died.

Winds in the vicinity of the chamber reached 300 miles per hour. Hoses, electrical lines, cables, and appurtenances on the bubble chamber went up in flames.[154] Firemen arrived at 3:38 A.M., two minutes after the first call. Eight minutes later, there was a second explosion, of longer duration than the first, probably involving other gases; everything ignited. The fire grew into a wave, rose to-

153. AEC, New York Operations Office, "Investigation of Explosion and Fire, Experimental Hall, Cambridge Electron Accelerator, Cambridge, Massachusetts, July 5, 1965," typescript [1966], 24, 27, CEAP.

154. AEC, New York Operations Office, "Investigation of Explosion and Fire, Experimental Hall, Cambridge Electron Accelerator, Cambridge, Massachusetts, July 5, 1965," typescript [1966], 27, CEAP.

Figure 5.14 Aftermath of CEA fire (1965). Source: LRL Bubble Chamber 1551, LBL. Courtesy of Lawrence Berkeley Laboratory, University of California.

ward the ceiling, and spread. One graduate student had managed to crawl into a space between the bubble chamber electronics room and the south wall. Unable to escape further because of his injuries, he remained there until the fire seemed to be closing in. Radioing an ambulance to the east exit, the deputy fire chief, an engineer, a cryogenics expert, and some firemen hacked their way to him and brought him out on a stretcher.[155]

An hour and a half after the explosion, the Cambridge Fire Department declared the fire under control. It took 17 pieces of fire apparatus. The damage estimate was $1.5 million (see figure 5.14). Every major laboratory in the United States—and many abroad—felt the impact of the explosion on their day-to-day procedures. Soon after the disaster, the AEC established a group to set out "Safety Guidelines for High Energy Accelerator Facilities," bringing together representatives from Los Alamos, the AEC, Brookhaven, Argonne, MIT,

155. On Huld's ordeal, see AEC, New York Operations Office, "Investigation of Explosion and Fire, Experimental Hall, Cambridge Electron Accelerator, Cambridge, Massachusetts, July 5, 1965," typescript [1966], 13, 24–25, 31–32, CEAP.

the CEA, the Radiation Laboratory, SLAC, and the Princeton-Pennsylvania Accelerator.[156]

From the first page of the AEC report, it was clear that the flexibility CEA had enjoyed was over. From then on, every laboratory supported by the AEC was expected to have a general safety committee reporting to the director, manuals of procedures, and periodic reviews. To my knowledge, no senior person in the ambit of the AEC ever again argued the way Livingston had that he needed the freedom from "industrial type construction" based on safety procedures to work in the more "flexible" style appropriate to "research." No one after 5 July 1965 could relegate "production" methods to the dustbin of "arbitrary bureaucratic limitations" or maintain the absolute prerogative to modify components continuously. Danger, refracted through the history of the time, sealed that debate.

Beyond plans and organizations, the explosion and ensuing guidelines stressed the changes in daily life that were to be established to promote safety. Such "daily procedures" were to be "applicable to all personnel, including experimenters and visitors." Industrial and standard laboratory practices needed to be adopted, even amplified, given the special dangers of ever-changing equipment. The procedures also applied to communication, electrical power, transfer of equipment, hazardous materials, hand tools, clothing, and monitoring equipment. Allocation of space had to be made according to rules designed to isolate flammables and suffocants. The AEC asked for control over who could enter the experimental areas: "[U]ncontrolled access to accelerator and experimental areas by the public is not only undesirable but is potentially dangerous." This was "men's work." "Wives and children of accelerator or experimental group personnel" were singled out, as was the public that had frequently and problematically been escorted on site.[157] "Proper housekeeping," the guide added, "is essential to the conduct of a safe experimental program. Housekeeping is the responsibility of everyone. Management has the responsibility for assuring the development and continuance of proper attitudes and attention with regard to this subject by laboratory personnel."[158] Over and over again one reads in the safety guides that constant training is necessary to establish and maintain a style of work appropriate to the new scope, complexity, and danger of science on the industrial scale.

Before 5 July 1965 there had been several models of how to live the life of physics, even within the constraints of bubble chambers and accelerators. One

156. National Accelerator Safety Committee, "Safety Guidelines for High Energy Accelerator Facilities," TID-23992 [1967], v, CEAP.

157. National Accelerator Safety Committee, "Safety Guidelines for High Energy Accelerator Facilities," TID-23992 [1967], 27–29, on 27, CEAP.

158. National Accelerator Safety Committee, "Safety Guidelines for High Energy Accelerator Facilities," TID-23992 [1967], 29, CEAP.

was avowedly patterned on industry, the military, and large-scale engineering. Berkeley embodied this style from the start. The other, exemplified by Livingston's insistence on the unique character of the research environment and his absolute resistance to "assembly line" and "industrial" scientific work, tried to continue with the more autonomous work style that had characterized smaller scale research. I do not mean to imply that it was inevitable that the small-scale experiment style writ large would encounter disaster. Nor would I argue that industrial-style structure offers an iron-clad assurance of safety. Rather, what needs to be emphasized historically is that given the CEA's insistent, explicit, and written determination *not* to work on industrial prototypes, and the equally vehement attempts by the AEC to enforce "production style" work, once the accident occurred the events were read as an object lesson in the price of deviating from centralized, industrialized workplace.

Accidents in complex systems rarely, if ever, lead to unambiguous etiological accounts. Catastrophic technological failures, of airplanes, ships, spacecraft, or bridges, can be (and are) glossed in a myriad of ways: as stories about material failure, as stories about organizational failings, as stories of individual lapses or malfeasance, or as stories of a flawed cultural ethos. In the case of the CEA failure one detects many of these: some pointed an accusing finger at the designer of the window, at the properties of the window itself, at the head of the laboratory, or at lack of supervision of the personnel filling the chamber. In principle, the overdetermined nature of any accident leaves explanation open, but what strikes me above all in the ascription of meaning to this tragedy is that meaning was primarily vested in the ethos of the laboratory.

Danger, as Mary Douglas has argued, is handled through the imposition of ritual.[159] Safe procedures, enforced by checklists and dictating behavior, would enforce "proper housekeeping," and enforced orderliness would breed habits that would secure against the fearful fate of anyone near a ruptured tank of liquid hydrogen. For Douglas, a highly structured social system *creates* categories of danger in order to enforce lines of authority. In high energy physics laboratories like the CEA, danger (or rather the response to it) was in part constitutive of the highly ordered social structure that became characteristic of bubble chamber groups. It had already been so in the work space of Berkeley and the NBS; with the CEA detonation, that way of work became normative. The meaning of laboratory danger crystallized an image of an orderly industrial work space, and by doing so helped stamp out an older space and style of experimentation.

159. Douglas, *Purity and Danger* (1966), e.g., 94ff. Douglas emphasizes the association of power and danger, a theme that can be tracked through the physicists' experience of World War II and its aftermath. It goes without saying that the production of war matériel brought physicists into the corridors of power. More subtly, the dangers of the laboratory itself created new reasons for restricting access to the laboratory and casting the experience of being an experimenter in a new light. Contrast this modality of exclusion, e.g., with exclusion based on questions of reliable (gentlemanly) witnesses of seventeenth-century Gresham House. See Shapin, "Experiment," *Isis* 79 (1988): 373–404.

The drive to industrialization, it appears, came from several sides at once. From the example of the war projects and the civilian scientific enterprises that grew out of them, new laboratories constructed a positive guiding image. In the example of the conflagration at the CEA, new laboratories and research groups saw what was broadly acknowledged as a cautionary tale of what might happen in the absence of industrial training and organization that limited and structured the prerogatives of the individual investigator.

At Berkeley (even before the CEA disaster), the increasing complexity and danger inherent in the large bubble chambers led to the initiation of check-list procedures for the laboratory floor.[160] Such routinization "eliminat[ed] operating mistakes and improv[ed] the safety." It also gave "all crew members the same objectives and permit[ted] them to anticipate the next step of the operation." "[C]onfusion," Hernandez concluded, was brought down "to a lower level because routine operating decisions [had] been eliminated."[161] Typical safety checkoff entries were the following: "Barriers and warning tape installed definitely outlining danger area. Explosion-proof flashlights in area. Personnel wearing conductive shoes. Liquid nitrogen jacket filled well in advance to reduce thermal shock upon introduction of liquid hydrogen. Continuous explosimeter in proper operating condition."[162] As Hernandez put it to the Pittsburgh Naval Reactor Office, the "check-off approach" is "used in the sense of an attitude between safety personnel and the user."[163] After the CEA disaster, industrial regulations like these soon spread around the world; an often intrusive policy sought to alter that elusive quality "attitude"[164] and with it the experimenter's "personality (in physics)"—as Hofstadter had put it a decade before. Physical danger may not seem to have a place in the ethereal realm of epistemology. To believe that would be a mistake. The threat posed by these vats of hydrogen loomed large to practitioners of the 1950s and 1960s, and it reshaped the material culture surrounding experimental physics.

5.6 Expansion

By the time that Berkeley Radiation Laboratory was ready to inaugurate its 10-inch hydrogen chamber, the scale, tone, and orientation of the project could not have contrasted more with what Glaser had wanted. Whereas Glaser sought to save cosmic ray physics, Alvarez wanted to beat it. On the opening page of

160. UCRL Engineering Note 4311-17, file M6, cited in Hernandez, "Safety" (1960), 2:348.

161. Hernandez, "Safety" (1960), 2:348.

162. "Check-off List," 4 November 1955, used for operation of 10-inch chamber, RL-1353-5, LBL.

163. Hernandez to Earhart, Pittsburgh Naval Reactor Office, 19 October 1967, box 1, book 117, HHP.

164. One mode of transfer was through the AEC itself, a process that began at least as far back as 20 October 1958, when Hernandez answered a request from the AEC for internal documentation of reports on hydrogen safety. See Hernandez to Weinstein, 20 October 1958, which included a discussion of burning, venting, and explosion questions for chambers including the planned 72-inch detector, box 1, book 114, HHP.

his April 1955 proposal for a new, very large bubble chamber (which was to culminate in the 72-inch chamber), Alvarez said bluntly that "the accelerator physicist is often like the hare, in competition with the tortoise-like cosmic ray physicist."[165] As long as the accelerator physicist was limited using variants of the cosmic ray physicist's detectors (cloud chambers, emulsions, and counters), the contest was more or less equal. Even though the accelerator produces a much higher flux density than the cosmos, the extra particles could not be detected. From Alvarez's perspective, only fast scintillation and Cerenkov counters could keep up with the accelerators, and they had very poor track resolution. (Not everyone agreed with Alvarez—important discoveries such as that of the antiproton led several Berkeley physicists to defend the counter tradition staunchly.) By pushing the cloud chamber to high pressures, the accelerator physicists could get about 50 times more tracks than the cosmic ray physicists could even on mountain peaks. Since the cosmic ray particles have a higher energy maximum and thus interact more, the accelerator physicists actually held only a tenfold advantage, however. "This is an important increase," Alvarez asserted, "but not enough to justify much of the expenditure of ten million dollars, which went into the Bevatron."[166] Such a justification, Alvarez continues, could be based only on a fast-recycling high-density detection device such as the bubble chamber. Cloud chambers could, by 1955, be made to operate at pressures of over 30 atmospheres,[167] but the bubble chamber could perform at a density that would be the equivalent of 700–1,000 atmospheres. Since the number of events per length of traversed chamber is proportional to the density, this difference represented an enormous improvement. Even better, instead of cycling every 15 minutes (the characteristic time for a high-pressure cloud chamber), the bubble chamber was ready in five seconds. In several ways, the Bevatron therefore called for a new detection device, and the bubble chamber fit the bill. Physicists too played a role in shaping the plans for making a larger bubble chamber, primarily through Alvarez's consideration of the reaction[168]

$$\begin{array}{llll} \pi^- p \rightarrow \Lambda^0 & + & \Theta^0 & \\ \quad\;\; \hookrightarrow p + \pi^- & & \hookrightarrow \pi^+ + \pi^- & \end{array} \tag{5.5}$$

(The identification $\Theta^0 \equiv K^0 \equiv \tau$ would not be made until the downfall of parity in 1957.)[169] Since the Θ^0 and Λ^0 are neutral, they do not leave tracks in the chamber. In order for them to be studied, the active volume must be large enough for

165. Alvarez, "The Bubble Chamber Program at UCRL," 18 April 1955, 1, LAP.

166. Alvarez, "The Bubble Chamber Program at UCRL," 18 April 1955, 4, LAP.

167. E.g., Elliott et al., "Thirty-Six," *Rev. Sci. Inst.* 26 (1955): 696–97; the authors are Wilson Powell's group at Berkeley.

168. The above and following discussions are based on Alvarez, "The Bubble Chamber Program at UCRL," 18 April 1955, 10–12, LAP.

169. For an excellent history of this episode, see Franklin, "Discovery and Nondiscovery," *Stud. Hist. Phil. Sci.* 10 (1979): 201–57.

the neutral mesons to decay and for their decay products to leave measurable tracks. How large a chamber would one need to observe this double decay pattern? In the center-of-mass frame the π^- and the proton have equal and opposite momenta, so the outgoing Θ^0 and Λ^0 must have equal values of momentum. Solving for the momentum p is easy using energy conservation; knowing p, we deduce immediately that the speed of the Θ^0 will be twice that of the Λ^0, since the Λ^0 is almost twice as heavy as the Θ^0.

Since the Λ^0 lives twice as long as the Θ^0, however, the two particles travel approximately equal distances (in the center-of-mass frame) before decaying. For different π^- energies, one can plot the distance traveled before a typical particle will decay. In the center-of-mass frame, this will give a circle whose radius will grow for higher initial energy. The inverse Lorentz transformation that takes us back to the laboratory frame from the center-of-mass frame expands the axis along which the pion is traveling, so that the circle becomes an ellipse. Plots of the ellipses along which Θ^0 and Λ^0 will decay (figure 5.15) were the prize exhibit of the Alvarez proposal. Adding some extra volume of hydrogen to serve as a target before the decay, some to observe the decay products, and some to accommodate the finite size of the beam itself, one had (plus or minus a little hand waving) 50 × 20 × 20 inches. "This," Alvarez conceded, "is such a large extrapolation from the present 4-inch diameter chamber, that it would be quite foolhardy to make the jump in one stage. We therefore have well under construction a chamber 10 inches in diameter."[170] Without much ado, the length was soon expanded to 72 inches on the basis of the greater interaction lengths of other strange particles.[171]

When Alvarez voiced concern that the Bevatron was not being fully exploited, his words apparently fell on receptive ears. George Kolstad (chief of the Physics and Mathematics Branch of the AEC Division of Research) drafted a June 1955 memorandum to T. H. Johnson (director of the AEC Division of Research) arguing for more money: "With the increased emphasis on peacetime applications of atomic energy, the Russian publications and the great interest of U.S. physicists in this field, it would appear that a substantial increase in funds for high energy physics should be provided as soon as possible."[172] Kolstad had in mind $750,000 for bubble chamber equipment at the Radiation Laboratory and nearly another $2 million for associated facilities. Principal among the expenses were some $200,000 for a power supply, $370,000 for the chamber, and $150,000 for engineering. Some funds came directly through the AEC, others by transfer to the NBS facility in Boulder. By the time a reexamination

170. Alvarez, "The Bubble Chamber Program at UCRL," 18 April 1955, 9, LAP.

171. Alvarez, "Liquid Hydrogen Bubble Chambers" [1956], UCRL-3367, 4, LAP.

172. Kolstad to Johnson, memorandum, "Expanded Operating Costs for High Energy Physics," 7 June 1955, LAP. Also see letters in DC files numbered DC 55-180, 183, 226, 227, 530, 744, LBL.

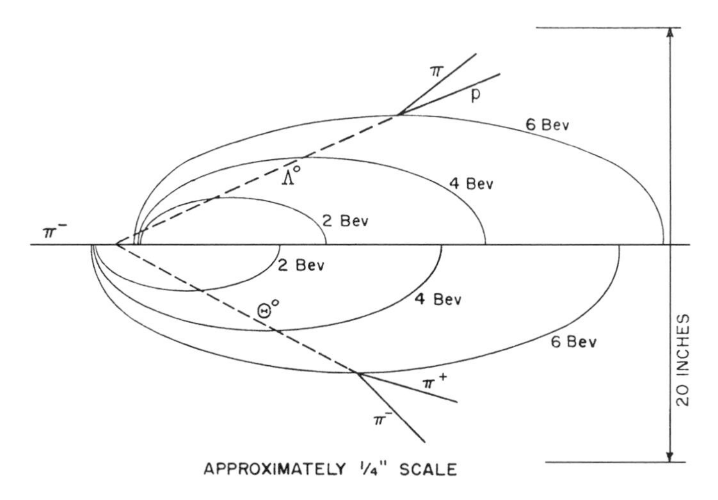

Figure 5.15 Decay ellipse, physics motivation for 72-inch chamber (1955). Produced in a π^--proton collision, the Λ^0 and the θ^0 decay by the weak interactions over a length that depends on their energy. The ellipses mark the typical point to which a Λ^0 or a θ of given energy (in billion electron volts) will travel; the dashed lines give the mean angle from the beam (horizontal axis). This diagram summarized the physics justification for the size of the big chamber. Soon, however, the physicists working with the chamber turned their attention much more to the strong than to the weak interactions. Source: Alvarez, "The Bubble Chamber Program at UCRL," typescript, 18 April 1955, LAP. Courtesy of Lawrence Berkeley Laboratory, University of California.

of these figures had been undertaken in late 1955, the cost of building the monster bubble chamber had increased to $1.25 million.[173] Safety features, data-processing equipment, and the magnet were to blame.

With these additional funds, construction of the 72-inch chamber advanced furiously; the laboratory first produced pictures in the early months of 1959 (see figure 5.16). Dollars poured into the project and physics results poured out, both at an unprecedented rate. To Alvarez's competitors, both outside LRL and within it, the ratio of productivity to dollars expended became a controversial issue. Provoked in particular by comments from Robert L. Thornton, also at LRL, Alvarez responded with a singular piece of sociophysics. In recent conversations, Alvarez asserted, "[y]ou have made statements of the following nature: that it is certainly true that our group does a majority of the high energy physics in the laboratory, but this isn't surprising in view of the fact that we are the largest group

173. Cooksey to Johnson, 21 November 1955, DC 55-1313, EMP. My thanks to R. Seidel for pointing this item out to me.

Figure 5.16 Seventy-two–inch LRL bubble chamber (1959). Visible here are the hydraulic feet used for moving the device, the magnets (*left-hand side, just above lower walkway*), and the instrumentation for refrigeration (*upper walkway*). The actual chamber is obscured from view, and the compressors that pumped in pressurized hydrogen gas were located in a separate room. Operation began in March 1959. Source: LRL Bubble Chamber 720, LBL. Courtesy of Lawrence Berkeley Laboratory, University of California.

and spend the most money. You often conclude such discussions by suggesting that other groups get more physics done per dollar expended, or per person in the group." Striking back, Alvarez asserted that nothing of the sort was true, as one could see by counting references over the last 27 months (1 July 1960–

Table 5.1
Alvarez's Dollars per Reference

Physics Group	Total References	References per Ph.D.	Research Budget for Past 27 Months	Dollars per Reference
Alvarez	215.5	10.75	*5,280,000	24,500
Segrè-Chamberlain	3	.27	2,432,000	816,667
Moyer	23	2.3	2,379,000	103,435
Powell-Birge	25.5	3.64	***1,575,000	62,000
Lofgren	31.5	3.93	1,410,000	44,760
Thornton (Crowe)	21	5.25	1,323,300	63,000
Trilling-Goldhaber	19.5	3.9	**1,005,000	51,540
Barkas	4	1.3	692,500	173,000

Source: Luis W. Alvarez to R. L. Thornton, 7 November 1962, appended, EMP.
*Includes 75% of the 72-inch bubble chamber operating budget plus 100% of the 15-inch operating budget for the period of interest ($1,250,000).
**Includes the transfer of funds (estimate) from the Segrè and Lofgren groups to the Trilling-Goldhaber group ($280,000).
***Includes 35% of the operating budget for the 30-inch propane chamber, which was $60,000 for this period.

1 October 1962). Counting publications alone might measure a group's own feelings about its work. "But references to group work have the built in evaluation factor, as well as a reasonable relaxation time." Older work of continuing importance would continue to be cited in the target period, whereas work of "lesser importance" would "relax" into obscurity more quickly. Collaborative work with other institutions would be credited on a prorated scheme, and references to software would be omitted (which penalized the Alvarez group more than any other group). Data on manpower and costs would come from Thornton's own files or those of the centralized budget office, and Bill Nolan from the budget office would resolve any financial or personnel questions in ways that would aid other groups and be counted to the detriment of the Alvarez group. Alvarez then attached his data (table 5.1) to argue that the $5.25 million spent by his group between 1960 and 1962 had yielded a per citation cost of a mere $24,500 compared with the $63,000 cost of a citation to Thornton's group—and the $816,667 dispensed to get a citation for the Segrè-Chamberlain group. Playing on the work itself, Alvarez taunted his competitors to come by his office to examine the 150 cards that carried the data—anyone could "make spot checks on them for 'scanning efficiency.'"[174]

As this struggle over LBL funding indicated, competition between individuals for resources was now repositioned as a tough struggle between groups—identified by their leaders' names—for vastly larger amounts of money. Everyone could recognize the identity of these new entities: the Alvarez group, the

174. Alvarez to Thornton, 7 November 1962, EMP.

Segrè-Chamberlain group, the Moyer group, the Powell-Birge group, the Lofgren group, the Thornton (Crowe) group, the Trilling-Goldhaber group, and the Barkas group. But with the Alvarez budget nearly four times that of its nearest competitor and references in *Physical Review Letters* occurring at a rate of nearly 100 per year, there was little question that this group alone had come to dominate physics at the laboratory and beyond. Across the country and overseas, it was the Alvarez laboratory more than any other that garnered attention through its domination of elementary particle physics and its new life of experiment.

Spreading the LRL approach to bubble chamber physics took several forms. Physicists of course gave lectures and corresponded with a wide variety of other colleagues, but one of the salient features of the LRL program was that engineers began their own network of links between laboratories, and a great deal of the information transfer between bubble chamber programs passed through the less visible stratum of engineers. One Oxford physicist, Dennis Shaw, wrote Hernandez to ask how quality control was imposed on film as it emerged from the chamber. To this query, Hernandez responded with a brief outline of LRL practices: processing the film immediately on removal from the chamber, scanning immediately upon drying, and instantaneous feedback on chamber operation via closed circuit television, augmented by pictures put directly onto Polaroid film.[175] Other questions had a more detailed and specific character, such as these, from 1961, about the all-important inflatable seals that bound glass to metal: "[I]s the gasket-chamber seal different from the gasket-glass seal?" "Have any materials other than indium and lead been tried for the actual seal?" "What is the relationship between inflation pressure and sealing force per linear inch?"[176] To questions of this sort, written responses sufficed: "The relationship between the inflation pressure and our sealing force is based on the glass compressive stress limit. We used a design force of about 1,400 pounds per lineal inch on the 72-inch gasket and actually operate with a force of about 900 pounds per lineal inch." To other queries, such as a list of questions from CERN in 1960, the engineers answered with pictures rather than words: "Drawing 7Q1784C should answer your requests (a) and (b). The hydraulic system panel 7Q1833 is also included." Supply information, such as sources (the hydraulic pump and sump tank from the Rucker Company in Emeryville, California, to give one example), was often as valuable as bits of physics or engineering.[177]

Perhaps the most important mode of information transfer was the engineering note, which left the laboratory in enormous quantity. In 1957, Pierre Amiot at

175. Shaw to Hernandez, 5 November 1962; and Hernandez to Shaw, 12 November 1962, both from box 1, book 113, HHP.

176. McKenzie to Hernandez, 2 June 1961, box 1, book 113, HHP.

177. Hernandez to Ackermann, CERN, 28 March 1960, box 1, book 113, HHP.

CERN was sent 38 drawings, 150 engineering notes, 49 bubble chamber pictures, and 7 pictures from the propane chamber, supplemented by 22 graphs, booklets, bulletins, catalogs, and books. Two years later, CERN requested and received over a hundred bubble chamber engineering notes, in addition to 14 drawings and 15 photographs.[178] Others of the tens of laboratories that received massive bundles of these documents included the Imperial College of Science and Technology (London), the Lab. v. Massaspectrografie (Amsterdam), and the Nuclear Physics Research Laboratory (Liverpool).[179] Still, these representational forms—photographs, letters, engineering notes, and drawings—were highly reduced from the full context in which they were originally used. Ultimately, the form of work that the laboratory had developed could not be conveyed simply in print or charts, and it is for this reason, at least in part, that physicists and engineers flocked to Berkeley to observe at first hand the mode of construction and design that had become the hallmark of Radiation Laboratory design.

Percy Bowles, chief engineer at the Rutherford High Energy Laboratory in Harwell, wrote to Berkeley reporting that a number of his scientific staff had already visited the laboratory, but that this had not been enough. "I feel it would be of most value if I made my visit complementary to theirs by looking more at the organisational side of the laboratory with particular reference to the engineering support that is given to the projects." Above all, Bowles avowed, "I believe you have a great ability for knowing when to be exact and when to be approximate."[180] "Good enough engineering," as the locals referred to it, was characteristic of Lawrence's laboratory and, more generally, of American postwar physics. This industrial pragmatism counted more in big physics than it had in the craft world of blowing cloud chambers, where *Fingerspitzengefühl* had been the key.[181] Unlike so much of the bubble chamber's technological scientific core, which could be and was transmitted by engineering note, this knowledge of when to cut refinement short demanded immersion in the Berkeley subculture. All the subtle shifts in planning, building, and scaling up that had made "radar philosophy" and the "Los Alamos man" had come to the heartland of experimentation; Berkeley had become a powerful factory of physics.

178. List for Pierre Amiot at CERN, HHP; and Hernandez to Vilain, CERN, 29 June 1959, box 1, book 113, HHP.

179. Hernandez to C. C. Butler (London), 14 May 1957; Hernandez to J. Kistemaker (Amsterdam), 18 July 1955; Hernandez to W. H. Evans (Liverpool), 18 December 1957, all from box 1, book 113, HHP.

180. Bowles to Gordon, LBL, 11 May 1962, box 1, book 113, HHP.

181. The locus classicus for discussions of face-to-face transfer of knowledge is in Polanyi's *Personal Knowledge* (1958), e.g., 206–8, and *Knowing and Being* (1969), e.g., 142–48, work that has been given a sociological orientation in the influential writings of Collins, especially on the TEA laser: Collins, *Changing Order* (1985), chap. 3; and Shapin and Schaffer, *Leviathan* (1985). These issues will be taken up in greater depth in chapter 6.

PART II: Building Data

5.7 Data and Reading Regimes

From 1954 to 1968, the material culture of particle physics was wrenched into an entirely new configuration. The hardware and its legions of cryogenic, mechanical, and structural engineers permanently altered the laboratory landscape and shaped the life that developed inside this new environment. From physicists' daily submission to safety checklists to their long-term isolation from the shop floor, they had to come to terms with a new concept of what it meant to be an experimenter. The infiltration of loudspeakers, control rooms, and engineering into the inner laboratory was, however, just the first of two stunning blows to the autonomous self-image of the experimenter. For as the new bubble chamber factories began to spew kilometers of 70 mm film, the advances in understanding symmetry principles and nuclear forces came at an increasing cost. Already inflated by construction outlays, the price of doing particle physics surged ever higher as "armies" of technicians and scanners joined physicists in sorting through staggering quantities of data. But the burden of photographic plenitude was not simply economic: physicists struggled to define the right relation between scientist and data, between physicist and technician, and between human and machine. At stake, the physicists argued, was not simply a matter of new technologies: each solution to the picture problem immediately joined a continuing argument about the boundary of what it meant to be a physicist, a scientist, and a human being.[182]

All of these concerns were embodied in a novel class of devices designed to "read" photographs, often bearing names like "Rapid Reader" or "Spiral Reader." It is to these "reading machines" that this section is devoted. The nomenclature itself forces a twist on Michael Mahoney's important study, "Reading Machines."[183] For in the case at hand we must work in two directions simultaneously: we must "read" the machines for the cultural assumptions built into them, and simultaneously, we must reconsider the nature of reading implied by the textualization of photographs. In the gerundive usage of "reading," we need to parse the use of the machines, both hardware and software. In the adjectival sense of "reading" machines, we must ask what stands behind the project of treat-

182. See Hermann et al., *CERN II* (1990), esp. chaps. 6, 8, 9; chap. 9, subsec. 6, includes the best history of CERN's efforts in developing picture-handling facilities. I have profited greatly from this and from Krige's preliminary version in the preprint CERN History Series, CHS-20.

183. As far as I know, the first systematic exploration of the concept of reading machines as text is in an unpublished paper of Michael Mahoney's, "Reading A Machine" (1983); more recent relevant work on this subject can be found in Carlson and Gorman, "Understanding Invention," *Soc. Stud. Sci.* 20 (1990): 387–430, which focuses on the inventor's use of mental models in the construction of new devices. Also of interest in the "reading" of images are the essays in Lynch and Woolgar, *Representation* (1990).

ing photographic evidence as text. What do the builders (are they authors?) of these devices have in mind when they decompose "reading" into constituent parts such as scanning and measuring? Reading machines (machines that read) offer us an opportunity to approach the problem of interpretation from the ground up. Instead of beginning with a high-level theory about reading, rhetoric, and writing and then applying it to the analysis of an instrument, I want to begin with the object and the day-to-day operations of economics, work, gender, and science in which it participates. The difficulties of reading, interpretation, and discovery then emerge embedded in the machine design and use. To see how these broader issues are entangled with the practice of data reduction, we must enter the struggle between two strategies of reading.

Alvarez, leader of the crash bubble chamber program at LBL, stood at one pole. In a long series of hardware and software developments, he and his group defended what I will call an "interactionist" view; they held fast to the position that human beings, by virtue of their peculiar capacities, had to remain central to the processing of track pictures. Machines would *aid* the human, but the technology would revolve around the "good enough" engineering that eschewed any attempt to supplant the intrinsically human gift of recognizing patterns and seizing on the striking or unusual.

At the other pole was Lew Kowarski, an early participant in, propagandist for, and organizer of CERN. From the beginning of the reading machine industry, Kowarski defended what might be called the "segregationist" view: that, provisionally, humans would do what preparatory work they had to do before the machines would take over, but that, ultimately, machines would put people out of the photo-reading business. As Kowarski put it while chairing a session at a 1960 data-processing conference: "The evolution is towards the elimination of humans, function by function."[184] The machine he hoped would usher in the posthuman age was invented by Paul Hough, a Michigan physicist, and Brian Powell, a staff physicist at CERN. From their first publication in 1960, Hough and Powell saw their partially automated scanner, the Hough-Powell Device (HPD), as but a first step in the complete elimination of human beings from the process of reading photographs. In Kowarski's later thinking, the ultimate goal transcended not only high energy physics but physics more generally. The reading machine was to "give the computer an eye" by solving the pattern recognition problem, eventually transforming other scientific technologies, from the tools of cellular biology to those of aerial reconnaissance.

Each side imposed what I will call a *reading regimen:* a specification of who would see the pictures, what they would search for, and how the information would be recorded. In the process of exploring the dynamics of each regi-

184. Kowarski, "Introduction" (1961), 223.

men, we can see more than mere data for particle resonances. Embodied in the technology of reading is a social order of the workplace, an epistemological stance toward discovery, and a vision of the relation between physics and the engineering arts. The task of this section is to trace the development of these competing strategies and to situate them in a wider universe of beliefs ranging from the social structure of the laboratory to conflicting visions of human nature itself.

Given the deluge of photographs, it was clear from the first months of Berkeley bubble chamber life that extracting data from pictures was an essential part of any research program. Alvarez ended his 1955 prolegomenon to future big machines by pointing to the Achilles' heel of all three visual detectors—the cloud chamber, the nuclear emulsion, and the bubble chamber. All, he suggested, "suffer from a common difficulty which is not present in counter experiments. Each event must be studied and measured individually."[185] For cloud chambers, the stereo pictures were typically "reprojected" through lenses set at the same angles from which the pictures were taken. By rotating a gimbaled frosted glass until the images came into focus, one could find the plane of interaction. Then the curvature could be measured with templates of known curvature. The process was slow and painful. With emulsions the situation was worse. Bubble chambers would provide in a single day enough data to "keep a group of cloud chamber analysts busy for a year." "If this situation could not be improved," Alvarez continued, "the bubble chamber would be nothing but an expensive toy."[186] Instead of a physicist delicately adjusting a movable glass screen, Alvarez pictured a "relatively untrained person" following the tracks by turning a steering wheel. At selected intervals, the coordinates of the crosshairs would be automatically recorded. A computer could then interpolate the three-dimensional path the charged particle had followed. Summaries of events could be put on IBM cards, and the fantasy was that future "experiments" would involve a physicist sitting happily behind a computer console, sifting automatically through boxloads of cards.

The fantasy was not very far from reality. Computers had been used for several essential tasks in the Manhattan Project, and after the war Alvarez had seen the MANIAC at work at Los Alamos.[187] This recent history, in conjunction with his war work on automatic target tracking and automatic data readout (in the radar program), made Alvarez optimistic about data reduction. Alvarez wanted a device that could employ nonphysicists to digitize individual stereo views and then reconstruct the spatial track. After all, while physicists and engi-

185. Alvarez, "The Bubble Chamber Program at UCRL," 18 April 1955, 18, LAP. At the time Alvarez's "Bubble Program" was written, there was a heated debate over whether to build high energy and low flux machines or high flux and low energy machines. Bubble chambers were appropriate for the former and counters for the latter; Alvarez's "Bubble Program" was part of this continuing controversy.

186. Alvarez, "The Bubble Chamber Program at UCRL," 18 April 1955, 18, LAP.

187. Alvarez, "Round Table" (1966), 271.

neers had designed radar equipment, they certainly were not the ones who, once the machines were in production, would sit hour after hour along banks of glowing cathode ray tubes.

The radar model (which Alvarez cited over and over) was above all a *system:* transmitters, receivers, switching mechanisms, and antiaircraft guns, for example, all had to be designed, built, and tested together. In just this way he believed that both institutionally and conceptually, the mechanical and cryogenic problems of big bubble chambers had to be solved of a piece. And data analysis, like the other components, had to be integrated into the system as well.[188]

The first of the reading machines drew its name from one of its inventors, the LBL engineer Jack V. Franck. Physicist Arthur Rosenfeld was sufficiently horrified by its jerry-built illumination stand, microscope measuring engine, and other features to label Franck's contraption a monstrosity: "Franckenstein" (see figure 5.17).[189] The name stuck, even while the hardware evolved. By 1957, Franck and Hugh Bradner succeeded in producing a Franckenstein that could automatically follow and measure tracks.[190] A typical measurement proceeded in five steps:

1. *Scanning:* A nonphysicist would look for possible interesting events.
2. *Sketching:* A physicist would record rough data, including tentative identifications, and instructions for measurement.
3. *Measuring:* A nonphysicist would measure *x-y* coordinates of fiducial (calibration) marks and sample points on the tracks by "driving along the track" in the automatic following mode, using a steering wheel for rough guidance and pressing a pedal to produce an IBM punch card on which the coordinates would be encoded.
4. *Computing:* The computer would fit the tracks to a spatial curve.
5. *Checking:* A specialist would accept or reject the computer output (e.g., offering a new particle hypothesis).[191]

Two years later, after building an improved "Mark II Franckenstein," the measurement group was ready to take on the early output of the 72-inch bubble chamber. Despite their ambition, it was clear that even the grandest plans for extending the Franckensteins could not cope with the flood of pictures that were expected from the large bubble chamber.[192]

From the start, the Berkeley team integrated physicists with engineers,

188. On the Franckenstein as integrated into the full analysis system, see Bradner, "Analysis of Bubble Chamber Photographs," UCRL-9104, 29 January 1960, LBL. Also, Franck, interview by the author, 4 September 1991.

189. Franck, interview by the author, 4 September 1991.

190. Alvarez, "Recent Developments" (1972), 267.

191. Bradner, "Capabilities" (1961), 225.

192. Bradner, "Capabilities" (1961), 225.

Figure 5.17 Franckenstein. Named after one of its inventors, Jack V. Franck, the Franckenstein became the prototype for a generation of track-measuring "reading machines," not only in the United States but in Europe and the Soviet Union as well. Source: LRL Bubble Chamber 764, LBL. Courtesy of Lawrence Berkeley Laboratory, University of California.

experimenters with computer programmers: Alvarez, for example, was involved in many ways with engineers and technicians in work on scanning and measuring devices, just as many of the earlier bubble chamber articles were written by technicians and engineers. In constant interaction with plans for new bubble chambers, the scanning tables and measuring machines went through a long series of new models: Measuring Projector (MP) I, MPIa, b, c, d, e, and f, followed by larger scale devices for the 46 mm film needed for the bigger chambers. By contrast, at CERN efforts to produce a Franckenstein-like device—the Instrument for the Evaluation of Photographs (IEP)—were relegated to the Scientific and Technical Services (STS) division, renamed *Données et Documents* in 1961, and headed by Kowarski until 1963.[193] I argue that this institutional con-

193. Krige, "Development of Techniques" (1987), introduction.

trast between the Alvarez group and the CERN approach went hand in hand with differences in the design of hardware, and with much wider differences in the relation of physics to engineering. Indeed, in many respects the reading machines simultaneously captured, and then helped to fashion, the work life that went with experimentation.

For example, the practice of nuclear physics in the two decades after World War II was largely a male preserve, and this included the design of reading machines. But the actual reading of photographs was, from the start, "women's work"—the operators of these "male" machines were almost exclusively female. For years, women had been poring over astronomical star plates[194] and scrutinizing scintillation screens.[195] Since the 1940s, as we saw in chapter 3, women had been the first to examine and record the nuclear and particle tracks on photographic emulsions. At the same time, women had served as "computers" (as they were called) who calculated numerical solutions to the tangle of differential and integrodifferential equations that arose in various war projects, most prominently in the nuclear weapons work at Los Alamos during and after World War II. (When the electronic calculators began to execute this work more quickly, they took on the name "computer," and women were rapidly assimilated into the job of programming their functional namesakes.) Thus, when it came time to define who was likely to fit the job of measuring tracks on bubble chamber film using a computer-aided device, it was inevitable that the occupation should appear "naturally" to be women's work (see figure 5.18).

Even when physicists disagreed about how to configure the data analysis system, there was general agreement that there be a gendered division of labor in whatever process they supported. The self-evidence of this appears in a myriad of places. Here is one: At the end of the first track analysis conference at CERN in 1962, Kowarski rose to summarize the fundamentally different approaches of the Alvarez system and the HPD; in so doing, he pointed to many axes of disagreement. But ultimately, "both [approaches] pursue the same aim—to solve the problem of man versus machine (or, rather, the scanning and measuring girl versus machine)."[196] In a similar vein, Alvarez reported in 1966 that, absent innovations, Franckensteins would have demanded a staff of a thousand to measure a million events per year—then the current production quota for a single bubble chamber. By Alvarez's lights, it was a situation not unlike the one that faced the telephone company half a century before; the industry had predicted that "if everyone was going to own a telephone, about half of all the women [in

194. On women's work in the analysis of star charts, see Mack, "Straying from Their Orbits" (1990); and see Rossiter's discussion of women's work in *Women Scientists in America* (1982), esp. chap. 3.

195. In nuclear physics, women often worked as the observers of scintillation screens. See Stuewer, "Artificial Disintegration" (1985).

196. Kowarski, "Concluding Remarks" (1962).

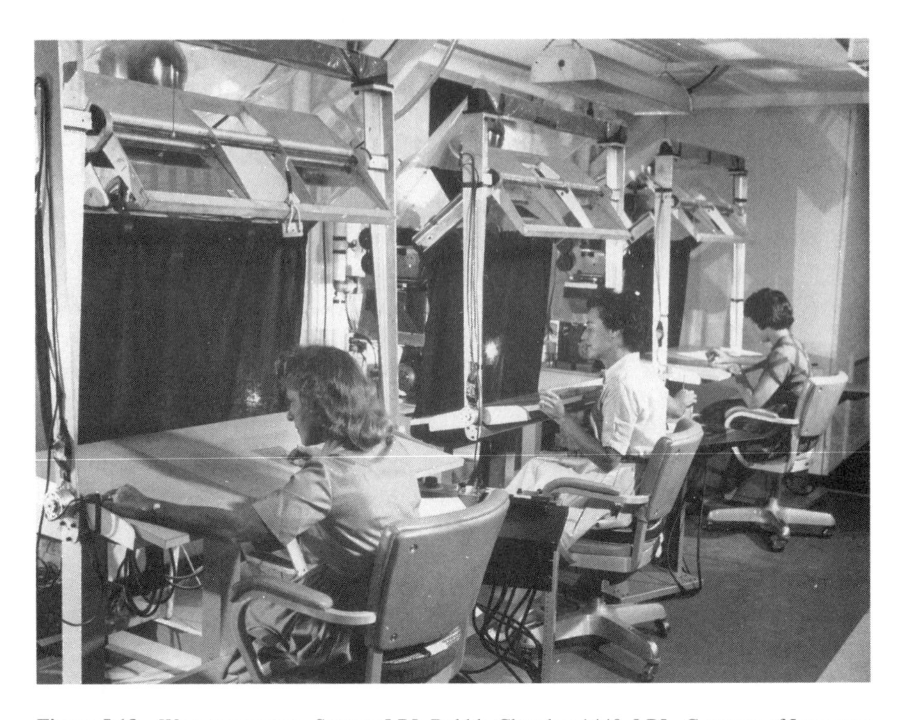

Figure 5.18 Women scanners. Source: LRL Bubble Chamber 1448, LBL. Courtesy of Lawrence Berkeley Laboratory, University of California.

the United States] would be required as telephone operators. [The company] concluded that the efficiency of each operator had to be increased enormously; the dial system is the result of that engineering analysis."[197] A similar engineering analysis was the order of the day in the 1960s: how to build on and augment the productivity of everyone within the system, from the "scanning girl" to the physicist.

The point here is not that the scanning of bubble chamber images was intrinsically unskilled labor and was therefore cast to women. Rather, while scanning had routine aspects, it was in fact decidedly *skilled* labor, an activity located where "discovery" had been a few decades earlier. In Powell's Bristol laboratory—as we have seen—this shift had already begun. Now, with the massive intervention of computer-aided data processing coupled to a factory-style organization, an even greater number of tasks previously "owned" by the physicists moved to the scanning shop floor. But to characterize the practices involved in scanning and measuring, to understand properly the change in ex-

197. Alvarez, "Round Table" (1966), 276.

perimental work, we must examine who did what in the analysis of data. Out of these details of the reading regimen will emerge a picture of the changing status of what counted as making a discovery, running an experiment, and being an experimenter.

We would not know the details of the work of reading were it not for an extraordinary set of training documents that were written at Berkeley and used for years in laboratories all over the world. Until such manuals became part of the routine, track-reading skills were passed orally from physicist to division leader to scanning supervisor to scanner. As the procedures expanded and became routinized in the early 1960s, however, the Alvarez group began to prepare training documents.[198] Starting in 1961, and established firmly by the mid-1960s, a rather formal "course" was in place to instruct the new scanner. In the most popular of these courses, the scanner learned while seated at the scanning table, working through the instructions frame by frame. Projected onto the table, just as bubble chamber film would be, the training film covered a distinctive admixture of physics, scanning tricks and procedures, and the broader culture of the Radiation Laboratory. After her training the new apprentice could become a routine scanner, able to follow the classificatory demands of the physicists, or she could rise to the level of advanced scanner, capable of disambiguating complex events that had baffled the computer, the physicist, or both.

The Alvarez Group Scanning Training Memo (written in 1961, revised in 1964 and again in 1968) opens with a presentation of the basics of accelerator design: how the bending magnets isolate particles with a chosen momentum, how velocity spectrometers separate just those particles with a specific velocity (and therefore mass), and how focusing magnets collimate the beam. At the same time, the apprentice scanner scrolled through pictures not only of the chamber and its photographs but of the wider accelerator laboratory itself. If the environment appeared unlovely, this was no accident—rather it was presented as a part of the world the scanner was entering: "The elements of the beam described above are sometimes sophisticated. Designing such a beam and putting it together in working order requires great skill. Nevertheless, as is so often the case with brilliant scientists, their contraptions are rarely beautiful. Consider the photo below."[199]

The vision of physics thus passed to the scanners by the physicists was both immensely complex and highly simplified. It was simplified insofar as a large fraction of the entities and the laws governing them were deleted in the presentation. For example, of the neutrino the guide has this to say: "The

198. E.g., Stevenson, "Elementary Particles," LRL Physics Note 327, 28 August 1961, LBL.

199. Hoedemaker, "Alvarez Group Scanning Training Film," UCRL Physics Note 595, October 1968, 11. For an earlier manual, see Stevenson, "Elementary Particles," LRL Physics Note 327, 28 August 1961; and Stevenson, "Reaction Dynamics for Scanners," LRL Physics Note 300, 16 June 1961, all from LBL.

neutrino . . . can pass through a block of concrete the size of the earth without realizing it was there. You will never see one or suspect that one was involved in the interactions you find while scanning. For our purposes, neutrinos may be forgotten."[200] Or the manual informs the scanners that the η will not be discussed because, like other resonances, it decays too quickly to leave a track. "None [of these resonances] are significant to scanners because, like the η, there is no way to tell whether or not they are there." No way, that is, for the *scanner* to tell. And again, now referring to the classification of forces: "The scanning instructions on the experiment for which you'll be scanning will very probably say nothing about strong, electromagnetic and weak interactions. Scanning is mostly a function of topology, and a knowledge of what is occurring inside the particular nucleus involved will rarely affect the proper identification of the event."[201]

But if the physics is truncated in some places to a kind of "foreigner talk" or pidgin, in others, in which it leaves the visible tracks projected on the scanning table, it is subtle, difficult, and anything but routinized. For a start, the scanner learned five basic tricks, and a myriad of delicate variations on them:

1. Scanning two views at a time (superimposing)
2. Reconstructing invisible paths of neutral particles (deducing lines of flight)
3. Matching paths of tracks to known trajectories (deploying stopping templates)
4. Measuring curvatures of track (using curvature templates)
5. Counting bubble density (consulting ionization charts)

At the simplest level, superimposing meant taking two of the three stereo views and positioning a known point seen on both images one on top of the other. For example, if a particle stopped in the interior of the chamber, the endpoint of the track (such as point p) could easily be identified and the two views superimposed—as in figure 5.19. As an aid to the determination of points inside the liquid hydrogen, fiducial points (surveying marks) are inscribed on the interior of the top bubble chamber glass and on the bottom of the chamber. If the point p is near the top of the chamber, when the two views of p are made to sit on top of one another the top fiducial points will also appear to be superimposed. If p is deep in the chamber, the top fiducial marks will appear to be far apart. A much more difficult exercise can be undertaken as the scanner becomes more experienced; she can take two views of track and "run" along it, superimposing points further and further along the track. By watching the relative position of the two

200. Hoedemaker, "Alvarez Group Scanning Training Film," UCRL Physics Note 595, October 1968, 28, LBL.

201. Hoedemaker, "Alvarez Group Scanning Training Film," UCRL Physics Note 595, October 1968, 128, LBL.

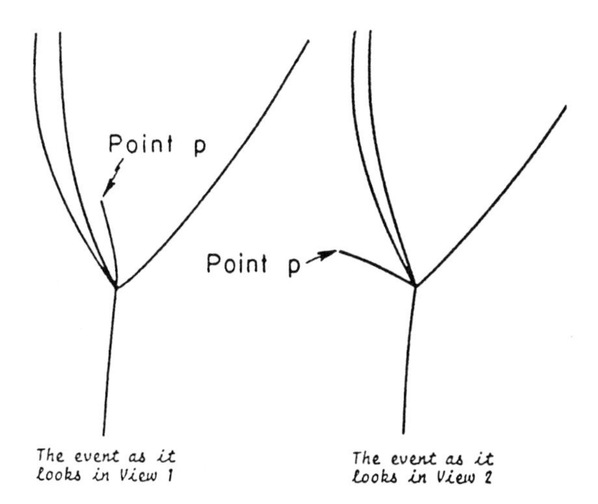

Figure 5.19 Superimposition (1968). Bubble chamber photographs (like cloud chamber pictures) were taken in stereo, and these different perspectives were combined to deduce the full spatial trajectory. Though the final cartesian coordinates were calculated by computer, the initial classification and identification had to be executed by "scanners" whose job it was to locate "interesting" phenomena. This skilled task demanded the ability to slide the stereo projections over one another in order to create an abstract mental "movie" of the unfolding microphysical interactions. Source: Hoedemaker, "Alvarez Group Scanning Training Film," UCRL Physics Note 595, October 1968 (PH 19-5-50), LBL. Courtesy of Lawrence Berkeley Laboratory, University of California.

pictures of the fiducial points, she can determine whether the track is dipping or rising in three-dimensional space. What begins as an intellectual exercise becomes three-dimensional visual intuition on which many of the other operations will build.

In particular, many of the scanner's tricks work only for an event that occurs roughly in the horizontal plane; superimposing points can help the scanner establish that the event neither dipped nor rose. For example, one trick for reconstructing the path of a neutral particle is to mark off tangents to the visible particles of a V—an event with two visible prongs. Using curvature templates, the scanner measures the approximate radius of the left-moving particle and marks off a length on the tangent line equal to that radius; the same is done for the right-moving particle (see figure 5.20). The two line segments marked along the tangents then form half of a parallelogram whose diagonal gives the line of flight of the unseen neutral. If the reconstructed neutral has a path leading back to a plausible interaction at which the neutral might have originated, the scanner

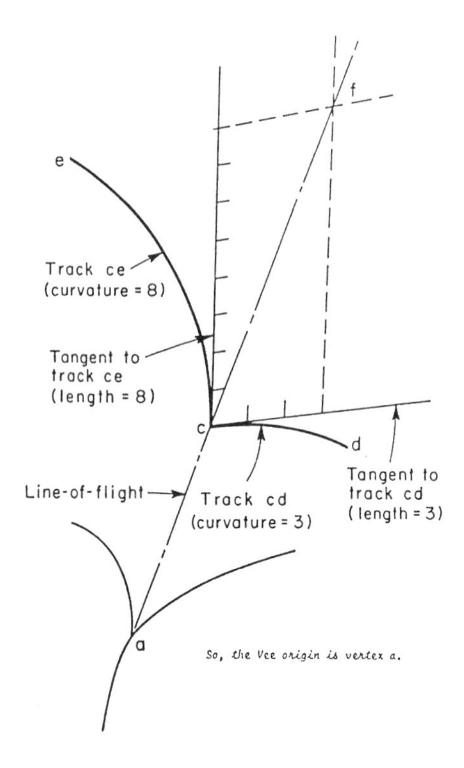

Figure 5.20 Tangent method for reconstruction of neutral line-of-flight (1968). Scanners learned to "see" the unseen by reconstructing the paths of trackless neutral particles. Tangent lines to a V (two diverging tracks from a vertex) gave their separate momenta. By completing the parallelogram and drawing the diagonal, the scanner could see whether there was a candidate source for a neutral particle that could have given rise to the charged (visible) particles of the V. Nothing *guarantees* the inference. There is no way, for example, to be absolutely sure that the decay is truly only a two-body process. Here, as elsewhere, scanners were expected to exercise judgment where algorithmic reasoning could not tread. Source: Hoedemaker, "Alvarez Group Scanning Training Film," UCRL Physics Note 595, October 1968 (PH 19-5-50), LBL. Courtesy of Lawrence Berkeley Laboratory, University of California.

will often conclude that she has completed a valid reconstruction. Implicit in the method, but unsaid, is the conservation of momentum: by classical electromagnetism, the radius of the track is proportional to the momentum, and assuming the neutral decays into two (and only two) particles, the vector sum of these momenta must equal the momentum of the invisible neutral. Pidgin particle physics for scanners is not a theory-free observation language; it is a structured set of practices that is partly separated from the main body of physical theory.

For the Alvarez group, it was clear that these scanning methods could not be completely routinized; this belief was repeated over and over again in the scanners' instructions—for example, in the discussion of neutral particle reconstruction: "Remember," the scanners would read, "that these methods are only approximate. They are not foolproof." From experiment to experiment, the protocol changed: "You should always follow the scanning instructions for your experiment to *the letter*. The mis-use of the above technique could be detrimental to the final data."[202] Even within a single experiment, judgment was required—

202. Hoedemaker, "Alvarez Group Scanning Training Film," UCRL Physics Note 595, October 1968, 66, LBL.

this was not an algorithmic activity, an assembly line procedure in which action could be specified fully by rules. For example, if the V dipped or rose, if there was a third (neutral) particle emitted, or if any of a number of other conditions held, the method would fail. Measuring errors were far less dangerous than misidentifications, and the waiting computers were nearly immobilized by wrong qualitative judgments made by the scanners.

Tricks multiplied. The scanner learned to eliminate certain particles as candidates. The track of a particle that stopped in the visible volume of the chamber left a distinctive curve. No particle of greater mass could ever curve more than that. Take, for example, a curvature template for a stopping proton; the training memo explained that if a positive track of unknown origin curves more than that, it cannot possibly be a proton and is likely to be (by an eliminative process) a stopping positive pion.

Momentum, a further aid to particle identification, could be determined from a particle's curvature, which was fixed by trying out different curvature templates. This method would work, however, only if the track lay on a relatively level plane. In this circumstance, the scanner could use the curvature and the magnetic field (printed on the film) to determine the momentum of the particle measured in millions of electron volts (MeV). For all of the techniques, the scanner spent equal time being instructed in when *not* to use the tricks at her disposal. If, for example, the three stereoscopic views gave curvatures differing by more than a specified inverse length, the orientation of the track was such that the scanner simply could not determine its curvature. Later, the computer would make the reckoning.[203]

Alvarez expounded frequently on the virtues of human intervention as a methodological precept. But the force of his view was inscribed not in these pronouncements, but on the scanning tables. However deft the Alvarez electronic computer routines became, there were many points at which the scanner could advance when the machine could not. One was the density of ions left by a particle, a quantity measured by the density of bubbles. No technical means existed for the computer to gain access to the density of bubbles on a photographic track; it was at best a qualitative guess. But by making such an "eyeball" estimate and using superposition to find the dip angle, the scanner could examine a chart on which was printed the curve of ionization as a function of dip angle for each particle. Sometimes a rough estimate could eliminate a candidate particle or reaction, even where a precision measurement could not be made.[204] For example,

203. Hoedemaker, "Alvarez Group Scanning Training Film," UCRL Physics Note 595, October 1968, 71, LBL.

204. Hoedemaker, "Alvarez Group Scanning Training Film," UCRL Physics Note 595, October 1968, 75–77, LBL. Late in the bubble chamber era, Brian Powell at CERN and Dick Strand at Brookhaven did succeed at getting HPD bubble density data. Given that the track coordinate problem had still not been solved, even that success failed to win over many converts to the new method; Hough to author, e-mail, 11 December 1995.

the track might be so dark or so light that an interpretation had to be thrown out: a very heavily ionizing track thought to be a proton for other reasons probably is; its best impersonator, the positive pion, is lightly ionizing. As the Alvarez group emphasized in every section, ionization estimates, indeed scanning in general, was explicitly taught as *defying* any attempt to reduce it to a set of rules: "As you have seen, ionization, or track density, can help you to identify particles. As with the other scanning techniques, it is approximate and can only be relied upon as such. Experienced scanners will rarely, if ever, say 'I *know* that track was made by a π.' What they will more likely say is 'I bet it is a π,' or 'it is most likely a π.' One should always use track density information with the awareness that it is not foolproof."[205] How does this work in practice? After an initial scan, measurement, and computer run, certain events are rejected because they fail to represent "the expected hypotheses." These are then sent back to the scanning table "for further 'eyeballing.'" Though the computer does not itself obtain any ionization information from the photograph, it can use its reconstruction of energy and momentum and particle type to tell the scanner:

> If this event is, as you suppose, [$K^- p \rightarrow \Lambda\pi^-\pi^+$, with the Λ going to $\pi^- + p$] and track 3 is a π^+, then track 3 should be 1.2 times minimum. It will give similar information for the other tracks. The experienced scanner can then look at track 3 and if it is quite dark, and clearly not 1.2 times minimum, conclude that it is not a π^+ as supposed. . . . He [the scanner] is in a better position to solve the problem than the computer because he can see how dark the tracks are on the scan table. He also has an advantage over the scanner who originally found the event because he has additional information from the computer.[206]

From these procedures themselves—not from grand methodological pronouncements—emerge deeply embedded assumptions about what goes into the reading of a photograph. First, the marks (the tracks) are allusive: they point elsewhere. A delta ray signals the passage of a heavy charged particle, the e^+e^- pair with no opening angle heralds a photon, a highly ionizing track suggests a proton, and the list goes on. Second, the trained scanner has acquired a set of conventions. There are conventions conveyed by instantiation—as in the cornucopia of exemplary photographs used to illustrate *K*'s, electrons, Dalitz pairs, conversion pairs, protons, pions, sigmas, and π-μ-e chains (see figure 5.21). And there are conventions of procedure: rules about throwing out angle measurements when views disagree by more than a fixed curvature, rules about when the tangent method can be applied to find a neutral's line of flight. Finally, and perhaps most important, there is caution before the text, a kind of herme-

205. Hoedemaker, "Alvarez Group Scanning Training Film," UCRL Physics Note 595, October 1968, 78, LBL.

206. Hoedemaker, "Alvarez Group Scanning Training Film," UCRL Physics Note 595, October 1968, 78, LBL.

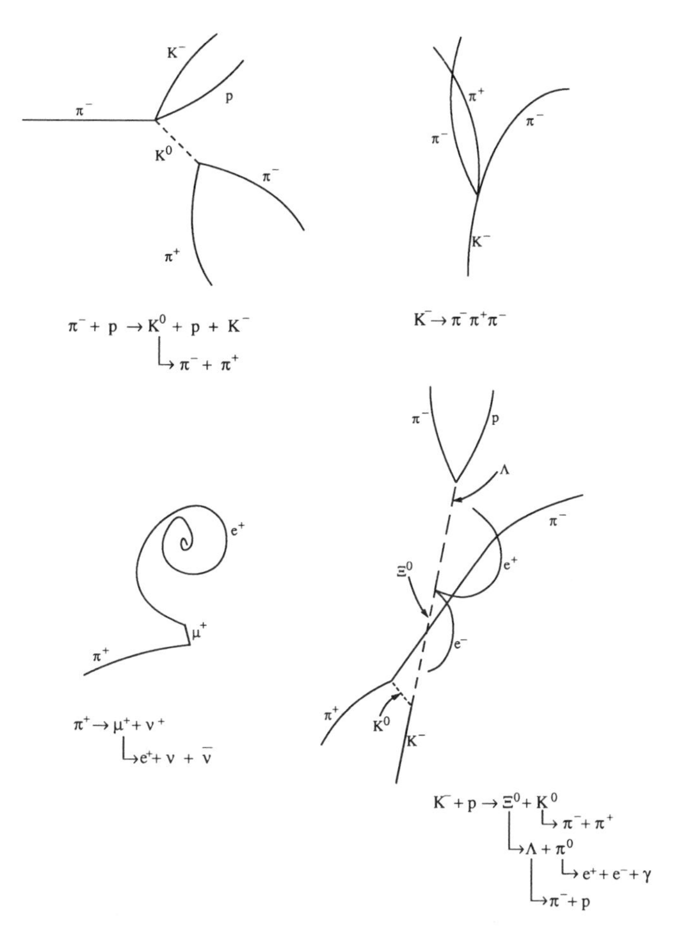

Figure 5.21 Exemplary tracks (1968). These "cleaned up" tracks (with their interpretations) were reproduced in the training memoranda to instruct initiates into the art of classification. Trainees would learn to find not only these particular processes but also classes of related processes in which elements of these interactions were found. Given the complexity of the interactions, along with the variety of optical perspectives and distorting mechanisms, classifying events was never entirely routine. Source: "Alvarez Group Scanning Training Film," UCRL Physics Note 595, October 1968 (PH 19-5-50), LBL. Courtesy of Lawrence Berkeley Laboratory, University of California.

neutical hesitancy that explicitly forbids the reader to go too far toward the claim of an exclusive interpretation. Judgment was inherent in the Alvarez reading regimen: the thorough integration of "scanning girl," physicist, and electronic computer *was* the system, not a prelude to it.

The Alvarez reading regimen relied on a social and technical infrastructure that supported constant interaction between humans and machines, and between technicians and physicists. Along with this "interactionism," work at the Franckensteins and Scanning-Measuring Projectors (SMPs) went hand in glove with a particular reading strategy in which the human readers learned from the start to exercise caution in the application of rules at every step of their interpretive work. None of these aspects of the regimen were dictated by "technical requirements," as the very different evolution of CERN's data processing attested. At CERN (and at its allied groups at Berkeley and Brookhaven), the HPD pointed like St. John toward a new world in which computer-driven artificial intelligence would read photographs from the first pixel to the final analysis. Engineering would obviate judgment, and "scanning girls" would tackle more specific and less sophisticated tasks until every last human gesture was absorbed into automation. It is to this alternative vision of data analysis that we now turn.

5.8 HPD versus SMP

In August 1960 Paul Hough, then on a sabbatical visit to CERN from the University of Michigan, began his efforts to automate scanning in what Yves Goldschmidt-Clermont called his "crazy ideas seminar." Hough's hope was to use his leave to get out of the nuclear emulsion business he had been in at Michigan, and to find a way make the move into bubble-chamber–based high energy physics. Collaborating with Hough was the British physicist Brian W. Powell, and the two developed a radically new automated method of reading bubble chamber photographs, intended to replace the human being entirely. Their goal was to develop a track analysis system that would stand in a natural sequence of devices leading, ultimately, to a fully automatic pattern-recognizing instrument. To this end, their scheme was divided into four parts. First, they would eliminate mechanical track-following systems, instead building hardware that would cast a spot of light smaller than the mean bubble diameter and flash this spot over the film. When the spot hit a darkened bubble image, the intensity of transmitted light dropped, and a photomultiplier recorded this event as a halt in its electrical output. By shining their beam over several thousand parallel lines, they could digitize the positions of individual bubble images to within 2 microns; in only 10 seconds, the device (see figure 5.22) could reduce a 70 mm photographic image from the 72-inch bubble chamber to numerical information. Hardware was, however, only part of the new reading regimen.

As they unveiled the device, Hough and Powell introduced their second idea, "parallel human guidance," which specified the position people would occupy in reading. In particular, as a necessary (but all hoped provisional) concession, a *person* would intervene before the machine began its work, help it begin the task, then cede control entirely to the technology. In the Alvarez reading

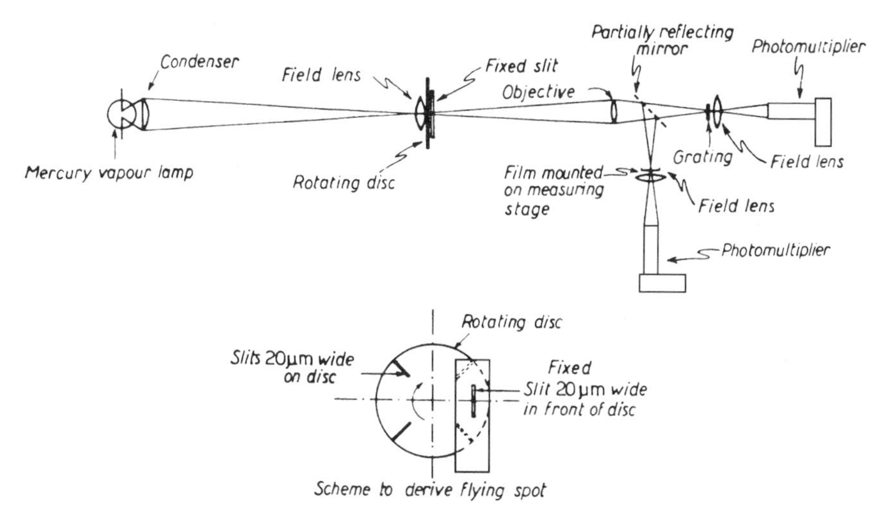

Figure 5.22 Schematic HPD (1960). A mercury vapor lamp illuminates a mechanical flying spot generator. The generator works (as indicated in the blown-up insert) when the rotating slit crosses the stationary vertical slit. This spot then passes through the objective lens and is split by a half-silvered mirror: one image continues through a field lens and is captured by the photomultiplier at the upper right-hand portion of the schematic diagram. The other image is reflected down through the bubble chamber film and onto a second photomultiplier. By subtracting the signal from the first photomultiplier from that of the second, only the effects of the film image remain. A dark spot (a bubble) then would cause a spike in the electronic output. These spikes serve as the basis for the computer analysis, which Hough and Powell hoped would eventually eliminate human scanners altogether. Source: Hough and Powell, "Faster Analysis," *Nuovo Cimento* 18 (1960): 1184–91, on 1186.

regimen, the goal was to integrate the human and the automatic at every stage (computer and human alternated back and forth many times in the process of measurement). In the HPD reading regimen, the ambition was the opposite: human and machine were to be segregated (all human intervention would occur at the outset, and even that involvement would eventually be replaced by an automatic process). If you will pardon the abuse of Freud, the HPD philosophy announced: where human judgment was there FORTRAN would be. Reading à la HPD obviated the need to insert human beings serially between each measurement. Instead, they would give the computer a leg up on its task by applying judgment in parallel to the whole lot of tracks by laying out "roads," crude demarcations of the rough trajectories in which the actual particle track was to be found. At first, Hough and Powell imagined the operator marking out masks around the track to keep the computer from needlessly taking time on irrelevancies or succumbing to the ambiguities of which tracks were relevant, or even what constituted a track. Soon, at Kowarski's suggestion, Hough and Powell simplified road engineering by having the operator merely indicate the location and identity of a track by a few rough designations on or near it.[207] Specifying roads in this way accelerated computer image-processing calculations, since tracks were now unambiguously identified and vast areas of the film were stricken from computer consideration as "uninteresting."

Third, Hough and Powell determined that they needed to extract far less information from a bubble chamber photograph than others had expected. Instead of the order of 10^8 or 10^9 bits, they would only need to register about 10^6 bits. Fourth and finally, the system involved a transfer mechanism from the flying spot device itself to a computer. "A measuring system of this kind," the authors repeatedly emphasized, "has the advantage that it can be developed into one capable of varying degrees of pattern recognition."[208] For the Franckenstein designers, such ambitions were irrelevant; their appointed task was the design of engineering solutions to facilitate the human processing of images, not to advance artificial intelligence per se. Alvarez group philosophy sought to aid the human scanner—not to replace her.

What was the appeal of (eventual) full automation? Answers varied. One participant on the HPD project, the Brookhaven mathematician J. W. Calkin, conceded that the goals were not purely practical and spotlighted his fascination with the general problem of automatic pattern recognition: "The great argument for [a full-fledged flying spot device] is of course the elimination of human intervention. The importance of this varies from place to place and is, in effect, a function of the available labor market. I take it, nevertheless, that on the whole we regard it as desirable, uncertainties in the labor supply for such work being what they are. In fairness, I should add that our motivations in pursuing this work at Brookhaven were mixed—certainly, anyway, insofar as I and others in the Mathematics Department at Brookhaven are concerned."[209] Calkin and his colleagues wanted what they called "blue sky" projects, endeavors leading to lofty ambitions in artificial intelligence far beyond the physicists' merely terrestrial concerns with controlling costs and "scanning girls" in a difficult labor market. Calkin's framework for such open-ended pattern recognition work also included four stages, beginning with *acquisition*—the accumulation of raw data from the photograph in numerical form. Second, Calkin stressed, was *erasure*—the striking out of information deemed unworthy ("irrelevant" or "uninteresting") now and forever more. From the surviving information, the goal was to extract "physics," which for Calkin meant effecting a *translation* from information to what he called the "Cartesian-Newtonian-Einsteinian language [that] we speak scientifically."[210] Then, once they were in the language of physics, there

207. Powell and Hough, "Faster Analysis" (1961), 245.

208. Hough and Powell, "Faster Analysis," *Nuovo Cimento* 18 (1960): 1184–91, on 1186.

209. Calkin, "Mathematician" (1963), 193.

210. Calkin, "Mathematician" (1963), 192.

would be a *division* of experimental results either into the domain of understood theory or into the challenging category of new physics.

Success with the HPD was striking in the acquisition stage, where rough digitization of data could be effected at an extraordinarily rapid rate. It was therefore clear almost immediately that the Franckenstein, with its dependence on slow human vision and eye-hand control, had little chance against the new device. Scanning at a rate of merely tens of thousands of events per year, the Alvarez group had to invent or capitulate. On 8 November 1960, Alvarez finished a 27-page Physics Note 223 to be distributed inside the Radiation Laboratory. It began by setting out by analogy what Alvarez called a "basic design philosophy" for the new machine: "If one wishes to know the coordinates of his home relative to the appropriate fiducial mark, which is just outside the backyard of the White House, in Washington, he doesn't hire a party of surveyors to run a survey line between his house and the fiducial mark; he makes use of an extensive grid of 'bench marks,' which was set down by the Coast and Geodesic Survey many years ago."[211] Instead of subsidizing the surveyors, the homeowner lays a rough measurement from his house to a nearby benchmark and then consults a map to provide the distance between this benchmark and the one outside 1600 Pennsylvania Avenue. The resulting uncertainty in this shrewd procedure is surprisingly small: the error in the rough local measurement is far outweighed by the precision of the much greater distance between the local benchmark and the White House. (The reader is presumed to live somewhere far from Washington, D.C.—Berkeley, California, e.g.)

The Alvarez group fastened on the control of labor costs, the expense of paying for thousands of scanner hours. But as soon as the physicists began to take apart the act of photographic reading, it became apparent that the purely economic could not be isolated from the empyrean realms of pure physics. It was obvious, Alvarez pointed out, that the full cost of analysis would depend on the fraction of "interesting events" that "needed" to be scanned. Less evidently, as Alvarez saw it, the number of interesting events that merited scanning depended on cost: "We have fallen into the habit of speaking of 'interesting events,' as though there was an absolute measure of 'interest.' If we look at the situation more realistically, we find that our definition of 'interest' is exactly tailored to the measuring capability we either have, or expect to have in the near future! This is not at all surprising, since in planning an experiment, a physicist must decide what events he should measure."[212] Nothing mattered more than a crucial ratio: the rate of picture production from the bubble chamber (per year)

211. Alvarez, "Rapid Measurement of Bubble Chamber Film," LRL Physics Note 223, 8 November 1960, 3, LBL.

212. Alvarez, "Rapid Measurement of Bubble Chamber Film," LRL Physics Note 223, 8 November 1960, 20, LBL.

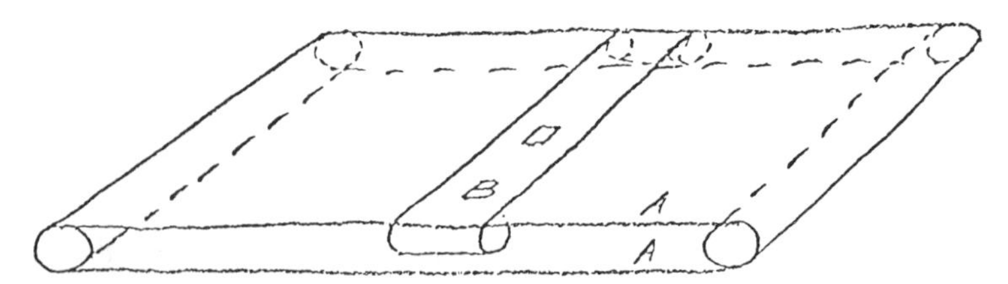

Figure 5.23 Design sketch of Scanning and Measuring Projector (SMP) (1960). Alvarez's first hand-sketch of the SMP. Visible is the sliding "curtain" with its window that moves longitudinally and rolls from side to side. The bubble chamber image is projected, and the section to be measured appears in the window. Beneath the window, photomultipliers identify a fiducial mark (giving a global measurement); the glass vibrates automatically until the fiducial mark crosses the track. At that moment, from the position of the glass relative to its "rest" position, the computer can calculate the "local" distance of the track relative to the fiducial point. Source: Alvarez, "Rapid Measurement of Bubble Chamber Film," LRL Physics Note 223, 8 November 1960, LBL. Courtesy of Lawrence Berkeley Laboratory, University of California.

divided by the rate of measurement of these pictures (per year). At the end of 1960, this ratio was about 60 : 1 for the 72-inch chamber; when all of the Franckensteins went on-line, this ratio would fall to about 15 : 1. Alvarez wrote: "Since N is 60, we can only measure one event per 60 frames now, but we hope to measure 1 event in 15 frames, when 'the experiment is finished'; we therefore define an interesting event as one which occurs every 20 frames!"[213] Pragmatic concerns (scanning capacity) could not be separated from physics concerns.

For example, two different event types give information on the interaction of pion-pion interactions in which a beam of negative pions (π^-) was shot into the hydrogen bubble chamber:

$$\pi^- p \rightarrow \pi^- p \pi^0, \tag{5.6}$$

$$\pi^- p \rightarrow \pi^- n \pi^+. \tag{5.7}$$

While the first set of reactions had been extensively studied, the second had not. Purely because of the lack of scanning tables, an entire set of events, according to Alvarez, had been rendered "uninteresting."[214]

Alvarez turned to the material aspects of the reading machine. Following the reasoning of his survey metaphor of calculating the distance from Berkeley to Washington, he described the Alvarez group machine as proceeding in two steps. First, as indicated in figure 5.23, a small (1 cm) window in a spring-loaded

213. Alvarez, "Rapid Measurement of Bubble Chamber Film," LRL Physics Note 223, 8 November 1960, 21, LBL.

214. Alvarez, "Rapid Measurement of Bubble Chamber Film," LRL Physics Note 223, 8 November 1960, 21, LBL.

shutter was placed over the desired section of the projected bubble chamber image. (The projected image falls on a glass plate; underneath the plate is a photomultiplier.) Reading the position of the window automatically, the computer stored the identity of the fiducial mark visible in the window. Next the "local measurement" (distance of a point on the track to the fiducial mark) took place by vibrating the glass plate on which the fiducial point is etched. When the benchmark crosses the image of the track, the light transmitted to the photocell diminishes, and a circuit signals the position of the moving benchmark to the computer.[215]

Practically, what happens is this: the operator moves the window along the track and, as often as desired, sets the machine to work at measuring the precise position of the track (see figure 5.24). Since measuring and scanning take place in a single motion of the window over the track, the human and machine are in constant joint action. The method lends itself to modification in which the machine asks for a remeasure (e.g., if, upon computer analysis, a track is physically "impossible" the computer can ask the operator to check again to see whether perhaps what was thought to be one track was in fact a juxtaposition of two). But since the human is necessarily active at every stage, the method participates only awkwardly in a broader program of full automation. This did not disturb Alvarez: "Much of the thinking in the past has been flavored by thoughts of thousands of technicians. No doubt automatic pattern recognition will be one of the most important technological developments of the 20th century, but it is hard to see just how the bubble chamber fraternity could be justified in spending appreciable sums of money to bring it into being at a date much earlier than it will appear for other important reasons."[216]

215. More specifically, the huge Berkeley bubble chamber was 72 inches long and was typically photographed on film with a useful area of 12 cm × 3 cm. An accuracy of 4 microns ($4\ \mu m = 4 \times 10^{-4}$ cm) on the film was equivalent to a 60 micron uncertainty in the chamber. In the long direction, this amounted to a fractional error of 4/120,000 or 1/30,000. Measuring three parts in 10^5 is no mean feat, and at root it was this difficulty that led to the high cost of the Franckenstein. Put into binary form, 30,000 is 2^{15} or 15 bits; since 3 cm is a factor of 2^2 smaller than 12 cm, 13 bits of information were required in the x-direction. Franckenstein projectors were commonly available with a magnification of 10 times; it was therefore practical to adapt the projectors for use with the new reader. The dimensions of the projected image (120 cm × 30 cm) corresponded to a required accuracy of 40 microns. The analogue of the Coast and Geodetic Survey benchmarks of Alvarez's metaphor would be 300-micron-diameter frosted circles etched on the otherwise clear glass stage. (The glass serves both to lay out the grid of benchmarks and to act as a light guide, which ushers light from the bubble chamber image and moves it by internal reflection to the light-sensitive cell.) Since the grid consists of benchmarks in a square array with interbenchmark distance of 1 cm, approximately 7 bits of information are needed to specify the "name" of the x-benchmark in binary (the long distance is 120 cm, which is nearly $2^7 = 128$ dots or 128 cm). So although it first appears that one needs a measurement of 1 in 2^{15}, the bubble position measurement in the x-direction then reduces to two measurements. First, one provides the "name" of the x-benchmark that has come into view (e.g., 0100 110); then, the other provides a fraction of the distance between benchmarks that corresponds to $40\ \mu m \approx 1/256$ cm, or 8 bits (e.g., 0011 1101). Alvarez, "Rapid Measurement of Bubble Chamber Film," LRL Physics Note 223, 8 November 1960, 21, LBL.

216. Alvarez, "Rapid Measurement of Bubble Chamber Film," LRL Physics Note 223, 8 November 1960, 24, LBL.

Figure 5.24 Operational SMP. The SMP—and indeed all of Alvarez's scanning devices—never eliminated human intervention in the establishment of "interesting" physics. If the CERN attempt was to solve the pattern recognition problem in general, Alvarez's motto was always (roughly speaking) to give to the machine just that which was machinelike. Source: LBL Bubble Chamber 1531, LBL. Courtesy of Lawrence Berkeley Laboratory, University of California.

If Alvarez's Physics Note was the vision of interactionism in hardware, J. N. Snyder's of 25 August 1961 did the same for software. For hardware alone could not abstract the "interesting" bubbles from the prosaic ones in real-world film. Noise, in the form of dirt, had to be filtered out; spurious tracks from beam particles had to be eliminated; and the confusion resulting from track crossings had to be resolved. Just as Alvarez resisted pushing too far into the general problem of pattern recognition in the construction of hardware, so Snyder held back in software design. "Progress," he asserted, "is never made in such an uncharted area [as pattern recognition] by over-generalizing in a vacuum of ignorance; rather one first picks some specific problems and tries to solve them. The things learned can then be extended. An attempt to create an automated, computer-controlled, scanning, measuring, and analysis system for bubble chamber film is one example of such a specific problem."[217] Participants in SMP development

217. Snyder, "Some Remarks on a Data Analysis System Based upon the Scanning-Measuring Projector (SMP)," UCRL Physics Note 326, 25 August 1961, 2–3, LBL.

held a "great personal interest" in particle physics, "hence it has been chosen as a representative of the larger field." But no attempt would be made to leapfrog directly into the general problem of nonnumerical information processing.[218] Cryogenics, optics, data-processing hardware, and data-processing software all were saturated with the philosophy of a "good enough" engineering that resisted "blue sky" projects and seized the resolution of "specific problems" to be solved.

Alvarez and his team wanted the "human operator" to be "the black box pattern recognizer." According to Snyder, "from this point of view one does not regard the system as a human scanner and an SMP with an on-line computer to do the computational part of the work on request, but rather views it as a comprehensive analysis program in the computer with an on-line human-and-SMP which can be interrogated when necessary for those visual and pattern recognizing tasks which are not yet automated and which a human does so well."[219] The competition (the HPD) by contrast detached the human from the machine processing; and once the machine had hold of the information, recourse to the human became a more difficult and time-consuming affair. While the CERN fantasy was total automation, the Berkeley imagination clasped a highly mechanized but still interactive process of reading pictures. One is a classical robot, the other a cyborg.[220] The output from a future Alvarez group analysis might go to an oscilloscope, asking the operator for more information. As Snyder reasoned: "For example, the following type of question might be asked: 'This track (arrow points to one of the tracks in the scope reconstruction) gave a sloppy fit, re-search it for a kink'; or, 'This V fitted badly, (arrow) search for a recoil from a possible neutral scatter in this (arrow or box) area.' etc."[221] While this scenario might seem far-fetched, Snyder maintained that it was not wildly beyond current air traffic control or teaching machines and would eventually find its place under the broader rubric of pattern recognition. As these remarks suggest, the Alvarez team sought to maintain human intervention at several key points in the reading. Nonetheless, the die of automation was cast. The first function to go was "sketching." The computer took over many of the decisions of the "sketchers," such as

218. In Snyder's analysis, the goal of pattern recognition lay behind two great barriers. First, one needed a "front end" pattern recognizer that, given the location of points, would decide where the points constituted line segments. Second, given the line segments, there were the equally daunting sequential tasks of putting these segments into particle tracks, eliminating noise, discarding "uninteresting tracks," recording the topological features of the interesting ones, executing the geometrical reconstruction of the tracks in space, and performing the kinematic analysis (determination of the particles' energy and momentum). For Snyder, even if a "front end" pattern recognizer existed as a black box, the project of full automation was far from complete. Snyder, "Some Remarks on a Data Analysis System Based upon the Scanning-Measuring Projector (SMP)," UCRL Physics Note 326, 25 August 1961, LRL Physics Note 326 (1961), 3–4, LBL.

219. Snyder, "Some Remarks on a Data Analysis System Based upon the Scanning-Measuring Projector (SMP)," UCRL Physics Note 326, 25 August 1961, 4, LBL.

220. On cyborgs, see Haraway, e.g., *Primate Visions* (1989).

221. Snyder, "Some Remarks on a Data Analysis System Based upon the Scanning-Measuring Projector (SMP)," UCRL Physics Note 326, 25 August 1961, 7, LBL.

which two of three views to use, what number to assign each track, and whether a track stopped on the film.[222] Programs that supplanted human labor bore such names as PANG, which reconstructed tracks from measured points (a later version that fit a helix to the measured point was called TVGP). Next KICK took the track data and performed a kinematical analysis by assigning a mass to each track emerging from a vertex and executing a least squares fitting subroutine subject to the constraints of energy and momentum conservation. On the output tape of KICK was a χ^2 test for the mass interpretations used on a particular vertex, the momentum, azimuth, and dip angle for each track at the vertex, and a matrix of errors for every measurement. Variants of the program were widely distributed—CERN had one called GRIND, for example. Then, stringing the tape back up on the IBM 7094 or equivalent, the bubble chamber team would run EXAMIN, a program that put together all the vertex information from KICK and picked the most probable interpretation of the event as a whole. The output of EXAMIN became the group's library of events, known then and for decades afterward as Data Summary Tapes. Finally, SUMX summarized the data of many points,[223] producing graphical displays so easily that it became evident that the physicist would "no longer [be] rewarded for his ability in deciding what histograms he should tediously plot and then examine."[224]

Distributed to laboratories throughout the United States, Europe, and beyond, these prepackaged programs took over functions that previously were the "high level" part of experimentation, and in so doing took another piece out of the scientific author. In some collaborations, the programs would be run by a group that had never been at the bubble chamber and never met "coauthors" from other participating universities. Think back to the myriad of plots carefully drawn in 1953 at Bagnères de Bigorre. Now, in the heart of analysis—the sanctum sanctorum of the physicist's physics—part of what had been a physicist's task was isolated, routinized, and severed from the activities that had previously defined experimentation. Dalitz plots in 1965 were no longer the master summary of hand-scanned emulsions. They were the automatic output of a computer, flashed on a computer screen and photographed.

Software automation did not deter the Berkeley group from its "interactive" philosophy. It did lead it to invert the prior (Franckenstein) relationship of human operator to on-line computer. Before the introduction of the SMP, the operator "[determined] the sequence of operations, calling upon the on-line computer to carry out arithmetic tasks quickly and to present the results of these calculations for inspection in order to aid in the human decision process." With

222. Rosenfeld, "Current Performance," *Nucl. Inst. Meth.* 20 (1963): 422–34, on 422.

223. Rosenfeld, "Current Performance," *Nucl. Inst. Meth.* 20 (1963), 422–34, on 424–30; the previous discussion also draws on Alston, Franck, and Kerth, "Data Processing" (1967), esp. 2:76–84.

224. Alvarez, "Recent Developments" (1972), 267.

the SMP, the machine set the sequence, "calling upon the human operator via the typewriter for those tasks such as pattern recognition, guidance along tracks, etc., at which a human is so adept."[225] Millions of times the computer would ask and the operator would respond in a dialogue dubbed a "script." Here are typical exchanges, with the operator responses underlined (T2 means track 2, and π indicates measurement completed): "Roll 6748; Meas. Y; Reject N; T1 π; T2 π; T3 π." Not exactly Shavian, but every such exchange saved enormous amounts of time over the course of an experiment. Soon it was said that "the operators even learn to distinguish the patterns of sounds made by the typewriter [and] to recognize the messages which are transmitted without looking, as the vocabulary of these systems is rather limited."[226]

Further advances could be expected, one programmer wrote, when "a careful 'human engineering' analysis of the entire measuring sequence" was undertaken.[227] Meanwhile, production was gradually speeding up for many reasons. Trivial mistakes by the operator were being corrected by the computer. Moreover, the operator knew that his mistakes would be caught, so he could "push his speed to the limit" since it was now "possible for him to determine what that limit is." Where a more automated device was placed near an older machine, the speed of the new device was "communicated by contagion and emulation to the operators of the non-automated Franckensteins of the same laboratory" in "an interesting example of psychology in man-machine relationships" (an example, the reader might observe, of aggressive production management as well).[228]

But Alvarez's interactive cyborg was still too slow. No matter how cleverly the scripts were rewritten or the scanners' psychology manipulated, the exchanges between computer and operator seemed doomed to produce no more than an event every few minutes, or tens of thousands per year. Since the big bubble chambers would soon be churning out millions of events per year, a more completely automated device that could analyze pictures in seconds was needed. In laboratories across the United States, Europe, and the Soviet Union, groups began exploring the possibility that a fully automated process could be perfected. At Berkeley, the Alvarez group continued to build according to the "interactive" philosophy. Their work culminated in the Spiral Reader (figure 5.25).[229] Using this device, the operator identified the "interesting" vertex in the field of view, put a cursor over it, and then let the machine take over numbering, following,

225. Snyder et al., "Bubble Chamber" (1964), 243.

226. Hulsizer, Munson, and Snyder, "System," *Meth. Comp. Phys.* 5 (1966): 157–211, on 176; this paper also includes a good chronology of SMP development. Goldschmidt-Clermont, "Progress" (1966), 445.

227. Snyder et al., "Bubble Chamber" (1964), 243.

228. Taft and Martin, "On-Line" (1966), 392; Goldschmidt-Clermont, "Progress" (1969), 443.

229. See McCormick and Innes, "Spiral Reader" (1961). Compare the results of the working version in Alvarez, "Round Table" (1966).

Figure 5.25 Spiral reader. Unlike the SMP, which moved the glass and kept the photomultiplier stationary, the spiral reader spiraled the photomultiplier out from the center while keeping the glass fixed. By knowing the precise position of the photomultiplier, the machine could determine the position of a track segment as it crossed its optical field. Source: LBL CBB 681-366, LBL. Courtesy of Lawrence Berkeley Laboratory, University of California.

and measuring the tracks that emanated outward; as queries arose, the operator responded. As its name might suggest, the Spiral Reader was based on a photomultiplier tube that moved out from the vertex along a spiral beneath the projected bubble image. In this way both the radial and the angular position of points along the tracks could be determined. One of the most ambitious features of the system was the use of a filtering program to eliminate spots and scratches on the film. In 1967 the Spiral Reader would, it was estimated, measure over a million events per year, whereas in 1957 the Franckenstein had performed a mere 13,000.[230]

The competition between human-centered and fully automatic readers roused passion. Kowarski, summarizing the state of analysis in 1960, insisted that the choice of automation was unambiguous: "Clearly the problem is that of

230. Alvarez, "Round Table" (1966), 288.

speed, and since human attention and action introduce a rock-bottom bottleneck, speed can be achieved either by pouring in parallel through many bottlenecks, or by eliminating them altogether. Either vast armies of slaves armed with templates and desk calculators—maybe even strings of beads—or few people operating a lot of discriminating and thinking machinery. The evolution is towards the elimination of humans, function by function."[231] Franckensteins meant slavery; the HPD was the harbinger of "discriminating and thinking machinery." If, for Kowarski, liberty meant freedom from human involvement in reading pictures, then it should hardly be surprising that he would consider Alvarez's invention of the SMP to have left slavery intact. Indeed, three years later, in 1963, Kowarski summarized advances in data analysis that had occurred over the previous year, and the contrast between the Alvarez approach and the HPD was as strong as ever. At issue, he believed, was the fundamental division between humans and machines, a division that had evolved in four stages. First, Kowarski contended, human observers, aided by nothing but an inch rule, had explored physics amid the slow output of the grand old cloud chambers. Workers proceeded with no speed constraint whatsoever. A second stage had followed in 1956 in Berkeley under the watchful eye of Alvarez. This was the age of the Franckenstein (and CERN's equivalent device, the IEP), which drove production rates to the order of 100,000 pictures per year. Only recently had the third epoch begun, the age of a million events, ushered in on one side by Alvarez's SMP and on the other by Paul Hough with his innovative hardware. As Kowarski put it, though both still required humans, their philosophies were utterly at variance: "On one end, the Hough-Powell system segregates the machine from the human; it puts the human operation in its allotted corner and lets most of the machine operation work without the human hampering it. On the contrary, in the Alvarez system, the human operator still is in the center of things but the machine is arranged so that it speeds up the human operation, and corrects its inherent lack of accuracy."[232] For Kowarski, the HPD's tentative steps toward total automation were just what was needed to reach the fourth stage, a coming generation of experiments—presumably involving tens of millions of events—that would *only* be accessible to an inhuman reader. Here too the road of development would bifurcate. On one side, HPD-like devices would be readied with a "philosophy" of "tricky hardware": a computer would drive the spot in ways that reflected the physics interest of the experiment. On the other side, the tricky software solution was to write selectivity into computer programs that would extract information from the experiment without ever producing a photograph.[233]

231. Kowarski, "Introduction" (1961), 223.
232. Kowarski, "Introduction" (1963), 2–3.
233. Kowarski, "Introduction" (1963), 4.

To the Alvarez group, Kowarski's longed-for "fourth stage" was a dystopia. For as the Alvarez group wrote software and crafted hardware, its members hailed the interactive feature of their "design philosophy" as a virtue, not a vice. If the "interrogation" of the human by the machine got too complex for the scanner, then so much the better. As four Berkeley software writers pointed out, when that happened the physicists could be part of the reading (of) machines:

> [I]f no hypothesis contained in the KICK [kinematic analysis] programs sufficed to fit the event, then additional hypotheses to try could either be generated automatically or requested of the operator. However, if the dialogue demands too involved decisions or too much physical understanding from the operator then it will be necessary to use more highly trained personnel on the SMP tables. This in turn opens a whole new vista of using the SMP and the dialogue concept as tool for understanding and processing very complicated or very recalcitrant events by manning such tables with highly trained experimental physicists.[234]

Nothing like these remarks can be found among the programmatic statements of the HPD enthusiasts. If "scanning girls" were necessary in the early stages of road preparation, HPD advocates would reluctantly and provisionally make that concession. But everything in the CERN program (and its Brookhaven and non-Alvarez Berkeley links) was designed to obviate the need for a physicist to sit behind a computer. Reading in the HPD mode was entirely without the hermeneutic pause desired by Alvarez's group. Alvarez, by contrast, saw the "scanning girl's" pattern recognition as a capability unmatched (and probably unmatchable) by electronics. Working physicists into the computer exploration of "recalcitrant" images was nothing but an improvement. For at every level of analysis, these authors argued, reliability would be enhanced, not reduced, by the human-computer interaction. One group member commented: "Immediate feedback of results to the SMP operator enhances the operating efficiency and reliability of the analysis system."[235]

The differences ran deep. For the Alvarez group, as for many of the physicists who emerged from the American war laboratories, engineering and physics entered together. Alvarez was perfectly willing to write to a colleague that he had been working "as an engineer," or elsewhere "I am wearing my engineering hat." Technicians and engineers published articles on bubble chambers, and physicists recognized the need to incorporate many of the engineering strategies into the construction and operation of the chambers. In such a world, it was reasonable to expect physicists to sit occasionally behind scanning tables, and to train scanners in at least some elements of particle physics—enough to give the scanners judgment about events and interpretations. Alvarez himself took it as

234. Snyder et al., "Bubble Chamber" (1964), 243–44.

235. Hulsizer, Munson, and Snyder, "System," *Meth. Comp. Phys.* 5 (1966): 157–211, on 159.

necessary—even obvious—that scanning technique should be treated from his first proposal; he was himself continuously involved in the details of the mechanics of the hardware. At CERN, such engineering was handled, as mentioned earlier, in a division separate from the one in which the leading bubble chamber experimenters were working. None of this should be mistaken for a claim that European engineering was in any way inferior to American engineering. In fact, it was the vaulting ambition of the HPD device to solve the pattern recognition problem more generally that distinguished it from the pragmatic goals of Alvarez's computer experts like Snyder.

The commitment of the Alvarez group to the symbiosis of computer and human extended all the way from the identification of tracks to the legitimation of a histogram for publication. At every level, the team wrote computer routines to test both humans and machines. One program titled FAKE used a Monte Carlo routine to generate simulated bubble chamber tracks "faking" real events. Such collections of simulated images could then be used to test the analysis and track reconstruction programs against a set of events with specified characteristics and statistical distributions. Along with this event simulator, the Berkeley team developed a second program, GAME, that produced artificial histograms—reduced data in precisely the form that would normally appear in *Physical Review*. Such phony "final" data had a variety of applications, one of which was to test the reliability of the human eye as it gazed, not on individual events, but on the laboratory's end product. Here is what one of the lead programmers, Arthur Rosenfeld, had to say about FAKE in 1963: "As an example of its application let us consider [a particular histogram] and ask the following questions. 'Is the peak above [one of the bins] a resonance or a statistical fluctuation?' 'What is the Poisson probability that these adjacent bins will be over populated in so striking a way?' It is difficult to formulate a definition of 'striking' so as to answer this question with a chi-squared tail (thus χ^2 does not distinguish between under- and over-population)."[236] To probe the robustness of the physicists' judgment, the programming team would draw a curve through the real data and feed it to GAME. The program then produced 99 fake distributions of this smooth curve modified by Poisson fluctuations. These 100 printed histograms, the real curve and the 99 fakes, were then interleaved and presented to the physicists, who were asked to rank them from the one they considered to most forcefully indicate a bump (new particle) to the one that least smacked of new physics. If the real McCoy was not the top contender, then the team agreed that an equally "striking" peak would present itself in 100 runs by Poisson statistics alone. "GAME," Rosenfeld concluded, "has protected us from publishing statistical fluctuations instead of resonances."[237]

236. Rosenfeld, "Current Performance," *Nucl. Inst. Meth.* 20 (1963): 422–34.
237. Rosenfeld, "Current Performance," *Nucl. Inst. Meth.* 20 (1963): 422–34, on 433.

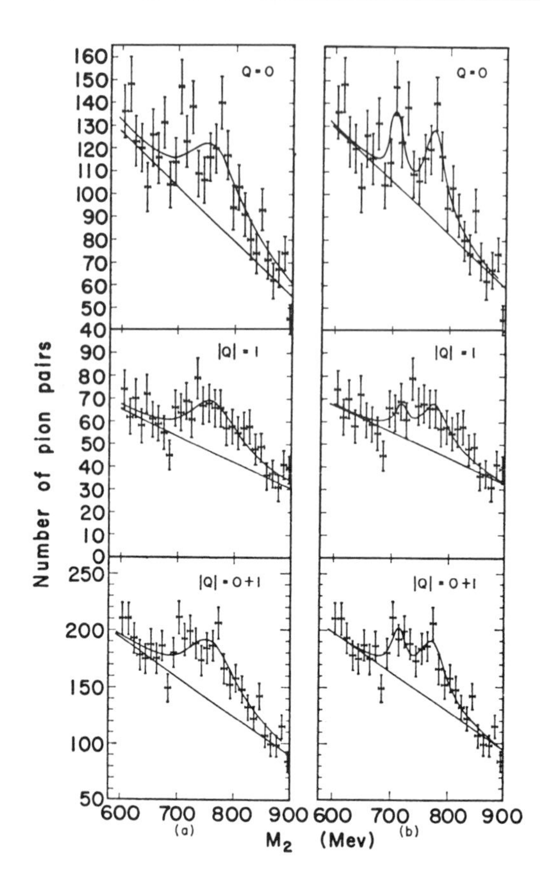

Figure 5.26 FAKE/GAME (1962). In contrast to the developers of the flying spot device program, the Alvarez group remained staunchly committed to a central role for human judgment in the assessment of data. For example, when the data seemed ambiguous between "two peaks" (*column b*) and "one peak" (*column a*), the Alvarez group generated Monte Carlo data displaying statistical fluctuations from a distribution with only one "true" peak. These pictures were then interspersed with real data, and physicists were asked to rank the images in "two-peakedness," from the most two-peak-like to the least. When the real data failed to emerge as the clear victor, the physicists withheld the accolade of "reality" from the second peak. Source: Reprinted from Button et al., "Pion-Pion Interaction," *Phys. Rev.* 126 (1962): 1858–63, on 1861. Copyright 1962 The American Physical Society.

An example from the published literature can be found in the pages of *Physical Review* from 1962, where the Alvarez group reported on pion-pion interactions, resulting from the collision of antiprotons with protons, in particular the reaction that produced two positive pions, two negative pions, and n neutral ones,

$$\bar{p} + p \rightarrow 2\pi^+ + 2\pi^- + n\pi^0. \tag{5.8}$$

Earlier studies had shown there to be a resonant phenomenon in the pion-pion interaction by showing that if one took pairs of pions produced in pion-nucleon scattering, there was a peak in the number of events around the invariant mass near 750 MeV. Typically one took pairs of pions with a given charge (0 or 1) and plotted the number of events with different values of $M_2 = [(E_1 - E_2)^2 - (\mathbf{P}_1 - \mathbf{P}_2)^2]^{1/2}$, with E_1, E_2, and $\mathbf{P}_1$, $\mathbf{P}_2$ representing the measured energies and momenta of the two pions. On examining their data, the Alvarez bubble chamber experimenters were divided about whether their results indicated the existence of one new particle or two. All hinged on the interpretation of figure 5.26—the upper left-hand plot ran a single-peak hypothesis curve through the data for pion pairs with a charge sum of zero, and the upper right-hand plot indicated the double-peak hypothesis on the same data. Again making use of the FAKE/GAME procedure, 20 experimental physicists were shown 72 Monte Carlo–faked histograms of the invariant mass distribution for the pions, with the real data included in the sample. Each physicist was asked to identify the one "most suggestive of a double peak." In response, the experimenters chose the real data second most frequently; six picked it as "most suggestive" of a double peak. "Although such a test is an extremely subjective one," the authors surmised, "we can conclude from it that there is about a 1 to 3% chance of obtaining data . . . which are as suggestive of a double peak as are our data when only a single resonance exists."[238] Humans checked computers with their pattern-recognizing capability; computers checked humans' inclination to see patterns where there were none.

Film, kilometers of it, continued to spin from the bubble chamber cameras, and the debate over reading regimens raged. In 1963, Alvarez joined in a discussion of the HPD to comment on the relative merits of that system and the SMP. Both machines, he observed, cost on the order of $35,000. In terms of speed, he confessed, the HPD held an advantage, as its proponents had repeatedly pointed out, taking 15 seconds per event, whereas an event occupied the SMP and its operator for 4 minutes. In part this disparity was due to the radically different procedures. Alvarez:

> On the SMP a cursor goes along the whole track, whereas on the FSD [Flying Spot Device, another name for the HPD] only three points are measured, but the three points are measured more accurately. . . . At the end of the measuring, the event is completely measured and analysed on the SMP. With the FSD it now goes onto tape, and a day or so later it goes to a $250,000 machine where it is then measured. I think with this method of comparison that the FSD is slower and more expensive. However, when a pattern recogniser is built into the FSD, everything I have said must be reexamined, and the FSD clearly has a potential advantage. One must then reconsider the question, taking into account economic factors, and one's basic philosophy about the interaction of physicists with their own data.[239]

Alvarez's caveats were two. First, as he had six years earlier, Alvarez stated his pessimism about even the medium-term prospects for pattern recognition: "I do not think that automatic scanning of bubble chamber film is going to come soon enough to contribute to our present campaign in particle physics. Perhaps it will help in some renaissance, twenty-five years from now, just as the laser has brought new vigor into the field of optical spectroscopy, after it had lain dormant for

238. Button et al., "Pion-Pion Interaction," *Phys. Rev.* 126 (1962): 1858–63, on 1863.

239. Alvarez, comment from discussion following Macleod, "Development," *Nucl. Inst. Meth.* 20 (1963): 367–83, on 382.

almost twenty-five years."[240] But his real objection lay in the loaded phrase "taking into account . . . one's basic philosophy about the interaction of physicists with their own data." Alvarez saw in the two reading regimens far more than this or that projector design: he saw competing definitions of the nature of discovery and of what it meant to be a physicist.

5.9 Track Reading and the Nature of Discovery

Computers were in no position to discover the fluke event, the tracks that fit no prior hypotheses and conformed to no previously known entity. Already in 1960, the Alvarez group was insisting on the importance of the search for the individual occurrence: "As in cloud chamber work, it was considered very important to examine and keep a record of each event. . . . A large part of the interest was in studying rare, new, or unexpected events. . . . [T]his is still considered very important today, though some experiments involve so many thousand events that individual scrutiny is almost impossible."[241] These rare but exemplary beasts dubbed "zoo-ons" to symbolize their oddness were prized by Alvarez and his group: "I should like to say a few things about zoo-ons because I think that they are exceedingly important. At present I believe we miss many of them because physicists no longer do the scanning. There is a hope that this will not be the case when the SMP's are in operation, where the computer will be able to help the scanner or the physicist in interpreting an odd event."[242] Here we see the confluence of the various streams of Alvarez's reading regimen. The attachment to discoveries by means of individual events runs deep—Alvarez was as proud of his group's discovery of the cascade zero and the muon-catalyzed fusion as of anything in the history of his bubble chamber program (more on both of these below). Even the possibility of such discoveries would be jeopardized by what Alvarez saw as the destructive segregation of the physicists from the scanning procedure. Hardware—that is, the SMP itself—would open the way to a reintegration of physicists into a more intimate relation with their data and at least create the conditions in which the golden event could be found.

One hundred and eighty degrees separated Alvarez's perspective from Kowarski's. The HPD, at least in its full-fledged automatic form, would focus on reducing masses of data without any human intervention. Kowarski granted Alvarez's point about golden events: "These devices [such as the HPD] are therefore promising mostly in statistical experiments in which the chamber acts

240. Alvarez, "Round Table" (1966), 294.

241. Bradner, "Capabilities" (1961), 225.

242. Alvarez, comment following the discussion after Macleod, "Development," *Nucl. Inst. Meth.* 20 (1963), 367–83, on 383.

as a 'seeing counter'; for rarer and more sophisticated events, in which human search and evaluation have a greater role, their promise is less immediate."[243] Thus, even within the enterprise of reading bubble chamber photographs, we find the clash between the visual and the logic traditions. Now, though, it is the statistical form of the HPD search that accords pictures the role more usually played by Geiger counters and coincidence circuits. The strategy of automating the reading process bypasses the epistemic role of the human search for the golden event, and in so doing has thrown one style of picture analysis out of the world of images and into that of logical counts by electronic gear. In Kowarski's words, the HPD style of reading has effectively changed the reader of bubble chamber pictures into a "seeing counter." As we will see many times throughout the following chapters, it was a continuing goal of both image and logic experimenters to reproduce the epistemic virtues of their opponents within their own traditions.

Anticipating demurral from an audience well schooled in the great photographic triumphs of the past, Kowarski then moved to undermine the unwritten supposition that the golden event exceeded the statistical discovery in importance or value. One might take it as obvious that rare events—in physics or elsewhere—were more precious than those appearing in multitudes. This, Kowarski insisted, was prejudice: "[I]f a 100,000-picture run contains information on several kinds of physical processes, one of which will require a measurement on only one picture out of ten, and another will call for measuring them all it will be only human to find that the first process is more interesting and rewarding than the second. The physics of rare events thus gets, for purely material reasons, preference over the physics of high-number statistics."[244] This preference had led to the conclusion that physics should be left to physicists; they were the ones who could fully extricate the golden events from those of fool's gold. This was specious reasoning, Kowarski maintained. In Kowarski's book, three arguments militated against the supervaluation of the extraordinary event. First, physicists often found themselves in situations in which "there are millions of events to be processed, and where zoology plays only a small role." Second, physics did not exist in isolation, and the techniques of automated analysis, far more than the interactive SMP, could help advance the cause, for example, of microbiology. Third and finally, automated pattern recognition was so manifestly important to defense analysts (e.g., in the analysis of aerial reconnaissance) that the military would surely be developing techniques relevant to the search for a fully automated flying spot. And "after a suitable time lag to let the secrecy abate, it may well happen that fundamental science will yet profit from gadgets

243. Kowarski, "Introduction" (1961), 224.
244. Kowarski, "General Survey" (1964), 26.

developed for defense purposes." One such "gadget," not yet a full pattern recognition device, had already come on the market, he reported.[245]

CERN's ambition, however, hampered its progress. As historians John Krige and Dominique Pestre have pointed out,[246] many commentators have wrongly censured CERN for not engineering enough to keep up with the Americans. The opposite was the case: if CERN lagged behind the United States (especially Brookhaven and Berkeley), it was principally because they *over*designed apparatus. Data analysis was no exception. As Kowarski quipped: "So Berkeley lacks universality, Brookhaven lacks automaticity, and CERN which proudly refuses to sacrifice either universality or automaticity, so far lacks successful achievement in bubble chamber physics."[247]

Even if the glory of new discoveries eluded them for the moment, CERN's drive for ever larger bubble chambers and more sophisticated and faster analysis machines pressed ahead. The next year (1964) at the Karlsruhe meeting on automatic data acquisition, Kowarski repeatedly extolled the virtue of gathering massive quantities of data and then analyzing them with various automated techniques. His audience, clearly disquieted, responded with a mixture of shock, dismay, and resistance to the notion of discovery that seemed implicit in the new experimental order. The following dense exchange is transcribed from the taped discussion that followed the talk:

> K. EKBERG: I would like to ask a question of principle which touched on the point that one might wish to do the first scanning of these pictures automatically. Now it is surely so that many important discoveries have been made because the scientists have noticed something which they did not expect. Something new which they could not explain from their previous knowledge. Now we surely can't programme computers and automatic devices for this kind of thing. Do you think there is any danger here?
>
> KOWARSKI: Well, first of all you must know what you want. You obviously cannot produce in a year several million or several tens of million pictures and at the same time explore each picture in an adventurous way as Blackett used to treat his cloud chamber pictures back in 1934. It just cannot be done. These millions will be taken mainly for kinds of experiments in which you know more or less in advance what kind of information you want to extract. In some intermediate cases there might perhaps be not millions but shall we say a hundred thousand pictures, which are worth to look at not in a statistical way but in a more enquiring mood. . . .
>
> H. SCHOPPER: This point of view frightens me a little because it would mean that in a few years if one wants to do a high energy experiment one would not go to start a new experiment but one would just go into the archives, get a few

245. Kowarski, "Concluding Remarks" (1963).
246. Pestre and Krige, "Some Thoughts" (1992).
247. Kowarski, "Concluding Remarks" (1963), 238.

> magnetic tapes[,] and start to scan the tapes from a new point of view[—]that would be the experiment.
>
> KOWARSKI: If you want to measure something in . . . space, for example about ionized belts or something like that you do not immediately jump and go up there. You send a satellite, usually an unmanned one. You take the information collected by the satellite and then you process the information in your laboratory and I think that this kind of attitude will more and more prevail in high energy physics. . . .
>
> W. W. HAVENS: I would like to point out that looking for the unexpected and hidden is not characteristic for these times. . . . It is important for the physicist to examine his data to look for the unexpected. If the results come only exactly as planned very little is learned.[248]

In the midst of a conference on data reduction, something has broken the tone and rhythm of a quite technical discussion. What up to then has been implicit becomes explicit: reading photographic evidence one way rather than another entrains a scientific life altogether at odds with everything physicists have known. This is why Ekberg brings to the floor "a question of principle." This is why Schopper is "frightened." This is why Havens disparages the character of the times, times that compare poorly with a lost and better world that preceded them, earlier times when the search for the unexpected and hidden was the rule. And it is why Kowarski himself half laments and half celebrates the passing of the "adventurous" era of Blackett, the old days of viewing pictures in an "enquiring mood."

5.10 FORTRAN and Human Nature

If Kowarski thought that CERN had moved away from the "enquiring mood" of an earlier age, Alvarez's headlong plunge into engineering practice would have shocked him still further. At the beginning of a major roundtable discussion at SLAC in 1966, Alvarez suggested that the participants would do well to be "more concerned with engineering and production" than with instrumentation. And engineering and production meant economically informed activities. Citing an inaugural lecture at the United States Academy of Engineering the previous year, Alvarez paraphrased: "Engineering of necessity deals with economics, and any engineering-like activity that does not deal with economic realities is, in fact, not engineering at all." Instrumentation, which Alvarez referred to as that "delightful activity" without pecuniary considerations, did not qualify. Student shops, department shops, laboratory shops—even professional instrument-making shops—were not engineering. In the problem at hand, thinking like an

248. Kowarski, "General Survey" (1964), 39–40.

engineer meant making certain choices that improved the number of "events per dollar" rather than perfecting a piece of hardware or software that would marginally assist in the analysis of a handful of events.[249]

By 1966, data reduction had come of age, and nostalgia for the student shop or even the departmental shop was irrelevant. "We operate a very large business," Alvarez noted, and he went on to outline just how large. In the United States alone the thriving bubble chamber industry owned \$15 million worth of scanning and measuring devices and spent about \$13 million per year, of which \$8 million was dedicated to technicians' salaries and \$5 million to computer analysis. With the stakes this high, competition for the development of data analysis techniques became fierce. There was no time for the luxury of tinkering with a system to produce marginal improvements. For example, certain "fitting programs" straightened out a track degraded by thermal gradients or turbulent hydrogen flow within the chamber. These, Alvarez contended, cleaned up the data, but at the cost of slowing down the analysis far too much. The juggernaut could not stop.[250]

To make his analysis more systematic, Alvarez then turned to the techniques of operations analysis. Operations analysis was a new field, inaugurated during the Second World War to recommend battle strategies for a variety of objectives, most notably the location and destruction of enemy submarines, and the most effective bombing runs. After the war, many of the key figures in operations analysis turned to economic as well as military matters. Key to all their work, and to Alvarez's use of it, was the breakdown of a desired outcome into its constituent parts, each part describable in terms of partial derivatives. In the case at hand, the goal was the efficient production of scanned and measured pictures, and the key constituent quantities were R, the specific scanning rate per hour for an individual scanner, and r, the scanning rate per person averaged over the whole group (many employees are supervising, accounting, etc., and not scanning):[251]

$$\partial R/R \equiv \text{fractional increase in individual scanning rate,}$$
$$\partial r/r \equiv \text{fractional increase in group scanning rate arising from } \partial R/R.$$

In 1966, $R \approx 100$ per hour, and $r \approx 5$ per hour. The crucial quantity then becomes the ratio k of these two dimensionless fractions, that is,

$$\partial r/r \equiv k\, \partial R/R. \tag{5.9}$$

Alvarez argued that the fact that k is near unity implies that the group rate depends almost entirely on the individual scanning rate. Other factors were negligible.[252]

249. Alvarez, "Round Table" (1966), 272.
250. Alvarez, "Round Table" (1966), 272.
251. Alvarez, "Round Table" (1966), 272.
252. Alvarez, "Round Table" (1966), 290–91.

Following this operations analysis led the group, for example, to focus attention on variables, X, that were leading to the greatest change in the ratio of $\partial X/X$ to $\partial R/R$. These included disambiguating tracks that crossed at a small angle, locating the fiducial marks, and measuring the scanning stage to mark the end of short tracks. To Alvarez, the HPD users' focus on the rate, f, at which events are digitized is irrelevant, since it has nothing to do with the more fundamental quantities r (which, in the HPD case, depends only on the number of road makers) and R. In particular, $\partial r/\partial f = 0$. Alvarez conceded that this did not prove that the HPD rate r was not higher than his group's r, but he did conclude that CERN was not using the same operations analytical approach to the problem.[253]

The transition from instrument crafting to industrial engineering and operations analysis affected attitudes as well as specific designs. Early in the 1960s, it had seemed that measuring systems that "corrected" the tracks to ensure their conformity to conservation laws might prove worthwhile. Now it seemed that such systems were not cost effective, given the errors of turbulence, multiple scattering, and film distortion. A cheaper, practical analysis giving the most "events per dollar" was good enough. "The idea of 'good enough,'" Alvarez remarked, "comes from engineering, and although it may upset instrument designers, I feel it is a concept we must embrace."[254] This was certainly the approach taken in the building of the Berkeley accelerators and detectors; now it could be applied to the analysis of results as well. Engineering and industry offered helpful models in management as well as in instrumentation, as Alvarez noted: "[T]he head of our scanning and measuring group spends a large fraction of his time in a role that would be called, in industry, production management." For example, he would prepare daily, weekly, monthly, and yearly reports for himself and Alvarez, indicating such output data as the number of vertices scanned per operator hour, the comparative outputs of operators, machines, and shifts, and the allocation of time to measuring, maintenance, instruction, and programming.[255]

The transition to corporate industrial practice was not painless. As Alvarez reminded an audience of physicists, it was well known among industrialists that little companies often go bankrupt trying to pass from an organization in which individuals handled many jobs to one in which high efficiency comes from "production line operation with very expensive production tooling."[256] The main difficulty is inefficient use of supervisory personnel; if this pitfall can be avoided, the company operates at a vastly higher level of productivity. As in business, so in the analysis of physical data.

253. Alvarez, "Round Table" (1966), 292–93.
254. Alvarez, "Round Table" (1966), 272.
255. Alvarez, "Round Table" (1966), 277–88.
256. Alvarez, "Round Table" (1966), 289.

In these remarks, Alvarez identified a business-historical trend that Alfred Chandler has singled out as the essential feature of modern corporations: the introduction of middle management.[257] Each box in a corporate flowchart stands for an organizational group that can function independently. By the mid-1960s, the Alvarez group clearly had such middle-level groups, such as bubble chamber operation and development under James Gow and data analysis development that at various times was under Hugh Bradner and Frank Solmitz.[258] Indicative of the relative autonomy of the subgroups are the many reports, conference talks, and even publications that issued from cryogenic, optical, and data-processing collaborations. Once the large structure of scanners, engineers, and physicists became integrated, production did indeed speed up considerably. But Alvarez, knowing that his business methods for accelerating data analysis would startle his audience, reiterated the importance of a pragmatic approach: "For those of you who may be horrified to hear a scientist setting such unscientific goals, let me remind you that I have my 'engineering hat' on at the moment so I have no apologies."[259] Although the value of scientific goals could not be defined in production terms, the preconditions for important work could.

On first glance, the Alvarez group seems dominated by pragmatic concerns (events per hour, operations analysis of production, commonsense definition of "interesting physics"). But despite the attention to supervisors, accounting, and engineering; despite the reams of paper devoted to production per operator hour, machine hour, and team hour, I suspect the real animus behind their opposition to the HPD had little to do with partial derivatives of production. We have seen hints of this earlier—when Alvarez said in 1963 that even if the HPD speed increased there would be differences in philosophy about how physicists should interact with their data. Now, this deep concern emerged more strongly than ever: "[M]ore important than [my] negative reaction to the versatile pattern recognition abilities of digital computers is my strong positive feeling that human beings have remarkable inherent scanning abilities. I believe these abilities should be used because they are better than anything that can be built into a computer."[260]

Alvarez expresses a belief in a peculiarly human capacity to unravel pictures that is, I would argue, at the root of the development of his reading regimen in all its many facets. It is, for example, the basis for making all of the apparatus interactive, rather than segregated. And Alvarez's sense about our special human capacities lies behind all the talk of zoo-ons, and his vision of discovery.

To support his position, Alvarez told a cautionary tale. An astronomer had

257. Chandler, *Visible Hand* (1977), 1–3.

258. On Gow and Bradner, see Heilbron and Seidel, *Lawrence* (1989), 95.

259. Alvarez, "Round Table" (1966), 262. See Alvarez, "Recent Developments" (1972), 266ff., for a discussion on who did what in the development of computer data analysis.

260. Alvarez, "Round Table" (1966), 294.

found Pluto by using a blink comparator. Flicking into focus first one picture of a star field and then another taken sometime later, the discoverer had noticed a single spot of light that had changed. What, Alvarez rhetorically asked, would have been the outcome if astronomers had searched with full automation? "One could certainly scan a star field with an FSD-like device, and store the coordinates and intensities of all stars in some memory system. Then he could repeat this measurement on a star plate taken at a different time, and again store the coordinates and intensities. And finally, he could perform the arithmetic comparisons between the two lists, and throw out all stars that had duplicate images on the other plate to within some 'least count.'"[261] "Certainly," he concludes, it would have been a "mistake" to try "to beat the human with a computer." Everything, even evolutionary theory, pointed to the "extraordinary" capacities of the eye-brain system, a system not easily duplicated by machines, however complex. Pluto was a warning to those hoping to push the human eye aside. "I hope that we never go so far in automatic scanning that we do in a similar tedious and expensive way something that can be done so easily by a human scanner. I'm sure I can't discourage anyone who really wants to replace scanners by machines. But I don't think that such an effort is usefully related to bubble chamber physics."[262]

To maintain the humanity of reading, Alvarez was willing to impose industrial production as a work regime. By contrast, the HPD advocates called the style of work imposed by the interactive mode "slavery" and, to avoid it, were willing to sacrifice the human being as part of the regimen of discovery. K. Gottstein held out the hope that the HPD would alleviate the oppression of routinized round-the-clock labor. Perhaps physics could be restored to physicists. At the 1967 Munich meeting on HPDs, Gottstein sketched a Dickensian picture of film reading in the age of million-photograph experiments:

> We are now still used to a situation in which the accelerators and their bubble and spark chambers turn out data faster than the physicists and their assistants can analyze them. As a consequence, many laboratories which analyze . . . bubble chamber film, work around the clock, in shifts, day and night and weekends, like a blast-furnace or the fire-brigade, and the lights never go out. Now the day may not be so far on which with help of computers, flying spots, etc. the data can be analysed faster than they are produced. This may mean that our laboratories could return to a normal day-time routine of operation which would certainly be more appropriate to the natural rhythm of human life.[263]

Here was a dream of physics as pure ideation without labor and all its dirty blast furnace associations. Like Leprince-Ringuet a decade before, Gottstein was repelled by the work ethos of the new American laboratory.

261. Alvarez, "Round Table" (1966), 294–95.
262. Alvarez, "Round Table" (1966), 295.
263. Gottstein, "Introductory Remarks" (1967), 3.

To Kowarski, the struggle against the human was one marked, as we saw, by an almost aesthetic pleasure in isolating the machine from the human being. With palpable frustration he ended a 1964 conference with the impression that "no matter how automatic any given system claims to be, the human element comes galloping back and the system begins to incorporate again some sort of human guidance." Brookhaven's "gadget" for spark chamber pictures might aspire to the "completely inhuman" but ended up employing "quite a bit of human guidance"; even the HPD spark chamber analyzer which momentarily "achieved complete inhumanity" fell back upon human intervention.[264]

Summarizing the views of one of his HPD associates at their 1964 conference, Kowarski noted that the term "interface" had been used as but "another word for humanity's ugly face showing up again in some way, so that the system is not fully automatic after all."[265] Perhaps discouraged by the eternal return of repressed humanity, Kowarski concluded the last of the dedicated HPD meetings with the concession that "[t]he real purpose of the whole HPD development, including its recent cathode-ray tube sequels, may turn out to be definable not exclusively in relation to a certain class of problems arising from high-energy physics, but rather in terms of a simpler and wider ambition to provide a computer with an eye."[266]

In a sense, the reading machine debate of the 1960s pitted two equal, though differently situated, ambitions against one another. The Alvarez group at Berkeley aimed at the discovery of novelty, classically defined in terms of a startling golden event. And they did find such objects, indeed several of them. Muon-catalyzed fusion was one, in which a muon caused two hydrogen atoms to fuse at liquid hydrogen temperatures. (For this fleeting moment, long before the great Utah Brouhaha of the 1980s, the excitement of cold fusion swept into physics.) Another golden event inaugurated the discovery of the cascade zero, a single remarkable image that only painstaking analysis could transform into an argument. Only later–seven years after Stevenson opened the first Berkeley bubble chamber notebook–did a second kind of analysis begin to emerge. By amassing great storehouses of data, and automating what not long before had been sophisticated state-of-the-art theoretical interpretation, the team began to explore resonances, excited states of baryons and mesons. Joined together, the bubble chamber and the Alvarez style of interactive data processing made possible both the golden event and the resonance work of the 1960s. The Alvarez group aimed to combine industrial engineering with an irreducibly "human" authorship at every stage.

At CERN, the data-processing ambition was equally grand, to do new

264. Kowarski, "Concluding Remarks" (1965), 264–65.

265. Kowarski, "Concluding Remarks" (1965), 264.

266. Kowarski, "Concluding Remarks," in Powell and Seyboth, *Programming* (1967), 409–16, on 415.

image tradition physics with the massive statistical sweep previously possessed only by the competing tradition of electronic counters. More than that, the CERN physicists (and their American allies at Brookhaven and elsewhere) hoped their new device would find application far beyond particle physics. Some imagined automatic scanning in applications from chromosomal analysis to aerial photography—the dream was to build a device that would read photographic evidence with the ease that ordinary computers processed punchcards. At one level, the HPD advocates were astonishingly successful: the national high energy physics laboratories of practically every technically advanced country and many universities launched HPD efforts. Leaving aside the two hand-built HPDs (both of which resided at Brookhaven), 26 commercially produced Sogenique HPD II's went to laboratories, three to the Soviet Union. Amsterdam used its machines for data reduction, as did the Berkeley group and others at CERN led by Brian Powell himself. But even after all this effort, the device never produced the kind of results that emerged from Alvarez's laboratory. In fact, Hough ended up establishing a "manual rescue procedure" (OFFMIF for Off-Line Manual Integral Filter) that "scooped up the correct points on tracks that failed geometrical reconstruction, using human recognition (women, naturally) and a light pen."[267] Data produced this way could compete with Alvarez's Franckensteins, but by importing the human, the dream of full automation had come to a halt.

To a certain extent some of the hopes and most of the fears of both sides were realized in the coming decades of the 1970s and 1980s. Experiments grew vastly larger, leaving teams of Alvarez size (20, 30, or 40 Ph.D.'s) looking minuscule by comparison. Electronic devices took over, making the individual scrutiny of events by human beings less and less feasible. Yet at the same time, the pattern recognition device, like the automatic language translator and so much else in the burgeoning field of artificial intelligence, compromised its goals in order to realize them.

As he surveyed the field for the last time in 1967, Kowarski had to admit that "the goal of complete automation is slightly receding." And if neither elegant electromechanical devices nor fancy programming could blot out the human as a reader, then Kowarski would bring back the reader in the form she had originally taken. Alluding to his oft-repeated dichotomy between the two paths reading might take, he now added (or returned) to a third: "Should we add to tricky hardware and tricky software also tricky girlware?"[268] At the end of the day, both Berkeley and Geneva had teams of physicists with vastly revised daily practices, working alongside engineers of many stripes, and bands of women, linked to computers, tracing particle trajectories across the photographs.

Changes in the organization of work and in the evolving man-machine

267. Hough to author, e-mail, 11 December 1995.

268. Kowarski, "Concluding Remarks," in Powell and Seyboth, *Programming* (1967), 409–416, on 414.

relationship deeply affected the day-to-day running of an experiment. Indeed, the nature of experimentation had reached a watershed. Kowarski prophesied in 1964 that "today's idea of a physical experiment is concerned mainly with the setting-up and operation (in the accelerator's real time) of the beam-defining and detecting apparatus; in the future, the 'runs' of such an apparatus may become more comprehensive, and the individuality of an 'experiment' may shift to the dialogue between the physicist and the data-processing device."[269] That transformation had already begun.

The process by which the tasks of a complex technological workplace are fragmented into smaller units has been studied in a number of professions (though more by labor historians than historians of science). For our purposes, a particularly interesting example is computer programming. During and immediately after World War II, each task assigned to a computer had to be individually prepared at the level of machine language. Stored-program computers, introduced in the United States in 1950, revolutionized the relation of computer to user on every level, from the training of specialists to the types of problems the computer could solve.[270] Almost immediately a huge demand for programmers was created by the American government, which wanted to double the number of programmers from 2,000 to 4,000 in order to prepare the early-warning radar network code-named SAGE (Semi-Automatic Ground Environment). In addition to facilitating improvements in hardware, SAGE required a new staff structure. Early in the project's history the Systems Development Corporation was created to produce the programming workforce. The large scale of the project necessitated restructuring of software training and work organization from an apprenticeship system (largely within electrical engineering) to a more specialized division of labor. Systems analysts, for example, were trained to supervise programmers, whose programming objectives and methods were highly specific. As this trend continued over the next decades, programming was further broken down into a hierarchy of skill levels. By the late 1970s a chief programming team would make large-scale architectural decisions, less skilled programmers would fill in the modules outlined by the chief programming team, and coders would employ canned programs with minor changes for specialized uses.[271]

This shift in programming work organization is relevant on two levels. In general terms, the specialization of tasks and the isolation of less skilled from more skilled tasks parallels the evolution of laboratory life surrounding the bubble chamber. More specifically, we can see in the history of programming the outlines of the changing role of the computer at the Lawrence Berkeley Laboratory. The first important computer program written for the bubble chamber

269. Kowarski, "General Survey" (1964), 36.
270. Kraft, *Programmers* (1977).
271. Kraft, *Programmers* (1977).

group was PANG, the track reconstruction software, which had to be prepared entirely in machine language. By the time KICK and SUMX were assembled (principally by Frank Solmitz and Arthur Rosenfeld), FORTRAN was available, enormously simplifying a still difficult job. Later, as this software was copied and distributed, its use became easier, with smaller changes needed for adaptation to specific experiments. Exemplary among such cases are the discoveries at LBL of the strange resonances and the ω and η particles. Both made fundamental use of the new strategies of reading photographs. Data reduction and discovery had become inseparable in the experimental workplace.

5.11 Physics Results

Programming, data reduction, engineering, and physics converged with the discovery of a wealth of new unstable particles in the late 1950s and the early 1960s. Already in 1957 the startling success of the bubble chamber program led Alvarez to doubt the future of competing detectors. Writing to Edwin McMillan, he elaborated: "For one to appreciate such an assertion the way I feel it, he would have to look at tens of thousands of bubble chamber pictures, the way I have done. When one sees the way in which an interaction of charged particles gives rise to neutral particles which go some distance and then decay into charged secondaries which then may decay into other charged secondaries, or undergo charge exchange reactions, and reappear as neutral particles and so on, he can be as discouraged as I am about the future of counters."[272] Just such a long string of decays would be needed to find the cascade zero, predicted to exist by Murray Gell-Mann and Kazuhiko Nishijima.[273] For theorists, the cascade zero fit into a broader structure of group theory and concerns with the evolving quantum field theory. For experimenters, the cascade zero was a problem of a very different sort, one that involved sorting and analyzing vast numbers of bubble chamber pictures.

From over 10,000 K mesons passing through the chamber, Alvarez hoped to find some of the elusive Ξ^0's. Writing to W. Libby, commissioner of the AEC, Alvarez announced in December 1958 that his group had such a candidate. With the help of the computer's kinematical analysis, the complicated production and decay of the cascade zero could be deduced from a single specimen:

$$\begin{array}{llll} K^- p \rightarrow K^0 & + & \Xi^0 & \\ \quad\;\; \hookrightarrow \pi^+\pi^- & & \hookrightarrow \Lambda^0 + \pi^0 & \\ & & \quad\;\; \hookrightarrow \pi^- p & \end{array} \tag{5.10}$$

272. Alvarez to McMillan, 17 September 1957, widely circulated. Office of W. A. Wenzel, Lawrence Berkeley Laboratory, WWP.

273. Gell-Mann, "Interpretation," *Nuovo Cimento*, Suppl. 4 (1956): 848–66; and Nishijima, "Independence Theory," *Prog. Theor. Phys.* 13 (1955): 285–304.

Alvarez reflected: "We, of course, hope to get several more of these particles before the run is finished, but even if we don't, I feel we will publish this event without reservations."[274] No more cascade zeros showed up despite a ferocious search, so the singular event was reported, along with its characteristic photograph.[275]

Exhibition of such golden events was one way that the bubble chamber could be used to construct an experimental demonstration. Another, equally important, use of the bubble chamber was to gather a statistical sample of many decays. During 1960–61, the data reduction program at Berkeley used such statistical evidence to educe the existence of three particles, the strange resonances: $Y_1^*(1380)$, $K^*(890)$, and $Y_0^*(1405)$.[276] Illustrative of the mode of their discovery is the case of the $Y_1^*(1380)$. In this experiment, led by two graduate students, Stan Wojcicki and Bill Graziano, a monoenergetic beam of K^- particles was incident on the 15-inch hydrogen bubble chamber in which the K^- particles interacted. Forty-nine events of the type $K^- p \rightarrow \Lambda^0 \pi^+ \pi^-$ were isolated, after the data had been scanned for events with a V (decay products of the lambda zero) and the two prongs that were candidates for the pion pair. Using the computer and a half-automated Monte Carlo simulation, Wojcicki and Graziano concluded that another 92 could have been either $\Lambda^0 \pi^+ \pi^-$ or $\Sigma^0 \pi^+ \pi^-$, as these looked kinematically quite similar.

Intriguingly, when Wojcicki and Graziano plotted the energy of the final-state positive and negative pions, each displayed a peak at about 285 MeV (see figure 5.27). This energy corresponded to a mass for the resonant system of lambda and the other pion of about 1380 MeV (i.e., the decay looked as if it were two-body, with the available energy split between a pion, e.g., the π^+, and a single resonant entity, e.g., $\Lambda^0 \pi^-$). Though some of the local theorists protested, Harold Ticho (visiting from UCLA) and Alvarez himself backed the students' explanation that the reaction was proceeding not directly, but by way of an intermediary, a new particle Y_1^* (later known as Σ^*) of mass near 1380 MeV:[277]

$$
\begin{aligned}
K^- p \rightarrow\ & Y_1^{*+/-} + \pi^{-/+} \\
& \hookrightarrow \Lambda^0 \pi^{+/-}
\end{aligned}
\tag{5.11}
$$

Were this the actual production process, a second pion peak (between 58 and

274. Alvarez to Libby, 12 December 1958, LAP.

275. Alvarez et al., "Neutral Cascade," *Phys Rev. Lett.* 2 (1959): 215–19.

276. Alston et al., "Resonance in the Lambda-π System," *Phys. Rev. Lett.* 5 (1960): 520–24; Alston et al., "Resonance in the K-π System," *Phys. Rev. Lett.* 6 (1961): 300–302; and Alston et al., "Study of Resonances," *Phys. Rev. Lett.* 6 (1961): 698–705.

277. Wojcicki, "Alvarez Group" (1987).

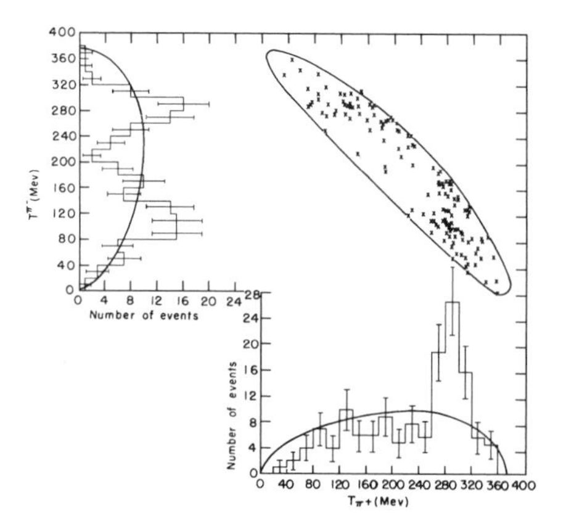

Figure 5.27 Y* resonance (1960). One form of demonstration, unanticipated even at the time Alvarez proposed the 72-inch chamber in 1955, was a statistical use of bubble chamber data to find excited states of strongly bound particles. Here the collaboration examined data from a negative kaon beam hitting a proton target and producing a visible Λ, π^+, and π^-. The claim was that this proceeded by way of an intermediate state:

$$K^- p \rightarrow Y^{*+/-}\pi^{-/+} \rightarrow \Lambda^0\pi^+\pi^-.$$

In this figure, the histogram of π^- energy is plotted on the upper left, the π^+ energy on the bottom. Solid lines over histograms indicate the phase-space curves—the curves that would be followed if there were no intermediate state. Without such an intermediary, the Dalitz plot (*oblong*) would be uniformly populated, whereas its strikingly clustered appearance signaled the new resonance: the Y*(1380). This demonstration launched a long epoch of "bump hunting." Source: Alston et al., "Resonance in the Lambda-π," *Phys. Rev. Lett.* 5 (1960): 520–24.

175 MeV, cleanly separated from the 285 MeV pions) should be observable when the Y_1* broke up into a lambda and a pion—some spread is expected, due to the width of the Y_1*. The additional structure of a new particle is apparent in figure 5.27, where the Dalitz plot exhibits strong deviation from phase-space homogeneity: a vertical band is apparent near 285 MeV for the positive pion, and a horizontal band near 285 MeV for the negative pion. Were there no intermediate Y_1*, one would expect an even spread of Dalitz points over the part of the plane allowed by energy conservation (the oblate region in figure 5.27). Instead, and unmistakably, the points huddled together. Further analysis of the data indicated the spin and the isotopic spin of the particle, and similar analyses were used in the unveiling of the *K**(885) and Y_0*(1405). This work capped an eminently successful hardware and software effort, firmly establishing the Alvarez group as the world leader in particle physics.

In addition to the success of the in-house physicists, one of the salient features of the new kind of experimentation was the distribution of the bubble chamber film to outside groups. For example, in the summer of 1959, a visitor to the laboratory, Bogdan Maglić, began a search for the ω, an isospin-zero neutral meson.[278] Since charge is conserved when an ω decays into pions, it must go into neutral combinations such as $\pi^+\pi^-\pi^0$. (The symmetry of the wave function discourages the ω from decaying into two pions.) Maglić then reasoned as follows: In a proton-antiproton collision the ω could be produced along with a $\pi^+\pi^-$ pair. The ω would then decay into three pions, leaving a total of five pions in the final state. Finding the event candidates was no mean feat. Maglić exploited the standard, Alvarez-group KICK program to identify the three-pion decays and to calculate the momentum of the unseen neutral one. With the energies and momenta of the pions, he could plot (by hand, since SUMX did not yet exist) the invariant mass of neutral three-pion combinations. A peak confirming the existence of the ω stood out in the region of 800 MeV. That no similar peak could be found for the charged three-pion combinations, such as $\pi^+\pi^+\pi^-$, indicated that the particle was indeed without isospin partners.

Soon afterward, another group led by Aihud Pevsner from Johns Hopkins began examining bubble chamber film shipped from Berkeley documenting the effects of a 1.23 GeV beam of positive pions incident on deuterons in the Berkeley 72-inch chamber.[279] Like Maglić, Pevsner's Johns Hopkins–Northwestern collaboration was searching for a neutral meson decaying to $\pi^+\pi^-\pi^0$. Here, the team would look for $\pi^+d \rightarrow ppX^0$, where X^0 (the searched-for meson) would then decay via $X^0 \rightarrow \pi^+\pi^-\pi^0$. The hunt, therefore, was on to find events of the type: $\pi^+d \rightarrow pp\pi^+\pi^-\pi^0$. Again, PANG reconstructed the tracks—all except the trace of the neutral pion, which of course left no footprint. But by plotting the missing mass on the tracks that were visible, the collaboration could show that the missing-mass peak fell on the mass of a neutral pion. Confidence rose. KICK performed the kinematical analysis, doing its statistical sorting. Events had to satisfy three statistical tests: $\chi^2 \leq 6$ for the hypothesis $\pi^+d \rightarrow pp\pi^+\pi^-\pi^0$, and at the same time KICK enforced a criterion that made would-be imitators unlikely: it demanded $\chi^2 \geq 25$ for the hypothesis $\pi^+d \rightarrow pp\pi^+\pi^-$ (no neutral pion) and imposed a final χ^2 test to eliminate events in which protons and pions were not clearly distinguishable. So vetted by the computer program, the data could then be plotted to show the effective mass of the unseen neutral meson. Clearly visible was the Maglić group's omega meson at about 770 MeV, and alongside it, the new particle, called η, weighing in at 550 MeV (see figure 5.28). As the η and ω discoveries indicate, by the early 1960s not only had the Bevatron been opened to collaborations with other laboratories, so had the bubble chamber and the data reduction machinery. Conducting an "experiment" (and therefore being a scientific author) meant something different

278. Maglić et al., "Evidence," *Phys. Rev. Lett.* 7 (1961): 178–82. The actual data reduction was performed by Maglić alone, according to Alvarez's Nobel Lecture, "Recent Developments" (1972), 241–90. The distribution of exposed photographs for analysis was not unique to bubble chambers; emulsion groups had done this also.

279. Pevsner et al., "Evidence," *Phys. Rev. Lett.* 7 (1961): 421–23.

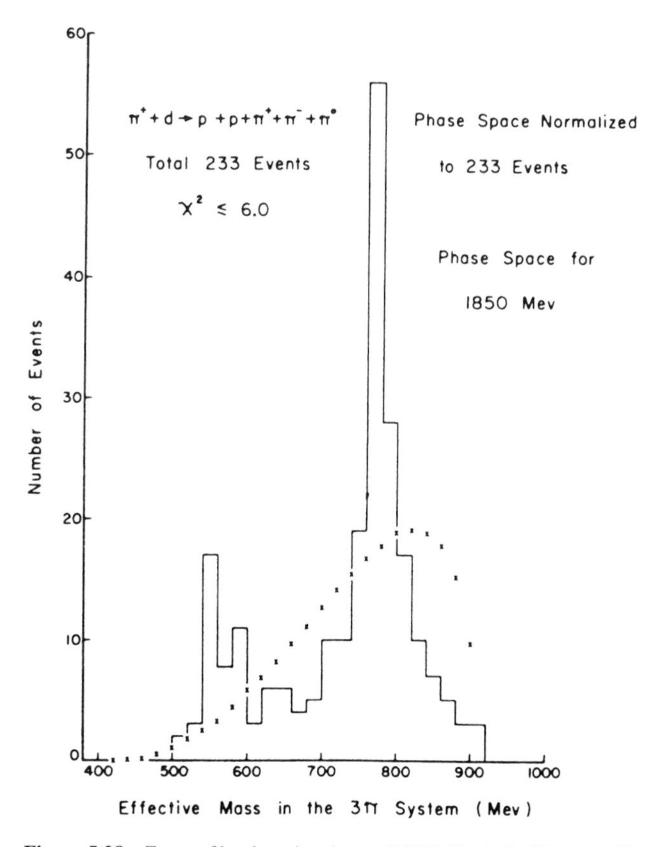

Figure 5.28 Export film, locating the eta (1961). Typical of "exported" film analysis, the Johns Hopkins–led group took the film, ran the various LBL-packaged film analysis programs, and deduced that a meson, dubbed the eta (η), existed at a mass of 550 MeV. Note that now it is possible to author a discovery principally through involvement in analysis. At the same time, the "site" of experimentation diffuses, now that construction, manipulation, and analysis are no longer located in the same place. Source: Pevsner et al., "Three-Pion Resonance," *Phys. Rev. Lett.* 7 (1961): 421–23, on 422.

when both the film and much of the data reduction machinery traveled from laboratory to laboratory.

Evidence for the centrality of data reduction to the new definition of experimentation comes from another quarter as well. Emulsion groups all over the world had grown expert at the painstaking reconstruction of tracks and their corresponding kinematics. When the Alvarez group began to distribute some unprocessed film, many emulsion collaborations were eager to participate. Werner

Heisenberg, on behalf of his colleagues at Göttingen, queried Alvarez about the possibility of obtaining film for the relatively large photographic plate group of Gottstein.[280]

For bubble and emulsion groups, the huge new Berkeley bubble chamber provided hundreds of events where other methods of particle detection could not even find one. Larger data sets in the ω and the η discoveries aided in identifying the peak of the effective mass histograms; equally important, the larger data sets allowed the determination of ΔE, the width or uncertainty in mass. Heisenberg's uncertainty principle tells us that $\Delta(T)(\Delta E) = \hbar$, where $\Delta(T)$ is the uncertainty in time. Therefore, good mass measurements also determine a particle's lifetime. Lifetime measurements on strongly interacting particles were an unexpected benefit of the bubble chamber, since originally the 72-inch device was designed to measure lifetimes of the Λ^0 and θ^0, which lived to the ripe old age of 10^{-10} seconds; the strongly decaying mesons survived but 10^{-23} seconds.

Over the early and mid-1960s, a great many more particles were discovered in bubble chambers at Berkeley, Argonne, Brookhaven, CERN, and elsewhere. Prominent among these finds were the $\Xi^*(1530)$ (independently discovered by a collaboration using the 72-inch bubble chamber[281] and by a group working with the 20-inch chamber at the Brookhaven AGS)[282] and the $\Sigma^*(1385)$ (discovered by the Alvarez group). Indeed, when the $\Xi^*(1530)$ was announced on 5 July 1962, at the CERN International Conference on High Energy Physics, it fit neatly into the recently developed SU(3) particle classification scheme that Gell-Mann and Ne'eman had recently developed. Five days later, at a plenary session of the CERN meeting, Gell-Mann had pointed out in a conference discussion that the SU(3) scheme could account for an equal mass spacing between the $\Delta(1238)$ and the $\Sigma^*(1385)$, between the $\Sigma^*(1385)$ and the $\Xi^*(1530)$, and between the $\Xi^*(1530)$ and a hitherto unseen particle. Perhaps, he concluded, "our speculation might have some value and we should look for the last particle, called, say, Ω^-." This particle would have strangeness 3, isospin 0, and mass 1685 MeV. Its discovery at Brookhaven in 1964 dramatically confirmed the SU(3) symmetry.[283] Pursuing this symmetry would lead physicists toward the quarks and a new generation of physics (see figure 5.29).

By the time the Ω^- was found, plans were being set at laboratories around the world to exploit the bubble chamber for the study of neutrino interactions. It was a direction taken by many physicists who previously had been engaged with the strongly interacting resonance physics, including several from the Alvarez

280. Heisenberg to Alvarez, 21 September 1956, LAP. Gottstein had visited Berkeley to learn about bubble chamber technique.

281. Pjerrou et al., "Resonance" (1962).

282. Bertanza et al., "Possible Resonances," *Phys. Rev. Lett.* 9 (1962): 180–83.

283. Gell-Mann, comment in the discussion following Snow, "Strong Interactions" (1962), 805. Barnes et al., "Observation," *Phys. Rev. Lett.* 12 (1964): 204–6.

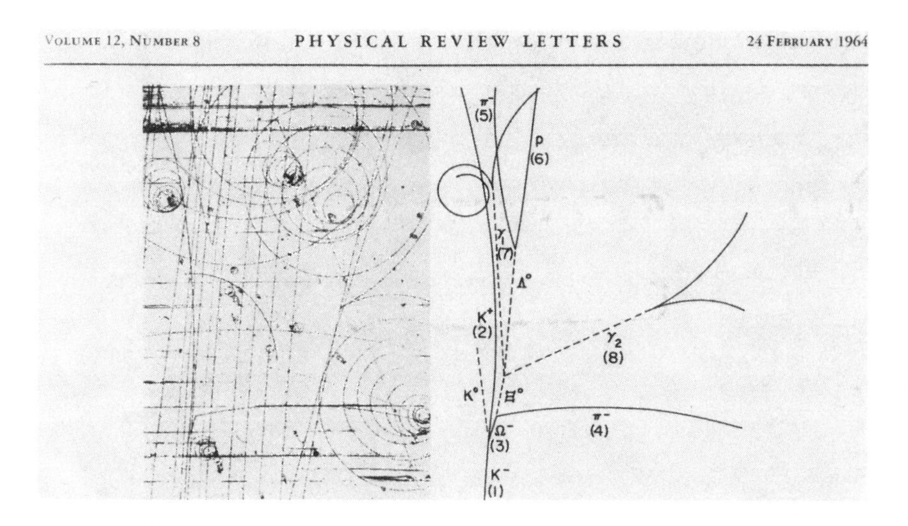

Figure 5.29 Omega minus (1964). Of all the golden event photographs prized from the bubble chamber, the omega minus (Ω^-) is perhaps the most often reproduced. The discovery of this event (*left,* photograph; *right,* schematic) by the Brookhaven group was a great competitive victory over the ever-powerful team at Berkeley. Source: Courtesy of Brookhaven National Laboratory.

group, such as Lynn Stevenson. Of the many remarkable successes of neutrino bubble chamber physics, the most influential was undoubtedly the discovery at CERN of weak neutral currents (see figure 5.30). As a confirmation of the electroweak theory it was remarkable; it also marked a great resurgence in European experimental particle physics.

As important as the physics was, the rapid growth of detectors and data reduction deeply affected the nature of laboratory life. Physicists were hopeful about the new possibilities for experiments, but worried as well. W. Jentschke (later to become director general of CERN) voiced concern that the growth of detectors and data analysis equipment was driving more and more physicists and engineers into administration.[284] Similarly, at the Dubna conference Goldschmidt-Clermont regretted "[t]he necessity to learn and use extensively the art of programming, or to enlist the help of professional programmers, at the price—amongst others—of a feeling of remoteness from the actual workings of the experiment."[285]

For students, the danger was even greater, for while R. J. Spinrad took "the mechanization and acceleration of the more elemental tasks" to be "a great boon to the experienced worker," he added that this same automation might work to

284. Jentschke, "Invited Summary," *Nucl. Inst. Meth.* 20 (1963): 507–12, on 512.
285. Goldschmidt-Clermont, "Progress" (1969), 441.

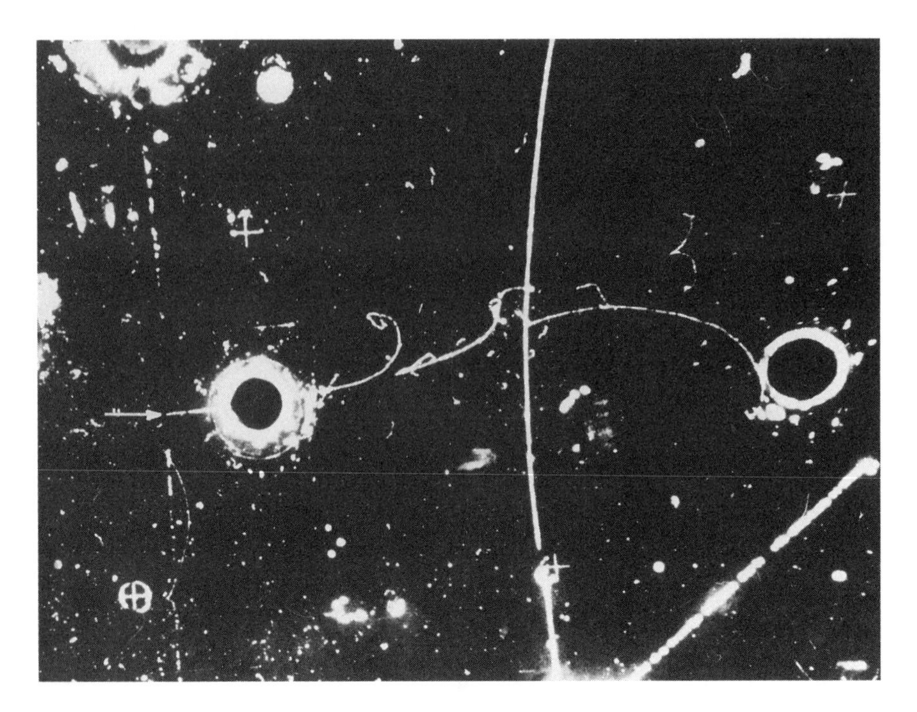

Figure 5.30 Neutral current (1973). This was, as one leading Gargamelle physicist called it, a "picture-book event," a single-electron event found in Aachen in early January 1973. It was a candidate for the purely leptonic scattering of a muon-neutrino from an electron. Many theorists and experimenters found this event, and a handful of others like it discovered in the next months, particularly compelling because their analysis did not require any assumptions about the interior of protons and neutrons and they were particularly simple to analyze. The electron's trajectory goes from left to right, beginning at the arrow's end, where it ostensibly was hit by a right-moving neutrino. Haloed black circles are lights illuminating the bubble chamber liquid. Source: Hasert et al., "Muon-Neutrino," *Phys. Lett. B* 46 (1973): 121–24, on 122.

the detriment of the student for whom the task might not be "elemental" by virtue of his never having performed it. Students deprived of a chance to "get [their] hands dirty" would find themselves cut off from the "realities of the experiment."[286]

5.12 The Rationalizer Rationalized

The industrialization of laboratory life began to wear even on the originators of the new methods. During the spring of 1959, Berkeley approached Glaser, hoping

286. Spinrad, "Digital Systems," *Prog. Nucl. Tech. Inst.* 1 (1965): 221–46, on 245.

to lure him to the Bay Area with the promise of his own group. Glaser accepted, bringing several of his own: George Trilling, who had done his degree with Carl Anderson at Caltech; John Brown, a Michigan Ph.D. who was already expert with computers and data handling; and John Kadyk, who specialized in optics and electronics that could supplement the bubble chamber.[287] From the start, indeed before he came, Glaser told McMillan that he expected to keep his group structure relatively weak in cohesion, allowing individuals to separate if common interests no longer bound them together:

> A number of other wilder ideas have been proposed and discussed among us. If we can all agree on the best next step for us to take after the most interesting xenon experiments are finished, it is my plan that we continue to work together. If no joint project fires our enthusiasm sufficiently, then it is my intention that each of us will choose what work interests him most and what collaborators in the laboratory he most prefers to work with. I am not fully familiar with the detailed operation of the Research Group concept at the laboratory, but have firm ideas about how I would like to run my own group.
>
> At the beginning we would work with the xenon chamber. . . . I am violently opposed to the notion that there is any "group allegiance" or commitment that goes far beyond this present work. I think that it is in the best interests of the development of the individual scientist and of the quality of scientific work, that there should be wide opportunity for work in many directions and with flexible possibilities for collaboration.[288]

What for Glaser was the necessary preservation of an individual scientist's "flexible possibilities" was for Alvarez a necessary sacrifice. Within his hydrogen bubble chamber group, Alvarez practiced a tough-minded managerial style; he believed production methods demanded them. And as he looked at other instrument programs, such as the electronic counters, he saw only inadequate alsorans. Faced with Alvarez's discouragement of their development of heavy liquid chambers, T. G. Pickavance wrote to Glaser in July 1959, saying that University College London hoped to build a heavy liquid chamber as an "adjunct" to the hydrogen chamber at the NIMROD cyclotron: "Alvarez visited us recently and, not unexpectedly, was not inclined to favour propane or heavy liquid chambers."[289]

Glaser responded by thoroughly backing the move toward a propane chamber—it had good gamma detection efficiency, it offered the possibility of exploring cooperative phenomena inside nuclei as they affected particle production and interaction, and its stopping power would aid in particle identification; stopping power would help too in the analysis of rare events such as double or triple

287. Glaser to McMillan, 17 March 1959, box 9, folder 112, NA-SB, EMP.
288. Glaser to McMillan, 17 March 1959, box 9, folder 112, NA-SB, EMP.
289. Pickavance to Glaser, 7 July 1959, box 9, folder 112, NA-SB, EMP.

scattering or in the conversions of the kaon system. But beyond these specific advantages, Glaser was keen to encourage diversity:

> To me it has always seemed important in high energy physics to have available a great diversity of techniques so that new theoretical ideas may be tested, and experimental uncertainties resolved with the greatest possible range of conditions of observation. Only with such flexibility can the creative imagination of the experimenter attain its best results. . . . To limit oneself to one or several techniques, except in case of dire economic necessity, is to fly in the face of experience in experimental physics and to deny the worth of inventive ingenuity. I have been a bit long-winded and perhaps allowed myself an excess of philosophy. Please excuse me for this.[290]

Given his predilection for "wilder ideas" and "inventive ingenuity," along with individual rather than group allegiance, and a multiplicity of techniques, it is perhaps not too surprising that Glaser and Alvarez did not frequently see eye to eye at the Radiation Laboratory during Glaser's first academic year there, 1960–61. In October 1960, Glaser was awarded the Nobel Prize for Physics, specifically for his invention of the bubble chamber. But by the end of that year Glaser had begun to distance himself from high energy physics and began to turn his attention toward biology. From Europe, he wrote to McMillan, in whom he found a sympathetic spirit, and expressed delight at the new world: "I am learning to use pipettes, centrifuges, refrigerators, agar petri dishes, anion resin exchange columns, spectrophotometers and all the rest of the paraphernalia. The techniques seem childishly simple and in general, so do the general theoretical ideas one is trying to test. The real complexity and virtuosity lies, as you once told me, in knowing the detailed biochemistry and accompanying technical tricks."[291]

In a sense, the agar petri dish and centrifuge were far closer in work style to the vacuum pumps and 2 cc bubble chambers than such chambers were to Alvarez's 72-inch monster. In its scale and in its potential for individual exploration and authorship, microbiology in 1961 was the natural continuation of a life world, unknown in the Radiation Laboratory of Lawrence and Alvarez, but one that had existed in Caltech physics under Millikan and Anderson. It was a world in which an individual scientific worker could refine a technique, explore an interpretation, and write a paper. As he turned to reflect on the by-then distant world at Berkeley, Glaser appended this to his letter to McMillan: "I most sincerely hope that things are calmer with Luis [Alvarez] than they were last year. His behaviour put everyone's nerves on edge and could do much to erode the morale and scientific quality of the work at the Rad Lab. It is so important that

290. Glaser to Pickavance, 23 July 1959, box 9, folder 112, NA-SB, EMP.
291. Glaser to McMillan, 3 April 1961, box 9, folder 112, NA-SB, EMP.

individuals and scientific ideas be always considered as the most important ingredients of our work, and that the teams and industrial organization are no more than a necessary means to an end."[292]

What smaller scale work did survive in the bubble chamber business did so by means of the heavy liquid chambers, devices Alvarez always thought of as poor relations to the more "fundamental" hydrogen chambers, which could function without the nuclear complexity of hydrocarbons. Thus, when Glaser's group (Brown, Kadyk, and Trilling, along with Gidal and Powell) proposed a new high-magnetic-field propane chamber, Glaser sent a letter of support, adding that the proposal writers had neglected to mention neutrino physics as a possible application. "I think it is important and will be productive to have available as large a number of varied experimental techniques for whatever ideas become crucial as well as a number of separate groups of people who are interested in these different techniques and will be able to expertly apply them as the problems arise."[293]

Glaser's one-year venture into biology extended to two as he took a visiting professorship at MIT. Eventually, he left physics entirely to pursue a career in biophysics. One of his first important contributions was the adaptation of bubble chamber data analysis and handling techniques to the monitoring of cell culture growth in petri dishes; one technique was implemented with Livermore's flying spot scanner, another with an HPD built originally for spark chambers at Berkeley.[294] Along with the equipment, Glaser moved technicians; Ronald Baker, for example, had come with Glaser from Michigan and now accompanied the data analysis devices into the Virus Laboratory.[295] While he assured the NSF that he was learning new biological techniques, for the moment he would employ "techniques borrowed from physics." These included not simply the FSDs, but also computer facilities, photographic processing methods, multichannel analyzers (to do photometric studies on centrifuged samples), and much else.[296] In Glaser's move to biophysics we have an extraordinary example of the discontinuity in subject that could accompany continuity in material culture.

In his choices, Glaser had adapted some of the techniques of particle physics while shedding the work structure of the physics factory. The material culture of the laboratory could travel—even across disciplinary lines as long

292. Glaser to McMillan, 3 April 1961, box 9, folder 112, NA-SB, EMP.

293. Glaser to McMillan, 5 June 1961, box 9, folder 112, NA-SB, EMP.

294. Glaser to James Liverman, Biology Branch, AEC, 13 February 1964, box 9, folder 112, NA-SB, EMP.

295. Glaser to McMillan, 11 May 1964, box 9, folder 112, NA-SB, EMP.

296. See Glaser, proposal to NSF, "Study of Mechanisms of Genetic Control of Cell Physiology and Structure," 1 May 1964–30 April 1965, Donald Glaser, Principal Investigator, box 9, folder 112, EMP. On the borrowing of specific resources from physics for his biological work, see Glaser to McMillan, 4 June 1964, box 9, folder 112, EMP.

and deep as those separating a virus from a cascade zero. Together the image analysis techniques served as a wordless creole that made it possible for Glaser to move between what otherwise would seem unbridgeable portions of a career in science.

By February 1967, Alvarez too had begun to tire of the factory world that he had done so much to create; he too began to devote almost all his time to other projects, principally his balloon work on cosmic rays. Asked at that time about experimental nuclear physics, he responded by labeling it "just a little dull":

> [S]o much of the work can be done by technicians. . . . You have technicians who run alpha particle spectrometers and beta ray spectrometers and gamma ray coincidence circuits. And the people working in the field are doing very much what our graduate students are doing, they are putting things into computers and analyzing the print-out, and they are pretty well disconnected from the experimental side of it, in the same way that we are. I can't complain because our people don't go down and look at the bubble chamber very often or at the bevatron. They ask the bubble chamber operators to expose a certain number of millions of frames of film, and then they ask somebody else to measure them, and then run them through computer programs, and then they start with computer program output and process this data.[297]

This sense of an oppressive routinization weighed increasingly on Alvarez, and in one of the great twists of physics history, he began to turn toward cosmic rays. Cosmic rays—the subject Glaser had invented the bubble chamber to preserve and which his invention had destroyed; the subject Alvarez had dismissed a decade and a half before—now, in the form of high-altitude balloons, beckoned like Leprince-Ringuet's alpine peak high above the laboratory-turned-factory. Writing to McMillan, Alvarez rehearsed his trajectory over the past years in a plea to be allowed the freedom either to shift his group toward the skies or, if that were not possible, to extricate himself from the strict and ordered infrastructure he had built.

Lawrence, Alvarez told McMillan, had not favored the 4- to 72-inch bubble chamber expansion, but he had believed "in me," supporting the move on condition that Alvarez stick with it, a reasonable enough request, Alvarez reckoned, since he had indeed jumped from "nuclear physics to radar to atomic energy, to accelerator building, and back to nuclear physics, in a twenty year period." Now, 13 years later, Alvarez believed he had done the job, by measuring over a million events per year (the 1968 rate was 1.5 million per year). His program of data analysis remained unchallenged (pace the FSD), and the 72-inch chamber

297. Alvarez, transcript of interview by Charles Weiner and Barry Richman, 14–15 February 1967, AIP. I thank L. Alvarez for permission to see the transcript.

was productively harbored at SLAC. Rather hyperbolically, he added in March 1968, "Like most everyone else in the high energy physics world, I believe that bubble chamber physics has passed its prime. In its prime, it stood out as by far the most productive technique of experimentation available to accelerator users. And even though it has passed its prime, it is my considered judgment that no other experimental technique has risen to take its place as the most versatile and productive tool in experimental high energy physics: In my opinion, this simply means that experimental high energy physics is not as interesting a field now as it was a few years ago."[298] In the place of accelerator-based bubble chamber physics, Alvarez wanted a balloon-based cosmic ray physics. Already, he argued, J. Warren Keuffel (whose work had inspired the pre–bubble chamber Glaser used to explore sparking as a means to a new detector) might have in hand the first evidence of the intermediate vector boson from mine work in Utah. Forgetting perhaps that he had been ready to cut his electronic colleagues out of laboratory support 10 years before, Alvarez now chastised McMillan for denying Bruce Cork AEC funding channeled through the laboratory. Rubbing his favorite salt in McMillan's least favorite wound, Alvarez added, "Certainly Ernest Lawrence would never have adopted such a course of action in his laboratory."[299]

Unable to get funding for his balloon-based cosmic ray group, Alvarez withdrew from his group (Group A), the group that for years had been the symbol of the new, industrial-scale physics and had been known everywhere by his name: "Since I believe that the principal function of a group leader is to lead a group, and since I have learned that AEC-supplied funds can only be used for things that used to interest me (but don't any more) and can't be used to support the things I believe are now more important, I reluctantly conclude that I have in fact lost my long-time status as a group leader. And so to make a de facto situation de jure, I am by this letter resigning my group leadership." No longer would he call the team the "Alvarez group," no longer would he serve on the senior staff appointment committee composed of group leaders, and no longer would the Monday night seminars—started nearly a decade before—be held in his home.[300]

Seven months later, in October 1968, the Nobel Committee telegraphed Alvarez to announce that he had received the 1968 prize for his "decisive contributions to elementary particle physics in particular the discovery of a large number of resonance states made possible through your development of the technique of using hydrogen bubble chamber and data analysis."[301] Surely the recognition

298. Alvarez to McMillan, box 1, folder 9, NA-SB, EMP.
299. Alvarez to McMillan, 18 March 1968, box 1, folder 9, NA-SB, EMP.
300. Alvarez to McMillan, 18 March 1968, box 1, folder 9, NA-SB, EMP.
301. Telegram to Alvarez, 30 October 1968, box 1, folder 9, NA-SB, EMP.

afforded by the Nobel Prize was welcome, especially so soon after the machines that Alvarez had created had now chased not only Glaser but Alvarez himself from the field he had pursued for more than 30 years.

5.13 Summary and Conclusion

5.13.1 Instruments, Physics, and Work Organization

Interwoven in this chapter are three levels of history. The first narrative attends to the specific physics questions that motivated the creation of a new instrument. Concerns such as the nature of strange particles, their decay properties, and their spectra were certainly on the minds of the Berkeley physicists. Interactions of strange particles were also of interest; later, hyperon decays and short-lived resonances became an essential part of the physics program. In the earliest stages of his exploratory work on new track detectors, Glaser was particularly concerned with the upper end of the cosmic ray muon spectrum and the properties of the "pothooks." Summing up the contribution of bubble chamber physicists in the 1950s and the early 1960s, one might say that above all a classification scheme was established—the SU(3) scheme of Gell-Mann and Ne'eman. Later work explored the weak interaction through neutrino experiments in even larger chambers.[302]

This level of narrative, the history of results, is essential. Experimental results are used by theorists for inspiration, confirmation, and refutation of ideas, and by engineers and experimenters in the design of equipment. Ought one to conclude that physics goals constitute the motor for instrumental innovation? Should one argue that the discovery of the bubble chamber was purely an outcome of the search for a better way of measuring the properties of muons and the new unstable particles? Did Alvarez's bubble chamber program issue simply from the desire to explore *K* decays? There is some truth in answering yes to these questions, but clearly not the whole truth. It is not only experiments but instruments that have a life of their own in the multiple senses I explored in chapter 1.[303]

It is the life stories of these instruments, and even more important the human lives that interact with them, that constitute the second layer of the history of experimentation. In Glaser's case, cloud chamber experiments carried over in a myriad of ways to the early bubble chamber. For example, the later rejected but historically essential electrostatic theory of bubble formation is of thoroughbred cloud chamber extraction. The bubble chamber itself, with its creation of a metastable thermodynamic state precipitated by a sudden drop of pressure,

302. See Galison, *Experiments* (1987), chaps. 4–6.

303. Hacking, *Representing* (1983), xiii.

holds a natural parallel with the cloud chamber. Finally, the product of both apparatuses is the same: detailed photographic tracks in which one can see the interaction images so prized that they gave rise to golden events and a burning desire for "computers that see." Looking ahead to the contrasting tradition of electronic logic detectors discussed in chapters 6 and 7, we might also say of the two devices what they are not. Neither the cloud chamber nor the bubble chamber employs complicated electronics, neither depends on high-voltage technology, and neither produces a logically selected, high-volume output of statistical clicks. Where the image detectors produce fine-grained photographs that often had to prove their worth through a single event, the logic detectors deliver coarse-grained but plentiful tracks that convince through overwhelming enumeration. Unable to build the kind of trigger that would combine the two, a vast array of data reduction techniques aimed to recover for the bubble chamber the control over statistics that would at least step toward a synthesis of image and logic.

Before 1953, Berkeley housed a diverse blend of both image and logic detectors, with neither type dominating the other. Some groups exploited emulsions; others applied cloud chambers or counters. With the arrival of Alvarez and his group, a new, massive experimental program exclusively devoted to images suddenly towered over the other groups. Even before the bubble chamber was invented, there was frustration with the clumsiness of counter, emulsion, and cloud chamber work at the Bevatron. The bubble chamber offered more and better pictures than the cloud chamber or nuclear emulsion ever could, but on two conditions: The transition to liquid hydrogen had to be effected, and the pictures had to be processed quickly and effectively. Together these demands resulted in the integration of three new cadres of workers into the process of experimentation: structural and cryogenic engineers, computer programmers, and nonphysicist scanners. Each group developed its own methods and standards that were brought to bear on the construction and operation of the bubble chamber, fundamentally altering the nature of instrumentation and experimental life. Cryogenic engineering, for example, drew on a tradition of safety testing established in industry and in the nuclear weapons program. Computer programming, including the central notion of the simulation, had roots outside fundamental physics in any narrow sense. Scanning was in some respects a new skill, but in other respects its gendered aspect had links back to astronomical, calculational, and counting functions inside the laboratory and observatory.

We have seen in some detail how particular tasks were distributed as the building and running of accelerators passed increasingly into the control of engineers, not physicists. It has even been possible to observe the separation of skills in the analysis of data. If, as Kowarski put it, the evolution was toward "the elimination of humans, function by function," that dehumanization was fought over, line of code by line of code, and task by task. The deliberate cycle of analyzing tasks, breaking them down to simple routines, and finally automating them was in no way a purely technical procedure; behind the division and deskilling of labor lay differing visions of what it was to do physics.

These three historical categories—the history of physics questions and results, the history of instrumentation, and the history of work organization—afford us a particularly stark contrast between the early invention of the bubble chamber by Glaser and its subsequent exploitation by the Alvarez group. Moreover, we see how closely the three levels are bound together. For example, the heavy-nucleus bubble liquid has high stopping power and remains liquid at room temperature. Glaser hoped that he could use such a device in small cosmic ray collaborations. Hydrogen is suited for studying many simple particle interactions at the accelerator but demands a complex reorganization of the research team to fully exploit the apparatus. To ignore the experimental organization is to risk a grave misunderstanding of what experimental life was like in the 1960s, and why it was Alvarez's group that was positioned to do much of the experimental physics leading to SU(3). Said one more time for emphasis: The organization of specialized subgroups and the integration of engineers, programmers, and scanners was as much a component of the changing experimental physics of the late 1950s and the early 1960s as were the bubble chambers themselves.

5.13.2 Instruments and Arguments

In scale, purpose, and use, the bubble chamber had thus changed drastically from its invention in 1952 to its productive exploitation in the late 1950s through the 1960s. Making use of this detector system, experimenters began to develop new strategies for the demonstration of novel phenomena. The unexpected possibility of using the invariant mass plot and the uncertainty principle to identify the decay times of short-lived particles was more than just another technique. "Bump hunting" surely drew on the tradition of visual understanding that the bubble chamber fostered; and quite rapidly these bumps became more than the sign of a resonance, they became signs of a new particle in nature. To physicists like Alvarez, the analysis of rare decays and long chains of inferences about particle identities, momenta, and interaction patterns became part of what it meant to do high energy physics. Mere statistical information, not backed by pictures, looked hopelessly impoverished as a guide to the subatomic world, as Alvarez had told McMillan in 1957.

In many ways, the bubble chamber was the culmination of the photographic visual tradition that began with Wilson's photography of clouds half a century before. The promise of the homomorphic representation, the ability to mimic nature through nature's own inscription, marked particle physics profoundly. It was, in the words of one of photography's founders, a pencil of nature, with all that implies for an objectivity exteriorized from our own intentions and desires. What is singular in this story of the bubble chamber is an element that was not present in the cloud chamber or nuclear emulsion, and that is the industrial scale of the enterprise. Let us explore for a moment how the changing character of work around the bubble chamber altered the criteria for an "adequate" demonstration: How did the development of the large-scale bubble chamber affect its users' reasons for believing in the validity of its data? In other words, how did the new experimental procedures affect the manner in which real phenomena were extracted from the artifactual?

A particularly insightful parallel approach is provided by Hacking's analysis of our basis for belief in the images offered by various types of microscopes.[304] His grounds are three. First, Hacking points to the *similarity of results* obtained by apparatus of different types, based on different physical principles, or applied to markedly different situations. For Glaser, the bubble chamber's first promise of success came from the correspondence between the physical explosion of superheated diethyl ether and the known frequency of cosmic rays as measured by computer experiments. When the tracks were eventually seen, the correspondence with counters, cloud chambers, and emulsions became that much more vivid. Second, our beliefs are strengthened by the possibility of *intervening;* the predictable manipulation of a phenomenon gives added credence to its reality. In a microscope, our success in observing a man-made calibration grid gives us faith in the verisimilitude of the enlarged images of hitherto unseen objects. This provides us with an appropriate way of describing the radiation sensitivity experiments in which Glaser could instantly and deliberately induce the violent boiling of superheated diethyl ether by exposure to radioactive cobalt. Finally, an understanding of the *physical principles* behind the apparatus inspires further confidence in the reports we glean from it. Often the description of the process can be provided in terms of older, well-understood theories, or even by low-level empirical generalizations and approximations. Glaser's work provides a twist, since his electrostatic repulsion theory (which aided in the construction of the detector) was later abandoned. Each of these factors certainly played a historical role in establishing the bubble chamber as a reliable detector of the paths of charged particles.

Thus far the discussion has centered on the certification of the first prototype chambers by Glaser, and the process bears a strong structural similarity to the legitimation of earlier desk-sized instruments such as Hacking's microscope. But the legitimation of instrumentation is itself a dynamical process; it too has a history. We can ask, we must ask: How did the problem of extracting a real effect change when the device was brought to a military or industrial scale and in the process utterly transformed by the Alvarez group?

First, new problems arise in the upscaling of the size of the machine—especially problems of *image distortion.* The most troublesome of these aberrations

304. Hacking, *Representing* (1983), 186–209.

came from setting errors on the film, lens distortion, optical distortion from the windows, thermal turbulence in the chamber, and displacement of tracks by liquid motion.[305] (It took the periscope builder for the French nuclear submarine fleet to construct the optics for one of the larger French chambers in the late 1960s.)[306] Each of these distorting factors became the subject of working subgroups within various bubble chamber collaborations. Only by understanding and controlling these effects could real phenomena be extracted from machine artifacts.[307] Second, equipment failures have their parallels in human faults. During bubble chamber runs employing many scanners, it was common to calculate scanning efficiencies for each individual. From their coefficients, experimental errors and limits on certain processes could be adjusted. Though this sort of compensation for *personal error* was hardly new with the advent of bubble chambers,[308] it became prominent when the scale of experimentation grew. More surprising, perhaps, were tests of human error at a higher, interpretive level. As we saw, the Berkeley group developed the GAME program, which faked histograms resembling the final reduced data of a particular experiment. With only random bumps as structure, a physicist would be asked to pluck the real—putatively significant—signal of his experiment from a pile of histogram impostors. Alvarez wrote that "one can appreciate how many retractions of discovery claims have been avoided in our group by the liberal use of [this] program."[309] With apologies to the psychologists, we could label this tendency to project unwarranted structure on random information a "Rorschach effect." (As far as I know, the bubble chamber project at Berkeley was the first in particle physics to attempt systematically to compensate for the human proclivity, individually and in groups, to see patterns.)

The third and most important innovation in the extraction of reliable signals comes through the extensive development of *data reduction.* As has been stressed throughout this chapter, the computer played a crucial role in this regard, taking over an increasing number of the intermediate steps between particle interactions and published histograms. One by one, scanning, measuring, track reconstruction, kinematic analysis, and experiment analysis were transformed through the exploitation of computers. First, by speeding up scanning, the computer made possible experiments that could not have been contemplated otherwise. Second, by statistically fitting a curve to the measured data points, the computer could produce a track more accurate than the photograph would naively

305. Derrick, "Bubble Chambers" (1966), 449–52.

306. Galison, *Experiments* (1987), 146.

307. See Gargamelle Construction Group, "Gargamelle" (1967); or Dykes and Bachy, "Bubble Chamber Liquid" (1967).

308. Personal errors were often discussed in standard textbooks on experiments in the late nineteenth and early twentieth centuries. See, e.g., Palmer, *Measurements* (1912). For more on the role of personal errors in (earlier) physics, see Schaffer, "Astronomers Mark Time," *Sci. Cont.* 2 (1988): 115–45.

309. Alvarez, "Recent Developments" (1972), 267.

reveal. Third, by resolving ambiguities of particle identification, the computer could "interpret" events where the human eye could not. This was not an insignificant advance, since, as one physicist put it, "everyone has a favorite bump which he thinks can be explained by event misidentification."[310] Finally, the automated and routinized reduction of data to usable form (such as invariant mass and Dalitz plots) expanded the class of possible demonstrations by combining many events into a single, visualizable representation. And these visual data reduction displays were themselves rapidly integrated into the automatic processing of data. Dalitz plots began to appear on the cathode ray tube along with the data themselves.

Together, the variety of automated tasks gave data reduction a far more prominent role in the construction of experimental demonstration than ever before in the history of physics. Indeed, its introduction permanently altered both the organization of work in particle physics laboratories and the nature of experimental argumentation. For in the bubble chamber, the physicists' control over the apparatus was almost entirely lost—Alvarez himself could be (and was) ordered out of the control room if the engineers judged it necessary for operational safety. And for most of the bubble chamber physicists, the production end of data was truly in another world. In these circumstances, data analysis became not an auxiliary part of doing an experiment or being an experimenter: *data analysis was the experiment.* Pevsner's η and Maglić's ω experiments were precisely coextensive with their analysis of film delivered to them from the 72-inch chamber. Indeed, as control over the physical apparatus slid from the physicists' grasp, they sought ever more sophisticated forms of control over the data through ever more complex forms of software manipulation.

In sum: Our grounds for faith in our instruments' reports are manifold. In addition to tests by correlation among diverse instruments, our ability to intervene, and our understanding of underlying physical principles, we have seen many new methods of avoiding misreadings arise with the growing scale of particle physics. These include the development of subfields for the study and control of distortion, the understanding of personal errors, and the avoidance of spurious ascription of patterns. But, above all, the mark of the new large-scale physics is the creation of data reduction as an integral part of the experiment. All these techniques figured prominently in the increasingly subtle severance of real effects from the merely artifactual.

5.13.3 The Control of Objectivity

Underlying these various strands of history is a dominant concern with control: who (or what) dictates what can be altered in the process of experimentation. Seen one way, control appears as a standard part of labor history: the laboratory

310. Derrick, "Bubble Chambers" (1966), 452.

is, after all, a workplace, and it would be surprising if those who worked there did not care about maintaining control over the many faces of their work. But at the same time, control is an essential component of any ideological defense of the objectivity of physical knowledge: the ability to manipulate the conditions of inquiry, to intervene and establish the grounds for belief, is a central feature of what has been understood as scientific experimentation since the time of Bacon, if not before. My suspicion is that these two images of control are, for the experimenter, inextricable: control over the instruments of physics is so vexed precisely because the very category of the experimenter is at issue.

When Glaser struggled to prevent cosmic ray particle physics from being subjugated to accelerator-based physics, when he sought to adapt the bubble chamber to a triggering mechanism, or when he tried over and over to wrest a place for heavy-liquid bubble chambers out of the empire of hydrogen, he was at the same time attempting to preserve a way of working and a form of demonstration. For Glaser, the routinized hierarchy squelched creativity; "teams" and "industrial organization" threatened to prevent potentially valuable alternative sets of experimental conditions from being explored.

For Alvarez, the progressive automation of scanning tasks threatened to eliminate human judgment from the interpretive process, and by doing so could cut off the very possibility of discovering something altogether unexpected, the "zoo-on." As cases in point, he emphasized the importance of individual discoveries such as the cascade zero and muon-catalyzed fusion (which he compared to the discovery of Pluto). Despite his efforts to hold on to the human, routinization gradually edged the physicist not only out of the bubble chamber control room but also out of the scanning rooms, a catastrophic development (by Alvarez's lights) that made it even less likely that they would find new events. But Alvarez's view of human judgment was ambivalent. While full automation threatened the discovery of a zoo-on, he knew that full faith in judgment would lead to the "projection" of patterns even where there were none. Humans would check the idiocy of rule-bound machines, but machines would rein in the unbridled fantasy of the human eye.

Objectivity, for the Alvarez group, thus stood in a precarious interstitial zone, not quite in the tradition of those who felt that expert judgment constituted our best shot at objective knowledge, but not quite in the mechanistic tradition that saw artificial intelligence as a goal that was well on its way to replacing the merely natural. What this meant, concretely, was that the subject of experimentation, the experimenter, was now located inside a complex system of hardware and software. Scanners and physicists acted inside a network of mechanical and programmatic reading regimens. Objectivity was defined as the product of the human eye interacting with the products of liquid hydrogen, compressors, photographic apparatus, FAKE, GAME, PANG, and SUMX.

Alvarez, it seems to me, had it right in his letter of resignation to McMillan. By the time he asked to remove his name from the group title, he was changing what was de facto to de jure. It is no accident, I would argue, that the Alvarez group then became known simply as "Group A"; the era in which the group could be seen as the extension of a single individual had long since passed. In a very deep sense the individual had—reluctantly—handed over the concept of the experimenter to the group. In 1967, Brookhaven's Alan Thorndike, then leading one of the most accomplished bubble chamber groups in the world, reflected on the meaning of this elusive term in the quotation that serves as the epigraph for this book. "The experimenter," Thorndike judged, "is not one person, but a composite." Dispersed over institutions and around the world, the experimenter could be permanent or fleeting. "He is a social phenomenon, varied in form and impossible to define precisely. One thing, however, he certainly is not. He is not . . . a cloistered scientist working in isolation at the laboratory bench."[311] The author of microphysical experimentation had come to reside in the interaction between chamber, computer, and collaboration.

311. Thorndike, "Summary" (1967), 299–300.

6 The Electronic Image: Iconoclasm and the New Icons

6.1 Instruments and History

Reflecting on a lifetime of experiments with image-making devices, Louis Leprince-Ringuet remarked in 1985: "I always wanted to see things—I am a painter. One never knew what was going on with counters. I always loved photography—since I was eight or nine years old. I always loved to have tracks."[1] Images presented a fullness, a completed form of knowledge about the world—no gaps, no processes hidden behind the discrete data points provided by counters. More than just an aesthetic, his commitment to images was a system of belief; like many of his colleagues in pictures, Leprince-Ringuet himself suggested the existence of a new kind of particle based on a handful of cloud chamber photographs.[2]

Decades after the demise of the bubble chamber, its pictures adorn the covers of physics texts, histories, popularizations, and conference proceedings. Cloud chamber images have stood for over half a century, as the guidepost into the debate over the reality of the unseen. Indeed, when it comes time to point to the most secure of evidence, physicists, historians, philosophers, and sociologists can rally around the microphysical image of gently arcing tracks to say that *this* is as persuasive as evidence can get. How can one argue for a skepticism against the extraordinary detail of these frozen interactions? As Wittgenstein once said in a different context, "Everything speaks for it and nothing against."

But there were those who would speak against it, those possessed of an antivisual impulse as strong as Leprince-Ringuet's visual one, iconoclasts persuaded that demonstration should avoid slavish devotion to the image. One experimenter,

1. Leprince-Ringuet, interview by the author, 14 May 1986.
2. Leprince-Ringuet and Lhéritier, "Existence probable," *Comptes Rendus* 219 (1944): 618–20.

Curry Street, remarked about the golden event (in perfect opposition to Leprince-Ringuet): "Anything can happen once"—he wanted numbers, large numbers of events. More generally, the logic experimenter wanted overwhelming statistical significance and the ability to alter at will the experiment. To the physicist metaphorically (and sometimes literally) growing up with radio tubes and circuits instead of cameras and paint, knowledge of the microphysical world was not a matter of seizing upon a perfect specimen of an *X* or *Y* decay. It was the ability to manipulate the world at the finest level, to marshall statistically powerful numbers of events on demand, to alter the apparatus and see an effect vanish—then return the apparatus to its original state and find the phenomenon again. Experimentation in the aniconic tradition of the logic experimenter was getting results as they occurred, not waiting weeks for a photolab to return ten thousand pictures, and obtaining results by way of a scanning squadron. Experimentation and demonstration for the "logicians" meant shattering the hierarchical, specialized tasks into which the bubble chamber factories had fossilized. Data for the bubble chamber groups were manufactured, stored, and shipped; experimentation meant processing these data at distant sites. Data for the logic groups were just part of a cycle of work, in which the physicist could still build, manipulate, take data, and rearrange the device itself.

At the center of the iconoclastic drive stood the issue of control in the dual sense I have emphasized previously. Control to the many electronic "logic tradition" experimenters meant a kind of Baconian control, an activity contrasted with the passivity of photographic registration. Logicians selected events, culled the interesting phenomena from the background, manipulated the apparatus—in some experiments they acted back on nature and through that back-action demonstrated what was so. To the electronic experimenters, those they derisively called "bubblers" were at the mercy of their machines, unable to alter already-built factories. Logic experimenters did not want to receive their data in the mail, and they certainly did not want to be ordered out of the experimental site by an engineer or safety officer. As we will see, these two faces of control are really one: the manifold attempts to recapture particle physics with a plethora of devices was at one and the same time both a Baconian effort to recover the epistemic purchase on the world through manipulation and a workplace desire to recover control over their own laboratory lives.

Though long, the narrative of the previous chapters is, at the level of hardware, fairly simple: Wilson's cloud chamber was reshaped in Glaser's hands into the bubble chamber, and in Powell's hands into the nuclear emulsion. Cloud chamber physicists went to work on emulsions and then on bubble chambers; emulsion physicists migrated in great numbers to the bubble chamber. Among the logic instruments, our narrative will necessarily be far more ramified, and less easy to schematize. The devices themselves splinter into an endless parade of variants: Geiger-Müller counter, spark counter, proportional counter, spark chamber, sonic chamber, current-division chamber, magnetostrictive chamber, multiwire proportional counter, drift chamber, and many more. In fact, the number of devices is so great that we might well ask whether the notion of an experimental or instrumental "tradition" has any meaning across this variety.

From the Latin for "handing down" or "delivery," the term *tradition* had, by the Renaissance, an interlinked set of connotations. First was the handing over of material, as in the delivery of goods.[3] But there is also an early notion of tradition as a set of practices, customary behaviors that may or may not have the force of law. The handing down of practices and goods came together in the notion of tradition as teachings and codes of regulations: for the Jews those received from Moses and textualized in the Mishnah, for the Catholics the teachings derived from Christ and the Apostles institutionalized by the Pauline church, and for the (Sunni) Mohammedans the sayings of Mohammed not contained in the Koran. Custom, religious practices, and material embodiment flowed together in the doctrine of the "tradition of the instruments," the handing to a new priest of the paten and chalice. "Handing down" in the material culture of the laboratory has, I have argued, taken many forms—as we have seen repeatedly in chapters 4 and 5. It can be as literal as the passing on of laboratory equipment: roomfuls of machines passed to the universities from war work in the United States and Europe. It can be partial, as in the "cannibalization" of instrument parts from one experiment to another. But the transfer can also be of a more subtle kind. To a large degree this chapter is an exploration of the heterogeneous ways in which things migrate from radars, computers, televisions, and warships into the laboratory, to be soldered together into entirely "new" instruments. This productive bricolage comes at a price. In an iconoclastic culture of experimentation that valued control over the experimental apparatus and experiment, there was a reluctance to allow the task of experimentation to fall to others, such as cryogenic engineers, chemists, or scanners.

As the "tradition of the instruments" makes clear, objects travel clothed in culture and human interactions. Objects (like the paten and chalice) are encumbered, covered with meanings, symbolisms, power, and the ability to represent but also to preserve specific elements of continuity. Yet precisely because things come dressed with meaning, it is essential not to picture the handing down as occurring without alteration. There are no purely neutral exchanges or donations, no "technological transfer" that is isolable from the contexts of origin and destination. When Alvarez imported the hydrogen liquefier from the H-bomb project, it was not an isolated piece of hardware—it came with a human infrastructure of cryogenic engineers from the National Bureau of Standards, a particular relationship with the Atomic Energy Commission, a social organization of experimental research that had both a military and an industrial cast. This is

3. See *Oxford English Dictionary,* 2d ed., s.v. "tradition."

not to say that Alvarez's laboratory was simply a replication of the Eniwetok bomb test—far from it. As we saw, Alvarez's bubble chamber system was built around much more than hardware; it embraced thoroughgoing commitment to the necessity of human intervention at every stage from the scanner to the group leader. It included a devotion to the rare and unusual event—the zoo-on—as well as the accumulation of statistics. Similarly, Marietta Blau came to her nuclear emulsions using X-ray dental plates; while it may be fruitful to track the continuities of the penetrating filmic gaze from molars to muons, it does not follow that the practices of plate analysis in 1911 and 1938 are entirely overlapping. So while it would be an error to suppose that machines can be plucked cleanly from their context, it would be equally distorting to assume that objects carry the totality of their cultural embedding with them. Clothes wear away, dyes fade, and meanings themselves shift over time. One of the central arguments of this book is that there is a partial peeling away, an (incomplete) *dis*encumberance of meaning that is associated with the transfer of objects. Radar klystrons become accelerator klystrons, nuclear weapons isotopes become nuclear interaction targets, and gun mounts become spectrometer stands. Our histories must be dense and specific enough to understand the limits of the malleability of objects and meanings as they travel from domain to domain.

Objects draw together clusters of cultural practices the way pidgins and creoles bind languages. The wordless creole of film tracks, for example, ties together the practice clusters of bubble chamber, nuclear emulsion, and cloud chamber work. Not entirely attached to any one of these arenas of experimental physics, the skills of track recognition, film processing, event sorting, and background calculations allow practitioners to move between the domains and to find confidence in the procedures. When tracks in the early bubble chamber are classed as pion decays, they build on an already-established process of sufficient similarity to the cloud chamber and emulsion that legitimation of both device and event does not have to begin from the beginning. To explore the material culture of the logic tradition, we will necessarily immerse ourselves in a wide spectrum of devices, their physical principles, their electronic functions, and the way their data were interpreted. For only through this sometimes daunting specificity is it possible to characterize the laboratory moves, historical associations, and conceptual structures that made it possible for several generations of instrument builders to move back and forth from vidicons to computer memory, and from Geiger counters to drift chambers.

As in the image tradition, the logic tradition is defined by continuity at three levels: a continuity of skills and technology, a continuity of personnel, and a continuity of the kind of evidence produced with the machinery at hand. Both traditions began near the turn of the century, and each was inaugurated with the invention of a new kind of instrument for atomic physics. The image tradition "began" around 1911 in the sense that while image making with instruments is old, single-event demonstration of physical phenomena is, as I argued in chapter 2, a break with more general uses of scientific photography. Similarly, the coincidence-arrayed counter of 1930 built statistics into physical experimentation in a fundamentally new way.

C. T. R. Wilson's cloud chamber rendered the paths of particles visible and so made subatomic processes "real" for generations of physicists. So well resolved were these trajectories that individual pictures could, and did, serve as evidence for novel phenomena in the 1930s and 1940s. While the use of statistical arguments, based on scores of photographed events, was of course an option for the cloud chamber (image) physicists, for the counter (logic) physicist of the 1930s it was a necessity. No individual click of a Geiger-Müller counter could serve as evidence; the electronic devices could persuade only by *accumulating* coincidences or anticoincidences. For example, to show that charged particles could penetrate thick lead plates, the counter physicist had to demonstrate that the joint counting rate of one device above the plate and another below the plate was higher than that expected on the basis of chance.

Over the years since the first coincidence and anticoincidence experiments, the two traditions competed. Each had advantages, each had weaknesses. On the one hand, the image devices provided detail but often were vulnerable to the charge that "anything can happen once"; some unexplained fluke might have occurred in the device that could fool the experimenter. On the other hand, the electronic logic devices typically produced ample statistics but remained open to the objection that they recorded only a very partial description of any single subatomic process. Unlike the image devices, the electronic experiment, being "blind," might miss some crucial feature of the phenomenon being described. It is not a story where one side is right and the other wrong. Rather it is a story of competing subcultures of physics and the at-first tentative, then powerful trading zone they built around computers, statistics, electronics, and images.

In his study of iconoclasm in art from antiquity to the twentieth century, David Freedberg remarks that there never are truly aniconic societies—even iconoclasts alternately need, fear, and revere images.[4] I would add that there never was a purely iconic culture within physics—even the cloud, emulsion, and bubble iconodules needed, at certain points, to transform their pictures into the statistical arguments of Dalitz and invariant mass plots. Advocates of each of the two ideals of experimentation, while frequently at odds, persistently tried to incorporate the epistemic virtues of the other, until, as we will see in this and the following chapter, the boundary between number and picture blurred in the mid-1970s. While the golden event is a long way from the golden calf, we will find over and again that our iconoclasts, like their statue-bashing forebears, were unwilling to forgo the image. But instead of adopting the photograph, they

4. Freedberg, *Images* (1989), esp. chap. 4.

replaced it. The complex, multiple, and nonlinear story of the new icon is a principal subject of this and the next two chapters.

6.2 Coincidence

Rutherford's first experiments on uranium rays, in 1898, involved the study of their penetration through different materials. These studies led to his classification of the rays into alpha and beta, the former penetrating more effectively than the latter. Using magnets, Rutherford and others were able, by 1903, to convince their colleagues that the beta rays were fast electrons. Alpha particles appeared undeflectable by magnets and so remained problematic, especially since they seemed to slow down like charged particles. Bragg concluded that the alpha was just an ionized atom. According to conventional wisdom, an atom consisted of equal and very large numbers of positive and negative electrons. An ion was simply an atom missing one or several electrons.

From various experiments that deflected alpha particles with electric and magnetic fields, the ratio of the alpha's charge to its mass could be determined. On the basis of these efforts, Rutherford had concluded by 1907 that the alpha particle was a doubly ionized helium atom. Now he needed to prove it. His idea was to count the number of particles that deposited a measurable charge Q on a detecting device. The ratio of total charge to total number of particles would give the charge per particle.

After some unsuccessful attempts at McGill University in Montreal, Rutherford returned to Manchester where he teamed up with a visiting research assistant, Hans Geiger. Together they built a cylindrical capacitor, which they intended to function as a counter registering a signal proportional to the amount of charge that entered. The idea was that when a charged particle passed through the gas, it would ionize some of the particles. The more charge on the particle, the greater the current of positive ions that would wander toward the chamber's negative walls. According to John S. Townsend's collision theory, the actual current across the cylindrical capacitor could be increased by making the ions travel with such speed that they would ionize other atoms, creating an avalanche. To increase the sensitivity, all one needed to do was to turn up the voltage between the inner and outer cylindrical surfaces. With the new device, Rutherford and Geiger's original hypothesis of a doubly charged alpha particle seemed, by 1909, on the way to confirmation.[5]

Unexpectedly, the mica window at the entrance to the Rutherford-Geiger counter scattered the alpha particles much more than the two physicists had expected. Although it had not been part of their original interest, scattering

5. Rutherford and Geiger, "α-Particle," *Proc. Roy. Soc. London A* 81 (1908): 162–73. See Pais, *Inward Bound* (1986), 61, for further references.

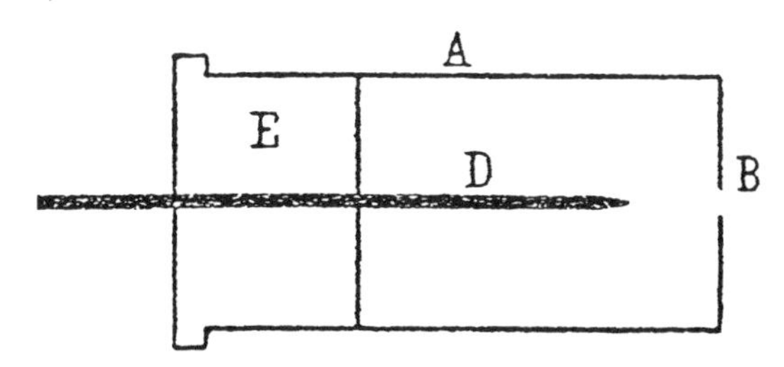

Figure 6.1 Geiger's *Spitzenzähler* (1913). Geiger's pointed center needle created a counting tube of considerable sensitivity, one that became the standard for radioactive experiments for a number of years. Source: Geiger, "Einfache Methode," *Verhand. Deut. Phys. Gesell.* 15 (1913): 534–39, on 535.

became so now. Geiger and a 20-year-old undergraduate, Ernest Marsden, abandoned the electrical counter, replacing it with a scintillator screen that flashed when and where the alpha struck. Though the details of Rutherford's exploitation of the Geiger-Marsden results is a fascinating story—intermeshed with Rutherford's 1911 nuclear model of the atom—I want to maintain our focus on the material culture of the laboratory.[6]

The problem with the electrical counter was that, when the sensitivity of the device was increased by bringing the voltage near the gas breakdown point, "wild collisions" (*wilde Stösse*) would set in—even in the complete absence of radioactive sources. Something seemed faulty in the detector itself. In 1912, Geiger became director of radium research at the Physikalisch-Technische Reichsanstalt, where he once again turned to his extraordinarily sensitive but misbehaving detector. In the course of experimenting with the shape and dimension of the old (1908) Rutherford-Geiger counter, Geiger came across a device considerably more sensitive than the one he had worked with earlier (see figure 6.1).[7] Using a negatively charged, sharply pointed needle (with its powerful electric field near the sharp tip), Geiger had a device that could detect beta particles as well as alphas. Because of the powerful field, when a potential difference between the needle and counter wall exceeded about 1,000 volts, the *Spitzenzähler* could precipitate an avalanche of charge with a much smaller number of initiating electrons. In fact, the device was so sensitive that photons could be detected indirectly by the electrons they knocked out of atoms.

Still, the *Spitzenzähler* exhibited the same "natural disturbances" as the old device. If the problem was a radioactive contamination of the gas, as the pressure was lowered the effect ought to have vanished. It did not. This led Geiger and others to conclude that the difficulty must reside in the counter's walls or

6. On the scattering work by Rutherford, Geiger, and Marsden and the theoretical work by Rutherford in the creation of the nuclear model of the atom, see Pais, *Inward Bound* (1986), 188–93.

7. In the following discussion of the Rutherford-Geiger and Geiger-Müller counter, I have benefited from discussions with Thaddeus Trenn, whose articles should be consulted for further details: see, e.g., Trenn, "Geiger-Müller-Counter," *Ann. Sci.* 43 (1986): 111–35; and Trenn, "Geiger-Müller-Zählrohres," *Deut. Mus. Abh. Ber.* 44 (1976): 54–64. On the experiments on alpha and beta rays, see also Heilbron, "Atomic Structure" (1964).

central wire. Consequently, he and others launched a great effort between 1926 and 1927, to vary, clean, and coat the wire and walls, but to no avail. Finally, in 1928, Walther Maria Max Müller, a postdoctoral research assistant to Geiger at the University of Kiel, began searching, systematically, for the source of the spontaneous discharge. Speculating that the unwanted discharges might be coming from outside the detector, he tried, as a last resort, isolating the counter from the outside world with thick blocks. The wild collisions ceased. Müller thereby unearthed, quite literally, the source of the background. No longer an emanation from the earth or a contaminant in his counter, he now attributed the wild collisions to the cosmic rays Victor Hess had found in his balloon experiments a quarter-century earlier.[8]

The consequences of Müller's unearthing were twofold. First, assigning the wild collisions to cosmic rays retrospectively legitimated the original Rutherford-Geiger device of 1908. The Rutherford-Geiger detector, though it had been used by physicists, had long carried a reputation for unreliability. Now the instrument appeared exculpated, retrieved from the dustbin of history, and even celebrated for its ultrasensitivity. Second, with the wild collisions now attributable to cosmic rays, the detector brought a fundamentally new tool to the analysis of these "new" entities, one vastly more flexible and sensitive than the electrometers of an earlier time. As much as any hardware change, as much as the shift from point detector (*Spitzenzähler*) to electron tube (*Elektronenrohr*), the transformation of the instrument was conceptual. By shifting the wild collisions from the gas to cosmic ray detection events, the Geiger-Müller counter went, without altering a single screw, from a device with a fundamental, poorly understood, and bothersome limit of sensitivity to the most sensitive instrument in the cosmic ray physicist's toolkit.

"Coincidence counting," the registration of a count only if two detectors fired simultaneously, began with Walther Bothe's famous experimental devastation of the Bohr-Kramers-Slater (BKS) paper in 1924. BKS's theoretical salvo was a last-gasp attempt to preserve a nonquantized theory of light, in which the electromagnetic field was to be described as continuously changing in energy, but the atomic processes of emission and absorption would be described discontinuously. Causality and conservation of energy would be obeyed only in the statistical long run. What about the already-demonstrated Compton effect, in which the frequency of light shifted and electrons were scattered? BKS answered: The Compton effect was only known in the statistical average. No individual process at the atomic level had been shown to obey energy conservation laws. Bothe and Geiger, using two independent counters, shot this down, showing that when a photon hits an electron, the arrival of the secondary photon at

8. Trenn, "Geiger-Müller-Counter," *Ann. Sci.* 43 (1986): 111–35; and Trenn, "Geiger-Müller-Zählrohres," *Deut. Mus. Abh. Ber.* 44 (1976): 54–64.

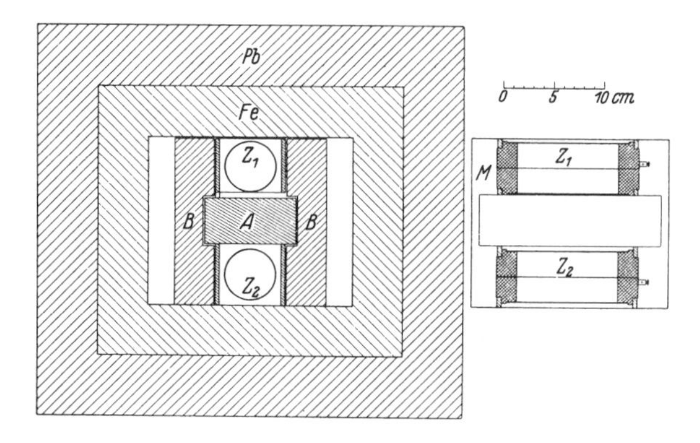

Figure 6.2 Bothe-Kolhörster coincidence setup (1929). *Left,* End-on view of the two counting tubes, Z_1 and Z_2, both with inner diameters of 5 centimeters, both 10 centimeters in length; the bodies of the cylinders were fabricated from zinc, and the ends from ebony through which the central wire was pulled. Here, and throughout the preparation of the counters, the two experimenters followed Geiger and Müller closely—the counters were filled with dry, "emanation-free" air of about 4–6 centimeters of mercury pressure. The output of the circuit moved a delicately supported mirror that reflected a beam of light; the reflected light registered on a strip of photographic film that then had to be developed. Pulses from the circuit (coincidences) thus appeared as spikes on the developed film strips. Source: Bothe and Kolhörster, "Höhenstrahlung," *Z. Phys.* 56 (1929): 751–77, on 754. © 1929 Springer-Verlag.

one counter and the "knock-on" electron at the other were occurring within 10^{-3} seconds. The BKS theory dropped dead.[9]

Very soon after Müller's rehabilitation of the counter (1929), Bothe (who had been one of Geiger's assistants) and Werner Kolhörster used the Geiger-Müller device to probe the nature of cosmic rays. Their idea was to place one Geiger-Müller counter above, and one below, a gold block 4.1 centimeters thick (see figures 6.2 and 6.3). Both counters were wired to a single chart recorder. As in the anti-Bohr-Kramers-Slater experiment, Bothe and Kolhörster could demonstrate that there was a high coincidence rate between the discharge of the two counters, some 76% of what they found without the gold block, and definitely in excess of the rate expected from "accidental" coincidences.[10] There were charged cosmic ray particles capable of traversing that much gold.

9. Wasserman, "Bohr-Kramers-Slater" (1981); and Pais, *Subtle* (1982), chap. 22.

10. "Von den so ermittelten Koinzidenzen mußten noch die 'zufälligen' in Abzug gebracht werden, um die gesuchte Häufigkeit der 'systematischen' Koinzidenzen zu erhalten" (Bothe and Kolhörster, "Höhenstrahlung," *Z. Phys.* 56 [1929]: 751–77, on 755).

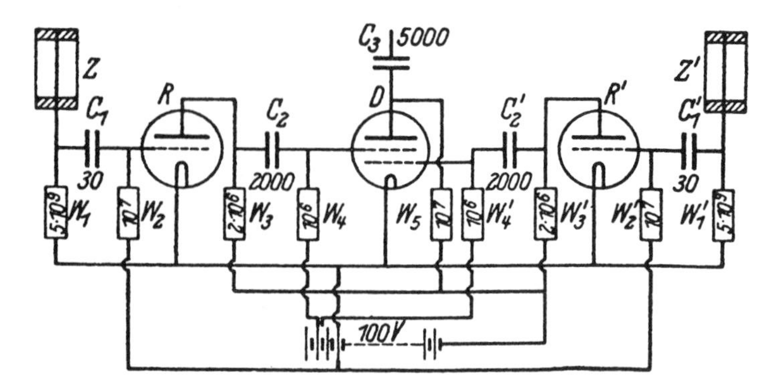

Figure 6.3 Bothe-Kolhörster circuit (1930). In order to eliminate the dreaded accumulation of film strips and the tedious task of examining them, Bothe designed this circuit, which would pulse when the two counters (*Z* and *Z′*) fired in quick succession. Clever as this circuit was, it was designed specifically for the setup of figure 6.2 (just two counters) and lent itself in no obvious way to generalization. Resistances (*W*) are given in ohms, and capacitances (*C*) in centimeters. Source: Bothe, "Vereinfachung," *Z. Phys.* 59 (1930): 1–5, on 2. © 1929 Springer-Verlag.

The threefold confluence of the Geiger-Müller counter, the coincidence method, and the conceptual-technical separation of cosmic rays from wild collisions opened a new chapter in physical research. Natural radioactivity from below ceded to rays from above. Bothe's institute became a center for work on the nature of cosmic rays. Among others, the young Italian physicist Bruno Rossi came to work there in the late 1920s. Deeply impressed by Bothe's coincidence experiments with Geiger and with Kolhörster, Rossi spent the year developing an electronic circuit that would detect the coincidental discharge of two, three, even an arbitrary number of Geiger-Müller tubes (see figures 6.4 and 6.5).

In Rossi's device, when no tube is firing, there is no voltage on the grids of the vacuum tubes, T_1, T_2, and T_3; consequently a voltage produced by a battery, P, induces a current through the vacuum tubes and a consequent voltage drop across the large resistor, R_4. This drop puts the voltage at A (the far side of the resistor R_4) at approximately the potential of the ground. If one of the Geiger-Müller tubes, say GM_2, were to discharge, then the grid of T_2 would charge up with electrons, and current could no longer flow across it. This would have no effect on the voltage at A because current could still easily flow through T_1 and T_3. But should all three Geiger-Müller tubes fire simultaneously, no current could flow through the circuit and the voltage at A would simply become the voltage of the positive end of the battery, B. This sudden increase in voltage at A from ground (zero) to the battery voltage could then be amplified by using the voltage at A to set the grid potential of an amplifying tube that Rossi wired to a

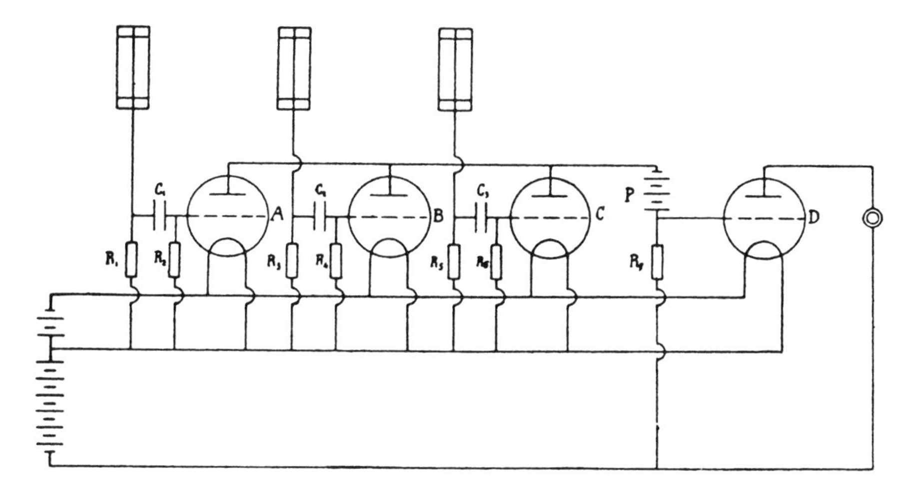

Figure 6.4 Rossi circuit (1930). The Rossi circuit is to the logic tradition what the cloud chamber was to the image tradition. Here was a device of quite general applicability, easily suited to an arbitrary number of tubes and even—as was done shortly thereafter—to anticoincidences as well. Source: Rossi, "Method," *Nature* 125 (1930): 636.

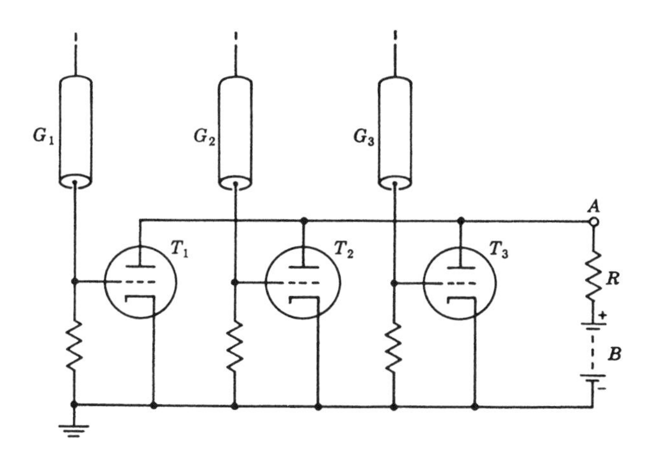

Figure 6.5 Rossi circuit, simplified (1964). The firing of all three tubes would charge up the grids of the three vacuum tubes T_1, T_2, and T_3 simultaneously, effectively blocking the flow of electricity from A to B—thus increasing the voltage difference from zero to the voltage of the battery. Such a voltage spike could be amplified and used to drive a scaler (a counting mechanism that would record only increments of, say, 10). Source: Rossi, *Cosmic Rays* (1964), 44. Reproduced with permission of McGraw-Hill, Inc.

telephone receiver (shown in Rossi's original circuit diagram, figure 6.4, but not in the simplified figure 6.5).[11] Indeed, unlike Bothe's circuit (figure 6.3), Rossi's was easily generalized—the elementary "cells" of coincidence circuitry could be extended, *ad desideratum,* by simply adding similar arrangements of tubes and wires.

Bothe recalled that his laboratory had "resembled an industrial laundry" of hanging graph paper (even with just two counters),[12] so Rossi's device was an immense practical improvement over the simple photography of counter output. But the coincidence circuit was more than a simplifying device. In fact, Rossi's idea is easily extended to record *A* or *B*, as well as *anti*coincidences[13] (e.g., *A* and *not B*), and thus any logical combination could be written into wires and tubes (see figure 6.6*a*–6.6*f*). Tubes, circuits, and electronic counters (scalers) soon gained an ordered form of experimentation, one distinct from the practices of the image-making tradition.[14]

Rossi, for example, could demonstrate the shower-making capabilities of cosmic rays by arranging a cluster of tubes (*A* and *B* and *C*) just underneath a lead plate (see figure 6.6*a*): because the placement of the tubes was not collinear, only a shower could trigger a statistical excess of coincidences in all three counters.[15] Reasoning similarly, Thomas Johnson (of the Bartol Research Foundation at the Franklin Institute) and Curry Street (from Harvard) showed, in 1932, that the arrangement in figure 6.6*b* (placing lead slabs at *A* and *B*) *increased* the number of coincidences in the counters. Such a result was not to be expected if there was only one kind of charged particle involved, since no straight line could be drawn through a plate and both counters. This, they argued, led to the conclusion that a shower of secondary particles was being produced in the plates, and it was this secondary shower that then went on to trigger the two counters.[16] Indicative of the further work that could be pursued in the analysis of collision dynamics, Street and Johnson introduced a more complex array (figure 6.6*c*) in which they hoped to deduce the properties of the particle causing a

11. Rossi, "Method," *Nature* 125 (1930): 636.

12. Bothe, "Coincidence" (1964), 271–79, on 272.

13. Anticoincidence circuits were so widely used from the mid-1930s on that only a fraction of those exploiting the technique published specifically on their particular circuit variations. For a sample of these, see Herzog, "Circuit," *Rev. Sci. Instr.* 11 (1940): 84–85; Jánossy and Rossi, "Photon," *Proc. Roy. Soc. London A* 175 (1940): 88–100; Swann, "Report," *J. Franklin Inst.* 222 (1936): 647–714, reporting on work of W. E. Ramsey, 702–3; Swann and Locher, "Variation," *J. Franklin Inst.* 221 (1936): 275–89; Street and Stevenson, "New Evidence," *Phys. Rev.* 52 (1937): 1003–4; and Herzog, "Search," *Phys. Rev.* 55 (1939): 1266.

14. After the Rossi circuit, the method of electronic coincidence counting rapidly spread. See, e.g., Eckart and Shonka, "Accidental," *Phys. Rev.* 53 (1938): 752–56; Fussell and Johnson, "Vacuum," *J. Franklin Inst.* 217 (1934): 517–24; Johnson and Street, "Circuit," *J. Franklin Inst.* 215 (1933): 239–46; Johnson, "Cosmic-Ray Intensity," *Rev. Mod. Phys.* 10 (1938): 193–244; Mott-Smith, "Attempt," *Phys. Rev.* 39 (1932): 403–14; Mouzon, "Discrimination," *Rev. Sci. Inst.* 7 (1936): 467–70; and Tuve, "Multiple," *Phys. Rev.* 35 (1930): 651–52.

15. Rossi, "Sekundärstrahlung," *Phys. Z.* 33 (1932): 304–5.

16. Johnson, "Cosmic Rays," *J. Franklin Inst.* 214 (1932): 665–89, on 682.

shower. First they showed that there was a statistically significant number of triple coincidences—since the three tubes were nowhere collinear, more than one particle must be at work. They then sought to discriminate among three hypotheses: (1) that a nonionizing ray knocked several ionizing particles out of the lead, hitting the bottom two counters; (2) that a nonionizing ray hit one nucleus and emitted one ionizing particle, then hit a second nucleus and again released an ionizing particle; and (3) that an ionizing ray released at least two ionizing particles from a nucleus. By showing that the number of double coincidences among the bottom two chambers was approximately the same as the number of triple coincidences, the two authors argued that ionizing particles, arriving from above, were responsible for the majority of showers below. While arguably not definitive, the experiment strongly suggested hypothesis 3: that the primary shower-producing radiation was charged.

A fourth example of logic circuitry comes in the frequently used "cosmic ray telescope" (figure 6.6*d*), where Johnson strung three tubes together in coincidence and used them to register only penetrating particles coming from a restricted solid angle of the sky. By pointing this coincidence telescope at different portions of the sky, he could demonstrate the statistical asymmetry between the number of particles coming from the east and the number arriving from the west.[17] If cosmic rays were photons, there should be no such "east-west" discrepancy, but if primary cosmic rays were charged and preponderantly (say) positive, then the earth's magnetic field would ship them preferentially from one direction.[18]

Illustrative of the joint use of coincidence and anticoincidence circuits is the work of W. E. Ramsey illustrated in the photograph and schematic of figure 6.6*e*. Here again is an attempt to sort out the nature of cosmic radiation: *S* is an array of Geiger-Müller counters, *A* is a block of lead in which the bursts are supposed to originate, *B* is another array of counters, as is *C*. The geometry is arranged so that no ray can traverse *A* and *C* without penetrating *S*. How to test the efficiency of *S*? Removing the lead, the coincidence rate of *B* and *C* is measured for 15 minutes; then *B* and *C* are put in anticoincidence with *S*. If the rate plummets from, say, 100 per 15 minutes to 2 per 15 minutes when *S* is turned on, then *S* is 98% efficient at registering the passage of particles. Now the lead is restored and the array *C* wired to count only if more than one counter of *C* fires. If a single ray penetrates the experiment, no count is made: *S* is in anticoincidence so it would rule out the count, and so a single ray cannot elicit a count from *C*. If a single ray goes through *S*, makes a shower in the lead (*A*), and therefore covers several counters in *C*, the experiment still will not count since a

17. See Johnson, "Comparison," *Phys. Rev.* 43 (1933): 307–10, figure on 308.

18. See, e.g., Johnson, "Asymmetry," *Phys. Rev.* 43 (1933): 834–35; and for a more extensive discussion of this period of cosmic ray experimentation, see Galison, *Experiments* (1987), esp. chap. 3 and references therein.

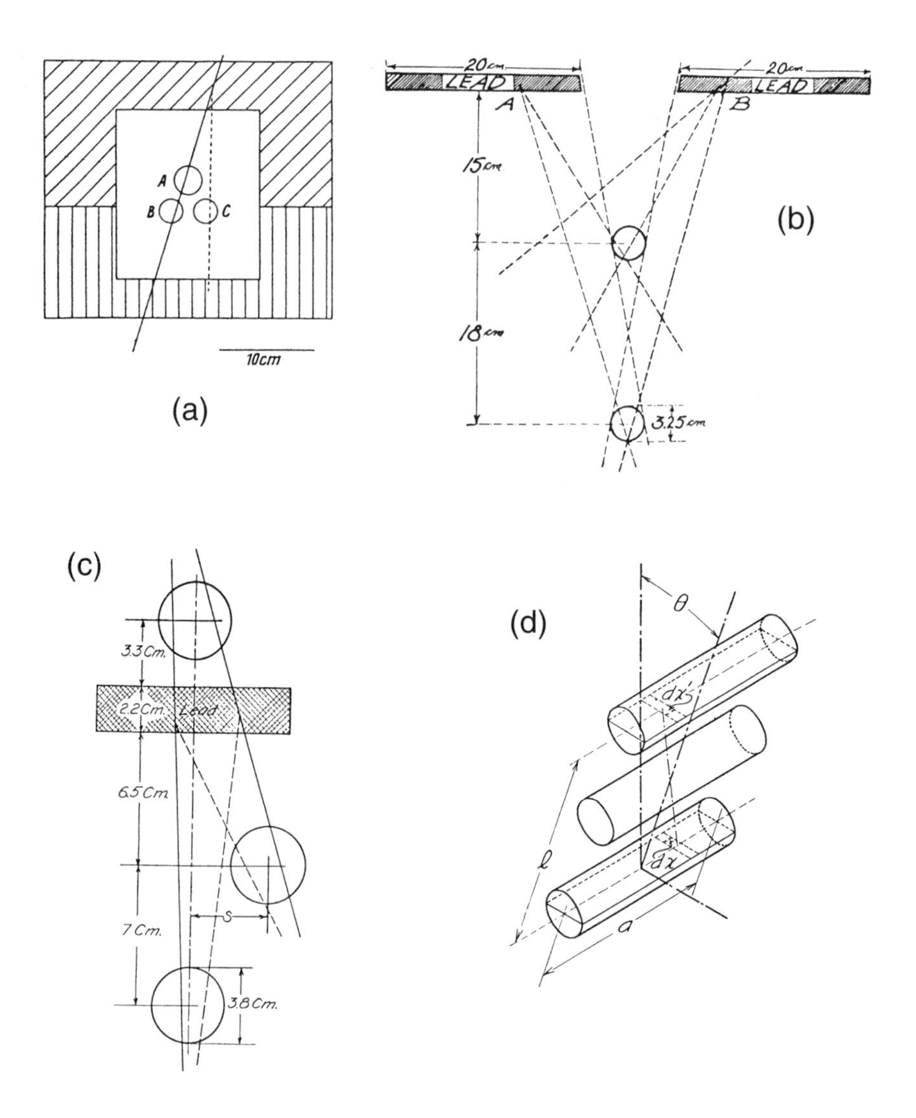

Figure 6.6 Logic tubes (1932–36). To understand the new language that electronic logic afforded experimental physics, nothing is quite so useful as a glance at a representative sample of coincidence and anticoincidence experiments. In each case, no single event could ever be called "golden"—every argument was predicated on a particular electronic architecture coupled to a statistical contention that the number of outcomes exceeded that to be expected under background or random discharges. See the text for descriptions of the particular experiments. (*a*) Rossi, shower tubes (1932). Source: Rossi, "Sekundärstrahlung," *Phys. Z.* 33 (1932): 304–5. (*b*) Johnson and Street, secondary showers (1932). Source: Johnson, "Cosmic Rays," *J. Franklin Inst.* 214 (1932): 665–88, on 682. (*c*) Street and John-

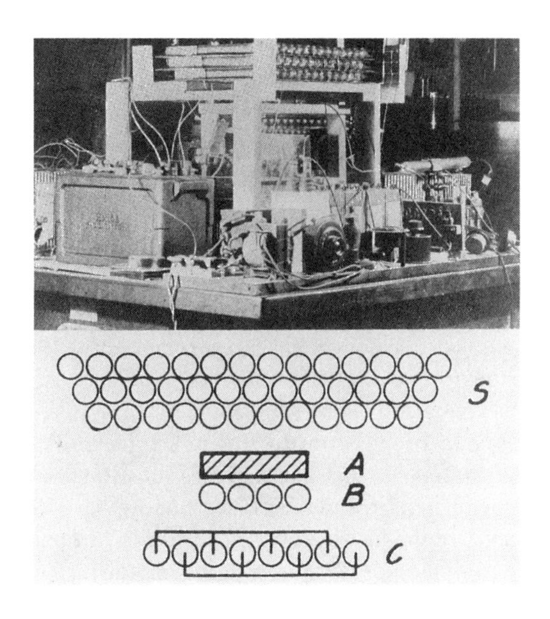

Barometric Pressure Elevation		76 cm 60 meters			64 cm 1620 meters			52 cm 3250 meters			44 cm 4300 meters		
1	2	3	4	5	6	7	8	9	10	11	12	13	14
0.00	0.00	0.014	0.0017	0.0010	0.066	0.007	0.004	0.138	0.005	0.006	0.206	0.012	0.008
.63	.00				.39	.004	.012				1.28	.017	.033
.95	.00							1.23	.007	.005			
1.27	.00	.320	.0062	.0048	.75	.019	.014				2.82	.037	.035
1.59	.00	.388	.0064	.0054	.89	.017	.012	1.84	.032	.028	3.35	.020	.040
1.91	.00	.398	.0072	.0051	.91	.011	.012	1.98	.040	.025	3.38	.034	.027
2.22	.00	.398	.0065	.0064	.85	.011	.012	2.00	.026	.025	3.44	.038	.030
2.54	.00	.345	.0075	.0061				1.97	.019	.027	3.35	.040	.044
2.85	.00				.72	.018	.013						
3.18	.00							1.59	.017	.034	2.86	.054	.037
4.77	.00	.197	.0037	.0047	.43	.014	.012	0.93	.007	.028	1.84	.039	.030
7.30	.00										1.02	.017	.034
9.85	.00	.102	.0021	.0036	.20	.006	.007	.46	.030	.019	0.79	.009	.018
12.40	.00										.77	.007	.030
0.00	.63				.043		.006						
.00	5.08	.011		.0017	.030		.006	.071		.009	.105		.008
.63	.32										.62		.024
.63	1.59										.39		.021
.63	5.08										.19		.010
1.91	.16	.259		.0057	.60		.015						
1.91	.32	.161		.0042	.37		.012	.85		.028	1.54		.042
1.91	.63	.115		.0047	.19		.008	.60		.029			
1.91	1.11	.069		.0042									
1.91	1.59	.063		.0033	.14		.007	.39		.024	.63		.041
1.91	5.08	.013		.0015	.035		.004	.17		.010	.29		.016

(f)

son, mechanism of secondary showers (1932). Source: Johnson, "Cosmic Rays," *J. Franklin Inst.* 214 (1932): 665–88, on 684. (*d*) Johnson, cosmic ray telescope (1933). Reprinted from Johnson, "Comparison," *Phys. Rev.* 43 (1933): 307–10, on 308. (*e*) Ramsey, photograph and schematic (1936). Source: Swann, "Report," *J. Franklin Inst.* 222 (1936): 647–714, on 702. (*f*) Woodward, counts and statistics (1936). Source: Woodward, "Coincidence," *Phys. Rev.* 49 (1936): 711–18, on 715. Parts (*b*) and (*c*) copyright 1932, with permission from Pergamon Press, Ltd., Headintgon Hill Hall, Oxford OX3 0BW, UK; (*d*) copyright 1933 The American Physical Society.

traversal of *S* will rule out the count. Therefore, only nonionizing radiation through *S* (such as a photon) and a shower through *C* will be registered. Reasoning this way Ramsey, could show that two-thirds of showers are produced by nonionizing radiation such as photons, and one-third by ionizing particles like electrons.[19] Data from counter experiments were expressed, not pictorially, but in coincidences and probabilities. In the typical example of figure 6.6*f*, Richard H. Woodward, a counter physicist at Harvard, exhibited the count rate and statistical errors for a quadruple coincidence experiment.[20]

By 1934, at the nuclear physics conference in London, speakers could display the disposition of logically connected counters without any reference to the details of wiring or the particular construction of Geiger-Müller counters. Counters, logic circuits, counting circuits, and statistics constituted a sufficiently stable and well-known set of tools for details to be unnecessary. Such experiments could be used to analyze the lifetimes and absorbability of particles, to sort out the complexities and properties of shower particles, for further inquiry into the east-west effect, and for many other applications.[21]

What was the working theory of the Geiger-Müller counter, the practical knowledge of how it operated at a microscopic level? This is a question easily asked, less easily answered. In one sense, much of post-1900 physics concerning the electron, discharges, diffusion, and atomic theory is relevant and a full answer would involve a complete rehearsal of electron physics in its classical, semiclassical, and quantum theoretical forms. But more relevantly, there was a widely held semiquantitative sense of how the counter worked; it is the explanation one would find in Ph.D. dissertations such as Woodward's and learned treatises such as Montgomery and Montgomery's famous 1941 article on Geiger-Müller counters or Leonard Loeb's Berkeley lectures published in 1939:[22] Many electrons (roughly 100 per centimeter of wire) are released by ionization somewhere within the volume enclosed by the copper cylinder. Driven by the voltage difference between wire and cylinder, positive ions migrate (relatively slowly) toward the cylinder and electrons liberate further electrons by ionization as they accelerate (rapidly) toward the central wire (anode). From oscillograph studies, the discharge avalanche builds up exceedingly quickly (within a millionth of a second). The current passing through the circuit drops the potential between copper cylinder and anode until the ionization by collision comes to an end.

By 1941, counter lore could be codified into the often-reproduced chart

19. Ramsey's experiment as reported by Swann, "Report," *J. Franklin Inst.* 222 (1936): 647–714, on 702–3.

20. Woodward, "Interaction" (1935); this example from Woodward, "Coincidence," *Phys. Rev.* 49 (1936): 711–18, on 715.

21. Many of these papers are discussed in Galison, *Experiments* (1987), chap. 3.

22. Loeb, *Fundamental* (1939); Montgomery and Montgomery, "Geiger-Müller Counters," *J. Franklin Inst.* 231 (1941): 447–67, 509–45.

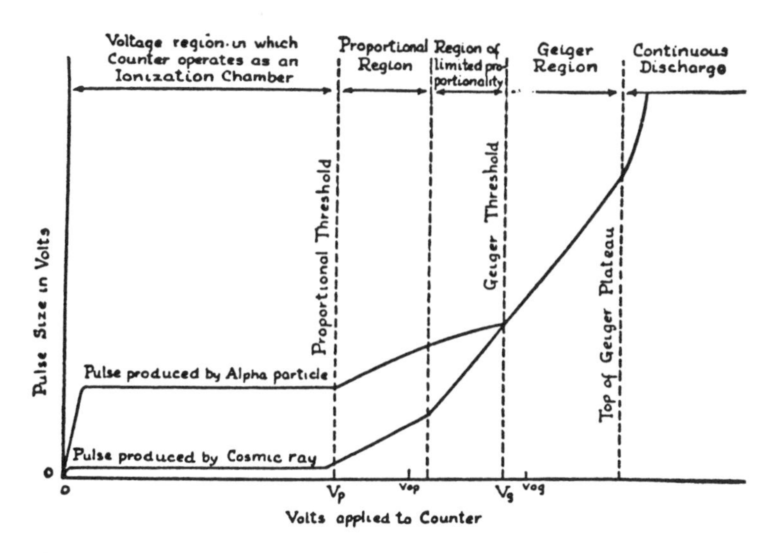

Figure 6.7 Korff, Montgomery-Montgomery diagram (1955). This diagram became the standard characterization of the behavior of Geiger-Müller tubes. The top curve shows the pulse resulting from the passage of an alpha particle, which left some 10,000 along its trajectory through the counter. A typical fast cosmic ray (represented by the bottom curve) would leave only 30 ions along its path. In the low-voltage region (between 0 and V_p) the pulse is simply the result of the accumulation of these residual ions. Between V_p and V_g, the deposited ions are accelerated by the field and hit other atoms, ionizing them, and so deliver a pulse proportional to the original number but considerably greater. Finally at the Geiger threshold, V_g, any ionization causes a general discharge the full length of the center wire—the counter fully discharges, and so no distinction is made between alphas, betas, or any other charged particles moving at any speed. Source: Korff, *Counters* (1955), fig. 1.2 on 13.

drawn by Montgomery and Montgomery (figure 6.7). With a low voltage between the wire and cylinder, the counter functions as an ionization chamber: the charge ranging from 30 or so electrons liberated by a fast cosmic ray particle to 10,000 electrons left in the wake of an alpha particle—these charges are collected on the central wire without any secondary ion production and independent of voltage. Moving upward in voltage, the counter enters the proportional region, in which (close to the central wire, where the field is strongest) the Townsend avalanches amplify the signal by a factor ranging from 1 to 100,000.[23] Pumping up the voltage even higher, one comes to the Geiger region, in which radiation from the ionization processes causes further ionization up and down the central wire, making the discharge independent of the original charge.

23. The maximum factor varies from text to text; I follow here Kleinknecht, *Detectors* (1986), 48.

(During the 1960s, various groups explored "streamer" effects within the Geiger region—special combinations of gases were used in concert with fast, large pulses to make discharges visible along the particle track with less "dead time" between events.)[24] Finally, in the continuous (actually a pseudocontinuous) discharge region multiple discharges occur with each passing particle. In a very crude sense, the vast spectrum of electronic detectors all fall into the categories set out by this simple chart.

The theory of the instrument enters into the design of the counter in a variety of ways. For example, various authors speculated that as the electrons move through the discharge they emit soft X rays. This radiation causes the photochemical emission of further electrons on the cylinder surface, and this conjecture was tied (in a hand-waving way) to treatments aimed at modifying the emissive capacity of the cylinder walls by baking the tube in air or adding free iodine. But most of the practical imperatives of manufacturing tubes did not even attempt to tie construction to the detailed theory of the instrument. Heating the wire to incandescence was thought to eliminate false discharges. Surface leakage along the glass could cause trouble, too—it was to be combatted by dipping the counters in ceresin. But of all the home-grown remedies, it was thought that cleansing the surface interior with HNO_3 and H_2SO_4 was the most essential.[25]

Some of the experts on gas discharge physics (such as Loeb) were not leaders in cosmic ray experimentation with counters; but among the counter physicists Rossi and the Montgomerys were among the most concerned with the details. The Montgomerys published widely used articles in 1940 and 1941 in which they outlined the mechanism of Geiger-Müller discharge. Additional considerations include the detailed structure of the space charge produced by the cloud of slow-moving ions left behind when the electrons began their avalanching migration toward the anode—how did this swarm of ions reshape the field experienced by electrons as they drifted toward the central wire? What happens when the ions hit the cathode? Are further electrons released? How do different gases alter the properties of the cathode? How do the gases affect recombination and photon production? How precisely do photons emitted in the avalanche cause photoelectrons from the cathode, which induce further avalanches all along the length of the counter? Questions like these became part of the counter physicist's stock-in-trade, and utterly irrelevant to the life or happiness of an image maker in cloud chambers, bubble chambers, or nuclear emulsions.

In particular, Montgomery and Montgomery enumerated what they considered to be the five ideal properties of a counter: high efficiency (registers every

24. See, e.g., Rice-Evans, *Chambers* (1974), chap. 6.

25. Woodward, "Interaction" (1935), 27ff.

ray), large pulses (no problem recording a discharge), no spurious count (minimize counts without ray passage), fast pulses (change of anode potential should be rapid), and small lags (time interval between ray passage and potential change should be minimal). In each case, the Montgomerys outlined how the theory of the device could be used to understand the way that each characteristic could be improved, and in some cases what the physical limits of the device might be. Everywhere people built counters after the mid- to late 1930s one finds a complex amalgam of physically understood processes of gas discharge physics, experimental procedure, and very rough rules of thumb.[26]

Like the Bothe-Kolhörster experiment, these explorations were inherently and inalienably statistical. Estimation of probable errors and the statistical excess over background is not a side issue in these detectors—it is central to the possibility of any demonstration at all.[27] While statistics could certainly be used within the image tradition, it was by no means necessary for most applications. Indeed, as we have seen, even as late as the early 1950s arguments for the existence of new particles based on purely statistical analyses of cloud chamber and emulsion data met with great resistance. Recall from chapter 3 Blackett's plaintive call for an end to all the statistical argumentation being invoked at Bagnères de Bigorre in support of new particles. Quoting the venerable Lord Rutherford, Blackett had urged his fellow image physicists, "Do, goodness sake, forget about the theory of error and go back into the laboratory and do the experiment again!"

The status of statistics could not have been more different in the logic tradition: statistics could not be eschewed, even the quintessentially simple application of the coincidence method. Anyone picking up a textbook on counters in the 1940s or 1950s, such as Curran and Craggs's *Counting Tubes: Theory and Applications* (1949) or Korff's widely used *Electron and Nuclear Counters: Theory and Use* (1955), would have found thorough discussions of statistical tests of counters and statistical methods for evaluating data.[28] Even more general monographs on experimental method—such as Braddick's *Physics of Experimental Method* (1954)—typically followed the discussion of Rossi circuits with one on "Errors and Statistics in Counting." In the same textbook, discussion of the visual methods (cloud chambers and emulsions) abstained from statistical reasoning—even though, as we saw in chapter 5, bubble chamberites invoked statistics for a variety of purposes, including lifetime determinations and Dalitz chart analysis.[29]

Statistics appeared inscribed into the very concept of logic instruments. And though both image and logic traditions exploited statistics, they did so in

26. Montgomery and Montgomery, "Geiger-Mueller Counters," *J. Franklin Inst.* 231 (1941): 447–67, 509–45.

27. See the estimation of probable error in Johnson, "Comparison," *Phys. Rev.* 43 (1933): 307–10, on 308.

28. Curran and Craggs, *Counting* (1949), 82–108; and Korff, *Counters* (1955), 231–64.

29. Braddick, *Experimental* (1954), 378–83.

utterly different ways. Logic experimenters regularly used statistical tests to certify the instrument itself. Statistics granted the instrument its license to function, as is evident from a standard description of how to proceed. Begin with the standard statistical fluctuation given by the Poisson distribution. (The Poisson distribution gives the probability P_n of observing n counts in a time t when the number of counts per time t [averaged over many observations] is $\langle n \rangle$: $P_n = (\langle n \rangle^n / n!) e^{-\langle n \rangle}$.) Standard statistical arguments relate this quantity to the Gaussian "normal" error law for large $\langle n \rangle$. Then the standard deviation of $(n - \langle n \rangle)$ becomes $D = \sqrt{\langle n \rangle}$, and the standard deviation as a percentage of the average value is $D_p = 100/\sqrt{\langle n \rangle}$. Thus, one popular test of whether a counter was "behaving properly" was to compare the observed standard deviation in counts over some period of time (e.g., measuring the number of counts per minute, repeated each minute over a 30-minute period) with the square root of the mean number of counts measured per minute. If D (observed in the 30-minute test) differed from D (computed from the square root rule), one suspected the counter of malfeasance.[30]

A second, even more popular application of statistics occurred within the realm of hardware: one compared an uncertified instrument with ones that were already among the elect. Street and Woodward introduced a prototype of such tests in 1934 when they placed the candidate counter as an interior member in a line of at least three counters (see the cosmic ray telescope in figure 6.6*d*). With the candidate counter disconnected, and the others wired in coincidence, the count rate C_{off} is measured. Then the candidate counter is turned on, and the coincidence count rate for the whole array C_{on} measured. The ratio C_{on}/C_{off} therefore defines the total efficiency of the counter—if the new counter only fires half of the times its colleagues do, its total efficiency would be 0.5; if it fires every time, its total efficiency would be 1. Street and Woodward also used statistics to characterize more subtle aspects of a counter's behavior: the efficiency of a counter would generally increase if the number of rays coming into it was smaller, and one could measure this improvement. The efficiency increases because part of the inefficiency of a counter is caused by its complete insensitivity for a certain "dead time" as it recovers from a discharge. If rays are infrequent, the counter is rarely in its dead time when a ray arrives, and so the counter will be relatively accurate. Woodward and Street's statistical model led to the following formula for the efficiency of the counter: $E_{total} = e^{-\sigma n} E_\rho$, where σ is the dead time, n is the counting rate, and E_ρ is the efficiency of the counter when not in its dead time. This formula would then tell an experimenter at what rates n the counter would be useful.[31]

For all these reasons, counters considered individually were statistical

30. Korff, *Counters* (1955), 236–51.

31. Street and Woodward, "Counter Calibration," *Phys. Rev.* 46 (1934): 1029–34.

devices. Then, over and above their individual behavior, coincidence counting provided a second, irreducibly statistical feature of the logic tradition. When four, five, even six counters fired within a short time, there was always the possibility that they had been triggered, not by a single penetrating ray, but by the separate particles of a widespread avalanche—or even by coincidental spurious discharge. Any argument for the penetration of a single particle (as opposed to a shower) or for a shower (as opposed to a single particle) had to be of the form: one hypothesis is more probable than another. Consequently, a single event, whether it was the click of a Geiger counter or the pulse from a complex array of counters, was meaningless in itself. Data in the logic tradition became persuasive only in their statistical aggregation.

Did the image tradition use statistics? Of course—we have seen statistics applied extensively to many features of cloud chamber, nuclear emulsion, and bubble chamber physics. Emulsion analysts carefully probed fluctuations in the grain density of a track; bubble chamber physicists measured the scanning efficiencies of their staffs and gave statistical significance to the results they obtained when "bump hunting" in the glory days of the new hadron physics; Dalitz plots were irreducibly statistical. Examples abound. Nonetheless, throughout the history of the image tradition, from Wilson's first golden event to the discovery of the omega minus, there stood an abiding faith in the power of the individual image. Scanners might miss a track, tracks might need to be considered in aggregate to make a point about particle lifetimes, but the fundamental authority of the individual event stood, sometimes explicitly, sometimes implicitly, at the epistemic foundation of the edifice of argument. Alvarez could write that he was perfectly willing to go to press with the existence of a new entity, such as the cascade zero, on the basis of a single, well-articulated event. Not so in the logic tradition. As the logic physicists interpreted the data, statistics had to enter—there was not and could not be any equivalent of the golden event—no click, no set of coincidences, no pulse from the complicated amalgam of circuitry could ever persuade anyone of anything without the essential validation of statistical significance.

These objects (coincidence and anticoincidence circuits, gases, scaling circuits, Geiger-Müller tubes, high-voltage sources, and gases), rules of combination (into telescopes, hodoscopes, and other experimental apparatus), and forms of reasoning (statistical use of errors for significance and background calculations) constitute a perfect example of what we have been calling a *wordless creole.* This concatenation of objects and practices forms a creole because, although it is tied to particular sets of experiments (such as those designed to probe cosmic rays), the quasi autonomy of the patterns of circuitry and rules of combination make the use of these logic circuits easily transferable to other domains, across the gamut of electronics. Engineers borrowed extensively from this materialized sublanguage as the logic circuits and scalers played an important role in the creation of the computer. The link is direct: On 3 August 1941, John W. Mauchly, one of the builders of the first electronic general purpose computer, the ENIAC, recorded in his notes his first explorations of the digital electronic calculator. That same day, he jotted down references to the physicist Thomas Johnson's 1938 *Review of Scientific Instruments* cosmic ray article, "Circuits for the Control of Geiger-Müller Counters and for Scaling and Recording Their Impulses,"[32] a move made without the slightest visible interest in cosmic rays, east, west, or otherwise. But the use of the logic creole extended far beyond even these two domains—it became a lingua franca among instrument makers using scintillator tubes, Cerenkov counters, spark chambers, and the plethora of wire chambers put together in the 1960s, 1970s, and 1980s.

6.3 Counting Light

With the war over, scintillation and Cerenkov counters exploded into use. Both techniques for producing light from fast-moving charged particles had roots in prewar experimentation. Zinc sulfide scintillation screens, of course, went back to Rutherford's early experiments as he and his assistants watched in darkened rooms for flashes that would signal alpha particles caroming off nuclei. That light was produced by the passage of radioactive emanations through liquids was also known early in the century, though the fact had never been exploited in an experimental instrument.[33] By the mid-1930s, a trio of Moscow physicists (Pavel Čerenkov, Igor Tamm, and Il'ja Frank) had produced, analyzed, and largely explained the eerie bluish glow that accompanied the passage of radioactive particles through liquids. Before the war, the visual use of scintillation screens had acquired a reputation for unreliability, and for that reason Rutherford had forsaken them for electrical devices. Even at the end of the war, Cerenkov radiation remained principally a curious laboratory effect to be studied with film. What transformed the scintillator's flash and Cerenkov's glow into basic building blocks of the logic tradition was the electronic revolution begun during the war. When attached to the new high-gain photomultiplier tubes and strung into the array of amplifiers, pulse-height analyzers, and scalers that emerged from the Rad Lab and Los Alamos, then and only then did the scintillator and Cerenkov radiation become part of the material culture of postwar physics.

When peace came, it did not take long for the new devices to spread. Physicist Hartmut Kallmann had worked before the war at the Kaiser Wilhelm Institut in Berlin on various aspects of atomic and nuclear physics. After emerg-

32. Burks and Burks, *Electronic Computer* (1988), 160. The specific reference is to Johnson, "Circuits," *Rev. Sci. Inst.* 9 (1938): 218–22.

33. Mallet exhibited the bluish-white light produced when gamma radiation penetrated water. For references, see Jelley, *Čerenkov Radiation* (1958), 7–8.

ing from hiding (he had spent the war sequestered in a Berlin basement), Kallmann explored the combined use of napthalene crystals to produce light and photomultiplier tubes to amplify the signal. Following hard on the heels of Kallmann's 1947 publication, Robert Hofstadter—straight from the logic tradition—showed the even more powerful scintillating effects of sodium iodide crystals (see figure 6.8).[34] On the Cerenkov side, the Harvard physicist Ivan Getting played a role similar to Kallmann's. Thoroughly versed in prewar logic circuits, Getting had used coincidence and anticoincidence circuits to study cosmic rays and in the 1930s had invented a fast new scaler to aid in the counting of electrical pulses that issued from these circuits. During the war, he continued his electronics at the MIT Rad Lab, where he worked on fire control radar. He was therefore well positioned in late 1946 to muster the electronics needed to use Lucite and Plexiglas Cerenkov light producers with photomultipliers. Getting's work was paralleled by that of Robert Dicke, who also drew on his electronic war work in his 1947 publication: like Getting, Dicke fed Cerenkov light into the new photomultipliers, went on to plug his counter into a Los Alamos 501 amplifier, and from there into a war-originated pulse-height analyzer.[35]

Cerenkov counters are based on the simple but important observation that charged particles radiate at a characteristic angle as they pass through matter. An easy analogy is this: A ripple glides through still water at a given speed. If a boat travels faster than that speed, there is a fixed angle at which the V-shaped bow wave opens, an angle related to the speed at which the many ripples inter-

34. Among the early postwar papers on scintillation, the following are among the most important: In a combined Manhattan Project–Berkeley Radiation Laboratory project in 1944, Curran and Baker, "Photoelectric Alpha-Particle," *Rev. Sci. Inst.* 19 (1948): 116, used silver-sensitized zinc sulfide combined with an RCA photomultiplier tube to count alpha particles; Blau and Dreyfus, "Multiplier Photo-Tube," *Rev. Sci. Inst.* 16 (1945): 245–48, were principally interested in using zinc sulfide with photomultipliers to measure alpha intensity. The most important papers were those of Broser and Kallmann, "Anregung," *Z. Naturforsch.* 2a (1947): 439–440; and Kallmann, "Quantitative Measurements," *Phys. Rev.* 75 (1949): 623–26, which demonstrated napthalene's sensitivity to beta and gamma radiation. In quick succession, Deutsch, "High Efficiency," *Phys. Rev.* 73 (1948): 1240, focused on thicker phosphor screens that increased gamma sensitivity; Bell, "Use of Anthracene," *Phys. Rev.* 73 (1948): 1405–6, demonstrated the superior scintillating characteristics of anthracene; and Hofstadter, "Alkali Halide," *Phys. Rev.* 74 (1948): 100–101, launched the program he pursued over many years of using Na I as a detector. Scintillator counters became Hofstadter's workhorse detector, as he went on to studies of electron scattering and nucleonic structure.

35. For the original papers, see Čerenkov, "Vidimoje," *C.R. Acad. Sci. URSS* (1934): 451–54; Frank and Tamm, "Coherent Radiation," *C.R. Acad. Sci. URSS* 14 (1937): 109–14; Collins and Reiling, "Radiation," *Phys. Rev.* 54 (1938): 499–503; all of these prewar studies used photographic means to record the light. Postwar electronic Cerenkov detectors were launched by Getting, "Proposed Detector," *Phys. Rev.* 71 (1947): 123–24, about which see Getting, *Lifetime* (1989); Furry, "Measuring Masses," *Phys Rev.* 72 (1947): 171; Dicke, "Čerenkov Counter," *Phys. Rev.* 71 (1947): 737; and Jelley, "Čerenkov Effect," *Phys. Soc. London., Proc A* 64 (1951): 82–87. Both Cerenkov and scintillation counters were classified as counters from the start: e.g., Samuel Crowe Curran and John D. Craggs's *Counting Tubes* (1949) moved smoothly from Geiger-Müller tubes to scintillation counters in their exposition, as did J. B. Birks's *Scintillation Counting* (1964), when he divided detectors into "track visualization instruments" on one side and the various "counters" including those exploiting Cerenkov and scintillation effects on the other.

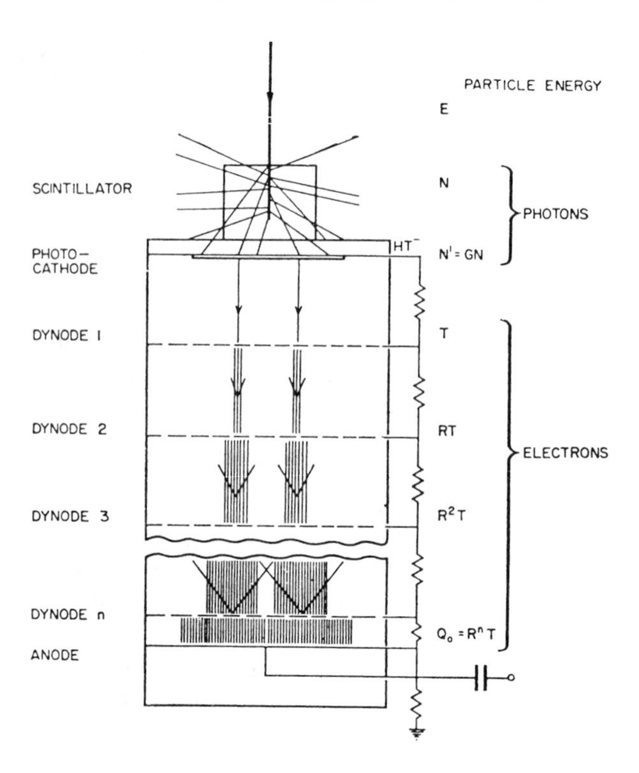

Figure 6.8 Scintillation counter (1964). In this schematic diagram, a particle of energy *E* (MeV) hits the scintillator, ionizing and exciting molecules. Some of this energy emerges as *N* photons, a fraction of which (*GN*) strike the photocathode. On impact, some of these *GN* photons produce photoelectrons out of the photocathode. The photoelectrons are then accelerated through a potential, so that T (= pGN) photoelectrons hit a second electrode (Dynode 1), producing *RT* electrons at Dynode 2. This process is iterated, each time multiplying by *R*, until a substantial current of electrons (R^nT of them) collects at the anode—a number that depends directly on the energy *E* of the original charged particle. Source: Birks, *Scintillation Counting* (1964), 11.

fere constructively (see figures 6.9 and 6.10). The same phenomenon occurs with faster-than-sound travel through air, resounding with a sonic boom that trails like a V behind the plane. In the case of Cerenkov radiation, when a charged particle moves faster than light in a medium, it too radiates a "bow wave," along a trailing V with opening angle $\cos \theta = 1/n\beta$, where $\beta = v/c$, v is the speed of

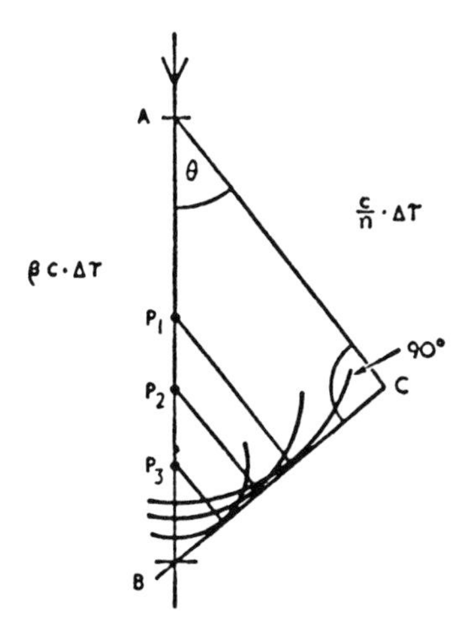

Figure 6.9 Cerenkov radiation (1958). A charged particle traveling from A to B at velocity βc (c is the speed of light in vacuo) emits radiation at every point: Consider the wavelets emitted at arbitrary points P_1, P_2, and P_3. Now there is some angle θ such that the line BC runs through a coherent set of wavelet phases. What is that angle? If $\Delta\tau$ is the time the charged particle takes to get from A to B, then $AC = (c/n)\Delta\tau$ is the distance that the radiated light travels in the medium, during $\Delta\tau$. Since AC is perpendicular to BC, and AB and AC are given, θ is fixed: $\cos\theta = 1/\beta n$. Source: Jelley, *Čerenkov Radiation* (1958), 5.

the particle, and n is the refractive index of the medium.[36] Since the angle of the radiation depends on speed, not momentum or energy, Cerenkov radiation can be used in combination with the usual magnetic deflection measurement of momentum to determine the mass of the fast-moving particle.[37]

In the case of Cerenkov radiation, the Tamm-Frank theory did give an account of light emission; the same could certainly not be said of the new scintillators. Six years after their introduction one of the leading researchers wrote: "The theory does not tell us what substances will be good scintillators or what combinations will perform best. Our working knowledge has come almost solely from trial and error. Physicists ransacked the chemists' shelves for substances that could be tested as scintillators."[38] There were questions open about the electronic excitations in the organic compounds, about the transport of energy to impurities within the scintillator, and about the final emission process; even the photomultiplier tubes involved a certain trial and error of the plates to minimize random emission of electrons and maximize their response to electrons.[39] Faith in the reliability of the new counters did not come from a microphysical

36. Tamm's theory went further, predicting that the total visible radiation (in quanta per unit path length) would be $dN/dL = (z^2e^2/hlc^2)(1 - 1/n^2\beta^2)d\omega$, where ze is the particle charge, ω is 2π times the frequency of the emitted quanta, and the other quantities are as previously. See the articles cited earlier and, e.g., Mather, "Proton Velocity," *Phys. Rev.* 84 (1951): 181–90.

37. By 1951, it was possible to use Cerenkov radiation to measure the velocity of individual protons (not just electrons) and to design velocity discriminators—strategically placed photomultiplier tubes that would pick up the flash of light only when it emerged from the radiator at a specific angle. Marshall, "Fast Electrons," *Phys. Rev.* 81 (1951): 275–76; and Mather, "Proton Velocity," *Phys. Rev.* 84 (1951): 181–90.

38. Collins, "Scintillation Counters," *Sci. Amer.* 189 (November 1953): 36–41, on 38–39.

39. Birks, *Scintillation Counters* (1953), chap. 3.

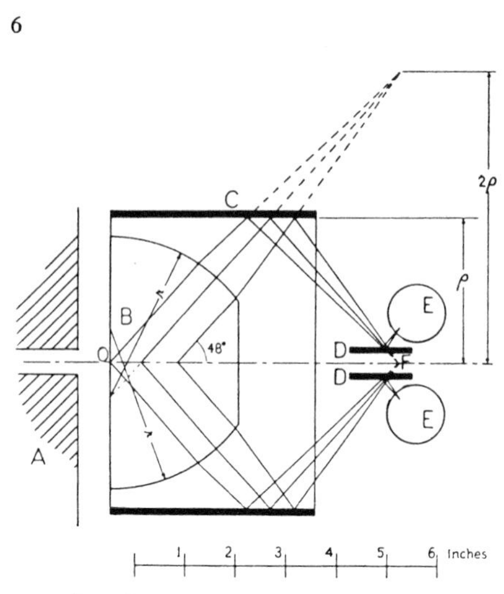

Figure 6.10 Marshall, Cerenkov velocity counter (1951). Originally designed for the University of Chicago proton synchrocyclotron, Marshall exploited the characteristic angle of Cerenkov radiation to design a detector that would only be fired by passing particles of a certain velocity. In the diagram, *A* is a lead collimator through which electrons pass; *B*, the radiator, is a figure of revolution with radius *r* designed to focus the Cerenkov radiation into a sharp ring of radius 2ρ, where ρ is the radius of a cylindrical reflector, *C*. *C* focuses the radiation at *F*, from which it is reflected into two photomultipliers *E* wired in coincidence (to escape noise). By moving the position of the photomultipliers until they catch the light, the Cerenkov radiation can be determined, and with it the velocity of the electron. Note that logic methods were used inside the velocity counter, and the counter was wired inside logic circuits (see figure 6.11). Source: Marshall, "Fast Electrons," *Phys. Rev.* 81 (1951): 271–76, on 276.

theory of the device; by and large reliability issued from tests against other instruments. Some of the earliest certification trials for the scintillator were made by comparing it to ionization chambers (zinc sulfide with a wartime phototube turned out to yield an efficiency of 80%–100%); Cerenkov counting achieved the same approval after it passed a test in which the new counter competed with Geiger-Müller tubes when both were wired into a coincidence circuit.[40] The point is one already familiar to us from the minor role thermodynamic theory played in the design of the cloud chamber, from the nugatory role emulsion chemistry or the quantum theory of silver bromide played in the development of nuclear emulsions, and from the indirect role that nucleation theory played in the design of the bubble chamber.

The logic tradition figures therefore in a variety of ways: the physicists inventing the new counters were following a logic tradition trajectory; the devices themselves were tested using logic circuitry; and both scintillator and Cerenkov counters were instantly and permanently classified as counters alongside the Geiger-Müller tube and its cousins. But perhaps most important, scintillation

40. See, e.g., Kallmann, "Quantitative Measurements," *Phys Rev.* 75 (1949): 623–26; and Jelley, "Čerenkov Effect," *Phys. Soc. London, Proc. A* 64 (1951): 82–87.

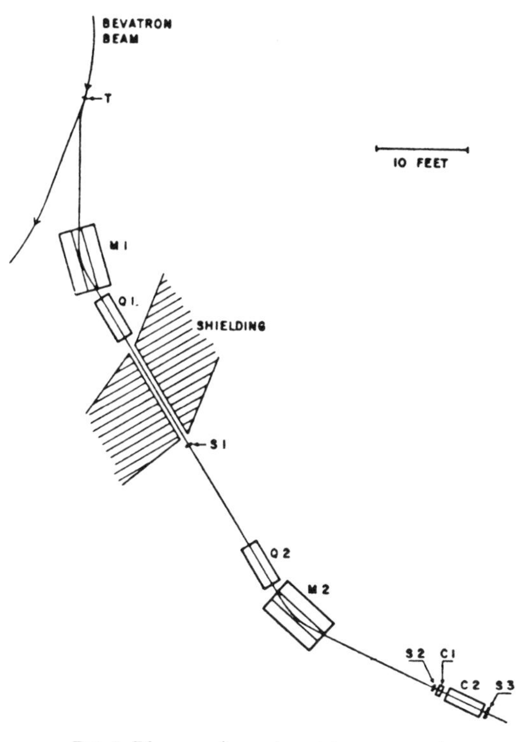

Fig. 1. Diagram of experimental arrangement.
For details see Table I.

Figure 6.11 Antiproton logic (1955). Using a variant of Marshall's Cerenkov velocity counter (figure 6.10) along with various older forms of logic circuitry, Chamberlain and his colleagues were able to pick out one antiproton candidate per 62,000 negative pions. Though soon confirmed by visual methods, the case for the antiproton was widely seen as a triumph for the then-beleaguered logic tradition. Source: Chamberlain et al., "Observation," *Phys. Rev.* 100 (1955): 947–50, on 947.

and Cerenkov counters figured as fundamentally statistical devices deployed in logical arrays.

A paradigmatic application of the new counters within the logic tradition was made in the mid-1950s by the Segrè-Chamberlain group at LBL. From the Dirac equation, many theorists assumed that the proton would have a negative counterpart, just as the positron stood opposite the electron. But that assumption was not accepted as an obvious feature of the world by most experimenters. The Berkeley antiproton experiment wired the new light-based counters in a sophisticated configuration (see figure 6.11), though the principle of using statistical sorting based on coincidence counts had been a venerable tradition of logic experiments since long before the war. From the Bevatron, a beam of protons hit a copper target *T,* after which the resultant negative particles had an energy of 1.19 GeV; they were then focused by magnets *M1* and *Q1.* With *C* denoting Cerenkov counters, and *S* scintillation counters, the beam passed through *S1,* was refocused by magnets *Q2* and *M2,* and finally flew through *S2, C1, C2,* and

S3 in that order. The experimenters worried that the copiously produced negative pions might pass for antiprotons, so they had designed the logic sequence to exclude that eventuality. Since mesons are lighter than protons, 1.19 GeV pions would be traveling significantly faster ($\beta = v/c = 0.99$) than antiprotons of the same energy ($\beta = 0.78$ at *S2* and $\beta = 0.765$ after the energy loss suffered in *C1* and *C2*). *C1* served to ferret out the antiprotons by producing a count if the candidate particle had a velocity greater than $\beta = 0.79$ and was therefore lighter than an antiproton. *C2* (modeled on Marshall's velocity discriminator shown in figure 6.10) applied the even more exacting criterion that $0.75 < \beta < 0.78$, a condition that selected particles having a mass near that of the antiproton. As a third check, the logic circuitry would count a particle as a bona fide antiproton only if the time of flight between *S1* and *S2* clocked between 40 and 51 millimicroseconds (10^{-9} seconds), a time that would *include* a particle of mass near the proton and *exclude* the wily negative pion. Finally, just to make sure that there was no large-angle scattering in *C2,* which might slow a pion down until it looked like an antiproton, the experimenters set a third scintillator, *S3,* so that it had to fire in coincidence with *S1* and *S2*. The experiment produced a count if and only if all the triggers fired: $S1 \cap S2 \cap S3 \cap$ (not *C1*) $\cap$ *C2*. Each particular piece of this logical combination could have been duplicated by a spurious effect (such as nuclear scattering in *C2*). But the superimposed demand that all criteria be satisfied left only one putative antiproton per 62,000 negative pions. Finally, by altering the strength of the magnetic field, Chamberlain, Segrè, et al. could vary the momentum of the particles sent into the detector. If everything else remained the same, particles of a new momentum running through the same velocity selector would yield counts for particles of a different mass. With all the counters running smoothly, they declared that a "test for the reality of the newly-detected negative particles"[41] would be a pronounced peak in counts when they tuned the apparatus to the mass of the proton/antiproton. When that peak turned up in their scalers, the authors declared that they had a persuasive case for the new particle. No single count meant anything; the existence argument was irreducibly statistical.

Another previously unseen particle, the neutrino, also turned up in a logic experiment. This one, launched by Frederick Reines and Clyde Cowan, had its beginning in the nuclear weapons tests of the Cold War. Reines, though trained in cosmic ray physics, had become a "Los Alamos man" in every sense. During the war, he worked in Feynman's group, especially in measuring the blast wave of an atomic explosion. After writing a (classified) Ph.D. dissertation, Reines decided at war's end to continue working at the laboratory, where he held positions of increasing authority in the instrumentation of blast effects, including

41. Chamberlain et al., "Observations," *Phys. Rev.* 100 (1955): 947–50, on 949.

X-ray measurement, gamma-ray determination, and the measurement of α, a key blast parameter. The scale of the enterprise, the ability to do physics with what he called "infinite resources," held an appeal that was hard to resist. Of the George test in May 1951 that mixed fusion with fission, Reines recalled: "George was a man-made star. [We] could get those conditions on earth. I thought it was pretty fascinating. To have your own private star. Gave a scientist such a different way of looking at things, at trying to understand things."[42] Before plunging into preparations for the Mike test of the first full thermonuclear detonation in October 1952, Reines requested a leave so he could stop thinking about H-bombs and start thinking about physics.

Instead, during the summer of 1951, Reines began an inquiry into whether he could do both: could a nuclear explosion produce detectable neutrinos, the elusive particles postulated two decades earlier yet never directly detected? The underlying physics process would be an inverse beta decay, in which an antineutrino hit a proton, producing a positron and a neutron:

$$\overline{\nu} + p \rightarrow e^{+} + n\,.$$

A bomb would produce huge numbers of neutrons, and their decay a wealth of neutrinos. But as plans developed, Reines and Cowan decided to turn to a somewhat calmer neutrino source: the nuclear reactor. After inconclusive results at Hanford, they moved their ever-growing experiment to the AEC's more powerful Savannah River plant. By then, their goal was to take huge vats of scintillator—some 1,400 gallons—in which the characteristic signal from the electron-positron annihilation might be observed. In order to reject cosmic rays and electronic noise, the two Los Alamos physicists added a second condition. Once released in the scintillator, the neutron would itself be captured by cadmium, and the resulting gamma rays, released a few microseconds later, would make the "real" neutrino events quite distinct (see figures 6.12 and 6.13). As in the antiproton experiment, the neutrino search built its argument into hardware, truckloads of it, that included Los Alamos–designed preamplifiers, coincidence circuits, scalers, and high-voltage power supplies.[43] With this formidable electronic arsenal, Reines and Cowan could show that the number of allowed antineutrino candidates varied with the power setting of the reactor in a way consistent with theory. Next, they demonstrated that the prompt gamma radiation was likely to be honest-to-god positrons by trying the system with known positron emitters, and that the delayed signal from the capture of a neutron by cadmium varied with the amount of cadmium they put in the scintillator. Finally, by

42. Reines, interview by the author, 15 August 1988.

43. I would like to thank Robert Seidel for making available to me his unpublished paper, "The Hunting of the Neutrino" (n.d.), which tracks the Los Alamos origins of some of the electronics used in Reines and Cowan's experiments.

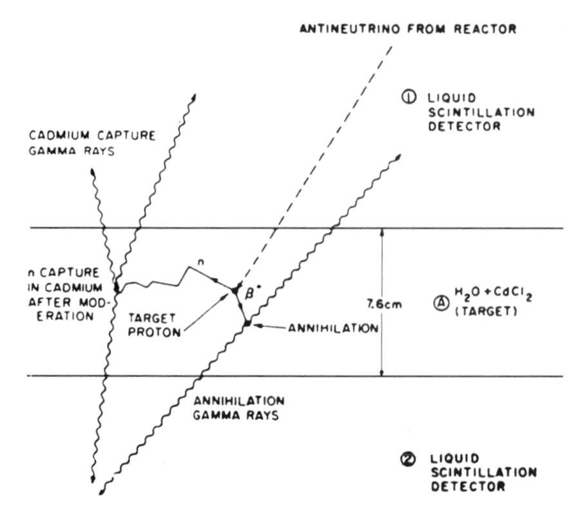

Figure 6.12 Mark of the neutrino (1960). An antineutrino from a fission reaction in the Savannah River reactor hits the water target in which cadmium chloride is dissolved. Inverse beta decay ($\bar{\nu} + p \rightarrow e^+ + n$) yields two particles to be detected. The positron annihilates an electron, producing the characteristic two 0.5 MeV gammas. These were observed in coincidence by two scintillation detectors. The neutron, moderated by the water, took longer before being captured by a cadmium nucleus and releasing multiple gamma rays. Together, the prompt and delayed signals constituted the mark of the antineutrino. Source: Reines et al., "Free Antineutrino," *Phys. Rev.* 117 (1960): 159–73, on 159.

inserting shielding of different thicknesses, they could cut the number of neutron and gamma rays that would reach the detector.[44]

We need to see these two experiments in a double light. As demonstrations of long-conjectured entities, the antiproton and the antineutrino, they are exceptional, one-of-a-kind instances of the laboratory manifestation of theoretical entities. (It should be said that both won Nobel Prizes.) But at the same time we ought to see these experiments as sophisticated instantiations of a longer lived logic tradition, in which unlike the image tradition much of the interpretive work is designed into the device itself. Backgrounds to the neutrino are wired out of existence; backgrounds to the antiproton are killed by the electronics.

Epistemically, these are not open-ended explorations. Neither experiment was set up in such a way as to exhibit a range of previously unknown phenomena the way a bubble chamber or nuclear emulsion might be. Both these electronic wonders aimed at showing that something the experimenters believed to be the case was, in fact, demonstrable, by displaying how the reasonable alternatives could be statistically excluded. These experiments aimed at head→world reliability. If there were other phenomena nearby, they would have to be brought

44. Reines, "Neutrino," *Ann. Rev. Nucl. Sci.* 10 (1960): 1–26; Reines, "Neutrinos," *J. Phys.* 43 (1982): C8-237–C8-260; Reines and Cowan, "Free Neutrinos," *Phys. Rev.* 92 (1953): 830–31; Cowan et al., "Detection," *Science* 124 (1956): 103–4.

Figure 6.13 Electronic neutrino. Reines and Cowan's electronics for the Savannah River experiment. Picking the signal from the background by means of pulse height, coincidence, and anticoincidence was the central feature of the experiment. Source: Courtesy F. Reines.

to the experimenter in some other way. These experiments and the larger tradition for which I have them stand were not exploratory pictures. There would be no meticulous scanning in the hopes of finding a wild golden event, undreamed of previously. World→head reliability was not the name of this game.

6.4 A Democratic Detector

It is all too easy to rhapsodize about visualization, to give it a triumphalist history as a self-evident desideratum: Schrödinger over Heisenberg, Feynman over Schwinger, or geometrical over field-theoretical representations of general relativity. In experiment, too, the cloud or bubble chamber image of particle physics has come to stand, metonymically, for the branch of physics or even physics itself. Vision and visuality have come to be culturally supervalued, not only but markedly in the history and philosophy of science. But the nonvisual, even occasionally antivisual side of experimentation that corresponds to the logic tradition is a powerful part of the history of experimental physics. And with the establishment in the 1930s of a material creole of electronic logic, we can begin to track an iconoclastic branch of empirical inquiry.

Out of the established electronic logic tradition of counters came a technology that struggled to compete with the bubble chamber, the cloud chamber, and nuclear emulsion that were its image predecessors in generating evidence for new particles. The new devices, spark chambers and their many cousins, are both easier and harder to analyze historically. They are easier to study because in their earliest days spark chambers were child's play to build, in contrast with the imposing cryogenic facilities surrounding the hydrogen bubble chambers. Yet precisely this ease of construction makes the historian's task vastly more difficult. Whereas there were only a handful of costly, centralized bubble chamber installations, spark chambers soon proliferated everywhere. Whereas the bubble chambers demanded teams of 20, 30, or 40 physicists, technicians, engineers, and scanners to service, run, and analyze data, the early spark chamber groups rarely comprised more than 10 physicists. Whereas bubble chamber development can be recounted in a fairly linear fashion, the spark chamber grew up sporadically, in multiple versions, and in sites scattered across the United States, England, Germany, Japan, and the Soviet Union. This democratic, even anarchic aspect of the various counters and spark chambers will prove important in understanding their cultivation as instruments of physics, but their multiple origins confound a tidy narrative scheme.

The spark chamber is related to counters in many of the same ways that the bubble chamber is a descendant of the cloud chamber. As in the transition from cloud to bubble chamber, there is continuity of personnel, there is continuity of technique, and there is continuity in the style of experimental demonstration. But because the spark chamber sprouted from at least three (probably more) independent roots, a narration of its evolution must reflect its ramified branches, where a few growths end abruptly, and some split into a multitude of smaller stems. As with so many of the new instruments of the postwar period, to grasp the origin of the spark chamber it is essential to understand the new reservoir of technical knowledge available after the war came to an end.

As we saw in chapter 4, both of the big American World War II weapons projects, the atomic bomb and radar, had to produce a new breed of sophisticated electronic equipment to accomplish their tasks. At Los Alamos, physicists needed fast timing circuits for many purposes. Among other applications, Manhattan Project scientists required them to instrument the implosion process in the plutonium bomb and to analyze neutron production and multiplication. Seth

Neddermeyer led one of the most sophisticated timing projects, the "chronotron," which determined the region of superposition of two pulses traveling along transmission lines in opposite directions.[45] Ultimately, Neddermeyer's goal was a determination of the time between the firing of counters and therefore a measurement of the velocity of fast charged particles. By the summer of 1945 Neddermeyer's team had in hand a device that could measure time differences to an accuracy of 3×10^{-10} seconds. Potentially, it was a remarkable contribution to the logic tradition, but it suffered from a troublesome gap.

Missing from Neddermeyer's system was a counter with an accuracy commensurate with that of the timing system.[46] Lamentably, the conventional Geiger-Müller tubes to be used with the chronotron had an inherent uncertainty of about 10^{-7} seconds.[47] At root, the counter's precision was understood to be limited because electrons, liberated by the passage of the charged particle, had to drift from somewhere inside the tube toward the central wire before they encountered a field strong enough to initiate an avalanche. Experimenters had to overcome this weak link in the instrumental chain if they were going to determine the lifetimes of the new particles.

Jack Warren Keuffel, a graduate student working with Henry Neher at Caltech, seized the opportunity presented by Neddermeyer's "timer in search of a detector." Keuffel realized that since charged parallel plates create a constant field throughout the volume between them, spark counters—essentially flattened Geiger counters—would allow avalanches to begin *anywhere* in the sensitive volume. Technically, Keuffel's problem was to modify the known parallel-plate counters so that he could rapidly eliminate ions left over from earlier discharges. A perfected spark counter that could sweep the old ions away would complete Neddermeyer's otherwise powerful measuring system.[48]

Keuffel's thesis advisor, Neher, was an expert on counters, and from him Keuffel must have learned the magic tricks of counter preparation: how to bake and bevel electrodes, how to seal and outgas containers, and how to handle the vagaries of electronic logic circuits and high-voltage apparatus. For years, counter building had been thought "a kind of witchcraft," as one of its practicing

45. Upon leaving Los Alamos, Neddermeyer had the equipment transferred to him at Washington State University, where he continued to refine the technique. See Hawkins, Truslow, and Smith, *Project Y* (1983), 320–21; and Neddermeyer et al., "Measurement," *Rev. Sci. Inst.* 18 (1947): 488–96.

46. "The difficulties in the achievement of [building a fast counter] are probably great but may not be fundamental" (Neddermeyer et al., "Measurement," *Rev. Sci. Inst.* 18 [1947]: 488–96, on 488).

47. See, e.g., Sherwin, "Short Time Delays," *Rev. Sci. Inst.* 19 (1948): 111–15; for an example of the difficulty of using Geiger counters for microsecond physics, see Rossi and Nereson, "Experimental," *Phys. Rev.* 62 (1942): 417–22.

48. Workers *at the time* thought of the parallel-plate counter as a material analogue of the Geiger counter; this is not a retrospective connection. E.g., Pidd and Madansky wrote in "Properties I," *Phys. Rev.* 75 (1949): 1175–80, on 1175: "The parallel plate counter is an adaptation of a gas counter in which the electrodes have plane parallel symmetry."

magicians, Bruno Rossi, later put it.[49] Even the dry prose of physics textbooks could not hide the mystery behind counter manufacturing. In one of the most widely used instructional texts in prewar experimental physics, *Procedures in Experimental Physics* (1938), Neher described the rudiments of the trade: coat the copper tube with six normal nitric acid, then cleanse with a 0.1 nitric acid bath. The acolyte should then rinse the assembly (at least) 10 times with distilled water, dry the electrodes, heat them over a flame until the copper turns brownish black, seal the counter, heat the equipment for several hours until the copper turns bright red with cupric oxide, evacuate, admit dry NO_2 until the copper turns dark and velvety, pump out the gas, and then admit argon bubbled through xylene. Then seal it off. "All the above steps may not be necessary in all cases," Neher confessed, "yet this procedure has been found to give very satisfactory counters" with reaction times less than 10^{-5} seconds.[50] If it works, use it—even when you don't know why.

Just as Glaser was able to take the skills he had honed on the cloud chamber to the bubble chamber, Keuffel could transfer the skills and techniques he learned as Neher's counter student to a novel instrument. But while the tradition of counter physics could open possibilities, it served at the same time to constrain the options Keuffel anticipated for the new instrument. Most important, Keuffel only gave passing mention in his 1948 thesis to what Glaser had found most intriguing in his work, the possibility of using the spark counter to *follow* a charged particle: "[T]he discharge is localized, presumably, in the neighborhood of the initiating ion, with the streamer channel plainly visible. This has obvious possibilities as a means of determining the path of a particle."[51] Keuffel then let the subject drop.

Though spatial localization later proved of immense importance, we can understand why Keuffel was not particularly interested: the tradition from which he emerged, a tradition of fast timers and logic circuits, valued first and overwhelmingly the binary yes/no answers of the counter array. Keuffel wanted to fill a technological gap by providing a fast detector for a potentially successful timing circuit. The technical tradition and accepted experimental context defined a "need" for temporal, not spatial, localization. With its missing counter, the chronotron system illustrates perfectly the simultaneously productive and restrictive functions of technical constraints, defining the meaning of the object as a new spark counter and *not* a proto–spark chamber.

49. Rossi, "Arcetri" (1985), 56.

50. Neher, "Geiger Counters" (1938), 270–71.

51. Keuffel, "Parallel-Plate" (1948), 8; see also Keuffel, "Parallel-Plate," *Phys. Rev.* 73 (1948): 531; and Keuffel, "Parallel-Plate," *Rev. Sci. Inst.* 20 (1949): 202–8. Similarly, three other counter workers, F. Bella, Carlo Franzinetti, and D. W. Lee, maintained that "the main use of this type of counter is in the measurements of very short time intervals" on the order of one billionth of a second; "Spark Counters," *Nuovo Cimento* 10 (1953): 1338–40; see also Bella and Franzinetti, "Spark Counter," *Nuovo Cimento* 10 (1953): 1461–79.

Physicists studied Keuffel's work, abroad as well as in the United States. One avid reader was Erich Bagge, who had also worked on a nuclear fission project during the war, but for the Germans. Enlisted to the task in September 1939, his efforts had been primarily directed toward the development of an isotope separation device on the principle of thermal diffusion. Heating the uranium source, he set in motion spinning disks timed so that the faster moving, lighter nuclei of U^{235} would fly through the slits while the U^{238} would be caught.[52] By June 1944, the sluice was producing enriched uranium hexafluoride.[53] Like several of his colleagues on the German project, Bagge also found time to pursue cosmic ray physics, researching the problem of nuclear fragmentation in the lower atmosphere.[54]

As the Allied occupation forces advanced through Germany, Samuel Goudsmit's Alsos team swept up Bagge, along with others including Otto Hahn, Werner Heisenberg, Walther Gerlach, and Carl Friedrich von Weizsäcker, and brought them as privileged prisoners to Farm Hall in England. There the Anglo-American intelligence teams attempted to learn how far the Germans had advanced in their quest for the bomb; and there, as we saw in chapter 4, the German physicists learned for the first time that their American counterparts had succeeded.[55] Released in early 1946, Bagge explored issues related both to fission and to cosmic rays. In 1947, for example, he published a theoretical paper intended to account for the asymmetric fragments released in spontaneous nuclear fission.[56] Both nuclear and cosmic ray physics demanded more precise and sophisticated measuring tools; Bagge turned to an examination of the rapidity of droplet growth in cloud chambers.[57] Responding to Keuffel's paper in *Physical Review,* Bagge set his student Jens Christiansen the task of reproducing and refining the American work on parallel-plate spark counters.[58]

Bagge and Christiansen went to press with their improved spark counter in 1952, whereupon Bagge promptly presented another student, Paul-Gerhard Henning, the thesis problem of investigating the space behavior of the spark in counters like Christiansen's. What prompted Bagge to press the possibility of track localization where Keuffel had not? Bagge's prior interest in cloud chambers no doubt catalyzed his attempt to transform the spark counter into a track-following device, a transformation that would yield pictures where there had

52. Irving, *German Atomic Bomb* (1967), 43–49, 88–89; and Walker, *German National Socialism* (1989), 33, 53, 126, 133.

53. Walker, *German National Socialism* (1989), 133–34.

54. Bagge, "Nuclear Disruptions" (1946), 128–43.

55. See Frank, *Operation Epsilon* (1993); Walker, *German National Socialism* (1989), 153–65; and Bagge, Diebner, and Jay, "Uransfaltung" (1957), 42–72.

56. Bagge, "Massen-Häufigkeitsverteilung," *Z. Naturforsch.* 2a (1947): 565–68.

57. Bagge, Becker, and Bekow, "Bildungsgeschwindigkeit," *Z. angew. Phys.* 3 (1951): 201–9.

58. Bagge and Christiansen, "Parallelplattenzähler," *Naturwissenschaften* 39 (1952): 298.

Figure 6.14 Allkofer group, spark counters (ca. 1953). Paul-Gerhard Henning's spark counters were the first to be used specifically to illuminate the trajectory of a passing particle. Previously, spark counters had been entirely assimilated in the cylindrical counter; both types of counters had been conceived as devices to determine if (and when) an event had occurred within its volume. Source: OAP.

only been counts. To someone immersed in the cluster of practices surrounding the image tradition, finding tracks was natural. Keuffel, by contrast, had come from a scientific heritage of pure counts—no prior work with either emulsions or cloud chambers. In the logic tradition, the culture of images was at best irrelevant; at worst, images were positively to be avoided. Over the next three years Bagge had Henning investigate the properties of a system of three spark counters set one on top of another. To Keuffel's design Henning added a light intensifier (*Aufhellverstärker*) that discharged a capacitor through the spark after the chamber had fired (see figure 6.14). The added current brightened the sparks of just those events in which a charged particle plunged through the whole apparatus. Thus, by setting the camera so that only bright events would register, Henning effectively recorded only events in which coincidences occurred.

From March until October 1953, Henning designed and redesigned his electronics, calculating pulsing circuits, time constants, and coincidence circuits. Finally, on Saturday 17 October 1953, he began to analyze film, recording the frames in his notebook one by one: "*künstl. Funken*" (spurious sparks), then "*nichts gesehen*" (nothing seen) for picture after picture. More spurious sparking . . . then something: "*Koinzidenz gesehen?*"[59] Another spurious spark, where-

59. Henning, "Labor-tagebuch," 17 October 1953, 48–50, PHP.

upon the film switched to exposures of one minute instead of 15 seconds. In rapid succession the excited comments follow: "*Mit Sicherheit Koinz. gesehen!*" (coincidences seen with certainty!), "*vermütlich zwei Koinzidenzen,*" and finally, "*mit Sicherheit Koinz. gesehen!*"[60] From these remarks and the ones that follow in the notebook it is clear that for Henning the persuasive experiments were these first double coincidences.

Simply showing double coincidences would not, however, persuade other physicists: it is always possible to draw a line between two points. Henning was obliged to extend the method to *threefold coincidence,* which was recorded in his notebook, without fanfare, almost a half-year later (28 April 1954).[61] It was this demonstration that the Bagge group took as the conclusive exhibit for their case that spark counters could follow the path of a penetrating particle and describe it in spatial terms. It was a first, highly tentative step toward incorporating the capacity of picturing within the logic tradition.

The struggle to extend the two to threefold coincidences illustrates the gap that separates "private" from "public" evidence. But this epistemological separation is not necessarily binary: in more complex experiments involving many collaborators, the "circle of belief" expands slowly outward, from small subgroup, to wider divisions within an experiment, through the larger collaboration, and then outward to the wider community.[62]

Henning set his detector to photograph sparks left by the passage of multiple scattering of muons and reported his results to the Deutsche Physikalische Gesellschaft in Wiesbaden. Soon after completing his dissertation in 1955 Henning left physics for industry, and his work languished on the shelf. Bagge himself took his group to Kiel in 1957 and soon thereafter turned his attention to the installation of a nuclear reactor in Geesthact (Schleswig-Holstein). With both Henning and Bagge out of the picture, Henning's dissertation (and its published single-paragraph summary) passed into obscurity—even among his Kiel colleagues "nobody felt responsible to push the publication."[63] And when the work was finally ready, Bagge insisted that it appear in a largely ignored new journal, *Atomkern-Energie,* for which he was an advisor.[64] Thus the Kiel team's work, while innovative in its use of stereophotography, track-following sparks, multiple plates, and selective recording and its cosmic ray applications, had no

60. Henning, "Labor-tagebuch," 17 October 1953, 48–50, PHP.

61. Henning, "Labor-tagebuch," 17 October 1953, 48–50, PHP; Henning, "Labor-tagebuch," 28 April 1954, 82–83, PHP.

62. On the expanding circle of belief in a large high energy physics experiment, see Galison, *Experiments* (1987), chap. 4.

63. See Allkofer et al., "Ortsbestimmung," *Phys. Verh.* 6 (1955): 166. A paper that is almost identical to the thesis was published by Henning in March 1957: "Ortsbestimmung," *Atomkern-Energie* 2 (1957): 81–88. Allkofer, *Spark Chambers* (1969), 1–5, contains a brief history of the Kiel work, as does Allkofer to Conversi, 11 October 1972, OAP.

64. Allkofer, interview by the author, 24 April 1983.

immediate effect outside the confines of that port city.[65] With Keuffel's work it is easy in retrospect to see an "anticipatory" move toward the spark chamber; historically, the work led elsewhere. Similarly with the Kiel work: it too illustrates how spark detectors grew up at multiple sites, while failing to connect with the wider world of microphysical research.

A third and final example of an evanescent spark discovery occurred in Paris, where Georges Charpak settled after being liberated from a German concentration camp. Upon finishing his studies at the École des Mines de Paris in 1954, he entered the Nuclear Chemistry Laboratory at the Collège de France under the supervision of Frédéric Joliot-Curie. Virtually all of their equipment had to be built from scratch, and Charpak turned his attention to the measurement of the excitation of atomic shells following beta decay, using especially large angle Geiger counters.[66] This experience not only formed in Charpak the habit of building his own electronic equipment, it also gave him detailed knowledge about the tools of the logic tradition's trade: the choice of gases used in counters, the theory of sparks, the techniques of counter construction, and the assembly of electronic filters, coincidence circuits, and amplifiers.

By July 1957, Charpak had shifted his attention to the sparking process itself in order to create a device that would follow the tracks of charged particles. His idea was to provide a very fast high-voltage pulse in order to precipitate small (~1 millimeter) but visible avalanches around each electron liberated by the charged particle.[67] For the task he could use a gas combination in his new counter similar to that exploited in his thesis work (9:1 argon to alcohol; 8:1 argon to alcohol in his thesis work) and adapt his skills with high-voltage pulses and fast electronics. Reporting his results to Joliot-Curie in late 1957 gained him a tentative nod from his old thesis advisor: "I see that you have worked a lot, and already obtained some interesting results with your spark chamber. I think, like you, that this work ought be of substantial interest for physicists working on theories of the spark." Concurring with Charpak's own assessment, Joliot-Curie hoped to set him on a more productive path: "It would be desirable to continue this kind of research but I completely agree that it would be more worthwhile to return to your first work consisting of using the avalanche effect to amplify the effect of the liberation of electrons by particles in the gas."[68] As in the Keuffel work, Joliot-Curie's comment suggests that sparks and electronic devices fit more easily within the field of discharge physics than they did in the enterprise of

65. Research continued at Kiel by Allkofer and colleagues; e.g., Allkofer, "Ansprechvermögen" (1956). Additional references from the Kiel group may be found in Allkofer, *Spark Chambers* (1969).

66. Charpak, "Phénomènes atomiques" (1954); and Charpak, curriculum vitae, 9 December 1981, 1, GCP.

67. Charpak, "Nouveau détecteur," *J. Phys.* 18 (1957): 539–40. A longer version was presented at the International Conference on Mesons and Recently Discovered Particles, Venice, 28 September 1957: Charpak, "Principe d'un détecteur," 1957, GCP.

68. Joliot to Charpak, 29 November 1957, GCP.

following particle tracks. The wordless creole of logic work, with its attention to sparks, gases, and fast electronics, had little vocabulary for spatial–pictorial–reconstruction.

Like Charpak, the Italian experimenter Adriano Gozzini had been imprisoned by the Nazis during World War II. When he returned to his laboratory in Pisa he found a setting far more destitute than postwar Paris. His laboratory building had been ransacked by the German occupation forces. While holding the building they had destroyed or stolen all the important equipment, as well as the library collection and the laboratory's prized possession, a signed Roland grating. Research opportunities at the end of the war seemed almost nil, restricted essentially to experiments that used improvised materials. One good source of electronic hardware was the contingent of American troops pulling out from the woods outside the city and divesting themselves of technological detritus. From these troops, in 1950, Gozzini purchased a magnetron that had been designed for radar use and set about converting the war relic into an instrument of microwave physics.[69]

With his refurbished microwave source, late in 1954 Gozzini devised a technique for detecting small impurities in chemical samples. His idea was this: if the molecules of the sample had a different electric moment from impurities located in the sample, then the microwave should heat the two differently. Hotter and cooler liquid regions refract light differently, so a shining light could produce a visual display of the liquid's purity. To accentuate the effect, Gozzini chose for his liquid cyclohexane, a chemical that has almost no electric moment. Nothing happened.[70]

Thinking that his surplus magnetron was probably broken, Gozzini tested it—in a routine way—by holding the magnetron next to a tube of neon gas. When working properly, a magnetron causes such tubes to emit light. To see the tube more clearly, the Italian physicist brought the whole apparatus into a darkened room. The tube remained dark. When the room lights were turned on, however, the neon tube clearly, if faintly, glowed. Gozzini decided to consult a veteran counter physicist, Marcello Conversi. Discussing the recalcitrant behavior, the two physicists concluded that only free electrons, accelerated by the microwave's electromagnetic pulse, could ionize and excite atoms, producing a breakdown of the gas. Ionizing particles excite atoms; when the atoms deexcite they radiate

69. A. Gozzini, interview by the author, 12 July 1984. See, e.g., Gozzini, "Costante dielettrica," *Nuovo Cimento* 8 (1951): 361–68; and Gozzini, "Sull'effetto Faraday," *Nuovo Cimento* 8 (1951): 928–35. For historical recollections, see Conversi, "Development," *J. Phys.* 43 (1982): C8-91–C8-99; and the especially useful typescript, Conversi and Gozzini, "Electrically Pulsed Track Chambers and the Origin of the Spark Chamber Technique," July 1971, AGP. Note that independently of the Pisa work, Tyapkin and his group developed a pulsed-fed hodoscope system of Geiger-Müller counters for use with accelerators; see, e.g., Vishnyakov and Typakin, "Investigations," *Sov. J. Atomic Energy* 3 (1957): 1103–13.

70. The original chemical sample that Gozzini used was still in Pisa in Gozzini's possession in 1983.

light. (Later instrument designers concluded that photons emitted in deexcitation played an important role in propagating the discharge throughout the tube.) In the dark, so Conversi and Gozzini reasoned, there were no free electrons with which the microwave could start an avalanche. Consequently, the neon tube would not glow.[71]

From this rough and ready qualitative explanation of the phenomenon, the two physicists deduced that they could build a detector by exploiting the fact that only gases with free electrons would glow. Like Glaser's multiple charge theory or Wilson's ion theory, *some* theory played an essential enabling function in the production of the new device. But for none of these instruments was the underlying theory mathematically articulated, experimentally secured—the sense of "theory" usually invoked in philosophical references to the "theory ladenness" of instruments. Conversi and Gozzini reasoned this way: If a charged particle traversed the tube, it would leave a trail of ions and free electrons. When a microwave pulse bathed the apparatus, those (and only those) tubes would glow that had been penetrated by a charged particle. As a precaution against one tube's light triggering another tube, the experimenters carefully wrapped each in a protective jacket of black paper.[72] On 25 March 1955, a pulsed electric field activated soda glass tubes of pure argon and, for the first time, produced a straight line of flashes in coincidence with counters. Soon, the Italian team switched to narrower, neon-filled tubes and began recording events (see figures 6.15 and 6.16).[73]

An essential feature of the flash tubes was that they were only sensitive for 10^{-5} seconds prior to the pulse—any track older than that would no longer be made visible, since the free electrons would have diffused to the glass walls where they recombined with ions. The Conversi-Gozzini tubes thus offered the possibility of tracking cosmic ray particles as they scattered and showered within the chamber. Just like the Geiger-Müller tubes on which they were patterned, the tubes provided a two-dimensional projection of the particles' precise trajectories, though, as in the case of crossed Geiger-Müller tubes, it was possible to recover some of the track's three-dimensionality by alternating layers of

71. Conversi, "Development," *J. Phys.* 43 (1982): C8-91–C8-99, on C8-91–C8-92.

72. Conversi and Gozzini, "'Hodoscope Chamber,'" *Nuovo Cimento* 2 (1955): 189–91, on 189. A longer paper was given at the 1955 Pisa International Conference on Elementary Particles: Conversi et al., "Hodoscope," *Nuovo Cimento* 4 Suppl. (1956): 234–37. Franzinetti brought to the collaboration his expertise with the theory and operation of spark counters. See Bella and Franzinetti, "On Spark Counters," *Nuovo Cimento* 10 (1953): 1461–79; Bella and Franzinetti, "Spark Counter," *Nuovo Cimento* 10 (1953): 1335–37; and Bella, Franzinetti, and Lee, "Spark Counters," *Nuovo Cimento* 10 (1953): 1338–40. Note that Franzinetti and his collaborators built on and cited the earlier work by Keuffel and by Pidd and Madansky; see nn. 20 and 23 above. Wolfendale and his colleagues at Durham were the first to incorporate flash tubes in a usable detector; see, e.g., Gardener et al., "Flash Tube," *Proc. Phys. Soc. B* 70 (1957): 687–99.

73. Conversi, "Development," *J. Phys.* 43 (1982): C8-91–C8-99, on C8-92, C8-93.

Figure 6.15 Conversi and Gozzini, hodoscope device (1955). Conversi and Gozzini's "hodoscope chamber," which first operated on 25 March 1955, consisted of glass tubes wrapped in black paper. Though modeled on stacks of logic-combined Geiger-Müller counters, Conversi and Gozzini here partially hybridized the logic and track-following ideals. Source: Conversi and Gozzini, "Electrically Pulsed Track Chambers and the Origin of the Spark Chamber Technique," July 1971, AGP.

tubes pointing one way, with layers aligned perpendicularly.[74] Once again, a hesitant step had been taken toward the production of spatially specific track data from within the logic tradition.

Unlike some of the other moves toward a spark-based track detector, the Italian work caught the attention of the logic community. Recognizing the value of the triggered pulse system used by Conversi and Gozzini, two physicists, T. E. Cranshaw and J. F. De Beer, working in Harwell sought to combine the spark counter (of Keuffel et al.) with the Italians' pulsing technique. That is, they tried to alleviate the problem of spurious discharges in the spark counter by pulsing the plates with high voltage only when a particle had entered the system.

74. Indeed, a triggering system altogether analogous to the Conversi-Gozzini system was used with Geiger tubes by Piccioni in 1948: Piccioni, "Search," *Phys. Rev.* 74 (1948): 1754–58. This is not to imply that the flash tubes were always replaceable by Geiger-Müller tubes: Geiger-Müller tubes have a notoriously strong field gradient around their central wire; flash tubes operate in a uniform electric field, which makes their response uniform.

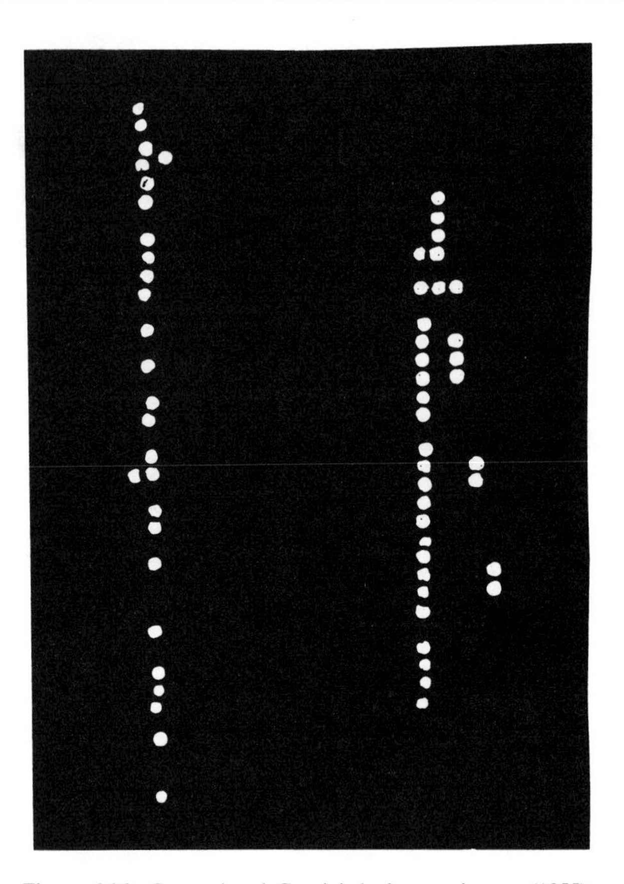

Figure 6.16 Conversi and Gozzini, hodoscope images (1955). Representative photographs of track pictures taken in the hodoscope chamber showing a single and double track event. Source: Conversi and Gozzini, "Electrically Pulsed Track Chambers and the Origin of the Spark Chamber Technique," July 1971, AGP.

Not following the Italians, the Harwell physicists used air, not a noble gas, and contemporaries ascribed their difficulties to that choice.[75]

Once more a temporary halt in one part of the world was the starting point for new modifications in another. A Japanese physicist, Shuji Fukui, was well prepared to reproduce the Pisa physicists' neon-filled device and, after 1957, to exploit Cranshaw and De Beer's more sophisticated triggering electronics. As an undergraduate, Fukui had extensive training in glasswork for vacuum experi-

75. Cranshaw and de Beer, "Triggered Spark," *Nuovo Cimento* 5 (1957): 1107–16.

ments, gaseous discharges, and classical atomic spectra. At the same time the financial constraints of Japanese physics pointed to cosmic ray research as the only economically viable way to participate in particle physics. War damage left only limited research funds, and the emergency demands of the Korean War left only partly restored industrial resources. Moreover, in November 1945 American occupation forces had destroyed five Japanese cyclotrons: two at the Institute for Physical and Chemical Research in Tokyo, two at the Osaka Imperial University, and one at Kyoto Imperial University.[76]

Acting within these limits, in February 1956 a Japanese workshop on "Physics of Super-High-Energy Interactions" set a twofold program for the cosmic ray division of the recently completed Institute for Nuclear Study (INS) in Kyoto. First, physicists would expose nuclear emulsions on mountain peaks and send films into the upper atmosphere in high-altitude balloons. Second, they would exploit Conversi and Gozzini's recently invented detector to examine the particle distribution in the core of extensive cosmic ray showers. In pursuit of this second goal, Fukui, by then a research associate (*joshu*), wrote to Conversi requesting reprints of his and Gozzini's *Nuovo Cimento* articles.[77]

Fukui and a doctoral student, Sigenori Miyamoto,[78] planned to assemble 5,000 small glass balls filled with neon. They would then stack the delicate containers on the laboratory floor in a 7-square-meter array designed to detect the particle density and lateral distribution of extensive air showers. At the beginning of their study, little was available from commercial sources: the neon and argon gases, the hydrogen thyratron tubes (to generate the pulse), the camera with automatic film advance, the high-speed film, and miscellaneous high-voltage equipment—all had to be specially imported. Gozzini and Conversi had used their American war-surplus pulse-forming network; Fukui and Miyamoto, in contrast, had to wire together their own pulser, since they could not find a Japanese manufacturer capable of producing a square-pulse generator. Consequently, they had a cruder circuit in which a capacitor, discharging through a resistor, produced a pulse with a steep rise and exponential decay, rather than the more elegant square wave that would have issued from a commercial product. In a moment we will see how the home-grown character of Fukui and Miyamoto's generator was central to their success.[79] This was not big physics.

76. Groves, *Manhattan Project* (1983), 367–72.

77. Fukui, "Chronological Review" (1983), 1–2. At the same time, the INS set its nuclear physics division the task of building a cyclotron. Miyamoto, letter to the author, 21 October 1986.

78. Miyamoto was a doctoral student of Hushimi's "for form's sake" but was "actually" a graduate student of Yuzuru Watase's. Similarly, Fukui was formally Professor Ogata's *joshu*, but his actual affiliation was with Watase's laboratory. The Osaka laboratory was effectively a small branch of Watase's laboratory. Miyamoto, letter to the author, 21 October 1986.

79. Fukui, "Chronological Review" (1983), 1–2.

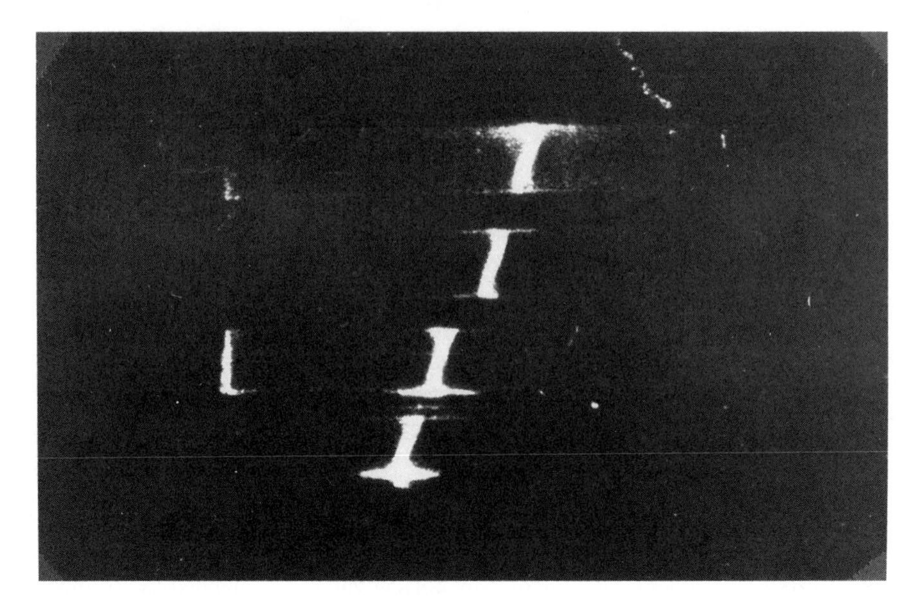

Figure 6.17 Fukui and Miyamoto, tracks (1957–58). Once they realized that the discharge was localized in the Conversi tubes, Fukui and Miyamoto could turn to a system of glass boxes and photograph tracks, even somewhat oblique tracks (as shown here). The angular restrictions on tracks, however, meant that the chamber was only sensitive to *some* of the tracks passing through—the Achilles' heel of electronic imaging for many years. Source: Fukui and Miyamoto, "'Discharge Chamber,'" *Nuovo Cimento* 11 (1959): 113–15, pl. following 114.

On receiving the Italian papers, Fukui and Miyamoto began tests to determine whether the flash tubes could be fired by their steeply rising, exponentially decaying pulse. During fall 1956 they obtained their first positive results with a few tubes and immediately turned to the task of assembling the extensive device foreseen at the INS workshop. By December 1957, the Japanese experimenters had amassed a significant collection of gas-filled glass tubes. According to Conversi's instructions, the next step was to cover the tubes with dark paper, leaving one side open. Stuck in the laboratory without any technical assistance, they foresaw nothing but tedium in the prospect of wrapping 5,000 little glass bulbs in black paper. To lighten their burden, the physicists piled a few bulbs, unwrapped, with gas of varying pressures on top of one another. They turned the pulser on and the lights off. Out of the darkness came a remarkable image that Conversi and Gozzini could not have seen because of the dark paper: inside at least some of the tubes, *tracks* were forming (see figure 6.17). Instead of the crude two-dimensional mosaic that had appeared to Conversi and Gozzini, before the Japanese physicists stood a well-defined three-dimensional image.[80] Though limited in precision, the Japanese device represented one more advance by the logic tradition into the image-producing capability of their competitors at the rich centers of bubble chamber physics.

On the basis of images like the one recorded in figure 6.17, Fukui and Miyamoto realized that they could build a new kind of detection chamber with much greater spatial resolution than had been possible using counters. But clear tracks appeared in only some of the tubes—those with higher pressure. This phenomenon puzzled them at first, since they anticipated that the discharges in gases would be the same for two tubes at different pressures as long as the pulse height was greater than the critical minimum level for each tube. Since they thought that they were exposing all of the tubes to a field well above even the highest tube threshold, the tubes' uneven responses puzzled them.

Finally, they understood. Fortune had smiled on their poverty: the differing responses of the tubes with different pressures was traceable to their jury-rigged pulsing device, which produced pulses that were not square. Consider (in figure 6.18*a*) the case of a square wave such that its peak field exceeded the critical electric field minimum of all the tubes; for example, the maximum would be greater than the minimum E_1 needed for a high-pressure tube, and also greater than the minimum needed for a low-pressure tube with a minimum $E_2 < E_1$. In this square wave case, all the tubes received a track-producing pulse for the same amount of time—the time t_1 during which E is above E_1 would be equal to the time t_2 during which E was above E_2. This is what Fukui and Miyamoto expected. But because their pulse had a shape closer to that depicted in figure 6.18*b*, the situation was different: the time t_1 was considerable *shorter* than the time t_2. The two physicists were therefore exposing the high-pressure tubes to a shorter effective pulse than the low-pressure tubes. Inadvertently, they had discovered optimal conditions for producing track-following discharges: short pulses on high-pressure tubes.[81]

Above all, Fukui and Miyamoto's observation of tracks meant that they no longer needed to model the detector on the binary response of Geiger counters. Suddenly they were free to alter the geometry from hundreds of stacked cylinders to a few glass boxes made with electrically conductive surfaces.[82] Within

80. Fukui and Miyamoto, "Hodoscope Chamber" (1957); and Fukui, "Chronological Review" (1983), 3–4. Miyamoto remembers the pressure tests as independent of the tube-wrapping problem. Miyamoto, letter to the author, 21 October 1986.

81. The discharge could be improved in quality by reducing the high-voltage pulse to 10^{-7} seconds from the 10^{-3} seconds used previously. Fukui and Miyamoto, "Hodoscope Chamber II" (1957), 4. Note that following Conversi and Gozzini's lead, the Japanese physicists had, from the start, used neon as their primary gas; by studying reference tables they discovered that they could operate at a lower electric field by mixing in a small amount of argon.

82. Fukui, "Chronological Review" (1983), 5.

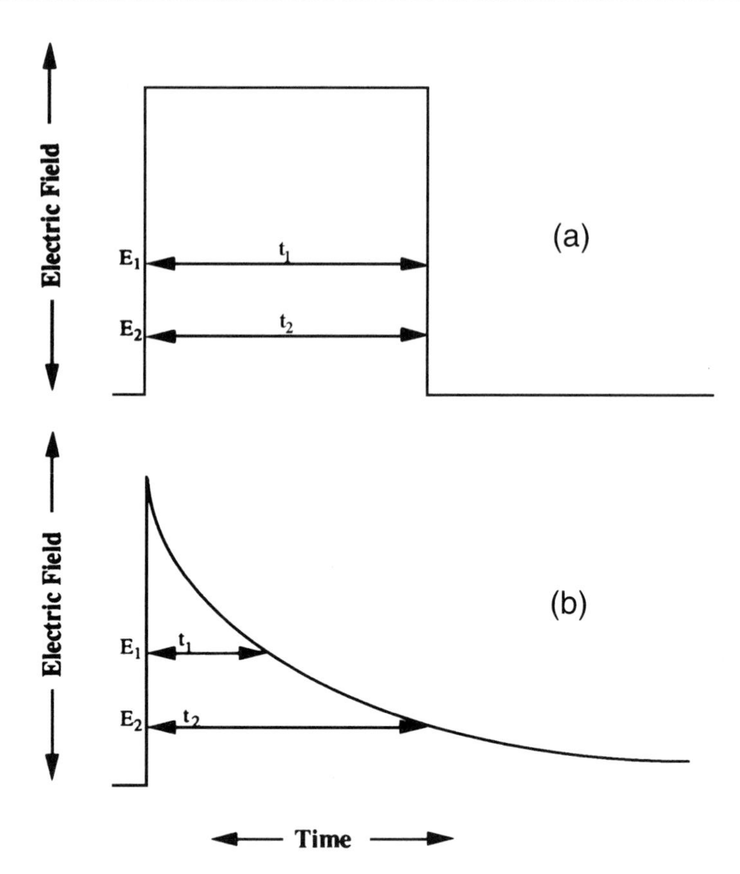

Figure 6.18 Fukui and Miyamoto, pulse shapes (1959). (*a*) Square-wave pulse; (*b*) non–square-wave pulse. With Fukui and Miyamoto's non-square-wave pulses, the higher the voltage threshold, the shorter the pulse time. Thus, accidentally, they discovered the advantage of using steeply rising, short pulses on higher pressure tubes (a legacy of its Geiger-Müller antecedents) and promptly dispensed with the tubular geometry in favor of glass boxes.

months they had used the new apparatus to produce clear pictures of particle tracks, and by September 1959, the two physicists had composed their contribution to *Nuovo Cimento* (see figure 6.19).[83] Endowing their now-enclosed device with the title "discharge *chamber,*" Fukui and Miyamoto had completed their transformation of the instrument. What had begun as a stack of criss-crossed counters deployed in the style of Conversi and Gozzini had now been circumscribed, literally and figuratively. The assortment of counters had become *an*

83. Fukui and Miyamoto, "'Discharge Chamber,'" *Nuovo Cimento* 11 (1959): 113–15.

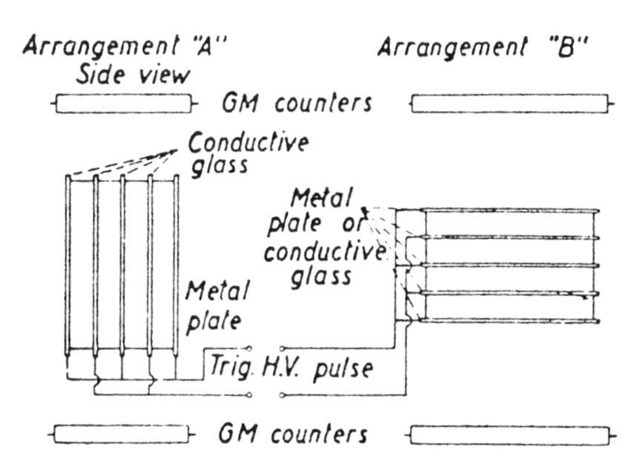

Figure 6.19 Fukui and Miyamoto, discharge chamber (1959). Chamber used to produce the tracking picture of figure 6.17. It was this device that reanimated the logic tradition at LBL in 1960, then reeling from the combination of bubble chamber success and Alvarez's blistering 1957 attack on the future of nonvisual experimentation. Source: Fukui and Miyamoto, "'Discharge Chamber,'" *Nuovo Cimento* 11 (1959): 113–15, on 114.

instrument in its own right. It was an instrument that produced a logic-based image, one manifestly schematic in its rendition of a trajectory, but—unlike the untriggerable emulsion and bubble chamber—at the same time amenable to the *epistemic control* afforded by the logic of adjustable triggering and small enough to remain under the *workplace control* of a handful of experimenters.

6.5 The Hegemony of Bubble Chambers

The triumphs of the bubble chamber at American accelerator laboratories made the remaining American logic experimenters a ready audience for the Japanese device. Two Bevatron physicists, William A. Wenzel and Bruce Cork, along with two visiting colleagues, James Cronin from Princeton and Rodney L. Cool, had pushed counter experiments to the limit of complexity in an effort to capture the complicated topology of multiple interactions and decays. The prognosis for the logic tradition appeared grim. With each passing day it seemed more obvious that bubble chambers in general, and Alvarez's group in particular, would outdistance all electronic competitors. Laboratory policy and scientific planning were intertwined; soon the belief that electronic apparatus was obsolete burst into the open.

During 1957, a decision had to be reached about future accelerator building at the Berkeley Radiation Laboratory. Alvarez led the charge for big bubble chamber physics, pressing for a high energy low-intensity accelerator (good for bubble chamber physics) while offering a dim estimate of the future contributions of counter experiments using high-intensity machines. In a letter of 17 September 1957 to the laboratory director, Edwin McMillan, Alvarez acknowledged that while high-intensity beams and counters may have made contributions in nuclear and medium energy pion physics, these were domains in which elastic scattering was common.[84] For such experiments, he granted, visual detectors, cloud chambers, and emulsions had been truly only "exploratory tools of low statistical accuracy." He wrote: "I am just as familiar with those historical facts as anyone else, but I think they have very little to do with the future of high energy physics."[85]

New-style high energy physics, Alvarez contended, was in no way analogous to old-day medium energy nuclear physics. When particles interact *elastically,* as is the rule in many aspects of classical nuclear physics and medium energy pion physics, particle collisions do not produce a multitude of other particles. The counter physicist may be able to measure total cross sections by simply seeing how many events take place, or by measuring the energy and angle of the scattered particles; the complication sometimes arises of charge exchange between neutral and charged pions, but this is not a major obstacle. But history was not destiny; now, Alvarez argued, high energy processes were the subject of real interest, and in high energy physics collisions were almost invariably *inelastic.* Multiple production of particles in collisions was the rule, not the exception. In this new world, the counter physicist "can hardly get started in a serious investigation of the much more common inelastic processes if he does not have a visual detector, and preferably a bubble chamber." If a high energy negative pion interacts with a proton, a variety of particles are produced, including many neutrals. "In order to find out what happened, one must measure the angles and momenta of the charged prongs and then calculate by relativistic kinematics the number, the nature, the energies, and the directions of the invisible prongs."[86] Counters, Alvarez concluded, were simply not up to the task of this detailed reconstruction of events, a task that had to be visual and had to be done precisely, track by track.[87]

Some of Cork and Wenzel's 1957 work figured in Alvarez's letter as an example of the inadequacy of counter work at high energies. The two countermen had devised an electronic experiment to measure scattered protons from target protons at 2, 4, and 6 GeV. From Alvarez's perspective, the elastic cross section that Cork and Wenzel measured was not of primary theoretical concern because "the elastic part of the cross section is a small part of the whole affair. Some day the whole job will be done in a hydrogen filled bubble chamber, and in the process the elastic cross section will be measured again, so the theorists will have the real details of the interaction."[88] With the exception of work on low energy *K* mesons, Alvarez felt, in short, that "the counter is on the way out

84. Alvarez to McMillan, 17 September 1957, ELP.
85. Alvarez to McMillan, 17 September 1957, 4, ELP.
86. Alvarez to McMillan, 17 September 1957, 5, ELP.
87. Alvarez to McMillan, 17 September 1957, 5, ELP.
88. Alvarez to McMillan, 17 September 1957, 5, ELP.

as a precision high-energy physics tool."[89] Without the details, the identity, momenta, and spatial arrangement of tracks—in short the image of the event—theory could not proceed.

Even the counter physicists' greatest pride—statistical power—came in for Alvarez's attack. As far as Alvarez was concerned, bubble chambers were ahead there as well. In recent weeks his group had taken some 200,000 pictures and scanned, measured, and analyzed them within days. By contrast, the primitive state of counter work left one experiment after another in need of analysis two to three years after the physicists had left the laboratory floor. "One is accustomed to believe that at the end of a day one has written down in his notebook all the results of a counter experiment, but that he will have to wait for a long period of time to find out what his [bubble or cloud] chamber has to tell him. It is certainly strange to find that this situation can be reversed."[90] Here was another highly targeted attack, for the ability to achieve rapid response from the experimental apparatus was central to the counter physicist's self-definition.

This all-out struggle over the nature of experimentation formed the context of Alvarez's polemic against counters that I cited in chapter 5; we can now examine it more fully:

> I believe that in the high energy range, counter techniques will never be able to do the job that has been done so well in the lower energy ranges. For one to appreciate such an assertion the way I feel it, he would have to look at tens of thousands of bubble chamber pictures, the way I have done. When one sees the way in which an interaction of charged particles gives rise to neutral particles which go some distance and then decay into charged secondaries which then may decay into other charged secondaries, or undergo charge exchange reactions, and reappear as neutral particles and so on, he can be as discouraged as I am about the future of counters.[91]

Sure, he conceded, there were specialized niches left over for his colleagues in counters: they could slow particles produced in high energy accelerators and study them. But this was not truly high energy physics—it was the physics of particles produced in high energy machines. Counters were, he reiterated, on the way out.

Bubble chambers were not. After rehearsing the advantages of the new generation of bubble chamber and data analysis facilities, Alvarez attacked again: "After reading what I have just dictated, you will probably say as many others have said to me in the past two years, 'You really believe that the bubble chamber is going to be the source of most of the high energy physics information in the future, don't you?' My answer is 'Yes. . . .' Two years ago neither

89. Alvarez to McMillan, 17 September 1957, 9, ELP.
90. Alvarez to McMillan, 17 September 1957, 8, ELP.
91. Alvarez to McMillan, 17 September 1957, 8, ELP.

emulsion people nor counter people in general were reading the handwriting on the wall the same way I was, but now the situation is quite different, at least insofar as the emulsion people are concerned."[92] According to Alvarez, emulsion groups all over the world were now vying for bubble chamber film to analyze; even members of the LBL group—in front of McMillan himself—were exhausted after grappling with the latest batch of propane bubble chamber data that had landed on their desks. The emulsion technique might not be "dead"; Alvarez allowed that its practitioners could still scurry to specialized corners. But as a technique it had been replaced in high energy physics. Counter physicists, Alvarez judged, were in the same situation: if they did not yet recognize their own demise they soon would; their only hope lay in conversion or in retreating to unoccupied niches. Leaving no monument standing, Alvarez went after the counter physicists' greatest triumph—the discovery of the antiproton in 1955.[93] This discovery had not been made by the bubble chamber group only because the big chamber had been "in its infancy," "too young to vote" at the time of the experiment.

Challenged to find a single example—any example—of an experiment that could not be done with the bubble chamber, only Wolfgang "Pief" Panofsky (a leading LBL physicist, later director of SLAC) had succeeded in finding a solitary example: large angle elastic scattering of charged particles from protons. The bubble chamber, Alvarez concluded, had become the "universal detector" of fundamental physics, and every non–bubble chamber experimenter had better get used to the new *Pax Physica.* Copies of Alvarez's anticounter report went not only throughout LBL but also to the highest scientific advisory group to the AEC, the prestigious General Advisory Committee. Alvarez's message was clear: the image tradition, specifically the bubble chamber, had won—decisively.

With Alvarez ringing the death knell for their electronic livelihood, it was small wonder that the logic physicists Cork and Wenzel felt under siege. Appealing to McMillan just a few days after Alvarez's onslaught, Wenzel conceded that he agreed with "most of Luis' observations" yet hastened to add that "counters are not on the way out as precision instruments."[94] First, he suggested, the counter is *flexible,* "it comes in small inexpensive units, easily moved and altered as to size and type on short notice." In high energy experiments the counter's fast time resolution could be used for particle velocities above those in which "emulsions, cloud chambers and (potentially) bubble chambers" were useful. For example, Cerenkov counters and devices that were selected according to time of flight offered an extremely efficient way of separating K^- particles from antiprotons.

92. Alvarez to McMillan, 17 September 1957, 13, ELP.

93. Chamberlain et al., "Observation," *Phys. Rev.* 100 (1955): 947–50.

94. Wenzel to McMillan, 27 September 1957, 1, ELP.

Exploiting the efficiency of counters in the study of antiprotons, Wenzel and Cork could plan experiments that would determine the angular distribution and total cross section of proton-antiproton scattering at low energies. Theorists had made predictions about this process even though they had no detailed account to offer of the annihilation process itself. For such an experiment, counters could generate many more data, and do it faster, than the bubble chamber. In short, Wenzel granted Alvarez's claim that the bubble chamber would probably be the most useful single tool of high energy physics. Nonetheless, Wenzel continued, there was "no reason to regard it as the universal detector. While it may be true, as Luis says, that the bubble chamber does complete experiments, it is also true that completion may take a long time."[95] By offering rapid answers to theoretical predictions that dealt with "a few details of the complete experiment," counters could stake a provisional claim in territory that might eventually be deeded to the big bubble chambers.

Under the circumstances, Wenzel's qualified defense was probably as strong an endorsement of the electronic tradition as could be made. Caught in such straits, Wenzel, Cork, and Cronin greeted Fukui and Miyamoto's reprint with understandable enthusiasm. The three physicists submitted a counter paper on 26 May 1960 using their old techniques. But by 6 June 1960, Wenzel and Cork had drafted a proposal to build a spark chamber and within weeks had a machine yielding clear track photographs.[96] Immediately upon his arrival back east, Cronin too began constructing a modified Fukui-Miyamoto chamber, cannibalizing Keuffel's old spark counters for spark chamber plates and exploiting Keuffel's old data for a design.[97] To say that the spark counter lay at the core of the spark chamber is to speak both literally and figuratively. And with that appropriation (should we say ingestion?) came an attitude about pictures, statistics, and laboratory control.

The new spark chamber proved itself an ideal match for high energy proton accelerators: the sensitive time of the chamber was on the order of several microseconds, and by applying a clearing field between the positive and negative spark chamber plates the experimenters could remove unwanted tracks.[98]

95. Wenzel to McMillan, 27 September 1957, 3, ELP. In a separate letter, Cork seconded Wenzel's arguments, adding a list of promising extensions of the counter technique—better phototubes, better coincidence circuits, dense glass scintillators, faster oscilloscopes, and improved data-recording methods—along with a series of new experiments they could undertake. "The chief point of this note is that even though many high energy physics experiments can be done very well with bubble chambers and with other detectors, the fast counter techniques still have a very important role" (Cork to McMillan, 27 September 1957, 3, ELP).

96. Cork, "Charged-Particle Detector" (1960); this typescript was subsequently revised by Beall (13 June 1960), then by Murphy (14 June 1960), and put in revised form as Beall et al., "Spark Chamber" (1960); some months later it was published as Beall et al., "Spark Chamber," *Nuovo Cimento* 20 (1961): 502–8.

97. Cronin, interview by the author, 4 January 1983. Keuffel went to Princeton as a postdoc after receiving his Ph.D. from Caltech.

98. Beall et al., "Spark Chamber" (1960), 1–2.

As Cronin reported to Cork in late 1959 in the flush of excitement over the new detector, "I am finally building a [high-voltage] pulser! Have you done more with your spark chamber?"[99]

6.6 Sparking the Revolt

High-statistics electronic machines were back in the running. Cronin, at Princeton, consulted with engineers to find the best thyratrons to produce large electrical pulses,[100] and by March 1960 Cronin could already report (even while completing an older model counter experiment) that he had successfully tested an eight-gap chamber with plates 4 inches by 5 inches. Enthusiasm for the new device leapt ahead of specific projects: "We are now building a larger chamber with decent optics. It will have a volume of 6″ × 6″ × 12″ with 18 gaps. The purpose will be to test larger spark chambers. No particular experiment is in mind."[101] Unlike the bubble chambers, the new spark chambers could be prepared in short order. Within a few weeks the Princetonians hoped to test the device at the Cosmotron and then the following summer at Berkeley.

Word spread quickly. Letters of inquiry arrived from laboratories around the world—from the Università degla Studi–Roma to the University of Maryland.[102] Here was an easily assembled, affordable device for institutions that could not dream of competing with the great hydrogen bubble chamber efforts at Berkeley or Brookhaven. As Cork put it, echoing Wenzel's defense against Alvarez, "The first impressive thing about . . . spark chamber[s] is the fact that they are extremely easy to build, are inexpensive, and there seems to be no difficulty in making them go. Anyone can build one in the basement and make it work."[103] This was no exaggeration. The first model used by Gerard O'Neill's Princeton group "was largely the work of college sophomores majoring in physics,"[104] a far cry from the massive engineering and scientific efforts necessary for the dangerous construction of bubble chambers.

As the new device entered the consciousness of the experimental community, applications proliferated. Cork, Wenzel, Cronin, and George Renninger (a graduate student) brought the device to an accelerator for the first time to study the scattering of polarized protons and spin-spin correlations in proton-proton scattering (see figures 6.20 and 6.21).[105] In November 1959, just a few months

99. Cronin to Cork, 24 November 1959, JCP.

100. Brady, Kuthe Laboratories Inc., to Cronin, 17 February 1960, JCP.

101. Cronin to Cork, 22 March 1960, JCP. The first public report of the Cronin-Renninger accelerator work was made in Cronin and Renninger, "Studies" (1961), 271–75.

102. E.g., Bernardini to Cronin, 1 April 1960, JCP; and Burnstein to Cronin, 18 January 1961, JCP.

103. Remark by Cork in discussion at the Argonne National Laboratory, 7 February 1961, in "Spark Chamber Symposium," *Rev. Sci. Inst.* 32 (1961): 480–98, on 486.

104. O'Neill, "Spark Chamber," *Sci. Amer.* 207 (August 1962): 36–43, on 43.

105. See "Spark Chamber Symposium," *Rev. Sci. Inst.* 32 (1961): 480–98, on 487–89.

Figure 6.20 Cronin and Renninger, device (1961). William Wenzel, Bruce Cork, and James Cronin received Fukui and Miyamoto's reprint; within weeks they had a working spark chamber. This device and pictures taken with it of scattering protons at the Berkeley Bevatron received wide attention at the 1960 conference on instrumentation and encouraged many other groups to deploy spark chambers at accelerators. After years of marginalization, the logic tradition was back. Source: Cronin and Renninger, "Studies" (1961), 273.

after the first of these experiments, students and faculty gathered around T. D. Lee at Columbia to debate how to test weak-interaction theory at high energies. Melvin Schwartz realized that it might be done with neutrinos, a thought that occurred almost simultaneously to Bruno Pontecorvo, a physicist working in the Soviet Union after having been exposed as a Soviet spy.[106]

The second type of neutrino, one conjectured to be produced in reactions that made muons instead of electrons, was equally elusive. As Schwartz soon discovered, the difficulty was that banks of Geiger-Müller counters and stacks of neon-filled tubes all had poor spatial resolution. And hydrogen bubble chambers—even the 72-inch Berkeley chamber—were not a massive enough target

106. Schwartz, "Feasibility," *Phys. Rev. Lett.* 4 (1960): 306–7; Pontecorvo, "Neutrinos," *Sov. Phys. JETP* 37 (1960): 1236–40; Schwartz, "Discovery," *Adventures Exp. Phys.* 1 (1972): 81–100; and Schwartz, interview by the author, 20 October 1983.

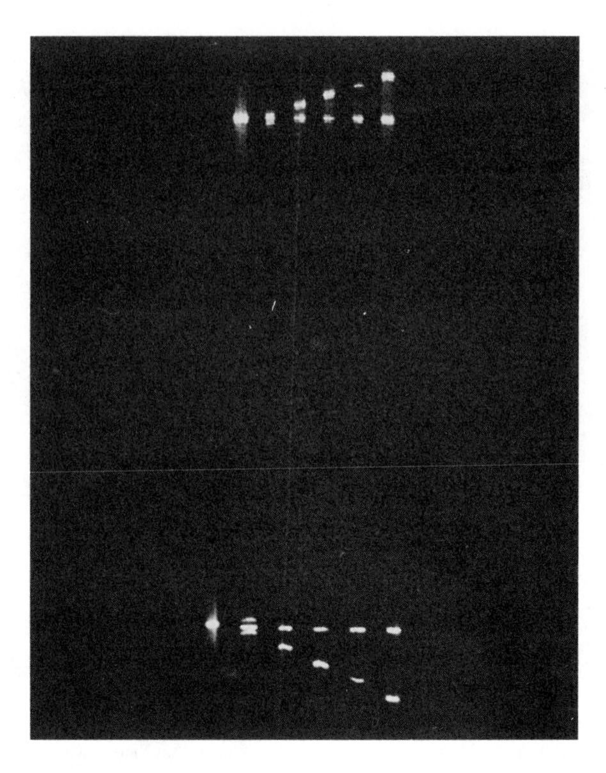

Figure 6.21 Cork, Cronin, and Renninger, tracks (1961). Proton scattering in the Cronin and Renninger chamber depicted in figure 6.20. Source: Courtesy of Bruce Cork.

to use to study neutrino interactions. During the early summer of 1960, Irwin Pless from MIT reported to Schwartz and Leon Lederman about Cronin's working desktop spark chamber. In pursuit of a better target for neutrinos to hit, the two physicists raced to Princeton to look at it.[107] Again, as in the early stages of the "dirty" bubble chamber that the technician Wood had assembled in Berkeley, the key was to transform a desktop device into a large engineering-grade structure.

With money from the navy, the AEC, Columbia, and Brookhaven, the hunt for a second neutrino was set in motion. Engineering salaries and electronics came from Brookhaven, and money from Columbia's Nevis Laboratory. The navy supplied surplus cruiser deck plates, weighing between two and three thousand tons, gratis. By setting an example for a generation of spark chamber

107. Schwartz, interview by the author, 20 October 1983.

Figure 6.22 Schwartz with two-neutrino detector. Of all the experiments conducted with spark chambers, perhaps the most startling was the one run by Schwartz, Steinberger, and Lederman using the device depicted here. Not only did this device explore the properties of a neutrino beam (extraordinary if one recalls that the neutrino itself was only detected a few years earlier), Schwartz, Steinberger, and Lederman demonstrated that there were actually two kinds of neutrino, one associated with the production of an electron and the other with the production of the muon. Immediately, the experiment became the prototype for many others, including the Fermilab-based work of the collaboration E1A that codiscovered neutral currents in the mid-1970s. Source: 1–644–63, Brookhaven National Laboratory Archives.

neutrino experiments, the Columbia-Brookhaven machine yanked the spark chamber from the tabletop into the engineering age of physics. Only muons emerged from the neutrino events, rapidly convincing almost everyone that this second neutrino was very different from the electron-making neutrino known theoretically since Pauli and Fermi in the 1930s and experimentally since Reines and Cowan. The two-neutrino hypothesis entered physics with force (see figure 6.22).[108]

108. Danby et al., "High Energy Neutrino," *Phys. Rev. Lett.* 9 (1962): 36–44.

Spark chambers prospered during the next few years. They were built in every conceivable size and shape. Experimenters devoted specialized meetings and major sessions at more general gatherings to the noisy detectors,[109] drawing participants from laboratories around the world. By its relative ease of construction, the device satisfied a number of felt needs, offering a fast, triggerable detector for accelerator physics and restoring to small groups and less wealthy laboratories the opportunity to experiment in particle physics without joining the monster bubble chambers that were big enough to be working at the frontiers of the science. Simultaneously, the spark chamber renewed in the logic physicists (who overwhelmingly were the ones to use it) the enthusiasm that had begun to fade as the bubble chambers had risen to absolute domination.

Excitement flashed through the electronic community at the possibility of competing again on something like an equal footing with the image groups. Cronin drafted a letter in February 1961 to Thomas J. Watson, Jr., president of IBM, pleading for faster delivery of his company's card punch. The sooner it arrived, the sooner the spark group could link it to their encoding system and beat the competition to a resolution of certain ambiguities in nuclear forces: "[W]e have employed the newly-developed spark chamber technique and it will undoubtedly be a triumph for this technique if we are able to arrive at decisive results before groups in Russia and England, doing similar experiments, succeed with their more antiquated techniques."[110]

As the logic machines grew both in size and sophistication, they produced ever more pictures. Soon it was the *counter* physicists who were clamoring for better optics, faster film-advance mechanisms, more scanners (mostly women) to analyze events,[111] and ever more subtle computer programs written to perform kinematical and particle identification analyses. As Cronin put it, in many spark chamber collaborations, experimenters were now finding themselves "in a situation similar to that of our bubble chamber colleagues."[112] The flexibility and small scale of the traditional counter system was yielding to the large scale of the contemporary image-producing devices. Absorbing so much of the material culture of the bubble chamber laboratory, the logic laboratory in the 1960s came increasingly to resemble the site it had so long opposed.

109. E.g., the symposium on spark chambers held at Argonne National Laboratory on 7 February 1961 ("Spark Chamber Symposium," *Rev. Sci. Inst.* 32 [1961]: 480–98); Session 5, "Spark Chambers," of the 1962 Conference on Instrumentation for High-Energy Physics, CERN, 16–18 July 1962 ("Spark Chambers," *Nucl. Instr. Meth.* 20 [1963]: 143–219); and the session "The Latest Advances in Spark and Luminescent Chambers and Counters Method" at the International Conference on High-Energy Physics held at Dubna in 1964 ("Latest Advances," in Smorodinskii et al., *XII International High-Energy Physics* [1966], 301–77).

110. Cronin and Engels to Watson, 27 February 1961, JCP.

111. Roberts, "Properties" (1964), 368: "Need for film scanning may require manpower (womanpower)"; Rosenfeld, "Current Performance," *Nucl. Inst. Meth.* 20 (1963): 422–34, on 422: "Next, the scan cards are keypunched and verified by two full-time girls (currently we are scanning at about 200,000 events per year)."

112. Cronin, "Present Status," *Nucl. Inst. Meth.* 20 (1963): 143–51, on 150.

6.7 Computers, Physics, and the Nature of the Laboratory

Paradoxically, success brought the counter physicists into exactly the world they had derided—among the large-scale image techniques and work practices that were anathema to them. But as spark chambers had grown, the "logicians" had been forced to supplement the technical core of their tradition, electronics, with the photographic skills and procedures of the "bubblers." Many soon yearned for a return to the purity and control of circuits and counters. Wenzel: "I'd always been interested in electronics and after a while I got tired of the business of scanning and reconstructing the film. . . . [The experiments were] a little tedious because I'd always done electronic experiments before. I started to think: what were other ways to record the information?"[113] By 1964, Wenzel's was a discontent shared widely among the displaced counter physicists. In response, like-minded counter physicists convoked in March 1964 what they thought would be a small, "informal meeting" at CERN on "Film-less Spark Chamber Techniques and Associated Computer Use."[114]

Originally, the organizers planned the assembly as an intimate colloquium for 20 people to set the future direction of the CERN program. But the idea resonated so forcefully with the concerns of the logic physicists that, when the opening session began, over two hundred physicists had arrived to participate.[115] Introducing the meeting, P. Preiswerk welcomed the larger than expected crowd, expressing a hope he obviously shared with many in the audience that "with computers on-line the physicist will receive certain answers during the running of the experiment in the experimental halls, [which] might give him back the pleasure of being an experimenter and not only an operator, who is able to act and to put new questions on the grounds of this information during the running of the experiment." Prophesying the eventual triumph of the imageless devices, Preiswerk concluded, "The high cost of the new techniques might damp the speed of the development you have initiated, but not withhold it."[116]

G. R. Macleod then outlined the field, stressing that the technical basis for the anti-image movement came from rapid advances in electronics and computers. Just as hydrogen bomb work had proved important in laying technical prerequisites for the first large-scale bubble chamber work, other military developments proved useful to the counter scientists. Macleod noted that the on-line exploitation of computers was a "fairly widely used technique" in the guidance of satellites and missiles and the deployment of advanced radar systems.[117] Indeed, even a cursory reading of military literature from the 1950s reveals the

113. Wenzel, interview by the author, 7 March 1983.

114. Macleod and Maglić, *Film-Less* (1964).

115. Preiswerk, "Introduction" (1964), 1.

116. Preiswerk, "Introduction" (1964).

117. Macleod, "On-Line Computers" (1964), 4.

importance of computer-aided machine feedback for military applications. To take one example, the Air Force's 1960 edition of its Air Training Command textbook *Fundamentals of Guided Missiles* has several chapters that stress the importance of on-line computers for telemetric feedback. In missile delivery, many channels of information need to be recorded quickly, including the position of control surfaces, airspeed, pitch, yaw, roll, temperature, acceleration, altitude, and ordnance functions. Typically, each such on-board instrument was connected to a transducer that produced an audio frequency signal. This signal then modulated an FM carrier. Using a distributor, the system sampled each of these transducers and sent the data back to the tracking station. If guidance information was to be fed back to the missile in time for in-flight correction, computers had to process and retransmit coded information quickly.[118]

Many of the problems that arose later in elementary particle physics were present earlier in such telemetry applications. For example, telemetric information often arrived from the missile faster than it could be processed, so the Air Force had to develop systems for rapidly recording the data on magnetic tape. New technology was needed to separate the various data channels and, further, to avoid interference between channels. Similar technological work continued in other research areas, especially in advanced radar technology, and in the oil, chemical, and communication industries. Closer to home, particle physicists could borrow from the multichannel pulse-height analyzers used frequently by experimental low energy nuclear physicists.[119] Together these various arenas of electronic applications provided a ready reservoir of techniques from which the counter physicists could draw.

For the 10 years preceding the 1964 CERN meeting, computers had performed vital tasks in bubble chamber research. In that context the computers were designed to take over ever more of the tedious and repetitious work involved in sorting and analyzing film, and then in collating the data into useful forms. Perhaps, one spark chamber physicist ruminated, the "well known fact that all elementary particle physics in the last few years has been done in bubble chambers may not be unconnected with the fact that the bubble chamber physicists have had well developed data handling facilities." With a new class of logic instruments ready to defend the cause of control over experimentation, this physicist noted, "we should listen to the reports [of those] experimental groups who have been working very hard to try and change this lamentable situation [the poverty of the logic tradition's data processing]."[120]

Spark chamber workers had additional tasks in mind for the new electronic brains, ones that went beyond reducing photographic tracks. First, the logic

118. U.S. Air Force, Air Training Command and Technical Staff, *Guided Missiles* (1960), esp. 497–569.
119. Macleod, "On-Line Computers" (1964), 4.
120. Macleod, "On-Line Computers" (1964), 9.

experimenters hoped to use computers for *data acquisition.* As data are produced by the detector, the electronic signals could be recorded on magnetic tape by a computer for processing at a later time. Second, computers could perform a *check and control* function by monitoring experimental parameters, magnet currents, counting rates, voltage levels, beam intensities, and so forth. By doing so, the computer could partially obviate gross errors by alerting experimenters to malfunctions in the apparatus. Such work was comparable to the upkeep of a logbook, recording "the sort of things which nowadays tend to be written (or worse, not written) in note books in illegible pencil writing at three o'clock in the morning by harassed physicists . . . [could be] done automatically by the computer with rather more consistency and reliability."[121]

A third function of the computer would be to perform *sample computations.* Suppose one knew that the beam particles entered at a certain point in the chamber. One might have the computer check whether many particles were produced or only a few. Thus, the computer functioned as a variable logic element in the detecting system and could be used to determine which events to record. Unlike the bubble chamber with its post hoc use of the computer, the spark chamber brought the computer on-line, in a role that directly extended Rossi's use of logical discriminators. Computers became part of the instrument itself, with the logic circuits black-boxed in the calculation device; no longer would custom-built coincidence and anticoincidence circuits need to be wired for each experiment. This was an assimilation altogether unavailable in the chemical and thermodynamic processes underlying emulsions and bubble chambers.

After comparing notes, the various groups assembled at CERN for the 1964 meeting agreed that for a typical particle physics experiment, programming took "several man-months." Not all of this work was equally hard. Macleod noted that "[t]he thinking is part of the programmer's job." Once the physicists in charge of the project did that thinking, they could "give [the detailed computer work] to a very junior coder for example."[122] In all, by 1964 programming was occupying 10%–20% of a spark chamber group's work. To M. G. N. Hine, such an investment of time seemed quite reasonable: "The design of the programmes has quite properly occupied most of the physicists in the groups concerned as this is nowadays really another name for the design of the experiment itself."[123] Once again, the understanding of what it meant to be an experimenter was in flux, and with new understanding came a new version of the meaning of experiment: computer program qua experimental design. But whereas in bubble chamber experimentation computer programming defined the experiment through data extraction, in the logic tradition programming entered into all

121. Macleod, "On-Line Computer" (1964), 5.

122. Macleod, discussion in Macleod and Maglić, *Film-Less* (1964), 310.

123. Hine, "Concluding Remarks" (1964), 374.

phases of data production—from the design of the apparatus to the recording of results.

As always, bubble chambers were the competition. At each stage of the development of electronic techniques, workers compared their machines to the great liquid detector that had produced so much physics during the 1950s and early 1960s. One obvious advantage of the bubble chamber was its ability to record immensely complicated interactions and decay patterns. Often these decays would be quite rare. Nonetheless, because of the detail of the bubble pictures, evidence from just a few events could be persuasive. For the electronic experimenters, persuasive evidence came more from high statistics than from such golden events. The rarity of the fully analyzed bubble events struck Iowa State physicist Arthur Roberts—a vocal advocate of the spark chamber—as one of the difficulties with the bubble chamber. Consider weak interactions:

> [I]f we talk about decays of various particles there is nothing inherently different about the leptonic decay of the omega and the leptonic decay of the neutron, and the number of neutron decays that have been observed is numbered in the hundreds of thousands. To do an equivalent job on the decay of the omega which is equally interesting would require the same number of events. We are used to thinking of 50 Λ decays as a large sample not because there is anything in the physics which says that the Λ decay is any different from the neutron decay, but just because they are expensive.[124]

It was becoming clear that experimentation would be computer limited, not simply because computers were needed for data analysis, but because computers now constituted part of the detectors themselves. Roberts's comment about the lambdas captured the long-standing split between the image tradition's willingness to sacrifice number for detail and the logic tradition's reluctance to settle for the rare when massive statistics could be extracted electronically. In the search for the power to sort and extract more data, one conference participant, Samuel J. Lindenbaum, only half-jokingly contended that his group "could easily use up all the computers at CERN, Brookhaven and the whole East coast."[125]

Lindenbaum was a quintessential logic physicist. Where an Alvarez or a Leprince-Ringuet could revel in the fine points of visualizable data, Lindenbaum bluntly said, "I am not the visual type." From early in his career counters were his instrument of choice; a brief, not particularly happy excursion into emulsion physics left him persuaded that "emulsions were a dinosaur, dying." He added, "I thought we should connect instrumentation to where the big bucks

124. Roberts, discussion of Roberts, "Some Reflections" (1964), in Macleod and Maglić, *Film-Less* (1964), 298.

125. Lindenbaum, discussion of Roberts, "Some Reflections" (1964), in Macleod and Maglić, *Film-Less* (1964), 298.

were," and in the late 1950s that meant binding his experiment to computers like the MANIAC III. Deliberately surrounding himself with colleagues who had electronic backgrounds—not bubble chamber experience—he wanted a "new departure," one "best prosecuted by people that had not been prejudiced by the bubble chamber approach. We would start all over." From the fall of 1962, Lindenbaum and his collaborators began harvesting data, and his enthusiasm for the new style of physics was boundless.[126]

From 1962 on, Lindenbaum began article after article reciting the by-then classical division between gas discharge counters and their cognates on one side and the "visual techniques" such as cloud and bubble chambers on the other. Among their other failings, counters could not detect in all directions at once and could not handle multiple production of particles; visual methods, by contrast, could not handle large numbers of events. One needed, as Lindenbaum stressed, some way of combining the strengths of the two traditions in order to overcome their weaknesses. Simply multiplying the number of "counters, coincidence electronics, and scalers" would not only overwhelm any laboratory, the resulting data reduction problem would overwhelm the physicist if the hardware did not. It was the on-line computer, Lindenbaum emphasized, that was the key.[127]

One advantage of the computer was its speed in returning information about data and curves: "Because of the almost immediate on-line computer data processing feature, this complicated counter system which would normally be inherently blind is given a remarkable degree of vision. One can now see almost instantaneously the progress of the experiment and make standard checks whenever desired."[128] Here we have a multiple meaning of the restoration of sight. At one level, the blindness of counters refers to their inability to deliver a visual record; they cannot produce the images of a cloud or bubble chamber. But at the same time, Lindenbaum's ascription of vision is to the logic experimenters themselves. As never before the experimenters could "see" the progress of their experiments—where "seeing" meant being able to visualize what was happening *as it happened* and to exert control over the events proceeding in front of them (see figure 6.23). Once again we find an attempt to combine vision and manipulability by combining features of image and logic.

What made this vision and consequent control possible was the radical break between the production of data and their evaluation. Language—a language that permitted dialogue between experimenter and machine—was essential in bridging the gap. Lindenbaum:

126. Lindenbaum, interview by the author, 21 December 1984.

127. Lindenbaum, "On-Line Computer," *Ann. Rev. Nucl. Sci.* 16 (1966): 619–42.

128. Lindenbaum, "Hodoscope System," *Nucl. Inst. Meth.* 20 (1963): 297–302, on 300.

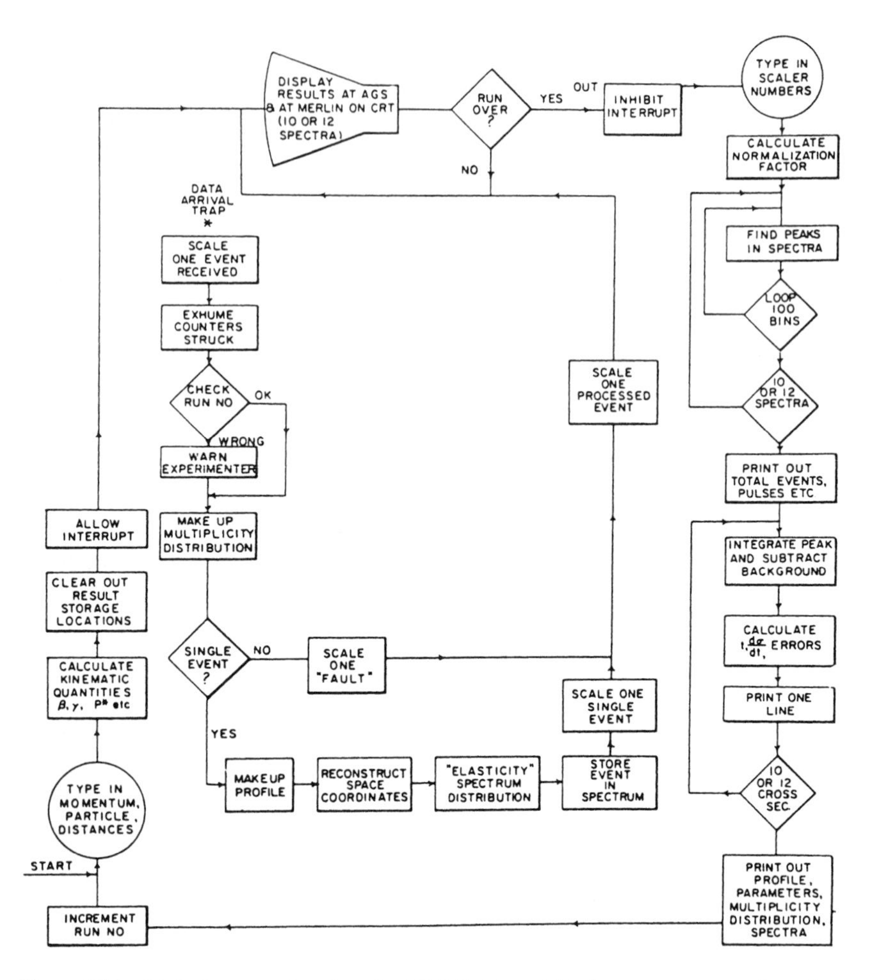

Figure 6.23 Lindenbaum, on-line flowchart (1964). Samuel Lindenbaum was one of the first physicists to integrate the computer into the experimental apparatus itself. In so doing, he brought the logic tradition full circle: logic circuits of cosmic ray physics became the basic unit of computation; now the computer became part of the instrument, not simply a fast way of reducing data after the fact. In this flowchart from 1964, Lindenbaum's collaboration sketches out how a newly sophisticated on-line computer program would function. Source: Foley et al., "Computer System," *Nucl. Inst. Meth.* 30 (1964): 45–60, fig. 10 on 55.

> Language can be a varied subject ranging from primitive man's sign signals to a powerful modern language with rich vocabulary, written as well as vocal. . . . Language communications can range from exchange of letters to a rapidly changing dialogue between individuals or public-meeting type of interactive debate. Thus, although there is no doubt that computing is by now a language of physics for all

> disciplines, I believe that in the field of Data Processing for Electronic Techniques in High-Energy Physics, computing has reached its most advanced and prolific level as "a language of physics."

According to Lindenbaum, only in the interaction "between the physicist and his rather sophisticated detector devices" is the possibility of a truly "Interactive Language" realized.[129] Among other tasks, Lindenbaum addressed a variety of problems in high energy scattering in nucleon-nucleon collisions, pion-nucleon collisions, and kaon-nucleon systems. These bore directly on predictions made by the then-current Regge pole model, which related, phenomenologically, the energy and spin dependence of reaction cross sections.

In the decade from 1962 to 1972, Lindenbaum's interactive use of the computer was impressive, both as a method of returning the experimenter to the experiment and as a way, at least partially, of removing images from the image makers. So astonishing was this new kind of experimentation that in 1963 when the theorist Abdus Salam was called upon to conclude a Stanford conference on nucleon structure, he projected a slide of Lindenbaum to the assembly. "Both in Cambridge and here what came through strongly [from Lindenbaum's presentation] was the might of American physics—I mean experimental physics. For me the use of on-line computation in such an effective manner seems like magic. I wish to point next to my favorite illustration of this conference [figure 6.24]. I think that the only thing needed to complete Lindenbaum's felicity would be a theoretician in place of that stool, biting the dust off the floor."[130] If only for a brief moment, even the theoreticians were impressed by the power of this new and more fundamental use of the computer, coupled to spark and wire chambers, in experimentation.

For many at the conferences of 1962 and 1963, the spark chamber's fast (and inexpensive) response to complicated phenomena was its primary advantage; indeed the possibility of eliciting meaningful data on the spot from the hardware was a salient feature of the logic physicists' mode of work. Computers coupled to imageless detectors allowed the counter physicist to regain this capacity, which had been lost as the spark chamber had begun to drift toward the technology and hierarchical work organization of the bubble chamber. Even among bubble chamber physicists there was a growing resistance to the reliance on large groups to sort, scan, and measure the photographs. As we saw in the last chapter, Glaser left physics for biology in the early 1960s in large measure because the structure of the bubble chamber workplace seemed to leave no room for solitary or small-group experimentation. By the mid-1960s, the malaise even

129. Lindenbaum, "Data Processing" (1972), 209.
130. Salam, "Summary" (1964), 397–414.

Figure 6.24 Lindenbaum, computer, and stool-theory (1963). View from inside the data trailer; present too are the digital electronics that recorded data automatically on tape and transmitted them to an on-line computer in scope-displayed computed data. Lindenbaum is to the right. Source: Courtesy of S. J. Lindenbaum.

reached Alvarez, who (as we saw) had begun having second thoughts about the structure of laboratory life he had done so much to create.

Logic physicists also wanted a style of work in which they could respond to experimental conditions *during* the experiment—indeed within a few minutes of the run. It was a desire, at once about knowledge and work, that motivated a great deal of the instrumental design then under consideration. As Macleod reminded an audience, with the computer's aid

> [o]ne can see rather rapidly the effect of changing experimental parameters; one can make the setting up on an experiment a rather less laborious and less random affair; one can actually see that the data being recorded is the data one wants to

> record; and one can plan the experiment on the basis of analysed data rather than having to make intuitive guesses based on what we did last time. All this I think makes for the physicist having a much better *control* over what is going on and gives him the information on which to base considered decisions on changes to be made during the experiment.[131]

Macleod's hope for "control over what is going on" is a refrain heard time and again within what I have called the "logic" tradition. Logic physicists were acutely aware of the changing organization of the experimental workplace that often forced experimenters into supervisory jobs, unable to interact directly with their tools and the objects of their investigation. Concern about the passivity of the bubble chamber joined with the inability of the physicist to find control in the bubble chamber workplace and drove many of the electronic logic experimenters toward the development of imageless, on-line devices. For as the "visual" spark chamber came to resemble the bubble chamber—with scanners and stacks of unprocessed data—the logic experimenters ran once again toward electronics, this time toward the computer. It was not an unproblematic move.

Some opponents of an increased use of computation contended that computers would make experimentation automatic, and so block discovery. It was a debate we already glimpsed in the bubble chamber world, as the Alvarez group fought the attempt to introduce pattern-recognizing computers. Macleod had no truck with these fears of the automatic, insisting: "This is about as different as it possibly could be from the real situation. . . . In using an on-line computer . . . the physicist's role of decision making during an experiment is enhanced by having the possibility to make judgments on the basis of real live information, not just on intuition. He is in a much better position to be really aware of what is going on at any given time, what has gone on, and what he should do in the future in running his experiment."[132] Control lies central both to the work life of experimentation and to the security of the experimental results. Epistemic and workplace control come in together.

In addition, the design of the instruments was intimately connected to the type of experimental problem that could be investigated. One can see this in one of the discussion sessions at the CERN conference, where an experimenter commented that computers ought to be used to *exclude* certain events from consideration. Otherwise the primary purpose of spark chambers would be lost—that is, the experimenter would sacrifice the possibility of presetting the logic for the type of event desired. If the computer was to record *everything,* the physicist would do better with a bubble chamber.[133] Making pictures, surrendering control—this would have been a betrayal of the guiding values of the logic tradition.

131. Macleod, "On-Line Computers" (1964), 6, emphasis added.
132. Macleod, "On-Line Computers" (1964), 8.
133. Maglić, discussion in Macleod and Maglić, *Film-Less* (1964), 305.

6.8 Image of Logic; Logic of Image

One way for the logicians to eliminate the photograph was to pass directly from the flash of the spark chamber into the electronic eye of a television camera. Inspired by a remark of Kowarski's to the effect that images were not really necessary for the spark chamber, H. Gelernter, a visitor to CERN from IBM, tried television registration in 1961. In particular, he wanted to capitalize on the spark chamber's rapid repetition rate and its production of luminous tracks.[134] Over the next few years, several groups took up the idea, including teams from LBL, Princeton, Chicago, and Lund/CERN. There was even a proposal to use the device in an earth satellite.[135] The theory of the *vidicon* was fairly simple and could be borrowed wholesale from the broader technological base that, by 1964, supported commercial television: The sensitive element was a photoconductive layer deposited on a transparent conducting surface. This photoconductive layer is charged to a homogeneous potential by an electron beam. When light strikes this layer, it discharges the layer at the points of illumination; upon recharging by the electron beam a current is generated, and it is this current that can be used to transmit a signal for analysis or reproduction (see figure 6.25). This is not to say that the technology fell easily into the language of a high energy physics laboratory. Indeed, the documentation for the vidicon devices regularly proved irrelevant once the vidicons were divorced from their original television context. Here, in a remark by LBL's Victor Perez-Mendez, is an example of the dissociation between the original (in this case, commercial) context and the laboratory context; it is from an exchange during an instrumentation conference in 1965:

> LEBOY: Can you tell us on these new vidicons, the new sensitive ones, what the spacial resolution and accuracy is?
>
> PEREZ-MENDEZ: Well, the first thing I would say is that whatever I read in the RCA's bulletins never conveys that much information because the RCA bulletins tell you the things concerning the commercial use of the vidicons.[136]

In particular, the spatial resolution of the vidicon tube was limited by the size of scanning spot, a characteristic simply irrelevant for the tube's intended (commercial) users. Or consider the following interaction between Gelernter and Perez-Mendez at the CERN meeting. The problem being considered on the surface appeared rather elementary: how to erase a piece of tape by modifying the signal coming out of the video tube:

134. Gelernter, "Automatic Collection," *Nuovo Cimento* 22 (1961): 631–42.

135. See the reports in Macleod and Maglić, *Film-Less* (1964): Vernon, "Spark Chamber Vidicon" (Princeton); Andreae et al., "Automatic Digitization" (LBL); Anderson et al., "Vidicon System" (Chicago); Fazio, "Vidicon Spark Chamber"; Dardel and Jarlskog, "Vidicon Development" (Lund); see also the *Proceedings of the Purdue Conference on Instrumentation for High-Energy Physics, IEEE Trans. Nucl. Sci.*, NS-12, no. 4 (1965); and Dardel, Jarlskog, and Henriksson, "Status Report," *IEEE Trans. Nucl. Sci.*, NS-12, no. 4 (1965): 78–82 (Lund).

136. Discussion of Perez-Mendez, "Film-Less" (1965), in *IEEE Trans. Nucl. Sci.*, NS-12, no. 4 (1965), 16.

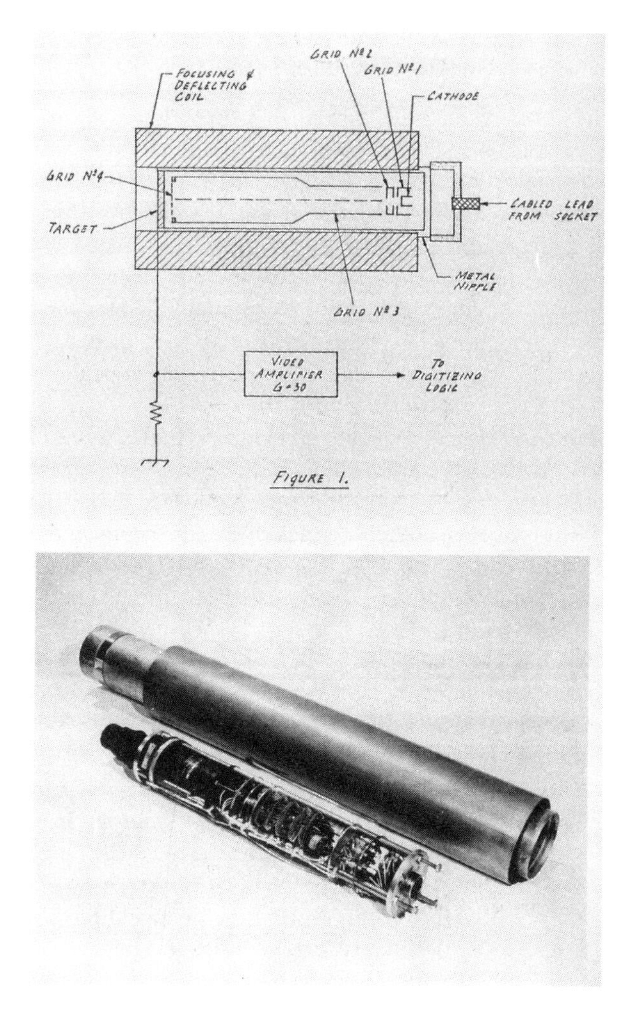

Figure 6.25 Vidicon tubes (1964). Source: Andreae et al., "Automatic Digitization" (1964), plate following 78.

GELERNTER: Is there any reason why in order to erase the previous event you can't increase the beam current in the vidicon and also increase the sweep speed [rate at which the electron beam sweeps across the sensitive layer]?

PEREZ-MENDEZ: . . . [RCA told him that this wouldn't help.] Now I don't understand the physics of this phenomenon but they assure me that this is so, and that it doesn't really matter at what rate—I am quoting the RCA people—they claim that it doesn't really matter at what rate you erase providing you go through a number of erase cycles. . . .

> GELERNTER: I would guess that they don't understand the physics of the situation either or else they would have solved the problem.[137]

Again, the physicists wanted the "physics" of the device, an understanding of certain underlying mechanisms responsible for the functioning of the instrument in a situation distant from commercial television. But (not too surprisingly) these particular physicists' questions intersected only marginally with the concerns of a commercial firm like RCA. Consequently, even in a piece of equipment as common as a video tube, it was possible for the physicists and the RCA engineers to raise such different sets of questions that the standard manuals were simply superfluous in places like CERN or Brookhaven—even when the device itself functioned perfectly in both places. In the trading zone of devices, objects could travel where words could not.

Despite the uselessness of technical manuals, the vidicon soon flourished. In high energy physics laboratories the video recorder, like the spark chamber itself, seemed so obviously useful for recording events, and so cheap to build, that none of the groups even bothered to acknowledge the inventor of the method (Gelernter). By 1966, however, there was only one paper on the method at the SLAC instrumentation conference.[138] It became one more stump branch on the tree of imageless detectors.

Another iconoclastic technology that branched out further was the sonic chamber—a spark chamber with microphones positioned at its edges. When the spark crackled through the air it caused a loud retort. Picking up this "bang," the microphones sent an electrical pulse. By clocking the times of arrival, an electronic circuit could then reconstruct the spark's position—at least if no ambiguities arose from multiple sparks—once again without the hated image (see figure 6.26).[139] Other methods sprouted up around a wide range of available technological systems; most were never heard of again. One such lost innovation took advantage of the tape recorder. By passing tape directly under a spark chamber, the magnetic field of the spark would leave a pattern of grains polarized on the tape (see figure 6.27). Playing the tape back through a reader and into an oscilloscope would then yield pulses at particular positions on the tape, and therefore positions in the original chamber. No icon would ever need to be reproduced. Here, as elsewhere, the collected audience at CERN was ever vigilant for ways to combine elements of various existing technologies, always hoping to remain clear of the image. When the inventor of the direct tape method

137. Discussion of Andreae et al., "Automatic Digitization" (1964), in Macleod and Maglić, *Film-Less* (1964), 70.

138. The paper on vidicon spark chambers presented at SLAC 1966 was by the Chicago group Hincks et al., "Spectrometer" (1966).

139. Fulbright and Kohler, University of Rochester Report NYO 9560 (1961), cited in Charpak, "The Evolution of Filmless Spark Chambers," 15 April 1970, 4, GCP; Maglić and Kirsten, "Acoustic," *Nucl. Inst. Meth.* 17 (1962): 49–59; Bardon et al., "Sonic" (1964); and Blieden et al., "System" (1964).

Figure 6.26 Sonic spark chamber (1964). Source: Bardon et al., "Sonic" (1964), plate following 46.

reported that one particularly nasty spark had blown a hole clear through his tape, another responded: "I was already thinking about this possibility. Why not use paper tape? The sparks would make holes in the paper leaving a permanent record. Computers of course like to read holes in paper tape."[140] Here was a festival of devices: tape drives from tape recorders, paper tape readers from computers, microphones from audio equipment, vidicon tubes from commercial television—each wired into a bricolage detector.

Even the office xerox machine was enrolled, as Charpak speculated that electrostatic photography might be useful in tracking particles. He and his group noted that the standard account of the xerox went this way: A precharged layer (usually amorphous selenium) was deposited on a conducting surface. Light alters this charge pattern, and charged powder particles, when dispersed on the surface, differentially sort themselves into the image. These powder particles could then be fixed by heat treatment and, with luck, read out by a machine. In particular, Charpak and his associates hoped to modify the usual copying method to include *magnetic* particles—these could then be read by the standard check-

140. Farley, discussion of Quercigh, "Direct Recording" (1964), in Macleod and Maglić, *Film-Less* (1964), 349.

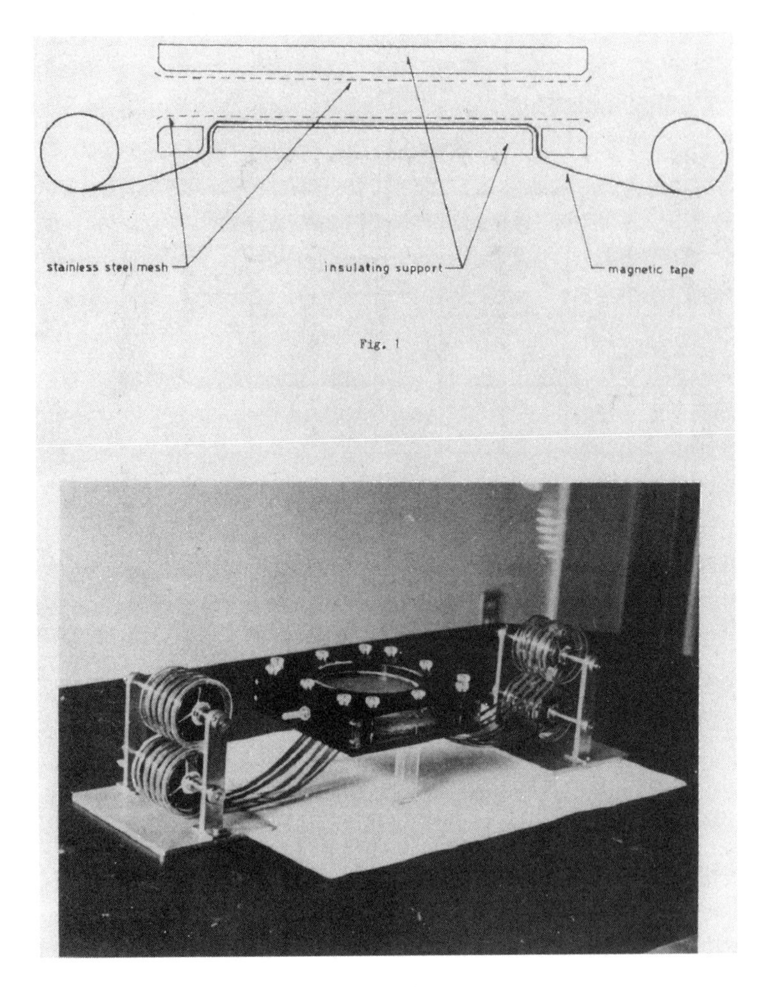

Figure 6.27 Burning in the tape (1964). Source: Quercigh, "Direct Recording" (1964), plate following 348.

reading machines used by every bank. Like Perez-Mendez, Charpak hoped to "suppress" (his word) the "servitudes of film and delayed processing." If xerographic registration worked (the method had not yet been fully tried), the CERN team hoped to preserve the control afforded by real-time results and "yet present the advantages of visualisation and permanent record of film spark chambers."[141] Here was photography without light, and images without the passivity and loss of control that accompanied the image laboratory.

141. Charpak et al., "Electrostatic" (1964), 341–42.

Charpak's equivocal bow to permanent imagery was the exception to a lifelong career of abolishing pictures. Nonetheless, it is perhaps testimony to the power of sight that drove him to follow this xerographic technique. Following the presentation of Charpak's idea, one dubious conference participant challenged its originator to explain any conceivable advantage of the xerox method over the vidicon (which also used electrostatic imaging but had a built-in readout). Charpak responded, "[I]f you have the picture you can make a prescanning by eye, that is an important feature. This was in fact the main feature that attracted us."[142] Even the most successful iconoclast of the entire logic tradition felt the seduction of the icon.

Xeroxing (indeed, imaging in general) was not Charpak's main interest; and the price of creating and scanning a permanent visual record generally seemed too high. Instead, virtually all of the myriad of devices he created over several decades deliberately avoided the intermediate production of pictures. Characteristically, Charpak would use some property of the spark itself to try to locate it electronically. In one attempt, he used the electrostatic charge that was deposited on a wire.[143] In another, he let the spark function as a radio transmission (à la Hertz) and used loops as receivers. Since the strength of the received signal declined quickly with distance, this too could give rise to a method of distance determination.[144] But the most promising of Charpak's methods used a supremely simple principle: the current going through a wire would be inversely proportional to the resistance, and the resistance would be proportional to distance (see figure 6.28).[145] Suppose a charged particle (indicated in 6.28 by the dashed line) discharges a capacitor, depositing a charge on the lower plate, from which two wires issue, one from each side of the plate. The division of the charge between the two paths will be inversely proportional to the impedance from the spark site to each wire. Since the impedance will be proportional to the distance, the ratio Q_1 to Q_2 will be inversely proportional to the ratio of distances from the spark site to the edges of the plate. These two charges (Q_1 and Q_2) are then routed through a ferrite transformer in opposite directions (so that any signal engrossing the whole apparatus will cancel out). If Q_1 is equal to Q_2, no net current will traverse the ferrite coil; the greater the discrepancy between Q_1 and Q_2, the greater the current and therefore the greater the induced current in the secondary circuit in the wires of the transformer. This signal can then be taken out and measured, on an oscilloscope, for example.

Together these bits and pieces—logic circuits, vidicons, sonic sensors, current dividers, tape drives and readers, core memories,[146] magnetostrictive delay

142. Discussion of Charpak et al., "Electrostatic" (1964), in Macleod and Maglić, *Film-Less* (1964), 344.

143. Charpak, "Location," *Nucl. Inst. Meth.* 15 (1962): 318–22.

144. Charpak, "Localization," *Nucl. Inst. Meth.* 48 (1967): 151–83.

145. Charpak, Favier, and Massonnet, "Method," *Nucl. Inst. Meth.* 24 (1963): 501–3.

146. Krienen, "Digitized," *Nucl. Inst. Meth.* 20 (1963): 168–70.

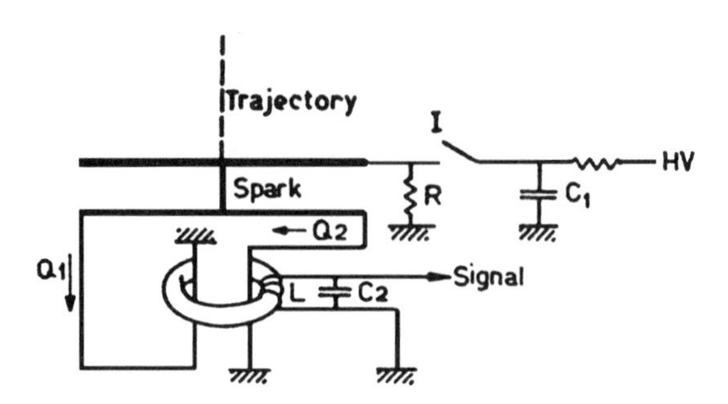

Figure 6.28 Charpak, current division (1963). Source: Charpak, Favier, and Massonnet, "New Method," *Nucl. Inst. Meth.* 24 (1963): 501, 503, on 501.

lines—formed a kind of *object pidgin,* a set of procedures and objects ripped from their original functions and now forming the basis for a new class of combinations. The xerox machine would make an instant image of a spark, not reproduce the surface of a lighted page. Struggling to compete with the overwhelmingly powerful bubble chamber, experimenters in the logic tradition moved restlessly among these different component parts, struggling to find the right language, the right hardware pidgin, with which they could express the empirical structure of physics. Even the variety of concoctions described here only begins to tap the myriad of devices and schemes proposed as a way to proceed without pictures.

Though distinct in their original technical and physical locations, these various practices now formed part of the expanded logic tradition: a cluster of moves, objects, and heuristics that were linked through shared local bits of practice drawn from the industries, weapons, and communication technologies of the postwar decades. It would be hopelessly involved to trace back each of these elements; instead, take one—the magnetostrictive delay line—and let us explore a bit how it came to figure as a prominent component in this technological patois of the detector-building business. Bear with me a moment then, as we detour into the history of the delay line to explore its roots in two much broader technological systems and how it was deracinated from both to form an important component in the new era of particle detection. At stake, I submit, is an example of a general process by which largely delimited components and procedures retain some aspects of their original meanings, gain new ones, and lose others as they pass from one set of technical uses to another.

6.9 Disencumbering Memory

At the height of World War II, Allied radar laboratories began struggling with a persistent and deadly difficulty. Marching across the mountainous terrain of China and Burma, Allied ground troops found themselves vulnerable to Japanese attack from the air. They desperately needed a long-range warning against aerial assault, and a shorter-term warning against low-flying strafing runs. Both types of strike were all too often lost to defensive radar amid the clutter of stationary objects on the surrounding hills and valleys. The Moving Target Indicator program was supposed to remedy this failure by producing a display image in which all objects not in motion would be deleted, leaving only the images of enemy planes (see figure 6.29).[147]

Schematically, the solution was fairly straightforward. A radio frequency pulse at frequency f is sent out and upon return mixed with a copy of the original signal. A moving airplane Doppler shifts the signal to a frequency f', and the beat frequency $(f - f')$ could be used to produce an audible signal (see figure 6.30). This signal gives warning of an airplane; range can be obtained by pulsing the signal in the usual way.

Now the Moving Target Indicator group had to solve the problem of getting rid of the image of stationary objects so that an operator could stare at a polar position indicator and see only the moving objects. It was here that the *delay line* became crucial: Suppose successive sweeps yielded patterns 1 to 4 in figure 6.31. If the first sweep could be temporarily stored, it could be subtracted from the second; the second could be subtracted from the third, and so on. To do this, the receiver had to be able to delay a signal by the period between pulses. This done, the old signal could be electronically inverted and added to the new signal, yielding the "canceled signals" at the bottom of figure 6.32.

Functionally, the delay line stored a microwave radar pulse to be compared with a later pulse. Physically, it worked by electrically pulsing a quartz crystal that (by the piezoelectric effect) vibrated, setting off a supersonic vibration in a tank of liquid. The vibration propagated across the tank, where it was picked up by a second quartz crystal, this time converting vibration into an electric signal. For the time of transit, the signal was "memorized." Invented by William Shockley, the first liquid was a mixture of water and ethylene glycol, though this was soon replaced by mercury in the hands of John Presper Eckert, Jr. Eckert at this time (1942) was working on radar applications and introduced mercury into the delay line because it could handle higher frequencies and was more reliable.[148]

Memory soon became important to Eckert in a very different context. In 1943, Eckert and John William Mauchly famously launched their effort to built the first general purpose electronic computer, the ENIAC. Their goal was to produce, as rapidly as possible, a device that would expedite the ballistics calculations needed for the war effort. In the interest of speed, memory and the

147. Guerlac, *Radar* (1987), 616.

148. William B. Shockley built the first (water/ethylene glycol) model at Bell Laboratories; the second model was built by Eckert's group at the Moore School for the MIT Radiation Laboratory. See Emslie et al., "Ultrasonic," *J. Franklin Inst.* 245 (1948): 101–15; Burks and Burks, *Electronic Computer* (1988), 285; and Goldstine, *Computer* (1972), 188–89.

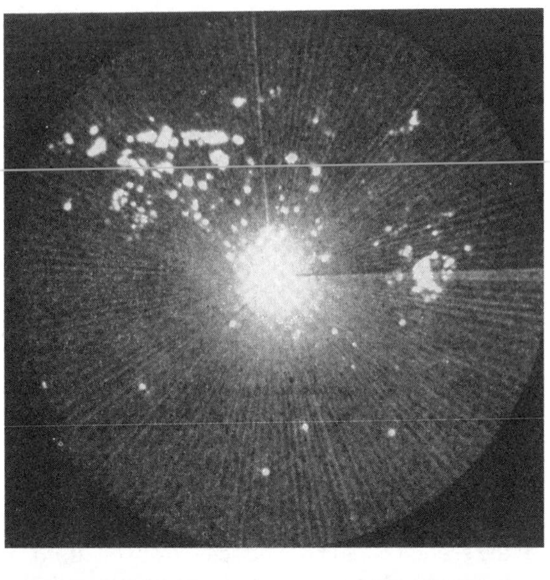

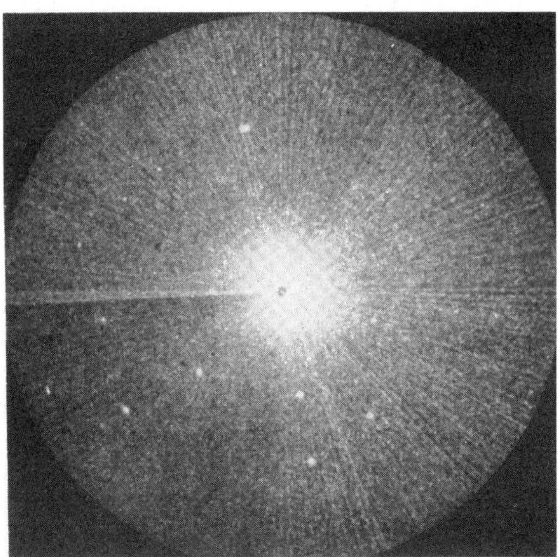

Figure 6.29 Moving Target Indicator (1947). The first use of the memory tube was to subtract out nonmoving "ground clutter" to reveal low-level Japanese fighters. *Top,* Image with clutter; *bottom,* image after the stationary objects are removed. Source: Emslie and McConnell, "Moving-Target" (1947), 626.

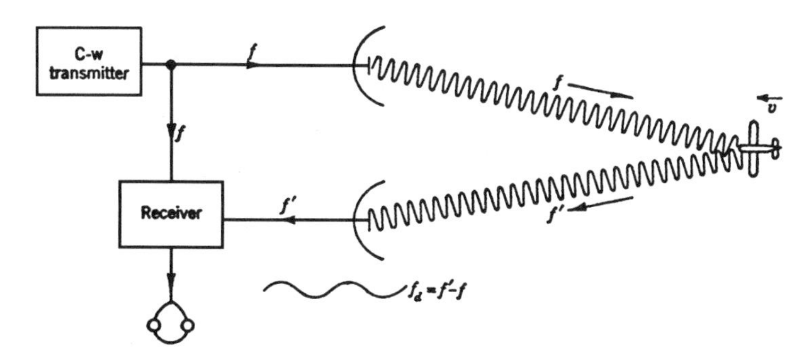

Figure 6.30 Low-frequency Doppler effect (1947). Source: Emslie and McConnell, "Moving-Target" (1947), 629.

capability of storing programs were set aside.[149] But in 1944, when the Moore School team began to consider the next computer, the EDVAC, memory became a crucial feature of the device. Indeed, when John von Neumann wrote his highly influential 1945 "First Draft of a Report on the EDVAC," in which he set out a logical (as opposed to an electrical engineering) description of the machine, he patterned five "organs" on faculties of the human mind: input, output, central arithmetic, central control, and memory. Memory functions as a feedback loop in which a delay line stores information during its transit time and then reinjects the data into the line (see figure 6.33). It was here that Eckert's familiarity with the delay tank provided the link between radar and computer. For those like Eckert who had to deal with the concrete analogue of von Neumann's "memory organ," creating memory meant using pulses, 0.5 microseconds long, to stand for a "1" and no pulse to stand for "0." By recirculating the signal through the mercury delay tank, information could be stored for seconds at a time, and the EDVAC was to have a memory of 2,000 words, a hundredfold increase over its predecessor, the ENIAC. By 1951, the mercury delay line was built into the EDSAC at Cambridge University, the Pilot ACE, the National Bureau of Standards' SEAC, and Eckert and Mauchly's BINAC.[150]

The delay line therefore had its roots in two important (and historically connected) *systems*—radar and the computer. In a strong sense its technological "meaning" was coupled to both, structurally and functionally. Structurally, the detailed physical and electronic structure of the radar delay tube was carried over

149. The origin of the stored-program concept in computers is one of the most vexed and controversial in this vexed and controversial field. Readers are referred to the following sources, which contradict one another about the distribution of credit among Eckert, Mauchly, von Neumann, and others: Goldstine, *Computer* (1972), e.g., 191–92, 253–60; Stern, *ENIAC to UNIVAC* (1981), e.g., 28, 58, 74–86, 168–73; and Burks and Burks, *Electronic Computer* (1988), 195–255, 285–87.

150. Stern, *ENIAC to UNIVAC* (1981), 58–61, 151.

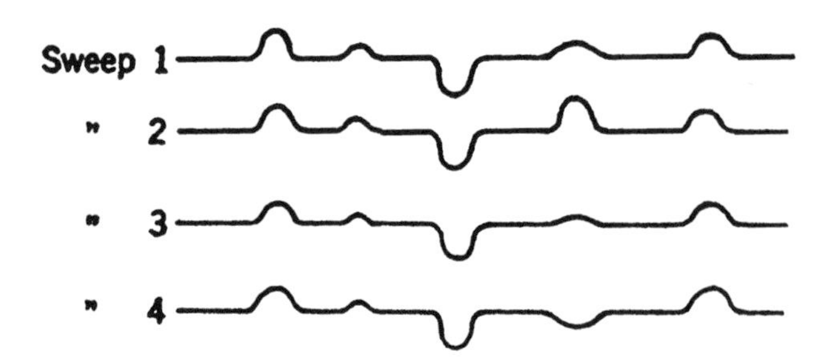

Figure 6.31 Successive sweeps (1947). Source: Emslie and McConnell, "Moving-Target" (1947), 631.

in its entirety by Eckert from radar to computer. Functionally, the delay line had essentially the same abstract representation in both radar and the computer—in the schematics that characterized one level of design work at the Rad Lab and in von Neumann's logical coding system for the computer. Given the emphasis on function within both systems, it is not surprising that a raft of physical structures succeeded one another as instantiating the abstract role of the device.

Electromagnetic delay lines were considered as one of many alternatives to mercury lines for use in computers in 1950.[151] In 1953, when Eckert reviewed the field of memory devices, several other means were added, including magnetostriction, which were being fabricated by the Hazeltine Electronics Corporation and the English company Elliott Brothers (see figure 6.34).[152]

Magnetostriction used a current pulse through a coil to produce a magnetic field; the magnetic field in turn constricted a wire (made of, e.g., nickel) that then propagated an acoustic pulse longitudinally until it reached a second coil in which the inverse effect took place: the mechanical compression was converted back into a magnetic field that induced a current in the read-out coil. During the time in which the acoustic pulse propagated along the wire, the signal was stored.

Radar engineers had used the delay line to cancel a reflected signal; computer designers had made it into more abstract recyclable memory. Now particle physicists wrenched the magnetostrictive delay lines out of the context of *memory* altogether. Ripped from these other technological contexts, the delay line was given a starkly reduced purpose: determining the position of a spark that had hit the wire by the transit time it took the vibration to reach a pick-up at the end of the wire. Instead of feeding highly complex digitized information into a

151. See, e.g., Engineering Research Associates, *High-Speed* (1950), 341–54.

152. Eckert, "Survey," *Proc. IRE* 41 (1953): 1393–1406, on 1396.

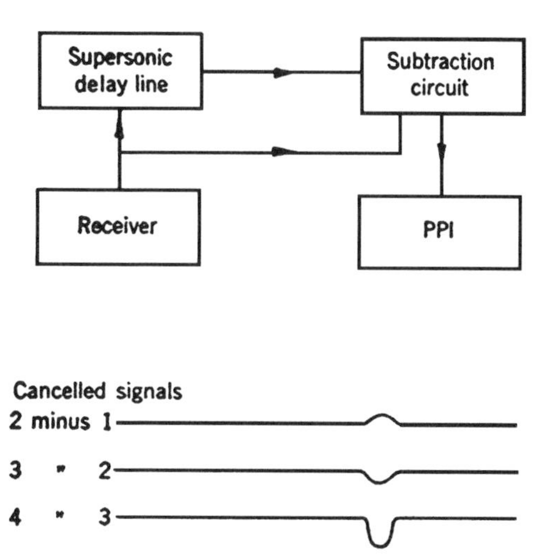

Figure 6.32 Delay line schematic and pulse-to-pulse cancellation (1947). Source: Emslie and McConnell, "Moving-Target" (1947), 631.

line and then recirculating it as memory, the physicists simply used the line to hold one (or at most several) pulses for a time proportional to the length of wire between the spark and the pick-up coil. The first such application appears to have been made by an electrical engineer, G. Giannelli (University of Bari), who in 1963 measured the time it took between spark and pulse arrival (see figure 6.35).[153]

Victor Perez-Mendez at LBL heard about Giannelli's work at the 1964 CERN conference and immediately set to work on a modification of the device in which the magnetostrictive wires were distinct from but adjacent to the electrodes.[154] Self-consciously plucking pieces of technology from the computer, he counseled his fellow instrument makers to do the same. What follows are taped comments from a 1965 instrumentation conference at Purdue:

> PEREZ-MENDEZ: [Another] thing I want to say is that the old computer books, especially from the period of English books around 1950, which deal with the work of Fer[r]anti [Ltd.] and Manchester University, are great sources of information on

153. Giannelli, "Magnetostriction Method," *Nucl. Inst. Meth.* 31 (1964): 29–34, which cites "Digitalizzazione," paper delivered at Frascati Congress, 6 May 1963, report LNF/63/54; see also Giannelli, "Magnetostriction" (1964), 325–331.

154. See Perez-Mendez and Pfab, "Magnetostrictive Readout," *Nucl. Inst. Meth.* 33 (1965): 141–46.

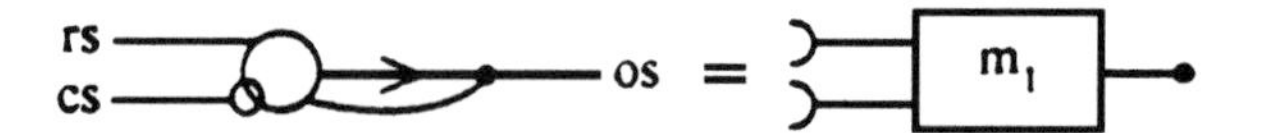

Figure 6.33 von Neumann, abstract memory (1947). A striking example of the disencumbrance of the technical "meaning" of a piece of technology can be found in the history of memory. For von Neumann, the crucial point was to analyze the computer, especially the stored-program computer, without overattention to the particular electrical engineering modality by which the components were to be realized. As part of this process, he treated memory as an "organ" of the computer. Source: von Neumann, "First Draft" (1981), 199.

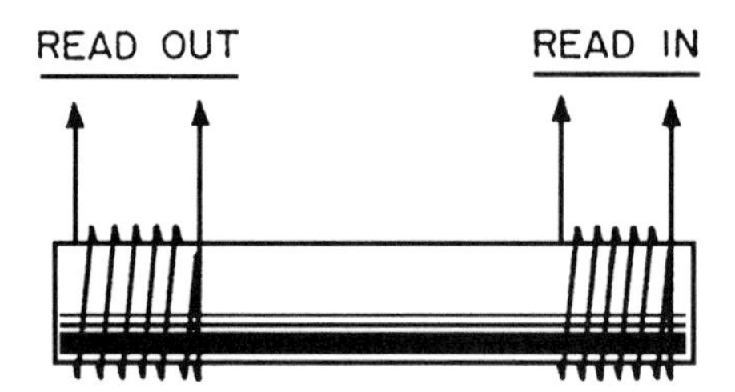

Figure 6.34 Magnetostriction as computer memory (1953). It was all very well to abstract à la von Neumann, but for engineers like Eckert the question was precisely how a signal could be stored. Source: Eckert, "Survey," *Proc. IRE* 41 (1953): 1393–1406, on 1396. © 1953 IRE (now IEEE).

> things that could be useful to high energy physics and which the computer industry has decided to forget because cores were the cheapest components which were available. I found out a lot of things by reading them. I looked up the papers by the Fer[r]anti engineers, and they discussed all the details of magnetostriction and condensor storages, electrostatic storages, and things that they had discarded later on in favor of the cores.[155]

And once the instrument makers had their hands on the magnetostrictive lines, they could be combined in many ways with the other components of their systems. For example, Perez-Mendez, commenting on a wire chamber talk, reminded his audience that once they had a wire chamber they could use it as a proportional counter and "then make use of the ionization information. . . . You can pick up these signals magnetostrictively or otherwise."[156]

155. Perez-Mendez, discussion of Higinbotham, "Wire Spark" (1965), in *IEEE Trans. Nucl. Sci.*, NS-12, no. 4 (1965): 199–205, on 203.

156. Perez-Mendez, discussion of Higinbotham, "Wire Spark" (1965), in *IEEE Trans. Nucl. Sci.*, NS-12, no. 4 (1965): 199–205, on 203.

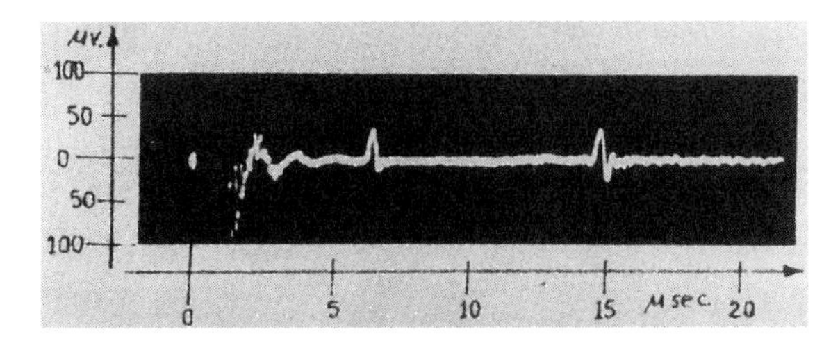

Figure 6.35 Magnetostriction as locator (1964). Finally, Giannelli and other physicists remove the memory device entirely from its computational context, and the storage of a signal becomes secondary to the use of time delay as a way of locating the original disturbance: the more time it takes for the signal to reach the end of the device, the further the disturbance was from the end. Source: Giannelli, "Magnetostriction Method" (1964), plate following 328.

Though the specific path of magnetostriction is tortuous, the physicists' acquisition of the lowly delay line epitomizes their adoption of many such devices. As much as anything else, the process of borrowing is at root one of *disencumbrance,* stripping away the structural and functional links that originally defined a piece of the material culture in order to readapt it elsewhere. One could, for example, pursue other "organs" of the computer through equally complex systems and strings of systems. Core memory too was "cannibalized" from the computer and deployed within the new detectors, just as the vidicon was stripped from the more complex system of full-scale television systems. These bits of hardware, often sophisticated pieces of electronic gear, thus became elements of a materialized pidgin in which many local, internal relationships were understood in the same way by all who used them—the radar engineer, computer builder, and high energy physicist all shared input-output electronic knowledge—but the global function of the device was understood altogether differently in the original and adopting cultures. Local structure could travel where global relations could not.

The flowering of imageless devices in the mid-1960s culminated in two electronic systems that became the standards of high energy physics for the next several decades: the multiwire proportional chamber and the drift chamber. Charpak's laboratory, working on variants of his earlier track-following electronic chambers, produced the first multiwire proportional chamber. For years people had dreamed of using a wire chamber in the proportional mode but imagined the end result too much like a three-dimensional array of individual proportional counters, in which each "cell" of the array would have to be isolated from the others. The reason for this requirement is fairly straightforward. Suppose in a plane of wires a negative pulse is sent along one wire. By capacitive coupling, it would be expected that a positive pulse would appear on adjacent wires—and this is, in fact, observed in the laboratory if one sends a negative pulse through the wire from an external source. Consequently, anyone trying to use several wires in the proportional mode would try to insulate the wires one from the other with at least one intermediate "field" wire. This extra wire doubled the space required between working wires and therefore drastically limited the spatial resolution possible.

What Charpak and his technician R. Bouclier discovered was that an avalanche of electrons cascading toward an anode wire L induced a pulse on the adjacent anode wire A, but that the cloud of ions left behind in the formation of the avalanche induced a pulse of exactly opposite strength and sign on wire A as well. The two induced pulses canceled, making the two anode wires L and A act essentially as independent cells—without the need for any intermediate field-shaping wires. This independence very much simplified the design and reduced the distance needed between anode wires. From the Charpak notebooks it is clear that by September 1967, they had abandoned the field wires and within a year had reduced the separation between anode wires from 2 centimeters to 2–3 millimeters.[157] Justifying the argument for wire independence, Charpak traced equipotential lines on conductive paper (see figure 6.36). Inaccurate as this analog method was, it clearly indicated concentric potentials around the wires and planar potentials far from the wires: a stack of proportional counters had been integrated into a single, unified detector. The new chamber promised efficiencies close to 100%, good amplification requiring less than $2 per wire of further electronic processing before sending the signal to logic circuits, good localization even when the wires were very near one another, a high counting rate, and operation in a high magnetic field.

Again from Charpak's notebooks, one sees the smooth transition from the multiwire proportional chamber to the drift chamber before the end of 1968. Because of the (approximately) flat potential lines distant from the wires, an electron released at some distance from the anode would drift toward it at a constant rate (see figure 6.37). That is, if the multiwire proportional chamber was laid on its side, and one measured the time between the particle's passage and the arrival of the electrons at the sense wire, one had in effect created a new instrument: the drift chamber. In fact, the first such experiments were actually done with one of

157. Charpak, interview by the author, 6 July 1984; Bouclier, interview by the author, 9 July 1984; notebooks in GCP. First results of the multiwire proportional chamber appear in Charpak et al., "Multiwire," *Nucl. Inst. Meth.* 62 (1968): 262–68.

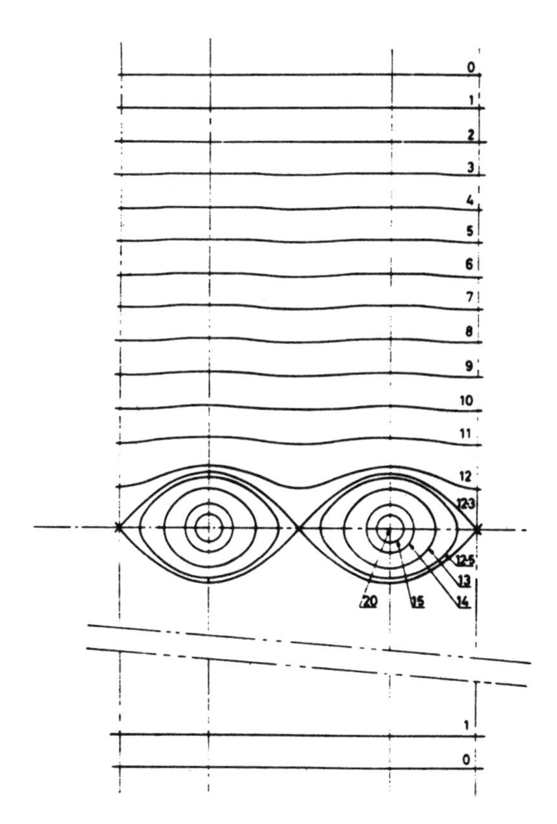

Figure 6.36 Charpak, equipotential lines (1968). This traced picture of the equipotential lines showed them to be circular around each of the wires of the multiwire proportional chamber. Thus, the unshielded wires were in essence a set of independent proportional chambers. Suddenly, the old technology dating back to Geiger and Müller could be multiplied many times over—limited only by the expense of amplifying the signals from each wire. And that expense was dropping precipitously in the late 1960s. Source: Charpak et al., "Multiwire," *Nucl. Inst. Meth.* 62 (1968): 262–268, fig. 2 on 262.

the prototype multiwire proportional chambers. The significant differences were in the way the data were processed and in the introduction of a timer set to begin when a fast counter registered the passage of the particle and set to stop when the electron pulse reached the anode. Charpak and his colleagues first presented their work on drift chambers in 1969 and published it in 1970 under the name of its precursor the multiwire proportional chamber. Indeed, as we have seen so many times before, one device was used *physically* as well as *analogically* as the basis for another: the drift chamber is a reformulated multiwire proportional chamber. In the 1970 article "Some Developments in the Operation of Multiwire Proportional Chambers," "Drift Chamber" is included as a subsection after a section on multigrid structures and before the treatment of multiwire proportional chambers in magnetic fields.[158] Heidelberg physicist A. H. Walenta also

158. Charpak, Rahm, and Steiner, "Some Developments," *Nucl. Inst. Meth.* 80 (1970): 13–34.

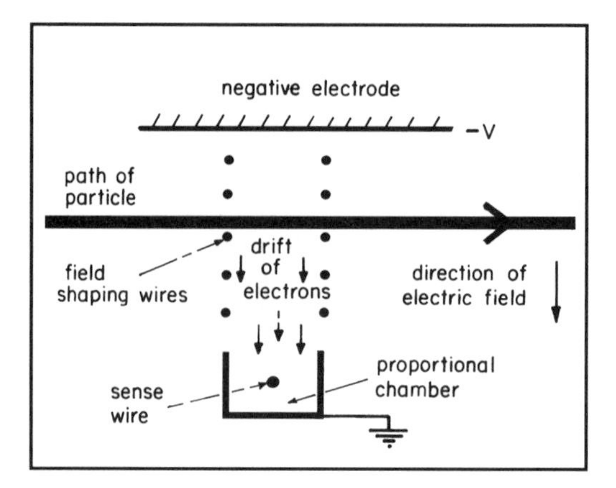

Figure 6.37 Drift chamber principle (1974). Field-shaping wires maintain a uniform electric field throughout the chamber so that electrons knocked free of their atoms by the passing particle (indicated by the dark arrow) will move to the sense wire at a constant average velocity of about 5 centimeters per microsecond. Consequently, the time between the particle's passage (registered by a fast counter) and the arrival of an avalanche at the sense wire yields the "height" of the particle track. Source: Rice-Evans, *Chambers* (1974), fig. 10.1.0.1 on 360.

boosted the proportional chamber into a drift chamber capable of exploiting time-based measurements to determine position. Equalizing the electric field using field wires, reducing the electronics, and joining the output to a computer, he and his colleagues achieved a location accuracy of just 0.47 millimeters. The new technology spread rapidly as laboratories around the world achieved ever more linear relations between drift length and time, rendering the device one of the workhorses of colliding beam physics as CERN, Fermilab, DESY, and the once but never SSC.

For all the many roots and stunted branches of the logic tradition, we can now characterize its underlying practices. The Geiger-Müller counter and the other electrical counters of the 1920s shared a statistical interpretation, a theory of the instrument based on the diffusion of charged particles in a gas and the complexities of discharge phenomena. Out of the Geiger-Müller counter, quite literally, came the spark counter, the Cerenkov counter, and the scintillation counter; all were promptly classified as counters, sharing the Geiger-Müller's integration into the electronic environment and many of its original physics uses in timing and counting. Indeed, the spark counter was, materially, little more than a flattened Geiger-Müller counter. Out of the spark counter, in turn, came the spark chamber (a spark counter with a visible spark); out of the spark chamber, the sparking wire chamber (a spark chamber with wires instead of plates); and out of wire chamber work by Charpak and others, the multiwire proportional chamber, and its alter ego, the drift chamber (a multiwire proportional chamber on its side). People moved freely back and forth between the counters, in part because much of the technical vocabulary was common to the different devices: gas mixtures, quenching circuits, amplifiers, logic circuits, high-voltage technologies. But throughout the 1960s, physicists committed to the logic tradition were struggling not only *for* the control over the experiment that the logic tradition afforded, they were struggling as well *against* the life they saw associated with the big bubble chambers. The logic tradition had always promised and demanded control—control over the physical process and control over the experimental life that surrounded these sparking and pulsing instruments.

6.10 Pure and Postmodern Machines

6.10.1 Hybrids

In 1957 Luis Alvarez had been ready to celebrate, perhaps even hasten, the demise of the logic tradition. Bubble chambers were producing more data, more discoveries, and more precise information about the subatomic world than their counter counterparts could even imagine. The world of visualization seemed triumphant. But by the early 1970s the bubble era was drawing to a close, and massive detectors had begun to shut down in laboratory after laboratory. In a more than symbolic act, Alvarez's old 72-inch chamber, expanded when it was sent to SLAC, was ceremoniously shut down at 3:30 P.M. on 16 November 1973, and the extinction process proceeded at laboratories around the world, driven by the new triumph of electronics.[159] Indeed, the last great hurrah for the boiling liquid was the discovery of weak neutral currents in Gargamelle, a heavy-liquid bubble chamber at CERN. The establishment of the weak neutral current was a contentious process, one that lasted quite some time but was more or less complete by 1973. But if bubble chamber physicists knew they were at the end of the line, the epistemic ideal embodied in these detectors lived on: somehow, within the sphere of logic electronics, rare, often complex events would have to be extracted with the indelible tracks that had previously been the purview of the image tradition. This section, on one particular detector, begins an exploration of the hybrid that continues in chapter 7: hybrid teams, hybrid equipment, and hybrid modes of demonstration.[160] This hybridization took place only with difficulty and, as I have indicated, after a myriad of failed attempts to capture the image without the world of the photograph.

On one side of the image-logic divide stood the pure forms of image makers: the cloud chamber, the bubble chamber, and the nuclear emulsion. On the other stood the pure electronic world of the Geiger-Müller counter, the spark

159. Invitation to shutting down ceremonies, 16 November 1973, vol. 2, GGP.

160. See Galison, *Experiments* (1987), chaps. 4–6.

chamber, and the wire chambers. A single physical process is implicated in each: the cloud chamber was predicated on the transformation from a vapor to a liquid, the Geiger-Müller counter and its cognates from a gas to a plasma, the bubble chamber from a liquid to a gas, and the emulsion from silver halide to its ionic components. And while the instruments in some cases could be used in conjunction with one another—one thinks immediately of the counter-controlled cloud chamber—as a broad generalization one can reasonably characterize experimentation before the late-1960s as grounded on relatively "pure" devices. Bubble chambers were never triggerable nor was film. Indeed, because of the background problem of old tracks superimposed on new ones, and the difficulties associated with fading tracks, the lack of time control on films remained a problem throughout its usage. And electronic devices, despite hundreds of attempts, never achieved the detailed, omnidirectional responsiveness that the bubble chamber had vaunted. In the 1960s and early 1970s, however, the purity of the iconoclast and iconodule was lost as hybrid instruments became the standard. This shift would be a powerful alteration of the material culture of physics.

Facing off against one another, the iconoclasts and iconodules each tried to incorporate the other: imageless devices would create the simulacrum of a photographic image using electronics—everywhere the "electronic bubble chamber" became a holy grail. There were streamer chambers—gas-filled parallel-plate chambers in which emitted light from the recombination of ions and electrons would leave tracks that could be photographed. There were wide-gap spark chambers that had improved track-following capabilities. Among the image physicists, there was an endless series of attempts to restore control over the experiment by creating the ever elusive triggerable bubble chamber. Image makers wanted the control and high statistics of the logicians; the logicians wanted the powerful ability to be able to rely on the data, event by detailed event. As Peter Rice-Evans, author of a compendious book covering the full spectrum of detectors, wrote some 20 years after Glaser's invention: "Every undergraduate knows that the bubble chamber has the grave disadvantage that it cannot be triggered by external electronic counters set to register a certain combination of emerging particles. Thus, in normal operation, the expansion cycle of the chamber is synchronized with the burst of particles from an accelerator; in a series of exposures, desired events may occur as a matter of chance—uninteresting photographs are rejected."[161] In one of the last chapters of his massive review of detector technologies, Rice-Evans hoped that people would reconsider the problem in order to reconstruct the bubble chamber into something useful for cosmic rays and for a new accelerator being built at CERN. Even Charpak, arguably the most creative instrument maker of the postwar era, was unable to find a niche for his ideas

161. Rice-Evans, *Chambers* (1974), 312.

in triggerable liquid detectors.[162] Pictures proved a poorer starting point than counts.

Triggered bubble chambers never succeeded in entering the mainstream of physics. Image and logic physicists, their devices and their strategies of demonstration, met elsewhere: in the computer.

6.10.2 Mark I and the Psi

The halting move toward a fusion of image and logic traditions took much more than beefing up hodoscope arrays of counters with computers, more than photographing track-following spark chambers, and much more than the abortive attempts to make triggerable bubble chambers. For accompanying (and indeed in some ways precipitating) the shift there was a concomitant alteration in the social structure of the experimental group and a change in the mode of demonstration. Since this shift is the primary theme of the next chapter (on the Time Projection Chamber), I want here merely to introduce these concepts by discussing what is arguably the single most important instrument that high energy physics ever produced: the detector known by the appallingly uneuphonious name of the SLAC-LBL Solenoidal Magnetic Detector, at the colliding beam facility SPEAR. For it was with this device, which I will call by its later name, Mark I, that the psi (ψ) particle was codiscovered.[163] Along with its role in ushering in the quark theory of matter through the psi, the Mark I was also used in arguing for the existence of the tau (τ) lepton and for the production of the D^0 meson that gave credence to the claim that the psi was a bound state of charmed mesons. More than any single device, the Mark I became a prototype for the next several decades of high energy colliding beam physics at laboratories around the world, including Fermilab, CERN, and DESY.

The introduction of colliding beam facilities radically altered instrumentation in several ways. In a fixed target facility, much of the beam's energy is never converted into new particles—it is simply carried forward along the line of motion set by the projectile. Colliding beam facilities, by contrast, smash a particle and antiparticle together with equal and opposite momenta. Since the two objects annihilate one another the energy is fully utilized and the resulting new particles can emerge in any direction—not just along the line of the projectiles' trajectories.[164] Even a cursory glance at figure 6.38 reveals the geometri-

162. Charpak, interview by the author, 6 July 1984.

163. My purpose here is to characterize the emerging nature of hybrid colliding beam detectors and not to give a comprehensive history of the *J*/psi discovery, which would surely give equal weight to the MIT-Brookhaven experiment led by Samuel Ting. For more on the latter, see Riordan, *Hunting* (1987): 262–321; Pickering, *Quarks* (1984), 258–73; and Ting, Goldhaber, and Richter, "Discovery," *Adventures Exp. Phys.* α (1976): 114–49.

164. In the case of proton-antiproton collisions, the quark–anti-quark annihilation may have its center of mass in motion with respect to the accelerator because the quarks are not at rest inside the proton and anti-

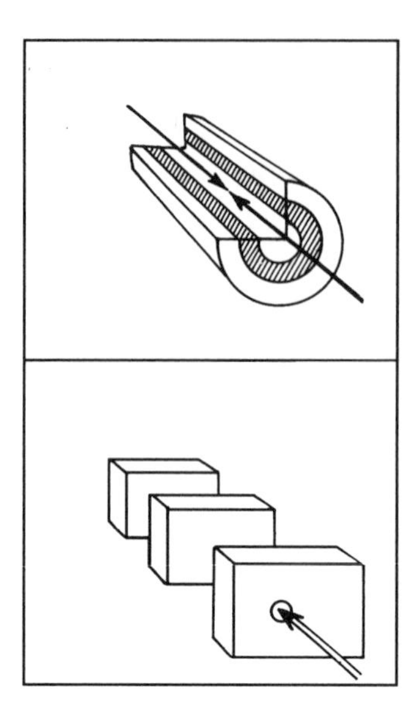

Figure 6.38 Drawing of schematics. *Bottom,* Typical detector geometry for fixed target experiments. The beam hits a target, and the resulting particles are then measured by a sequence of parallel slabs which may consist of counter arrays, calorimeters, spark chambers, wire chambers, filters, etc. *Top,* Characteristic cylindrical geometry typical of colliding beam experiments. Especially when equipped with "end cap" detectors, this detector setup can be sensitive to nearly all the secondary particles, which can fly in any direction.

cal alteration that detectors underwent in the passage from fixed targets to colliding beams. Instead of being responsible for a fairly narrow cone in the direction of the (fixed target) beam, the colliding beam detector has to search over as much of a full sphere as possible. Here was a transformation in the geometry of detectors, but at the same time, as we will see, an alteration in the sociology of experimentation as great as the shift from tabletop to bubble chamber.

The production of a triggerable detector with nearly 4π acceptance was only the latest attempt to join the logic and the image tradition, an attempt to join the control we saw emphasized in the logic tradition to the possibility of picture-like representation of the data. Building a 4π detector would be the basis of the imaging. Control, however, would not be ceded as it was in a bubble chamber. Burton Richter:

> Part of my background made me want to know what was happening in the experiment while it was happening. I felt that it was much better when exploring the unknown to know where you are so as to better plan where you can go, rather than after the journey to only know where you have been. [And it's different in] the bubble chamber business, the physicists took hundreds of thousands of pictures which then had to be developed and scanned, and didn't know anything about what

proton. I will not review the long and important history of colliding beam facilities here, but see Paris, "Building" (1991). She effectively reviews work at the CEA, ADONE, and elsewhere, concentrating on SPEAR.

> was happening in the experiment until a year or two after the experiment was already over.[165]

Given the views expressed from the late 1950s forward, Richter's remarks are hardly surprising. Wenzel, Charpak, Preiswerk, Macleod, Hine, and others too numerous to cite had emphasized time and again how abhorrent they found the dissociation of data taking from analysis. In every aspect of instrument construction, the logic tradition experimenters continued to express in hardware their demand for results they could use "while they happened." This insistence on the priority of on-line results remained central to the Mark I collaboration, reflected most clearly in the extraordinary lengths to which the team went to provide single-event displays in real time.

The hybrid nature of the effort is as visible in the social structure of the collaboration as it was in the material structure of its component parts. In particular, three teams joined to form the Mark I collaboration: One group, from LBL, brought with it a long tradition of expertise in bubble chamber track analysis. The other two, from SLAC, carried with them an equally long history of electronic experimentation. Look for a moment at the Berkeley side of the venture. George Trilling was, by the early 1970s, one of LBL's most senior bubble chamber experimenters—he had come with Glaser to Berkeley. Gerson Goldhaber had begun scientific life in the world of emulsions, culminating in his registration of the first track of an antiproton in 1956. By 1959, he had turned to bubble chambers, first with Wilson Powell and subsequently with Trilling. Together, Trilling and Goldhaber had directed a large and successful LBL bubble chamber group, one of the main competitors to Alvarez's own LBL outfit. Goldhaber also helped recruit William Chinowsky to Berkeley, first as a visitor and then permanently. Though Chinowsky's Ph.D. research was done with counters, he too had turned, by 1954, entirely to track experiments, and he worked for many years with bubble chambers.[166]

Like the other LBL members of the team, John Kadyk and Gerry Abrams were both bubble chamber veterans; the LBL graduate students, John Zipse, Robert Hollebeek, and Scott Whitaker, filled out the group. Asked later if the bubblers and electronics people felt they lived in different communities, Chinowsky laughed, recalling the competition between the two: "The people doing electronics tended to look down on us in the bubble chambers; on the other hand, it was the bubble chamber that got all the results. I guess, to some extent, it didn't seem fair that they should work so much harder and not get a return."[167] Chinowsky went on to recall the search for one particle, the ρ, in which

165. Richter, interview by the author, 18 March 1991, 34; addendum, Richter to the author, 11 November 1995.

166. Goldhaber, interview by the author, 14 August 1991, 10–11.

167. Chinowsky, interview by the author, 14 August 1991, 11.

the electronic team had mounted a vast array of counters, but the Brookhaven bubble chamberites won the race handily with invariant mass plot analysis.[168] As we saw in chapter 5, bubble chamber work—certainly once the chamber was up and running—revolved almost entirely around analysis: programming for the reconstruction of tracks and the identification of particles and processes. It was such skills—the elaborate deciphering of track structures—that would carry over to the annihilation of colliding electrons and positrons. For example, one track program used in the Mark I exploited the "road" idea from the HPD that the Goldhaber-Trilling group had used at LBL in its bubble chamber film analysis. Another program was lifted virtually intact from bubble chamber analysis to the Mark I.[169] As an indication of just how powerful the highly honed track analysis skills were, Abrams, Goldhaber, and Kadyk spent many hours poring over photographs of computer-reconstructed tracks that were printed from the magnetic tape.[170] This was a process altogether outside the logic tradition, but in part constitutive of the Berkeley physicists' credence in (or doubts about) any putative new effect.[171]

There were other differences between the Berkeley and SLAC groups. During the 1960s, the Berkeley team had made its mark by its analysis of hadron dynamics, the meson-nucleon resonances, and the myriad of new particles that came out of that program. At SLAC, the emphasis had been more on electron-nucleon interactions used to explore nucleon structure (Hofstadter's program) and on many tests of quantum electrodynamics within the interactions of electrons, muons, and photons. LBL necessarily looked at events with fairly high cross sections because the bubble chamber could not be selective; at SLAC, picking rare events out of the noise electronically was standard fare. For all these reasons, combining the two subcultures was not trivial.

The one piece of electronic hardware that Berkeley undertook was the "shower counter," a sandwiched structure of plastic scintillator (which measured the energy of electrons and photons by way of the light deposited in it) and lead plates (which converted the photons into electron-positron pairs). In light of the LBL team's prior experience, it is perhaps understandable that the software and tracking programs were a spectacular success and the shower counter was occasionally a source of friction with their cross-bay, logic tradition

168. Chinowsky, interview by the author, 14 August 1991.

169. For a discussion of the use of "roads" in the PASS 1 filter program, used for primary data analysis, see Hollebeek, "Inclusive" (1975), 21–22; see also Goldhaber, interview by the author, 14 August 1991, 18. Richter: "This was going to be a 4 pi tracking detector, and I felt that the experience of the bubble chamber people in data analysis would be extremely useful in designing the programs that were needed to turn these spark chamber bits . . . into physics" (Richter, interview by the author, 18 March 1991).

170. Goldhaber, interview by the author, 14 August 1991.

171. Jonathan Treitel has made use of the image-logic distinction in his dissertation, "Structural Analysis" (1986), which was then published as an article, "Confirmation," *Centaurus* 30 (1987): 140–80.

electronic collaborators.[172] By December 1972, it was clear that the scintillators had been badly scratched in fabrication, and a crash program began to try to salvage what could be rescued. Passing a steam jet over the surface failed; applying the manufacturer's polish did not work; spraying the sheets with acrylic helped not; and heating to 50°C for 24 hours did not aid the cause. What did help (somewhat) was giving the scintillators a coat of Johnson's Glo-Coat floor wax and standing them up to dry. In the end forgoing this ad hoc solution, the detector did achieve its design specifications and even was used in the detector trigger, an unanticipated bonus.[173] Software-embedded image reconstruction and analysis was the Berkeley forte, and aside from the flap over scratched plastic, collaboration between the two institutions worked without much friction.

One link between the Berkeley image and the Stanford logic traditions was made through Martin Perl, who invited Chinowsky to join the collaboration. Perl had been a student of I. I. Rabi's and in the early 1950s had gone to Michigan to work with Donald Glaser on his newly minted bubble chamber. Perl's fascination had been for quite some time the muon. Heavier than the electron but otherwise apparently indistinguishable from it, this particle was a standing rebuke to any motivated understanding of particle physics. Rabi was most adamant on the point, repeating often and insistently his refrain about this heavy new particle: "Who ordered that?" After Glaser left Michigan for Berkeley, Perl moved to Stanford and began a six- or seven-year struggle to answer Rabi's question by trying to find some difference other than mass between the electron and the muon. Unable to explain who ordered the peculiar second course of lepton, Perl began ordering more . . . and more. For years, Perl and his collaborators in SLAC Group E (including J. Dakin, G. Feldman, and F. Martin) had used electronic means to explore possible distinctions between the muon and the electron, and they pushed hard to add muon detectors to the Mark I to allow the collaboration to search for even heavier leptons that might be within reach. So it was that this new detector, while it might have looked like a hadron finder to the LBL team and to Richter's Group C at SLAC, was to Perl an instrument for uncovering the "next" in a sequence of heavy leptons. One other Stanford experimenter, David Fryberger, came from SLAC's Experimental Facilities Department, formally outside of the structure of Groups C and E.

172. The problem LBL had was that the surfaces of the delicate plastic scintillators had many microscopic scratches due to incorrect wiping to remove excess glue during fabrication. This altered the light-transmitting properties of the plastic, reducing the attenuation length for light transmission. See, e.g., Hollebeek, "Inclusive" (1975), 11–12. Indeed as late as August 1973, the team debated about replacing the plastic or trying to patch over the difficulty: "[N]ew scintillator vs. Johnson's Wax was debated, without decision" (Lynch to SP-1 Physicists, 10 August 1973, vol. 2, GGP). In the end, neither was done. Kadyk to the author, 4 December 1995.

173. Whitaker to SPEAR Detector Distribution, "Rejuvenation of Pilot F Scintillator," TN-191, 5 December 1972, MBP; see also Goldhaber, Abrams, and Kadyk to the author, 4 December 1995.

The majority of the SLAC contingent, however, was affiliated directly with Richter's long-standing research in electron physics. One of Richter's interests had been in photoproduction—using photons produced in electron collisions to make new particles. Out of this effort came one half of Richter's group, including Adam Boyarski. Boyarski was the computer wizard; it was his job to ensure that the data produced by the various detector elements came in with compatible formats, that the detector had a reliable on-line monitoring system, and that a final data tape was produced that could be accessed by anyone within the large collaboration. Two SLAC researchers, Roy Schwitters (who had been an MIT graduate student with Louis Osborne, working with Richter's group at SLAC's End Station A) and Martin Breidenbach (also an MIT alumnus, who had worked at SLAC and then CERN before returning to SLAC), also moved into the Mark I effort out of this earlier enterprise. On the other side of Richter's (pre-SPEAR) physics pursuits was a series of spark chamber experiments that included SLAC staff physicists Rudolf Larsen and Harvey Lynch. Larsen became the spokesman for the Mark I collaboration, and he and Lynch contributed their extensive experience with spark chambers, including magnetostrictive readouts and other electronic lore. Some members of the Group C team were entirely devoted to machine building—not to making the detector. John Rees, for example, originally moved to SLAC from the Cambridge Electron Accelerator in 1965–66, bringing expertise in accelerators and storage rings, especially on new ways to handle the RF (radio frequency) facilities and beam-focusing equipment. Gerry Fischer managed the magnet systems for SPEAR.[174]

From the outset, in his design and in the composition of Group C itself, Richter conceived of the experiment as a colliding beam machine *and* a detector. His intention was evident on the ground, where, unlike virtually any other experiment in high energy physics after 1970, Mark I and SPEAR shared a single control room. Even in 1971 this integration was unusual: laboratories such as Brookhaven and CERN had long separated the production of particles from their consumption.

The search for full coverage underlay the design of the Mark I and distinguished it from all previous electronic detectors (see figure 6.39). By immersing the whole in a magnetic field, the Mark I collaboration hoped to be able to identify all of the resulting particles. Aside from the scintillator (in the shower counters), there were three main subdetectors within the whole: the cylindrical wire spark chambers that would be the primary track detector, the beam pipe counter that monitored the beam and collision times, and outer muon spark chambers that identified muons, distinguishing them from strongly interacting particles of similar mass (pions, largely).

174. Rees, interview by the author, 21 June 1991; Schwitters, interview by the author, 17 June 1991, 13–14.

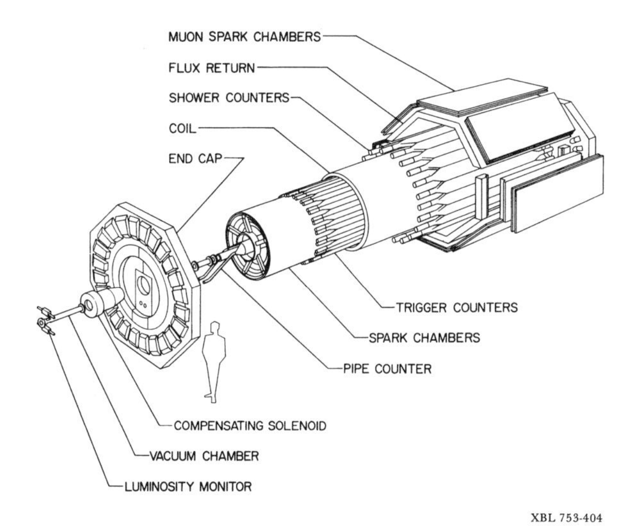

Figure 6.39 Schematic Mark I. In this exploded schematic, the cylindrical layers of the Mark I are exposed—this instrument in many ways inaugurated the era of colliding beam detectors. Note how each of the elements—proportional chambers, spark chambers, trigger counters, shower counters, and muon spark chambers—derives from the electronic logic tradition. But taken together, and integrated with a new order of computing power, these various devices produced the nonphotographically derived "pictures" of figure 6.40. Source: XBL-404, LBL. Courtesy of Lawrence Berkeley Laboratory, University of California.

The centermost detector, the pipe counter, was designed to exclude cosmic rays by triggering the registration of data only when an electron-positron annihilation took place. Effectively, this selection reduced the number of cosmic ray events from about a thousand per second to about one per second—dramatic improvement. Assembled from two semicylindrical sheets of 90-cm–long scintillator, the pipe counter wrapped directly over the vacuum chamber in which the electron and positron bunches traveled—as is visible in figure 6.39. From each end, each semicylinder was linked to a Lucite pipe that brought the light collected outside the detector to where a phototube measured the output; the phototube could then be linked to the logic circuits of the trigger.[175]

Moving outward from the beam pipe, the next and most important detector was a cylindrical wire spark chamber, built to lie concentric with the beam

175. Note that a new pipe counter was installed in 1974; see Larsen to SP-17 Experimenters, "New Pipe Counter for Magnetic Detector," 23 September 1974, MBP.

line. This instrument was divided into four concentric components, each optically isolated from the others but sharing a single gas volume. As an indication of scale, the outermost chamber had a radius of 53 inches, a length of 106 inches, and 31,900 wires at 1/24-inch intervals. To measure the efficiency of these chambers, the following test was established: Any track that left a space point in all four chambers counted as a success for each one. Every track that left three space points (one chamber not firing) was tallied as a failure for the missing chamber. "Efficiency" could then be defined as the ratio of successes to successes plus failures, and this quantity could be computed as a function of angle and distance down the beam line.[176]

Built into the notion of "efficiency" and the programs that computed it is a continuation of the fundamentally statistical nature of the logic tradition. For even as the computer reconstructed a track, the absence of a spark in any particular chamber would be excused; no individual spark location was deemed essential to the reconstructed trajectory.

Magnetostriction, as we saw, was by the early 1970s hardly a new technology for track construction. As a mode of detection, it was no longer in question. The problem was that no one had any idea how to make such a chamber function in the magnetic field that engulfed the Mark I, and the magnetic field was necessary to bend the particle tracks so that each particle's momentum (and thus identity) could be determined. Here is the issue: the standard geometry (which put wires in rings around that cylindrical detector shown in figure 6.39) would have put the magnetostrictive readout wires perpendicular to the magnetic field (which pointed along the cylinder axis). Fatim Bulos had the idea of spiraling the readout wires around the cylinder axis (which gave the wires some magnetic field along their length), and Schwitters determined that the spiral angle should be 30° or so from the ambient field.[177]

Muon detectors formed the outermost layer of the cocoon, insulated from the rest of the apparatus by a thick layer of concrete that would stop a large percentage of any strongly interacting particles. Though they were necessarily larger, this fifth set of magnetostrictive wire spark chambers would ferret out the muons. These, like the inner spark chambers, had their own set of tests to pass before they could be trusted. How would they respond to pulses of different voltages? How frequently would the chambers fire "accidentally" (in the absence of any known signal)? Could "correlated" firings of the chamber (coincident and

176. Hollebeek, "Inclusive" (1975), 6–10; see also Schwitters to SPEAR Detector Distribution, "Inner Spark Chamber Configuration," 16 March 1972, MBP; and Schwitters to SPEAR Detector Distribution, "Inner Spark Chambers," 2 August 1972, MBP.

177. Schwitters to SPEAR Detector Distribution, "Magnetostrictive Wand Orientation in the Magnetic Field of the SPEAR Magnetic Detector," 31 January 1972, vol. 2, GGP; see also Hollebeek, "Inclusive" (1975), 7; and Richter to the author, 11 November 1995.

therefore easily confused with a bona fide muon event) be avoided by the introduction of alcohol in the chamber?[178]

Separately, then, each component detector had to meet a certain standard, achieve a certain measured efficiency. But separate functioning was not enough. Nothing in the epistemological structure of experimental high energy physics is as important to understand as the coordination between subgroups. For it is this mutual alignment—at once social and epistemic—that both restructures the function of a scientific author and undergirds the demonstration of a new effect.[179] Again and again through the earlier chapters of this book I have stressed how local coordination between diverse approaches to a problem is central to understanding the building of an argument and the cohesion of the larger scientific community. As experimentation grows in scale, this coordinative function is exhibited more and more often within the construction of the instrument itself. It is a theme that will become central in chapter 7 when we move from groups the size of the Mark I collaboration (20–30 physicists) to the vastly larger collaborations of the Time Projection Chamber at SLAC, CERN's Large Electron Positron collider, and the planned but never built Superconducting Supercollider. One element that holds the components together is the language of computation, for it was ultimately the computer that had to mesh together the output signals from the beam pipe detector, the inner track chamber, the lead scintillator sandwich, and the outer muon detector.

I quite intentionally foreground the linguistic character of this synchronization. As one physicist put it after listing various problems, "Assuming that all the foregoing problems could be solved, we are still on the path to Babel unless positive steps are taken to recognize the reality that we are a large group of people who depend upon each other."[180] First priority: computer programs had to give correct answers, and the programs therefore had to be checked by someone other than the author. This demand for *correctness* was quickly followed by demands for *reliability* (the programs could not crash) and *intelligibility* (documentation both external and internal to the code had to be clear and well directed). A fourth priority was the constraint of *efficiency;* scarce resources both in core memory and CPU time could be swallowed up by an inefficient piece of code, and off-line work, while it needed less core, would still tie up crucial CPU time. Fifth and finally, "The programs must be easy to use. . . . Since we are a group where people use and depend upon the programs of others, considerable attention should be applied to the '*human engineering*' aspect."[181] Technique and

178. Dakin to SPEAR Detector Distribution, "μ Chamber Performance," 14 August 1972, file "Detector," MBP; see also Dakin to SPEAR Detector Distribution, "Muon Chambers," 18 September 1973, vol. 2, GGP, which describes the muon chambers with cross-sectional view.

179. This is the main thesis of work elsewhere; see Galison, *Experiments* (1987), chaps. 4–6.

180. Lynch to SPEAR Physicists, 21 February 1973, "Computer Programming," GGP.

181. Lynch to SPEAR Physicists, 21 February 1973, "Computer Programming," GGP, emphasis added.

social structure were here, as throughout the history of experimentation, inextricably bound.

Even if Babel could be averted, there remained the coordination of pieces in data reduction. On 8 November 1973, for example, Larsen issued a memo that set out the basic problem: "What we all want to do is to extract the physics from our data tapes, right? Right! . . . To date, we have not been remarkably successful in this preparation. Many problems have arisen: duplication of effort, conflict over use and status of software routines, lack of definition of problems. . . . While many of the problems can be attributed to the 'lack-of-communication' cliché, the most important void is the absence of a structure that everyone understands and within which we can work."[182] There were two parts to the process of coordination, event identification and a full characterization of the four fundamental components—cylindrical wire spark chamber, muon wire spark chamber, trigger counters, and shower counters. As Larsen emphasized, most of the collaborators had been affiliated with the production of only one piece of hardware and its attached software for extracting quantitative information. Now that isolation had to end.

Again Larsen: "It is necessary to formalize this existing situation so that all know whom to turn to when they need information; there is clearly a good deal of cross-talk between the various hardware components."[183] As this remark made clear, structural architecture and social architecture must move together. "Cross talk" between the wire spark chamber and the shower counter depended on links between the appropriate software, and that meant coordinating the wire spark chamber group with the shower counter group. So the collaboration was subdivided once more, this time into "software" components:[184]

Cylindrical WSC	Muon WSC	Triggers-Pipe	Showers	Public Analysis Program
Lynch	Bulos	Moorhouse	Kadyk	Boyarski
Hollebeek	Lyon	Feldman	Feldman	Breidenbach
Zipse	Dakin	Larsen	Whitaker	Hanson
Schwitters	Pun		Friedberg	Abrams
Augustin			Perl	Chinowsky
Breidenbach				Goldhaber
Chinowsky				

182. Larsen to SP-1, SP-2 Experimenters, "Data Analysis," 8 November 1973, MBP.

183. Larsen to SP-1, SP-2 Experimenters, "Data Analysis," 8 November 1973, MBP.

184. Larsen to SP-1, SP-2 Experimenters, "Data Analysis," 8 November 1973, MBP. According to Goldhaber, in the end, Perl and Feldman joined the Muon WSC group and Abrams joined the Showers group. Goldhaber to the author, 4 December 1995.

Significantly, the segregation of LBL and SLAC, present in the building of hardware, had been breached. The cylindrical wire spark chamber now brought Chinowsky and two Berkeley students, Hollebeek and Zipse, into the collaboration with the builders, Schwitters and Lynch; the shower counter, similarly, now introduced Perl and Feldman, from the hardware (muon chamber) side of SLAC, to the original LBL shower counter builders, Kadyk, Whitaker, and C. E. Friedberg.

But over and above these divisions had to be a geographically representative supergroup that would create the integrative "analysis" program. This software would take output from the component software and weave it into a coherent data set with a clear and consistent set of calibrations. "While everyone is free to maintain his own file, we can't have everyone altering 'the' primary analysis routines at his will."[185]

The next month (December 1973), Richter reported on the Data Analysis Steering Committee's deliberations. There had to be consensus on the two first computer data-crunching programs: PASS 1 and PASS 2. PASS 1 "filtered" out events that were, as one participant wrote, "the most obvious sources of background events."[186] Events had to satisfy certain PASS 1 constraints before they would even be recorded as basic data: A minimum time lapse had to have passed between the moment of electron-positron annihilation and the detection of tracks; this guaranteed that the colocation of tracks was minimally plausible as a physical occurrence. There had to be a prima facie case against the process's being of cosmic origin. There had to be enough (at least four) points in the wire spark chambers for a track to be considered viable. And, finally, there had to be at least one "road," that is, a football-shaped area formed by two tracks of opposite charge and a minimum amount of energy deposited perpendicular to the beam line. PASS 2 then took the filtered tapes produced by PASS 1 and filtered them once more: this time using spark information to determine the "hits" in space, then grabbing these space points to make tracks, and finally sorting the tracks into particle types.[187]

By examining photographs of the sparks, the Mark I physicists determined that sparks tended to develop perpendicular to the cylindrical surfaces, no matter what the actual inclination of the track. This "hypothesis" (as it was referred to) was then built into the computer program. Given a hit in the innermost chamber (chamber 1), the software would search within a specified area of chamber 2 for another hit; if the routine located a hit in chamber 2, then the search would continue in chamber 3 and, mutatis mutandis, in chamber 4. By convention, at most one chamber could be missing a point for a putative track to

185. Larsen to SP-1, SP-2 Experimenters, "Data Analysis," 8 November 1973, MBP.
186. Hollebeek, "Inclusive" (1975), 21.
187. Hollebeek, "Inclusive" (1975), 22–49.

be certified as "real." And only real tracks would be used by the computer as it attempted to draw a helix through the space points in question. This "fitting" routine determined the least squares fit to the helix and, when done, rechecked the fit with the crossing points through each of the cylindrical spark chambers and discarded any point falling outside of three standard deviations. As one collaborator wrote: "The purpose of the point deletion is to eliminate crossings which, though they fall within the large tolerances of the recognition programs, do not properly belong to the rest of the track segment."[188]

Even after the tracks were certified and statistically idealized to a helix there was more filtering. For example, there was a rule: no two tracks were allowed to share a point in space unless (1) they were oppositely charged and (2) the shared point was in the chamber closest to the beam. The reasoning here was that a photon from the annihilation might convert into an electron-positron pair in the material before the first chamber. But aside from this possibility, the collaborators calculated the probability of two independent tracks crossing within a 0.04 cubic centimeter volume to be less than one part in 10 billion.[189] The sorting and filtering process continued: one routine probabilistically assigned vertices; another sorted the particles into hadrons, electrons, muons, and "junk."

Analyzed tracks (certified, idealized, and identified) became the "raw data" of the experiment, to be modified, resorted, and hand-scanned in whatever way either of the laboratories judged best. "From the PASS-2 level on, each laboratory can go its own way and the final cross check will be whether or not the physics results are the same."[190]

But what are these "raw data"? In what sense "raw?" "Data analysis" is a relative concept, relative to the extent to which the data have already been analyzed. Just where the boundary is drawn between data and analysis must be decided: in this case after PASS 2. On the business end (the output) of PASS 2, physicists from LBL could ply their trade. It was fine, according to the Data Analysis Steering Committee, for the Berkeley group to hand-scan the tracks, but it was agreed that the PASS 2 tapes would not be modified—the steering committee mandated that a new PASS 3 tape would henceforth electronically summarize the categorization of events based on the hand scan. But this whole issue would never have arisen had not the possibility been seriously entertained of using the visual "hand scans" to rewrite the PASS 2 tape. You do not put up a "No Trespassing" sign unless you suspect trespassers.

Hand scans themselves had their own routines, modeled on the bubble chamber procedures that had long been established at LBL (see figure 6.40). Kadyk, Abrams, and Goldhaber put together just the kind of scan protocols

188. Hollebeek, "Inclusive" (1975), 35.

189. Hollebeek, "Inclusive" (1975), 38.

190. Richter to Distribution, SP1,2, "Meeting of the Data Analysis Steering Committee of 4 December," 5 December 1973, 3, vol. 2, GGP.

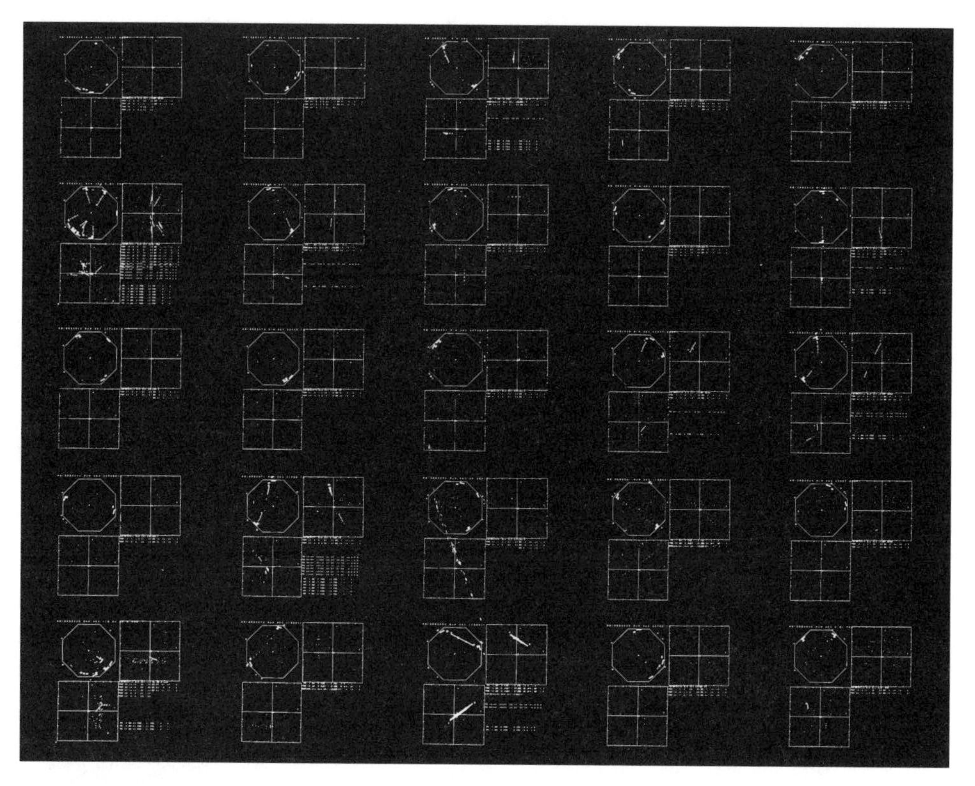

Figure 6.40 Hand scans of electronic images (1973). The hybrid practices joining the image and logic traditions become apparent in the uses to which this figure—and thousands like it—were put. Though these images, taken from the CRT computer output, are the product of a purely electronic device, Goldhaber and his LBL colleagues used them as the basis for scanning procedures modeled on the long history of bubble chamber practices. It was one more step in the joint image-logic epistemic project of creating a controllable image. Source: Microfiche detail, GGP.

that we saw in the bubble chamber manuals, described in chapter 5. One difference between 1954 and 1974 was that pictures would not, in 1974, be photographs from the bubble chamber, but rather microfiche printed with computer-reconstructed tracks (the output of PASS 2). In addition to the tracks, the microfiche would include information about timing, the number of nanoseconds (billionths of a second) from the origin of the annihilation to the counter. If the computer judged the time of flight to be incompatible with a "real" event, an "F" (for false) would be imprinted on the microfiche by the appropriate counter. Similarly, the shower counters, labeled on the microfiche 1 through 24, would be stamped with a number indicating the amount of energy deposited in each. The scanner's job was to use the time-of-flight, energy, and track pattern to sort the events into categories: Bhabhas, muon pairs, cosmic rays, hadrons, and "junk." Take the first, all-important category: Bhabhas. These "simple" electron-positron scattering events (named after Homi Bhabha, an Indian physicist) gave the collaboration a calibrating fix on the total number of annihilations occurring in the collider. Characteristics that the scanners had to bring to bear on the microfiche pictures included: (1) the timing will be about 5 or 6 nanoseconds; the pulse should be 40 units of energy in the shower counter and, unless the particle slithered through a crack between two shower counters, usually much higher than that; (2) the two outgoing particles should have a difference in azimuthal angle of zero; (3) the sum of the dip angles (measured relative to the plane defined by $z = 0$, where z is the beam axis) should be near zero; (4) by conservation of momentum, the event will look collinear—that is, in all three views the two tracks should appear to make up a single line; (5) fiducial volume (coordinates of the tracks) should be close to zero in x and y since the origin should be at the center of the interaction region; and (6) the two track-quality indicators XSQ (chi-squared) and DF (degrees of freedom) should be zero since no vertex can be assigned to a collinear event (a line with no kink).[191]

Scanners were to sort muon pairs more or less by the same criteria as the Bhabhas, but this time they would look for a lower number in the shower counters and a muon registration in the muon detector (circle or square) just beyond the registered hit on the shower counter. The real action, however, according to the scanning guide, was elsewhere: "*Hadrons.* This is what the experiment is all about." To the particle physicist, hadrons were defined by their susceptibility to the strong nuclear force; for the scanners, a hadron would be "any particle except an electron, muon, and gamma."[192] Clearly, these two definitions were not equivalent in principle (e.g., neutrinos were not hadrons by the physicists' definition but were other than "an electron, muon, and gamma"). As in bubble chamber or emulsion scanning, the classification was made operationally: the scanner only "needed" to know about particles that left tracks, and neutrinos left no footprints.

Scanners were taught to look for a timing of about 5 or 6 nanoseconds, though they were instructed to be alert for slower particles. There would typically be a low shower counter pulse, less than 30 units of energy deposited, but the counter could cause a cascade, precipitating a much higher pulse. "Obviously, the data is not 'black and white,'" the guide cautioned, "so questions will be the best way to get the feel of what is happening."[193] It is precisely this "feel"

191. Sieh, Kadyk, Abrams, and Goldhaber, "Beginning SPEAR Scanning," TN-194, 27 November 1973, vol. 2, GGP.

192. Sieh, Kadyk, Abrams, and Goldhaber, "Beginning SPEAR Scanning," TN-194, 27 November 1973, vol. 2, on 4, GGP.

193. Sieh, Kadyk, Abrams, and Goldhaber, "Beginning SPEAR Scanning," TN-194, 27 November 1973, vol. 2, GGP.

that lay behind the demand for hand scanning in the first place, a sense that a human being could pick out events from the thousands recorded that raised specific or systematic questions about the automatic sorting routines. But would it make sense to call the human classification "interpretation" and the computer-based operation merely data "provision"? I think not.

Look at the category "junk." "This category," Kadyk, Abrams, and Goldhaber wrote, "requires special emphasis and discussion, since there are several different possibilities, and a misjudgment could easily lead to a good event being missed." Here it is explicitly, though everywhere implicitly: judgment. The hand scanning, by both physicists and scanners, was designed to reintroduce the human faculty of assessment as a check on the facility of algorithmic procedure. One species of junk was, phenomenologically, the activation of a large number of triggers and counters—more than, say, 20 triggers plus shower counters. This would be type A junk. Type B junk was the polar opposite, a mere one track visible, counting as a "track" any smooth set of points even where the computer had not actually drawn a tracklike line. Now type C junk was trickier—events with a time spread of more than 7 nanoseconds. Such an event might be junk, then again it might be hadrons, the very object of the search. Finally, type D junk consisted of background hadron events in which only one beam was involved, since these events occurred outside the fiducial interaction region. Keep them tallied, the scanners were told: the computer would filter them later based on their coordinates and would use the information to sift out unwanted events. "Please keep all events when in doubt, and mark those on which you would like to ask the advice of a physicist."[194] Judgment enters at every stage, whether explicitly, in open debates about the status of a particular event, or implicitly, in the programs, counters, and analysis programs that separated wheat from chaff. Interestingly enough, the statutory convergence of LBL and SLAC subgroups through PASS 2 stopped on completion of that particular data run. Indeed, after the establishment of the statutorily designated "raw data," the two groups would go their separate ways, and then only cross-check their final "physics results" as the steering committee members put it.[195] This internal quasi independence was a nontrivial part of the group's ability to make a persuasive argument, as had already been made clear a few weeks before—"It was pointed out by Goldhaber that if these [analysis] programs were not identical but did give the same physics results, it would greatly increase the confidence of the group that our results were correct."[196] On 5 December that agreement became apparent: "Everyone

194. Sieh, Kadyk, Abrams, and Goldhaber, "Beginning SPEAR Scanning," TN-194, 27 November 1973, vol. 2, GGP.

195. Richter to Distribution, SP1,2, "Meeting of the Data Analysis Steering Committee of 4 December," 5 December 1973, vol. 2, GGP.

196. Richter to Distribution, SP1,2 Experimenters, "Meeting of the Data Analysis Steering Committee of 21 November," 26 November 1973, vol. 2, GGP.

was extremely gratified to know that the total cross sections as derived at the [SLAC and LBL] laboratories agree to within about 15%, well within the 30% errors that will be assigned to these cross sections in preliminary presentations to the physicists at the two labs."[197]

Bit by bit, the two cultures came together. While at first it was news that the SLAC and LBL groups were getting similar results, the months of software design eventually persuaded both sides that computer programs could bridge the gap. To Richter, the hand scans were temporary means to verify that the programs were assigning events to the correct categories and that the tracking programs were including the proper points. To Abrams, the goal was to "tur[n] the logic tradition on its ear" by producing programs that would "match the pattern recognition capabilities of the eye-brain." And in the end, both sides did see the synthesis as successful: the SLAC team saw the Mark I as a full-bore electronic device with sensitivity in nearly every direction; and the LBL group saw the device as the realization of an electronic bubble chamber.[198]

With the twice-calculated cross sections in hand, the LBL and SLAC experimenters found common ground, but it was still not entirely clear what these numbers meant to the theorists. I put it this way rather than "what these numbers meant theoretically" because theory had already entered, virtually every step of the way. But whether the experimental results could be aligned with *specific models* of the elementary particles remained to be seen. Could, for example, the data be found compatible with the parton model? With vector dominance? These were theorists' theories, not experimenters' theories, so to speak.

To establish a trading zone with the SLAC theorists, the Mark I collaboration scheduled a meeting for Monday 12 November 1973, in SLAC's "Green Room." "Subject," the memo read: "What Does Inclusive Include? Performers: Various Theorists."[199] This was contact within the laboratory, although outside the experimental group. Typical of internal communication of this sort was a 1972 memo from B. J. (James D. Bjorken) and Helen Quinn to Richter and Gerry Fischer, in which the theorists took various theoretical ideas (such as Weinberg's Z^0, Georgi and Glashow's $J = 0$ particle, and Ne'eman's "fifth force") and calculated the likelihood and signature of neutral resonant states that SPEAR might detect in Mark I.[200]

A different and sociologically deeper divide had to be crossed with the decision to take results outside the participating laboratories. Indeed, one of the

197. Richter to Distribution, SP1,2, "Meeting of the Data Analysis Steering Committee of 4 December," 5 December 1973, vol. 2, GGP.

198. Richter to the author, 11 November 1995; Kadyk, Abrams, and Goldhaber to the author, 4 December 1995.

199. Richter to SP-1, SP-2 Experimenters, "Special Seminar," 7 November 1973, vol. 2, GGP.

200. Bjorken and Quinn to Richter and Fischer, "Search for Neutral Resonant States at SPEAR," 14 June 1972, MBP.

defining elements of any collaboration is the boundary crossed when results become "public." I put the word "public" in scare marks because in a world of large collaborations, replete with collaboration meetings, informal talks, e-mail, conference proceedings, faxes, and published "physics" letters, it is problematic to decide just where the private-public divide lies. The decision about when to publish is inseparably coupled to standards of demonstration, standards that have to be set. In early August 1973, Goldhaber had made some rough estimations of the cross sections and wondered whether it was worth making a statement at an upcoming physics conference. Larsen wrote Richter on 3 August 1973, horrified at the idea that any disclosure might be forthcoming: "My position is (and I'd like to have it read verbatim): Until we are ready to quote cross sections at the 10% level, we say nothing to anyone, anytime, anyplace. . . . I think any premature statements are likely to be wrong; they would compromise the ultimate potential of the experiment and, as for the argument that many are awaiting the results, I don't think any theoretician worthy of the title is awaiting any more factor-of-two or is-consistent-with statements."[201]

According to minutes of a 10 August 1973 collaboration meeting, a "spirited" discussion about disclosure then followed. "A strong case was made that no public statement of any kind be made until we can confidently quote [cross section] total to 10%. Anything less convincing is to be publicly 'denied' to exist. An attempt was made to lower the standard to 20%, but this was quashed."[202] Such discussions were not new. Already in the days of the bubble chamber, "spirited discussions" reigned, as participants struggled over the significance of a bump in a mass plot or teams divided over whether a bump was more like a two-humped bactrian or a simple, dromedarian curve.

The first and most important physics problem for the Mark I collaboration was to march SPEAR through its energy range from 2.4 to 5.0 GeV at 200 MeV intervals and to measure the likelihood that an electron-positron interaction would occur at each energy. More specifically, the team wanted to see how the ratio (R) of two crucial quantities would vary with energy. The numerator of R is the rate of production of hadrons (strongly interacting particles such as the proton or the pion) in e^+e^- annihilation; the denominator is the rate of production of muons in e^+e^- annihilation. The relatively recent parton model (which represented hadrons as composed of essentially noninteracting point constituents) militated for an R that would be constant with energy; other, older theories suggested declining values of R as a function of energy.

The first half of 1974 was occupied with understanding the new results from this series of SPEAR runs. Here and there physicists in the collaboration

201. [Larsen] to [Richter], "Miscellaneous Issues," 3 August 1973, in SPEAR FY 74, no. 1 "Miscellaneous memos," 1 July 1973 through 31 December 1973, Richter, Group C leader [91014 box 4], SLAC.

202. Lynch to SP-1 Physicists, "Minutes of Meeting of 10 August [1973]," vol. 2, GGP.

pointed to oddities in the data; some quickly dissolved, others persisted. John Kadyk at Berkeley hand-scanned the data in January 1974 and found an anomalous 30% excess of hadron events (defined as events that appeared to have three or more prongs) at 3.2 GeV. Preoccupied as the physicists were with other matters, the excess receded into the background. One more pressing concern was that not enough of the energy was showing up in the shower counters, and the collaboration was unsure whether the counter was registering the neutral pions as expected or whether energy was slipping by in unexpected ways. Were neutrinos exiting the detector, produced in some new form of weak decays? Yet another level of uncertainty was reached when the results were compared to data generated elsewhere: Frascati and CEA both seemed to be producing large numbers of hadrons at higher energies. In June 1974, Marty Breidenbach from SLAC reopened the case and gathered more data on R at 3.1, 3.2, and 3.3 GeV.[203]

Nothing showed up, and the reason is revealing. As part of an effort to understand the production and energy of gamma-ray photons, a "converter" (essentially a thin steel cylindrical can) was inserted over the interaction region of the beam pipe in which electrons hit positrons.[204] When a photon emerged from the collision it would hit the side of the "can" and convert into an electron-positron pair that could then be detected in the Mark I. To compensate for this increase in electron-positron pair production, the analysis program was modified to take pairs of oppositely charged particles and classify them as electron-positron pairs (so-called ECODE 3 events) so they would not be confused with bona fide hadrons (ECODE 5 events).[205] The gamma measurement done, the can was removed, *but its compensatory software remained in place.* From that moment on, any pair of hadrons produced in the annihilation was reclassified by the computer as an electron-positron pair. Silently, in the heart of the computer, ECODE 5 events became ECODE 3 events. It was impossible to observe the production of any hadron pair at 3.1 GeV. Had any evidence for a resonance arisen, the computer would have instantly killed it.[206] (Shortly afterward, the conversion computer code was corrected.)

Roy Schwitters, one of the SLAC physicists, took on the task of drafting the "total cross section paper" and, working with Willy Chinowsky, Gary Feldman, and Harvey Lynch, prepared a version that was ready for collaboration critique on 5 July 1974. While numbers were still needed on detector efficiencies and other experimental errors, the conclusion was clear: "The total cross section is a rather smooth function of C.M. [center of mass] energy over the range

203. Goldhaber, in Ting, Goldhaber, and Richter, "Discovery," *Adventures Exp. Phys. α* (1976): 114–49, on 132.

204. On the converter, see Larsen to Distribution, "Inner Package for May–June 74 Cycle," 29 January 1974, vol. 2, GGP.

205. A list of the ECODES can be found in Hollebeek, "Inclusive" (1975), 47.

206. Feldman, interview by the author, 21 January 1994; Schwitters, interview by the author, 17 June 1991; and Goldhaber, in Ting, Goldhaber, and Richter, "Discovery," *Adventures Exp. Phys. α* (1976): 114–49, on 135.

covered in this experiment. There is no strong evidence for resonance peaks or production thresholds. In strong contradiction to the predictions of asymptotic scale invariance, the cross section is essentially constant for C.M. energies between 3 GeV and 5 GeV. As yet, no generally satisfactory theoretical framework encompassing these results has emerged."[207] No "satisfactory theoretical framework" was putting it rather mildly. When the results were presented at the London Conference later in July, theorist John Ellis declared that "there is no consensus among theoreticians working on electron-positron annihilation, not even about such basic questions as . . . whether or not to use parton ideas." The quantity *R*, Ellis concluded, could be variously deduced to be anywhere from 0.36 to 70,383. Go figure. Richter then presented the SLAC-LBL results, which were "in violent disagreement" with the quark model. As the experimenter Michael Riordon nicely put it, "[E]xperimenters thought Theory was pretty confused, and theorists—at least those who trafficked in gauge field theory—felt Experiment was the one befuddled."[208]

Beginning in July 1974 SPEAR was shut down for three months to ready it for higher energies. During that time, Schwitters and Scott Whitaker reanalyzed the data and found that measurements at 1.6 GeV/beam (3.2 GeV total) indicated a 30% higher cross section and that the measurement at 1.55 GeV/beam (3.1 GeV total) also looked larger than expected. More peculiar yet, in mid-October 1974, Schwitters returned to the logbooks to study an odd set of eight runs the team had conducted a few months before. On 29 June 1974, the notebook indicated that the energy scan had been boosted to 3.1 GeV. "Beam dumped, ready for 1.55 [GeV/beam = 3.1 GeV in the center of mass]." "YOU MAY NOTICE," the shift inscribed in self-congratulation, "WE ARE EXACTLY ON TIME." "Unfortunate fill, tortuous beam configuration." Run 1381 failed miserably: "Aargh! Some Power Supply has developed a massive leak. Dump, end run." Run 1383 went fairly normally. Run 1384: "Luminosity is rather disappointing. The boys are studying the situation." So it progressed. Run 1387 "is miserable. Fill is beginning with .7 × 10*30 and we expected [approximately] 1.2 × 10*30. Furthermore lifetime only about 1 hour. Furthermore it takes a full hour to fill. This STINKS. Our lead is gone and we sink into the morass." A few hours later, Run 1389: "CONDITIONS STABLE. . . . KEEP ON TRUCKIN!"[209] And at 14:10 on 30 June, the 1.55 run came to an end: "STOP Run 1389 with 171 [mu]'s logged in at 1.55; on to 1.60. With time precious and running uncertain we defer a background run at each energy. We will try to make a complete scan at 1.6 [GeV/beam]."[210]

207. Schwitters to SPEAR Friends, "Draft Paper," 5 July 1974, MBP.

208. Riordan, *Hunting* (1987), 259, 261, quoting Ellis, "$e^+e^- \to$ Hadrons", and Richter, "Plenary Report," both in *Proceedings 1974 PEP Summer Study* (1974), IV-30, 37, 41, 54.

209. SLAC logbook, 29 June 1974.

210. SLAC logbook, 30 June 1974.

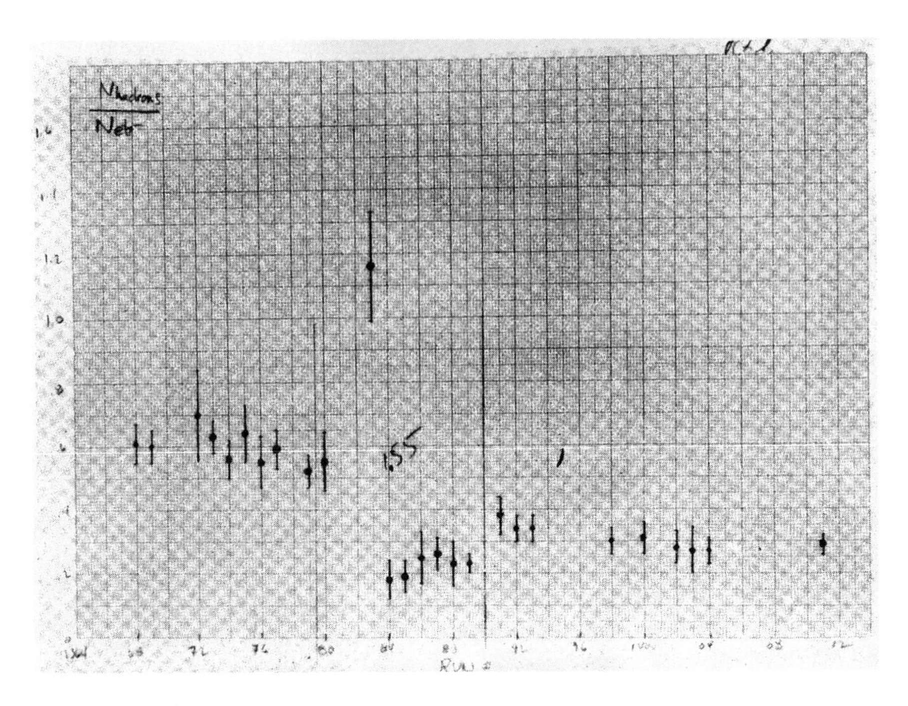

Figure 6.41 Schwitter, reanalysis of runs showing excess of hadrons (1974). In this hand-drawn plot, Schwitter summed up the data recorded over the previous months. Something had appeared to be amiss, and here the oddity stood out: an extreme deviation at 1.55 GeV/beam (equivalent to a total center-of-mass energy of 3.1 GeV). It was this peculiarity, along with an apparent excess of kaons, that prompted a systematic rerun of the experiment. Source: Ting, Goldhaber, and Richter, "Discovery," *Adventures Exp. Phys.* α (1976): 131–49, on 134.

Remarkably and inexplicably two of the eight usable runs (1380 and 1383) at 1.55 GeV/beam yielded an excessive number of hadrons: Run 1383 yielded a full five times the harvest of hadrons the team expected (see figure 6.41).[211] On 22 October, Schwitters asked Goldhaber and Abrams to look again at these odd runs, which they did, frame by frame.[212] Partly because of the excess hadrons, and partly to clean up the manifestly inconsistent set of measurements at 1.55 GeV/beam, Goldhaber began on Monday 4 November 1974 to lobby hard for revisiting the anomalous energy region around 3.1 GeV. If there was something going on near 1.55 GeV/beam, the varying results from earlier runs suggested they might need all the resolution in energy they could get—Abrams and Gold-

211. Goldhaber, in Ting, Goldhaber, and Richter, "Discovery," *Adventures Exp. Phys.* α (1976): 114–49, on 135.

212. Goldhaber, Kadyk, and Abrams to the author, 4 December 1995; Riordan, *Hunting* (1987), 272.

haber pressed Richter to determine just how precisely the accelerator energies were known. There were, however, countercurrents. Other physicists, including the senior Stanford experimenter Bob Hofstadter, wanted to push ahead in the unexplored new regions of energy in order to test the validity of quantum electrodynamics. Since Richter was both running the Mark I and in charge of the accelerator, he could not easily set aside Hofstadter's program. Finally, persuaded by a seeming excess of strange particles that Goldhaber and Whitaker had noticed on the microfiches, Richter relented and the machine again began taking data near 3.1 GeV.[213]

In the predawn hours of 9 November 1974, new data began to come in. Harvey Lynch, watching the events as they crossed the CRT, put pen to logbook: "The man was tired, for he had diligently worked the area for weeks. He stooped low over the pan at the creek and saw two small glittering yellow lumps. 'Eureka!' he cried, and stood up to examine the pan's contents more carefully. Others rushed to see, and in the confusion the pan and its content fell into the creek. Were those lumps gold or pyrite? He began to sift through the silt once again." Sifting silt meant first of all establishing that the machine was functioning correctly, functioning for known regions of energy (such as 2.4 GeV) as it had previously.

> I. Our first priority is to be sure that the detector is alive, and that the normal analysis program functions properly on the triplex [the three-CPU computer used to take data]. We should log ≥ 100 μ pair equivalents at a beam energy of 2.4 GeV. These results should reproduce our previous result of [the total cross section] $\sigma \sim$ 18nb, with a detection efficiency of 0.63.
>
> II. Having completed the "checkout" phase we can begin the energy scan from 1.5 GeV to 1.6 GeV in 0.01 GeV steps.[214]

This scan would take roughly three hours per energy step, and Lynch expected roughly 10 events as a baseline, 40 or so if there was a "good 'bump.'" For each run, the team would plot the ratio of ECODE 5 events (hadrons) divided by the number of Bhabhas, indicating the luminosity (to normalize the hadron production to the total number of electron-positron annihilations). Events would pop up on the CRT screens, one by one.

At 08:00 on 9 November 1974, the log records: "Filled ring with $E_0 =$ 1.5[5?] GeV. Watch data on one-event display. The table below gives the result of the hand scan. A total of 22 hadron candidates were found along with 37 Bhabha events. . . . If all this makes sense this means a *trigger* cross section of ~72 nb! Now if the signal just 'disappears' when we run at 1.50 we will be happy."[215] After a hand tally, James (Ewan) Paterson and Roy Schwitters

213. Richter, interview by the author, 18 March 1991; and Goldhaber, interview by the author, 14 August 1991.

214. SLAC logbook, 9 November 1974.

215. SLAC logbook, 9 November 1974.

formalized their astonishment, "We the undersigned certify . . . J[ohn] S[cott] W[hitaker]'s above count to be a valid representation of the data." Back to checking cosmic rays; the shower counter efficiency still "looks o.k." Then, at 15:40, "Frustration! We have had no colliding beams since the 1.56 GeV beams went away. Is there no hope?" More hand scans, back to normal levels of hadron production away from the mystery region of 1.56 GeV. At 1:47 on the morning of 10 November, the crew completed the 1.56 GeV run and moved on to 1.57 GeV. At 10:05 that morning, as the crew tried to zero in even closer to the peak of this new resonance, they set up Run 1460 at 1.555 GeV, a sequence that ended 56 minutes later. Schwitters scribbled: "This past fill has been incredible. While running 1.55 we saw essentially the baseline value of τ_T (the total cross section). During the middle of the fill, we bumped the energy to 1.555 and the events starting [*sic*] pouring in. The visual scan had 61 hadrons in 87 Bhabhas. This is a remarkable resonance indeed!"[216]

On 13 November, the collaboration submitted its article "Discovery of a Narrow Resonance in Annihilation" to *Physical Review Letters.* The word "charm" never appears; perhaps the closest to it is the widely cast remark at the end of the article: "It is difficult to understand how, without involving new quantum numbers or selection rules, a resonance in this state which decays to hadrons could be so narrow" (see figure 6.42).[217]

Charm did, of course, appear elsewhere in the physics community. As Andrew Pickering has nicely shown, there were several alternative explanations that prospered within the theory community, all vying for pride of place in explaining the new peak. For the charm theorists, the psi was a bound state of two quarks, a charmed quark and anti–charmed quark. What allowed theoretical calculations to proceed was that, as David Politzer and Tom Applequist radically contended, the force that tied quarks together got weak at small distances. This doctrine—asymptotic freedom—was a fundamental part of what the gauge theorists meant when they referred to the psi as a bound quark-antiquark pair.

Not so the experimenters.

6.10.3 No Raw Data

Physicists, first theorists and now experimenters as well, have come to speak of the events beginning the second week of November 1974 as the November Revolution. Half tongue-in-cheek, the allusion to the storming of the Winter Palace nearly six decades earlier evokes a break, a radical discontinuity in physics that accompanied the discovery of the *J*/psi. In part, this language of a rupture fit the theorists' image of their subculture at a time of uncertainty. It is reflected in talk of an "*R* crisis" and in the heated ways in which theory was discussed. Theorist

216. SLAC logbook, 10 November 1974.

217. Augustin et al., "Narrow Resonance," *Phys. Rev. Lett.* 33 (1974): 1406–8.

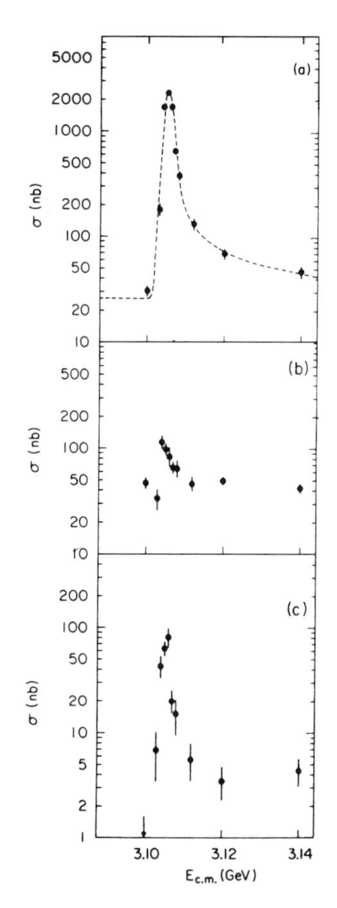

Figure 6.42 Spike at 3.1 GeV (1974). These results, received by *Physical Review Letters* on 13 November 1974, contained the astonishing peak at 3.1 GeV. (*a*) Cross section in nanobarns of the production of multihadron states from e^+e^- collisions as a function of energy (in GeV). (*b*) Cross section to e^+e^- in the final state. (*c*) Cross section for production of pairs of pions, kaons, or e^+e^- (since the collaboration's muon detector was not working, no distinction could be made among these different results). The extraordinary height of the hadron curve (*a*) (rising by a factor of 100) and the exceptional narrowness was like nothing known in physics. "It is difficult to understand how, without involving new quantum numbers or selection rules, a resonance in this state which decays to hadrons could be so narrow." Indeed, discussion then turned to possible quantum numbers including—but not restricted to—charm. Source: Augustin et al., "Narrow Resonance," *Phys. Rev. Lett.* 33 (1974): 1406–8, on 1407.

John Ellis, for example, referred in the summer of 1974 to the shocking blow inflicted by the experimenters at SPEAR and elsewhere to theorists, who had been "almost unanimous" in their expectation that R would be constant. The nonconstancy of R, in particular its rise with energy, was nothing short of a "theoretical debacle,"[218] "R crisis," or "mini-R crisis."

Such crisis talk, made popular in the 1970s by the oil industry's "energy crisis," permeated the world of high energy theory—but only theory. And indeed, there seems to be little reason to doubt that the theorists experienced their inability to account for experimental results as a crisis. What is problematic is the extension of such a break to the instrument makers and experimenters. To

218. Ellis, "$e^+e^- \to$ Hadrons," in *Proceedings 1974 PEP Summer Study* (1974), IV-20–35, on 20, cited in Pickering, *Quarks* (1984), 256.

my knowledge there is not a single reference, among experimenters, to a crisis that refers to the reliability or efficacy of the methods of *their* subculture in this period. By the early days of November 1974, the accelerator at SPEAR had been functioning for a year and a half. The Mark I detector had a yearlong track record, including a series of conference papers on the total cross section that had flown in the face of prevailing theoretical work in particle theory. Did judgment and theory enter into the certification of the detector and its functioning? Of course: from the first pretrigger filter to the hand scans on the CRT. Does this mean that the prediction of charm and the associated development of the "new physics" of quantum chromodynamics played a crucial or even a significant role in the spike of figure 6.42? Absolutely not.

In his insightful book *Constructing Quarks,* Andrew Pickering argues along very different lines. His is an intriguing interpretation of the events at Mark I, one worth pursuing not just for its own sake but because it nicely illustrates a line of reasoning grounded firmly in the highly influential antipositivist philosophy of science that began in the early 1960s with the work of Kuhn, among others. "Monitoring the beam energy very precisely, the experimenters obtained curves like that [of cross section versus energy in figure 6.42]. These they regarded as manifestations of a genuine phenomenon, and accordingly put aside their worries about detector performance. They had not proved that the detector was working perfectly; they assumed this because it produced credible evidence."[219] What characterized this credible evidence? It became credible just because it fit the preexisting theoretical framework: "[T]heory is the means of conceptualisation of natural phenomena, and provides the framework in which empirical facts are stabilised."[220] Adopting Kuhn's argument that "each theory would appear tenable in its own phenomenal domain, but false or irrelevant outside it,"[221] Pickering then turns to his example of the new physics and old physics: "Each phenomenological world was, then, part of a self-contained, self-referential package of theoretical and experimental practice. To attempt to choose between old- and new-physics theories on the basis of a common set of phenomena was impossible: the theories were integral parts of different worlds, and they were incommensurable."[222] It is the point of this chapter that the practices of colliding beam physics at Mark I were *not* contained entirely within the "self-referential package" of either the old physics or the new.

Pickering's argument, as I understand it, has two parts, each containing a general thesis and a specific claim about the psi:

1a. Generally, an experimenter's "theoretical construct" precedes and picks out a set of experimental results by means of "tuning"—experi-

219. Pickering, *Quarks* (1984), 274.
220. Pickering, *Quarks* (1984), 407.
221. Pickering, *Quarks* (1984), 409.
222. Pickering, *Quarks* (1984), 411.

menters adjust their techniques "according to their success in displaying phenomena of interest."[223]

1b. Specifically, the assumption of charm led to a theoretical prediction about the value of R, and this result preceded and determined the experimental result (the narrow resonance at 3.1 GeV depicted in figure 6.42): "[T]he discovery of the psi can be seen as an instance of the 'tuning' of experimental techniques . . . to credible phenomena."[224]

2a. Generally, experimenters do not establish the proper functioning of their instruments prior to certifying an effect; credibility of the instrument is an outcome of finding the effect that is expected.

2b. Specifically, the SLAC-LBL collaboration at Mark I had not "proved" that its detector was functioning properly ("working perfectly"); the SLAC-LBL physicists *assumed* this *because* they found the resonance at 3.1 GeV that they expected.[225]

Why are theses 1a,b and 2a,b important? First, they serve to undermine a view (which Pickering calls the "scientist's account") that takes theory to be the inductive limit of a series of prior experimental observations. Second, and more important, if experiments are tuned to theory and theory suffers a discontinuity, then the picture of physics is rent all the way down through experiment itself, and the "world" splits into two incommensurable parts. (As Pickering put it: "The old and new physics constituted, in Kuhn's sense, distinct and disjoint worlds.")[226] Philosophically, this amounts to a case for incommensurability; historically, it is predicated on a block periodization; and sociologically, it is the statement that the cultures of experimentation and theory are sufficiently interwoven to function as a single paradigm.

The evidence for claim 1b is that in mid-October 1974, Goldhaber took the excess number of kaons as evidence for an interesting new phenomenon, possibly charm, at 3.1 GeV. Goldhaber's view, however, was by no means universally accepted; I can find not a single endorsement of the charm hypothesis in any published or unpublished document by the Mark I collaboration prior to 11 November. Stranger still, even in the days after 11 November such traces do not appear in the experimental literature. Of course it is possible that emphasizing historical documents rather than interviews systematically omits material that would be more evident from oral histories. Here again, however, in a systematic set of interviews with the main actors in the Mark I collaboration, I see little support for the "tuning" of data to confirm a preestablished conviction that charm existed. Schwitters put it this way:

> Charm for us never really meant [narrow] resonances. We didn't really think about it—it wasn't in our psyche as I say ["We pronounced it that weekend . . .

223. Pickering, *Quarks* (1984), 14.
224. Pickering, *Quarks* (1984), 273.
225. Pickering, *Quarks* (1984), 274.
226. Pickering, *Quarks* (1984), 409.

> sort of joking with people . . . the phi-c's {feces}."] I didn't even really understand [charm] until that Gaillard, [Lee,] and Rosner paper about the strangeness-changing currents. I really wasn't that up on those things. I think that's a fair description of how most of us viewed it. It was much more of an experimental issue of getting in there and understanding things experimentally and then being somewhat intrigued by the possibility that there was something really strange with a constant total cross section. Of course it sounds like such nonsense now—it's embarrassing to mention it! But one does get into that kind of mindset."[227]

Breidenbach recalled that his and his colleagues' main concern in going back to the 3.1 GeV region was to clean up their data—to find out whether something was wrong with their procedures. "There was never the feeling that this [narrow resonance] was just what the charm theorists ordered. It was nothing like that."[228] Whitaker: "We didn't know this had anything to do with charm until after we'd established the weird behavior and the theorists said o.k. you've discovered charmonium."[229] To Chinowsky, Schwitter's plot of mid-October seemed downright impossible: "Roy showed me this plot, he said 'Oh, look what I did,' . . . we looked at each other and I said, 'You know this can't be happening,' and he said 'I know it can't [be] happening.' I said, 'What should we do?' He said, 'We'll run [again at that energy].'"[230]

Downplaying the importance of the charm hypothesis in November 1974, Lynch added this: "Charm was not part of my thinking, not the slightest. We didn't pay a lot of attention to that—we were an iconoclastic bunch. As you know the total cross section we had found [earlier] was so completely different from what everyone was saying that we didn't pay attention to the theorists." As for the hours following 08:00 on 9 November 1974, Lynch had this to say: "It was a qualitative difference. You could simply watch them on the CRT. They were very fast and very clean: click, click, click, every few seconds. Pief was in the control room pacing back and forth saying, 'Oh my God, Oh my God'; he was pounding his head in his hands: 'I hope we're not making a mistake,' 'I hope we are not making a mistake.' Finally, he said, 'there's no mistake.' You looked and there was simply no question that it was real. *We had no idea what it was, but it was real.* That's how we all felt."[231] Even after the enormous resonance emerged, Lynch remained dubious about charm. Similarly, Rees, Fryberger, and other physicists on the experiment had essentially no stake in the charm hypothesis before the narrow resonance was established on the days after 9 November 1974.

227. Schwitters, interview by the author, 17 June 1991, 20.
228. Breidenbach, interview by the author, 14 February 1994.
229. Whitaker, interview by the author, 24 February 1994.
230. Chinowsky, interview by the author, 14 August 1991.
231. Lynch, interview by the author, 8 March 1994.

The underlying difficulty in the Pickering analysis is the assumed opposition between an algorithmic move to "prove" that the detector was "working perfectly" and a theory-laden "tuning" of the experiment on the basis of a "theoretical construct." This sort of dichotomy, characteristic of both positivist and antipositivist philosophy of science, has obscured a richer, subtler spectrum of registers in which experimental argumentation proceeds.

Antipositivist discussions about experiment-theory relations typically contrast "observation" and "theory" and insist upon theory's penetration into the heart of experiment. One can sympathize with the antipositivist's impatience with his imagined foe asserting the long-discarded doctrine that there are observations that are "independent" of all theory. In the extreme limit lies the notion that observations lead to theory through an inexorable inductive sequence. Here, the more recent sociology of science (Collins, Pinch, Pickering, and others) has drawn frequently on the venerable work of Kuhn, Hanson, Hesse, Feyerabend, and others, who blasted the naïveté of believing that science was cumulative data codified into a whole.[232] In doing so the antipositivists stress the psychological and epistemic shaping that occurs as data are melded by "conceptual frameworks" into self-reinforcing wholes.

I do not see how even to start an analysis of an experiment like the Mark I with such an abstracted notion of "observation." Like its cousins "raw data" and "theory," the dichotomy between theory and observation obscures the central phenomenon that I find most interesting. This way of speaking suggests a set of untrammeled data and a more or less coherent theoretical agenda. On the old positivist view, data were hard and theory ephemeral. On the view espoused by the antipositivists, data are "tunable" and theory powerful and controlling. Neither view seems adequate.

Data are always already interpreted. But "interpreted" does not mean shaped by a governing high-level theory. The notion that quantum field theories led to asymptotic freedom, and that asymptotic freedom coupled with charm dictated the tuning of the Mark I, flies in the face of the continuity of experimental and instrumental practices. Data are already interpreted in the PASS 3 visual scan tape; they are already interpreted in the event identification portions of the PASS 2 filter program. So we go back further in search of the raw, but the data still are not in that Edenic state in the early sections of PASS 2, where the program minimizes the least squares reconstruction of the trajectory helix. So perhaps the raw data were untouched one stage earlier. But here we find the space points adjusted, discarded, confirmed in the softly grinding machinery of the PASS 1 filter. Back up again. We find the beam trigger counters excluding cosmic ray events, calculating flux, checking the timing of hits in the outer detector relative to the electron-positron annihilation. There are no original, pure,

232. See Galison, "Contexts" (1995).

and unblemished data. Instead, there are judgments, some embodied in the hard-wired machinery, some delicately encoded into the software. Some judgments enter by way of scanners and physicists peering over CRTs at event displays, others at histograms, and others at microfiche reproductions. Interpretation and judgments go all the way down. But to call these moves of interpretation "theories" is to grossly misread the nature of experimental culture.

What follows? One might conclude from the saturation of interpretation that there are no bedrock data, that the physics conclusions drawn from the experiment rest on a bottomless and shifting sea of sand. This seems to me backward. It is rather the picture of rigid observation and arbitrary interpretation that lies behind a whole class of claims for relativism. What we see here is more like the intercalated picture of partial continuities and discontinuities.[233] Take the spark chamber. It draws its certification, if you will, from a long sequence of device types running back at least as far as the Geiger-Müller counter. In Conversi and Gozzini's radar flash tubes, in Fukui and Miyamoto's discharge chamber, in the German spark counter group, in the prototype devices of Wenzel and Cronin, in the massive neutrino detector of Schwartz, Steinberger, and Lederman, the long-term legitimacy of the chamber found its ground. What licenses the PASS 1 filter program to interpolate a "missing" point in a track when one spark chamber fails to fire? It is underwritten by a mesh of processes working simultaneously at a multitude of timescales. Behind it is the aggregated force of decades of work within the logic tradition. On the longest timescale lies the logic physicists' commitment to statistical argumentation: PASS 1 simply does automatically what Bothe and Kolhörster had done self-consciously and dramatically in 1932. In both cases, reasoning follows probabilistic lines, from the unlikelihood of independent hits to the conclusion that a single particle had passed through each chamber.

It might be useful to think of the establishment of the logic tradition in general as but one in a series of certifications: Validation of the spark chamber certifies the genus, of the magnetostrictive wire chamber the species, and of the particular wire chambers sitting in the Mark I the individual. Each level has its own set of certifying processes; their starting points may be intercalated in complex ways. Delay line memory has its roots in techniques of wartime radar, and magnetostriction issues from efforts to extend radar memory to computer memory. Perez-Mendez's use of magnetostrictive memory preserved the manufacture but altered the function of stored pulses; no longer employed to store information, the wires now used the delay time to locate the original spark. So to ask the question: Why accept the results of a magnetostrictive wire spark chamber? The answer reaches back in all these directions at once. What sustains commit-

233. On the periodization question in physics, see Galison, "History," *Sci. Cont.* 2 (1988): 197–212; and Galison, "Contexts" (1995).

ment to the psi measurement is a multiplication of commitments across the different subsystems: pipe counters, scintillators, muon chambers. Then there are the myriad of cross-talk and coordinative moves that bind the detector into a whole. Does this intercalated history of certification render the whole system immune to any possible skeptical challenge? Of course not. As I have stressed, vesting reliability in an instrument does not always come from any single mode of reasoning. There are techniques supported by detailed theoretical justifications, and there are techniques adopted without any theoretical understanding whatsoever. There are pieces of machine lifted bodily from one machine and planted lock, stock, and barrel into another, and there are technologies that emerged for this particular machine in this particular application. And importantly, as we have seen briefly here and will discuss more extensively in chapter 7, experimental groups began in the 1970s to take the antithetical image and logic traditions and to compact them, with the computer, into a new and hybrid whole.

6.11 Instruments and Experimental Work

The renaissance in instrumentation between 1952 and 1964 matches any in the history of the physical sciences. During this period physicists invented or improved high-pressure cloud chambers, hydrogen, hydrocarbon, and xenon bubble chambers, discharge chambers, spark chambers, wire chambers, flash chambers, acoustic chambers, current-division chambers, video chambers, and new kinds of scintillators. There were new timers, pulse-height analyzers, new computers, and a myriad of combinations and variations of these devices.

To understand this renaissance in the material culture of physics, our discussion had to proceed on several levels. On one plane there was the story of physics questions and physics answers. When Glaser set about building a new chamber, the problem of strange particles and the origin of cosmic rays had his part of the physics world puzzled. The old generation of detectors was neither dense enough to produce sufficient numbers of interactions nor well enough resolved in space and time to yield the detail needed to advance the field. For Fukui and Miyamoto cosmic ray air showers were similarly inaccessible—the cloud chamber simply could not handle phenomena that took place over several square meters. Cork, Cronin, and Wenzel also had a physics agenda—studying the decays and interactions of the strange particles first seen in cloud chambers and emulsions and then, more recently, produced in accelerators. They asked, for example, about the polarization properties of the lambda zero.

New instruments thus were partly motivated by various programmatic goals within physics. I say *programmatic* deliberately, to distinguish the physics ambitions of the experimental tradition from the suggestion that they were "simply" trying to resolve specific theoretical issues. Of course there are examples of theorists' "ordering" a particle—famously in the case of the omega minus. In general, though, the physics questions that precipitated instrumental innovation were not nearly as specific as the properties of a single particle. More commonly, experimenters responded to classes of phenomena: weak decays, nuclear lifetimes, and extensive air showers were typical of the broad-based physics concerns that drove the reformulation of equipment.

Closely coupled to such strategic, rather than tactical, physical goals were even broader issues about the direction that experimental physics should take. One such cut separated what I have called the "image" from the "logic" tradition. We saw how Glaser drew on skills developed within the visual detector tradition as he built prototype instruments culminating in the bubble chamber. Once laboratories around the world established bubble chamber programs, they rapidly drew together groups of physicists experienced in other image-type experiments, either cloud chambers or nuclear emulsions. For it was these scientists particularly who could most easily transfer their skills to the new technology. Similarly, it was from the resources of the logic tradition that physicists like Charpak, F. Bella, Gozzini, D. C. Allkofer, Henning, Bagge, Carlo Franzinetti, Keuffel, Cork, Cronin, Richter, Schwitters, and others primarily drew.

As argued in chapter 4, there are several different levels of continuity hiding under the widely used term "technological transfer," and it is useful to distinguish among them. One strand of historical continuity embraces the laboratory skills of the experimenters. As Neher described in his rough textbook sketch, there were a myriad of craft operations, many of them poorly understood, which were needed to build a working counter. Keuffel knew those skills and applied them in his important contribution to the spark counter. It is obviously not always possible to work by first principles in experimentation; often one has to imitate what worked if the device and experiment are to function properly. Charpak put it eloquently in 1962: "You will see one physicist state that after filling his chamber he waits several days for the gas to get dirty—he says it works better then. Some other physicist will scrupulously purify his gas with a calcium oven to remove oxygen from it. Yet another adds a 'quenching gas' while another adds nothing at all. In the end people do what's wise—they do whatever those are doing whose chambers work well—without trying to understand why!"[234] Thus, part of the experimenter's commitment to one or another of the traditions is practical—in one domain the physicist knows how to make devices work reliably, even if theory comes up short. After a talk in 1964, Karl Strauch was asked about the difference in spark registration time between spark chambers and the new streamer chambers. He began with a theoretical response: To make the spark visible in the spark chamber, massive amounts of energy must be pumped into the ionic path; a later particle traversing the gap has less energy available to it, since the discharge is nearly complete. In the streamer

234. Charpak, "Chambre à Etincelles," *Indus. Atom.* 6 (1962): 63–71, on 68.

mode, however, this is not the case, since there is no sudden discharge. Later particles might still be visible. But then, his experimenter hat back on, Strauch continued: "Let me emphasize that the theory is a good guide but *only the experiment should really be believed,* and I'm unaware of any experiments that have studied this point carefully."[235] In another place, after citing some detailed calculations, Strauch added, "[I]t is clear that the avalanche, streamer and spark formation mechanisms are not well understood, certainly not in detail, and we must not keep from trying out things 'that cannot work.' After all, why did it take so long to try a track spark chamber?"[236] The message was clear: Instrument making should not wait on theory, even the midlevel applied theory that purports to explain ionization, cascades, and recombination.

Sometimes the technological transfer can involve not just skills but a material connection. Not only did the personnel from the NBS/AEC facility bring their knowledge to Alvarez's laboratory, the government handed over the hydrogen compressor itself. On a much smaller scale, Gozzini's work was made possible by the pulse-forming bricolage he constructed from radar equipment bought from the American army. Frequently, as in Cronin's use of Keuffel's Princeton spark plates, one experiment is actually cannibalized for another. The spark counter is thus contained in the spark chamber in this most literal sense.

There is a third category of technological transfer that encompasses a carryover of whole *structural systems* from one device to the next. In the 1930s, physicists in the logic tradition designed electronic circuits to record selectively the firings of their Geiger-Müller counters. Two decades later, Conversi and Gozzini adapted such logic circuits to activate their flash tubes. And spark and wire chamber experimenters later took analogous though more sophisticated circuits to sensitize their detectors at precisely the time at which particles were to arrive from the accelerator. Such systemic similarities in the generation and analysis of data in the logic tradition are paralleled on the image side. There, techniques and skills that developed around the processing of pictures were handed back and forth between cloud chambers, bubble chambers, and nuclear emulsions.

Along with the transfer of skills and hardware that links devices within a tradition, there are broader exchanges involving the nature of experimental work that condition the development of modern instrumentation. Competition was stiff in the mid-1950s between accelerator and cosmic ray physics. As Glaser put it, he wanted the bubble chamber to "save" cosmic ray experimentation as a viable enterprise within physics. Alvarez, by contrast, believed that the huge investment the government had put into the Bevatron and other big machines would be justified by producing a new and significantly better kind of physics

235. Strauch, "Innovations," *IEEE Trans. Nucl. Sci.*, NS-12, no. 4 (1965): 1–12, on 7, emphasis added.
236. Strauch, "Innovations," *IEEE Trans. Nucl. Sci.*, NS-12, no. 4 (1965): 1–12, on 7.

using the bubble chamber. Thus, Glaser's attempt to preserve cosmic ray physics transmuted into Alvarez's successful large-scale projects that effectively killed cosmic ray research in high energy experimentation. Similarly, on the electronic side, many experimenters, including the Japanese, the Italian, and the German groups, intended their small-scale new detectors to be used to study cosmic rays. The Americans, by contrast, from the very start found the new flash chamber interesting *because* it was adaptable to accelerators and brought it into their large laboratories, reinforcing—across the image-logic divide—the trend begun by Alvarez.

The debate over the appropriate site for the emerging field of high energy physics was only the most outward sign of a deeper division. For not far beneath the rift between accelerator and cosmic ray research was a very human concern with the nature of laboratory work. Glaser's selection of cosmic ray physics was not just a choice of field, it was simultaneously a choice of work style, one removed from "the factory environment of the big machines." When Alvarez chose to develop the engineering-scientific team, he did so intentionally—his was a decision to further the integration of the two cultures he had witnessed during the war. Hierarchically structured, specialized "line work," he hoped, would prove as successful in advancing high energy physics as it had in expediting military enterprises.

The effect of Alvarez's transformation of the experimental workplace was not lost on his contemporaries. Even before the invention of the spark chamber, physicists defended the electronic counter tradition by citing its flexibility. Just a few years after the construction of the first spark chambers, experimental workers around the world tried to build into the device their vision of the ideal workplace. This is the sense of Macleod's remarks at the CERN conference to the effect that experimental physicists were going to build devices that would "give . . . back the pleasure of being an experimenter and not only an operator," or that physicists needed instruments that would give the experimenters back "control" over the apparatus, and so provide them with answers to their questions "in the experimental halls."

Throughout the fifties, experimenters often expressed their optimism that new techniques would transform the life of their laboratories. There was a sense that almost any technological system could be turned to use in the search for the very small: hydrogen liquefiers, radar, computer cores, television, stored-program computers. Not only would these marvels serve as useful prototypes, physicists could draw on the industrial and military experience of the war and Cold War years for models of the successful integration of engineering and physics. At the same time there were real worries about how education and collaboration would proceed.

Such issues even brought into the question the nature of the experiment and the experimenter's place in the laboratory, as Kowarski stressed in the Karlsruhe debate over the nature of experimentation that we reviewed extensively in the last chapter:

> High-energy experiments give us glimpses of a world very different from our everyday universe with which we can maintain a brief contact only at a very high cost. In this respect, high energy physics is directly similar to space research or to deep oceanography. It would hardly pay to send down a bathyscaphe to hunt for a few rare specimens of a definite species, or to send up a space ship to gather a special kind of cosmic dust; collecting excursions of a more eclectic kind seem to make more sense in these two categories of exploration.[237]

Like these sciences of acquisition, physicists would dredge now and look later. Such a future was, for Kowarski, hardly utopian. Rather it was a "vision of a Brave New World . . . offered not as a happy anticipation, but as a logical consequence of the ever-growing cost of accelerating and detecting apparatus in high energy physics."[238] Work would proceed at fewer accelerators running, like factories, around the clock.

Inundated by data, the image and logic traditions responded differently, though paradoxically the outcome was similar. On the image side, the European drive toward full automation continued. As the HPD devices scanned the data, and physicists were pulled further and further from the "raw" photographs, it became less and less likely for the unexpected to crop up in the routinized course of events. Within the logic tradition, a similar filtration occurred even earlier in the data acquisition process: events that did not fit the electronic trigger selection never even left their pulsed footprints on the magnetic tape. The loss of even the possibility of such serendipitous discovery disturbed many physicists, among them several in Kowarski's Karlsruhe audience. Ekberg bemoaned the danger of automation foreclosing the unexpected;[239] Havens feared the loss of "exploratory" experimentation. The "late" discovery of parity nonconservation, Havens insisted, was a warning that, by working only according to plan, experimenters could have donned blinders. "Looking for the unexpected and hidden," Havens sadly mused, "is not characteristic for these times."[240] When Kowarski riposted that one could still "roll out the film to look again if you don't throw it out," he spoke of a possibility of course foreclosed to the imageless device.[241]

237. Kowarski, "General Survey" (1964), 35.

238. Kowarski, "General Survey" (1964), 36.

239. Ekberg, discussion of Kowarski, "General Survey" (1964), in Beckurts, Gläser, and Krüger, *Automatic Acquisition* (1964), 39.

240. Havens, discussion of Kowarski, "General Survey," in Beckurts, Gläser, and Krüger, *Automatic Acquisition* (1964), 40.

241. Kowarski, "General Survey" (1964), 39–40.

Kowarski's vision of archival resurrection was small consolation to many physicists. For though Kowarski seemed resigned to the new laboratory order, his reluctant fascination with the charms of high technology rattled his audience. In his bathyscaphic model lay the dispossession of physicists from their experimental workplace and the elimination of the idea of an experiment that had drawn many of them to their work.

Thus, in addition to finding missing links between elements of experimental or theoretical systems there, we must treat the structure of the laboratory workplace as a dynamic variable shaping the construction of instruments. Just as the big bubble chambers embodied one form of work, so the early on-line spark chambers instantiated another. As physicists began to emphasize, the design of computer programs for the acquisition and processing of data became the same thing as the design of the experiment.

Doubts arose too about the style of technology being employed. To some observers the systems of the 1950s seemed awkwardly juxtaposed, a high-technology bricolage: Oscilloscopes scanned photographs. In one popular device, the acoustic chamber, transducers converted sound waves into electrical pulses, then the computer sorted the signals. In another detector, a television camera swept its eye across the spark chamber, digitizing the location of discharges. Bubble chamber physicists invented a myriad of new instruments to photograph and reduce the tracks of boiling hydrogen with the help of hired hands. A bit of unease with the kludgelike quality of some of the instruments emerged in M. G. N. Hine's concluding words to the CERN conference of 1964:

> I do not put my money on the acoustic chamber. I think it is too like the bubble chamber in having a 19th century steam-age feel about it. Especially I think that, apart from this esthetic disadvantage, its inability to cope with multiple tracks is going to be a practical reason why it will not be able to keep its present predominance. I would also vote against any photographic system or any vidicon system mainly because I think the difficulty of actually optically looking into a spark chamber is going to become more and more of a nuisance as time goes on.[242]

Hine preferred Charpak's current-division method, which used the current produced by the spark to fix the spark's location. Solid state detectors, too, might satisfy Hine. Here we catch a glimpse of an esthetics of detector building that only partly overlaps with the pragmatic exigencies of experimentation. It is a subject discussed much less than the "beauty" of group-theoretic symmetries in quantum field theory—perhaps only because experimenters have written fewer rhapsodies about their discipline. By the time the Mark I opened for business in 1973, the transition away from visual spark chambers was thoroughly underway.

242. Hine, "Concluding Remarks" (1964), 374–75.

The renaissance of instrumentation in the postwar years is thus a story of many beginnings. It is a story that can be told in terms of theory—"explained," in a certain sense, by the necessity of clearing up the dynamics of strong resonances and weak interactions. It is equally a story about clusters of experimental skills propagated from one instrument to the next: from the cloud chamber to the bubble chamber, from the Geiger-Müller counter to the hodoscope chamber or the spark counter, from the spark counter to the spark chamber, and from the spark chamber to the many species of imageless electronic detectors—magnetostrictive, acoustic, core, video. In a related sense, it is part of a historical epistemology—an account of competing ways of constructing demonstrations about the natural world. So too is the instrument renaissance linked inextricably to the history of wartime and postwar technology. Without the fast timing devices, cryogenics, pulse-forming circuitry, and computer systems developed for the military and industry, neither the bubble chamber nor the more advanced spark chambers would have amounted in the 1950s to more than curious prototypes for accelerator physics. Finally, the story of new instruments is one of revolution in the laboratory. For experimenters, instruments could hardly be more important: they constitute the stage for all of laboratory life. Through the design and use of instruments experimenters continuously struggled to set the terms by which they would interact with machines, collaborators, technicians, scanners, computers, and finally, nature.

The quarter-century following the acoustic chamber did not end the debate over the character of experimental work. Among many other examples, it will be instructive in the next chapter to investigate the ways that the heterogeneous detectors of the 1970s and 1980s concatenated the various electronic mechanisms to form a community of communities, a cluster of clusters of physicists and apparatus. In addition to the problems of division of labor raised by emulsions, the challenges posed by the new role created for engineers by the bubble chamber, and the unpredictability brought by the new menagerie of electronic instruments, integration took on a new importance. With a myriad of laboratories participating in the design, operation, and interpretation of a hybrid device, the twin problems of efficient governance and unified argumentation became central.

If attempts to contain the ever-growing scale of physical experimentation made little headway, the effort to draw together the image and logic traditions fared well, as the psi/tau discoveries indicated. The 1960s, 1970s, and 1980s brought a profound technological shift in the domain of microelectronics, one that welded together diverse tools of high energy physics. For with the introduction of integrated circuitry that could time, sort, and amplify tens of thousands of independent channels, there emerged a system that could synthesize images electronically. By the 1980s, for the first time, the old dream came true—golden events like the one that certified the omega minus could be produced by electronics, fished by a computational net out of the ocean of microphysical debris.

But as the scale jumped higher, from the size of the Mark I to the next generation larger, a new dynamic emerged, and with it the meaning of experimentation changed once again. In the Time Projection Chamber, which we explore next, just holding the machine and collaboration together emerges as an extraordinarily difficult and revealing task.

7 Time Projection Chambers: An Image Falling through Space

7.1 The Floating Image

My view is this: as an analytic term, "big physics" is about as helpful to the historian of science as "big building" would be to a historian of architecture. What Ernest Lawrence did for accelerators, or what Luis Alvarez did for the bubble chamber, is surely big physics. In the case of Alvarez, we have seen how engineers from across the country gathered to assemble the massive chamber, how experimenters from several continents imported data from the Bevatron to run their physics programs, and how huge teams of bubble chamber scanners strained their eyes in the darkness of graveyard-shift projector rooms. To many commentators, and indeed to many participants at the time, these developments, and the accompanying projects to build ever larger devices, signaled the beginning of "large-scale experimentation" the way Robert Boyle (or Galileo or Francis Bacon) marked the beginning of experimentation *simpliciter* three centuries earlier. In this chapter, I want to press the architectural metaphor further and suggest, somewhat speculatively, that we view the transition of big physics from the phase of spark chamber, counter array, emulsion stack, or bubble chamber to the vastly more complex phase of multicomponent electronic detectors as emblematic of a shift from the pure to the hybrid. The reverberations of this shift are felt in the definition of the experimenter, in the structure of demonstration, and in the self-conscious reevaluation of what it means to be an author in modern physics. Collectively, these changes might usefully be designated as a shift from the "modern" to the "postmodern" laboratory. Already hinted at in our discussion of the Mark I in chapter 6, the full impact of the social, technical, and epistemic hybrid instrument becomes apparent in the building of the massive colliding beam detectors of the 1970s and 1980s.

In particle physics, this union of the image and logic traditions was sealed with an extraordinary detector installed at SLAC, the Time Projection Chamber (TPC). It took each one of the 20-odd tracks in an electron-positron annihilation and manipulated it electronically: identifying the particle that made each track, coding it by its momentum and energy, determining its precise trajectory, and feeding that information back instantaneously to the operators while allowing decisions about triggering and storage to be made in real time. Simultaneously, the TPC borrowed heavily from the representational image tradition of the cloud chamber, the emulsion, and the bubble chamber. As the charged particles emerged from the annihilation event, they left ionized tracks in a gas. But this time, the idea was not to intervene physically at the site of each bit of track. The sensitive volume of the chamber would not be used to create a region of phase instability that would be catalyzed by the ions. There would be no superheated hydrogen that would burst out boiling around visible tracks; no unstable chemical medium from which silver iodide could be precipitated in film developer; and no supersaturated gas in which to precipitate visible droplets. Nor would the sensitive volume be electrified to the threshold of sparking, or even of proportional cascades of other ions. What then was the sensitive volume to contain? Nothing but gas.

The TPC was emptied of everything but argon and methane. When a particle passed through the sensitive volume, an electric field would transport the tracks toward an array of electronics at the end caps, where the arrival location of each ionic track fragment would reveal the radial and azimuthal position of that portion of the original particle trajectory. The arrival time—this was the real innovation—would determine the longitudinal position of that track fragment with remarkable precision. Key to the concept of the TPC was the presence of a magnetic field parallel (rather than perpendicular) to the electric field. This field would drive the drifting electrons around in tight helices as they approached the end cap, keeping a track from diffusing into an electronic haze. By holding the web of tracks together as it was shifted to the measuring apparatus, the TPC had, in effect, become an electronic realization of Glaser's imaginary three-dimensional "christmas tree" of solid tracks that could be measured with calipers. In the TPC, the entire subatomic image fell through space, intact, to a waiting electronic eye. In its austere transport of an entire image, in all its complexity, the TPC embodied the fusion of image and logic.

Complex as it was, the TPC did not travel alone. Linked fundamentally to a hybrid detector, the TPC was accompanied by a phalanx of other devices: muon detectors, calorimeters, and drift chambers, among others. Why think of the construction of a bubble chamber as "modern" and the TPC detector as "postmodern"? First, the architect of high modernism reached for a unity and simplicity of form: hard-edged and geometrical, these were forms that built, paradigmatically, on the simplicity of the cube and solid rectangle.[1] Like these pure forms, the bubble chamber was essentially a pure or monolithic technology. These vats of superheated liquid hydrogen seemed to repel hybridization, refusing to mesh with electronic counters or other devices. From Donald Glaser's first hesitant and unsuccessful attempts to graft counters to his tiny chamber, through the massive bubble chamber complexes of the late 1960s, physicists and engineers could produce only awkward agglomerates. Some groups inserted a heavy-liquid chamber inside a hydrogen chamber; occasionally, physicists mounted external muon chambers. But in spite of determined efforts to combine the bubble chamber with electronic instruments, the results never amounted to a triggerable chamber. Conversely, as we saw in chapter 6, adherents of the electronic tradition wanted nothing more than to expunge every aspect of the filmic laboratory, from the "armies" of scanners, to the reliance on individual golden events, to the much derided passivity of the image experimenter with respect to his photographic instrument. In the place of the photographic, the logicians wanted to restore the integrity of a thoroughly electronic instrument. In both architecture and physics, utilitarian justifications of these pure forms merged inextricably with a visceral distaste for the impure, eclectic construction that mounted a photographic camera on a spark chamber. To the modernist sensibility, the eclectic was imperfection on the path to the simple.

Second, the modernist construction of *authorship* survived into the epoch of big bubble chambers with 10 or 20 individuals in a way that it did not when one had 10 or 20 *institutions* participating in a collaboration. Alvarez borrowed freely and deeply both from centralized corporate management techniques and from the military chain of command—as his use of a "chief of staff," a "production manager," and operations research methods testify. During the mid-1950s to mid-1960s, groups grew in size, but they remained centralized around individual physicists. "Alvarez group" was more than simply a name, for all authority ultimately did lead back to him. Decisions about when and what to publish, engineering judgments, planning schemes, and even styles of data reduction converged on a single person. And while it is certainly true that other bubble chamber collaborations were less authoritarian than Alvarez's, their magnetic lines of force led back to a single pole: the Thorndike group, the (Wilson) Powell group, and the Glaser group (to cite but a few).

Viewed from outside, each group carried its own mode of work and could stand for a relatively well defined style of action. Technical experts visiting the Alvarez group from abroad could, as we saw, clearly recognize his stamp of

1. There are, of course, many sources on the transition between modern and postmodern architecture. Of particular help are Klotz, *Postmodern Architecture* (1988); and Jencks, *Language of Post-Modern Architecture* (1984).

"good-enough engineering," and the team he assembled—programmers, cryogenic experts, and analysis supervisors—formed an entity that was relatively stable both in the personnel composition of the group and in the division of labor. The characteristic "modernist" problem was one of competition with other, similarly constituted groups, and the expansion (through export of film, analysis techniques, and hardware) of their style of work. Groups competed under the name of the leader, and the dynamic within the experimental community replicated, in a sense, the dynamic of individuals competing in an earlier time.

Emblematic of the transition in authorship that I am pointing to is a change from the "X" group (named after and directed by X) to collaborations in which an elected spokesperson "stood" for his or her colleagues for a term of office. It is not just that as Alvarez retires his group's name becomes Group A; it is that in a myriad of ways, participants and outsiders stop thinking of collaborations as individuals writ large. No one ever questioned that the 72-inch bubble chamber belonged to the Alvarez group (though legal ownership was vested in the AEC), but the original operating team of the TPC, which we will be examining here (or of the detectors at Fermilab, CERN, and the ill-starred Superconducting Supercollider detectors), did not necessarily have presumptive control forever over the detectors they built. Perhaps here it is more useful to think of shifts in other cultural fields (in literary and art theory, e.g.) than in architecture. Foucault and Barthes famously have pointed to the fall of the autonomous author in the years after World War II. Together their work underlines how an analysis of the use of literary productions fixes the notion of "author." Instead of starting with the supposition that there is one and only one notion of an author (for example, as an isolated, purely creative individual), Foucault proposed that we begin with the works. By focusing on the literary productions themselves, he suggested, we can see how in a specific culture works are valued (or disvalued), how they are circulated (or held secret), how they are attributed (or kept anonymous), and we can then use the status of the texts to show us the function of the category "author" at a given historical moment. Among artists of the second half of the twentieth century, there was a simultaneous questioning of the meaning and use of the unexamined category "author." As Caroline Jones argues, the industrialization and mechanization of the studio engendered a profound ambivalence about whether artists should seize authorship or abandon it. She quotes Andy Warhol as saying that with the use of certain techniques, "I don't have to work on my objects at all. One of my assistants or any one else, for that matter, can reproduce the design as well as I could."[2]

No such theatrical loss of authorship marked the heyday of the Berkeley bubble chambers in the 1960s, despite their industrialization; that came, as we

2. On the death of the author in literature, see, e.g., Barthes, *S/Z* (1974); and Foucault, "What Is an Author?" (1984). On Warhol and Stella and their challenges to the concept of the author, see Jones, "Warhol's Factory," *Sci. Con.* 4 (Spring 1991): 101–31; and Jones, *Machine in the Studio* (1997).

will see, in the 1980s. During the Alvarez period, authorship was reinforced by a well-articulated system that simultaneously allowed significant latitude to technicians and engineers and enforced a strong hierarchical structure. So while technicians like Pete Schwemin and his colleagues could more or less autonomously produce their own small bubble chambers, there never was any doubt that the authority ran from Alvarez through his chief of staff down through the managers and lead engineers to the technicians. This division of labor was reflected, especially in the larger chambers, in a well-defined "front" (in the sense of weather or war) between physics and engineering. Certainly, by the time of the 72-inch chamber, the ranks of the physicists and technicians were supplemented (indeed, to a certain extent supplanted) by engineers. The whole liquefaction and hydrogen transport system was, as we saw, designed, fabricated, and run by mechanical engineers. Sometimes the separation between engineering and physics was spatial: safety testing and work on the storage, transport, and thermodynamic characteristics of hydrogen and its containers was to a great extent executed in Boulder, Colorado, by cryogenic engineers at the NBS. Even where both engineers and physicists worked at the same site, activities fell more or less squarely into broad blocks of what the participants recognized as one dominion or the other. When physicists such as Alvarez or Lynn Stevenson intervened, they were self-consciously acting as engineers, wearing, as they said, their "engineers' hats."

Similar fronts developed along the radical disjunctions between film production and film analysis, and between data collection and interpretation. From the beginning, Alvarez (and most bubble chamber leaders) assembled two teams: one to work on the reduction of data and the other to produce the data. The former was a technology of projectors, computers, and analysis programs; the second was a technology of cryogenic and photographic equipment. Construction, maintenance, and operation of the bubble chamber itself had practically nothing to do with the construction, maintenance, and operation of the analysis "factories" that sprung up not only at Berkeley but around the world. Analysis was separated in space, time, and (through the export of film) even institutionally from the production site of data.

One result of this separation, already apparent from the days of nuclear emulsions, was that the "experimenter" became increasingly identified with analysis, rather than production. In no small measure, as we saw in the last chapter, it was because of this isolation that the spark chamber boosters shunned the bubblers and their prefabricated data. Though they succeeded for a while, the spark chamber could remain a small, craft-built machine only a very few years before it too expanded to industrial scale, restoring the division between production and consumption.

To these fronts between physicists and engineers, production and consumption, must be added a third: the boundary between experiment and theory.

While the experimenters demonstrated the existence of a wealth of mesons and baryons, the theorists took these entities and manipulated them in a variety of ways: group-theoretical classifications, Regge pole theory, and so on. However complex the sociology of big physics (bubble chamber style), there appeared to be segmented lines of production. At one end, the experimenters manufactured exposed bubble chamber film. Next came computer punchcards replete with scanning and measurement information. At the final stop on the production line, experimenters (and occasionally computers) took scanning information and transformed it into analyzed data in the form of histograms and Dalitz plots, culminating either disappointingly in the garbage bin of artifacts or triumphantly in the line entries of the *Particle Data Book*. These entries were the bottom line of experimentation: the cross sections, resonant states, decays, lifetimes, spins, masses, parities, and strangeness of the elementary particles and their interactions. Then these subatomic entities were ready to be embodied in the myriad of theoretical schema, ranging from rules of thumb to grand overarching interpretations.

Now contrast the bubble and spark chambers with the massive hybrids that followed. In the great majority of detectors before 1960, the technologies were relatively homogeneous: emulsions, spark chambers, bubble chambers, or counter arrays. Where the different technologies met, as in counter-triggered cloud chambers, the counter electronics were simple compared with work in the mainstream of the logic tradition. The great experts at cloud chamber work through the 1950s, C. T. R. Wilson, Carl Anderson, R. W. Thompson, Louis Leprince-Ringuet, and others, had used electronic triggers, of course, but their central concerns revolved around the manipulation of the chamber itself, exploiting craft tricks (such as temperature gradients and gas mixtures) to increase resolution and eliminate distortion. By the 1970s, the situation had changed. Almost no detector functioned by itself, and the typical experiment was founded on a complex superposition of a host of subassemblies, as was evident from experiments like the Mark I at SLAC. Moreover, with this *material* hybridization came a *social* and *epistemic* hybridization that changed the way both arguments and subgroups were assembled. Each of anywhere from five to fifty institutions would build separate component parts—a scintillator detector, a calorimeter, a muon detector, a multiwire proportional chamber—and each would become an authority on its particular fraction of the whole. One institution would design and build a muon detector, for example. Graduate students and postdocs of that institution would supervise the maintenance of the muon detector during runs, and their Ph.D. hopefuls would write theses grounded in an analysis of muon events. Ultimately, it would fall to the builders and maintainers of the muon chambers to vouch for the reliability of their device and so contribute to the security of any discovery claim involving that detector. In short, any global argument about a new phenomenon would have to be built, like the apparatus itself, out of the integrated fragments of these subsystems.

It is to an exploration of this hybridization that I now turn. For it is my argument that the shift from monolith to hybrid embodies a deep alteration in the experimenter's relation with machines, engineers, and theorists. As we will see in detail, the *macro*coordination between physicists and engineers became a *micro*coordinative problem in a way that significantly differed from the "use" of engineers by bubble chamber physicists. Each component detector required both engineers and physicists. But more than that, each aspect of the subassemblies *within* each component demanded such integration. Aside from a few subcontracted items, engineering and physics were thoroughly intercalated.

In a certain sense, the sacrifices in autonomy that came with the hierarchical structure of the 72-inch chamber, or even of the Mark I, paled before the hybrid big science heralded by the image-producing detectors that were launched in the mid-1980s. In the 72-inch chamber, physicists had to confront their need for engineers—scale, safety, and cryogenics all made it almost impossible to proceed without them. But the sense of physicists as clearly dominating the engineers, as ultimately if not proximately in control of construction, could still be maintained. In the TPC any illusions on that score were shattered. Coordination between physicists and engineers at every level became essential; painfully, over some 10 years, the physicists had to learn how to speak in the language of engineering—how to use critical path analyses, how to freeze designs, how to make systematic cost reviews, how to track hundreds of subtasks, and how to integrate a raft of different subassemblies built by groups with cultures as different as physics and engineering.

Perhaps it was because the physicists thought they *knew* "big science" that they took this last blow to their autonomy hard. Big (modernist) apparatus as a centrally structured hierarchy became big (postmodernist) apparatus as locally synchronized, globally dispersed work. In the former, Alvarez-style science, the physicists could order components from the engineers, and occasionally they could and did contribute to the design of specific seals, light reflectors, or bellows. But the coordination between physicists and engineers in the TPC and its descendant devices was of an entirely different order: it was a dialogue that had to occur at every stage from preliminary conceptual design to final blueprints, from prototype manufacture to full-scale production. Big science, in other words, is not simply what happens when a lot of money is spent on a single device. In this chapter we will explore the difference between what I will call macrointegrated big science and microintegrated big science. To understand the turning point from one to the other, both in instrument making and in the definition of experimentation, we need to turn first to the invention and then to the unprecedented expansion of the TPC.

7.2 Time into Space

David Nygren "grew up" entirely within the eclectic regime of late-1960s logic detectors. In that great instrument renaissance, practically every piece of the detector was modified, transposed, and replaced. Spark chambers and wire chambers of every description blossomed; readout systems were cut out (often literally) from the postwar technologies of computers, televisions, and radars and soldered into place in the laboratory. Nygren's thesis experiment, completed in 1968 at the University of Washington, had examined neutron-neutron interactions through an electronic array of neutron and gamma ray detectors.[3] Afterward, in postdoctoral work with Jack Steinberger at Columbia, he used spark chambers to study muon scattering. While Nygren was building detectors at Columbia, George Charpak published his papers on multiwire proportional chambers, which Nygren, along with many others, immediately adopted and began expanding and improving.

Not long after he came to LBL in 1972, Nygren became the first "Division Fellow," and he used his freedom from day-to-day responsibilities to explore novel detector techniques, including the assembly of a sufficiently large number of drift proportional wires to measure curvature, and with enough ionization measurements to determine particle type. The problems with variants of the multiwire proportional chamber were twofold: First, the position at which the particle crossed the wire could not be determined accurately. Second, the relative orientation of the electric and magnetic fields made it much harder to deduce longitudinal position from the drift time. This needs explaining.[4]

For decades, it had been standard procedure to set spark chamber plates and wire planes *perpendicular* to the beam direction. After all, the impact of the beam particles on the target would send reaction products flying down the beam line, and the standard procedure was to bend the trajectories of the reaction products to measure their momenta. At the same time, it had always made sense to set an electric field *parallel* to the beam direction: the plates or wires would set up such a field to clear the intervening spaces of charge, or in the case of drift chambers the electric field would draw ions to the nearest plate or wire. Many early chambers lost efficiency if the plates or wire planes were not perpendicular to the tracks. These "natural" assumptions meant that the electric and magnetic field were perpendicular to each other. And because the E and B fields were perpendicular, when electrons drifted in the electric field, they crossed magnetic field lines and were deflected: *since the drift trajectories were curved in complicated ways, it was extremely difficult to deduce position from drift time.*

With the electron-positron collider SPEAR, collision products were emerging from annihilation over a far wider set of angles, and for some time the name

3. Nygren, "Neutron-Neutron Scattering Length" (1968).

4. Marx and Nygren, "Time Projection Chamber," *Phys. Today* 31 (1978): 46–53, on 49.

of the game had been to "cover" as much of the sphere as possible. What would happen, Nygren wondered, if the electric field were parallel instead of perpendicular to the magnetic field? After all, since one was looking for an isotropic chamber the naturalness associated with preferring the beam direction had become unnatural. Two things followed: First, with E and B parallel, the drift time would be strictly proportional to the distance drifted. Second, and this was more subtle, Nygren remembered some experiments at the University of Washington that showed a pronounced but unexplained sharpening of sparks when a strong magnetic field had been set parallel to them. Perhaps, he speculated, the magnetic field was collimating the drifting electrons by sending the particles in a spiral and therefore suppressing diffusion.[5] To go any further, he would need a quantitative treatment of this diffusion-suppressing effect.

The place to turn was more or less evident: Sir John Townsend had written one of the most widely read texts on the motion of charge, his 1915 *Electricity in Gases.* Like its successor, *Electrons in Gases* (1947), it was a virtual bible of the logic tradition, a standard source for designers of counters and electrical detectors of every sort. *Electrons in Gases* is a supremely simple book with far-reaching results, exploiting over and again a few elementary mathematical principles and recording the results of straightforward tabletop experiments involving pressurized vessels, different gases, and electric and magnetic fields. Nothing places the TPC more squarely within the logic tradition than Nygren's choice of Townsend's *Electrons* as his principal reference for his first, crucial memorandum. Here he found a ready-to-use treatment of the antidiffusion capacity of a magnetic field.[6]

A few days after Nygren realized that diffusion would be suppressed by a magnetic field parallel to the electric one, he sat down to write out the first informal report on the proposed detector. Dated 22 February 1974, it began: "Consider . . . the experimental difficulties confronting the physicist who wishes to detect in entirety an event occurring in PEP [Positron-Electron Project]. He must operate in high backgrounds, have very good spatial resolution in order to measure momenta[,] . . . be able to reconstruct *many* tracks occurring over 4π [i.e., to detect in all directions] unambiguously, identify particle types, as well as measure the neutrons, K's and γ's."[7] The central claim of Nygren's note was that it was possible to radically suppress the diffusion constant normally encountered for a cloud of electrons $\sigma_z = (2DT)^{1/2}$, where σ_z gives the resolution (the square root of the mean squared distance) of electrons along the z-axis (i.e., the

5. Nygren, interview by R. Chandler, 3 December 1990; and Nygren, interview by the author, 28 January 1992.

6. Nygren cites two sources: Townsend, *Electrons in Gases* (1947); and Brown, *Electrical Discharges* (1966).

7. Nygren, "Proposal to Investigate the Feasibility of a Novel Concept in Particle Detection," 22 February 1974, 1, photocopied typescript, DNyP I.

drift direction), D is the diffusion constant, and T is the total drift time elapsed since the creation of the electron cloud. In particular, in the presence of a magnetic field collinear with the electric field, D would be suppressed by a factor dependent on the cyclotron frequency $\omega = eB/mc$, where B is the magnetic field, m the electron mass, and c the speed of light. Though this effect came as something of a shock to many detector builders, in other branches of physics it was standard knowledge. The derivation of these quantities was among the first topics addressed by Townsend.[8] "Ordinary" diffusion of particles, and this

8. Suppose we have n_0 electrons starting from the y-z plane, and let n be the number of electrons that arrive at a point x from the plane without suffering any collisions. Let dn be the number that collide between x and $x + dx$:

$$dn = -\theta n\, dx \tag{7.1}$$

or $n = n_0 e^{-\theta x}$. The sum of all the free paths of lengths between x and $x + dx$ is therefore

$$\int_0^\infty x\, dn,$$

which integrates by parts to

$$[xn]_0^\infty - \int_0^\infty n\, dx.$$

The first term vanishes since n is zero at infinity, and the second can be solved by substituting $-dn/\theta$ for ndx, yielding n_0/θ as the sum of a large number of free paths. Define l to be the mean free path. We then immediately have another expression for the sum of all the free paths: $n_0 l$, from which we can deduce that $l = 1/\theta$.

This expression proves useful in deriving a second quantity, the mean of the squares of a large number of free paths. The sum of all squares of paths is

$$\int_0^\infty x^2\, dn.$$

Integrating by parts once again, we have $2n_0 l^2$, or a mean free path squared of $2l^2$. With this result in hand, Townsend goes on to derive the rate of diffusion. Consider the one-dimensional case first: Electrons of density $\rho(x, y, z)$ are located between two planes A and B normal to the x-axis. The total number of electrons between the two planes is therefore q:

$$q = \iint n\, dy\, dz.$$

It is useful to consider the diffusion of electrons as if the whole population simply moved with an average drift velocity of (u,v,w) in the x-, y-, and z-directions even though these are really quite small compared to the agitation velocity of the particles as they ricochet off one another. In one dimension, we then define the diffusion constant, D, to be the constant of proportionality tying the difference in density from one region to the next to the net flow of particles: $\rho u = -D\, d\rho/dx$, or integrating out the y- and z-components, $qu = -D\, dq/dx$. The continuity equation then simply expresses the fact that the change in the number of particles between the planes $(dq/dt)dx$ is equal to the difference between the number that flow through plane A (qu) and the number that flow through plane B $(qu + [d(qu)/dx]dx)$. This difference is $d(qu)/dx$ and so $dq/dt + d(qu)/dx = 0$, or (using the expression $qu = -Ddq/dx$)

$$dq/dt = D\, d^2q/dx^2. \tag{7.2}$$

What is the rate of diffusion, i.e., the time rate of change of $\langle x^2 \rangle$? The mean value of the squared distance from the origin for a group of N electrons between x and $x + dx$ is

$$(1/N)\int x^2 q\, dx = \langle x^2 \rangle,$$

which has a time derivative of

$$(1/N)\int x^2 (dq/dt)\, dx = \frac{d\langle x^2 \rangle}{dt}.$$

includes electrons, was of course first derived by Einstein in his first, famous 1905 paper on Brownian motion,[9] and Townsend adapted Einstein's work more or less verbatim to the case of electrons, $\sigma_z = (2DT)^{1/2}$, now standing for the root mean square of the electron position.

Nygren's main idea, announced in his draft paper on the TPC, amounts to this: a magnetic field parallel to the drift direction opposes the electron's tendency to diffuse. Thus, the diffusion factor in the perpendicular direction, D_{perp}, effectively is transformed into

$$\mathrm{D} \rightarrow \frac{\mathrm{D}}{(1 + \omega^2\tau^2)}, \tag{7.5}$$

where ω is the cyclotron frequency $eB/mc = 1.76 \times 10^7$ rad/sec-gauss and τ is the mean time between electron-gas collisions. The effective D_{perp}, therefore, can be suppressed for a sufficiently large value of the magnetic field or a sufficiently large mean time between collisions; more specifically, the job is to make $\omega\tau \gg 1$.[10]

With dispersion squelched, the detector would determine the azimuthal and radial positions by arrival spot on the end cap, and depth in the chamber would be given by the amount of time (since the creation of the track) that the

---

By the continuity equation (7.2), we have

$$(D/N) \int x^2 (d^2q/dx^2)\, dx = \frac{d\langle x^2\rangle}{dt}.$$

Integrating by parts and discarding the surface term leaves

$$-2\,(D/N) \int x\, dq = \frac{d\langle x^2\rangle}{dt};$$

again integrating by parts we get

$$2\,(D/N) \int q\, dx = \frac{d\langle x^2\rangle}{dt},$$

which is simply $2D$ since $\int q\, dx = N$. Integrating with respect to time, $\langle x^2\rangle = 2DT$, and by the definition of the root mean squared:

$$\sigma_x = \sqrt{2DT}, \tag{7.3}$$

which we wanted to show.

Now suppose that a group of electrons is released at a point in space, P, and that upon colliding with a molecule, the electrons are equally probable to scatter in any direction. After a large number of collisions, the electrons are symmetrically distributed around P. The intervals between collisions are t_1, t_2, etc., corresponding to free paths divided by velocity, l_1/V, l_2/V, etc. Once the electrons have scattered through two free paths in a time $t_1 + t_2$, their mean square distance from P is $l_1^2 + l_2^2$. Using our previous result (eq. [7.3]), after a time cl/V (c is a large number) the mean square of the distance from P is $2cl^2$. So the rate of change of the mean square distance $d\langle R^2\rangle$/dt is therefore $2cl^2/cl/V$ or $2lV$. Invoking $d\langle R^2\rangle/dt = 6D$ gives us

$$D = lV/3. \tag{7.4}$$

9. Einstein, "Bewegung der Teilchen" (Movement of Small Particles), *Ann. Phys.* 17 (1905): 549–56, repr. in Einstein, *Brownian Movement* (1956), 1–18.

10. Nygren, "Proposal to Investigate the Feasibility of a Novel Concept in Particle Detection," 22 February 1974, 2, DNyP I.

electrons took to arrive at the end cap. The task then was how to get a high value of $\omega\tau$. A magnetic field of 56 kilogauss yields a cyclotron frequency of about 170 gigahertz; Nygren also assumed a value for D determined by another experimental group (0.2 m^2/sec). Using the theory of diffusion we just described, $D = Vl/3$, where V is the speed of electrons (not the drift speed, but the speed between collisions) and l is the mean free path. In a gas in which the electrons are in thermal equilibrium, V would be 1×10^7 cm/sec. Therefore,

$$\tau = l/V = 3D/V^2 = 6 \times 10^{-11} \text{ sec} \qquad (7.6)$$

or $\omega\tau = 60$. This calculation implied that dispersion would be suppressed by the extraordinarily large factor of 60. Concretely, it meant that even after drifting a meter, the dispersion in the x-direction would be a mere 14 microns.[11]

Nygren's new detector appeared attractive at LBL (and beyond) on several counts. There was the possibility of unambiguous reconstruction in space, high spatial and temporal resolution, the ability to handle high-multiplicity (many-particle) events, and the simplicity of a sensitive volume filled merely with gas. Unlike most other detectors, there was a complete absence of wire planes and other objects in the sensitive volume of the detector. Electronics at the end cap emptied the machine's center. The simultaneous drop in diffusion and the removal of ambiguities promised an enormous improvement in resolution.

Under the section heading "Program," Nygren proposed to enlist physicists to collaborate and criticize, to search the literature for information on electron-molecule interactions, to build a device to measure what he called the "entrainment factor," $(1 + \omega^2\tau^2)$, in a strong magnetic field, and finally, to "study" the problem of the end cap detector. This last question was pressing. As Nygren worried, "[T]he quality of information reaching the end caps may be very high, [but] is there any way to obtain it?" One way was to exploit wire devices, many of which had come cascading out of Charpak's laboratory: current division, induced pulses, or the much newer technology of the charged coupled device (about which much more in a moment). Much simpler than any of these was the possibility of exploiting what he called an "ideal honeycomb."[12]

The honeycomb was an array of wire needles, capped with a platinum ball to avoid the sharp point that might cause gas discharge. Around each wire would be walls (whence the term "honeycomb") held at a high negative potential (see figure 7.1). The ball-wire assembly occupied Nygren's research during the first half of 1974, when he worked mainly with two other physicists: Peter Robrish (from LBL) and Marcel Urbain (visiting from the École Polytechnique). The three were joined not long afterward by Jay Marx, a Yale associate professor

11. Nygren, "Proposal to Investigate the Feasibility of a Novel Concept in Particle Detection," 22 February 1974, 3–4, DNyP I.

12. Nygren, "Proposal to Investigate the Feasibility of a Novel Concept in Particle Detection," 22 February 1974, 6–8, DNyP I.

Figure 7.1 Honeycomb and ball-wire detector (1974). The TPC idea depended on being able to collect the liberated electrons as they drifted toward the end planes. At first, Nygren proposed that this collecting mechanism be a ball mounted on a very thin wire—depicted here is a prototype ball-wire collector, photographed under enormous magnification. Source: LBL XBB 743 1836, LBL. Courtesy of Lawrence Berkeley Laboratory, University of California.

visiting LBL, with whom Nygren had worked when Nygren had been a postdoc and Marx a graduate student.

In April 1974, LBL and SLAC submitted a joint proposal to the AEC to build the Positron-Electron Project (PEP), and the call went out for ideas on how this replacement for the colliding ring SPEAR could best be exploited. The first part of SLAC's 1974 Summer School was devoted to educating the community about PEP, and the second half to preparing experiments for the new collider.[13]

13. Strauch, "Introductory Remarks" (1974).

Nygren's presentation to the 1974 summer school rehearsed his February discussion of the entrainment factor and went on to provide some results of his exploration into the ball-wire detector. With an appropriate potential, the wires acted in the proportional mode. One problem was that the time an electron took to arrive at the ball was affected by two distinct processes: the time to drift down to the ball and, *if* the electron was not directly aligned with the ball, the time it took the particle to spiral in to it. There was thus a fundamental ambiguity about any single time measurement; it was impossible to disentangle the straight drift time and the time taken by the electron on its spiral into the platinum globe. Here the computer entered: Nygren hoped that by repeating the measurement of a track some 40 times, it could resolve the ambiguity by fitting a plausible track to a path between the various measuring points.[14]

Three other features of the device then had to be addressed. First, the incipient TPC group had to demonstrate that the magnetic diffusion suppression worked. To this end they designed a cylindrical chamber with a collimating entry tube at the top and a radioactive source on top. As the electrons entered the cylinder they were aligned along a very thin vertical column; by moving a probe along the radius below, the physicists could measure the radial dispersion (see figure 7.2). In many ways, this test device fit perfectly into an older technical tradition, within which Townsend had been a master—it was an experiment that would have found a natural home in any of several studies of the motion of electrons in gases, in 1912 as easily as in 1974. Even the protocols of the experiment, such as varying gas type, pressure, and magnetic field, were just the sort of manipulation that a Cavendish physicist in the time of Wilson would have pursued.

Dispersion was surely the enemy of track reconstruction. It was equally the bane of any attempt to extract information on the identity of the particles. In the days of cloud chamber, emulsion, or bubble chamber pictures, particle type often had to be assessed by the subtle exercise of judgment we saw discussed in Alvarez's training manuals. In the electronic tradition, such judgments went against the grain of automation, and a variety of devices, such as Cerenkov and transition radiation detectors, could do a rough job of automating the analysis (by measuring emitted light) when isolated particles came sailing through. Unfortunately, the environment of colliding beam detectors left particles anything but isolated: the 10 to 20 objects would overwhelm any Cerenkov detector, and the transition radiation detectors only worked for electrons if the momenta of the particles were much higher than those expected at the next generation of machines.[15] Instead, people working on the new multiwire proportional chambers began to explore the possibility of using the chambers themselves to measure ionization density and exploiting this information to identify particles.

14. Nygren, "Time-Projection Chamber" (1974).

15. Allison et al., "Ionisation Sampling," *Nucl. Inst. Meth.* 119 (1974): 499–507, on 499.

Figure 7.2 Dispersion device (1974). To test the basic principle of TPC operation, Nygren constructed this tiny test device; using a radioactive source, he released electrons at the top and measured their dispersion in a longitudinal electric and magnetic field. If the TPC idea worked, the electrons should stay clustered together without significant dispersion. Source: LBL BBC 744 2492, LBL. Courtesy of Lawrence Berkeley Laboratory, University of California.

Although the details of the transfer of energy from passing charged particles to matter are extraordinarily subtle (Hans Bethe considered his 1932 paper on the subject to be his greatest work), the basic physical ideas are straightforward, and many of the detector papers simply refer to the standard electrodynamics textbook by J. D. Jackson.[16] In particular, for nonrelativistic particles ($v/c \ll 1$) the energy transferred to an atom declines with speed because there is less time during which the passing particles are close enough to an electron to

16. Jackson, *Classical Electrodynamics* (1975), esp. chap. 13.

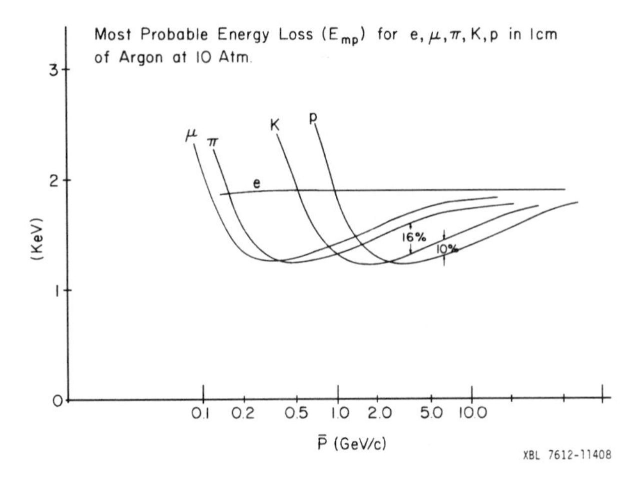

Figure 7.3 Specific ionization loss. In order to be able to distinguish between particles of different types, the TPC designers had to be able to sort out the specific energy losses of different particles. This chart and many others like it plotted the loss per centimeter in kilo–electron volts, as a function of momentum (in GeV/*c*) for muons, pions, electrons, kaons, and protons. Three features were immediately apparent: First, the TPC was extraordinarily good at distinguishing types of particles. Second, at certain critical energies the lines cross, which meant that ionization per centimeter could not distinguish those particles at the crossing energy. Third, the distinction between pions and muons—critical for many experiments—was not very large at any energy. Source: LBL XBL 7612–11408, LBL. Courtesy of Lawrence Berkeley Laboratory, University of California.

transfer momentum. Since the time spent in the interaction region is inversely proportional to the velocity, the momentum transfer falls with v ($\Delta E = (\Delta p^2)/2m$, so the energy transfer falls with v^2). Once the projectile reaches relativistic speeds, however, the field lines begin to intensify around the direction perpendicular to the motion, and this intensification causes a net *increase* in the energy transfer. In the momentum range from a gamma of about five to a few hundred MeV/*c*, the differences between the masses of muons, pions, kaons, and protons mean that at the same momentum, the particles have significantly different velocities and so different specific energy losses to ionization (see figure 7.3).[17]

17. Finally, at extremely high energies, the charged particles have such large transverse fields that they begin to polarize the medium, and at such energies (upward of hundreds of GeV) pions, protons, and electrons all begin to behave indistinguishably. See Jackson, *Classical Electrodynamics* (1974), chap. 13.

Perhaps the most salient of the attempts to use specific ionization to determine particle type had been made in 1974 by Wade Allison and his group at Oxford using a type of drift chamber. They designed a chamber that would allow the electrons released by a passing particle to drift up to 2 meters before they were captured by a sense wire. The time of arrival would fix the position of the track, and the integrated height of the pulse would provide an estimate of the primary ionization. But because any individual particle could lose a wide spectrum of energy per centimeter, it was necessary to sample each track many times—certainly more than 100 times in order to achieve a statistically significant result. The problem remained diffusion. As the electrons diffused perpendicularly to the drift direction, they caused confusion between the different sense wires (cross-talk), and as they diffused in the drift direction they caused ambiguity between closely spaced tracks.[18] How could the drift chamber be altered?

For many physicists, including the TPC designers, the TPC *was* a drift chamber[19] in the strong sense that spark counters were flattened Geiger-Müller counters and drift chambers modified multiwire proportional chambers. Charpak, for example, could write in 1978 that "[o]ne of the most sophisticated drift chambers is described . . . by Jay Marx and David Nygren."[20] At the same time, however, there was a sense that something had changed in the shift from the drift chambers of the late 1960s and early 1970s to the TPC. One review article from 1981 identified "[t]wo generations of drift chambers," the first of which included the "classical drift chambers" in which 20 or so points are measured per track. Typically, 10 layers of wire would be strung along the direction of the beam (and the magnetic field), and these would be used for pattern recognition—that is, to see whether a sufficient number of adjacent wires were hit to justify counting the hits as a genuine particle. The other 10 or so planes of wire would be offset by about 3° and would provide the "stereo" information needed to reconstruct the z-coordinate of a space point, since the longitudinal wires only gave information on radius and angle.[21] Just because these "first generation" chambers were deployed for statistical rather than pictorial purposes, they were "classical" instantiations of the logic tradition.

A second generation almost immediately stood out as distinct from the first. These newer devices were "*pictorial or imaging drift chambers* since the aim was to record a complete picture of particle trajectories even in the case of

18. Allison et al., "Ionisation Sampling," *Nucl. Inst. Meth.* 119 (1974): 499–507. Of considerable historical significance are the first drift chambers with large drift distances: e.g., Saudinos, "Large Drift Length Chambers" (1973).

19. Clark et al., "Proposal for a PEP Facility Based on the Time Projection Chamber," 30 December 1976, box 2, 18, DNyP II.

20. Charpak, "Multiwire and Drift Proportional Chambers," *Phys. Today* 31 (October 1978): 23–30, on 27.

21. Wagner, "Central Detectors," *Phys. Scripta* 23 (1981): 446–58, on 447.

complex events." Instead of sampling 15–20 times, 40–200 samples per track were made. Where the classical generation projected the space point onto two planes and then deduced the position, the imaging generation used time to find the "true space points." And where the classical generation had to use external devices to determine particle type, the pictorial generation could exploit its many samples to find the particle type.[22] Here a quantitative alteration becomes qualitative; somewhere, in the passage from 15–20 to 40–200, counts became "pictorial" and we moved away from the classical. Again, image and logic had fused, not dramatically at first, but out of the very practical demand for more information to effect particle identification. Silently, below the dramatic features of discovery claims, an epistemic shift had occurred: without photography, electronics had gone visual.

The fusion of the image and logic traditions occurred on several levels simultaneously. For the first time, the output of an electronic detector was recognized as "pictorial," with a detail previously available only in the image tradition. But to achieve that refinement of representation, the electronics and the accompanying computer processing introduced automatic statistical inference into the identification of particles.

If the TPC was going to work, the end caps at which the charges would arrive had to be able to extract the information. The ball and honeycomb solution was awkward from the first. Just as Nygren was completing his first paper (February 1974), *Scientific American*'s issue for the month arrived. It was there that he learned about a new technology—charged coupled devices (CCDs)—that the author (Gilbert Amelio of Fairchild Camera and Instrument Corporation) asserted put us "at the dawn of a revolution" that would alter forever the way information could be processed and stored. Built with the same materials and the same quantum electronic principles as the transistor, the new device was conceptually different: "[I]t is a functional concept," Amelio argued, "that focuses on the manipulation of information rather than an active concept that focuses on the modulation of electric currents."[23]

This passing remark captures a sea change in the nature of material culture during the late 1970s and 1980s. As Nygren and his collaborators insisted over and again, their hope was to carve up space virtually—with information—rather than with physical detector elements. Thus, for example, in the 1976 proposal they write that the cylindrical volume (on the order of 6 meters cubed) is segmented into three billion cells of 2 cubic millimeters each "by the use of modern electronic techniques rather than by physical detectors within the fiducial volume itself."[24] Even the invocation of the "modern" here signals the

22. Wagner, "Central Detectors," *Phys. Scripta* 23 (1981): 446–58, on 446, 448.

23. Amelio, "Charge-Coupled Devices," *Sci. Amer.* 230 (February 1974): 22–31, on 23.

24. Clark et al., "Proposal for a PEP Facility Based on the Time Projection Chamber," 30 December 1976, box 2, 15–16, DNyP II.

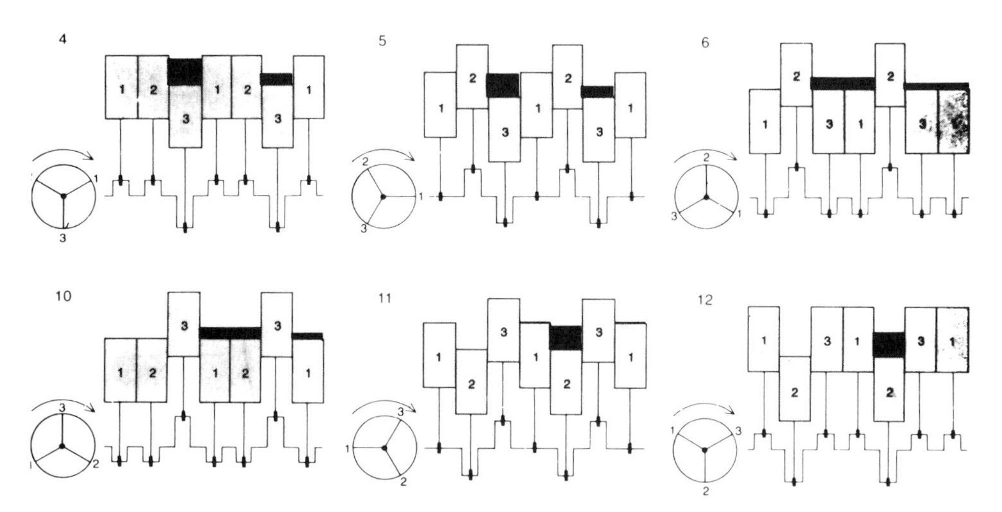

Figure 7.4 CCD mechanical analogue (1974). CCDs came to Nygren's attention through an article in *Scientific American.* This figure is an excerpt from the mechanical analogue presented there: buckets on a crankshaft systematically move volumes of water from left to right, indicating, by analogy, the movement of bits of information (charge) through the registers of the CCD. Source: Amelio, "Charge-Coupled Devices," *Sci. Amer.* 230 (1974): 22–31, on 22. Copyright © 1974 by Scientific American, Inc. All rights reserved.

perceived break with the past; one that goes, I suspect, far beyond the boundaries of high energy physics, suggesting a kind of postmodern break between the "physical" and the information-based detector.

Central to the dephysicalized partition of the machine was the use of the new electronics of the CCD. The basic idea was simple. Information, in the form of a given number of electrons, arrives at the device. By means of an externally manipulated voltage, the collection of charges is transferred to a similar storage element adjacent to the first, then to a third station, and so on. Mechanically, a striking analogy exists—as Amelio stressed. Imagine a crankshaft attached to a battery of pistons (see figure 7.4). As the crankshaft turns, the liquid resting on top of piston number 3 sloshes over the tops of 1 and 2, then covers only number 2, and so on as the changing potential energy slides the liquid progressively toward the right. Electronically, the equivalent of the liquid is charge in the form of a number of electrons; the changing gravitational potentials in the mechanical case have, as their analogical elements, the externally applied voltages. The crankshaft and connecting rods stand for the driving voltage on the CCD, and the rate at which the crank is turned corresponds to the timing (or clock frequency) that fixes how often the cycle of voltages repeats itself, driving packets of electrons through the device. Already by the time Nygren read the *Scientific American* article, engineers were working on two distinct functions for the new device: it could serve as an electro-optical element converting visible light into electrical signals (for television cameras), or it could serve as a memory device, since the information is preserved as it shifts from cell to cell. In a sense, Nygren and his coworkers borrowed from both. Because the CCDs could be arrayed in huge numbers over very small areas (by 1974 it was possible to construct 100 × 100 CCDs on a 0.12-inch by 0.16-inch chip),[25] vast numbers of independent channels could be processed to reconstruct an image of a very complicated annihilation event. But whereas a television camera exploits Einstein's photoelectric effect in silicon to convert light into a pool of electrons, the "signal" in the TPC is already a floating cloud of electrons gathered up through a wire or pad.

It would be more precise, however, to describe the function of the charged coupled device in the TPC as that of an analog shift register, an electronic device that stores analog information. Nygren's hope was that the entire history of an event from each of the output channels of the end cap would be stored on a CCD, only to be actually "read out" if the event satisfied certain specified criteria. During 1975, the ball-wire scheme began to look increasingly problematic, and by March 1976, Marx and Nygren were telling potential collaborators that the readout would "most likely [be] CCD delay lines."[26] Uncertainty faded over the next nine months, and the proposal to the PEP leadership (December 1976) asserted categorically "that these CCD delay lines are not a visionary technology but are current shelf items whose price should drop in the future."[27] Specifically, Fairchild was producing a CCD (model 321) for about $50, which had 455 cells, or "buckets," each one of which would be filled with a packet of electrons and then passed on to the next after 50 nanoseconds (50×10^{-9} seconds). Stringing the buckets end to end, 23 microseconds (23×10^{-6} seconds) of information could be stored, which was more than the drift time of an electron across the whole chamber. In other words, *all* the information from an entire event could reside in the CCD (see figure 7.5).[28]

To go much further into the design and construction of the TPC would be to get ahead of ourselves. For most of 1974 and 1975 the TPC remained a long way from the industrial-scale facility to which speculative papers alluded. Instead, this was a time of tabletop prototype manufacture and of experiments (varying the magnetic field) to explore transverse diffusion with a long menu of

25. Amelio, "Charge-Coupled Devices," *Sci. Amer.* 230 (February 1974): 22–31, on 22.

26. [Marx and Nygren], "A PEP Facility Based on the Time Projection Chamber," distributed for recruitment meeting, 15 March 1976, file "TPC Politics-Internal," JMP.

27. Clark et al., "Proposal for a PEP Facility Based on the Time Projection Chamber," 30 December 1976, box 2, 15–16, DNyP II.

28. Clark et al., "Proposal for a PEP Facility Based on the Time Projection Chamber," 30 December 1976, box 2, 15–16, DNyP II.

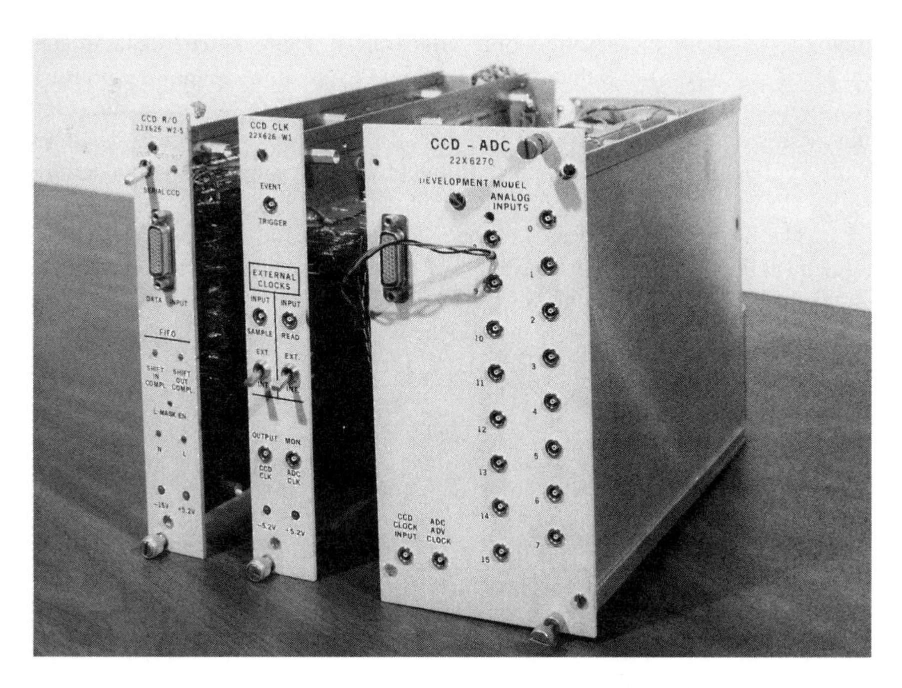

Figure 7.5 Fairchild CCD 321 (1978). Ball-wire charge collectors proved inadequate on several counts. Most dramatically, the curvature of the ball created an inhomogeneous field that pulled the drifting electrons into highly variable trajectories depending on where they arrived relative to the center of the ball. Because they took different paths, the arrival time of drifting electron clusters was no longer proportional to the drift distance from the site of ionization, and the TPC's guiding principle was destroyed. CCDs promised to remedy this situation by providing flat electrodes with an electronic processing system that could count charges and effectively forward this information to a computer. Depicted here are the prototype Fairchild units that were first used in 1978. Source: LBL BBC 787 9075, LBL. Courtesy of Lawrence Berkeley Laboratory, University of California.

gas mixtures. Aside from the electronics of the end plane, Nygren and his colleagues, especially Peter Robrish, looked at the distortion that would be imposed on tracks if the electric and magnetic fields were misaligned, and as they had not yet procured sample CCDs, they continued their exploration of the ball-wire detectors under a variety of conditions. How quickly, for example, could the ball-wire recover after receiving a pulse of electrons?[29]

But as PEP moved toward its selection of a detector facility, Nygren and his little group increasingly saw themselves at a crossroads: they could compete or drop out. Either they would continue their tinkering, or they would restructure

29. See, e.g., Nygren, "Time-Projection Chamber-1975" (1975).

their efforts around the assembly of a collaboration. Their choice was facilitated by Marx's arrival—more than the others Marx threw himself into the political fray of mounting a huge hybrid detector. Almost immediately, Marx began printing organizational memos that, up to then, had been entirely absent. Together with Nygren, Marx issued an appeal to the physics division staff, announcing a series of TPC-related seminars, many of which would be given by theorists.

The first such topic was "Lepton Detection and the Search for New Phenomena at PEP," the collaborative SLAC-LBL colliding beam facility that would supplant SPEAR, proposed in April 1974 and estimated at the time to cost some $78 million. Other topics would include the importance of particle identification at PEP, weak interactions at PEP, and a variety of more specific theoretical probes such as tests of quantum electrodynamics. But most of the topics chosen were set at a more inclusive level of phenomenology; rather than ask after the particular consequences of, say, the Glashow-Weinberg-Salam theory, the colloquium would explore gross features of multihadron final states, such as how spherical the end products would be, how many particles were likely to emerge from the colliding beams, and the correlations between products expelled in different directions. Instruments provided a third category of seminar topics: exploring drift chambers, the TPC, identification by dE/dx, superconducting magnets, calorimeters exploiting liquid argon, and new species of Cerenkov detectors. Then a fourth slate of seminars encompassed PEP politics, the "posture" LBL would take toward the user community, and intermediate topics such as "small experiments," which was at once an inquiry into specific physics questions and, as we saw in chapter 6, a search for a way of working.[30]

7.3 Collaboration

Worries about "posture" and scale were very much to the point for the incipient collaboration. In one sense, LBL was codesigning PEP; in another, PEP would redesign LBL. For as the scale of physics expanded, the option of continuing life even in the Alvarez mode ceased to exist. Writing to Robert Birge in early November 1975, Marx suggested that the LBL staff had previously functioned much the way university groups had, an arrangement that presupposed that no individual group would swallow a dominant fraction of the division's budget. Now PEP threatened to do just that, and proposals by other groups within LBL were essentially competing among themselves for survival. In the age of Alvarez, other groups could argue that Alvarez received too many resources from

30. Marx and Nygren to Physics Division research staff, 26 September 1975, file "LBL PEP Seminars," box 2, JMP.

the laboratory; but even in the heyday of the hydrogen bubble chamber five or six other groups had survived. Now, by Marx's lights, there would only be one: "Without the fortuitous situation of a brilliant idea that all can rally around, the decisions of priority can be accomplished in one of two ways: the dictatorial or the quasidemocratic (all votes are never equal). I sense that you are reluctant to choose the first alternative and I agree with that point of view. There are few philosopher kings and still fewer who are so recognized by their subjects." On Jay Marx's view, only such an informed "quasi-democratic" entity—such as one composed of the director, division head, and group leaders—could enforce a coherent effort at PEP from LBL, and without that unity, the "impact of LBL [would] be diluted."[31]

It did not take long for the explosive growth to begin. In early 1976, Marx and Nygren were focusing their group on the design of a major PEP detector with the TPC at its heart, and by March 1976 they were rallying other physicists to the project. One physicist, Mark Strovink, left a particularly clear trace of his deliberations in the form of a decision tree, which he mailed to Nygren (see figure 7.6). Most strikingly, Strovink had seized on the dilemma of whether to pursue an "experiment" oriented toward a more narrowly defined set of physics questions, or toward a "facility": "The facility [as opposed to an experiment] may do more physics, though it is not at all the style of physics in which I have worked up to now." If he was going to work on a facility, it should emphasize particle identification; at a minimum, it had to be able to distinguish pions from kaons at the level of a few parts in a thousand. As branch point 4 indicated, this might be done by Cerenkov radiation (in which case Strovink would be choosing not to work on the TPC). Or, it might be done with dE/dx, in which case (as branch point 5 indicated) he could either opt for drift or proportional chambers—the competition—or choose the TPC. Ultimately, his choice would rest on the ability of the TPC advocates to show that the dE/dx would work properly, that diffusion over long drift distances would not cause problems, and that the long sensitive time would not allow so many background tracks into the picture that extracting the signal would become impossible. It was therefore the collaboration that had to up the ante: "Obviously my participation in PEP planning during this period will be meager, compared to yours. Whether it is concentrated on the TPC depends on branch points (4) and (5). The advantages of the TPC *apart* from possible particle identification in my view are not sufficient to warrant a commitment to the TPC at this point."[32] Over the next few months, Nygren and his core group plunged further into the question of particle resolution. Strovink signed on.

31. Marx to Birge, 7 November 1975, file "LBL Politics and Memos," box 1, JMP.
32. Strovink to Nygren, 11 March 1976, file "TPC Politics-Internal," box 2, JMP.

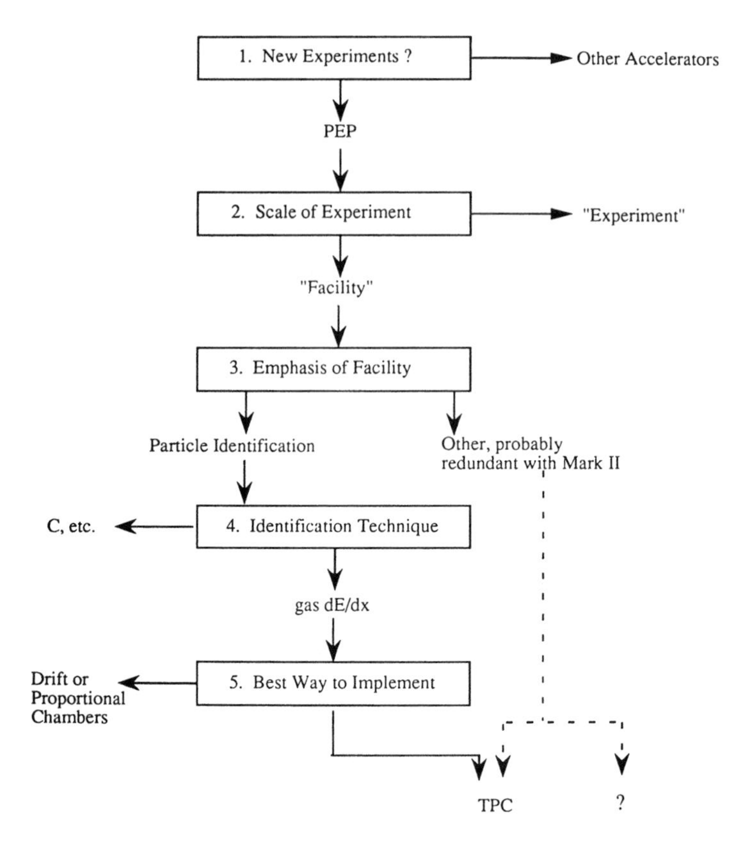

Figure 7.6 To participate or not? Strovink's decision tree (1976). Source: Modified from Strovink to Nygren, 11 March 1976, file "TPC Politics-Internal," box 2, JMP. Courtesy of Lawrence Berkeley Laboratory, University of California.

But before Strovink put his name on the roster, the recruiters opened a 15 March 1976 meeting, specifically to help enlist their colleagues to the TPC project: "We need able, creative collaborators with any and all interests. There is much to be done and more than enough room for many people to make decisive impacts on the physics and instrumentation." Like special deals and unusual sales, this "opportunity" (the flyer proclaimed) would not be available for long; December's deadline for proposals set a limit on how long LBL could procrastinate. The "physics" advertised fell into four categories: the 4π detection of electrons and muons, the 4π tracking coupled with an intrinsic three-dimensional readout, the identification of particles with *dE/dx* and time-of-flight measurements, and the ability to reconstruct neutral *V*'s. In the aftermath of the staggering success of the Mark I, which found the psi by its decay to e^+e^- and $\mu^+\mu^-$

pairs, "[w]e are guessing that leptons are the key to seeing new phenomena" and the proposed detector—a TPC with CCD readout, surrounded by a magnet and muon detector—promised to do that splendidly.[33]

In manuscript notes, Marx jotted down his audience's response: "Carithers—Is not sure of longterm commitment to . . . Mark II. . . . Told him he is welcome to join anytime up to 1981." To succeed, the collaboration would clearly need to win over large numbers of people from existing groups at LBL; it was therefore welcome news that Group A, the old Alvarez group, might be interested: On 16 March 1976, Marx talked with Morris Pripstein, who reported that "others in group A are quite interested *if dE/dx* works. Have heavy commitments now. Would become more involved after proposal. He will be supportive e.g. will try to get 1 group A physicist + 1 programmer + possibly Lynn Stevenson to work on Monte Carlo." A few days later Lynn Stevenson was "interested" but had neutrino commitments for the short term; when he was free for PEP work, he would design a Monte Carlo for charm decay and add it to an already extant PEP Monte Carlo. Philippe Eberhard (*the* LBL expert on magnets) "says our stuff is best going at LBL" but hesitated to commit beyond consulting on the superconducting magnet. He would, however, calculate the effect on a particle's trajectory of misaligned E and B fields. Orin Dahl was willing to work 50% time on the TPC, preferably on software, but with the caveat that he wanted "to do physics before 1981."[34] Dahl and Strovink's hesitancy was understandable. To commit oneself to a project of this scale was to invest at least five years before any results would arrive, and after that another several years—at minimum—would be needed to take and process data. In the course of a 30-year career in physics a scientist could only incur a handful of such 7- to 10-year obligations. One mistake could cost a physicist dearly.

As many of these and other comments indicated, recruiting in the early days was difficult and obviously an admixture of sociology and physics. To expand their numbers, the kernel of TPC advocates had, above all, to persuade a wider universe of allies that the new device could deliver the goods: it had to be able to identify particles. It is not surprising then that the remainder of March and April 1976 was devoted exclusively to *dE/dx* resolution studies followed by a month and a half on tracking resolution studies. Only then, at the end of June, would they begin the four-month process of writing a letter of intent, followed by the more intricate task of preparing the proposal, which was to include budgets, logistics, Monte Carlo simulations, and designs for the all-important end cap.[35]

33. "Meeting to Recruit Help with TPC," and handwritten notes; and [Marx and Nygren], "A PEP Facility Based on the Time Projection Chamber," 15 March 1976; both in file "TPC Politics-Internal," box 2, JMP (present were Birge, Eberhard, Carithers, Groves, Loken, Shapiro, Strovink, Pripstein, Dahl, Wenzel, Stevenson, Marx, Nygren, Robrish, Urban, and Steiner).

34. Marx, handwritten notes, 16, 18, 21, and 23 March 1976, file "TPC Politics-Internal," box 2, JMP.

35. [Marx], timeline, 15 March 1976, file "TPC Politics-Internal," box 2, JMP.

Having secured a core of LBL physicists, Marx and Nygren began trying to roust collaborators out of other laboratories. They drew up a master chart on 15 May 1976 in which they sorted the leading experimental physicists of the time into three groups: "happy to work with," "willing to work with," and the last category "unwilling to work with."[36]

Only collaborations can write the proposals needed to fund collaborations. No individual institution—let alone an individual—could possibly develop the hardware, software, and theory necessary to pry loose the tens of millions of dollars needed to run an experiment of the kind that became the norm by the 1980s. By May 1976, therefore, Nygren and Marx were already organizing the collaboration in order to write a formal proposal to the Department of Energy. Needed would be a physics argument—the facility would plumb the "new charm-like spectroscopy à la Gell Mann et al.," as well as leptonic and hadronic decays and the conversion of photons into lepton pairs. Someone would have to draft this into much more specific language, while others would formulate sections on weak interactions, new phenomena, total and differential cross sections, and jets. Additional subgroups would be assigned to write up aspects of the detector and to compile long appendices on subjects ranging from technical studies to cost estimates and management strength.[37] Even the fairly large experiments of a few years before—experiments like the Mark I or the 72-inch bubble chamber—had nowhere near this complexity at the proposal stage.

To produce a document of this scope required a major collaborative effort in which the experimenters splintered into "short term TPC working groups." Their task included answering questions about the expected resolution of the facility, and they addressed the problem both with theoretical studies using Monte Carlos and with experimental tests using already operational multiwire proportional chambers. Similarly, the working groups took on the problem of positive ions that might accumulate in a region of the chamber in which they could generate a local fluctuation in the ambient electric field. Such "space charges" would accelerate and decelerate the drifting electrons in irregular ways as they floated through this field aberration, endangering the vaunted accuracy of the chamber. On the theoretical side, the working group undertook to study the limits of "tolerable space charge." On the experimental side, a subgroup initiated empirical tests on the prototype TPC that would be tried out in the Bevatron. Another subgroup examined ambiguities that might prevent the energy loss measurements (dE/dx) that would unilaterally identify a particular particle species; yet others

36. Marx and Nygren, handwritten chart, 15 May 1976, file "TPC Collaboration," box 7, JMP. There were four in this last category, every one of which was a name that would be universally recognized across the profession, including two Nobel Prize winners.

37. Marx to TPC Proposal Distribution List, 24 May 1976, with enclosure "TPC Proposal Outline #2," 20 May 1976, file "TPC Politics-Internal," box 2, JMP.

Figure 7.7 End plane sector, *dE*/*dx* and position wires (1981). Source: LBL BBC 813 2223, LBL. Courtesy of Lawrence Berkeley Laboratory, University of California.

explored time-of-flight measurements and recounted the state of play in liquid-argon calorimeters that might complement the TPC in the full-blown detector.[38]

Out of these discussion groups came a consensus that the heart of the TPC would be an end plane sector outfitted with two sorts of electronics. First, there would be a plane of 192 wires, forming a multiwire proportional chamber that would measure the ionization produced by the passing particle in order to determine *dE*/*dx*. To increase the number of collisions and therefore the statistical significance of the ionization samples, the chamber would be pressurized to 10 atmospheres. In addition, there would be 12 wires dedicated to the determination of position. More specifically, the electrons would drift from the track toward these position wires, and as the avalanche arrived, a strip of electrodes (cathodes) located below these wires would determine the time of arrival and therefore the z-position of the measured track segment (see figures 7.7 and 7.8).

Already, then, even before the device was formally proposed or approved (much less built), the problem of disengaging signal from background was

38. Marx to TPC Group, "Short Term Working Groups," 6 August 1976, file "TPC Collaboration," box 7, JMP.

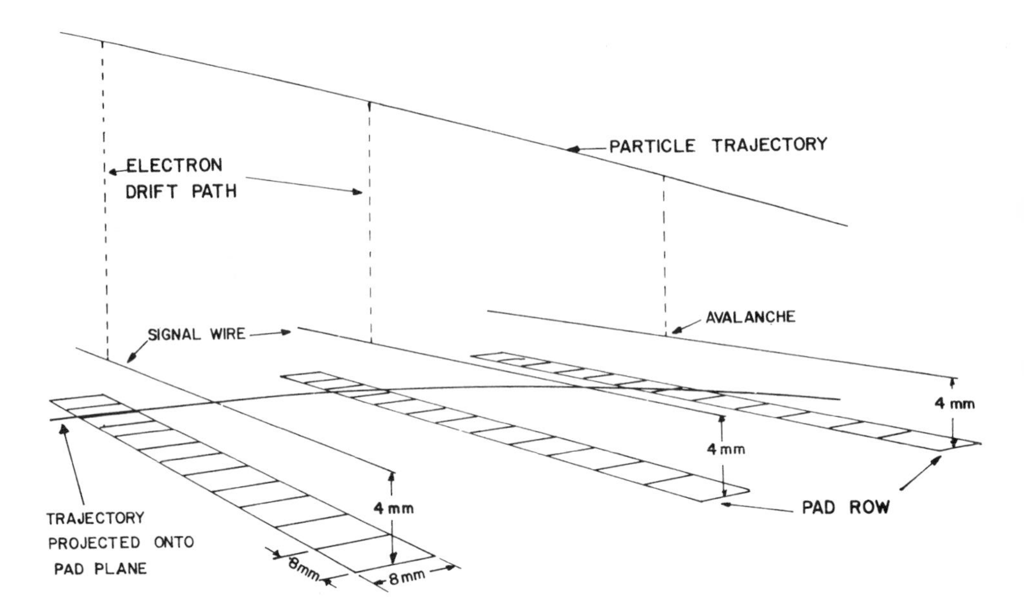

Figure 7.8 Passing particle and wires (1979). Here one sees a schematic representation of the TPC. The time of the liberated electrons' arrival (relative to the time of beam passage) is determined by the CCD cathode pads, which are arrayed in rows of 8 millimeter squares: the time of drift is proportional to the distance in the *z*-direction, which is fixed to within 350 microns. Above these cathode pad rows lies a plane of wires: 183 of these wires are dedicated to the determination of the ionization loss per centimeter of each passing particle (*dE/dx*) which identifies the particle; the other 15 sense wires per sector (one over each pad row) fix the *x-y* position in the cathode plane to within 185 microns. Source: Fancher et al., "Chamber," *Nucl. Inst. Meth.* 161 (1979): 383–90, fig. 4 on 385.

central. Where the bubble chamber physicists designed techniques of particle identification to be used on the raw data, in the massive electronic device that lay at the heart of this new detector the raw data would be, to a striking degree, already analyzed. Or perhaps one should say it this way: the always subtle distinction between fact and interpretation was driven ever further inside the apparatus, to the point where it became virtually impossible to speak of "raw data" in the sense that this phrase had in the earlier days of cloud chambers, emulsions, and bubble chambers.

In June 1976, the PEP Conference offered a venue at which the now expanding collaboration could present a status report. Motivations continued to include the original goals of studying high-multiplicity events and particle identification over a wide solid angle. In the first version of the proposal (3 June 1976), Marx insisted "that the most exciting physics is probably, at the present time, unanticipated." There was indirect (and disputed) evidence for a fifth quark in data from Fermilab. Coming on the heels of the epochal work of late 1974 and 1975, the TPC physicists looked forward to nothing so much as a return to the halcyon days of Mark I, in which the fourth quark had generated so much excitement. There would be "a new ψ-like spectroscopy and hypercharm." Other new particles such as vector mesons, Higgs scalars, gluons, and heavy leptons all demanded a similar sort of detector. Weak interactions were relegated to a secondary position, quantum electrodynamics to a two-sentence paragraph, and last came a passing reference to collisions with two gamma rays in the final state.[39]

For particle identification to work, the detector would have to sample ionization at least 200 times; this in turn implied pressurizing the TPC gas to give more electrons per centimeter of track. The good news was that 8 atmospheres of pressure yielded over 400 electrons per centimeter; the bad news was that this same pressure meant that $\omega\tau \approx 1$, and so the entraining factor would be not 60 but 2. To make up for this undesired spread of the clump of electrons, Nygren and his allies now used the electronics to extract the information: by sampling the number of electrons arriving in short units of time, the computer could deduce the center of gravity of the resulting histogram and therefore pinpoint the spatial origin of the bump with far greater precision than might otherwise have been imagined.[40]

Having grown to a substantial size by summer 1976—LBL had now been joined by Yale, Johns Hopkins, and UCLA—even the proposal-writing collaboration needed a political structure, and the collaboration began debating its shape. Dubbed the "commissariat," the representative body would function as the "central decision making body within the collaboration" and would communicate decisions to the "home groups." According to one proposal, each group would send two representatives, one of whom must be present at each monthly meeting: each representative would report on the progress of his group, and each group would have one vote. This august entity would be responsible for setting criteria for new collaborators, assessing design options for experiments, setting "logistical," "strategic," and "political" plans, allocating tasks within the experiment, deciding who would give which talks, and being final arbiter on who would and who would not count as an author. Finally, the commissariat would choose the experimental "spokesperson" from its body, and this spokesperson would represent the TPC collaboration in negotiations with PEP management.[41] (Nygren became spokesperson and Marx deputy spokesperson.) At the commissariat's first meeting, it would debate the assignment of tasks to

39. Marx, "PEP Proposal-I," 3 June 1976, TPC-LBL-76-24.

40. Nygren, "Time Projection Chamber," talk at 1976 PEP Conference, 23–25 June 1976, xeroxed transparencies, in Notebook QCD 191, SLAC.

41. Marx to Members of TPC Collaboration, "Communications and Decision Making," 20 July 1976, file "TPC Collaboration," box 7, JMP.

groups, allocate manpower, review both theoretical and experimental progress, and establish either a "Senate" or "House" scheme to represent the constituent groups. Then it would rename itself, "commissariat" evidently being a bit too Stalinist for public consumption.[42] The "TPC Executive Board" evidently had a suitably corporate ring.[43]

Who else should join? At its first meeting of 16–17 August 1976, the board considered Harvard University, the University of Arizona, the University of Massachusetts, and the University of Colorado—all seemed "weak." None were considered further. In addition to the quality of a group, the board was desperately concerned about putting warm bodies into the collaboration. Without guaranteed resources, especially definite commitments by physicists, a new group would prove a liability rather than an asset. The University of Maryland, for example, had inadequate support in the short term, whereas the University of Washington and the University of California, Riverside, both had other gaps in their resources. Nygren asked these last three groups for written projections of their commitment to TPC work on the basis of which a final decision would be made.[44] With these in hand at the next board meeting (in September), Washington's paltry 0.5 full-time equivalent (FTE) knocked it out of the running. Johns Hopkins unsuccessfully defended its neighbor the University of Maryland, protesting that even though Maryland's short-term abilities were limited, it would aid the collaboration in the long run. Riverside (also in the University of California system) looked more promising to LBL, even though some at the laboratory had doubts. Riverside's ability to field 2.5 FTEs immediately was an enormous attraction, and on this basis the Executive Board invited U.C.-Riverside to join the core of the collaboration.[45]

As the group was assembled, so was the conceptual design. One working group reported to the board in September 1976 on schemes for resolving the *dE*/*dx* ambiguity—that is, the possibility of confusing two particles that tended to lose similar amounts of energy at similar momenta. Another worry raising its ugly head again was that the positive ions' space charge would wreck the carefully controlled electric field in the TPC. With an electronics design in hand, Mike Zeller (from Yale) provided "an existence proof" that the CCD data could be processed adequately for transmission to the computer. For its part, the Johns Hopkins group issued a plea for more Monte Carlo simulations to help with its

42. Marx to TPC Collaborators, "The First Meeting of the TPC Commissariat," 27 July 1976; and Marx to TPC Collaborators, "Agenda for Meeting of TPC Commissariat," 4 August 1976; both in file, "TPC Collaboration and Intragroup Organization," box 7, JMP.

43. "The TPC Executive Board," 18 August 1976, file "TPC Charter," DNyP.

44. Minutes of the TPC Executive Board Meeting, 16–17 August 1976, file "TPC Executive Board," DNyP II.

45. Minutes of the TPC Executive Board Meeting, 21–22 September 1976, file "TPC Executive Board," DNyP II.

design of the muon detector. The interim working group reports left much undetermined, and the group acknowledged that many crucial questions, including the detector's ability to ferret out π^0's by measuring their decay products (two photons), were still up in the air. Other problems remained as well: Would a liquid argon calorimeter be scientifically justified? What neutral particles really needed to be measured?[46] As the team struggled to bring coherence to reliability estimates for different instruments, to link different subgroups, and to coordinate various directions of research, it soon found the constraints of physics, sociology, and politics leaching into one another.

That their work would be conditioned by politics, as well as by physics and sociology, became clear when the experimenters faced the immediate consumer of their collaboration's wares: the directorate of the proposed colliding ring, PEP, and more specifically the joint leadership of PEP in the persons of SLAC Director Pief Panofsky and LBL Director Andrew Sessler. From long experience, both men were familiar with political pressure to transform the increasingly large and centralized accelerators of high energy physics from the possessions of individual laboratories into nationwide "facilities." Indeed, by 1965 when the Joint Committee on Atomic Energy was considering the future of accelerator research, the necessity of opening facilities to outside groups was a topic of sustained discussion, and the laboratories were analyzed one by one for their percentage of outside users. At that time, the term "facility" was an attribute more or less applicable to accelerator laboratories like Argonne National Laboratory, Brookhaven National Laboratory, the Cambridge Electron Accelerator, Lawrence Radiation Laboratory, and the Princeton-Pennsylvania Accelerator.[47] With the TPC, however, there was a category shift. From a political tag attached to accelerators, "facility" became something that could be applied to the detector itself, the sanctum sanctorum of the experimenter. In a 1976 directive, Panofsky and Sessler explained what they wanted from the experimental groups, and in the process they articulated the joint physics and political constraints that resulted in a reconceptualization of the terms "experiment" and "facility": "The concept of the 'PEP Facility' was devised to solve a very real problem: large general detectors may have useful lifetimes of many years but it is undesirable to commit the use of PEP for such long periods of time to the initiating groups. Therefore, we decided that we needed to make the large general detectors available for use by physicists other than the original builders after an initial period, called the 'Priority Period.'" This period would run somewhere between 12 and 18 months, during which time the builders would be allowed exclusive rights to harvest what data they could: "The priority period is long enough to guarantee

46. Minutes of the TPC Executive Board Meeting, 21–22 September 1976, file "TPC Executive Board," DNyP II.

47. U.S. Congress, *Hearings*, 89th Cong., 1st sess., 1965, 377–78.

the builders an opportunity to make early discoveries but it is relatively short compared to the projected lifetime of the detectors."[48]

Nomenclature—the designation as "facility" or "experiment"—was therefore a political-scientific issue of real import. Whoever classified the device would exercise considerable power over who would take data, who would control the machine, and for how long. Not surprisingly, the PEP administration added a clause that jealously guarded the right to make the decision about just what it was that they were building: "[W]hile the proponent takes the initiative in calling the detector an experiment or a facility, the Experimental Program Committee (EPC) will ultimately recommend which detectors should be developed as facilities." Part of the initiating group's duty would then be to plan for future support staff after they were gone, to determine whether support staff would come from the laboratory or from the world outside. One solution, and whether this or another would be used would have to be negotiated, was that cryogenic, vacuum, magnet, power supply, and electronic support might be pooled to service several facilities.[49]

Working up to the wire, 36 participants filed their proposal for a TPC facility with the SLAC authorities on 30 December 1976. By then, everything at SLAC—and elsewhere in the high energy physics community—revolved around the *J*/psi spectroscopy of the 1974 November Revolution. No longer would the experimenters identify as their goal, as Nygren had two years earlier, "covering" the phenomena produced at PEP. Now there was a definite set of phenomena under inquiry, and the proposal opened with an homage to it: "The spectacular sequence of discoveries made at presently operating e^+e^- colliding beam facilities has established this class of reactions as a major avenue of progress in experimental high energy physics." Jet events pointed toward the reaction production of two quarks, which then hadronized into a panoply of pions, kaons, and heavier particles. There was the determination of R (the ratio of hadron to muon pair production in e^+e^- annihilation) that had been key in the discovery of the *J*/psi just two years earlier, and there was the astonishing demonstration of the new lepton, the tau.[50]

To probe and perhaps add to these new phenomena, the PEP-4 collaboration proposed a detector with four major subsystems: the TPC itself; a superconducting magnet; an electromagnetic calorimeter, which would measure the total energy deposited by charged particles; and the muon detector, which

48. [Sessler and Panofsky], "Guidelines for What Is Meant by a PEP Facility," n.d. [late 1976], file "PEP Politics," box 2, JMP.

49. [Sessler and Panofsky], "Guidelines for What Is Meant by a PEP Facility," n.d. [late 1976], file "PEP Politics," box 2, JMP. A slight variant of this item is found in the "Experimental Program Advisory Committee Charter," attachment A, 12 September 1978, file EPAC I, box G014-F, SLAC.

50. Clark et al., "Proposal for a PEP Facility Based on the Time Projection Chamber," 30 December 1976, box 2, 1, DNyP II.

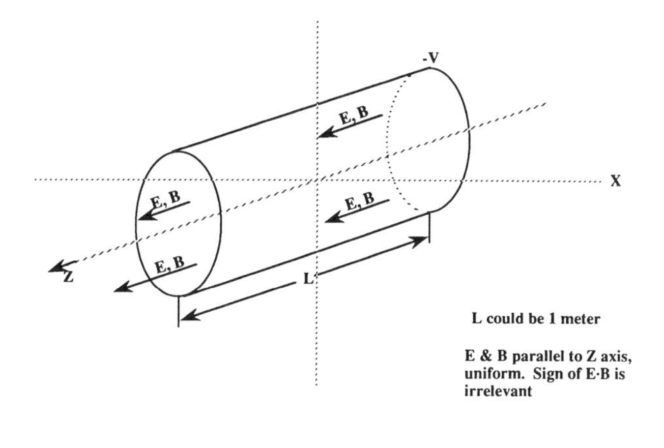

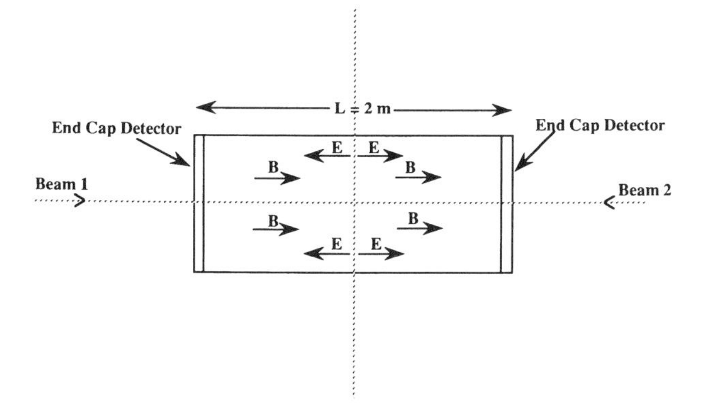

Figure 7.9 Nygren, first TPC sketch (1974). Redrawn from a fading pencil sketch by Nygren, this figure is the first to depict the basic idea of the full-scale TPC. Source: Nygren, "Proposal to Investigate the Feasibility of a Novel Concept in Particle Detection," 22 February 1974, 2, DNyP I. Courtesy of Lawrence Berkeley Laboratory, University of California.

at the extreme outside of the detector, would call a muon any particle penetrating several hundred grams of matter. Here it is instructive to watch, in pictures, the unfolding of a facility from the first notebook sketches of February 1974 to full-fledged multiexperiment operation a few years later (see figures 7.9 to 7.13).

Nothing in the new environment of theory, nor among the astonishing experimental discoveries at the Mark I in SPEAR, shifted the basic design of the TPC. Rather, the collaboration exploited these developments to reformulate the "physics objectives," emphasizing those features that were uniquely or at least

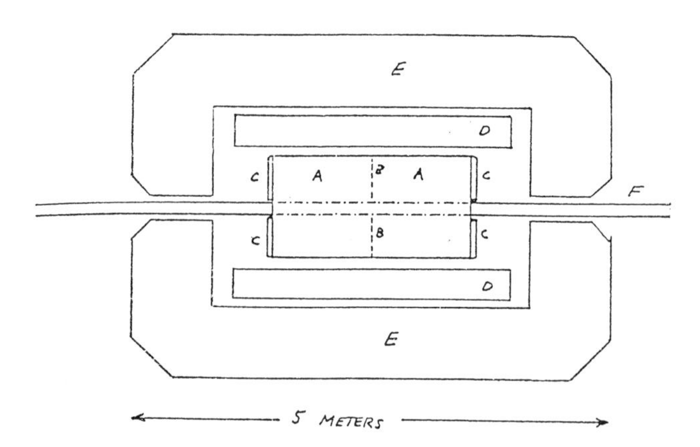

Figure 7.10 TPC sketch (1974). Five months after the preliminary sketch of figure 7.9, Nygren began to outline the other elements of the hybrid TPC detector. Elements of the device were (in summer 1974) given by: *A*, Methane-filled region approximately 1 meter in diameter, 2 meters in length. *B*, Screen or foil to establish *E* field. *C*, End cap detectors. *D*, Superconducting solenoid (3.33 tesla). *E*, Iron return yoke for *B* field. *F*, Beam vacuum pipe. Not shown are trigger scintillators, compensations, luminosity monitors, etc. Source: Nygren, "Time Projection Chamber" (1974), fig. 2 on PEP-144-18. Courtesy of Lawrence Berkeley Laboratory, University of California.

well suited to the device. Similarly, the 1973–74 discovery of neutral currents at Fermilab and CERN made further, more precise studies of the Glashow-Weinberg-Salam model pressing, but again, the new electroweak unified field theory left the basic structure of the experiment intact. If one read the discovery of neutral currents as evidence for the neutral weak boson Z^0, then it was to be expected that PEP-4 could explore the interference effects between

$$e^+e^- \to \gamma \to l^+l^- \quad \text{and} \quad e^+e^- \to Z^0 \to l^+l^-. \tag{7.7}$$

Such purely leptonic phenomena were particularly appealing to theorists because they were free of the complicating features of strongly interacting particles.

If the material culture of the laboratory did not change with the unveiling of hidden charm, that did not mean the experimenters were indifferent to the new discoveries. On the contrary, the new physics discovered at SPEAR was so staggeringly interesting that many in the TPC team fantasized about a repeat performance. Should PEP produce a new set of objects analogous to the *J*/psi and its excited states, the TPC collaboration contended, the TPC was particularly well positioned to find them. Their detector could see the gamma rays

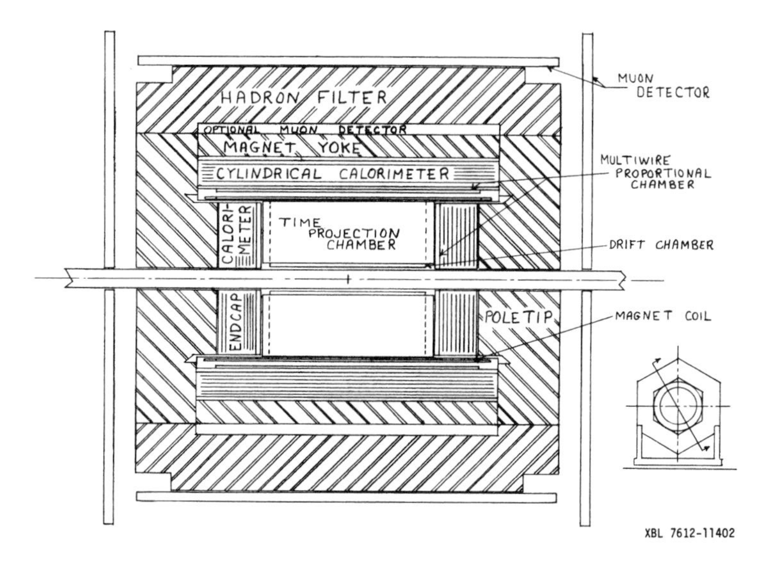

Figure 7.11 Proposal image of PEP-4 facility (1976). As the collaboration came together, both the device and the sociological structure of the collaboration ramified. In this cross-sectional diagram from 1976, the device reveals several elements not present in figure 7.10, including end cap calorimeters, cylindrical calorimeter, and a muon detector. Source: LBL XBL 7612–11402, LBL. Courtesy of Lawrence Berkeley Laboratory, University of California.

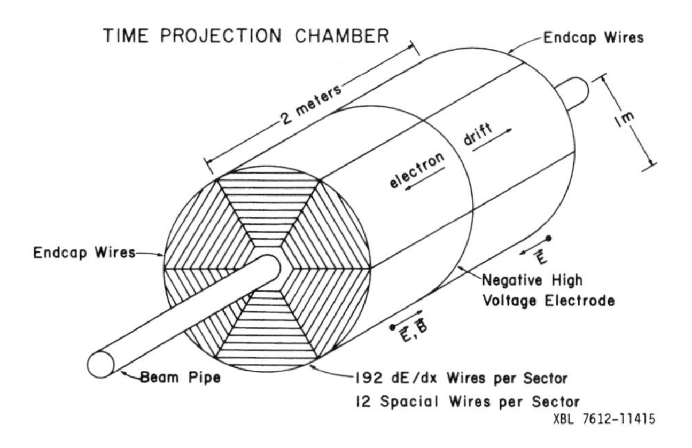

Figure 7.12 Proposal image of TPC (1976). By the time of the TPC proposal to the PEP facility at SLAC, the TPC itself had been doubled in size from Nygren's first sketch and had gained significant complexity in its wire plane. Source: LBL XBL 7612–11415, LBL. Courtesy of Lawrence Berkeley Laboratory, University of California.

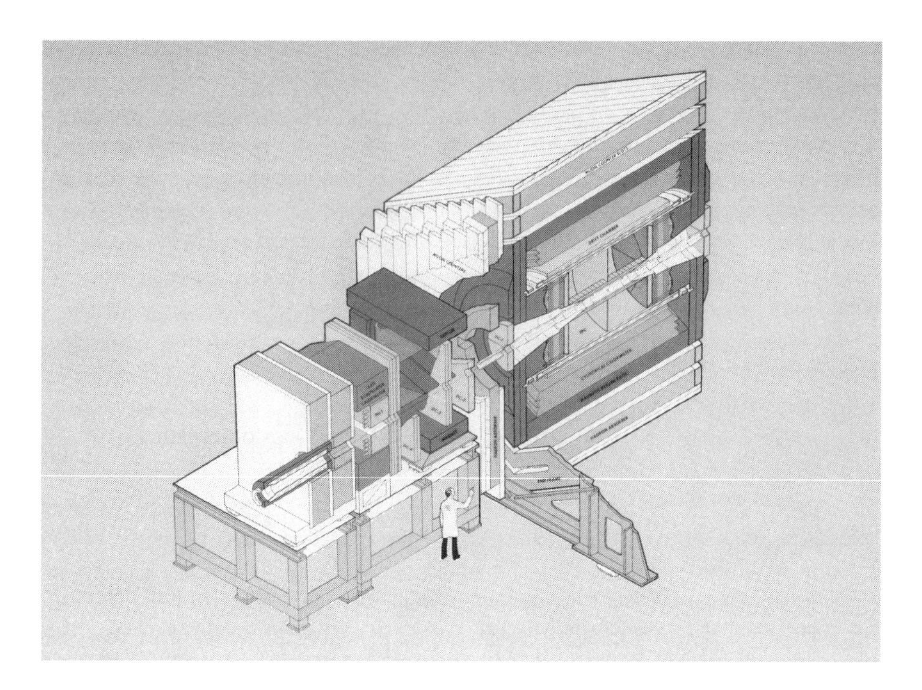

Figure 7.13 TPC/PEP-4/PEP-9 Facility (1980). Finally, in 1980 one sees not only the full TPC/PEP-4 facility but its "other half," PEP-9, the experimental collaboration that joined with PEP-4 to form (sometimes uneasily) a facility twice composite: first, each "experiment" had its own hybrid hardware, software, and sociology, and second, the two experiments were administratively bound at the top. PEP-4 is the larger diameter piece of the apparatus to the right of the "end frame." Source: LBL BBC 794 4269, LBL. Courtesy of Lawrence Berkeley Laboratory, University of California.

produced in decays from excited states over all angles. And with the discovery of the tau, there was a widely accepted theoretical argument for the existence of a new generation of quarks (a top quark and a bottom quark) that, if they came in with an accessible energy range, would "provide a fascinating and experimentally challenging kaleidoscopic richness." Just as the Mark I had found a surge in the ratio *R*, there would be a new leap in *R* that the TPC collaboration would be sure to catch because it was designed precisely to capture charged particles as they flew off from the interaction region, no matter what their direction.[51]

But the TPC's experimental advantages had social consequences. By studying a single wedge of solid angles around the interaction region, fixed target experiments could specialize their apparatus and maintain a kind of technological purity: a relatively simple array of spark chambers, for example. In the

51. Clark et al., "Proposal for a PEP Facility Based on the Time Projection Chamber," 30 December 1976, box 2, 9–11, on 9, DNyP II.

jargon of the time, each of the solid angle sectors defined a piece of "real estate," and by positioning their apparatus within a particular angular zone, each of several groups could pick out its very own piece of physics. Fixed target detectors therefore more easily fit into the category "experiment." Colliding beams and their concomitant centralization of available physics into a few interaction regions created a new social as well as physical world in which experimenters had to work. Big, expensive, and comprehensive, colliding beam detectors disrupted the relation of "private" ownership by any single group. Even the architecture of the colliding beam detector revealed and guided a new intergroup dynamic in which no group could operate alone. Cylindrically symmetric, this concentric structure of plastics, metals, gases, and chips embodied a new social structure among laboratories, physicists, and physics argumentation. But the transition from "experiment" to "facility" was not unproblematic.

The "facility" style of collaboration was manifested in the juxtaposition of devices: the electromagnetic calorimeter, the muon detector, the drift chambers, the superconducting magnet, and of course, the TPC itself. Each of these was the visible presence of each of the constituent laboratory groups: UCLA built the pole tip calorimeters, Johns Hopkins took on the muon detectors, Riverside adopted the drift chambers, Yale designed data acquisition, and LBL undertook the construction of the magnet and the TPC. Despite their quasi autonomy, these different subsystems did not sit next to each other in total independence. At the level of pure construction, the different subsystems shaped each other by constraining each other's form and function. The physical configuration of one device hemmed the shape of its neighbors, while the temperature and gaseous environment of the TPC conditioned the working conditions of all its surrounding subassemblies. Across the *physical* interface between subdevices lay an *electronic* one. Beyond the reliability of any particular component, the apparatus as a whole had to function in a global electronic network that communicated both through signals and through the elaborate coordination of software. Since any given university group was typically attached to a particular subassembly of the detector, each team was responsible not only for its own piece of the machine but for the interaction of its software with others'. Put a different way, the sociological, institutional, hardware, and software architectures had to run in coordination; they had to be integrated, not simply amalgamated.

Along with the proposal itself, the PEP authorities demanded a proposal defense, and the TPC Executive Board had to orchestrate this part of the campaign in its jointly political and scientific aspects. At its 17–18 December 1976 meeting, for example, the board began parceling out tasks. Within the domain of computer simulations alone, several studies were needed: One group would determine the spectra for the detection of π, K, P, e, μ, η^0, K^0, and Λ^0. Another would demonstrate the "physics gain" netted by using a cylindrical calorimeter rather than the competition's planar design. A third would explore one-gamma and two-gamma processes to demonstrate the virtue of being able to measure over 4π steradians. And a fourth would simulate processes involving charm decay and the hoped-for discovery of a top quark. Other groups would prepare "defense kits" to guard against attacks on the TPC's ability to handle field nonuniformities, large backgrounds, positive ions, and the overall precision they had "advertised." Buried as a subcategory was the assignment of Wenzel to "construction schedule," a topic that would become the collaboration's bane for the next decade.[52] Each person or subgroup assigned to a topic was to prepare a summary no more than a page long. According to the minutes, these were to "be like staff briefs for a politician's news conference."[53]

Readying these briefs was only part of the political positioning necessary to get the proposal through the review process. Other members of the collaboration would attempt to meet with individual members of the Experimental Program Committee, later renamed the Experimental Program Advisory Committee (EPAC), at SLAC to try to sway those members assigned to the TPC. Others would go on the offensive: "Readers" would prepare attacks on each of the competing proposals. One team member indicated, for example, that he thought magnet costs were "considerably underestimated" in the Barry Barish–Richard Taylor proposal.[54] Much more delicate was how to handle their potential ally, PEP-9, also known as the 2γ collaboration. This group had, from early on in the competition, proposed to put a detector in the one spatial region in which the TPC was not particularly good at capturing particles—along the beam line. Here the TPC team faced a conundrum: EPAC would probably value an explicit collaboration of collaborations, and therefore it was in the interests of PEP-4 to appear welcoming to PEP-9. At the same time, to collaborate meant to accept yet another set of technical, political, funding, and ultimately, physics constraints that might restrict both the physics moves available to and the possible credit reaped by the TPC collaboration. One sees in the minutes of the Executive Board traces of this tension, a distinction made, for example, between a "public position" that would cede 120 milliradians around the beam pipe to the 2γ collaboration and a private investigation of increasing this to 200 milliradians. In a somewhat tepid endorsement, the TPC board concluded that it "sees no objection to sharing a pit with [George] Masek [the PEP-9 spokesperson], if necessary."[55]

52. Minutes of the TPC Executive Board Meeting, 17–18 December 1976, file "TPC Executive Board," box 3, DNyP II.

53. Minutes of the TPC Executive Board Meeting, 4–5 February 1977, file "TPC Executive Board," box 3, DNyP II.

54. Minutes of the TPC Executive Board Meeting, 4–5 February 1977, file "TPC Executive Board," box 3, DNyP II.

55. Minutes of the TPC Executive Board Meeting, 4–5 February 1977, file "TPC Executive Board," box 3, DNyP II.

Figure 7.14 Prototype (1978). Before construction on the TPC could begin—or the device could even successfully compete for funds and authorization from PEP—its supporters had to build and operate a small-scale prototype that would demonstrate its capacity to perform precision measurements. Source: LBL CBB 787 9080, LBL. Courtesy of Lawrence Berkeley Laboratory, University of California.

At the defense (in February 1974), the TPC/PEP-4 collaboration (as it became known) had a crucial item that it had lacked in the original proposal: a prototype that worked (see figures 7.14 and 7.15). On 15 February 1977, the collaboration operated this TPC prototype at the Berkeley Bevatron, and these results, even more than the original proposal, soon became their most persuasive display. Essentially a mockup of a portion of the TPC as it would be built, the prototype yielded a resolution for *dE*/*dx* that was within a few percent of theoretical predictions and a spatial resolution of 220 microns from the cathode pads. Along with this new information, the collaboration rushed through a Monte Carlo simulation of how the liquid argon calorimeter would respond to jets of hadrons containing large numbers of photons that would be released in

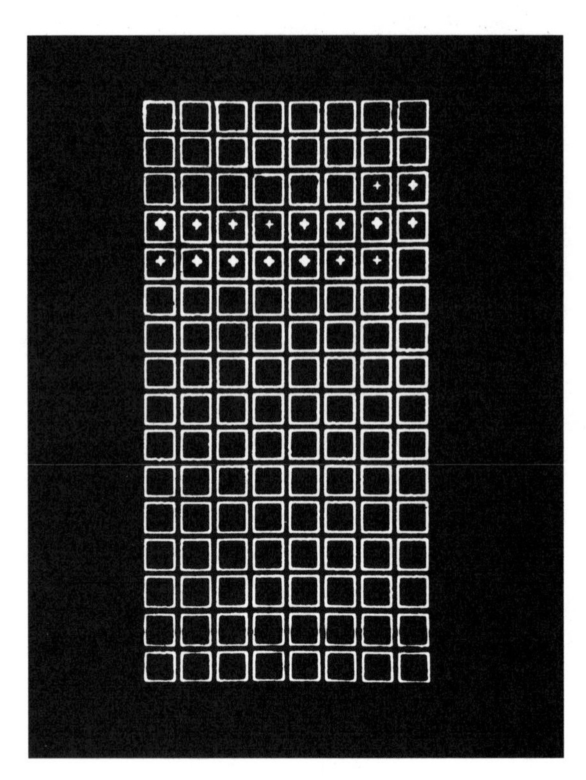

Figure 7.15*a* Pad hits, CCD prototype (1974). This picture, illustrating the pad "hits" beneath a track in the prototype TPC, was one of the principal instrumental results that persuaded the PEP committee to adopt the novel device as the main detector for PEP. Source: LBL XBB 783 2951, LBL. Courtesy of Lawrence Berkeley Laboratory, University of California.

the decay $\pi^0 \rightarrow \gamma\gamma$. In the absence of a nuts-and-bolts calorimeter prototype that could be tested in the demanding environment of an annihilation event producing large numbers of such pions, the Monte Carlo was designed to persuade the PEP hierarchy that the calorimeter could disentangle this myriad of photons and assign them properly to their particles of origin. As a final part of their written defense kit, Eberhard and Michael Green composed a memo dated 24 February 1977 defending the stability of the superconducting magnet design and arguing that it would survive a "quench" (being suddenly turned off) without generating forces and fields that would destroy it.[56]

Just days before the official "Supplement" went to PEP, the collaboration reported on its results. On 24 February 1977, Nygren and Marx received official notification that the PEP-4 proposal was approved by the directors of LBL and SLAC, and they were instructed to begin, post haste, negotiations with their

56. Proposal Updates Submitted to Experimental Program Committee, 26 February 1977, TPC-LBL-77-3, including the following: "Supplement to PEP-4 Report on Preliminary Results from the Pressurized TPC Prototype," 25 February 1974; Hauptman and Ticho, "Calorimeter Monte Carlo Program," 23 February 1977; and Eberhard and Green, "Intrinsic Stability," 24 February 1977.

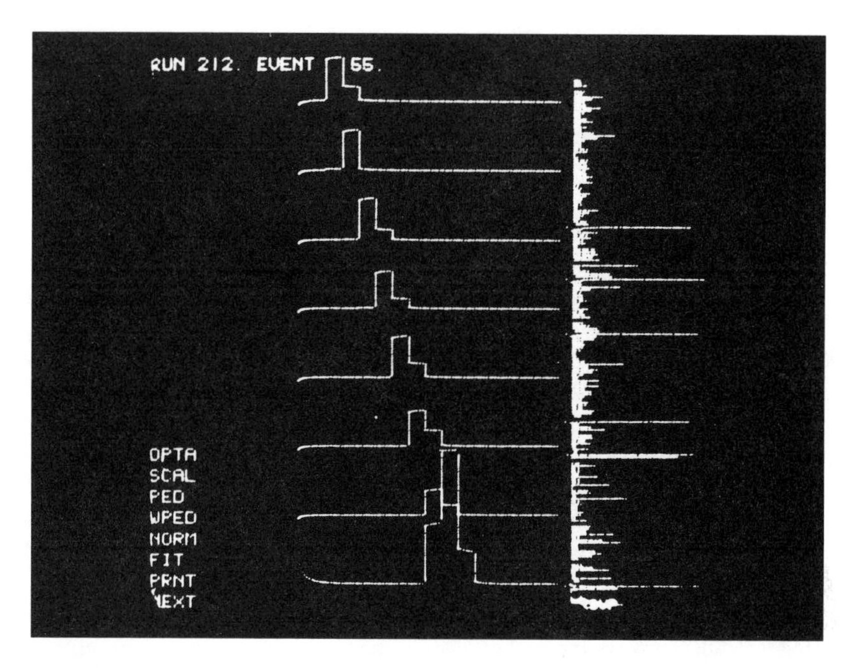

Figure 7.15*b* Timing data, CCD prototype (1974). Along with figure 7.15*a*, this display (from the prototype TPC) showed that the timing mechanism of the CCDs could effectively register the time delay of electrons as they drifted from their liberation from atoms along the particle trajectory. Source: LBL XBB 772 1328, LBL. Courtesy of Lawrence Berkeley Laboratory, University of California.

participating groups, funding agencies, the two laboratories, and the PEP leadership to establish the dates of their priority access to the facility. Without elaboration, the directors effectively merged PEP-4 with PEP-9, and the two facilities were instructed to "integrat[e]" their detectors.[57]

7.4 Intercalation and Coordination

"Integration," that refrain of the hybrid experiment, was more easily sung than done. In a series of meetings the PEP-4 and PEP-9 groups hammered at several questions, the most important of which involved the opening angle (θ_{max}) available to the 2γ group. The issue was this: For the PEP-9 facility to get pictures, there had to be a "tunnel" along the beam line through the PEP-4. How wide should that tunnel be? On one hand, if the opening got too large, the geometry

57. Fischer, secretary of the Experimental Program Committee, to Nygren and Marx, 18 April 1977, file "PEP Politics," box 2, JMP.

of restrictions on the TPC end caps made it impossible for the TPC electric field to be as uniform as was necessary to make precise determinations of particle positions; in short, it endangered the very feature that made the TPC attractive in the first place. In addition, the physical presence of the 2γ detector could interfere with access to the apparatus and might interfere with the TPC's active volume. On the other hand, if the opening was too small, the 2γ collaboration would lose too many photons or charged particles from reactions involving two photons for *their* experiment to be worthwhile.

Negotiations during March 1977 proceeded simultaneously on the plane of physics and on the plane of politics. Nygren tried a compromise that set $\theta_{max} =$ 160 milliradians. Then the TPC collaboration would work to correct the field distortions with a software massage off-line. If this proved feasible, θ_{max} would be upped to 190 milliradians. Flexibility of this kind, however, was vetoed by both collaborations. The TPC could not afford to wait because it needed to freeze the magnet design, which depended crucially on the opening, and the 2γ experiment had to have an answer immediately because it too needed to complete its designs. Bottom line: the TPC refused to go above a θ_{max} of 200 milliradians because it was sure its device would become unworkable, and the 2γ contingent refused to go under 170 milliradians because its experiment would be rendered useless. By the end of March, the two groups were negotiating over this 30 milliradian territory, the steradian analogue of Taba on the Israel-Egypt border.

In further testimony to the inseparable admixture of categories, the minutes labeled one section "Physics/Politics/Sociology." Here the issue was control of information. Who would get the data? There were two schemes as to how to divide the goods. One split it by the "physics": that is, events with a one-gamma intermediate state would go to the TPC/PEP-4 collaboration, and those with a two-gamma intermediate state would go to the 2γ collaboration. This division was rejected because for many of the goals of the TPC group, it would need two-gamma events, and if the 2γ collaboration had not been present, the TPC would have gone quite some way (down to 80 milliradians) toward capturing two-gamma events on its own. Another possibility was that the data be divided by trigger, but the problem was that a one-gamma trigger would get most of the two-gamma events, but not vice versa. The minutes concluded: "The general feeling is that each group will get all the data obtained by their own trigger and [in those events in which] their own detectors play [a] major role. . . . As to overlapping triggers, maybe each group should have and be gentlemen about it."[58]

Gentlemen or not, having won the approval of PEP the group now had to produce the most complex high energy physics facility ever built. A new tone

58. Chien, "On the Question of Compatibility between PEP-4 and PEP-9," 25 March 1977, file "TPC Executive Board," box 3, 6, DNyP II.

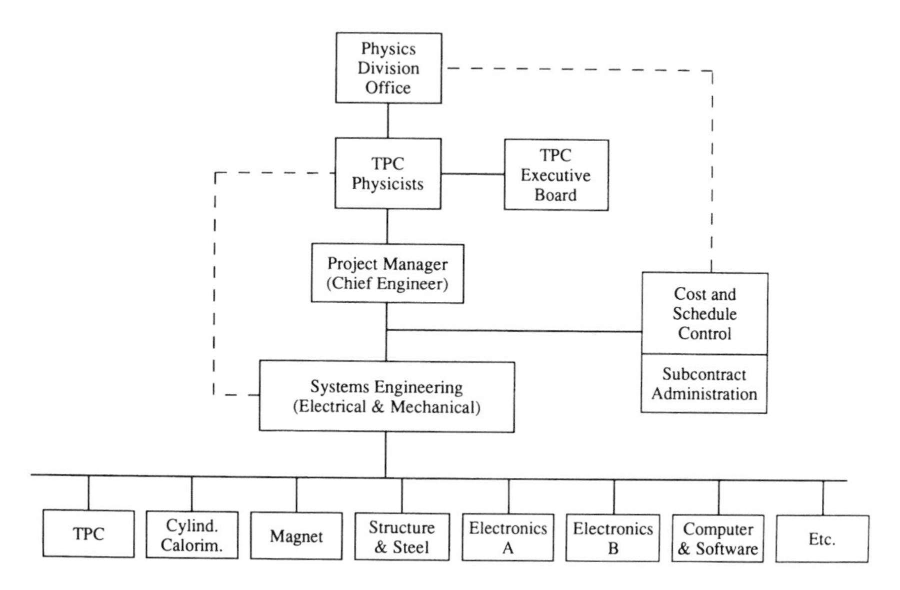

Figure 7.16 Proposed TPC management structure (1977). Although this early (1977) management structure was simplicity itself when contrasted with later organizational charts (see figures 7.24–7.27), it already departs from the relatively centralized work structure of the Alvarez laboratory. No one person is central here—the group TPC Physicists represents one pinnacle of power with the quasi-democratic TPC Executive Board confederated with it. Source: Marx to Birge, "Management of the TPC Project," 21 March 1977, with enclosure, "Management Structure for the TPC Facility," file "TPC Politics-Internal," box 2, JMP. Courtesy of Lawrence Berkeley Laboratory, University of California.

entered the memoranda, and a vocabulary including such terms as "management," "interfaces," "control," "critical path," and "PERT diagrams" joined the more traditional lexicon of physics. "My goals," Marx wrote to Birge on 21 March 1977, "are . . . to keep control in the hands of the Physics Division[,] . . . to allow engineering to manage the engineering effort[,] to [minimiz]e the amount of bureaucracy[,] to avoid . . . cost overruns and schedule delays," to "interface" the subsystems, and to create milestones and review points to facilitate the interlocking design and construction of the whole. To master the management, Marx would create a Systems Engineering Group, which would have "primary day-to-day control on engineering decisions"; it was this entity that would provide "interface definition and control."[59]

The administrative structure (see figure 7.16) would involve a project manager (who would also be chief engineer), the Systems Engineering Group

59. Marx to Birge, "Management of the TPC Project," 21 March 1977, with enclosure, "Management Structure for the TPC Facility," file "TPC Politics-Internal," box 2, JMP.

(which would have one full-time engineer and several physicist advisors), and the Cost and Schedule Control Group (which would be made up of one manager and one administrative assistant). The TPC Executive Board (the old "commissariat") would be the physicists' chief decision-making body.[60]

Physicists—that is, those not serving in the Systems Engineering Group—would have two avenues of access to the construction: by directing their ideas to the project manager or by appealing to the Systems Engineering Group. "Lastly," Marx added, "physicists will interact with engineers involved in the various subsystems as they have traditionally done—sharing their ideas and insights with the engineering staff whose job it is to evolve feasible designs and construction techniques." Engineers would vouchsafe the "feasibility of the construction," and physicists would guarantee the "integrity of the science." This carefully orchestrated pairing would assure the phased development that would be needed in individual subsystems (such as the liquid argon calorimeter or the high-voltage system) as well as in the system writ large.

> Each phase must include clear intermediate milestones, goals, objectives, and periodic design reviews. After certain, specified time periods, designs in a given area must be fixed, and we proceed to the next phase of the development program. This process must be coordinated so that indecision in one area doesn't adversely affect another. *At times, we may have to live with designs that will work well even though they are not our newest, most clever, ideas in order to meet our schedule and budget.* This is a primary motivation for a control of systems changes within the Systems Engineering Group.[61]

The culture of engineering, rather than that of physics, lay behind the four "phases" that would be enforced: a Feasibility Study that would produce proposal and prototype test results; a Definition Phase that would produce preliminary design reviews including system requirements, constraints, trade-offs, and definitions; and a Design and Development Phase that would terminate with final design decisions and a "critical design review" by outside, "objective" experts as well as inside experts. Finally, there would be the Construction Phase, which embraced purchasing, fabrication, assembly, and testing for "a July 1980 turn-on of the facility." In a revealing coda to the management plan, Marx captured the leadership's vision of what physicists and engineers wanted: "well-defined decision making, control and communication responsibilities with a minimum of bureaucracy for the physicists, engineers and others working 'in the trenches.'" Everyone wanted that. In addition, though, physicists would

60. Marx to Birge, "Management of the TPC Project," 21 March 1977, with enclosure, "Management Structure for the TPC Facility," file "TPC Politics-Internal," box 2, JMP.

61. Marx to Birge, "Management of the TPC Project," 21 March 1977, with enclosure, "Management Structure for the TPC Facility," file "TPC Politics-Internal," box 2, JMP.

maintain "ultimate scientific control," while engineers would be able to "work within a structured environment" under well-defined constraints.[62]

Problems were numerous and pressing. At the 6 June 1977 TPC Executive Board meeting, the issues at stake filled several pages. "How do we assure *E* field uniformity? Guard ring construction and resistor selection. Is there another way? How are *E* and *B* aligned? What tolerances are needed in construction of field shaping cage?" To create the electric fields needed to cause the ionized electrons to drift down the TPC, a powerful high-voltage system was needed. "What's the maximum HV [high voltage] we want? How much above that do we want to test materials? What's already known about HV properties of useful dielectrics (e.g., Kapton and Mylar)?" Typical of a wide set of questions concerned with the interaction between the environmental conditions presented by other aspects of the apparatus was the following: "What's known about the degradation of HV strength in a radiation environment? What tests do we do to lessen our ignorance?" Then there were questions that related directly to the discriminatory capabilities of the detector itself—how accurate would it be: "How do we string sense wires and field wires to minimize dead spots in chamber? . . . How does our sensitivity and energy resolution vary as we approach the wire termination?" Finally, calibrating the chamber presented its own difficulties: would a single source of test radiation (such as radioactive Fe^{55}) be sufficient "or do we want several sources of different energy?" After tens of such queries, the author ended with the daunting remark "I'm sure I've left out 20 pages of questions."[63]

With so many unknowns in so many areas of the experiment, nothing was more important than dividing responsibility effectively. This division was an elaborate dance, partly political, partly economic, partly pragmatic. Take, for example, the calorimeter, essentially a huge vat of very low temperature liquid argon that was instrumented to produce ions when traversed by charged particles in such a way as to register their energy. In divvying up this subsystem of PEP-4 in June 1977, the collaboration reasoned as follows: Cryogenics, including cooling systems, storage, and so forth, had been an LBL specialty since the days of the hydrogen bubble chamber. These tasks, the team concluded, ought to remain based at LBL. The cryoenclosures (which contained the liquid argon), both thermal and structural, would interact fundamentally with so many aspects of the TPC itself that LBL argued it should take that too. But the guts of the calorimeter, the sensitive volume, was more easily parted from the rest of the TPC, and this

62. Marx to Birge, "Management of the TPC Project," 21 March 1977, with enclosure, "Management Structure for the TPC Facility," file "TPC Politics-Internal," box 2, JMP. It is interesting to note that the 15 March 1977 draft of this document left out the clause that specified that engineers ought be in charge of engineering.

63. [Marx ?], "TPC Problems," for 6 June [1977] TPC Executive Board Meeting, file "Job Lists for 6/6 TPC Board," box 7, JMP.

was appropriately farmed out to UCLA. Electronics presented a more troublesome case. Of that part of the electronics that interacted with the TPC, LBL would take charge, consulting closely with UCLA. But the "special electronics," that part tied most closely to the sensitive volume of the calorimeter, logically would remain with UCLA. To tie the effort at UCLA and LBL together a human link was needed, and LBL planned to send an engineer to its Southern Californian associates from August 1977 to March 1978.[64] These sorts of negotiations were repeated for each of the many subsystems that made up the composite detector.

All was not proceeding according to plan, however. Overruns in the cost of moving the Mark II detector from SPEAR to PEP threatened to gobble up funds previously earmarked for PEP-4.[65] By October 1977, the collaboration had already begun to slip back against its projected internal timeline. Sy Horowitz, a mechanical engineer with administrative responsibilities, titled a sharply worded memorandum to his colleagues: "Do We Mean What We Say in Our Schedules?" All too foolishly, he believed, the collaboration was letting crucial milestones, such as the 15 October 1977 systems definition, "gracefully . . . slip by." Instead, the team should be working hard to finish tasks early and "banking" time: they would need it later. "The Mark II project was allowed to float in its early stages for what then seemed to be good reasons. Plans were treated casually then and advice on the results of continual change went unheeded. . . . It may sometimes seem fashionable to malign the Mark II project but in the words of Santayana, 'Those who do not learn from history are condemned to relive history.'"[66]

The reasons for the slippage were many; the most painful was the necessary, but often tense, coordination between physicists and engineers, managers and managed, and electrical and mechanical engineering. There were small details that would have been insignificant had the groups been working in isolation; now that they were building an integrated system, the misalignment was highly disruptive. For example, the schedules between the electrical and mechanical engineering groups were set differently. The electrical engineers used the fiscal year, whereas the mechanical engineers followed the calendar year. This opened chasms in the scheduling of end cap wiring development (the end cap terminated the gaseous volume and contained the electrical transfer from the wires in the chamber to the readout electronics of the CCDs); it disrupted planning for the installation of the electronic house, the pressure vessel, the TPC wiring sectors, and the final assembly of the whole machine. In his distinctive

64. [No author], "Logistics of TPC-LAD Design and Construction," 9 June 1977, book 71, HHP.

65. Marx to Sessler and Panofsky, 21 June 1977, file "Notes to/from LBL Directors," box 7, JMP.

66. Horowitz to Distribution, "Do We Mean What We Say in Our Schedules?," 17 October 1977, file "John Hada," box 8, JMP.

administrative voice, John Hada (a former military man brought into the TPC as a project administrator) informed the leadership that these were "conditions that anathematize effective cooperation and coordination among physicists, electrical and mechanical engineers and collaborators." Hada complained that the lead electrical engineer had "[i]nsulat[ed]" his colleagues from the collaboration, and that the electrical engineer resented having a mechanical engineer as project manager. "I responded rather brusquely . . . that it doesn't make any difference whether the Project Manager was a Gynecologist, [he, the electrical engineer] is obliged to support the Project Manager. . . . Disloyalty is a character deficiency I simply cannot tolerate in any person."[67]

At the same time, the working relationship between physicist and engineer remained very much in flux. I suspect that at the root of much of the early tension in the experiment lies the transitional nature of the TPC/PEP-4 facility's sociology. In earlier experimental apparatus, even devices as large as the 72-inch hydrogen bubble chamber or the Bevatron, there was a nonnegotiable subordination of engineer to physicist. While engineers were certainly central contributors to all the major devices in the Berkeley hills (the role of a William Brobeck can hardly be overestimated at the Bevatron, e.g.), power lay entirely in the hands of an Ernest Lawrence or a Luis Alvarez. With the TPC, the level of instrumental complexity, for the first time, evidently exceeded the capability of any individual to master. Strictly speaking, no "one" was in charge; and the project consequently had to proceed in a collaborative mode that required detailed, constant, and powerful intervention from the engineers at each step. These changes came hard, both for the physicists and for the engineers.

In an anonymous letter to Nygren and Marx from late 1977, one sees the manifest discomfort of one participant, probably (and one might at first think surprisingly) an older technician or engineer:

> I have heard of a large corporation where the engineers are the bosses and the physicists are the consultants—and flit from job to job at the beck and call of their superiors—offering advice which is often ignored. I hope Ernest never heard of this!
>
> In *this* lab the physicist (chemist, metallurgist, etc.) is the BOSS and the engineer is the HANDMAIDEN. I don't know how you guys got started in this business but that's how it all got started out here and that's how it works out BEST. The physicist does not ask the engineers what they ought to be doing—the physicist does not even tell the engineers what they ought to be doing—the physicist DEMANDS that the engineers do thus and such.
>
> When you are running a 13 megabuck project you do not think of "How to Win Friends and Influence People." . . . You have to be the boss—you have to step

67. Hada to Marx, "Conditions that Anathematize Effective Cooperation and Coordination among Physicists, Electrical and Mechanical Engineers and Collaborators," file "John Hada," box 8, JMP.

> on toes—you have to be a bastard—you have to be a prick—you have to be a sonofabitch. . . . If you guys really object to adopting this "altered state of consciousness" consider the alternative. You remain "nice guys"—you accept glib promises—and you wind up with another Mark II fiasco.
>
> You physicists somehow think of these ideas for new hardware—you beget the concept—you go off somewhere and then you come back home with all this money to pay our salaries. Who then should run the project—you or the engineers?"[68]

This rather raw report was striking in its invocation of gender and power imagery but still characteristic in its warning that the violation of the social order between physicist and engineer was leading to chaos. More generally, there was a pervasive concern with attaining the right relationship between engineering and physics. Along these lines were the more sober (though less startling) remarks by Michael Zeller, when he reported to Marx on the situation in electronics. Some aspects of electronics development were going well, Zeller contended; he had "absolute confidence" in several of the engineers, especially in Mitch Nakamura's group, which Zeller reported as doing an outstanding job of testing the new CCD chips, the eyes of the device. Nakamura and Kai Lee had successfully built a board with 16 channels that could serve as a prototype for the tens of thousands of channels that would be needed in the full apparatus. Combined on one board would be the CCDs themselves, the clock mechanism that set the time intervals between sampling, the device that shaped the output signals, and the electronics that permitted many channels to be kept apart. Not only would this prototype test crucial elements of the system, it would give the collaboration a better sense of the expected cost of this part of the readout. But despite the success of Nakamura and Kai, the task "of evaluation and prototyping should not be left in the hands of the engineers alone. We [physicists] need intimate familiarity with the operation of this system and thus should have a physicist in close liaison with this group." Best of all, Zeller judged, would be if Nygren himself took on the role of liaison, but others could do it. "Mitch is very concerned about physicist incompetence causing interference with Kai's work. Kai has shown unhappiness when physicists screw up and he has to fix it, so you have to be careful about this. Nonetheless we must be involved in this program."[69] Evidently not all the nonphysicists were as keen to be ordered around as the author of the Boss/Handmaiden memorandum would have it.

Indeed, one clear alternative to the plea for the return of the physicist-patriarch came from those who wanted a rule-governed assembly, in which the dead hand of efficiency alone would determine who would do what when. "Cost

68. Anonymous to Nygren and Marx, n.d. [probably November or December 1977], file "LBL Politics & Memos," box 1, JMP.

69. Zeller to Marx, 10 January 1978, file "TPC Politics-Internal," box 2, JMP.

effectiveness studies" was one possible scheme, and its considerable influence at the Pentagon seemed to make it a promising candidate for adoption in big physics. Hada, as project administrator, was asked to report on whether it would be appropriate as a management tool for PEP-4. Though he had used it extensively at the Department of Defense, Hada now advised against it because, he argued, cost-effectiveness only works well as a criterion when the type of output is constant and the need is to maximize other quantities. Given that future technologies relevant to the TPC were in flux and physicists set extensive restrictions on engineering variations, cost-benefit studies struck him as nearly useless.[70]

If cost management could not provide a neutral, impersonal key to management, other schemes still offered that hope. Most people put their bets on "critical path management," in which an administrator would lay out "milestones" (key accomplishments) and the necessary prerequisites for them. Then, either by hand or by computer, the interleaving strands would be woven together to generate a "critical path," the master list of tasks to be accomplished by specified dates. This form of management demanded detailed reporting at every level, from the project manager, through the group leaders, and down to subtasks on such items as software, the calorimeter, and the end caps of the TPC. To maintain an up-to-date path, the project administration issued reporting requirements in January 1978—it would take months to put into effect—that included "problems," "downstream" impact on the project, corrective action proposed, person responsible for implementation of the correction, person reviewing and approving the problem analysis of report, along with cost, manpower, and any variance from the original specification.[71] These microscale exercises in authority were supposed to align all tasks with the critical path. But in the end, however much power diffused in Foucauldian "capillary" form into the limbs and fingers,[72] the functions of the sovereign high leadership, especially the project manager, remained matters of the utmost importance.

One suggestion that was entertained was to divide the functions of the project manager—Paul Hernandez, who had been Alvarez's right-hand engineer on the bigger bubble chambers—into an electronic and a mechanical part. To at least one senior engineer, Sy Horowitz, this course was doomed from the start: "If the TPC Project Manager's planning functions are split into M[echanical] E[ngineering] and E[lectrical] E[ngineering] parts, it seems much like asking a

70. Hada to Nygren and Marx, and Hada to Hernandez, "Cost Effectiveness Approach, Its Values and Dangers," file "John Hada," box 8, JMP.

71. Hada to PEP-4 TPC Project Reporting Entities, 12 January 1978; and Marx to Distribution, 16 January 1978; both in book 70, box 1, HHP.

72. The classic references to Foucault on the capillarity of power can be found in Foucault, *Power/Knowledge* (1980), chap. 2, "Prison Talk," esp. 37–39; on these themes also see Foucault, *Discipline* (1979), esp. chapters "Panopticism" and "The Carceral."

carpenter to build a house. Then his hammer is split in two and the halves are given away to electricians to do the job." Suppose the functions were divided, Horowitz added, who then would lead the way in developing shared standards and cost planning? Who would integrate these two interlocked sides of the project? Who would coordinate manpower, the project plan, the schedule, and changes in design?[73]

In the event, far from clipping the project manager's power, Marx wanted to boost it. He wanted Hernandez to be responsible for keeping the TPC on schedule and at cost, meeting specifications and quality requirements, and maintaining an up-to-date project plan that would both guide participants and inform the scientific spokesman. All reports from below would reach him, and these would then culminate in a monthly account to the scientific spokesman. On a draft of a memo, Marx scribbled to Birge, "Do you think this will bruise any egos or cause trouble between Divisions at the Lab. Here's your chance to head me off before I cause ruffled feathers." Birge replied: "Jay, perhaps this should be a 'discussion paper' rather than a final action."[74]

In January 1978, the team began preparing for a major design review, which involved surveying all of the subsystems. On one overhead transparency from the time, Hernandez gave the project status his (weak) green light: "In general TPC schedule & cost OK[;] problems but no panics[;] everyone seems slightly behind[;] aggravation about equally distributed; overall rating about C− but catching up."[75] By this time, Marx had attended "management school" (one more sign of the distance from the hierarchical but still improvised management structure of the Alvarez era), and he had strong views about what the manager should do.[76] To Nygren on 21 February 1978, Marx wrote that because of Hada's problems in finding direction in the project, because of his own experience at project management school, "and especially as a result of my own deep concerns and instincts, I feel the TPC Project is *doomed* to failure unless we have a functioning Project Manager soon (<1 month from now)." This permanent position would have to be responsible to the "Scientific Leadership" and, crucially, "[h]ave his finger on the pulse of the project in *all* areas including Mechanical Engineering, Electrical Engineering, and collaborators." Though Hernandez was project manager, Marx argued that the authority of the position was not extensive enough to hold the center. Solving the problem of authority would

73. Horowitz to Hernandez, "Should Planning be Centralized or Split Into Electronic and Mechanical Engineering Parts," n.d. [mid-January 1978], book 70, box 1, HHP.

74. Marx to Nygren, "Project Management Responsibilities," file "Project Management Responsibilities and Authorities in the TPC Project," box 7, JMP.

75. Hernandez, photocopy of transparency shown at TPC Executive Board Meeting, 13 January 1978, file "TPC Executive Board," box 3, DNyP II.

76. Marx later considered the two-day course to have been too much focused on the management of manufacturing projects. Marx to the author, December 1995.

require enlisting the Physics Division head (Birge), the associate director of the Engineering and Technical Systems Division (Hartsough), and the LBL director (Sessler). At stake were their own futures. "We have both worked hard for many years to gain the reputations we now have in the Physics Community. These reputations are now on the line. Speaking for me, I don't intend to see mine destroyed because we are too 'nice' to *demand* proper management for this project."[77]

As the project advanced, and the leadership found itself advancing farther from the known shores of previous endeavors, Nygren and Marx continued to hope that the structure of the project would soon come to seem more familiar. Review procedures had demanded an assessment of progress during 1977, and in early 1978 Nygren and Marx wrote the following: "[T]he further along on the R&D spectrum (from basic research to operational development) a project appears, the less the uncertainty and hence, the less risk it exhibits. . . . [T]he differences between R&D management [and more conventional management] are most striking at the basic research and exploratory end of the spectrum. As the nature of the effort approaches the development phase, however, management will become more hierarchical, applying orthodox controls."[78] In reassuring the funding agencies—and perhaps by way of reassuring themselves—the collaboration emphasized its orthodoxy, and in particular its intention to return to the familiar hierarchical organization of most earlier experiments. This struggle between the reality of a largely dispersed quasi-democratic organization and the comfort of an autocratic structure is one that appears in the earliest rhetoric of the experiment and lasts to its final moments in PEP 18 years later. To explain the pressures under which they were operating, the participants cited the twin "forces of 'technology-push' and 'requirements-pull'" that issued from the different orientations of the physicists and the engineers. On one side, the "engineers sought to design material[s] to take advantage of new discoveries," while "the physicists initiated the 'requirements-pull' by insisting on advanced items to satisfy current or future needs." To keep both in check, the leadership promised to follow a "technological building blocks theory" that would defer technological development until after the technology had been fully tested at the prototype stage. This, they assured their Department of Energy reviewers, would "reduce complexity" and "increase reliability" of the final PEP-4 apparatus. "Reliability" was not merely a property of electronic components; it had to be built into managerial structures as well.[79]

77. Marx to Nygren, "The Project Manager Problem," 21 February 1978, file "Project Management," box 7, JMP.

78. [Marx and Nygren?], "Review and Appraisal of PEP-4 During FY 1977," file "Review and Appraisal of PEP-4 During FY 1977," box 1, DNyP II.

79. [Marx and Nygren?], "Review and Appraisal of PEP-4 During FY 1977," file "Review and Appraisal of PEP-4 During FY 1977," box 1, DNyP II.

The redefinition of managerial responsibilities in early 1978 immediately affected the spokesperson, deputy spokesperson, and project manager. But soon, the ranks below were also touched by structural questions—in particular, the rank Marx, Nygren, and Hernandez dubbed the "prime physics contractors." This term was not, as the name might at first suggest, a reference to outside industrial contractors. Prime physics contractors were the *physicists* in charge of significant sections of the project. They were "contractors" insofar as they were contracting out tasks to engineering and technical personnel. In the new management scheme, they would set specifications after consulting both with the physics leadership and with other physicists on the project. More important, each prime physics contractor would work with engineers to get specifications and definitions worked into the system: "This process involves communications with engineers on a frequent basis." Presumably, the process of implementing these requirements would present engineering alternatives. It was the physics contractor's job to vet these proposed trade-offs in budget, time, and specifications: "[R]elationships should be established between engineers and physicists that encourage free and open communication." In other words, the physicists had to make their requirements clear to the engineers, and the engineers had to let the physicists know what was possible, and what it would cost in time, space, parameters, and money.[80]

To the engineers, Nygren and Marx wrote a separate note, correcting "prime physics contractor" to the less confusing (and perhaps less industrial-sounding) "physics contact personnel." Simultaneously, they created a category "responsible engineer" that became the engineering analogue of the physics contact.[81] Thus, each major task had a pair of leaders, one physicist and one engineer. Here are some examples:

Task	Responsible Engineer	Physics Contact
Readout System	J. Meng	S. Loken
Compensation Coils	G. Miner	L. Stevenson
Safety	E. McLaughlin	D. Nygren
Trigger Electronics	B. Jackson	M. Zeller
TPC Field Cages	F. Jansen	P. Robrish and D. Nygren

Boosting the status of the engineers in this way would only be meaningful if the engineers spoke up, if they confronted the physicists with the consequences of

80. Nygren, Marx, and Hernandez to TPC Management and Engineering Leadership, "Responsibilities of Prime Physics Contractors," 5 April 1978, file "PEP-4 Project Management," box 7, JMP.

81. Nygren and Marx to Distribution (ME [Mechanical Engineering]), 11 April 1978, book 70, HHP; for an explicit discussion of the vocabulary change, see Marx and Nygren to Distribution, 25 April 1978, book 70, box 1, HHP.

their choices and outlined alternatives. Sometime in mid-April 1978, Nygren and Marx appealed to the engineers to do just that in a memorandum entitled "Responsibility of Engineers." Using the specifications and requirements provided by the physicists, the engineers were directed first to establish the feasibility of *exact* compliance with the physicists' plans and to review these assessments with their physics contact. Then, it was up to the engineers to take a more active role, suggesting "trade-offs," "de-scoping," or "schedule adjustments" and indicating the consequence of such "evasive maneuvers." "During this phase," the leaders insisted, "close and open teamwork between engineering and physics is imperative." Over and again, Nygren and Marx stressed that many of the physicists' proposals and specifications were "tentative" or "in the early stages." Physicists *needed* engineers to guide them with layouts and calculations that would articulate options and ways to proceed.[82]

With the responsible engineers, physics contacts, project manager, and project administrator all in place, and a reporting system established, a work structure had begun to emerge. But coordination of the various component entities was still problematic, and once again the leadership turned to formalized managerial schemes. In July 1978, a group of mechanical engineers met to figure out how to exploit more fully and to computerize the widely hailed "critical path network" (CPN), a system designed to coordinate a large number of tasks by sequencing them in a network that would display just what was required for each to take place. D. Ohmen, an engineer, gave the pitch. The CPN would be flexible, allowing changes to be easily entered into the computer. Money would be practically visible as it flowed through the system, and the shops could integrate their acquisition of parts with the rest of the network. Since other graphical management techniques were obviously not doing the job—the TPC was falling ever further behind—Ohmen went on to criticize them. Gantt charts of the early 1950s, for example, merely displayed progress (they are essentially "percent completed" bar graphs) and revealed nothing about sequencing.[83]

As the CPN advocate recounted to the TPC engineers, critical path methods had begun as a planning procedure for large military and industrial projects during the later 1950s, when several groups undertook the development of network methods of management control. One, the Critical Path Method (1957), was produced by Du Pont and the Sperry Rand Corporation to structure design, construction, and plant maintenance projects. The following year, the Navy Special Projects Office produced another network system to manage the Fleet Ballistic Missile Program, responsible for producing the Polaris, the first submarine-to-ground missile. The success of the navy's contribution, PERT

82. Draft of Nygren and Marx to Distribution, n.d. [probably mid-April 1978], book 70, box 1, HHP.
83. Brown, "Meeting Notes," 18 July 1978, book 79, box 2, HHP.

(Program Evaluation and Reporting Technique), brought huge popularity to network management systems as it was credited with harnessing the efforts of some 3,000 contractors and driving them to complete the Polaris two years early.[84] That success story left a lasting legacy at Lawrence Livermore Laboratory (LLL), the weapons branch of the Lawrence Laboratories. For it was it that had designed the most crucial feature of the system, the Polaris warhead, a lightweight high-yield thermonuclear bomb capable of being loaded on a submarine-launched ballistic missile, building not only in coordination with other parts of the AEC but in an integrated fashion with the vast rocket program at the Department of Defense.[85]

Roughly speaking, both PERT and critical path management worked by identifying milestones—key events in the chain of development—and determining what tasks had to be accomplished for each milestone to be realized. More specifically, a *node,* or *event,* is defined as the time at which a particular task is to be finished and is represented graphically by a triangle. Activities are processes that take time; these take the form of arrows. When an event depends logically on another event, but no time-consuming activity is required between them, a dashed arrow is drawn as a "dummy" logical place holder. With a complete network diagram in hand, managers could identify which tasks set the pace of the whole project and which parts of the effort could be shortened or lengthened in time without affecting the completion date of the whole (see figure 7.17).

LBL had already used critical path management during the rebuilding of the Bevatron in the early 1970s. Perhaps this is not surprising, given the close affiliation of LBL with Lawrence Livermore. But the attachment was deeper than one might at first guess. In fact, the particular computer program—Critical Path Management Program, revision G (CPMG)—was actually set up as a joint LBL/LLL system by a single programmer, Ruth Hankins. Thus, the carryover between the institutional base of the old Polaris Program and the institutional base of the TPC was direct.[86] Adopting critical path management was symbolic. It meant that the individual experiment now carried the complexity, budget, and institutional weight of the projects to which the method had previously been applied: an accelerator, the Polaris Program, and a host of other major weapons development projects.

To implement critical path management in the TPC project, the engineers had to specify the nodes (their milestones) and indicate whether they had necessary predecessors and what these preconditions were. If several antecedent tasks could be completed in parallel, this too was to be indicated. Long activities were

84. Horowitz, *Critical Path Scheduling* (1967), 5. At the same time, network management schemes were devised by British operations research staff for power station rebuilding, and by the U.S. Air Force under the code name PEP; see Lockyer, *Critical Path Analysis* (1964), 1–2.

85. Hansen, *U.S. Nuclear Weapons* (1988), 203–5.

86. Brown, "Meeting Notes," 18 July 1978, book 79, box 2, HHP.

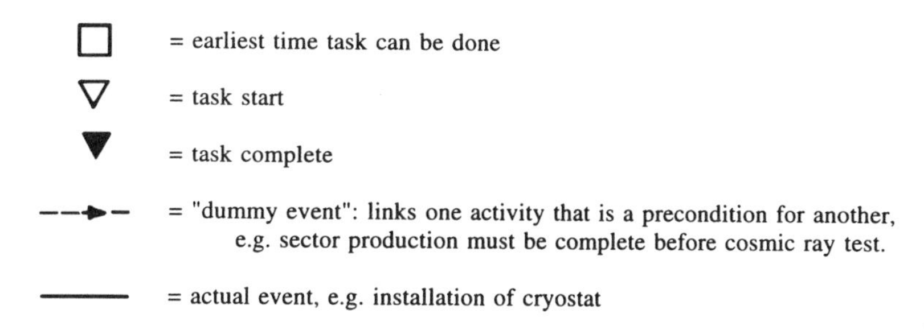

Figure 7.17 Elements of critical path analysis. These are some of the symbols commonly used in critical path diagrams, many of which appear in the highly abbreviated critical path sketch of figure 7.18. See, e.g., Barnetson, *Planning* (1970).

to be subdivided into identifiable components, and each activity was to be classified by craft code, name of task, number of people required, time demanded, and cost of procurement, and then submitted to the computer, which produced a critical path (see figure 7.18). As new information emerged the computer would revise the critical path.[87]

To Hernandez, the need for this guidance was practically self-evident, and he immediately called on the engineers to reduce all mechanical tasks for the TPC to critical path form: "For those of you who recall the situation of the Mark II project [previously the largest detector built for use at SLAC], you remember the near chaos we experienced in scheduling and rescheduling work activities as we tried to stay coordinated and get the modules shipped to SLAC on time. This project is about 8 times larger than Mark II, and we will rely on the computer and the programming help from Sy [Horowitz]'s staff for fast turn around." The critical path would guard the project's budget, and "you will readily know the last dates by which you can order materials, assign people to do a task, and get your job orders in to the Shops." In an environment in which slippage would cost $7,800 per day, to be paid out of the almost inflexible PEP budget, the group had to regain control of the schedule.[88]

Formalized control mechanisms alone, however, did not seem to provide as much managerial authority as Marx wanted. While Hernandez was away in August 1978, Marx took over as acting project manager; from that position he gave voice to his frustration about the central leadership of the project. Hernandez, he felt, was trying too hard to make "the physicists happy," and decisions

87. Brown, "Meeting Notes," 18 July 1978, book 79, box 2, HHP.

88. Hernandez and Brown to Mechanical Engineers, "Network Planning and Cost Estimating," 10 July 1978, file "PEP-4 Project Management," box 7, JMP. Some 12 years later, Hernandez judged the method a failure: "This critical path system did not work as we were not mentally set to spend as much time and money as required. The program was too complex, we did not have personnel with the computer expertise and dedication that CPM required. To obtain the weekly status of work accomplished was very difficult to keep current" (from description of book 79, 11 December 1990, box 2, HHP).

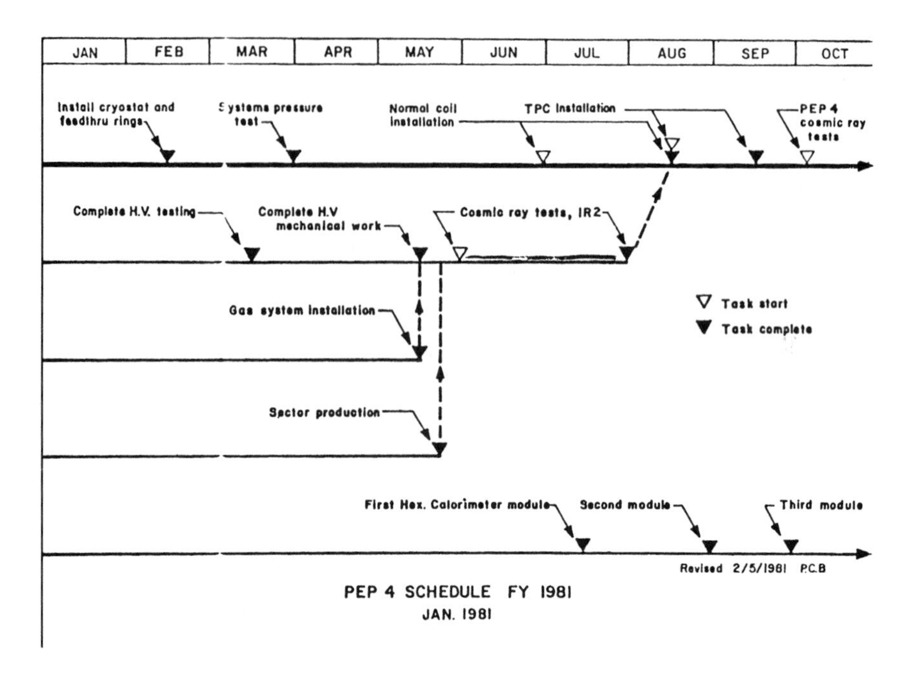

Figure 7.18 Critical path (1978). This small sample of critical path management indicates the kinds of tasks that had to be coordinated. A full critical path diagram was many times this size and vastly more complex. Source: Marx, "PEP-4 Status Report," TPC-LBL-81-23, talk given at SLAC Department of Energy review, 2 April 1981, JMP. Courtesy of Lawrence Berkeley Laboratory, University of California.

that were right in the long term but caused "unhappiness" in the short term were not being made. Lead engineers had to be strengthened; the drift chamber needed additional engineering help; one engineer desperately had to have assistance, as he was doing double duty as chief mechanical engineer and lead engineer for the support structure; the lead calorimeter team needed more engineering staff; and the high-voltage system was falling woefully behind schedule. Worst of all, the heart and soul of the project, the TPC proper, "is threatened with failure due to insufficient leadership." A "TPC Czar" was needed, and absent that Marx argued that Nygren himself should take over. "We are not doing well, and the flaws remain in the system. The project needs aggressive leadership. . . . You asked that I level with you. I have done so."[89]

Less than two months later, Marx himself was asked to take over as project manager of the now troubled enterprise. It was a job, Marx asserted, that he

89. Marx to Nygren, "View of PEP-4 from Prospective of the Acting Project Manager," 14 August 1978, file "PEP-4 Project Management," box 7, JMP.

neither wanted nor was "trained to perform." With the project behind schedule and running over budget, "I am being placed in the difficult position of having to reverse a long-standing trend towards loss of time and cost overruns." But the main effect of the "promotion" would be a separation from the "type of work I enjoy and expected to do as a staff physicist" to relieve a crying absence of leadership from the senior personnel.[90]

7.5 Group Structure

As much as the management duties weighed on the upper echelons, the pressure clearly had its effect on younger and mid-level physicists as well. In one recommendation for a TPC physicist, written at the end of 1978, Nygren commented, "I have sometimes thought that the group structure at LBL may stifle initiative. My own feeling is that collaborative association with a particular group of physicists tends to become unrewarding after a period of five to ten years, and that vitality and enthusiasm is promoted by forming new research organizations on this . . . time scale."[91] Of course he hoped the TPC would encourage greater leadership than usual, but the constraints of coordinated work and the consequent loss of control over physics activities weighed on him.

By May 1978, Nygren was concerned enough by the effects of group structure to write to Birge, complaining that a "network of self-interest, loyalties, and the inertia of habit" undergirded the laboratory's historical group structure. There were too many small "university-sized groups" with no clear purpose; SLAC was more successful than LBL because the linear accelerator had better leadership. Unlike other LBL entities, Nygren stressed, the TPC did not have a permanent staff; it "is project-oriented, and is unique in that I explicitly recognize and support the notion that people can vote with their feet as the TPC reaches the end of its fruitful period." But if it was laudable that people could walk away from the project, it was problematic that the project "lacks senior staff." By this Nygren meant, though he did not say so explicitly, that people and resources stuck in the present inertia at LBL could not, at least not easily, "vote with their feet" with a stroll *into* the TPC. On one side, the traditional group structure created an impasse for projects like the TPC because it was too rigid and confining; on the other side, a purely project-oriented system might prove too "fluid" and burden people with a mixture of projects. Nygren therefore proposed establishing a committee on group structure, heavily weighted with physicists from the TPC.[92] "I have always said [that group structure] is only an administrative problem," began Birge's reply, but in the end he conceded that some

90. Marx to Birge, 9 October 1978, file "LBL Politics and Memos," box 1, JMP.

91. Nygren to Jackson, 22 December 1978, file "Correspondence D. Nygren," box 2, DNyP I.

92. Nygren to Birge, 31 May 1979, file "Correspondence D. Nygren," box 2, DNyP I.

rethinking was needed to attract postdocs and, more generally, to generate new ideas.[93] Marx was appointed to the committee, and he began ruminating on the problem: under "*Model 1:* Project Based Task force," he wondered whether one would assign technicians to a project or pool them on particular jobs. Other staff and space could be organized around projects. Under "*Model 2:* Current Situation—Groups," he indicated that the group system left project leadership "not influential enough," and that "staffing is patchwork" (the TPC obviously the referent here). But the worst feature of the current system was that "Group survival distorts things," especially by generating small activities simply to "survive as groups."[94] "*Model 3,*" the liberal compromise, killed the bad groups and reallocated the resources of the victims to projects.

As the committee members focused their thoughts, the possibility of a projects-based system appeared more and more attractive. Organized around a physics experiment or facility, the task force would "create a dynamic structure which reflects the current physics direction of the Laboratory." In a sense, then, the task force was positioned as the "modern" (or should one say "postmodern") structure that would reflect changing realities of affiliation, rather than the "traditional" or "historical" associations that had characterized the laboratory's early years. At the same time, the committee contended that the rearrangement of space that would follow from an administrative reshuffling would be of enormous help. "It encourages people working closely on a particular project to share office space and lab space in one [area] so that vital day to day communication is facilitated."[95] That its personnel could not work at close quarters was lamented throughout the TPC's development and was often blamed for the lack of coordination between project manager and director, or between electrical and mechanical engineering.

While the TPC's fortunes ebbed and flowed (actually ebbing more than flowing), the committee saw the reform of LBL's group structure as ever more necessary. By the time of its final report in February 1980, the committee viewed the task force solution as a means of attracting postdocs and clearing up lines of authority and responsibility, healing the split between the scientific structure (which now lay to a large measure in projects) and the administrative and staffing structure (which still lay in the groups). Finally, the task force structure would foster innovation because even a small group could start a project by recruiting a small nucleus of participants. Anyone not part of a task force would, by default, be put into a "New Projects Group," and postdocs and technicians would work under the supervision of individual physicists. Immediately, the committee

93. Birge to Distribution, 4 June 1979, file "Group Structure," box 1, JMP.

94. Marx, "Problems of Task Force," handwritten notes, 15 June 1979, file "Group Structure," box 1, JMP.

95. Loken, "Interim Thoughts on Division Structure," 17 August 1979, file "Group Structure," box 1, JMP.

concluded, LBL should formally recognize the Mark II, the Free Quark Search, and the TPC as its three task forces.[96]

The upshot then was this: The TPC collaboration began largely inside LBL. By early 1980, not only was TPC/PEP-4 absorbing a major fraction of LBL's resources, it had left a permanent stamp on the organizational structure of the parent laboratory. And the stamp of TPC/PEP-4 reflected its own dynamics. As a technological and sociological hybrid, a concatenation of laboratories, types of engineering, and management, the task force was becoming a prototype for future collaborations.

7.6 The Japanese Connection

Cost overruns at the TPC came early and stayed late. Among the many ideas the collaboration put forward to make ends meet was the recruitment of additional resources from outside the collaboration. One pitch went to the oil-producing countries of the Middle East, who were seen, in 1977 at least, as potential donors. Writing to one potential Middle Eastern source of funds, the collaborators began with a tried and true argument for the funding of high energy physics: its potential future applicability to matters more relevant to running cars and building machines:

> As you know, every advance in fundamental knowledge of the properties of new forms of matter has been invariabl[y] followed by new concepts and products in chemistry, electronics, and energy management and production. . . . The scarcity of funds threatens to delay and may even maim the [TPC] project and we need help. It would be nice if you could pique the interest of your Arab friends in the energy picture to the point of them pitching in a few million dollars on this unique project. A gesture like that would be sure to draw attention and the publicity would underscore their serious concern (as you told me) about the energy situation after oil runs out.[97]

Oil magnates did not arrive in great numbers. More promising was an approach to the Japanese physics community. Much impressed by the new quark discovery at Mark I, along with the new tau lepton, the Japanese physicists indicated that they might be interested in participating in TPC developments when they too were approached in a preliminary way in late 1977, and then more formally in October 1978. At one of the first of these meetings, Marx proposed that the Japanese contingent could make important contributions to many areas, including the cylindrical calorimeters, software, electronics, pole tip calorimeter, and

96. Carithers, Loken, Marx, Oddone, and Strovink, "Report of the Committee on Physics Division Structure," 2 January 1980, file "Group Structure," box 1, JMP. Final submission is Carithers, Loken, Marx, Oddone, and Strovink to Birge, 1 February 1980, file "Group Structure," box 1, JMP.

97. Fred [Jansen] to Bratheri Man, 20 December 1977, file "Arabs and TPC," box 2, JMP.

so forth, though the LBL group rejected assistance with the TPC itself. On the Japanese side, the prospective collaborators surrounded a University of Tokyo professor, Tadao Fujii; Fujii was accompanied by Tuneyoshi Kamae, an associate professor, along with two research associates and six to eight graduate students.[98] By the time the two groups (one from the Institute for Nuclear Study, the other from the Department of Physics) from the University of Tokyo had a formal proposal, they could specify physics goals (new leptons, new quarks, lifetime of the b-quark, quark-gluon system, etc.), a collaboration goal (building the cylindrical calorimeter to detect photons and electrons), and finally "technical goals" (the establishment of a strong detector group in Japan).[99]

When the Americans looked at the Japanese proposal, they obviously saw the possibility of financial support at a time when the project was careening over budget. "One might worry," one collaborator wrote Marx, "that the possible strings accompanying such a contribution might lead to intolerable compromises in the operation of our organization." But the letter writer dismissed this as completely unwarranted. His own experience with Japanese collaborators had persuaded him that the fiscal and "other" arrangements were "totally independent." "They [the Japanese physicists] provided manpower when and where it was most advantageous . . . and have worked diligently in whatever capacity necessary." Now came a rather two-sided compliment: "On the question of how the Japanese are as collaborators, they are, from an American's point of view, ideal. They seldom have their egos involved in their work and thus do not waste time in arguing simply to def[e]nd a position. This cannot be said of many groups of people in general and especially of physicists. They do, however, work very hard, meticulously, and accurately." In the context of large-group physics, the attributes ascribed to the overseas physicists squared perfectly with American perceptions of the physics culture needed in new colleagues: "If there is a fault in their work," this author concluded, "it is their lack of imagination. They seem to rely very strongly on authority and seldom deviate from prescribed procedures. For us this attribute will probably enhance our collaboration rather than detract from it." It was, as far as this advocate of collaboration was concerned, as good an opportunity as could be imagined. "The Japanese collaboration brings us both the funding and a group of competent, hard working, and cooperative colleagues. I strongly support it."[100]

98. Marx to Distribution, "Proposed Collaboration with Japanese," file "Japanese Collaboration," box 1, JMP; see also Tuneyoshi Kamae to Nygren, 27 November 1978, file "Japanese Collaboration," box 1, JMP.

99. T. Fujii, T. Kamae, K. Nakamura, H. Fujii, J. Chiba, H. Aihara, M. Yamauchi, and H. Okuno, "Participation in Electron-Positron Colliding Experiment (PEP-4) at SLAC-PEP and Development of New Detection and Data Handling Technology," typescript, n.d. [1978], file "Japanese Collaboration," box 1, JMP. All except Okuno came from the Department of Physics, University of Tokyo, and Okuno was from the Institute for Nuclear Study, University of Tokyo.

100. Zeller to Marx, 4 January 1979, box 1, JMP.

In negotiations throughout the project, it was almost always necessary for individuals and groups to be able to point to material evidence of their participation. The material culture of the laboratory *stood for* the groups, it was their totemic representation to their supporting agencies, their colleagues, and themselves. Both the PEP administration and the Japanese needed such visible tokens. In one document, for example, Marx explained to Birge and Nygren in February 1979 that if the Japanese joined the TPC by building the cylindrical calorimeter with their money, it would help them by giving them "an 'identifiable' area which is good politically." (Indeed, in the records, one sees pictures of the Japanese calorimeter being displayed for visiting officials.) At the same time, by bracketing off its construction during initial planning, Pief Panofsky could "hold the cylindrical calorimeter for security against a substantial overrun on PEP-4." Continuing the metaphor, Marx added that if the collaboration could raise additional moneys later on, these "could be used to 'ransom' some or all of the calorimeter."[101] Now the collaboration had the opportunity to "ransom" its calorimeter, and the Japanese collaborators had the chance to control and display a material manifestation of their efforts.

If we ask, as is perfectly natural, "*Why* is there a calorimeter in the TPC?" the answer must come on several levels. From a "pure physics" perspective, the calorimeter is there because a true 4π inclusive detector capable of measuring all particles must be able to measure neutrals as well as charged particles, and the neutral pion could be detected by its decay into photons, which deposited their energy in the calorimeter. But the physics goals of the enterprise were not absolute. For the TPC, the calorimeter was also a bargaining chip. It could be provisionally sacrificed to the PEP leadership as a symbol of the leaders' fiscal responsibility. At the same time, the (now redesigned hexagonal) calorimeter was a tangible sign of participation that the Tokyo group could present to its scientific administrators.

Detector components like the "hex" were anthropological tokens, and it was necessary that they be quasi-autonomous from the rest of the detector. The identifiability of the entity, that it could be built, moved, run, and even mortgaged (sacrificed) to fiscal necessity, was an essential feature of the collaboration. At a review of the Japanese/American Agreement some years later, the speaker's transparencies immediately focused on the calorimeter, explaining its function, its workings, and its success in reconstructing photons and electrons in tests. Everything pointed to the concrete traces of the investment: a slide of a Japanese physicist wiring together a hex section, accompanied by two non-Japanese collaborators; next slide, a zoom in on the same scene now including only the Japanese experimenter; next slide, the hex-based theses of five Univer-

101. Marx to Birge and Nygren, 5 February 1979, box 1, JMP.

sity of Tokyo graduate students. In different ways, these theatrical performances would be repeated at each of the participating institutions: Johns Hopkins with its muon detectors, Riverside with the drift chamber, and so on.[102] External politics, internal sociology, and the details of physics and engineering all converged in the hybrid structure of the postmodern detector.

Horizontal coordination, coordination among institutions, was just one dimension of the network. Another, at least as difficult, was the vertical trade between the cultures of physics and engineering.

7.7 Physicists' Dreams, Engineers' Reality

The recurring nightmare of the engineer was the physicist's revision ("recycling") of plans to seize the latest technology or brainstorm. By the beginning of July 1979, Marx was all too aware of the demoralizing effect these rewritten scripts were having on the mechanical and electrical engineers who had to play the parts. To Nygren, Marx wrote perhaps the most sharply worded memo he ever directed to his boss: "I was shocked to sit at Tuesday's meeting on TPC motherboards and hear a complete change of philosophy and redesign of the motherboards and sector artwork being discussed. Let me remind you of the obvious—the building 80 tests are scheduled to begin in three months. A major change at this time is intolerable to our schedule and budget."[103] Switching the motherboards would mean at least a six-month delay and probably a quarter of a million dollars in additional expense. Worst of all, the alteration would have "[a] devastating impact on the morale of people who are busting their ass to stick to the schedule in other areas of the project, and an inevitable drain of precious technical manpower from other areas to help implement these changes. I urge you to take the following point of view: if the TPC will work with the existing design, *do not authorize any changes,* no matter how much the design is improved. If the TPC will not work with the existing design (which I doubt), *authorize the minimum number* of changes to fix whatever problems fundamentally jeopardize the TPC."[104] Indeed, if a major redesign of this magnitude were allowed, for anything less than "life threatening" problems, Marx declared, "I feel that I can no longer be responsible for keeping the project on schedule or budget. I also will have no credibility with the engineering and technical staff in being able to keep them working to our schedule." Instead, he suggested completing the machine as planned and improving it later, if at all.[105] One source of the engineers' discomfort was their own engineering hierarchy, which, accord-

102. "TPC Review Japan/U.S. Agreement," TPC-LBL-87-05.
103. Marx to Nygren, 3 July 1979, file "PEP-4 Project Management," box 7, JMP.
104. Marx to Nygren, 3 July 1979, file "PEP-4 Project Management," box 7, JMP.
105. Marx to Nygren, 3 July 1979, file "PEP-4 Project Management," box 7, JMP.

ing to Marx's handwritten interview notes, they found without "leadership," "direction," or "management." They wanted decisions by fixed dates and were not getting them; they called for "concepts" that would allow them "to carry on." One was on the verge of quitting, telling Marx, "I refuse to compromise my professional integrity any longer," while a chorus of others found too much "recycling." Physicists did not look much better than the weak engineering leadership; "young physicists—Nobel Chasers—want [the] impossible[; they have] no appreciation of practical limits." And it was clearly just such practical limits that the engineers felt too weak to present. It had become impossible, one engineer reported, for the engineers in the trenches to "stand up for themselves."[106]

As Marx worked over the results of his interviews, he handwrote his conclusions on a scrap of paper. Among them was an extraordinary phrase, from which we can read volumes:

> Role of Eng[ineer] vs Physicist.—keep us real[107]

At one level, Marx was simply summing up his view that the project needed practical boundaries on costs and time schedules. But at another level, the tension that surfaces at every turn in this massive endeavor reveals a forced retreat for the physicists from their traditional understanding that they could intervene at every point, that they could position the engineers as tools toward ends only physicists could set. For despite the vast increase in size that accompanied earlier experimental apparatus, the physicist could still retain the self-image of a jack-of-all-trades, able to fill an engineer's shoes one day and a computer programmer's the next. With a project the size of the TPC, which had already lost its sheen, this self-image was now shattered. Limits to the physicists' control came not by slamming up against the powerful array of funding sources; limits came in confronting a social and technical world in which the physicists' own decisions, when made without the structuring reaction of the engineers, were driving the project in circles.

Across the wide spectrum of tasks, the physicists were having a rough time learning when to intervene and when to withdraw. On a quite different front, the high-voltage system (which fueled the migration of electrons to the end plates) was heading for disaster. Marx pleaded with Sessler to grant the high-voltage project priority in the laboratory shops, accelerated purchasing capacity, and access to resident high-voltage experts, along with overtime and double shifts. But money and laboratory facilitation were not enough. Physicists

106. Marx, handwritten notes, 10 July 1979, file "PEP-4 Project Management," box 7, JMP.

107. Marx, handwritten notes, 10 July 1979, file "PEP-4 Project Management," box 7, JMP. Obviously, this is not to say that logic physicists constructed everything ab initio—vacuum tubes and counters already existed before the war, and standardized electronic logic circuitry became increasingly common during the late 1960s and 1970s. New were the metaphorical "off-limit" markers separating the logic physicists from their machines.

here, as in the motherboard case, had to stop interfering with tasks that the project leadership deemed better accomplished by engineers. It is illustrative of the delicacy of managing more senior physicists that Marx turned to lab director Sessler when he wanted Owen Chamberlain and Ronald Madaras to delegate responsibility for field cages and insulators to the engineers. As physicists, Marx wrote, they and their colleagues should focus on the high-voltage cable, power supply resistor box, and tests. It was in these areas that the physicists could, presumably, make contributions that the engineers could not. Even more specifically, during the field cage assembly the physicists should get their hands off the device: "soldering and small tool work must be done by trained technicians, not by physicists." Here again was a stunning blow; everything in the logic tradition spoke to the possibility, no, the necessity of the physicists' hands-on, down-and-dirty involvement in the very core of their instruments. Yet in these tasks as in many others, the technicians' work was both more efficient and more reliable, just as the fabrication of insulators was better left to the engineers. In contrast, as the engineers defined prototypes and monitored fabrication, they had to coordinate with physicists at every stage of design changes to make sure the alterations would not diminish the capacity of the system to run at 200,000 volts.[108]

To create the 200,000-volt potential drop that would cause electrons to drift adequately, the collaboration had to link the flat electrode plane of the TPC with a high-voltage source outside. Because the cable would necessarily cut across several other subassemblies, it precipitated a major coordination crisis that epitomized the difficulties the collaboration faced. For example, if the "inner radius" solution were taken (see figure 7.19), then the cable would penetrate UCLA's pole tip calorimeter. It would also (possibly) penetrate PEP-9 and one of the drift chambers. These were political as well as construction problems. There were also possible physics difficulties: more background would cloud the data because the "mask" (one of the shields around the sensitive volume) might be penetrated, and the cable, by presenting extra material up-beam of the TPC, could itself be the source of additional particle interactions that could complicate the physics analysis. This was not to say that the alternatives were free of complications. The "outer radius" solution demanded specially constructed end plane sectors that would cause disruption in the fabrication process, and "Val Reznik's solution" had its own downside. As this example illustrates, integration of the hybrid elements of the machine almost always involved an inhomogeneous mixture of political, engineering, and physics constraints (see figure 7.19).[109]

108. Marx to Sessler, 12 October 1979, and Marx, Chamberlain, Madaras, Brown, and Horowitz, 8 October 1979, "Clarification of the Role of Physicists and Engineers in the Fabrication of the TPC High-Voltage System Components," file "Notes to/from LBL Directors," box 7, JMP.

109. Chamberlain, Gorn, and Madaras, "TPC High Voltage Cable Penetration Location," 10 July 1978, file "High Voltage," box 8, DNyP II.

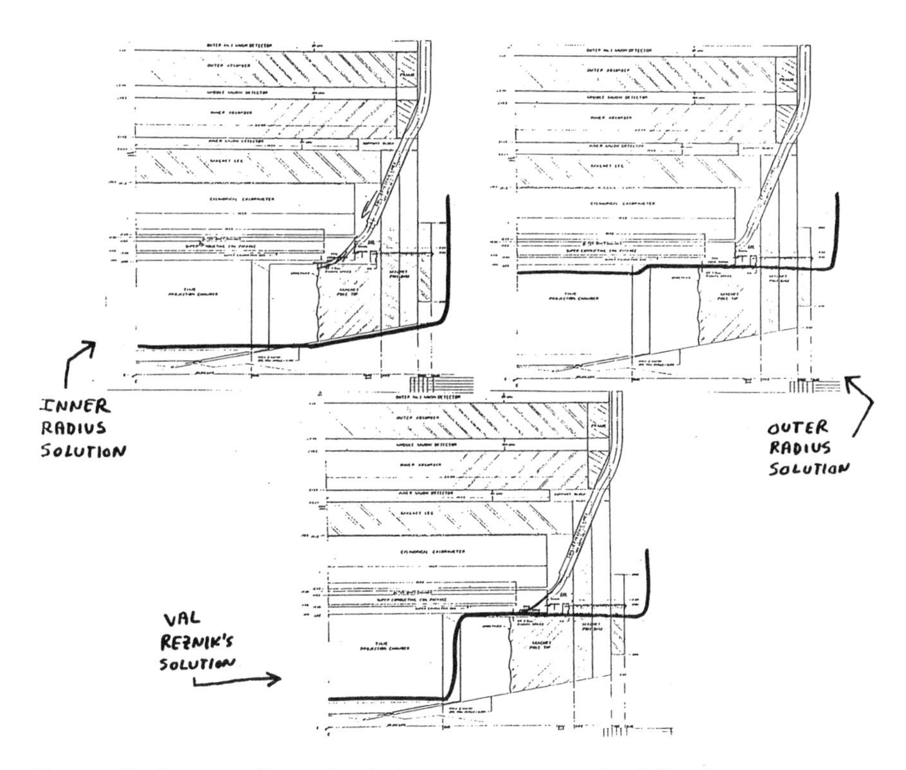

Figure 7.19 Problems of integration, high-voltage cable penetration (1978). This memo indicates both how difficult it was to coordinate the many different subgroups of the collaboration and how difficult it was (politically and scientifically) to bore a cable passage through the many different groups' proprietary electronic and mechanical space. Source: Chamberlain, Gorn, and Madaras, "TPC High Voltage Cable Penetration Location," 10 July 1978, file "High Voltage," box 8, DNyP II. Courtesy of Lawrence Berkeley Laboratory, University of California.

On 17 July 1979, Marx took up these concerns with Nygren, again addressing the central and precarious relation between physics and engineering:

> The role of engineering personnel vis-à-vis physicists in reaching design specifications, design concepts and in the execution of detail[ed] engineering design has not been satisfactory. In my view, the physicist's proper role is to push the technical design as far as he is able in the direction of maximizing the scientific output of the device being designed. His job is to demand specifications and determined by scientific goals, even if these specifications are at the "state of the art" level. The engineer, on the other hand, must be more conservative and realistic. His proper role is to advise on what can be done at what cost, in what period of time. He should advise the physicist as to the impact of specifications on cost, schedule, reliability, as well as feasibility, and especially as to the impact of varying specifications—the effect of the first derivative. The engineer's job is to inject reality into

> the physicists'. . . dreams and then to help to translate their mutually acceptable specifications into a working piece of hardware.[110]

Nothing would succeed without this dialogue between experimenters and engineers, and the engineers "must be willing and able to engage physicists in such a technical give and take. The engineering leadership must set a tone where such dialogue can take place and where individual engineers feel support from their leaders in entering into these technical debates." On their own, Marx clearly believed, engineers on the floor were unlikely to face down the physicists, so they had to be supported by their superiors: "Such technical discussion between engineers and physicists requires confident engineers who understand what this system demands as well as strong, confident leadership in the engineering department. Without such debate, the engineer is left with the frustrating task of designing to unreasonable specifications . . . and then having to recycle the design when, at some later time, the physicist himself embraces more realistic specifications. The breakdown of this process of dialogue has been the largest symptom of failure on the part of the engineering leadership."[111]

The constant recycling of designs, the ignoring of milestones, the tinkering on technical matters that treated cost as no object was, in Marx's view, leading the project straight to destruction. An end to the whole scheme was not unimaginable. Just in case the experimenters were not worried by their problems with the motherboards, the TPC, the lead calorimeter, and the high-voltage system, the Physics Division Review Committee (which included various physicists from across the country) issued an April 1980 report that, quite naturally, focused on the division's biggest project, the TPC/PEP-4 detector. The authors reviewed the slowness of some construction, especially of the TPC and the superconducting magnet, neither of which were finished: "In general we feel that the experimenters should switch their emphasis from improvement of the TPC subsystems (perhaps at the expense of perfecting every one) to construction, testing, and assembly of the full device. Realistic milestones should be set up and met. If necessary, priority should be given to the TPC if manpower help is needed."[112]

Enclosing a copy of the report, the laboratory director wrote Birge, commenting that the September 1980 completion date for the 12 end sectors "strains the credulity of even the most optimistic among us." He added that he was similarly uneasy about the possibility of testing and certifying the superconducting

110. Marx to Nygren, "Endemic Problems with PEP-4 Mechanical Engineering," 17 July 1979, file "Project Management," box 1, DNyP II.

111. Marx to Nygren, "Endemic Problems with PEP-4 Mechanical Engineering," 17 July 1979, file "Project Management," box 1, DNyP II.

112. Brodsky, Cox, Derrick, Gittleman, Quigg, Smith, and Wojcicki to Sessler, 24 April 1980, file "Notes to/from LBL Directors," box 7, JMP.

magnet on time, and asked Birge to meet with Jay Marx.[113] Anxiously, Marx issued a long memorandum entitled "Major Problems in Maintaining Schedule/Budget of PEP-4," and another labeled "Major Technical Concerns (or — Jay's nightmares)." These "major problems" included uncertainties in the state of various high-tech arts, control over the interface between the various subgroups, and the division of electronic, mechanical, and physics personnel into separate buildings. "Next time," he insisted, "build a circus tent to house everyone." Then the geometry turned out to be too tight, and a long parade of specification changes arose at the time the R&D phase was supposed to be transformed into the engineering phase and finally into the phase of fabrication. Compounding these troubles, the project ran into delays in the availability of the PEP intersection ring, and late deliveries by Sheldahl Company of the TPC insulators and by U.S. Steel of the field cage (see figure 7.20).[114]

But the underlying difficulty remained the awkward and perennially frustrating one of coordinating two cultures as different as those of physics and engineering. Part of the problem, Marx contended, was the inexperience of the engineers and designers in "working hand in hand with physicists": "*This inexperience hindered the development of a strong dialogue between physicists and engineers.* Such a dialogue is essential early in the project so that the engineers understand the scientific requirements and the specifications which can and cannot be varied to save time and cost. The physicists also need this dialogue to educate them as to what specifications can be met at a given cost and schedule, and . . . the gain in cost and schedule that can be made if certain specifications are relaxed."[115] Instead of real understanding between the two groups, the dialogue remained "sporadic" and "diffuse." Marx wanted an engineering community that would stand up to the physicists and protest vigorously when their scientific associates made impractical demands, or set specifications that admitted simpler, cheaper, and faster alternatives: "*Our engineers were not sufficiently combative.*" But the fault was clearly not all on the side of the engineers: "On their part, the physicists lacked experience in working in a 'big project' environment. There was *insufficient discipline in making design decisions* in a manner consistent with the schedule needs; an *inability to take prudent risks* to improve schedule, and to lessen cost. Finally, *physicists were unable to resist making design changes late in the game to 'improve things.'*"[116] The engineers were insufficiently strong to resist untimely changes; the physicists lacked the will to

113. Shirley to Birge, 30 April 1980, file "Notes to/from LBL Directors," box 7, JMP.

114. Marx, "Major Problems in Maintaining Schedule/Budget of PEP-4," 14 May 1980, file "PEP-4 Project Management," box 7, JMP.

115. Marx, "Major Problems in Maintaining Schedule/Budget of PEP-4," 14 May 1980, file "PEP-4 Project Management," box 7, JMP, Marx's emphasis.

116. Marx, "Major Problems in Maintaining Schedule/Budget of PEP-4," 14 May 1980, file "PEP-4 Project Management," box 7, JMP, Marx's emphasis.

Figure 7.20 TPC insulators from Sheldahl (1980). As detectors increased in cost and complexity, physicists turned to outside contractors. Here is one such piece, the TPC insulator, that had to be produced outside the sphere of university or even national laboratory shops. Source: LBL BBC 807 9049, LBL. Courtesy of Lawrence Berkeley Laboratory, University of California.

stop making them. Acting as if they were tinkering on a bench top instead of building on the scale of a factory, physicists (even those with memories of the big bubble chambers) resisted the microintegration that pressed on them from all sides.

Whatever their causes, Marx's "nightmares" of mid-May 1980 had specific names and addresses. "Will the cryogenic magnet work? . . . Can we produce 1 TPC sector/week[?]" Would there be excessive cross-talk between the thousands of electronic channels? ("We don't expect these problems at a serious level because we've been careful, but until the whole system is run . . .") "Will the TPC high voltage work . . . without causing excessive noise in electronics[?]" "[U]ntil whole H.V. system is assembled and run we don't know for sure." And "Will Sheldahl succeed in fabricating the large H.V. insulator before the profit motive forces them to drop the job?"[117] In the case of the Sheldahl

117. Marx, "Major Technical Concerns (or — Jay's Nightmares)," 14 May 1980, file "Physics TPC Nygren Group," box 1, DNyP I.

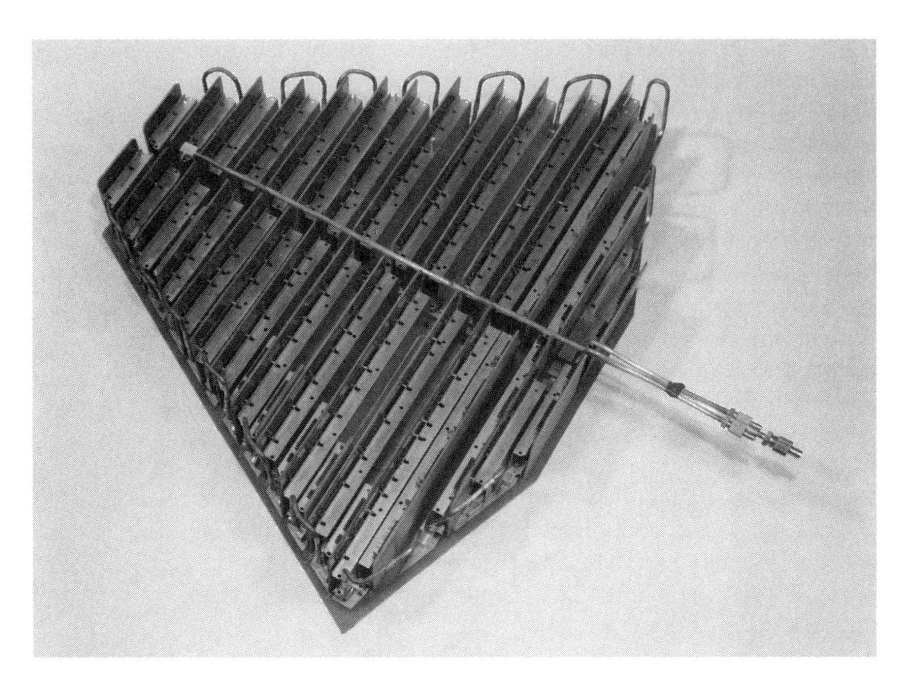

Figure 7.21 Cooling kluge in the end sector (1980). Coordinating physicists and engineers was only one aspect of the synchronization that became such a difficult problem in the TPC project. Here one sees the awkward cooling tubes inserted after the fact when it was discovered that the electrical and mechanical engineers had misunderstood each other's efforts, leaving the end plane producing much too much heat for the effective operation of this crucial aspect of the TPC. Source: LBL BBC 806 7183, LBL. Courtesy of Lawrence Berkeley Laboratory, University of California.

Company, the anxiety was awakened by the complete lack of control the collaboration exerted over an external contractor and its precarious economic situation. Characteristic of most of these fears was their *systemic* character; it was the interaction among electronic channels rather than the functioning of a preamplifier that caused sleeplessness. It was the high-voltage effect on the delicate electronics rather than the specific production of 200 kilovolts that was worrisome. And the production of the TPC sectors was not a problem in principle; it was a problem of detailed coordination among a great number of sophisticated subtasks. Illustrative of the interpretive problems was the mechanical and the electrical engineers' separate consideration of the sector end planes. Because the two groups had worked apart, the cooling of the electronics had not been considered during either the mechanical or the electrical design phase. Consequently, an extraordinarily clumsy hand-sculpted tubing system had to be built to snake through the sector, cooling by the transport of cold distilled water (see figure 7.21). Like the high-voltage cables, these copper pipes are the material patchwork that was required to bind together (barely and locally) two cultures that otherwise seemed determined to remain fiercely independent.

The project was therefore by no means on secure footing when Marx went a few days later (19 May 1980) to Isabelle, the troubled massive high-luminosity collider under construction at Brookhaven. The Brookhaven group had hardships of its own, chiefly with its own superconducting magnet, and therefore must have come with some interest to Marx's talk on large-scale experimentation. "How Large is Large?" he began by asking. In the case of PEP-4, the answer was (by then) about $18 million, not including physicists' salaries, but comprising approximately 100 engineers, designers, draftsmen technicians, shop workers, secretaries, and administrators. Large also meant complex; the complexity entered through the cryogenic magnet, through the 200-kilovolt high-voltage system, and through the low-noise electronics. Moreover, the electronics had to function in a high-pressure environment (some 10 atmospheres) that was secure, given that there would be significant quantities of poison gas (Ar and Br). And complexity issued from the necessity of holding the electronic and gas systems in mechanical structures of 2 meters crafted to a tolerance of 4 mil. Finally, the whole had to produce some 25,000 channels of electronic information.[118]

Projects like the TPC (and here the Brookhaven audience would have recognized its own experience with Isabelle) began with summer studies, passed through stages of small-group exploration, then to approval of the project, and then to the dreadful moment when "[y]ou regret having reduced contingencies to make your detector look more financially reasonable—you must live with your optimism. You begin the political process of pushing your home institution (and those of your collaborators) to contribute resources. You have fantasies of collaborating with Texas oil barons, Arab sheiks or your rich uncle."[119]

Back home, the problem was staffing the project, and here no one appeared to Marx, at least in May 1980, to be more crucial than the lead engineers. Not only did they need to be brought on board during the proposal-writing stage, they needed to contribute clear ideas from the beginning about what was rigid and what flexible in the project design. Only then (but definitely then) should the physicists insist on realistic cost and schedule estimates. Marx continued his confessional lesson to those assembled at Brookhaven: "Teach the lead engineers to read your mind, to argue with you and to be honest with you about what can/cannot be done at a given cost. . . . THE INTERACTION BETWEEN ENGINEERS AND PHYSICISTS IS CRUCIAL. Leave an adequate R&D phase in the beginning where specifications and cost tradeoffs can

118. Marx, transparencies of talk at Isabelle, 19 May 1980, file "Physics TPC Nygren Group," box 1, DNyP I. Note: There is a penciled date "1981" on this document that is incorrect.

119. Marx, transparencies of talk at Isabelle, 19 May 1980, file "Physics TPC Nygren Group," box 1, DNyP I.

be argued by physicists and engineers. These debates should be aggressive and forthright."[120]

After the debates had raged, and by Marx's lights they had to, it was crucial to freeze the specifications and work as hard as possible to make the various tasks "interface." Physicists should develop test facilities and procedures, they should monitor the engineers to make sure the device met specifications, but they should not otherwise interfere: "[L]et the engineers do their job their way. . . . Trust your engineers or don't hire them!" Under "miscellaneous . . . but important," Marx added: "Maintaining morale over 6 years without physics (how to keep full timers from becoming part timers)."[121]

Magnets soon crushed the morale of the speaker, his colleagues, and his audience. On 27 August 1980, during a quench test of the TPC/PEP-4 magnet, a shard of iron that had lodged in the insulation caused a catastrophic short, completely ruining the hugely expensive device (see figures 7.22 and 7.23).[122] And in October 1982, the Department of Energy shelved the Isabelle project, in large part because its magnet development program had grown hugely expensive and was not working well.

Just a few days after the magnet disaster, recognizing that PEP-4 was in trouble, the LBL director reshuffled the project administration. LBL engineer Kenneth Mirk took over as project manager from an exhausted Marx, who had been serving both as interim project manager and as deputy spokesman. Mirk had begun his career as a test engineer with Pacific Gas and Electric, specializing in electric automatic control equipment and boiler load tests. In 1951 he had joined LBL, where he worked on the Bevatron construction, especially vacuum piping and pumping refrigeration. He had done the conceptual design for gas handling at the Livermore-run Nevada nuclear weapons tests during the Rover Program and then returned to LBL to do the gas pumping and control systems for Alvarez's 72-inch chamber. Like so many of his colleagues, the joint experience of large-scale weapons and civilian accelerator work had initiated him into a new style of work.[123] Thanking Marx for the effort he had put into the project (Marx continued as deputy spokesman), Mirk promised his colleagues that he would stay faithful to the project's schedule and scientific goals.[124] Mirk was joined in the new project directorate by Lee Wagner, a senior electrical engineer,

120. Marx, transparencies of talk at Isabelle, 19 May 1980, file "Physics TPC Nygren Group," box 1, DNyP I.

121. Marx, transparencies of talk at Isabelle, 19 May 1980, file "Physics TPC Nygren Group," box 1, DNyP I.

122. Isabelle ended in stages between April and October 1982, and in a short continuance under the acronym CBA (for Colliding Beam Accelerator), it lasted until December 1983.

123. Mirk, curriculum vitae, file "Project Management," box 1, DNyP II.

124. Shirley to Division and Department Heads of Physics, Computer Science, and Mathematics Division and of Engineering and Technical Services Division, "TPC Project Management," 11 August 1980, file "PEP-4 Project Management," box 7, JMP.

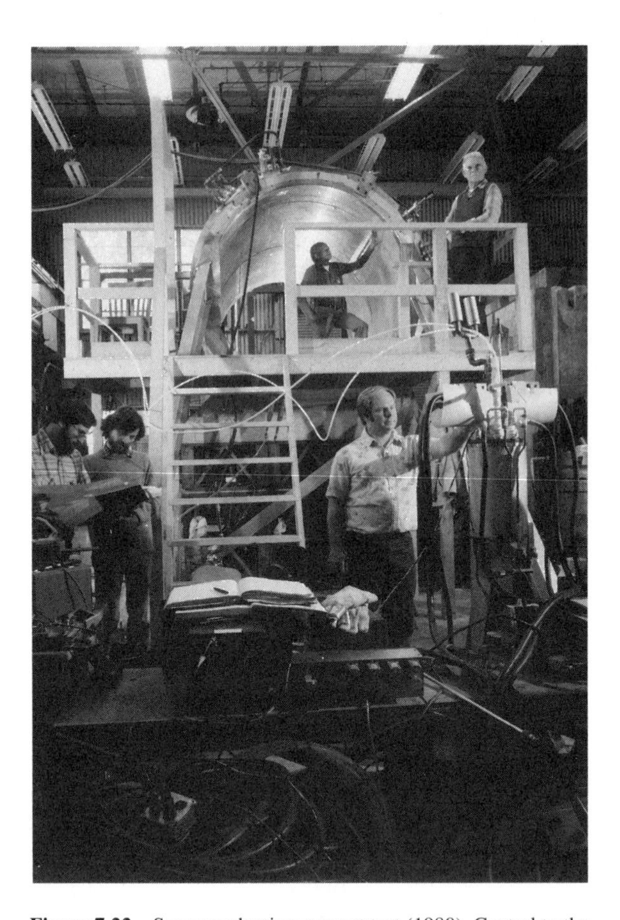

Figure 7.22 Superconducting magnet test (1980). Central to the TPC operation was the effective production of a powerful magnetic field. This picture depicts the test of the superconducting magnet, moments before the catastrophic failure that almost derailed the entire effort. Source: LBL BBC 803 3252, LBL. Courtesy of Lawrence Berkeley Laboratory, University of California.

and Ed Hodemacher, a budget and planning specialist. Marx himself returned to a more specifically scientific administrative role along with Nygren.[125]

But the magnet setback threw even the stalwarts into confusion, and the TPC Executive Board issued a "Dear Colleague" letter: "Delays in the past have severely damaged our credibility and have seriously complicated our future access to the PEP beams. The reputation of each of us and of LBL is at stake. We

125. Marx to the author, December 1995.

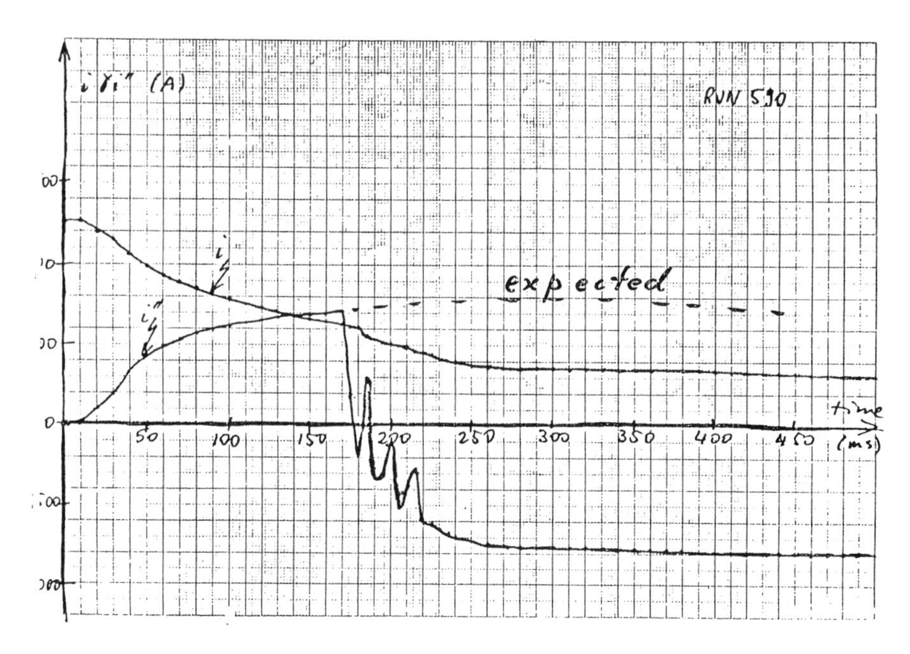

Figure 7.23 Great magnet catastrophe (1980). At turn-on plus 160 milliseconds, the superconducting magnet experienced a sudden and devastating failure. Later attributed to a short circuit in the coil from a tiny impurity in the coil, this disaster almost terminated the experiment. Source: LBL file set III EPAC, 28 October 1980, file "TPC Executive Board," box 3, DNyP II. Courtesy of Lawrence Berkeley Laboratory, University of California.

must complete and deliver the entire PEP-4 system to PEP at the earliest possible date." There would be no refuge from the rush to completion: "*Every subsystem is now in the critical path.*" More collaborators would be needed "in the trenches"; the collaboration would institute double shifts and take on any new ideas to speed up production. "It has been 4 years since we started our collaboration and we are approaching the final phase of the construction of our detector. Let us make an all out effort now."[126]

While the board threatened the loss of beamtime and reputation, Marx, though no longer project manager, offered gentler persuasion. Trying in late October 1980 to raise morale, he proposed brown bag lunches, informal get-togethers, and even town meetings: "Finally, let me say to those of you who have spent the past few years working hard in the trenches to make PEP-4 happen that I deeply appreciate your efforts. I recognize how it is sometimes hard to keep your spirits up in spite of the many technical problems and the long years

126. TPC Executive Board to Dear Colleague, 15 October 1980, file "Physics TPC Nygren Group," box 1, DNyP I.

with perhaps little recognition." In one more year, he assured the faltering troops, they would "have the best detector ever built."[127]

Others were less sure. On 28–29 October 1980, SLAC's EPAC convened to discuss the fate of PEP-4. In a carefully worded status report, Nygren put the best light on developments. The prototype electrostatic system worked above operating standards, and the high-voltage electrostatic structure would be ready for electrical testing in December 1980. Conceding that work on the readout planes had not been as smooth as desired, Nygren nonetheless hoped for the production line to be in operation in early November, at which point it would churn out one of the 12 planes per week. The gas system (which kept the TPC filled with the right quantity of purified gas) and the data acquisition system (which converted the 25,000 channels to digital form and shipped them to computers) were both on track. Software development, which had concentrated on the generation of simulated "raw data," was now turning more and more toward the analysis of results from the various detectors, and diagnostics on the dead superconducting coil were advancing in an effort to provide a complete accounting to the EPAC. Finally, with the appointments of Mirk and Wagner, LBL hoped to "improve our record in meeting project objectives."[128]

Along with Nygren's precirculated "status report," Panofsky sent to EPAC an "issues paper" that began by rehearsing the history of the sometimes ill-starred project. The original 1976 proposal had been approved in part because of its promise to exceed the features of its competitors at a cost only moderately higher. And the superconducting coil had actually been seen as an advantage, since the TPC (by its smallness) could fit inside the coil and therefore would not be "paced" by the magnet's development. Now, with skyrocketing costs and a catastrophe in magnet development, the program was in deep trouble. Moreover, the PEP-9 collaboration, which was going to use PEP-4 as its central tracker, was paralyzed without it.[129]

Because it was blocked by the PEP-4 difficulties, PEP-9 now pressed for a surrogate central tracker with a conventional magnet, a temporary replacement for the TPC. The substitution would bring it "on-line" and give it a chance to do physics. For the TPC side of the collaboration (here one must read between the lines), the disadvantages must have seemed equally obvious. Its members would be further delayed in getting to the physics, possibly losing the chance to discover new phenomena. But given its predicament, and its consequences for

127. Marx to Dear Colleague, 22 October 1980, file "Physics TPC Nygren Group," box 1, DNyP I.

128. Kirk (acting secretary) to Experimental Program Advisory Committee, "Resume of EPAC meeting, 28–29 October 1980," 3 November 1980, attachment Nygren, "PEP-4 Status Report," file "EPAC I (1975–1980)," box G014-F, SLAC.

129. Kirk to Experimental Program Advisory Committee, 3 November 1980, attachment B, Panofsky, "Issues Paper on TPC," file "EPAC I (1975–1980)," box G014-F, SLAC.

others, the PEP-4 collaboration was in a position to object only weakly; by the time of the meeting, the PEP-4 representatives "had no objection." So, Panofsky concluded, EPAC had to decide what to do about the coil—wait for the superconducting replacement or install a conventional magnet. Next, it had to decide whether to authorize a replacement central tracker and, if so, which one. Finally, and most important, it had to reauthorize (or block or partly block) further development of PEP-4.[130]

Philippe Eberhard from PEP-4 argued to the committee that the experimenters could have the superconducting coil repaired by about May 1981; in the words of the minutes, "The Committee appeared to be skeptical." Armed with the show-and-tell of a completed TPC sector, Nygren then assured the committee that the TPC would be ready for installation within a few months of November 1981. Once more, according to the minutes, "[t]he committee . . . was . . . rather skeptical about the LBL schedule forecasts." After Marx, Nygren, Masek, and the other proponents had left, the EPAC members continued their deliberations into the evening. The superconducting coil looked like it was in deep trouble, and the TPC was not in a position to assure delivery. Yet the alternative of providing a surrogate central tracker met with only "mixed" reviews, and the committee called back a series of witnesses including the new managers, Mirk and Wagner, to understand "in more detail their roles in the new PEP-4 organization." In the end, the committee voted to keep the TPC with no alternate central tracker (though they agreed to reconsider a more detailed proposal for one) and to proceed with a conventional magnet.[131] On 1 July 1981, some six years after planning began, the TPC saw cosmic rays for the first time. "The champagne was excellent," one memo boasted: "TPC ran stably. It was turned on and off each day with no problems."[132] Still no new physics, but at least the eye of the machine had opened and taken a look around.

The prospect of results precipitated a sudden reappraisal of who could speak where on what. At the TPC Executive Board meeting of 9 October 1981, a new entity came into existence, its job to regulate output: SCAT, the Standing Committee for Assignment of Talks. At stake, quite explicitly, was the partitioning of prestige at several levels, among the institutions (LBL, Johns Hopkins, Tokyo, UCLA, U.C.-Riverside, and Yale), among the various classes (students, postdocs, etc.), and among individuals.[133]

130. Kirk to Experimental Program Advisory Committee, 3 November 1980, section III, and attachment B, Panofsky, "Issues Paper on TPC," file "EPAC I (1975–1980)," box G014-F, SLAC.

131. Kirk to Experimental Program Advisory Committee, 3 November 1980, file "EPAC I (1975–1980)," box G014-F, SLAC.

132. Marx, "Cosmic Ray Test of TPC," 15 September 1981, file "Originals J. Marx Correspondence," box 4, DNyP I.

133. "Procedure for Assignments of PEP-4 Talks: TPC Board Draft," 9 October 1981, file "Minutes of TPC Executive Board Meeting," box 4, DNyP II.

There was consensus about the typology of prestige. Class I occasions were the "top conferences and meetings for high energy physics": these prestigious events included invited talks at the famed Rochester meetings in High Energy Physics that had begun after the war, the European High Energy Physics conference, and the New York and Washington meetings of the American Physical Society. Class II talks were those at other conferences or colloquia at "major laboratories"—those were the elite set of Argonne, Brookhaven, DESY, Fermilab, LBL, Rutherford, and SLAC. Fine distinctions mattered in matters of protocol: a "contributed" paper (one simply submitted into a regular session) to a Class I event would only count as Class II. Finally, Class III encompassed essentially everything else: talks at departmental colloquia and seminars, hardware or software talks, and all student talks. The only exception to the "hardware" clause was a talk that constituted a "PEP-4 review-type" presentation, which would fall in a contribution of a higher class. In short, the status of a presentation depended on the speaker's status, the site, and the nature of the contribution.[134]

Based on this classification, SCAT would enforce a set of authorship rules, which, I would suggest, carve out three boundaries to expression: power, propriety, and proof. The pressures behind the rules were, of course, not new—each set of rules made visible invisible forces that had long governed the intricacies of credit and demonstration. But now that such extraordinary resources were under one dominion, the dynamics of the community were written into code, not simply absorbed through socialization.

Rule 1: No member of the collaboration could deliver a Class I or Class II talk on any subject without authorization from SCAT. This was the assertion of the power of the collaboration as a whole, through its representatives to the board, and through the Board's appointed entity, SCAT, over the individual. Rule 2: "No special or exclusive rights to the presentation of particular physics results or analysis will belong to any TPC member or members." Here the point was to destabilize any dominance a particular subgroup might claim over specific physics. The fear behind Rule 2 is widespread in the era of very large groups. It arises, often contentiously, because after years of work building an apparatus, it is quite common that the final physics analysis (say the plotting of an invariant mass histogram or the calculation of a cross section) falls to a handful of people. To those in the collaboration who did not add to the physics analysis, the situation resembles the building of a house: should the roofer who puts on the last shingle be lauded as the true and main builder of the edifice? Interdicting this kind of last touch possessiveness is part and parcel of the maintenance of the

134. "Procedure for Assignments of PEP-4 Talks: TPC Board Draft," 9 October 1981, file "Minutes of TPC Executive Board Meeting," box 4, DNyP II.

collaboration as such; without it, the social bond that keeps someone working on the calibration of a muon detector or rewiring the TPC would collapse into a Hobbesian dash for the analysis routines. Rule 3: "[N]ew experimental results must be reviewed by the board before presentation." Without such a clause the epistemic commitment of the collaboration as a whole would seem to stand behind any individual's claim to have found a new effect or entity. Since to the outside world the collaboration appears as a pseudoindividual, any inconsistency or retraction would debilitate the collaboration as a whole. Thus, proof or adequacy of demonstration had to come from the full collaboration (or more precisely the Senate-like entity that the collaboration chose to represent it).[135]

Are these rules part of the physics? Are they merely "external sociology"? I would argue that here—as so often in the understanding of scientific argumentation—we have a false dichotomy. Among other strictures, these rules constructed what it meant to be an author, insisting that the design, building, and testing of apparatus count as a sufficient part of the physics demonstration to merit inclusion—even the person in question never touched the analysis. These rules, maintaining group control, had a dual aspect: at once acting to preserve the social cohesion of the collaboration and to vouchsafe the reliability of the physics results.

Further guidelines mixed the mechanical distribution of goods with a need-based political pragmatism. For example, Class I talks would rotate among the PEP-4 institutions, with a double weighting for LBL. Class II talks, by contrast, would strictly follow the numerical weight of each institution's membership: 13 for LBL, 3 for Johns Hopkins, Tokyo, UCLA, and U.C.-Riverside, and 1 for Yale. Finally, Class III talk assignments could be had for the asking—so long as SCAT was properly notified. These criteria would bend, however, when they had to: "Due consideration will be given to persons without tenure." Just to indicate the scale of SCAT's control functions, there were some 108 Class I and II presentations *already* scheduled.[136]

Combatting the centrifugal spin of such a massive organization had other effects. Ceaselessly, PEP-4 and PEP-9 struggled to navigate their way between autonomy and coordination. In May 1982, for example, PEP-9 wanted fast data from PEP-4 as part of its triggering mechanism, and PEP-4 agreed to support a test to that end. At the same time, PEP-4 made it clear that *it* was not reciprocally interested in data from the PEP-9 trigger and so would not assign a person to help the two-photon group analyze the data tapes. If PEP-9 wanted to send an emissary to PEP-4 to learn the ropes, that would be acceptable. A cold

135. "Procedure for Assignments of PEP-4 Talks: TPC Board Draft," 9 October 1981, file "Minutes of TPC Executive Board Meeting," box 4, DNyP II.

136. "Procedure for Assignments of PEP-4 Talks: TPC Board Draft," 9 October 1981, file "Minutes of TPC Executive Board Meeting," box 4, DNyP II.

peace reigned between the two facilities-in-a-facility, a collaboration coordinated only insofar as the wider world forced detente upon it.[137]

Even within PEP-4, the struggle to hold the myriad of groups and subgroups together was a constant preoccupation. One sees this, for example, in the many-fold iterations of documents on the role of the "spokesman," who stood both as the human face of the collaboration in the wider community and as the leader, and whose task it was to have the interest of the whole rather than the parts in mind. In one May 1982 draft (by Harold Ticho and Jay Marx), the authors reflected on the problem: "In a large collaboration, individual collaborators will, in the main, be concerned with that portion of the effort which is their immediate responsibility. There is a real risk that the totality of these individual efforts will not remain focussed on a common goal and that important aspects of an overall plan will be neglected because they are not clearly the responsibility of one or another individual." This was the spokesman's job. He ("spokesman" became "spokesperson" not too long afterward, here and in many other experiments) would define the short- and medium-range goals of the experiment, and do so by creating a document that would be revised as needed. Of course, the spokesman would consult with the board and the TPC collaborators, but as the collaboration expanded, the need for an individual figure of responsibility grew stronger. As the document emphasized, "[T]he collaboration must be secure in its knowledge that there is a designated individual who regards overall issues [as] his primary responsibility."[138]

The issue of leadership was timely. For a variety of reasons, Nygren wanted to pull back from his position in the directorate of the collaboration. And Jay Marx, who had assumed the role of spokesman, began exploring other career moves. In August 1982, Marx wrote to the TPC board to let it know that he wanted to leave the project for a year and serve in Washington at the Department of Energy, contributing to the national program in high energy physics.[139] It was a move initiated by LBL Deputy Director Herman Grunder, who hoped the assignment would advance the laboratory's fortunes.[140] Though the Executive Board immediately began deliberations on its choice of a new spokesman, the diffusion of authority in what was already perceived as an overly democratic project had clearly begun to worry Panofsky and others both at SLAC and at the Energy Department.

It might be thought that a project some nine years old with a budget that

137. Minutes of TPC Executive Board Meeting, 3 May 1982, file "TPC Executive Board Meeting," box 4, DNyP II.

138. [Ticho and Marx], untitled draft statement on role of spokesman, file "Minutes of TPC Executive Board Meeting," box 4, DNyP II.

139. Marx to TPC Board Members, 5 August 1982, file "Minutes of TPC Executive Board Meeting," box 4, DNyP II.

140. Marx to the author, December 1995.

was, by 1982, edging past $40 million would have a bureaucratic inertia that exempted it from the threat of cancellation.[141] This would be an error. At every level, from its capability of effectively registering quark decays to the economic support of the detector, the collaboration faced extinction. Indeed, after two years of desperate work to get the detector on-line, EPAC issued yet another death sentence on 14 May 1983. Once again, the project stood on the edge of the abyss: "On the basis of the presentation made to us, the EPAC recommends that the Director should not approve any additional running for the PEP-4 (TPC) collaboration at this time. We are not confident that the collaboration has the leadership, organization, and manpower which is necessary to conduct a program all the way through to final physics results."[142] Panofsky immediately notified the new spokesman, Benjamin C. Shen of Riverside, that there would be no beam for PEP-4 in January 1984 unless these problems were "remedied."[143] The collaboration tried to rally a defense at once. On 27 May 1983, Shen wrote to Panofsky, arguing for a reprieve: "Both Roy Kerth [the deputy spokesman] and I have explained to the collaboration the negative impressions the EPAC had of our efforts, as you conveyed them to us: that we have not produced physics in a timely manner, that our computing and data analysis are not organized, that we tend to do things in series and do not have sufficient advance planning, that we do not have adequate leadership and management organization and seem too democratic to be effective. We are determined to erase these impressions but we maintain they are largely unfounded."[144] Arguing on physics grounds, Shen contended that, in fact, the new antidistortion upgrades, including refined *dE/dx* calibrations, were on the verge of yielding important results and that the spring harvest of data would soon be in Data Summary Tape form. On more human grounds, he pleaded that suspending approval would seriously damage the morale of young people in the collaboration, make it harder to recruit physicists and graduate students, scare off the funding agencies, and diminish support in collaborating institutions.[145] Intriguingly, what had been a virtue (not submitting to any philosopher-king) had now become a liability for the collaboration, as it was perceived as "too democratic."

It was time to rally the allies, and J. D. Jackson, a senior theoretical physicist at LBL and now head of its Physics Division, came to the collaboration's

141. Costs estimated different ways often differ significantly. But a minimum estimate for 1982 can be assessed by taking the $28.32 million that was to complete the project (Panofsky and Sessler to Wallenmeyer, 26 February 1976, file "PEP Politics," box 2, JMP) and adding it to the supplementary estimate of $11.046 million for FY 1981 and 1982 ("Status of TPC," November 1989, file "Notes to/from LBL Directors," box 7, JMP). Note that these amounts do not include physicists' salaries and much else.

142. Experiment Program Advisory Committee to Panofsky, "PEP-4 Running Time," 14 May 1983, DNyP.

143. Panofsky to Shen, 16 May 1983, DNyP.

144. Shen to Panofsky, 27 May 1983, DNyP.

145. Shen to Panofsky, 27 May 1983, DNyP.

defense. Writing to each member of EPAC individually, he first apprised them of the 17 May 1983 cooldown and activation of the superconducting magnet. The magnet had withstood its full current (over 2,200 amperes) and therefore promised to deliver the full 1.6 tesla magnetic field for which the TPC had been designed almost a decade earlier. Then, the blast:

> The brief brutal memorandum [from EPAC] to Panofsky is so broad in its condemnation and so unspecific as to be of little or no value to the collaboration. The message to the outside world was, however, very clear. Within a few hours the message that the TPC was finished at PEP reached literally every place where there are high-energy physicists. One had the impression that the news reached some places with a speed faster than light or even via the backward light cone. You may say that the audience has oversimplified and under-interpreted your recommendation, but the damage has been done.[146]

Jackson went on to object both to the generalized accusation of incompetence and the specific grounds for it. "I am reminded of those British boards of examiners who, on the basis of an intimidating half-hour interview, award their beta minuses and so determine some poor soul's whole career." Instead of discouraging the experimenters, and stigmatizing both the spokesman and the collaboration, he enjoined both EPAC and the director to support the TPC in every way they could.[147]

The same day that Jackson intervened (31 May 1983), Panofsky appointed Abraham Seiden, a physicist from U.C.–Santa Cruz, to lead a systematic investigation into the status of the project and report to EPAC: "I would like the committee to examine in substantial detail the past achievements of [PEP-4 and PEP-9] in respect to software activities leading to Data Summary Tapes and from these to extraction of physics results. I would like to gain understanding as to the performance of the groups in terms of adequate planning for software work and the prospects of extracting physics in a timely fashion."[148] The committee, Panofsky insisted, would need to find a mechanism to establish an "early warning system" to find and correct problems as they arose. As director of the laboratory, Panofsky was especially worried since the Department of Energy (the source of all the laboratory's funds) saw both the TPC/PEP-4 and the PEP-9 as SLAC projects, and with both in trouble the laboratory's own future was in danger. A copy of Panofsky's letter reached Jackson and clearly went some way toward assuring him that a more serious inquiry into the TPC would be undertaken.[149]

146. Jackson to members of Experimental Program Advisory Committee, 31 May 1983, DNyP.
147. Jackson to members of Experimental Program Advisory Committee, 31 May 1983, DNyP.
148. Panofsky to Seiden, 31 May 1983, DNyP.
149. Panofsky to Seiden, May 1983; and Jackson to Seiden, 13 June 1983; both in DNyP.

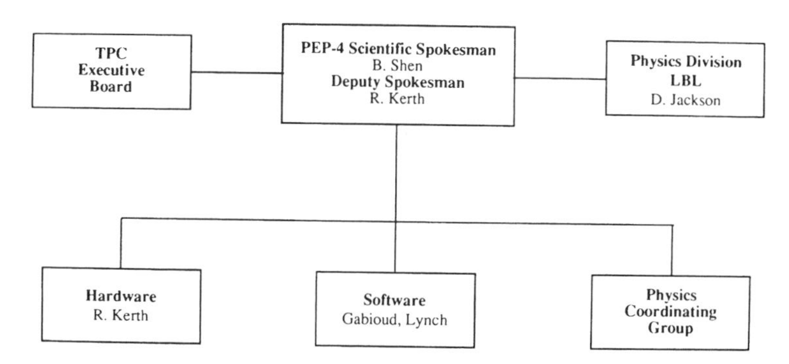

Figure 7.24 Overall management structure (1983). Its organizational structure under attack, the TPC collaboration in 1983 revised and elaborated a scheme they expected would be both effective and reassuring to the Department of Energy and SLAC. Note the designation of a scientific spokesman and the tripartite division between hardware, software, and physics. Source: Shen to members of PEP-4, 17 June 1983, DNyP. Courtesy of Lawrence Berkeley Laboratory, University of California.

Seiden let PEP-4 know on 16 June 1983 that his committee would be concentrating on three areas. First, it wanted to know what the first *physics* results would be, how soon they could be expected, and how many people were at work on the necessary software and physics analysis. Second, it wanted to inquire more closely into the precise nature of the two-photon physics: what would the relation be between TPC/PEP-4 and PEP-9? Finally, the committee wanted to know who was responsible for finding hardware and software problems, and how authority functioned in achieving a resolution of difficulties once they were identified.[150]

Nothing concentrates the mind like an impending execution. Shen began assembling detailed organizational charts, summaries of data runs, descriptions of software and analysis routines, physics topics, and projections of future computing needs and capabilities. Under the TPC/PEP-4 scientific spokesman (Shen) and his deputy (Kerth) stood three organizational structures of roughly equal complexity: Hardware, Software, and the Physics Coordinating Group (see figures 7.24–7.27).

By e-mail, Shen had established the Physics Coordinating Group (PCG) a few days earlier. It was a move, Shen argued, that was necessary to avoid duplication of efforts among the various groups. The group would define the "topics of interest." Under pressure to deliver a plan for the provision of physics results, the collaboration began treating analysis as an activity not unlike any other part of the collaboration. There were scarce resources and finite time; no longer

150. Seiden to Shen, 16 June 1983, DNyP.

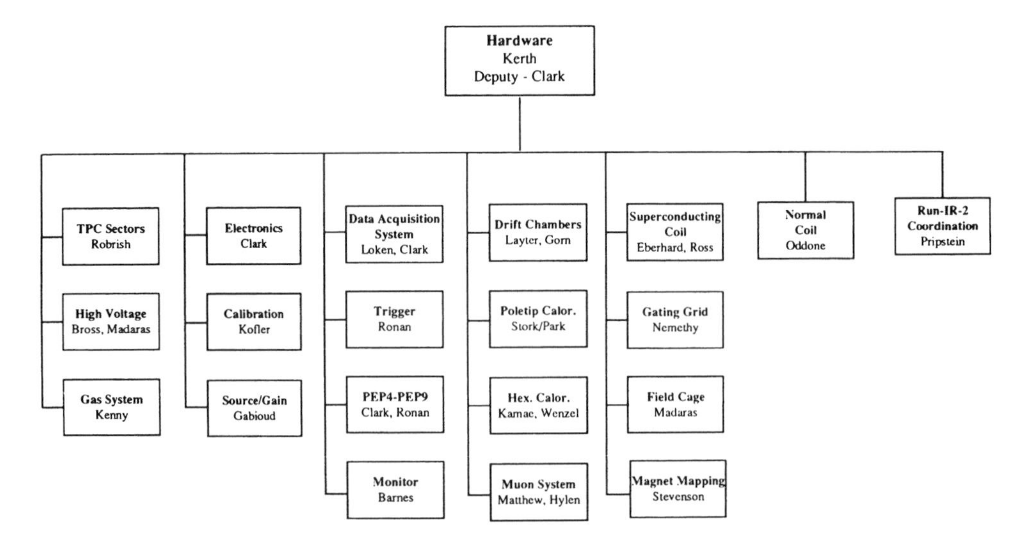

Figure 7.25 Hardware organization (1983). Note in contrasting figures 7.25–7.27 that the same person may appear in positions of varying authority in each of the three "organizations": hardware, software, and physics. Source: Shen to members of PEP-4, 17 June 1983, DNyP. Courtesy of Lawrence Berkeley Laboratory, University of California.

would it be possible to work in the redundant, free-wheeling mode of an earlier time. The PCG's tasks would include the "critical review" of *every* plan for the examination of *every* physics topic, evaluating significance, feasibility, manpower, Monte Carlo, computing load, and timescale. Doing "physics" no longer was exempted from the general approach brought to bear on a drift chamber or a data buffer—the idea was to break down tasks into modules and determine the most efficient path of attack. Here, for example, is one of the directives by Shen to the PCG: "Coordinate the physics topics among the various groups to avoid conflicts, duplications, overlaps; coordinate efforts which are commonly required by more than one physics topic or group, e.g. luminosity measurement, Monte Carlo etc." Tracking progress, forecasting problems, and setting agendas for physics all fell under the purview of the PCG. There was no firebreak in the rationalization of work between equipment building and physics analysis.[151]

The Monte Carlo makers, now performing an essential part of the physics analysis, also came in for microcoordination. Earlier, the group had conducted a "software workshop" in which various participants had presented their programs. Now, in part under the watchful eye of a highly critical EPAC, Shen

151. Shen to Cahn, Galtieri, Hofmann, Madaras, Maruyama, Matthews, and Stork, e-mail, 5 June 1983 at 23:44, file "Minutes of the TPC Executive Board Meetings," box 4, DNyP II.

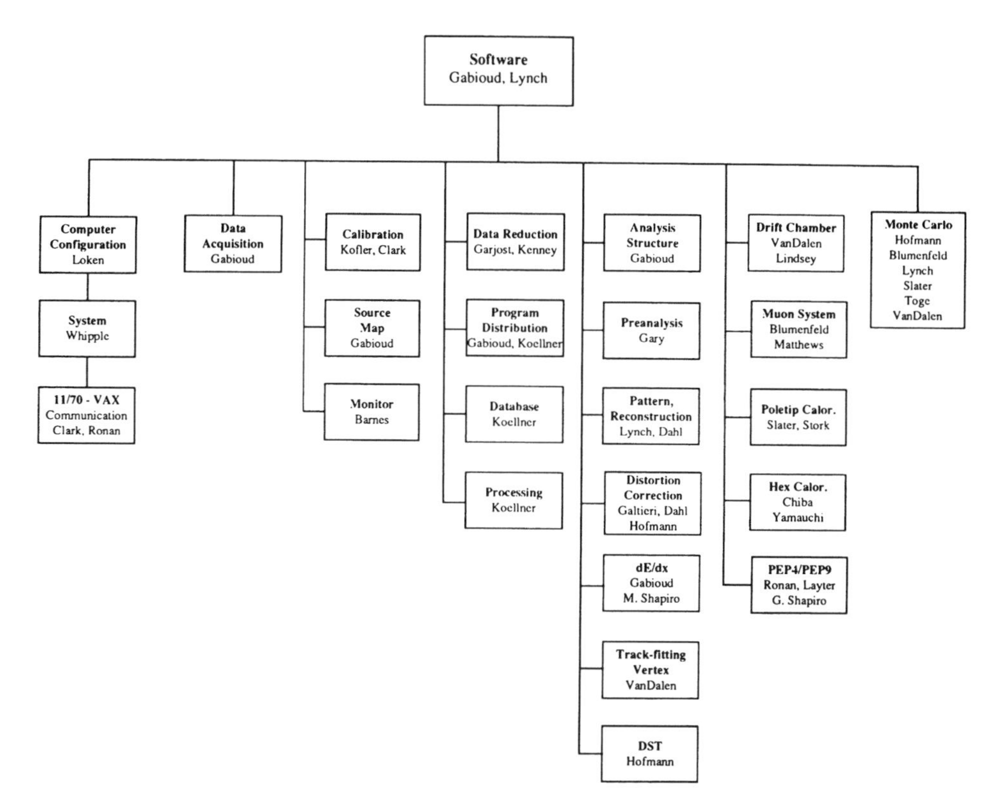

Figure 7.26 Software organization (1983). Source: Shen to members of PEP-4, 17 June 1983, DNyP. Courtesy of Lawrence Berkeley Laboratory, University of California.

began to restructure the simulations through the establishment of another formal subgroup: "[S]ince the workshop, the efforts [on Monte Carlos] were still individual efforts motivated by people's own interests and needs. There is a lack of communication and coordination among these efforts." Just as in the PCG, the Monte Carlo Subgroup would enforce a nonduplicative work division and, in addition, standardize software and its accompanying documentation.[152]

It is not trivial that data analysis had become comparable to hardware in its complexity and in the effort needed to build and maintain it: this is a characteristic feature of the new mode of hybrid experimental particle physics, and increasingly of fields such as plasma physics, astrophysics, and condensed matter physics. Before, all the hierarchies culminated in a single leader (such as

152. Shen to Blumenfeld, Hofmann, Lynch, Slater, Toge, and VanDalen, e-mail, 5 June 1983 at 20:04, file "Minutes of TPC Executive Board Meeting," box 4, DNyp II.

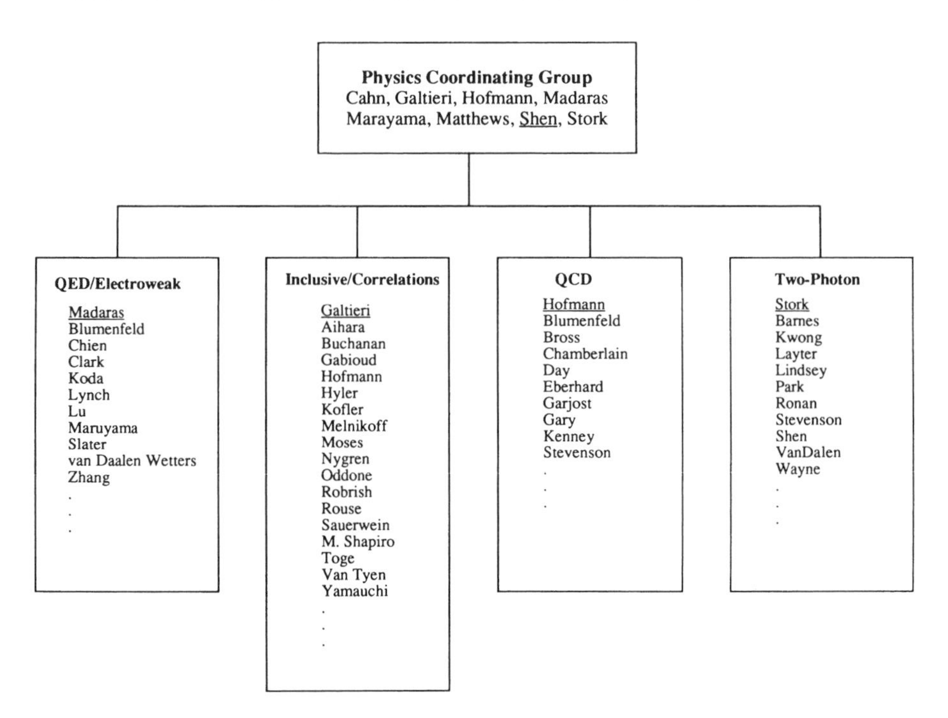

Figure 7.27 Physics Coordinating Group (1983). Source: Shen to members of PEP-4, 17 June 1983, DNyP. Courtesy of Lawrence Berkeley Laboratory, University of California.

Alvarez); now, entirely new structures came into existence with no single person (or even single institution) in charge of all of them. Physics analysis culminated in one place, simulations in another, detector components, detector coordination, and data acquisition each in its own pinnacle.

What is involved in processing data, for example? Physically, the data tape is passed through the analysis five times, for several reasons. The size of the program simply exceeded the capacity of the available computers, the results of the initial passes had to be understood before passing them to the next stage, and finally operations could be accelerated if some analysis could be performed while subsequent steps were in preparation. Figure 7.28 indicates the basic sequence. In Pass 1 (done on-line), the trigger rates were monitored, and a program called PREANALYSIS rejected a large number of triggers, cutting out events that were calculated to have occurred away from the beam-crossing position. These illegitimate collisions might come from cosmic rays or from the beam's hitting gas in the pipe—as a result, nearly a third of events were summarily removed, from roughly 7,000 triggers per tape to some 5,000. A second pass, off-line, took about an hour per tape, during which three-dimensional

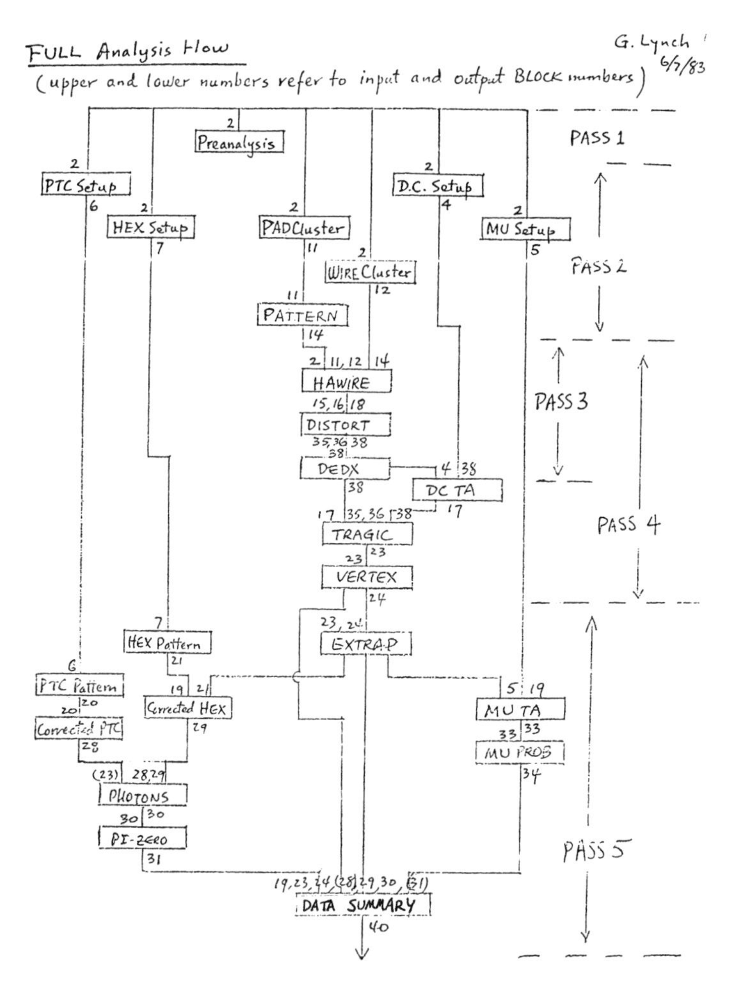

Figure 7.28 Full analysis flow (1983). As indicated in the text, these programs governed and coordinated the data produced by the various segments of the hybrid detector. After this massive restructuring of the data, a Data Summary Tape is produced, and it is this that becomes the new "raw data" of experimentation. Even the data one might consider most "primitive," the space points that determine the track, have already been processed many times to compensate for inhomogeneities in the field and to find the statistical center of the cluster of migrating electrons. Source: Shen to members of PEP-4, 17 June 1983, DNyP. Courtesy of Lawrence Berkeley Laboratory, University of California.

points in the TPC were located by examining the outputs from the 14,000 cathode pads (program: PAD Cluster). Another program constructed tracks out of the pad points (PATTERN) and calibrated the two calorimeters: the hexagonal calorimeter (HEX Setup) and the pole tip calorimeters (PTC Setup). Then using this information about the disposition of tracks and the energy deposited, the computer could eliminate many more triggers, cutting the total from 5,000 to 2,000. Finally, auxiliary programs calibrated the drift chamber (DC Setup), the proportional wires (WIRE Cluster), and the muon chambers (MU Setup). No further information was needed for the computer to find the drift velocity, the vertex position, and so forth, and these parameters were then available for subsequent passes.

In Pass 3 (about 45 minutes on the computer), the tape was monitored to determine variations from run to run in dE/dx and in track distortions. Wire data (which are used to determine the specific energy loss) were associated with the tracks that were reconstructed by cathode pad output in Pass 2 (HAWIRE). Variations between the wire data and the computer-massaged pad-generated tracks then gave a measure of the distortions present in the chamber, as well as an improved measure of the drift velocity.

In Pass 4, the computer compared the wire response to its response to known calibration sources and used this to give a definite value of dE/dx (DEDX), which along with time-of-flight information yielded a particle identification. From previous measurements of the fields, the computer could then correct for inhomogeneities in E and B, along with the more transitory effects of positive ions that vary from event to event (DISTORT). It also pieced together the tracks to find the primary vertex (VERTEX) and associated points in the two drift chambers with the tracks. Finally, Pass 5 extrapolated the TPC tracks to the calorimeters and muon detector (EXTRAP), located photons and π^0's, and executed the final event reduction, such as locating candidates for jets, highly collimated bursts of strongly interacting particles that were attributed to the production of quark-antiquark pairs, where the quark and antiquark shot off in opposite directions (there were about 30 of these jet candidates per tape).[153]

In a formal response from Shen to Seiden, the TPC assured the committee that it would have the equivalent of 8 full-time physicists, drawn from among 16 people who would be working on physics, *strictu sensu.* Of highest priority in the short term would be the determination of the inclusive cross section (that is, any combination of reaction products) for all charged particles, for π^0 and γ, and for muon production. These numbers were bread-and-butter physics; numbers that would be useful in checking a wide variety of theories but that were not accessible in any other accelerator except PETRA in Hamburg, Germany. Along the same lines, the collaboration could determine the production rates of several known particles: ρ, $K^*(890)$, and $\phi(1020)$. All of these are questions that could have been anticipated at the time of the original design of the TPC in

153. Gabioud, "TPC Data Analysis Organization for Data between Nov 82 and Jun 83," 16 June 1983, DNyP.

1974. The same could not have been said of quantum chromodynamics, the gauge theory of the color force that made its appearance in the mid-1970s.[154] By 1983, however, tests of QCD were fairly routine, and the specific physics topics that PEP-4 proposed to study included flavor tagging, that is, using decay products to identify what kinds of quarks (up, down, strange, charm, etc.) were produced in the interaction and with what momenta and frequency. Electrons and muons could be tracked, for example, to study the distribution of momentum in the transverse direction (perpendicular to the beam line). Determining fragmentation functions (fixing the distribution of quarks within a proton or neutron), identifying the kaon and baryon contents of jets, and computing the correlation between electrons and kaons all promised to reveal more about the inner composition and dynamics of quark-gluon interactions. Similarly the group would examine flavor tagging with gammas to study QCD by comparing light quark jets with heavy quark jets. A nonstandard avenue available to the team was the setting of limits on the production of isolated quarks: charged particles with charge $4/3e$ or $2/3e$. Finally, they could look at a novel combination of quarks in the previously unseen meson, F^*, that would be registered in two-gamma–producing interactions.[155]

Integration, therefore, was necessary within and between each of many planes of activity: electronic, mechanical, electromagnetic, structural, software, and physics. And once again, but not for the last time, PEP-4 and PEP-9 faced each other with mutual suspicion. Ordered to reassess this relationship once again by Panofsky and his EPAC, the leaders of the two "experiments" came together on 17 June 1983. Now the dispute was not over the value of θ_{max}, it was over the division of data itself. In an e-mail report to the collaboration, one member reported his impressions of the peace talks:

1. On the total collaboration of one-gamma and two-gamma physics
   PEP9 said most likely yes
   PEP4 said most likely no
2. On collaboration of all two-gamma physics
   PEP9 said most likely no
   PEP4 said most likely yes
3. All two-gamma and those topics requiring both detectors
   PEP9 said most likely no
   PEP4 said maybe yes
4. Single-tag and no-tag part of the two-gamma
   PEP9 gave more than 50%
   PEP4 said maybe

154. On the history of QCD, see Pickering, *Quarks* (1984).

155. Shen to Seiden, n.d. [on internal grounds shortly after 20 June 1983 Department of Energy review], DNyP.

5. Single-tag of two-gamma only
   PEP9 gave less than 50%
   PEP4 gave more than 50%[156]

Here one sees in full dress the interconnected nature of data and prestige. Take Category 1. For the PEP-9 collaboration, sharing in the full spectrum of one- and two-gamma physics was a worthwhile move; while it would no longer have a proprietary piece of physics, its expanded territory would encompass more than what could be extracted near the beam. To PEP-4, however, such a total collaboration must have appeared a poor bargain indeed: in exchange for access to a larger portion of the two-gamma events (it had some of its own), it would be sharing other physics it could do quite well by itself. Category 2 is similarly understood. For PEP-9, sharing *all* two-gamma physics meant giving up its unique claim to fame in exchange for the marginal additional two-gamma physics available through the PEP-4 data. For precisely that reason, PEP-4 thought it a swell deal. A week later, Shen had to dodge the issue when he wrote Seiden, mentioning only that "we are in the midst of discussions which could lead to a more formal collaborative arrangement."[157] Nine years later, in 1992, even as the collaboration of collaborations rushed headlong toward termination, these negotiations were still underway—traces of their quasi autonomy visible even in the software of data analysis (see figure 7.29).[158]

By 12 August 1983, the ad hoc committee on the status of PEP-4 and PEP-9 was, on the whole, convinced. The principal failing of PEP-4, the committee concluded, was in the small number of people actually devoted to physics analysis, and this weakness was probably inevitable given the immense amount of effort needed to work out the distortion corrections. For example, the committee pointed out that dE/dx resolution was only about $\sigma = 3.6\%$, which, "while impressive," yielded the capacity to distinguish π's from K's only at the level of 3σ, and K's from p's only at the level of 2σ. As far as hardware was concerned, Seiden's committee was reasonably optimistic. The superconducting magnet would cut the momentum resolution by a factor of 3.5, and a further factor of 4 by the elimination of distortions within the TPC. Even if this combined factor of 14 was too high by a factor of 2, the committee thought the device would be "competitive."[159] Panofsky agreed, signaling to Shen that while as laboratory director he could not say anything official before the November EPAC meeting, as far as he was concerned, "the necessary steps . . . have been taken."[160]

156. [Shen], e-mail, [17 June 1983], file "Minutes of TPC Executive Board Meeting," box 4, DNyP II.
157. Shen to Seiden, 24 June 1983, file "Minutes of TPC Executive Board Meeting," box 4, DNyP II.
158. Elliott Bloom, interview by the author, 27 February 1992.
159. Seiden to Panofsky, 12 August 1983, DNyP.
160. Panofsky to Shen, 22 August 1983, DNyP.

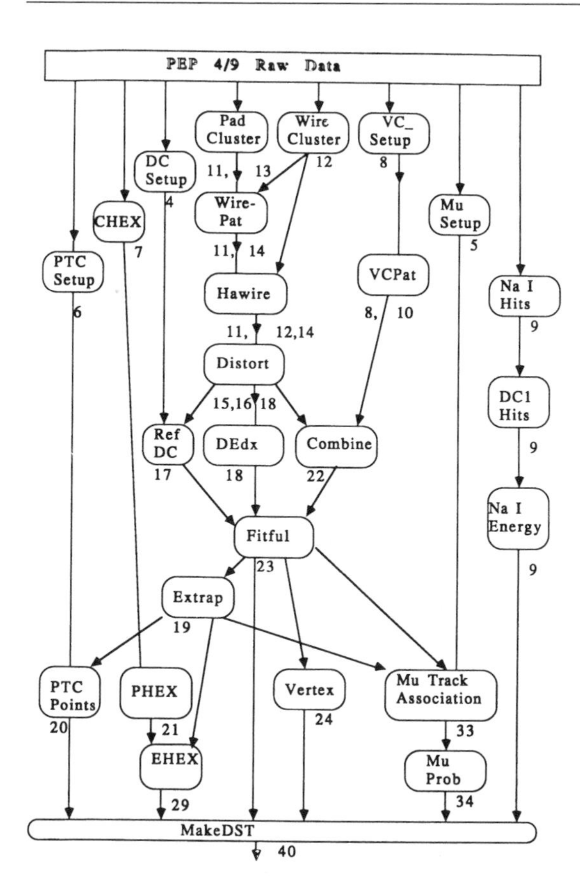

Figure 7.29 Data analysis of PEP-4/PEP-9 (1992). Contrasting the data analysis routine of 1992 with that of 1983 indicates how efforts to increase coordination among sectors of the experiment reached all the way into the production of the Data Summary Tape. Note, for example, that the EXTRAP program (which extrapolated tracks from the TPC into the calorimeters and muon detectors in 1983) began, by 1992, to reach into the pole tip calorimeters as well. But most important, note that PEP-9 figures not at all in the flowchart of 1983—and, while listed in 1992, slides data from "raw data" to Data Summary Tape without linkage to the software "guts" of PEP-4. Source: Courtesy of Michael Ronan, LBL, 1992.

7.8 Coordinating Theory and Data

As difficult as it was to mesh LBL with the other laboratories, or the physicists with the engineers in TPC/PEP-4, coordinating the culture of the experimenters with that of the theorists presented problems of an entirely different order. Our interest here is in the dynamics of this contact. Specifically, I want to explore the work done in the trading zone between the two domains. In this interstitial territory are the "phenomenological" physicists—modelers drawn from both camps who acted as mediators between those with solder in their hands and those with chalk. There is a sociohistorical or sociological facet of the question: who were the model builders and what was the relation between their professional activity and the more established traditions of experimentation and theory? There is also a linguistic or rhetorical side: how do the model builders talk to their more traditionally situated colleagues? And finally, there is a philosophical dimension to be understood: what is the relation between the entities, laws, and grounds for belief assumed by the theorists, experimenters, and phenomenologists?

The gap between a cathode pad and the Lagrangian that defined the theory of strong interactions, QCD, was unprecedented in the history of physics. If one looks for antecedents, only general relativity, with all its calculational and observational difficulties, offered a comparable gulf between theory and experiment, and even general relativity came with Einstein's three quantitative predictions in 1915. Thus, having probed in some detail the complex building-up process that takes data from a centerpoint in a cathode "hit" through the various distortion, field reconstruction, and other analysis programs up to particle identification, we now need to view movement from the top down. How did QCD look from above in the late 1970s and early 1980s, and how did theorists and the experimenters of TPC/PEP-4 assemble a middle ground, a trading zone in which the two cultures could find local, if not global, agreement?

The theory of quarks had its own complex history and traditions, passing through three distinct phases. Beginning in 1964, Murray Gell-Mann and George Zweig introduced the notion of a quark to be used as a classificatory tool much the way atoms had functioned within the periodic table during the nineteenth century. By 1967 or 1968 the idea of the quark and its mathematical representation in the group SU(3) was taken by most particle physicists to be altogether as important as Dmitri Mendeleev's table had been. And just as the nineteenth-century chemist retained a studied agnosticism toward the reality of physical atoms, so many particle physicists of the 1960s refrained from commitment to the reality of quarks. Such ontological abstinence was challenged in 1969 when a team at SLAC produced a new and higher energy set of data on the scattering of electrons off the interiors of neutrons and protons. For several years the common interpretation had been that the inside of a neutron, for example, was a complicated and continuous distribution of charge. But within months of the SLAC work, Richard Feynman had shown that the new data were consistent with a picture in which the electrons were scattering off pointlike objects located within neutrons and protons. Others, led by SLAC theorist Bjorken, went on to identify the "partons," as Feynman called them, with a now *physicalized* set of quarks. If the first, classificatory stage of quark theorizing could be dubbed the "Mendeleev quark" period, the quark-parton identification ushered in the age of "Rutherford quarks." Then, between 1969 and 1974, in an explosive period of particle physics, the partons themselves came to be treated dynamically, that is, integrated into a quantum field theory in which the old "strong force" was replaced by a gauged quantum field theory, QCD, modeled explicitly on the successful quantum theory of electrodynamics, QED.[161]

161. For an excellent discussion of the development of the quark idea, see Pickering, *Quarks* (1984). Pickering has much to say about the recycling of ideas from earlier theories, including scattering and resonance.

By the early 1980s, QCD was a standard part of any particle theorist's education; review articles and textbooks abounded. For theorists, the theory drew its strength not from quantitative links to specific experiments, but from a combination of qualitative explanations and the intratheoretic links it provided between different domains of phenomena. When Oxford theorist C. H. Llewellyn Smith gave the summary talk of the 1981 Rencontre de Moriond, he introduced his discussion of QCD with the words "Why I Believe." Faith, it seemed, had come to him not through any detailed quantitative comparison of experiment with theory, but through what he called "a priori" (theoretical) reasons. For example, without color, the electroweak theory would be unrenormalizable. Nothing an experimental physicist could do or say could *prove* that the theory had to be such that a finite number of parameters would suffice for calculations of arbitrary accuracy. But for theorists, renormalizability had by the 1980s become a touchstone of truth, and by bolstering it, color looked ever more attractive to the theorists. Another theoretical consideration came from a symmetry that the theory exhibited. If the masses of the up, down, and strange quarks were small compared to the typical energy scale of QCD (about 300 GeV), then the theory exhibited what was called a chiral symmetry: substitutions among these three quarks in a wide variety of ways left the theory unchanged. Chiral symmetry, in turn, led to a number of predictions that were roughly consistent with past experiments. For the theorists, chiral symmetry became appealing in its own right, a metatheoretical constraint on any putative theory of the strong interactions. Any such candidate had to satisfy, at least approximately, the chiral symmetry—and QCD did. In addition to these a priori reasons, Llewellyn Smith cited the explanatory function of QCD in accounting for the success of the naive parton model and even, occasionally, leading to corrections to it.[162]

Other bits of evidence could be culled from new experiments and old. The rate of neutral pion decay into two gammas had, for quite some time, been observed to occur about three times more frequently than theory predicted. With the assumption that quarks could come in three colors, it followed that the predicted rate of decay should be three times what had originally been predicted; the theorists recouped their missing factor of three and could match experiment. Similarly, R (the ratio of the rate at which e^+e^- produced hadrons divided by that at which e^+e^- produced muons) exhibited a similar "need" for an additional factor of three that could be supplied by the assumption of three colors. Even the late-1970s discovery of three-jet events at DESY found a simple qualitative explanation in QCD: out of an e^+e^- annihilation, a virtual photon emerged. The photon then produced a quark-antiquark pair (each of which makes a jet), and in addition, one of the two particles radiated a gluon providing the third jet.[163]

162. Llewellyn Smith, "Summary Talk" (1981), 432–35.
163. Llewellyn Smith, "Summary Talk" (1981), 432–35.

Conceding that none of the processes that purportedly tested QCD did so quantitatively, Llewellyn Smith defended the theory on the grounds that it worked "as well as [could] be expected." One couldn't, after all, hope for too much when perturbation theory, the nearly universal tool by which predictions were made, had to be used in conditions in which the perturbative series did not converge. "However," he argued, "the qualitative success of the theory in all cases together with the other nice features of the colour force and the 'a priori' arguments . . . convince me that QCD is correct."[164]

Harald Fritzsch was one of the theorists at the center of these theoretical developments. In 1979 he, like Llewellyn Smith, endorsed the theory more for its exhibition of theoretical virtues than because of its close numerical correspondence with experiment:

> At present most of the particle theorists have great confidence that the electromagnetic and weak interactions can be described by a system of non-Abelian gauge fields, and that the masses of the *W* and *Z* bosons are generated by a spontaneous breakdown of the gauge symmetry. Furthermore the prospects have become very good that the strong interactions are a manifestation of a pure, unbroken non-Abelian gauge theory, the theory of colored quarks and gluons (QCD). This is all very satisfying, in particular in view of the fact that the theory which is most simple and beautiful in its structure, Einstein's theory of General Relativity, is nothing other than a non-Abelian gauge theory, employing the symmetry of space and time as the underlying gauge symmetry.[165]

To an experimenter, resting belief on the possible gauge reformulation of general relativity would have seemed utterly alien—even more so than the metaphysical assumption that true theories should be renormalizable. Indeed, when experimenters reviewed QCD in their own lectures, the post hoc reformulation of the Einstein equation into a spin-2 massless quantum field theory did not even arise. But for Fritzsch, not only did QCD seem meritorious in its own right, it appeared as a harbinger of the grand unified theory "whose completion can be considered as the ultimate goal in physics and would imply the end of the development in fundamental theoretical physics." Though far from his and his colleagues' immediate goals, Fritzsch advertised the method by claiming that "we have arrived at a stage at which the construction of an ultimate theory of all interactions has become thinkable."[166]

"Ultimate theories," chiral symmetries, non-Abelian mathematics linking gravity and particle physics—these were a very long way from the ionization measurements, magnets, and data bases that had preoccupied the TPC/PEP-4

164. Llewellyn Smith, "Summary Talk" (1981), 435.
165. Fritzsch, "Chromo and Flavour Dynamics" (1980), 279.
166. Fritzsch, "Chromo and Flavour Dynamics" (1980), 317.

experimenters for almost a decade. Now, however, the TPC was finally working—it was about to go on-line at SLAC—and there was no avoiding the difficulties of sorting out how the physics would be coordinated between these two communities. Retiring to Asilomar, a conference center near Monterey, California, experimenters from the collaborating institutions gathered to exchange views among themselves and with a select group of experimentally minded theorists on how to close the gap with theory. One subgroup centered its attention on possible tests of QED and electroweak physics, another explored the physics of two photons. A third team focused its attention on the correlations between particles found among the reaction products, hoping to gain insight into the dynamics of jet production. Finally, in what rapidly became the most prominent of the collaboration's efforts, a QCD subgroup assembled around Werner Hofmann, a German physicist who had come to LBL in mid-1982.

Hofmann had completed his doctoral thesis on proton-proton interactions at the CERN Intersecting Storage Ring (ISR), in Karlsruhe, in 1977. In his post-doctoral years at the ISR, he continued this work, using jets as a probe of the motion of quarks. By 1981 he was sufficiently immersed in the subject to have written a book on it (*Jets of Hadrons*).[167] To Hofmann, the TPC was a perfect device with which to continue these studies. Unlike traditional collider detectors, the TPC could identify in detail the nature of each of the reaction products, enabling the experimenter to examine the detailed correlations between final state particles. The quantum number "strangeness" is conserved in the strong interactions. So if a strange particle appeared in one place among the final state particles, the TPC collaboration could search its data for (and presumably find) the anti-strange quark that must lie somewhere else in the debris. By sifting through these correlations, Hofmann hoped to shed light on the dynamics by which the initial quarks produced in the collision made other quarks and combined to form the observable entities such as pions, protons, kaons—but no "free" quarks. This elaborate recombination of quarks and gluons, known as "hadronization" (figure 7.30), would take place as a two-part process. First, in a short distance scale (by the uncertainty principle, a region of high momentum) the two quarks would behave like two electrons making an electron-positron pair in QED (when momentum was large, the interaction strength of QCD was weaker, and perturbation theory held good). Then came the hard part: on a longer scale of distance (lower momentum involved) the quarks would bind up together with an interaction strength so strong that the simplifying assumptions of perturbation techniques were sure to fail. In short, the process of hadronization appeared trebly difficult—first, because QCD appeared as a new and largely untested theory; second, because the hadronization process itself involved far

167. Hofmann, *Jets of Hadrons* (1981).

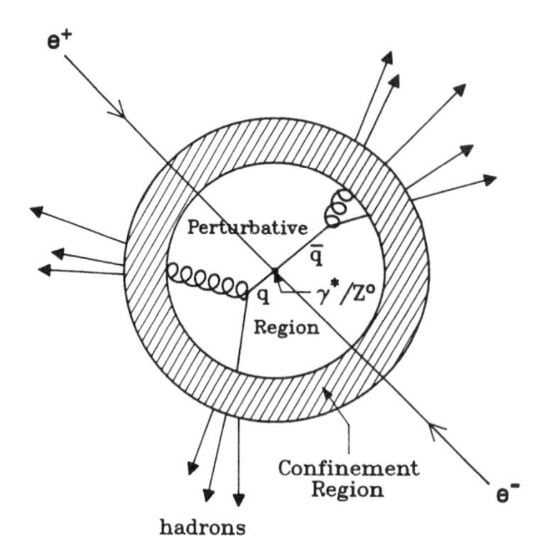

Figure 7.30 Hadronization. Schematic representation of hadron production: electron and positron annihilate one another at the center, producing a virtual photon or a virtual neutral vector boson (γ^* or Z^0) that in turn materializes as a quark-antiquark pair (labeled q and $\bar{q}$). As indicated, these outgoing quarks can radiate gluons (*curly lines*). In the high-momentum sector of the spacetime picture (close to the annihilation event, labeled "perturbative region"), the coupling constant is small and processes can be treated by methods analogous to those used in QED. When momentum falls (as distance from the event increases), the coupling becomes large and perturbation theory fails. For this reason the quark and gluon recombination process is not simple and surely not to be understood by perturbative methods alone (this troublesome sector is shaded here and labeled "confinement region.") At the periphery, black outgoing arrows ("hadrons") are the observed entities—particles like the pion, lambda, or kaon. Source: Gary, "Tests of Models," LBL-20638 (1985).

more physics than could be extracted in any straightforward way from QCD; and third, because the collaboration had spent practically no time on physics during the crisis years of construction.[168]

One of the theorists who came down from LBL to meet with the TPC collaboration at Asilomar was Ian Hinchcliffe, who early in his presentation to the experimenters put up a transparency saying "QCD cannot predict how quarks ⇒ hadrons. We can only make models." Warning of the dangers inherent in the vagaries of Monte Carlo simulations, and the precarious dependence of the whole notion of jets on how one defined these objects, he sketched out the "straight" QCD part of the theoretical picture. If Hinchcliffe reached out to make contact with the experimenters, Hofmann was the experimenter who wanted to meet him halfway. Hofmann's talk began by setting out three goals. First, he took a rhetorical stance in which he reasserted his identity as an experimenter, reassuring a worried audience: "NO THEORY! (+ NO MATH EXCEPT FOR +, −, ×, ÷)," and carried on from there.[169]

168. Hofmann, interview by the author, 16 June 1992.

169. Hinchcliffe, "Theory," and Hofmann, "Experimental," in "QCD Studies," for TPC Asilomar Workshop, 4–6 February 1983, TPC-LBL-83-13.

Immediately the QCD subgroup faced problems of a complexity that exceeded their grasp. In QED, the coupling constant that characterizes the strength of the electric force is defined as $\alpha = 1/137$. Because α is so small, it is possible to expand the basic theory into a perturbation expansion in which each term is smaller by a factor of 1/137 than the preceding one. It therefore becomes a fairly straightforward matter to calculate quantities to an arbitrary degree of accuracy: one simply chooses how far out in the expansion to go. In QCD, alas, the analogous coupling constant α_s is on the order of 1, and so terms are not typically getting much smaller as one goes further out in the expansion. Thus, in QED, processes that involve the radiation of four photons, for example, are much less likely (about 137^2 times less likely) to occur than processes involving the radiation of two. In QCD, however, the likelihood of radiating four gluons is *not* particularly smaller than the likelihood of radiating two. In addition, once the quarks and gluons were blasted apart in a collision, QCD predicted that only bound states of quarks, not individual ones, would ever be seen. The process of recombination into mesons and baryons was also not understood and clearly would need to be before any direct test of QCD could be made. The QCD subgroup members thus stood on the far side of a brick wall from their theoretical colleagues. According to a summary of their deliberations written by Hofmann in February 1983, these two considerations—higher order effects and hadronization—stood squarely between theory and experiment:

> From this it is obvious, that considerable theoretical input and interaction with theory is required; the main effort is *not* to solve technical problems, but instead to understand the physical impact and differences of the various models; obviously such a topic is not a short-range task, and [it] is not clear if the integrated knowledge of the present PEP-4 group is sufficient to attack these problems. On the other hand, there seems to be general agreement that on a long-term basis understanding all these things is a necessary ingredient for almost any kind of jet physics.[170]

By the time their meeting had ended, the group assigned itself a series of tasks, all of which involved grappling with a growing subfield that hoped to mediate between experiment and theory. It was, in a sense, the descendant of Feynman's original parton model over a decade earlier.

Specifically, the collaboration knew it had to begin by understanding the different "models" of jet production and what they implied both at the level of partons and at the level of observable particles. Next, it had to get hold of the relevant Monte Carlo simulations and explore them enough to "find out which differences are basic." For only with a detailed computer simulation could the models produce graphs commensurate with the results of the experiment. Finally,

170. Nygren and Sub-Group Organizers to PEP-4 Collaborators, "Summary of Asilomar Physics Workshop," 7 February 1983, TPC-LBL-83-5.

the TPC/PEP-4 subgroup would have to equip itself with computer algorithms that would meet this amalgam of simulation and model halfway: the computer would be needed to find and characterize jets from among the scattered debris registered on the data tapes.

At a very general level all the models were designed to bind theory to experiment. But more specifically, we need to know how this purported binding was supposed to take place. Were the models provisional heuristics useful only as guiding posts on the path to a fundamental truth? Were the models themselves candidates for a True Theory of the World? Were they merely empirical summaries the way a curve-fitting algorithm might be used to find a polynomial that would neatly arc through a set of data points? Were the models deductions from the "fundamental" theory, QCD?

Feynman and Field offered a good beginning point for the experimenters' inquiry. Some 30 years before, Feynman had been instrumental in creating the theory of quantum electrodynamics, a theory that, from the outset, was designed to give an account of electrodynamic interactions involving the structureless entities the electron and the photon. By the 1970s, QED was considered by most theorists to be the very prototype of a "fundamental" theory. Because Feynman had played such a formative role in the development of QED and because the electroweak and chromodynamic theories were so explicitly modeled on QED, many physicists were surprised at his deep-seated suspicion of the new gauge physics. It was far better, he believed, to reason closer to the phenomena. It was no accident that Feynman was not among those who insisted vehemently on the identification of partons with quarks: silence on the further attribution of properties to these scattering centers was more his style. In 1972, for example, when he concluded his book on photons and hadrons, Feynman held back from the ascription of physical reality to QCD-style quarks or partons-as-quarks. For even if the identification of quarks at low energy with partons at high energy could be made, there remained the possibility that one had discovered algebraic regularities rather than objects that "exist." "From this point of view the partons would appear as an unnecessary scaffolding that was used in building our house of cards." As such, the importance of quarks-as-partons might be seen as purely "psychological." Yet for Feynman to call something psychologically useful was not, in and of itself, a condemnation. It was possible that the assumption of quarks would lead people to come to other "valid expectations" and that at the end of the day quarks would "become 'real,' possibly as real as any other theoretical structure invented to describe Nature."[171] The ontological fate of quarks remained very much open.

When it was discovered not long afterward that neutrinos could scatter intact from quarks and leptons (weak neutral currents), high energy physicists

171. Feynman, *Photon-Hadron* (1972), 269–70.

celebrated the news as powerful evidence for the Glashow-Weinberg-Salam electroweak theory and its applicability to partons-as-quarks. Not Feynman. He insisted in 1974 that it was far more appropriate to restrict discussion to what followed more closely from the experiments.[172] On similar grounds, his view of jet phenomena was entirely consistent with his earlier ontological hesitancy. Feynman did not buy the idea that the collimated blasts of strongly interacting particles known as jets were the manifestation of pairs of quarks producing hadrons out of the vacuum by emitting gluons, which in turn made more quarks that then bonded into quark-antiquark pairs (mesons) or three-quark combinations (nucleons and other baryons). Over and again in their joint paper, Feynman and his collaborator Richard Field shied away from such overarching claims about the status of the independent fragmentation model as a theory: "Although the model is probably not a true description of the physical mechanism responsible for quark jets, many predictions of the model seem quite reasonable, possibly much like real quark jets. . . . The purpose of this work is to provide a model useful in the design of experiments . . . and further to provide a standard to facilitate the comparison of lepton-generated jets with the high-$P_\perp$ jets found in hadron collisions."[173] ($P_\perp$ is the momentum orthogonal to the beam.)

On one level, Feynman and Field were taking a stance not dissimilar from that of the philosopher Bas van Fraassen, who has argued strenuously against the view that the unification of different domains of phenomena *forces* one to interpret a theory realistically. Instead, van Fraassen claims, the virtue of bringing diverse domains under one theoretical roof is pragmatic.[174] Unification, according to this position, is to be sought as vigorously by an instrumentalist as by a realist. While there are many other features of van Fraassen's work, I want only to highlight the particular argument that unification need not imply realism. Here, Feynman and Field emphasized throughout their paper that their primary purpose was to *link* the dynamics of jets produced in proton collisions to the jets observed in electron-positron collisions at SLAC, but they certainly did not follow that goal with the conclusion that the model must therefore be interpreted realistically: "[W]e think of our jet model, not as an interesting theory to be checked by experiment, but rather as a possibly reliable guide as to what general properties might be expected experimentally. In particular, it can assist in the program of comparing hadron [high-$P_\perp$] jets to lepton-generated jets."[175] I would suggest that we see the Feynman and Field model not in the image of

172. Feynman encouraged the team of Barish et al. to shy away from the neutral-current search in large part because he had doubts about the Glashow-Weinberg-Salam theory. When the existence of neutral currents came to be indisputable, Feynman emphasized that the phenomenon should be discussed in its own right—not simply as a test of the electroweak theory. See Galison, *Experiments* (1987), 238–40.

173. Field and Feynman, "Parametrization," *Nucl. Phys. B* 136 (1978): 1–76, on 1.

174. Van Fraassen, *Scientific* (1980).

175. Field and Feynman, "Parametrization," *Nucl. Phys. B* 136 (1978): 1–76, on 74.

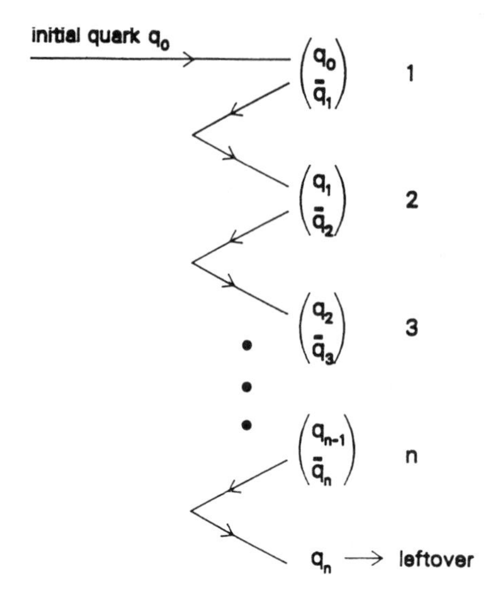

Figure 7.31 Feynman-Field independent fragmentation model. Designed to characterize and standardize the new "jet" phenomena, the independent fragmentation model deliberately abstained from any commitment to "fundamental" theory. The initial quark q_0 fragments into two further quarks q_1 and $\bar{q}_1$, q_0 and $\bar{q}_1$ form a meson, and so on. To "tune" the model, a parameter is chosen that fixes the width of the transverse momentum of the mesons (crosswise to the jet direction) and another parameter governs the parabolic distribution of momentum in the direction of the jet. Other parameters fix the ratio of quark types. Not quite "theory" (*strictu sensu*) and not quite "experiment," such modeling served as a common meeting ground for theorists on one side and electron-positron and proton-antiproton colliding beam experimenters on the other. Source: Gary, "Tests of Models," LBL-20638 (1985).

theory exemplified by Newton's universal theory of gravity, Einstein's general relativity, or even Feynman's QED, but rather in the tradition of standardization that lay behind the establishment of nineteenth-century assessments of electrical resistance, for example.[176] The authors explain: "We thought it might prove useful to have some easy-to-analyze 'standard' jet structure to compare to. Thus, a hadron experiment could say 'the real jets differ from the "standard" in such and such a way,' and the lepton experiment could then see whether they deviated from the same 'standard' in a similar way."[177] Precisely because the Feynman-Field model invoked a series of rather ad hoc assumptions in the enforcement of conservation laws, Feynman could not and did not imagine that their "independent fragmentation" model was *true*.

The model worked this way. Feynman and Field assumed that a parton, q_0, emerges with momentum W_0 and direction w. It creates (somehow) a pair of other quarks, labeled anti-q_1 and q_1, such that a parton of type q_0 combines with a parton of type anti-q_1 to form a meson. Further pairs are similarly created out of the vacuum, for example, anti-q_2 and q_2, with the q_1 parton combining with the anti-q_2 to form a second meson. The process continues (see figure 7.31, in which anti-q_2 is written $\bar{q}_2$) according to a probability function in which $f(\eta)d(\eta)$ is the probability that the primary meson (q_1 and anti-q_2) leaves a fraction η of the original jet momentum to the remaining cascade; each such frag-

176. See Schaffer, "Manufactory of Ohms," unpublished paper, presented at conference "Mediums of Exchange," UCLA, 2 December 1989.

177. Field and Feynman, "Parametrization," *Nucl. Phys. B* 136 (1978): 1–76, on 2.

mentation is independent of the previous one, determined by a new roll of the dice, so to speak. (Note that the distribution is given for the center-of-mass frame only—it is not intended that the distribution hold good in other frames of reference.) Eventually a single quark is left over, and the momentum of the beam is insufficient to make any more parton pairs. Now comes a brute empirical fitting question. Could the authors choose the function $f(\eta)$, along with three additional parameters (the mean momentum of the mesons transverse to the jet, the ratio of the different species of quarks, and the spin of the mesons), such that the resulting jets would fit the experimental data? It seemed so, with choices such as fixing $f(\eta)$ to be a parabola with one adjustable parameter, setting the mean transverse momentum at 330 MeV, allowing the creation of half as many *s* anti-*s* pairs as *u* anti-*u* pairs, and demanding that vector mesons come into existence about as often as pseudovector mesons. These basic assumptions could then be written into a computer-run simulation that would produce jets, and calculate such experimentally observable quantities as correlations between certain pairs of mesons, or the total charge of a jet. But what is this fragmentation scheme? If one asks whether the assumptions lying behind it necessarily emerge from QCD, the answer is a resounding "no." "The predictions of the model are reasonable enough physically that we expect it may be close enough to reality to be useful in designing future experiments and to serve as a reasonable approximation to compare to data. We do not think of the model as a sound physical theory."[178]

If one asks whether the model is entirely fictitious, the answer is also "no." Feynman departs from a run-of-the-mill philosophical instrumentalism in that he does not deny the possibility of a theory deeper than such "fitting" schemes. While he made it abundantly clear that the independent fragmentation model (as it came to be called) held no pretensions to being an "ultimately" correct theory, this did not foreclose the possibility of some such theory being found. QCD was not the savior for which he was waiting, and he remained dubious about the ability of the whole class of phenomena subsumed under jets ever to reveal the inner secrets of quark dynamics:

> [Q]uark jets are often investigated not just to be used as a tool to investigate hadron collisions, but rather as a subject of interest in itself. How are quark jets actually generated? From what we have learned, we think it will tax the ingenuity of experimenters to see behind those properties which are overshadowed by the effects of resonance decay [in which mesons in excited states decay to other mesons] in order to study effects more intrinsically related to the process of jet formation. In this connection, the most fundamental experimental question is whether lepton-induced jets really have a quark origin at all. Even here we find difficulties.[179]

Those difficulties stemmed from the inevitable averaging necessary to extract data about the interactions; the inability of jet physics to focus on individuated

178. Field and Feynman, "Parametrization," *Nucl. Phys. B* 136 (1978): 1–76, on 2.

179. Field and Feynman, "Parametrization," *Nucl. Phys. B* 136 (1978): 1–76, on 74.

phenomena left Feynman pessimistic about the physicists' ability to extract fundamental information about the underlying processes.

More generally, Feynman conceived of the fragmentation model neither as a theory in its own right nor as a consequence of QCD. It was not a full-fledged theory in and of itself because the model did not conserve energy or momentum: a massless initial quark could be transformed into a number of massive mesons. There was no mechanism for making higher spin states, and no way to produce baryons (three-quark combinations like protons or neutrons). And by leaving one quark floating free at the end of the chain (see figure 7.31), quantum numbers were not conserved. (Over time various pieces of the model were fixed up: two jets could mate their free quarks into a meson, e.g.; and individual partons could be artificially boosted to more or less conserve energy and momentum.) However problematic the model was, Feynman certainly did not take it for granted that the independent fragmentation scheme was a poor "approximative" cousin to a "true" QCD. Not only did he harbor doubts about QCD, he considered it entirely unestablished that the jets seen at SLAC or Hamburg's DESY had anything at all to do with quarks, and problematic that the experiments could be used to resolve the question. What the independent fragmentation model could do was to establish a point of reference for the TPC/PEP-4 collaboration that would throw deviations from that standard into relief.

On the other side of the Atlantic, a group of theorists at Lund approached QCD and hadronization very differently from Feynman. Instead of looking at QCD askance, they accepted QCD, and searched for ways to use their knowledge of the "basic" theory as "inspiration" for their model building. And while Feynman thought that jets would ultimately have little to offer the quark model, the Lund group had the highest expectation that the dialectic between model building and experimental jet production would lead inexorably to an ever richer description of the quarks and their chromodynamic interactions. As Bo Andersson and his collaborators argued in 1983, fragmentation models were necessary to compare experiment to "basic theory" (QCD): "Such models may on the one hand be looked upon solely as phenomenological parametrizations and rules of thumb in order to obtain a translation from one language to another. As such they are useful for analysis of experiment as well as for the planning. On the other hand one may as always in connection with phenomenology try to obtain a dynamical framework that serves as a motivation and a generalizing principle for the constructions."[180]

Feynman would have liked neither of these two hands. For Lund, the "language" of experimentation was that of hadrons, with observable decays—the descriptive vocabulary that included the old standbys of 1960s bubble chamber physics: lambdas, pions, kaons, protons, and sigmas embedded in the dynamics

180. Andersson et al., "String Dynamics," *Phys. Rep.* 97 (1983): 33–145, on 34.

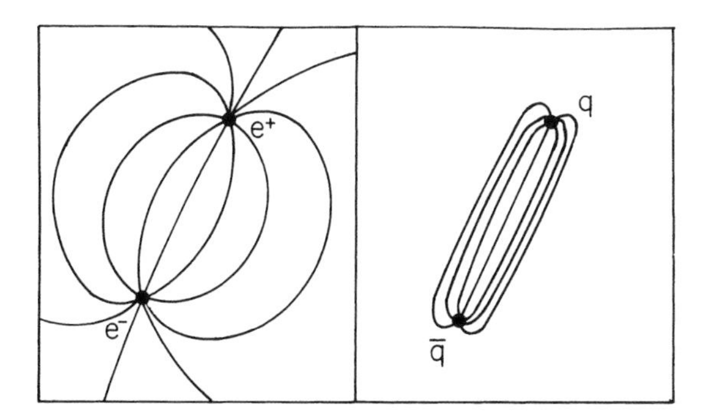

Figure 7.32 Field lines in QCD. The string model of QCD is only loosely based on first principles of quantum field theory. But very roughly, it is helpful to contrast the field lines of electrodynamics, which all eventually get from, say, a positron to an electron (but do so over a very large volume of space) with those of QCD, in which the field lines between a quark and an antiquark are restricted to a thin "flux tube." In a sense, the dispersion of the lines in QED is due to the fact that the photons that carry the electrodynamic force are not attracted to each other; the confinement of the flux lines in QCD results from the mutual attraction of the gluons that carry that force. (Photons are electrically neutral, but gluons carry color. Color is the charge in QCD that functions as the analogue of electrical charge in electrodynamics.)

that described their production and transformation. Read as a "phenomenological parametrization" or "rule of thumb," the Lund model was proffered as a means of linking this experimental language with the language of "basic theory"—quarks, gluon, and their interactions. Since Feynman had little faith in QCD as the "true" theory, he was hardly likely to be interested in effecting such a "translation." Nor, for similar reasons, would Feynman have wanted to search for QCD principles by the light of which he could understand the building blocks of his own fragmentation model.

The guiding idea of the Lund group was that the color field between two quarks would act as a string (see figure 7.32). Since gluons themselves carry "color charge," gluons are attracted to gluons; it is as if in QED photons carried charge themselves instead of simply reacting to it. Because of this, the field lines in QCD collapse around a single line between the two quarks instead of venturing out in all possible directions, as is the case in electrodynamics. This "string" then serves to hold a pair of quarks together. In figure 7.33*a*, the antiquark heads left and the quark heads right until they are drawn back together by their string. Yo-yoing in this fashion, the quarks shown in this picture depict a meson at rest. In another frame of reference, the yo-yoing quarks form a meson moving to the

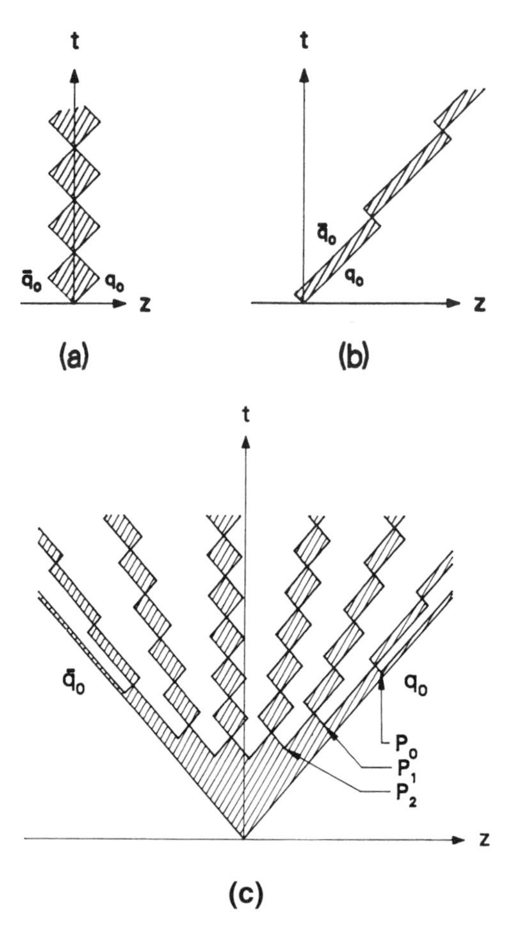

Figure 7.33 Strings, yo-yos, and mesons. (*a*) The string model, represented in a spacetime diagram, shows a quark-antiquark pair (a meson) oscillating with its center of mass at rest. (*b*) That same meson is viewed from a moving frame of reference. In both cases the shaded region indicates the spacetime region in which the field is nonvanishing (i.e., where the string can be found). (*c*) The Lund string model allows the quark-antiquark pair to produce further q-$\bar{q}$ pairs, obeying the conservation of energy and momentum. Note that the scheme is very different from that of Feynman-Field. Feynman-Field partons fragment one by one; in the Lund model it is the string system as a whole that fragments. Furthermore, in the Feynman-Field model, particle number and Lorentz invariance are violated if one moves to other frames of reference; in the Lund string scheme the number of mesons and the conservation laws are preserved. Source: Gary, "Tests of Models," LBL-20638 (1985), 49–50.

right (figure 7.33*b*). Quite unlike the gauge theorists before, the Swedish model builders could not buttress their vision of the world with the abstract virtues of theory. Here, there were no anticipations of a final unification, no talk of the structural similarity with gravity, and little place for abstract rumination about the beauty of theory. While gauge theorists could offer renormalization as a signal of a true theory, or point at the Glashow-Weinberg-Salam theory as a golden example of (partial) unification, these were not avenues open to phenomenologists trying to tie something from the QCD to something from the Data Summary Tape. Nor could the phenomenologists retreat to the aesthetics of starting with symmetries and deriving force laws, as the gauge theorists were wont to do—the whole point of their endeavor was to forsake the intratheoretical arguments in favor of concepts that tied more directly to the experimental. They were not writing a classical literature in a world language—they could function neither as gauge theorists nor as experimenters—instead, they were composing a pidgin that would mediate, locally, between them. But having lost the intrinsic appeal of a purely theoretical grounding, the warrant for phenomenological models was harder to establish. The Lund group (rather altering the chancellor of the exchequer's words): "It should be kept in mind, however, that there are no easily available measures of the success of such a venture. As Bacon told us a long time ago, it is actually only possible to learn that one is wrong by a comparison between model calculations and experimental findings. If the prediction agrees there is no reassurance that one is even working in the right basic direction (although there is evidently a possible reason to feel some confidence!)."[181]

To achieve this confidence, the Lund group had to remain in extremely close contact with experimental groups in a way that the gauge theorists did not. Bo Andersson, for example, came to the TPC to deliver a series of lectures on the string model in late November 1983. Again, in 1984, one of the Lund group spoke at a TPC-organized conference at UCLA. Whereas Hofmann had somewhat despairingly worried about the "integrated competence" of the TPC group to handle the link to theory in 1983, by 1986 contact between the experimenters and the phenomenological theorists had grown to the point where Andersson and Hofmann could coauthor a *Physics Letter.*

Feynman and Field had begun their modeling with what they took to be concepts "prior" (epistemically) to QCD—foremost among them, the parton. Andersson and colleagues took a highly simplified physical model (the string) and looked for results that this model held in common with QCD on one side, and with experiment on the other. Yet a third line of reasoning grafted a good deal of QCD onto some purely experimental results: this was the strategy that Richard Field (of Feynman and Field) pursued with Stephen Wolfram.

These "cluster models" by Field and Wolfram took seriously the claim of QCD to be a valid perturbative theory when the quarks have high momenta (see figure 7.34). So for the initial stage of the interaction, the cluster model simply included calculations based on perturbative QCD, yielding a detailed picture of the incipient features of jet formation, including gluon radiation. Then, as time passed in this cluster model, the characteristic momenta of the quarks dropped, and QCD no longer could be used validly as a perturbative theory. At this point, the cluster model is incapable of providing insight into the formation of hadrons, and one (or rather the computer) simply groups the quarks into high-mass "clusters" that then are allowed to decay into the hadrons we know and love—sigmas, lambdas, protons, and so on. The ratios of these final state particles are put in by hand to match observation, but their energy and momentum distributions emerge from the motion and composition of the cluster from which they spring.

181. Andersson et al., "String Dynamics," *Phys. Rep.* 97 (1983): 33–145, on 34.

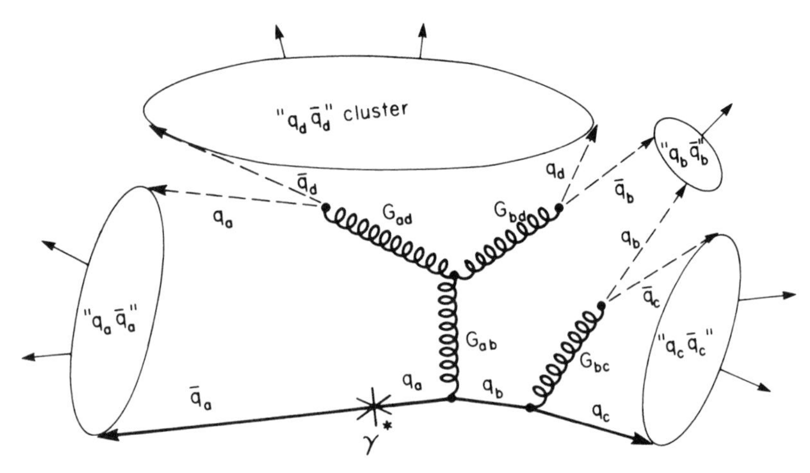

Figure 7.34 Field and Wolfram QCD model (1983). Schematic illustration of the procedure used to isolate color singlet clusters. Each parton line carries a spinor color index (one for quarks, two for gluons). Final gluons are forcibly split into collinear q-$\bar{q}$ pairs (*dashed lines*). The quark and antiquark at the ends of each group-theoretical "string" are combined into a color singlet (color neutral) cluster. The clusters thus produced are taken to decay independently into hadrons. Each cluster decay is assumed isotropic in the cluster rest frame, and the final state is determined from a simple phase-space model. Source: Field and Wolfram, "QCD Model," *Nucl. Phys. B* 213 (1983): 65–84, on 73.

One contributor to the cluster scheme, Thomas Gottschalk, then a postdoctoral fellow in theory at Caltech, propagandized this way at the June 1984 TPC UCLA workshop:

DOGMA

1). QCD (Whatever L_{QCD} Really Means) Does Describe Hadronic Physics
2). At Short Distances (Early Times) Perturbative Theory Provides a "Reasonable" Approx. To QCD
3). At Later Times, the Character of QCD Somehow Changes. Dominant Phenomenon Becomes (Linear) Confinement—*Not* Perturbation Theory
4). Hadrons are Colorless
5). Hadrons Form Primarily At Late Times In An Event . . .
6). (Unknown) Mechanism of Hadron Formation Is Characterized by
 (i) Locality
 (ii) Universality
7). Pions are Not Mass-Shell Yo-Yo's: Direct Hadron-Parton $\leftrightarrows$ Connections Are Suspect

ANATHEMA

Independent Fragmentation ie the Feynman-Field Model Or Anything Like It.[182]

182. Gottschalk, "QCD Cluster Modeling of Fragmentation," talk presented at TPC QQ Analysis Workshop, Berkeley, 1–2 June 1984, xeroxed transparencies, MRP.

Like schismatics, the clusterites now left the mother church, the Feynman-Field model. Field now threw his weight behind the Wolfram-Field model, and Gottschalk (borrowing from Wolfram-Field) called the doctrine of the church fathers anathema. In a sense the religious terms should not surprise us. At stake were what participants held to be the fundamental entities of nature, and these appeared in utterly different forms. In QCD, the basic objects are quarks and gluons; they are embedded in a relativistic quantum field theory, and the full set of theoretical assumptions that went with it: vacuum fluctuations, relativistic invariance, renormalization, and a myriad of conservation laws. Feynman and Field's quarks hold only certain very specific features in common with those of QCD. There is no dynamical connection between quarks and gluons, for example. Even in principle the independent fragmentation model cannot conserve momentum and energy simultaneously. And the lack of relativistic invariance was manifest: in one frame of reference the electron-positron collision could spew out 15 particles, in another 23. Feynman and Field had produced a phenomenological model that was isolated from QCD in all these and other ways.

By 1983–84, many of the younger theorists, like Gottschalk, had grown up using QCD; it was the first recitation line of the dogma. The audience for Feynman and Field had not been Feynman's usual audience of theoretical physicists; not surprisingly it was the experimenters who embraced it. For what it did was to offer a way out of the localized context of particular experimental schemes. Suddenly, an experiment colliding protons and antiprotons could be likened to an experiment colliding electrons and positrons. Even if Feynman and Field had violated every dearly held theory in the book from energy-momentum conservation to relativistic invariance, the scheme could coordinate these very different experimental activities. Feynman and Field's model served as *an intraexperimental pidgin language*. It, not QCD, served the experimenters by delocalizing their results, allowing competition, communication, and the concatenation of results. Secondarily, and only partly, Feynman and Field's independent fragmentation scheme offered the beginnings of a trading zone between QCD and experiment.

Each of these models had its own motivations, along with different objects and distinct laws of combination. "[T]he main goals behind the physics of jets," Hofmann wrote in 1987, "are . . . to test techniques developed in perturbative QCD, and to derive a deeper knowledge as well as phenomenological models of the nonperturbative regime."[183] But within the strictures of this "fundamentalist" orientation, radical disagreement persisted. The string modelers invoked one-dimensional QED to bolster the plausibility of the string itself—their fundamental entity. Cluster theorists rejected the direct passage from partons to hadrons; they insisted that some higher mass clusters form first and only then decay into hadrons.

183. Hofmann, "Jet Physics PEP and PETRA," SLAC Summer Institute 1987, LBL-24086.

On the block-relativist view, these models would pass each other like ships in the night. "Quark," "jet," "fragmentation" refer so differently in the different pictures of independent fragmentation, strings, clusters, and QCD that it would be a fluke, a pun, to identify the "quark" of Field and Feynman with the "quark" of the clusterites, the "quark" of the Swedish stringmen, or the QCD "quark" in which Fritzsch or Llewellyn Smith "believed." But is it true that by virtue of their different theoretical embeddings these phenomenologists could no longer put one theory against a common standard? No. Had these systems of representation become incommensurable in the old sense of 1960s philosophy of science? Certainly not.

After the many meetings and collaborations with theorists and model builders, a set of site-specific ties had been worked out, a local consensus about how to connect TPC output to the models and, thereby, indirectly to the high theory of fundamental physics. Specifically, the TPC collaboration looked at three-jet events and tuned each of the three basic schemes (independent fragmentation, string fragmentation, and cluster fragmentation) to work as well as it could to account for global features of the data such as the number of particles per jet. They then tested all three against an observed feature: the number of particles observed to emerge between the jets. Consider the three schemes, each set up to account for a gluon and a quark-antiquark pair to emerge from the e^+e^- annihilation. The independent fragmentation models produced mesons in all directions at about the same rate and so led to no great asymmetry, and none of the "fix-ups" to the model seemed to help (figure 7.35*a*). String models, by contrast, have one string running from the quark to the gluon, and another string from the gluon to the antiquark (i.e., a gluon appears as a kink in the string). If the gluon is heading off to the right, then any mesons emerging out of the string begin with a "boost" to the right. Special relativity then tells us that while production may be isotropic in the rest frame of the string (in fact, the process is essentially the same as the independent fragmentation scheme), in the laboratory frame there will be a depletion of particles behind the string. Particles therefore appear depleted in the region between the quark and antiquark (figure 7.35*b*). Finally, in the cluster scheme, there are three sources of the depletion, and each could be examined separately in the computer simulations (figure 7.35*c*). First, perturbative QCD allows virtual gluons to be emitted from quark or antiquark, and these interfere destructively in the region between the quark and antiquark, causing a depletion in that sector. Second, because the particles that are observed in the laboratory do not have color, the quark and gluon color indices must cancel. A quark carries one index (red, blue, or green), while gluons carry two—they are supposed to link quarks (or quark-antiquark pairs) into color-neutral combinations. So an anti-red/green gluon might bind together a red-antired pair of quarks, turning the pair into a green-antigreen pair. Since the clusters are required to be color-neutral, more clusters end up between the quark and the gluon than end up

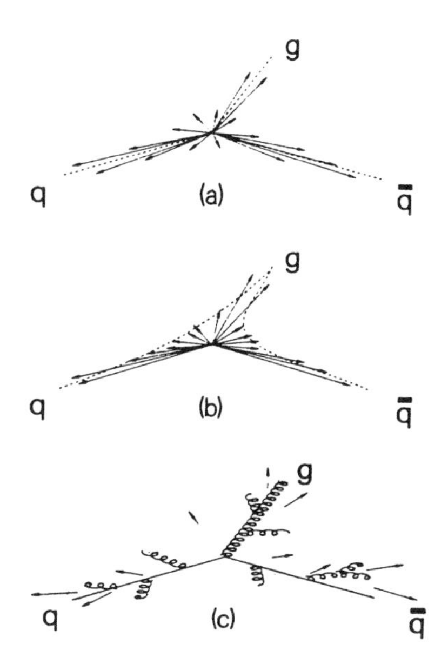

Figure 7.35 Three stories of depletion. In three-jet events (events attributed to a quark-antiquark pair in which one of the two radiated a gluon) it was observed that there was a depletion in the number of observed particles emerging between the quark jet and the antiquark jet. (*a*) This depletion was inexplicable on the basis of Feynman-Field independent fragmentation—the model itself presupposed isotropic particle production. (*b*) In the Lund string scheme, a gluon corresponds to a kink in the string. The ensuing motion of the kink then preferentially tends to create particles in the direction that the kink is traveling ("boost effect"). (*c*) The cluster model explains the depletion differently: low-momentum radiated gluons from the quark and antiquark interfere destructively, tending to suppress particle production in that sector. Source: Aihara et al., "Parton," *Phys. Rev. Lett.* 54 (1985): 270–73, on 271.

associating the quark to the antiquark. (Mutatis mutandis, there are more clusters in the antiquark-to-gluon sector than between the quark and antiquark.) Last, the clusters tying the gluon to the quark or antiquark are in motion, so (just as in the string model) the boosted origin of the observed hadron leaves a depletion in the region behind the gluon.[184]

Coming "down" from QCD to a prediction proved to be anything but automatic. And this is not a problem of a "difficult" calculation: it is "difficult" to reckon QED to 10 significant digits, a task for which great cleverness and endurance is required. That is not the problem here. QCD does not coherently admit a prediction of hadronization using perturbation theory, a state of affairs recognized across the board. In QCD, the strong coupling constant rises as momentum falls; when the quarks begin to bind with one another and with gluons, then the perturbation expansion yields nonsense in almost any calculation. In the face of this in-principle incalculability, enormous efforts went into the creation of models, each with its own putative links (or nonlinks) with the fundamental theory. Models made it possible to continue. With each of the three models in place, Hofmann and the phenomenologists could begin to make predictions, assertions that, at last, would intersect with the output of the TPC.

184. The analysis of the three models by the TPC/PEP-4/PEP-9 collaboration was presented in Aihara et al., "Quark and Gluon," *Z. Phys. C* 28 (1985): 31–44; Aihara et al., "Parton," *Phys. Rev. Lett.* 54 (1985): 270–73; and Gary, "Tests of Models," LBL-20638 (1985). The conclusion that cluster and string models beat independent fragmentation models confirmed work by the JADE collaboration: Bartel et al., "Particle Distribution," *Z. Phys. C* 21 (1983): 37–52.

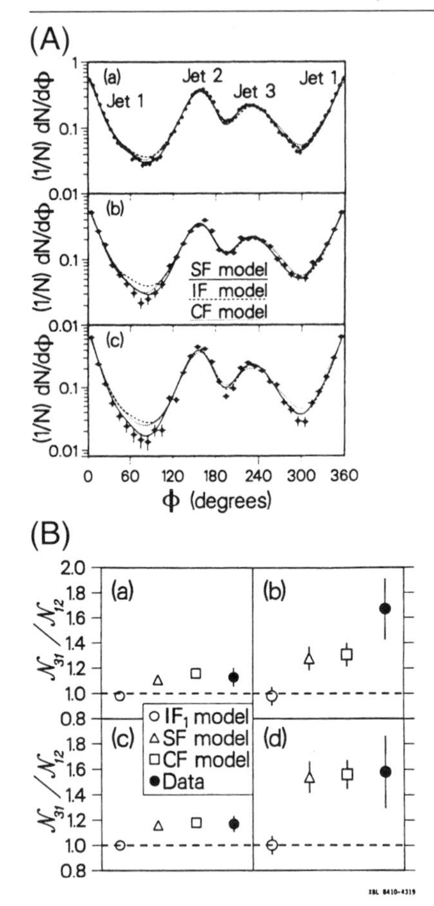

Figure 7.36. Models and data, (*A*) heavy particles versus angle and (*B*) pions and depletion. (*A*) As a further test of the various fragmentation schemes, the azimuthal density of particles in the three-jet sample, $(1/N)\, dN/d\phi$, is plotted as a function of the angle in the event plane ϕ between the direction of a particle and jet 1. Because the TPC could make quite accurate particle identifications, it was possible to plot the discrepancy between models and experiment for different species—here for kaons, protons, and lambdas. As expected, the Feynman-Field model overpredicted the number of particles in the "depletion" zone between jets 1 and 2 (roughly 40°–120°); strings beat the secondbest cluster model. To further explore the "valleys" between jets, the collaboration then divided the region between jets *i* and *j* into a region parameterized by the values 0 to 1; N_{ij} is defined to be the number of particles between 0.3 and 0.7. (*B*) Plots of N_{31}/N_{12}, the surplus, so to speak, of particles in the quark-gluon region over the quark-antiquark region. Independent fragmentation predicts isotropy, $N_{31}/N_{12} = 1$. Top row shows the different models for different pion momenta; bottom left sums all pion momenta; bottom right is a replotting of the heavy-particle data from (*A*). Conclusion: string and cluster models do about equally well, definitely better than independent fragmentation. Source: Aihara et al., "Parton," *Phys. Rev. Lett.* 54 (1985): 270–73, on 272; Gary, "Tests of Models," LBL-20638 (1985), 202.

That output was itself highly processed. After the long journey of figure 7.28, the drift electrons had been transformed into the contents of the Data Summary Tape. Then this tape ran through the collaboration's "jet finder" program, which scooped out events that were highly aspherical and had axes not lying along paths at which detector acceptance was poor. With a final selection of events having 1.5 GeV per candidate jet and at least two particles, the resulting output was a data set of 3,022 "three-jet" events. It was a long way from QCD to the ions, and here, in figure 7.36, we see, at last, the confrontation, but it is a confrontation not between QCD and raw data, but rather, necessarily, between the series of models and the highly processed data of the PEP-4/PEP-9. First conclusion: independent fragmentation models looked dead in the water. No ad hoc energy conservation, no retuning of the parameters, could lift its depletion ratio off the ground. As for the other two models (cluster and string), two features seemed irresistible in accounting for the depletion effect: one was an event early in the process that suppressed particle production between the quark and antiquark. In one picture, this emerged as the result of virtual gluons destructively interfering with one another; in the other picture, the kink itself preferentially produced particles in the direction of the kink's motion. Second conclusion: something resembling QCD color seemed to play a role. In the string picture, the color flow again corresponds to the kink, enforcing the string binding; in the cluster picture, color flow explicitly ties quarks, gluons, and antiquarks together more on the side of the primary gluon. Never would the authors claim to have "proved" QCD. What they had done was to produce, from top down and bottom up, a scheme that gave features of that theory a local and experimentally consequential meaning.

With these data in hand, Werner Hofmann returned to the issue of hadronization models in 1987, offering what he called "A Shoot-Out of Fragmentation Models." For purposes of comparison, all talk of the underlying mismatches between a model that violates relativity and one that does not fades away, as do discussions of exact or approximate energy conservation, or the precise interaction between quark and gluon. In the midst of the "shoot-out" only one thing mattered: what do each of these various conceptualizations of QCD have to say about the structure of jets—how planar they were, for example. At the end of the day, when the gunsmoke cleared, only the Lund modelers appeared to walk away relatively unscathed. Eighteen distributions were brought in, with a total of 450 data points. The standard statistical χ^2 test now served as arbitrator. The Lund string model walked off with a χ^2 of 960, where the Webber cluster model shuffled in with a χ^2 of 2,870 and the Caltech model staggered under a χ^2 of 6,830. "Data," Hofmann concluded, "seem to indicate a preference for string models with normal mesons and baryons (instead of heavy clusters) as primary hadrons."[185] Here was a place for the TPC/PEP-4 results to weigh in, though not perhaps in as dramatic a form as the old Mark I—the TPC collaboration's discovery of the F^* was hardly the stuff of a November Revolution. Nonetheless, adjudication among these different models, the formation of a trading zone despite their differing claims to fundamentality or instrumentality, was a crucial part of any understanding of QCD, not only for the late 1980s but for any future experiment that would have to know what QCD said about the data. It is easy to dismiss the decade of the 1980s as "merely confirming" the standard model. That, in my view, would be an enormous mistake. In the face of the epistemic gap between a theory like QCD and the Data Summary Tapes, nothing could be learned. While theorists increasingly took QCD for granted and used it to build grand unified theories or the even more ambitious string models of the 1980s and 1990s, QCD was anything but transparent in the laboratory. With an indefinitely bounded background there is not, and could never be, a foreground.

185. Hofmann, "Jet Physics PEP and PETRA," SLAC Summer Institute 1987, LBL-24086, 14.

This was a brute experimental obstacle, and nothing but a combination of precision measurement and model building could get around it.

One response to the χ^2 significance test of different fragmentation and string models could be that it was not truly a via media between the various models; it was really only a form of instrumentalism—a comparison among uninterpreted calculi. Now it might be logically possible to reduce the quark models to mere symbolic manipulation, but the participants clearly did not treat them that way. This is manifest, for example, in the often made distinction between "artificial" and "natural" aspects of the models. Everyone recognized that the radical distinction between the perturbative and the nonperturbative phase of the analysis could hardly correspond to anything in "nature." This "highly artificial" element would "have to be overcome in any real theory of parton fragmentation."[186] And it was just this coordination that the string community wanted.

The search for a more natural, philosophically defensible model of hadronization is visible in the work of Charles Buchanan, a TPC collaborator from UCLA. In a 1987 *Physical Review Letter,* Buchanan began a program of *explaining* the underlying dynamics of the Lund model.[187] To defend his view, in a January 1989 talk at LBL Buchanan enlisted the support of a philosophico-historical conceit, a table of correspondences between the development of planetary motion and that of hadronization (see figure 7.37). Early data were on the Planetary Motion side to be identified with Copernicus et al., the analogue to PETRA and PEP's first runs at 30 GeV in electron-positron collisions. Building on these still primitive observational results, all that could be expected was the "many-parameter" phenomenology of epicycle physics up to 1620, and these, Buchanan asserted, corresponded to the independent fragmentation model (Feynman-Field) of 1978. Copernicus and Feynman had, according to Buchanan, more or less abandoned hope of linking their results with what he elsewhere called the "actual production mechanism."[188] Instead, they chose to describe the world in terms driven by observation (experiment) alone.

Access to a "fundamental" phenomenology came later, according to Buchanan. The year 1620 (or thereabouts) marked the beginnings of Kepler's three laws, elliptical orbits, constant real velocity, and the harmonic law (which set the period squared proportional to the radius cubed). What made Kepler's phenomenology "fundamental" was (according to Buchanan) that it revealed "Important regularities. Simple, elegant, compelling explanations of how major features work. *'Trustable'* predictions." Part of Kepler's break with the era of epicycles was the reduction in the number of parameters. In hadrons this same

186. Hofmann, "Jet Physics PEP and PETRA," SLAC Summer Institute 1987, LBL-24086, 8.

187. Buchanan and Chun, "Predictive Model," *Phys. Rev. Lett.* 59 (1987): 1997–2000; and Andersson and Hofmann, "Correlations," *Phys. Lett. B* 169 (1986): 364–68.

188. See Buchanan and Chun, "Predictive Model," *Phys Rev. Lett.* 59 (1987): 1997–2000, on 1997.

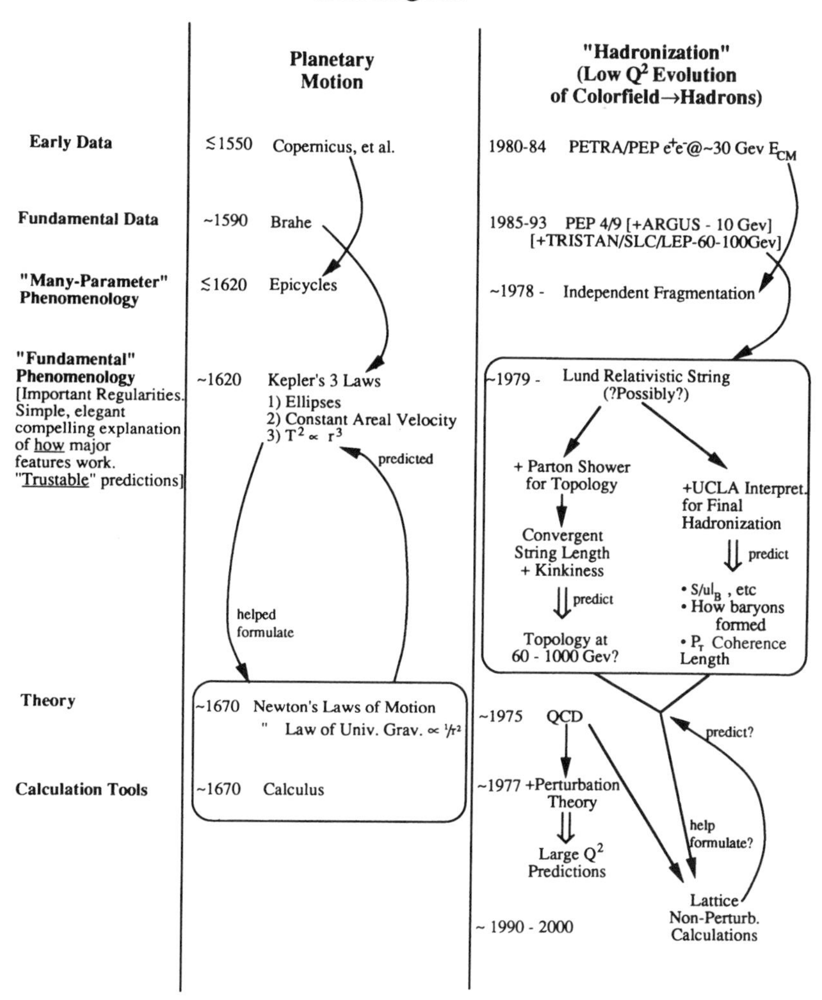

Figure 7.37 Kepler and string theory, according to Buchanan (1989). Redrawn from a transparency made by Buchanan in 1989, this chart must be read on several levels. First, the much-disputed phenomenological analysis of hadron physics is identified with the most venerated episodes in the history of the physical sciences: the story that opens with Copernicus's *De Revolutionibus* and closes with Newton's *Principia Mathematica.* Second, it breaks out of the story, and underlines, a category that might strike some readers as oxymoronic: "fundamental" phenomenology. It is this category alone that is elaborated in the left-hand column ("Important Regularities. Simple, elegant compelling explanation of *how* major features work. *'Trustable'* predictions"); it is here, at the modern equivalent of Kepler's three laws, that Buchanan is trying to establish a conceptual and partly explanatory space for a domain of theory that could mediate between "theory" and the data pouring out of hadron experimentation. Buchanan, "The Lund Symmetric Fragmentation Function as a Complete Description of 'Color Field' Hadronization: Progress in the UCLA (Hadron-Level) Interpretation," talk presented at LBL, 6 January 1989, xeroxed transparencies, MRP.

historical evolution was supposed to have taken place. Whereas the Feynman-Field model had to invoke many parameters that had no "natural" interpretation, and the original Lund model had maintained 10 to 12 such parameters, the UCLA model would have only three variables with three "natural" parameters that could be obtained from observable physics. Finally, theory (Newton's laws of motion and the inverse square law) would correspond to QCD. Perturbative QCD and lattice nonperturbative calculations might then both follow from some of the UCLA-grounded string models and, in turn, help predict them. Thus, in the long run, Buchanan hoped some scheme like the following might bind the output of the TPC to the elusive realm of QCD:[189]

$$\text{QCD} \Leftrightarrow \text{Lattice Calculations} \Leftrightarrow \text{UCLA "Fundamental" Phenomenology} \Leftrightarrow \text{TPC Data}$$

Between the standardizing Ansatz of Feynman and the "explanatory" fundamental phenomenology of Buchanan lay vast philosophical differences: global differences in meaning ran riot. That the various practitioners could agree to the local coordinative mission of a "shoot-out" is the sign of the long-sought and often elusive coordination. It was commensurability bought at the price of a vast amount of work.

Taken together the various hadronization models constituted a new occupation, a subfield sufficiently rich to precipitate conferences and a myriad of publications. One could say they were "like" the many approximative models used, for example, in the many nuclear models discussed in the 1930s. But to do so, it seems to me, is to miss the point: here we have a full-blown theory, QCD, that for many theorists was an exemplary instance of what they wanted in an account of nature and at the same time an enormous production of data. The fragmentation models became an intermediate domain in every sense—a domain that professionally absorbed both theorists (like Andersson) and experimenters (like Buchanan), a domain that coordinated elements of the "fundamental" theory such as gluons and color flow with endlessly reprocessed ionization tracks, modulated through a skein of soft- and hardware. From the days when theorists could pore over a bubble chamber picture, or the time when experimenters could draw up an interpretation on their own, the world had turned. I would argue that in this new phenomenological zone something more than lengthy calculations were taking place: quietly, for over a decade, these modelers were struggling to define the meaning of both QCD and Data Summary Tapes by mapping uncertain paths in the territory between them.

189. C. Buchanan, "The Lund Symmetric Fragmentation Function as a Complete Description of 'Color Field' Hadronization: Progress in the UCLA (Hadron-Level) Interpretation," talk presented at LBL, 6 January 1989, xeroxed transparencies, MRP. For a comparison of the Webber, Lund, and UCLA models, see Buchanan and Chun, "A Simple Powerful Model which Describes Meson and Baryon Formation in e^+e^- Annihilations," May 1990, UCLA-HEP-90-003, MRP.

In addition to the shoot-out, the TPC collaboration explored the hadronization process through an examination of the quantum (Bose-Einstein) correlations in the production of pions. The idea was this. When particles are produced in an interaction, their quantum-mechanical wave functions can interact constructively, neutrally, or destructively. Decades earlier (already in 1956), R. Hanbury-Brown and R. Q. Twiss had used this correlation to study photons emitted from stars; by looking at the correlations between two photons, they could estimate the diameter of the emitting body. In particle physics, Gerson and Shula Goldhaber extended this analysis in 1959 to same-sign pairs of pions emerging from proton-antiproton annihilation. These same-sign pairs would preferentially have small opening angles when compared with opposite-sign pion pairs. Why? From the early days of quantum theory, it was known that bosons (even half-spin particles) would have symmetric wave functions when identical particles were exchanged, whereas fermions (odd half-spin particles) would have quantum state functions that were antisymmetric under exchange. This meant that the amplitude for the production of two identical bosons in the same state could be written in the form $A(p_1,p_2) + A(p_2,p_1)$. Where $p_1 \sim p_2$ are the (nearly identical) momenta of the two particles, the direct amplitude $A(p_1,p_2)$ would interfere constructively with the exchanged amplitude $A(p_2,p_1)$. In the case of small opening angles, the two same-sign pions would be identical particles with similar momenta, and the Bose-Einstein interference would cause an enhancement of their number. For opposite-sign pions, the particles would be distinguishable and no such correlation would occur, whence the effect that was observed in the late 1950s.[190]

Flashing forward to the 1980s, the TPC physicists wanted to use the Bose-Einstein correlations to explore the production mechanism of hadrons, and so to illuminate the difficult-to-articulate domain between the original event in which e^+e^- annihilations produced a quark-antiquark pair and, eventually, the output of a jet of hadrons. In particular, the correlation of pions is going to depend on the distribution of pion sources, so the correlations could help get at the messy, multistage process in which the original quark-antiquark pair emitted gluons, further pairs of quark and antiquarks, and so on. In the low energy nuclear physics of the late 1950s, it was reasonable to assume that the sources of the pions were static; in the very high energy experiments of the 1980s at SLAC, the correlations revealed that deviations from the static source model, in accord with the relativistic models of the hadronization process, were many. There were all the usual TPC-related hardships of compensating for distortions in the

190. On the 1956 work by Hanbury-Brown and Twiss, see "Stellar Interferometer," *Nature* 178 (1956): 1046–48; the particle physics applications came from Goldhaber et al., "Pion-Pion Correlations," *Phys. Rev. Lett.* 3 (1959): 181–83; and Goldhaber et al., "Influence of Bose-Einstein Statistics," *Phys. Rev.* 120 (1960): 300–312. An excellent discussion of the correlation work is to be found in Avery, "Bose-Einstein Correlations" (1989), which has full references to the series of TPC/PEP-4/PEP-9 collaboration publications by Aihara et al.

field and in the registration and identification of particles; there were competing physical effects (such as pions produced by the decay of strong interaction resonances); and there were the difficulties of articulating the theoretical picture amid competing models for the hadronization process itself. By 1989, the collaboration had, through simulations and further testing, begun to remove the real correlation both between two pions and among three pions. Removing the other physical effects that contributed to pion correlations, and coming to unique conclusions about the various models, was more difficult.[191] These were tasks that continued into the 1990s, long after the TPC facility was gathering dust off the beam line.

As an indication of the immense complexity of data reduction, much of the Bose-Einstein analysis was conducted during the two-and-a-half-year shutdown that spanned part of 1986, 1987, and into the fall of 1988. After this long hiatus, the TPC/PEP-4 collaboration finally came back into action during the fall of 1988. But by 1988, Burton Richter (then director of SLAC) was hesitant to inject particles into the PEP/SPEAR rings until the Stanford Linear Collider (SLC), the troublesome new machine just coming on-line, "is operating satisfactorily for physics."[192] As the TPC collaboration saw it, Richter's particle parsimony threatened recruitment, funding, and the competitive edge of the already troubled collaboration. A case in point was the growing restiveness of U.C.-Riverside. With very few data on hand, the senior Riverside physicists had little to provide their students. Shen was pulling back his involvement, and others grumbled about withdrawing from the experiment. Writing to Mike Ronan, Shen concluded: "I would like to be able to say that we could devote a lot of our effort to TPC, but I could not without knowing what to expect. . . . [W]hether we will be part of the new TPC is up to you."[193]

Problems at the SLC continued to rain heavily on the TPC/PEP-4/PEP-9 parade. The collaboration would protest, and Richter would respond sympathetically but insistently that SLAC's race against CERN to determine the number of particle "generations" (how many neutrinos there were, e.g.) had to have laboratory priority. Key to this measurement was the production of *Z*'s—and lots of them. CERN was gearing up to make tens of thousands of the *Z*'s, and the Stanford machine was running far behind schedule. In April 1989, Richter shot an e-mail to Piermaria Oddone, an LBL colleague who had been on the TPC and was now director of the LBL Physics Division: "[W]e are not going to start the TPC up until the SLC is settled down and producing physics. You know well that five *Z*'s do not make a physics program, and we have considerably more

191. See conclusion to Avery, "Bose-Einstein Correlations" (1989).

192. Bloom et al. to Richter, 22 December 1988, citing letter of 9 November 1988 from Richter to the collaboration, in Bloom notebook "TPC Notebook 1989–A," EBP.

193. Shen to Ronan, e-mail, 30 January 1989 at 17:04:10, file "TPC Notebook 1989–A," EBP.

work to do to get the machine in good running shape." If they could produce more *Z*'s—a lot more *Z*'s—then perhaps the laboratory could bring the TPC on in July or August 1989. "[I]f you feel that you must do your planning on worst-case assumptions [a nine-month delay], then so be it." Tension was high everywhere in the laboratory, and Richter concluded by hoping "that we can get out of our crisis mode of living in the next few months."[194]

The only hope the TPC/PEP-4/PEP-9 collaboration had was to introduce a switch that would alternate running between the *Z*-producing SLC and PEP, which fed particles to the TPC. In July 1989, Oddone wrote Richter setting out the terms that LBL wanted met for its continued involvement in the TPC program. These included a successful test of the PEP/SLC switch during the summer of 1989, a decent number of electron-positron collisions delivered to it in the rest of 1989 and 1990 (350–400 pb^{-1}), and sufficient support from the Department of Energy for LBL's other activities. If the TPC was going to run physics, it would have to be possible for SLC to go from colliding beam conditions, to the conditions necessary for injection into PEP, and then quickly back to operating conditions at SLC. Oddone doubted that this technical tour de force could be executed and concluded by saying that if the switch did not work he (and LBL) would be just as happy to renounce an immediate physics program in favor of the long-term possibility of running an R&D program and perhaps a facility that one day, in the not too distant future, would probe the *B* mesons.[195] Hopes for a TPC renaissance proved short-lived.

By October 1989, Oddone and Richter had signed an agreement stating the need for an adequate number of interactions and acknowledging the pressure LBL faced to reallocate resources from the TPC to newer projects (such as Superconducting Supercollider detector development). Under these circumstances a minimum of 200 pb^{-1} had to be delivered to PEP during 1990, and some 400 pb^{-1} during 1991 and 1992, if the project was to be worthwhile. Finally, LBL wanted SLAC to assume the equivalent of $250,000–$300,000 of operating expenses for the TPC. Should any of these conditions fail, LBL expected to discontinue its participation in the TPC.[196]

SLC, however, continued to be an unstable, finicky device and SLAC was under enormous pressure to produce Z^0's at a sufficient rate to determine that particle's width—essentially the only parameter that SLAC could hope to snag in its widely publicized race with CERN. On the afternoon of 6 September 1990, Richter faxed Oddone informing him that although the PEP/SLC switch seemed to be working, SLC conditions were not stable and that even if a further extension were granted during 1990, it would add only 10 pb^{-1} to the 50 pb^{-1}

194. Richter to Oddone, e-mail, 26 April 1989 at 16:43:55, file "TPC Notebook 1989–A," EBP.
195. Oddone to Richter, 16 July 1989, DNyP.
196. Oddone and Richter, "Memorandum of Understanding on TPC Operations," 25 October 1989, DNyP.

already accumulated. So, with the hope that he could provide significant luminosity to PEP starting in February 1991, Richter canceled PEP operations, and with them the TPC, for the remainder of 1990.[197]

The next day, Oddone scheduled an "urgent meeting on the fate of the TPC" for 10 September 1990. The decision was made for LBL to withdraw from the TPC. Sixteen years after its inception, the TPC at SLAC had finally succumbed.

7.9 Conclusion: Control and Diffusion

7.9.1 The Final Expansion

Was the TPC typical of high energy experimentation? Once again, we run up against the inadequacy of "typicality." The expansion of detector facilities begun with the TPC in the 1970s continued in the 1980s and 1990s with the construction and operation of four major facilities at the CERN colliding beam LEP (OPAL, L3, ALEPH, and DELPHI). Each had hundreds of collaborators and ate up budgets in the hundreds of millions of dollars. Two—ALEPH and DELPHI—exploited the time projection chamber ideas developed in the Californian TPC, while the others used a combination of the wire chambers and other image-producing electronic detectors discussed earlier. At Fermilab two other proton colliding beam detectors drew physicists from around the world: the so-called D0 and its competitor the Colliding Detector Facility (CDF). These six detectors, joined to a certain extent by work at SLAC and DESY, made up the high energy physics world of the mid-1990s. In a universe of six objects it is difficult to say what is typical and what not.

Still it is clear that while many of the TPC/PEP-4's crises were site specific (magnet coils blew up nowhere else), the tensions created by the building of that detector were indeed endemic to the construction of this new generation of hybrid machines. For example, the association of universities and laboratory groups with individual component parts of the detector is a nearly universal practice. DELPHI distributed its property this way: microstrip vertex detector, CERN/Milan/Saclay/Rutherford; inner wire detector tied to the triggering apparatus, Cracow/Amsterdam; time projection chamber, CERN/ Paris/ Lund/Orsay/Saclay; ring imaging Cerenkov counter, Amsterdam/Athens/ CERN/ Orsay/Paris/Strasbourg/Uppsala/ Wuppertal; electromagnetic calorimeter, Ames/ Bologna/CERN/Genoa/Karlsruhe/Milan/Rome/Stockholm/Warsaw; and soon.[198] Each of the other big detectors divided into similar groupings. Each had massive Monte Carlo simulation projects designed to create virtual reenactments of the full detector as part of the planning, construction, and analysis phases of the

197. Richter to Bloom, Ronan, and Sheppard, fax, 6 September 1990 at 15:53, DNyP.

198. "DELPHI," *CERN Courier* 24 (July–August 1984): 227–29.

machine. Each had to grapple with the tremendous problem of integrating the data, processing them in stages, and preparing them for analysis. And each had to combine physicists and engineers, while accommodating the ever-increasing quantity of data.

As TPC/PEP-4/PEP-9 suggested, governance in the era of colliding beam detectors was a difficult task. CDF was surrounded by a fluctuating cloud of some 400 members when it went to press with a claim for the existence of the top quark in 1994. D0 sported 424 members from 40 institutions as of 1994. ALEPH had hosted 475 people by 1990, representing 30 institutions.[199] Each worked out its own form of decision making; some, like L3, had a highly centralized directorship around a single leader (Samuel Ting), while others tried to preserve a more democratic structure.

D0, for example, presents us with a governance structure comparable to, though more formalized than that of, TPC/PEP-4/PEP-9. Like TPC/PEP-4/PEP-9, D0 had an Executive Committee, though it was stripped of certain purely institutional functions (such as the admission of new groups). These tasks were borne by an Institutional Board. Choice of the "cospokesmen" was through balloting by the full collaboration, and like several other collider collaborations, a Spokesman Nomination Committee set out the candidates for the ballot. As "usual," there were physics groups headed by two "physics conveners," who directed each of the five physics groups that shaped the output of the experiment, allocated computer resources, and advised on issues such as experimental triggers. The other "boards" included:

Trigger Certification (responsible for detailed execution of trigger decisions)
Off-Line Computing Policy Board (responsible for overseeing the processing of events and simulations)
Speakers Bureau (coordinated who spoke where)
Analysis Computing Planning Board (recommended long-term changes in computing environment)

Along with these, there were groups in charge of present operations, upgrading operations, and designing triggers and data acquisition, groups that handled algorithms (including one for particular particle subdetectors), and groups concerned with specific objects (electrons, muons, taus) and several higher level phenomena (jets and missing energy).

In D0, TPC/PEP-4/PEP-9, and all the other collider collaborations, *authorship* became a troublesome issue. On one side lay the possibility of including anyone and everyone who had ever worked on the design, construction, acquisition, or analysis of the data. The obvious risk was that the meaning of

199. ALEPH data from Decamp et al., "Aleph: A Detector," *Nucl. Inst. Meth. A* 294 (1990): 121–78.

individual authorship would evaporate. On the other side lay exclusivity, and with it all the problems that any criterion of choice immediately raised. D0, for example, began in 1991 with a definition of "author" that demanded several preconditions. First, to join the list of authors on a paper, the collaboration insisted that the would-be author be a "current member" of the collaboration, namely, someone who had devoted at least one year of time to the experiment and left the fold less than a year before submission of the relevant paper. Second, the putative author had to have made a significant contribution *both* to shift work *and* to either the apparatus or the analysis of results. "Shift work" in this context meant working at least one-half of the average number of shifts per person during the relevant run. "Contribution to the detector" meant either a year's worth of effort on hardware or the same amount of time on software used for data acquisition; "contributing to the analysis" meant making a major contribution to the analysis of the data from which the paper was derived. Any exceptions to these rules (such as nonphysicist authorship, authorship by a junior graduate student, or authorship by a computer programmer) required the approval of the Executive Committee. In 1994 the task of adjudicating such matters was shipped to a Committee on Authorship.

D0's Committee on Authorship decided whose names graced the cover pages of *Physical Review Letters;* it did not determine what went in the paper. How do 400 physicists write anything at all? With the help of an Editorial Board. This entity was usually composed of the person who wrote a particular piece of analysis, and four others: The advisor (also known in D0 parlance as the "godparent") was a collaboration-based expert chosen to watch over the preparation of the paper. The godparent was joined by a person associated with the physics or algorithm group out of which the paper emerged, and then two other people, at least one of whom did *not* come from the originating group. When the Editorial Board gave its *nihil obstat,* the paper went "up" on the electronic bulletin board for criticism by the rest of the collaboration, followed by a "public reading" open to all and sundry members.

All these rules were established before data taking began; when it did many members felt that a revision was in order. In particular, there was a widely shared view that the emphasis on shift work was disproportionate. Other, equally significant tasks might take its place, and the collaboration drew up a new set of rules in June 1994 that defined the bottom-line contribution to be 12 hours per week devoted to "service tasks." Now, however, "service tasks" included alternatives to data taking:

1. Design construction, or debugging of detector components for D0 or its test facilities
2. Design, implementation, or debugging of software for simulations, data acquisition, or data processing

3. Processing of Monte Carlo events or events obtained at test facilities
4. Shift work
5. Personnel, management, or administration of grants; work as physics convener or coordinator of technical topics such as identification of electrons

The point of the original rule set from 1991 was clear: the collaboration wanted every author to pitch in to the nitty-gritty of the experiment. Everyone had to sit by the instruments of the control room, be on site at least some number of hours, and do something about extracting data from the machine. As the experiment progressed, however, it became evident that in some cases the diversity of tasks ought to include efforts far removed from any hardware at all. Software development, the massive programming effort needed to extract, coordinate, assemble, and analyze data, was now an essential part of being an experimenter.[200] One could go through each of the major collider experiments and extract either explicit or implicit conventions that address the same point. Everywhere, however, one finds of necessity an expanded (or perhaps one should say contracted) conception of what counts as an author.

Given the spiraling size of detectors, from tabletop to bench, from the individual worker like C. T. R. Wilson or Donald Glaser to the cottage industry of Cecil Powell, from the physics factory of Luis Alvarez to the national and international TPC/PEP-4/PEP-9, LEP, and Fermilab machines, one can be forgiven for asking whether there is an end to the process of expansion. There is. While the cloud chamber cost thousands of dollars and bubble chambers a few million, the TPC and its siblings at LEP and Fermilab ran up bills of over a hundred million dollars. The end seemed to be reached in the early 1990s with the billion-dollar detectors that would have sat underground at the two interaction regions of the Superconducting Supercollider (SSC) in Waxahachie, Texas. Never built, the SSC detectors are the asymptote of scientific instrumentation, the limiting case of particle energy, collaboration size, budgetary expenditure, political commitment, and brute square mileage. In the SSC one had an instrument of research that was projected to cost $12 billion, an amount large enough to pass from a congressional subcommittee hearing to a national issue.[201] In collaboration size, the assembled physicists would have numbered, not the 500 or so of Fermilab or LEP, but over 1,000. With 2,500 or so experimental high

200. [Name deleted], "Rules on Authorship of DZero Publications, 2 June 1994," from Authorship Rules, D0NEWS, e-mail, ca. May–June 1994. Interestingly, one participant sent an e-mail in which he emphasized the need for flexibility in defining an author: "[I]t may not make sense to [enforce more service on new members] because of the different blend of talents in any individual. . . . [I]f some bright young person we hire is specifically interested in D0 because of some physics analysis. . . . I would consider it foolhardy to prevent that individual from doing that research."

201. On the political debate over the SSC, see Kevles, *The Physicists* (1994), preface.

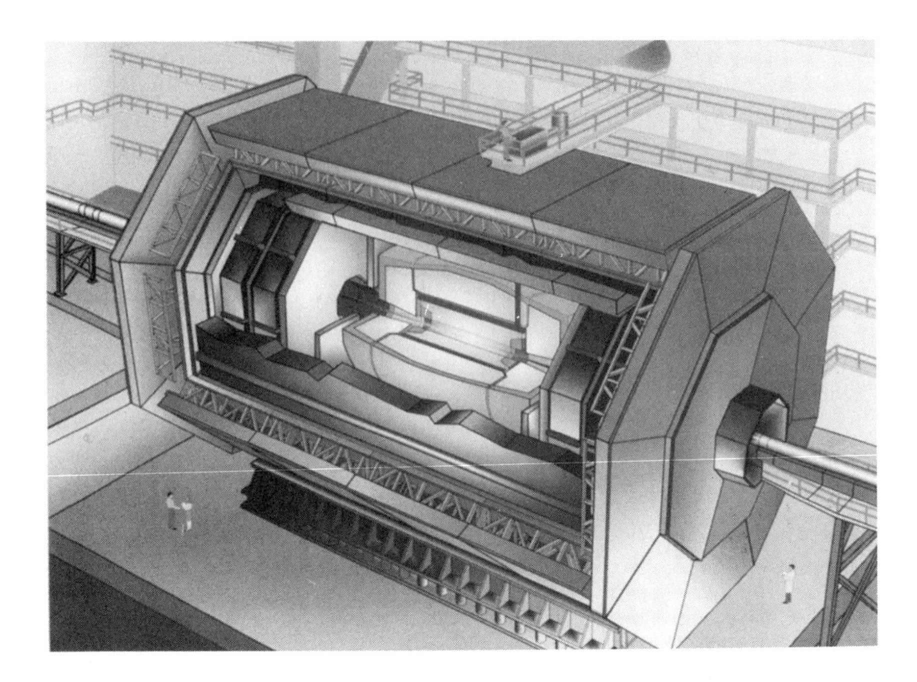

Figure 7.38 Solenoidal Detector Facility (1992). Source: Solenoidal Detector Collaboration, *Design* (1992), cover figure.

energy physicists in the United States and roughly the same number in the rest of the world, it is clear that there are no further factors of a hundred to be had—a trillion dollars is the order of magnitude of the U.S. national budget. There will not be projects of 100,000 scientists occupying sites a hundred times the two-county acreage of the SSC. So in the unbuilt SSC detectors we can see the dreams of a final experiment (to paraphrase Steven Weinberg), an intensified image of the ideals that animated so much of the large-scale physics of the late twentieth century.

Two detectors were approved for the SSC: one directed by the Solenoidal Detector Collaboration (SDC) and the other by the Gammas, Electrons, and Muons Collaboration (GEM). Their physics goal was to explore the Achilles' heel of the Standard Model (see figure 7.38). For however successful the Standard (electroweak) Model had been—defeating essentially all attempts by Fermilab and CERN to find a hole in its phenomenological predictions—the mechanism of the symmetry-breaking process that separated the various forces of nature was clearly inadequate. In the Standard Model at high energies, the electromagnetic and weak interactions are indistinguishable, mediated by force-carrying particles that made no distinction between the two. At the much cooler temperatures at which we live, however, the forces are self-evidently different. Weak interactions (governing, e.g., the decay of a neutron into a proton, electron, and antineutrino) are, according to the Standard Model, transmitted through a short-range force that acts on distances no greater than a nucleus. Electromagnetism knows no such bounds; as we look to the stars we receive photons from objects far beyond the outer rim of our galaxy. According to the Standard Model, this bifurcation in the forces, this breaking of symmetry, is due to the effects of a postulated "Higgs" particle, named after the British physicist Peter Higgs. But however helpful the Higgs mechanism was in formulating the Standard Model, it cannot, on theoretical grounds of self-consistency, be right if the mass of the Higgs particle is too great. For too high a Higgs mass, the theory stops making sense; so by the theory's own lights, something interesting had to happen below the SSC design energy of about 40 TeV, and it is there that the Supercollider was designed to annihilate particles. Advocates argued that either the Higgs would be found or some other mechanism would surface that would account for symmetry breaking. Either way, current theory would reveal its flaw.

At SSC energies, the number of particles produced in a proton-antiproton annihilation grows enormously, and the particle identification software and hardware had to expand to match. Within a year of commencing their data harvest, the SSC physicists building GEM and SDC estimated, even after throwing out all but a fraction of the observed interactions, that the SSC would have to store a petabyte of information: 50 times the information content of the 20-million–volume Library of Congress. The ring itself, to be located on prairie land, would be 54 miles around, enclosing a town and spreading a laboratory across distances several times larger than the venerable laboratories of SLAC, Fermilab, and CERN. Hiring the architect Moshe Safdie, the directorate of the SSC hoped to plan the laboratory rather than proceed helter-skelter, as had so many earlier particle physics installations. Safdie went to other laboratories to explore, with the physicists, how they thought about the space and community they would need.

Built into the computer architecture of the SDC, from the start, was a dispersion of computation among a series of "regional centers" located in Japan, in Europe, and at least two sites within the United States. Each of these centers would have computing power equivalent to some 10,000 VAXs and would be able to support multiple users interacting with large batches of stored data and performing repeated computational runs.[202] And at the level of controlling the SSC itself, it was a stated goal in the SDC technical design report that "high level run control" was to be "location independent."[203] Videoconferencing would be built in from the start, and when technically feasible, wide area networks and multimedia workstations would "hold the promise of 'personal

202. Solenoidal Detector Collaboration, *Design* (1992), 10–16 on 17.
203. Solenoidal Detector Collaboration, *Design* (1992), 9–1.

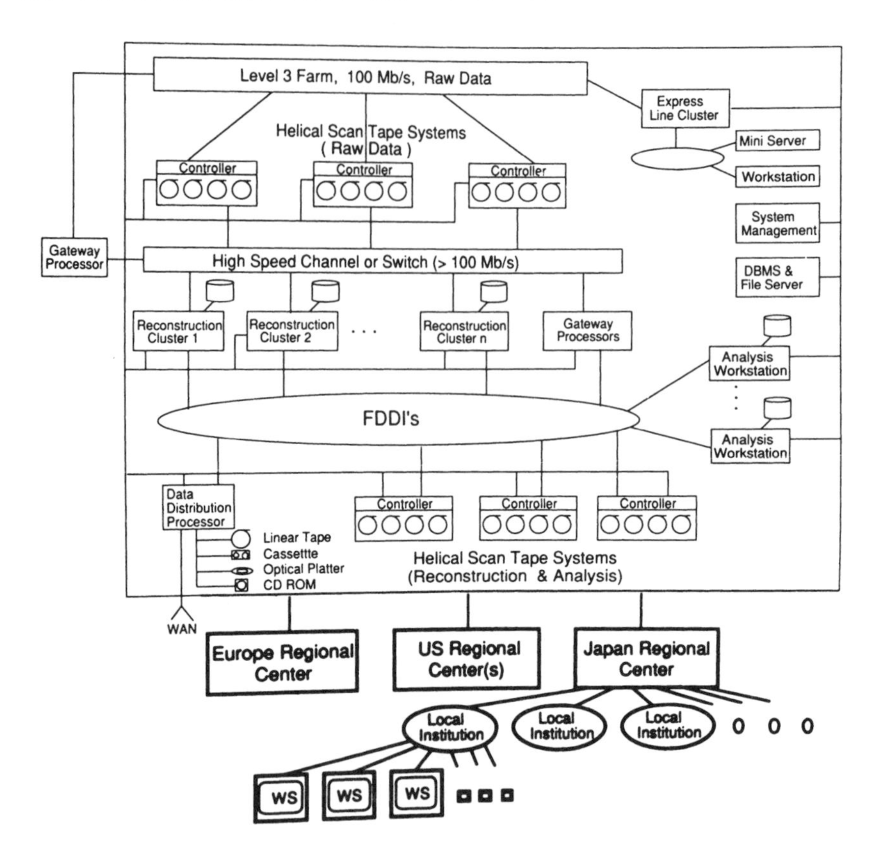

Figure 7.39 Delocalized control, SDC (1992). Control rooms operated remotely would, according to the unrealized plans of the SDC, make it possible for physicists across several continents to run shifts without coming to the "site" of the SSC. Source: Solenoidal Detector Collaboration, *Design* (1992), 10–15, 10–17.

video' conferencing."[204] But beyond conferencing, the high-speed networking might allow the establishment of a remote control center in Japan; Japanese physicists could then run data shifts 12,000 miles from the annihilation event. And if that center were successful, further remote control sites were contemplated elsewhere, from the central campus of the SSC to Europe (see figure 7.39).[205]

But if the international character of SDC militated for maximal dispersion of data taking, sorting, and analysis, the force of the local had not entirely dissipated. In October 1992 Melissa Franklin of CDF reported to Safdie her grave doubts about the plan to move the experimenters "temporarily" to the eastern

204. Solenoidal Detector Collaboration, *Design* (1992), 10–8.

205. Solenoidal Detector Collaboration, *Design* (1992), 9–4.

part of the campus, a half hour by car from the control room and center for accelerator physics. Her fear was that de facto, if not de jure, while the theorists and accelerator physicists resided in luxury at the central west site, the experimenters would end up "squatting in an industrial facility and makeshift arrangements. This would lead to a strengthening of the schism and separation between accelerator people, detector people, and theoretical physicists." Or, and this was no improvement as Franklin saw it, the physicists would move to the west campus leaving "only technicians and 'lesser' physicists . . . to service the detectors, thus downplaying the significance of detector physicists in the process. The issues are those of status in the delicate hierarchy." Already there was a crying absence of community among the accelerator physicists, detector physicists, and theoreticians—"not to mention the other hierarchy of technicians and support people."[206] When Safdie asked about the experience at CERN, where people do go back and forth, Franklin replied that CERN, a much wealthier organization than most American laboratories, ordered more work off-site and that American physicists had to do more of the maintenance, construction, and reconstruction on-site. Roy Schwitters agreed, though he pointed to a style of work—"American inclination for a hands-on work with the machines, with the European inclination to accept the distance and operate via electronic controls." Though the issue was fraught, Safdie concluded, "My sense is that we are beginning to zero in on the fundamental master-plan strategy issues."[207]

With the issue of centralization front and center, Safdie scheduled a trip to CERN specifically to address the way the Europeans were handling the relation between the central facility and the decentered experiments. Of Hans Hoffmann at CERN, Safdie asked in November 1992, "How often do [the CERN physicists] move from one part of the complex to the other? Do the physicists all have offices in the central complex and how often do they move to other places within CERN?"[208] More generally, he wanted to know about the work practices in offices, in laboratories, and between one and the other, and about the movement of people to and from the Central Laboratory. From his visit, Safdie emerged with reinforced conviction that the "nucleus of the campus" served as a social, intellectual, and administrative center point, even if the campuses' "home bases" were as large (and as distant) as OPAL or ALEPH (see figures 7.40 and 7.41). Safdie continued, "All those interviewed at CERN emphasized the importance of having a vital campus available early in the development" at the SSC. Later construction of a center might be too late: habits might not be reversible.[209]

206. Moshe Safdie to Superconduction Supercollider, "SSC Master Plan," 26 October 1992, file "Conversation with Melissa Franklin," MSP.

207. Moshe Safdie to Superconduction Supercollider, "SSC Master Plan," 26 October 1992, file "Conversation with Melissa Franklin," MSP.

208. Safdie to Hoffmann, 4 November 1992, MSP.

209. Safdie et al., *SSC Laboratory* (1993).

Figure 7.40 Model of SSC campus center (1993). According to the master plan, the four structures forming the bridge to the "H" of the central facility would be the dead center of the campus. The second module from the left would be the operations center, the third from the left would be the library, and the fourth, the auditorium. In front of the left two modules stood the cafeteria, and in front of the right two modules was the conference center. Source: Safdie et al., *SSC Laboratory* (1993). Photograph by Michal Safdie.

Scale, however, differentiated CERN from the SSC, and many, like Franklin, thought it anything but obvious that a center would hold. Barry Barish, a senior experimenter at UCLA and spokesperson for GEM, was not at all comfortable with the idea of a centralized campus, especially if the center sat over the accelerator control center and not near the interaction region—called by Lew Kowarski "the site of the kill." Still reeling from a miserable situation in which the laboratory was split into two geographically distinct sites (making up the temporary laboratory), he acceded to the idea of a unified laboratory. "For our GEM group, and particularly for the many experimenters . . . who come for short visits, there will be a tremendous pull to spend their time at or near our I[nteraction] R[egion]. It is no secret that a central campus on the east side would be far more natural for us to use, and I have serious questions as to whether the experimental groups will use a west campus site during this period."[210]

In these exchanges we see the inherent tensions between the local and the dispersed aspects of experimental work. Only on the scene can the social hierarchies be negotiated between theorists, engineers, technicians, experimenters, and accelerator physicists. Only on site can one fix things and alter the experi-

210. Barish to Schwitters and Safdie, 10 January 1993, MSP.

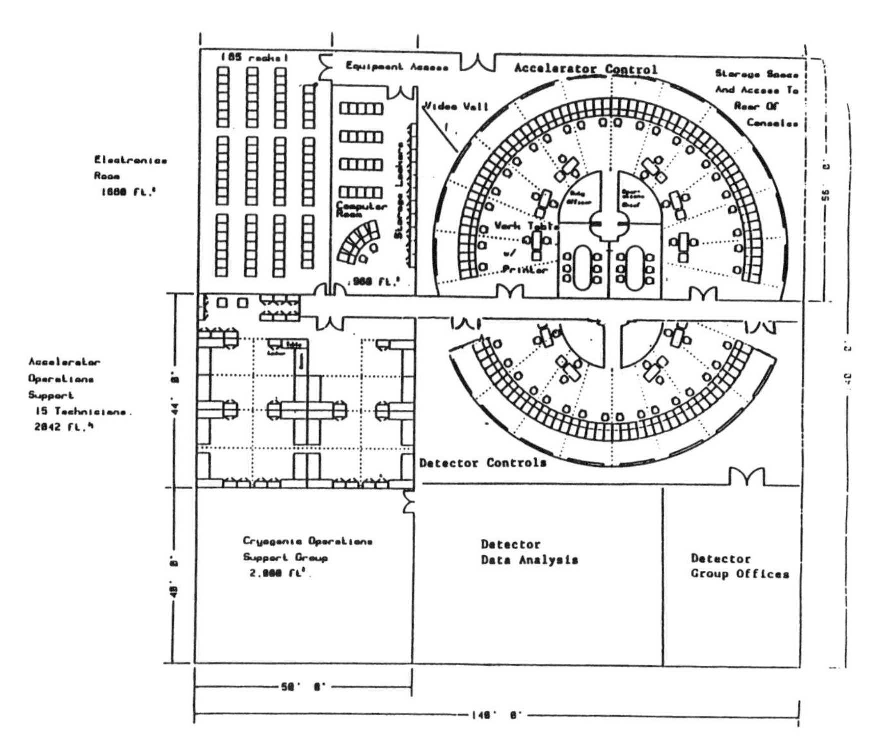

Figure 7.41 Blueprint of SSC control center (1993). The twin poles of the post-1970s laboratory are dispersion and centralization. In the SSC control center one sees powerful centripetal forces at work, for here would be located not only all the detector controls but also injection accelerator control, test beam control, collider control, and a host of monitoring/control systems including water, power, communications, fire, and radiation. Operated around the clock, this complex was to have been the "nerve center" of the entire laboratory. At the same time, plans were being laid for the introduction of detector control centers on several different continents to accommodate the dispersion of control and authorship that accompanied the thoroughly internationalized collaboration. Source: Timothy E. Toohig to Distribution, "Preliminary Building Specification Proposed SSC Central Control Center," 3 February 1993, MSP. Photograph by Michal Safdie.

ment in what Schwitters took to be a characteristically American fashion. At the same time the call of globalization was clear. This was a collaboration drawing heavily on resources—human, technical, and financial—from many places. Here was multiculturalism not as an abstraction, but as computer architecture. Regional centers came into the planning across three continents. Simulations in California, shifts in Japan, particle collisions in Waxahachie, and analysis spread over four European countries: Where, we can ask, as we have already with the TPC/PEP-4/PEP-9, is the experiment? Author committees, godparents, editorial committees, town meetings, and virtual preprints complicate (self-consciously) the category of authorship. "Who is an author?" becomes an issue not of hermeneutic contemplation but of billion-dollar contracts with the Department of Energy. In the end, an author might well be a postdoc working on a piece of simulation software for a component of a device that he has never seen. Amid this unprecedented hybridity, the difficulty was the continued necessity of holding it all together.[211] The powers of dispersion and cohesion remained in tension, as was evident throughout the history of TPC/PEP-4/PEP-9.

7.9.2 The Danger of Disaggregation

At practically every moment, from 1974 to 1985, the TPC/PEP-4/PEP-9 facility threatened to disaggregate. The risks lurked everywhere: the space image of the track would diffuse into the gas, the magnet wires would burn, the magnet would rip from its stand, the gas would destroy the calorimeter, the high-voltage system would destabilize, the software would not interface, the university groups would pull out, PEP would cancel the program, the Department of Energy would withdraw, collaboration between physicists and engineers would shred, and the experimental data would fail to make contact with theory. And the list goes on: Could the data acquisition system mesh the output of one subassembly with the others? Could PEP-9 be integrated with PEP-4? Could PEP itself, in the final years, be effectively brought on-line through a switching system with the newer particle-accelerating system SLC? At every moment of this enormous project the issue was whether quasi-autonomous parts could be integrated into a coordinated, if not homogenous, whole. The coordination problem makes the history of this epoch of physics at once a social history (in the coordination of people and institutions), a machine history (in the coordination of detector components), and an epistemological history (in the coordination of data with models and theories). In each of these histories, the centrifugal forces at work threatened to end the TPC. Arrayed against these shattering impulses were a multiplicity of local binding forces. There were moves to codify a task-oriented group structure, a critical path management system, an externally imposed review process, a pairwise coordination of experimenter and engineer. And there was a new pidginlike modeling system that stood like a shifting sandbar between the ocean of QCD and the sea of TPC data.

One does not need Tocqueville to see the footprints of American political culture in the organization of this facility, the particular political solution participants used to address the problem of disaggregation. From the time of the

211. In the preamble to their e-mailed document on authorship, the D0 Executive Committee reminded its 400 members that they all were expected to maintain their names on the header of all physics papers: "Withdrawal of individual names due to lack of close involvement in a particular topic will tend to undercut the impact of the paper and is strongly discouraged" (Executive Committee, D0, "Criteria of Authorship of D0 Physics and Technical Papers," e-mail to the author, original last revised 14 March 1991).

founding documents of the collaboration that set out a choice between a "House of Representatives" or a "Senate," a specific rhetoric saturated the organizational structure. Nor is the rhetoric "mere rhetoric." Reflecting on the TPC project not long after its demise, Marx wrote, "In some sense we [Nygren and Marx] were products of the 60s political and cultural environment. That was a time of great distrust of authority, and skepticism about hierarchical organizations. I believe that we shared these views and naturally this influenced our approach to organizing the collaboration."[212] But the attempt to apply democratic principles (albeit hierarchical ones) did not stop with Marx and Nygren, it shaped decision making all the way from the TPC Executive Committee to the designation of speakers at the hundreds of conferences at which the experiment was to be represented. Indeed, the collaboration was so insistent on preserving its democratic structure—against all the pressures to centralize, streamline, and govern—that democracy itself became a debated issue in several of the reviews by the PEP administration. Shen, Nygren, Marx, and others repeatedly had to defend the project against what outsiders saw as excesses of governance by the many.

We can read the history of the collaboration as a chapter in the oldest of all political philosophical debates, that between individual authority and democratic assent. When Jay Marx protested to Birge, "[T]here are few philosopher kings and still fewer who are so recognized by their subjects," his tone was half-ironic, but his intention—to democratize the setting of physics priorities—was serious. He, as much as anyone on the collaboration, was an architect of the democratic decision-making process. But as the project advanced, even Marx came to press for a more centralized and directed leadership while at the same time struggling to exert authority over a group that included physicists considerably his seniors. These discussions erupted violently over and again: over the project manager's function in the early days: Did he adequately command developments in the electrical sector of the experiment? Could he enforce scheduling and cost constraints? Later, similar tension erupted into a battle over leadership within both the physics and the engineering sides of the experiment, and more specifically over the lack of direction from various spokespersons and deputy spokespersons. Oscillating sometimes violently between the demand for democratic representation and the desire for an enlightened despot, the collaboration struggled to coordinate not just an assembly of individuals and institutions but also a host of different subcultures, from electrical engineering and computer programming to high energy theory. While these debates over governance (in the collaboration, in the TPC Executive Committee, and in SLAC's Experimental Program Advisory Committee) may have fallen short of the eloquence of Plato's *Republic,* it was not for want of passion.

Over budget and out of time, the TPC lurched from crisis to crisis. As

212. Marx to the author, December 1995.

everyone recognized, the stakes could not have been higher: reputations were on the line, as were years of work by well over a hundred people. More specifically, questions of authority entered in several ways. One was the problem of autonomy and control in the workplace. Would the electrical engineers (for example) be able to guard the pace and structure of their work against what they perceived as inappropriate intervention by mechanical engineers and physicists? Would both mechanical and electrical engineers find themselves in a position to "stand up" to the physicists, with the right to point out that a given "improvement" would compromise cost, schedule, or reliability? Who would—who *could*—demand that the different components of the master simulation program fit together? Who had the power to insist on the electronic compatibility of PEP-4 and PEP-9? Who could compel the team to produce simulated data and process Data Summary Tapes in a form commensurate with the fragmentation models and ultimately with QCD? Who would judge that a discovery could be made public?

By contrast, the simpler power structure of a bubble chamber collaboration displayed authority that climbed pyramidally toward the top. And in such a structure, the idea of deskilling—of the routinization of tasks—was useful in understanding the application of time-and-motion studies and "factory floor" methods to some aspects of bubble chamber work. Putting nearby workers in competition with one another, breaking down measuring tasks into component parts, automating even the production of histograms—all of these moves fit well into Alvarez's and others' use of classical factory managerial strategy. But even within the bubble chamber regime, no single system of work prevailed. As we saw, there were striking differences between the work practices associated with the Alvarez-guided interactive technologies (which used machines to assist people) and the robotic work model of the Flying Spot Device (which used people to compensate for the technical inadequacies of picture-reading machines). Control over the workplace had to be understood in the context of very different understandings of the nature of scientific discovery and, ultimately, different images of what human beings were good at doing.

Nonetheless, the various bubble chamber groups had in common a centralized, well-structured hierarchy. They were *centered systems,*[213] with well-defined leadership structures. By the time one gets to the TPC, that pyramidal structure has become a *decentered network.* This is not to say it is a democracy of peers. The university is one of the most hierarchical social systems ever devised, and the physics team from each university was no exception: full professor, associate professor, on down through the assistant professor, postdoctoral

213. The literature on systems and systems engineering is, of course, vast. Of particular interest is the work of Hughes, whose *Networks of Power* (1983) and related articles offer ways of thinking about innovation within complex technological systems.

fellow, and graduate student. Impeding the establishment of any "meta" hierarchy was the brute economic-political relationship that existed between the various participating institutions. Each needed the others and no group could afford to lose the political and financial support provided by the others. Facing the selection committee at PEP, or the reviewers at the Energy Department, meant presenting a unified and equitable relationship among self-sustaining groups. In such a system no university team could afford to enter into a collaboration as anything but an equal (or nearly equal) participating entity.

Consequently, within the collaboration no single group could dominate absolutely. While LBL could lay claim more than most to the title of "center," even it had to work alongside the other university groups from Riverside, Yale, Tokyo, Johns Hopkins, and UCLA. *Intra*university structures were perfectly adapted to hierarchy; *inter*university collaborations amalgamated different hierarchies and so were bound to generate resistance to any hierarchy of hierarchies.

One way this resistance to central authority manifested itself is in what one might call the collaboration's multifocal character. There was a center for TPC design, and it clearly lay at the desk of David Nygren. The center for motherboard electronics development lay in good measure with the electrical engineers at LBL, with physicists more peripheral. And when it came to doing string modeling to analyze jet physics, the relevant analysis orbited around Werner Hofmann and the Swedish string group. Rather than say that the center has disappeared, one should probably say that centers have proliferated.

It is certainly possible to imagine large-scale collaborations with purely centralized structures (in fact, one can point to a successful and well-known group at CERN that maintains an extremely hierarchical structure). Nonetheless, the pressures are many, probably more in the United States than in Europe, though in Europe too, toward the kind of democratic amalgamation of hierarchies that I have sketched here. Ultimately, the high energy physics community in the United States draws its constituency from the university groups. These are the entities that in the end "vote with their feet" to put resources into the experiment. It is through the universities that graduate students and postdocs are recruited, and it is only by assembling a politically acceptable consortium of participating institutions that the facility committees (be they at SLAC, Fermilab, CERN, or DESY) can muster a hundred or five hundred million dollars. By the 1980s, there simply was no longer even the possibility of an Alvarez bargaining more or less directly and personally with the AEC to fund a major piece of equipment.

Given the network structure of the collaboration, new modes of governance were necessary. The Alvarez organization was born from a cross between the big business of the 1940s and the military command structure of the war. It had a head, a chief of staff, and a host of other deputies directing various "departments." By contrast, the TPC absorbed many of its structural features from the huge coordinated consortia of companies that aligned their work on structures such as the Polaris missile. The critical path method was one attempt to distribute resources in a way that would rationalize the assignment of priority and to guide various roughly equal subgroups toward a common goal. A second institutional structure that reflected the shift from centralized hierarchy to dispersed network involved the critique of the old LBL group structure. Where the laboratory had earlier relied almost entirely on the guidance of individuals in the "Goldhaber-Trilling group," the "Alvarez group," or the "Powell group," members of the TPC collaboration demanded a structure revolving around tasks, not the dead weight of historical leaders. A third mode of coordination was the computer network, originally set up to bind Stanford to Berkeley and eventually arcing over the entire high energy physics community. These microwave links made it possible for each of the groups not only to access, monitor, and manipulate data but also actually to direct the on-line behavior of the TPC facility. All three structures—the critical path method, task forces, and computer networking—thus served double functions. In the first instance, each simply mirrored the federalist configuration of constituent institutions. At the same time, however, the advocates of each hoped to use these social technologies as a means of integrating the multiplicity of local cultures and hierarchies without forcing them into homogeneity. In a very real sense, these social and managerial technologies enabled new trading zones to emerge.

The dispersal of responsibility away from a single center was not, as we have seen, without conflict. At one level of analysis, the PEP-4 facility was a sometimes painful coordination of subassemblies, each standing for an institution. Coordinating the muon detectors with the drift chambers was at once a technical and sociological act: it meant aligning teams from U.C.-Riverside and Johns Hopkins who had designed, built, maintained, programmed, and analyzed their respective devices. Similarly, integrating data from PEP-9 with that from PEP-4 was never separable from the tension between the two institutional collaborations. Top-down structures like the critical path method or the task force were not bringing individuals into face-to-face contact, and problems such as the cooling system on the motherboards or the cabling fiasco were the result. Though it was a hard lesson for the participating physicists, over the course of building the TPC a different conception of the relation between physicists and engineers was reached, a relation I have called "postmodern" in recognition of its decentered and heterogeneous structure. These hybrid collaborations are in many respects experiments in dispersal, attempts to create affiliations without annihilating distinct identities.

In an essential way, the TPC was a collaboration between professional subcultures—including electrical engineers, mechanical engineers, theorists, and phenomenologists. The forced marriage of "physics contacts" and "responsible engineers" was one attempt to coordinate two cultures that, more than once, came to the verge of tearing apart. Recounting these struggles in which the physicist confronted the dangers of his own unchecked power, I am reminded of Hegel's parable of the master and the slave. A master at first appears to have everything: his intentions and desires seem immediately realized, since someone else (the slave) does the work. To the master, the world therefore seems to offer no opposition; everything around him appears infinitely pliable. The slave, by contrast, continually encounters the world as independent, as a stubborn, recalcitrant place, where he cannot arbitrarily impose his will; he "merely works on [the world]." Eventually, though, the master, who at first seemed to have nothing but pure enjoyment, suffers the fate of perceiving everything around him as ephemeral: his only experience of the world is of its changeability. The servant, however, because the world will not conform easily to his desires, has the experience of permanence and objectivity. His work leaves things modified in a lasting fashion. In this "Robinsonade" as Hegel calls it, there is a moral as well as a philosophical judgment. "Labour," according to Hegel, "is desire restrained and checked, it is the ephemeral postponed." It is the master's lack of self-control that lies behind his fall.[214]

This moral injunction to restrain, check, and channel desire is manifest in the increasingly desperate pleas by Jay Marx, who found in his fellow physicists *"insufficient discipline in making design decisions,"* on top of an *"inability to take prudent risks."* Instead, as a group, they were *"unable to resist making design changes."* By invoking a language of discipline, prudence, and the inability to resist, Marx was speaking of purely technical matters—how to modify readout planes, for example. But in choosing this mode of speech, Marx (and many of his contemporaries in the experiment) had also begun to criticize deeper flaws within the physicists' culture. Conversely, perhaps because the physicists were used to their position of mastery, Marx found that *"our engineers were not sufficiently combative,"* that the team needed more "confident engineers" and a tougher engineering leadership. Such questioning had to be disturbing to both sides. The anonymous engineer's letter to Marx and Nygren was an explicit demand for the return of a lost style of leadership, in which "Ernest" imposed the authority of the physicist (the Boss) on the engineer (the Handmaiden).

If Marx was continually frustrated by his inability to foster the kind of "dialogue" or "debate" that he wanted between physicists and engineers, it was in part because he was working against an entrenched cultural arrangement. Too many physicists were willing to dictate the terms of design, and engineers were used to papering over their objections to avoid confrontation. Marx and others introduced a series of exchange procedures developed between planner and builder: "feasibility phase," "definition phase," "design and development phase," and "critical design review." These categories served to break the "front" between

214. See, e.g., Lukács, *The Young Hegel* (1975), 326–28.

the cultures of physics and engineering—to infiltrate the experimenter into the engineering and construction problems, and conversely to insert engineering culture into that of physics. In a sense, by demanding that physicists and engineers meet face to face, Marx was trying to avoid the nightmare of Hegel's master who, because of his disconnection from the fabrication of things, becomes in the end completely ineffective. By putting the engineer's knowledge on the table, even when it opposed the physicist's first inclination, the physicist began to confront the reality of large-scale construction. The existence of the TPC—over budget and late, but working nonetheless—was a final testimony to the joint aspiration of injecting "reality into the physicists' . . . dreams and then . . . translating it," as Marx had put it, "into a 'working piece of hardware.'"

Supplementing face-to-face contact with electronics, the collaboration addressed the coordination problem with a technical solution and, in the process, reflected in the experiment the decentering that already existed institutionally. In a first stage of dispersion, data could be analyzed at different laboratories. The transport of magnetic tape was, after all, not different in any important way from the export of bubble chamber film in the 1950s and 1960s. Delocalization of the experiment progressed a significant step further when the collaboration established a microwave computer link between LBL and SLAC, followed by a full-scale network among the participating institutions. With this change it became possible for the collaborating physicists to access data from remote sites, to modify analysis programs, and to maintain an interactive pool of shared knowledge. Finally, completing the despatialization of the experiment, it became possible for distant collaborators to watch data come out of the TPC in real time and to modify detector settings as seemed appropriate. At this stage, one can reasonably ask: *Where* is the experiment? If people are doing everything from observing to altering the experimental setup from Los Angeles, New Haven, and Tokyo, is it still the case that the experiment "really" is in Palo Alto?

7.9.3 Modern and Postmodern: The Pure and the Hybrid

Architects and architectural historians have found it useful to distinguish between modernism and postmodernism in the structure of buildings. High modernism, exemplified by the pristine works of the Dessau Bauhaus architects such as Hannes Meyer, reveled in the purity of geometric form. The quintessential modernist building of the 1920s contrasts starkly with the eclectic pastiche of a structure by Frank Gehry, who entered architecture in the 1970s and 1980s. The definitive contrast is between Mies's slogan "Less is more" and Robert Venturi's sardonic rejoinder "Less is a bore."[215] Without pressing the point too strongly, something like this characterized the transition between the idealized detector of the 1950s—the emulsion stack or the vat of liquid hydrogen—and the fantastic

215. Klotz, *Postmodern Architecture* (1988), 142.

assemblage of traditions, groups, specialties, and components that marked the detectors of the 1970s and 1980s.

What kind of detector is TPC/PEP-4/PEP-9? It is a hybrid entity without even a single name, the physical manifestation of an extraordinary diversity of functions and origins. It is a Čerenkov ring, a drift chamber based on technology from the 1960s, a gas calorimeter, a lead calorimeter, a muon detector, and of course, the novel Time Projection Chamber. The scientific principles behind these component parts are varied and draw on a plethora of physical principles from Lorentzian electrodynamics to quantum transitions. Furthermore, the institutional origins of the component parts are diverse. While Group A under Alvarez controlled the entirety of the hydrogen bubble chamber, that same group dominated only pieces of the TPC. It was inconceivable—politically, financially, or indeed scientifically—that all parts of a machine of this magnitude could be constructed by a single institution. Finally, the roles of engineers and physicists (indeed, the kinds of engineers and the kinds of physicists) diverged. Cryogenic, electrical, mechanical, and structural engineers all had to bring their expertise to the machine. On the physicists' side, the profession fragmented into specialties twice, once around the hardware and again around the integrative computer programs that pulled the hardware together. Each component separately and the organization as a whole depended on software. The machine existed twice—first in circuitry and then in the virtual reality of software.

In the monolithic ("modernist") machines of the past that we have mapped in the image and logic traditions (bubble chambers, nuclear emulsions, Geiger arrays, spark chambers), the coordination of the various groups was always an important feature, but for the most part, specialties were well defined and responsibilities clear. NBS engineers could study safety matters, and Hernandez and the mechanical engineers could use their results to design a reshaped chamber. Now, in the intensely hybridized detector, negotiating these contacts became a primary concern of the experiment.

Take theory. Perhaps the emblematic moment of the bubble chamber epoch was the meeting at which the Alvarez group announced its discovery of the $\Xi(1530)$. There, from the floor of the auditorium, on 5 July 1962, Gell-Mann could exploit the SU(3) theory to predict the existence of the final member of the octet, the Ω^-. The unambiguous experimental result, the equally unequivocal theoretical interpretation, and the decisive experimental rejoinder—these were hallmarks of an era of physics in which the categories of experiment and theory, as well as experimenter and theorist, had well-defined boundaries and more or less clear modes of interaction. As we have seen, the era of the TPC could not be more different. By the time the TPC group assembled at Asilomar to discuss the physics it would do, it had become evident that linking its work with the theorists was more than daunting; it might well lie beyond the group's resources. In an effort to bridge that chasm, physicists on both sides had to transform themselves. From the theorists' side, people like Wolfram, Fox, Gottschalk, and Andersson had to plunge themselves into the details of event simulations and specific experimental conditions that took them a long way from the clear ethereal world of renormalization, chiral symmetries, and grand unification. From the experimenters' side, it meant that Werner Hofmann and Charles Buchanan began coauthoring articles on the underlying basis of the string model. And it meant that Buchanan and his collaborators began seeking a "fundamental" phenomenology no longer focused on the more traditionally conceived experimenter's role of gathering and presenting data. The role of Wilson loops in the symmetric Lund fragmentation function was a long way from cooling the motherboards, building a calorimeter, or running a data check.

The experimenter who faced the TPC as a variant of the wire chambers in 1974 is not the same experimenter who faced the multinational collaboration in 1990. When the construction of the TPC/PEP-4 facility began with the assembly of the full-scale proposal in 1976, LBL's Group A existed as a legacy of the age of Alvarez. Experimenters knew engineers were necessary for the building of big equipment—there is no doubt of that. But the sense of necessity was predicated on a vision of the engineer as executing tasks set by the physicists. "Will" resided in the physicists, and instrumentality in the engineers. Similarly, the experimenters could make use of theorists' work, but even in the work at Mark I there was no need for the minor industry that arose around hadronization modeling. By 1990, role definitions had changed—physicists were rewriting their own part in the laboratory script. Activities that had been unrelated in the age of bubble chambers now had to be coordinated at every level. In the age of Alvarez, cryogenic engineers did not need to worry about data reduction and simulation—that was a different department. Experimenters planning the 72-inch chamber essentially planned on the basis of theory no more elaborate than the decay distance of a lambda. Those clear, sharply defined fronts between practices had become vastly more complex. The computer, for example, now entered everywhere. Computers simulated detectors during the design process; computers sorted data before, during, and after the information was written to tape; simulations produced events to test models and calculate backgrounds; simulations mocked up the detectors to see how events would be "seen." At every stage, activities thought to be polar opposites now interlocked in ever more elaborate forms:

design::execution
data::interpretation
physics::engineering
hardware::software
experiment::theory
reality::simulation

These changes began to reshape the way students were trained, evaluated, and placed in jobs after their stint at the TPC. Of one of the collaboration's 75 or so doctoral students, Nygren remarked in a letter of recommendation that he worked rapidly and had shown considerable interest in apparatus development, test setups, data acquisition, and microcomputers. "[E]xperience with TPC has taught [this student] much about the special character of large scale enterprises and has improved [his] interpersonal skills." These "interpersonal skills" obviously become much more important working in a team of a hundred than they were when one could work individually or in a small group of one's choosing. Nygren went on with a qualified endorsement that seemed almost rueful—what the TPC had produced was a physicist for the new generation of big physics: "[W]hen . . . goals are clearly defined and well thought out, [he] is enormously productive. . . . I believe he would make a valuable team member for many of the challenging aspects of [your accelerator laboratory] program."[216]

If the high energy physics community painfully adopted a new picture of what a young "team-playing" physicist was or ought to be, the anxiety did not stop there. Even senior physicists were concerned with the long hiatus between the initiation of a project and the time when they would be "doing physics."[217] Indeed, just months after the project proposal left LBL in 1976, Jay Marx wrote Birge indicating just how threatening a stint at administration might be: "Because the TPC is a long term project, I will not have done any 'physics' research during my term at LBL. In addition, my role in TPC R&D is that of a generalist who fills whatever cracks need filling. I will not have a particular hardware or software effort that I can point to as mine. This will work to my disadvantage in the inevitable 'show and tell' that must be a part of staff committee deliberations."[218] For Marx, the relation of administration to "physics" loomed large. For others, the tensions pulled elsewhere, but across the field people were asking, what would count as experimenting? Detector design? Computer simulation? Hadronic model building? Shift running? Data analysis? Authorship—the very idea—crystallized these anxieties in the ever more elaborate criteria invoked to reflect the splintered actions that together make a machine, an argument, and in the end, an experiment.

In these and other ways, the traditional career-defining boundaries had mutated along with the material culture of experimentation. Engineer and physicist, data and interpretation, hardware and software, experimenter and theorist, reality and simulation—all began to spill across the hard divisions that had once

216. Nygren to [recipient and date deleted for confidentiality], file "Miscellaneous Memos," box 2, DNyP II.

217. Nygren to [recipient and date deleted for confidentiality], file "Miscellaneous Memos," box 2, DNyP II.

218. Marx to Birge, 2 November 1977, file "LBL Politics & Memos," box 1, JMP.

defined them. In a field of particle physics that had devolved to a handful of megalithic machines, each detector was a world in itself. Like the largest multinational companies, the detector collaboration was concerned not so much with competition as with the simple task of remaining intact against the forces of disaggregation.

Pulled a thousand ways by the dynamics of its various subcultures, the Time Projection Chamber is emblematic not just of high energy physics but of our culture in a time when the legitimation of centralized authority has lost much of its power to compel. The TPC was an experiment in postmodernism.

8 Monte Carlo Simulations: Artificial Reality

8.1 Introduction: Simulations

When we look at a machine like the Time Projection Chamber—or for that matter almost any large-scale high-tech device of the late twentieth century—we are seeing double. We see the object in its materiality, layer upon layer of large-scale integrated circuits, cryogenic magnets, gases, scintillating plastics, and wires. But then, as if we could dimly make out Ryle's ghost in the machine, we begin to uncover various superposed virtual machines. If we stare at a drift chamber manufactured in the last decades of the century, we can be sure there was before it a simulation; before the scintillation counter was its simulation; before their interaction was the simulation of their interface. And when the machine is turned on, yet other simulations kick into action, mimicking the microphysical events to extract, process, clean, and interpret data. Sociologically (as even a cursory glance at the work flowcharts of chapter 7 makes clear) the experiment existed at least twice—the "soft" calorimeter had to function for the "hard" one to yield anything at all. Without the computer-based simulation, detectors like the TPC were deaf, blind, and dumb: they could not acquire data, process them, or produce results. This could be put in still stronger terms: without the computer-based simulation, the material culture of late-twentieth-century microphysics is not merely inconvenienced—it does not exist. Nor is this true only of particle detectors—machines including the huge plasma-heating Tokamaks, the complex fission-fusion-fission nuclear weapons, the guidance systems of rockets are inseparable from their virtual counterparts—all are bound to simulations. But what are artificial versions of complex processes, these indispensable procedures that are not quite experiment and yet not quite theory?

To get at this quasi-material dimension of material culture, I will focus on what appears at first to be a chaotic assemblage of disciplines and activities: thermonuclear weapons, enhanced A-bombs, poison gas, weather prediction, pion-nucleon interactions, number theory, probability theory, industrial chemistry, and quantum mechanics. No entities bind them together; they fall into no clear framework or paradigm; they have no single history that can be narrated smoothly across time. Yet the practice of these activities was congruent enough in the years just after World War II for Enrico Fermi, John von Neumann, Stanislaw Ulam, and others to move back and forth between these widely divergent domains. They did not share commonly interpreted laws, and most certainly not a common ontology. They did hold a new cluster of skills in common, a new mode of producing scientific knowledge that was rich enough to coordinate highly diverse subject matter.

Their common activity centered on the computer. More precisely, nuclear weapons theorists transformed the nascent "calculating machine," and in the process created alternative realities to which both theory and experiment bore uneasy ties. Grounded in statistics, game theory, sampling, and computer coding, these simulations constituted what I have been calling a "trading zone," an arena in which radically different activities could be locally, but not globally, coordinated.

To workers in these various domains, the most astonishing feature of these simulations was that they worked as well as they did. The basic idea exploited randomness in a way that with the aid of a computer was supremely simple to implement. Suppose one wanted to estimate π to arbitrary accuracy. Inscribe a circle of radius r in a square (see figure 8.1). Then take a random collection of points generated in the square and count the ratio of points falling within the circle to the total number of points generated inside the square. In the limit of large numbers of points, this ratio will approach the ratio of the area of the circle (πr^2) to the area of the square ($4r^2$), yielding $\pi/4$. Analogues of this Monte Carlo process (named for the Monacan gambling mecca) could be invoked not only to solve vastly more complicated mathematical systems but also to imitate physical processes that defied analytical solution.

Reaching back into the annals of the history of mathematics, many of the elements of the Monte Carlo method can be exhibited from particular endeavors. There were techniques like the Monte Carlo used to find π, for example, and a rich legacy of numerical methods for estimating areas extending back at least as far as Archimedes. But only during and shortly after World War II, with the advent of the computer, was a generalized mode of inquiry created to address the wide class of problems too complex for theory and too remote for experiment. By using random numbers (chosen *à la roulette*), nuclear weapons designers could simulate stochastic processes too difficult to calculate in full

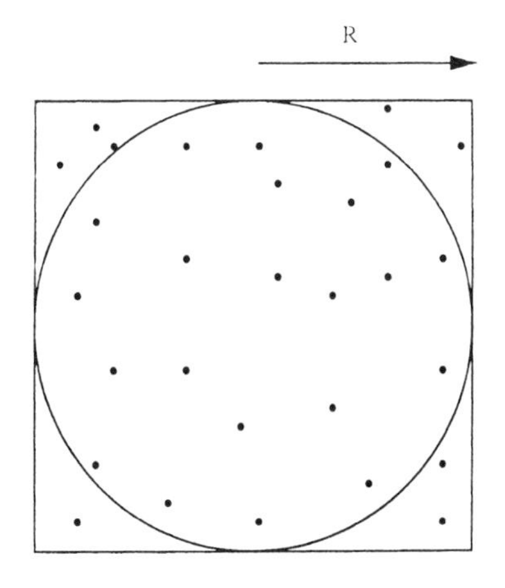

Figure 8.1 Monte Carlo estimation of π. Long before the electronic computer and the introduction of Monte Carlo simulations as a routine part of physics, there were a few instances of techniques by which random events could be used to solve nonstochastic problems. Here, for example, randomly chosen points within a square can be used to estimate the ratio of the area of the inscribed circle to the square as a whole: The ratio of the number of points in the circle to the number of points in the square approaches $\pi/4$ as the sample size gets larger.

analytic glory.[1] But physicists and engineers soon raised the Monte Carlo above the lowly status of a mere numerical calculation scheme; it came to constitute an alternate reality—in some cases a preferred one—in which "experimentation" could be conducted. Proven on the most complex physical problem that had ever been undertaken in the history of science—the design of the first hydrogen bomb—the Monte Carlo ushered physics into a place dislocated from the traditional sociointellectual poles of experiment and theory. Monte Carlos formed a tertium quid, a simulated reality that borrowed from both experimental and theoretical domains, fused these borrowings together, and used the resulting amalgam to stake out a netherland at once nowhere and everywhere on the usual methodological map. As they struggled to find a place in the traditional categories of experiment and theory, the simulators both altered and helped define what it meant to be an experimenter or a theorist in the decades following the Second World War. Faced with several years of extraordinary successes and the hopes for many more, A. N. Marshall opened a "Monte Carlo" conference in March 1954, with astonishment: "one seems to be getting something for nothing." Somehow, the disparate members of the ad hoc community formed from a myriad of specialties had to explain to themselves how it was that "everything

1. Almost all of the historical material on the Monte Carlo method appears as subsections within broader histories of the computer. Of these, several are particularly helpful. All subsequent histories of the computer are greatly indebted to Goldstine, *Computer* (1972), on the Monte Carlo method see 295–97; the best history of von Neumann's scientific work on the computer is Aspray, *von Neumann* (1990), on the Monte Carlo method see 110–14; the most reliable source on nineteenth-century stochastic simulations is Stigler, "Simulation," *Stat. Sci.* 6 (1991): 89–97. Here Stigler looks in particular at work by De Forst, G. H. Darwin, and Galton. See also Stigler, *History of Statistics* (1986), 265–99.

comes out all right in the end; the efficiency of the methods in particular cases seems unbelievable. The results quite literally have to be seen, and seen through, to be believed."[2] In this chapter, we will track the history of the quest for this *characteristica universalis,* with its promise to decode fields from all across the disciplinary map.

At Los Alamos during the war, physicists soon recognized that the central problem was to understand the process by which neutrons fission, scatter, and join uranium nuclei deep in the fissile core of a nuclear weapon. Experiment could not probe the critical mass with sufficient detail; theory led rapidly to unsolvable integrodifferential equations. With such problems, the artificial reality of the Monte Carlo was the only solution—the sampling method could "recreate" such processes by modeling a sequence of random scatterings on a computer. Simulations advanced the design (more particularly, the refinement) of fission weapons, but remained somewhat auxiliary to the A-bomb theorists. When, at war's end, nuclear weapons work turned to the thermonuclear bomb, Monte Carlos became essential. For there was no analogue of Fermi's Metallurgical Laboratory nuclear reactor—no controlled instance of fusion that physicists could study in peacetime. Instead, Stanislaw Ulam, Enrico Fermi, John von Neumann, Nicholas Metropolis, and a host of their colleagues built an artificial world in which "experiments" (their term) could take place. This artificial reality existed not on the bench, but in the vacuum tube computers—the JONIAC, the ENIAC and, as Metropolis aptly named it, the MANIAC.

This story proceeds on several intersecting planes. It is a tale in the history of epistemology: a new method for extracting information from physical measurements and equations. It is a narrative of the history of metaphysics: a scheme of representation that presupposed a nature composed of discrete entities in irreducibly stochastic interaction. It is a workplace history: traditional professional categories of experimenter and theorist challenged by an increasingly large and vocal cadre of electrical engineers, and later computer programmers. Overall, it is an account of fundamental physics inextricably tied to the development of the superbomb, the weapon with no limit to its potential destructive power, and a description of the calculating machine from computer-as-tool to computer-as-nature.

8.2 Random Reality

The computer began as an instrument, a tool like other tools around the laboratory. Von Neumann could refer to it in precisely the way that he would discuss a

2. Marshall, "Introductory Note" (1956), 14.

battery or a neutron generator. To Julian Huxley in 1946, for example, von Neumann would write: "We want to use this machine in the same sense, in which a cyclotron is used in a physics laboratory."[3] As has been rehearsed endlessly in the histories of the computer, there were two fairly distinct traditions of computer building. One sought to create machines that would function in an analog fashion, recreating physical relations in media or at scales different from their natural occurrence. Among these could be counted harmonic analyzers, gunnery computers, and network analyzers, as well as analog models including model ship basins or wind tunnels.[4] Such models have long histories; one thinks for example of nineteenth-century British electricians, as they put together pulleys, springs, and rotors to recreate the relations embodied in electromagnetism. But it was the second tradition, devoted to digital information processing, that interests us here. Indeed, my concern is considerably narrower still: the attempt to use the emerging digital devices of the late war and postwar periods to simulate nature in its complexity.

With the Japanese surrender, the laboratory at Los Alamos began to disperse. But before it did, scientists convened to collect and record their knowledge, should it later prove necessary to return to a major effort in nuclear weapons. One such meeting was called for mid-April 1946, to discuss the still hypothetical fusion weapon, a device that the staid, classified report regarded with some awe—even after the devastation of Hiroshima and Nagasaki: "Thermonuclear explosions can be foreseen which are not to be compared with the effects of the fission bomb, so much as to natural events like the eruption of Krakatoa . . . values like 10^{25} ergs for the San Francisco earthquake may be easily obtained."[5] No longer content to mimic Krakatoa in vitro (as had the Victorian natural philosophers of chapter 2), the physicists of Los Alamos now contemplated its imitation in vivo.

If imagining the destructive power of the H-bomb was a matter of no great difficulty, designing it was. Though the early bomb builders had thought that heavy hydrogen could be set into self-propagating fusion with relative ease, work during the war had indicated that the problem was much more complicated. Elsewhere in this same report, the Los Alamos physicists commented that "[t]he nuclear properties of the key materials are still fundamental in that the energy is supplied by nuclear reactions; but control and understanding of the phenomenon involves purely atomic consideration to a much greater extent than was the case in the fission bomb. What the reaction depends on is the complex behav-

3. von Neumann to Huxley, 28 March 1946, JNP.

4. See, e.g., Goldstine, *Computer* (1972); Aspray, "Mathematical Reception" (1987), 168; Aspray, *von Neumann* (1990); Burks and Burks, *Electronic Computer* (1988), 74–87, 106–109; and Stern, *ENIAC to UNIVAC* (1981), 8–14.

5. Bretscher et al., "Super," LA-575, 16 February 1950, 4.

iour of matter at extremely high temperatures. For prediction, then, the primary requisite is a deep insight into the general properties of matter and radiation derived from the whole theoretical structure of modern physics."[6] When the authors said that they would need the "whole theoretical structure of modern physics," they were not exaggerating. Not only were the nuclear physics of hydrides, the diffusion of hard and soft radiation, and the hydrodynamics of explosion difficult in themselves, they had to be analyzed simultaneously and in shock at temperatures approaching that of the sun's core. How was energy lost, what was the spatial distribution of temperature, how did deuterium-deuterium and deuterium-tritium reactions proceed? How did the resultant helium nuclei deposit their energy? These and other problems could not be solved by analytical means, nor did they lend themselves to "similarity" treatments by analogue devices. Experiments appeared impossible. A hundred million degrees kelvin put the laboratory out of the picture; there was no thermonuclear equivalent to Fermi's reactor, no slow approach to criticality obtained by assembling bricks of active material. Where theory and experiment failed, some kind of numerical modeling was necessary, and here nothing could better the prototype computer just coming into operation in late 1945: the ENIAC (Electronic Numerical Integrator and Calculator).[7]

It is no mere accident that the first problem on the first computer was the thermonuclear bomb. For three years the Manhattan Project had grown to staggering size, and commanded, by war's end, the full range of new technology. A key player in these developments was John von Neumann, a Hungarian refugee mathematician and physicist, who had been thinking about automatic calculation throughout the war. From his early work as a consultant on ballistics to the Aberdeen Proving Ground (and on the Scientific Advisory Committee beginning in 1940), von Neumann had been concerned with the adaptation of complex physics problems to the computer. By 1941, he was widely recognized as an expert on shock and detonation—skills he brought that year to the National Defense Research Council (NDRC), and from 1943 through 1945 to the Manhattan Project.[8] Each of these projects had its own complexity, but perhaps the most difficult problems surrounded the calculation of implosion necessary to drive plutonium together fast enough and precisely enough to ensure an efficient nuclear detonation.[9] What were the hydrodynamics of the neutron reflector that surrounded the plutonium? How would the hydrodynamics of the fissile core shift when hit by the converging blast wave? Amid the complicated motion of plutonium and reflector, how would the blast wave itself behave? These were calculations that had no chance of being solved by hand, much less in closed form,

6. Bretscher et al., "Super," LA-575, 16 February 1950, 3.
7. Bretscher et al., "Super," LA-575, 16 February 1950, 20.
8. Aspray, "Reception" (1987), 171.
9. See Hoddeson, "Mission Change" (1992).

and von Neumann was driven even further toward an understanding of how to transform coupled differential equations into difference equations which, in turn, could be translated into language the computer could understand.[10]

Here is an example from von Neumann's 1944 work on numerical methods, designed for the digital processing of a thermodynamic model of hydrodynamical shocks.[11] It illustrates how the underlying objects in a physical model are often conditioned by the very unphilosophical need to be able to *calculate.* In question is a compressible gas or liquid in which heat conduction and viscosity are negligible. In the Lagrangian form, the equations of motion are constructed by labeling the elementary volume of substance by a quantity, a, which is located by its position x at the time t. We characterize the gas by its caloric equation of state, which gives its internal energy, $U = U(V,S)$, where U is a function of specific volume, $V = \partial x/\partial a$, and specific entropy, S. Normalized mass units made the density equal unity, so the amount of substance located between x and $x + dx$ is just da. It follows that the gas density is the inverse of its specific volume, and pressure, p, and temperature, T, are given by the usual thermodynamical equations,

$$p = -\frac{\partial U}{\partial V}, \qquad T = \frac{\partial U}{\partial S}. \tag{8.1}$$

The equations of motion are designed to provide x as a function of a and t. Conservation of momentum ($dp/dt = 0$) has two parts: a time derivative of $m\partial x/\partial t$ and a term corresponding to the change of pressure as one moves along in space:

$$\frac{\partial^2 x}{\partial t^2} = -V \cdot \left(\frac{\partial p}{\partial x}\right)_{t\,=\,\text{constant}}. \tag{8.2}$$

Substituting $V = \partial x/\partial a$ leads to

$$\frac{\partial^2 x}{\partial t^2} = -\frac{\partial}{\partial a} p\left(\frac{\partial x}{\partial a}, S\right). \tag{8.3}$$

The conservation of total energy (internal plus external) is just

$$dE = \frac{\partial U}{\partial V} dV + \frac{\partial U}{\partial S} dS + \left(\frac{\partial^2 x}{\partial t^2}\right) dx = 0, \tag{8.4}$$

recalling that mass units are normalized to unity. When combined with the thermodynamic relations (8.1), equation (8.4) yields

10. See Bethe, "Introduction" (1970), 3.
11. von Neumann, "Numerical Method" ([1944] 1963), 6:367.

$$dE = -pdV + TdS + \left(\frac{\partial^2 x}{\partial t^2}\right)dx = 0. \tag{8.5}$$

From equation (8.3), we can substitute $-\partial p/\partial a$ for $\partial^2 x/\partial t^2$. After integration by parts, the first and third terms on the right-hand side of equation (8.5) cancel, yielding

$$\frac{\partial S}{\partial t} = 0. \tag{8.6}$$

This is a crucial part of von Neumann's argument: because of the result that entropy is conserved not as a condition but as a consequence of energy conservation, the solution to equation (8.3) can be made assuming p is a function of V alone. Analytic and numerical techniques work on this well-understood hyperbolic differential equation perfectly well. If, however, a shock is introduced in the system, the entropy changes as our bit of matter crosses the shock. Since changes of entropy (and therefore of the coefficients in the equation of motion [8.3]) depend on the trajectory of the shock, the problem is rendered vastly more complicated. Instead of a hyperbolic differential equation, one has a hyperbolic differential equation with variable coefficients, which in general cannot be solved either analytically or numerically.

Von Neumann's central idea was to recapture the simplicity of equation (8.3) even for the shock case (where S is not constant) by assuming that the internal energy could be divided into two noninteracting parts: $U(V,S) = U_*(V) + U_{**}(S)$. Then only $U_*(V)$ contributes to p in the thermodynamical equation (8.1), and this quantity, which we can label p_0, is as free of entropy dependence as it was in the nonshock case. In plain English, von Neumann's assumption that the energy $U(V,S)$ divides up this way amounts to the postulation of an "interactionless" substance that can be precisely modeled by beads of definite mass on a chain of springs. Shocks can still propagate, but entropy does not contribute to the pressure.

Letting a run over the integers from negative to positive infinity and

$$x = x(a,t) \equiv x_a(t), \tag{8.7}$$

we can reduce differential equation (8.3) to an approximate system of difference equations, well suited for machine coding:[12]

$$\frac{d^2 x_a}{dt^2} = p_0(x_a - x_{a-1}) - p_0(x_{a+1} - x_a). \tag{8.8}$$

Once on the machine, the simulation could produce a graphical representation of the shock wave as it propagated (see figure 8.2).

12. von Neumann, "Numerical Method" ([1944] 1963), 6:367.

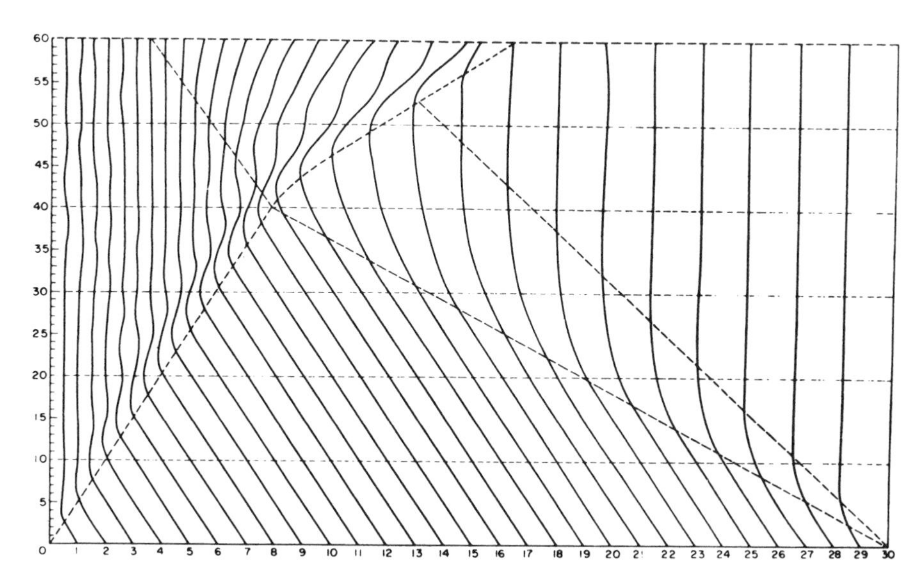

Figure 8.2. Shock simulation (1944). At the Aberdeen Ballistic Research Laboratory, von Neumann put his "beads and springs" shock problem on the punch-card calculator. Here the y-axis marks time and the x-axis labels the "molecules" and their original positions. The solid lines are the "worldlines" of individual beads. For a case where the exact hydrodynamic equation can be solved, the shock wave front is shown as the dashed line from the lower left; the two dashed lines from lower right are the front and back of the "rarefaction wave" that moves in advance of the shock. Impressive for the closeness of the numerical results to the exact solution—and for the additional detail given by the numerical work—these investigations launched von Neumann's interest in what became "computer simulations," even where exact solutions did not exist. Source: von Neumann, "Numerical Method" ([1944] 1963), fig. 2 on 377, copyright 1963, with permission from Pergamon Press Ltd., Headington Hill Hall, Oxford OX3 0BW, UK.

Here is what has happened. For purely computational reasons (i.e., in order to process the shock problem on an electromechanical calculator), von Neumann redefined the goals of his work to include the elimination of entropy dependence. Until then, the only way to do this was to cancel the shocks; his new approach was to alter the model, introducing the beads-and-springs. Thus, the beads-and-springs is a model of the real gas that has been constructed to conform to the new calculational apparatus. It was this model, calculable only by machine, that would simulate a physical system even when analytic techniques led nowhere. Von Neumann then had to defend the validity of his machine-directed representation, and he did so with a long string of plausibility claims. First, he turned the tables on the visual "representational" argument for differential equations. In particular, he pointed out that it is the hydrodynamical equation (8.3) that distorts the world, by using a continuum to describe what in "reality" (von Neumann's term) is a discrete set of molecules. By contrast the beads-and-springs "corresponds" to the "quasi-molecular description of this substance." Of course, as von Neumann acknowledged, no device is about to process 6×10^{23} of these beads, but then again even the hydrodynamical equation never explicitly demands such vast numbers.

Somewhat wistfully, von Neumann concluded: "The actual $N \sim 6 \times 10^{23}$ is certainly great, but much smaller numbers N may already be sufficiently great. Thus there is a chance that $N \sim 10^2$ will suffice."[13] If one thinks of a dynamical theory as usually treating one or two particles (*microscopic physics*) and statistical mechanics or experiment as treating 10^{23} particles (*macroscopic physics*), then what von Neumann is doing here is trying to carve out a via media, a zone of what one might call *mesoscopic physics* perched precariously between the macroscopic and the microscopic. Because this intermediate sphere lacks the traditional justification of either of the two extremes, von Neumann needed other arguments to show that this representation was a reasonable stand-in for the "true" state of affairs. These further arguments contained a defense of the simplified intermolecular forces, the process of energy degradation between forms of energy, and so on. On these and other grounds, von Neumann prophesied success for more complicated equations of state and more spatial dimensions—especially for the case of spherical symmetry. Inward-bound spherical shock waves were precisely what was needed to blast plutonium toward its supercritical assembly at the heart of an atomic bomb.

As von Neumann knew, shock calculations like those of 1944 were only a small fraction of the mathematical-physical difficulties confronting bomb builders at the end of the war. For if the plutonium bomb presented more difficult calculational problems than the simple "gun assembly" uranium bomb, none of the atomic bomb problems were even remotely as difficult as the problem of building a thermonuclear weapon.[14]

Imagined before the creation of Los Alamos, the hydrogen weapon had been on weapons designers' minds since the summer of 1942. When the H-bomb was pushed aside in favor of the A-bomb, only a small group around Edward Teller continued superbomb work (as it was called) during the war. And of all the outstanding problems left by the war effort in nuclear weapons, the hydrogen bomb was at once the most difficult and potentially the most important.

Building on the success of machine-based hydrodynamical models like those just described, von Neumann in 1945 invited Stanley Frankel and Nicholas Metropolis to try their vastly more complex problem (the thermonuclear weapon) on the new ENIAC.[15] Just as with the beads-and-springs shock model, one difficulty would be to present a compelling case that the difference equations would

13. von Neumann, "Numerical Method" ([1944] 1963), 6:368.

14. For more on the hydrogen bomb, see Galison and Bernstein, "In Any Light," *Hist. Stud. Phys. Sci.* 19 (1989): 267–347, and references therein.

15. Goldstine, *Computer* (1972), 214.

indeed replicate the essential features of the "underlying" differential equation; another problem would be to demonstrate that the restricted number of particles allowed by the ENIAC would properly "correspond" to the reality of a more complex world of 10^{23} particles.

Before the calculation could begin, Metropolis and his collaborators wrote out a handful of differential equations giving a simplified description of the deposition of energy by photons and nuclei, the release of energy by nuclear fusion, and the hydrodynamics of the simulated bomb. Then, after they were translated into difference equations and hardwired into the computer, the ENIAC began to spit out punchcards at about one per second. For six weeks during December 1945 and January 1946, the runs continued. Even when all the equipment worked perfectly, the simulation took several days to study one particular configuration of deuterium and tritium. As each card emerged as output, it was fed back in as input for the next time step. From this work came the temperature and configuration of the fusionable material as a function of time.[16]

Metropolis and Frankel presented their results to the April 1946 superbomb meeting. Present were all the members of Teller's wartime group, along with many others, including the theoretical physicist Robert Serber who had greatly contributed to the A-bomb, the wartime head of the Los Alamos theory group, Hans Bethe, the mathematician Stanislaw Ulam, and to the Americans' later chagrin, the Soviet spy Klaus Fuchs. Ulam, just five days before, had completed a speculative Los Alamos paper with J. Tuck on how thermonuclear reactions might be initiated in deuterium by the intrusion of jets from a fission device. Now, at the meeting, Ulam heard Frankel and Metropolis's results, fresh from the ENIAC. Having processed over a million IBM punchcards, they reported that the program had run successfully, leaving them provisionally optimistic (if one should use that term) that the super would detonate as designed.[17] But even with cybernetic assistance, it was apparent that the super problem far exceeded the ENIAC's capacity to model, and Ulam began to search for a more efficient way to exploit the capabilities of the new machine.

Somewhere between reflections on thermonuclear weapons, probability, neutron multiplication, mechanics, and the card game of solitaire, Ulam began to sketch the Monte Carlo. The public face of the Monte Carlo first appeared in a short abstract by Ulam and von Neumann, received at the American Mathematical Society on 3 September 1947. Saying only that they had a new computational procedure for studying differential equations, the authors emphasized that the technique built on their earlier work combining statistical and stochastic processes. "This procedure," they wrote, "is analogous to the playing of a

16. Metropolis, interview by the author, 21 May 1991; see also Aspray, *von Neumann* (1990), 47.

17. Ulam [and Tuck?], draft of "Preliminary Jet Experiments of Interest for Thermonuclear Study," SUP; handwritten note indicates that this document is a draft of LA-560 (Ulam and Tuck, "Possibility of Initiating a Thermonuclear Reaction").

series of 'solitaire' card games and is performed on a computing machine. It requires . . . 'random' numbers with a given distribution." "Random" appeared in quotation marks because the generating process itself could be deterministic. For example, the function $f(x) = 4x(1 - x)$, for $0 < x < 1$, would yield another number between 0 and 1; that in turn could be fed back into $f(x)$. The authors continued: "By playing suitable *games* with numbers 'drawn' in this fashion, one can obtain various other distributions." Many of the themes of the remainder of this chapter will explore elements already present in this early, and telegraphic, communication: the nature of the "random," "games," "model," and "analogy."[18]

The name "Monte Carlo" was coined by Metropolis, and was first printed in the formal contribution by Metropolis and Ulam in the September 1949 volume of the *Journal of the American Statistical Association.* Clearly following von Neumann's lead in his hydrodynamic computations, Ulam and Metropolis began by pointing to the uncharted region of mechanics between the terra firma of the classical mechanician, in which only very few bodies could be treated, and the newer world of the statistical mechanician, in which $N = 10^{23}$. Something of this mesoscopic terra incognita existed as well in combinatoric analysis, where the numbers were too large to calculate and not yet large enough to call in the law of large numbers. Ulam and Metropolis staked their claim on the new turf, announcing the key to their method: "To calculate the probability of a successful outcome of a game of solitaire (we understand here only such games where skill plays no role) is a completely intractable task. On the other hand, the laws of large numbers and the asymptotic theorems of the theory of probabilities will not throw much light even on qualitative questions concerning such probabilities. Obviously the practical procedure is to produce a large number of examples of any given game and then to examine the relative proportion of successes."[19] Sampling could apply even when no stochastic element was present, as we saw in the "computation" of π above, or as in the evaluation of a finite volume within a many (say 20) dimensional space. The volume could be defined, for example, by specifying a series of inequalities along each of the axes. Dividing up the unit cube into 10 parts along each of the 20 axes would produce 10^{20} cubes. Systematically checking each box to see whether there is a point in it would be, to put it mildly, impractical. Instead, the authors suggest, one could simply sample the cubes, say 10^4 times and find the ratio of points falling inside the required volume to the total number of points (10^4). More physically, the same sort of difficulty arises in understanding cosmic ray showers in which a high energy proton collides with a nucleus and produces a shower of other particles; these in turn give rise to more; and the chain continues in an extraordinarily complex fashion. Although each step of the process is played out

18. Ulam and von Neumann, "Stochastic and Deterministic," *Bull. Amer. Math. Soc.* 53 (1947): 1120.
19. Metropolis and Ulam, "Monte Carlo Method," *Amer. Stat. Assoc.* 44 (1949): 335–41, on 336.

according to previously given probabilities and is therefore calculable through a specifiable algorithm, obtaining the full solution is beyond the reach of analytic methods.

In each of these cases (solitaire, volume determinations, particle cascades) the authors proposed *sampling* particular realizations of a process from among the astronomically large number of possibilities. It was then possible to study features of this sample set of "experiments" for their statistical properties. By focusing on these instantiations of a general process, the Monte Carlo appeared experimental; by eschewing the bench, it appeared theoretical. The categories themselves had begun to slip, and in what must have appeared provocatively oxymoronic, Ulam and Metropolis declared: "These experiments will of course be performed not with any physical apparatus, but theoretically."[20] At stake was the boundary of the category "experiment," and as subsequent events will show, the widening of that concept did not go unopposed.

Such philosophical niceties took a back seat to the primary business to be expected from Metropolis and Ulam's return address: nuclear weapons. For fission, the problem was to calculate the diffusion of neutrons through an active substance such as plutonium. In such cases a neutron can scatter from the nucleus (inelastically or elastically), be absorbed by the nucleus, or cause the nucleus to fission (with the emission of n neutrons). Each process has predetermined probabilities, and a sampling of some finite number of neutrons is far simpler than trying to apply the integrodifferential equations of Boltzmann to groups of neutrons at given energies. Unmentioned, but undoubtedly coequal with the fission was fusion, on which Ulam was then hard at work.

For the Monte Carlo method to work, Ulam and von Neumann needed a large collection of random numbers, preferably generated by computer, with which to sample (perform experiments on) the infinite world of scatterings, fissions, and fusions. Borrowing thousands of "true" random digits (random numbers generated from "truly random" physical processes such as radioactive decay) from books was thoroughly impractical, so the two physicists set about fabricating "pseudorandom" numbers. In one of their earliest postwar formulations, von Neumann's scheme was to take some 8- or 10-digit "seed" number n to start the process; a sequence of further numbers would emerge by the iterative application of a function (typically a polynomial) $f(n)$. Not surprisingly, this procedure left people puzzled at how the Monte Carlo players could expect a definite function, applied iteratively, to produce a truly random number. The short answer is that he and Ulam did not expect $f(n)$ to do anything of the kind; unable to generate random numbers in the classical sense, they expanded the meaning of the term.

Von Neumann's attitude at the time emerges in a January 1948 exchange

20. Metropolis and Ulam, "Monte Carlo Method," *Amer. Stat. Assoc.* 44 (1949): 335–41, on 337.

with Alston Householder, a physicist at the Oak Ridge laboratory, who in puzzlement asked, "Is it the case that one starts with a single number, randomly selected, and generates therefrom an infinite sequence of random numbers?"[21] Work at Oak Ridge on Monte Carlos would stop, Householder added, pending the master's reply. Far from being infinite random number generators, Neumann responded, a finite machine sequentially applying $f(n)$ would, in the fullness of time, repeat a finite cycle of numbers over and over again ad infinitum. Suppose, for example, that the machine had a 10-digit storage capacity. Then, even if every 10-digit number were produced before repeating, after 10^{10} numbers $f(n)$ would necessarily produce a number already generated, causing the entire cycle to repeat precisely. "Consequently," von Neumann acknowledged, "no function $f(n)$ can under these conditions produce a sequence which has the essential attributes of a random sequence, if it is continued long enough." His and Ulam's hope was more modest. They wanted their computer to fabricate somewhere "between 1000 and 10,000 numbers . . . which up to this point look reasonably 'random.'" Such sequences need be only "random for practical purposes."[22]

By "random for practical purposes," von Neumann meant that the numbers would satisfy two conditions. First, the individual digits would be as equidistributed as one would expect for a random sample of the same size. Second, the correlation of *k*th neighbors would be zero for $1 \leq k \leq 8$ (the number of digits in the original seed number). Procedurally, the generative process worked this way: The first n_0 would be a randomly chosen eight-digit number. This would then be inserted into, say $n_1 = n_0^2$, and the middle eight digits extracted for reinsertion into $n_2 = f(n_1)$.[23] Perhaps acknowledging the legacy of his training in "pure" mathematics, von Neumann confessed: "Any one who considers arithmetical methods of producing random digits is, of course, in a state of sin. For, as has been pointed out several times, there is no such thing as a random number—there are only methods to produce random numbers, and a strict arithmetic procedure of course is not such a method."[24]

Over the next few years, von Neumann's simple generation scheme and correlation test gave way to others,[25] and a lengthy debate ensued over just what

21. Householder to von Neumann, 15 January 1948, JNP.

22. von Neumann to Householder, 3 February 1948, JNP.

23. See Householder to von Neumann, 15 January 1948; and von Neumann to Householder, 3 February 1948; both in JNP. Ulam also endorsed this "reasonably random" notion in his presentation to a conference in September 1949: "It may seem strange that the machine can simulate the production of a series of random numbers, but this is indeed possible. In fact, it suffices to produce a sequence of numbers between 0 and 1 which have a uniform distribution in this interval but otherwise are uncorrelated" (Metropolis and Ulam, "Monte Carlo Method," *Amer. Stat. Assoc.* 44 [1949]: 335–41, on 339).

24. von Neumann, "Various Techniques" ([1951] 1961).

25. It was first suspected on empirical grounds, and later on rigorous mathematical grounds, that von Neumann's original generative scheme did not provide a sufficiently reliable set of pseudorandom numbers. See Tocher, "Application," *J. Roy. Stat. Soc. B* 16 (1954): 39–61, esp. 46–47.

was meant by "random." But to the users of the Monte Carlo, the abstract mathematical concept moved to the periphery. At a January 1954 Monte Carlo symposium, there was an extended discussion of the detailed mathematics of various generative schemes for pseudorandoms, with only an occasional mention of the use of "true" physical generators such as thermal noise fluctuation in a vacuum tube circuit or the decay rates of radioactive materials. While ceding the economic utility of replacing true random numbers with their pseudorandom counterparts, one exasperated auditor, E. C. Fieller, declared: "I have to confess, however, that in spite of the elegant investigations that have been made into [the] properties [of pseudorandom numbers] my own attitude to all deterministic processes remains at best one of grudging and puzzled acceptance. I do not feel that I am getting random numbers; I just wonder whether I am applying the right set of tests of randomness—if indeed the search for a 'right set' is not itself illusory." Clearly, it was the idea of a definition "by test" that troubled him; for he continued, "What I much prefer are numbers, specifically generated as in [a specific device at Manchester] by a process that has what I can intuitively accept as random characteristics." Then, and only then, could tests be offered, not as *constituting* randomness, but as *monitors* of possible malfunction ("quality control," in his words).[26]

In response to Fieller, two of the main contributors to the conference, J. M. Hammersley (who held the lectureship in the Design and Analysis of Scientific Experiment at Oxford and served as a consultant at Harwell) and K. W. Morton (also at Harwell), submitted a barrage of rhetorical questions: "[D]o [random numbers] even exist; can they be produced to order and, if so, how; can they be recognized and can we test that they are not impostors?" Pragmatism, not rigor, was needed: "These are diverting philosophic speculations; but the applied mathematician must regard them as beside the point. He knows that random numbers are not really necessary in Monte Carlo work: approximately random numbers, pseudo-random, and approximately pseudo-random numbers will do instead; and the only remaining (but nevertheless formidable) question for him is how pseudo and how approximate."[27] Pseudo and approximate enough to be sufficient unto the day, was their response. Each problem would have its own demands. To gather up every test and pile all such criteria one on top of the other would be disastrous: "it is nigh impossible to construct anything positive when bound by a mass of negative principles." All that mattered was to get the "right answer" to the problem at hand; getting this right answer meant mainly not having any regularity in the "random" number sequence correspond to any structural feature of the problem itself. Indeed, for some integration problems, the sequence 1, 2, 3, 4, 5, 6, 7, 8, 9, 0, 1, 2, 3, 4, 5, . . . would work just fine.[28]

26. Fieller et al., "Discussion," *J. Roy. Stat. Soc. B* 16 (1954): 61–75, on 62.

27. Hammersley and Morton, in Fieller et al., *J. Roy. Stat. Soc. B* 16 (1954): 61–75, on 73.

28. Hammersley and Morton, in Fieller et al., *J. Roy. Stat. Soc. B* 16 (1954): 61–75, on 73.

These arguments about the adequacy of the pseudorandom generators were not idle. One author referred to the 1960s as the "dark ages" of pseudorandom generators. On the basis of various "tests" authors tried to adjudicate among the different schemes. But despite much spilled ink, the random generator built into the queen of computers, the IBM 360's RANDU, was notoriously unrandom, sending unsuspecting authors into disaster as late as 1980. F. James caustically summarized the pseudorandom philosophy: "if a pseudo-random number generator has passed a certain number of tests, then it will pass the next one, where the next one is the answer to our problem."[29] There was no reason, of course, that luck should hold in this way. In fact, the old IBM 709 generator had glided through its tests, but the remarkably high fluctuations these randoms produced when inserted into a Monte Carlo left one 1962 high energy physicist (Joseph Lach) deeply suspicious. Taking triplets of sequential points (x, y, and z), he considered the subset of points for which $z < 0.1$ and had the computer plot x and y to the screen. In a most shocking display of organization, the computer flashed a series of slanted bands with nothing between them (see figure 8.3). Why? At the time, Lach could not say, but such manifest structure showed that the "random" generator was clearly and profoundly flawed. At root, the difficulty could be explained by a number-theoretical analysis that revealed that all such "multiplicative congruential generators" produced bands. (A multiplicative congruential generator is one of the type $r_i = ar_{i-1} + b \bmod m$, where m would typically be 2 raised to power of the word size of the computer in use. So an elementary example for a 5-bit number sequence would be to set $a = 21$, $b = 1$, $m = 32$, and $r_0 = 13$; this choice would yield $21 \times 13 + 1 \bmod 32$, or the remainder of 274/32, which is 18. Now insert 18 where 13 was, and in this way get the series 13, 18, 27, 24, 25, 14, ...) Lach's analysis also suggested that by choosing a different multiplier (the number a) it was possible to make those bands "go away." The lesson many people drew was that tests alone had their limits: better plot and look than pile arbitrary tests one on the other.[30]

In an elegant application of Minkowski's geometry of numbers published in 1968, George Marsaglia, of the Mathematics Research Laboratory within Boeing Scientific Labs, showed that Lach's bands were inevitable. It would always be the case that any multiplicative congruential generator would send points into hyperplanes; change the multiplier (one learned from Marsaglia's paper) and you could alter the hyperplanes, but they would not vanish. This "defect," Marsaglia warned, could not be remedied by adjusting the starting value, multiplier, or modulus. Any way you did it there would be many systems of parallel hyperplanes containing all the points. This was bad: any application of a Monte Carlo in which some physical feature of the system coincided with the hyperplanes

29. James, "Monte Carlo" ([1980] 1987), 656.

30. James, "Monte Carlo" ([1980] 1987), 657, including reference to J. Lach, unpublished (1962); Warnock, "Random-Number Generators," *Los Alamos Science* 15 (1987): 137–41.

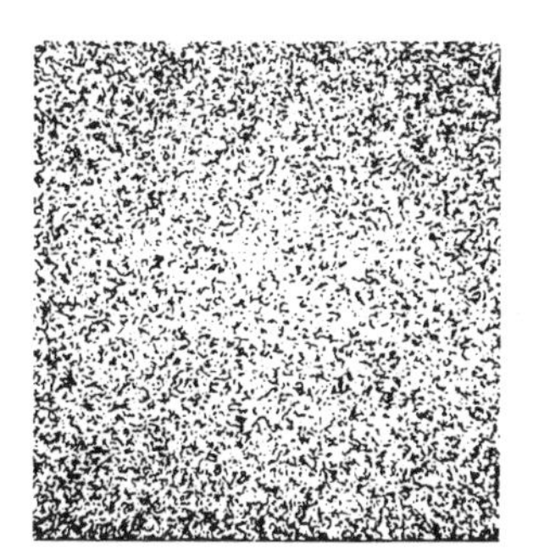
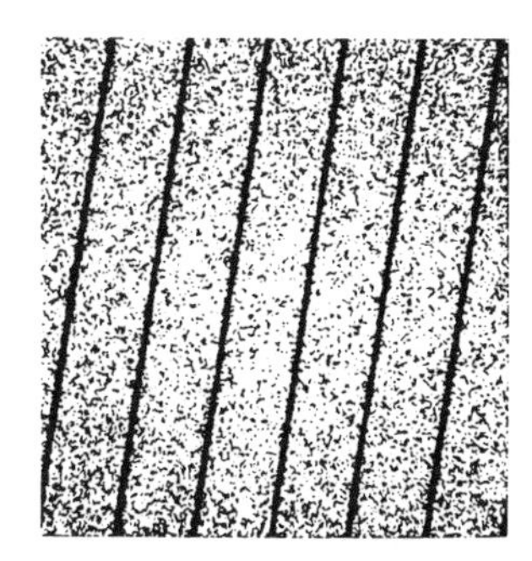

Figure 8.3 Unrandom randomness (1963). The simplest test of a random number generator is to check that numbers are equidistributed in the unit interval. Next, one can check that pairs of successive points are uncorrelated when plotted on the unit square. Two early 1960s generators GAS1 and RANNO passed both tests. But when particular *triplets* of points were considered (triplets such that the first element z was less than 0.1 and the next two points were taken to be x and y), RANNO produced the figure on the right (failing the test), whereas GAS1 on the left seems to pass. Source: J. Lach, Yale Computer Center Memorandum Number 27, 29 October 1963.

would produce misleading results. Worse, admonished Marsaglia, "for the past 20 years such regularity might have produced bad, but unrecognized, results in Monte Carlo studies" using these random generators, which sat in practically every major computer in the world.[31]

With this deeper mathematical grasp of the situation, the maximum number of planes could be increased, and the orientation and spacing of the planes improved for Monte Carlo purposes. (One review article commended the McGill University "Super-duper" that would give a repetition period of 2^{64}.) But it was also plain as day that whatever else this type of random generator was, it was not "random" no matter how many correlation or distribution tests it passed. As one author for *Los Alamos Science* put it: "Calling these criteria 'tests of randomness' is misleading because one is testing a hypothesis known to be false." Failing a test was, of course, a warning that ought be heeded. Nevertheless, "passing all such tests may not . . . be enough to make a generator work for a given problem, but it makes the programmers setting up the generator feel better."[32]

31. Specifically, Marsaglia used Minkowski's famous "geometry of numbers" result that showed that a symmetric, convex set of volume 2^n in n-space must contain a point (other than the origin) with integral coordinates. With Minkowski's theorem, Marsaglia could demonstrate that for any modulus m, used in a multiplicative congruential generator to produce n-tuples of numbers in the unit n-cube, there would be no more than $(n!/m)^{1/n}$ parallel hyperplanes containing all the points. Marsaglia, "Random in Planes," *Proc. Nat. Acad. Sci.* 61 (1968): 25–28.

32. James, "Monte Carlo" ([1980] 1987), 662; the quotation is from Warnock, "Random-Number Generators," *Los Alamos Science* 15 (1987): 137–41, on 140.

Not everyone felt better. S. K. Zaremba, working at the Mathematics Research Center, United States Army, at the University of Wisconsin–Madison (bombed in late August 1970 by radicals), thought that the whole philosophy of test-based, or even physically based, randomness was utterly misguided: "The term 'random numbers' is likely to lead to misunderstandings, because, in general parlance, randomness is a property of a process, and applying it to determinate sequences has been the origin of a mistaken belief that such sequences have some properties which, in fact, can apply only to stochastic processes."[33] As Zaremba saw it, probabilistic justifications of random or pseudorandom sequences in Monte Carlo calculations were simply beside the point. The issue—the only issue according to Zaremba—was how effectively the sampling procedure approximates the integral in question. Specifically (according to the guiding principle of the *quasi Monte Carlo*) our eyes ought to be fixed not on the philosophical status of the sampling points but on minimizing the error, E, between some integral and an approximation based on sampling over the best possible set of points x_k:

$$E = \left| \int f(x)\, dx - N^{-1}\Sigma_k f(x_k) \right| .$$

Down to business: take a set of points, S, in the unit cube, Q^d, of dimension d, with each point x to have coordinates between 0 and 1. Let the function $v(x)$ stand for the total number of points from S located in the rectangular volume from the origin to the point x. Let x_i stand for the ith coordinate of x, and let N be the total number of S points in the cube. The volume from the origin to x is obviously given by the product of the coordinates: in three dimensions, for example, $x_1x_2x_3$ yields the volume of a solid rectangle. In the spirit of a result that dates back to Hermann Weyl in 1916,[34] we now define the *local discrepancy* $g(x)$ as the difference between the number of points one would expect to have in the solid rectangular volume from the origin to x (if the points were distributed randomly in the cube) and the number of points actually found in that volume:

$$g(x) = v(x)/N - x_1x_2x_3 \ldots x_d.$$

Based on the local discrepancy, Zaremba then defined a *global discrepancy* for the cube as a whole. One such measure he dubbed the *extreme discrepancy* measure: the absolute value of the maximum value of $g(x)$ for all x in the set S. In three dimensions:

$$D(S) = \sup \{|g(x)|\} \text{ for } x \text{ in } Q^3.$$

33. Zaremba, "Quasi-Monte Carlo," *Stud. App. Math.* 3 (1969): 1–12, on 2.

34. Weyl, "Gleichverteilung," *Math. Ann.* 77 (1916): 313–53.

Another perfectly good measure of global discrepancy is the mean square discrepancy given by

$$T(S) = \sqrt{\int (g(x))^2 \, dx}.$$

Summarizing a powerful series of mathematical results by Hlawka, Koksma, Smirnov, and Kolmogorov, it was possible to bound the error E in terms of the extreme and mean square discrepancy.[35] For random or pseudorandom numbers the error decreased with the number of points sampled, n, as $1/\sqrt{n}$.

But now, Zaremba reported, the philosophy of random number generation ought to change: the goal should be to minimize the error E by finding sets S with a smaller global discrepancy (i.e., sets that were less clumped) than random sets. In particular, it was possible to show that for a set of points more evenly distributed than a random sequence, the error decreased much faster than it would for random or pseudorandom numbers; perhaps, in the limit, E would decrease with n as $1/n$ instead of $1/\sqrt{n}$. This was an enormous improvement; it meant that such a "quasi Monte Carlo" could get a result with 1,000 points that would take a million points with physical or pseudo–Monte Carlos. Could anyone resist the lure of the low discrepancy? Some could.

Far from leaping from the pseudo to the quasi, many workers in the field shied away. Zaremba lamented:

> It is curious that statisticians are occasionally warned against using sequences with too low discrepancy. For instance [the Soviet applied mathematician] Golenko [dismissed extreme discrepancies less than those given by random sequences as] '. . . most undesirable.' No reason for this warning is given; is there perhaps, behind it, a wish to pretend that the numbers were taken at random even when they are known to be obtained by a determinate process. . . . Instead of clinging to vague concepts of randomness, it might be better to aim at working with sequences making no pretense of random origin, but so devised as to give the best possible guarantee of accuracy in computation.[36]

These best guarantees came from the quasi Monte Carlos. One such sequence of quasi randoms is that generated by the "radical-inverse function" that takes a number n and converts it into a new number $\phi(n,b)$, where b is a base. It works this way:

1. Choose a starting number n_0 (e.g., $n_0 = 14$).
2. Choose a base b (e.g., $b = 3$), and express n in terms of b (here $14 = 112$).

35. See, e.g., Hlawka, "Funktionen," *Ann. Mat. Pura Appl. (4)* 54 (1961): 325–34; and articles by Koksma and others, cited in Zaremba, "Quasi-Monte Carlo," *Stud. App. Math.* 3 (1969): 1–12.

36. Zaremba, "Quasi-Monte Carlo," *Stud. App. Math.* 3 (1969): 1–12, on 6. Golenko's words are quoted from Schreider, *Monte Carlo Method* (1966), 278.

3. Reverse the digits (here 211), and write the result as a fraction of 1, so that $\phi(14,3) = 211/1000 = 0.211$.
   In a similar manner a whole sequence $\phi(1,b)$, $\phi(2,b)$, $\phi(3,b)$, . . . , $\phi(N,b)$ is quasi-randomly given by the input 1, 2, 3, . . . , N and b.[37]

Tension over the meaning of the Monte Carlo is visible in the literature. F. James's widely reprinted article on Monte Carlos for high energy physicists put it this way: "Since we have now dropped all pretense of randomness, the reader may object at this point to retaining the name Monte Carlo. Strictly speaking he is right, but it is probably more justified to enlarge the concept of Monte Carlo to include the use of quasi-random sequences."[38] Quasi Monte Carlos, he argued, continued to be applicable to high-dimensional spaces, to perform more or less equally well independent of dimension, and to be robust with respect to the continuity properties of the function to be integrated. In this sense quasi Monte Carlo was still Monte Carlo. Importantly, too, the theory of the quasi Monte Carlo was far closer to that of the Monte Carlo than to that of ordinary quadrature. For example, the trapezoidal rule divides a one-dimensional integral into n subintervals and estimates the integral in a particular subinterval by the area of a trapezoid inscribed under the curve to be integrated. Think of a curve's Taylor expansion, with a big constant term, smaller first derivative, smaller-still second derivative, and so forth. Then the trapezoid takes care of the constant and first-derivative term, locating the error mainly in the second derivative term. That term is proportional to the interval squared, and of course the interval decreases as $1/n$, for n subdivisions. Therefore, the error decreases as $1/n^2$, which is much faster than the Monte Carlo's $1/\sqrt{n}$. Unfortunately for the trapezoid rule, the number of points needed to sample a higher dimensional integral increases like n^d, which leads to an error in d dimensions of $n^{-2/d}$; the Monte Carlo converges as $1/\sqrt{n}$ no matter what the dimension—and independence of dimension holds good for physical, pseudo, or quasi Monte Carlo. While other quadrature rules did better than the simpleminded trapezoid technique, all demanded regular sampling, all got worse with increasing dimension, and all eventually yielded errors larger than the Monte Carlo error for sufficiently high dimension.[39] Because the various types of Monte Carlo had much in common, and because they differed in this respect from the broad and historically rich class of quadrature techniques, James advocated a meaning of "Monte Carlo" that ignored differences about the ontology of randoms and captured common theory and practice.

In the fixing of meaning, local practices trump global issues of ontology. This is one of our recurring themes from the cloud chamber to the TPC. Here it

37. Warnock, "Random-Number Generator," *Los Alamos Science* 15 (1987): 137–41, on 141.

38. James, "Monte Carlo" ([1980] 1987), 662.

39. James, "Monte Carlo" ([1980] 1987), esp. 646ff. For the Gauss rule, one of the best of the quadrature methods, a 10-point Gauss rule converges more slowly than the Monte Carlo only in 38 dimensions or more.

is again. Dig down in the coding of a Monte Carlo and you find random numbers, perhaps just a line of code, "random x," calling up such a list. But there is, within that superficially simple line of FORTRAN, some consequential philosophy about the very meaning of the Monte Carlo, for "random" meant different things to different groups of workers. To some the term designated a purely physical process such as electronic noise, cosmic ray arrival times, or alpha decay. Others saw "random" as a title, conferred on a pseudorandom series upon its successful passage through a battery of tests designed to probe distributions and correlations. Yet others rejected the tests: In randomness, they contended, what you see is what you get. Better graph and look than trust a set of algorithms that would always remain inadequate. Finally, there were those like Zaremba who set their stock by the post hoc usefulness of a set of numbers as they worked (or failed to work) in a given application, a philosophy that rejoiced in the abandonment of "illusions" of randomness and aimed at justifiable equidistribution (quasi randomness) for a given case that would get the job of approximation done, quickly. We will return to this set of global disagreements about the "true nature" of the random, but for now let it suffice that, however they were imagined, random number generators were furiously pumping digits into a burgeoning simulations technology: bombs, earths, airplanes, neutrinos, gluons, and galaxies all seemed to fall within its grasp.

Effectively, the Monte Carlo therefore hinged on a twofold simulation. First, the computer would simulate the mathematics of randomness, then the Monte Carlo would use these simulated random numbers to prosecute the simulation of physics. If the former already rankled some mathematicians, statisticians, and physical scientists concerned about the nature of the random, the latter precipitated an equally powerful debate about the meaning of experimenter and experiment.

8.3 The Artificial Bomb

Thus far we have seen only the public side of the Monte Carlo. In secret, by the time Metropolis and Ulam's article appeared in 1949, an intense and sophisticated program of Monte Carlo analysis had already been under way for over two years. One of the earliest classified applications of a Monte Carlo to a physical system came on 11 March 1947, when von Neumann laid out his and Ulam's scheme to simulate nuclear fission. The proposal was made in a classified letter to Robert Richtmyer, Richtmyer having succeeded Bethe as head of the theory division at Los Alamos.

At issue was the challenge of representing the enormous complexity of neutrons diffusing through various components of a weapon, a process made even more difficult by the fact that the whole assembly would be exploding. This was, without exaggeration, the most pressing theoretical issue facing Los

Alamos, for the result would answer two crucial questions. What was the critical mass? (Critical mass determined how much material had to be produced and how large the bomb would have to be.) How efficiently would the "active" material be used? (The efficiency would determine how much energy would be released.) One set of methods built on Boltzmann's transport equation, an integrodifferential equation that tracked in detail the temporal evolution of particles in any system, spherically symmetric or not. It was hopeless to solve such an equation for a realistic version of a bomb with its manifold nuclear reactions, so the theorists made do with the simplifying assumptions that all neutrons had the same energy and that all scattering was isotropic. By combining such one-velocity solutions for different velocities, they could piece together a crude model of neutronic history. Eventually, during the summer months of 1944, Feynman and his group were able to work out a method that gave a solution for several velocity groups in terms of a single group, taking into account a more realistic picture of scattering in the tamper. Only then did Bethe and a group of theorists consider the critical mass calculations to be reliable. Efficiencies were even tougher to calculate analytically, as no one had any idea how to combine the diffusion problem with the explosive hydrodynamical equations in a system that had gone critical; those estimates were a combination of guesses, empirical interpolations, and other approximations.[40]

By the end of the war, a variety of refinements improved criticality and efficiency calculations, but these improvements still left many questions unanswered. If the new Monte Carlo could actually follow a sizable set of neutrons as they ricocheted and fissioned their way through the core and tamper, it would be indispensable to the Los Alamos scientists who remained on the scene in the hopes of producing smaller and more powerful nuclear weapons.

Von Neumann and Ulam were therefore addressing the ground zero of nuclear weapons design when they worked out the neutron multiplication method using the Monte Carlo. In his letter to Richtmyer, von Neumann wrote, "[I]n accordance with the principle suggested by Stan Ulam" it would be possible to simulate a critical assembly (presumably either a bomb or a reactor) by using a computer to "follow" a sample of neutrons through their peregrinations. In the simplest case, von Neumann considered a spherical arrangement of matter: active material (abbreviated A) to fission, usually plutonium or enriched uranium, tamper material (abbreviated T) such as beryllium to reflect neutrons back into the fissile center, and a slower-down material (labeled S), typically some form of hydrogen, to remove energy from the fission neutrons and thereby increase their chance of causing fission. Because of its spherical symmetry, the problem could be treated as if it occurred along a single spatial dimension.

40. On diffusion methods, see Hoddeson et al., *Critical Assembly* (1993), 179–84; also Weinberg and Wigner, *Chain Reactors* (1958), chap. 8.

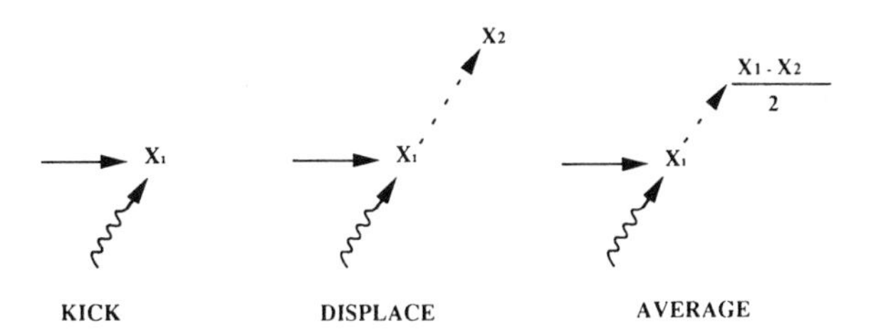

Figure 8.4 Kick, displace, and average. Monte Carlo simulations typically proceeded in steps, iterated over and over. In one of the first such simulations, the neutrons would move according to some predetermined velocity distribution; a particular neutron would then kick a nucleus and both the nucleus and the neutron would recoil for a further unit of time. Taking the average of the nucleus position after the kick and its original position, the history of the neutrons would be repeated. This cycle continued until the hundred neutrons and the matter they were in came to a self-consistent state.

Many other parameters of the problem were fixed and could simply be inserted from prior experiment and theory; these included some, but not all, of the velocity-dependent cross sections for the scattering or capture of neutrons by nuclei and for the emission of new neutrons by fission. Von Neumann assumed, for simplicity, that all neutrons produced were launched isotropically. The statistical character of the Monte Carlo revealed itself in several ways: through the selection of the number of fission-produced neutrons (say two, three, or four), through the velocity of neutrons chosen in a weighted but random way from a preprogrammed velocity distribution, and most important, through the random free path the neutron took between interactions. As formulated, the problem was, von Neumann acknowledged, inert—it did not take account of the hydrodynamics that would move materials around as time advanced and energy was released.[41]

Von Neumann went on to specify the way the simulation of this "inert criticality" could be expanded to one that would take into account material transfer as well. First, 100 neutrons would proceed through a short time interval, Δt, and the energy and momentum they transferred to ambient matter would be calculated. With this "kick" from the neutrons, the matter would be displaced. Assuming that the matter was in the middle position between the displaced position and the original position, one would then recalculate the history of the 100 original neutrons (see figure 8.4). This iteration would then repeat

41. von Neumann to Richtmyer, 11 March 1947, in von Neumann, *Collected Works* (1961), 5:751–62, on 751–52.

"until a 'self-consistent system'" of neutron histories and matter displacement was obtained. The computer would then use this end state as the basis for the next interval of time, Δt. Photons could be treated in the same way, or if the simplification was not plausible because of photon-matter interactions, light could be handled through standard diffusion methods designed for isotropic, blackbody radiation.[42]

Because this was the first worked-out Monte Carlo scheme, and because it captures the crucial features of the thousands of subsequent simulations from the earth sciences to airplane design, it is worth following in some detail. Bracketing off hydrodynamics, for a first attempt von Neumann introduced a spherically symmetric geometry of total radius R, divided into N concentric homogenous shells or zones, where zone number i is defined as those points of distances r from the center such that $r_{i-1} \leq r \leq r_i$ and $0 = r_0 < r_1 < r_2 < \cdots r_{N-1} < r_N = R$. In zone i let there be fractions of A, T, and S of x_i, y_i, and z_i. Now von Neumann defined a series of velocity-dependent functions that gave the cross section per cubic centimeter for neutron absorption (a), neutron scattering (s), and neutron-induced fission (f) in each of the three materials for a neutron of incoming velocity v. A cross section (Σ) gives the likelihood of an interaction and can be imagined as the area facing an incoming neutron within which the given interaction would occur. Here those areas are expressed per cubic centimeter. There are three possible outcomes to an interaction.

1. Absorption in A, T, S:

$$\sum_{aA}(v), \quad \sum_{aT}(v), \quad \sum_{aS}(v). \tag{8.9}$$

If the neutron is absorbed, that is the end of the story. No other neutrons emerge.

2. Scattering in A, T, S:

$$\sum_{sA}(v), \quad \sum_{sT}(v), \quad \sum_{sS}(v). \tag{8.10}$$

If scattering takes place, then a "new" neutron emerges with a velocity v', where the new velocity is given as a product of the old velocity and a function appropriate to the medium ($\phi_A(\rho)$ for the active medium, $\phi_T(\rho)$ for the tamper, and $\phi_S(\rho)$ for the slower-down material). Each function takes on values between 1 and 0, as a function of a random parameter ρ, which also lies between 0 and 1. We therefore have: $v' = \phi_A(\rho)v$, $v' = \phi_T(\rho)v$, or $v' = \phi_S(\rho)v$ depending on the material in which the neutron interacts (A, T, or S).

Fission, which only occurs in the active material A, produces (in this scheme) two, three, or four neutrons (as indicated in eq. 8.11), and von Neumann

42. von Neumann to Richtmyer, 11 March 1947, in von Neumann, *Collected Works* (1961), 5:753.

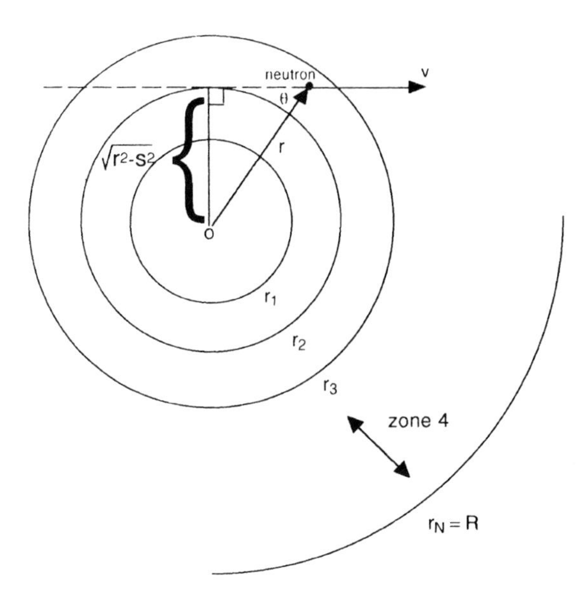

Figure 8.5 von Neumann's simulated neutron. In a first application of the Monte Carlo, von Neumann gave detailed instructions on how to model a fissioning system (presumably an atomic bomb) by a series of concentric spheres, each with its own relative proportion of fissioning, scattering, and slower-down material. As described in the text, the Monte Carlo generated a series of "artificial" neutrons and the computer followed their fate, allowing a pseudorandom number generator to govern how and whether each scattered, fissioned, and was absorbed. Though this first Monte Carlo kept the material stationary, more complex simulations allowed the bomb to explode as the neutrons diffused.

assumed that these all emerge with velocity v_0. These probabilities are given by three expressions.

3. Fission in A, T, and S:

$$\sum_{fA}^{(2)} (v), \quad \sum_{fA}^{(3)} (v), \quad \sum_{fA}^{(4)} (v). \tag{8.11}$$

Now a neutron can be characterized by its position (r), its velocity (v), and the angle it makes with the radius (θ)—see figure 8.5. For convenience, von Neumann reparameterized θ, writing $s = r\cos\theta$, so that $(r^2 - s^2)^{1/2}$ is the point of closest approach between the linear extension of the path and the center of the system. For isotropic scattering, all angles are equally likely, and s lies equidistributed between $-r$ and r. Von Neumann let a random variable σ fix the particular new value of s. Finally, the path of a given neutron's excursion and its velocity provide us with t, the time of flight; and depending on whether the neutron is moving in or out, the new zone is labeled by $i+1$ or $i-1$. Each neutron is then fully described by these five quantities: i, r, s, v, and t.

What are the probabilities of the various fates in store for a neutron? There are seven possibilities: absorption in one of A, T, or S; scattering by A, by T, or by S; fission by A with two, with three, or with four outgoing neutrons. It is simple to list those relative probabilities in terms of the sigmas, with $f1$ to $f6$ defined through the following relations:

1. Absorption in A, T, or S: $f1 = \sum_{aA} (v)x_i + \sum_{aT} (v)y_i + \sum_{aS} (v)z_i$
2. Scattering by A: $f2 - f1 = \sum_{sA} (v)x_i$
3. Scattering by T: $f3 - f2 = \sum_{sT} (v)y_i$
4. Scattering by S: $f4 - f3 = \sum_{sS} (v)z_i$
5. Fission with two outgoing neutrons: $f5 - f4 = \sum_{fA}^{(2)} (v)x_i$
6. Fission with three outgoing neutrons: $f6 - f5 = \sum_{fA}^{(3)} (v)x_i$
7. Fission with four outgoing neutrons: $f - f6 = \sum_{fA}^{(4)} (v)x_i$

The intervals $(0, f1)$, $(f1, f2)$, $(f2, f3)$, $(f3, f4)$, $(f4, f5)$, and $(f6, f)$ are therefore proportional in length to each of the seven relative probabilities: the length $(f4, f5)$ is proportional to the probability of a fission in the tamper with the emission of two neutrons. A number chosen at random between 0 and f can therefore be imagined to be a random dart thrown on the line element $(0, f)$. A hit between $f4$ and $f5$, for example, would be a fission with two outgoing neutrons.

$$0 \leftarrow\!\!\!-\!\!-\, f1 - f2 -\!\!-\, f3 -\!\!-\!\!-\!\!-\!\!-\, f4 - f5 -\!\!-\, f6 -\!\!-\, f.$$

A random (or pseudorandom) "dart" is thrown by choosing a random μ between 0 and 1, putting μf between 0 and f. Our set of random variables thus acquires μ as one more member.

With these assumptions, we need only further specify the stochastic element in these calculations that sets the path length of the neutron. Define $\lambda = 10^{-fd1}$ as the probability that a neutron will travel a distance d^1, through zone i (in which a fraction x_i is active, y_i is tamper, and z_i is slower-down material) without colliding, and where f is given by:

$$f = \left(\sum_{aA} (v) + \sum_{sA} (v) + \sum_{fA}^{(2)} (v) + \sum_{fA}^{(3)} (v) + \sum_{fA}^{(4)} (v) \right) x_i + \left(\sum_{aT} (v) + \sum_{sT} (v) \right) y_i + \left(\sum_{aS} (v) + \sum_{sS} (v) \right) z_i \tag{8.12}$$

(f has units of cm^{-1}). Somehow, by throwing dice, by consulting a table, or by creating a computer-generated algorithm, the probability λ must be chosen from an equidistributed set of points on the unit interval. Each such choice specifies d^1 for one neutron of velocity v making one excursion through the bomb or reactor by the relation $d^1 = -(\log_{10}\lambda)/f$.[43]

The procedure for the ENIAC is then clear. A card records the values of i, s, r, v, and t for a single neutron; the necessary random numbers (such as ρ, σ, μ, or λ) are entered separately to launch the next generation of neutrons. With a hum of the vacuum tubes, and a shuffle of the cards, the ENIAC would then take those values and send the neutron through its first interaction and spit out another card C′ with i', s', r', v', and t' (or perhaps up to four new cards C′, C″, C‴, C⁗ in the event of a four-neutron fission). If a neutron escaped ($i \geq N + 1$) or was absorbed ($v' = 0$), the computer would sort the card out of the pack. All remaining cards would then be run through the computer again, and the cycle repeated. Estimating that 100 neutrons cycled 100 times each, von Neumann predicted a run time per "assembly" of about five hours. In the course of such a calculation, the machine could not only compute the standard goals of bomb designers—such as efficiency and critical mass—but provide unheard-of detail about neutron statistics.

After years of struggling to get even the crudest estimates of criticality and efficiency using complicated approximations to the Boltzmann equation, the new method must have seemed miraculous. Microsecond by microsecond, they could watch the detonation occur. How many neutrons at each instant? Count the cards. Do you want the spatial or velocity distributions at a particular time? The ENIAC could tell that too. It was as if Laplace's dream machine had come into existence. Every particle could be tracked.

Richtmyer responded (2 April 1947) in ways that might be expected given his position at Los Alamos. For the nuclear weapons under discussion, there was no need for slower-down material of the sort characteristically employed in a reactor (or the problematic "hydride gadget," where the light atoms in uranium hydride or plutonium hydride would act to slow the neutrons). Under these

43. von Neumann to Richtmyer, 11 March 1947, in von Neumann, *Collected Works* (1961), 5:753.

circumstances "[m]aterial S could, of course, be omitted for systems of [interest to us]." Similarly, in the bomb, the tamper itself typically would contain fission, especially in the new enhanced weapons then under consideration—there would be a "49 core" (plutonium) and a fissionable "tuballoy tamper" (uranium).[44] Richtmyer added that the Los Alamos designers were not likely to get data distinguishing between the velocity dependence of core fissions producing two, three, or four new neutrons; he suggested using a single function with a random variable fixing the number of emitted neutrons. Finally, the simple formal substitutions,

$$\sum_{fT} (v) \neq 0, \qquad \sum_{a,sS} (v) = 0,$$

along with a few other modifications made von Neumann's model into a more up-to-date bomb design problem. The basic Monte Carlo strategy remained the same, and with the ENIAC still not ready after having been moved, a Los Alamos colleague (Bengt Carlson) had begun running a hand calculation for a simplified case.

Ulam too began to use the method, though his efforts were even more consistently directed toward the superbomb than von Neumann's. In response to a lost letter from Ulam that probably indicated his tentative efforts to execute a Monte Carlo by hand for the super (it is clear, on other grounds, that this is what he was doing), von Neumann sent his enthusiastic endorsement in December 1947: "I am waiting with great expectations to see the details of the manual Monte-Carlo procedure. What you tell me about it is intriguing, and in any case, this is the first large scale application of the statistical method to a deep-lying problem of continuum-physics, so it must be very instructive."[45] To von Neumann, the calculation was at one and the same time a contribution to the super and an exploration of the applicability of discontinuous physics to continuous phenomena of hydrodynamics.

As Los Alamos geared up for a major new simulation of the super, this time with the Monte Carlo, von Neumann wrote Ulam in March 1949: "I did, however, work on the 'S[uper,]' and about a week ago I finished the discussion of the non-M[onte] C[arlo] steps that are involved."[46] This "discussion" meant setting out the logical structure of the calculation: the equations, storage, and logical steps that would be needed. He needed to calculate what he called the "time economy" of the calculation: how it would run on different machines, what ought to be calculated precisely, and through which Monte Carlo. These

44. Richtmyer to von Neumann, 2 April 1947, in von Neumann, *Collected Works* (1961), 5:763–64.
45. von Neumann to Ulam, 17 December 1947, SUP.
46. von Neumann to Ulam, 28 March 1949, JNP.

decisions were not resolved unambiguously, and he reported a certain amount of "demoralization" because "several, mutually interdependent and yet not mutually determining, choices as to procedure are both possible and important." Hardest of all would be the treatment of photons, because they could traverse the greatest number of zones in an interval Δt. Neutral massive particles (neutrons) and charged particles cut across fewer (usually only one or two zones) per Δt and so demanded less computer time.[47]

To make the calculation, von Neumann set up a spacetime division in which Δt was 1/10 shake (shakes are 10^{-8} seconds in bomb builder jargon), and there were 100 such intervals. The hydrodynamics he referred to in his initial fission Monte Carlo was now essential rather than optional. Space (in radius r) was to be divided according to:

$$0 = r_0 < r_1 < \cdots < r_{19} < r_{20} \sim 100, \qquad r_{i+1} - r_i = aq^i \quad (i = 1, \ldots, 19), \qquad (8.13)$$

where the simulated superbomb was described as a series of these 20 concentric zones. If one assumes that the zones grow geometrically in radius, then q must be 1.1 because $r_{20} = 100$, and the radius of the initial zone is given by $\alpha = 1.75$.[48] (I assume the spatial units here continue to be in centimeters.)

Many of the simulation steps did not involve Monte Carlo methods; they built on the sorts of hydrodynamical calculations at which von Neumann had become expert. These deterministic operations used less than 100 multiplying steps (designated 100(*M*), where (*M*) is a multiplying step with all accompanying operations and *M* indicates a multiplying step without this complement of operations) per unit time Δt and per zone Δr. There are 20 zones and 100 time intervals. Estimating a need for about four iterations per step, this gave $20 \times 100 \times 4 = 8{,}000$ repetitions. Multiplying the number of steps by the multiplications per step gives $8{,}000 \times 100(M)$; in other words, the non–Monte Carlo part of a super calculation would take 0.8 mega(*M*). Additional multiplications are needed to check and run the Monte Carlo, but these 0.8 mega(*M*) give a rough guide to a comparison of "hand" computing with various computers.

If the time for a hand multiplication $M = 10$ seconds, then a hand $(M) = 8M = 80$ seconds and mega$(M) = 8 \times 10^7$ "man" seconds $= 2.2 \times 10^4$ "man" hours.[49] I should pause here to point out that von Neumann put "man" in scare

47. Von Neumann added: "I think that I have now come near to a procedure, which is reasonably economic at least as a first try: I will treat the Ph[otons?] by the old 'blackjack' method, which seems very well suited to the peculiarities of this species in this milieu; the other species [presumably neutrons] by the normal MC, and, of course, one of them [presumably charged particles such as helium nuclei] by 'local deposition'" (That is, all their energy was deposited at once; von Neumann to Ulam, 28 March 1949, JNP). I do not know (nor does Metropolis remember) what the "blackjack method" was, though I suspect it was a simple method of generating random collisions.

48. von Neumann to Ulam, 28 March 1949, JNP.

49. von Neumann to Ulam, 28 March 1949, JNP.

Figure 8.6 von Neumann with Oppenheimer at the IAS machine, Princeton (1952). Source: Faculty File, "Oppenheimer, von Neumann, and computer," 1952, PUA. Courtesy of Princeton University Libraries.

quotes, presumably because the people doing these millions of calculations were all women. So von Neumann's "man" = woman. Moreover, von Neumann's wife, Klara von Neumann; Herman Goldstine's wife, Adele Goldstine; Foster Evans's wife, Cedra Evans; Teller's wife, Augusta Teller; John William Mauchly's wife, Kathleen McNulty Mauchly, along with many others were all programming these early computers. (In fact, the word "computer" designated the usually female calculators; only later did the term shift from the woman to the device.) All this is background to von Neumann's peculiar quotational terminology in which a "man" year = 50 "man" weeks = 50×40 "man" hours = 2×10^3 "man" hours. A single mega(M), therefore, would take 11 "man" years to do by hand, and so the deterministic part of the super simulation would demand 8.8 years for a single computing person to reckon.

The estimate of 0.8 mega(M) enabled von Neumann to contrast the way a "super" simulation would be calculated by the ENIAC, by the SSEC IBM computer in New York, and by "our future machine" to be built at the Princeton Institute for Advanced Study (see figure 8.6).

For the ENIAC, the time of multiplication, M, was a mere 5 milliseconds, so $(M) = 40$ milliseconds and mega$(M) = 4 \times 10^4$ seconds = 11 hours. Therefore 0.8 mega$(M) = 8.8$ hours, which when compounded with typical running efficiency could be accomplished in one 16-hour day. However, this estimate was academic, as von Neumann knew. Because the ENIAC lacked sufficient memory (it could store a mere 20 words), the problem simply could not be executed. For the SSEC IBM machine in New York, however, there was enough memory, and it was faster as well: it ran multiplication about 25% more slowly than the ENIAC but made up the speed in other ways. And the future machine could run the same operation fast enough to simulate the deterministic part of the same exploding superbomb in about 15 minutes (the Los Alamos machine, the MANIAC [see figure 8.7], was roughly equivalent to the machine at Princeton's Institute for Advanced Study).[50]

Von Neumann next described the effect of adding in the Monte Carlo (MC) steps which had so far been ignored. If one defines a as the ratio

$$a = \text{(time for MC steps)/(time for non-MC steps)}, \tag{8.14}$$

then the total time would be the non-MC time multiplied by $1 + a$. The Monte Carlo steps are such that a single photon would take less than $60(M)$, while any other particle needs less than $40(M)$. There are 300 photons and 700 particles of other species. This yields $(300 \times 60 + 700 \times 40)(M) = 46{,}000(M)$. While this appears to be much greater than the $100(M)$ required for the non–Monte Carlo step, there are 20 zones, each of which must be examined each time for the non–Monte Carlo steps (the hydrodynamics), whereas the Monte Carlo step (a photon interaction, e.g.) would take place in only one zone. So $a = 23$, and the total amount of time needed for the super simulation would be 24 times the non–Monte Carlo totals.[51] This meant that, were the calculation assigned to a single computer (woman), a given thermonuclear bomb simulation would take her 211.2 years. In a sense, this number was a crucial factor in the historical decision to devote vast resources to replace the human computer with an automatic one.

Working with these highly schematized Monte Carlo simulations, over the course of 1949 Ulam and C. J. Everett managed to put a crude version of the super problem on the computer, but they had to invoke assumptions and simplifications at every stage to accommodate the limited capacity of the ENIAC. Foster Evans wrote von Neumann in September of 1949, reporting on changes to the hydrodynamic expressions, the energy production terms, and the rates of reaction. With the Soviet "Joe 1" explosion of August still reverberating, Evans reported that at the lab "the revelation of a 'nuclear explosion' in Russia" had

50. von Neumann to Ulam, 28 March 1949, JNP.
51. von Neumann to Ulam, 28 March 1949, JNP.

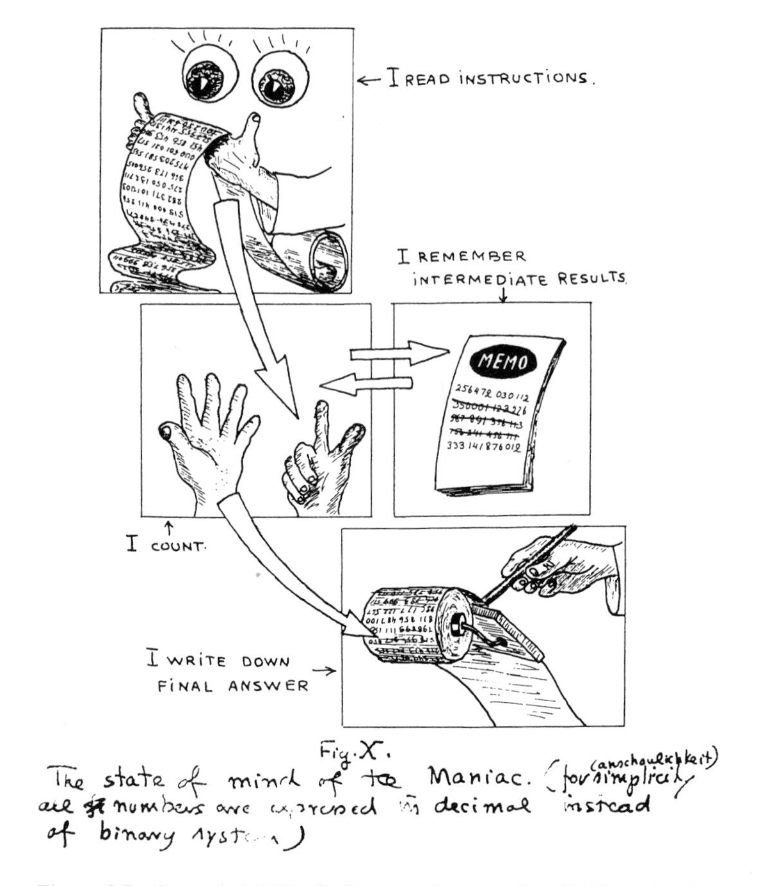

Figure 8.7 Gamow's MANIAC. Gamow, always ready with his cartoonist's pen, sketched this shortly after the MANIAC was installed. Source: Ulam, handwritten diaries, SUP.

provoked "considerable interest." "Perhaps the present problem [i.e., the super] will lead again to an 'unbalance' of power," the previous unbalance presumably being the fission monopoly the United States had enjoyed since 1945.[52]

At stake in these initial calculations (destined for the Aberdeen ENIAC) was the determination of the amount of tritium needed to initiate the superbomb in a spherically symmetric geometry. In particular, the designers wanted to study the reaction $D + T \rightarrow {}^4He + n$(at 14.1 MeV) + 17.6 MeV—the Monte Carlo would track the dynamics of these 14 MeV fusion neutrons and some of the

52. Evans to von Neumann, 23 September 1949, FEP.

charged nuclei.[53] How would the neutrons deposit their energy? When hit by these fast neutrons, with what angle would the deuterons recoil? Would only the first neutron collision be significant, or would subsequent ones need to be included? These and a host of questions like them occupied much of 1950.[54] With the program finally installed, Ulam wrote von Neumann in March 1950: "Everett managed to formalize everything so completely that it can be worked on by a computer. . . . It still has to be based on *guesses* and I begin to feel like the man I know in Poland who posed as a chess champion to earn money, gave nine 'simultaneous' exhibitions in a small town playing 20 opponents, was *losing* all 20 games and had to escape through the window!"[55]

When the ENIAC began computing the super problem, it looked as if the chessmasters would indeed have to defenestrate themselves: even the "5% problem" (with more tritium) was running cold. Here was the problem: When the many low energy photons were taken into account, they cooled the fast-moving electrons. In the "usual" laboratory Compton effect, a high energy photon hits an electron so that the electron gets a boost and the photon loses energy; the opposite was true here. Low energy photons would gain energy in collision with high energy electrons, and the electrons, consequently, would be cooled. As Evans put it to von Neumann, "[I]t is not clear that under these circumstances the 5% mixture would have ignited."[56] Results like these were not at all promising for the classical superbomb. At the least, another, much more complex Monte Carlo would be needed, one that took the photons properly into account. And that calculation was not done until after the bomb was radically redesigned by Ulam and Teller.

In the midst of one of the super simulations (an early one done by hand, it seems) Ulam sent a message to von Neumann in late January 1950: Initial results were apparently not promising for detonation, and Ulam judged that earlier fears that hydrodynamical instabilities would quench the reaction were misplaced: "Hydrodynamics, so far at least, far from being a *danger,* is the *only hope* that the thing will go!" Interwoven with his cautious technical optimism came a similar political forecast: "I think that in matters of 'politics' a victory of unbelievable proportions is preparing." He was right: within four days, President Truman delivered his decision to proceed with an intensified effort on the H-bomb. Whether Ulam had inside information to this effect, I do not know. It certainly seems so.[57] In any case, von Neumann was delighted with the out-

53. Evans to the author, 22 March 1991.

54. von Neumann to Evans, 11 February 1950; Evans to von Neumann, 21 February 1950; von Neumann to Evans, 28 February 1950; Evans to von Neumann, 6 March 1950; Evans to von Neumann, 14 April 1950; von Neumann to Evans, 18 April 1950; Evans to von Neumann, 16 May 1950; all from FEP.

55. Ulam to von Neumann, 17 March 1950, box 22, JNP.

56. Evans to von Neumann, 23 August 1950, FEP.

57. Ulam to von Neumann, 27 January 1960, box 22, JNP.

come: "I need not tell you how I feel about the 'victory.' There are, however, plenty of problems left, and not trivial ones either."[58] Mixing the political and technical inextricably, "hope," "victory," and a "go" all stood on the side of a simulated detonation; "danger" and "problems" lurked in a computational dud.

In public, both Ulam and von Neumann kept low profiles in what had become a raging debate over whether the United States should build the hydrogen bomb. Albert Einstein appeared on television to warn that the weapon was bringing the world closer to total annihilation. Hans Bethe and Victor Weisskopf made public appearances arguing that the level of destruction contemplated with thermonuclear weapons was practically genocidal; in addition, Bethe made no secret of his hope that the bomb would prove scientifically impossible to build. But Harold Urey, the chemist, insisted in the *Bulletin of Atomic Scientists* that, should the Russians get the H-bomb first, they would immediately issue ultimatums as they sought to usher in the "millennium of communism." Only overwhelming political, commercial, ideological, and military power could counter the threat: "I am very unhappy to conclude that the hydrogen bomb should be developed and built. I do not think we should intentionally lose the armaments race; to do this will be to lose our liberties, and, with Patrick Henry, I value my liberties more than I do my life."[59]

Bethe struck back in the same 1950 volume of the *Bulletin:*

> I believe the most important question is the moral one: Can we, who have always insisted on morality and human decency between nations as well as inside our own country, introduce this weapon of total annihilation into the world?. . . . It is argued that it would be better for us to lose our lives than our liberty; and this I personally agree with. But I believe that this is not the question; I believe that we would lose far more than our lives in a war fought with hydrogen bombs, that we would in fact lose all our liberties and human values at the same time, and so thoroughly that we would not recover them for an unforeseeably long time.[60]

Bethe's strongest opponent was Edward Teller who, in his contribution to pages of the *Bulletin* ("Back to the Laboratories"), blasted his fellow scientists for having "been out on a honeymoon with mesons. The holiday is over. Hydrogen bombs will not produce themselves."[61] Domestic responsibility lay in Los Alamos.

Ulam dutifully read these tracts, but from amid the immensely complex calculations that had been proceeding more or less unaffected by the outside world, this contretemps seemed laughably ill informed, even irrelevant. To von Neumann, he joked with great detachment on 24 March 1950:

58. von Neumann to Ulam, 7 February 1950, SUP.
59. Urey, "America," *Bull. Atom. Sci.* 6 (1950): 72–73, on 73.
60. Bethe, "Hydrogen Bomb," *Bull. Atom. Sci.* 6 (1950): 99–104, on 102, 125.
61. Teller, "Laboratories," *Bull. Atom. Sci.* 6 (1950): 71–72, on 72.

> Read with constant amusement a whole series of new articles in the press about hydrogen bomb—the statements by Zacharias, Millikan, Urey, Einstein and Edward in the Bulletin of A[tomic] S[cientists] in Chicago each contribute[s] a share of merriment.
>
> I propose to you hereby to write jointly a "definitive" article on this subject. It will be signed by fictitious names of two foreign born scientists, "key men" in various projects, *not* atomic scientists *but* experts on the hydrogen bomb, former scientists on radar and submarine detection work etc. The first paragraph will say how secrets and lack of free exchange thwarts progress in basic (science) and prevents development of new ideas. The next [paragraph] will document it by pointing out how there are surely no secrets in nature & how *any* scientist can figure out these secrets by himself in 5 minutes. Then next [paragraph] how hydrogen bomb is too big & very immoral, the next one that it is too big to be useful, but not really big enough to be decisive.
>
> The next that it is not clear to anyone whether it can be built at all; after that that Russia probably has it already but let us all pray that it is mathematically impossible.
>
> After that, like Edward's, a pitiful plea for all scien[tists] to work on [it], spending 7/8 of the time on the projects and 3/4 on [other topics].[62]

In these paragraphs Ulam stands aloof from the debate, casting the moral discourse between quotation marks and voiding it of the immediacy it carried for Teller, Bethe, Einstein, and others. He had, in effect, created a simulation of the debate, an artifice that stood for it, but not in the language of feasibility, overkill, and scientific imperatives.

It was not that Ulam thought the bomb was necessarily possible, however. In one of their "hand" studies during 1950, Ulam and Everett showed that the initiation of fusion using deuterium and tritium would demand prohibitive quantities of tritium, a fantastically expensive hydrogen isotope. Then, during the summer of 1950, the two authors used the Monte Carlo method to study the behavior of thermonuclear reactions in the mass of deuterium. In the still-classified report LA-1158, Ulam, Fermi, and Plank invoked the hydrodynamics of the motion of the material, the interaction of radiation energy with matter, and various reactions between nuclei (presumably deuterium-deuterium and deuterium-tritium) that were dependent on temperature, density, and geometry. Though the size of the calculations was small (they were still able to use desk computers and slide rules), their conclusion was that the reaction would *not* propagate in a volume of deuterium. Together, the two pieces of work appeared to sound the death knell for the super: it would neither light nor burn. The slightly later, massive Monte Carlo simulations on electronic computers discussed above (by von Neumann, Foster Evans, and Cedra Evans) confirmed their assessment.[63]

62. Ulam to von Neumann, 24 March 1950, SUP.
63. Ulam, Introduction to "Studies" (1965), 2:978.

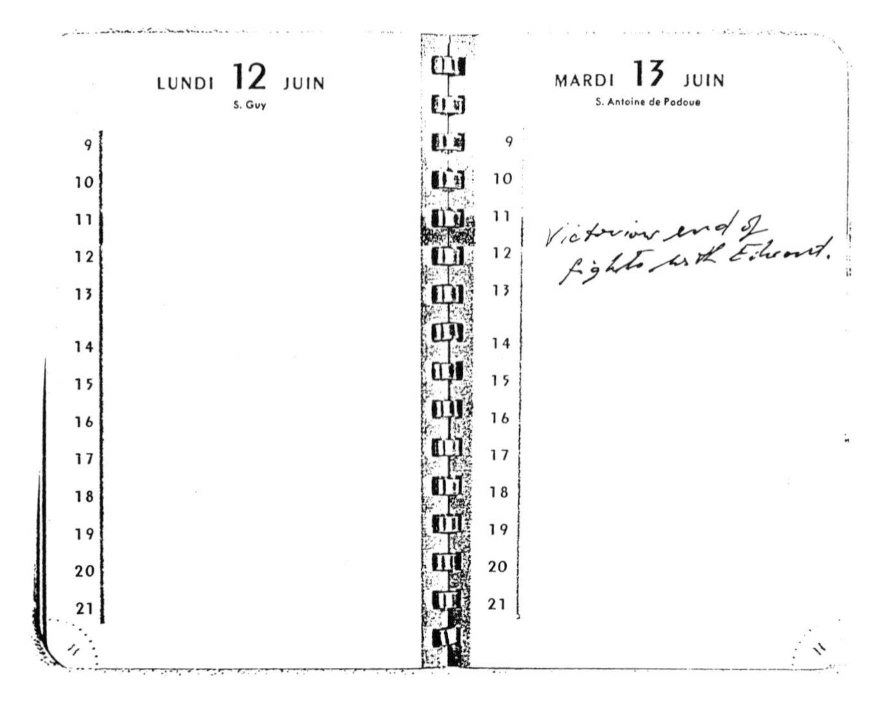

Figure 8.8 Ulam's diary, "Victorious end" (13 June 1950). Source: Ulam, handwritten diaries, SUP.

By all accounts, Teller took Ulam's fizzling bomb news hard. Ulam's pocket diary for 10 May 1950 reads, "Fights with Edward." By 13 June 1950, with more data in hand, Ulam relished vanquishing Teller's greatest project: "Victorious end of fights with Edward" (see figure 8.8). The terse entries carry multiple meanings. On the larger scale, Ulam was preoccupied with a "victory" of national politics, Truman's endorsement of the H-bomb; here, "victory" means triumph in Ulam's personal struggle with Teller over the design of the bomb. Given the extraordinary publicity and resources already devoted to the project, it is not too surprising that Ulam's dim reports ended neither the technical nor the moral battles, which continued throughout the second half of 1950. By January 1951, almost exactly a year after the presidential order, there was little prospect of technical success with the weapon, and tempers were wearing thin. On Thursday 18 January 1951 Ulam recorded one encounter between Bethe (who hoped and argued that the bomb was impossible) and Teller, "Amusing fights: Hans-Edward," or a few days later, "Big fight fairly amusing." Then, sometime between 18 and 25 January 1951, Ulam realized that a radically different configuration of the hydrogen bomb might make it possible. Instead of trying to create

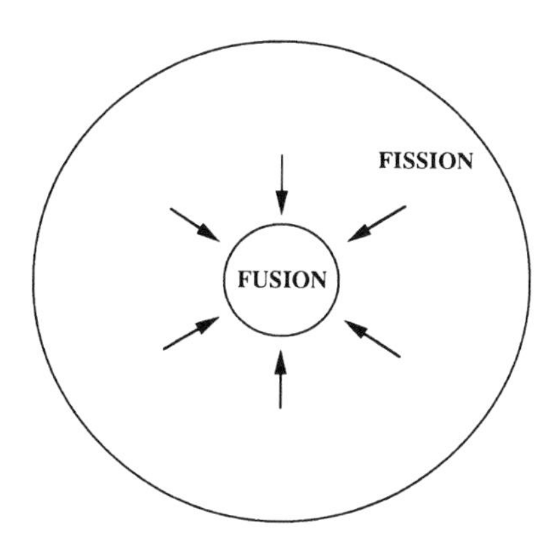

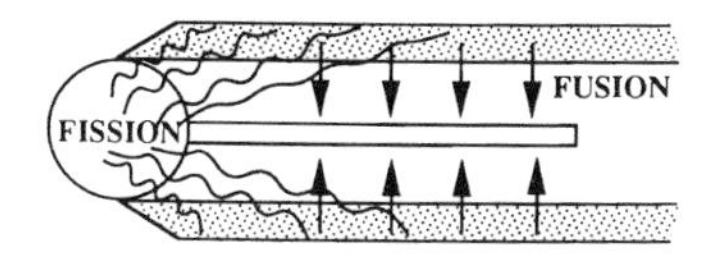

Figure 8.9 Radiation implosion. In the older (pre–Teller-Ulam) fusion experiments (*top*) large amounts of fissioning material were used to cause fusion in a limited, centrally located region. In radiation implosion (*bottom*) an atomic (fission) bomb produces X rays, which ionize material surrounding the body of hydrogen and its isotopes. The expanding plasma then drives in the hydrogen fuel, compressing it to very high pressures, allowing detonation at lower temperatures. At the center of this oversimplified sketch is further fissionable material, which detonates outward. The idea of using an external atomic bomb to drive the thermonuclear weapon was due to the joint efforts of Ulam and Teller.

enormous temperatures with a fission bomb, the A-bomb could be used to compress the fusionable material; high pressure would enable the reaction to proceed with much less heat.[64] On 25 January 1951, the tone of his entries abruptly changed: "Discussion with Edward on '2 bombs.'" Apparently, Ulam had brought the idea of compression with neutrons to Teller, and Teller had added his own idea—that the same effect could be obtained more easily by "radiation implosion," the compression of cylindrically arranged fusion fuel by an expanding plasma created by X rays from the fission bomb (see figure 8.9). (There were therefore "two bombs" in question, both involving compression.)

The fights ended. Ulam recorded on 26 January 1951: "Discussion with Bethe[,] Evanses[, and] Carson [Mark] on the set-up for the cylinder propagation. Write up dis[cussion] with Johnny [von Neumann] & write to Hans [Bethe]." Over the next few weeks Ulam finished his part of the paper and sketched the introduction. Its somewhat unwieldy title appeared the next month on 15 February—"Wrote Lenses. (jointly with Teller) Heterocatalytic deton[ation]' Radiation lenses and hydro[genous] lenses"—and was formally issued on 9 March 1951.[65] In a cartoon that would have been comprehensible only to those cleared to know about this report, Gamow invited Ulam to post his victory (see figure 8.10).

64. Handwritten diaries in SUP.
65. Handwritten diaries in SUP.

Figure 8.10 Gamow's tortoise and hare (ca. 1951). Once again Gamow's wicked pen strikes: here, Teller—working step-by-step as a tortoise—is soundly beaten to the hydrogen bomb by Ulam, who pulls the carrot out of the hat. Source: Ulam, handwritten diaries, SUP.

With this classified Los Alamos report, debate "inside the fence" over the hydrogen bomb virtually ended among physicists privy to the classified breakthrough. In June 1951, faced with new simulations done on the National Bureau of Standards' SEAC and elsewhere, the General Advisory Committee, whose moral opposition to the weapon had stunned the secret community, retracted its earlier statement and endorsed a full-tilt effort to build and test the bomb. Even Bethe, who had long been the strongest public scientific voice against it, ceased to object. On 30 October 1952, the United States detonated a "Teller-Ulam" (heterocatalytic) hydrogen bomb on the South Pacific island of Eniwetok. The force of the 10.2 megaton explosion, 500 times that of the Hiroshima weapon, wiped the island from the face of the earth.

Over the next few years, thermonuclear bombs proliferated. The United States tested a deliverable H-bomb in 1954, and the Soviet Union followed suit shortly thereafter. Von Neumann saw these developments as inevitable and constructive, if sometimes fraught with instability. In a 1955 article, "Can We Survive Technology?" he sketched a future in which reactors would "grow up" to use not fissionable heavy elements, but fusionable light ones ("fission is not nature's normal way of releasing nuclear energy").[66] Directly following this discussion is a section on controlled climate in which he prophesied a time when deliberately lofted stratospheric dust would alter the world's weather system just as Krakatoa had in the previous century. Given the continuing rhetoric linking nuclear weaponry to Krakatoa and the placement of this discussion directly after his explanation that fusion is natural, one can only read von Neumann as implicitly suggesting the possibility of altering climate by hydrogen weapons. "What could be done, of course, is no index to what should be done; to make a new ice age in order to annoy others, or a new tropical, "interglacial" age in order to please everybody, is not necessarily a rational program." Yet von Neumann harbored no doubt that the analyses necessary to predict such consequences would be possible, presumably building on the twin simulations of hydrogen weapons and meteorological analyses he had been conducting during the past few years (see figure 8.11).

Having watched the evolution of the fission and fusion bombs, the exploitation of computers, and all the other developments in technological warfare that had come helter-skelter out of World War II and its aftermath, von Neumann leaned hard on a technological determinism: "Present awful possibilities of nuclear warfare may give way to others even more awful. After global climate control becomes possible, perhaps all our present involvements will seem simple. We should not deceive ourselves: once such possibilities become actual, they will be exploited."[67] If some unspecified political impediment to technological warfare did not intervene, von Neumann saw only a long series of ever more severe conflagrations.

Thermonuclear and enhanced fission weaponry had crystallized a new mode of doing scientific work. During 1948 and 1949 statisticians, mathematicians, electrical engineers, and a restricted few others joined physicists in the creation of this new trading zone in which Monte Carlos, games, pseudorandom numbers, variance reduction, and flowcharts were common parlance. Now the style of work grew outward and through a series of conferences passed beyond the domain of the secret and into the wider community of researchers. As it crossed that fence, the debate spread about who might count as an experimenter or as a theorist, and what as an experiment or theory.

8.4 Lifestyle of the Tertium Quid

Already by 1948, frustration was building outside the restricted community about the hidden world of Monte Carlo. The mathematician George Forsythe,

66. von Neumann, "Can We Survive," *Fortune* 51 (1955): 106–8, 151–52, repr. in von Neumann, *Collected Works* (1963), 6:508.

67. von Neumann, "Can We Survive," *Fortune* 51 (1955): 106–8, 151–52, repr. in von Neumann, *Collected Works* (1963), 6:519.

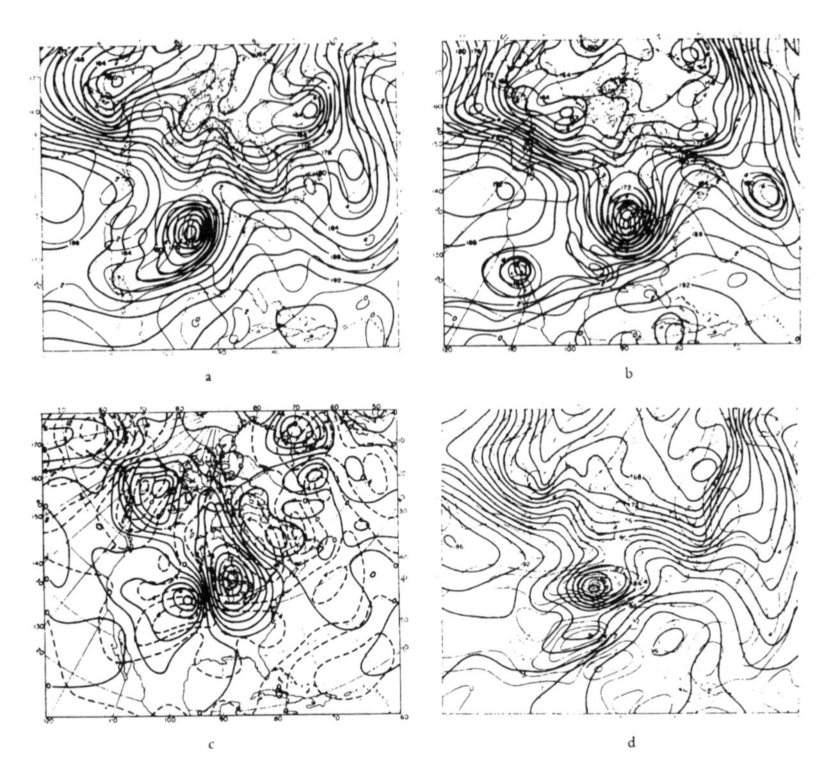

Figure 8.11 Numerical weather prediction (1949). Along with the Monte Carlo, other simulations became ever more popular. Here, in work by von Neumann and colleagues, is a simulated rendition of the weather, headed "Forecast of January 5, 1949, 0300 GMT." Source: Charney, Fjörtoft, and von Neumann, "Numerical Method" (1963), fig. 2 on 6:422, copyright 1963, with permission from Pergamon Press Ltd., Headington Hill Hall, Oxford OX3 0BW, UK.

who later helped found computer science as a discipline, scribbled a note indicating that he knew that von Neumann and Richtmyer had conducted a neutron propagation study with the Monte Carlo but that one had to have a Q-clearance (nuclear weapons) to see it.[68]

At the time, Forsythe was at the Institute for Numerical Analysis, the Los Angeles unit of the National Applied Mathematics Laboratories of the National Bureau of Standards. In the fall of 1948, he along with others in the Los Angeles unit created a Monte Carlo Project to investigate the new technique but rapidly discovered that progress was impeded by the inaccessibility of the clas-

68. Forsythe, notes entitled "Learned on Trip," SC 98, box 8, folder 33, GFP.

sified material. Forsythe's frustration was shared by his colleagues elsewhere. Blocked by the AEC's embargo on applications, the only information leaking out was highly abstract, and as the head of the National Applied Mathematics Laboratories, John H. Curtiss, put it, "a fruitful development of the theory of a computation technique must be oriented in the practice." In December 1948, the mathematicians Cuthbert Hurd and Alston Householder (both from Oak Ridge) joined forces with two statisticians from the RAND Corporation (Hallett H. Germond and Theodore E. Harris) to arrange the first open, unclassified meeting on Monte Carlos, to be scheduled at the end of June 1949 in Los Angeles. The organizing committee was augmented by John H. Curtiss and Raymond P. Peterson, also of the National Bureau of Standards.[69] From the start, the intention was to mix physicists (including the experimenter Robert R. Wilson and the theorists Herman Kahn and Wendell C. De Marcus) with chemists (most prominently Gilbert King), mathematicians (especially Ulam and von Neumann), and statisticians (including Harris and Curtiss).[70]

Shortly after the Los Angeles meeting, Hurd (now at IBM) began to convene another set of meetings, this one to take place in Endicott, New York, in November and December 1949, as part of the IBM Seminar on Scientific Computation. They too would "center around a discussion of problems which are of mutual interest to chemists and physicists; problems such as atomic structure, molecular structure, equilibrium calculations, reaction rates, resonance energy calculations, shielding calculations, Monte Carlo calculations and the fitting of decay curves."[71] In this regard, the Monte Carlo played as important an integrative function as the computer itself; the Monte Carlo became a lingua franca, whether the participants were discussing Kolmogorov's results in mathematical probability theory, the engineering details of the IBM 402 Card-Programmed Electronic Calculator, or the physical significance of the eigenvalues of the Schrödinger equation. Enthusiasm generated for the Monte Carlo and its applications continued to grow, culminating in a third major Monte Carlo meeting on 16 and 17 March 1954 at the University of Florida at Gainesville.

All of these conferences assembled groups of participants that were heterogeneous not only in their specializations (mathematics, statistics, chemistry, physics, engineering, design specialists, biometrology) but also in the institutions from which they came. At the IBM meetings, for example, scientists and engineers gathered from the Bureau of Mines, Northrop Aircraft, Westinghouse, Corning Glass Works, Consolidated Vultee Aircraft Corporation, Shell Oil, IBM, General Electric, Fairchild Aircraft, Watson Scientific Computing Laboratory, Bell Telephone Laboratories, General Aniline and Film Corporation, Aberdeen

69. Curtiss, "Preface" (1951), iii.

70. Hurd to Ulam, 26 April 1949, SUP.

71. Hurd to Ulam, 22 September 1949, SUP.

Proving Grounds, Underwater Ordnance Department, Air Weather Service, and many others. For this limited purpose, the head of the Flutter and Vibrations group at Fairchild Aircraft had to and could talk to an engineer with the Phillips Petroleum Corporation and a physicist from the RAND Corporation. Their language was that of the Monte Carlo simulation.

But what *was* this Monte Carlo? How did it fit into the universally recognized division between experiment and theory—a taxonomic separation as obvious to the product designer at Dow Chemical as it was to the mathematician at Cornell? As we saw in chapter 4, World War II had greatly enhanced the status of theorists and theoretical groups in many American universities, and in the new plans for regional and national laboratories. An advertisement recruiting for a faculty job would clearly specify one or the other; students would declare themselves to be theorists or experimenters, not both. Soon degree requirements would be different for theoretical and experimental students. Meetings began to partition along the theoretical-experimental divide; indeed, by the mid-1950s, Fermi's youthful adhesion to both camps had become the stuff of legend, not a model for everyday life. How, then, did physicists understand this new set of simulation practices that was not quite one or the other?

From the earliest efforts of Ulam, Metropolis, and von Neumann, some Monte Carlo workers saw simulations as borrowing essential features from the character of theoretical physics, while for others simulations were self-evidently a part of experimentation. Still others took them to be a hybrid, which I suggest we think of as a tertium quid between the traditional epistemic poles of bench and blackboard. Even 15 years later, the anxiety behind this puzzle in classification persisted. "Computational physics," the physicist Keith V. Roberts wrote, "combines some of the features of both theory and experiment. Like theoretical physics it is *position-free* and *scale-free,* and it can survey phenomena in phase-space just as easily as real space. It is *symbolic* in the sense that a program, like an algebraic formula, can handle any number of actual calculations, but each individual calculation is more nearly analogous to a single experiment or observation and provides only numerical or graphical results."[72] But if the frankly symbolic character of labor resembled techniques of the blackboard, it was still true that the error analysis and problem shooting held more in common with techniques of the bench. Roberts continued: "Diagnostic measurements are relatively easy compared to their counterparts in experiments. This enables one to obtain many-particle correlations, for example, which can be checked against theory. On the other hand, there must be a constant search for 'computational errors' introduced by finite mesh sizes, finite time steps, etc., and it is preferable to think of a large scale calculation as a *numerical experiment,* with the program as the

72. Roberts, "Computers and Physics" (1972), 7.

apparatus, and to employ all the methodology which has previously been established for real experiments (notebooks, control experiments, error estimates and so on)."[73]

In short, the daily practice of error tracking bound the Monte Carlo practitioner to the experimenter. A physicist using a spark chamber would know that there were limits to the spatial resolution of a spark, and this would filter down to an eventual uncertainty in the momentum or energy of a particle. Computational errors were of uncertainties of this sort; they had their origin in the resolution of the device, but the "device" was now a piece of software.

The simulator was joined to the experimenter by other activities as well. In experimental practice it is routine to use the stability of an experimental result as a sign of its robustness: does it vary as one repeats it or shift depending on parameters that ought (on prior grounds) to be irrelevant to the outcome? From the inception of the Monte Carlo, its practitioners were equally aware that their nightmare would be the production of results without constancy. In Ulam's 1949 paper with Metropolis, the authors made this clear: a "procedure is repeated as many times as required for the duration of the real process or else, in problems where we believe a stationary distribution exists, until our 'experimental' distributions do not show significant changes from one step to the next."[74] He should know about stability: many of the early weapons calculations had been anything but that. Problems of locality, replicability, stability, and error tracking constitute some of the reasons (I will come back to other, deeper ones) that simulators came to identify their work as fundamentally experimental.

At the same time, however, other practice clusters tied the simulator to the mathematician and theoretical physicist. In 1964, J. M. Hammersley and D. C. Hanscomb asserted that the linking of mathematics to the computer demanded a fundamentally new classification of mathematics, one that cut across the old typology of "pure" and "applied" mathematics:

> A relatively recent dichotomy contrasts the theoretical mathematician with the experimental mathematician. These designations are like those commonly used for theoretical and experimental physicists, say; they are independent of whether the objectives are pure or applied, and they do not presuppose that the theoretician sits in a bare room before a blank sheet of paper while the experimentalist fiddles with expensive apparatus in a laboratory. Although certain complicated mathematical experiments demand electronic computers, others call for no more than paper and pencil. The essential difference is that theoreticians deduce conclusions from postulates, whereas experimentalists infer conclusions from observations. It is the difference between deduction and induction.[75]

73. Roberts, "Computers and Physics" (1972), 7.

74. Metropolis and Ulam, "Monte Carlo Method," *Amer. Stat. Assoc.* 44 (1949): 335–41, on 339.

75. Hammersley and Handscomb, *Monte Carlo Methods* (1964), 1.

While this remark may be naive in its split of experiment and theory into induction and deduction, it is important as an indicator of the powerful identification of the simulator with the experimenter. This language of "theoretical experiments," or "mathematical experiments," saturates the literature. In the pages of *Physical Review* from the early to mid-1950s one finds tens of such examples. Of the albedo of 1 MeV photons, two authors write: "It occurred to us that a more reliable estimate of this quantity, on which there really existed no information, could be obtained by a 'theoretical experiment' using the Monte Carlo technique."[76]

Caught between a machine life and a symbol life, computer programmers in physics risked becoming pariahs in the sense that many university physics departments found them neither fish nor fowl: they could successfully apply neither for theory positions nor for experimental jobs. At the same time the simulators were also irreplaceable intermediaries between experiment and theory, principally at the major national and international laboratories such as Brookhaven, CERN, Fermilab, and SLAC. In this dual role as marginal and necessary, they served a precarious transactional function known to border peoples on every continent. Over the course of the 1960s, as we saw in chapter 5, the computer transformed particle physics; among the changes it brought was the creation of a category of action (data analysis) that was as all-embracing a career as accelerator building or field theory. The novelty of this situation was not lost on the physics community, as Lew Kowarski made clear in the summer of 1971: "As scientists get used not only to writing their own programs, but also to sitting on-line to an operating computer, as the new kind of nuclear scientist develops—neither a theoretician, nor a data-taker, but a data-processor specialized in using computers—they become too impatient to sit and wait while their job is being attended to by computer managers and operators."[77] Quoting approvingly from a colleague at New York University, Kowarski presented mathematics as the analogue of mining diamonds—finding extraordinary theorems among the dross of uninteresting observations. Computers, on the other hand, sought truth the way one mines coal, with the massive, everyday labor that methodically moves earth from pits to furnaces. "This analogy," Kowarski observed, "illustrates the difference between the spirit of mathematics and that of computer science and helps us to realize that being a computational physicist, or a computational nuclear chemist, or what not, is not at all the same thing as being a mathematical physicist and so on, so that, in fact, a new way of life in nuclear science has been opened."[78]

For Kowarski, the *Lebensform* of the computer console was one more akin to that of coal mining, oceanography, selenology, and archaeology than it was to

76. Hayward and Hubbell, "Albedo," *Phys. Rev.* 93 (1954): 955–56, on 955.

77. Kowarski, "Impact of Computers" (1972), 35.

78. Kowarski, "Impact of Computers" (1972), 29.

physics as it had been previously understood. "There will be a lot of attempts to judge such new situations by old value criteria. What is a physicist? What is an experimenter? Is simulation an experiment? Is the man who accumulates printouts of solved equations a mathematical physicist? And the ultimate worry: are we not going to use computers as a substitute for thinking?"[79] The anxiety over identity was both social and cognitive. The fear that "thinking" might be destroyed must be glossed as a fear that the pleasure (and status) of controlling the activities previously associated with experimenting would be lost.

Once again, anxiety rises as the bounds of the experimenter's self-definition blur. We saw this tension earlier as the experimental physicist had to work out a mode of demonstration (and authorship) with the scanner, the chemist, the engineer, the big group, and to a certain degree the theorist. Now it is the last refuge, the materiality of the laboratory that appears in jeopardy. At issue is the sense of "demonstration" used by the simulator—is it theoretical, experimental, or something entirely new? Just asking the question throws the category "experiment" into relief. Recall the shifting criteria for experimental authorship that we tracked in chapter 7, as experimenters debated whether simulation writing would count.

Were simulations "like" experiment or "like" theory? Take replicability. For theoretical work, the recreation of argumentation is generally considered to be unproblematic. Derivations are relatively easy to repeat, if not to believe. Experimental efforts, by contrast, present notorious difficulties, as many historians and sociologists of science have effectively illustrated: air pumps, prisms, lasers, all were initially immensely difficult to transfer to new locations and new contexts.[80] From the start, simulations presented a hybrid problem. The work was unattached to physical objects and so appeared to be as transportable as Einstein's derivation of the A and B coefficients for quantum emission and absorption. But in practice, this was not the case. K. V. Roberts, among many others, bemoaned this fact and with considerable effort began to delocalize simulations, and computer analysis programs more generally. Opening the "bottleneck" (as it was commonly called) required three concurrent efforts. First, programs had to be openly published.[81] While desirable, this proved in the fifties (and ever since, I should add) a wistful hope since most interesting programs were far too large to allow distribution on the printed page. Second, Roberts pressed for what he called "portability": the use of universal languages and the physical distribution of data tapes.[82] While the former encountered difficulties associated with variations between machine types and local programming customs, the latter ran into

79. Kowarski, "Impact of Computers" (1972), 36.

80. On replicability, see Collins, *Changing Order* (1985); Shapin and Schaffer, *Leviathan* (1985); and Schaffer, "Glass Works" (1989).

81. Roberts, "Computers and Physics" (1972), 17.

82. Roberts, "Computers and Physics" (1972), 17.

a host of property right difficulties. For example, in the 1980s questions arose as to whether the distribution of data tapes could be considered analogous to the distribution of cell samples in biology. Finally, Roberts insisted that "modularity" ought to be a goal for programmers, analogous to the routine elements of practice in theoretical physics such as Laplace's equation, group theory, vector algebra, or the tensor calculus.[83]

Each of these programmatic responses—advocacy of publication, portability, and modularity—was partial; none could truly universalize a set of practices that bore the deep stamp of its localized creation. Lamenting this state of affairs, Roberts commented: "There are many good programs that can only be used in one or two major laboratories (notably the Los Alamos hydrodynamics codes), and others which have gone out of use because their originators moved on to other work."[84] In more recent times this phenomenon has become known as "program rot"—people and machines move on, leaving older programs dysfunctional, often irretrievably so.

At the same time that physicists struggled to uproot the world of simulated realities from a particular place, others applauded the deracination already achieved. Kowarski, for example, began to speak about a simultaneous "liberation in space" and "liberation in time" afforded by the new modality of computer-aided research. "Perhaps, when links as comprehensive as those used in television become available at long distance, there will be even less reason for the user to spend a lot of his time on the site where his physical events are being produced. This may even abolish the kind of snobbery which decrees today that only those may be considered as physicists who are bodily present at the kill, that is at the place and time when the particle is actually coming out of the accelerator and hitting the detector."[85]

Kowarski had in mind events originally encoded from particle collisions at a centralized laboratory, usually by multiinstitutional bubble chamber teams. But just as events produced at Brookhaven or Berkeley could be issued as a stack of cards or a spool of tape, so too could events created by simulations. In both cases, analysis, still the central stage of an experiment, was removed from any single place.

Liberation in time similarly made the experiment fit more easily into the life of the experimenter. Kowarski extolled the fact that the computer expanded the timescale of events—often 10^{-9} seconds—to the scale of the minutes, hours, and days in which we live. Second, with the computer's help, ordinary, unidirectional time became repeatable time as physicists reprocessed the same set of

83. Roberts, "Computers and Physics" (1972), 19.

84. Roberts, "Computers and Physics" (1972), 20.

85. Kowarski, "Impact of Computers" (1972), 35.

events in ever different ways to reveal different patterns of order. Finally, the computer allowed the physicists to break the ties between the accelerator "beamtime" and their own time, time to live the lives of university-based scholars with teaching, departmental, and familial duties. It therefore linked, in yet another way, the different worlds of beamtimes and lifetimes (to borrow the evocative title of Sharon Traweek's anthropological study of SLAC).[86]

8.5 Expanding "Experiment"

8.5.1 Data, Error, and Analysis

As Monte Carlo simulations developed, it became clear that their practitioners shared a great deal with experimenters—I gave the examples of a shared concern with error tracking, locality, replicability, and stability. But the self-representation of Monte Carlo users as experimenters is so pervasive that I now want to zero in on this notion in two different ways in an effort to uncover the practices underlying this talk of "experiment" done on keyboards rather than lab benches.

The first point is that the simulator spends time processing as well as generating "data"—the scare quotes denoting that the term has now been expanded to include that generated from pseudorandom numbers in Monte Carlo simulations. These practices, as the Monte Carlo folks immediately recognized, held more in common with experimental than with theoretical activity. As much as anyone, Herman Kahn of the RAND Corporation (later famous for his 1960 *On Thermonuclear War*)[87] continually emphasized that the simulator had no business simply reporting a probability, say the probability that a neutron will penetrate the concrete shielding wall of a nuclear reactor. Instead, the only meaningful statement would be a probability p along with a certainty (error estimate) m. "This situation is clearly not unknown to the experimental physicist, as the results of measurement are in this form." If the problem is surprising, Kahn insisted, it is because the need to reduce variance is not a situation theorists typically encounter. Variance reduction arises in the context of Monte Carlos because "one is not carrying out a mathematical computation in the usual (analytic) sense, but is carrying out a mathematical experiment with the aid of tables of random digits."[88]

With a limited sample of "particles" that the computer can track, it is frequently the case that the interesting phenomena occur so rarely that uncertainty

86. Kowarski, "Impact of Computers" (1972), 36. (Sharon Traweek has put this nicely in her *Beamtimes* [1988], 158, where she writes that detectors are where "two kinds of cosmological time converge," one of which is the time of particle interactions and the other "experiential" time.)

87. Kahn, *Thermonuclear War* (1960).

88. Goertzel and Kahn, "Shield Computation," ORNL 429, 19 December 1949, 10.

runs riot. If, for example, a thousand neutrons are sent out from a simulated reactor and only 10 penetrate the barrier, then the accuracy of statements about these 10 particles will be slight. Kahn and others (following on early work by Ulam and von Neumann) particularly pressed three strategies for reducing this uncertainty: splitting, statistical estimation, and importance sampling. He argued in 1954 that these three methods had particular applicability to the Monte Carlo method because they were well suited to a situation in which (unlike the physical experimenter) the experimenter had "complete control" of the objects under inquiry: "If for example [the Monte Carlo operator] wanted a green-eyed pig with curly hair and six toes and if this event had a nonzero probability, then the Monte Carlo experimenter, unlike the agriculturist, could immediately produce the animal."[89] Beyond the ability to make green-eyed pigs, there were other, more traditional methods drawn from the established theory of statistical sampling—these included correlation and regression, systematic sampling, and stratified or quota sampling.[90]

Here is a simple, worked example from Kahn. Suppose one wants to know the probability of tossing a total of 3 on two fair dice. Analytically, the solution is easily obtained: Each face of a die has a probability of 1/6; each permutation of two dice has a probability of 1/36; and since there are two ways to get a total of 3 (1,2 and 2,1) the total probability, p, of a toss of 3 is just twice 1/36 or $p = 1/18$. The Monte Carlo works by "experimenting" N times and taking the ratio of the number of successes, n (here a success is a 3), to the total number of throws, N:

$$\hat{p} = \frac{n}{N}. \tag{8.15}$$

After N trials, $\hat{p}$ will usually differ from p, and the statistical error can be measured by σ, the standard deviation:

$$\sigma = \sqrt{\frac{p(1-p)}{N}}, \tag{8.16}$$

which can be expressed as a percentage error by:

$$\frac{100\sigma}{p} = 100\sqrt{\frac{1-p}{Np}}. \tag{8.17}$$

89. Kahn, "Sampling Techniques" (1956), 147.

90. During the early days of Monte Carlo work, variance reduction techniques were explicated in many different ways. This particular organization of the techniques into three novel and three traditional modes follows Kahn, "Sampling Techniques" (1956), 146–90.

Evidently, the variance can be reduced by increasing N, but this works slowly since the percentage error will decrease only as the square root of N. All of the variance techniques were designed to squelch the variance by other means.[91]

"Importance sampling" is one of the most crucial of such methods. If we could weight the dice so that a 1 or a 2 would come up twice as often as usual, then the probability p of getting a 3 would be quadrupled: $p = 2/9$. From equation (8.17), it is plain that the percentage of error is consequently reduced by a factor just over two. Naturally, the Monte Carlo estimate for $\hat{p}$ is not simply n/N but must now be corrected to account for the lead shot we sequestered in the dice: $\hat{p} = (1/4)n/N$, where the 1/4, known as the weighting factor, compensates for the biased sampling. In general, the idea of importance sampling is to augment the examination of a particular region of phenomena of interest, get a reduced variance, and then compensate for this bias in the final estimate of the probability. For example, neutrons might be given an extra large probability of scattering toward the outside of a reactor shield (where the people are) in an effort to reduce the error associated with the estimate of neutrons penetrating the concrete barrier.[92]

A second technique is known as "splitting and Russian Roulette." The idea is this: In a typical Monte Carlo, many sequences of events can be identified as uninteresting based on early stages of their evolution. In our dice case, a 3, 4, 5, or 6 on the first die makes irrelevant whatever happens with the second throw. We therefore eliminate 2/3 of throws of the second die and need to make 1/3 fewer tosses than we would have done naively. More generally, it is possible to decrease the variance in two ways. First, split the cases in which the first stage is interesting. For example, if a neutron made it most of the way out of a reactor, it might be worthwhile to spend a large amount of computer time studying possible paths it might take from there on. Conversely, if the neutron has burrowed far away from the exterior, it may not be worthwhile to spend any computer time studying its subsequent evolution. Therefore, in some percentage of cases, it is economical to end this neutron's simulation. As Kahn puts it (note metaphor): "The 'killing off' is done by a supplementary game of chance. If the supplementary game is lost the sample is killed; if it is won the sample is counted with an extra weight to make up for the fact that some other samples have been killed. The game has a certain similarity to the Russian game of chance played with revolvers and foreheads—whence the name."[93]

Less brutal in its imagery, the final variance reduction technique, the "method of expected values," works by combining Monte Carlo and analytic

91. Kahn, "Sampling Techniques" (1956), 148–49.

92. Kahn, "Sampling Techniques" (1956), 149–50. Richtmyer also did fundamental work on variance reduction.

93. Kahn, "Sampling Techniques" (1956), 151.

techniques. In our example, after the first die toss, it is a simple matter to calculate the probabilities for the remaining die. After a 1, the probability is 1/6 of getting the requisite 2; after a 2, the probability is 1/6 of getting the requisite 1; and after anything but a 1 or 2, the probability of getting a winning number is zero. The average of these expectations is an estimate for *p*: 1/3 divided by 6 to get 1/18 in this case. As in splitting, this both reduces the variance and cuts the number of throws needed.[94]

All of these methods are characterized by an effort, not to get a different estimate of the central value of the "answer," but instead to reduce the scatter around that estimate. In this respect they resemble the strategies of experimenters as they strive to reduce random error in their conclusions. And, as in the experimenter's case, it is here that the simulator must insert prior physical knowledge to narrow the gaze of inquiry.

I conclude from these variance reduction techniques, and from the earlier discussion of error tracking, that there are two sides to the simulators' concern with error. The first (error tracking) bears on the ability of the Monte Carlo to get the "correct" expectation values; it is experienced by the simulators as akin to the experimenters' search for accuracy by quashing systematic errors in the apparatus itself. The second (variance reduction techniques) is the direct analogue of the experimenters' attempts to increase the precision of their examinations. Taken together, these day-to-day commonalities between the practices of the simulator and those of the bench experimenter tended to press the two groups together in their self-identification. Some simulators went further, arguing that, because they could control the precise conditions of their runs, they in fact had an edge on experimenters; in other words, it was the simulator not the experimenter who should be seen as the central figure in balancing theory. "The Monte Carlo methods," Kahn concluded, "are more useful . . . than experiments, since there exists the certainty that the comparison of Monte Carlo and analytic results are based on the same physical data and assumptions."[95] Simulators, not bench workers, could make the green-eyed, six-toed curly-haired pigs that science now demanded.

8.5.2 Stochasticism and Its Enemies

All of the forms of assimilation of Monte Carlos to experimentation that I have presented so far (stability, error tracking, variance reduction, replicability, and so on) have been fundamentally epistemic. That is, they are all means by which the researchers can argue toward the validity and robustness of their conclusions. Now I want to turn in a different direction, toward what amounted to a *metaphysical* case for the validity of Monte Carlos as a form of natural philo-

94. Kahn, "Sampling Techniques" (1956), 151–52.

95. Goertzel and Kahn, "Shield Computation," ORNL 429, 19 December 1949, 8.

sophical inquiry. The argument, as it was presented by a variety of people (including occasionally Ulam), was based on a purportedly fundamental affinity between the Monte Carlo and the statistical underpinnings of the world itself. In other words, because both Monte Carlo and nature were stochastic, the method could offer a glimpse of reality previously hidden to the analytically minded mathematician. As Ulam once put it, his and von Neumann's hunt for the Monte Carlo had been a quest for a *homomorphic image* of a physical problem in which the particles would be represented by fictitious "particles" in computation.[96]

Gilbert King, a chemist at Arthur D. Little who had been in operations analysis at the Office of Scientific Research and Development (OSRD) during the war, is a good spokesman for this simulacrum interpretation of the Monte Carlo. Already in late 1949 at the IBM Seminar on Computation, he insisted that "[f]rom the viewpoint of a physicist or a chemist, there really is no differential equation connected with the problem. That is just an abstraction."[97] Two years later he amplified on these comments, arguing that the computer should "not be considered as a glorified slide rule" but as an "organism" that could treat a problem in an entirely new way. For King, the directness of the Monte Carlo gave it a role vastly more important than just another approximation method: "Classical mathematics is only a tool for engineers and physicists and is not inherent in the realities with which they attempt to deal. It has been customary to idealize and simplify the mechanisms of the physical world in the form of differential and other types of equations of classical mathematics, because solutions or methods of attack have been discovered during the last few hundred years with means generally available—namely, pencil, paper, and logarithm tables."[98] Engineering was far too complex for such traditional paper-and-pencil solutions. As a result, engineers substituted difference equations for the differential equations and sought approximate solutions by numerical methods. For the classical mathematical physicist, such crude methods were a poor man's solution. King's worldview inverted the mathematical physicist's epistemic hierarchy.

The platonic view of partial differential equations was espoused by some of the leading theoretical physicists of midcentury, including Einstein and P. A. M. Dirac. Dirac valued mathematical beauty above all other theoretical virtues and insisted, even in quantum mechanics, that the fundamental differential equations capturing causality still held good: one must, however, understand that the differential equations are "retained in symbolic form" as they govern the change in quantum states.[99] While Einstein did not subordinate experiment to beauty quite so thoroughly as Dirac, he did identify the crucial break point

96. Ulam, "von Neumann," *Bull. Amer. Math. Soc.* 64 (1958): 1–49, on 34.

97. King, "Further Remarks" (1951), 92.

98. King, "Diffusion Problems," *Ind. Eng. Chem.* 43 (1951): 2475–78, on 2475.

99. Dirac, cited in Kragh, *Dirac* (1990), 81, on mathematical beauty see chap. 14.

between Newtonian and modern physics in the differential equation. In a well-known homage to Maxwell, Einstein wrote, "After Maxwell [physicists] conceived physical reality as represented by continuous fields, not mechanically explicable, which are subject to partial differential equations." Since that time, quantum mechanics had called the Maxwellian scheme into question by identifying theoretical quantities with probabilistic notions. This "indirect" representation offended Einstein, and he felt sure that, in the fullness of time, physics would "return to the attempt to carry out the program which may be described properly as the Maxwellian—namely, the description of physical reality in terms of fields which satisfy partial differential equations without singularities."[100]

Refusing the mathematical physicist's invitation to the sacred realm of partial differential equations, King argued that such expressions refracted the world through a distorting prism; he insisted that the engineer's tools mapped directly onto something deeper: "[T]here is no fundamental reason to pass through the abstraction of the differential equation. Any model of an engineering or physical process involves certain assumptions and idealizations which are more or less openly implied in setting up the mathematical equation. By making other simplifications, sometimes less stringent, the situation to be studied can be put directly to the computing machines, and a more realistic model is obtained than is permissible in the medium of differential or integral equations."[101] King's claim is radical. For, contrary to a long tradition of supervaluing differential and integral equations as reflecting a platonic metaphysics hidden behind appearances, King believed that it is the engineer—not the mathematical physicist—who has something to say about reality. This "more realistic model" is so because *nature is at root statistical,* and representative schemes such as integrodifferential equations that eschew the statistical are bound to fail.[102]

Consider the diffusion equation, the bread and butter of industrial chemists like King:

$$\frac{\partial \mu}{\partial t} = D\frac{\partial^2 \mu}{\partial x^2}, \tag{8.18}$$

where t is time, μ is the dye concentration, x is the spatial dimension, and D is the diffusion constant. For the specific, simple case of a capillary tube and dye released at its center point, the solution is known in closed form. According to the solution, the dye molecules will move a specifiable mean distance Δx in a

100. Einstein, "Maxwell's Influence" (1954), 269–70.

101. King, "Diffusion Problems," *Ind. Eng. Chem.* 43 (1951): 2475–78, on 2476.

102. Since the differential equation itself presupposes a continuous rather than quantized spacetime, it is possible to continue the argument against differential equations in other ways. King, however, was arguing against deterministic continuous *processes* not against the spacetime continuum itself.

time Δt. This process can be simulated by a simple Monte Carlo conducted with a coin: we increase x by Δx if the coin comes up heads and decrease x by Δx if the coin reads tails. A random number generator in a computer can proceed in a similar manner by using even numbers as the basis for a positive increment and odd numbers for a decrement. In this way, repeated over many runs, one obtains a distribution. King celebrated this sequential process of random events: "The mathematical solution [that is, the analytic solution] of the diffusion equation is an *approximation* of the distribution. The mathematical solution of the diffusion equation applies to the ideal situation of infinitely small steps. In setting up the diffusion equation, more assumptions were used than were put directly, in an elementary fashion, into the computing machines, and a solution has been reached by an entirely different computing scheme from any that would be used by hand."[103] The most concise formulation of King's view emerged in an animated exchange between King and the New York University mathematician Eugene Isaacson, at one of the earliest meetings on the Monte Carlo method. King had just spoken on the problems of applying Monte Carlo methods to quantum mechanics.

> MR. ISAACSON: Isn't it true that when you start out on your analysis of the physical problem and you have a complicated finite difference process, you then look at a continuous differential equation and approximate that by a simpler finite difference process and so cut down some of your work?
>
> DR. KING: I think one can dodge a good deal of differential equations by getting back of the physics of the problem by the stochastic method.[104]

King's view—that the Monte Carlo method corresponded to nature (got "back of the physics of the problem") as no deterministic differential equation ever could—I will call *stochasticism* (see figure 8.12). It appears in a myriad of the early uses of the Monte Carlo, and clearly contributed to its creation. In 1949, the physicist Robert Wilson took cosmic ray physics as a perfect instantiation of the Monte Carlo just because the physical system itself contained a random element: "The present application has exhibited how easy it is to apply the Monte Carlo method to a stochastic problem and to achieve without excessive labor an accuracy of about ten percent."[105] And elsewhere: "The shower problem is inherently a stochastic one and lends itself naturally to a straightforward treatment by the Monte Carlo method."[106]

Two, radically different metaphysical pictures of how simulations relate to nature are illustrated in figure 8.12. On the platonic view, "physical reality" was or ought to be captured by partial differential equations. In the particular

103. King, "Diffusion Problems," *Ind. Eng. Chem.* 43 (1951): 2475–78, on 2476, emphasis added.

104. Discussion of King, "Stochastic Methods," in Hurd, *Computation Seminar* (1950), 48.

105. Wilson, "Showers" (1951), 3.

106. Wilson, "Monte Carlo," *Phys. Rev.* 86 (1952): 261–69, on 261.

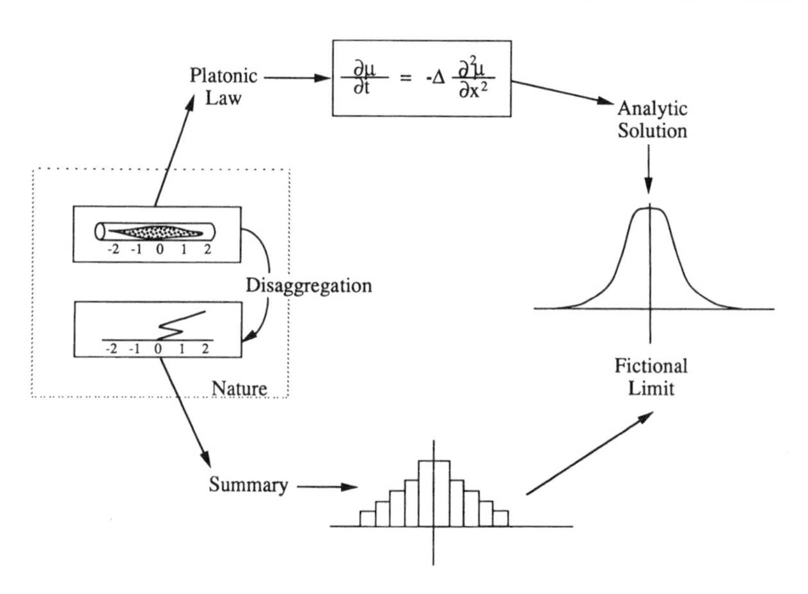

Figure 8.12 Monte Carlo metaphysics. One way of viewing the Monte Carlo was simply as another form of numerical approximation. On this view, what we want (ultimately) are the partial differential equations of a theory like Maxwell's account of electrodynamics—or like classical thermodynamics. Even if later physics showed that there were underlying discontinuities in the physics, Einstein considered thermodynamics to reflect some fundamental truth about the world. Gilbert King saw things differently. As far as he was concerned, it was the partial differential equations that were crutches, not the Monte Carlo. It was only the limitations of our calculating ability that had forced an overappreciation of *continuum* physics, and he saw the computer-driven Monte Carlo as a remarkable remedy. Now, with the Monte Carlo and the electronic computer, we could truly recreate the underlying, *discontinuous* reality that surrounds us.

case of diffusion, the physical reality was the spread of red dye in a thin capillary tube. It was "represented," as it rightfully should be, by a differential equation. This equation would, in any particular case, have a solution, as illustrated at the right. In what sense were the equations of thermodynamics true for Einstein? Not, obviously, because they were irreducible. As one of the founders of statistical mechanics and in particular the author of a famous series of papers on diffusion, Einstein certainly did not believe that the thermodynamic laws were absolute. Instead Einstein argued throughout his work that he wanted to proceed from compelling physical principles—such as the impossibility of a perpetuum mobile—and from such principles derive laws, expressible in differential equations, that were true. As he put it in his autobiographical essay (in the context of his derivation of the equations of special relativity), "[T]he longer and more despairingly I tried, the more I came to the conviction that only the discovery of a universal formal principle could lead us to assured results. The example I saw before me was thermodynamics."[107] The covariance of physical law under transformations of inertial frames of reference was a principle likened to the impossibility of perpetual motion machines.

Maxwell proceeded to his equations in the end without a detailed building-up from first mechanical principles; thermodynamics had been grasped without being deduced from detailed theories of matter; and special relativity emerged without detailed knowledge of the structure of the electron and ether dynamics. It is in this sense that equations like the diffusion equation or expressions relating entropy to other thermodynamical variables could be true and yet not irreducible. On the platonic view, which I will label in a necessarily schematic way the "Einstein-Dirac view" of physical theory, a Monte Carlo simulation would seem always an approximation, a poor man's way of arriving at prediction without identifying the underlying principles that found expression through differential equations.

On the stochasticist view (dare one call it the "King-Kahn view"?), the walk of a counter to the right and to the left in the simulation was *like* the diffusion. Both were predicated on what at root was a random process. Thus, for the stochasticist, the simulation was of a piece with the natural event. The Monte Carlo was a phenomenon of the same type as diffusion, but disaggregated into the motion of individual entities. Each particle ends at a particular point on the x-axis; when we collect these ending points into a histogram, we get the summary indicated at the bottom. Smoothing this distribution to the curve at the right is a fictional limit, obtained when one (imaginatively) extrapolates to arbitrarily many "runs" of the experiment. To the platonist, the stochasticist has merely developed another approximative method, useful perhaps but not more. To the stochasticist, the platonist has interposed an unnecessary conceptual entity (the analytic continuum equation) between our understanding and nature—the Monte Carlo, the King-Kahn view holds, offers a direct gaze into the face of nature.

As I have stressed, the problem of neutron diffusion was at the heart of the creation of the method, and it continued to provide difficult applications, mostly in weapons design and effects, but also in related civilian problems of reactor design and safety. Alston S. Householder, writing on neutron age calculations in "Water, Graphite, and Tissue," contended that "[t]he study of the diffusion of heavy particles through matter provides an ideal setting for the Monte Carlo approach. Here we are not forced to construct an artificial model to fit a given functional equation, but can go *directly* to the physical model and, in fact, need never think about the functional equations unless we choose."[108] This notion of

107. Einstein, "Autobiographical Notes" (1970), 53.

108. Householder, "Neutron Age" (1951), 8, emphasis added.

directness is central; it underlines the deep philosophical commitment to the mimetic power of the Monte Carlo method. This same faith in the metaphysical replication of process and representation appears in "health physics," where one finds a prominent Oak Ridge contributor, Nancy Dismuke, writing of neutron propagation through tissue: "The type of problem I am concerned with is a *natural* Monte Carlo problem in the sense that the physical model suffices as the model for proceeding with the calculation. An experiment is carried out (on a computing machine) which at every stage resembles closely the true physical situation. Whenever a random selection must be made in our experiment, the corresponding physical situation seems to be a matter of random choice."[109] A. W. Marshall spoke of the Monte Carlo as being most productive when it was applied to "nature's model" of a process such as diffusion, avoiding the differential equations.[110]

Unlike the authors just cited, some advocates of "naturalness," "directness," or mimesis of "nature's model" were people unwilling to endorse the program on abstract grounds of correspondence; instead, they argued on purely pragmatic grounds that it was the set of stochastic processes that were most effectively modeled by Monte Carlos. Listen to the comments of a programmer (a Dr. Howlett) at the British atomic weapons research center at Harwell during a conference in the 1950s: "I am responsible for the computing service at Harwell and represent the user of the Monte Carlo technique who views it, professionally at any rate, as just another numerical weapon with which to attack the problems he is asked to solve. . . . I have to answer the questions, what are the problems to which Monte Carlo is best suited, how good a method is it—how much better than conventional numerical analysis—and how can it be improved?"[111] In a sense this is the hardest-line pragmatic view possible: simulated reality would be accorded just that degree of credence that it earned in competition with other numerical processing packages. A slightly softer version of the pragmatic approach emerged from those who wanted to marginalize Monte Carlos to a heuristic function, a kind of suggestive scaffolding to be kicked aside at the earliest possible moment. This attitude is evident in the following two statements from the December 1949 IBM Seminar, the first by Princeton mathematician John Tukey: "[O]ne point of view for the use of Monte Carlo in the problem is to quit using Monte Carlo after a while. That, I think, was the conclusion that people came to. . . . After you play Monte Carlo a while, you find out what really goes on in the problem, and then you don't play Monte Carlo on that problem any more." A bit later, the Cornell mathematician Mark Kac commented on Robert Wilson's use of Monte Carlos to examine cosmic ray showers in a

109. Dismuke, "Monte Carlo Computations" (1956), 52, emphasis added.

110. Marshall, "An Introductory Note" (1956), 4–5.

111. Howelett, in Fieller et al., "Discussion," *J. Roy. Stat. Soc. B* 16 (1954): 61–75, on 63.

similar way: "They [Wilson and his collaborators] found the Monte Carlo Method most valuable because it showed them what goes on. I mean the accuracy was relatively unimportant. The five per cent or the seven per cent accuracy they obtained could be considered low; but all of a sudden they got a certain analytic picture from which various guesses could be formulated, some of them of a purely analytical nature, which later on turned out to verify very well . . . one of the purposes of Monte Carlo is to get some idea of what is going on, and then use bigger and better things."[112] In recent years some authors have combined both metaphysical and pragmatic views. As late as 1987, a textbook on experimental particle physics could stress the *special* relation between stochastic objects (in the world) and their Monte Carlo representations: "Historically, the first large-scale calculations to make use of the Monte Carlo method were studies of neutron scattering and absorption, random processes for which it is quite natural to employ random numbers. Such calculations, a subset of Monte Carlo calculations, are known as *direct* simulation, since the 'hypothetical population' . . . corresponds *directly* to the real population being studied."[113]

This more contemporary view wanted it both ways—metaphysical justification for simulations of stochastic processes and a pragmatic validation for the modeling of deterministic processes. But not everyone was so sanguine about Monte Carlos, and it is worth giving a representative hostile response. For some scientists could barely conceal their fury at the widespread adoption of the Monte Carlo method in physics. Monte Carlos, as far as the applied mathematicians John Hammersley and Keith Morton were concerned, ought to be expurgated line by line from the discipline. In January 1954, they put it this way:

> We feel that the Monte Carlo method is a last-ditch resource, to be used only when all else has failed; and even then to be used as sparingly as possible by restricting the intrinsic random processes to bare essentials and diluting them wherever possible with analytic devices. . . . We cannot emphasize too strongly that the random element of Monte Carlo work is a nuisance and a necessary evil, to be controlled, restricted, excised, and annihilated whenever circumstances allow; and this applied mathematician's attitude, of seeing how much can be done without appealing to random processes, we contrast with the pure mathematician's attitude, of seeing how much can be done by appeal to random processes."[114]

To these two authors, the battle line was drawn between the applied mathematicians (presumably like themselves) and the pure mathematicians (including Ulam, von Neumann, and Everett). Why? To these applied mathematicians, the extraction of physical laws from the world demanded analytic techniques

112. Both quotations from the discussion following Donsker and Kac, "Monte Carlo Method," in Hurd, *Computation Seminar* (1951), 81.

113. James, "Monte Carlo" ([1980] 1987), 628, emphasis added.

114. Hammersley and Morton, in Fieller et al., "Discussion," *J. Roy. Stat. Soc. B* 16 (1954): 61–75, on 74.

because they led to reliable results. On this view of the world, pure mathematicians were merely playing with curiosities when they spent their efforts evaluating simple problems like linear differential and integral equations in only one or two variables, for these were easily tackled with everyday numerical methods. Such frivolous mathematical activities would be better replaced by serious inquiry into the true challenges, such as nonlinear integral equations in several unknowns.

Two separate attacks on Monte Carlo enthusiasm had thus been made by the early 1950s. There were those who, on principle, remained deeply suspicious of the artificiality of pseudorandom numbers and the simulations themselves. And there were those, like Hammersley and Morton, who for pragmatic reasons doubted the new method's reliability. Either way, the status of the enterprise was in question. Not only was the broader community split over whether these new simulations were legitimate, the camps of the defenders and the attackers were themselves divided. Viewed "globally" the enterprise of simulating nature was on the shakiest of grounds. In the absolute absence of any agreed-upon interpretation of simulations, it might be thought that the whole enterprise would collapse. It did not. Practice proceeded while interpretation splintered.

8.5.3 The Practice of Simulations

For many demonstrations across a wide variety of fields, Monte Carlo simulations were not optional: without them there was no demonstration at all. To ground the more abstract discussion about the nature of Monte Carlo and to underscore the variety of ways in which the Monte Carlo realigned the boundaries of experiment and theory, it is useful to consider some exemplary ways in which the new instrument was used.

Consider, for example, those cases in which the Monte Carlo was deployed, not as the model itself, but as a way of choosing from among a universe of models. Geophysics presents such a striking instance. In 1968, Frank Press, a geophysicist from MIT and later science advisor to President Carter, took available geophysical data (the mass of the earth, the dimensionless moment of inertia of the earth, travel times of the measured seismic phases at six distances, and a variety of eigenperiods for the free oscillation of the planet). He then let his computer explore a huge variety of earth models, checking each against the geophysical data. More specifically, Press examined some five million models in which, within certain broad bounds, the program randomly chose a density distribution of the earth. Each earth possessed a compressional velocity, a shear velocity, and a density distribution in the mantle all randomly distributed within certain bounds; each earth's core radius and density were also chosen at random, again within bounds. When the computer then checked the predicted geophysical data against measurements for our earth, only six of the five million earths passed. The goal of the exercise was to see how well the surviving models agreed—that is, how constraining the available geophysical data were on any potential model of the earth. Sure enough, all six "passing" models had certain features in common including an increased core radius, with an iron-nickel solid component and an iron-silicon alloy for the fluid core. Interestingly enough, Frank also saw the Monte Carlo procedure as liberating, by freeing the scientific community from "bias stemming from 'initial' models or other preconceived notions of the Earth's structure."[115] How should one class this type of data-to-model (*inversion*) simulation? As experimental theory? Theoretical experiment? Is it a case of induction from data? Deduction from theory? Each such attempt to force the argumentative form back into the older categories strikes me as awkward, a rearguard action unable to capture the novelty of procedure. Once again, I would suggest that the Monte Carlo is best seen as expanding the spectrum of persuasive evidence, a tertium quid.

The Monte Carlo is sometimes the phenomenon, not simply the model. By this I mean the following. In high energy physics we saw several examples of an effect's being certified as solid only insofar as it differed from random variation. Take the FAKE/GAME routines run at Berkeley, in which the experimenters had to rank 100 smooth curves run through some data, from the most "two bumpy" to the most "one bumpy." (Or similarly let the Monte Carlo program simulate a Dalitz plot and ask whether the measured Dalitz plot is more clustered.) By agreement, the group would only publish a claim to a new discovery if it ranked the actual bumpiness above those generated by Monte Carlo variation. Similar practices were used in many different fields. In work launched in 1969, P. J. E. Peebles and his collaborators sought to resolve the much disputed question: was there superclustering of galaxies? Some astronomers (including G. O. Abell) took the standard Abell galaxy cluster catalog to show such superclustering, others challenged the claim. The arguments of Peebles and his subsequent collaborators relied, in an essential way, on the contrast between statistical measures applied to the "real" Abell catalog and those same functions put to work on "artificial Monte Carlo catalog[s]." Without the stochastic background, galactic "clumpiness," like Berkeley's "bumpiness," had no meaning.[116]

115. Press, "Density Distribution," *Science* 160 (1968): 1218–21; see also Keilis-Borok and Yanovskaja, "Inverse Problems," *Geophys. J. Roy. Astr. Soc.* 13 (1967): 223–34, for a review of Soviet work on inversion Monte Carlos of this sort (from data-to-model instead of model-to-data); also the earlier work by Press and S. Biehler, "Inferences," *J. Geophys. Res.* 69 (1964): 2979–95.

116. See Yu and Peebles, "Superclusters?" *Astr. J.* 158 (1969): 103–13; Peebles, "Statistical Analysis I," *Astr. J.* 185 (1973): 413–40; Hauser and Peebles, "Statistical Analysis II," *Astr. J.* 185 (1973): 757–85; Peebles and Hauser, "Statistical Analysis III," *Astr. J. Suppl.* 253 (1974): 19–36; Peebles, "Distribution," *Astr. Astr.* 32 (1974): 197–202; Peebles and Groth, "Statistical Analysis V," *Astr. J.* 196 (1975): 1–11; Peebles, "Statistical Analysis VI," *Astr. J.* 196 (1975): 647–52; and Groth and Peebles, "Statistical Analysis VII," *Astr. J.* 217 (1977): 385–405.

Biologists too used the Monte Carlo to explore clustering of certain phenomena, including the possible periodicity of mass extinctions.[117] Each of these examples, in different ways, illustrates what one might call an argument by excess granularity: the Monte Carlo presents the random background, and the observed world is or is not seen to possess more graininess. But again—as in the argument by inversion—as each of these investigators asks whether our world is less random than a Monte Carlo world, it seems ever more peculiar to call such comparative calculations of correlations either "theory" or "experiment."

A third form of Monte Carlo demonstration proceeded this way: Using known or established physics, physicists would simulate the data that would emerge from a particular experimental setup. Frequently, these simulated results could not be obtained through any other analytical means. Then if the measured data indicated deviations from the simulated results, they would be counted as evidence for a new effect or particle. As we saw in chapter 5, one of the early applications of Monte Carlos to particle physics came in 1960 when Stanley Wojcicki and Bill Graziano, working with Alvarez, wanted to know whether their claimed observation of the production of $\Lambda\pi\pi$ was legitimate since $\Sigma\pi\pi$ events might imitate their quarry.[118] How many of their $\Lambda\pi\pi$ candidates could be accounted for by the impostors? Only a Monte Carlo could tell. Not knowing about pseudorandom generators, they had a scanner laboriously type random numbers from a book onto IBM punchcards for insertion into their magnetic-core IBM 704 machine. Incomplete, inaccurate as the data may have been, the 704 spat out the result that the $\Sigma\pi\pi$ events would be few in number. The Λ's were therefore the real thing. But the central idea was already there in this first reckoning: calculate a known effect—just the way the Los Alamos people were tracking neutron flux—and see whether the result could account for the whole of an observed experimental signal.

This signal-background use of the Monte Carlo is such an accepted component of many high energy physics experiments that numerous physicists have become specialists in it. But the use of Monte Carlos is not straightforward. Far from it. One of the most stunning (and initially contentious) such uses of Monte Carlo was in the weak neutral-current experiments of the early 1970s.[119]

117. On mass extinction and Monte Carlos, see Raup and Sepkoski, "Periodicity," *Proc. Natl. Acad. Sci.* 81 (1984): 801–5. There are, of course, many other uses in biology of the Monte Carlo. E.g., in house mice (*mus musculus*) there are certain fatal and semifatal *t* alleles; interestingly enough, the standard deterministic model predicts a too-high value (larger than observed) for the maintenance of this allele in a population. Using the Monte Carlo, Lewontin and Dunn demonstrated that the stochasticity of Monte Carlo, coupled with the finite size of populations, accounted better for the actual percentage. Monte Carlo "experimental populations," especially small populations, exhibited a genetic drift effect in which the allele was lost. See Lewontin and Dunn, "Evolutionary Dynamics," *Genetics* 45 (1960): 706–22. For an interesting discussion of Monte Carlos in biology viewed as a type of experiment, see Dietrich, "Defense of Diffusion" (n.d.).

118. Wojcicki, "First Days," in Trower, *Discovering Alvarez* (1987), 168.

119. A much more detailed account of these experiments can be found in Galison, *Experiments* (1987), chaps. 4–6; the Monte Carlos are discussed extensively.

Prompted in part by the then-new electroweak unified theories, the experimentally sought effect was a scattering of muon neutrinos from quarks or electrons in which the neutrino did *not* change into a muon. For, while "classical" weak interaction lore held that such events would hardly ever take place (virtually all muon-neutrino interactions were supposed to produce muons), the unified electroweak theory of Weinberg, Glashow, and Salam of 1967–68 said otherwise: muonless events should be legion. Even a decade or more after Wojcicki and Graziano's crude estimates, even after the development of highly sophisticated Monte Carlos, the technique was still viewed with considerable apprehension by many experimenters. Often quite persuasively, they worried aloud that the simulations could not or did not accurately recreate the phenomena they purported to represent.

At CERN, a multinational bubble chamber team was mounting a large-scale exploration of various physical processes using its new heavy-liquid bubble chamber, Gargamelle. Over the course of the early 1970s, the existence of neutral currents rose to the top of its physics priorities, but the most frequent manifestation of the effect (neutrinos scattering from inside protons and neutrons) had an imitative background that was exceedingly hard to quantify—and at the time seemed potentially large enough to explain the entire set of several hundred neutral current candidates. As we have seen in gory detail, bubble chambers were superb at catching charged particles—their omnidirectional acceptance and complete registration was, after all, their most celebrated virtue. *Uncharged* particles were a different matter, and the issue was this: could incoming neutrinos hit nuclei in the walls, floors, and shielding surrounding the chamber in such a way that the neutrons they released would carom into the chamber, producing events that looked like true neutrino events (blasted nucleus, no muon) but were not (see figure 8.13)? Here again, the Monte Carlo was essential, as only an excess of muonless events over and above the number produced by the Monte Carlo would give firm evidence for the new physics. As I have shown elsewhere, the battles over the Monte Carlos were long and hard: Were they sufficiently like the real world to be reliable? Could one emitted neutron precipitate a cascade of others, extending their dangerous radius? Were the simulators realistically assessing the length even a single neutron could travel in the chamber? On the answers to these questions, and ones like them, rode everything, since no matter how many muonless events the bubble chamber team saw, it would mean nothing unless the signal exceeded the background expected from "ordinary" physics. Indeed, for several key members of the Gargamelle collaboration the establishment of the Monte Carlo as a robust simulation was precisely coextensive with the establishment of the existence of neutral currents. No background, no signal; no Monte Carlo, no demonstration.

On the American side a Harvard-Wisconsin-Pennsylvania-Fermilab (HWPF) collaboration used spark chambers and calorimeters in its competi-

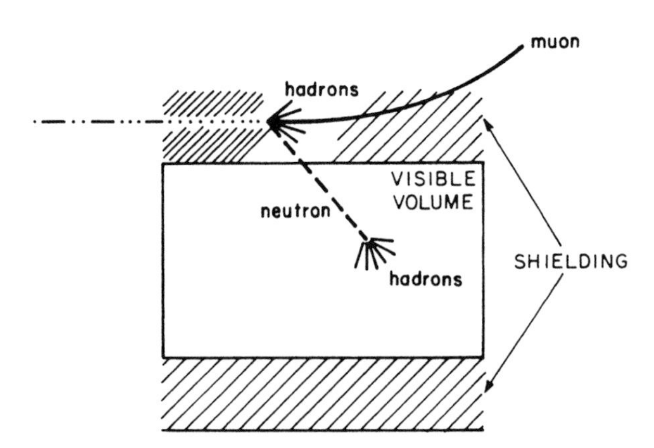

Figure 8.13 Monte Carlo acceptance. For the Gargamelle collaboration, wide-angle muons were no problem: bubble chambers saw all. Their nightmare was rather that neutrons, created outside the chamber in old-physics neutrino interactions, might diffuse into the chamber and mimic neutrino-induced muonless events. To evaluate this background, the team simulated the production of neutrons and tracked their activity inside the chamber. (In many ways this calculation recreated the original neutron diffusion problem set out years before by von Neumann.) Only by showing that the number of observed muonless events exceeded the background could the discovery claim for neutral currents be established.

tive hunt for the neutral currents. But HWPF's exploitation of the Monte Carlo was somewhat different. At the high energies then available at Fermilab, neutron background to muonless events was not particularly important, but another danger was. The HWPF terror was that an incoming neutrino could blast apart a proton or neutron producing an altogether unexciting muon, and that the muon might then leave the detector without being snagged by the muon detector (see figure 8.14). Under these circumstances, the detector would record the event as "muonless," but that would be the apparatus talking, not the processes in question. Such events were observed, and in great number. Again, the question arose: signal or background? In this respect, the two teams were after similar results, and as at CERN, there was much debate about the resemblance of the Monte Carlo to the processes it purported to depict.

There was, however, a distinction. The Americans were not so much worried about two similar-looking effects (e.g., $\Lambda\pi\pi$ mimicking $\Sigma\pi\pi$, neutron background mimicking a neutrino event). Instead, they were concerned about the geometrical design of their detector itself: what, they were asking, was the acceptance of their muon detector for the wide-angle muons? Just as the imita-

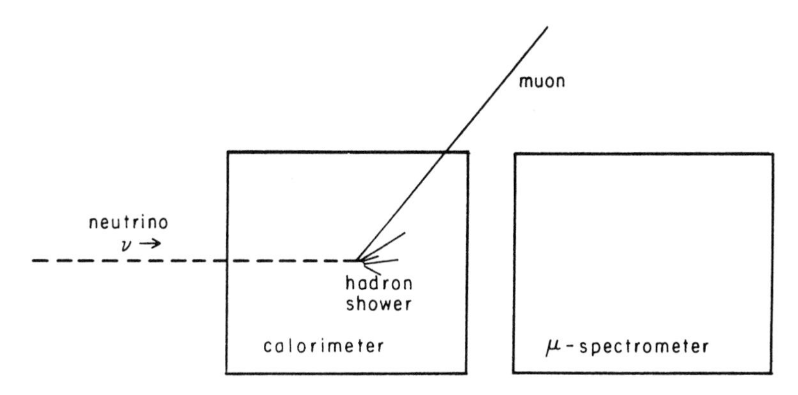

Figure 8.14 Simulating fake neutral currents. At issue in the Harvard-Wisconsin-Pennsylvania-Fermilab collaboration, E1A, was whether the muonless events they were recording were, in fact, neutrino-induced weak interactions without muons. Could the muons be escaping at large angles? To address this question, the team took the known physics (the old weak-interaction theory in which muon neutrinos always made muons when they interacted). They then let the Monte Carlo simulate old-physics events, calculating how many muons would miss the muon detector and so be counted (wrongly) as "muonless." Only by comparing the simulated muonless events with the observed could the excess, if any, be calculated. "Discovery" in such cases demands a contrast with background, and Monte Carlos formed part of that background calculation in growing numbers of experiments.

tive effect problem became a dominant concern in many experiments in the 1970s, 1980s, and 1990s, so too did the problem of Monte Carlo–calculated acceptance for electronic detectors over the same period. Thousands of physicist-years were spent on the design of the Superconducting Supercollider detectors alone, many of which were devoted to such simulations.

At the time they were conducted, the neutral-current simulations were extraordinarily difficult, in no small measure because not everyone in the collaborations felt comfortable with the ability of the Monte Carlos to generate reliable backgrounds. But in one sense, the experiments were simple, or at least simpler than the hadronization experiments discussed in chapter 7. In the neutral-current experiments, the relevant theory was clear: the old weak interaction theory gave fairly straightforward predictions about how neutrinos should interact. The problem was in implementing the Monte Carlo: how, for example, would the neutrons scatter and sometimes induce cascades of hadrons? When we considered the various hadronic jet models in section 7.8, much more than that was up for grabs. For theorists reaching toward experiment, the Monte Carlo was not merely necessary to accommodate the complications produced by the detector: the QCD Lagrangian itself allowed very little to be calculated about the way quark and gluon dynamics led to their amalgamation into observed particles. Consequently, the scientific-philosophical status of the simulations was much more problematic.

Some theorists, like Feynman and Field, used Monte Carlo methods even while eschewing any commitment to the "fundamental" theory of QCD. At the same time, from the experimental side, Monte Carlos served an equally important role in the design and testing of detectors. Ultimately, though, it was between the two poles of experiment and theory that the jet simulation business grew up, achieving a distinct professional identity for its practitioners and an epistemic function that stood outside both theory and experiment. Some chose "clusters," some chose independent fragmentation, and others settled on "strings." Yet others began the difficult task of lifting the Monte Carlo out of its merely instrumental role in reproducing data. Physicists like Buchanan began to use the simulations to get at an explanation for the phenomena in question: perhaps, they argued, they could use features of QCD considered in a lattice instead of the spacetime continuum as a way of tying specific features of QCD to experiment strongly enough that they would *explain* the data. But there was little consensus. With names like ISAJET, PYTHIA, FIELDAJET, COJET, EUROJET, among many others, each Monte Carlo embodied different assumptions about the fundamental interactions (the "hard" physics of quark-gluon interactions) and the poorly understood, lower energy hadronization. But one thing was clear by the late 1980s, QCD (and any higher energy theory that included QCD) would have to pass through a net of simulations or remain forever isolated from the laboratory world. Neither theory nor experiment, simulations were here for the long run.

8.6 FORTRAN Creole, Thermonuclear Patois

Pure mathematician, applied mathematician, physicist, bomb builder, statistician, numerical analyst, industrial chemist, numerical meteorologist, and fluid dynamicist: the Monte Carlo was something different for each. To the pure mathematician, the Monte Carlo was a measure defined on the space of infinite graphs, for a coupled set of Markov and deterministic equations. For statisticians, the method was another sampling technique, with special application to physical processes; they considered such techniques well known and as a result were at first hesitant to join the plethora of postwar conferences, discussions, and research efforts. To the numerical analyst such as Wishart, the method was one more numerical tool for the solution of integrodifferential equations. To the industrial chemist such as King, the stochasticist view made his subject amenable to a form of direct representation never before possible. Bomb builders saw a modeling technique for a physical process so complicated by radiation transport and hydrodynamics in hot media that it defeated all their usual means, both experimental and theoretical. The symbols and procedures of the method therefore sat variously in each domain, and connected differently to the terms, theorems, and style of each discipline.

Yet for all this diversity, it is clear from our discussion thus far that at meeting after meeting, beginning with the June 1949 Institute for Numerical Analysis conference in Los Angeles, and continuing through the 1949 IBM Scientific Seminars and the 1954 Gainesville meeting, representatives of these professions could and did find common cause. Moreover, individuals, including Ulam, King, Curtiss, and Householder, could switch back and forth between problem domains without difficulty. Von Neumann passed as easily to meteorology as he could to the superbomb, to reactor problems, or to fluid dynamical shock calculations. A chemical engineer could speak to a nuclear physicist in this shared arena of simulation; the diffusion equation for chemical mixing and the Schrödinger equation were, for purposes of this discussion, practically indistinguishable. Yet the common ground of the Monte Carlo implied no further alignment; if anything, with the growth of nuclear engineering, cyclotron design, and meson theory, nuclear physicists were growing ever further from their chemist colleagues.

But in the heat of the moment, the diverse assemblage of technical workers staked out a trading zone by forging a pidgin in which techniques were partly (but not totally) divorced from their broader signification. Everyone came to learn how to create and assess pseudorandom numbers—without necessarily sharing views about the relation of these pseudorandoms to the "essence" of being "random." It became a matter of common ability to transform pseudorandom sequences into the particular weighted distributions needed for a particular problem. Everyone learned the techniques of variance reduction.

It may be tempting to think of the Monte Carlo as "mere" technique—that is to say, formalism without content or meaning. But, as we have seen, the metaphysical debate over the nature of randomness, or about the link between simulation and a stochastic world, belies this interpretation. Part of the Monte Carlo pidgin involved the establishment of a locally shared vocabulary, a retinue of guiding concepts and mathematical manipulations. To illustrate this process, consider the subtle family of meanings associated with the notion of "game."

The term "game" permeates Ulam's and von Neumann's writing. In Ulam's early work we saw how the Monte Carlo was thought of as a "game" binding solitaire to mathematical and physical problems. This was no passing flourish. Analogies built on analogies. In late August of 1950, Ulam attended a meeting of the American Mathematical Society, where he again exploited the concept of Monte Carlo as a game to unify his wide-ranging interests. For example, long before World War II, Ulam had been concerned with set theory and mathematical logic. Now, looking back on the prewar work of Kurt Gödel, Ulam declared that logic could be considered a branch of combinatorial analysis. "Metamathematics," on his conception, "introduces a class of games—'solitaires'—to be played with symbols according to given rules. One sense of Gödel's theorem is that some properties of these games can be ascertained only by playing them."[120] A variety of terms, "game" among them, serves to form part of the exchange language that binds these diverse formal discourses. Each was an "analogue" to the other.

In the continuous, orderly realm of multidimensional integration, Ulam again saw an opportunity to bet at Monte Carlo, this time, as mentioned earlier, by sampling from within a known volume that contained the volume to be integrated. "[B]y playing *a game of chance* (producing the points at random) we may obtain quantitative estimates of numbers defined by strictly deterministic rule." Here, Ulam invokes a statistics applicable to sampling within a platonic mathematical space. As he puts it, the multidimensional integral is just an instance of the more general problem of exploring "geometric probabilities."[121]

One "analogous situation" must have been startling: Ulam took Monte Carlo gaming to the heart of logical proof. For some time, Ulam and a number of other logicians had explored the representation of logic as an algebra definable in geometric terms. The idea was this: A formal system in mathematics involves the Boolean operations of logic or set theory (union and intersection of sets of points) and in addition must make use of the existential quantifier, $\exists$, and the universal quantifier, $\forall$. These quantifiers admit of geometric interpretations leading to "cylindrical" or "projective" algebras. Each variable corresponds to a dimension of an n-dimensional space, where a point $(x_1, x_2, \ldots, x_n)$ is an assignment of values for each of the variables x_1 through x_n. The existential quantifier can then be thought of as an orthogonal projection of a set of points in n dimensions onto a subspace of fewer than n dimensions, and the universal quantifier can be defined in terms of $\exists$. Any proposition of metamathematics can in this way be formulated in terms of these algebraically defined geometrical sets. If a (closed) proposition is true, the set corresponding to its negation will be vacuous; if it is false, the set will encompass the whole space over the allowed values of the variables. The link with the Monte Carlo then becomes clear: we sample randomly to show heuristically (in a probabilistic sense) that the relevant set is vacuous (that the proposition is true). In other words, we choose random points from the space; with each successive one that turns out *not* to satisfy the algebraic conditions for the relevant set, we are that much more certain that the set is empty (see figure 8.15): we say that the set is empty with probability p. This meant (rather startlingly) that the corresponding proposition in logic was known to be true with just that probability. While such a demonstration necessarily remained what Ulam called "heuristic," it would be possible to

120. Ulam, "Transformations" (1952), 2:266.
121. Ulam, "Transformations" (1952), 2:267, emphasis added.

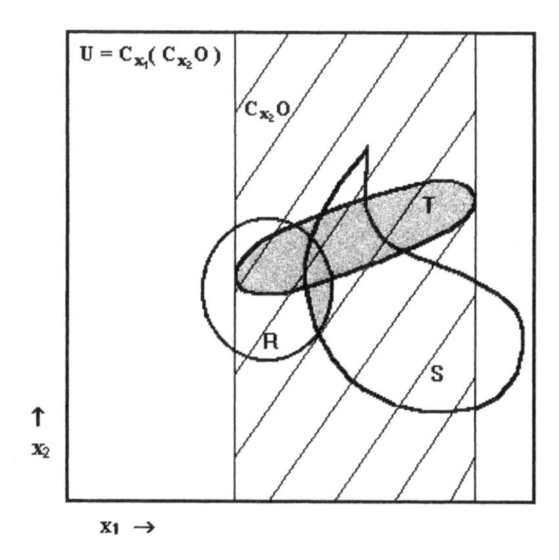

Figure 8.15 Monte Carlo logic. Consider the following example, where the proposition $Q = \exists x_1 \exists x_2$ $[T(x_1,x_2) \bigvee (R(x_1,x_2) \bigwedge S(x_1,x_2))]$ is represented "geometrically." The relations *R*, *S*, and *T* having been arbitrarily defined as pictured, the gray-shaded area, call it *O*, then identifies all points satisfying the square-bracketed relation (crudely, [*T* or (*R* and *S*)]). To existentially quantify this relation with respect to x_2, we take the cylinder with respect to x_2 of the area *O*, and get the region $C_{x2}O$ indicated with stripes. Existentially quantifying with respect to x_1 amounts to taking the x_1 cylinder of $C_{x2}O$, and yields the whole space: $C_{x1}(C_{x2}O) = U$. This is as it should be since, with *T*, *R*, and *S* as given, there do exist points that satisfy the relation [*T* or (*R* and *S*)], so the proposition is true, and true (closed) propositions are to be represented by the whole space, their negations by the empty set (via set complementation). The set corresponding to the negation of the proposition in question here is obviously vacuous, by inspection, but visual inspection is not in general possible. However, that negation set will, in general, be governed by a set of equations. Using the Monte Carlo method to estimate the truth or falsity of a proposition involves taking random points from the space *U* and checking each one to see whether it satisfies the equations governing the negation set of the proposition of interest. (This is precisely analogous to sampling a rectangle to find out whether $\int f(x)|\, dx = 0$: the more points that do not satisfy $x_i < f(x_i)$, the higher the probability is that $f(x) = 0$.) Here, after taking sample points each of which does not satisfy the negation of *q*, we may say: "The region is vacuous with a certain probability," or equivalently, "*Q* is true with that probability." Source: Courtesy Sherri Roush.

render these Monte Carlo "proofs" as probabilistically persuasive as one desired (or had machine time to execute). The Monte Carlo method had rendered this corner of logic a science of probabilities.[122]

With integrodifferential equations that ostensibly had a more direct physical reference, such as Boltzmann's equations describing the motion of a fluid, the continuum expressions have traditionally been interpreted as being the

122. Ulam, "On the Monte" (1951), 211; also see Henkin, Monk, and Tarski, *Cylindric Algebras* (1971).

limits of more fundamental motions of random walks (among molecules, e.g.). As we have seen, Ulam thought of the Monte Carlo as an inversion of this procedure in order to "construct models of suitable games and obtain distributions or solutions of the corresponding equations by playing these *games,* i.e., by experiment." In particular, with cosmic ray showers, one can "'produce' a large number of these showers by playing *a game of chance* with assumed probabilities and examine the resulting distributions statistically."[123] "Experiment" here glosses the term "game," emphasizing the notion that trials are involved, the outcome of which cannot be forecast deductively. Such "experiments" apply as much to games of geometric (platonic) probabilities as they do to physical systems.

The analysis grows more challenging in the problem of neutron propagation in a complicated mixture of fissionable and nonfissionable materials, since the mathematics of continuously distributed velocities is formidable. Following his earlier work with Metropolis, Ulam advocated treating the evolution of the neutrons as the iteration of a linear transformation instead, pretending that space and time were divided into a grid of finite cell size. In von Neumann's formulation (and I will come back to the links between Ulam and von Neumann in a moment) a "probabilistic game" is defined through the elements a_{ij} of a matrix that correspond to the transition probabilities of particles of type i into type j. Further, the a_{ij} would sum in any given row to unity, and each term would have to be nonnegative. Ulam suggested that this can be generalized in three ways. First, each a_{ij} could be considered as the expected value of a transition probability, leaving it open just what probability distribution actually obtained. Second, he notes that for an arbitrary real matrix, it is possible to write down a nonnegative matrix with the same algebraic relations, for example, $(-1)^2 = 1$ and $(1)(-1) = -1$, if the following identifications are made:

$$1 \approx \begin{vmatrix} 1 & 0 \\ 0 & 1 \end{vmatrix}, \qquad -1 \approx \begin{vmatrix} 0 & 1 \\ 1 & 0 \end{vmatrix}. \tag{8.19}$$

Yet a third generalization is possible (by identifying ordered pairs of real numbers $\mathbb{R}^2$ with the complex numbers $\mathbb{C}$); this identification makes it possible to write down a nonnegative real matrix equivalent (isomorphic) to a matrix with complex entries. In this way, any process representable in terms of a generating function, with arbitrary complex entries, could be written as a real, nonnegative matrix, and therefore be seen as a version of a von Neumann probability game.[124]

Referring to nonclassical physics, Ulam reiterated his earlier insistence (from his original 1949 paper with Metropolis) that the Schrödinger equation

123. Ulam, "Transformations" (1952), 2:267–68, emphasis added.
124. Ulam, "Transformations" (1952), 2:269, emphasis added.

can be studied by means of a Monte Carlo random walk: "we study the behavior of such a system empirically by *playing a game* with these particles according to prescribed chances of transitions." While the random walk model of the Schrödinger equation will not give precise answers, it will provide an answer within ϵ of the analytic solution with probability $1 - \eta$. Immediately thereafter, Ulam repeated that "in reality" the integral or differential equations only express averages of physical quantities. Consequently, "[t]he results of a *probability game* will reflect, to some extent, the deviation of such quantities from their average values. That is to say, the fluctuations unavoidably present as a result of the random processes performed may not be purely mathematical but may reflect, to some extent, the physical reality."[125]

Just this sort of remark about the "reflection" of "physical reality" makes it impossible to read Ulam's vision of the Monte Carlo as a "merely" instrumental or calculational tool. While the world referred to may be platonic, classical, or quantum mechanical, the Monte Carlo gives us access where purely deductive means fail.

One of the most striking, and consequential, extensions of the simulation qua experiment came a few months later (1953) as Ulam, collaborating with John Pasta and Enrico Fermi, began exploring nonlinear dynamics with computational methods. At one level, the problem of their paper "Studies of Nonlinear Problems" is easy to state. If beads on a string are connected by springs of given potential $v(x_i)$, where x_i is the displacement of the ith point from its original position, how much energy will go into the different modes of oscillation as a function of time? More specifically, with a linear potential, the beads set up stationary modes of oscillation with calculable frequencies. Fermi, Pasta, and Ulam expected that if they added a nonlinear term, the modes would disperse, and after a sufficient amount of time, the system would have "thermalized": that is, the system would have distributed its energy equally over all available frequencies. To their astonishment, this did not happen: "[T]he results of our computations show features which were, from the beginning, surprising to us. Instead of a gradual, continuous flow of energy from the first mode to the higher modes, all of the problems show an entirely different behavior."[126] For example, in figure 8.16, the MANIAC produced a result that showed if all the energy began in low-frequency mode 1, early data seemed normal: mode 2 starts to grow, followed by mode 3, mode 4, and mode 5. But all is not normal. "For example, mode 2 decides, as it were, to increase rather rapidly at the cost of all other modes. . . . At one time, it has more energy than all the others put together!" (around $t = 13{,}000$ cycles). At the end of the day ($t = 30{,}000$), amazingly enough, almost all the energy was back in mode 1. "It is, therefore, very hard to

125. Ulam, "Transformations" (1952), 2:270, emphasis added.

126. Fermi, Pasta, and Ulam, "Nonlinear," LA-1940, 1955; repr. in Ulam, *Analogies* (1990), 142.

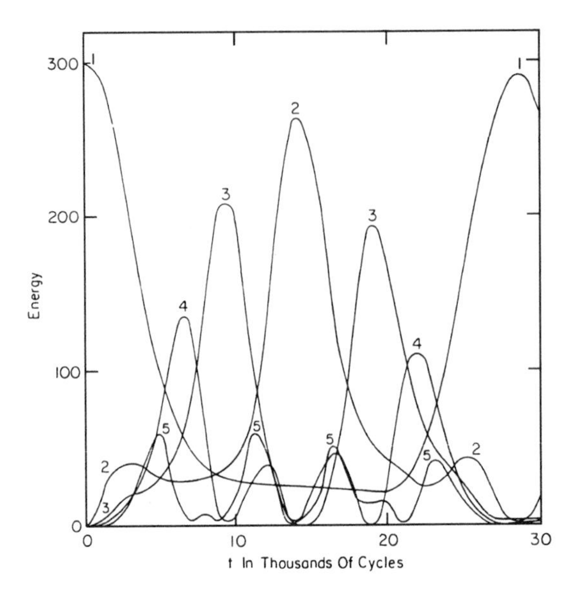

Figure 8.16 Thermalization undone (1955). Source: Fermi, Pasta, and Ulam, "Nonlinear," LA-1940 (1955), repr. in Ulam, *Analogies* (1990), 146.

observe the rate of 'thermalization' or mixing in our problem, and this was the initial purpose of the calculation."[127] It was an extraordinary piece of "experimental work on a computing machine" (Ulam's own characterization)—as if a bowl of hot tea cooled down in a room (unexceptional), and then the room cooled to warm the tea (unbelievable).[128] It was not supposed to happen, as the authors' exclamation point made clear.

The discovery of such nonthermalizing phenomena was crucial to the development of chaotic and turbulent dynamics, as Thomas Weissert has shown in some detail.[129] And one could—as Ulam and others have suggested—trace the concern back to Fermi's longstanding interest in the old Boltzmannian problem of explaining irreversibility on the basis of fundamentally reversible systems. But we are now in a position to say something more specific about the immediate (Los Alamos) site of the calculation.

In particular, sometime around May 1952 planning for the cylindrical (Teller-Ulam) simulation was more or less ready for implementation on the machine just being finished at the Institute for Advanced Study.[130] By September 1953, the Institute team was reporting its results in kilotons of energy released graphed against the cycle number of the simulation, where one cycle corre-

127. Fermi, Pasta, and Ulam, "Nonlinear," LA-1940, 1955; repr. in Ulam, *Analogies* (1990), 142.
128. Ulam, Introduction to "Studies" (1965), 977.
129. See Weissert, *Fermi Pasta Ulam* (1997).
130. Evans to von Neumann, 7 May 1952, FEP.

sponded to 3×10^{-8} seconds in real time.[131] In addition to the radical new bomb design, the simulation itself was far more sophisticated than before and was set to account not only for the nuclear dynamics of the DD and DT interactions but also for the Monte Carlo treatment of photons and electrons as the reaction proceeded. Hydrodynamics became more and more complicated; Lagrangian methods (following the position of the masses) were used for the longitudinal direction in which the motion was not too great, and Eulerian methods (which specified flow as a function of position and time) were used to characterize the enormous flows in the radial direction during the compression. As the bomb began to detonate, its constituents became plasmas of different densities and velocities. Fluid-dynamical instabilities arose as the now "liquid" materials mixed, merged, and sloughed off one another: Rayleigh-Taylor instabilities developed as dense plasmas accelerated sparse ones, and Kelvin-Helmholtz instabilities grew when streams of plasma at one velocity sheared past streams of plasma moving faster or slower. At one point, Evans and his collaborators at the Institute hooked up a speaker to the computer in order to "play" tones corresponding to the different excited modes in the fluids; Monte Carlo's produced a white noise of static. "Lovely tunes," Evans reported much later, "they would go for a couple of minutes in each mode" and eventually cycle back.[132] Kepler's spheres had their celestial harmonies, thermonuclear weapons their own.

In the midst of these hydrodynamical calculations that were set up to explore the Teller-Ulam weapon, Ulam and Pasta wrote their 1953 paper on the "heuristic" studies of mathematical physics using high-speed computation.[133] Their stated goal was to launch a series of papers exploring the computer's capacity to perform "'mental experiments' on mathematical theories and methods of calculation," from gases to star clusters. What distinguished their approach was the focus it put on functionals (functions of functions) that characterized the process without following its constituents in microphysical detail. For example, they would pursue the degree of mixing of two gases, not the microscopic progress of each bit of matter; or they would follow the rate of energy shift from simple modes to higher frequency modes. In this way they were able to exhibit larger scale phenomena such as the billowing of smoke as it expanded. Turning to gravitationally bound star systems, they determined the angular momenta of randomly chosen subsystems (chosen à la Monte Carlo), not the motion of each star. Again, Ulam and Pasta aimed to characterize the system as a whole—in case of stars, the authors aimed to let the cluster settle to a stable size, after which they could calculate its properties, such as the number of binary and triple stars.

131. Evans to von Neumann, 24 September 1953, FEP.

132. Evans, interview by the author, 2 March 1991.

133. Ulam and Pasta, "Heuristic Studies," LA-1557, 1953, repr. in Ulam, *Analogies* (1990).

With the H-bomb and Ulam-Pasta's "heuristic studies" as background, we can view the famous Fermi-Pasta-Ulam paper lit from behind, so to speak. The problematic that informs the paper—from the MANIAC modeling of two-dimensional mixing processes to the tracking of energy modes, from the back-and-forth between stochastic and deterministic systems of the Pasta-Ulam paper to the broader notion of "mental experiments"—was rooted in the H-bomb work at Los Alamos from 1947 to 1953. This new style of inquiry had already begun to reshape research across the disciplinary map, in part by creating a new and shared vocabulary for this hybrid between theory and experiment.

Together these alterations in the cluster of concepts including "randomness," "proof," and "experiment" signal a profound shift in the notion of a demonstration in physics. "Randomness" has been transformed from a platonic idea to a practical problem in the correlation of computer-generated digits; "proof," the very glue of reason, has been approximated in logic; "experiment" has become a series of computer-simulated runs, whether of a physical system, a logical proposition, or a mathematical proof. All circulate together and gather around the image of a game, now cut loose from its moorings in everyday life.

It is in part through the concept of a game that I want to situate Ulam's work in the postwar environment. More specifically, I want to sketch the links between Ulam's concept of a game and the specific sense attributed to it in the first instance by von Neumann but, more broadly, in the amalgamated context of game theory and operations analysis during and just after the Second World War. As an assistant to the mathematician David Hilbert, von Neumann became an ardent defender of formalism, the mathematical movement that attempted to build mathematics up from an axiomatic foundation and divorce its content from any reference to a platonic or physical world of mathematical entities. As part of this process, Hilbert came to speak about the rule-governed pursuit of higher mathematics as a kind of game. In 1928 (and in more detail in the early 1940s), von Neumann developed a rigorous theory of games that included not only parlor games but a vast array of competitive activities among n participants.[134]

In particular, von Neumann and Morgenstern delimited the terms associated with a game, including the word itself, along with the associated notions of plays, choices, and strategies. In his 1944 book with Morgenstern, he put it this way: "[T]he use of [these terms] in everyday language is highly ambiguous. The words which describe them are used sometimes in one sense, sometimes in another, and occasionally—worst of all—as if they were synonyms." The two authors refined these loose definitions into technical ones, which they sketched informally at first by describing a *game* as nothing else but the "totality of rules which describe it." A given worked-out instance of the game is a *play,* and a *move* is the "occasion" on which a *choice* is made; for example, the fifth move in

134. von Neumann, "Gesellschaftsspiele" ([1928] 1963).

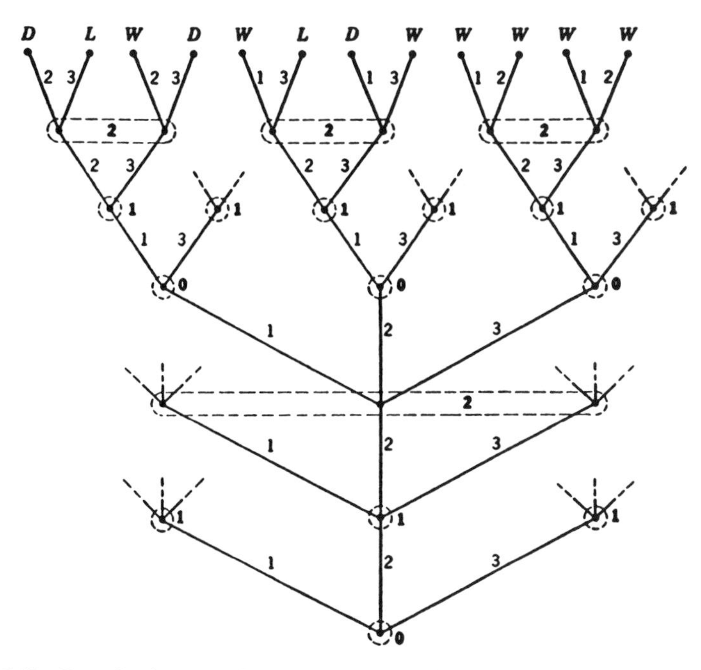

Figure 8.17 Extensive form, von Neumann game. In this simple game two players and a (non-playing) house are each dealt three cards, with face values of 1, 2, and 3. The house reveals its first card (shown as three equiprobable branches from node 0 at bottom). Then both players choose and simultaneously show a card—high card gets points equal to the value of the house card. Illustratively, player 1 picks a "2" (at circled node 1 above circled node 0), and player 2's three possible moves are each shown (as branches from node in long dashed oblong). The house then displays its second card (indicated by branches from the three 0 nodes representing a "1" and a "3"); play repeats. Continuing up the graph, D(raw), L(oss), and W(in) mark each play's outcome for player one. Von Neumann showed how all games can be represented using graphs like these to stand for moves of strategic, calculational, and random choice. He and Ulam thought of Monte Carlos as random-choice games. Source: Luce and Raiffa, *Games and Decisions* (1957), 46.

a chess game designates all the choices present at a given point in the game. In this sense, a play is composed of choices, and a game is composed of moves. Last, a *strategy* is the principle on which a player makes choices.[135]

The representation of a game in terms of a decision tree indicates all possible choices at each point at which decisions can be taken (see figure 8.17). This kind of representation (technically known as the extensive form of a pure game) rapidly grows cumbersome, and it was one of von Neumann's contributions to re-represent a game in terms of a matrix. In the simple case of a two-person game, the m rows might be labeled by an index alpha i corresponding to each of player 1's m strategies. Similarly, the n columns would be designated by

135. von Neumann and Morgenstern, *Games* (1953), 48–49.

beta *j* corresponding to player 2's *n* strategies. Entries in the matrix would then designate the "payoff," in the simplest case the amount of money to be transferred from player 1 to player 2.

During World War II, while von Neumann was returning to his game theory, the notion of a game became widely current through the efforts of "operations analysts." Operations research began as part of an interdisciplinary effort to solve specific problems: How, with finite resources, to locate the maximum number of enemy submarines at sea? How to estimate bomb damage given the random fluctuations of bomb trajectories?[136] One "game" played frequently by operations researchers was to distribute points on the map of a target, draw ellipses of destruction around each point, and calculate the nonintersecting area to estimate the damage inflicted. Trained in such methods, someone like Gilbert King could write in 1952 that the method of random walks (which he took to be key to Monte Carlo) was "revived during the war to investigate military problems in operations research" and at the same time put forward by von Neumann and Ulam as the basis for the Monte Carlo. Indeed, for King, operations research was defined by the "setting up of a working model," and Monte Carlo methods simply added to this strategy the flexibility to move back and forth between deterministic and stochastic rules as appropriate.[137] On this reading the Monte Carlo was, as King dubbed it, the "natural mode of expression" for operations research,[138] especially with the much greater calculational speed afforded by the computer.

The war games played by the operations researchers drew, frequently and explicitly, on von Neumann's earlier game theoretical work. For example, one strategist in 1950 analyzed the Battle of the Bismarck Sea during World War II. As the fight over New Guinea raged, intelligence reports showed that the Japanese would shift a troop and supply convoy from Rabaul to Lae. The convoy could travel by a northern route, where visibility was poor, or by a southern route, where it would be clear. Either way, the convoy would travel for three days. As soon as General Kenney's reconnaissance planes spotted the convoy, bombing could commence, but Kenney had to choose to send his spy planes on one route or the other. In terms of their respective choices, this led the strategist to block out the matrix in figure 8.18, in which the Allied northern reconnaissance strategy yields two days of bombing against a Japanese move south or north, and an Allied southern strategy yields one day bombing against a Japanese decision to go north but three days of bombing against a Japanese decision to deploy to south. Note that in this "normal form" of a game, each "player" does not consider each move or choice as in the extensive form; rather, each side

136. I am indebted to Robin Rider for sharing with me two unpublished pieces on the history of operations research: "Early Development" (1988) and "Capsule History" (forthcoming).

137. King, "Natural Mode," *J. Oper. Res. Soc. Amer.* 1 (1952–53): 46–51, on 49.

138. King, "Natural Mode," *J. Oper. Res. Soc. Amer.* 1 (1952–53): 46–51.

	Japanese Strategies	
	Northern Route	Southern Route
Kenney's Strategies:		
Northern Route	2	2
Southern Route	1	3

Figure 8.18 Game theory tactics. Source: Luce and Raiffa, *Games and Decisions* (1957).

merely chooses a strategy ($i = 1 =$ north, e.g.), and this completely specifies the play and its outcome. Together, the two strategies possible for each commander specified the game. (These strategies are called "pure strategies" because each one fully specifies what happens in every circumstance.) In this example, an equilibrium point exists for a north-north decision. The allies seek to maximize their bombing, knowing that the Japanese will seek a strategy designed to minimize their exposure. Conversely, from the Japanese perspective, their decision to go north is at least as good as their decision to go south *no matter what strategy the Allies deploy.* As it turned out, the Japanese did go north and suffered terrible losses.[139]

Game theory gets much more complicated. Von Neumann and Morgenstern explored games with more than two players, and games that involved collaboration as well as strict competition. Importantly, von Neumann had also explored the conditions of equilibrium not only for pure strategies such as the one illustrated in the example of the Bismarck Sea but also for "mixed strategies," strategies that involved a probability distribution over pure strategies. (In the above example, this might mean flipping a coin to decide between north and south.) Operations research groups leapt at the new technology: one such team (at Johns Hopkins) went so far as to conclude, "A war . . . is composed of many battles, and each battle separately may be considered as a game in which the friendly commander and the enemy commander each has to choose between several strategies." Here, as almost everywhere else, "strategy" was defined "in the usual manner" by direct reference to von Neumann and Morgenstern's *Theory of Games and Economic Behavior.*[140]

I now bring this excursion back to Monte Carlos. Von Neumann had more or less set aside game theoretical work in 1928 and returned to it in 1940–41, bringing *Theory of Games* to completion in January 1943, for publication in 1944. Significantly, the notion of a game does not appear in Ulam's writing during the war, even when he wrote on subjects (such as the mathematics of

139. Haywood, "Military," *J. Oper. Res. Soc. Amer.* 2 (1954): 365–85, on 369, cited in Luce and Raiffa, *Games and Decisions* (1957), 65.

140. Smith et al., "Theory of Value," *J. Oper. Res. Soc. Amer.* 1 (1952–53): 103–13, on 112.

neutron diffusion) in which he later would deploy the concept.[141] But when Ulam and von Neumann wrote the first open publication on the Monte Carlo in September 1947, they introduced games twice in an abstract of only one paragraph. First, they referred to the Monte Carlo procedure as "analogous to the playing of a series of 'solitaire' card games" on a computer. Then, more generally, the method is identified with the "playing [of] suitable *games* with numbers 'drawn' [from a sequence of pseudorandoms]."[142] From this point on, games saturate both Ulam's and von Neumann's writing on the subject.

We can now understand more formally what von Neumann and Ulam meant by the term "game." With Morgenstern, von Neumann had described solitaire as a one-person, non–zero-sum game. In the normalized form, they defined solitaire as the choice of a number $\tau = 1, \ldots, \beta$, following which the one and only player receives a payoff, $H(\tau)$.[143] It is important to remember that each choice of τ specifies not a single "move" but an entire strategy—that is, a determination of what the player will do in every possible situation. In a Monte Carlo, the player's moves are determined by the plucking of a random (or pseudorandom) number and using that number in the algorithmic execution of a step. The "strategy" of the Monte Carlo player is therefore, technically, a "mixed strategy." We can therefore specify the sense in which the evaluation of a multidimensional integral inside a unit cube by a computerized Monte Carlo is a game: such a Monte Carlo is a one person, non–zero-sum von Neumann formal game for mixed strategies $\tau = 1, \ldots, \beta$, where each τ gives a sequence of N n-dimensional points. The "payoff" $H(\tau)$ is the sum of all the "hits" (points in the domain of integration) divided by N.

We can now close the circle between game theory, Monte Carlo, and operations research. Game theory shaped operations research in two ways. First, it led to a specific formulation of deterministic strategic problems (such as the analysis of the Battle of the Bismarck Sea), giving warriors, business leaders, and other analysts a mathematical theory in which the notion of strategy held a formal and restricted meaning. At the same time, operations research began using Monte Carlo simulations as the "natural mode of expression" for the incipient discipline. Philip Morse of MIT, the retiring president of the Operations Research Society of America, emphasized both aspects of "game" in his farewell address in 1952. Of the Monte Carlo, Morse insisted that if one was willing to "play the game" often enough, it would constitute a "random experiment, carried out on paper, so to speak." And this kind of simulated experiment could

141. Hawkins and Ulam, "Multiplicative Processes," LA-171, 14 November 1944, repr. Ulam, *Analogies* (1990).

142. Ulam and von Neumann, "Stochastic and Deterministic," *Bull. Amer. Math. Soc.* 53 (1947): 1120. The paper was first presented by Ulam during the 53rd summer meeting of the American Mathematical Society, New Haven (Yale), 2–5 September 1947.

143. von Neumann and Morgenstern, *Games* (1953), 85–86.

then be combined with human actors playing submarine captains, provided with electrical contacts and charts, battling against other humans simulating the "air player." Together, the human competition and Monte Carlo constituted "gaming technique" or "simulated operational experiment."[144] Though Morse shied away from identifying the game theoretic and Monte Carlo meanings of "game" in his address, in practice the concepts slipped back and forth.

As the concept of game circulated, so too did the scientists. At the 1954 Gainesville conference, Ulam contributed a formal game theoretical paper entitled "Applications of Monte Carlo Methods to Tactical Games," binding the Monte Carlo to formal game theory by combining the two senses of von Neumann's game—formal theory and probabilistic trial.[145] Morse linked the Monte Carlo to games of human competition in his air-sea battles. Participants in this amalgam of activities even merged their concerns with the ordinary sense of game, as when George Gamow took time off from his work on the hydrogen bomb and operations research to design a board game representing a battle between tanks (see figure 8.19). Combining skill of movement and a random element (a "kill" is decided by a flip of a coin), the game might well "have commercial possibilities in the next Christmas gift season," according to the director of the Operations Research Office, Thornton L. Page.[146]

For the interlinked group of Monte Carlo advocates, "game" was a powerful term that bound together apparently extraordinarily diverse activities. It embraced characteristics as diverse as human competition, sampling, probabilistic trials, stochastic physical systems, computer-based simulations, and even, occasionally, entertainment. It joined domains of application—set theory, geometry, molecular physics, nuclear weapons, partial differential equations. As the examples I have given indicate, any combination of these traits could be found together, and none could be called upon as essential. To a philosopher such a state of affairs cannot but be amusing; for Wittgenstein, beginning with the *Blue Book,* the *Brown Book,* and *Philosophical Investigations,* used the concept of a game as the centerpiece of his argument against an essentialist understanding of what it means to be a concept:

> Consider for example the proceedings that we call "games." I mean board-games, card-games, ball-games, Olympic games, and so on. What is common to them all? Don't say: "There *must* be something common, or they would not be called 'games'" but *look and see* whether there is anything common to all. . . . Are they all "amusing"? Compare chess with noughts and crosses. Or is there always winning and losing, or competition between players? Think of patience. In ball games there is winning and losing; but when a child throws his ball at the wall and

144. Morse, "Trends," *J. Oper Res. Soc. Amer.* 1 (1952–53): 159–65, on 163–64.

145. Ulam, "Tactical Games" (1956), 63.

146. Page, "Tank Battle," *J. Oper Res. Soc. Amer.* 1 (1952–53): 85–86, on 85.

Figure 8.19 Gamow's tank strategy game (1952–53). The tank battle game designed by George Gamow is a variant of the Kriegspiel form of chess. The game is played with three boards of hexagonal spaces, one for each player and a third for a referee, with the light areas signifying open fields and the darkened areas signifying woodlands. Neither player can see the other's board; each must infer the opponent's moves from play. The referee records all the moves. A "move" occurs when one player moves any number of his or her tanks into any of the adjacent hexagons (the shape was selected since it allows for a wider choice of movement than a simple square format). Moves alternate until a tank enters a space adjacent to an opposing tank, at which point a "battle" is called by the referee, which is decided by a coin toss. If a tank on a light square adjoins one on a darkened square, the latter automatically wins, and if a tank comes into contact with more than one opposing tank, the outnumbered tank must first defeat one tank in a coin toss, and then move on to those remaining. "Battles" can only take place in the woods (*dark squares*) if a tank tries to move onto the same square as an opposing tank. The game is won when one player's tanks are wiped off the board. Source: Reprinted with permission from Page, "Tank Battle," *J. Oper. Res. Soc. Amer.* 1 (1952–53): 85–86, on 86, copyright 1952, Opertions Research Society of America. No further reproduction permitted without the consent of the copyright owner.

> catches it again, this feature has disappeared. Look at the parts played by skill and luck. . . . And the result of this examination is: we see a complicated network of similarities overlapping and criss-crossing: sometimes overall similarities, sometimes similarities of detail.[147]

In perhaps his most famous analogy, Wittgenstein went on to characterize these similarities as "family resemblances" where traits criss-cross but no single trait is held by all members. "And I shall say: 'games' form a family."[148]

What Wittgenstein says is surely true of the notion of game as sketched above—that is, there is no essential feature included in everything from Gamow's board game to the evaluation of n-dimensional integrals, chemical diffusion, the Battle of the Bismarck Sea, and economic theory. Yet Wittgenstein's metaphor is grounded in the extraordinarily dispersed sense of "game" used in ordinary language. For such a wide domain it is no doubt appropriate to bring to mind an image of a family tree that rapidly disperses as it moves up (or down) the generational line.

In the example examined here, however, there is a density to the meanings of "game" that is not captured by the abstraction of Wittgenstein's analysis. In part this is due to the historically specific state of affairs in which the concept is deployed. Von Neumann and Ulam, along with Gilbert King, Haywood, and others, engaged in mission-directed activity during and just following the Second World War. Not surprisingly, scientists at war shared a way of speaking that, while not completely isolated from the general linguistic community, was partly so. Technical skills and the exclusiveness of formal language served to separate "game" from its everyday sense. And this tendency was then reinforced by the fence of secrecy that tended, at least during the initial and formative stages of the concept, to isolate meaning as well as people.

Considerations like these suggest a modification of Wittgenstein's notion of family resemblances. For quarantined communities, or even partly isolated technical communities, we may find it more useful to think of meanings as bound into local knots rather than dispersed in family trees (see figures 8.20 and 8.21). These knots are not isolated—like a knot in a spider's web, here and there a string leads out. Nonetheless, there is a strong sense in which the local, dense, interconnected knot of technical meaning carries a certain autonomy: movement in the knot is partly autonomous of motions in the web as a whole. Gamow's board game could well have been marketed that Christmas; if it had been it would have been recognizable to the wider community of people as a game not unlike other board games. But for most of the uses discussed here, and as von

147. Wittgenstein, *Philosophical Investigations* (1958), par. 66.
148. Wittgenstein, *Philosophical Investigations* (1958), par. 67.

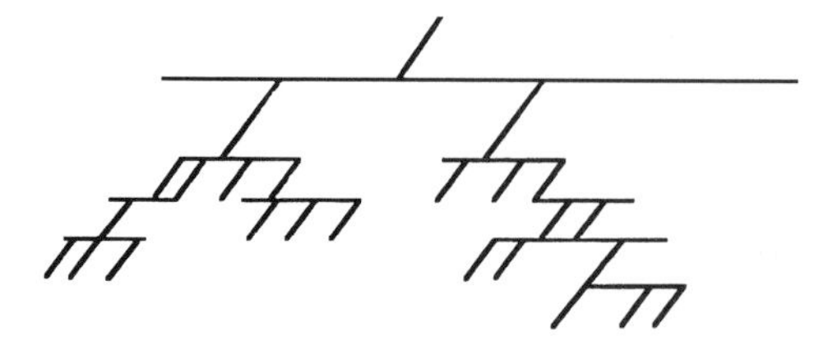

Figure 8.20 Family tree of meaning. In the family tree metaphor, a set of attributes resides with each individual and nearby relatives share some of those attributes; but while each set of close relations has some attributes in common, distant relatives may share no specific attributes. Extended to the often-cited notion of a semantic web, the nodes correspond to statements and the connecting lines to the "semantically salient relations." If one takes seriously the notion that the meaning of a term like "electron" could only be given by the full specification of its location in this web, then only identical theories could have the same set of entailments. (See the critical assessment of holism in Fodor and Lepore, *Holism* [1992], 42.)

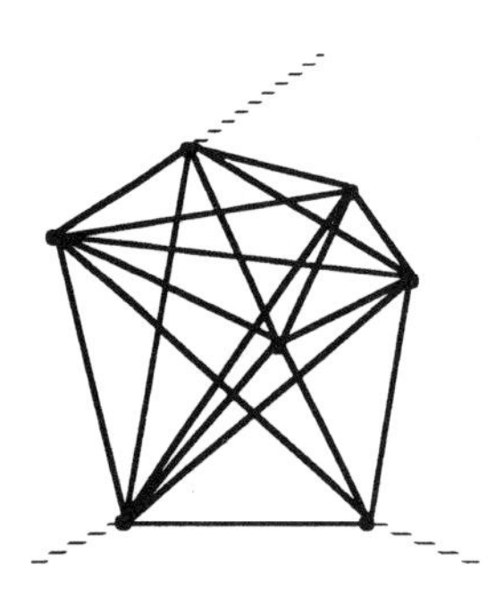

Figure 8.21 Knot of meaning. The "network" remains a useful metaphor for meaning, but it is suggested here that the overly rigid interpretation of the metaphor leads to the unhelpful conclusion that no meaning can be shared among different theories (or languages) if the networks differ in any respect. Instead of such a hard-line view, it is useful to think of real mechanical nets in which complex pieces of the net may move all of a piece (as in a knot), or electrical networks in which the function of a subnetwork may be quasi-independent of its relation to other parts of the network. Such partial independence makes it possible for the subnetwork or knot to specify local meaning relationships more or less without reference to the whole.

Neumann and Morgenstern self-consciously indicated, "game" had to be divorced from everyday usage (see figure 8.22).

Taken together, terms and procedures like "sampling," "game," "pseudorandomness," "importance sampling," "splitting," "Russian Roulette" began to capture a way of approaching problems from a wide variety of fields. The turn of mind grasped by this way of thinking became distinctive, perhaps most strikingly so to outsiders. One British visitor, Dr. Wishart from Cambridge University, stood up after the 1949 Los Angeles conference on Monte Carlos to remark: "As a visitor to this country, I have been impressed, both here and at Oak Ridge, by the knowledgeable way in which speakers who are primarily physicists, or, if not, machinery users, speak of the operation of statistics. I doubt if quite the same mastery of the subject would be found in a similar gathering in England

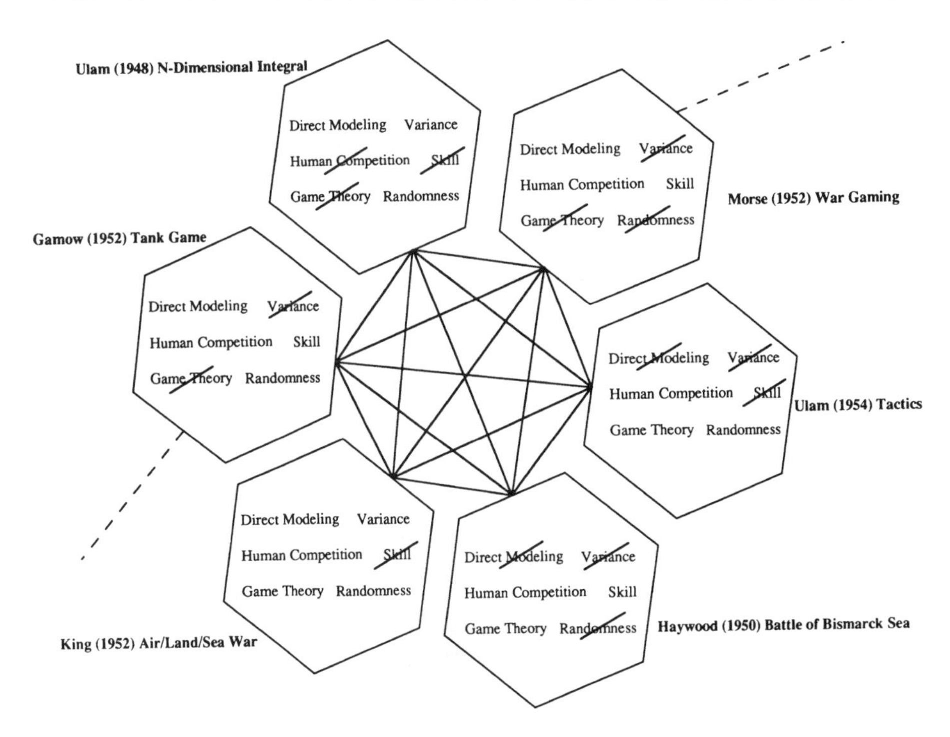

Figure 8.22 Knot meaning for "game." Surely Wittgenstein is right in his famous dissection of the term "game." There are no necessary and sufficient criteria that pick out all and only the things we know as games. Nonetheless, in ordinary games as in the complex of actions and beliefs surrounding game theory and its related structures, certain features do come to the fore—and others recede into the background. Some of the interconnections that were salient to von Neumann, Ulam, and their peers are sketched in this figure.

today, even although times are changing. I find, for example, that the Cavendish Laboratory has become much more statistically minded since the war than it was before."[149] Slowly, this locally worked out language began to build into something more than a matter of provisional utility. Research in Monte Carlos, and into their inner structure, began to gather momentum; theorems postulated and proved, rules of thumb generated, computer programs exchanged. Though the Monte Carlo began as a few crude mediating moves binding different fields of work together, it now began to acquire its own journals, its own experts. By the 1960s, what had been a pidgin had become a full-fledged creole: the language of a self-supporting subculture with enough structure and interest to support a research life without being an annex of another discipline, without needing

149. Wishart from Germond's summary, "Round Table" (1951), 40.

translation into a "mother tongue." The principal site of creation for this hybrid language was Los Alamos. For it was there, in the search for improved nuclear weapons, that a new mode of coordinating activities was built, and there that scientists from different disciplines (different practice and language groups) could trade.

Of course not everyone shared all of the skills of this new "trading zone." Some focused on the game theoretical aspect, others more on variance reduction or convergence problems. There were formalists and practical-minded researchers, workers interested in special methods for particular problems such as inverting matrices and others content to exploit general approaches. For a few years, between 1944 and 1948, the exploration of sampling and stochastic processes proceeded in the hothouse of the weapons laboratory, nurtured with the endless resources of the AEC and the devoted efforts of von Neumann and Ulam. Already during this time, a shared proficiency developed that made the translation of problems into the pidgin of Monte Carlos inviting, almost required. Perhaps because secrecy had at first bound together and isolated the originators of the method, when Monte Carlos broke outside of the AEC enclave they exerted an enormous fascination over the great uncleared; it was an attraction so intense that it sparked warnings about overenthusiasm.

Linguists who study trading languages distinguish among foreigner talk, pidgins, and creoles. Foreigner talk, as its name suggests, requests no reply. It is the language used with a linguistic outsider, in which the speaker reduces structure syntactically, lexically, and phonetically. More nuanced is the pidgin, in which groups are exchanging goods. Highly specific to its exchange function, the pidgin nonetheless functions in a much more varied way than foreigner talk. As the various linguistic functions increase, a pidgin may be pressed into service as the primary language for children as they grow up. At that point, great ranges of linguistic behavior are demanded of the language, such as metalinguistic reflections, irony, humor, and the uncounted conventions that go with daily life. The pidgin becomes a creole.

I will argue at greater length in the next chapter that the language of intertheoretical and interdisciplinary exchange is usefully considered as following such a dynamic. The computer simulation began as an auxiliary function, in which highly reduced versions of physical problems were handed over to the computer (first the woman, then the machine) who executed the calculation. Gradually, the site of simulations expanded; through conferences and collaborative publication, mathematicians, statisticians, physicists, and others carved out a language sufficient to express their joint concerns and to allow the exchange of theorems, algorithms, programs, and tricks. The creole stage was reached when people, graduate students and more advanced scholars, could make a living in the trading zone—when it made sense to become a Monte Carlo computer specialist, neither experimenter nor theorist. Their professional identity might be as physicist, or it might be as a computer scientist. Their domain of inquiry was artificial reality.

Fiction writers have recognized (more clearly than philosophers or historians) the powerful and problematic temptations of artificial reality. In William Gibson's remarkable trilogy,[150] cyberspace—the unlocated computer-driven reality outside of physical existence—holds as much significance as our daily three-dimensional space. Simulations become physical and the physical becomes simulation. Human personalities "jack" into a universal "matrix" of computers, and data protection has the inviolability of a deadly minefield. Indeed, the entire trilogy is structured, in large part, through an unresolved tension between a plot on the plane of physicality and a plot taking place in the matrix. When, at one point, the protagonist stares down through his console to see himself standing in cyberspace, neither he nor we can know where his true self lies. Physical and computer-generated reality have lost the well-understood duality of reality and representation.

The seductiveness of cyberspace as an alternative to coping with the harsh edges of the everyday was apparent to those who worked with simulations. Years after working on simulations of the first H-bomb, a physicist told me this about the time when he was a young postdoc: "I had a strange attitude toward the reality of hardware and the reality of explosions, which it's hard for me to explain now. But [it] was intense and real at the time. I didn't want to see the actual hardware of an atomic bomb in the laboratory . . . in the machine shop, in the metallurgical facility. And I didn't want to see a nuclear explosion."[151] The alternative world of simulation, even in its earliest days, held enough structure to captivate its practitioners. And in their fascination they learned a new way of work and a new set of skills that marked them for a long time to come.

The Monte Carlo and the computers that gave it power did much to bring cohesion to the scientific community, but only by establishing trading zones between otherwise highly disparate groups. It was a cohesion based on the establishment of locally shared culture, a cohesion far from a Comtian dream of a hierarchical tree of knowledge.

8.7 Conclusion

8.7.1 Origins and Interdisciplinarity

For many "native speakers" of mathematics and statistics, the claims by physicists to originality in Monte Carlo work were exaggerated, sometimes even offensive. John Curtiss, a mathematician and statistician trained at Harvard, and in

150. Gibson's trilogy consists of *Neuromancer* (1984), *Count Zero* (1986), and *Mona Lisa Overdrive* (1988).

151. Ford, H-Bomb interview by P. Galison and P. Hogan, 1987.

1949 chief of the National Applied Mathematics Laboratories at the NBS, can be taken as illustrative. He vigorously objected to the new fashion of applying "a rather picturesque name, the 'Monte Carlo Method,' to any procedure which involves the use of sampling devices based on probabilities to approximate the solution of mathematical or physical problems." While granting that Metropolis and Ulam had, in their *Journal of the American Statistical Association* paper, made "a stimulating philosophical introduction," he regretted that they, along with many others, had omitted reference to "background literature." As a result (he diplomatically added), "misconceptions regarding the novelty of the method have arisen from time to time among persons interested in the applications rather than the history."[152]

Probability and functional equations, Curtiss pointed out, had been linked for ages, certainly since the time of de Moivre, Lagrange, and Laplace. In theoretical physics, such connections were a main concern not only of Einstein, but of Smoluchowski, Lord Rayleigh, Langevin, and others. True, Curtiss allowed that Ulam and von Neumann had taken a new point of view by inverting the procedure and using probability to solve deterministic equations. But even here, Curtiss insisted, "the Monte Carlo method is not at all novel to statisticians." In particular, he argued that: "For more than fifty years, when statisticians have been confronted with a difficult problem in distribution theory, they have resorted to what they have sometimes called 'model sampling.' This process consists of setting up some sort of urn model or system, or drawings from a table of random numbers, whereby the statistic whose distribution is sought can be observed over and over again and the distribution estimated empirically.[153] Theoretically, this amounted to the evaluation of a multiple integral over a multi-dimensional region, in short (according to Curtiss), "'model sampling' is clearly a Monte Carlo method of numerical integration." William Sealy Gosset's (a.k.a. Student's) *t*-distribution, he adds, was determined in precisely this way.[154]

Physicists and mathematicians working as physicists typically opened their conferences with exactly the opposite declaration. For example, in the course of introducing the collected papers of one of the early and important meetings on the Monte Carlo method at the University of Florida in 1954, A. W. Marshall (a physicist) defended the priority of Ulam and von Neumann. He conceded that statisticians had for decades been interested in randomized test procedures to investigate the effects of nonnormality on statistical methods devised for normally distributed populations. But "in any case, the statisticians did not have the analogue idea and this is what got Monte Carlo in its current form started." That is, the statisticians were concerned with statistical sampling from a probabilistic

152. Curtiss, "Sampling Methods" (1950), 87–88.
153. Curtiss, "Sampling Methods" (1950), 88.
154. Curtiss, "Sampling Methods" (1950), 88.

population, whereas Ulam and von Neumann wanted to take deterministic equations for a probabilistic analogue model and sample from the latter. Furthermore, Marshall contended, the statisticians did not combine their (separate) variance reduction with problems in the domain of sampling.[155]

Student's t-distribution had, by 1954, become a touchstone. For later statisticians Student was emblematic of the priority their discipline enjoyed. Marshall, not surprisingly, sought to topple Student by dealing him left-handed praise: "As to the [purported] invention of Monte Carlo by Student, it is a strange case. Student did something slightly different, but much better than most Monte Carlo calculations." According to Marshall, Student had come to the t-distribution by purely analytic (if unrigorous) means and wanted to be surer of his conclusion. Taking an exploratory sample, he computed t and tested the sample distribution against his hypothesis using a χ^2, but this was a test, not a generative part of the theory. For both t and r, Student did make guesses, based on prior samples, and then checked them by comparing samples with analytic formulae. "All of this may or may not be Monte Carlo; it is different from most applications, and in any case an isolated instance of first rate use of sampling for statistical purposes." In the same section, Marshall made clear the significance of his removing Student from the lineage of the method: "The development of Monte Carlo independently of the statisticians and particularly their poor showing in applying sampling techniques to their own problems may be surprising to some; especially to whose who think Student invented Monte Carlo."[156]

In sum, Marshall argued, there were three possible definitions of "Monte Carlo." The first identified the method with sampling methods applied to deterministic problems; the second said a Monte Carlo was in use when there was *any* sampling at all (either deterministic *or* probabilistic problems); and the third definition said that the identifying feature of Monte Carlo was the use of sampling methods with variance reduction techniques. Under this last definition (which he preferred) the case was open and shut, since the statisticians had not combined variance reduction with sampling as a means of solving otherwise intractable problems: "It is always hard to decide questions of priority and direction of influence, but in the case of the current development of Monte Carlo the situation is relatively clear."[157] Marshall concluded that the Monte Carlo was born of mathematical physics, and more particularly of the work of Ulam and von Neumann.

Back and forth, the statisticians and the mathematical physicists traded insults and grabbed priority. After Kahn presented a variance reduction talk at the November 1949 IBM Seminar, the mathematician John Tukey riposted: "It seems to me that the physicists are growing up! They are beginning to tackle the

155. Marshall, "Introductory Note" (1956), 3.
156. Marshall, "Introductory Note" (1956), 6–7.
157. Marshall, "Introductory Note" (1956), 3.

hard problems and they are starting in to use the techniques that have been used in the other sciences all along."[158] Kahn had his revenge at the Gainesville meeting of 1954, where he concluded by remarking that there was considerable overlap between certain standard sampling techniques from statistics and the variance reduction techniques he had discussed. But importance sampling and Russian Roulette and splitting had not been treated by the statistical textbooks. "For this reason, it is very valuable to have professional statistical help in designing these calculations. However, if one has to choose between a person who is mainly interested in statistics and one who is mainly interested in the problem itself, experience has shown that, in this field at least, the latter is preferable. This last remark is not intended as a slur on statisticians."[159] Kahn's intention, he said, was to argue that detailed knowledge of a problem is worth far more than the "routine application of general principles." In passing, I would suggest that one general, routinely applicable principle is that when someone says no slur is intended, a slur is usually intended.

Elsewhere, one mathematician went so far as to identify an event from the seventeenth century as the first recorded instance of what might be regarded as an application of the Monte Carlo method, while other statisticians, with some annoyance, presented genealogies of precedents and antecedents to show that sampling methods were nothing new to the statisticians.[160] A few days after the Gainesville conference, Herman Kahn, then writing a brief "historical" introduction that would set Ulam's place in this now-extended pantheon of Monte Carlo progenitors, asked Ulam how to set the scene. Ulam responded:

> It seems to me that while it is true that cavemen have already used divination and the Roman priests have tried to prophesy the future from the interiors of birds, there was not anything in literature about solving differential and integral equations by means of suitable stochastic processes. In fact the idea of Monte Carlo seems to me to consist mainly in inverting the procedures used before, that is to solve problems in probability by reducing them to certain special differential equations. Of course, sampling processes were used in statistics but the idea of using probabilistic schemes to solve problems in physics or pure mathematics was not used before Johnny v. N. and myself were trying it out.[161]

Instead of trying to adjudicate the correct definition of "Monte Carlo" and award credit to this or that contributing individual or field, it is surely more valuable to step back from the debate in order to find out what in the postwar scene fostered the vast expansion of these sorts of calculations and then, more

158. Discussion of Kahn, "Modification" (1950), in Hurd, *Computation Seminar* (1950), 26.
159. Kahn, "Sampling Techniques" (1956), 190.
160. Germond, "Round Table" (1951), esp. 39–40.
161. Ulam to Kahn, 23 March 1954, SUP.

abstractly, to ask why the Monte Carlo had become such disputed disciplinary territory.

Simulations and their growth cannot be understood merely as the sum of a small set of ideas in the statistics of sampling. Instead, the proliferation of the Monte Carlo in the postwar years grew out of distinct elements, some of which came from rather abstract mathematics and some from the details of physics, while others were institutional, technological, even philosophical.

It is impossible to separate simulations from World War II. First, the military demand for shock thermodynamics arose not only for the uranium and plutonium atom bomb projects but also for a variety of other armaments. Von Neumann's involvement with these projects led him deeply, already in 1944, into the creation of models and mechanical calculational devices that would facilitate these simulations. Ulam too was captivated by the nuclear weapons problems, most specifically neutron propagation in fission. It was in this context that early studies of multiplicative processes emerged, allowing approaches to analytic solutions when stochastic processes were at play. Second, the late war and postwar periods were characterized by the installation of high-power computers. It was a two-way street, with simulations driving the increase in computing power and the increased computing power allowing the exercise of ever more complex simulations. This dialectic permitted the redesign of fission weapons with larger and smaller yields, with different radioactive products, and in smaller sizes. The computer facilitated the improvement of nuclear reactors, the abortive (but expensive) development effort of a nuclear airplane, nuclear shielding, and the feasibility studies of the fusion bomb. With both a domain of application (nuclear engineering) and computers with which to calculate, scientists working with the Monte Carlo had virtually limitless resources. Furthermore, the extensive mixing of disciplines in scientific war work encouraged continuing alliances after the war. Whether chemists, physicists, electrical engineers, and mathematicians had worked together on the bomb, radar, or operations research, they had needed to form ways of speaking to one another. And the Monte Carlo techniques did just that, bringing the industrial chemist, meteorologist, number theorist, statistician, and nuclear physicist into the same computer rooms and conference halls with a set of procedures relatively free of ties to any single discipline.

But this partial amalgamation was not without friction. Each group ascribed meanings to the Monte Carlo method that went beyond the region of local assent. And in those exterior regions, there was no consensus. (Here, I would argue, lies the origin of the priority dispute.) In particular, there was little agreement about what the success of Monte Carlos said about nature and indeed no consensus on the boundaries of what would count as a Monte Carlo. Some mathematicians remained uncomfortable with the pragmatic definition of pseudorandomness, and statisticians were never fully satisfied with what many Monte

Carlo workers understood by sampling techniques. But within the more localized region of Monte Carlo simulations, a partial autonomy underwrote at first a set of practices, and later a new domain of inquiry, not quite experimental and not quite theoretical.

8.7.2 The Status of Simulations

When it came time to justify the new technique, the Monte Carlo community's responses were varied. If the borrowed error analysis methods of experimentation provided a first, epistemic argument for calling Monte Carlo runs "experiments," a second metaphysical argument came close on its heels: Like nature itself, and unlike differential equations, the computer-based Monte Carlo proceeds by a series of (simulated) random, finite occurrences. In this strong sense, the early Monte Carlo applications to the diffusion of gases, the scattering of neutrons, and the production of cosmic ray showers *simulated* nature. On this view (the view of many early practitioners of Monte Carlos), the Monte Carlo offered a relation of resemblance between sign and signified in a pre-sixteenth-century semiotic sense. As the Monte Carlo became a standard tool for the resolution of problems with no stochastic elements, the vision of simulations as offering a uniquely privileged vantage point began to dissolve. Nonetheless, the sense of direct access to a problem "as nature poses it," "behind" the equations never quite left the players of Monte Carlo.

Others challenged the legitimacy of knowledge based on simulations. For many theorists, an analytic solution continued to have a cachet inaccessible to the electrical engineers and their approximation methods; analytic solutions got at why something happens in a way that approximations and simulations never could. Indeed, if any theoretical representations stood in Plato's heaven, they were the delicate hypersurfaces of differential equations—not batch-generated random numbers and the endless shuffle of magnetic tape. For experimenters, Monte Carlos never came to occupy a position of "true" experimentation, as exemplified in debates that continued decades later over the legitimacy of according doctorates to students who had "only" simulated experiments. This left the engineers and their successors, the computer programmers, in a peculiar position. They spoke an intermediate language, a kind of formalized creole: the language of computer simulations was understood both by theorists and by experimenters. (It was no accident that conferences flourished with names like "Computing as a Language of Physics.") As such the simulators became indispensable as links between high theory and the gritty details of beam physics and particle collisions. But just as they occupied an essential role in this *delocalized trading zone,* they also found themselves marginalized at both the experimental and theoretical ends of particle physics.

I have argued here that the atomic bomb served both as the subject and metaphorical generator of the Monte Carlo technique. Simulations were essential in enhanced fission weapon work and in the basic design of the thermonuclear bomb. Not only did the weapons projects provide the resources, they yielded the prototype problems on which virtually all early thinking about simulations was predicated: neutron transport, stability analysis, radiation diffusion, pseudorandom generators, and hydrodynamics. Everywhere, from von Neumann's early work on mounting the Monte Carlo on computers to Ulam's original reflections on the method, from Herman Kahn's variance reduction methods to a myriad of applications in air bursts, tissue penetration, and the diffusion of gamma rays, one sees the hand of weapons design. There were, naturally, exceptions: Robert Wilson continued his work on cosmic ray showers with the Monte Carlo, and Enrico Fermi used the computer for calculations on pion-proton resonances. And as the method leapt the fence to evolutionary biology, star clustering, and earth sciences, its Los Alamos origins faded.

But the early world of simulations was steeped in weapons considerations; nuclear bombs saturated every aspect of these early discussions from the language to the self-representation of the simulators. In this respect, I close with a paraphrased summary of Jerzy Neyman, a statistician at the Statistical Laboratory (Berkeley), who opened a 1951 roundtable discussion on the Monte Carlo with an analysis of origins of the new method: "Speaking first of the history of science, he [Neyman] observed that it seemed rather general that an idea begins to explode with sporadic, disconnected events. Each of these may trigger off further explosions in turn, as we imagine the events to occur in a chain reaction. In time these explosions of ideas occur more frequently and we eventually have what might be likened to a mass explosion in which the ideas blossom out and become common knowledge."[162] Perhaps. Looked at from a microscopic point of view, an atomic bomb does proceed in this way: disconnected fissions, scatterings, emissions—a Markov universe plummeting into detonation. But from outside the bomb, we see a mission-oriented laboratory, an extraordinary group of scientists allied with a military infrastructure, struggling to create particular weapons and the intellectual superstructure to facilitate their design and implementation. Have we witnessed the cascade of the participants' narrative of history into the narrative of the physicists' simulations?

In baldest possible form: the computer began as a "tool," an object for the manipulation of machines, objects, and equations. But bit by bit (byte by byte), computer designers deconstructed the notion of a tool itself as the computer came to stand not for a tool, but for nature. In the process, discrete scientific fields were linked by strategies of practice that had previously been separated by object of inquiry. Scientists came together who previously would have lived lives apart, and a new subfield came to occupy the boundary area. Notions of simplicity were stood on their heads: where compact differential equations

162. Germond, "Round Table" (1951), 39.

previously appeared as the essence of simplicity, and numerical approximations looked complex, now the machine-readable became simple and differential equations complicated. Where the partial differential equation had appeared as the exalted furniture decorating Plato's heaven, now Monte Carlo methods appeared to truly represent the deeply acausal structure of the world.

Considerations like these lead us to a clash between two ways of thinking about language and scientific practice. One approach, long familiar, is to take meaning as given implicitly in the relations defined by use. Famously this is what Hilbert and his followers advocated when they let "line," "point," and "plane" be defined only through the propositions of Euclidean geometry. Implicit definitions of this sort appealed to the Vienna Circle, and something like it has often been invoked in the postpositivist philosophy of science.

But the practice of physics is not so homogeneous as Euclidean geometry, and at the dynamic edge of working knowledge, the model of an axiomatized branch of mathematics serves us poorly. I suspect that behind the picture of block relativism lies this notion of meaning: for a given conceptual scheme (such as Einsteinian mechanics or Newtonian mechanics), usage picks out the basic ontological structure of the theory. "Time," "length," "mass," get their meaning from this usage, by analogy with Hilbert's implicit definition of "line," "point," and "plane," in Euclidean geometry. The argument then continues: since "length" (for example) has different properties in Einsteinian mechanics than it does in Newtonian mechanics, physicists working within the two "frameworks" speak "past one another," and in the strongest formulation "live in different worlds," since the basic objects are picked out differently.

In the case of Monte Carlos, this type of argument would shatter the practice of simulation into a thousand pieces. "Game," "experiment," "random," even the notion of a "Monte Carlo" would splinter. "Random," for example, as used by one group of practitioners, including E. C. Fieller, would pick out the instantiation of a process "that has what I can intuitively accept as random characteristics": electrical noise, alpha decay, cosmic ray arrival times. But others, including von Neumann, had no time for such a strict physical notion of random process; "random" designated a series of numbers that was "random for practical purposes." Since von Neumann's pseudorandoms were algorithmically generated, they were surely not compatible with Fieller's usage. And quasi randoms, extraordinary as they were in getting good answers fast, were guaranteed to fail the anticorrelation tests that, for pseudorandom advocates, were the price of admission to the tables of Monte Carlo. On the block-relativist view, we could, with some justification, say that these Monte Carlos had little to do with one another, predicated as they were on different notions of randomness.

Proceeding in this way, we would be led inexorably to split frameworks. Stan Ulam consistently used "experiment" in ways unfamiliar, even shocking, to many working at the bench. Runs of his Monte Carlos became "experiments" in physics done on a computer, and there was worse. Experiments showed up not only in simulated nuclear weapons but in the calculation of nonlinear differential equations, and in the extreme case in logic itself. Similarly, the term "game" applied by von Neumann and Ulam to simulations had long left the parlor. Now this concept took on meanings that evolved over time. There were games with one player and games with several, games with stochastic elements and games without, games played against an enemy nation and games against nature. If we fix our ideas rigidly on the picture of Euclidean geometry with its consistent propositions and unchanging meanings, we can only describe these early, heady days of simulation as the confrontation of a hundred frameworks, each with its own basic entities, each aligned skew to the others. Instead, I would suggest, the growth of Monte Carlo ought intrigue us *because* meanings were in flux, *because* statisticians, mathematicians, weaponeers, quantum mechanicians, and aerodynamical engineers were hammering out meanings they could share at the local intersection of their concerns.

My sympathies are frankly with F. James who, as we saw, looked at the varieties of "randomness," recognized their striking differences, weighed those differences against the common practices of usage and proof, and argued for keeping a shared, practice-based notion of Monte Carlo. True, he argued, quasi Monte Carlo had no stochastic element, and in this respect split significantly from the guiding notion of the original Monte Carlo. Still, the quasi Monte Carlo was used far more like a "regular" Monte Carlo than it was like an eighteenth-century quadrature. Fragile at first, Monte Carlos gradually gained such lore, including techniques of error reduction, inversion, proof structure, and tests. In domain after domain, the simulation came to occupy an irreplaceable crossing point between experiment and theory, the subject of individual expertise, conferences, and textbooks, a pillar in the edifice of early computer science. As was plain in our study of quantum chromodynamics at the Time Projection Chamber, by the 1980s it had become as unimaginable to design, run, and interpret a high energy experiment without Monte Carlos as it would have been to proceed without the physical detector.

At each moment of transition in the meaning of laboratory work, there have been those who resisted, who saw in a vivid if sometimes imaginary past a more complete experiment and more integral experimenter. Even in the age of emulsions, some physicists regretted the passing of the experimenter who did "everything" from building the instrument to reducing the data and writing the paper. Then there were those who accepted the division of labor, but who—facing collaborations of 500—bemoaned the loss of "small" collaborations of 10, 20, or 50. Think too of those senior physicists of the 1960s who looked askance at their young bubble chamber colleagues whose data tape analysis sheltered them from pistons, solder, and wires. Others regretted the loss of equipment building to industry or engineers, discovery to scanners, or authorship to committees. How easy it is to see Monte Carlo coding—the formation of unmaterial experiments—as one fateful last step away from the "reality" afforded by inquiry in an bygone era of "real" experiment.

I have exactly the opposite view. It is precisely because physics has been so deeply enmeshed in the twentieth century, not only in philosophy but also in industry, in war, in technology, and in other sciences, that it has embraced new modes of demonstration: golden events, statistical experiments, simulations. Behind the felt coherence, continuity, and strength of physics lies immense heterogeneity: understanding the coordination of these subcultures is the task of the next and final chapter.

9 The Trading Zone: Coordinating Action and Belief

PART I: Intercalation

Throughout the preceding chapters, and in many different ways, I have tried to convey a sense of the extraordinary diversity of scientific cultures that participate in the production of data. There are long-term traditions of image and logic, but also the differently situated pieces of the laboratory, from Wilson and Powell's engagement with Victorian themes of clouds, volcanoes, and photography through the often uneasy relations between engineers and physicists in the weapons laboratories of World War II. It is a diversity that takes us from the physics factories of radar and A-bombs to the hybrid, dispersed computer-linked laboratories of the late twentieth century. Over and again, I have tried to emphasize both the distinctness of the identities of these different groups and the complex dynamics by which common cause is made between and among them. In this final chapter, I want to turn, more intensively than before, to some of the philosophical and methodological issues that surround the tension between the twin poles of autonomy and interconnection.

9.1 Introduction: The Many Cultures of Physics

I will argue this: science is disunified, and—against our first intuitions—it is precisely the disunification of science that brings strength and stability. This argument stands in opposition to the tenets of two well-established philosophical movements: the logical positivists of the 1920s and 1930s, who argued that unification underlies the coherence and stability of the sciences, and the antipositivists of the 1950s and 1960s, who contended that disunification implies instability. In the previous chapters I have tried to bring out just how partial a theory-centered, single-culture view of physics must be. Forms of work, modes of demonstration, ontological commitments—all differ among the many traditions that compose physics at any given time in the twentieth century. In this chapter, synthesizing previous discussions, and drawing on related work in the history and philosophy of science, I will argue that even specialties within physics cannot be considered homogeneous communities. Returning to the intuition sketched in the introduction, I want to reflect at greater length on a description of physics that would neither be unified nor splintered into isolated fragments. I will call this polycultural history of the development of physics *intercalated* because the many traditions coordinate with one another without homogenization. Different traditions of theorizing, experimenting, instrument making, and engineering meet—even transform one another—but for all that, they do not lose their separate identities and practices.

To oversimplify one might say the following: the logical positivists took the unification project to involve the identification of a "basis" language of observation that would be foundational across all theory. Antipositivists conclusively (in my view) demolished the possibility of such a hard and fast line between experiment and theory and concluded (rightly) that no such "protocol language" could exist. But their argument went further, to a vision of science in which theory and experiment not only became inextricable from one another but so lost their separate dynamics that it did not make sense to think about breaks in one sphere of activity without assuming concomitant shifts in the other. There is another (logical/historiographical/philosophical) alternative: Invert the quantifiers. Agree that there is no observation language valid across every theory change, but at least leave open the possibility that for each change of theory (or experiment or instrumentation) there is a sphere of practice that continues unbroken. The burden of this chapter is to explore both historiographically and philosophically what it would mean to have such an intercalated history.

There are different ways to describe the partial autonomy of experiment from theory. Ian Hacking came to the split between theory and experiment through his argument for a restricted realism, in which the epistemic grounds for belief in an entity come from the ability to manipulate it, not from the entity's role within a grand theory.[1] Beyond the realism question, as we have seen time after time, experimenters have their own continuing concerns and systems of belief that cut across even extreme changes in theory. Cloud chamber, bubble chamber, and spark chamber physicists all worked with near-perfect continuity across major dislocations of high theory—and in many cases even across dislocations in the theory of the instrument itself. Wilson and his cloud chamber and Glaser and his bubble chamber both show that instrument making can even cut continuously across radically separated domains of experimental inquiry.

1. Hacking, *Representing* (1983).

Wilson passed smoothly back and forth from meteorology to atomic physics, while Glaser rode his visualizing machines from cloud chamber particle physics to bubble chamber particle physics to microbiology.

My original hope (which I sketch in part I of this chapter) was that such a laminated description of the larger community would do two things at once: it would underline the heterogeneity of practice within the wider physics community while allowing continuities on one level to bridge discontinuities on another. Instrument practices would continue unbroken when theory split; theories would carry on when new technologies transformed the laboratory; experimental practices might go on even when instruments altered; and in this way a macrocontinuity would coexist with local breaks. Physicists' own experience of physics as maintaining a certain continuity even across conceptual breaks might, on this account, be ascribed to the local existence of continuity in the not purely conceptual arenas of practice.

But the more I pressed the laminated picture of intercalated practices (part II of this chapter), the more it seemed to delaminate. The criteria that divided the practitioners of theory, experiment, and instrumentation—different meetings, different preprint exchange, different journals—were the classic sociological dividers Kuhn and many others since have invoked to identify distinct communities. Moreover, the experimenters and theorists often disagreed as to what entities there were, how they were classified, and how one demonstrated their existence—just the criteria Kuhn used to identify incommensurable systems of belief. With distinct communities and incommensurable beliefs, the layers seem to fall apart like decaying plywood; if they are significantly disconnected—if there are distinct communities using terms like "mass" and "energy" in significantly different ways—then the continuity of one level would hardly bolster discontinuity at another.

These considerations so exacerbated the problem that it seemed as if any two cultures (groups with very different systems of symbols and procedures for their manipulation) would be condemned to pass one another without any possibility of significant interaction. But here we can learn from the anthropologists who regularly study unlike cultures that do interact, most notably by trade. Two groups can agree on rules of exchange even if they ascribe utterly different significance to the objects being exchanged; they may even disagree on the meaning of the exchange process itself. Nonetheless, the trading partners can hammer out a *local* coordination despite vast *global* differences. In an even more sophisticated way, cultures in interaction frequently establish contact languages, systems of discourse that can vary from the most function-specific jargons, through semispecific pidgins, to full-fledged creoles rich enough to support activities as complex as poetry and metalinguistic reflection. The anthropological picture is relevant here. For in focusing on local coordination, rather than global meaning, one can understand the way engineers, experimenters, and theorists interact. At last I come to the connection between place, exchange, and knowledge production. But instead of looking at laboratories simply as the places at which experimental information and strategies are generated, my concern is with the site—partly symbolic and partly spatial—at which the local coordination between beliefs and action takes place. It is a domain I call the trading zone.

9.2 Logical Positivism: Reduction to Experience

Early in this century, the logical positivists sought to ground knowledge on the bedrock of experience. Rudolf Carnap's masterwork, *Der Logische Aufbau der Welt,* is usually translated as *The Logical Structure of the World* but might better be construed as *The Logical Construction of the World.*[2] For it is a construction, a building-up from the elementary bits of individual sensory experience to physics, then to individual psychology, and eventually to the totality of all social and natural sciences. To secure the foundations of this construction, both Carnap and Otto Neurath argued at length that some form of "protocol statements" and their manipulation through logic would form a language that would guarantee the validity of complex inferences constructed with them. "We assumed," Carnap recalled later, "that there was a certain rock bottom of knowledge, the knowledge of the immediately given, which was indubitable. Every other kind of knowledge was supposed to be firmly supported by this basis and therefore likewise decidable with certainty. This was the picture which I had given in the *Logischer Aufbau.*"[3] Carnap had a picture of knowledge raised like a building, from a firm foundation of observation, through the upper stories of physical theory, and up from there to the autopsychological, the heteropsychological, and the cultural.[4]

The picture in figure 9.1 might be helpful, encapsulating what I will call the positivists' "central metaphor." Historians begin any investigation, implicitly or explicitly, with a periodization—a methodological commitment that prescribes the breaks and continuities appropriate to the domain under study. By fastening on reports of experience as the basis and the unifier of all science, the positivists committed themselves to an unbroken, cumulative language of observation. For Carnap, theories carried no such guarantee. Only as long as they could account in a shorthand way for the results of experience would they stay. Theories come and go, but protocol statements remain.

2. On the notion of *Aufbau,* see Galison, "Cultural Meaning of *Aufbau*" (1993); an interpretation of Carnap's *Aufbau* that looks forward to Carnap's later work in *Logical Syntax* can be found in the excellent article by Michael Friedman, "Carnap's *Aufbau* Reconsidered," *Noûs* 21 (1987): 521–45.

3. Carnap, "Intellectual Autobiography" (1963), 57.

4. While Carnap himself enriched this view considerably with varying forms of "structuralism" and conventionalism (as discussed in the introduction), it was this "bedrock" view that was more widely associated with his work among both historians and philosophers.

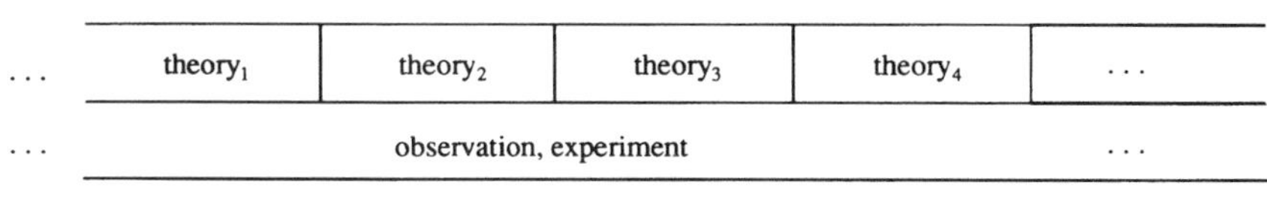

Figure 9.1 Positivist periodization. In the positivist periodization, the continuity and strength of the scientific picture issue from the accumulation of empirical results. Theory can and does change dramatically as needed to accommodate the new data.

In some respects physicists early in the century found themselves comfortable with the positivists' representation of their discipline. The empiricist (Machist) temper is clear in Einstein's early work and writings, and in the operationalism of Percy Bridgman.[5] Indeed, the American predilection for experiment before theory was so entrenched that the physics department at Harvard refused to grant Edwin Kemble his 1917 Ph.D. for work on quantum theory until he produced an experimental result.[6] On the Continent, where theoretical institutes did exist (led by theorists like Planck, Lorentz, Riecke, Sommerfeld, and others), they were often physically separate (in their own buildings or in separated sections of the same buildings) from their experimental counterparts.[7]

During the 1930s, theoretical physics in the United States came of age with a first generation of American quantum theoreticians and an influx of eminent refugees. The growing subculture of theorists, however, did not entirely displace the prevailing positivist orientation, which remained visible, even after the war, in the building of physics facilities. For example, it was typically the case that, in mixed-discipline science buildings, chemistry was put above physics, and laboratories above shops. One might think that this was for practical reasons, and indeed practical reasons were often given: physics experiments must be free from vibrations and so must be lower down; chemistry labs produce fumes and so must be higher up to vent them. But architects point out that isolation from vibration can be ensured on upper floors and chemical fumes are as often heavier than air as lighter.[8] I suspect, therefore, that in the floor plans we are seeing far more than pragmatically situated air ducts; we are witnessing a physicalized architecture of knowledge. Consider the Brandeis science building in figure 9.2: the ground floor is primarily shops and instrumentation, the first

5. Holton, "Mach, Einstein, and the Search for Reality" (1988), 237–78. On Bridgman, see Walter, *Bridgman* (1990).

6. Kemble, interview by the author, 1977. For more examples of this proexperimental rule, see Sopka, *Quantum Physics* (1988), 23–25.

7. See Jungnickel and McCormmach, *Intellectual Mastery* (1986), e.g., Göttingen, 2:115; Leipzig, 2:181; Munich, 2:183, 274, 281–85; Berlin, 2:51–52, 254–55, 277. Often precision measurement was associated with theory and the combination of theory and precision measurement was differentiated from experiment.

8. Palmer and Rice, *Modern Physics Buildings* (1961), 28.

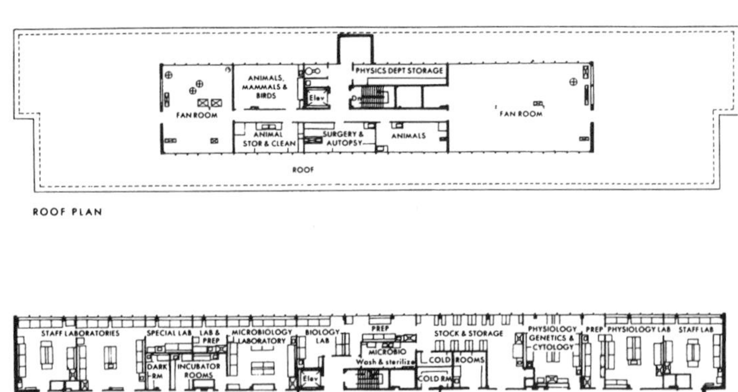

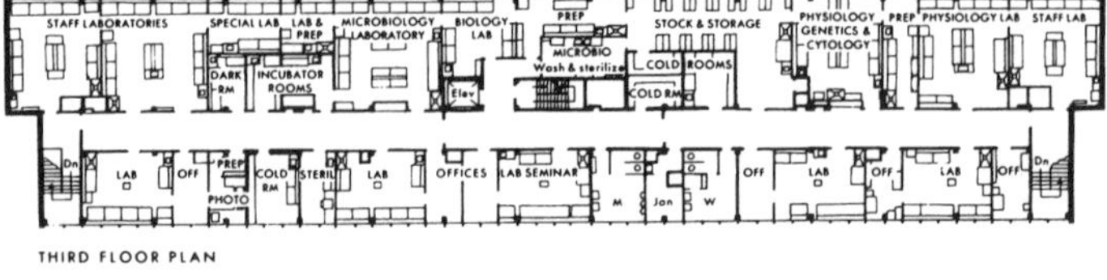

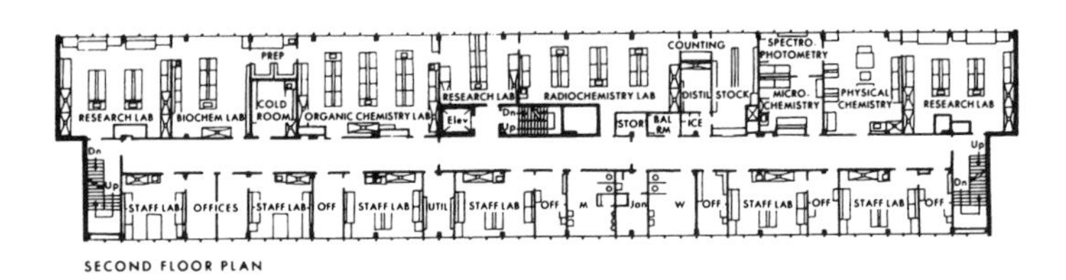

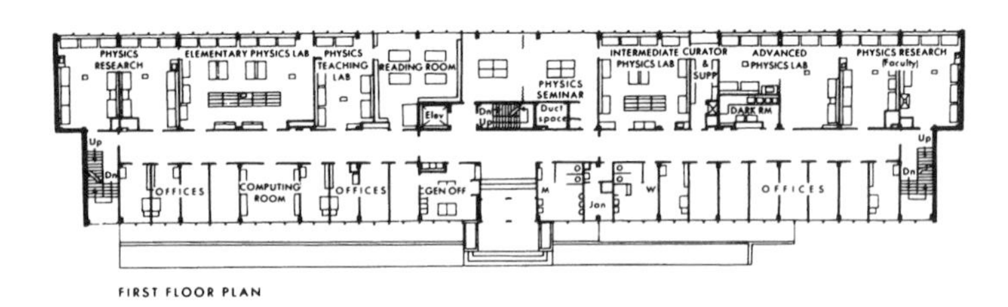

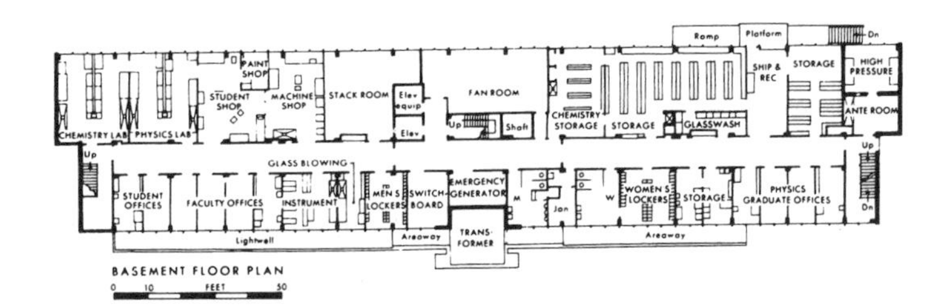

Figure 9.2 Positivist disciplinary architecture (1961). Kalman Science Building, Brandeis University, Waltham, Massachusetts. Source: Palmer and Rice, *Modern Physics Buildings* (1961), 91.

floor physics, the second floor chemistry, the third floor microbiology, and the roof for animals, birds, and physiology. Not surprisingly, when a building is dedicated to physics the theory group occupies the top (or next to top) floor, as in figure 9.3. It has become so routine to put the theorist on the top floor that the architect for the Liverpool physics tower (an eight-story building with experiments all the way up) commented in the late 1950s: "Usually a theoretical physicist will require services, a blackboard and in some cases a bed to lie on to help his deliberations. . . . [W]e have put him in an 'ivory tower' overlooking the city in the vain hope that the view over Liverpool may inspire him."[9]

Historians of science participated in the positivist movement of the philosophers and scientists. It is no accident that the justly famous *Harvard Case Histories in Experimental Science* chronicled *experimental* triumphs: Robert Boyle's uncovering of the gas law, Pasteur's inquiry into fermentation, and Lavoisier's overthrow of the idea of phlogiston.[10] As the laboratory workers marched onward, it came as no surprise to the positivists, or to their historian counterparts, that theory fractured. If the equation $PV = nRT$ better accommodated observation, let it stand; if oxygen organized the facts in the laboratory better than phlogiston, then leave phlogiston by the way. The unification of science occurred at the level of observation/experiment (no sharp distinction being made between them); and the stability of the scientific enterprise rested on the belief that this continuous, unified "physicalist" language provided a continuous, progressive narrative through the history of that science.

9.3 Antipositivism: Reduction to Theory

The 1950s and 1960s saw a sharp reaction in both history and in philosophy of science against the positivist picture with the infusion of the newly popularized notion of conceptual schemes. The notion of a conceptual scheme was itself built on a long line of thought. As Patrick Gardiner and Klaus Köhnke (among others) have pointed out, the late-nineteenth-century neo-Kantian tradition removed the universalist element from Kant's concepts and categories and emphasized the possibility of multiple schemes through which questions could be framed.[11] Certainly long before World War II, Pierre Duhem, Henri Poincaré, Otto Neurath,[12] and to a certain degree Carnap[13] were all taken with the possi-

9. Spence, "The Architect and Physics" (1959), 14.

10. Conant and Nash, *Harvard Case Histories* (1950–54).

11. Gardiner, "German," *Monist* 64 (1981): 138–54. See also the helpful introduction to Krausz and Meiland, eds., *Relativism* (1982); and Köhnke, *Neo-Kantianism* (1991).

12. On Duhem, Poincaré, and the holism of Neurath, see the essays by Uebel, Haller, and others in Uebel, *Vienna Circle* (1991). As Haller notes, Neurath emphasizes that he rejects "the expression that a statement is compared with 'reality', and the more so, since for us 'reality' is replaced by several totalities of statements that are consistent in themselves but not with each other" (Neurath, "Physicalism" [1983], 102).

13. One of the most significant recent essays in the history of logical positivism is Friedman's "Carnap's *Aufbau* Reconsidered," *Noûs* 21 (1987): 521–45. Opposing a foundationalist view of the *Aufbau*, Friedman argues

bility of different structures of scientific belief. By 1936 Quine was explicitly using the term "conceptual scheme" in "Truth By Convention," though he used it in the singular: "*our* conceptual scheme."[14] As usage pushed conventionalism ever further toward an explicit relativism, the plural usage gained ground, so much so that after World War II, its historical track of "conceptual schemes" is too ramified to be traced. James Bryant Conant puts the term on the pages of his *Harvard Case Histories in Experimental Science* (1950) over 30 times—and that is just in his short introduction and first chapter. There were Toricellian, Boylean, Daltonian, even nuclear structure conceptual schemes—conceptual schemes were to be studied in their competition and in the dynamics by which one supplanted the other.[15] Philipp Frank, having moved from prominence in the Vienna Circle to Harvard before World War II, already by the early 1950s strongly advocated the creation of a sociology of science that would explore how the shifts in the "conceptual environment" would transform a well-known effect, such as the elongation of the spectrum, into the basis for a "new revolutionary theory." This new sociology of science would take as a central goal "the resistance of scientists to new conceptual schemes: a) sources of the resistance, b) techniques employed in resisting."[16] Benjamin Whorf's widely read writings on the Hopi exploited the idea copiously,[17] and the conceptual scheme rapidly became such a pervasive idea in anglophone philosophy of science that it is hard to find a programmatic statement about science in the 1950s or 1960s without it.

One of the most powerful philosophical statements of the 1950s was Rudolf Carnap's essay "Empiricism, Semantics, and Ontology," defending the view that a given linguistic framework entirely determined the objects that existed for that framework. No questions about existence outside the framework were even

---

that the *Aufbau* should be taken as but one of many possible such structures, the purpose of which was to secure objectivity by locating every concept in a fully articulated place.

14. Quine, "Truth by Convention," *Paradox* (1976), 102: "There are statements which we choose to surrender last, if at all, in the course of revamping our sciences in the face of new discoveries; and among these there are some which we will not surrender at all, so basic are they to our whole conceptual scheme. Among the latter are to be counted the so-called truths of logic and mathematics, regardless of what further we may have to say of their status in the course of a subsequent sophisticated philosophy."

15. Conant and Nash, *Harvard Case Histories* (1950–54), x, 8, 9, 10, 11, 25, 50, 58, 59, and 62. Conant refers to Daltonian, Toricellian, Boylean, and nuclear structure conceptual schemes. For Conant, the term embraces something broader than narrow, specific beliefs. The closest he comes to a definition is: "The word 'theory' is commonly used to mean either a working hypothesis or a well-accepted conceptual scheme. Because of the resulting ambiguities, we prefer to use the phrases 'working hypothesis on a grand scale' or 'broad working hypothesis' for a new idea in its initial phases. As soon as the deductions from such a hypothesis have been confirmed by experimental test and the hypothesis is accepted by several scientists, it is convenient to speak of it as a conceptual scheme" (66).

16. Philipp Frank, "Possible Research Topics: Sociology of Science," [1953] mimeographed document distributed in the context of the Institute for the Unity of Science, RF RG 1.1 100 Unity of Science, 1952–1956; Box 35, Folder 285. RFP.

17. Whorf, *Language* (1956).

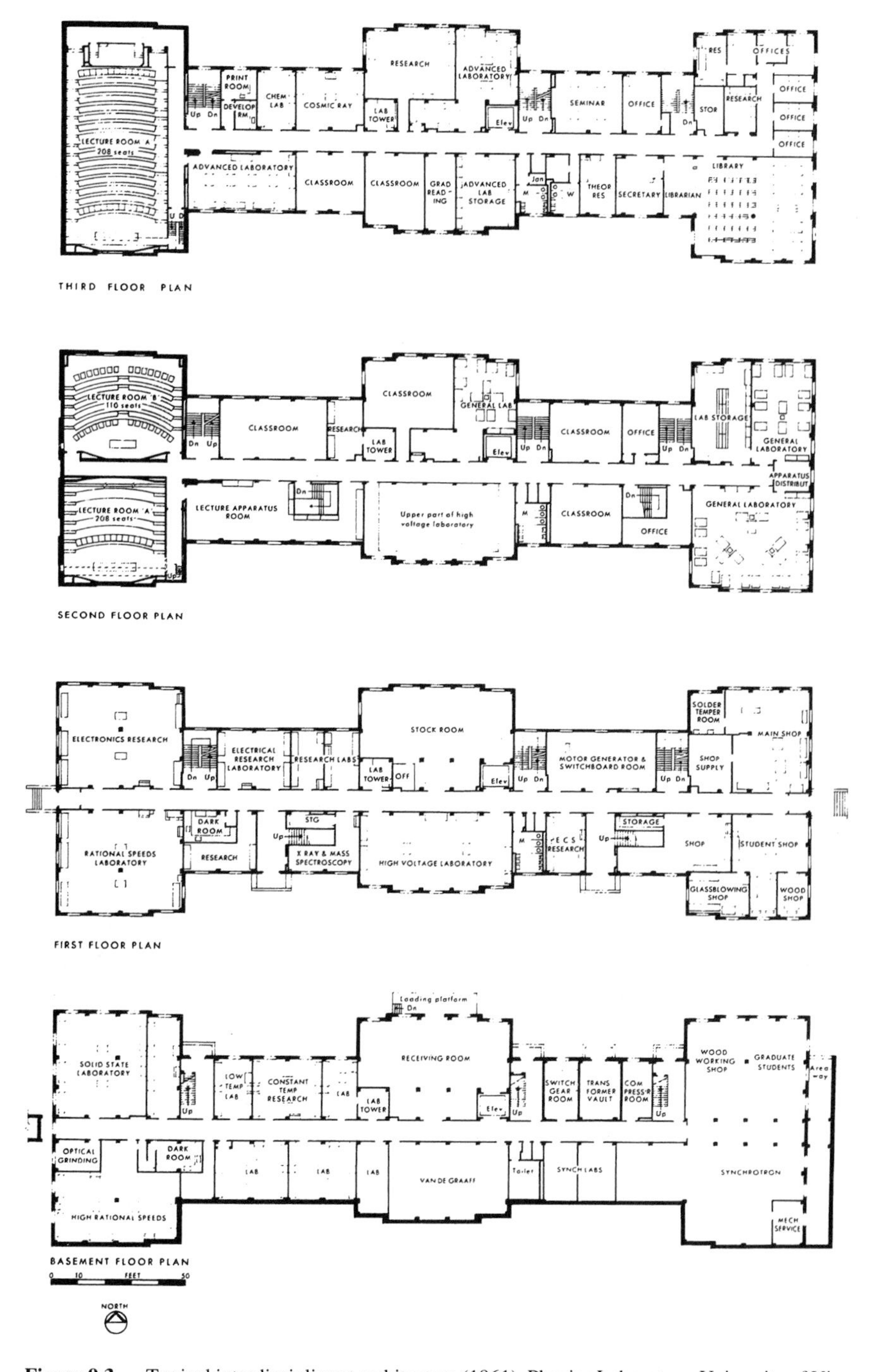

Figure 9.3 Typical intradisciplinary architecture (1961). Physics Laboratory, University of Virginia, Charlottesville, Virginia. Source: Palmer and Rice, *Modern Physics Buildings* (1961), 85.

interpretable as cognitively meaningful utterances.[18] Thomas Kuhn's paradigms served many functions (e.g., historical and sociological explanation) that bore no relation whatsoever to Carnap's syntactical or semantical account of ontology. But Kuhn's picture of different worlds of meanings and objects, located within blocks of knowledge (and language) incomparable with one another, ran continuously from Philipp Frank's "conceptual schemes" or Carnap's "frameworks" to "paradigms." It is in this new notion of a paradigm, and related ideas of the early 1960s, that the block representation of theory-plus-experiment gained the historiographical, sociological, and philosophical force it has held in differing forms for several decades.

Most important, the antipositivists insisted that no Carnapian protocol language (understood in the foundationalist sense) could exist even in principle, a condition sometimes referred to as "theory contamination" or "theory ladenness." Following the philosophers' lead—more than they sometimes might care to admit—historians of biology, chemistry, and physics adduced example after example in which theory changed first—and experiments then conformed to fit the mold.

Some of the leading antipositivists—including Kuhn and Russell Hanson—continued the positivists' fascination with early-twentieth-century Gestalt psychology and put it to new use. They now argued that theoretical and linguistic changes of science shifted with the abruptness and totality of a gestalt switch.[19] Just as the duck became a rabbit, experiments showing the absence of phlogiston now became experiments displaying the presence of oxygen. Theory shifts *forced* changes all the way through experience, leaving no bit of language, theory, or perception unaffected.

Paul Feyerabend spelled out his antipathy for the positivists' central metaphor: "[My] thesis can be read as a philosophical thesis about the influence of theories on our observations. It then asserts that observations . . . are not merely theory-*laden* . . . but fully *theoretical* (observation statements have no 'observational core'). But the thesis can also be read as a historical thesis concerning the use of theoretical terms by scientists. In this case it asserts that scientists often use theories to restructure abstract matters *as well as* phenomena, and that no

18. Carnap, "Empiricism" (1952), 219. It does not make sense, Carnap insists, to ask whether electrons exist independently of a framework: "[T]he acceptance [of new linguistic forms] cannot be judged as being either true or false because it is not an assertion. It can only be judged as being more or less expedient, fruitful, conducive to the aims for which the language is intended." While Quine protested against Carnap's use of the analytic/synthetic distinction, he saw no problem identifying frameworks with conceptual schemes: "Carnap maintains that ontological questions, and likewise questions of logical or mathematical principle, are questions not of fact but of choosing a convenient conceptual scheme or framework for science." Quine's caveat was that the framework could not be restricted to only part of the whole of science. Quine, "Carnap's Views," *Paradox* (1976), 211.

19. Kuhn, *Scientific Revolutions* (1970), chap. 10; and Hanson, *Patterns of Discovery* (1958), chaps. 1–4.

part of the phenomena is exempt from the possibility of being restructured in this way."[20]

For Feyerabend, the distinction between theoretical and observation terms was "purely psychological" (as opposed to the privileged role that observation held for the Vienna Circle). Through his own historical examples from the time of Galileo and classical antiquity, and via allusions to the wider historical and sociological literature, he contended that "[w]e may even say that what is regarded as 'nature' at a particular time is our own product in the sense that all the features ascribed to it have first been invented by us and then used for bringing order into our surroundings." In a doctrine he linked to Kant, Feyerabend insisted on the "all-pervasive character of basic theory."[21] And while Feyerabend allows that in certain particular cases, there may be facts held in common for different competing theories, in general that is not so: "Experimental evidence does not consist of facts pure and simple, but of facts analyzed, modelled, and manufactured according to some theory."[22] Sometimes theories shape the scientific community's treatment of error, sometimes theory fashions the criteria of data selection, and even more pervasively theory is used to express the data. As an epigram for his views Feyerabend chose a morsel of Goethe: "Das Höchste zu begreifen wäre, dass alles Faktische schon Theorie ist."[23]

Kuhn's view was similarly grounded in a thoroughgoing attack on the possibility of a sense-data language: "The point-by-point comparison of two successive theories demands a language into which at least the empirical consequences of both can be translated without loss or change. . . . Ideally the primitive vocabulary of such a language would consist of pure sense-datum terms plus syntactic connectives. Philosophers have now abandoned hope of achieving any such ideal, but many of them continue to assume that theories can be compared by recourse to a basic vocabulary consisting entirely of words which are attached to nature in ways that are unproblematic and, to the extent necessary, independent of theory."[24] As Donna Haraway put it in 1976, "Operations and manipulations [Kuhn] feels, are determined by the paradigm and nothing could be practically done in a laboratory without one. A pure observation language as the basis of science exactly inverts the order of things."[25] This was the enemy: a neutral, unproblematic Archimedean point outside of a theoretical structure.

20. Feyerabend, *Realism* (1981), x.

21. Feyerabend, *Realism* (1981), 45, 118. The link to Kant occurs on 45: "As is well known, it was Kant who most forcefully stated and investigated this all-pervasive character of theoretical assumptions."

22. Feyerabend, *Realism* (1981), 61.

23. Feyerabend, *Realism* (1981), x.

24. Kuhn, "Reflections" (1970).

25. Haraway, *Crystals* (1976), 7–8.

Central to the antipositivist image of science was the idea that the positivists' dream of correlating bits of experience with theoretical propositions would never succeed. Theory had to be accorded pride of place. Imre Lakatos modeled the basic unit of scientific progress, the program, by a series of concentric rings—not too different in spirit from Quine's battened-down fabric. In the center (where else?) lay the "hard core" of a theory; for example, in the Newtonian program, the dynamical laws and the inverse square law of gravitation lay in this hard core. Surrounding and insulating this theoretical core was the "protective belt" composed of auxiliary assumptions added to the program to rescue the hard-core beliefs against refutation. As long as the introduction of auxiliary assumptions led to fruitful new discoveries and explanations, the program was progressive; when the auxiliary assumptions contributed only marginally to the advancement of learning, the program "degenerated" and was discarded.[26]

Of interest in the Lakatosian model is, once again, the *primacy* of theory: theoretical assumptions lie in the hard core and survive all but the most sustained attack against the program as a whole. When the basic elements of theory do eventually crumble, the results are catastrophic (or liberating according to one's perspective). The totality of the program collapses, hard core along with the myriad of lesser, "lower level" assumptions linking high theory to the phenomenal world. Such a view was intended to break irrevocably with the positivists' position.

An example makes the contrast between positivist and antipositivist sharp. For a logical empiricist like Carl Hempel, both philosophy and history had shown the gradual build-up from experiment, to phenomenal laws, and ultimately to theory. Thus, according to Hempel, the archetypal progression from fact to theory was well illustrated in atomic physics: Anders Jonas Ångström mapped the spectral lines of hydrogen, J. J. Balmer codified those relations in an empirical formula, and Niels Bohr explained the formula on the basis of the old quantum theory.[27] By contrast, in Lakatos's scheme, experiment and empirical laws play a negligible role. Of Balmer, Lakatos has this to say: "[T]he progress of science would hardly have been delayed had we lacked the laudable trials and errors of the ingenious Swiss school-teacher: the speculative mainline of science, carried forward by the bold speculation of Planck, Rutherford, Einstein and Bohr would have produced Balmer's results deductively, as test-statements of their theories, without Balmer's so-called 'pioneering.'"[28] For Lakatos, as for many of his antipositivist contemporaries, theory was the engine of scientific change. Theory advanced of its own accord ("In the rational reconstruction of science there is

26. Lakatos, "Methodology" (1970).
27. Hempel, *Natural Science* (1966), 37–39.
28. Lakatos, "Methodology" (1970), 147.

little regard for the pains of the discoverer of 'naive conjectures.'"),[29] and the interesting history was the history of this domain of speculative exploration. Because theory was so central, when theory itself fragmented the whole cloth of scientific activity was rent, effectively torn into unrelatable bits. Instead of the image conveyed in figure 9.1, the antipositivists chose as their central metaphor a picture of scientific change that was grounded in theory.

Antipositivist historical examples multiplied. Hanson argued that before the Dirac theory in 1932, cloud chamber workers like Dimitry Skobeltsyn simply could not "see" certain particle tracks as positrons.[30] In his elegant work on Millikan's oil drop experiments, Gerald Holton contended that "theoretical presuppositions" shaped the ways in which Millikan accepted and rejected data in his notebooks.[31] John Heilbron took the Leyden jar as a perfect example of the ways in which theory caused an instrument to be reformulated.[32] As was evident from Kuhn's use of the psychological works of Albert Hastorf, Jerome Bruner, Leo Postman, and John Rodrigues,[33] it was perceptual psychology that offered a model for the attack on neutral sense-data.

The positivist central metaphor was upended: now theory had primacy over experiment/observation; phenomena were no longer exempt from breaks. When theory changed, the rupture tore through the whole fabric of physics—including experiment/observation. Over such fissures between the plates of science nothing could cross. A new central metaphor replaced the old (see figure 9.4).

The antipositivists' central metaphor has been extraordinarily fruitful. It has precipitated new philosophical debates on meaning and reference, and novel historical insight into the practice of science. No longer can science be described by the fantasy in which observation was simply cumulative, and in which theory was isolated from philosophical commitments, reduced to a mere shorthand for logical strings of protocol statements.

The positivist and antipositivist periodizations have a grandeur to them: they both sought and found a single narrative line that would sustain the whole of science—in observation for the positivists and in theory for the antipositivists. Both agreed that language was the linchpin of science—though the positivists looked for a language of experience, and the antipositivists located the key terms in theory. The positivists concluded that the common foundation of all specialties in basic observations guaranteed the unity of science. By denying the possibility of this foundation antipositivists, preeminently Kuhn, split even the single discipline of physics into a myriad of noncommunicating parts separated

29. Lakatos, "Methodology" (1970), 147.

30. Hanson, *Positron* (1963), 136–39.

31. Holton, *Scientific Imagination* (1978), 25–83.

32. Heilbron, "Leyden Jar," *Isis* 57 (1966): 264–67.

33. Kuhn, *Scientific Revolutions* (1970), 113.

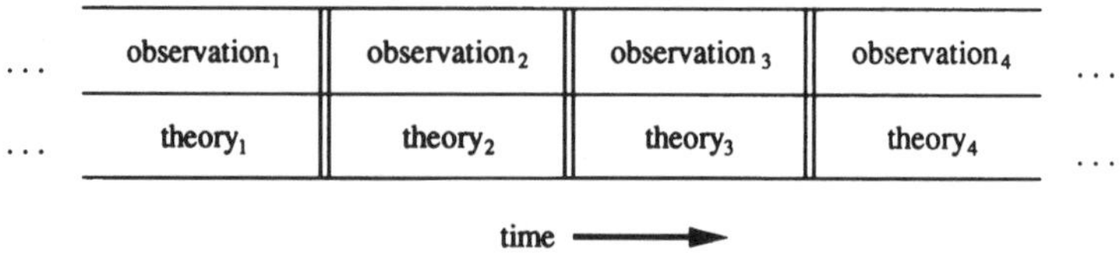

...	observation$_1$	observation$_2$	observation$_3$	observation$_4$	...
...	theory$_1$	theory$_2$	theory$_3$	theory$_4$	...

time ⟶

Figure 9.4 Antipositivist periodization. The antipositivist periodization inverts the positivist one in this respect: theory comes first. But now an additional assumption enters the picture, as theory and observation are assumed to be coperiodized—every powerful change in theory carries with it a concomitant shift in the standards of observation.

by "microrevolutions." All was tied to the language and reference of theory, and theory was multiply torn.

To enforce the idea that the shift was a gestalt switch, it was necessary to insist that the moment of theory change was also the moment of empirical fracture. I have tried to capture this image in figure 9.4, now with the breaks of periodization occurring simultaneously at the theoretical and experimental levels. Furthermore, epistemic primacy has shifted from the empirical to the theoretical. The statement that it is impossible to communicate across empirical gaps appears in this image as the totality of the rupture through all layers of scientific practice. Or, said another way (Kuhn's way), it is the absence of a continuous substratum of common practice across the break that underlies the image of "paradigm shifts" and "different worlds," in which there is no overarching notion of progress. This is the thesis that has generated so much controversy in the community of historians and philosophers of science since the early 1960s, reactivated in the 1980s by the neo-Kuhnian strand of social constructivism (think here of Andrew Pickering's important contribution "Against Putting the Phenomena First").[34]

The central metaphor of the antipositivists has much to recommend it. By their critique of the positivist vision of a simply progressive empirical domain, the antipositivists drew attention to the dynamic role that theory plays in experimental practice. This strategy created historiographical room to link theoretical concerns with the larger context of scientific work, including philosophical commitments, ideological assumptions, or national styles of science. A myriad of interesting historical studies have by now revealed how theoretical notions have significantly altered the construction, interpretation, and valuation of experimentally produced data. Moreover, there is no doubt—as the antipositivists persuasively argued—that there are breaks in the world of observation. The systematic study of the attraction and repulsion of rubbed objects does not contin-

34. Pickering, "Phenomena," *Stud. Hist. Phil. Sci.* 15 (1984): 85–117.

uously meld into later experimental investigations into electrostatics and then electrodynamics.[35]

There is an elegance to both the positivist and antipositivist pictures. Both have a grand scope as they set out to find universal patterns of scientific evolution, and both follow language as the guide to that understanding. In Carnap's words, "there is a *unity of language* in science, viz., a common reduction basis for the terms of all branches of science, this basis consisting in a very narrow and homogeneous class of terms of the physical thing-language."[36] Neurath put the same thought less technically when he insisted on the universality—the internationality—of the language of science which was to underlie the move toward unified science: "Unified science is therefore supported, in general, by the scientific attitude which is based on the internationality of the use of scientific language."[37] It is this sense of the linguistic nature of the science problem that underlay both the positivist credo of unification and the antipositivist departures from it. Thus, as Charles Morris wrote, "The degree of unity or disunity of science reveals itself here in the degree to which the sciences have or can have a common linguistic structure."[38]

When the antipositivists lashed out at the positivists they accepted Carnap's, Neurath's, and Morris's parameters of the debate—and answered that there was no common linguistic structure. Certainly this is a consequence of the later Carnap's linguistic frameworks or Quine's extraordinarily influential work on the indeterminacy of translation. It is also what I take to be the import of Kuhn's "meaning incommensurability," the inability of one language and its referential structure to translate fully into another language system.[39] Allied with that split are others mentioned earlier: for Kuhn there is no "protocol language" that would serve as a common referent for the two languages. For the young Carnap there was. For Kuhn, theory came epistemically first by setting out the classificatory boundaries of the language; for Carnap (or at least for a common reading of Carnap), the observation language came first, providing just that set of primitives from which the conventions of the higher order language would be built.[40] On the Kuhnian view there is a unity, a vertical or synchronic unity if you will, between observation and theory *at a given time* (within a paradigm). For the younger Carnap that unity came in the cumulative and continuous observation language—from below.

35. Heilbron, *Early Modern* (1982).

36. Carnap, "Logical Foundations" (1955), 61.

37. Neurath, "Unified Science" (1955), 23.

38. Morris, "Scientific Empiricism" (1955), 69.

39. Quine's indeterminacy of translation is, of course, quite different from Kuhn's meaning incommensurability. For Quine, translation is determinate with respect to a given manual of translation; the difficulty is that there are many possible manuals. For Kuhn, there is not even one fully adequate translation system between theories.

40. In recent years there has appeared a variety of sophisticated nonfoundationalist readings of the logical positivists. See, e.g., the work by Cat, Cartwright, and Chang, "Otto Neurath" (1996).

Despite their disagreements about the existence of an independent observation language, at the core of the views of both Kuhn and Carnap lies the assumption that the activity of science is principally to be understood as an unravelling of the difficulties of language and reference. Furthermore, antipositivists such as Kuhn joined the positivists in insisting that there is *a structure* to which the evolution of scientific propositions could be reduced. For the logical positivists there were various criteria that gave rationality to the substitution of one theory for another, criteria that they expounded and explored in verificationism and confirmation theory. On the sociological level, Frank—a "founding father" of the Vienna Circle—believed that the dynamic was to be worked out by the analysis of conceptual schemes and resistance to them. For Kuhn too there was a universal structure for the supplanting of one theory with another in his timeless cycle of epochs: normal science, crisis science, revolutionary science, and the return of normal science.[41] Like Polybius's cycle of governments, the Kuhnian structure transcended time and place. In a curious way it fulfilled one dream of the Encyclopedists by providing a unification of the sciences at the level of method. It was thus highly appropriate—more so than usually realized—that *The Structure of Scientific Revolutions* formed chapter 2 in volume 2 of the *International Encyclopedia of Unified Science,* edited by Carnap, Neurath, and Morris.[42]

Antipositivism and logical positivism thus share the search for a universal procedure of scientific advancement and a view that language and reference form the chief difficulty in the analysis of the experiment-theory relation. But the ties between positivist and antipositivist go much further. Both models have a well-established hierarchy that lends unity to the process of scientific work. True, they are flip sides of one another, but in their mirrored images there is a good deal of similarity. The central metaphor of figure 9.4 is an inversion of figure 9.1, with the special assumption, in Kuhn's case, that the important experimental and theoretical breaks occur contemporaneously. The unity of each account is, to a certain extent, enforced by the provision of a privileged vantage point, what Lyotard would call a "master narrative": in the case of the positivists it is from the "observational foundation"; in the case of the antipositivists it is from the theoretical "paradigm," "conceptual scheme," or "hard core" looking down and out.[43] This shared intuition that there are blocks of unified knowledge that float past each other without linking has been expressed in many places and many ways.

41. Of course, for some antipositivists, such as Feyerabend, there is no such universal schema; the exchange of one theory for another is anarchic.

42. The history of Kuhn's *Scientific Revolutions* in the *Encyclopedia* is discussed in Galison, "Contexts and Constraints" (1995).

43. On the notion of the master narrative, see, e.g., Lyotard, *Postmodern Condition* (1984), 27–41.

As compelling as this antipositivist picture is, recent historical and philosophical work on experimentation suggests that it needs revision. In the remainder of this chapter, I would like to expand on the alternative sketch of the relation of experiment, theory, and instruments developed in the previous chapters and to reflect on other recent studies of experimentation.

9.4 Intercalation and Antireductionism

Like Gaul, the practice of twentieth-century physics is divided into three parts. Indeed, precisely those criteria that Kuhn laid out some years ago as the symptoms of separate scientific communities apply to the groupings of experiment, theory, and instrumentation.[44] There are separate journals, such as *Nuclear Instruments and Methods* and *Reviews of Scientific Instruments,* for those physicists and physicist-engineers concerned with the design and implementation of particle detectors, accelerator technology, and computer data analysis systems. So too are there specifically theoretical publications, including *Theoretical and Mathematical Physics* and the *Journal of Theoretical Physics.* And there are specifically experimental journals, such as the eminent *Methods of Experimental Physics.* There are separate conferences on theoretical, experimental, and instrumental subjects. Furthermore, the invisible colleges defined by preprint and reprint exchange frequently fall within these stratifications. In recent decades, graduate students at many institutions have been accepted qua experimenter or qua theorist, and increasingly Ph.D.'s are awarded for contributions to instrumentation, considered as an area of research distinct from experimentation.[45] There are prominent workshops, conferences, and summer schools that segregate these different subcultures. Think of the Johns Hopkins Workshop on Current Problems in Particle Theory, which in a given year might focus on lattice gauge theory, supersymmetry, or grand unification; consider the World Conference of the International Nuclear Target Development Society (its members make beryllium plates, not ICBMs), or the Winter School of Theoretical Physics in Karpacz. Quite obviously there are national and international laboratories dedicated to experimental physics, some with significant and others with tiny theoretical groups. Less visible are the laboratories in industry, or at universities (and sometimes sections within larger laboratories), devoted solely to the development of instrumentation. Theorists have fewer places to themselves, but they are not insignificant: the Center for Theoretical Physics in Santa Barbara, the Institute for Theoretical Physics in Leningrad, the Institute for Advanced Study in

44. Kuhn, "Second Thoughts" (1974), 462.

45. The problem of the awarding of Ph.D.'s in physics for *purely* instrumental research has been the subject of much contention within the physics community. See, e.g., U. S. Department of Energy, *Future Modes* (1988), 33, 53.

Princeton, and the International Center for Theoretical Physics in Trieste, to name a few. Nor are such assemblies restricted to high energy or nuclear physics. Condensed matter theorists often convene without their experimental colleagues in order to discuss the theory of metals or many-body problems. Astronomers sometimes find it appropriate to meet about instrumental techniques in the radio or optical domains, and when the quantum gravity theorists meet there are few experimenters or instrumentalists. More recently, computation has arisen as a distinct arena from all of the above, and computer specialists regularly assemble for workshops such as "Computing for High Luminosity and High Intensity Facilities."[46]

While defections from one arena to another are possible, they are rare and discouraged. (Particle physicists like to point to the brilliant exception Enrico Fermi, who in his youth contributed powerfully both to theory and experiment; he is a physicist's hero precisely because he traversed a barrier that only a handful have crossed in the last 50 years.) For all these reasons, it has become awkward to treat physics and physicists as constituting a single, monolithic structure. As historians, we have become used to treating cultures as composed of subcultures with different dynamics. It is now a commonplace that the political dislocations of the French Revolution did not alter economics, social structure, politics, and cultural life in the same measure. Indeed, as Lynn Hunt has shown, even the political impact of the revolution was felt differently by workers concentrated in towns and textile workers dispersed over the countryside.[47] It is high time that we recognize that the physics community is no less complex. Experimenters—and one could make a similar statement about theorists or instrumentalists—do not march in lockstep with theory. For example, the practice of experimental physics in the quantum mechanical revolution of 1926–27 was not violently dislocated despite the startling realignment of theory: spectroscopy continued unabated, as did measurements of specific heat and blackbody radiation. And practitioners of these experimental arts continued undaunted their dialogue with theorists on both sides of the great theoretical divide. Each subculture has its own rhythms of change, each has its own standards of demonstration, and each is embedded differently in the wider culture of institutions, practices, inventions, and ideas.[48]

46. A tiny sample of such meetings are represented in the following volumes: Domokos and Kovesi-Domokos eds. *Johns Hopkins Workshop* (1983); and Jaklovsky, *Nuclear Targets* (1981).

47. Hunt, *Revolution* (1978).

48. As we search to locate scientific activities in their context, it is extremely important to recognize that for many activities there is no single context. By viewing the collaboration among people in a laboratory, not as a melding of identities, but as a coordination among subcultures we can see the actors as separately embedded in their respective, wider worlds. E.g., in the huge bubble chamber laboratory at Berkeley one has to see the assembled workers as coming together from the AEC's secret world of nuclear weapons, and from an arcane theoretical culture of university physics. The culture they partially construct at the junction is what I have in mind by the "trading zone."

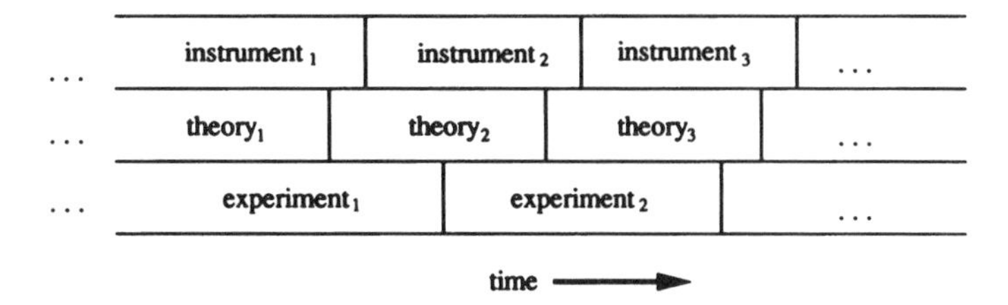

Figure 9.5 Intercalated periodization. An intercalated periodization drops the assumption of coperiodization and separates the subcultures of physics into (at least) the three quasi-independent groupings of theory, experiment, and instrument making. But there is nothing sacred about the tripartite division. Theory may well split *internally* into intercalated pieces of varying duration, as may instrument making or experimentation. The point is that breaks in one need not coincide with breaks in the others.

Thus, for historical reasons, instead of searching for a positivist central metaphor grounded in observation, or an antipositivist central metaphor grounded in theory, I suggest that we admit a wider class of periodization schemes, in which three (or more) levels are intercalated as in figure 9.5.

Different quasi-autonomous traditions carry their own periodizations. There are four facets of this open-ended model that merit attention. First, it is at least tripartite, granting the possibility of partial autonomy to instrumentation, experimentation, and theory. It is not necessary that each subculture be represented separately, as one can easily identify moments in the history of physics when the instrument makers and the experimenters (to give one example) are not truly distinct. Nor is it always the case that break points occur separately. And there are many times (as we have seen in the previous chapters) when there were competing experimental subcultures each working in the same domain (such as bubble chamber users and spark chamber users). Second, this class of metaphors incorporates one of the key insights of the antipositivists: *there is no absolutely continuous basis in observation.* Both the level of experimentation and the level of instrumentation have their break points, just as theory does. Third, the local continuities are intercalated—we do not expect to see the abrupt changes of theory, experimentation, and instrumentation occur simultaneously; in any case it is a matter of historical investigation to determine if they do line up. Indeed, there are good reasons to expect that at the moment one stratum splits, workers in the others will do what they can to deploy accepted procedures that allow them to study the split before and after. When a radically new theory is introduced, we would expect experimenters to use their best-established instruments, not their unproven ones. Fourth, we expect a rough *parity* among the strata—no level is privileged, no subculture is the arbiter of progress in the field or serves as the reduction basis (the intercalated strands should really be mapped in three dimensions to demonstrate that no one subculture is always "on top" and that each borders on the other two). Just as bricklayers would not stack-set the bricks for fear the whole building would collapse, researchers try to set breaks in one practice cluster against continuities in others. As a result of such local actions (not global planning), the community as a whole does not stack-periodize its practices.

Examples of the survival of experimental practices across theoretical breaks are now abundant in the new literature on experiment. For the first time there is a real interest in the dynamics of experiment outside the provision of data to induce, confirm, or refute specific theories. And among the philosophers, no one has done more than Hacking to articulate the variety of roles experiment plays in the production of knowledge, roles that go far beyond the merely confirmatory function usually assigned to "experiment" in overly abstract accounts of scientific research.[49] Surely, then, Hacking would grant experimentation and the creation of phenomena just the sort of partial autonomy I have in mind with this class of periodization models. He would also agree, I take it, that the experimental/phenomenal domain has its own break points.

Where I differ, perhaps, is in regard to parity among the subcultures. For while I am all for granting experimentation a life of its own, I do not think its life should come at the cost of poor theory's demise. More specifically, I read Hacking's work on the production of experimental entities this way: the possibility of intervening—making, moving, changing—is a way of imposing constraints on what can be the case. On Hacking's view, when it is possible to spray objects like positrons at will, these restrictions are so severe that there is nothing for it but to acknowledge their "reality."

Theory (or at least high theory), for Hacking, lacks the compulsive force of interventionist experimentation. For this reason he has defended an antirealism about theories and condemned those entities that theory alone demands—such as gravitational lenses or black holes.[50] But for many of the reasons Hacking originally defended the robustness of a quasi-autonomous experimental domain, I want to defend the robustness of theory and of instrumentation: there are quasi-autonomous constraints on each level. When Duhem talks about the many theories that can each account for the data, he often has in mind positional astronomy as his example;[51] but most theoretical physics—such as particle physics or condensed matter theory—is as far from models of positional astronomy as the

49. Franklin, *Neglect* (1986), e.g., 103ff.; Galison, *Experiments* (1987); Lenoir and Elkana, *Sci. Con.* 2 (1988): 3–212; Gooding, Pinch, and Schaffer, *Uses of Experiment* (1989); Achinstein and Hannaway, *Observation* (1985); special issue on artifact and experiment, Sturchio, *Isis* 79 (1988); Shapin and Schaffer, *Leviathan* (1985); and Hacking, *Representing* (1983).

50. Hacking, *Representing* (1983), 274–75; and Hacking, "Extragalactic Reality," *Phil. Sci.* 56 (1989): 555–81.

51. Duhem, *Aim and Structure* (1954), 168–73, 190–95.

determination of Snell's law is from an experiment at SLAC. The theorist is *not* free to admit any particle or effect in order to come into harmony with the experimenter.

Experimenters come to believe in an effect for various reasons; one is the *stability* of the phenomenon—you change samples, you shift the temperature, and still the effect remains. Another road to the closure of an experiment involves the increasing *directness* of our probing of the phenomenon. By increasing the power of a microscope, the energy of a particle beam, the sensitivity of the apparatus, or the amplification of a signal, one probes further into the causal processes linking phenomena together.[52]

The theorist's experience is not so different. You try adding a minus sign to a term—but cannot because the theory then violates parity; you try adding a term with more particles in it—forbidden because the theory now is nonrenormalizable and so demands an infinite number of parameters; you try leaving a particle out of the theory—now the law has uninterpretable probabilities; you subtract a different term—all your particles vanish into the vacuum; you split a term in two—now charge is not conserved; and you still have to satisfy conservation laws of angular momentum, linear momentum, energy, lepton number, and baryon number. Such constraints do not all issue axiomatically from a single, governing theory. Rather, they are the sum total of a myriad of interpenetrating commitments of practice: Some, such as the conservation of energy, are over a century old. Others, such as the conservation of parity, survived for a very long time before being discarded. And yet others, such as the demand for naturalness—that all free parameters arise in ratios on the order of unity—have their origin in more recent memory. Some are taken by the research community to present nearly insuperable barriers to violation, while others merely flash a yellow cautionary light on being pushed aside. But taken together, the superposition of such constraints makes some phenomena virtually impossible to posit, and others (such as black holes) almost impossible to avoid.

Indeed, the astonishing thing about black holes is that they form (theoretically) in the face of enormous variations in the basic structure of our theory of matter. They do not depend on the details of this or that theory of the strong, the weak, or the electromagnetic force; and to remain consistent with other observations there is practically nothing one can do with the theory of gravity that would get in the way of the formation of black holes. The situation is similar with antiparticles. If one accepts special relativity, quantum mechanics, and local causality (the notion that cause and effect should be by near action, not action at a distance), then changes in the charges of particles, the number of particles, the nature of forces, the existence or nonexistence of unification schemes all

52. See Galison, *Experiments* (1987), sec. 5.6. On the role of causal explanations and realism about phenomena, see Cartwright, *Laws* (1983).

leave the basic symmetry intact: for every particle there is an antiparticle. This stubbornness against variation is the theoretical analogue of stability, and it is the experience of this stability that eventually brings theorists to accept such objects, come (almost) what may from their experimental colleagues.

In experiment, the search for directness is a search to measure quantities previously only deduced—for example, the production of the Z^0, after its effects had been measured in scattering experiments. Theory has its analogue of such increasing directness as well. André Marie Ampère searched for force laws directly linking one microscopic current element to another, without specifying how they were to be linked. Gradually, over the course of the middle decades of the nineteenth century, Faraday, Thomson, and Maxwell spelled out those intermediate causal elements while (by and large) preserving the Ampèrian force relations. In more modern times, the search for a quantum field theory of electrodynamics in the 1940s was an analogous attempt to provide the intermediate steps that were elided in the point interactions of Fermi. The demand for stability and directness thus has representation in theory and experiment and heavily constrains both activities. Each develops a partial autonomy, and in the old sense—the sense of Kant—each becomes, in part, a lawgiver to itself. It is this self-regulation coupled with the development of partially independent systems of symbols that fills out the notion of theory and experiment as subcultures of the wider discipline of physics. Nowhere is this quasi autonomy better illustrated than in superstring theory, in which many of the leading practitioners see mathematical and mathematical-physical constraints as the ultimate test of the theory. As Michael Green, John Schwartz, and Edward Witten put it in 1987: "Quantum gravity has always been a theorist's puzzle *par excellence.* Experiment offers little guidance. . . . The characteristic mass scale in quantum gravity is the Planck mass [10^{19} GeV]. This is so far out of experimental reach that barring an unforeseeable stroke of good luck . . . we can hardly hope for direct experimental tests of a theory of quantum gravity. The real hope for testing quantum gravity has always been that in the course of learning how to make a consistent theory of quantum gravity one might learn how gravity must be unified with other forces."[53] In superstring theories, Weinberg argued similarly, there are "almost no string theories at all." And just because of that "rigidity" (Weinberg's term) "there is nothing you can tinker with in these theories; they are either right or wrong as they stand."[54]

My sense of the heavily constrained nature of theory is what underlies my discontent with the heavy emphasis on the "plasticity" of physics. Constraints at different levels allow theorists to come to beliefs about particles, interactions,

53. For this quote, and more on constraints in superstring theory, see Galison, "Theory Bound and Unbound" (1995), 385.

54. Galison, "Theory Bound and Unbound" (1995), 384–85.

electronic effects, stellar phenomena, black holes, and so on, even when their experimental colleagues disagree or remain silent. The enterprise as a whole is strong, on this view, not because the domains of action are so plastic, but because they are so robust—and yet, despite that, fit together. The process by which this fitting occurs is emphatically not a reduction to a protocol language or a mutual translation of the two finite traditions. This is the argument that motivates the historical material of the preceding chapters, and the understanding that informs these metahistorical reflections on it. My focus throughout has been on finite traditions with their own dynamics that are linked not by homogenization, but by *local coordination.*

PART II: The Trading Zone

9.5 The Locality of Exchange

In an effort to capture both the differences between the subcultures and the felt possibility of communication and joint action, we consider again the picture of intercalated periodizations discussed earlier (figure 9.5), but now we need to focus on—and to expand—the boundaries between the strata. To characterize the interaction between the subcultures of instrumentation, experiment, and theory I want to pursue the idea that these really are subcultures of the larger culture of physics. Like two cultures distinct but living near enough to trade, they can share some activities while diverging on many others. What is crucial is that in the local context of the trading zone, *despite* the differences in classification, significance, and standards of demonstration, the two groups can collaborate. They can come to a consensus about the procedure of exchange, about the mechanisms to determine when goods are "equal" to one another. They can even both understand that the continuation of exchange is a prerequisite to the survival of the larger culture of which they are part.

I intend the term "trading zone" to be taken seriously, as a social, material, and intellectual mortar binding together the disunified traditions of experimenting, theorizing, and instrument building. Anthropologists are familiar with different cultures encountering one another through trade, even when the significance of the objects traded—and of the trade itself—may be utterly different for the two sides. And with the anthropologists, it is crucial to note that nothing in the notion of trade presupposes some universal notion of a neutral currency. Quite the opposite, much of the interest of the category of trade is that things can be coordinated (what goes with what) without reference to some external gauge.

For example, in the southern Cauco valley in Colombia, the peasants, mostly descended from slaves, maintain a rich culture permeated with magical cycles, sorcery, and curing. They are also in constant contact with the powerful forces of the landowning classes: some of the peasants run shops, others work on the vast sugarcane farms. Daily life includes many levels of exchange between the two groups, in the purchase of goods, the payment of rent, and the disbursement of wages. And within this trading zone both sides are perfectly capable of working within established behavioral patterns. But the *understanding* each side has of the exchange of money is different. For the landowners, money is "neutral" and has a variety of natural properties; for example, it can accumulate into capital—money begets money. For the peasants, funds obtained in certain ways have intention, purpose, and moral properties, though perhaps none more striking than the practice of the secret baptism of money. In this ritual, a godparent-to-be hides a peso note in his or her hand while holding the child as the Catholic priest baptizes the infant. According to local belief, the peso bill—rather than the child—is consequently baptized, the bill acquires the child's name, and the godparent-to-be becomes the godparent of the bill. While putting the bill into circulation, the owner quietly calls it by its name three times; the faithful pesos will then return to the owner, accompanied by their kin, usually from the pocket of the recipient. So, when we narrow our gaze to the peasant buying eggs in a landowner's shop we may see two people harmoniously exchanging items. They depend on the exchange for survival. Out of our narrow view, however, are two vastly different symbolic and cultural systems, embedding two incompatible valuations and understandings of the objects exchanged.[55]

I invoke this anthropological scene for the specific purpose of illustrating how sharply different global meanings can nonetheless come to (even very complex) coordination in specific contexts. Such a partial sharing of meanings became salient at many points in the history of laboratory practice and the material culture of physics. Condensation physics marked a set of procedures and interpretations aimed at revealing the production of droplets around ions. To Victorian meteorologists, here was a miniaturization of the natural world of rain, fog, and thunderstorm. To the Cavendish ion physicists, here was a world of ions-made-visible, in which the droplet itself was a tag not a prime subject of inquiry. Thunderstorms made small to one side, atoms made large to another—but in the middle, in the zone of exchange, was the cloud chamber and its associated laboratory moves. I would argue that we should forget the either-or dichotomization of the gestalt switch that says the worlds of big atoms and little storms were incompatible, incommensurable representations of the wispy cloud chamber tracks. For in Wilson's own practices, his manipulation of droplets in electric fields, there was no radical distinction between meteorological recreation and ionic experimentation. Even his words, "fog," "rain," "clouds," and "drizzle," were partially stripped of many of their macroworld connotations as they came to designate the tracks of his chamber. The coordination of Caven-

55. The example of the secret baptism of money is from Taussig, *Commodity Fetishism* (1980), chap. 7.

dish ion physics with Scottish natural philosophy took up space—the physical space of the chamber and the conceptual space of the pursuit of condensation phenomena.

In the structuralist vision of things, boundaries have no size and can occupy no such site. The argument of this book is the opposite: the point is precisely that the delimiting arena not only exists—it (the trading zone) was substantial enough to contain at least a decade of C. T. R. Wilson's work and to guide his student Cecil Powell from the cloud chamber to the steamship engine to the volcanic cloud. Both effected the trade coordination between the realm of their Cavendish ion colleagues and the wider world of geologists, meteorologists, and engineers.

There was nothing permanent about the trading zone of condensation physics; in fact, as a binding site for meteorology and the theory of matter, it rapidly fell apart. I want to emphasize this, to stress that there is no teleological drive toward ever greater cohesion. It is altogether possible that, at some moments, fields previously bound, fall apart. Just as some pidgins or creoles die out, so too can scientific interfields atrophy or mutate to the point of being unrecognizable. Eighteenth-century iatromechanics did not hold together any more than Einstein's attempt to directly unify electromagnetism and general relativity did a century and a half later. Like condensation physics, the trading zone of *ionographie* was of finite duration–the flourishing, productive trading zone Demers displayed in figure 3.12 eventually cracked like dried film.

As Demers imagined it, ionography was the study of materialized tracks in solids. To some practitioners of the art (those coming from photography) it was an exploration of the fundamental processes of atomic, crystalline, and gelatinous materials. To cosmic ray physicists ionography was a probe of deep space, to nuclear physicists it was a look into the basic bits of matter, to the geochronologist it was a probe of the unimaginably distant past. Like condensation physics, ionography was a substantive conceptual space, roomy enough to support conferences and collaborations, rich enough to support massive treatises explaining the ins and outs of emulsion fabrication, response, distortion, and analysis. It was the quasi autonomy of condensation physics and ionography that underwrote the making of track pictures (emulsion and cloud chamber) into data. And with the creation of this quasi-autonomous data came a lore of interpretation, a domain in which it became possible for experimenters and theorists to say things like "this is a pion decaying into a muon, and a muon decaying into an electron." Interpretations could conflict, or could come to consensus, but this intermediate set of linguistic and procedural practices bound together experimenters, instrument makers, and theorists in collaboration.

"Collaboration" as a term is helpful insofar as it indicates different individuals or groups aiming at certain shared goals, but we can and have gone further toward a specification of how the coordination takes place. Indeed, far from melting into a homogeneous entity, the different groups often maintain their distinctness, whether they are electrical engineers and mechanical engineers, or theorists and engineers, or theorists and experimenters. The point is that these distinct groups, with their different approaches to instruments and their characteristic forms of argumentation, can nonetheless coordinate their approaches around specific practices. Frequently, as we have seen, theorists work out a detailed, coordinative trade between experimental predictions and experimenters' results. Note that here, as in any exchange, the two subcultures may altogether disagree about the implications of the equivalencies established, the nature of the information exchanged, or the epistemic status of the coordination. Theorists may predict the existence of an entity with profound conviction because it is linked to central tenets of their practice—for example, group symmetry, naturalness, renormalizability, covariance, or unitarity. The experimenter may receive the prediction as something quite different, perhaps as no more than another curious hypothesis to try out on the next run of the data analysis program: look for a certain Higgs particle, a new heavy neutrino, a supersymmetric partner to the electron, a short-lived proton. Conversely, in the 1970s, two experimental collaborations from atomic physics used reactions in bismuth to argue that the much-celebrated neutral currents of Weinberg, Salam, and Glashow did not exist. In fairly short order, theorists began treating the experimental results as mistakes even without an account of the details.

Think back to our view of Dirac in chapter 2, with his deep suspicion of some of the new results, coming as they did on the heels of a failed experimental attempt to overthrow relativity. Writing to Blackett that he would wait a year before he, Dirac, would get worried about "unexpected experimental results," he articulated what theorists often suspected: anomalous experimental results were rife and often fleeting. It was a theme he repeated throughout his long life: "Let us now face the question, that a discrepancy has appeared, well confirmed and substantiated, between the theory and the observations. How should one react to it? How would Einstein himself have reacted to it? Should one then consider the theory to be basically wrong? I would say that the answer to the last question is emphatically no. Anyone who appreciates the fundamental harmony connecting the way nature runs and general mathematical principles must feel that a theory with the beauty and elegance of Einstein's theory *has* to be substantially correct."[56] Blackett could plead with Dirac to look at the photos, but it is clear that what was persuasive for Blackett the experimenter was not necessarily so for Dirac the theorist.

In the trading zone both sides impose constraints on the nature of the exchange. Over the course of this book, we have seen in detail how the instrument subculture of *image* and that of *logic* competed and ultimately joined. More

56. Cited in Kragh, *Dirac* (1990), 285–86.

accurately, the coming together of the two traditions was a halting coordinative effort that frequently ran aground on the technical obstacles. While casting aspersions on the other, each side insistently tried to acquire the virtues of its rival: logic had statistics and experimental control and wanted persuasive detail; image had the virtues of being fine-grained, visible, and inclusive but wanted the force of statistics and control over experimentation. Nonetheless, the "electronic bubble chamber" remained an elusive ideal for the image physicists, an imaginary instrument always desired, often anticipated, and no less frequently abandoned. Bubble chamber physicists tried over and again to rig control into their machines of production and their machines of analysis—but the triggerable bubble chamber would not, could not, be built until the last of the behemoths was dismantled. Full automation of pattern recognition slipped away time and again.

Despite the suspicion that "anything can happen once" among the electronic physicists, the trustworthy image remained a prize—in the photographic spark chamber, in the streamer chamber, in Glaser's imagined "solid Christmas tree of tracks" that could be extracted from a chemical soup, and in the many bits of bricolage that tried to glue images to electronic control. It was only in the deliberate construction of the SLAC-LBL collaboration, in the joint efforts of a leading image group and a leading logic one, that the Mark I detector could fabricate an image-producing, triggerable machine. There, as described in chapter 6, individual pictures could be scanned with full image expertise. Bubble chamber computer programmers infused their image subculture into the heart of the logicians' electronics, and scanners scoured the pictures imprinted on microfiche by calorimeters, scintillators, and computer-aided data acquisition. Physicists like Gerson Goldhaber—a veteran of emulsion photographs and bubble chamber prints—could carry on analyses of a logic experiment with the eye of an image physicist. In a sense, both traditions remained intact, preserved inside an encompassing collaboration; the coordination of exchange took place around the production and use of the image.

But even as the Mark I physicists exploited their pictures, they would not publish individual golden events as decisive evidence for a new entity. Rather like the Harvard-Wisconsin-Penn-Fermilab (E1A) collaboration that codiscovered the weak neutral current in the early 1970s, individual pictures could be mobilized *within* the collaboration to persuade individuals and widen the circle of belief. But such images could not be used to move the outside world.

With these laboratory moves toward electronic imagery in the mid-1970s, the pattern of particle physics experimentation had been set: experimental devices were becoming information machines, and the boundary between computation and experimentation blurred. The Time Projection Machine made this transformation explicit (to the extent that the inner workings of a high-tech machine are explicit): in the very heart of the detector there was only gas—everything went into the processing of information from the charged coupled detectors

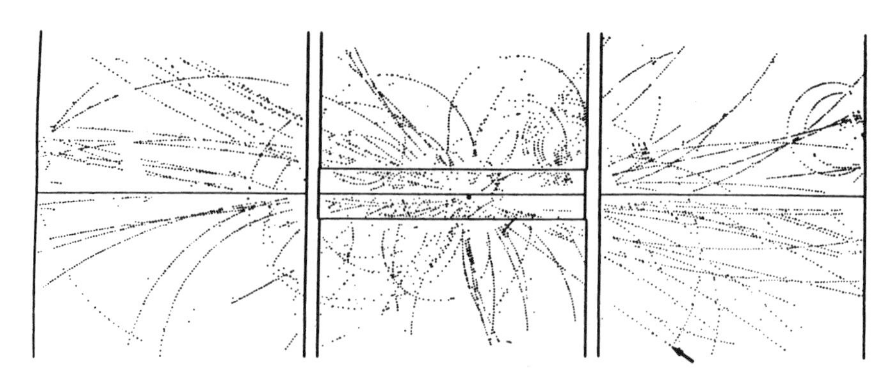

Figure 9.6 Golden *W* (1983). This picture was arguably the first purely electronic golden event ever found—that is, the first instance of a single electronic picture invoked, in the absence of a strong statistical argument, to demonstrate the existence of a new particle or effect. Each track corresponds to the trajectory of a charged particle, and the arrow marks the track that is the purported electron resulting from the decay of the *W* boson. Produced along with the electron (if the event was to follow theoretical prediction) was an unseen neutrino, the existence of which could be inferred from an energy deficit in the observed charged particles. As usual in any golden event argument, the collaboration had to defend against the possibility of a background event "mimicking" the *W*—no "lost" charged particles, for example. Source: Arnison et al., "Experimental," *Phys. Lett. B* 122 (1983): 103–16, on 112.

through to the microprocessors and the many passes of computer programming that worked and reworked the "hits" until they emerged as the highly interpreted data of experimental argumentation. Simulations created a detector before it was built, figured in the acquisition of data, and eventually were crucial in analyzing results.

In the physics of the very small, with the pictures produced by machines like the TPC/PEP-4 and the CERN, Fermilab, and (planned) SSC detectors, we come to a confluence of the two great epistemic rivers of image and logic that flow through a century of physics. And if one wanted a pair of images to represent this confluence, it would be that of the first candidate *W* and *Z* events, put forward by the 150 members of the CERN-based UA1 collaboration in January 1983. Out of the hybrid assemblage of drift chambers, calorimeters, scintillators, and muon chambers came a handful of electronically produced pictures, and hundreds of physicists spent thousands of hours probing their every pixel. Here were golden events fabricated in a world of electronics, sorted with the control only available in the logic tradition, yet assigned a place in physical knowledge previously only granted to the photographic (see figures 9.6 and 9.7).

These pictures instantiated the working-out of the image and logic traditions, a hybrid epistemic basis for data that became the rule in the massive collider detectors that followed. It is familiar to characterize the broad history of

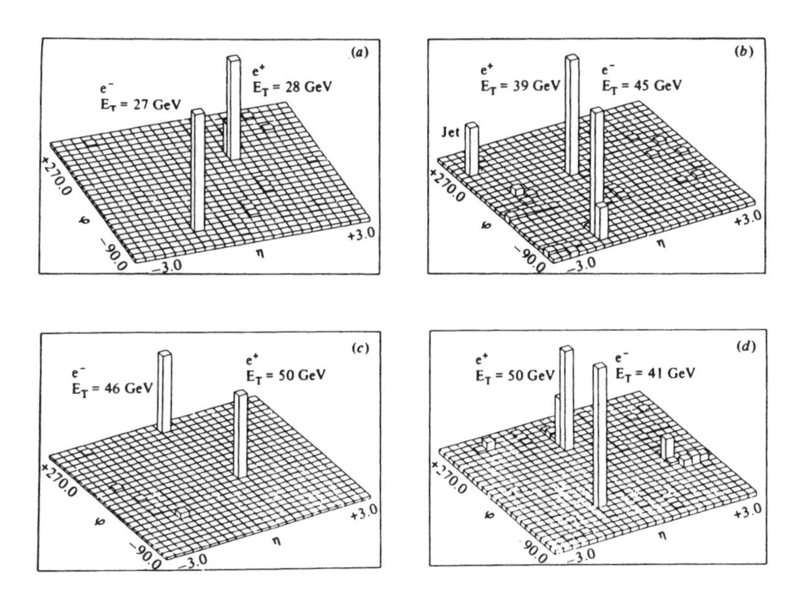

Figure 9.7 Golden Z, lego graph (1983). Unlike the *W*, the *Z* can decay into particles that are both charged: an electron and a positron. The first four *Z* candidates are displayed in this figure, but represented not as tracks but as processed data points in a "lego graph." Along the phi axis one sees the coordinates of the detector unrolled, as it were. The eta direction corresponds to the measuring cells along the length of the detector. These pictorial but nonrepresentational presentations of data make visible the absence of any other processes that might imitate *Z* production and exhibit a widening use of the computer for pictorial purposes divorced from a literal re-creation of tracks in space. Source: Arnison et al., "Experimental," *Phys. Lett. B* 126 (1983): 398–410, on 403.

physics by its theories: classical mechanics, classical electromagnetism, relativity, quantum theories, and theories of unified fields, chaos, and strings. Less dramatically, perhaps, certainly less publicly, there have been ages of data: Data of the nineteenth century, in which precision and standardization became watchwords in so many spheres of inquiry. Early-twentieth-century data on matter, with counts and cloud chamber pictures focusing on individual events, followed by the deluge of data that spewed from bubble chambers after the midcentury mark. If the late twentieth century had its typical form of data, perhaps it would be the controllable image. For in many domains beyond particle physics, one sees the binding of picturing with counts. X rays and the visual culture of the radiologist on one side combined with the nonvisual traditions to form computer-aided tomography (CAT) scans and nuclear magnetic resonance (NMR) imaging. Radio astronomers grafted their careful electronic deductions to visual astronomy with charged coupled detectors and computers—often borrowing, trading, and exchanging the new technologies with the particle physicists. Plasma physicists begin to form images, some representational, some not, out of the instabilities that are their stock in trade; geophysicists exploited the computer to combine traditionally statistical, linear data of seismology with morphological and structural concerns as the computer begins to image the surface of the earth's core. Across these many disciplines, the controllable image came to supplant the century-old ideal of objectivity that the chemical photograph had embodied. Knowing through the passive registration of the eye no longer stood apart from knowing through manipulation.

Less abstractly, one can look through the collaboration lists of the SSC, D0, ALEPH, and CDF and pick out the adherents of the image and logic traditions among the older physicists—this one worked in emulsions and then the long span of CERN bubble chambers; that one went from scintillators and counters to spark chambers and wire chambers. But increasingly, especially for the younger experimenters of the 1980s and 1990s, these are distinctions without meaning; the fragile trading zone of the 1970s between the two traditions has become the site of a new generation of experimentation. But the coordinative process continues beyond the conduct of experiment and the production of data: empirical claims—about the *W* and *Z*, for example—had to be meshed with the myriad of theoretical constraints. For theorists the central and most powerful reason for attending to the Glashow-Weinberg-Salam theory had been its ability to produce finite (renormalizable) results—and the production of *W*'s and *Z*'s in small numbers did not in and of itself test this feature. From the experimenter's point of view, there were other, different constraints. For the particle to be discussed jointly by theorists and experimenters, some subset of both sets of constraints had to hold. In such trading zones there were highly constrained coordinations and complex languages that bound the otherwise disparate subcultures together. It is worth turning to this coordination of theory and experiment now, in greater detail.

9.6 Trading between Theory and Experiment

The example of relativistic mass is an appropriate place to start because over the last 30 years it has become the locus classicus for discussions of meaning incommensurability. For Kuhn, the advent of Einsteinian dynamics was a prototype of revolutionary change and, he argued, only at low velocities could the two concepts of mass lead to measurements in a single experiment.[57] On this

57. "This need to change the meaning of established and familiar concepts is central to the revolutionary impact of Einstein's theory. . . . We may even come to see it as a prototype for revolutionary reorientations in the sciences" (Kuhn, *Scientific Revolutions* [1970], 102).

view, one would expect there to be no experimental mode of comparison of Einstein's concept of mass and the concepts of mass his theory displaced—those of H. A. Lorentz, Max Abraham, and Henri Poincaré, none of whom shared Einstein's view of an operationally defined space and time. Feyerabend simply says there is no single experiment: where it appears there is one measurement of mass there actually are several—one experiment for the classical mechanic and one for the relativist. Any scientists who think differently, according to Feyerabend, are instrumentalists not interested in interpretation at all, or "mistaken," or simply such remarkable translators that they "change back and forth between these theories with such speed that they seem to remain within a single domain of discourse."[58] None of these alternatives seems to capture what goes on between theorists and experimenters.

There is no doubt that the term "mass" was used differently by the different participants in what was referred to as the physics of the electron. Max Abraham and H. A. Lorentz both believed that electrons' masses originated purely from their interaction with their own electromagnetic fields. Since Abraham and Lorentz also took electrons to be the basic building blocks of matter, the "electromagnetic mass" of the electron was the basis of a worldview in which mechanical mass was a derivative concept, and electricity the primary substance of nature. But while Abraham took the electron to be a rigid sphere with a uniform surface charge, Lorentz postulated, in addition, that electrons were flattened as they moved through the ether, and he used this hypothesis to explain the Michelson-Morley experiment. Soon afterward, Henri Poincaré introduced a modified version of Lorentz's theory, adding a nonelectromagnetic force to keep the deformable electron from blowing apart under the stresses of its deformation.[59]

These theories differ significantly from one another about the meaning of mass. And as radical as these theories might have seemed at the time, Einstein's was surely as shocking. Einstein abandoned the attempt to embed his notion of mass in the grand scheme of the electromagnetic world picture and founded his theory on a positivist critique of the metaphysical categories of space and time, replacing them with clocks and rulers. Kuhn's claim is that prerelativistic and relativistic uses of the term "mass" make comparison impossible: "Only at low relative velocities may the [Newtonian and Einsteinian masses] be measured in the same way and even then they must not be conceived to be the same."[60] In

58. Feyerabend, *Problems of Empiricism* (1981), 159: "It is no good insisting that scientists act as if the situation were much less complicated. If they act that way, then they are either instrumentalists . . . or mistaken: many scientists are nowadays interested in *formulae* while we are discussing *interpretations*."

59. In the massive literature on Einstein's special theory of relativity, the best historical book is Miller, *Special Theory* (1981). I have drawn liberally on this source for the following discussion of early experimental evidence on transverse electron mass experiments.

60. Kuhn, *Scientific Revolutions* (1970), 102.

fact, there was a rich experimental subculture preoccupied precisely with comparing these different theories—and not at low velocities. With Max Kaufmann and Alfred Bucherer leading the way, these physicists produced experiment after experiment using magnetic and electric fields to measure what was called the "transverse mass" of the electron. Moreover, their efforts were clearly understood by all four of the relevant theorists (Poincaré, Lorentz, Abraham, and Einstein) to arbitrate among theories. Lorentz recognized the relevance of one such set to his work and immediately conceded defeat: "Unfortunately my hypothesis [explaining mass by] the flattening of electrons is in contradiction with Kaufmann's results, and I must abandon it. I am, therefore, at the end of my Latin." These are not the words of someone for whom the experiment was irrelevant or incomprehensible. Only slightly less despairingly, Poincaré conceded that at "this moment the entire theory may well be threatened" by Kaufmann's data.[61] Einstein was more confident of his theory and doubted the execution of Kaufmann's work; he did not challenge the relevance *in principle* of the results. Quite the contrary: Einstein went to considerable pains to produce predictions for the transverse mass of the electron so that Kaufmann and Bucherer could use their experimental methods to study the theory; he constructed a detailed analysis of Kaufmann's data; and he even designed his own modification of the electron deflection experiments that he hoped someone would execute.[62] For the participants in the fast electron experiments, there does not seem to be a problem in talking about the experiment or its proximate significance.

Feyerabend suggests that should scientists not acknowledge the existence of two (or presumably more) experiments lurking behind the apparent existence of just one, there were three possibilities. They could be instrumentalists. At least in the present case, that would seem to be a hard position to defend. Einstein is famous for his insistence that his goal was to discover how much choice God had in the design of the universe. And while acknowledging that the axiomatic basis of theoretical physics could not be inferred from experience, Einstein consistently maintained a deep-seated optimism about theoretical representations: "Can we hope to be guided safely by experience at all when there exist theories (such as classical mechanics) which to a large extent do justice to experience, without getting to the root of the matter? I answer without hesitation that there is, in my opinion, a right way, and that we are capable of finding it." He goes on to say that experience may suggest theoretical ideas in the formal structure of a theory, and experience surely must be the standard against which physical theories are certified. "But the creative principle resides in mathematics. In a certain sense, therefore, I hold it true that pure thought can grasp reality, as the ancients dreamed."[63] These are not the words of an instrumentalist.

61. See Miller, *Special Theory* (1981), 334–35.
62. See Miller, *Special Theory* (1981), 341–45.
63. Einstein, *Ideas and Opinions* ([1954] 1982), 274.

Could it be that Einstein, Lorentz, Poincaré, and Abraham were superfast translators and so could appear (but only appear) to remain in "a single domain of discourse"? Presumably one would look for instances in which Einstein switched into the language and calculational practices of the adherents of the electromagnetic worldview. Such evidence might be reflections on the details of the charge distribution within or on the surface of the electron, or dynamical explorations of the means by which the electron might resist electrostatic self-destruction, or methodological statements advocating electromagnetism as the starting point of physical theory. As far as I know there are no such examples of this kind of work in the published or unpublished record. On the side of Lorentz (or Poincaré or Abraham) one would look for the opposite: indications, perhaps in private, that these theorists alternated their calculations with ones beginning with Einstein's heuristic starting point. Even if direct methodological statements were not forthcoming, we would expect at least some calculations that began with simple mechanical reflections and set aside the structure of matter. Again, even among the unpublished papers I know of no such indications. The third and last alternative that Feyerabend put forward was that a scientist who denied the "two-experiments-in-one" interpretation was just plain "mistaken." Lorentz might simply not recognize that Einstein had a different programmatic commitment. But Lorentz once remarked that Einstein "simply postulates what we have deduced."[64] Conversely, Einstein explicitly argued that he did not believe that mechanics could be reduced to electromagnetism. Each side recognized the gap that existed between their orientations and that this gap was central to the present and future development of physical theory.

The lesson I want to draw from this example is this: despite the "global" differences in the way "mass" classifies phenomena in the Lorentzian, Abrahamian, and Einsteinian theories, there remains a localized zone of activity in which a restricted set of actions and beliefs is deployed. In Kaufmann's and Bucherer's laboratories, in the arena of photographic plates, copper tubes, and electric fields, and in the capacity of hot wires to emit electrons, experimenters and theorists worked out an effective, though limited, coordination between beliefs and actions. What they worked out is, emphatically, not a protocol language—there is far too much theory woven into the joint experimental-theoretical action for that. Second, there is nothing universal in the establishment of jointly accepted procedures and arguments. And third, the laboratory coordination does not fully define the term "mass," since beyond this localized context the theories diverge in a myriad of ways.

We have seen (in chapter 7) a similarly local coordination between experiment and theory at the opposite end of our time period—in the extended border zone lodged between the Time Projection Chamber, model builders, and the

64. Lorentz, *Theory of Electrons* (1909), 230.

full-blown theory of quantum chromodynamics. Here, once again, in the attempt to understand hadronization (the recombination of quarks or partons into jets of observable particles like pions) we have accounts of phenomena in which terms (jets, quarks, partons, gluons, hadronization) are used in such heterogeneous ways that on the face of it carry such different meanings that we might expect to locate them in different and incommensurable conceptual schemes or paradigms. And yet, once again, we have a site at which the actors worked furiously to coordinate and adjudicate among alternatives. As in the philosophical charge of incommensurability leveled against Einstein, Abraham, Lorentz, and Poincaré, we can well ask whether these differences in meanings rendered their communication illusory. Restructuring and summarizing the material of chapter 7, we pose Feyerabend's three alternatives: Do we here have an instance of "mere instrumentalists" seeming to communicate but ultimately uninterested in the physical meaning of the terms in question? Are we faced with superfast translators who flick back and forth between different conceptual schemes? Or are the physicists in question "mistaken," unable to recognize that their adversaries are after different ends with different means? I contend that none of these three "illusion" accounts hold good. Something else, something at once more substantial and more interesting, is occurring in the interstitial zone.

Recall how Llewellyn Smith launched his 1981 discussion of QCD with the words "Why I Believe," staking out through such religious imagery a stance about as far from a desiccated instrumentalism as can be imagined. Citing the "a priori" (theoretical) motivations more than any detailed comparison with experiment, he put faith in the theory's satisfaction of certain conditions. "Color" was needed to keep the electroweak theory renormalizable; chiral symmetry to offer a guarantee of the rough equality of certain observed processes—these became, for Llewellyn Smith, touchstones of truth, not guidebook summaries of data. To Harald Fritzsch, reasons for belief were similarly theoretical—the charms of QCD included the circumstance that the account of the strong interactions was "pure, unbroken non-Abelian gauge theory," and in virtue of that fact, the theory resembled nothing so much as that "most simple and beautiful" structure of general relativity. Because of this aesthetic-theoretical framework, QCD was in his view a step on the road to the "ultimate goal in physics . . . the end of the development in fundamental theoretical physics." We saw in chapter 7 that even model builders like Thomas Gottschalk consistently invoked a physically interpreted conception of the processes they were sketching on paper and keyboard—he wanted eventually, but above all, to tie his modeling to the "dogma" of QCD while relegating the competition (independent fragmentation models) with no "true" gauge basis to the fires of "anathema." Charles Buchanan labeled his scheme "fundamental phenomenology," explicitly (figure 7.37), tying the development of hadronization models to the ultimate realist conceit in the history of science: the progression from Ptolemy through Copernicus and

Kepler to Newton. No, whatever these accounts were, they were not "merely instrumentalist."

Now as Feyerabend suggested, it is (logically) possible that when speakers of different languages appear to be lodged in "a single domain of discourse" they are actually translating so quickly that the mismatch is not noticeable. Were our model builders, experimenters, and theorists really all doing QCD, for example? Again, there is nothing either in the published or in the unpublished papers to suggest it. Feynman polemicized against taking the details of QCD too seriously; everything he did with jet physics in the mid-1980s eschewed the colored gluons and quarks embedded in a renormalizable Lagrangian. Bo Andersson characterized his "use" of QCD by saying that he turned to the "basic" theory as an "inspiration" to model building—but nothing in the string model of hadrons presupposed the detailed acceptance of QCD. There was nothing in the string model that corresponded to the gluon, no reference to Lagrangian field theory at all. Indeed, a detailed "translation" of QCD into the string model would have gone nowhere.

Finally, could the participants in these debates over hadronization simply have had a "mistaken" understanding about what their opposite numbers were saying? Might they have so misconstrued their opponents that Andersson or Gottschalk heard Feynman say "parton" and understood him to mean a QCD quark? Not a shred of evidence supports such a view. "Pure" quantum chromodynamicists had nothing to say about experimental data emerging from the TPC at all—it was not computable. And since no one could "derive" any of the models from QCD, it was surely the case that no Fritzsch or Llewellyn Smith could "mistake" a string, cluster, or independent fragmentation model for QCD itself. From the modelers' perspective, there was if anything a maximization of difference between their accounts: Buchanan's figure 7.37 explicitly contrasted the status of the different models; numerous authors derided the early Feynman-Field model precisely because it did not consistently conserve energy and momentum. The TPC collaboration itself argued in print that one could distinguish between particular physical features of the different models by comparing them with data. Whatever their differences, none of the participants in these debates had any illusion that they and their interlocutors were "really speaking" through the same concepts of quarks and jets.

These negative conclusions lead to a final, positive conclusion. First, bringing different accounts together—pulling them into what Werner Hofmann called a "shoot-out"—was of central importance to the TPC collaboration, as well as to the theoretical model builders. Second, for even the most phenomenologically oriented participants, the goal was always to understand the physical principles at work in the production of the observable hadrons. There was a *physically based* story to be told about these quarks or partons and their recombination into pions and protons, and everyone involved wanted, however partially, to contribute to it. Third, we see consistently the attempt to develop an interlanguage that could serve to bring various theoretical commitments into contact—and into contact with experiment. That is why Andersson and his collaborators insisted in 1983 that the models had "on the one hand" a role as phenomenology able to serve as an intermediary, "in order to obtain a translation from one language to another." "On the other hand" Andersson was always searching for a link to "a dynamical framework" that would serve to motivate and generalize physical principles. It is also why Feynman designed his model with Field to form a "standard" by means of which proton-antiproton collisions could be talked about in the same breath as electron-positron annihilations. Did Feynman hold the Feynman-Field model to be "true"? Never. Feynman was as clear as day: "we do not think of the model as a sound physical theory." But were these models unrelated to physical reality? No, again. The model might well be "close enough to reality" to design future experiments. Buchanan too aimed at such partiality. Likening his and his colleagues' work to the establishment of Kepler's laws—midway between Copernicus's pure phenomenology and the full dynamical theory of Newton—Buchanan aimed for explanations of how major features of the phenomena work and for trustable predictions. He—like Feynman, Andersson, and others—was under no illusion that this interstitial work between data tape and QCD was to be inscribed on the parchment of eternal law. It would rather be a way of establishing a place in practice where mediating concepts, understandable by otherwise skew approaches, could meet. "Quarks," "jets," and "hadronization," along with the simulations that embedded them, could form an interlanguage in a new trading zone.

Theorists and experimenters are not miraculous instantaneous translators and they are not "mere" instrumentalists uninterested in interpretation. They are traders, strategically coordinating parts of interpreted systems against parts of others. The holism that W. V. O. Quine advocated in the years after World War II is compelling. It is hard to imagine ever trying to resurrect a demarcation criterion that would sever the observable from the theoretical. Yet perhaps we could say this: in the trading zone, where two webs meet, there are knots, local and dense sets of quasi-rigid connections that can be identified with partially autonomous clusters of actions and beliefs.

9.7 The Place of Trading

Trading between theorists and experimenters in the heyday of electron theories was done by mail; given the separation of theoretical and experimental institutes on the Continent this is hardly surprising. In the United States and Britain, this geographical isolation was not as marked. When American universities began to acquire theorists in the 1930s they were intellectually and spatially closer to their experimental colleagues. But it would be a distortion to talk about these communities as if they were coequal: only in the Oppenheimer group at Berkeley was there a strong prewar contingent of theorists. Elsewhere, a Wendell Furry, a John Van Vleck, or a John Slater was distinctly in the minority.

For many reasons World War II changed this relation. Obviously, J. Robert Oppenheimer's performance as director of Los Alamos brought theory into prominence. But more important, as we saw in chapter 4, theorists, experimenters, and engineers were forced to work with one another in the large wartime projects. They emerged with nearly five years' experience of each other's ways of approaching problems, and an enduring faith that postwar science had to exploit the collaborative efforts that they credited with the atomic bomb and radar. In large part, collaboration consisted of establishing a place where ideas, data, and equipment could be passed back and forth between groups—constituting a trading zone.

The Rad Lab, as it came to be called, was established in late 1940, around the British invention of a device that could produce microwaves of the right frequency for an effective radar system. Lee DuBridge agreed to head the project on 16 October 1940; by late October a core group had established itself in Room 4-133 at MIT. At first, the divisional structure of the laboratory was designed to replicate the five-part electronic structure of radar, as if the laboratory were a small business: the modulator delivered pulses of power to the magnetron, the magnetron transmitter delivered microwave signals, the antenna emitted and collected these signals, the receiver sorted signals from noise, and the indicator displayed an image on a cathode ray tube (see figure 9.8). As can be seen in figure 9.9, the physical architecture closely matched the electronic architecture of figure 9.8. These three architectures—physical, electronic, and administrative—did not respect distinctions between engineers and physicists. William Tuller, for example, was an electrical engineer with a desk adjacent to that of Henry Neher, a physicist trained in experimental cosmic ray investigations. William Hall, who had been an electrical engineer working for Metro-Goldwyn-Mayer doing sound recording, now shared the indicator corner of 4-133 with A. J. Allen, a physicist and electrical engineer, and Ernest C. Pollard, a physicist who had taken his B.A. and Ph.D. at Cambridge and in 1940 was an assistant professor at Yale. At first, theoretical physicists had no physical location in the laboratory—they were consultants, appearing from time to time very much the way they would visit a prewar cosmic ray, spectrographic, or magnetic laboratory. Face-to-face contact—literally so, as is evident from figure 9.6—counted for much. As one experimental physicist put it at the time: "It is not enough that the discoveries and experiences of one group be occasionally presented in seminars or regular written reports. The former seldom go into sufficient detail to mean much, while the latter are either too detailed or simply

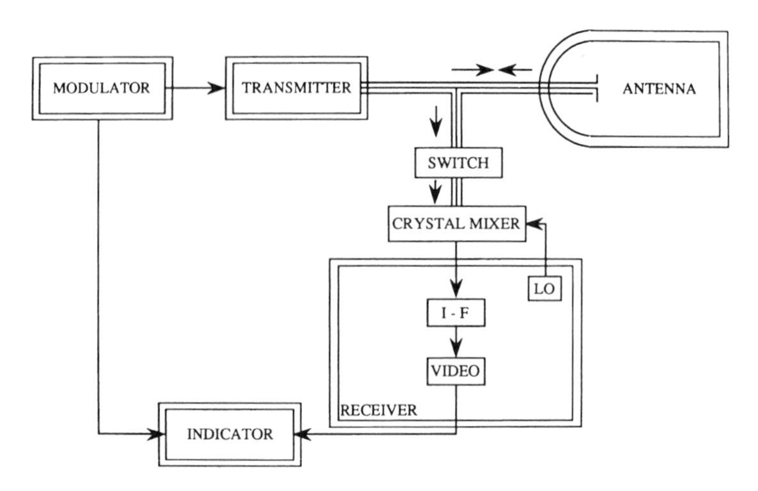

Figure 9.8 Electronic architecture of radar. Schematically, radar could be divided into several components.

unread." Instead, he suggested, the physicists needed to work physically in the same group. It was "[a] far swifter and more painless method of spreading new circuits and general Radar philosophy."[65]

But after Pearl Harbor, military needs pressed the laboratory ever harder to restructure around the production of finished projects, not merely "breadboard" models unready for manufacture. On the other side, many of the physicists tried to guard the "horizontal" autonomy of their component groups. The resulting reorganization of March 1942 was a compromise, providing a mixture of divisions based on components (transmitters, receivers) and divisions based on end products (ground systems, ship systems, airborne systems).[66] While this decision to impose a vertical integration may appear to be "merely" administrative,[67] it was hotly debated, as we saw in chapter 4, because it challenged the boundaries of the subculture of the physicists, forcing them to trade directly with the production-oriented engineers.

In early 1942, DuBridge established a formal division for research, including a theoretical group. To illustrate the kind of trading that went on between theorists and instrument designers, I will focus on the transfer of a research style from the instrument design of electrical engineers to the arcane, sometimes metaphysical world of particle physics. In this instance, the connection is not the

65. White, "A Proposal for Laboratory Organization," file "Reorganization," box 59, RLP.

66. Guerlac, *Radar* (1987), 292–95.

67. The vertical integration of component parts is not just a part of the creation of large-scale business, it is (according to Chandler) the single most salient structural change that signals the transition from small to big business. Chandler, *Visible Hand* (1977).

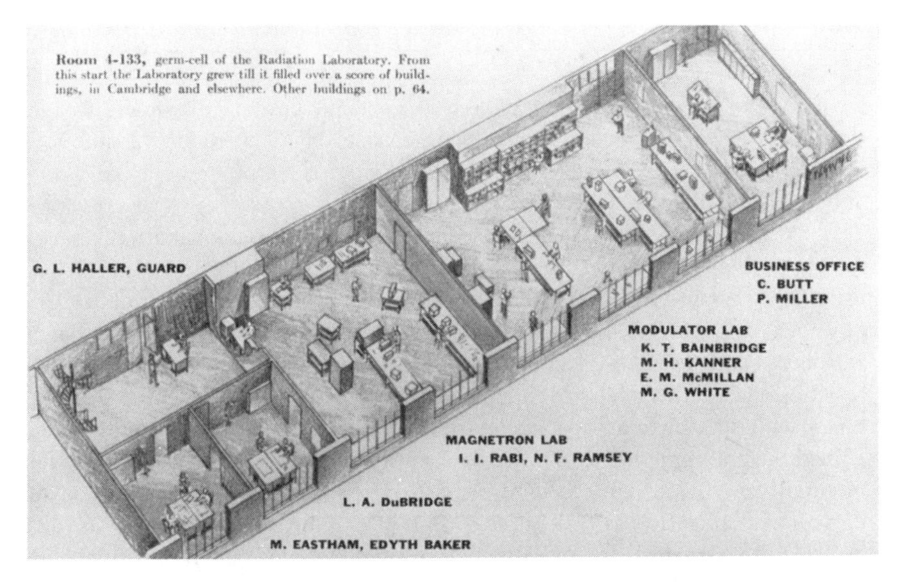

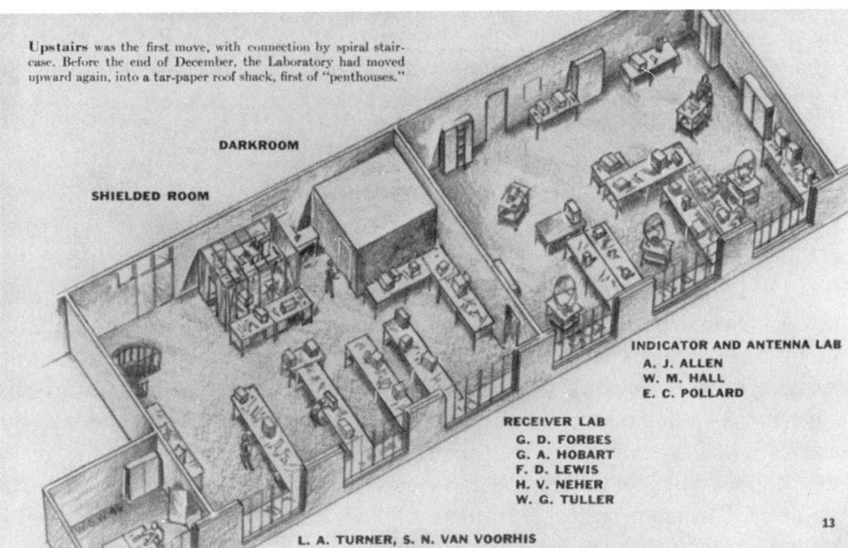

Figure 9.9 Physical architecture parallels electronic architecture (1946). One contemporary Rad Lab publication called Room 4-133 the "germ cell of the Radiation Laboratory." From there it pushed upstairs, then up to a tarpaper roof shack, and eventually into over 20 buildings stretching across the continent and into Europe. In this early electronic functional division of space, engineers and physicists structured their work around components rather than polarizing efforts around "pure" and "applied" science. Working out a common language became the order of the day. Source: Massachusetts Institute of Technology, *Five Years* (1947), 13.

trivial one in which theory is applied to the specific case of instrument design. Rather, the instrument makers—here the electrical engineers of the wartime MIT Radiation Laboratory—had a characteristic way of learning about the world, and it was the theorists who came to adopt it.

At first glance, the war would seem to have made no contribution whatsoever to such an abstruse and abstract subject as quantum electrodynamics. The usual story about QED runs roughly as follows: during the 1920s and 1930s physicists interested in the subject, including Victor Weisskopf, H. A. Kramers, J. Robert Oppenheimer, Niels Bohr, and Julian Schwinger, made halting progress in understanding how the quantum theory of the electron could be combined with special relativity. For reasons of war work, all those living in the United States supposedly broke off their efforts during World War II to do their required (but "irrelevant" to pure physics) work on engineering and then returned, triumphantly, to QED in the second half of the 1940s. The story is false on at least two levels. First, as Silvan Schweber has pointed out, the developments in QED were catalyzed in part by the results of wartime microwave technology that made possible the precision measurements of Willis Lamb, R. C. Retherford, Henry Foley, J. M. B. Kellogg, and P. Kusch et al. in I. I. Rabi's laboratory and the work of Dicke at Princeton.[68] These were extraordinary experiments, but the impact of the war went even deeper. Radar work reconfigured the strategy by which Schwinger approached physical problems. Julian Schwinger himself argued that his radar work had a strong impact on his postwar thinking; in what follows I will expand on his remarks, making use of his actual work in radar to complete the picture.

Here as in previous chapters let us attend to practice, not results. During the war, Schwinger worked in the theoretical section of the MIT Rad Lab; his group had the task of developing a usable, general account of microwave networks. Ordinary network theory—the theory of radio waves in resistors and capacitors—failed because the wavelength of microwaves is comparable to the size of ordinary electrical components. In ordinary components such as resistors, copper wires, or cylindrical capacitors, the microwave energy would radiate away. This made useless the full set of older calculational tools available for electronic circuits. Schwinger began with Maxwell's equations and, with the help of his coworkers, derived a set of rules by which engineers and physicists could make practical network calculations.[69]

68. Schweber, "Quantum Field Theory" (1984), 163.

69. Schwinger and others at the MIT Radiation Laboratory struggled with the problem of determining the input and output characteristics of radiation to and from waveguide junctions. Working in parallel (or perhaps antiparallel) the theorist Sin-itiro Tomonaga was at work on radar for the Japanese. During the first part of the war, each of the two physicists approached the difficulty using classical "physicists' techniques"—exploiting relations among the amplitudes of waves arriving at and leaving the junction. When put into a scattering "S"-matrix these quantities obeyed certain symmetries and, when the process was lossless, unitarity. Because of these symmetries, the S-matrix formalism facilitated the solution of problems (such as the split of one waveguide into two) by side-

As the war progressed and Schwinger assimilated more of the "good enough" and input-output engineering culture of the Rad Lab, he began to abandon the physicists' abstract scattering theory of electromagnetism and to search for the microwave analogue of the electrical engineers' more practical representations: simple "equivalent circuits" that imitated just the relevant aspects of the components. It was an old technique among electrical engineers, who were used to treating certain systems, such as loudspeakers, not according to their real electrical, mechanical, or electromechanical properties, but as if the loudspeaker were a circuit of purely electrical components. In other words, they (symbolically) put the complicated physics of the loudspeaker's electromechanically generated sound into a "black box" and replaced it in their calculations with "equivalent" electrical components. Similarly, the conducting hollow pipes and cavities of microwave circuits could be replaced (symbolically) by ordinary electrical components in order to make the cavities amenable to algebraic manipulation—without entering each time into the details of complex boundary-value problems for Maxwell's equations. As the postwar Rad Lab "Waveguide Handbook" put it, the adoption of equivalent circuits "serves the purpose of casting the results of field calculations in a conventional engineering mold from which information can be derived by standard engineering calculations."[70] It is just this process of appropriation, this "casting" into an "engineering mold," that intrigues me. In the detachment of field calculations from their original context, the full meaning of the terms is cut short. Nor is the physics meaning suddenly and of a piece brought into engineering lore: microwave frequencies did not allow any simpleminded identification of electrical properties with the well-known categories of voltages, currents, and resistances.

The most difficult cases to recast into the engineering pidgin involved the determination of equivalent circuits for waveguides (long hollow metal boxes) involving discontinuities (protrusions, gaps, dividers; see figure 9.10). To treat such devices as a prewar physicist would have involved intractable calculations of the currents around these protrusions and of the immensely complicated fields generated around them. One either had to measure wave amplitudes emerging from an already built circuit (which was hardly feasible if one was trying to design hundreds of components) or else devise—as Schwinger did—theoretical methods to circumvent the difficulties of such geometries. Schwinger's solution was predicated on localizing the difficulty around the discontinuity and using variational methods to determine the equivalent circuit for that part of the waveguide. With this and other methods, Schwinger and his collaborators calculated

stepping the vast amount of information contained in the full electromagnetic field description and representing only the relation of the *measurable* quantities coming in and out of the junction. Perhaps surprisingly, then, two of the founders of postwar particle physics built their theoretical scattering theory on their wartime radar work. Schwinger, *Shakers* (1980), 14–16.

70. Marcuvitz, *Waveguide Handbook* (1986).

Figure 9.10 Complex waveguide (1945). One of the extraordinary difficulties of microwave work was the theoretical prediction of how a complex waveguide would work. The usual radio frequency techniques foundered in the high-frequency domain, and when complex combinations of waveguides were needed, as in this "balanced duplexer," rigorous treatment by Maxwell's equations and Green's functions became completely intractable. Source: RL-53-409, RLP.

example after example of equivalent circuits. Since equivalent circuits for continuous transmission lines were well known, it became a matter of routine algebra to combine equivalent circuit elements in the building of novel microwave circuits and to derive the practical quantities called for by the engineers: relations among input voltages, output voltages, and currents.

To depict Schwinger's transformation as a shift to "mere approximation" would be to miss the point. There are many forms of approximation, including those that would have yielded approximate solutions to fields deep within the waveguides. This procedure, the procedure of concentrating uniquely on input-output relations, was specific and unlike other approximation schemes—Schwinger's labor had as its goal the production of a kind of simplified jargon (or pidgin) binding elements of the language of field theory with elements of engineering equivalent circuit talk (see figure 9.11). This was an attempt to restrict and modify the usual physicists' language of Maxwellian theory in order to make it algebraically (syntactically) compatible with routine computational practices of the engineers. Schwinger had manufactured a meeting point that both physicists and engineers could understand and that both could link to their larger concerns—on one side to the concepts of Maxwellian field theory, on the other to the practices of radio engineering.

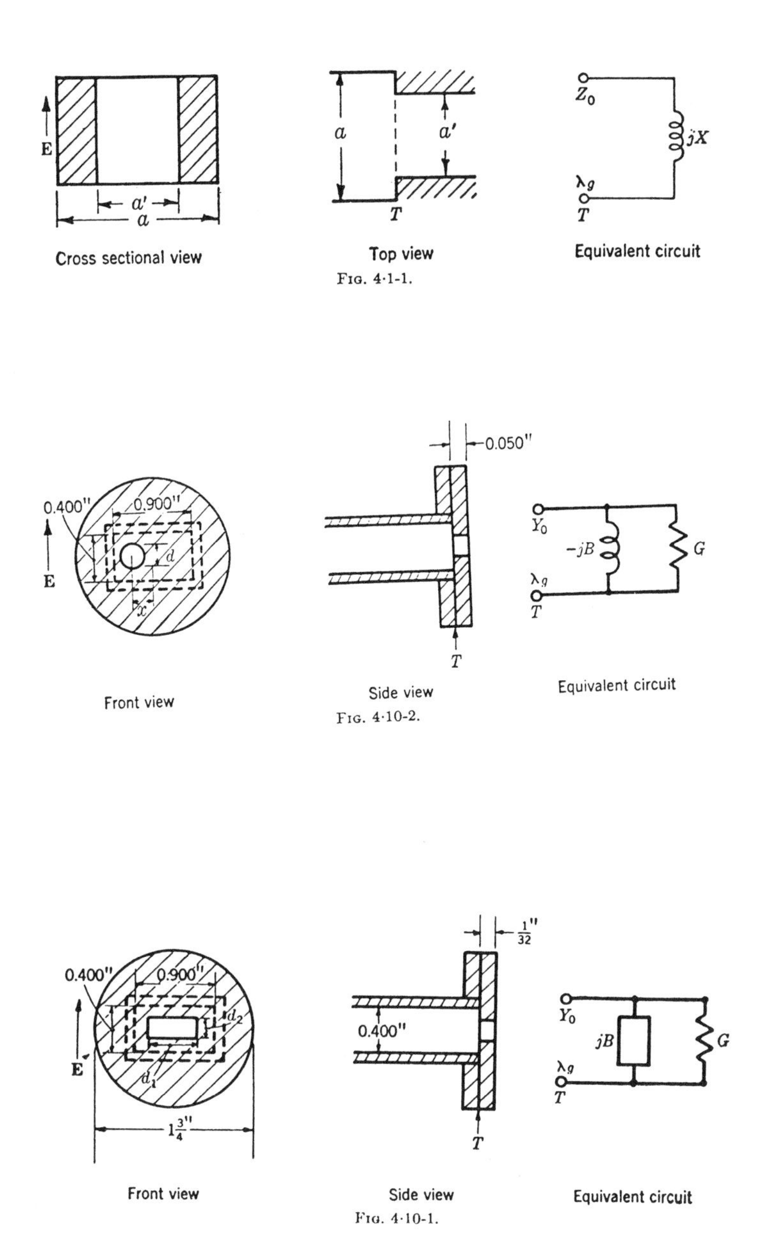

Figure 9.11 Physical waveguides and their equivalent circuits (World War II, published 1951). Schwinger's wartime task was to determine theoretically an electric circuit equivalent for microwave components. These were then used in ordinary electrical engineering calculations. Source: Marcuvitz, *Waveguide Handbook* (1986), 31a.

Figure 9.12 Schwinger and equivalent circuit calculations at the MIT Radiation Laboratory (1946). Schwinger's daily calculation of equivalent circuits in the microwave regime deeply impressed on him the lesson that one could calculate the pragmatically useful—and ignore or integrate out those physical quantities that were not. It was a lesson he and others went on to apply at the very core of "fundamental" physics in the years immediately after the war. Source: Massachusetts Institute of Technology, *Five Years* (1947), 34.

In short, the war forced theoretical physicists—such as Schwinger—to spend day after day calculating things about devices and, through these material objects, linking their own prior language of field theory to the language and algebra of electrical engineering (see figure 9.12, depicting Schwinger staring at the equivalent circuit calculations). Modifying the theory, creating equivalent circuits for microwave radiation, solving new kinds of problems was not—and this is the crucial point—a form of translation. Even the "glossary" of figure 9.11 was identifying newly calculated theoretical elements with recently fabricated fragments of microwave circuitry; neither was part of the prior practice of either the theorists or the radio engineers. Boundaries are substantial, translation is absent, and gestalt shifts are nowhere in sight.

The enforced contact, even in matters of symbolic notation, is evident when Schwinger writes to the head of the theory division, George Uhlenbeck: "There is a question to be decided relative to sign. I have called the reactance negative since an inductive impedance is $-i\omega L = iX$; the engineer however writes $J\omega L = JX$ and hence calls it positive. Do we bow to the engineer?"[71] The answer, as one can see from Schwinger's subsequent notes, is a resounding yes. More important, equivalent circuits became the local coin of exchange: for the physicists they were linked to the larger symbolic system of field theory; for the engineers they were the natural extension of their old radio engineering toolkit. The impact of this trade had a lasting impact on Schwinger, as it did (in other ways) on his fellow theorists and on the experimenters' view of the role of the theorist. Here is a letter of 12 February 1944 from Rabi to Sam Allison:

> It appears that through a series of unforgivable errors we are joined in a controversy over the services of Dr. Julian Schwinger. . . . The reason for having you in as an innocent bystander in the battle of pygmies I can't quite follow, but I am willing to oblige. Historically speaking, Schwinger was released to the Metallurgical Laboratory during an absent-minded moment of Wheeler Loomis without consultation with me. . . . [B]e that as it was, he went to Chicago where I suppose he performed a Herculean task. What should have been known in the first place was subsequently discovered, namely that his job here was far from complete and that there was a big accumulation of results that he had obtained which only he could work up into a form useful for killing Germans. . . . *Meanwhile its really great importance has become clearer to the various experimental physicists and engineers around the Laboratory.* . . . Please give my regards to the other members of the Laboratory, and ask them to stop recruiting from us for sometime, at least until after the invasion of Europe.[72]

By stripping field theory down to the input-output relations of the equivalent circuit, Schwinger made many aspects of radar design predictable. It became possible to foresee how a variation in a junction box would affect signal strength—in advance of banging metal on the bench. And this only became conceivable by abandoning the physicist's natural inclination to solve for the field at all points inside the elaborate copper linkages. After the war, it was this philosophy of pursuing the experimentally useful (and isolating buried complexity) that guided Schwinger in this quest for a consistent quantum electrodynamics. The problem, as Schwinger put it in late 1947, was this: "Electrodynamics unquestionably requires revision at ultra-relativistic energies, but is presumably accurate at moderate relativistic energies. It would be desirable, therefore, to isolate those aspects of the current theory that essentially involve high energies, and are subject to modification by a more satisfactory theory, from those aspects that involve only moderate energies and are thus relatively trustworthy."[73]

In particular, Schwinger observed that very high energy processes only affected observable consequences of the theory weakly (logarithmically), and

71. Schwinger to Uhlenbeck, n.d., file "S," box 876, RLP.

72. Rabi to Allison, 12 February 1944, file "Personnel-Division 4," box 784, RLP, emphasis added.

73. Schwinger, "Quantum-Electrodynamics," *Phys. Rev.* 73 (1948): 416–17, on 416.

only in a few specifiable places: vacuum polarization and self-energy. In the case of the electron's self-energy, Schwinger noted that the mass of an electron consists of two parts, an intrinsic mechanical mass and an electromagnetic mass caused by the emission and reabsorption of a photon. But all we ever see in the laboratory is the experimental mass of the electron—never the mechanical mass or electromagnetic mass separately. Therefore, Schwinger contended, we must calculate the infinite electromagnetic mass of the electron but then renormalize the total mass to be that of the particle we know in the laboratory. This renormalization factor, and the corresponding one for charge, allows the calculation of a myriad of other observable processes such as the magnetic moment of the electron.[74] While Schwinger's subject matter changed radically between the war and the next few years, the theoretical orientation he acquired in the engineering halls of the Rad Lab left a lasting mark on his subsequent research.

Schwinger himself alluded to the link between the two seemingly unrelated domains of waveguides and renormalization. "[T]hose years of distraction" during the war were more than that: "The waveguide investigations showed the utility of organizing a theory to isolate those inner structural aspects that are not probed under the given experimental circumstances. . . . And it is this viewpoint that [led me] to the quantum electrodynamics concept of self-consistent subtraction or renormalization."[75] With an understanding of Schwinger's work in waveguide physics, we are now in a position to unpack this connection between the calculations of radar and renormalization.

In the microwave case, it was impossible to calculate fully the field and currents in the region of the discontinuity; in the quantum electrodynamics case, it was hopeless to try to pursue the details of arbitrarily high energy processes. To attack the microwave problem, Schwinger (wearing his engineering hat) isolated those features of the discontinuity regions' physics that were important for "the given experimental circumstances"—for example, the voltages and currents emerging far from the discontinuity. In order to isolate the interesting features, he dumped the unneeded details of the electrodynamics of the discontinuity region into the parameters of an equivalent circuit. Faced with the fundamental problem of quantum electrodynamics, Schwinger concluded in 1947 that he should proceed by analogy: one had to isolate those features of the physics of quantum electrodynamics that were important for the given experimental circumstances—for example, magnetic moments or scattering amplitudes. To separate these quantities from the dross, he dumped the unneeded details of high energy interactions into the renormalization parameters.

One lesson that theoretical physicists learned from their engineer colleagues during the war was, therefore, both simple and deep: concentrate on what

74. Schwinger, "Quantum-Electrodynamics," *Phys. Rev.* 73 (1948): 416–17, on 416.

75. Schwinger, *Shakers* (1980), 16. Philosophically, renormalization blurs "fundamental" and "phenomenological" laws; on the unexpected relation of these two concepts, cf. Cartwright, *Laws* (1983).

you actually *want to measure,* and design your theory so that it does not say more than it must to account for these particular quantities. The adoption of this pragmatic attitude toward theorizing was a sufficiently sharp break with earlier traditions of theory that some of Feynman's and Schwinger's contemporaries never accepted it. Even P. A. M. Dirac, one of the greatest of twentieth-century theorists, resisted the idea of renormalization until his death in the 1980s. But the idea rapidly took hold, altering for at least several decades theorists' attitudes toward the character of their description of nature.

To a certain extent, one could trace the new attitude back to the long-held, central European philosophical view that the unobservable should be excluded from physics. Mach's hesitancy before atomism, Einstein's rejection of the light medium, Bohr and Heisenberg's reluctance to say more than we should about electron orbits surely shared something of a positivist orientation. But the American pragmatic, engineering style of the war is not the same as this variety of positivist proclamation to go no further than the observable. One could, in principle (in some sense of "in principle") calculate the detailed electromagnetic fields inside a complex waveguide, but it is not necessary. One could, similarly, "in principle" (no one had any idea how) resolve Boltzmann's integrodifferential equations for fission-causing neutrons in a bomb. The American wartime search for bottom-line calculability was something else, not a philosophically driven hesitancy before in-principle observability. It was a hunt, born of close work with engineering, for a calculable bottom line: a frequency shift in cycles per second, a critical mass in kilograms, a power loss in watts per second.

9.8 The Coordination of Action and Belief

In this trading back and forth between traditions at the Rad Lab, one can see an interesting *dis*analogy to Foucault's gloss of Jeremy Bentham's "Panopticon." The Panopticon was a central tower in an "ideal" prison that could survey, and control what it could survey. While the particular spatial disposition of radar workers was important, the heterogeneous, self-consciously collaborative structure of laboratories like Room 4-133 at the Rad Lab set no single group in absolute authority.[76] For at MIT each of the different subcultures was forced to set aside its longer term and more general symbolic and practical modes of work in order to construct the hybrid of practices that all recognized as "radar philosophy." Under the gun, the various subcultures coordinated their actions and representations in ways that had seemed impossible in peacetime; thrown together they began to get on with the job of building radar.

As the architecture emerged in parallel with the expanding Radiation Laboratory, one can see the visible, architectural manifestations of the new modes

76. Foucault, *Discipline* (1979), 195–228.

of exchange. Rooms were established with movable walls as the interchange with industry began to shape the physicists' self-conception. The laboratory not only resembled a factory, as was emphasized in chapter 4, its integration into industrial structure was deliberate and thoroughgoing: by the end of the war almost $3 billion had been spent on radar, the Rad Lab had 3,900 people in its employment, and the laboratory with its "model shop" had delivered $25 million worth of equipment to the armed forces.[77] These developments had, as we have observed, a profound effect on the physics community's plan for a huge centralized laboratory on the East Coast (Brookhaven), one modeled explicitly on the Rad Lab. Recall Henry Smyth's remark near the end of the war that "[t]he [new central] laboratory, should be essentially of factory-type construction, capable of expansion and alteration. Partitions should be nonstructural." And to emphasize the ideological democracy of this new institution, he had added: "[p]anelled offices for the director or any one else should be avoided."[78] Democratization would not be followed by homogenization of the community; there is no question of eliminating the categories of theorist, experimenter, and engineer. Rather, the engineers, theorists, experimenters, and chemists had come to see coordination among their efforts as the central feature of the new physics, as ever newer kinds of devices reconfigured the material landscape of the laboratory.

The planners of the National Accelerator Laboratory (NAL, later Fermilab) also saw coordination as their main goal, although they recognized the enduring gap between the subcultures. Theorists, while necessary for the laboratory, would also need contact with colleagues at neighboring universities; toward this end, the laboratory inaugurated a weekly NAL "theory day,"[79] and more ambitiously a "Theoretical Physics Center" as well. The time, space, and institutional rubric were to give theorists the means to consolidate and reinforce their separate identity. At the same time, it is significant—and a direct outcome of the wartime discussions—that a new experimental laboratory insisted upon the establishment of a theoretical group within its walls. (Similar spatial and institutional accommodations for theorists were created at CERN, SLAC, DESY, and other high energy laboratories.)

While Fermilab attended to the instrumentation and experiments of particle physics, the theorists would work not only in strong interaction dynamics, field theory, symmetries and groups, axiomatics, and phenomenological studies—they would also immerse themselves in gravitation, general relativity, nuclear structure, astrophysics, quantum liquids, and statistical mechanics. "[I]t is understood, of course, that any individual theorist may move from one field to

77. Guerlac, *Radar* (1987), 4. On the military and factory models for postwar laboratories, see Galison, "Physics between War and Peace" (1988).

78. Smyth, "Proposal," 25 July 1944, revised 7 February 1945, 8, PUA.

79. Goldwasser, circular letter, 14 January 1969, uncataloged files, FLA.

another within particle physics and from particle physics to one of the 'peripheral' fields with complete freedom of choice."[80] Because of this recognized difference in the conceptual organization of experiment as distinct from theory, one would see breaks and continuities in theory that would be entirely distinct from those of experiment.

Homogenization of theory and experiment was not the order of the day. As the Fermilab organizers wrote: "Sophistication in mathematical reasoning and technology that has accompanied progress in particle physics no longer allows an ordinary mortal to pursue the science both in an experimental laboratory and in the quiet of a study, as in the good old days of Faraday, Cavendish and Rayleigh, or even in the more recent time of Enrico Fermi."[81] I take it to be no accident that the separation of these cultures is signaled by a separation of place: the "experimental laboratory" is no longer coincident with the "quiet of a study." The contrast between the *vita activa* and the *vita contemplativa* had now been recreated inside the laboratory itself, and in the minds of the Fermilab directorate the division demanded a spatial solution: "All members of the group engage in exchange of ideas and knowledge with users [experimenters from outside NAL who used the facilities] and experimentalists on the staff at Fermilab. These meetings of minds take place more formally in . . . [the] joint Experimental-Theoretical seminar which takes place every Friday [and] is an innovative approach to communication among theorists and experimentalists at Fermilab." More frequent are informal meetings "in offices on the third floor of the Central Laboratory and at the Cafeteria, Lounge and airports";[82] these sites become trading zones. Throughout such exchanges there is no attempt to make experimenters into theorists or vice versa. On the contrary, the concept of collaboration embraced by the physicists during the war involved a reinforcement of these subcultures and an emphasis on exchange in particular, local circumstances.

These various examples of trading between subcultures suggest a model of scientific practice (and the architecture of knowledge) as much at odds with the picture of pure plasticity of experiment and theory as with the rigidly segregated observation language of the early logical positivists. Or perhaps I should add that it has links to both. *Within* traditions, I want to emphasize the relatively constrained nature of scientific practice—hardly anything goes. But when radical changes do occur—and no subculture is immune to such alterations—it does not necessarily follow that the other subcultures break as well. Moreover, the relative rigidity and foreignness of one subculture with respect to another does not make cross-talk between the strata impossible; rather, it ensures that as

80. "Proposal for Theoretical Physics Center of the National Accelerator Laboratory," 19 August 1969, 2, uncataloged files, FLA.

81. Lee, "Theoretical Physics," *NALREP*, March 1975, 1.

82. Lee, "Theoretical Physics," *NALREP*, March 1975, 7–8.

the trading domains become established, the structure of the enterprise as a whole has a strength that the antipositivists denied.

The MIT Radiation Laboratory offers us a picture of the trading zone as an epistemic matter *and* as a physical location. Figure 9.9 (Room 4-133), showing the disposition of personnel in the early Rad Lab, indicates that engineers and physicists worked within sight of one another. Indeed, as a response to the antipositivists' pessimism about the possibility of translation, the MIT Radiation Laboratory strikes me as a perfect counterexample: its success was directly related to the creation of such common domains in which action could proceed even though the physicists and engineers entered into the exchange with radically different understandings of the machinery and techniques involved. Once made salient during the war, these spaces of interaction were designed into laboratories from Brookhaven to SLAC. Indeed, it was the absence of a common space that thwarted Jay Marx's attempts to draw together his constituent groups—electrical engineers, mechanical engineers, and physicists of various stripes. "[N]ext time," Marx had urged late in the TPC project, "build a circus tent to house everyone." When the theorists and experimenters wanted to link their worlds in the TPC/PEP-4 work, they held their own workshops in Asilomar: *place* helped establish the coordination that was so desperately needed.

Or recall our discussion of the early 1990s. Even when the Superconducting Supercollider remained an embattled hope, the laboratory's architectural master plan insisted that the "cross piece of the 'H' plays a critical role in the life and vitality of the campus. It is the 'bridge' that connects the scientific and the public sides of the campus. It is the crossroads of the life of the campus and 'the people interactive zone'" (see figures 7.35 and 9.13). By lining the "street" of the "H" with computer terminals, adding a post office, a cafeteria, and a travel office to the already centralized control facilities, auditorium, and library, the architects and physicists hoped to recoup what so much of their laboratory had undermined: a center. "This," the SSC master plan continued, "is the place where various elements of the physics community meet formally and informally."[83] It was to be, by design, a site where experimenter, accelerator physicist, and theorist would avoid the social and intellectual fragmentation they feared: a physicalized trading zone constructed in a city-sized laboratory-universe, linked through terminals to the wider world of the data-bearing net.

Moving away from the stack periodization typical of conceptual schemes, radical translations, gestalt switches, and paradigm shifts comes at a price: we lose the vivid metaphorical imagery of totalistic transformations. In the place of such all-or-nothing mutations we need some guidance in thinking about the local configurations that are produced when two complex sociological and symbolic systems confront one another. Anthropologists are familiar with such

83. Safdie et al., *SSC Laboratory* (1993), 54.

Figure 9.13 H-Street trading zone, SSC (1993). This architectural sketch depicts the "street-like" corridor that Safdie and Associates hoped would form a meeting point both for the "public" and "private" sectors of the laboratory, and for the different populations working within the laboratory (theorists, experimenters, engineers). It was to have been a physical "trading zone," or at least to have provided the needed architectural infrastructure. Source: Safdie et al., *SSC Laboratory* (1993), 68. Photograph by Michal Safdie.

exchanges, and one of the most interesting fields of investigation has been the anthropological linguistics of pidginization and creolization; I have used these ideas throughout the preceding chapters to characterize the border zones between different theoretical, experimental, and engineering cultures. Both terms—"pidgin" and "creole"—refer to languages at the boundary between groups. A pidgin usually designates a contact language constructed with the elements of at least two active languages; "pidginization" is the process of restriction by which a pidgin is produced. By convention, "pidgin" is not used to describe a language that is used even by a small group of people as their native tongue. A creole, by contrast, is a pidgin extended and complexified to the point where it can serve as a reasonably stable native language.[84]

Typically, pidgins arise when two or more groups need to establish trade or exchange, a need that can issue from the exigencies of weapons projects, or from the quieter joining of analytic and mimetic experimentation. One way that such languages arise is when a dominant but smaller group withholds its full language, either to preserve its cultural identity, or because its members believe that their social inferiors could not learn such a complex structure. To communicate, the dominant group produces a "foreigner talk" that is then elaborated as it is used in day-to-day trading. For example, "Police Motu" appears to have been produced this way. Originally, the Motu (of what is now Papua New Guinea) created a simplified version of their language (a foreigner talk) to use in their extensive trading network, for example, trading pots and sea products for game and bush products. William Foley, an anthropological linguist, speculates that at this stage the simplified Motu was not a distinct language from Motu itself. Beginning in the 1870s Europeans and later Chinese, Pacific Islanders, and Malay Indonesians arrived; they too acquired the foreigner talk version of Motu. When the British established colonial rule, they enforced their dominance with police, often not native speakers of Motu; the police slipped rather easily into the only lingua franca available, the simplified Motu, but now elaborated the language to make it serve its more complex function within colonial rule. As a more intricate and (forcibly) widespread language, Police Motu gained in significance. Diffusion was unintentionally aided, in part, by the circumstance that the criminals arrested by the police were often men of high social status in their villages (e.g., headhunters). When the incarcerated returned home they carried with them Police Motu, according it yet greater status.[85]

The contraction and alteration of a native language to a pidgin occurs on many axes.[86] Reduction in linguistic structure can occur *lexically,* through restriction of vocabulary or through monomorphemic words; it can occur *syntactically,* through the elimination of subordinate clauses, or the hardening of word order; *morphologically,* through the reduction in inflection or allomorph; or it can occur *phonologically,* through the elimination of consonant clusters and polysyllabic words. At first such pidgins may be unstable, varying according to the prior linguistic practices of each learner. But gradually, in at least some cases, the pidgin will stabilize; sometimes this will occur when learners of different

84. Much of the following is based closely on the excellent review of the literature on pidginization and creolization given by Foley, "Language" (1988). For an account of this particular development in Motu, see Dutton, *Police Motu* (1985).

85. Foley, "Language" (1988), 173–74.

86. See Ferguson, "Simplified Registers" (1982), 60.

linguistic backgrounds need to communicate among themselves. As the pidgin expands to cover a wider variety of events and objects, it comes to play a larger linguistic role than merely facilitating trade. Eventually, as children begin to grow up "in" the expanded pidgin, the language is no longer acquired merely to serve specific functions, but now must serve the full set of human demands. Linguists dub such a newly created "natural" language a creole, and the process leading up to it "creolization."

The pertinent theoretical point is that coordination of action occurs between languages in the absence of a full-blown translation. In a revealing and often-cited passage, Kuhn argued: "Why is translation, whether between theories or languages, so difficult? Because, as has often been remarked, languages cut up the world in different ways, and we have no access to a neutral sub-linguistic means of reporting."[87] True enough, there is no "sub-linguistic means"—a pidgin or creole is not by any means "below," or "above," language. But it lies among our linguistic abilities to create these mediating contact languages and to do so in a variety of registers.[88] It is this ability to restrict and localize symbolic systems for the purposes of coordinating them at the margins that is important to the linking of the many subcultures of the discipline of physics. *The physicists and engineers of Room 4-133 are not engaging in translation as they piece together their microwave circuits, and they are not producing "neutral' observation sentences: they are working out a powerful, locally understood language to coordinate their actions.*

I point to the dynamics of contact languages and their mechanisms of stabilization, organization, and expansion because they raise questions relevant to the confrontation of theorists with experimenters (or experimenters with instrument makers). For example, the process by which experimenters, theorists, and instrument makers simplify their practices for presentation to other subcultures needs examination. Can we articulate the process along lines similar to the lexical, syntactical, morphological, and phonological axes presented by the creole specialist C. Ferguson? Consider the following example from the heart of physics. In the early 1960s, two prominent theorists, Sidney Drell and James Bjorken, set out to write a book on quantum field theory. They soon realized that they had in fact written two distinct volumes: the first, directed at an audience outside the subculture of theorists, began with the calculational rules of the theory, and the second contained theoretical justifications and proofs of the Feynman techniques. The first book covered Feynman diagrams and the classical applications they made simple—bremsstrahlung (in which a charged particle emits a

87. Kuhn, "Reflections" (1970), 268.

88. Whence comes this ability to pidginize our language? The question is raised by Derek Bickerton in his *Roots* (1981)—he advocates the view that it is part of an innate "bioprogram," but this assertion leads directly into the extensive debate over the existence, extent, and source of linguistic ability more generally.

photon), Compton scattering (in which an electron deflects a photon), and pair annihilation (in which an electron and antielectron fuse and emerge as a pair of photons). In order to study higher order corrections to processes including these, the authors introduced the renormalization procedure without a systematic exposition. Their first volume was a book of *techniques:* it began with rules such as "For each internal meson line of spin zero with momentum q [, insert] a factor: $i/(q^2 - \mu^2 + i\epsilon)$," where μ is the meson mass and ϵ is a small positive number).[89] "Such a development [of theory], more direct and less formal—if less compelling—than a deductive field theoretic approach, should bring quantitative calculation, analysis, and understanding of Feynman graphs into the bag of tricks of a much larger community of physicists than the specialized narrow one of second quantized theorists. In particular, we have in mind our experimental colleagues and students interested in particle physics."[90]

"Second quantization" is the systematic development of quantum field theory, and the "specialized narrow" community of "second quantized theorists" designated the particular grouping of theorists for whom the practices of quantum field theory were "natural." As this smaller group sought to connect its work to the results of laboratories like SLAC, the fullness of theoretical representation would perforce be restricted (pidginized). Left out of the "experimental" volume was the framework in which the rules found their justificatory place. Also removed were the more general proofs, such as the demonstration that a calculation within quantum electrodynamics, to any order of accuracy, will remain finite.[91] As in Police Motu, the creation of a "foreigner" version of the symbolic system occurs on many fronts. There is an emphasis on plausible, heuristic argumentation rather than a more systematic demonstration, and there is more attention to the calculation of measurable quantities than to formal properties of the theory at some remove from experiment (such as symmetries and invariances). Perhaps more subtly, the "theoretical" version often links phenomena that are left merely associated for the experimenters. For example, in the experimental volume it is simply postulated that particles with half-integer spins (such as electrons) obey the Pauli exclusion principle, whereas in the theoretical volume, this contention is demonstrated for any local quantum field theory obeying Lorentz covariance and having a unique ground state.[92] This and other results are ultimately linked for the theorists to a different structure in which the basic entities are embedded. In particular, in the experimental volume the basic object—the field Ψ—stands for a wave function of a single particle. The experimenters learn to manipulate this function in various ways following the rule of

89. Bjorken and Drell, *Relativistic Quantum Mechanics* (1964), 286.
90. Bjorken and Drell, *Relativistic Quantum Mechanics* (1964), viii.
91. Bjorken and Drell, *Relativistic Quantum Fields* (1965), 330–44.
92. Bjorken and Drell, *Relativistic Quantum Fields* (1965), 170–72.

what is called "first" quantization: the position x and momentum p of classical physics are replaced by the operator x and the spatial derivative d/dx. The differential equations that result are solved and the dynamics of the particle's wave function thereby determined. For the theorists, Ψ does not stand for the wave function of a single particle; rather Ψ is considered to be an operator at each point in space and time. Instead of standing in for a single particle, it represents a field of operators capable of creating and annihilating particles at each space-time point.

Despite this radical difference in the ontology—the set of what there is—a meeting ground exists between the experimenters and the quantum field theorists. That trading zone forms around the description of the phenomenological world of particle physics: How do photons recoil from electrons? How do electrons scatter from positrons? How do photons create pairs of electrons and positrons in the near presence of a proton? How are magnetic moments calculated for the muon? For these and similar questions, the experimenters and theorists come to agreement about rules of representation, calculation, and local interpretation. Bjorken and Drell's volume 1 is an attempt to create a stable pidgin language that mediates between experimenter and theorist. Reduction of mathematical structure, suppression of exceptional cases, minimization of internal links between theoretical structures, simplified explanatory structure—these are all ways that the theorists prepare their subjects for the exchange with their experimental colleagues. I take these moves toward regularization to be the formal language analogues of phonetic, morphological, syntactical, and lexical reduction of natural languages.

By invoking pidgins and creoles, I do not mean to "reduce" the handling of machines to discourse. My intention is to *expand* the notion of contact languages to include structured symbolic systems that would not normally be included within the domain of "natural" language. On one side, this expansive attitude can be grounded by criticizing attempts to isolate natural languages; after all, even languages like English have been conditioned in part by intentional intervention. Language games, such as backslang or rhyming slang, have left grammatical traces within "purely natural" languages.[93] On the other side, "unnatural" (constructed) languages such as signing, FORTRAN, and even electronic circuit design can be used in such broadly expressive modes that any demarcation criterion seems bound to fail.

And indeed there is, not surprisingly, a corresponding foreigner talk that experimenters develop on their side. Just as theorists reduce complexity by suppressing the "endogenous" structure linking theory to theory, so experimenters, when addressing theorists, skip the connecting details by which experimental

93. See, e.g., Mühlhäusler, *Pidgin* (1986), 60.

procedures bind with one another. As for the instrument makers, I would suggest that it is the regularization of machine manipulation that constitutes much of their foreigner talk. C. T. R. Wilson stripped out of his published work references to vagaries of spherical glass blowing; nowhere do we find traces of his erratically performing chambers. In the history of the emulsion, the layers of filtration were many: the chemists withheld their chemical formulas; the experimenters did not publish their detailed procedures of development, drying, and cutting, even through all of these were necessary to produce consistent, distortion-free tracks. Occhialini only wrote of his assorted tricks for the preparation of Geiger-Müller tube preparation many years later. Proprietary reticence, as in the case of Ilford and Kodak, only goes so far to explain these silences. More important, I suspect, is that many of these practices were *endoreferential;* they connected one set of laboratory moves with another, rather than linking the laboratory procedures more directly to the trading zone of phenomena that the experimenters (and instrument makers) shared with theorists. At the same time, many of these techniques eluded regularization, at least early on, and as such could not form part of an interlaboratory experimental pidgin. But bit by bit, regularized versions of laboratory procedures do begin to separate themselves from their parent practices; they begin to form what I have called wordless creoles.

These "separable" bits of procedure can come as isolable fragments of craft or engineering knowledge, as when the Alvarez group introduced indium as the material by which to bind bubble chamber glass to the steel chassis. Between such localized wisdom and matériel lay computer programs such as PANG or KICK—or more recently the various neutron Monte Carlos or hadronizing jet makers. Their exchange not only regularized practices in the image tradition, the track analysis programs carried over as well into the logic tradition, serving in the long run to facilitate the coalescence of the two previously competing cultures. Finally, in many cases, such as the postwar distribution of Ilford emulsions, radar oscillators, multichannel analyzers, or spark chamber components, the medium of exchange can be physical. This suggests that the process of "black boxing" can be seen as the precise material analogue of the more linguistic forms of pidginization; just as terms like "electron" can acquire a decontextualized meaning, so *items* like a local oscillator, a charged coupled device, and a computer memory can function as binding elements between subcultures when stripped from their original contexts and coordinated with new ones. After all, it was the military censors' abiding confidence that (in isolation) these instruments would not reveal their function in nuclear weapons or radar development that led them to declassify virtually all electronic instrumentation in the several years after the end of World War II.

Beginning in Rad Lab seminars during the war, the techniques of circuit assembly, component coordination, and testing and general lore were codified into courses and, in the electronic boom after the war, into a universe of practices self-contained enough that students could grow up "in" microwave electronics, attached neither to field theory nor to traditional radio engineering. The pidgin became a creole. Similarly, the development of "particle phenomenology" as a subfield of theoretical physics was an expansion of a trading zone: pidgin particle physics was pressed outward, embracing an ever-widening domain of practices, some borrowed from the experimenters and some from the quantum field theorists. As befitted their boundary identity such physicists sometimes found themselves both in theoretical and in experimental groups. And Monte Carlo techniques, begun between mathematics, nuclear weapons design, statistics, and computer building, grew into a tertium quid joining experiment and theory, and more broadly into a pillar of the new field of computer science. Pidgins and creoles, in the way I have used them here, are defined by having enough autonomous structure to be self-sustaining while at the same time serving to link other, distinct and much larger bodies of practices. Whether these trading languages circulated around calculational methods (as in the Monte Carlo) or around laboratory moves (as in ionography or the repertoire of electronic counting technology), they served to bind the diverse subcultures of physics into a larger, intercalated, and more resilient whole.

What stabilizes a pidgin? What takes an ephemeral alliance of linguistic practices assembled and delimited for a specific purpose and allows it to endure and expand? One interesting conjecture is that the alignment of three or more languages ("tertiary hybridization") serves to prevent the reabsorption of the pidgin into one of the source languages.[94] Perhaps the most effective feature of the huge war laboratories was the imposed orchestration of the practices of theorists, experimenters, and instrument makers with those of electronic and mechanical engineers. It was the felt *difference* of this coordinated activity from any constituent group's prior experience that led White, among others, to speak of a distinctive "radar philosophy." New forms of encounter mark, too, the concatenation of engineering, experiment, and theory in the Alvarez era of bubble chamber experimenting—it was then that experimenters had to assimilate the techniques of computer-run data analysis, distant-site partitioning of data, and isolation of experimenter from machines. Now it was Alvarez who could get up, admonishing his audience to overcome its horror at production lines, events per dollar, and "good enough" engineering. A new kind of physics emerged between cryogenics, factory structure, and hyperon physics in the threatening environs of the hydrogen bubble chamber. Nor did the recombination of experimental subcultures stop in the late 1960s. As the political economy of collabora-

94. The term "tertiary hybridization" is attributed to K. Whinnom (1971) by Foley, "Language" (1988), 173; see also Todd, *Pidgins and Creoles* (1990), 48: "Although the process of pidginization is not limited to such areas, it seems likely that extended pidgins and creoles develop only in multilingual areas. Where the contact is between two languages only, one or both groups acquire the other language, either keeping or relinquishing their own in the process."

tion shifted in the 1970s and 1980s, it frequently came to pass that no one individual, not even one group could be the absolute leader of a multiinstitutional, often multinational collaboration. As we saw in chapter 7, in the absence of such leadership, hundreds of physicists had to find new ways to coordinate their efforts around the massive colliding beam detectors; there were new disciplinary affiliations and new trading zones. Computers entered as connecting links between participants, in the spread of data, and into the devices themselves. Images and logical manipulation of data fused, and "runs" of the experiment could be prosecuted from distant sites. Mixed technical and sociological issues in the design, construction, and running of a Time Projection Chamber or a Superconducting Supercollider raised fundamental questions about what it meant to be an "experimenter." Nothing reflected the dislocations of these new patterns of demonstration better than the ever more awkward, ever more explicit redefining of the scientific author.

Superficially, the handing of charts, tubes, and circuit boards back and forth across the various interexperimental cultural divides might look like a case of worlds crossing without meeting. This description, however, would do violence to the expressed experience of the participants. They are not without resources to communicate, but the communication takes place piecemeal, not in a global translation of cultures, and not through the establishment of a universal language based on sense-data. As it is within the heterogeneous world of the experimenter, so too is it between experimenter and theorist. Here is a summary slogan, necessarily limited: laboratory and theoretical work are not about translation, they are about coordination between action and belief.

That coordination takes place according to no fixed rules. To search for a permanent pattern of experiment, instrument, and theory is to chase a setting sun. The working-out of hybrid forms of practice is never done once and forever; there is no single fixed set of techniques for demonstrating things in the laboratory, no permanent formula for tying experiment to theory or experiment to engineering. How could there be? I take it to be a sign of vibrant life, not fragility, that the material culture of the laboratory is in flux through changing modes of collaboration, techniques, simulations, and disciplinary alliances. True enough, at every stage of its development, there have been laboratory scientists who have bemoaned the loss of a previous Edenic era: groups instead of individuals, networks instead of isolated laboratories, factories instead of workshops, computers instead of notebooks, engineers instead of technicians, multinational collaborations instead of personally acquainted teams. This loss strikes me as imaginary, nostalgic: no individual *ever* commanded the whole of physics. Would laboratory practice be more secure in the year 2000 if experiments were prosecuted as they were by Galileo or Maxwell? Would we be better off without computers, simulations, logic circuits, and golden events? When the norms, values, and standards of experimentation stagnate for 50 years, when they *stop* changing, then the discipline will be dead. Each age remakes the laboratory in its own image, and the laboratory reacts to change the world around it.

Before turning to some final, more broadly philosophical questions, there is an inevitable and inevitably difficult question to address. Frequently, as in the description of Room 4-133, the Superconducting Supercollider's H-Street, the Theory Meetings held by the TPC, or the Fermilab Experimental-Theoretical Seminar, I have emphasized that the *site* of trade is a central concern to the scientific actors themselves. More generally, the sociological separation of theorists and experimenters over the course of the century has been part of our story, as were the difficulties, necessity, and strategies of local conciliation. We can ask: is such a sociometric separation a *necessary* part of the term "subculture" used here? Is it useful to speak about stack and intercalated periodization in those instances in which an individual physicist moves back and forth between experiment and theory? This is a subtle, and in my view incompletely resolved, question, one that can ultimately only be settled through an epistemologically sensitive history. One can, however, move toward a more precise formulation of the problem.

At the epistemological level, the issue is whether, in a given time and place, the concepts and arguments associated with experimental practice are distinct from those associated with theoretical practice. These constraints might be said to differentiate the conditions of possible theoretical moves at a given time from the analogous conditions of possibility for experimentation or for instrumentation. Understood this way, such conditions may be periodized differently even where individual researchers were moving back and forth between theory, experiment, and instrument making. Wilson moved back and forth. But his mimetic experimentation (and its affiliation to meteorology and geology) had a historical dynamic not coperiodized with the contemporary theoretical ion physics that he used (and their affiliation with Cavendish analytic matter theories).

That said, the character of twentieth-century physics cannot be understood if one ignores the powerful practical and institutional forces that drove theorists, experimenters, and instrument makers into separate, interacting communities. That history is not over. Just to name a few borders in transition, we now see realignment between new sectors of the electronic industry and experimentation, between computer science and experimentation, and between string theory and algebraic geometry. As the language and practices of theory (and of experiment, instrumentation, and the trading zones between them) shift, theorists and experimenters create new identities, identities reinforced through the production of new workplaces, affiliations, technologies, and publications. Perhaps one could formulate the problem this way: understanding the historical topology of the constraint map means constantly pulling together an entangled mix of technical, sociological, practical, and abstract questions. Consequently, the nature of and relations between the "conditions of theoreticity," the "conditions of instrumentality," and the "conditions of experimentality" are never fixed in one permanent, deadened configuration. The relations among the subcultures are not, for that, *reduced* to sociology in the sense of being "nothing but" sociology.

9.9 Cables, Bricks, and Metaphysics

The view of science as an intercalated set of subcultures bound together through a complex of hard-won locally shared meanings fits awkwardly into the debate over relativism and realism. In one sense, this picture might be labeled "anti-antirealist" since it directly opposes attempts to disintegrate science into blocks of knowledge isolated from each other and self-contained.[95] But being anti-antirealist is not a defense of a metaphysical or transcendent realism. It is hard to see how one could ever find anything in the history of scientific work that will bear one way or another on whether there truly is a substance beyond the reach of physics that different theories "carve," "organize," or "structure" in different ways.

I find most congenial a view that might be called a historicized neo-Kantianism. That is, we have the totality of our theoretical, instrumental, and laboratory practices; using them we have assembled a historically evolving description of what there is and how it works. Those conclusions are, as Kant would say, to be taken as empirically realistic, not transcendentally realistic.[96] Transcendental realism, the view that there is something beyond, prior, or outside of any possible conception (*ausser uns*), leads according to Kant to some form of idealism. The world becomes a form of illusion, projection, just in the measure that we suppose something, over the horizon, that is real. But why treat kaons or black holes as less real than computers and books? About that which might lie beyond the sum total of what we can do and think, we have, Kant tells us, nothing to say. By contrast, we can make progress in unraveling the place of experiment, theory, and instrument making in the world, and their relations to one another. It is a joint project, at once historical, scientific, and philosophic.

Though manifestly more historical, the philosophical sensibility that underlies this book as a whole has something in common with the justly influential work of Donald Davidson and Hilary Putnam. Davidson's argument against frameworks, paradigms, and their ilk is set out in "On the Very Idea of a Con-

95. I have nothing to say about claims for a relativist metaphysics predicated on a global skepticism about the reliability of our senses; nothing to say about the "possibility" that the world is just the dreams of a brain in a vat; and nothing to offer about comparisons between the physical sciences writ large and domains of belief drawn from cultures largely distinct in time and place. If the frame of "framework" has dimensions so wide that it encompasses the totality of the physical sciences over centuries, then it no longer corresponds to the meaning of this and similar terms and its relatives used in defending framework relativism (within science).

96. My reading of Kant follows closely on that of Henry Allison in his great work *Kant's Transcendental Idealism* (1983).

ceptual Scheme." Its goal is to challenge the notion that different schemes or paradigms render mutual intelligibility impossible. All such scheme talk, Davidson insists, eventually rests on some form of dualism, which he rejects—Whorf wrote that language "organizes" experience, where experience is exterior to the conceptual scheme. According to Davidson, Kuhn denies the existence of sense-data languages but continues to speak about the differing ways in which schemes "attach to nature," while Feyerabend refers to "human experience" that is "outside" system or language. Such dualisms depict conceptual schemes that ("predict," "organize," "face," or "fit") experience (as "world," "nature," "reality," or "sensory promptings"). As Davidson sees it, the very possibility of calling two conceptual schemes different presupposes a substantial body of shared belief. He says: The principle of charity that we invoke in trying to understand another view is not a gratuitous act of generosity, it is the precondition for intelligibility at all.[97] This is what I think is going on when very different views about ions, bubbles, hadronization, or Monte Carlos, are at stake. A vast sea of concepts and practices underlay the distinct deployment of, say, string, cluster, and independent fragmentation pictures despite their radically different ontologies and laws of interaction. That the various advocates of one or another picture could conduct a "shoot-out" (as the participants called it) was precisely because they *nonetheless* shared other strata of belief about calculation and laboratory procedure. Our concepts of pions, kaons, partons, quarks, and jets leak out from the confines of any scheme, and persuasive arguments in physics are the stronger for it.

With the understanding that I want to insist on the further articulation of "shared" given by the distinction between local coordination and global meaning, I completely concur with Davidson when he concludes, "In giving up the dualism of scheme and world, we do not give up the world." Instead, as he puts it, we "reestablish unmediated touch with the familiar objects whose antics make our sentences and opinions true or false."[98] The world, in both its stubborn physicality and its mathematical abstraction, is where we live; what we make of the physical world is not a degraded approximation to a "Reality" ever out of reach.

Hilary Putnam, in defense of his "internal" or "pragmatic" realism, concurs with Davidson in one crucial respect. Both diagnose a metaphysical dualism at the heart of many arguments for relativism. In this respect, he too takes the frankly neo-Kantian position that we cannot do without concepts; we cannot

97. Davidson, "The Very Idea" (1982), 74. Ian Hacking modifies Davidson's idea in an important and interesting way. Instead of saying that there is a core of shared belief, Hacking suggests that we take there to be a substantial body of sentences that are "candidates for truth or falsity"; see Hacking, "Language, Truth and Reason" (1982).

98. Davidson, "The Very Idea" (1982), 79.

go far with a world of doughlike "substance" that is then "carved up" as if by a cookie-cutter into different schemes. Such a metaphysical reality, one "outside" all concepts, leads to a disparagement of certain kinds of things as less real than others. For Kant, Berkeley's dogmatic idealism made ideas real and objects "mere" illusion; and because it was always predicated on an invidious contrast of our ideas with something else, dogmatic idealism was not the enemy of transcendental realism, but its twin. Internal realism, says Putnam, is like Kant's empirical realism, "the notion of a 'thing in itself' makes no sense; and *not* because 'we cannot know the things in themselves.'" The problem with *Ding an sich* talk is not a failure of skill or even a failure of our species-specific capabilities; it is worse than that. When we talk about the *Ding an sich,* Putnam insists, "we don't know what we are talking about." Remove all possible concepts and there is nothing left with which to talk at all. This is not an epistemological limitation; it is not as if we have failed. "The thing in itself and the property the thing has 'in itself' belong to the same circle of ideas and it is time to admit that what the circle encloses is worthless territory."[99] Put more positively, we do have concepts—like that of an object—and can, do, and should talk about the full world of assertions we can make with them, whether about a pie or a pion.

Reference to Kant's language about realism and idealism might well suggest some positions I do not want to take. Is this work a contribution toward his transcendental idealism? Yes, no, and no. The term "transcendental" carries two features in the Kantian account. First, it points toward Kant's characteristic form of analysis in asking about the conditions that must be fulfilled for a concept or thing to be considered at all. I do like this mode of inquiry, and it is helpful, for example, to ask about the conditions that must be satisfied for a laboratory effort to count as a valid experiment, for a device to be an instrument, or for a more formal account to claim the status of theory. Second, however, Kant's "transcendental" suggests conditions that are beyond history. The entirety of the last eight chapters speaks against such an assumption: I am exactly interested in how the criteria for valid instrument, computer, and experimental argumentation have altered over time. Finally, the very term "idealism" (transcendental or otherwise) suggests an exclusion of the material. But whether Kant has that implication in mind or not (many of his followers surely did), it should be clear by now that the material culture of physics enters into the picture painted here—down to the canvas, pigment, and wood.

This picture does, in important ways, fit well with Putnam's antimetaphysical realism, with his sense that not anything goes, and with his abiding suspicion of transcendental realism. My concern about the way Putnam expresses

99. Putnam, *Many Faces* (1987), 36. For more on Putnam's internal realism see, e.g., the 1994 Dewey Lectures, Putnam (1994), esp. lecture 1.

his view is this: in the course of his argument against forms of scheme-reality dualism, he reimports talk about conceptual schemes. True, Putnam's arguments *do* depend on the choice of concepts (whether one admits merological sums matters in determining how many "objects" can be composed of three elements). But the term "conceptual scheme" (and its relatives, including "paradigm") often carries the implication that a vast theoretical structure is to be designated, such as Newtonian mechanics, Einsteinian mechanics, or phlogiston chemistry. When Putnam writes, "[W]hat is wrong with the notion of objects existing 'independently' of conceptual schemes is that there are no standards for the use of even the logical notions apart from conceptual choices,"[100] I would keep the narrower language of "conceptual choices" and lose the much broader connotation of "conceptual schemes."

At the same time, it seems eminently reasonable to follow Putnam's refusal of the cookie-cutter metaphor, and with Davidson and Putnam, I see no great use for the postulation of some metaphysical Substance or Absolute outside or prior to the "mundane reality" in which we live. The problem is the supposition that concepts come in schemes, with its suggestion of steel walls (or axiomatic systems) containing an extensive block of concepts, rigorously excluding all others. Historically, as we have seen, notions like that of Monte Carlo, randomness, game, hadronization, ion, kaon, or drift chamber are understood through powerful changes in their fields of applicability and defining attributes. In a sense this book is, through and through, an argument against the historical-scientific delimiting of concepts and practices into conceptual schemes excluding other blocks of concepts.

Let me conclude with a metaphor. For years physicists and engineers harbored a profound mistrust of disorder. They searched for reliability in crystals rather than disordered materials, and strength in pure substances rather than laminated ones. Suddenly, in the last few years, in a quiet upheaval, they discovered that the classical vision had it backward: the electronic properties of crystals were fine until—*because* of their order—they failed catastrophically. It was amorphous semiconductors, with their *disordered* atoms, that gave the consistent responses needed for the modern era of electronics. Structural engineers were slow to learn the same lesson. The strongest materials were not pure—they were laminated; when they failed microscopically, they held in bulk. To a different end, in 1868 Charles Sanders Peirce invoked the image of a cable. I find his use evocative in just the right way: "Philosophy ought to imitate the successful sciences in its methods. . . . [T]o trust . . . rather to the multitude and variety of its arguments than to the conclusiveness of any one. Its reasoning should not form a chain which is no stronger than its weakest link, but a cable whose

100. Putnam, *Many Faces* (1987), 35.

fibres may be ever so slender, provided they are sufficiently numerous and intimately connected."[101] With its intertwined strands, the cable gains its strength not by having a single, golden thread that winds its way through the whole. No one strand defines the whole. Rather, the great steel cables gripping the massive bridges of Peirce's time were made strong by the interleaving of many limited strands, no one of which held all the weight. Decades later, Wittgenstein used the same metaphor now cast in the image of thread, as he reflected on what it meant to have a concept. "We extend our concept of number as in spinning a thread we twist fibre on fibre. And the strength of the thread does not reside in the fact that some one fibre runs through its whole length, but in the overlapping of many fibres."[102] Concepts, practices, and arguments will not halt at the door of a conceptual scheme or its historical instantiation: they continue, piecewise.

These analogies cut deep. It is the *disorder* of the scientific community—the laminated, finite, partially independent strata supporting one another; it is the *dis*unification of science—the intercalation of *different* patterns of argument—that is responsible for its strength and coherence. It is an intercalation that extends even further down—even within the stratum of instruments we have seen mimetic and analytic traditions as separate and then combining, image and logic competing then merging. So too could we see divisions within theory—confrontational views about symmetries, field theory, S-matrix theory, for example—as one incompletely overlapped the other.

But ultimately the cable metaphor too takes itself apart, for Peirce insists that the strands not only be "sufficiently numerous" but also "intimately connected." In the cable, that connection is mere physical adjacency, a relation unhelpful in explicating the ties that bind concepts, arguments, instruments, and scientific subcultures. No mechanical analogy will ever be sufficient to do that because it is by coordinating different symbolic and material actions that people create the binding culture of science. All metaphors come to an end.

101. Peirce, "Four Incapacities" (1984), 2:213.
102. Wittgenstein, *Philosophical Investigations* (1958), par. 67.

Abbreviations for Archival Sources

AdR	Archiv der Republik, Vienna, Austria.
AEP	Albert Einstein papers, published with permission of Princeton University Libraries, Princeton, New Jersey.
AGP	Adriano Gozzini papers (private).
AIP	Niels Bohr Library, American Institute of Physics, College Park, Maryland.
ARI	Archives of the Royal Institution, Churchill College, London, England.
ATP	A. M. Thorndike personal papers, Brookhaven National Laboratory, Upton, New York.
BCL	Bubble Chamber logbooks, 1953–57, Archives and Records Office, Lawrence Berkeley Laboratory, Berkeley, California.
BMuP	Bundesministerium für Unterricht, Osterreiches Staatsarchiv, Vienna, Austria.
BNB	Earliest bubble chamber notebook, M. Lynn Stevenson, Bubble Chamber logbooks, 1953–57, Archives and Records Office, Lawrence Berkeley Laboratory, Berkeley, California.
CEAP	Cambridge Electron Accelerator papers, by permission of the Harvard University Archives, Cambridge, Massachusetts.
CPnb	Cecil Frank Powell notebooks, Contemporary Scientific Archives Centre, deposited at University Library, University of Bristol, Bristol, England.
CPP	Cecil Frank Powell papers, Contemporary Scientific Archives Centre, deposited at University Library, University of Bristol, Bristol, England.
CWnb	C. T. R. Wilson notebooks, Royal Society, London, England.
CWP	C. T. R. Wilson papers (private).
DCP	Donald Cooksey papers, Director's Office, R&D Administration Files, Archives and Records Office, Lawrence Berkeley Laboratory, Berkeley, California.

DGP	Donald Glaser papers (private).
DGnb	Donald Glaser notebooks in Donald Glaser papers; among these papers are two bound notebooks in Glaser's handwriting, which will be referred to by their original numbers DGnb1 and DGnb2. The first has no date; the second begins in October 1952.
DNP	Darragh E. Nagle papers (private).
DNyP	David Nygren papers (private).
DNyPI	David R. Nygren papers, TPC project files, 1974–83, Archives and Records Office, Lawrence Berkeley Laboratory, Berkeley, California.
DNyPII	David R. Nygren papers, TPC project files, 1974–83, Archives and Records Office, Lawrence Berkeley Laboratory, Berkeley, California.
DTP	Donald Tressider papers, Department of Special Collections, Stanford University Libraries, Stanford, California.
EAP	Edward Appleton papers, by permission of the Master and Fellows of the University of Edinburgh, Edinburgh, Scotland.
EBP	Elliott Bloom papers (private).
EGP	Eugene Gardner papers, Meson Experiment Files, 1945–50, Lawrence Radiation Laboratory (AEC RG 326), National Archives, Pacific Sierra Region, San Bruno, California.
ELP	Edward Lofgren records, Accelerator Division, R&D Records, 1953–83, Archives and Records Office, Lawrence Berkeley Laboratory, Berkeley, California.
EMP	Edwin McMillan papers, 1907–84, RG 434, Subgroup Lawrence Berkeley Laboratory, Series IV, National Archives, Pacific Sierra Region, San Bruno, California.
ERP	Ernest Rutherford papers, Niels Bohr Library, American Institute of Physics, College Park, Maryland.
EWP	Eugene Wigner papers, Department of Rare Books and Special Collections, Princeton University Libraries, Princeton, New Jersey.
FBP	Felix Bloch papers, Department of Special Collections, Stanford University Libraries, Stanford, California.
FEP	Foster Evans papers (private).
FLA	Fermilab Archives, Fermilab, Batavia, Illinois.
FTP	Frederick Terman papers, Department of Special Collections, Stanford University Libraries, Stanford, California.
GCP	Georges Charpak papers (private).
GFP	George E. Forsythe papers, Department of Special Collections, Stanford University Libraries, Stanford, California.
GGP	Gerson Goldhaber papers (private).
GRP	George Rochester papers (private).
HHP	H. Paul Hernandez papers (Time Projection Chamber files), 1976–81, Archives and Records Office, Lawrence Berkeley Laboratory, Berkeley, California.
HML	Hagley Museum and Library Collections, courtesy of Hagley Museum and Library, Wilmington, Delaware.

HUA	Harvard University Archives, by permission of the Harvard University Archives, Cambridge, Massachusetts.
IfR	Institut für Radiumforschung, Vienna, Austria.
JCP	James Cronin papers (private).
JMaP	Julian Ellis Mack papers, University of Wisconsin–Madison Archives, Madison, Wisconsin.
JMP	Jay Marx papers, TPC R&D Administrative Records, 1972–83, Archives and Records, Lawrence Berkeley Laboratory, Berkeley, California.
JNP	John von Neumann papers, Manuscript Division, Library of Congress, Washington, D.C.
JSP	John Clarke Slater papers, American Philosophical Society Library, Philadelphia, Pennsylvania.
LAA	Los Alamos National Laboratory Archives, Los Alamos, New Mexico.
LAP	Luis Alvarez papers, Scientist's Files I, 1936–88, and II, 1934–86, Lawrence Berkeley Laboratory, National Archives, Pacific Sierra Region, San Bruno, California.
LBL	Lawrence Berkeley Laboratory, Archives and Records Office, Berkeley, California.
LHP	Leopold Halpern papers (private).
LSP	Leonard Schiff papers, Department of Special Collections, Stanford University Libraries, Stanford, California.
LStP	M. Lynn Stevenson papers (private).
MBP	Marty Breidenbach papers (private).
MMPP	Michigan Memorial-Phoenix Project papers, University of Michigan Physics Department, Ann Arbor, Michigan.
MRP	Michael Ronan papers (private).
MSP	Moshe Safdie papers, Moshe Safdie Associates, Cambridge, Massachusetts.
NBA	Niels Bohr Archive, Niels Bohr Institute, Copenhagen, Denmark.
OAP	Otto Claus Allkofer papers (private).
OFP	Otto Frisch papers, by permission of the Master and Fellows of Trinity College, Cambridge, England.
PBP	P. M. S. Blackett papers, Royal Society, London, England.
PHP	Paul-Gerhard Henning papers (private).
PRO	Public Record Office, Kew Gardens, London, England.
PUA	Princeton University Archives, published with permission of Princeton University Libraries, Princeton, New Jersey.
RFP	Rockefeller Foundation Papers, Rockefeller Archive Center, North Tarrytown, New York.
RHP	Roger Hickman papers, Physics Department, Historical Records, by permission of the Harvard University Archives, Cambridge, Massachusetts.
RHilP	Roger Hildebrand papers (private).
RHofP	Robert Hofstadter papers (private).

RLE	MIT Research Laboratory of Electronics Papers, Massachusetts Institute of Technology, Cambridge, Massachusetts.
RLP	Radiation Laboratory papers (MIT), Office of Scientific and Research Development, National Archives, New England Region, Waltham, Massachusetts.
ROP	Robert Oppenheimer papers, Manuscript Division, Library of Congress, Washington, D.C.
RTP	R. W. Thompson papers (private).
SLAC	Stanford Linear Accelerator Center, Archives and History Office, Stanford, California.
SUA	Stanford University Archives, Department of Special Collections, Stanford, California.
SUP	Stanislaw Marcin Ulam papers, American Philosophical Society, Philadelphia, Pennsylvania.
UWA	University of Wisconsin–Madison Archives, Madison, Wisconsin.
WHP	W. W. Hansen papers, Department of Special Collections, Stanford University Libraries, Stanford, California.
WWP	W. A. Wenzel papers (private).

Bibliography

Achinstein, Peter, and Owen Hannaway, eds. 1985. *Observation, Experiment and Hypothesis in Modern Physical Science.* Cambridge, Mass.: MIT Press.

Ackermann, Robert J. 1985. *Data, Instruments, and Theory: A Dialectical Approach to Understanding.* Princeton, N.J.: Princeton University Press.

Aihara, H., et al. 1985. "Tests of Models for Parton Fragmentation by Means of Three-Jet Events in e^+e^- Annihilation at $\sqrt{s} = 29$ GeV." *Physical Review Letters* 54:270–73.

———. 1985. "Tests of Models for Quark and Gluon Fragmentation in e^+e^- Annihilation at $\sqrt{s} = 29$ GeV." *Zeitschrift für Physik C* 28:31–44.

Aitken, John. (1873) 1923. "Glacier Motion." In *Collected Scientific Papers,* edited by C. G. Knott, 4–6. Cambridge: Cambridge University Press. First published in *Nature* 7 (February 1873).

———. (1876–77) 1923. "On Ocean Circulation." In *Collected Scientific Papers,* edited by C. G. Knott, 25–29. Cambridge: Cambridge University Press. First published in *Proceedings of the Royal Society of Edinburgh* 9 (1876–77).

———. (1880–81) 1923. "On Dust, Fogs, and Clouds." In *Collected Scientific Papers,* edited by C. G. Knott, 34–68. Cambridge: Cambridge University Press. First published in *Transactions of the Royal Society of Edinburgh* 30 (1880–81).

———. (1883–84) 1923. "On the Formation of Small Clear Spaces in Dusty Air." In *Collected Scientific Papers,* edited by C. G. Knott, 84–113. Cambridge: Cambridge University Press. First published in *Transactions of the Royal Society of Edinburgh* 32 (1883–84).

———. (1883–84) 1923. "Second Note on the Remarkable Sunsets." In *Collected Scientific Papers,* edited by C. G. Knott, 123–33. Cambridge: Cambridge University Press. First published in *Proceedings of the Royal Society of Edinburgh* 12 (1883–84).

———. (1888) 1923. "On the Number of Dust Particles in the Atmosphere." In *Collected Scientific Papers,* edited by C. G. Knott, 187–206. Cambridge: Cambridge University Press. First published in *Transactions of the Royal Society of Edinburgh* 35 (1888).

———. (1889–90, 1892, 1894) 1923. "On the Number of Dust Particles in the Atmosphere of Certain Places in Great Britain and on the Continent, with Remarks on the Relation Between the Amount of Dust and Meteorological Phenomena." In *Collected Scientific Papers,* edited by C. G. Knott, 297–331 (Part I), 332–62 (Part II), and 363–434 (Part III). Cambridge: Cambridge University Press. First published as Part I, *Proceedings of the Royal Society of Edinburgh* 17 (1889–90); Part II, *Transactions of the Royal Society of Edinburgh* 37 (1892); Part III, *Transactions of the Royal Society of Edinburgh* 37 (1894).

———. (1892) 1923. "On Some Phenomena Connected with Cloudy Condensation." In *Collected Scientific Papers,* edited by C. G. Knott, 255–83. Cambridge: Cambridge University Press. First published in *Proceedings of the Royal Society of London* 51 (1892).

———. (1900–1901) 1923. "Notes on the Dynamics of Cyclones and Anticyclones." In *Collected Scientific Papers,* edited by C. G. Knott, 438–58. Cambridge: Cambridge University Press. First published in *Transactions of the Royal Society of Edinburgh* 40 (1900–1901).

———. (1915) 1923. "The Dynamics of Cyclones and Anticyclones." In *Collected Scientific Papers,* edited by C. G. Knott, 459–67. Cambridge: Cambridge University Press. First published in *Proceedings of the Royal Society of London* 36 (1915).

Allison, Henry E. 1983. *Kant's Transcendental Idealism.* New Haven, Conn.: Yale University Press.

Allison, W. W. M., C. B. Brooks, J. N. Bunch, J. H. Cobb, J. L. Lloyd, and R. W. Pleming. 1974. "The Identification of Secondary Particles by Ionisation Sampling (ISIS)." *Nuclear Instruments and Methods* 119:499–507.

Allkofer, Otto Claus. 1956. "Das Ansprechensvermögen von Parallel-Platten-Funkenzählern für die harte Komponente der kosmischen Ultrastrahlung." Diplom thesis, University of Hamburg.

———. 1969. *Spark Chambers.* Munich: Karl Thiemig.

Allkofer, O. C., E. Bagge, P. G. Henning, and L. Schmieder. 1955. "Die Ortsbestimmung geladener Teilchen mit Hilfe von Funkenzählern und ihre Anwendung auf die Messung der Vielfachstreuung von Mesonen in Blei." *Physikalische Verhandlungen* 6:166.

Alonso, J. L., and R. Tarrach, eds. 1980. *Quantum Chromodynamics.* Lecture Notes in Physics, no. 118. Berlin: Springer.

Alston, Margaret, Luis W. Alvarez, Philippe Eberhard, Myron L. Good, William Graziano, Harold K. Ticho, and Stanley G. Wojcicki. 1960. "Resonance in the Lambda-π System." *Physical Review Letters* 5:520–24.

———. 1961. "Resonance in the K-π System." *Physical Review Letters* 6:300–302.

———. 1961. "Study of Resonances of the Σ-π System." *Physical Review Letters* 6:698–705.

Alvarez, Luis W. 1966. "Round Table Discussion on Bubble Chambers." In *Proceedings of the 1966 International Conference on Instrumentation for High Energy Physics,* 271–95. Stanford, Calif.: Stanford University, Stanford Linear Accelerator.

———. 1972. "Recent Developments in Particle Physics." Nobel Lecture, 11 December 1968. In *Nobel Lectures Including Presentation Speeches and Laureates' Biographies: Physics 1963–1970,* 241–90. New York: Elsevier.

———. 1987. *Adventures of a Physicist.* New York: Basic Books.

———. 1989. "The Hydrogen Bubble Chamber and the Strange Resonances." In *Pions to Quarks: Particle Physics in the 1950s,* edited by L. M. Brown, M. Dresden, and L. Hoddeson. Cambridge: Cambridge University Press.

Alvarez, Luis W., Phillippe Eberhard, Myron L. Good, William Graziano, Harold K. Ticho, and Stanley G. Wojcicki. 1959. "Neutral Cascade Hyperon Event." *Physical Review Letters* 2:215–19.

Amaldi, E. 1956. "Report on the τ-Mesons." *Nuovo Cimento* 4 Suppl.: 179–215.

Amelio, Gilbert F. 1974. "Charge-Coupled Devices." *Scientific American* 230:22–31.

Anderson, H. L., E. Fermi, E. A. Long, R. Martin, and D. E. Nagle. 1952. "Total Cross Sections of Negative Pions of Hydrogen." *Physical Review* 85:934–35.

———. 1952. "Total Cross Sections of Positive Pions of Hydrogen." *Physical Review* 85:936.

Anderson, H. L., S. Fukui, R. Gabriel, C. Hargrove, E. P. Hincks, P. Kalmus, J. Lillberg, R. L. Martin, J. Michelassi, R. McKee, and R. Wilberg. 1964. "Vidicon System of Chicago." In *Proceedings of the Informal Meeting on Film-Less Spark Chamber Techniques and Associated Computer Use,* CERN "Yellow Report" 64-30, edited by G. R. Macleod and B. C. Maglić, 81–93. Geneva: CERN Data Handling Division.

Andersson, B., G. Gustafson, G. Ingelman, and T. Sjöstrand. 1983. "Parton Fragmentation and String Dynamics." *Physics Reports* 97:33–145.

Andersson, B., and Werner Hofmann. 1986. "Bose-Einstein Correlations and Color Strings." *Physical Review Letters B* 169:364–68.

Andrade, E. N. da C. 1923. *The Structure of the Atom.* London: G. Bell and Sons.

Andreae, S. W., F. Kirsten, T. A. Nunamaker, and V. Perez-Mendez. 1964. "Automatic Digitization of Spark Chamber Events by Vidicon Scanner." In *Proceedings of the Informal Meeting on Film-Less Spark Chamber Techniques and Associated Computer Use,* CERN "Yellow Report" 64-30, edited by G. R. Macleod and B. C. Maglić, 65–79. Geneva: CERN Data Handling Division.

Appadurai, Arjun, ed. 1986. *The Social Life of Things: Commodities in Cultural Perspective.* Cambridge: Cambridge University Press.

Arnison, G., A. Astbury, B. Aubert, C. Bacci, G. Bauer, A. Bézaguet, R. Böck, T. J. V. Bowcock, M. Calvetti, T. Carroll, P. Catz, P. Cennini, S. Centro, F. Ceradini, S. Cittolin, D. Cline, C. Cochet, J. Colas, M. Corden, D. Dallman, M. DeBeer, M. Della Negra, M. Demoulin, D. Denegri, A. Di Ciaccio, D. DiBitonto, L. Dobrzynski, J. D. Dowell, M. Edwards, K. Eggert, E. Eisenhandler, N. Ellis, P. Erhard, H. Faissner, G. Fontaine, R. Frey, R. Frühwirth, J. Garvey, S. Geer, C. Ghesquière, P. Ghez, K. L. Giboni, W. R. Gibson, Y. Giraud-héraud, A. Givernaud, A. Gonidec, G. Grayer, P. Gutierrez, T. Hansl-Kozanecka, W. J. Haynes, L. O. Hertzberger, C. Hodges, D. Hoffmann, H. Hoffmann, D. J. Holthuizen, R. J. Homer, A. Honma, W. Jank, G. Jorat, P. I. P. Kalmus, V. Karimäki, R. Keeler, I. Kenyon, A. Kernan, R. Kinnunen, H. Kowalski, W. Kozanecki, D. Kryn, F. Lacava, J. P. Laugier, J. P. Lees, H. Lehmann, K. Leuchs, A. Lévêque, D. Linglin, E. Locci, M. Loret, J. J. Malosse, T. Markiewicz, G. Maurin, T. McMahon, J. P. Mendiburu, M. N. Minard, M. Moricca, H. Muirhead, F. Muller, A. K. Nandi, L. Naumann, D. Niemand, A. Norton, A. Orkin-Lecourtois, L. Paoluzi, G. Petrucci, G. Piano Mortari, M. Pimiä, A. Placci, E. Radermacher, J. Ransdell, H. Reithler, J. P. Revol,

J. Rich, M. Rijssenbeek, C. Roberts, J. Rohlf, P. Rossi, C. Rubbia, B. Sadoulet, G. Sajot, G. Salvi, G. Salvini, J. Sass, J. Saudraix, A. Savoy-Navarro, D. Schinzel, W. Scott, T. P. Shah, M. Spiro, J. Strauss, K. Sumorok, F. Szoncso, D. Smith, C. Tao, G. Thompson, J. Timmer, E. Tsheslog, J. Tuominiemi, S. Van derMeer, J.-P. Vialle, J. Vrana, V. Vuillemin, H. D. Wahl, P. Watkins, J. Wilson, Y. G. Xie, M. Yvert, and E. Zurfluh. 1983. "Experimental Observation of Isolated Large Transverse Energy Electrons with Associated Missing Energy at $\sqrt{s}$ = 540 GeV." *Physics Letters B* 122:103–16.

Arnison, G., et al. 1983. "Experimental Observation of Lepton Pairs of Invariant Mass around 95 GeV/c^2 at the CERN SPS Collider." *Physics Letters B* 126:398–410.

Aspray, William. 1987. "The Mathematical Reception of the Modern Computer: John von Neumann and the Institute for Advanced Study Computer." In *Studies in the History of Mathematics,* edited by E. Phillips, 166–94. Washington, D.C.: Mathematical Association of America.

———. 1990. *John von Neumann and the Origins of Modern Computing.* Cambridge, Mass.: MIT Press.

Aubert, B., et al. 1974. "Further Observation of Muonless Neutrino-induced Inelastic Interactions," *Physical Review Letters* 32:1454–57.

Augustin, J. E., A. M. Boyarski, M. Breidenbach, F. Bulos, J. T. Dakin, G. J. Feldman, G. E. Fischer, D. Fryberger, G. Hanson, B. Jean-Marie, R. R. Larsen, V. Lüth, H. L. Lynch, D. Lyon, C. C. Morehouse, J. M. Paterson, M. L. Perl, B. Richter, P. Rapidis, R. F. Schwitters, W. M. Tanenbaum, F. Vannucci, G. S. Abrams, D. Briggs, W. Chinowsky, C. E. Friedberg, G. Goldhaber, R. J. Hollebeek, J. A. Radyk, B. Lulu, F. Pierre, G. H. Trilling, J. S. Whitaker, J. Wiss, and J. E. Zipse. 1974. "Discovery of a Narrow Resonance in e^+e^- Annihilation." *Physical Review Letters* 33:1406–8.

Avery, R. E. 1989. "Bose-Einstein Correlations of Pions in e^+e^- Annihilation at 29 GeV Center-of-Mass Energy," LBL-26593. Ph.D. dissertation, University of California, Berkeley.

Bachelard, Gaston. 1971. *Epistémologie: Textes Choisis par Dominique Lecourt.* Paris: Presses Universitaires de France.

Bacher, R. F. 1939. "Elastic Scattering of Fast Neutrons." *Physical Review* 55:679–80.

———. 1940. "The Elastic Scattering of Fast Neutrons." *Physical Review* 57:352.

Badt, Kurt. 1950. *John Constable's Clouds.* London: Routledge and Kegan Paul.

Bagge, E. 1946. "Nuclear Disruptions and Heavy Particles in Cosmic Radiation." In *Cosmic Radiation,* edited by W. Heisenberg, translated by T. H. Johnson, 128–43. New York: Dover.

———. 1947. "Zur Theorie der Massen-Häufigkeitsverteilung der Bruchstücke bei der spontanen Kernspaltung." *Zeitschrift für Naturforschung* 2a:565–68.

Bagge, E., F. Becker, and G. Bekow. 1951. "Die Bildungsgeschwindigkeit von Nebeltröpfen in der Wilsonkammer." *Zeitschrift für angewandte Physik* 3:201–9.

Bagge, E., and J. Christiansen. 1952. "Der Parallelplattenzähler als Teilchenmeßgerät." *Naturwissenschaften* 39:298.

Bagge, E., Kurt Diebner, and Kenneth Jay. 1957. "Von der Uransfaltung bis Calder Hall." *Rowohlts Deutsche Enzyklopädie* 41. Sachgebiet: Physik. Hamburg: Rowohlt.

Baggett, N. V., ed. 1980. *AGS 20th Anniversary Celebration.* Upton, N.Y.: Brookhaven National Laboratory.

Barboni, Edward J. 1977. "Functional Differentiation and Technological Specialization in a Specialty in High Energy Physics: The Case of Weak Interactions of Elementary Particles." Ph.D. dissertation, Cornell University.

Bardon, M., J. Lee, P. Norton, J. Peoples, and A. M. Sachs. 1964. "Sonic Spark Chamber System with On-Line Computer for Precision Measurement of Muon Decay Spectrum." 1964. In *Proceedings of the Informal Meeting on Film-Less Spark Chamber Techniques and Associated Computer Use,* CERN "Yellow Report" 64-30, edited by G. R. Macleod and B. C. Maglić, 40–48. Geneva: CERN Data Handling Division.

Barkas, Walter. 1965. "Data Handling in Emulsion Experiments," In *5th International Conference on Nuclear Photography,* CERN 65-4, edited by W. O. Lock, 67–75. Geneva: CERN.

Barnes, Barry. 1982. *T. S. Kuhn and Social Science.* New York: Columbia University Press.

Barnes, V. P., et al. 1961. "Observation of a Hyperon with Strangeness Minus Three." *Physical Review Letters* 12:204–6.

Barnetson, Paul. 1970. *Critical Path Planning.* New York: Brandon Systems Press.

Bartel, W., et al. 1983. "Particle Distribution in 3-Jet Events Produced by e^+e^- Annihilation." *Zeitschrift für Physik C* 21:37–52.

Barthes, Roland. 1974. *S/Z,* translated by Richard Miller. New York: Hill and Wang.

Bateson, Gregory. 1944. "Pidgin English and Cross-Cultural Communication." *Transactions of the New York Academy of Sciences II* 6:137–41.

Beall, E., B. Cork, P. G. Murphy, and W. A. Wenzel. 1960. "Properties of a Spark Chamber," Bev-527. Typescript. 1 July.

———. 1961. "Properties of a Spark Chamber." *Nuovo Cimento* 20 (1961): 502–8.

Beckurts, K. H., W. Gläser, and G. Krüger, eds. 1964. *Automatic Acquisition and Reduction of Nuclear Data.* Karlsruhe: Gesellschaft für Kernforschung.

Bell, P. R. 1948. "The Use of Anthracene as a Scintillation Counter." *Physical Review* 73:1405–6.

Bella, F., and C. Franzinetti. 1953. "On the Theory of the Spark Counter." *Nuovo Cimento* 10:1335–37.

Bella, F., C. Franzinetti, and D. W. Lee. 1953. "Spark Counters." *Nuovo Cimento* 10:1338–40.

———. 1953. "On Spark Counters." *Nuovo Cimento* 10:1461–79.

Ben Nevis and Fort William Observatories, Directors. 1903. "Memorandum by the Directors of the Observatories of Ben Nevis and at Fort William in Connection with Their Closures." *Journal of the Scottish Meteorological Society* 12:161–63.

Benoe, M., and B. Elliott. 1962. *Informal Meeting on Track Data Processing, Held at CERN on 19th July 1962,* CERN Report 62-37. Geneva: CERN Data Handling Division.

Benvenuti, A., et al. 1973. "Observation of Muonless Neutrino-induced Inelastic Interactions," *Physical Review Letters* 32:800–803.

Berriman, R. W. 1948. "Electron Tracks in Photographic Emulsions." *Nature* 161: 928–29.

———. 1948. "Recording of Charged Particles of Minimum Ionizing Power in Photographic Emulsions." *Nature* 162:432.

Bertanza, L., V. Brisson, P. Connolly, E. L. Hart, I. S. Mittra, G. C. Moneti, R. R. Rau,

N. P. Samios, I. O. Skillicorn, and S. S. Yamamoto. 1962. "Possible Resonances in the Xi-π and *K-K* Systems." *Physical Review Letters* 9:180–83.
Bertholot, André, ed. 1982. *Journal de Physique: International Colloquium on the History of Particle Physics* 43:C8-1–C8-493.
Bethe, Hans A. 1950. "The Hydrogen Bomb." *Bulletin of Atomic Scientists* 6:99–104.
———. 1970. "Introduction." In *Computers and Their Role in the Physical Sciences,* edited by S. Fernbach and A. Taub, 1–9. New York: Gordon and Breach.
Beyler, Richard H. 1994. "From Physics to Organicism: Pascual Jordan's Interpretation of Modern Physics in Cultural Context." Ph.D. dissertation, Harvard University.
———. 1996. "Targeting the Organism: The Scientific and Cultural Context of Pascual Jordan's Quantum Biology, 1932–47." *Isis* 87:248–73.
Bickerton, Derek. 1981. *Roots of Language.* Ann Arbor, Mich.: Karoma.
Birks, J. B. 1953. *Scintillation Counters.* New York: McGraw-Hill.
———. 1964. *The Theory and Practice of Scintillation Counting.* Oxford: Pergamon.
Bjorken, J., and S. Drell. 1964. *Relativistic Quantum Mechanics.* New York: McGraw-Hill.
———. 1965. *Relativistic Quantum Fields.* New York: McGraw-Hill.
Blackett, P. M. S. 1952. "Foreword." In *Cloud Chamber Photographs of the Cosmic Radiation,* by G. D. Rochester and J. G. Wilson, vii. New York: Academic Press.
———. 1953. "Closing Remark." In *Congrès International sur le Rayonnement Cosmique,* 290–91. Informal publication. Bagnères de Bigorre, July.
———. 1960. "Charles Thomas Rees Wilson 1869–1959." *Biographical Memoirs of Fellows of the Royal Society* 6:269–95.
———. 1964. "Cloud Chamber Researches in Nuclear Physics and Cosmic Radiation." In *Nobel Lectures including Presentation Speeches and Laureates' Biographies: Physics 1942–1962,* 97–119. New York: Elsevier.
Blackett, P. M. S., and G. Occhialini. 1932. "Photography of Penetrating Corpuscular Radiation." *Nature* 130:363.
Blau, Marietta. 1925. "Über die photographische Wirkung Natürlicher H-Strahlen." *Sitzungsberichte, Akademie der Wissenschaften in Wien, Mathematisch-naturwissenschaftliche Klasse, Abteilung IIa* 134:427–36.
———. 1925. "Die Photographische Wirkung von H-Strahlen aus Paraffin und Aluminium." *Zeitschrift für Physik* 34:285–95.
———. 1927. "Über die photographische Wirkung von H-Strahlen II." *Sitzungsberichte, Akademie der Wissenschaften in Wien, Mathematisch-naturwissenschaftliche Klasse, Abteilung IIa* 136:469–80.
———. 1928. "Über die photographische Wirkung von H-Strahlen aus Paraffin und Atomfragmenten." *Zeitschrift für Physik* 48:751–64.
———. 1928. "Über photographische Intensitätsmessungen von Poloniumpräparaten." *Sitzungsberichte, Akademie der Wissenschaften in Wien, Mathematisch-naturwissenschaftliche Klasse, Abteilung IIa* 137:259–68.
———. 1931. "Über das Abklingen des latentem Bildes bei Exposition mit α-Partikeln." *Sitzungsberichte, Akademie der Wissenschaften in Wien, Mathematisch-naturwissenschaftliche Klasse, Abteilung II* 140:623–28.
———. 1931. "Über photographische Untersuchungen mit radioaktiven Strahlungen."

In *Zehn Jahre Forschung auf dem Physikalisch-Medizinischen Grenzgebiet,* edited by F. Dessauer, 390–98. Leipzig: Georg Thieme.

———. 1934. "La Méthode photographiquè et les problèmes de désintégration artificielle des atomes." *Journal de Physique et le Radium,* 7th ser., 5:61–66.

———. 1949. "Grain Density in Photographic Tracks of Heavy Particles." *Physical Review* 75:279–82.

———. 1950. "Bericht über die Entdeckung der durch kosmische Strahlung erzeugten 'Sterne' in photographischen Emulsionen." *Sitzungsberichte, Akademie der Wissenschaften in Wien, Mathematisch-naturwissenschaftliche Klasse, Abteilung IIa* 159: 53–57.

Blau, Marietta, and K. Altenburger. 1924. "Über eine Methode zur Bestimmung des Streukoeffizienten und des reinen Absorptionskoeffizienten von Röntgenstrahlen." *Zeitschrift für Physik* 25:200–214.

Blau, Marietta, and M. Caulton. 1954. "Inelastic Scattering of 500-MeV Negative Pions in Emulsion Nuclei." *Physical Review* 96:150–60.

———. 1953. "Meson Production by 500-MeV Negative Pions." *Physical Review* 92:516–17.

Blau, Marietta, and J. A. De Felice. 1948. "Development of Thick Emulsions by a Two-Bath Method." *Physical Review* 74:1198.

Blau, Marietta, and B. Dreyfus. 1945. "The Multiplier Photo-Tube in Radioactive Measurements." *Review of Scientific Instruments* 16:245–48.

Blau, M., and I. Feuer. 1946. "Radioactive Light Sources." *Journal of the Optical Society of America* 36:576–80.

Blau, Marietta, and Elisabeth Kara-Michailova. 1931. "Über die durchdringende γ-Strahlung des Poloniums." *Sitzungsberichte, Akademie der Wissenschaften in Wien, Mathematisch-naturwissenschaftliche Klasse, Abteilung IIa* 140:615–22.

Blau, Marietta, and Elisabeth Rona. 1926. "Ionisation durch H-Strahlen." *Sitzungsberichte, Akademie der Wissenschaften in Wien, Mathematisch-naturwissenschaftliche Klasse, Abteilung IIa* 135:573–85.

———. 1929. "Weitere Beiträge zur Ionisation durch H-Partikeln." *Sitzungsberichte, Akademie der Wissenschaften in Wien, Mathematisch-naturwissenschaftliche Klasse, Abteilung IIa* 138:717–31.

———. 1930. "Anwendung der Chamié'schen photographischen Methode zur Prüfung des chemischen Verhaltens von Polonium." *Sitzungsberichte, Akademie der Wissenschaften in Wien, Mathematisch-naturwissenschaftliche Klasse, Abteilung IIa* 139: 275–79.

Blau, Marietta, and Hertha Wambacher. 1932. "Über das Verhalten einer kornlosen Emulsion gegenüber α-Partikeln." *Sitzungsberichte, Akademie der Wissenschaften in Wien, Mathematisch-naturwissenschaftliche Klasse, Abteilung IIa* 141:467–74.

———. 1932. "Über Versuche, durch Neutronen ausgelöste Protonen photographisch nachzuweisen, II." *Sitzungsberichte, Akademie der Wissenschaften in Wien, Mathematisch-naturwissenschaftliche Klasse, Abteilung IIa* 141:615–20.

———. 1932. "Über Versuche, durch Neutronen ausgelöste Protonen photographisch nachzuweisen." *Wiener Anzeiger, Akademie der Wissenschaften in Wien, Mathematisch-naturwissenschaftliche Klasse, Abteilung IIa* 9:180–81.

———. 1934. "Physikalische und chemische Untersuchungen zur Methode des photographischen Nachweises von H-Strahlen." *Sitzungsberichte, Akademie der Wissenschaften in Wien, Mathematisch-naturwissenschaftliche Klasse, Abteilung IIa* 143: 285–301.

———. 1937. Disintegration Processes by Cosmic Rays with the Simultaneous Emissions of Several Heavy Particles." *Nature* 140:585.

———. 1937. "Längenmessung von H-Strahlbahnen mit der photographischen Methode." *Sitzungsberichte, Akademie der Wissenschaften in Wien, Mathematisch-naturwissenschaftliche Klasse, Abteilung IIa* 146:259–72.

———. 1937. "II. Mitteilung über photographische Untersuchungen der schweren Teilchen in der kosmischen Strahlung." *Sitzungsberichte, Akademie der Wissenschaften in Wien, Mathematisch-naturwissenschaftliche Klasse, Abteilung IIa* 146:623–41.

———. 1937. "Vorläufiger Bericht über photographische Ultrastrahlenuntersuchungen nebst einigen Versuchen über die 'spontane Neutronenemission.'" *Sitzungsberichte, Akademie der Wissenschaften in Wien, Mathematisch-naturwissenschaftliche Klasse, Abteilung IIa* 146:469–77.

Blieden, H., D. Freytag, F. Iselin, F. Lefebres, B. Maglić, H. Slettenhaar, S. Almeida, and A. Lang. 1964. "System Consisting of Sonic Spark Chambers, Time-of-Flight, and Pulse-Height Counters ('Missing-Mass Spectrometer') with On-Line Computer." In *Proceedings of the Informal Meeting on Film-Less Spark Chamber Techniques and Associated Computer Use,* CERN "Yellow Report" 64-30, edited by G. R. Macleod and B. C. Maglić, 49–56. Geneva: CERN Data Handling Division.

Block, Ned, ed. 1981. *Imagery.* Cambridge, Mass.: MIT Press.

Bloom, Harold H. 1973. *The Anxiety of Influence: A Theory of Poetry.* London: Oxford University Press.

Boas, F. 1887. "The Study of Geography." *Science* 9:137–41.

Bonner, T. W., and W. M. Brubaker. 1936. "Disintegration of Beryllium, Boron and Carbon by Deuterons." *Physical Review* 50:308–14.

Born, Max. 1953. "Physical Reality." *Philosophical Quarterly* 3:139–49.

Bothe, W. 1930. "Zur Vereinfachung von Koinzidenzzählung." *Zeitschrift für Physik* 59:1–5.

———. 1964. "Coincidence." In *Nobel Lectures including Presentation Speeches and Laureates' Biographies: Physics 1942–1962,* 271–79. New York: Elsevier.

Bothe, W., and W. Kolhörster. 1929. "Das Wesen der Höhenstrahlung." *Zeitschrift für Physik* 56:751–77.

Braddick, Henry John James. 1954. *The Physics of Experimental Method.* New York: Wiley.

Bradner, Hugh. 1961. "Capabilities and Limitations of Present Data-Reduction Systems." In *Proceedings of an International Conference on Instrumentation for High-Energy Physics,* 225–28. New York: Wiley.

Bragg, W. H., and J. P. V. Madsen. 1908. "An Experimental Investigation of the Nature of the Gamma Rays." *Philosophical Magazine,* 6th ser., 15:663–75.

———. 1912. *Studies in Radioactivity.* London: Macmillan.

Bretscher, Egon, Stanley P. Frankel, Darol K. Froman, Nicholas C. Metropolis, Philip Morrison, L. W. Mordheim, Edward Teller, Anthony Turkovich, and John von Neumann. 1950. "Report of Conference on the Super," LA-575. 16 February.

Brickwedde, F. G. 1960. "A Few Remarks on the Beginnings of the NBS-AEC Cryogenic Laboratory." In *Advances in Cryogenic Engineering,* edited by K. D. Timmerhaus, 1:1–4. New York: Plenum.

Bridgman, P. W. 1927. *The Logic of Modern Physics.* New York: Macmillan.

———. 1952. *The Nature of Some of Our Physical Concepts.* New York: Philosophical Library.

Broser, Immanuel, and Hartmut Kallmann. 1947. "Über die Anregung von Leuchtstoffen durch schnelle Korpuskularteilchen I." *Zeitschrift für Naturforschung* 2a:439–40.

Brown, J. L., D. A. Glaser, and M. L. Perl. 1956. "Liquid Xenon Bubble Chamber." *Physical Review* 102:586–87.

Brown, Laurie M., Max Dresden, and Lillian Hoddeson, eds. 1989. *Pions to Quarks: Particle Physics in the 1950s.* Cambridge: Cambridge University Press.

Brown, Laurie, and Lillian Hoddeson, eds. 1983. *The Birth of Particle Physics.* Cambridge: Cambridge University Press.

Brown, R. H., U. Camerini, P. H. Fowler, H. Muirhead, C. F. Powell, and D. M. Ritson. 1949. "Observations with Electron-Sensitive Plates Exposed to Cosmic Radiation." *Nature* 163:47–51, 82–87. Reprinted in *Selected Papers of Cecil Frank Powell,* edited by E. H. S. Burhop, W. O. Lock, and M. G. K. Menon, 265–75. New York: North-Holland, 1972.

Brown, Sanborn C. 1966. *Introduction to Electrical Discharges in Gases.* Wiley Series in Plasma Physics. New York: Wiley.

Brush, S. 1978. "Planetary Science: From Underground to Underdog." *Scientia* 113: 771–87.

Brush, Steven, and Helmut E. Landsberg. 1985. *The History of Geophysics and Meteorology: An Annotated Bibliography.* Bibliographs of the History of Science and Technology, vol. 7. New York: Garland.

Buchanan, C. D., and S. B. Chun. 1987. "Simple Predictive Model for Flavor Production in Hadronization." *Physical Review Letters* 59:1997–2000.

Buchwald, Jed. 1985. *From Maxwell to Microphysics: Aspects of Electromagnetic Theory in the Last Quarter of the Nineteenth Century.* Chicago: University of Chicago Press.

———. 1994. *The Creation of Scientific Effects: Heinrich Hertz and Electric Waves.* Chicago: University of Chicago Press.

———. 1995. *Scientific Practice: Theories and Stories of Doing Physics.* Chicago: University of Chicago Press.

Burke, Peter, ed. 1991. *New Perspectives on Historical Writing.* Cambridge: Polity Press.

Burks, Alice R., and Arthur W. Burks. 1988. *The First Electronic Computer: The Atanasoff Story.* Ann Arbor: University of Michigan Press.

Button, J., G. Kalbfleisch, G. Lynch, B. Maglić, A. Rosenfeld, and M. L. Stevensen. 1962. "Pion-Pion Interaction in the Reaction $p + p \rightarrow 2\pi^{+} + 2\pi^{-} + n\pi^{0+}$." *Physical Review* 126:1858–63.

Cahan, David. 1989. *An Institute for an Empire: The Physikalisch-Technische Reichsanstalt, 1871–1918.* Cambridge: Cambridge University Press.

Cahan, Robert N., and Gerson Goldhaber. 1989. *Experimental Foundations of Particle Physics.* Cambridge: Cambridge University Press.

Calkin, J. 1963. "A Mathematician Looks at Bubble and Spark Chamber Data Processing." In *Programming for HPD and Other Flying Spot Devices,* CERN Report 63-34, edited by J. Howie, S. McCarroll, B. Powell, and A. Wilson, 21–23. Geneva: CERN Data Handling Division.

Callendar, H. L. 1915. "On the Steady Flow of Steam through a Nozzle or Throttle." *Proceedings of the Institution of Mechanical Engineers* 131:53–77.

Callendar, H. L., and J. T. Nicolson. 1898. "On the Law of Condensation of Steam Deduced from Measurements of Temperature-Cycles of the Walls and Steam in the Cylinder of a Steam Engine." *Minutes of Proceedings of the Institution of Civil Engineers* 131:147–268.

Callon, Michel. 1986. "Sociology of Translation." In *Power, Action, and Belief: A New Sociology of Knowledge?* edited by J. Law. London: Routledge and Kegan Paul.

Cambridge Scientific Instrument Co. 1913. *The Wilson Expansion Apparatus for Making Visible the Paths of Ionizing Particles.* List No. 117, June.

Camerini, U., H. Muirhead, C. F. Powell, and D. M. Ritson. 1948. "Observations on Slow Mesons of the Cosmic Radiation." *Nature* 162:433–38. Reprinted in *Selected Papers of Cecil Frank Powell,* edited by E. H. S. Burhop, W. O. Lock, and M. G. K. Menon, 259–64. New York: North-Holland, 1972.

Cannon, Susan Faye. 1978. *Science in Culture: The Early Victorian Period.* New York: Dawson.

Cannon, W. B. 1900. "The Case System in Medicine." *Boston Medical and Surgical Journal* 142:563–64.

Carlson, W. Bernard, and Michael E. Gorman. 1990. "Understanding Invention as a Cognitive Process: The Case of Thomas Edison and Early Motion Pictures, 1888–91." *Social Studies of Science* 20:387–430.

Caroe, G. M. 1978. *William Henry Bragg, 1862–1942: Man and Scientist.* Cambridge: Cambridge University Press.

Carnap, Rudolf. 1950. "Empricism, Semantics, and Ontology." *Revue Internationale de Philosophie* 11. Reprinted in *Semantics and the Philosophy of Language,* edited by L. Linsky, 208–28. Urbana: University of Illinois Press, 1952.

———. 1955. "Logical Foundations of the Unity of Science." In *International Encyclopedia of Unified Science,* ed. O. Neurath et al., vol. 1. Chicago: University of Chicago Press.

———. 1963. "Intellectual Autobiography." In *The Philosophy of Carnap,* edited by P. A. Schilpp, 3–84. Library of Living Philosophers, vol. 11. La Salle, Ill.: Open Court.

———. 1967. *The Logical Structure of the World and Pseudoproblems in Philosophy,* translated by Rolf George. Berkeley: University of California Press.

Cartwright, Nancy. 1983. *How the Laws of Physics Lie.* Oxford: Clarendon.

Cat, Jordi, Nancy Cartwright, and Hasok Chang. 1996. "Otto Neurath: Politics and the Unity of Science." In *The Disunity of Science,* edited by P. Galison and D. Stump. Stanford, Calif.: Stanford University Press.

Čerenkov, P. A. 1934. "Vidimoje Svečenije Čistyx Židkostej pod Dejstvijem γ-Radiacii." *Comptes Rendus de l'Académie des Sciences URSS* 2:451–54.

Chadwick, J., A. N. May, T. G. Pickavance, and C. F. Powell. 1944. "An Investigation of the Scattering of High-Energy Particles from the Cyclotron by the Photographic Method." *Proceedings of the Royal Society A* 183:1–25.

Chamberlain, Owen. 1989. "The Discovery of the Antiproton." In *Pions to Quarks: Par-*

ticle Physics in the 1950s, edited by L. M. Brown, M. Dresden, and L. Hoddeson, 273–84. Cambridge: Cambridge University Press.

Chamberlain, Owen, Emilio Segrè, Clyde Wiegand, and Thomas Ypsilantis. 1955. "Observation of Antiprotons." *Physical Review* 100:947–50.

Chandler, Alfred D., Jr. 1977. *The Visible Hand: The Managerial Revolution in American Business.* Cambridge, Mass.: Belknap.

Charney, J. G., R. Fjörtoft, and J. von Neumann. 1961–63. "Numerical Integration of the Barotropic Vorticity Equation." In *von Neumann: Collected Works,* edited by A. H. Taub, 6:413–30. London: Pergamon.

Charpak, G. 1954. "Etude de phénomènes atomiques de basse énergie liés à des désintégrations nucléaires; La diffusion élastique des rayons γ par les noyaux." D.sc. dissertation, University of Paris.

———. 1957. "Principe et essais préliminaires d'un nouveau détecteur permettant de photographier la trajectoire de particules ionisantes dans un gaz." *Journal de Physique* 18:539–40.

———. 1962. "La Chambre à Etincelles." *Industries Atomiques* 6:63–71.

———. 1962. "Location of the Position of a Spark in a Spark Chamber." *Nuclear Instruments and Methods* 15:318–22.

———. 1967. "Localization of the Position of Light Impact on the Photocathode of a Photomultiplier." *Nuclear Instruments and Methods* 48:151–83.

———. 1978. "Multiwire and Drift Proportional Chambers." *Physics Today* 31 (October): 23–30.

Charpak, G., R. Bouclier, T. Bressani, J. Favier, and Č. Zupančič. 1968. "The Use of Multiwire Proportional Counters to Select and Localize Charged Particles." *Nuclear Instruments and Methods* 62:262–68.

Charpak, G., P. Duteil, R. Meunier, M. Spighel, and J. P. Stroot. 1964. "Electrostatic Photography as a Means to Obtain Magnetic Records of Spark Chamber Pictures." In *Proceedings of the Informal Meeting on Film-Less Spark Chamber Techniques and Associated Computer Use,* CERN "Yellow Report" 64-30, edited by G. R. Macleod and B. C. Maglić, 341–44. Geneva: CERN Data Handling Division.

Charpak, G., J. Favier, and L. Massonnet. 1963. "A New Method for Determining the Position of a Spark in a Spark Chamber by Measurement of Currents." *Nuclear Instruments and Methods* 24:501–3.

Charpak, G., D. Rahm, and H. Steiner. 1970. "Some Developments in the Operation of Multiwire Proportional Chambers." *Nuclear Instruments and Methods* 80:13–34.

Christenson, J. H., J. W. Cronin, V. L. Fitch, and R. Turlay. 1964. "Evidence for the 2π Decay of the K_2^0 Meson." *Physical Review Letters* 13:138–40.

Churchland, Paul M., and Clifford A. Hooker, eds. 1985. *Images of Science: Essays on Realism and Empiricism, with a Reply from Bas C. van Fraassen.* Chicago: University of Chicago Press.

Clifford, James. 1988. *The Predicament of Culture: Twentieth Century Ethnography, Literature and Art.* Cambridge, Mass.: Harvard University Press.

Coleman, William, and Frederic L. Holmes, eds. 1988. *The Investigative Enterprise: Experimental Physiology in Nineteenth Century Medicine.* Berkeley and Los Angeles: University of California Press.

Collins, George B. 1953. "Scintillation Counters." *Scientific American* 189 (November): 36–41.

Collins, George B., and Victor G. Reiling. 1938. "Čerenkov Radiation." *Physical Review* 54:499–503.

Collins, H. M. 1974. "The TEA Set: Tacit Knowledge and Scientific Networks." *Science Studies* 4:165–86.

———. 1985. *Changing Order: Replication and Induction in Scientific Practice.* London: Sage.

Collins, H. M., and T. J. Pinch. 1982. *Frames of Meaning: The Social Construction of Extraordinary Science.* Boston: Routledge and Kegan Paul.

Colodny, Robert G., ed. 1970. *The Nature and Function of Scientific Theories: Essays in Contemporary Philosophy.* Pittsburgh: University of Pittsburgh Press.

Conant, James Bryant, and Leonard K. Nash, eds. 1950–54. *Harvard Case Histories in Experimental Science,* 8 vols. Cambridge, Mass.: Harvard University Press.

Congrès International sur le Rayonnement Cosmique. 1953. Organized by the University of Toulouse with the support of UIPPA and UNESCO. Informal publication. Bagnères de Bigorre, July.

Conversi, M. 1982. "The Development of the Flash and Spark Chambers in the 1950s." *Journal de Physique* 43:C8-91–C8-99.

Conversi, M., S. Focardi, C. Franzinetti, A. Gozzini, and P. Murtas. 1956. "A New Type of Hodoscope of High Spatial Resolution." *Nuovo Cimento* 4 Suppl.: 234–37.

Conversi, M., and A. Gozzini. 1955. "The 'Hodoscope Chamber': A New Instrument for Nuclear Research." *Nuovo Cimento* 2:189–91.

Conversi, M., E. Pancini, and O. Piccioni. 1945. "On the Decay Process of Positive and Negative Mesons." *Physical Review* 68:232.

———. 1947. "On the Disintegration of Negative Mesons." *Physical Review* 71: 209–10.

Coon, J. H., and R. A. Nobles. 1947. "Hydrogen Recoil Proportional Counter for Neutron Detection." *Review of Scientific Instruments* 18:44–47.

Cork, Bruce. 1960. "Charged Particle Detector." Unpublished typescript. 6 June.

Coulier, J. P. 1875. "Note sur une nouvelle propriété de l'air." *Journal de Pharmacie et de Chimie,* 4th ser., 22:165–73, 254–55.

Cowan, C. L., Jr., F. Reines, F. B. Harrison, H. W. Kruse, and A. D. McGuire. 1956. "Detection of the Free Neutrino: A Confirmation." *Science* 124:103–4.

Cowan, E. W. 1950. "A Continuously Sensitive Diffusion Cloud Chamber." *Review of Scientific Instruments* 21:991–96.

Crane, H. R., E. R. Gaerttner, and J. J. Turin. 1936. "A Cloud Chamber Study of the Compton Effect." *Physical Review* 50:302–8.

Cranshaw, T. E., and J. F. de Beer. 1957. "A Triggered Spark Counter." *Nuovo Cimento* 5:1107–16.

———. 1963. "Present Status of Spark Chambers." *Nuclear Instruments and Methods* 20:143–51.

Cronin, James W. 1963. "Present Status." *Nuclear Instruments and Methods* 20:143–51.

———. 1981. "CP Symmetry Violation: The Search for Its Origin." Nobel Lecture, 8 December 1980, with preceding biographical remarks. In *Les Prix Nobel 1980:*

Nobel Prizes, Presentations, Biographies and Lectures, 56–79. Stockholm: Almquist and Wiksell.

Cronin, James W., and George Renninger. 1961. "Studies of a Neon-Filled Spark Chamber." In *Proceedings of an International Conference on Instrumentation for High-Energy Physics,* 271–75. New York: Wiley.

Crowther, J. G. 1968. *Scientific Types.* London: Barrie and Rockliff.

———. 1974. *The Cavendish Laboratory, 1874–1974.* New York: Science History Publications.

Cüer, P. 1959. "Introduction. Où en est la Photographie corpusculaire?" In *Photographie Corpusculaire II,* edited by P. Demers, 9–18. Montreal: Les Presses Universitaires de Montréal.

Curran, S. C., and W. R. Baker. 1948. "Photoelectric Alpha-Particle Detector." *Review of Scientific Instruments* 19:116.

Curran, S. C., and J. D. Craggs. 1949. *Counting Tubes: Theory and Applications.* New York: Academic Press.

Curtiss, John H. 1950. "Sampling Methods Applied to Differential and Difference Equations." In *Computation Seminar November 1949,* edited by C. C. Hurd, 87–100. New York: International Business Machines Corporation.

———. 1951. "Preface." In *Monte Carlo Method,* edited by A. S. Householder, G. E. Forsythe, and H. H. Germond. National Bureau of Standards Applied Mathematics Series, no. 12. Washington, D.C.: U.S. Government Printing Office.

Dalitz, R. H. 1953. "On the Analysis of τ-Meson Data and the Nature of the τ-Meson." *Philosophical Magazine* 44:1068–80.

———. 1953. "The Modes of Decay of the τ-Meson." In *Congrès International sur le Rayonnement Cosmique,* 236–38. Informal publication. Bagnères de Bigorre, July.

———. 1954. "Decay of τ-Mesons of Known Charge." *Physical Review* 94:1046–51.

———. 1982. "Strange Particle Theory in the Cosmic Ray Period." *Journal de Physique* 43:C8-195–C8-205.

Danby, G., J. M. Gaillard, K. Goulianos, L. M. Lederman, N. Mistry, M. Schwartz, and J. Steinberger. 1962. "Observation of High Energy Neutrino Reactions and the Existence of Two Kinds of Neutrinos." *Physical Review Letters* 9:36–44.

Dardel, G. von, and G. Jarlskog. 1964. "Vidicon Development for the Lund Synchrotron." In *Proceedings of the Informal Meeting on Film-Less Spark Chamber Techniques and Associated Computer Use,* CERN "Yellow Report" 64-30, edited by G. R. Macleod and B. C. Maglić, 105–9. Geneva: CERN Data Handling Division.

Dardel, G. von, G. Jarlskog, and S. Henriksson. 1965. "Status Report on the Lund Vidicon System." *IEEE Transactions in Nuclear Science,* NS-12, no. 4: 65–79.

Daston, Lorraine J. 1986. "The Physicalist Tradition in Early Nineteenth Century French Geometry." *Studies in History and Philosophy of Science* 17:269–95.

Daston, Lorraine J., and Peter Galison. 1992. "The Image of Objectivity." *Representations* 40:81–128.

Davidson, Donald. 1982. "On the Very Idea of a Conceptual Scheme." In *Relativism: Cognitive and Moral,* edited by M. Krausz and J. W. Meiland, 66–80. Notre Dame, Ind.: University of Notre Dame Press.

Davis, Natalie Zemon. 1983. *The Return of Martin Guerre.* Cambridge, Mass.: Harvard University Press.

Decamp, D., et al. 1990. "Aleph: A Detector for Electron-Positron Annihilations at LEP." *Nuclear Instruments and Methods in Physics Research A* 294: 121–78.

Dee, P. I., and T. W. Wormell. 1963. "An Index to C. T. R. Wilson's Laboratory Records and Notebooks in the Library of the Royal Society." *Notes and Records of the Royal Society of London* 18: 54–66.

Demers, P. 1946. "New Photographic Emulsions Showing Improved Tracks of Ionizing Particles." *Physical Review* 70: 86.

———. 1954. "Cosmic Ray Phenomena at Minimum Ionization in a New Nuclear Emulsion Having a Fine Grain, Made in the Laboratory." *Canadian Journal of Physics* 32: 538–54.

———. 1958. *Ionographie.* Montreal: Presses Universitaires de Montréal.

Derrick, M. 1966. "Bubble Chambers 1964–1966." In *Proceedings of the 1966 International Conference on Instrumentation for High Energy Physics,* 431–84. Stanford, Calif.: Stanford University, Stanford Linear Accelerator.

The Design of Physics Research Laboratories. 1959. Symposium held by the London and Home Counties Branch of the Institute of Physics at the Royal Institution on 27 November 1957. London: Chapman and Hall.

Deutsch, Martin. 1948. "High Efficiency, High Speed Scintillation Counters for Beta- and Gamma-Rays." *Physical Review* 73: 1240.

DeWitt, Bryce, and Raymond Stora, eds. 1984. *Relativity, Groups, and Topology II.* Les Houches, Session 40. New York: North-Holland.

Dicke, R. H. 1947. "Čerenkov Radiation Counter." *Physical Review* 71: 737.

Dietrich, Michael R. n.d. "Computational Testing: Monte Carlo Experiments and the Defense of Diffusion Models in Molecular Population Genetics." Typescript.

Dilworth, C. C., G. P. S. Occhialini, and R. M. Payne. 1948. "Processing Thick Emulsions for Nuclear Research." *Nature* 162: 102–3.

Dingle, Herbert. 1951. "Philosophy of Physics: 1850–1950." *Nature* 168: 630–36.

"Discussion on Symposium on Monte Carlo Methods." 1954. Proceedings of symposium on Monte Carlo Methods held before the Research Section of the Royal Statistical Society on 20 January 1954. *Journal of Royal Statistical Society B* 16: 61–75.

Dismuke, Nancy M. 1956. "Monte Carlo Computations." In *Symposium on Monte Carlo Methods,* edited by H. A. Meyer, 52–62. New York: Wiley.

Domokos, G., and S. Kovesi-Domokos, eds. 1983. *Proceedings of the Johns Hopkins Workshop on Current Problems in Particle Theory, 7, Bonn, 1983.* Singapore: World Scientific.

Donham, Wallace B. 1922. "Business Teaching by the Case System." *American Economic Review* 12: 53–65.

Donsker, M. D., and Mark Kac. 1951. "The Monte Carlo Method and Its Applications." In *Computation Seminar December 1949,* edited by C. C. Hurd, 74–81. New York: International Business Machines Corporation.

Döring, W. 1937. "Berichtigung zu der Arbeit: Die Überhitzungsgrenze und Zerreissfestigkeit von Flüssigkeiten." *Zeitschift für physikalische Chemie B* 36: 292–94.

Douglas, Mary. 1966. *Purity and Danger: An Analysis of Concepts of Pollution and Taboo.* New York: Praeger.

Duhem, P. 1954. *The Aim and Structure of Physical Theory,* translated by Philip Wiener. Princeton, N.J.: Princeton University Press.

DuPree, Hunter. 1972. "The Great Instauration of 1940: The Organization of Scientific Research for War." In *The Twentieth Century Sciences: Studies in the Biography of Ideas,* edited by G. Holton, 443–67. New York: Norton.

Dutton, Thomas E. 1983. "Birds of a Feather: A Pair of Rare Pidgins from the Gulf of Papua." In *The Social Context of Creolization,* edited by E. Woolford and W. Washabaugh, 77–105. Ann Arbor, Mich.: Karoma.

———. 1985. *Police Motu: iena sivari* (Its story). Waigani: University of Papua New Guinea Press.

Dykes, M. S., and G. Bachy. 1967. "Vibration of Bubble Chamber Liquid during Expansion." In *Proceedings of International Colloquium on Bubble Chambers,* CERN Report 67-26, edited by H. Leutz, 349–69. Geneva: CERN Data Handling Division.

Eckart, Carl, and Francis R. Shonka. 1938. "Accidental Coincidences in Counter Circuits." *Physical Review* 53:752–56.

Eckert, J. P., Jr. 1953. "A Survey of Digital Computer Memory Systems." *Proceedings of the IRE* 41:1393–1406.

Eddington, Arthur. 1939. *The Philosophy of Physical Science.* New York: Macmillan.

Ehrmann, Stephen C. 1974. *Past, Present, and Future: A Study of the MIT Research Laboratory of Electronics.* Cambridge, Mass.: MIT Sloan School of Management, MIT Center for Policy Alternatives. Typescript, MIT Archives.

Einstein, Albert. 1905. "Die von der molekularkinetischen Theorie der Wärme geforderte Bewegung von in ruhenden Flüssigkeiten suspendierten Teilchen." *Annalen der Physik* 17:549–60. Translated and reprinted as "On the Movement of Small Particles Suspended in a Stationary Liquid Demanded by the Molecular-Kinetic Theory of Heat," in A. Einstein, *Investigations on the Theory of the Brownian Movement,* edited by R. Fürth, translated by A. P. Cooper, 1–18. New York: Dover, 1956.

———. 1931. "Maxwell's Influence on the Evolution of the Idea of a Physical Reality." In *James Clerk Maxwell: A Commemorative Volume, 1831–1931,* by Sir J. J. Thomson, Max Planck, Albert Einstein, et al. Cambridge: Cambridge University Press. Reprinted in A. Einstein, *Ideas and Opinions,* translated by S. Bargmann, 266–70. New York: Crown, 1954.

———. 1954. *Ideas and Opinions,* translated by S. Bargmann. New York: Crown.

———. 1956. *Investigations on the Theory of the Brownian Movement,* edited by R. Fürth, translated by A. P. Cooper. New York: Dover.

———. 1961. *Relativity: The Special and General Theory: A Popular Exposition.* New York: Crown.

———. 1970. "Autobiographical Notes." In *Albert Einstein: Philosopher-Scientist,* edited by P. A. Schilpp, 2–94. La Salle, Ill.: Open Court.

Eliot, C. W. 1900. "The Inductive Method Applied to Medicine." *Boston Medical and Surgical Journal* 142:557–71.

Elliott, John B., George Maenchen, Peter H. Moulthrop, Larry O. Oswald, Wilson M. Powell, and Robert Wright. 1955. "Thirty-Six-Atmosphere Diffusion Cloud Chamber." *Review of Scientific Instruments* 26:696–97.

Elmore, W. C. 1948. "Electronics for the Nuclear Physicist. I and II." *Nucleonics* 2, no. 2: 4–17; no. 3: 16–36; no. 4: 43–55; no. 5: 50–58.

Elmore, W. C., and Matthew Sands. 1949. *Electronics: Experimental Techniques.* New York: McGraw-Hill.

Emslie, A. G., and R. A. McConnell. 1947. "Moving-Target Indication." In *Radar System Engineering,* edited by L. N. Ridenour, 626–79. Radiation Laboratory Series, vol. 1. New York: McGraw-Hill.

Emslie, A. G., et al. 1948. "Ultrasonic Decay Lines II." *Journal of the Franklin Institute* 245:101–15.

Engineering Research Associates. 1950. *High-Speed Computing Devices.* New York: McGraw-Hill.

Fancher, D., H. J. Hilke, S. Loken, P. Martin, J. N. Marx, D. R. Nygren, P. Robrish, G. Shapiro, M. Urban, W. Wenzel, W. Gorn, and J. Layter. 1979. "Performance of a Time-Projection Chamber." *Nuclear Instruments and Methods* 161:383–90.

Fazio, G. G. 1964. "A Vidicon Spark Chamber System for Use in Artificial Earth Satellites." In *Proceedings of the Informal Meeting on Film-Less Spark Chamber Techniques and Associated Computer Use,* CERN "Yellow Report" 64-30, edited by G. R. Macleod and B. C. Maglić, 95–104. Geneva: CERN Data Handling Division.

Ferbel, Thomas, ed. 1987. *Experimental Techniques in High Energy Physics.* Menlo Park, Calif.: Addison-Wesley.

Ferguson, Charles. 1982. "Simplified Registers and Linguistic Theory." In *Exceptional Language and Linguistics,* edited by L. K. Obler and L. Menn, 49–66. New York: Academic Press.

Ferguson, Leland, ed. 1977. *Historical Archaeology and the Importance of Material Things.* Papers of the Thematic Symposium, 8th annual meeting of the Society for Historical Archaeology, Charleston, South Carolina, 7–11 January 1975. Lansing, Mich.: Society for Historical Archaeology.

Fermi, E., H. L. Anderson, A. Lundby, D. E. Nagle, and G. B. Yodh. 1952. "Ordinary and Exchange Scattering of Negative Pions by Hydrogen." *Physical Review* 85:935–36.

Fermi, E., J. Pasta, and S. Ulam. (1955) 1990. "Studies of Nonlinear Problems," LA-1940. Reprinted in S. M. Ulam, *Analogies between Analogies: The Mathematical Reports of S. M. Ulam and His Los Alamos Collaborators,* edited by A. R. Bednarek and F. Ulam, 139–54. Los Alamos Series in Basic and Applied Sciences, vol. 10. Berkeley and Los Angeles: University of California Press.

Feyerabend, Paul K. 1978. *Science in a Free Society.* London: NLB.

———. 1981. *Problems of Empiricism: Philosophical Papers,* vol. 2. Cambridge: Cambridge University Press.

———. 1981. *Realism, Rationalism and Scientific Method: Philosophical Papers,* vol. 1. Cambridge: Cambridge University Press.

Feynman, Richard P. 1949. "The Theory of Positrons." *Physical Review* 76:749–59.

———. 1972. *Photon-Hadron Interactions: Frontiers in Physics.* Reading, Mass.: Benjamin.

Feynman, Richard P., Robert B. Leighton, and Matthew Sands. 1963–65. *The Feynman Lectures on Physics,* 3 vols. Reading, Mass.: Addison-Wesley.

Field, Hartry. 1990. "'Narrow' Aspects of Intentionality and the Information-Theoretic Approach to Content." In *Information, Semantics and Epistemology,* edited by E. Villanueva, 102–16. Oxford: Blackwell.

Field, R. D., and R. P. Feynman. 1978. "A Parametrization of the Properties of Quark Jets." *Nuclear Physics B* 136:1–76.

Field, R. D., and S. Wolfram. 1983. "A QCD Model for e^+e^- Annihilation." *Nuclear Physics B* 213:65–84.

Fieller, E. C. 1954. "Discussion on Symposium on Monte Carlo Methods." *Journal of the Royal Statistical Society B* 16:61–75.

Fine, Arthur. 1986. *The Shaky Game: Einstein, Realism, and the Quantum Theory.* Chicago: University of Chicago Press.

Fitch, Val Logsdon. 1981. "The Discovery of Charge-Conjugation Parity Asymmetry." Nobel Lecture, 8 December, 1980, with preceding biographical remarks. In *Les Prix Nobel 1980: Nobel Prizes, Presentations, Biographies and Lectures,* 80–93. Stockholm: Almquist and Wiksell.

Fodor, Jerry, and Ernest Lepore. 1992. *Holism: A Shopper's Guide.* Oxford: Blackwell.

Foley, K. J., S. J. Lindenbaum, W. A. Love, S. Ozaki, J. J. Russell, and L. C. L. Yuan. 1964. "A Counter Hodoscope Digital Data Handling and On-Line Computer System Used in High Energy Scattering Experiments." *Nuclear Instruments and Methods* 30:45–60.

Foley, William A. 1988. "Language Birth: The Processes of Pidginization and Creolization." In *Language: The Sociocultural Context,* edited by F. J. Newmeyer, 162–83. Linguistics: The Cambridge Survey, vol. 4. Cambridge: Cambridge University Press.

Forman, P. 1987. "Behind Quantum Electronics: National Security as Basis for Physical Research in the United States, 1940–1960." *Historical Studies in the Physical and Biological Sciences* 18:149–229.

Forster, T. 1815. *Atmospheric Phaenomena.* London: Baldwin, Cradock, and Joy.

Foster, B., and P. Fowler, eds. 1988. *40 Years of Particle Physics: Proceedings of the International Conference to Celebrate the 40th Anniversary of the Discoveries of the π- and V-Particles, held at the University of Bristol, 22–24 July 1987.* Bristol: Adam Hilger.

Foucault, Michel. 1979. *Discipline and Punish: The Birth of the Prison,* translated by A. Sheridan. New York: Vintage.

———. 1980. *Power/Knowledge: Selected Interviews & Other Writings 1972–1977,* edited by C. Gordon. New York: Pantheon.

———. 1984. *The Foucault Reader,* edited by P. Rabinow. New York: Pantheon.

———. 1984. "What Is an Author?" In *The Foucault Reader,* edited by P. Rabinow, 101–20. New York: Pantheon.

Frank, Charles, ed. 1993. *Operation Epsilon: The Farm Hall Transcripts.* Berkeley and Los Angeles: University of California Press.

Frank, F. C., and D. H. Perkins. 1971. "Cecil Frank Powell, 1903–1969." *Biographical Memoirs of Fellows of the Royal Society* 17:541–55, with an appendix by A. M. Tyndall, 555–57.

Frank, I., and Ig. Tamm. 1937. "Coherent Visible Radiation of Fast Electrons Passing through Matter." *Comptes Rendus de l'Académie des Sciences URSS* 14:109–14.

Franklin, Allan D. 1979. "The Discovery and Nondiscovery of Parity Nonconservation." *Studies in the History and Philosophy of Science* 10:201–57.

———. 1981. "Millikan's Published and Unpublished Data on Oil Drops." *Historical Studies in the Physical Sciences* 11:185–201.

———. 1983. "The Discovery and Acceptance of CP Violation." *Historical Studies in the Physical Sciences* 13:207–38.

———. 1986. *The Neglect of Experiment.* Cambridge: Cambridge University Press.
———. 1992. *The Rise and Fall of the "Fifth Force": Discovery, Pursuit, and Justification in Modern Physics.* New York: American Institute of Physics.
Freedberg, David. 1989. *The Power of Images: Studies in the History and Theory of Response.* Chicago: University of Chicago Press.
Freeman, Eugene, ed. 1976. *The Abdication of Philosophy; Philosophy and the Public Good: Essays in Honor of Paul Arthur Schilpp.* La Salle, Ill.: Open Court.
Freundlich, H. F., E. P. Hincks, and W. J. Ozeroff. 1947. "A Pulse Analyser for Nuclear Research." *Review of Scientific Instruments* 18:90–100.
Friedman, Michael. 1987. "Carnap's *Aufbau* Reconsidered." *Noûs* 21:521–45.
Fritzsch, Harald. 1980. "Masses and Mass Generation in Chromo and Flavour Dynamics." In *Quantum Chromodynamics,* edited by J. L. Alonso and R. Tarrach, 278–319. Lecture Notes in Physics, no. 118. Berlin: Springer.
Fukui, Shuji. 1983. "Chronological Review on Development of Spark Chamber in Japan." Unpublished typescript. March. Courtesy of the author.
Fukui, Shuji, and Sigenori Miyamoto. 1957. "A Study of the Hodoscope Chamber," INS-TCA-10. Kyoto: Institute for Nuclear Study, Air Shower Project, 14 December.
———. 1958. "A Study of the Hodoscope Chamber II: A Preliminary Study of a New Device of a Particle Detector" (in Japanese), INS-TCA-11. Translated by Shuji Fukui. Kyoto: Institute for Nuclear Study, Air Shower Project, April.
———. 1959. "A New Type of Particle Detector: The 'Discharge Chamber.'" *Nuovo Cimento* 11:113–15.
Furry, W. H. 1947. "Discussion of a Possible Method for Measuring Masses of Cosmic-Ray Mesotrons." *Physical Review* 72:171.
Fussell, Lewis, Jr., and Thomas H. Johnson. 1934. "Vacuum Tube Characteristics in Relation to the Selection of Coincident Pulses from Cosmic Ray Counters." *Journal of the Franklin Institute* 217:517–24.
Fyfe, Gordon, and John Law, eds. 1988. *Picturing Power: Visual Depiction and Social Relations.* Sociological Review Monograph no. 35. London: Routledge and Kegan Paul.
Galilei, Galileo. 1970. *Dialogue Concerning the Two Chief World Systems—Ptolemaic and Copernican,* translated by S. Drake. Berkeley: University of California Press.
Galison, Peter. 1983. "The Discovery of the Muon and the Failed Revolution against Quantum Electrodynamics." *Centaurus* 26:262–316.
———. 1983. "How the First Neutral Current Experiments Ended." *Reviews of Modern Physics* 55:477–509.
———. 1983. "Rereading the Past from the End of Physics: Maxwell's Equations in Retrospect." In *Functions and Uses of Disciplinary Histories,* edited by L. Graham, W. Lepenies, and P. Weingart, 7:35–51. Hingham, Mass.: Kluwer.
———. 1985. "Bubble Chambers and the Experimental Workplace." In *Observation, Experiment and Hypothesis in Modern Physical Science,* edited by P. Achinstein and O. Hannaway, 309–373. Cambridge, Mass.: MIT Press.
———. 1987. *How Experiments End.* Chicago: University of Chicago Press.
———. 1988. "History, Philosophy, and the Central Metaphor." *Science in Context* 2:197–212.

———. 1988. "Philosophy in the Laboratory." *Journal of Philosophy* 185:525–27.

———. 1988. "Physics between War and Peace." In *Science, Technology, and the Military,* edited by E. Mendelsohn, M. R. Smith, and P. Weingart, 47–86. Sociology of the Sciences, vol. 1. Dordrecht: Kluwer.

———. 1989. "Bubbles, Sparks, and the Postwar Laboratory." In *Pions to Quarks: Particle Physics in the 1950s,* edited by L. M. Brown, M. Dresden, and L. Hoddeson, 213–51. Cambridge: Cambridge University Press.

———. 1989. "The Trading Zone: The Coordination of Action and Belief." Preprint for TECH-KNOW Workshops on Places of Knowledge, Their Technologies and Economies, UCLA Center for Cultural History of Science and Technology, Los Angeles.

———. 1993. "The Cultural Meaning of *Aufbau.*" In *Yearbook of the Institute Vienna Circle 1/93: Scientific Philosophy: Origins and Developments,* edited by F. Stadler, 75–93. Dordrecht: Kluwer.

———. 1995. "Contexts and Constraints." In *Scientific Practice: Theories and Stories of Doing Physics,* edited by J. Buchwald, 2–13. Chicago: University of Chicago Press.

———. 1995. "Theory Bound and Unbound: Superstrings and Experiments." In *Laws of Nature: Essays on Philosophical, Scientific and Historical Dimensions,* edited by F. Weinert. New York: de Gruyter.

Galison, Peter, and Barton Bernstein. 1989. "In Any Light: Scientists and the Decision to Build the Superbomb, 1942–1954." *Historical Studies in the Physical and Biological Sciences* 19:267–347.

Galison, Peter, and Bruce Hevly, eds. 1992. *Big Science: The Growth of Large-Scale Research.* Stanford, Calif.: Stanford University Press.

Galison, Peter, Bruce Hevly, and Rebecca Lowen. 1992. "Controlling the Monster: Stanford and the Growth of Physics Research, 1935–1962." In *Big Science: The Growth of Large-Scale Research,* edited by P. Galison and B. Hevly, 46–77. Stanford, Calif.: Stanford University Press.

Galison, Peter, and Caroline Jones. "Laboratory, Factory, and Studio: Dispersing Sites of Production." In *The Architecture of Science,* edited by P. Galison and E. Thompson. Cambridge, Mass.: MIT Press. Forthcoming.

Gardener, M., S. Kisdnasamy, E. Rössle, and A. W. Wolfendale. 1957. "The Neon Flash Tube as a Detector of Ionising Particles." *Proceedings of the Physical Society B* 70:687–99.

Gardiner, Patrick. 1981. "German Philosophy and the Rise of Relativism." *Monist* 64:138–54.

Gardner, E., and C. M. G. Lattes. 1948. "Production of Mesons by the 184-Inch Berkeley Cyclotron." *Science* 107:270–71.

Gargamelle Construction Group. 1967. "The Large Heavy Liquid Bubble Chamber 'Gargamelle': The Optics." In *Proceedings of International Colloquium on Bubble Chambers,* CERN Report 67-26, edited by H. Leutz, 295–311. Geneva: CERN Data Handling Division.

Gary, J. W. 1985. "Tests of Models for Parton Fragmentation in e^+e^- Annihilation," LBL-20638. Ph.D. dissertation, University of California, Berkeley.

Geertz, Clifford. 1973. *The Interpretation of Cultures: Selected Essays.* New York: Basic Books.

———. 1983. *Local Knowledge: Further Essays in Interpretive Anthropology.* New York: Basic Books.

———. 1995. "Culture War." *New York Review of Books* 62, no. 19 (November 30).

Geiger, H. 1913. "Über eine einfache Methode zur Zählung von α- und β-Strahlen." *Vehandlungen der Deutschen Physikalischen Gesellschaft* 15:534–39.

Geitel, Hans. 1900–1901. "Über die Elektrizitätszerstreuung in abgeschlossenen Luftmengen." *Physikalische Zeitschrift* 2:116–19.

Gelernter, H. 1961. "The Automatic Collection and Reduction of Data for Nuclear Spark Chambers." *Nuovo Cimento* 22:631–42.

Gell-Mann, M. 1956. "The Interpretation of the New Particles as Displaced Charge Multiplets." *Nuovo Cimento* 4 Suppl.: 848–66.

"General Discussion on On-Line Computer Use." 1964. In *Proceedings of the Informal Meeting on Film-Less Spark Chamber Techniques and Associated Computer Use,* CERN "Yellow Report" 64-30, edited by G. R. Macleod and B. C. Maglić, 299–312. Geneva: CERN Data Handling Division.

Gentner, W., H. Maier-Leibnitz, and W. Bothe. 1940. *Atlas typischer Nebelkammerbilder mit Einführung in die Wilsonsche Methode.* Berlin: Julius Springer.

———. 1954. *An Atlas of Typical Expansion Chamber Photographs.* New York: Wiley.

Germond, H. H. 1951. "Round Table Discussion: Summary." In *Monte Carlo Method,* edited by A. S. Householder, G. E. Forsythe, and H. H. Germond, 39–42. National Bureau of Standards Applied Mathematics Series, no. 12. Washington, D.C.: U.S. Government Printing Office.

Getting, I. A. 1947. "A Proposed Detector for High Energy Electrons and Mesons." *Physical Review* 71:123–24.

Getting, Ivan A. 1989. *All in a Lifetime: Science in the Defense of Democracy.* New York: Vantage.

Giannelli, G. 1964. "Magnetostriction." *Nuclear Instruments and Methods* 31:29–34.

———. 1964. "A Magnetostriction Method for Spark Localization." In *Proceedings of the Informal Meeting on Film-Less Spark Chamber Techniques and Associated Computer Use,* CERN "Yellow Report" 64-30, edited by G. R. Macleod and B. C. Maglić, 325–31. Geneva: CERN Data Handling Division.

Gibson, W. 1984. *Neuromancer.* New York: Ace.

———. 1986. *Count Zero.* New York: Arbor.

———. 1988. *Mona Lisa Overdrive.* New York: Bantam.

Giere, Ronald. 1988. *Explaining Science: A Cognitive Approach.* Chicago: University of Chicago Press.

Gillispie, Charles C., ed. 1981. *Dictionary of Scientific Biography.* New York: Scribners.

Ginzburg, Carlo. 1983. *The Cheese and the Worms: The Cosmos of a Sixteenth Century Miller,* translated by J. Tedeschi and A. Tedeschi. New York: Penguin.

Glaser, Donald A. 1950. "The Momentum Distribution of Charged Cosmic Ray Particles near Sea Level." Ph.D. dissertation, California Institute of Technology. Published as "Momentum Distribution of Charged Cosmic-Ray Particles at Sea Level." *Physical Review* 80 (1950): 625–30.

———. 1952. "Some Effects of Ionizing Radiation on the Formation of Bubbles in Liquids." *Physical Review* 87:665.

———. 1953. "Bubble Chamber Tracks of Penetrating Cosmic-Ray Particles." *Physical Review* 91:762–63.

———. 1953. "A Possible 'Bubble Chamber' for the Study of Ionizing Events." *Physical Review* 91:496.

———. 1954. "Progress Report on the Development of Bubble Chambers." *Nuovo Cimento* 11 Suppl.: 361–68.

———. 1955. "The Bubble Chamber." *Scientific American* 192 (February): 46–50.

———. 1964. "Elementary Particles and Bubble Chambers." Nobel Lecture, 12 December 1960. In *Nobel Lectures including Presentation Speeches and Laureates' Biographies: Physics 1942–1962,* 529–51. New York: Elsevier.

Glaser, Donald A., and David C. Rahm. 1953. "Characteristics of Bubble Chambers." *Physical Review* 97:474–79.

Glazebrook, Richard, ed. 1923. *A Dictionary of Applied Physics,* 5 vols. London: Macmillan.

Godfrey-Smith, Peter. 1991. "Signal, Decision, Action." *Journal of Phillosophy* 88: 709–22.

———. 1992. "Indication and Adaptation." *Synthese* 92:283–312.

Goertzel, G., and H. Kahn. 1949. "Monte Carlo Methods for Shield Computation," ORNL 429. Report from Oak Ridge National Laboratory. 19 December.

Goethe, Johann Wolfgang, 1960. *Schriften zur Geologie und Mineralogie; Schriften zur Meteorologie. Gesamtausgabe der Werke und Schriften in zweiundzwanzig Bänden.* Stuttgart: J. G. Cotta'sche Buchhandlung Nachfolger.

Goldhaber, Gerson. 1989. "Early Work at the Bevatron: A Personal Account." In *Pions to Quarks: Particle Physics in the 1950s,* edited by L. M. Brown, M. Dresden, and L. Hoddeson, 260–72. Cambridge: Cambridge University Press.

Goldhaber, G., et al. 1959. "Pion-Pion Correlations in Antiproton Annihilation Events." *Physical Review Letters* 3:181–83.

———. 1960. "Influence of Bose-Einstein Statistics on the Antiproton-Proton Annihilation Process." *Physical Review* 120:300–312.

Goldschmidt-Clermont, Y. 1966. "Progress in Data Handling for High Energy Physics." In *XII International Conference on High-Energy Physics,* edited by Y. A. Smorodinskii et al., 439–62. Moscow: Atomizdat.

Goldsmith, Maurice, and Edwin Shaw. 1977. *Europe's Giant Accelerator: The Story of the CERN 400-GeV Proton Synchrotron.* London: Taylor and Francis.

Goldstine, Herman H. 1972. *The Computer: From Pascal to von Neumann.* Princeton, N.J.: Princeton University Press.

Gooding, David. 1990. *Experiment and the Making of Meaning: Human Agency in Scientific Observation and Experiment.* Dordrecht: Kluwer.

Gooding, David, Trevor Pinch, and Simon Schaffer, eds. 1989. *The Uses of Experiment: Studies in the Natural Sciences.* Cambridge: Cambridge University Press.

Gottstein, K. 1967. "Introductory Remarks." In *Programming for Flying Spot Devices: A Conference Held at the Max-Planck-Institut für Physik und Astrophysik, Munich, on 18–20 January 1967,* edited by B. W. Powell and P. Seyboth, 1–4.

Gozzini, A. 1951. "La costante dielettrica dei gas nella regione delle microonde." *Nuovo Cimento* 8:361–68.

———. 1951. "Sull'effetto Faraday di sostanze paramagnetiche nella regione delle microonde." *Nuovo Cimento* 8:928–35.

Grandy, Richard E., ed. 1973. *Theories and Observation in Science.* Central Issues in Philosophy Series. Englewood Cliffs, N.J.: Prentice-Hall.

Grashey, Rudolf. 1905. *Atlas typischer Röntgenbilder vom normalen Menschen.* Munich: Lehmann.

———. 1908. *Atlas chirurgish-Pathologischer Röntgenbilder.* Munich: Lehmann.

Greene, Mott. 1982. *Geology in the Nineteenth Century: Changing Views of a Changing World.* Cornell History of Science Series. Ithaca, N.Y.: Cornell University Press.

Gregory, B. P. 1965. "The Future Perspectives for Emulsion Work at CERN: Closing Talk." In *5th International Conference on Nuclear Photography,* CERN 65-4, edited by W. O. Lock, 107–10. Geneva: CERN.

Groth, Edward J., and P. J. E. Peebles. 1977. "Statistical Analysis of Catalogs of Extragalactic Objects. VII, Two- and Three-Point Correlation Functions for the High-Resolution Shane-Wirtanen Catalog of Galaxies." *Astrophysical Journal* 217: 385–405.

Groves, Leslie M. 1983. *Now It Can Be Told: The Study of the Manhattan Project.* New York: Da Capo.

Guerlac, Henry E. 1987. *Radar in World War II.* Tomash Series in the History of Modern Physics 1800–1950, vol. 8. Los Angeles: Tomash.

Guilbaut, Serge. 1983. *How New York Stole the Idea of Modern Art: Abstract Expressionism, Freedom, and the Cold War.* Chicago: University of Chicago Press.

Gupta, Akhil, and Ferguson, James. 1992. "Beyond 'Culture': Space, Identity, and the Politics of Difference." *Cultural Anthropology* 7 (February): 6–23.

Gurney, R. W., and N. F. Mott. 1938. "The Theory of the Photolysis of Silver Bromide and the Photographic Latent Image." *Proceedings of the Royal Society London A* 164:151–67.

Hacking, Ian. 1983. *Representing and Intervening: Introductory Topics in the Philosophy of Natural Science.* Cambridge: Cambridge University Press.

———. 1989. "Extragalactic Reality: The Case of Gravitational Lensing." *Philosophy of Science* 56:555–81.

Halliday, E. C. 1970. "Some Memories of Prof. C. T. R. Wilson, English Pioneer in Work on Thunderstorms and Lightning." *Bulletin of the American Meteorological Society* 51:1133–35.

Hammersley, J. M., and D. C. Handscomb. 1964. *Monte Carlo Methods.* London: Methuen.

Hanbury-Brown, R., and Twiss, R. Q. 1956. "A Test of a New Type of Stellar Interferometer." *Nature* 178:1046–48.

Hannaway, Owen. 1986. "Laboratory Design and the Aim of Science: Andreas Libavius versus Tycho Brahe." *Isis* 77:585–610.

Hansen, Chuck. 1988. *U.S. Nuclear Weapons: The Secret History.* Arlington, Tex.: Aerofax.

Hanson, Norwood Russell. 1958. *Patterns of Discovery: An Inquiry into the Conceptual Foundations of Science.* Cambridge: Cambridge University Press.

———. 1963. *The Concept of the Positron: A Philosophical Analysis.* Cambridge: Cambridge University Press.

Haraway, D. 1976. *Crystals, Fabrics, and Fields: Metaphors of Organicism in Twentieth Century Developmental Biology.* New Haven, Conn.: Yale University Press.

———. 1989. *Primate Visions: Gender, Race and Nature in the World of Modern Science.* New York: Routledge and Kegan Paul.

Harman, P. M. 1982. *Energy, Force, and Matter: The Conceptual Development of Nineteenth-Century Physics.* Cambridge History of Science Series. Cambridge: Cambridge University Press.

Hasert, F. J., W. Faissner, W. Krenz, J. Von Krogh, D. Lanske, J. Morfin, K. Schultze, H. Weerts, G. H. Bertrand-Coremans, J. Lemmone, J. Sacton, W. Van Doninck, P. Vilain, C. Bactay, D. C. Cundy, D. Haidt, M. Jaffre, P. Musset, A. Pullia, S. Natali, J. B. M. Pattison, D. H. Perkins, A. Rousset, W. Venus, H. W. Wachsmuth, V. Brisson, B. Degrange, M. Haguenauer, L. Kluberg, U. Nguyen-Khac, P. Petiau, E. Bellotti, S. Bonetti, D. Cavalli, C. Conta, E. Fiorini, M. Rollier, B. Aubert, L. M. Chounet, J. McKenzie, A.G. Michette, G. Myatt, J. Pinfold, and W. G. Scott. 1973. "Search for Elastic Muon-Neutrino Electron Scattering." *Physics Letters B* 46:121–24.

Hauser, M. G., and P. J. E. Peebles. 1973. "Statistical Analysis of Catalogs of Extragalactic Objects. II, The Abell Catalog of Rich Clusters." *Astrophysical Journal* 185:757–85.

Hawes, Louis. 1969. "Constable's Sky Sketches." *Journal of the Warburg and Courtauld Institutes* 32:344–65.

Hawkins, David, Edith Truslow, and Ralph Carlisle Smith, eds. 1983. *Project Y: The Los Alamos Story.* Tomash Series in the History of Modern Physics, 1800–1950, vol. 2. Los Angeles: Tomash.

Hawkins, D., and S. Ulam. (1944) 1990. "Theory of Multiplicative Processes, I," LA-171. 14 November. Reprinted in S. M. Ulam, *Analogies between Analogies: The Mathematical Reports of S. M. Ulam and His Los Alamos Collaborators,* edited by A. R. Bednarek and F. Ulam, 1–15. Los Alamos Series in Basic and Applied Sciences, vol. 10. Berkeley and Los Angeles: University of California Press.

Hayward, Evans, and John Hubbell. 1954. "The Albedo of Various Materials for 1-MeV Photons." *Physical Review* 93:955–56.

Haywood, D. G., Jr. 1954. "Military Decision and Game Theory." *Journal of the Operations Research Society of America* 2:365–85.

Hazen, W. E., C. A. Randall, and O. L. Tiffany. 1949. "The Vertical Intensity at 10,000 Feet of Ionizing Particles That Produce Penetrating Showers." *Physical Review* 75: 694–95.

Heilbron, J. L. 1964. "A History of the Problem of Atomic Structure from the Discovery of the Electron to the Beginning of Quantum Mechanics." Ph.D. dissertation, University of California, Berkeley.

———. 1966. "G. M. Bose: The Prime Mover in the Invention of the Leyden Jar?" *Isis* 57:264–67.

———. 1977. "Lectures on the History of Atomic Physics 1900–1922." In *History of 20th Century Physics,* Proceedings of the International School of Physics "Enrico Fermi" Course 70, edited by C. Weiner, 40–108. New York: Academic Press.

———. 1982. *Elements of Early Modern Physics.* Berkeley and Los Angeles: University of California Press.

———. 1989. "An Historian's Interest in Particle Physics." In *Pions to Quarks: Particle Physics in the 1950s,* edited by L. M. Brown, M. Dresden, and L. Hoddeson. Cambridge: Cambridge University Press.

Heilbron, J. L., and R. W. Seidel. 1989. *Lawrence and His Laboratory: Nuclear Science at Berkeley, 1931–1961.* Berkeley: University of California, Office for History of Science and Technology.

Heisenberg, Werner. 1945. *Wandlungen in den Grundlagen der Naturwissenschaft,* 6th ed. Leipzig: Hirzel.

———. 1952. *Philosophical Problems of Nuclear Science,* translated by F. C. Hayes. New York: Pantheon.

Helmholtz, Robert von. 1886. "Untersuchungen über Dämpfe und Nebel, besonders über solche von Lösungen." *Annalen der Physik und Chemie* 27:508–43.

Hempel, Carl G. 1966. *Philosophy of Natural Science.* Foundations of Philosophy Series. Englewood Cliffs, N.J.: Prentice Hall.

Henkin, Leon, J. Donald Monk, and Alfred Tarski. 1971. *Cylindric Algebras,* 2 vols. Amsterdam: North-Holland.

Henning, Paul-Gerhard von. 1957. "Die Ortsbestimmung geladener Teilchen mit Hilfe von Funkenzählern." *Atomkern-Energie* 2:81–82.

Hermann, Armin, John Krige, Ulrike Mersits, and Dominique Pestre, with a contribution by Lanfranco Belloni. 1987. *History of CERN Volume I: Launching the European Organization for Nuclear Research.* New York: North-Holland.

Hermann, Armin, John Krige, Ulrike Mersits, and Dominique Pestre, with a contribution by Laura Weiss. 1990. *History of CERN Volume II: Building and Running the Laboratory, 1954–1965.* New York: North-Holland.

Hernandez, H. P. 1960. "Designing for Safety in Hydrogen Bubble Chambers," UCRL Engineering Note 4311-14 M 33. Reprinted in *Advances in Cryogenic Engineering,* edited by K. D. Timmerhaus, 2:336–50. New York: Plenum.

Herz, A. J. 1948. "Electron Tracks in Photographic Emulsions." *Nature* 161:928–29.

———. 1965. "Measurements in Nuclear Emulsions." In *5th International Conference on Nuclear Photography,* CERN 65-4, edited by W. O. Lock, 81–87. Geneva: CERN.

Herzog, G. 1939. "Search for Heavy Cosmic-Ray Particles with a Cloud Chamber." *Physical Review* 55:1266.

———. 1940. "Circuit for Anticoincidences with Geiger-Müller Counters." *Review of Scientific Instruments* 11:84–85.

Hesse, Mary. 1970. "Is There an Independent Observation Language?" In *The Nature and Function of Scientific Theories,* edited by R. G. Colodny, 35–78. Pittsburgh: University of Pittsburgh Press. Reprinted in M. Hesse, *The Structure of Scientific Interference.* London: Macmillan, 1974.

———. 1974. *The Structure of Scientific Interference.* London: Macmillan.

———. 1974. "Theory and Observation." In M. Hesse, *The Structure of Scientific Interference.* London: Macmillan.

Highfield, Arnold, and Albert Valdman. 1980. *Theoretical Observations in Creole Studies.* New York: Academic Press.

Higinbotham, W. A. 1965. "Wire Spark Chambers." *IEEE Transactions in Nuclear Science,* NS-12, no. 4: 199–205.

Higinbotham, W. A., J. Gallagher, and M. Sands. 1947. "The Model 200 Pulse Counter." *Review of Scientific Instruments* 18:706–15.

Hildebrand, Roger H., and Darragh E. Nagle. 1953. "Operation of a Glaser Bubble Chamber with Liquid Hydrogen." *Physical Review* 92:517–18.

Hincks, E. P., H. L. Anderson, H. J. Evans, S. Fukui, D. Kessler, K. A. Klare, J. W. Lillberg, M. V. Sherbrook, R. L. Martin, and P. I. P. Kalmus. 1966. "Spark Chamber Spectrometer with Automatic Vidicon Readout." In *Proceedings of the 1966 International Conference on Instrumentation for High Energy Physics,* 63–67. Stanford, Calif.: Stanford University, Stanford Linear Accelerator.

Hine, M. G. N. 1964. "Concluding Remarks." In *Proceedings of the Informal Meeting on Film-Less Spark Chamber Techniques and Associated Computer Use,* CERN "Yellow Report" 64-30, edited by G. R. Macleod and B. C. Maglić, 371–76. Geneva: CERN Data Handling Division.

A History of the Cavendish Laboratory 1871–1910. 1910. London: Longmans, Green.

Hitch, Charles J. 1966. *Decision-Making for Defense.* Berkeley: University of California Press.

Hlawka, Edmund. 1961. "Funktionen von beschränkter Variation in der Theorie der Gleichverteilung." *Annali di Mathematica Pura ed Applicata* 54:325–33.

Hoddeson, Lillian. 1983. "Establishing KEK in Japan and Fermilab in the United States: Internationalism and Nationalism in High Energy Accelerators." *Social Studies of Science* 13:1–48.

———. 1992. "Mission Change in the Large Laboratory: The Los Alamos Implosion Program, 1943–1945." In *Big Science: The Growth of Large-Scale Research,* edited by P. Galison and B. Hevly, 265–89. Stanford, Calif.: Stanford University Press.

Hoddeson, Lillian, Ernest Braun, Jürgen Teichmann, and Spencer Weart. 1982. *Out of the Crystal Maze: Chapters from the History of Solid-State Physics.* New York: Oxford University Press.

Hoddeson, Lillian, Paul W. Henriksen, Roger A. Meade, and Catherine Westfall. 1993. *Critical Assembly: A Technical History of Los Alamos during the Oppenheimer Years, 1943–1945.* New York: Cambridge University Press.

Hofmann, Werner. 1981. *Jets of Hadrons.* Springer Tracts in Modern Physics, vol. 90. Berlin: Springer.

Hofstadter, Robert. 1948. "Alkali Halide Scintillation Counters." *Physical Review* 74: 100–101.

Hohmann, C., and W. Patterson. 1960. "Cryogenic Systems as Auxiliary Power Sources for Aircraft and Missile Applications." In *Advances in Cryogenic Engineering,* edited by K. D. Timmerhaus, 4:184–95. New York: Plenum.

Hollebeek, Robert John. 1975. "Inclusive Momentum Distributions from Electron Positron Annihilation at $\sqrt{s}$ = 3.0, 3.8, and 4.8 GeV." Ph.D. dissertation, University of California, Berkeley.

Hollis, Martin, and Steven Lukes. 1982. *Rationality and Relativism.* Cambridge, Mass.: MIT Press.

Holmes, Frederic L. 1985. *Lavoisier and the Chemistry of Life: An Exploration of Scientific Creativity.* Madison: University of Wisconsin Press.

Holton, Gerald, ed. 1972. *The Twentieth Century Sciences: Studies in the Biography of Ideas.* New York: Norton.

———. 1978. *The Scientific Imagination: Case Studies.* Cambridge: Cambridge University Press.

———. 1978. "Subelectrons, Presuppositions, and the Millikan-Ehrenhaft Dispute." In G. Holton, *The Scientific Imagination: Case Studies,* 25–83. Cambridge: Cambridge University Press.

———. 1988. *Thematic Origins of Scientific Thought: Kepler to Einstein,* rev. ed. Cambridge, Mass.: Harvard University Press.

Horowitz, J. 1967. *Critical Path Scheduling: Management Control through CPM and PERT.* New York: Ronald Press.

Horwich, Paul, ed. 1993. *World Changes: Thomas Kuhn and the Nature of Science.* Cambridge, Mass.: MIT Press.

Hough, P. V. C., and B. W. Powell. 1960. "A Method for Faster Analysis of Bubble Chamber Photographs." *Nuovo Cimento* 18:1184–91.

Hounshell, David A. 1988. *Science and Corporate Strategy: DuPont R&D, 1902–1980.* Cambridge: Cambridge University Press.

———. 1992. "DuPont and the Management of Large-Scale Research and Development." In *Big Science: The Growth of Large-Scale Research,* edited by P. Galison and B. Hevly, 236–61. Stanford, Calif.: Stanford University Press.

Householder, A. S. 1951. "Neutron Age Calculations in Water, Graphite, and Tissue." In *Monte Carlo Method,* edited by A. S. Householder, G. E. Forsythe, and H. H. Germond, 6–8. National Bureau of Standards Applied Mathematics Series, no. 12. Washington, D.C.: U.S. Government Printing Office.

Householder, A. S., G. E. Forsythe, and H. H. Germond, eds. 1951. *Monte Carlo Method.* National Bureau of Standards Applied Mathematics Series, no. 12. Washington, D.C.: U.S. Government Printing Office.

Howard, Luke. 1803. "On the Modifications of Clouds, and on the Principles of their Production, Suspension, and Destruction: Being the Substance of an Essay Read before the Askesian Society in the Session 1802–1803." *Philosophical Magazine* 16: 97–107, 344–357; 17:5–11.

———. (1822) 1960. "Luke Howard an Goethe." In J. W. Goethe, *Schriften zur Geologie und Mineralogie; Schriften zur Meteorologie. Gesamtausgabe der Werke und Schriften in zweiundzwanzig Bänden.* Stuttgart: J. G. Cotta'sche Buchhandlung Nachfolger.

Howie, J., S. McCarroll, B. Powell, and A. Wilson, eds. 1963. *Programming for HPD and Other Flying Spot Devices: Held at the Collège de France, Paris, on 21–23 August 1963,* CERN Report 63-34. Geneva: CERN Data Handling Division.

Howelett, J. 1954. "Discussion on Symposium on Monte Carlo Methods." *Journal of the Royal Statistical Society B* 16:61–75.

Hughes, Thomas P. 1983. *Networks of Power: Electrification in Western Society, 1880–1930.* Baltimore: Johns Hopkins University Press.

Hulsizer, Robert I., John H. Munson, and James N. Snyder. 1966. "A System for the Analysis of Bubble Chamber Film Based upon the Scanning and Measuring Projector (SMP)." *Methods in Computational Physics* 5:157–211.

Hunt, Bruce J. 1994. *The Maxwellians.* Ithaca, N.Y.: Cornell University Press.

Hunt, Linda. 1985. "U.S. Coverup of Nazi Scientists." *Bulletin of the Atomic Scientists* 41 (April): 16–24.

Hunt, Lynn Avery. 1978. *Revolution and Urban Politics in Provincial France: Troyes and Reims, 1786–1790.* Stanford, Calif.: Stanford University Press.

Hurd, Cuthbert C., ed. 1950. *Computation Seminar November 1949.* New York: International Business Machines Corporation.

———. 1951. *Computation Seminar December 1949.* New York: International Business Machines Corporation.

Hyde, H. Montgomery. 1980. *The Atom Bomb Spies.* London: Hamilton.

Hymes, Dell, ed. 1971. *Pidginization and Creolization of Languages: Proceedings of a Conference Held at the University of the West Indies, Mona, Jamaica, April 1968.* Cambridge: Cambridge University Press.

Ihde, Don. 1991. *Instrumental Realism: The Interface between Philosophy of Science and Philosophy of Technology.* Bloomington: Indiana University Press.

International Center for Theoretical Physics, ed. 1972. *Computing as a Language of Physics: Lectures Presented at an International Seminar Course at Trieste from 2 to 20 August 1971.* Organized by the International Centre for Theoretical Physics, Trieste. Vienna: International Atomic Energy Agency.

Irving, David. 1967. *The German Atomic Bomb: The History of Nuclear Research in Nazi Germany.* New York: Da Capo.

Jackson, John D. 1975. *Classical Electrodynamics,* 2nd ed. New York: Wiley.

Jaklovsky, Jozef, ed. 1981. *Preparation of Nuclear Targets for Particle Accelerators.* New York: Plenum.

James, Frank, ed. 1989. *The Development of the Laboratory: Essays on the Place of Experiment in Industrial Civilization.* London: Macmillan.

James, F. 1987. "Monte Carlo Theory and Practice." In *Experimental Techniques,* edited by T. Ferbel, 627–677. Menlo Park, Calif.: Addison-Wesley. First published in *Reports on Progress in Physics* 43 (1980): 1145–89.

Jánossy, L., and B. Rossi. 1940. "On the Photon Component of Cosmic Radiation and Its Absorption Coefficient." *Proceedings of the Royal Society London A* 175:88–100.

Jardine, Nicholas. 1991. *The Scenes of Inquiry: On the Reality of Questions in the Sciences.* Oxford: Clarendon.

Jdanoff, A. 1935. "Les Traces des Particules H et α dans les Émulsions sensibles à la Lumière." *Le Journal de Physique et le Radium* 6:233–41.

Jelley, J. V. 1951. "Detection of μ-Mesons and other Fast Charged Particles in Cosmic Radiation, by the Čerenkov Effect in Distilled Water." *Physical Society of London, Proceedings A* 64:82–87.

———. 1958. *Čerenkov Radiation and Its Applications.* New York: Pergamon.

Jencks, Charles. 1984. *The Language of Post-Modern Architecture,* 4th ed. New York: Rizzoli.

Jentschke, W. 1963. "Invited Summary and Closing Speech." *Nuclear Instrumentation and Methods* 20:507–12.

Johnson, T. H. 1932. "Cosmic Rays–Theory and Experimentation." *Journal of the Franklin Institute* 214:665–88.

———. 1933. "The Azimuthal Asymmetry of the Cosmic Radiation." *Physical Review* 43:834–35.

———. 1933. "Comparison of the Angular Distributions of the Cosmic Radiation at Elevations 6280 ft. and 620 ft." *Physical Review* 43:307–10.

———. 1938. “Circuits for the Control of Geiger-Müller Counters and for Scaling and Recording Their Impulses.” *Review of Scientific Instruments* 9:218–22.

———. 1938. “Cosmic-Ray Intensity and Geomagnetic Effects.” *Reviews of Modern Physics* 10:193–244.

Johnson, T. H., and J. C. Street. 1993. “A Circuit for Recording Multiply-Coincident Discharges of Geiger-Müller Counters.” *Journal of the Franklin Institute* 215:239–46.

Jones, C. A. 1991. “Andy Warhol’s ‘Factory’: The Production Site, Its Context, and Its Impact on the Work of Art.” *Science in Context* 4:101–31.

———. 1997. *Machine in the Studio: Constructing the Postwar American Artist.* Chicago: University of Chicago Press.

Jordan, Pascual. 1941. *Die Physik und das Geheimnis des Organischen.* Lebens Braunschweig: Friedr. Vieweg und Sohn.

———. 1949. “On the Process of Measurement in Quantum Mechanics.” *Philosophy of Science* 16:269–78.

———. 1972. *Erkenntnis und Besinnung: Grenzbetrachtungen an naturwissenschaftlicher Sicht.* Oldenburg: Stalling.

Judd, J. N. 1888. “On the Volcanic Phenomena of the Eruption, and on the Nature and Distribution of the Ejected Materials.” In *The Eruption of Krakatoa and Subsequent Phenomena: Report of the Krakatoa Committee of the Royal Society,* edited by J. G. Symons, 1–46. London: Harrison.

Jungnickel, Christa, and Russell McCormmach. 1986. *Intellectual Mastery of Nature: Theoretical Physics from Ohm to Einstein. Volume I: The Torch of Mathematics, 1800–1870; Volume II: The Now Mighty Theoretical Physics, 1870–1925.* Chicago: University of Chicago Press.

Kahn, Herman. 1950. “Modification of the Monte Carlo Method.” In *Computation Seminar November 1949,* edited by C. C. Hurd, 20–27. New York: International Business Machines Corporation.

———. 1956. “Use of Different Monte Carlo Sampling Techniques.” In *Symposium on Monte Carlo Methods,* edited by H. A. Meyer, 146–90. New York: Wiley.

———. 1960. *On Thermonuclear War.* Princeton, N.J.: Princeton University Press.

Kallmann, Hartmut. 1949. “Quantitative Measurements with Scintillation Counters.” *Physical Review* 75:623–26.

Kant, Immanuel. 1929. *Critique of Pure Reason,* translated by N. K. Smith. New York: St. Martin’s.

Kargon, R. 1981. “Birth Cries of the Elements: Theory and Experiment along Millikan’s Route to Cosmic Rays.” In *The Analytic Spirit,* edited by H. Woolf, 309–25. Ithaca, N.Y.: Cornell University Press.

Kay, Lily E. 1988. “Laboratory Technology and Biological Knowledge; The Tiselius Electrophoresis Apparatus, 1930–1945.” *History and Philosophy of the Life Sciences* 10:51–72.

Keener, William A. 1892. “Methods of Legal Education.” *Yale Law Journal* 1:144, 147.

Keilis-Borok, V. I., and T. B. Yanovskaja. 1967. “Inverse Problems of Seismology.” *Geophysical Journal of the Royal Astronomical Society* 13:223–34.

Keller, A. 1983. *The Infancy of Atomic Physics: Hercules in His Cradle.* Oxford: Clarendon Press.

Kenrick, Frank B., C. S. Gilbert, and K. L. Wismer. 1924. "The Superheating of Liquids." *Journal of Physical Chemistry* 28:1297–1307.

Keuffel, J. Warren. 1948. "Parallel-Plate Counters and the Measurement of Very Short Time Intervals." Ph.D. dissertation, California Institute of Technology.

———. 1948. "Parallel-Plate Counters and the Measurement of Very Small Time Intervals," *Physical Review* 73:531.

———. 1949. "Parallel Plate Counters," *Review of Scientific Instruments* 20:202–8.

Kevles, Daniel Jerome. 1978. *The Physicists: The History of a Scientific Community in Modern America,* 4th ed. New York: Knopf.

———. 1994. "Perspective 1995. The Death of the Super Collider in the Life of American Physics." In D. J. Kevles, *The Physicists: The History of a Scientific Community in Modern America,* 1–6. Cambridge, Mass.: Harvard University Press, 1994.

King, Gilbert W. 1950. "Stochastic Methods in Quantum Mechanics." In *Computation Seminar November 1949,* edited by C. C. Hurd, 42–48. New York: International Business Machines Corporation.

———. 1951. "Further Remarks on Stochastic Methods in Quantum Mechanics." In *Computation Seminar December 1949,* edited by C. C. Hurd, 92–94. New York: International Business Machines Corporation.

———. 1951. "Monte Carlo Method for Solving Diffusion Problems." *Industrial and Engineering Chemistry* 43:2475–78.

———. 1952–53. "The Monte Carlo Method as a Natural Mode of Expression in Operations Research." *Journal of the Operations Research Society of America* 1:46–51.

Kinoshita, S. 1910. "The Photographic Action of the α-Particles Emitted from Radioactive Substances." *Proceedings of the Royal Society London A* 83:432–53.

———. 1915. "The Tracks of the α Particles in Sensitive Photographic Films." *Philosophical Magazine* 28:420–25.

Kirsch, G., and H. Wambacher. 1933. "Über die Geschwindigkeit der Neutronen aus Beryllium." *Sitzungsberichte, Akademie der Wissenschaften in Wien, Mathematisch-naturwissenschaftliche Klasse, Abteilung IIa* 142:241–49.

Kleinknecht, K., and T. D. Lee, eds. 1986. *Particles and Detectors: Festschrift for Jack Steinberger.* Berlin: Springer.

Klotz, Heinrich. 1988. *The History of Postmodern Architecture,* translated by R. Donnell. Cambridge, Mass.: MIT Press.

Knoepflmacher, U. C., and G. B. Tennyson, eds. 1977. *Nature and the Victorian Imagination.* Berkeley: University of California Press.

Knott, Cargill G. 1923. "Sketch of John Aitken's Life and Scientific Work." In *Collected Scientific Papers,* edited by C. G. Knott, vii–xiii. Cambridge: Cambridge University Press.

Köhnke, Klaus C. 1991. *The Rise of Neo-Kantianism: German Academic Philosophy between Idealism and Positivism.* Cambridge: Cambridge University Press.

Kohler, Robert E. 1982. *From Medical Chemistry to Biochemistry: The Making of a Biomedical Discipline.* Cambridge: Cambridge University Press.

Korff, Serge A. 1955. *Electron and Nuclear Counters.* Toronto: Van Nostrand.

Kowarski, L. 1961. "Introduction." In *Proceedings of an International Conference on Instrumentation for High-Energy Physics,* 223–24. New York: Wiley.

———. 1962. “Concluding Remarks.” In *Informal Meeting on Track Data Processing, Held at CERN on 19th July 1962,* CERN Report 62-37, edited by M. Benoe and B. Elliott, 99–100. Geneva: CERN Data Handling Division.

———. 1963. “Concluding Remarks.” In *Programming for HPD and Other Flying Spot Devices,* CERN Report 63-34, edited by J. Howie, S. McCarroll, B. Powell, and A. Wilson, 237–41. Geneva: CERN Data Handling Division.

———. 1963. “Introduction.” In *Programming for HPD and Other Flying Spot Devices,* CERN Report 63-34, edited by J. Howie, S. McCarroll, B. Powell, and A. Wilson. Geneva: CERN Data Handling Division.

———. 1964. “General Survey: Automatic Data Handling in High Energy Physics.” In *Automatic Acquisition and Reduction of Nuclear Data,* edited by K. H. Beckurts, W. Gläser, and G. Krüger, 26–40. Karlsruhe: Gesellschaft für Kernforschung.

———. 1965. “Concluding Remarks.” In *Programming for Flying Spot Devices: A Conference Held at the Centro Nazionale Analisi Fotogrammi, INFN, Bologna, on 7–9 October 1964,* CERN Report 65-11, edited by W. G. Moorhead and B. W. Powell. Geneva: CERN Data Handling Division.

———. 1967. “Concluding Remarks.” In *Programming for Flying Spot Devices: A Conference Held at the Max-Planck-Institut für Physik und Astrophysik, Munich, on 18–20 January 1967,* edited by B. W. Powell and P. Seyboth, 409–16.

———. 1972. “The Impact of Computers on Nuclear Science.” In *Computing as a Language of Physics: Lectures Presented at an International Seminar Course at Trieste from 2 to 20 August 1971,* edited by International Center for Theoretical Physics, 27–37. Vienna: International Atomic Energy Agency.

Kraft, Phillip. 1977. *Programmers and Managers: The Routinization of Computer Programming in the United States.* Berlin: Springer.

Kragh, Helge. 1990. *Dirac: A Scientific Biography.* New York: Cambridge University Press.

Krausz, Michael, and Jack W. Meiland, eds. 1982. *Relativism: Cognitive and Moral.* Notre Dame, Ind.: University of Notre Dame Press.

Krienen, R. 1963. “A Digitized Spark Chamber.” *Nuclear Instruments and Methods* 20:168–70.

Krige, J. 1987. “The Development of Techniques for the Analysis of Track-Chamber Pictures at CERN,” CERN Report CHS-20. Geneva: CERN Data Handling Division.

Kuhn, T. 1970. “Reflections on My Critics.” In *Criticism and the Growth of Knowledge,* edited by I. Lakatos and A. Musgrave, 231–78. Proceedings of the International Colloquium in the Philosophy of Science, vol. 4. Cambridge: Cambridge University Press.

———. 1970. *The Structure of Scientific Revolutions,* 2d ed. International Encyclopedia of Unified Science. Chicago: University of Chicago Press.

———. 1974. “Second Thoughts on Paradigms.” In *The Structure of Scientific Theories,* edited by F. Suppe, 459–82. Urbana: University of Illinois Press.

———. 1978. *Blackbody Theory and the Quantum Discontinuity, 1894–1912.* London: Oxford University Press.

Kuper, Adam. 1994. “Culture, Identity, and the Project of a Cosmopolitan Anthropology.” *Man, The Journal of the Royal Anthropological Institute* 9:537–54.

Ladurie, Emmanuel Le Roy. 1979. *Montaillou: The Promised Land of Error,* translated by B. Bray. New York: Vintage.

Lakatos, I. 1970. "Falsification and the Methodology of Scientific Research Programmes." In *Criticism and the Growth of Knowledge,* edited by I. Lakatos and A. Musgrave, 91–195. Proceedings of the International Colloquium in the Philosophy of Science, vol. 4. Cambridge: Cambridge University Press.

Lakatos, I., and A. Musgrave, eds. 1970. *Criticism and the Growth of Knowledge.* Proceedings of the International Colloquium in the Philosophy of Science, vol. 4. Cambridge: Cambridge University Press.

Lamb, Willis E. 1951. "Anomalous Fine Structure of Hydrogen and Singly Ionized Helium." *Reports on Progress in Physics* 14:19–63.

Lamb, Willis E., and R. C. Retherford. 1947. "Fine Structure of the Hydrogen Atom by a Microwave Method." *Physical Review* 72:241–43.

Langdell, C. C. 1979. *Selection of Cases on the Law of Contracts with a Summary of the Topics Covered by the Case,* 2d ed. Boston: Little, Brown.

Langsdorf, Alexander, Jr. 1939. "A Continuously Sensitive Diffusion Cloud Chamber." *Review of Scientific Instruments* 10:91–103.

Latour, Bruno. 1986. "Visualization and Cognition: Thinking with Eyes and Hands." *Knowledge and Society: Studies in the Sociology of Culture Past and Present* 6:1–40.

———. 1987. *Science in Action: How to Follow Scientists and Engineers through Society.* Cambridge, Mass.: Harvard University Press.

———. 1988. *The Pasteurization of France.* Cambridge, Mass.: Harvard University Press.

Latour, Bruno, and Steve Woolgar. 1987. *Laboratory Life: The Construction of Scientific Facts.* Cambridge, Mass.: Harvard University Press.

Lattes, C. M. G., H. Muirhead, G. P. S. Occhialini, and C. F. Powell. (1947) 1972. "Processes Involving Charged Mesons." In *Selected Papers of Cecil Frank Powell,* edited by E. H. S. Burhop, W. O. Lock, and M. G. K. Menon, 214–17. New York: North-Holland. First published in *Nature* 159 (1947): 694–97.

Lattes, C. M. G., G. P. S. Occhialini, and C. F. Powell. (1947) 1972. "Observations on the Tracks of Slow Mesons in Photographic Emulsions." In *Selected Papers of Cecil Frank Powell,* edited by E. H. S. Burhop, W. O. Lock, and M. G. K. Menon, 228–55. New York: North-Holland, 1972. First published in *Nature* 160 (1947): 453–56, 486–92.

Lawrence Berkeley Laboratory, Stanford Linear Accelerator Center, and SLAC-LBL Users Organization. 1974. *Proceedings of the 1974 PEP Summer Study,* LBL-4800, SLAC 190, PEP-178. Berkeley, Calif.: Lawrence Berkeley Laboratory.

———. 1975. *Proceedings of the 1975 PEP Summer Study,* LBL-4800, SLAC 190, PEP-178. Berkeley, Calif.: Lawrence Berkeley Laboratory.

Lawrence Radiation Laboratory. 1960. *LRL Detectors: The 72-Inch Bubble Chamber.* Publ. no. 31, 10 M, July 1960. Berkeley: University of California.

Lee, B. W. 1978. "Theoretical Physics at Fermilab." *NALREP: Monthly Report of the Fermi National Accelerator Laboratory,* March: 1–9.

Lenoir, Timothy. 1986. "Models and Instruments in the Development of Electrophysiology, 1845–1912. *Historical Studies in the Physical and Biological Sciences* 17:1–54.

———. 1988. "Practice, Reason, Context: The Dialogue between Theory and Experiment." *Science in Context* 2:3–22.

Lenoir, Timothy, and Yehuda Elkana, eds. 1988. *Science in Context* 2:3–212.

Leprince-Ringuet, L. 1953. "Discours de clôture." In *Congrès International sur le Rayonnement Cosmique,* 287–90. Informal publication. Bagnères de Bigorre, July.

Leprince-Ringuet, L., and Michel Lhéritier. 1944. "Existence probable d'une particule de masse 990 m_0 dans le rayonnement cosmique." *Comptes Rendus de l'Académie des Sciences* 219:618–20.

Les Prix Nobel 1980: Nobel Prizes, Presentations, Biographies and Lectures. 1981. Stockholm: Almquist and Wiksell.

Leslie, Stuart W. 1990. "Profit and Loss: The Military and MIT in the Postwar Era." *Historical Studies in the Physical and Biological Sciences* 21:59–85.

———. 1993. *The Cold War and American Science: The Military Industrial Academic Complex at MIT and Stanford.* New York: Columbia University Press.

Leslie, Stuart W., and B. Hevly. 1985. "Steeple Building at Stanford: Electrical Engineering, Physics, and Microwave Research." *Proceedings of the IEEE* 73:1169–80.

Leutz, H., ed. 1967. *Proceedings of International Colloquium on Bubble Chambers, Held at Heidelberg, 13–14 April 1967,* CERN Report 67-26, vols. 1 and 2. Geneva: CERN Data Handling Division.

Levere, Trevor H., and William R. Shea. *Nature, Experiment, and the Sciences: Essays in Honor of Stillman Drake.* Dordrecht: Kluwer.

Levi, Giovanni. 1991. "On Microhistory." In *New Perspectives on Historical Writing,* edited by P. Burke, 93–113. University Park, Pa.: Pennsylvania State University Press.

Lewontin, R. C., and L. C. Dunn. 1960. "The Evolutionary Dynamics of a Polymorphism in the House Mouse." *Genetics* 45:706–22.

Lieberman, P. 1960. "E.R.E.T.S. LOX Losses and Preventative Measures." In *Advances in Cryogenic Engineering,* edited by K. D. Timmerhaus, 2:225–42. New York: Plenum.

Lindenbaum, S. J. 1963. "A Counter Hodoscope System with Digital Data Handler and On-Line Computer for Elastic Scattering and Other Experiments at Brookhaven AGS." *Nuclear Instruments and Methods* 20:297–302.

———. 1966. "On-Line Computer Techniques in Nuclear Research." *Annual Review of Nuclear Science* 16:619–42.

———. 1972. "Data Processing for Electronic Techniques in High-Frequency Experiments." In *Computing as a Language of Physics: Lectures Presented at an International Seminar Course at Trieste from 2 to 20 August 1971,* edited by International Center for Theoretical Physics, 209–79. Vienna: International Atomic Energy Agency.

Livingston, M. S. 1980. "Early History of Particle Accelerators." *Advances in Electronics and Electron Physics* 50:1–88.

Llewellyn Smith, C. H. 1981. "Summary Talk." In *QCD and Lepton Physics.* Proceedings of the 16th Rencontre de Moriond, vol. 1, edited by J. Tran Thanh Van, 429–448. Les Ares, France.

Lock, W. O., ed. 1965. *5th International Conference on Nuclear Photography Held at CERN, Geneva, 15–18 September 1964,* CERN Report 65-4. Geneva: CERN.

Lockyer, K. G. 1964. *An Introduction to Critical Path Analysis.* New York: Pitman.

Loeb, Leonard B. 1939. *Fundamental Processes of Electrical Discharge in Gases.* New York: Wiley.

Lorentz, Hendrick Antoon. 1909. *The Theory of Electrons and its Applications to the Phenomena of Light and Radiant Heat.* New York: Stechert.

Lubar, Steven, and W. David Kingery, eds. 1993. *History from Things: Essays on Material Culture.* Washington, D.C.: Smithsonian Institution Press.

Luce, R. Duncan, and Howard Raiffa. 1957. *Games and Decisions: An Introduction and Critical Survey.* New York: Wiley.

Lukács, Georg. 1975. *The Young Hegel: Studies in the Relations between Dialectics and Economics,* translated by Rodney Livingston. London: Merlin.

Lynch, Michael. 1985. "Discipline and the Material Form of Images." *Social Studies of Science* 15:37–66.

Lynch, Michael, and Samuel Edgerton. 1988. "Aesthetics and Digital Image Processing: Representational Craft in Contemporary Astronomy." In *Picturing Power: Visual Depiction and Social Relations,* edited by G. Fyfe and J. Law, 184–220. Sociological Review Monograph no. 35. London: Routledge and Kegan Paul.

Lynch, Michael, and Steve Woolgar, eds. 1990. *Representation in Scientific Practice.* Cambridge, Mass.: MIT Press.

Lyotard, Jean-Francois. 1984. *The Postmodern Condition: A Report on Knowledge.* Theory and History of Literature, vol. 10, translated by Geoff Bennington and Brian Massumi. Minneapolis: University of Minnesota Press.

Mack, Pamela. 1990. "Straying from Their Orbits: Women in Astronomy in America." In *Women of Science: Righting the Record,* edited by G. Kass-Simon and P. Farnes, 72–116. Bloomington: Indiana University Press.

Macleod, G. R. 1963. "The Development of Data-Analysis Systems for Bubble Chambers, for Spark Chambers, and for Counter Experiments." *Nuclear Instruments and Methods* 20:367–83.

———. 1964. "On-Line Computers in Data Analysis Systems for High Energy Physics Experiments." In *Proceedings of the Informal Meeting on Film-Less Spark Chamber Techniques and Associated Computer Use,* CERN "Yellow Report" 64-30, edited by G. R. Macleod and B. C. Maglić, 3–9. Geneva: CERN Data Handling Division.

———. 1964. "Discussion." In *Proceedings of the Informal Meeting on Film-Less Spark Chamber Techniques and Associated Computer Use,* CERN "Yellow Report" 64-30, edited by G. R. Macleod and B. C. Maglić, 299–312. Geneva: CERN Data Handling Division.

Macleod, G. R. and B. C. Maglić, eds. 1964. *Proceedings of the Informal Meeting on Film-Less Spark Chamber Techniques and Associated Computer Use,* CERN "Yellow Report" 64-30. Geneva: CERN Data Handling Division.

Madansky, Leon, and R. W. Pidd. 1950. "Some Properties of the Parallel Plate Spark Counter II." *Review of Scientific Instruments* 21:407–10.

Maglić, B. C., L. W. Alvarez, A. H. Rosenfeld, and M. L. Stevensen. 1961. "Evidence for a $T = 0$ Three-Pion Resonance." *Physical Review Letters* 7:178–82.

Maglić, B. C., and F. A. Kirsten. 1962. "Acoustic Spark Chamber." *Nuclear Instruments and Methods* 17:49–59.

Mahoney, Michael. 1983. "Reading a Machine: The Products of Technology as Texts for Humanistic Study." Unpublished typescript.

Maienschein, Jane. 1983. "Experimental Biology in Transition: Harrison's Embryology, 1895–1910." *Studies in History of Biology* 6:107–27.

———. 1991. *Transforming Traditions in American Biology, 1880–1915.* Baltimore: Johns Hopkins University Press.

Marcuvitz, Nathan. 1986. *Waveguide Handbook.* IEE Electromagnetic Waves Series. London: Peregrinus.

Margeneau, Henry. 1950. *The Nature of Physical Reality: A Philosophy of Modern Physics.* New York: McGraw-Hill.

Marsaglia, George. 1968. "Random Numbers Fall Mainly in the Planes." *Proceedings of the National Academy of Science* 61:25–28.

Marshall, A. W. 1956. "An Introductory Note." In *Symposium on Monte Carlo Methods,* edited by H. A. Meyer, 1–14. New York: Wiley.

Marshall, John. 1951. "Čerenkov Radiation Counter for Fast Electrons." *Physical Review* 81:275–76.

Marx, Jay N. 1976. "PEP Proposal-I," TPC-LBL-76-24. 3 June.

Marx, Jay N., and David R. Nygren. 1978. "Time Projection Chamber." *Physics Today* 31:46–53.

Massachusetts Institute of Technology. 1947. *Five Years at the Radiation Laboratory.* Cambridge: Massachusetts Institute of Technology.

Mather, R. L. 1951. "Čerenkov Radiation from Protons and the Measurement of Proton Velocity and Kinetic Energy." *Physical Review* 84:181–90.

Mayr, E. 1982. *Growth of Biological Thought.* Cambridge, Mass.: Harvard University Press.

McCormick, Bruce H., and Daphne Innes. 1961. "The Spiral Reader Measuring Protector and Associated Filter Program." In *Proceedings of an International Conference on Instrumentation for High-Energy Physics,* 246–48. New York: Wiley.

McLaren, John, John Murray, Arthur Mitchell, and Alexander Buchan. 1903. "Memorandum by the Directors of the Observatories on Ben Nevis and at Fort-William in Connection with Their Closure." *Journal of the Scottish Meteorological Society* 12: 161–63.

Mendelsohn, Everett, Merritt Roe Smith, and Peter Weingart, eds. 1988. *Science, Technology, and the Military.* Sociology of the Sciences, vol. 1. Dordrecht: Kluwer.

Menon, M., and C. O'Ceallaigh. 1953. "Observations on the Mass and Energy of Secondary Particles Produced in the Decay of Heavy Mesons." In *Congrès International sur le Rayonnement Cosmique,* 118–24. Informal publication. Bagnères de Bigorre, July.

Merz, J. T. 1965. *A History of European Thought in the Nineteenth Century,* vol. 2. New York: Dover.

Metropolis, Nicholas, and S. Ulam. 1949. "The Monte Carlo Method." *Journal of the American Statistical Association* 44:335–41.

Meyer, Herbert A., ed. 1956. *Symposium on Monte Carlo Methods: Held at the University of Florida, 16 and 17 March 1954.* New York: Wiley.

Michel, Louis. 1953. "Absolute Selection Rules for Decay Processes." In *Congrès International sur le Rayonnement Cosmique,* 272–79. Informal publication. Bagnères de Bigorre, July.

Michl, W. 1912. "Über die Photographie der Bahnen einzelner α-Teilchen." *Sitzungsbe-*

richte, Akademie der Wissenschaften in Wien, Mathematisch-naturwissenschaftliche Klasse, Abteilung IIa 121:1431–47.

———. 1914. "Zur photographischen Wirkung der α-Teilchen." *Sitzungsberichte, Akademie der Wissenschaften in Wien, Mathematisch-naturwissenschaftliche Klasse, Abteilung IIa* 123:1955–63.

Millard, Charles. 1977. "Images of Nature: A Photo-Essay." In *Nature and the Victorian Imagination,* edited by U. C. Knoepflmacher and G. B. Tennyson, 3–26. Berkeley: University of California Press.

Miller, Arthur I. 1981. *Albert Einstein's Special Theory of Relativity: Emergence (1905) and Early Interpretations (1905–1911).* Reading, Mass.: Addison-Wesley.

Miller, D. H., E. C. Fowler, and R. P. Shutt. 1951. "Operation of a Diffusion Cloud Chamber with Hydrogen at Pressures up to 15 Atmospheres." *Review of Scientific Instruments* 22:280.

Montgomery, C. G., and D. D. Montgomery. 1941. "Geiger-Müller Counters." *Journal of the Franklin Institute* 231:447–67, 509–45.

Moore, Walter, J. 1988. *Schrödinger, Life and Thought.* Cambridge: Cambridge University Press.

Morgenau, Henry. 1950. *The Nature of Physical Reality: A Philosophy of Modern Physics.* New York: McGraw-Hill.

Morris, C. 1955. "Scientific Empiricism." In *International Encyclopedia of Unified Science,* edited by O. Neurath et al., 1:63–75. Chicago: University of Chicago Press.

Morse, Philip, M. 1952–53. "Trends in Operations Research." *Journal of the Operational Research Society of America* 1:159–65.

Mott-Smith, L. M. 1932. "On an Attempt to Deflect Magnetically the Cosmic-Ray Corpuscles." *Physical Review* 39:403–14.

Mouzon, J. C. 1936. "Discrimination between Partial and Total Coincidence Counts with Geiger-Müller Counters." *Review of Scientific Instruments* 7:467–70.

Mozley, Ann. 1971. "Change in Argonne National Laboratory: A Case Study." *Science* 173:30–38.

Mühlhäusler, Peter. 1986. *Pidgin and Creole Linguistics.* Language in Society, vol. 11. Oxford: Blackwell.

Musgrave, Alan. 1985. "Realism *versus* Constructive Empiricism." In *Images of Science: Essays on Realism and Empiricism, with a Reply from Bas C. van Fraassen,* edited by P. M. Churchland and C. A. Hooker, 197–221. Chicago: University of Chicago Press.

Nagel, Ernest. 1961. *The Structure of Science: Problems in the Logic of Scientific Explanation.* New York: Harcourt, Brace, and World.

Nagle, D. E., R. H. Hildebrand, and R. J. Plano. 1957. "Scattering of 10–30 Mev Negative Pions by Hydrogen." *Physical Review* 105:718–24.

National Academy of Sciences. 1972. *Physics in Perspective,* 2 vols. Washington, D.C.: National Academy of Sciences.

Neddermeyer, S. H., E. J. Althaus, W. Allison, and E. R. Schultz. 1947. "The Measurement of Ultra-Short Time Intervals." *Review of Scientific Instruments* 18:488–96.

Needell, Allan. 1983. "Nuclear Reactors and the Founding of Brookhaven National Laboratory." *Historical Studies in the Physical Sciences* 14:93–122.

Needels, T. S., and C. E. Nielsen. 1950. "A Continuously Sensitive Cloud Chamber." *Review of Scientific Instruments* 21:976–77.

Neher, H. Victor. 1938. "Geiger Counters." In *Procedures in Experimental Physics,* edited by J. Strong, 259–304. New York: Prentice-Hall.

Neurath, Otto. 1955. "Unified Science as Encyclopedic Integration." In *International Encyclopedia of Unified Science,* edited by O. Neurath et al., 1–27. Chicago: University of Chicago Press.

———. 1983. *Philosophical Papers 1913–1946,* edited by R. S. Cohen and M. Neurath. Dordrecht: Reidel.

———. 1983. "Sociology in the Framework of Physicalism." In O. Neurath, *Philosophical Papers 1913–1946,* edited by R. S. Cohen and M. Neurath, 58–90. Dordrecht: Reidel.

Neurath, Otto, Rudolf Carnap, and Charles Morris, eds. 1955. *International Encyclopedia of Unified Science,* vol. 1. Chicago: University of Chicago Press.

Nishijima, Kazuhiko. 1955. "Charge Independence Theory of *V* Particles." *Progress of Theoretical Physics* 13:285–304.

Nye, Mary Jo. 1972. *Molecular Reality: A Perspective on the Scientific Work of Jean Perrin.* New York: Elsevier.

———. 1984. *The Question of the Atom: From the Karlsruhe Congress to the First Solvay Conference, 1860–1911.* Los Angeles and San Francisco: Tomash.

Nygren, D. 1968. "A Measurement of the Neutron-Neutron Scattering Length." Ph.D. dissertation, University of Washington.

———. 1974. "The Time Projection Chamber: A New 4π Detector for Charged Particles," PEP-144. In LBL, SLAC, and SLAC-LBL Users Organization, *Proceedings 1974 PEP Summer Study.* Berkeley, Calif.: Lawrence Berkeley Laboratory.

———. 1975. "The Time Projection Chamber-1975," PEP-198. In LBL, SLAC, and SLAC-LBL Users Organization, *Proceedings 1975 PEP Summer Study.* Berkeley, Calif.: Lawrence Berkeley Laboratory.

Obeyesekere, Gananath. 1990. *The Work of Culture: Symbolic Transformation in Psychoanalysis and Anthropology.* Chicago: University of Chicago Press.

Occhialini, G. P. S., and C. F. Powell. (1947) 1972. "Nuclear Disintegrations Produced by Slow Charged Particles of Small Mass." In *Selected Papers of Cecil Frank Powell,* edited by E. H. S. Burhop, W. O. Lock, and M. G. K. Menon, 224–27. New York: North-Holland, 1972. First published in *Nature* 159 (1947): 186–89.

O'Ceallaigh, C. 1951. "Masses and Modes of Decay of Heavy Mesons." *Philosophical Magazine* 42:1032–39.

———. 1953. "Determination of the Mass of Slow Heavy Mesons." In *Congrès International sur le Rayonnement Cosmique,* 121–27. Informal publication. Bagnères de Bigorre, July.

———. 1982. "A Contribution to the History of C. F. Powell's Group in the University of Bristol 1949–65." *Journal de Physique* 43:C8-185–C8-189.

Olesko, Kathryn M. 1991. *Physics as a Calling: Discipline and Practice in the Königsberg Seminar for Physics.* Ithaca, N.Y.: Cornell University Press.

O'Neill, Gerard K. 1962. "The Spark Chamber." *Scientific American* 207:36–43.

Ophir, Adi, S. Shapin, and Simon Schaffer, eds. 1991. *Science in Context* 4:1–218, with an introduction, "The Place of Knowledge," by A. Ophir and S. Shapin, 3–22.

Ordway, Frederick I., and Mitchell R. Sharpe. 1982. *The Rocket Team: From the V-2 to the Saturn Moon Rocket.* Cambridge, Mass.: MIT Press.

Oreskes, Naomi. 1988. "The Rejection of Continental Drift." *Historical Studies in the Physical and Biological Sciences* 18:311–48.

Ortner, Gustav. 1940. "Über die durch Höhenstrahlung verursachten Kernzertrümmerungen in photographischen Schichten." *Sitzungsberichte, Akademie der Wissenschaften in Wein, Mathematisch-naturwissenschaftliche Klasse* 149:259–67.

———. 1950. "Dr. H. Wambacher" [Obituary]. *Nature* 166:135.

Ortner, G., and G. Stetter. 1923. "Über den elektrischen Nachweis einzelner Korpuskularstrahlen." *Zeitschrift für Physik* 54:449–70.

Owens, Larry. 1990. "MIT and the Federal 'Angel': Academic R&D and Federal-Private Cooperation before World War II." *Isis* 81:188–213.

———. 1992. "OSRD, Vannevar Bush, and the Struggle to Manage Science in the Second World War." Unpublished typescript.

Page, Thorton L. 1952–53. "A Tank Battle Game." *Journal of the Operations Research Society of America* 1:85–86.

Pais, Abraham. 1986. *Inward Bound: Of Matter and Forces in the Physical World.* Oxford: Oxford University Press.

———. 1982. *"Subtle Is the Lord—": The Science and Life of Albert Einstein.* Oxford: Clarendon.

Palmer, A. de Forest. 1912. *The Theory of Measurements.* New York: McGraw-Hill.

Palmer, Ronald R., and William M. Rice. 1961. *Modern Physics Buildings: Design and Function.* Progressive Architecture Library. New York: Reinhold.

Paradis, James, and Thomas Postlewait, eds. 1981. *Victorian Science and Victorian Values: Literary Perspectives.* Annals of the New York Academy of Sciences, vol. 360. New York: New York Academy of Sciences.

Paris, Elizabeth. 1991. "The Building of the Stanford Positron-Electron Asymmetric Ring: How Science Happens." Unpublished typescript.

Parkinson, W. C., and H. R. Crane. 1952. "Final Report." Ann Arbor: University of Michigan Cyclotron, Engineering Research Institute.

Parmentier, Douglass, Jr., and A. J. Schwemin. 1955. "Liquid Hydrogen Bubble Chambers." *Review of Scientific Instruments* 26:954–58.

Parmentier, Douglass, Jr., A. J. Schwemin, L. W. Alvarez, F. S. Crawford, Jr., and M. L. Stevensen. 1955. "Four-Inch Diameter Liquid Hydrogen Bubble Chamber." *Physical Review* 58:284.

Paton, James. 1954. "Ben Nevis Observatory 1883–1904." *Weather* 9:291–308.

Pearce, Susan M. 1989. *Museum Studies in Material Culture.* London: Leicester University Press.

Peck, R. A., Jr. 1947. "A Calibration for Eastman Proton Plates." *Physical Review* 72:1121.

Peebles, P. J. E. 1973. "Statistical Analysis of Catalogs of Extragalactic Objects. I, Theory." *Astrophysical Journal* 185:413–40.

———. 1974. "The Nature of the Distribution of Galaxies." *Astronomy and Astrophysics* 32:197–202.

———. 1975. "Statistical Analysis of Catalogs of Extragalactic Objects. VI, The Galaxy Distribution in the Jagellonian Field." *Astrophysical Journal* 196:647–52.

Peebles, P. J. E., and Edward Groth. 1975. "Statistical Analysis of Catalogs of Extragalactic Objects. V, Three-Point Correlation Function for the Galaxy Distribution in the Zwicky Catalog." *Astrophysical Journal* 196:1–11.

Peebles, P. J. E., and M. G. Hauser. 1974. "Statistical Analysis of Extragalactic Objects. III, The Shane-Wirtanen and Zwicky Catalogs." *Astrophysical Journal Supplement* 253:19–36.

Peirce, Charles Sanders. 1984. "Some Consequences of Four Incapacities." In *Writings of Charles Sanders Peirce, A Chronological Edition. Vol. 2, 1867–1871,* 211–42. Bloomington: Indiana University Press.

Perez-Mendez, V. 1965. "Review of Film-Less Spark Chamber Techniques: Acoustic and Vidicon." *IEEE Transactions on Nuclear Science,* NS-12, no. 4: 13–18.

Perez-Mendez, V., and J. M. Pfab. 1965. "Magnetostrictive Readout for Wire Spark Chambers." *Nuclear Instruments and Methods* 33:141–46.

Perkins, Donald H. 1947. "Nuclear Disintegration by Meson Capture." *Nature* 159: 126–27.

———. 1987. *Introduction to High Energy Physics,* 3rd ed. Menlo Park, Calif.: Addison-Wesley.

Perret, Frank A. 1935. *The Eruption of Mt. Pelée 1929–1932.* Baltimore: Waverly.

Pestre, D., and J. Krige. 1992. "Some Thoughts on the History of CERN." In *Big Science: The Growth of Large-Scale Research,* edited by P. Galison and B. Hevly, 78–99. Stanford, Calif.: Stanford University Press.

Pevsner, A., R. Kraemer, M. Nussbaum, C. Richardson, P. Schlein, R. Strand, T. Toohig, M. Block, A. Engler, R. Gessaroli, and C. Meltzer. 1961. "Evidence for a Three-Pion Resonance Near 550 Mev." *Physical Review Letters* 7:421–23.

Peyrou, C. 1967. "Bubble Chamber Principles." In *Bubble and Spark Chambers: Principles and Use,* edited by R. P. Shutt, 1:19–58. Pure and Applied Physics, vol. 27. New York: Academic Press.

Piccioni, Oreste. 1948. "Search for Photons from Meson-Capture." *Physical Review* 74:1754–58.

———. 1989. "On the Antiproton Discovery." In *Pions to Quarks: Particle Physics in the 1950s,* edited by L. M. Brown, M. Dresden, and L. Hoddeson, 285–95. Cambridge: Cambridge University Press.

Pickering, Andrew. 1981. "Constraints on Controversy: The Case of the Magnetic Monopole." *Social Studies of Science* 11:63–93.

———. 1981. "The Hunting of the Quark." *Isis* 72:216–36.

———. 1984. "Against Putting the Phenomena First: The Discovery of the Weak Neutral Current." *Studies in History and Philosophy of Science* 15:85–117.

———. 1984. *Constructing Quarks: A Sociological History of Particle Physics.* Chicago: University of Chicago Press.

Pidd, R. W., and Leon Madansky. 1949. "Some Properties of the Parallel Plate Spark Counter I." *Physical Review* 75:1175–80.

Pinch, Trevor. 1982. "Kuhn—The Conservative and Radical Interpretations: Are Some Mertonians 'Kuhnians' and Some Kuhnians Mertonians?" *4S Newsletter* 7:10–25.

Pjerrou, G. M., D. J. Prowse, P. Schlein, W. E. Slater, D. H. Stark, and H. K. Ticho. "A Resonance in the Xi-π System at 1.53 Gev." In *International Conference on High-Energy Physics at CERN,* edited by J. Prentki, 289–90. Geneva: CERN.

Plano, Richard J., and Irwin A. Pless. 1955. "Negative Pressure Pentane Bubble Chamber Used in High-Energy Experiments." *Physical Review* 99:639.

Pless, Irwin A. 1956. "Proton-Proton Scattering at 457 Mev in a Bubble Chamber." *Physical Review* 104:205–10.

Pless, Irwin A., and Richard J. Plano. 1955. "A Study of 456-MeV Proton Interactions with Hydrogen and Carbon." *Physical Review* 99:639–40.

Polanyi, Michael. 1958. *Personal Knowledge: Towards a Post-Critical Philosophy.* London: Routledge and Kegan Paul.

———. 1969. *Knowing and Being: Essays.* London: Routledge and Kegan Paul.

Pontecorvo, B. 1960. "Electron and Muon Neutrinos." *Soviet Physics JETP* 37:1236–40.

Powell, B. W., and Paul V. C. Hough. 1961. "A Method for Faster Analysis of Bubble Chamber Photographs." In *Proceedings of an International Conference on Instrumentation for High-Energy Physics,* 242–485. New York: Wiley, 1961.

Powell, B. W., and P. Seyboth, eds. 1967. *Programming for Flying Spot Devices: A Conference Held at the Max-Planck-Institut für Physik und Astrophysik, Munich, on 18–20 January 1967.*

Powell, C. F. 1912. "On an Expansion Apparatus for Making Visible the Tracks of Ionising Particles in Gases and Some Results Obtained by Its Use." *Proceedings of the Royal Society London A* 87:277–90.

———. (1925) 1972. "Supersaturation in Steam and Its Influence upon Some Problems in Steam Engineering." In *Selected Papers of Cecil Frank Powell,* edited by E. H. S. Burhop, W. O. Lock, and M. G. K. Menon, 37–48. New York: North-Holland, 1972. First published in *Engineering* 127 (1925): 229.

———. (1928) 1972. Condensation Phenomena at Different Temperatures." In *Selected Papers of Cecil Frank Powell,* edited by E. H. S. Burhop, W. O. Lock, and M. G. K. Menon, 49–73. New York: North-Holland, 1972. First published in *Proceedings of the Royal Society A* 119 (1928): 555–77.

———. (1937) 1972. "Royal Society Expedition to Montserrat, B.W.I. Preliminary Report on Seismic Observations." In *Selected Papers of Cecil Frank Powell,* edited by E. H. S. Burhop, W. O. Lock, and M. G. K. Menon, 131–48. New York: North-Holland, 1972. First published in *Proceedings of the Royal Society A* 158 (1937): 479–94.

———. (1940) 1972. "Further Applications of the Photographic Method in Nuclear Physics." In *Selected Papers of Cecil Frank Powell,* edited by E. H. S. Burhop, W. O. Lock, and M. G. K. Menon, 158–63. New York: North-Holland, 1972. First published in *Nature* 145 (1940): 155–57.

———. (1942) 1972. "The Photographic Plate in Nuclear Physics." In *Selected Papers of Cecil Frank Powell,* edited by E. H. S. Burhop, W. O. Lock, and M. G. K. Menon, 164–69. New York: North-Holland, 1972. First published in *Endeavor,* October 1942.

———. 1950. "Mesons." *Reports on Progress in Physics* 13:350–424.

———. 1953. "H-I. Photographic-Post-Discussion." In *Congrès International sur le Rayonnement Cosmique,* 233–35. Informal publication. Bagnères de Bigorre, July.

———. 1953. "Recapitulation et discussion sur les mesons lourds chargés." In *Congrès International sur le Rayonnement Cosmique,* 221–24. Informal publication. Bagnères de Bigorre, July.

———. 1954. "A Discussion on *V*-Particles and Heavy Mesons." *Proceedings of the Royal Society A* 221:277–420.

———. 1972. "C. T. R. Wilson, Biography by C. F. Powell." In *Selected Papers of Cecil Frank Powell,* edited by E. H. S. Burhop, W. O. Lock, and M. G. K. Menon, 357–68. New York: North-Holland.

———. 1972. "Fragments of Autobiography." In *Selected Papers of Cecil Frank Powell,* edited by E. H. S. Burhop, W. O. Lock, and M. G. K. Menon, 7–34. New York: North-Holland.

———. 1972. *Selected Papers of Cecil Frank Powell,* edited by E. H. S. Burhop, W. O. Lock, and M. G. K. Menon. New York: North-Holland.

Powell, C. F., and G. E. F. Fertel. (1939) 1972. "Energy of High-Velocity Neutrons by the Photographic Method." In *Selected Papers of Cecil Frank Powell,* edited by E. H. S. Burhop, W. O. Lock, and M. G. K. Menon, 151–54. New York: North-Holland, 1972. First published in *Nature* 144 (1939): 115–18, and reprinted in *The Study of Elementary Particles by the Photographic Method,* C. F. Powell, P. H. Fowler, and D. H. Perkins. London: Pergamon, 1959.

Powell, C. F., P. H. Fowler, and D. H. Perkins. 1959. *The Study of Elementary Particles by the Photographic Method.* London: Pergamon.

Powell, C. F., and G. P. S. Occhialini. 1947. "Appendix A: Method of Processing Ilford 'Nuclear Research' Emulsions." In C. Powell and G. P. S. Occhialini, *Nuclear Physics in Photographs: Tracks of Charged Particles in Photographic Emulsions.* Oxford: Clarendon.

Powell, C. F., and G. P. S. Occhialini. 1947. *Nuclear Physics in Photographs: Tracks of Charged Particles in Photographic Emulsions.* Oxford: Clarendon.

Powell, C. F., G. P. S. Occhialini, D. L. Livesey, and L. V. Chilton. (1946) 1972. "A New Photographic Emulsion for the Detection of Fast Charge Particles." In *Selected Papers of Cecil Frank Powell,* edited by E. H. S. Burhop, W. O. Lock, and M. G. K. Menon, 209–13. New York: North-Holland, 1972. First published in *Journal of Scientific Instruments* 23 (1946): 102–6.

Preiswerk, P. 1964. "Introduction." In *Proceedings of the Informal Meeting on Film-Less Spark Chamber Techniques and Associated Computer Use,* CERN "Yellow Report" 64-30, edited by G. R. Macleod and B. C. Maglić, 1–2. Geneva: CERN Data Handling Division.

Press, Frank. 1968. "Density Distribution in Earth." *Science* 160: 1218–21.

Press, Frank, and Shawn Biehler. 1964. "Inferences on Crustal Velocities and Densities from P Wave Delays and Gravity Anomalies." *Journal of Geophysical Research* 69: 2979–95.

Proceedings of an International Conference on Instrumentation for High-Energy Physics: Held at the Ernest O. Lawrence Radiation Laboratory, Berkeley, California, 12–14 September 1960. 1961. New York: Wiley.

Proceedings of the International Congress of Mathematicians: Cambridge, Massachusetts, 30 August–6 September 1952. 1952. Providence: American Mathematical Society.

Proceedings of the Purdue Conference on Instrumentation for High-Energy Physics. 1965. Published as *IEEE Transactions in Nuclear Science,* NS-12, no. 4.

Purcell, E. 1964. "Nuclear Physics without the Neutron: Clues and Contradictions." In *Proceedings of the 10th International Congress of the History of Science,* 121–133. Paris: Hermann.

Putnam, Hilary. 1987. *The Many Faces of Realism.* La Salle, Ill.: Open Court.
———. 1994. "The Dewey Lectures 1994." *Journal of Philosophy* 91, no. 9.
Quercigh, E. 1964. "Direct Recording on Magnetic Tape in Spark Chambers." In *Proceedings of the Informal Meeting on Film-Less Spark Chamber Techniques and Associated Computer Use,* CERN "Yellow Report" 64-30, edited by G. R. Macleod and B. C. Maglić, 345–49. Geneva: CERN Data Handling Division.
Quine, W. V. O. (1955) 1973. "Posits and Reality." In *Theories and Observation in Science,* edited by R. E. Grandy, 154–61. Englewood Cliffs, N.J.: Prentice-Hall. From original version, later superseded, of *Word and Object.*
———. 1960. *Word and Object.* Cambridge, Mass.: MIT Press.
———. 1976. *The Ways of Paradox and Other Essays,* rev. ed. Cambridge, Mass.: Harvard University Press.
Rabinow, Paul, ed. 1984. *The Foucault Reader.* New York: Pantheon.
Rahm, David, C. 1956. "Development of Hydrocarbon Bubble Chambers for Use in Nuclear Physics." Ph.D. dissertation, University of Michigan.
———. 1969. "Donald A. Glaser: Nobel Prize for Physics in 1960." Unpublished manuscript. American Institute of Physics.
Rankin, Angus. 1891. "Preliminary Notes on the Observations of Dust Particles at Ben Nevis Observatory." *Journal of the Scottish Meteorological Society* 9:125–32.
Rassetti, Franco. 1941. "Disintegration of Slow Mesotrons." *Physical Review* 60: 198–204.
Raup, David M., and J. John Sepkoski. 1994. "Periodicity of Extinctions in the Geologic Past." *Proceedings of the National Academy of Science* 81:801–5.
Reineke, John E. 1971. "Tây Bôi: Notes on the Pidgin French Spoken in Vietnam." In *Pidginization and Creolization of Languages,* edited by D. Hymes, 47–56. Cambridge: Cambridge University Press.
Reines, Frederick. 1960. "Neutrino Interactions." *Annual Review of Nuclear Science* 10:1–26.
———. 1982. "Neutrinos to 1960–Personal Recollection." *Journal de Physique* 43: C8-237–C8-260.
Reines, Frederick, and C. L. Cowan, Jr. 1953. "Detection of the Free Neutrino." *Physical Review* 92:830–31.
Reines, Frederick, C. L. Cowan, Jr., F. B. Harrison, A. D. McGuire, and H. W. Kruse. 1960. "Detection of the Free Antineutrino." *Physical Review* 117:159–73.
Reiter, Wolfgang. 1988. "Das Jahr 1938 und seine Folgen für die Naturwissenschaften an Österreichs Universitäten." In *Vertriebene Vernunft II: Emigration und Exil österreichischer Wissenschaft,* edited by F. Stadler. Vienna and Munich: Jugend und Volk.
Renneberg, Monika, and Mark Walker, eds. 1994. *Science, Technology, and National Socialism.* Cambridge: Cambridge University Press.
Reyer, E. 1892. *Geologische und geographische Experimente.* Leipzig: Wilhelm Engelmann.
Reynolds, Osborne. 1879. "On the Manner in Which Raindrops and Hailstones Are Formed." *Memoirs of the Literary and Philosophical Society of Manchester,* 3rd ser., 6:48–60.
Rice-Evans, Peter. 1974. *Spark, Streamer, Proportional and Drift Chambers.* London: Richelieu.

Rider, Robin E. 1988. "Early Development of Operations Research: British and American Contexts." Presented at the joint meeting of the British Society for History of Science and the History of Science Society, Manchester, 11–14 July.

———. "Capsule History of Operations Research." In *Encyclopaedia of the History and Philosophy of Mathematical Sciences,* edited by I. Grattan-Guinness. London: Routledge and Kegan Paul. Forthcoming.

Riordan, Michael. 1987. *The Hunting of the Quark: A True Story of Modern Physics.* New York: Simon and Schuster.

Roberts, A. 1964. "Properties of Conventional Camera-Film Data Acquisition Systems with Narrow-Gap Spark Chambers." In *Proceedings of the Informal Meeting on Film-Less Spark Chamber Techniques and Associated Computer Use,* CERN "Yellow Report" 64-30, edited by G. R. Macleod and B. C. Maglić, 367–69. Geneva: CERN Data Handling Division.

———. 1964. "Some Reflections on Systems for the Automatic Processing of Complex Spark Chamber Events." In *Proceedings of the Informal Meeting on Film-Less Spark Chamber Techniques and Associated Computer Use,* CERN "Yellow Report" 64-30, edited by G. R. Macleod and B. C. Maglić, 287–98. Geneva: CERN Data Handling Division.

Roberts, K. V. 1972. "Computers and Physics." In *Computing as a Language of Physics: Lectures Presented at an International Seminar Course at Trieste from 2 to 20 August 1971,* edited by International Center for Theoretical Physics, 3–26. Vienna: International Atomic Energy Agency.

Rochester, G. D., and C. C. Butler. 1947. "Evidence for the Existence of New Unstable Elementary Particles." *Nature* 160:855–57.

———. 1982. "The Development and Use of Nuclear Emulsions in England in the Years 1945–50." *Journal de Physique* 43:C8-89–C8-90.

Rochester, G. D., and J. G. Wilson. 1952. *Cloud Chamber Photographs of the Cosmic Radiation.* New York: Academic Press.

Rollo Russell, F. A., and E. Douglas Archibald. 1888. "On the Unusual Optical Phenomena of the Atmosphere, 1883–6, Including Twilight Effects, Coronal Appearances, Sky Haze, Coloured Suns, Moons, Etc." In *The Eruption of Krakatoa and Subsequent Phenomena: Report of the Krakatoa Committee of the Royal Society,* edited by J. G. Symons, 151–463. London: Harrison.

Romaine, Suzanne. 1988. *Pidgin and Creole Languages.* London: Longman.

Rosenfeld, A. H. 1963. "Current Performance of the Alvarez-Group Data Processing System." *Nuclear Instrumentation and Methods* 20:422–34.

Rossi, Bruno. 1930. "Method of Registering Multiple Simultaneous Impulses of Several Geiger's Counters." *Nature* 125:636.

———. 1932. "Nachweis einer Sekundärstrahlung der durchdringenden Korpuskularstrahlung." *Physikalische Zeitschrift* 33:304–5.

———. 1964. *Cosmic Rays.* New York: McGraw-Hill.

———. 1982. "Development of the Cosmic Ray Techniques." *Journal de Physique* 43:C8-69–C8-88.

———. 1985. "Arcetri, 1928–1932." In *Early History of Cosmic Ray Studies: Personal Reminiscences with Old Photographs,* edited by Y. Sekido and H. Elliot, 53–73. Dordrecht: Reidel.

Rossi, Bruno, and Norris Nereson. 1942. "Experimental Determination of the Disintegration Curve of Mesotrons." *Physical Review* 62:417–22.

Rossi, Bruno, and Hans Staub. 1949. *Ionization Chambers and Counters: Experimental Techniques.* New York: McGraw-Hill.

Rossiter, Margaret. 1982. *Women Scientists in America: Struggles and Strategies to 1940.* Baltimore: John Hopkins University Press.

Rotblat, J. 1950. "Photographic Emulsion Technique." *Progress in Nuclear Physics* 1:37–72.

Rudwick, Martin J. S. 1976. "The Emergence of a Visual Language for Geological Science, 1760–1840." *History of Science* 14:149–95.

———. 1985. *The Great Devonian Controversy: The Shaping of Scientific Knowledge among Gentlemanly Specialists.* Chicago: University of Chicago Press.

Rutherford, E., and Hans Geiger. 1908. "The Charge and Nature of the α-Particle." *Proceedings of the Royal Society London A* 81:162–73.

Safdie, Moshe, and Associates, Inc., Architects and Planners, with Peter Walker, William Johnson and Partners Landscape Architects, Inc. 1993. *SSC Laboratory. Dallas, Texas. Main Campus Development Plan.* Doc. no. Y1100029—rev. A. 31 May.

Sahlins, Marshall David. 1985. *Islands of History.* Chicago: University of Chicago Press.

———. 1993. "Goodbye to *Tristes Tropes:* Ethnography in the Context of Modern World History." *Journal of Modern History* 65:1–25.

———. 1995. *How "Natives" Think: About Captain Cook, for Example.* Chicago: University of Chicago Press.

———. 1995. "'Sentimental Pessimism' and Ethnographic Experience; or, Why Culture is Not a Disappearing 'Object.'" Typescript.

Sahlins, Peter. 1989. *Boundaries: The Making of France and Spain in the Pyrenees.* Berkeley and Los Angeles: University of California Press.

Salam, Abdus. 1964. "Summary of Conference Results." In *Nucleon Structure: Proceedings of the International Conference at Stanford University, 24–27 June 1963,* edited by R. Hofstadter and L. I. Schiff, 397–414. Stanford, Calif.: Stanford University Press.

Saudinos, J. 1973. "Operation of Large Drift Length Chambers." In *Proceedings of the 1973 International Conference on Instrumentation for High-Energy Physics, Frascati, Italy, 8–12 May 1973,* edited by S. Stipcich. Frascati: Laboratori Nazionali del CNEN, Servizio Documentazione.

Schaffer, Simon. 1988. "Astronomers Mark Time: Discipline and the Personal Equation." *Science in Context* 2:115–45.

———. 1989. "Glass Works: Newton's Prisms and the Uses of Experiment." In *The Uses of Experiment: Studies in the Natural Sciences,* edited by D. Gooding, T. Pinch, and S. Schaffer, 67–104. Cambridge: Cambridge University Press.

———. 1989. "A Manufactory of Ohms: The Integrity of Victorian Values." Presented at Mediums of Exchange: Building Systems and Networks in Science and Technology, UCLA, 2 December.

Schilpp, Paul A., ed. 1970. *Albert Einstein Philosopher-Scientist.* Library of Living Philosophers, vol. 7. La Salle, Ill.: Open Court.

Schlick, Moritz. 1979. *Philosophical Papers: Volume 1 (1909–1922),* edited by H. L. Mulder and B. E. B. van de Velde-Schlick, translated by P. Heath. Vienna Circle Collection, vol. 11. Dordrecht: Reidel.

———. 1987. *The Problems of Philosophy in Their Interconnections: Winter Semester Lectures 1933–1934,* edited by H. L. Mulder, A. J. Kox, and R. Hegselmann, translated by P. Heath. Dordrecht: Reidel.

Schreider, Yu A., ed. 1966. *The Monte Carlo Method: The Method of Statistical Trials,* translated by G. J. Tee. Oxford: Pergamon.

Schwartz, M. 1960. "Feasibility of Using High-Energy Neutrinos to Study the Weak Interactions." *Physical Review Letters* 4:306–7.

———. 1972. "Discovery of Two Kinds of Neutrinos." *Adventures in Experimental Physics* 1:81–100.

Schweber, S. S. 1984. "Some Chapters for a History of Quantum Field Theory: 1938–1952." In *Relativity, Groups, and Topology II,* Les Houches, Session 40, edited by B. DeWitt and R. Stora, 37–220. New York: North-Holland.

———. 1988. "The Mutual Embrace of Science and the Military ONR: The Growth of Physics in the United States after World War II." In *Science, Technology, and the Military,* edited by E. Mendelsohn, M. R. Smith, and P. Weingart, 3–45. Sociology of the Sciences, vol. 1. Dordrecht: Kluwer.

———. 1992. "Big Science in Context: Cornell and MIT." In *Big Science: The Growth of Large-Scale Research,* edited by P. Galison and B. Hevly, 149–83. Stanford, Calif.: Stanford University Press.

———. 1994. *QED and the Men Who Made It.* Princeton, N.J.: Princeton University Press.

Schwinger, Julian. 1948. "On Quantum-Electrodynamics and the Magnetic Moment of the Electron." *Physical Review* 73:416–17.

———. 1980. *Tomonaga Sin-itiro: A Memorial. Two Shakers of Physics.* [Japan]: Nishina Memorial Foundation.

Scottish Meteorological Society. 1884. "Report of the Council." *Journal of the Scottish Meteorological Society* 7:56–60.

Secord, James A. 1986. *Controversy in Victorian Geology: The Cambrian-Silurian Dispute.* Princeton, N.J.: Princeton University Press.

Seidel, Robert W. 1978. "Physics Research in California: The Rise of a Leading Sector in American Physics." Ph.D. dissertation, University of California, Berkeley.

———. 1983. "Accelerating Science: The Postwar Transformation of the Lawrence Radiation Laboratory." *Historical Studies in the Physical Sciences* 13:375–400.

———. 1986. "A Home for Big Science." *Historical Studies in the Physical Sciences* 16:135–75.

———. 1996. "The Hunting of the Neutrino." Typescript.

Seitz, F. 1958. "On the Theory of the Bubble Chamber." *Physics of Fluids* 1:2–13.

Sekido, Yataro, and Harry Elliot, eds. 1985. *Early History of Cosmic Ray Studies: Personal Reminiscences with Old Photographs.* Astrophysics and Space Science Library, vol. 118. Dordrecht: Reidel.

Seriff, A. J., R. B. Leighton, C. Hsiao, E. W. Cowan, and C. D. Anderson. 1950. "Cloud Chamber Observations of the New Unstable Cosmic-Ray Particles." *Physical Review* 78:290–91.

Servos, John W. 1990. *Physical Chemistry from Ostwald to Pauling: The Making of a Science in America.* Princeton, N.J.: Princeton University Press.

Shankland, Robert S. 1936. "An Apparent Failure of the Photon Theory of Scattering," *Physical Review* 49:8–13.

———. 1937. "The Compton Effect with Gamma-Rays," *Physical Review* 52:414–18.

Shapin, Steven. 1988. "The House of Experiment in Seventeenth Century England." *Isis* 79:373–404.

———. 1994. *A Social History of Truth: Civility and Science in Seventeenth-Century England.* Chicago: University of Chicago Press.

Shapin, S., and S. Schaffer. 1985. *Leviathan and the Air-Pump: Hobbes, Boyle, and the Experimental Life.* Princeton, N.J.: Princeton University Press.

Shaw, Napier. 1931. "A Century of Meteorology." *Nature* 128:925–26.

Sherry, Michael S. 1977. *Preparing for the Next War: American Plans for Postwar Defense, 1941–45.* New Haven, Conn.: Yale University Press.

Sherwin, Chalmers W. 1948. "Short Time Delays in Geiger Counters." *Review of Scientific Instruments* 19:111–15.

Shutt, R. P., ed. 1967. *Bubble and Spark Chambers: Principles and Use,* 2 vols. Pure and Applied Physics, vol. 27. New York: Academic Press.

Smart, J. J. C. (1956) 1973. "The Reality of Theoretical Entities." In *Theories and Observation in Science,* edited by R. E. Grandy, 93–103. Englewood Cliffs, N.J.: Prentice Hall. First published in *Australasian Journal of Philosophy* 34 (1956): 1–12.

Smith, Crosbie, and M. Norton Wise. 1989. *Energy and Empire: A Biographical Study of Lord Kelvin.* Cambridge: Cambridge University Press.

Smith, Nicholas M., Jr., Stanley S. Walters, Franklin C. Brooks, and David H. Blackwell. 1952–53. "The Theory of Value and the Science of Decision—A Summary." *Journal of the Operations Research Society of America* 1:103–13.

Smith, Robert W., and Joseph N. Tatarewicz. 1985. "Replacing a Technology: The Large Space Telescope and CCDs." *Proceedings of the IEEE* 73:1221–35.

Smorodinskii, Y. A., et al., eds. 1966. *XII International Conference on High-Energy Physics: Dubna, 5–15 August 1964,* 2 vols. Moscow: Atomizdat.

Snow, G. A. 1962. "Strong Interactions of Strange Particles [and discussion]." In *International Conference on High-Energy Physics at CERN,* edited by J. Prentki, 795–806. Geneva: CERN.

Snyder, J. N., R. Hulsizer, J. Munson, and A. Schneider. 1964. "Bubble Chamber Data Analysis Using a Scanning and Measuring Projector (SMP) On-Line to a Digital Computer." In *Automatic Acquisition and Reduction of Nuclear Data,* edited by K. H. Beckurts, W. Gläser, and G. Krüger, 239–48. Karlsruhe: Gesellschaft für Kernforschung.

Solenoidal Detector Collaboration. 1992. *Technical Design Report,* SDC-92-201, SSCL-SR-1215. Berkeley, Calif.: Lawrence Berkeley Laboratory, 1 April.

Sopka, Katherine Russell. 1988. *Quantum Physics in America: The Years through 1935.* Tomash Series in the History of Modern Physics 1800–1950, vol. 10. Los Angeles: Tomash.

"Spark Chambers." 1963. *Nuclear Instruments and Methods* 20:143–219.

"Spark Chamber Symposium." 1961. *Review of Scientific Instruments* 32:480–98.

Spence, Basil. 1959. "The Architect and Physics." In *The Design of Physics Research Laboratories,* 13–17. London: Chapman and Hall.

Spinrad, R. J. 1965. "Digital Systems for Data Handling." *Progress in Nuclear Technology and Instruments* 1:221–46.

Stadler, Friedrich, Hrsg. 1988. *Vertriebene Vernunft II: Emigration und Exil österreichischer Wissenschaft.* Vienna and Munich: Jugend und Volk.

Star, Susan Leigh, and James R. Griesemer. 1989. "Institutional Ecology, 'Translations' and Boundary Objects: Amateurs and Professionals in Berkeley's Museum of Vertebrate Zoology, 1907–39." *Social Studies of Science* 19:387–420.

Steinmaurer, Rudolf. 1985. "Erinnerungen an V. F. Hess, den Entdecker der kosmischen Strahlung, und an die ersten Jahre des Betriebes des Hafelekar-Labor." In *Early History of Cosmic Ray Studies: Personal Reminiscences with Old Photographs,* edited by Y. Sekido and H. Elliot, 17–31. Astrophysics and Space Science Library, vol. 118. Dordrecht: Reidel.

Steppan, Elvira. 1935. "Das Problem der Zertrümmerung von Aluminium, behandelt mit der photographischen Methode." *Sitzungsberichte, Akademie der Wissenschaften in Wien, Mathematisch-naturwissenschaftliche Klasse, Abteilung IIa* 144:455–74.

Stern, Nancy. 1981. *From ENIAC to UNIVAC: An Appraisal of the Early Eckert-Mauchly Computers.* Digital Press History of Computing Series. Bedford, Mass.: Digital Press.

Stetter, G., and H. Thirring. 1950. "Hertha Wambacher." *Acta Physica Austriaca* 4:318–20.

Stetter, G., and Hertha Wambacher. 1944. "Versuche zur Absorption der Höhenstrahlung nach der photographischen Methode I: Zertrümmerungssterne unter Blei-Absorption." *Sitztungsberichte, Akademie der Wissenschaften in Wien, Mathematisch-naturwissenschaftliche Klasse* 152:1–6.

Stewart, Irvin. 1948. *Organizing Scientific Research for War: The Administrative History of the Office of Scientific and Research Development.* Boston: Little, Brown.

Stigler, Stephen M. 1986. *The History of Statistics: The Measurement of Uncertainty before 1900.* Cambridge, Mass.: Harvard University Press.

———. 1991. "Stochastic Simulation in the Nineteenth Century." *Statistical Science* 6:89–97.

Strauch, K. 1965. "Innovations in Visual Spark Chamber Techniques." *IEEE Transactions in Nuclear Science,* NS-12, no. 4: 1–12.

———. 1974. "Introductory Remarks," PEP-138, September 1974. In LBL, SLAC, and SLAC-LBL Users Organization, *Proceedings 1974 PEP Summer Study.* Berkeley, Calif.: Lawrence Berkeley Laboratory.

Street, J. C., and E. C. Stevenson. 1937. "New Evidence for the Existence of a Particle of Mass Intermediate between the Proton and Electron." *Physical Review* 52:1003–4.

Street, J. C., and R. H. Woodward. 1934. "Counter Calibration and Cosmic-Ray Intensity." *Physical Review* 46:1029–34.

Strong, John. 1930. *Procedures in Experimental Physics.* New York: Prentice-Hall.

Stuewer, Roger. 1971. "William H. Bragg's Corpuscular Theory of X-Rays and γ-rays." *British Journal for the History of Science* 5:258–81.

———. 1975. *The Compton Effect: The Turning Point in Physics.* New York: Science History Publications.

———. 1985. "Artificial Disintegration and the Cambridge-Vienna Controversy." In *Observation, Experiment and Hypothesis in Modern Physical Science,* edited by P. Achinstein and O. Hannaway, 239–307. Cambridge, Mass.: MIT Press.

———. 1986. "Rutherford's Satellite Model of the Nucleus," *Historical Studies in the Physical Sciences* 16:321–52.

Sturchio, Jeffrey, ed. 1988. Special Issue on Artifact and Experiment. *Isis* 79:369–476.

Sullivan, Woodruff T., III. 1988. "Early Years of Australian Radio Astronomy." In *Australian Science in the Making,* edited by R. W. Home, 308–344. Cambridge: Cambridge University Press.

Super, R. H. 1977. "The Humanist at Bay: The Arnold-Huxley Debate." In *Nature and the Victorian Imagination,* edited by U. C. Knoepflmacher and G. B. Tennyson, 231–45. Berkeley: University of California Press.

Suppe, Frederick. 1974. *The Structure of Scientific Theories.* Urbana: University of Illinois Press.

Sutherland, Arthur E. 1967. *The Law at Harvard: A History of Ideas and Men, 1817–1967.* Cambridge, Mass.: Belknap.

Sviedrys, R. 1970. "The Rise of Physical Science at Victorian Cambridge," with commentary by Arnold Thackray. *Historical Studies in the Physical Sciences* 2:127–51.

Swann, W. F. G. 1936. "Report on the Work of the Bartol Research Foundation, 1935–1936." *Journal of the Franklin Institute* 222:647–714.

Swann, W. F. G., and G. L. Locher. 1936. "The Variation of Cosmic Ray Intensity with Direction in the Stratosphere." *Journal of the Franklin Institute* 221:275–89.

Symons, J. G., ed. 1888. *The Eruption of Krakatoa and Subsequent Phenomena: Report of the Krakatoa Committee of the Royal Society.* London: Harrison.

Taft, H. D., and P. J. Martin. 1966. "On-Line Monitoring of Bubble Chamber Measurements by Small Computers." In *XII International Conference on High-Energy Physics,* edited by Y. A. Smorodinskii et al., 390–92. Moscow: Atomizdat.

Taussig, M. 1980. *The Devil and Commodity Fetishism in South America.* Chapel Hill: University of North Carolina Press.

Taylor, H. J. 1935. "The Tracks of α-Particles and Protons in Photographic Emulsions." *Proceedings of the Royal Society London A* 150:382–94.

Taylor, H. J., and V. D. Dabholkar. 1936. "The Ranges of α Particles in Photographic Emulsions." *Proceedings of the Royal Society London A* 48:285–98.

Taylor, H. J., and M. Goldhaber. 1935. "Detection of Nuclear Disintegration in a Photographic Emulsion." *Nature* 2:341.

Teller, Edward. 1950. "Back to the Laboratories." *Bulletin of the Atomic Scientists* 6:71–72.

Thackray, Arnold. 1970. Commentary in "The Rise of Physical Science at Victorian Cambridge." *Historical Studies in the Physical Sciences* 2:127–51.

Thomas, Nicholas. 1991. "Against Ethnography." *Cultural Anthropology* 6:306–22.

———. 1991. *Entangled Objects: Exchanges, Material Culture, and Colonialism in the Pacific.* Cambridge, Mass.: Harvard University Press.

Thomson, J. J. (1846) 1989. "Introductory Lecture." In *Energy and Empire: A Biographical Study of Lord Kelvin,* edited by C. Smith and M. N. Wise, 121–22. Cambridge: Cambridge University Press.

———. 1886. "Some Experiments on the Electric Discharge in a Uniform Electrical Field, with Some Theoretical Considerations about the Passage of Electricity through Gases." *Proceedings of the Cambridge Philosophical Society* 5:391–409.

———. 1888. *Applications of Dynamics to Physics and Chemistry.* London: Macmillan.

———. 1893. "On the Effect of Electrification and Chemical Action on a Steam-Jet, and

of Water-Vapor on the Discharge of Electricity through Gases." *Philosophical Magazine,* 5th ser., 36:313–27.
———. 1897. "Cathode Rays." *Philosophical Magazine,* 5th ser., 44:293–316.
———. 1898. "On the Charge of Electricity Carried by the Ions Produced by Röntgen Rays." *Philosophical Magazine,* 5th ser., 46:528–45.
———. 1937. *Recollections and Reflections.* New York: Macmillan.
Thomson, J. J., and E. Rutherford. 1896. "On the Passage of Electricity through Gases Exposed to Röntgen Rays." *Philosophical Magazine,* 5th ser., 42:392–407.
Thorndike, Alan M. 1967. "Summary and Future Outlook." In *Bubble and Spark Chambers: Principles and Use,* edited by R. P. Shutt, 2:299–300. New York: Academic Press.
Thornes, John. 1979. "Constable's Clouds." *Burlington Magazine* 121:697–704.
Timmerhaus, K. D., ed. 1960–63. *Advances in Cryogenic Engineering,* 4 vols. New York: Plenum.
Ting, Samuel, Gerson Goldhaber, and Burton Richter. 1976. "Discovery of Massive Neutral Vector Mesons." *Adventures in Experimental Physics* 5:114–49.
Tocher, K. D. 1951. "The Application of Automatic Computers to Sampling Experiments." *Journal of the Royal Statistical Society B* 16:39–61.
Todd, Loreto. 1990. *Pidgins and Creoles.* London: Routledge and Kegan Paul.
Tomas, David, G. 1979. "Tradition, Context of Use, Style, and Function: Expansion Apparatuses Used at the Cavendish Laboratory during the Period 1895–1912." M.S. thesis, University of Montreal.
Tomonaga, S., and G. Araki. 1940. "Effect of the Nuclear Coulomb Field on the Capture of Slow Mesons." *Physical Review* 58:90–91.
Toulmin, Stephen. 1953. *The Philosophy of Science: An Introduction.* New York: Harper and Row.
Townsend, John Sealy Edward. 1915. *Electricity in Gases.* Oxford: Clarendon.
———. 1947. *Electrons in Gases.* London: Hutchinson.
Traweek, Sharon. 1988. *Beamtimes and Lifetimes: The World of High Energy Physicists.* Cambridge, Mass.: Harvard University Press.
Treitel, Jonathan, A. 1986. "A Structural Analysis of the History of Science: The Discovery of the Tau Lepton." Ph.D. dissertation, Stanford University.
———. 1987. "Confirmation with Technology: The Discovery of the Tau Lepton." *Centaurus* 30:140–80.
Trenn, Thaddeus. 1976. "Die Erfindung des Geiger-Müller-Zählrohres." *Deutsches Museum, Abhandlungen Berichte* 44:54–64.
———. 1986. "The Geiger-Müller Counter of 1928." *Annals of Science* 43:111–35.
Trower, W. Peter, ed. 1987. *Discovering Alvarez: Selected Works of Luis W. Alvarez, with Commentary by His Students and Colleagues.* Chicago: University of Chicago Press.
Turner, G. 1981. "Wilson, Charles Thomas Rees." *Dictionary of Scientific Biography* 14:420–23.
Tuve, M. A. 1930. "Multiple Coincidences of Geiger-Müller Tube-Counters." *Physical Review* 35:651–52.
Tye, Michael. 1991. *The Imagery Debate.* Cambridge, Mass.: MIT Press.
Tyndall, A. M. 1971. "Powell." Appendix to F. C. Frank and D. H. Perkins, "Cecil Frank

Powell, 1903–1969." *Biographical Memoirs of Fellows of the Royal Society* 17 (1971): 541–55.

Tyndall, A. M., and C. F. Powell. 1930. "The Mobility of Ions in Pure Gases." *Proceedings of the Royal Society A* 129:162–80. Reprinted in *Selected Papers of Cecil Frank Powell,* edited by E. H. S. Burhop, W. O. Lock, and M. G. K. Menon, 77–95. New York: North-Holland, 1972.

Uebel, Thomas C. 1991. *Rediscovering the Forgotten Vienna Circle: American Studies on Otto Neurath and the Vienna Circle.* Dordrecht: Kluwer.

Ulam, S. M. 1951. "On the Monte Carlo Method." In *Proceedings of a Second Symposium on Large-Scale Digital Calculating Machinery, 13–16 September 1949,* 207–212. Cambridge, Mass.: Harvard University Press.

———. 1952. "Random Processes and Transformations." In *Proceedings of the International Congress of Mathematicians: Cambridge, Massachusetts, 30 August–6 September 1950,* 2:264–75. Providence, R.I.: American Mathematical Society.

———. 1956. "Applications of Monte Carlo Methods to Tactical Games." In *Symposium on Monte Carlo Methods,* edited by H. A. Meyer. New York: Wiley.

———. 1958. "John von Neumann, 1903–1957." *Bulletin of the American Mathematical Society* 64:1–49.

———. 1965. Introduction to E. Fermi, J. Pasta, and S. Ulam. "Studies of Non Linear Problems," Paper no. 226. In *Enrico Fermi: Collected Papers. Volume 2, United States 1939–1954,* edited by E. Segrè, 977–978. Chicago: University of Chicago Press.

———. 1990. *Analogies between Analogies: The Mathematical Reports of S. M. Ulam and His Los Alamos Collaborators,* edited by A. R. Bednarek and F. Ulam. Los Alamos Series in Basic and Applied Sciences, vol. 10. Berkeley and Los Angeles: University of California Press.

Ulam, S. M., and J. Pasta. 1990. "Heuristic Studies in Problems of Mathematical Physics on High Speed Computing Machines." In S. M. Ulam, *Analogies between Analogies: The Mathematical Reports of S. M. Ulam and His Los Alamos Collaborators,* edited by A. R. Bednarek and F. Ulam, 121–38. Berkeley and Los Angeles: University of California Press.

Ulam, S. M., and L. D. Tuck. "Possibility of Initiating a Thermonuclear Reaction," LA-560.

Ulam, S. M., and von Neumann, J. 1947. "On Combination of Stochastic and Deterministic Processes: Preliminary Report" [abstract]. *Bulletin of the American Mathematical Society* 53:1120.

U.S. Air Force. Technical Staff and Air Training Command. 1960. *Fundamentals of Guided Missiles.* Los Angeles: Aero Publishers.

U.S. Atomic Energy Commission. 1966. Final Report. Part I. "Report on Investigation of Explosion and Fire: Experimental Hall, Cambridge Electron Accelerator, Cambridge, Massachusetts, July 5, 1965," TID-22594. Washington, D.C.: U.S. Government Printing Office.

U.S. Congress. Joint Session. 1965. *Hearings before the Subcommittee on Research, Development, and Radiation of the Joint Committee on Atomic Energy: Eighty-ninth Congress, First Session on High Energy Physics Research, 2–5 March 1965,* 377–378. 89th Cong., 1st sess., 1965. Washington, D.C.: U.S. Government Printing Office.

U.S. Department of Energy. Office of Energy Research. 1988. "Report of the HEPAP Subpanel on Future Modes of Experimental Research in High Energy Physics," DOE/ER-0380. Washington, D.C.: U.S. Government Printing Office.

Urey, Harold C. 1950. "Should America Build the H-Bomb?" *Bulletin of the Atomic Scientists* 6:72–73.

Valdman, Albert, ed. 1977. *Pidgin and Creole Linguistics.* Bloomington: Indiana University Press.

Valdman, Albert, and Arnold Highfield, ed. 1980. *International Conference on Theoretical Orientations in Creole Studies (1979, St. Thomas, V.I.).* New York: Academic Press.

Van Fraassen, Bas. 1980. *The Scientific Image.* Oxford: Clarendon.

Van Helden, Albert, and Thomas L. Hankins, eds. 1994. Special Issue on Instruments. *Osiris* 9:1–250.

Vernon, W. 1964. "Spark Chamber Vidicon Scanner with Discrete Scan." In *Proceedings of the Informal Meeting on Film-Less Spark Chamber Techniques and Associated Computer Use,* CERN "Yellow Report" 64-30, edited by G. R. Macleod and B. C. Maglić, 57–63. Geneva: CERN Data Handling Division.

Vishnyakov, V. V., and A. A. Typakin. 1957. "Investigations of the Performance of Gas-Discharge Counters with a Controlled Pulsed Power Supply." *Soviet Journal of Atomic Energy* 3:1103–13.

Volmer, Max. 1939. *Kinetik des Phasenbildung die Chemische Reaktion.* Dresden and Leipzig: Theodor Steinkopff.

von Neumann, John. (1928) 1963. "Zur Theorie der Gesellschaftsspiele." In *Von Neumann: Collected Works,* edited by A. H. Taub, 6:1–26. London: Pergamon. First published in *Mathematische Annalen* 100 (1928): 295–320.

———. (1944) 1963. "Proposal and Analysis of a New Numerical Method for the Treatment of Hydrodynamical Shock Problems." In *Von Neumann: Collected Works,* edited by A. H. Taub, 6:361–79. London: Pergamon. First published as OSRD-3617. 20 March.

———. (1951) 1961. "Various Techniques Used in Connection with Random Digits." In *Von Neumann: Collected Works,* edited by A. H. Taub, 5:768–70. London: Pergamon. First published in *Monte Carlo Method,* edited by A. S. Householder, G. E. Forsythe, and H. H. Germond, 36–38. National Bureau of Standards Applied Mathematics Series, no. 12. Washington, D.C.: U.S. Government Printing Office.

———. 1955. "Can We Survive Technology?" *Fortune* 51:106–8, 151–52. Reprinted in *Von Neumann: Collected Works,* edited by A. H. Taub, 6:504–19. London: Pergamon, 1963.

———. 1961–63. *Design of Computers, Theory of Automata and Numerical Analysis.* Vol. 5 of *Von Neumann: Collected Works,* edited by A. H. Taub. London: Pergamon.

———. 1961–63. *Theory of Games, Astrophysics, Hydrodynamics and Meteorology.* Vol. 6 of *Von Neumann: Collected Works,* edited by A. H. Taub. London: Pergamon.

———. 1961–63. *Von Neumann: Collected Works,* 6 vols., edited by A. H. Taub. London: Pergamon.

———. 1981. "First Draft of a Report on the EDVAC." In *From ENIAC to UNIVAC: An Appraisal of the Early Eckert-Mauchly Computers,* edited by N. Stern, 177–246. Digital Press History of Computing Series. Bedford, Mass.: Digital Press.

von Neumann, John, and Oskar Morgenstern. 1953. *Theory of Games and Economic Behavior,* 3d ed. Princeton, N.J.: Princeton University Press.

Wagner, A. 1981. "Central Detectors." *Physica Scripta* 23:446–58.

Walenta, A. H., J. Heintze, and B. Shürlein. 1971. "The Multiwire Drift Chamber: A New Type of Proportional Wire Chamber." *Nuclear Instruments and Methods* 92: 373–80.

Walker, Mark. 1989. *German National Socialism and the Quest for Nuclear Power.* Cambridge: Cambridge University Press.

Waller, C. 1988. "British Patent 580,504 and Ilford Nuclear Emulsions." In *40 Years of Particle Physics,* edited by B. Foster and P. Fowler, 55–58. Bristol: Adam Hilger.

Walter, Maila L. 1990. *Science and Cultural Crisis: An Intellectual Biography of Percy Williams Bridgman (1822–1961).* Stanford, Calif.: Stanford University Press.

Wambacher, Hertha. 1931. "Untersuchung der photographischen Wirkung radioaktiver Strahlungen auf mit Chromsäure und Pinakryptolgelb vorbehandelten Filmen und Platten." *Sitzungsberichte, Akademie der Wissenschaften in Wien, Mathematisch-naturwissenschaftliche Klasse, Abteilung IIa* 140:271–91.

———. 1940. "Kernzertrümmerung durch Höhenstrahlung in der photographischen Emulsion." *Sitzungsberichte, Akademie der Wissenschaften in Wien, Mathematisch-naturwissenschaftliche Klasse* 149:157–211.

———. 1949. "Mikroscopie und Kernphysik." *Mikroskopie: Zentralblatt für Mikroskopische Forschung und Methodik* 4:92–110.

Wambacher, Hertha, and Anton Widhalm. 1943. "Über die kurzen Bahnspuren in photographischen Schichten." *Sitzungsberichte, Akademie der Wissenschaften in Wien, Mathematisch-naturwissenschaftliche Klasse* 152:173–91.

Warnock, Tony. 1987. "Random Number Generators." *Los Alamos Science* 15:137–41. Special issue on Stanislav Ulam.

Warren, Mark D. (1907) 1987. *The Cunard Turbine-Driven Quadruple-Screw Atlantic Liner "Mauretania."* Wellingborough: Patrick Stevens. Primarily a facsimile reprint of a 1907 volume.

Wasserman, Neil H. 1981. "The Bohr-Kramers-Slater Paper and the Development of the Quantum Theory of Radiation in the Work of Niels Bohr." Ph.D. dissertation, Harvard University.

Wedderburn, E. 1948. "The Scottish Meteorological Society." *Quarterly Journal of the Royal Meteorological Society* 74:233–42.

Weimer, Albert. 1947. "The History of Psychology and Its Retrieval from Historiography II: Some Lessons for the Methodology of Scientific Research." *Science Studies* 4:367–96.

Weinberg, Alvin M., and Eugene P. Wigner. 1958. *The Physical Theory of Neutron Chain Reactors.* Chicago: University of Chicago Press.

Weinberg, S. 1980. "Conceptual Foundations of the Unified Theory of Weak and Electromagnetic Interactions." *Reviews of Modern Physics* 52:515–23.

Weiner, C., ed. 1977. *History of Twentieth Century Physics.* Proceedings of the International School of Physics "Enrico Fermi" Course 70. New York: Academic Press.

Weinert, Friedel, ed. 1995. *Laws of Nature: Essays on Philosophical, Scientific and Historical Dimensions.* Berlin: de Gruyter.

Weissert, Thomas. 1997. *Fermi, Pasta, Ulam: Genesis of Simulation in Dynamics.* New York: Springer.

Weisskopf, Victor. 1972. "Life and Work of Cecil Powell, A Tribute." In *Selected Papers of Cecil Frank Powell,* edited by E. H. S. Burhop, W. O. Lock, and M. G. K. Menon, 1–6. New York: North-Holland.

Welton, Theodore A. 1948. "Some Observable Effects of the Quantum-Mechanical Fluctuations of the Electromagnetic Field." *Physical Review* 74:1157–67.

Westfall, Catherine. 1989. "The Site Contest for Fermilab." *Physics Today* 42:44–52.

Westfall, Richard S. 1977. *The Construction of Modern Science: Mechanisms and Mechanics.* Cambridge: Cambridge University Press.

Weyl, Hermann. 1916. "Über die Gleichverteilung von Zahlen mod. Eins." *Mathematische Annalen* 77:313–53.

Wheaton, Bruce R. 1983. *The Tiger and the Shark: Empirical Roots of Wave-Particle Dualism.* Cambridge: Cambridge University Press.

Wheeler, John A. 1946. "Polyelectrons." *Annals of the New York Academy of Science* 48:219–38.

———. 1946. "Problems and Prospects in Elementary Particle Research." *Proceedings of the American Philosophical Society* 90:36–47.

———. 1947. "Elementary Particle Physics." *American Scientist* 35:177–93.

Whipple, F. J. W. 1923. "Meteorological Optics." In *A Dictionary of Applied Physics,* edited by R. Glazebrook, 3:518–33. London: Macmillan.

Whorf, Benjamin L. 1956. *Language, Thought, and Reality: Selected Writings,* edited by J. B. Carroll. Cambridge, Mass.: MIT Press.

Wilson, C. T. R. 1895. "On the Formation of Cloud in the Absence of Dust." *Proceedings of the Cambridge Philosophical Society* 8:306.

———. 1896. "The Effect of Röntgen's Rays on Cloudy Condensation." *Proceedings of the Royal Society London* 59:338–39.

———. 1897. "On the Action of Uranium Rays on the Condensation of Water Vapor." *Proceedings of the Cambridge Philosophical Society* 9:333–38.

———. 1887. "Condensation of Water Vapor in the Presence of Dust-Free Air and Other Gases." *Philosophical Transactions of the Royal Society London A* 189:265–307.

———. 1899. "On the Comparative Efficiency as Condensation Nuclei of Positively and Negatively Charged Ions." *Philosophical Transactions of the Royal Society London A* 193:289–308.

———. 1899. "On the Condensation Nuclei Produced in Gases by Röntgen Rays, Uranium Rays, Ultra-Violet Light, and Other Agents." *Proceedings of the Royal Society London* 64:127–29.

———. 1901. "On the Ionisation of Atmospheric Air." *Proceedings of the Royal Society London* 68:151–61.

———. 1902. "Further Experiments on Radio-Activity from Rain." *Proceedings of the Cambridge Philosophical Society* 12:17.

———. 1902. "On Radio-Active Rain." *Proceedings of the Cambridge Philosophical Society* 11:428–30.

———. 1903. "Atmospheric Electricity." *Nature* 68:102–4.

———. 1903. "On Radio-Activity from the Snow." *Proceedings of the Cambridge Philosophical Society* 12:85.

———. 1910. "1899–1902." In *History of the Cavendish Laboratory, 1871–1910,* 195–220. London: Longmans, Green.

———. 1911. "On a Method of Making Visible the Paths of Ionising Particles through a Gas." *Proceedings of the Royal Society London A* 85:285–88.

———. 1912. "On an Expansion Apparatus for Making Visible the Tracks of Ionising Particles in Gases and Some Results Obtained by Its Use." *Proceedings of the Royal Society London A* 87:277–90.

———. 1954. "Ben Nevis Sixty Years Ago." *Weather* 9:309–11.

———. 1960. "Reminiscenses of My Early Years." *Notes and Records of the Royal Society of London* 14:163–73.

———. 1965. "On the Cloud Method of Making Visible Ions and the Tracks of Ionising Particles." In *Nobel Lectures including Presentation Speeches and Laureates' Biographies: Physics 1963–1970,* 194–214. New York: Elsevier.

Wilson, David B. 1982. "Experimentalists among the Mathematicians: Physics in the Cambridge Natural Sciences Tripos, 1851–1900." *Historical Studies in the Physical Sciences* 12:325–71.

Wilson, J. G. 1951. *The Principles of Cloud-Chamber Technique.* Cambridge Monographs in Physics. Cambridge: Cambridge University Press.

Wilson, Robert R. 1951. "Showers Produced by Low Energy Electrons and Photons." In *Monte Carlo Method,* edited by A. S. Householder, G. E. Forsythe, and H. H. Germond, 1–3. Washington, D.C.: U.S. Government Printing Office.

———. 1952. "Monte Carlo Study of Shower Production." *Physical Review* 86:261–69.

———. 1972. "My Fight against Team Research." In *The Twentieth Century Sciences: Studies in the Biography of Ideas,* edited by G. Holton, 468–79. New York: Norton.

Wise, M. Norton. 1988. "Mediating Machines." *Science in Context* 2:77–113.

———. 1994. "Pascual Jordan: Quantum Mechanics, Psychology, National Socialism." In *Science, Technology, and National Socialism,* edited by M. Renneberg and M. Walker, 224–54. Cambridge: Cambridge University Press.

———, ed. 1995. *The Values of Precision.* Princeton, N.J.: Princeton University Press.

Wismer, K. L. 1922. "The Pressure-Volume Relation of Superheated Liquids." *Journal of Physical Chemistry* 26:301–15.

Wittgenstein, Ludwig. 1958. *Philosophical Investigations,* 2d ed., translated by G. E. M. Anscombe. Oxford: Blackwell.

Wojcicki, Stanley G. 1987. "My First Days in the Alvarez Group." In *Discovering Alvarez: Selected Works of Luis W. Alvarez, with Commentary by His Students and Colleagues,* edited by W. P. Trower, 163–70. Chicago: University of Chicago Press.

Wood, John G. 1954. "Bubble Tracks in a Hydrogen-Filled Glaser Chamber." *Physical Review* 94:731.

Woodward, R. H. 1935. "The Interaction of Cosmic Rays with Matter." Ph.D. dissertation, Harvard University.

———. 1936. "Coincidence Counter Studies of Cosmic Ray Showers." *Physical Review* 49:711–18.

Woolford, Ellen, and William Washabaugh, eds. 1983. *The Social Context of Creolization.* Ann Arbor, Mich.: Karoma.

Worthington, A. M. 1908. *A Study of Splashes.* London: Longmans, Green.
Wu, C. S., E. Ambler, R. W. Hayward, D. D. Hoppes, and R. P. Hudson. 1957. "Experimental Test of Parity Conservation in Beta Decay." *Physical Review* 105:1413–15.
Yu, J. T., and P. J. E. Peebles. 1969. "Superclusters of Galaxies?" *Astrophysical Journal* 158:103–13.
Zaniello, Thomas A. 1981. "The Spectacular English Sunsets of the 1880s." In *Victorian Science and Victorian Values: Literary Perspectives,* edited by J. Paradis and T. Postlewait, 247–267. New York: New York Academy of Sciences.
Zaremba, S. K. 1969. "The Mathematical Basis of Monte Carlo and Quasi-Monte Carlo Methods." *Studies in Applied Mathematics* 3:1–12.
Zhdanov, G. B. 1958. "Quelques problèmes méthodologiques présentant de l'intérêt aux grandes et très grandes énergies pour l'investigation de processus par la méthode des émulsions." In *Premier Colloque International de Photographie Corpusculaire,* 233–239. Paris: Centre National de la Recherche Scientifique.
Zila, Stefanie. 1936. "Beitrage zum Ausbau der photographischen Methode für Untersuchungen mit Protonenstrahlen." *Sitzungsberichte, Akademie der Wissenschaften in Wien, Mathematisch-naturwissenschaftliche Klasse, Abteilung IIa* 145:503–14.

Index

Italic numbers refer to page numbers of illustrations.

Bach
1960s Jacqueline Depre
#1 in G minor BW 207

6 Shapin + Schaffer

does he ever question whether
what is = what is right & good?
evaluations of weakness
failure
inefficiency

Galison is top of food chain
& shows it many times over
not least in taken-for-granted
identification with "heros" of physics
does he ever pay atten to technicians?
p422 dull

49 - consider idea of pidgin re
evolution of concepts of space
in 17th cent.

49 diachrony talk about evolution of language
pidgins in preface to Ars Conjectandi

57-8 James Bryant Conant